ADAPTIVE FILTER THEORY

PRENTICE HALL INFORMATION AND SYSTEM SCIENCES SERIES

Thomas Kailath, Editor

ANDERSON & MOORE	*Optimal Control: Linear Quadratic Methods*
ANDERSON & MOORE	*Optimal Filtering*
ASTROM & WITTENMARK	*Computer-Controlled Systems: Theory and Design, 2/E*
BASSEVILLE & NIKIROV	*Detection of Abrupt Changes: Theory & Application*
BOYD & BARRATT	*Linear Controller Design: Limits of Performance*
DICKINSON	*Systems: Analysis, Design and Computation*
FRIEDLAND	*Advanced Control System Design*
GARDNER	*Statistical Spectral Analysis: A Nonprobabilistic Theory*
GRAY & DAVISSON	*Random Processes: A Mathematical Approach for Engineers*
GREEN & LIMEBEER	*Linear Robust Control*
HAYKIN	*Adaptive Filter Theory*
HAYKIN	*Blind Deconvolution*
JAIN	*Fundamentals of Digital Image Processing*
JOHANSSON	*Modeling and System Identification*
JOHNSON	*Lectures on Adaptive Parameter Estimation*
KAILATH	*Linear Systems*
KUNG	*VLSI Array Processors*
KUNG, WHITEHOUSE, & KAILATH, EDS.	*VLSI and Modern Signal Processing*
KWAKERNAAK & SIVAN	*Signals and Systems*
LANDAU	*System Identification and Control Design Using P.I.M. + Software*
LJUNG	*System Identification: Theory for the User*
LJUNG & GLAD	*Modeling of Dynamic Systems*
MACOVSKI	*Medical Imaging Systems*
MOSCA	*Stochastic and Predictive Adaptive Control*
NARENDRA & ANNASWAMY	*Stable Adaptive Systems*
RUGH	*Linear System Theory, Second Edition*
SASTRY & BODSON	*Adaptive Control: Stability, Convergence, and Robustness*
SOLIMAN & SRINATH	*Continuous and Discrete-Time Signals and Systems*
SOLO & KONG	*Adaptive Signal Processing Algorithms: Stability & Performance*
SRINATH, RAJASEKARAN, & VISWANATHAN	*Introduction to Statistical Signal Processing with Applications*
VISWANADHAM & NARAHARI	*Performance Modeling of Automated Manufacturing Systems*
WILLIAMS	*Designing Digital Filters*

Adaptive Filter Theory
Third Edition

Simon Haykin

Communications Research Laboratory
McMaster University
Hamilton, Ontario, Canada

PRENTICE HALL, Upper Saddle River, New Jersey 07458

Library of Congress Cataloging-in-Publication Data

Haykin, Simon
    Adaptive filter theory / Simon Haykin. —3rd ed.
      p. cm.—(Prentice Hall Information and system sciences
series)
    Includes bibliographical references and index.
    ISBN 0-13-322760-X
    1. Adaptive filters. I. Title. II. Series.
TK7872.F5H368 1996
621.3815′324—dc20   95-11120
                          CIP

Acquisitions editor: **TOM ROBBINS**
Editorial/production supervision
  and interior design: **SHARYN VITRANO**
Copy editor: **DIANE LANGE**
Cover designer: **TOM NERY**
Manufacturing buyer: **DONNA SULLIVAN**
Editorial assistant: **PHYLLIS MORGAN**
Supplements editor: **BARBARA MURRAY**

Printed in the United States of America

10 9 8 7

0-13-322760-X

Prentice-Hall International (UK) Limited, *London*
Prentice-Hall of Australia Pty. Limited, *Sydney*
Prentice-Hall Canada Inc., *Toronto*
Prentice-Hall Hispanoamericana, S.A., *Mexico*
Prentice-Hall of India Private Limited, *New Delhi*
Prentice-Hall of Japan, Inc., *Tokyo*
Pearson Education Asia Pte. Ltd., *Singapore*
Editora Prentice-Hall do Brasil, Ltda., *Rio de Janeiro*

This book is dedicated to

- The many researchers whose significant contributions to adaptive filtering have made it possible to write the book.
- The many reviewers, readers, and students who have helped me to improve the book through insightful and critical comments.

Contents

Preface   **xiii**

Acknowledgments   **xvi**

Introduction   **1**

1. The Filtering Problem   1
2. Adaptive Filters   2
3. Linear Filter Structures   4
4. Approaches to the Development of Linear Adaptive Filtering Algorithms   9
5. Real and Complex Forms of Adaptive Filters   14
6. Nonlinear Adaptive Filters   15
7. Applications   18
8. Some Historical Notes   67

PART 1   BACKGROUND MATERIAL   78

Chapter 1   Discrete-Time Signal Processing   79

1.1 z-Transform   79
1.2 Linear Time-Invariant Filters   81
1.3 Minimum-Phase Filters   86
1.4 Discrete Fourier Transform   87
1.5 Implementing Convolutions Using the DFT   87
1.6 Discrete Cosine Transform   93
1.7 Summary and Discussion   94
Problems   95

Chapter 2   Stationary Processes and Models   96

2.1 Partial Characterization of a Discrete-Time Stochastic Process   97

2.2   Mean Ergodic Theorem   98
2.3   Correlation Matrix   100
2.4   Correlation Matrix of Sine Wave Plus Noise   106
2.5   Stochastic Models   108
2.6   Wold Decomposition   115
2.7   Asymptotic Stationarity of an Autoregressive Process   116
2.8   Yule-Walker Equations   118
2.9   Computer Experiment: Autoregressive Process of Order 2   120
2.10   Selecting the Model Order   128
2.11   Complex Gaussian Processes   130
2.12   Summary and Discussion   132
       Problems   133

Chapter 3   Spectrum Analysis   136

3.1   Power Spectral Density   136
3.2   Properties of Power Spectral Density   138
3.3   Transmission of a Stationary Process Through a Linear Filter   140
3.4   Cramér Spectral Representation for a Stationary Process   144
3.5   Power Spectrum Estimation   146
3.6   Other Statistical Characteristics of a Stochastic Process   149
3.7   Polyspectra   150
3.8   Spectral-Correlation Density   154
3.9   Summary and Discussion   157
       Problems   158

Chapter 4   Eigenanalysis   160

4.1   The Eigenvalue Problem   160
4.2   Properties of Eigenvalues and Eigenvectors   162
4.3   Low-Rank Modeling   176
4.4   Eigenfilters   181
4.5   Eigenvalue Computations   184
4.6   Summary and Discussion   187
       Problems   188

PART 2   LINEAR OPTIMUM FILTERING   193

Chapter 5   Wiener Filters   194

5.1   Linear Optimum Filtering: Problem Statement   194
5.2   Principle of Orthogonality   197
5.3   Minimum Mean-Squared Error   201
5.4   Wiener-Hopf Equations   203
5.5   Error-Performance Surface   206
5.6   Numerical Example   210
5.7   Channel Equalization   217
5.8   Linearly Constrained Minimum Variance Filter   220
5.9   Generalized Sidelobe Cancelers   227
5.10   Summary and Disussion   235
       Problems   236

Chapter 6   Linear Prediction   241

  6.1   Forward Linear Prediction   242
  6.2   Backward Linear Prediction   248
  6.3   Levinson-Durbin Algorithm   254
  6.4   Properties of Prediction-Error Filters   262
  6.5   Schur-Cohn Test   271
  6.6   Autoregressive Modeling of a Stationary Stochastic Process   273
  6.7   Cholesky Factorization   276
  6.8   Lattice Predictors   280
  6.9   Joint-Process Estimation   286
  6.10 Block Estimation   290
  6.11 Summary and Discussion   293
       Problems   295

Chapter 7   Kalman Filters   302

  7.1   Recursive Minimum Mean-Square Estimation for Scalar Random Variables   303
  7.2   Statement of the Kalman Filtering Problem   306
  7.3   The Innovations Process   307
  7.4   Estimation of the State using the Innovations Process   310
  7.5   Filtering   317
  7.6   Initial Conditions   320
  7.7   Summary of the Kalman Filter   320
  7.8   Variants of the Kalman Filter   322
  7.9   The Extended Kalman Filter   328
  7.10 Summary and Discussion   333
       Problems   334

PART 3   LINEAR ADAPTIVE FILTERING   338

Chapter 8   Method of Steepest Descent   339

  8.1   Some Preliminaries   339
  8.2   Steepest-Descent Algorithm   341
  8.3   Stability of the Steepest-Descent Algorithm   343
  8.4   Example   350
  8.5   Summary and Discussion   362
       Problems   362

Chapter 9   Least-Mean-Square Algorithm   365

  9.1   Overview of the Structure and Operation of the Least-Mean-Square Algorithm   365
  9.2   Least-Mean-Square Adaptation Algorithm   367
  9.3   Examples   372
  9.4   Stability and Performance Analysis of the LMS Algorithm   390
  9.5   Summary of the LMS Algorithm   405
  9.6   Computer Experiment on Adaptive Prediction   406
  9.7   Computer Experiment on Adaptive Equalization   412
  9.8   Computer Experiment on Minimum-Variance Distortionless Response Beamformer   421
  9.9   Directionality of Convergence of the LMS Algorithm for Non-White Inputs   425
  9.10 Robustness of the LMS Algorithm   427

9.11 Normalized LMS Algorithm   432
9.12 Summary and Discussion   438
      Problems   439

Chapter 10   Frequency-Domain Adaptive Filters   445

10.1 Block Adaptive Filters   446
10.2 Fast LMS Algorithm   451
10.3 Unconstrained Frequency-Domain Adaptive Filtering   457
10.4 Self-Orthogonalizing Adaptive Filters   458
10.5 Computer Experiment on Adaptive Equalization   469
10.6 Classification of Adaptive Filtering Algorithms   477
10.7 Summary and Discussion   478
      Problems   479

Chapter 11   Method of Least Squares   483

11.1   Statement of the Linear Least-Squares Estimation Problem   483
11.2   Data Windowing   486
11.3   Principle of Orthogonality (Revisited)   487
11.4   Minimum Sum of Error Squares   491
11.5   Normal Equations and Linear Least-Squares Filters   492
11.6   Time-Averaged Correlation Matrix   495
11.7   Reformulation of the Normal Equations in Terms of Data Matrices   497
11.8   Properties of Least-Squares Estimates   502
11.9   Parametric Spectrum Estimation   506
11.10 Singular Value Decomposition   516
11.11 Pseudoinverse   524
11.12 Interpretation of Singular Values and Singular Vectors   525
11.13 Minimum Norm Solution to the Linear Least-Squares Problem   526
11.14 Normalized LMS Algorithm Viewed as the Minimum-Norm Solution to an
       Underdetermined Least-Squares Estimation Problem   530
11.15 Summary and Discussion   532
       Problems   533

Chapter 12   Rotations and Reflections   536

12.1 Plane Rotations   537
12.2 Two-Sided Jacobi Algorithm   538
12.3 Cyclic Jacobi Algorithm   544
12.4 Householder Transformation   548
12.5 The QR Algorithm   551
12.6 Summary and Discussion   558
      Problems   560

Chapter 13   Recursive Least-Squares Algorithm   562

13.1 Some Preliminaries   563
13.2 The Matrix Inversion Lemma   565
13.3 The Exponentially Weighted Recursive Least-Squares Algorithm   566
13.4 Update Recursion for the Sum of Weighted Error Squares   571
13.5 Example: Single-Weight Adaptive Noise Canceler   572

13.6   Convergence Analysis of the RLS Algorithm   573
13.7   Computer Experiment on Adaptive Equalization   580
13.8   State-Space Formulation of the RLS Problem   583
13.9   Summary and Discussion   587
        Problems   587

Chapter 14   Square-Root Adaptive Filters   589

14.1   Square-Root Kalman Filters   589
14.2   Building Square-Root Adaptive Filtering Algorithms on their Kalman Filter
        Counterparts   597
14.3   QR–RLS Algorithm   598
14.4   Extended QR–RLS Algorithm   614
14.5   Adaptive Beamforming   617
14.6   Inverse QR–RLS Algorithm   624
14.7   Summary and Discussion   627
        Problems   628

Chapter 15   Order-Recursive Adaptive Filters   630

15.1   Adaptive Forward Linear Prediction   631
15.2   Adaptive Backward Linear Prediction   634
15.3   Conversion Factor   636
15.4   Least-Squares Lattice Predictor   640
15.5   Angle-Normalized Estimation Errors   653
15.6   First-Order State-Space Models for Lattice Filtering   655
15.7   QR-Decomposition-Based Least-Squares Lattice Filters   660
15.8   Fundamental Properties of the QRD–LSL Filter   667
15.9   Computer Experiment on Adaptive Equalization   672
15.10   Extended QRD–LSL Algorithm   677
15.11   Recursive Least-Squares Lattice Filters Using *A Posteriori* Estimation Errors   679
15.12   Recursive LSL Filters Using *A Priori* Estimation Errors with Error Feedback   683
15.13   Computation of the Least-Squares Weight Vector   686
15.14   Computer Experiment on Adaptive Prediction   691
15.15   Other Variants of Least-Squares Lattice Filters   693
15.16   Summary and Discussion   694
        Problems   696

Chapter 16   Tracking of Time-Varying Systems   701

16.1   Markov Model for System Identification   702
16.2   Degree of Nonstationarity   705
16.3   Criteria for Tracking Assessment   706
16.4   Tracking Performance of the LMS Algorithm   708
16.5   Tracking Performance of the RLS Algorithm   711
16.6   Comparison of the Tracking Performance of LMS and RLS Algorithms   716
16.7   Adaptive Recovery of a Chirped Sinusoid in Noise   719
16.8   How to Improve the Tracking Behavior of the RLS Algorithm   726
16.9   Computer Experiment on System Identification   729
16.10   Automatic Tuning of Adaptation Constants   731
16.11   Summary and Discussion   736
        Problems   737

Chapter 17   Finite-Precision Effects   738

17.1   Quantization Errors   739
17.2   Least-Mean-Square Algorithm   741
17.3   Recursive Least-Squares Algorithm   751
17.4   Square-Root Adaptive Filters   757
17.5   Order-Recursive Adaptive Filters   760
17.6   Fast Transversal Filters   763
17.7   Summary and Discussion   767
        Problems   769

PART 4   NONLINEAR ADAPTIVE FILTERING   771

Chapter 18   Blind Deconvolution   772

18.1   Theoretical and Practical Considerations   773
18.2   Bussgang Algorithm for Blind Equalization of Real Baseband Channels   776
18.3   Extension of Bussgang Algorithms to Complex Baseband Channels   791
18.4   Special Cases of the Bussgang Algorithm   792
18.5   Blind Channel Identification and Equalization Using Polyspectra   796
18.6   Advantages and Disadvantages of HOS-Based Deconvolution Algorithms   802
18.7   Channel Identifiability Using Cyclostationary Statistics   803
18.8   Subspace Decomposition for Fractionally-Spaced Blind Identification   804
18.9   Summary and Discussion   813
        Problems   814

Chapter 19   Back-Propagation Learning   817

19.1   Models of a Neuron   818
19.2   Multilayer Perceptron   822
19.3   Complex Back-Propagation Algorithm   824
19.4   Back-Propagation Algorithm for Real Parameters   837
19.5   Universal Approximation Theorem   838
19.6   Network Complexity   840
19.7   Filtering Applications   842
19.8   Summary and Discussion   852
        Problems   854

Chapter 20   Radial Basis Function Networks   855

20.1   Structure of RBF Networks   856
20.2   Radial-Basis Functions   858
20.3   Fixed Centers Selected at Random   859
20.4   Recursive Hybrid Learning Procedure   862
20.5   Stochastic Gradient Approach   863
20.6   Universal Approximation Theorem (Revisited)   865
20.7   Filtering Applications   866
20.8   Summary and Discussion   871
        Problems   873

Appendix A Complex Variables   875
Appendix B Differentiation with Respect to a Vector   890
Appendix C Method of Lagrange Multipliers   895

Appendix D Estimation Theory 899
Appendix E Maximum-Entropy Method 905
Appendix F Minimum-Variance Distortionless Response Spectrum 912
Appendix G Gradient Adaptive Lattice Algorithm 915
Appendix H Solution of the Difference Equation (9.75) 919
Appendix I Steady-State Analysis of the LMS Algorithm without Invoking the Independence Assumption 921
Appendix J The Complex Wishart Distribution 924
Glossary 928
Abbreviations 932
Principal Symbols 933
Bibliography 941
Index 978

Preface

The subject of adaptive filters constitutes an important part of statistical signal processing. Whenever there is a requirement to process signals that result from operation in an environment of unknown statistics, the use of an adaptive filter offers an attractive solution to the problem as it usually provides a significant improvement in performance over the use of a fixed filter designed by conventional methods. Furthermore, the use of adaptive filters provides new signal-processing capabilities that would not be possible otherwise. We thus find that adaptive filters are successfully applied in such diverse fields as communications, control, radar, sonar, seismology, and biomedical engineering.

Aims of the book

The primary aim of this book is to develop the mathematical theory of various realizations of *linear adaptive filters with finite-duration impulse response (FIR)* and do the development in a unified manner wherever possible. There is no unique solution to the linear adaptive filtering problem. Rather, we have a "kit of tools" represented by a variety of recursive algorithms, each of which offers desirable features of its own. This book provides such a kit.

Another aim of the book is to provide an introductory treatment of supervised *neural networks*, where the emphasis is on learning. Neural networks represent an emerging technology with a great deal of potential for solving difficult nonlinear adaptive filtering problems.

Organization of the book

The book begins with an introductory chapter, where the operation of adaptive filters and their practical applications are discussed in general terms. The chapter ends with *histori-*

cal notes, which are included to provide a source of motivation for the interested reader to plough through the rich history of the subject. The concepts and algorithms introduced in this chapter are explained in detail in subsequent parts of the book.

The remaining 20 chapters of the book are organized in four parts, as described here:

- *Background material*, consisting of Chapters 1 through 4. This part reviews fundamentals of discrete-time signal processing, characterization of discrete-time stochastic processes in both time and frequency domains, and eigenanalysis. The background material so provided helps the reader to develop a deep understanding of what adaptive filters are all about.
- *Linear optimum filters*, consisting of Chapters 5 through 7. Specifically, the basic theory of Wiener filters, linear prediction, and Kalman filters is developed in detail. Wiener filters and Kalman filters, in their own individual ways, provide the framework for the formulation of linear adaptive filters.
- *Linear adaptive filters*, consisting of Chapters 8 through 17. This third part, by far the largest in the book, presents a detailed treatment of the two important families of linear adaptive FIR filters:
  1. *Least mean-square (LMS) algorithms*, the time-domain and frequence-domain versions of which are covered in Chapters 9 and 10, respectively. The method of steepest descent, related to the Wiener filter and from which the standard LMS algorithm is derived, is discussed in Chapter 8.
  2. *Recursive-least-squares (RLS) algorithms*, the formulation of which is presented in a "unified" manner under the umbrella of Kalman filtering. Specifically, the standard, square-root, and order-recursive forms of the RLS algorithm are discussed in Chapters 13, 14, and 15, respectively. Other related issues, namely, the method of linear least squares for block least-squares filtering and unitary rotations and reflections, are covered in Chapters 11 and 12, respectively. This part of the book finishes with Chapter 16 on finite-precision effects and Chapter 17 on the tracking of linear time-varying systems.
- *Nonlinear adaptive filters,* covered in Chapters 18 through 20. In particular, Chapter 18 discusses the *blind deconvolution* problems and how it can be solved by nonlinear modifications to conventional adaptive filtering algorithms; the use of cyclostationarity for solving the blind equalization problem is also covered here. The remaining two chapters are devoted to *multilayer feedforward neural networks*, with Chapter 19 discussing the multilayer perceptron (MLP) trained with the back-propagation algorithm, and Chapter 20 discussing radial-basis function (RBF) networks.

Ancillary material

- A total of ten appendices are included on a variety of topics, to provide supporting material for the book.
- A Glossary is included, consisting of a list of definitions, notations and conventions, a list of abbreviations, and a list of principal symbols used in the book.

- All publications referred to in the text are compiled in the Bibliography. Each reference is identified in the text by the name(s) of the author(s) and the year of publication. The Bibliography also includes many other references that have been added for completeness.

Computer Experiments

The book includes many computer experiments that have been developed to illustrate the underlying theory and applications of the LMS and RLS algorithms. These experiments help the reader to compare the performances of different members of these two families of linear adaptive filtering algorithms.

The reader is invited to verify the results of the computer experiments and use them as a basis for further exploration as he or she sees fit.

Problems

Each chapter of the book, except for the introductory chapter, ends with problems that are designed to do two things:

- Help the reader to develop a deeper understanding of the material covered in the chapter
- Challenge the reader to extend some aspects of the theory discussed in the chapter

Manual

The book has an accompanying *manual* composed of two parts:

- The software for all the computer experiments described in this book has been written, using MATLAB. Part I presents the MATLAB code used for this purpose.
- Part II presents detailed solutions to all the problems at the end of Chapters 1 through 20 of the book.

A copy of the Manual can be obtained by instructors who have adopted the book for classroom use by writing directly to the publisher.

The book is written at a level suitable for use in graduate courses on adaptive signal processing. In this context, it is noteworthy that the organization of the material covered in the book offers a great deal of flexibility in the selection of a suitable list of topics for such a graduate course. It is hoped that the book will also be useful to engineers in industry working on problems relating to the theory and applications of adaptive filters.

Simon Haykin

Acknowledgments

I am truly indebted to Dr. Ali H. Sayed at the University of California, Santa Barbara, Dr. Barry Van Veen at the University of Wisconsin-Madison, and Dr. James R. Zeidler at the University of California, San Diego, for reading the manuscript at an early stage of development, for their many critical comments, and for helping me to write selected sections of the book. The inputs from anonymous reviewers of the second edition of the book are also appreciated.

Many others have been kind enough to critically read selected chapters/sections of the book. In alphabetical order, they are:

Dr. N. Balakrishnan, McMaster University, Canada
Dr. Hans Butterweck, Eindhoven University of Technology, The Netherlands
Dr. Francoise Beaufays, SRI International, California
Dr. Sandro Bellini, Politenico di Milano, Italy
Dr. Colin F.N. Cowan, Loughborough University of Technology, England
Dr. James Demmel, University of California, Berkeley
Dr. Zhi Ding, Auburn University, Alabama
Dr. Pierre Duhamel, Ecole National Supérieure de Telécommunications, Paris
Dr. William A. Gardner, University of California, Davis
Dr. Ming Gu, University of California, Berkeley
Dr. Dimitrios Hatzinakos, University of Toronto, Canada
Dr. C. Richard Johnson, Jr., Cornell University, New York
Dr. Harold J. Kushner, Brown University, Rhode Island
Dr. James P. LeBlanc, Cornell University, New York
Dr. David Lowe, Aston University, England
Dr. Odile Macchi, Laboratoire de Signaux et Systéms, France
Dr. Roy Mathias, College of William and Mary, Williamsburg, Virginia
Dr. Bernie Mulgrew, University of Edinburgh, Scotland

Dr. Philip Regalia, Institut National des Télécommunications, France

Dr. James P. Reilly, McMaster University, Canada

Dr. Louis Scharf, University of Colorado, Boulder

Dr. William A. Sethares, University of Wisconsin, Madison

Dr. John J. Shynk, University of California, Santa Barbara

Dr. Sanzheng Qiao, McMaster University, Canada

Dr. Andrew Webb, Defense Research Agency, Great Malvern, England

Dr. Bin Yang, Ruhr University Bochum, Germany

Dr. Patrick Yip, McMaster University, Canada

I thank them all for their help.

I am grateful to my graduate student Paul Yee for the computer experiments (using MATLAB) that are included in the book and for reading the two chapters on neural networks. I thank Dr. Paul Wei at the University of California, San Diego, for preparing two figures on tracking included in Chapter 16.

I thank (1) Dr. Geoffrey E. Hinton, University of Toronto, (2) the Institute of Electrical and Electronic Engineers, and (3) Prentice-Hall, for their kind permissions to reproduce certain figures acknowledged in the book.

I am grateful to Dr. Tom Kailath, Technical Editor of the Prentice-Hall Series on Information Systems in which this book is included, for helpful comments on the chapter on Kalman filtering.

I thank my editor Tom Robbins and Sharyn Vitrano at Prentice-Hall for their help in the production of the book.

I am indebted to Brigitte Maier and Peggy Findlay at the Science and Engineering Library, McMaster University, for their help in checking the accuracy of many of the references listed in the bibliography.

Last but by no means least, I am deeply grateful to my Secretary Lola Brooks for typing different versions of the manuscript for the book, and always doing it in a cheerful and meticulous way.

Simon Haykin

Introduction

1. THE FILTERING PROBLEM

The term *filter* is often used to describe a device in the form of a piece of physical hardware or software that is applied to a set of noisy data in order to extract information about a prescribed quantity of interest. The noise may arise from a variety of sources. For example, the data may have been derived by means of noisy sensors or may represent a useful signal component that has been corrupted by transmission through a communication channel. In any event, we may use a filter to perform three basic information-processing tasks:

1. *Filtering*, which means the extraction of information about a quantity of interest at time t by using data measured up to and including time t.

2. *Smoothing*, which differs from filtering in that information about the quantity of interest need not be available at time t, and data measured later than time t can be used in obtaining this information. This means that in the case of smoothing there is a *delay* in producing the result of interest. Since in the smoothing process we are able to use data obtained not only up to time t but also data obtained after time t, we would expect smoothing to be more accurate in some sense than filtering.

3. *Prediction*, which is the forecasting side of information processing. The aim here is to derive information about what the quantity of interest will be like at some time $t + \tau$ in the future, for some $\tau > 0$, by using data measured up to and including time t.

We may classify filters into linear and nonlinear. A filter is said to be *linear* if the filtered, smoothed, or predicted quantity at the output of the device is a *linear function of the observations applied to the filter input*. Otherwise, the filter is *nonlinear*.

1

In the statistical approach to the solution of the *linear filtering problem* as classified above, we assume the availability of certain statistical parameters (i.e., *mean and correlation functions*) of the useful signal and unwanted additive noise, and the requirement is to design a linear filter with the noisy data as input so as to minimize the effects of noise at the filter output according to some statistical criterion. A useful approach to this filter-optimization problem is to minimize the mean-square value of the *error signal* that is defined as the difference between some desired response and the actual filter output. For stationary inputs, the resulting solution is *commonly* known as the *Wiener filter*, which is said to be *optimum in the mean-square sense*. A plot of the mean-square value of the error signal versus the adjustable parameters of a linear filter is referred to as the *error-performance surface*. The minimum point of this surface represents the Wiener solution.

The Wiener filter is inadequate for dealing with situations in which *nonstationarity* of the signal and/or noise is intrinsic to the problem. In such situations, the optimum filter has to assume a *time-varying form*. A highly successful solution to this more difficult problem is found in the *Kalman filter*, a powerful device with a wide variety of engineering applications.

Linear filter theory, encompassing both Wiener and Kalman filters, has been developed fully in the literature for *continuous-time* as well as *discrete-time* signals. However, for technical reasons influenced by the wide availability of digital computers and the ever-increasing use of digital signal-processing devices, we find in practice that the discrete-time representation is often the preferred method. Accordingly, in subsequent chapters, we only consider the discrete-time version of Wiener and Kalman filters. In this method of representation, the input and output signals, as well as the characteristics of the filters themselves, are all defined at discrete instants of time. In any case, a continuous-time signal may always be represented by a *sequence of samples* that are derived by observing the signal at uniformly spaced instants of time. No loss of information is incurred during this conversion process provided, of course, we satisfy the well-known *sampling theorem*, according to which the sampling rate has to be greater than twice the highest frequency component of the continuous-time signal. We may thus represent a continuous-time signal $u(t)$ by the sequence $u(n)$, $n = 0, \pm 1, = \pm 2, \ldots$, where for convenience we have normalized the sampling period to unity, a practice that we follow throughout the book.

2. ADAPTIVE FILTERS

The design of a Wiener filter requires *a priori* information about the statistics of the data to be processed. The filter is optimum only when the statistical characteristics of the input data match the *a priori* information on which the design of the filter is based. When this information is not known completely, however, it may not be possible to design the Wiener filter or else the design may no longer be optimum. A straightforward approach that we may use in such situations is the "estimate and plug" procedure. This is a two-stage process whereby the filter first "estimates" the statistical parameters of the relevant signals and then "plugs" the results so obtained into a *nonrecursive* formula for computing the filter parameters. For *real-time* operation, this procedure has the disadvantage of requiring excessively elaborate and costly hardware. A more efficient method is to use an *adaptive filter*. By such a device we mean one that is *self-designing* in that the adaptive filter relies for its operation on a *recursive algorithm*, which makes it possible for the filter to perform satisfactorily in an environment where complete knowledge of the relevant signal characteristics is not available. The algorithm starts from some predetermined set of *initial conditions*, representing whatever we know about the environment. Yet, in a stationary environment, we find that after successive iterations of the algorithm it *converges* to the optimum Wiener solution in some statistical sense. In a nonstationary environment, the algorithm offers a *tracking* capability, in that it can track time variations in the statistics of the input data, provided that the variations are sufficiently slow.

As a direct consequence of the application of a recursive algorithm whereby the parameters of an adaptive filter are updated from one iteration to the next, the parameters become *data dependent*. This, therefore, means that an adaptive filter is in reality a *nonlinear device, in the sense that it does not obey the principle of superposition*. Notwithstanding this property, adaptive filters are commonly classified as linear or nonlinear. An adaptive filter is said to be *linear* if the estimate of a quantity of interest is computed adaptively (at the output of the filter) as a *linear combination of the available set of observations applied to the filter input*. Otherwise, the adaptive filter is said to be *nonlinear*.

A wide variety of recursive algorithms have been developed in the literature for the operation of linear adaptive filters. In the final analysis, the choice of one algorithm over another is determined by one or more of the following factors:

- *Rate of convergence.* This is defined as the number of iterations required for the algorithm, in response to stationary inputs, to converge "close enough" to the optimum Wiener solution in the mean-square sense. A fast rate of convergence allows the algorithm to adapt rapidly to a stationary environment of unknown statistics.

- *Misadjustment.* For an algorithm of interest, this parameter provides a quantitative measure of the amount by which the final value of the mean-squared error, averaged over an ensemble of adaptive filters, deviates from the minimum mean-squared error that is produced by the Wiener filter.

- *Tracking.* When an adaptive filtering algorithm operates in a nonstationary environment, the algorithm is required to *track* statistical variations in the environment. The tracking performance of the algorithm, however, is influenced by two contradictory features: (1) rate of convergence, and (b) steady-state fluctuation due to algorithm noise.

- *Robustness.* For an adaptive filter to be *robust*, small disturbances (i.e., disturbances with small energy) can only result in small estimation errors. The disturbances may arise from a variety of factors, internal or external to the filter.

- *Computational requirements.* Here the issues of concern include (a) the number of operations (i.e., multiplications, divisions, and additions/subtractions) required to make one complete iteration of the algorithm, (b) the size of memory locations required to store the data and the program, and (c) the investment required to program the algorithm on a computer.

- *Structure*. This refers to the structure of information flow in the algorithm, determining the manner in which it is implemented in hardware form. For example, an algorithm whose structure exhibits high modularity, parallelism, or concurrency is well suited for implementation using very large-scale integration (VLSI).[1]

- *Numerical properties*. When an algorithm is implemented numerically, inaccuracies are produced due to *quantization errors*. The quantization errors are due to analog-to-digital conversion of the input data and digital representation of internal calculations. Ordinarily, it is the latter source of quantization errors that poses a serious design problem. In particular, there are two basic issues of concern: numerical stability and numerical accuracy. *Numerical stability* is an inherent characteristic of an adaptive filtering algorithm. *Numerical accuracy*, on the other hand, is determined by the number of *bits* (i.e., *bi*nary digi*ts*) used in the numerical representation of data samples and filter coefficients. An adaptive filtering algorithm is said to be numerically robust when it is insensitive to variations in the wordlength used in its digital implementation.

These factors, in their own ways, also enter into the design of nonlinear adaptive filters, except for the fact that we now no longer have a well-defined frame of reference in the form of a Wiener filter. Rather, we speak of a nonlinear filtering algorithm that may converge to a local minimum or, hopefully, a global minimum on the error-performance surface.

In the sections that follow, we shall first discuss various aspects of linear adaptive filters. Discussion of nonlinear adaptive filters is deferred to a later section in the chapter.

3. LINEAR FILTER STRUCTURES

The operation of a linear adaptive filtering algorithm involves two basic processes: (1) a *filtering* process designed to produce an output in response to a sequence of input data, and (2) an *adaptive* process, the purpose of which is to provide a mechanism for the *adaptive control* of an *adjustable* set of parameters used in the filtering process. These two processes work interactively with each other. Naturally, the choice of a structure for the filtering process has a profound effect on the operation of the algorithm as a whole.

---

[1]VLSI technology favors the implementation of algorithms that possess high modularity, parallelism, or concurrency. We say that a structure is *modular* when it consists of similar stages connected in cascade. By *parallelism* we mean a large number of operations being performed side by side. By *concurrency* we mean a large number of *similar* computations being performed at the same time.

For a discussion of VLSI implementation of adaptive filters, see Shanbhag and Parhi (1994). This book emphasizes the use of *pipelining*, an architectural technique used for increasing the throughput of an adaptive filtering algorithm.

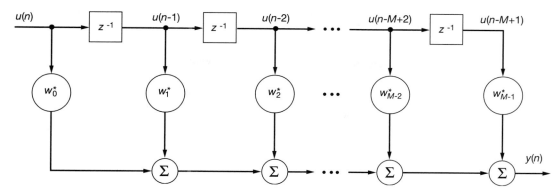

Figure 1   Transversal filter.

There are three types of filter structures that distinguish themselves in the context of an adaptive filter with *finite memory* or, equivalently, *finite-duration impulse response*. The three filter structures are as follows:

1. *Transversal filter.* The *transversal filter,*[2] also referred to as a *tapped-delay line filter*, consists of three basic elements, as depicted in Fig. 1: (a) *unit-delay element,* (b) *multiplier,* and (c) *adder.* The number of delay elements used in the filter determines the finite duration of its impulse response. The number of delay elements, shown as $M - 1$ in Fig. 1, is commonly referred to as the *filter order.* In this figure, the delay elements are each identified by the *unit-delay operator* z^{-1}. In particular, when z^{-1} operates on the input $u(n)$, the resulting output is $u(n - 1)$. The role of each multiplier in the filter is to multiply the *tap input* (to which it is connected) by a filter coefficient referred to as a *tap weight.* Thus a multiplier connected to the kth tap input $u(n - k)$ produces the scalar version of the *inner product,* $w_k^* u(n - k)$, where w_k is the respective tap weight and $k = 0, 1, \ldots, M - 1$. The asterisk denotes *complex conjugation,* which assumes that the tap inputs and therefore the tap weights are all *complex valued.* The combined role of the adders in the filter is to sum the individual multiplier outputs and produce an overall filter output. For the transversal filter described in Fig. 1, the filter output is given by

$$y(n) = \sum_{k=0}^{m-1} w_k^* u(n-k) \tag{1}$$

[2]The transversal filter was first described by Kallmann as a continuous-time device whose output is formed as a linear combination of voltages taken from uniformly spaced taps in a nondispersive delay line (Kallmann, 1940). In recent years, the transversal filter has been implemented using digital circuitry, charge-coupled devices, or surface-acoustic wave devices. Owing to its versatility and ease of implementation, the transversal filter has emerged as an essential signal-processing structure in a wide variety of applications.

Equation (1) is called a finite *convolution sum* in the sense that it *convolves* the finite-duration impulse response of the filter, w_n^*, with the filter input $u(n)$ to produce the filter output $y(n)$.

2. *Lattice predictor.* A *lattice predictor*[3] is *modular* in structure in that it consists of a number of individual stages, each of which has the appearance of a lattice, hence the name "lattice" as a structural descriptor. Figure 2 depicts a lattice predictor consisting of $M - 1$ stages; the number $M - 1$ is referred to as the *predictor order*. The mth stage of the lattice predictor in Fig. 2 is described by the pair of input–output relations (assuming the use of complex-valued, wide-sense stationary input data):

$$f_m(n) = f_{m-1}(n) + \kappa_m^* b_{m-1}(n-1) \tag{2}$$

$$b_m(n) = b_{m-1}(n-1) + \kappa_m f_{m-1}(n) \tag{3}$$

where $m = 1, 2, \ldots, M - 1$, and $M - 1$ is the *final* predictor order. The variable $f_m(n)$ is the mth *forward prediction error*, and $b_m(n)$ is the mth *backward prediction error*. The coefficient κ_m is called the mth *reflection coefficient*. The forward prediction error $f_m(n)$ is defined as the difference between the input $u(n)$ and its *one-step predicted* value; the latter is based on the set of m past inputs $u(n-1), \ldots, u(n-m)$. Correspondingly, the backward prediction error $b_m(n)$ is defined as the difference between the input $u(n-m)$ and its "backward" prediction based on the set of m "future" inputs $u(n), \ldots, u(n-m+1)$. Considering the conditions at the input of stage 1 in Fig. 2, we have

$$f_0(n) = b_0(n) = u(n) \tag{4}$$

where $u(n)$ is the lattice predictor input at time n. Thus, starting with the *initial conditions* of Eq. (4) and given the set of reflection coefficients $\kappa_1, \kappa_2, \ldots, \kappa_{M-1}$, we may determine the final pair of outputs $f_{M-1}(n)$ and $b_{M-1}(n)$ by moving through the lattice predictor, stage by stage.

For a *correlated* input sequence $u(n), u(n-1), \ldots, u(n-M+1)$ drawn from a stationary process, the backward prediction errors $b_0, b_1(n), \ldots, b_{M-1}(n)$ form a sequence of *uncorrelated* random variables. Moreover, there is a one-to-one correspondence between these two sequences of random variables in the sense that if we are given one of them, we may uniquely determine the other, and vice versa. Accordingly, a linear combination of the backward prediction errors $b_0(n), b_1(n), \ldots, b_{M-1}(n)$ may be used to provide an *estimate* of some desired response $d(n)$, as depicted in the lower half of Fig. 2. The arithmetic difference between $d(n)$ and the estimate so produced represents the estimation error $e(n)$. The process described herein is referred to as a *joint-process estimation*. Naturally, we may use the original input sequence $u(n), u(n-1), \ldots, u(n-M+1)$ to produce an estimate of the desired response $d(n)$ directly. The indirect method depicted in Fig. 2, however, has the advantage of simplifying the computation of the tap weights $h_0, h_1, \ldots, h_{M-1}$

---

[3]The development of the lattice predictor is credited to Itakura and Saito (1972).

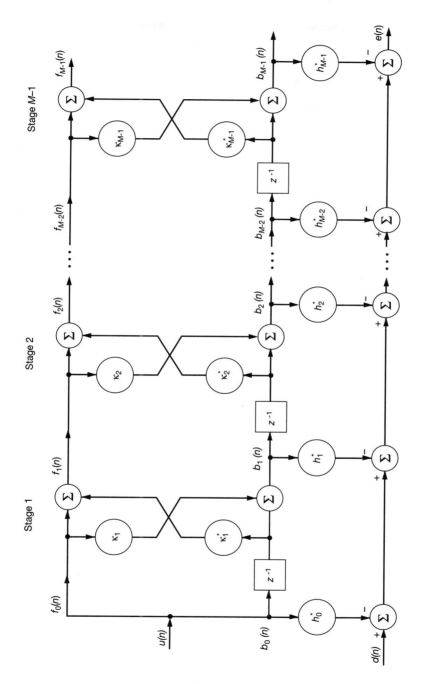

Figure 2 Multistage lattice filter.

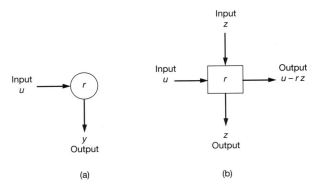

Figure 3 Two basic cells of a systolic array; (a) boundary cell; (b) internal cell.

by exploiting the uncorrelated nature of the corresponding backward prediction errors used in the estimation.

3. *Systolic array.* A *systolic array*[4] represents a *parallel computing* network ideally suited for *mapping* a number of important linear algebra computations, such as *matrix multiplication, triangularization,* and *back substitution.* Two basic types of processing elements may be distinguished in a systolic array: *boundary cells* and *internal cells.* Their functions are depicted in Figs. 3(a) and 3(b), respectively. In each case, the parameter r represents a value *stored* within the cell. The function of the boundary cell is to produce an output equal to the input u divided by the number r stored in the cell. The function of the internal cell is twofold: (a) to multiply the input z (coming in from the top) by the number r stored in the cell, subtract the product rz from the second input (coming in from the left), and thereby produce the difference $u - rz$ as an output from the right-hand side of the cell, and (b) to transmit the first input z downward without alteration.

Consider, for example, the 3-by-3 triangular array shown in Fig. 4. This systolic array involves a combination of boundary and internal cells. In this case, the triangular array computes an output vector $\mathbf{y}$ related to the input vector $\mathbf{u}$ as follows:

$$\mathbf{y} = \mathbf{R}^{-T}\mathbf{u} \tag{5}$$

where the $\mathbf{R}^{-T}$ is the *inverse* of the transposed matrix $\mathbf{R}^{T}$. The elements of $\mathbf{R}^{T}$ are the respective cell contents of the triangular array. The zeros added to the inputs of the array in Fig. 4 are intended to provide the delays necessary for pipelining the computation described in Eq. (5).

A systolic array architecture, as described herein, offers the desirable features of *modularity, local interconnections,* and highly *pipelined* and *synchronized* parallel processing; the synchronization is achieved by means of a global *clock.*

We note that the transversal filter of Fig. 1, the joint-process estimator of Fig. 2 based on a lattice predictor, and the triangular systolic array of Fig. 4 have a common

---

[4]The systolic array was pioneered by Kung and Leiserson (1978). In particular, the use of systolic arrays has made it possible to achieve a high throughput, which is required for many advanced signal processing algorithms to operate in *real time.*

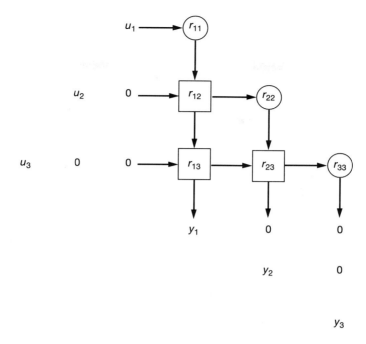

Figure 4 Triangular systolic array.

property: all three of them are characterized by an impulse response of finite duration. In other words, they are examples of a *finite-duration impulse response (FIR) filter*, whose structures contain *feedforward* paths only. On the other hand, the filter structure shown in Fig. 5 is an example of an *infinite-duration impulse response (IIR) filter*. The feature that distinguishes an IIR filter from an FIR filter is the inclusion of *feedback* paths. Indeed, it is the presence of feedback that makes the duration of the impulse response of an IIR filter infinitely long. Furthermore, the presence of feedback introduces a new problem, namely, that of *stability*. In particular, it is possible for an IIR filter to become unstable (i.e., break into oscillation), unless special precaution is taken in the choice of feedback coefficients. By contrast, an FIR filter is inherently *stable*. This explains the reason for the popular use of FIR filters, in one form or another, as the structural basis for the design of linear adaptive filters.

4. APPROACHES TO THE DEVELOPMENT OF LINEAR ADAPTIVE FILTERING ALGORITHMS

There is no unique solution to the linear adaptive filtering problem. Rather, we have a "kit of tools" represented by a variety of recursive algorithms, each of which offers desirable features of its own. The challenge facing the user of adaptive filtering is, first, to understand the capabilities and limitations of various adaptive filtering algorithms and, second,

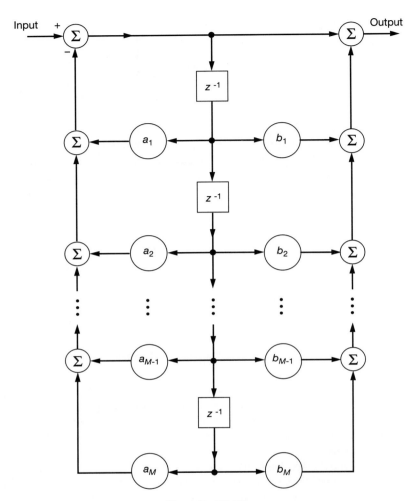

Figure 5   IIR filter.

to use this understanding in the selection of the appropriate algorithm for the application at hand.

Basically, we may identify two distinct approaches for deriving recursive algorithms for the operation of linear adaptive filters, as discussed next.

Stochastic Gradient Approach

Here we may use a tapped-delay line or transversal filter as the structural basis for implementing the linear adaptive filter. For the case of stationary inputs, the *cost function*,[5] also referred to as the *index of performance*, is defined as the *mean-squared error* (i.e., the mean-square value of the difference between the desired response and the transversal filter output). This cost function is precisely a second-order function of the tap weights in the transversal filter. The dependence of the mean-squared error on the unknown tap weights may be viewed to be in the form of a *multidimensional paraboloid* (i.e., punch bowl) with a uniquely defined bottom or *minimum point*. As mentioned previously, we refer to this paraboloid as the *error-performance surface*; the tap weights corresponding to the minimum point of the surface define the optimum Wiener solution.

To develop a recursive algorithm for updating the tap weights of the adaptive transversal filter, we proceed in two stages. We first modify the system of *Wiener—Hopf equations* (i.e., the matrix equation defining the optimum Wiener solution) through the use of the *method of steepest descent*, a well-known technique in optimization theory. This modification requires the use of a *gradient vector*, the value of which depends on two parameters: the *correlation matrix* of the tap inputs in the transversal filter, and the *cross-correlation vector* between the desired response and the same tap inputs. Next, we use instantaneous values for these correlations so as to derive an *estimate* for the gradient vector, making it assume a *stochastic* character in general. The resulting algorithm is widely known as the *least-mean-square (LMS) algorithm*, the essence of which may be described in words as follows for the case of a transversal filter operating on real-valued data:

$$\begin{pmatrix} \text{updated value} \\ \text{of tap-weight} \\ \text{vector} \end{pmatrix} = \begin{pmatrix} \text{old value} \\ \text{of tap-weight} \\ \text{vector} \end{pmatrix} + \begin{pmatrix} \text{learning-} \\ \text{rate} \\ \text{parameter} \end{pmatrix} \begin{pmatrix} \text{tap-} \\ \text{input} \\ \text{vector} \end{pmatrix} \begin{pmatrix} \text{error} \\ \text{signal} \end{pmatrix}$$

where the error signal is defined as the difference between some desired response and the actual response of the transversal filter produced by the tap-input vector.

The LMS algorithm is simple and yet capable of achieving satisfactory performance under the right conditions. Its major limitations are a relatively slow rate of convergence and a sensitivity to variations in the condition number of the correlation matrix of the tap inputs; the *condition number* of a Hermitian matrix is defined as the ratio of its largest

---

[5]In the general definition of a function, we speak of a transformation from a vector space into the space of real (or complex) scalars (Luenberger, 1969; Dorny, 1975). A cost function provides a quantitative measure for assessing the quality of performance; hence the restriction of it to a real scalar.

eigenvalue to its smallest eigenvalue. Nevertheless, the LMS algorithm is highly popular and widely used in a variety of applications.

In a nonstationary environment, the orientation of the error-performance surface varies continuously with time. In this case, the LMS algorithm has the added task of continually *tracking* the bottom of the error-performance surface. Indeed, tracking will occur provided that the input data vary slowly compared to the *learning rate* of the LMS algorithm.

The stochastic gradient approach may also be pursued in the context of a lattice structure. The resulting adaptive filtering algorithm is called the *gradient adaptive lattice (GAL) algorithm*. In their own individual ways, the LMS and GAL algorithms are just two members of the *stochastic gradient family* of linear adaptive filters, although it must be said that the LMS algorithm is by far the most popular member of this family.

Least-squares Estimation

The second approach to the development of linear adaptive filtering algorithms is based on the *method of least squares*. According to this method we minimize a cost function or index of performance that is defined as the *sum of weighted error squares*, where the *error* or *residual* is itself defined as the difference between some desired response and the actual filter output. The method of least squares may be formulated with *block estimation* or *recursive estimation* in mind. In block estimation the input data stream is arranged in the form of blocks of equal length (duration), and the filtering of input data proceeds on a block-by-block basis. In recursive estimation, on the other hand, the estimates of interest (e.g., tap weights of a transversal filter) are *updated* on a sample-by-sample basis. Ordinarily, a recursive estimator requires less storage than a block estimator, which is the reason for its much wider use in practice.

Recursive least-squares (RLS) estimation may be viewed as a special case of Kalman filtering. A distinguishing feature of the Kalman filter is the notion of *state*, which provides a measure of all the inputs applied to the filter up to a specific instant of time. Thus, at the heart of the Kalman filtering algorithm we have a recursion that may be described in words as follows:

$$\begin{pmatrix} \text{updated value} \\ \text{of the} \\ \text{state} \end{pmatrix} = \begin{pmatrix} \text{old value} \\ \text{of the} \\ \text{state} \end{pmatrix} + \begin{pmatrix} \text{Kalman} \\ \text{gain} \end{pmatrix} \begin{pmatrix} \text{innovation} \\ \text{vector} \end{pmatrix}$$

where the *innovation vector* represents new information put into the filtering process at the time of the computation. For the present, it suffices to say that there is indeed a one-to-one correspondence between the Kalman variables and RLS variables. This correspondence means that we can tap the vast literature on Kalman filters for the design of linear adaptive filters based on recursive least-squares estimation. Moreover, we may classify the *recursive least-squares family* of linear adaptive filtering algorithms into three distinct categories, depending on the approach taken:

1. *Standard RLS algorithm*, which assumes the use of a transversal filter as the structural basis of the linear adaptive filter. Derivation of the standard RLS algorithm relies on a basic result in linear algebra known as the *matrix inversion lemma*. Most importantly, it enjoys the same virtues and suffers from the same limitations as the standard Kalman filtering algorithm. The limitations include lack of numerical robustness and excessive computational complexity. Indeed, it is these two limitations that have prompted the development of the other two categories of RLS algorithms, described next.

2. *Square-root RLS algorithms*, which are based on *QR-decomposition* of the incoming data matrix. Two well-known techniques for performing this decomposition are the *Householder transformation* and the *Givens rotation*, both of which are data-adaptive transformations. At this point in the discussion, we need to merely say that RLS algorithms based on the Householder transformation or Givens rotation are numerically stable and robust. The resulting linear adaptive filters are referred to as *square-root adaptive filters*, because in a matrix sense they represent the square-foot forms of the standard RLS algorithm.

3. *Fast RLS algorithms*. The standard RLS algorithm and square-root RLS algorithms have a computational complexity that increases as the square of M, where M is the number of adjustable weights (i.e., the number of degrees of freedom) in the algorithm. Such algorithms are often referred to as $O(M^2)$ *algorithms*, where $O(\cdot)$ denotes "order of." By contrast, the LMS algorithm is an $O(M)$ algorithm, in that its computational complexity increases linearly with M. When M is large, the computational complexity of $O(M^2)$ algorithms may become objectionable from a hardware implementation point of view. There is therefore a strong motivation to modify the formulation of the RLS algorithm in such a way that the computational complexity assumes an $O(M)$ form. This objective is indeed achievable, in the case of temporal processing, first by virtue of the inherent *redundancy* in the *Toeplitz structure* of the input data matrix and, second, by exploiting this redundancy through the use of *linear least-squares prediction in both the forward and backward directions*. The resulting algorithms are known collectively as *fast RLS algorithms*; they combine the desirable characteristics of recursive linear least-squares estimation with an $O(M)$ computational complexity. Two types of fast RLS algorithms may be identified, depending on the filtering structure employed:

   • *Order-recursive adaptive filters*, which are based on a latticelike structure for making linear forward and backward predictions.
   • *Fast transversal filters*, in which the linear forward and backward predictions are performed using separate transversal filters.

Certain (but not all) realizations of order-recursive adaptive filters are known to be numerically stable, whereas fast transversal filters suffer from a numerical stability problem and therefore require some form of stabilization for them to be of practical use.

An introductory discussion of linear adaptive filters would be incomplete without saying something about their tracking behavior. In this context, we note that stochastic gradient algorithms such as the LMS algorithm are *model-independent*; generally speaking, we would expect them to exhibit good tracking behavior, which indeed they do. In contrast, RLS algorithms are *model-dependent*; this, in turn, means that their tracking behavior may be inferior to that of a member of the stochastic gradient family, unless care is taken to minimize the mismatch between the mathematical model on which they are based and the underlying physical process responsible for generating the input data.

How to Choose an Adaptive Filter

Given the wide variety of adaptive filters available to a system designer, how can a choice be made for an application of interest? Clearly, whatever the choice, it has to be *cost-effective*. With this goal in mind, we may identify three important issues that require attention: *computational cost, performance*, and *robustness*. The use of computer simulation provides a good first step in undertaking a detailed investigation of these issues. We may begin by using the LMS algorithm as an adaptive filtering tool for the study. The LMS algorithm is relatively simple to implement. Yet it is powerful enough to evaluate the practical benefits that may result from the application of adaptivity to the problem at hand. Moreover, it provides a practical frame of reference for assessing any further improvement that may be attained through the use of more sophisticated adaptive filtering algorithms. Finally, the study must include tests with real-life data, for which there is no substitute.

Practical applications of adaptive filtering are very diverse, with each application having peculiarities of its own. The solution for one application may not be suitable for another. Nevertheless, to be successful we have to develop a physical understanding of the environment in which the filter has to operate and thereby relate to the realities of the application of interest.

5. REAL AND COMPLEX FORMS OF ADAPTIVE FILTERS

In the development of adaptive filtering algorithms, regardless of their origin, it is customary to assume that the input data are in baseband form. The term "baseband" is used to designate the band of frequencies representing the original (message) signal as generated by the source of information.

In such applications as communications, radar, and sonar, the information-bearing signal component of the receiver input typically consists of a message signal *modulated* onto a carrier wave. The bandwidth of the message signal is usually small compared to the carrier frequency, which means that the modulated signal is a *narrow-band signal*. To obtain the baseband representation of a narrow-band signal, the signal is translated down in frequency in such a way that the effect of the carrier wave is completely removed, yet the information content of the message signal is fully preserved. In general, the baseband signal so obtained is *complex*. In other words, a sample $u(n)$ of the signal may be written as

$$u(n) = u_I(n) + ju_Q(n) \qquad (6)$$

where $u_I(n)$ is the *in-phase* (real) *component*, and $u_Q(n)$ is the *quadrature* (imaginary) *component*. Equivalently, we may express $u(n)$ as

$$u(n) = |u(n)|e^{j\phi(n)} \qquad (7)$$

where $|u(n)|$ is the *magnitude* and $\phi(n)$ is the *phase angle*.

Accordingly, the theory of adaptive filters (both linear and nonlinear) developed in subsequent chapters of the book assumes the use of complex signals. An adaptive filtering algorithm so developed is said to be in *complex form*. The important virtue of complex adaptive filters is that they preserve the mathematical formulation and elegant structure of complex signals encountered in the aforementioned areas of application.

If the signals to be processed are *real*, we naturally use the *real form* of the adaptive filtering algorithm of interest. Given the complex form of an adaptive filtering algorithm, it is straightforward to deduce the corresponding real form of the algorithm. Specifically, we do two things:

1. The operation of *complex conjugation*, wherever in the algorithm, is simply removed.
2. The operation of *Hermitian transposition* (i.e., conjugate transposition) of a matrix, wherever in the algorithm, is replaced by ordinary transposition.

Simply put, complex adaptive filters include real adaptive filters as special cases.

6. NONLINEAR ADAPTIVE FILTERS

The theory of linear optimum filters is based on the mean-square error criterion. The Wiener filter that results from the minimization of such a criterion, and which represents the goal of linear adaptive filtering for a stationary environment, can only relate to second-order statistics of the input data and no higher. This constraint limits the ability of a linear adaptive filter to extract information from input data that are non-Gaussian. Despite its theoretical importance, the existence of Gaussian noise is open to question (Johnson and Rao, 1990). Moreover, non-Gaussian processes are quite common in many signal processing applications encountered in practice. The use of a Wiener filter or a linear adaptive filter to extract signals of interest in the presence of such non-Gaussian processes will therefore yield suboptimal solutions. We may overcome this limitation by incorporating some form of *nonlinearity* in the structure of the adaptive filter to take care of higher-order statistics. Although by so doing, we no longer have the Wiener filter as a frame of reference and so complicate the mathematical analysis, we would expect to benefit in two significant ways: improving learning efficiency and a broadening of application areas.

Fundamentally, there are two types of nonlinear adaptive filters, as described next.

Volterra-based Nonlinear Adaptive Filters

In this type of a nonlinear adaptive filter, the nonlinearity is localized at the front end of the filter. It relies on the use of a *Volterra* series[6] that provides an attractive method for describing the input–output relationship of a nonlinear device with memory. This special form of a series derives its name from the fact that it was first studied by Vito Volterra around 1880 as a generalization of the Taylor series of a function. But Norbert Wiener (1958) was the first to use the Volterra series to model the input–output relationship of a nonlinear system.

Let the time series x_n denote the input of a nonlinear discrete-time system. We may then combine these input samples to define a set of *discrete Volterra kernels* as follows:

$$H_0 = \text{zero-order (dc) term}$$

$$H_1[x_n] = \text{first-order (linear) term}$$

$$= \sum_i h_i x_i$$

$$H_2[x_n] = \text{second-order (quadratic) term}$$

$$= \sum_i \sum_j h_{ij} x_i x_j$$

$$H_3[x_n] = \text{third-order (cubic) term}$$

$$= \sum_i \sum_j \sum_k h_{ijk} x_i x_j x_k$$

and so on for higher-order terms. Ordinarily, the nonlinear model coefficients, the h's, are fixed by analytical methods. We may thus decompose a nonlinear adaptive filter as follows:[7]

- A *nonlinear Volterra state expander* that combines the set of input values x_0, $x_1, \ldots, x_n$ to produce a larger set of outputs $u_0, u_1, \ldots, u_q$ for which q is larger than n. For example, the extension vector for a (3,2) system has the form

$$\mathbf{u} = [1, x_0, x_1, x_2, x_0^2, x_0 x_1, x_0 x_2, x_1 x_0, x_1^2, x_1 x_2, x_2 x_0, x_2 x_1, x_2^2]^T$$

- A *linear FIR adaptive filter* that operates on the u_k (i.e., elements of $\mathbf{u}$) as inputs to produce an estimate $\hat{d}_n$ of some desired response d_n.

[6]For a discussion of Volterra series, see the book by Schetzen (1981).
[7]The idea described herein is discussed in Rayner and Lynch (1989) and Lynch and Rayner (1989).

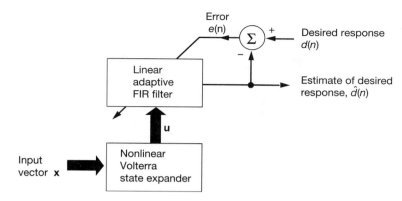

Figure 6 Volterra-based nonlinear adaptive filter.

The important thing to note here is that by using a scheme similar to that described in Fig. 6, we may expand the use of linear adaptive filters to include Volterra filters.

Neural Networks

An *artificial neural network*, or a *neural network* as it is commonly called, consists of the interconnection of a large number of nonlinear processing units called *neurons*; that is, the nonlinearity is distributed throughout the network. The development of neural networks, right from their inception, has been motivated by the way the human brain performs its operations; hence their name.

In this book, we are interested in a particular class of neural networks that *learn* about their environment in a *supervised manner*. In other words, as with the conventional form of a linear adaptive filter, we have a desired response that provides a target signal, which the neural network tries to approximate during the learning process. The approximation is achieved by adjusting a set of free parameters, called *synaptic weights*, in a systematic manner. In effect, the synaptic weights provide a mechanism for storing the information content of the input data.

In the context of adaptive signal processing applications, neural networks offer the following advantages:

- *Nonlinearity*, which makes it possible to account for the nonlinear behavior of physical phenomena responsible for generating the input data
- The ability to *approximate any prescribed input–output mapping* of a continuous nature
- *Weak statistical assumptions* about the environment, in which the network is embedded
- *Learning* capability, which is accomplished by undertaking a training session with input–output examples that are representative of the environment

- *Generalization*, which refers to the ability of the neural network to provide a satisfactory performance in response to *test data* never seen by the network before
- *Fault tolerance*, which means that the network continues to provide an acceptable performance despite the failure of some neurons in the network
- *VLSI implementability*, which exploits the massive parallelism built into the design of a neural network.

This is indeed an impressive list of attributes, which accounts for the widespread interest in the use of neural networks to solve signal-processing tasks that are too difficult for conventional (linear) adaptive filters.

7. APPLICATIONS

The ability of an adaptive filter to operate satisfactorily in an unknown environment and track time variations of input statistics make the adaptive filter a powerful device for signal-processing and control applications. Indeed, adaptive filters have been successfully applied in such diverse fields as communications, radar, sonar, seismology, and biomedical engineering. Although these applications are indeed quite different in nature, nevertheless, they have one basic common feature: an input vector and a desired response are used to compute an estimation error, which is in turn used to control the values of a set of adjustable filter coefficients. The adjustable coefficients may take the form of tap weights, reflection coefficients, rotation parameters, or synaptic weights, depending on the filter structure employed. However, the essential difference between the various applications of adaptive filtering arises in the manner in which the desired response is extracted. In this context, we may distinguish four basic classes of adaptive filtering applications, as depicted in Fig. 7. For convenience of presentation, the following notations are used in this figure:

$$u = \text{input applied to the adaptive filter}$$

$$y = \text{output of the adaptive filter}$$

$$d = \text{desired response}$$

$$e = d - y = \text{estimation error.}$$

The functions of the four basic classes of adaptive filtering applications depicted herein are as follows:

I. *Identification* [Fig. 7(a)]. The notion of a *mathematical model* is fundamental to sciences and engineering. In the class of applications dealing with identification, an adaptive filter is used to provide a linear model that represents the best fit (in some sense) to an *unknown plant*. The plant and the adaptive filter are driven by the same input. The plant output supplies the desired response for the adaptive filter. If the plant is dynamic in nature, the model will be time varying.

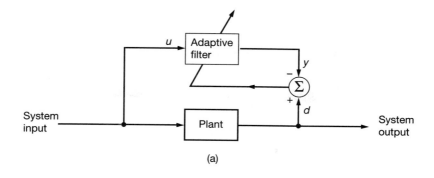

(a)

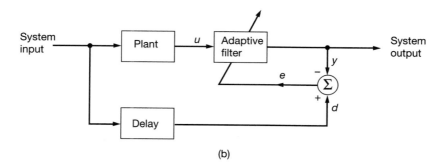

(b)

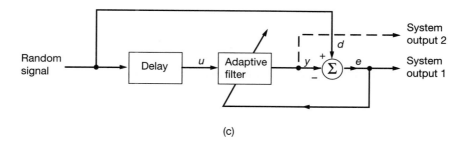

(c)

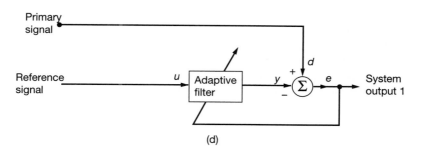

(d)

Figure 7 Four basic classes of adaptive filtering applications: (a) class I: identification;
(b) class II: inverse modeling; (c) class III: prediction; (d) class IV: interference canceling.

Introduction

II. *Inverse modeling* [Fig. 7(b)]. In this second class of applications, the function of the adaptive filter is to provide an *inverse model* that represents the best fit (in some sense) to an *unknown noisy plant*. Ideally, in the case of a linear system, the inverse model has a transfer function equal to the *reciprocal (inverse)* of the plant's transfer function, such that the combination of the two constitutes an ideal transmission medium. A delayed version of the plant (system) input constitutes the desired response for the adaptive filter. In some applications, the plant input is used without delay as the desired response.

III. *Prediction* [Fig. 7(c)]. Here the function of the adaptive filter is to provide the best *prediction* (in some sense) of the present value of a random signal. The present value of the signal thus serves the purpose of a desired response for the adaptive filter. Past values of the signal supply the input applied to the adaptive filter. Depending on the application of interest, the adaptive filter output or the estimation (prediction) error may serve as the system output. In the first case, the system operates as a *predictor*; in the latter case, it operates as a *prediction-error filter*.

IV. *Interference canceling* [Fig. 7(d)]. In this final class of applications, the adaptive filter is used to cancel *unknown interference* contained (alongside an information-bearing signal component) in a *primary signal*, with the cancelation being optimized in some sense. The primary signal serves as the desired response for the adaptive filter. A *reference (auxiliary) signal* is employed as the input to the adaptive filter. The reference signal is derived from a sensor or set of sensors located in relation to the sensor(s) supplying the primary signal in such a way that the information-bearing signal component is weak or essentially undetectable.

In Table 1 we have listed some applications that are illustrative of the four basic classes of adaptive filtering applications. These applications, totaling twelve, are drawn from the fields of control systems, seismology, electrocardiography, communications, and radar. They are described individually in the remainder of this section.

System Identification

System identification is the experimental approach to the modeling of a process or a plant (Goodwin and Payne, 1977; Ljung and Söderström, 1983; Ljung, 1987; Söderström and Stoica, 1988; Åström and Wittenmark, 1990). It involves the following steps: experimental planning, the selection of a model structure, parameter estimation, and model validation. The procedure of system identification, as pursued in practice, is iterative in nature in that we may have to go back and forth between these steps until a satisfactory model is built. Here we discuss briefly the idea of adaptive filtering algorithms for estimating the parameters of an unknown plant modeled as a transversal filter.

Suppose we have an unknown dynamic plant that is linear and time varying. The plant is characterized by a *real-valued* set of discrete-time measurements that describe the

TABLE 1   APPLICATIONS OF ADAPTIVE FILTERS

Class of adaptive filtering	Application
I. Identification	System identification
	Layered earth modeling
II. Inverse modeling	Predictive deconvolution
	Adaptive equalization
	Blind equalization
III. Prediction	Linear predictive coding
	Adaptive differential pulse-code modulation
	Autoregressive spectrum analysis
	Signal detection
IV. Interference canceling	Adaptive noise canceling
	Echo cancelation
	Adaptive beamforming

variation of the plant output in response to a known stationary input. The requirement is to develop an *on-line transversal filter model* for this plant, as illustrated in Fig. 8. The model consists of a finite number of unit-delay elements and a corresponding set of adjustable parameters (tap weights).

Let the available input signal at time n be denoted by the set of samples: $u(n)$, $u(n - 1), \ldots, u(n - M + 1)$, where M is the number of adjustable parameters in the model. This input signal is applied simultaneously to the plant and the model. Let their respective outputs be denoted by $d(n)$ and $y(n)$. The plant output $d(n)$ serves the purpose of a desired response for the adaptive filtering algorithm employed to adjust the model parameters. The model output is given by

$$y(n) = \sum_{k=0}^{M-1} \hat{w}_k(n) u(n - k) \tag{8}$$

where $\hat{w}_0(n)$, $\hat{w}_1(n), \ldots$, and $\hat{w}_{M-1}(n)$ are the estimated model parameters. The model output $y(n)$ is compared with the plant output $d(n)$. The difference between them, $d(n) - y(n)$, defines the *modeling (estimation) error*. Let this error be denoted by $e(n)$.

Typically, at time n, the modeling error $e(n)$ is nonzero, implying that the model deviates from the plant. In an attempt to account for this deviation, the error $e(n)$ is applied to an *adaptive control algorithm*. The samples of the input signal, $u(n)$, $u(n - 1), \ldots$, $u(n - M + 1)$, are also applied to the algorithm. The combination of the transversal filter and the adaptive control algorithm constitutes the adaptive filtering algorithm. The algorithm is designed to control the adjustments made in the values of the model parameters. As a result, the model parameters assume a new set of values for use on the next iteration. Thus, at time $n + 1$, a new model output is computed, and with it a new value for the modeling error. The operation described is then repeated. This process is continued for a sufficiently large number of iterations (starting from time $n = 0$), until the deviation of the

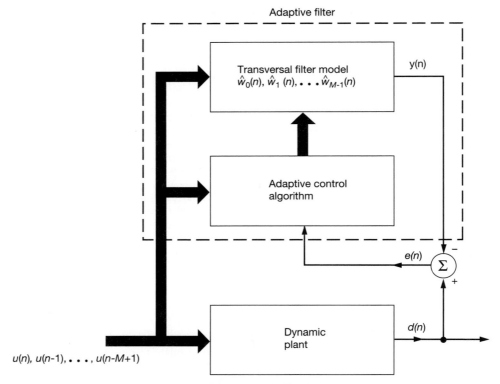

Figure 8 System identification.

model from the plant, measured by the magnitude of the modeling error $e(n)$, becomes sufficiently small in some statistical sense.

When the plant is time varying, the plant output is *nonstationary*, and so is the desired response presented to the adaptive filtering algorithm. In such a situation, the adaptive filtering algorithm has the task of not only keeping the modeling error small but also continually tracking the time variations in the dynamics of the plant.

Layered Earth Modeling

In *exploration seismology*, we usually think of a layered model of the earth (Robinson and Treitel, 1980; Justice, 1985; Mendel, 1986; Robinson and Durrani, 1986). In order to collect (record) seismic data for the purpose of characterizing such a model and thereby unraveling the complexities of the earth's surface, it is customary to use the *method of reflection seismology* that involves the following:

    1. A *source of seismic energy* (e.g., dynamite, air gun) that is typically activated on the surface of the earth.

2. *Propagation* of the seismic signal away from the source and deep into the earth's crust.

3. *Reflection* of seismic waves from the interfaces between the earth's geological layers.

4. *Picking up* and *recording* the seismic returns (i.e., reflections of seismic waves from the interfaces) that carry information about the subsurface structure. On land, *geophones* (consisting of small sensors implanted into the earth) are used to pick up the seismic returns.

The method of reflection seismology, combined with a lot of signal processing, is capable of supplying a two- or three-dimensional "picture" of the earth's subsurface, down to about 20,000 to 30,000 feet and with high enough accuracy and resolution. This picture is then examined by an "interpreter" to see if it is likely that the part of the earth's subsurface (under exploration) contains hydrocarbon (petroleum) reservoirs. Accordingly, a decision is made whether or not to drill a well, which (in the final analysis) is the only way of knowing if petroleum is actually present.

A seismic wave is similar in nature to an acoustic wave, except that the earth permits the propagation of shear waves as well as compressional waves. (In an acoustic medium, only compressional waves are supported.) The earth tends to act like an *elastic medium* for the propagation of seismic waves. The property of elasticity means that a fluid or solid body resists changes in size and shape due to the applications of an external force, and that the body is restored to its original size and shape upon removal of the force. It is this property that permits the propagation of seismic waves through the earth.

An important issue in exploration seismology is the interpretation of seismic returns from the different geological layers of the earth. This interpretation is fundamental to the *identification* of crusted regions such as depth rocks, sand layers, or sedimentary layers. The sedimentary layers are of particular interest because they may contain hydrocarbon reservoirs. The idea of a layered earth model plays a key role here.

The *layered-earth model* is based on the physical fact that seismic-wave motion in each layer is characterized by two components propagating in opposite directions (Robinson and Durrani, 1986). This phenomenon is illustrated in Fig. 9. To understand the interaction between downgoing and upgoing waves, we have reproduced a portion of this diagram in Fig. 10(a), which pertains to the kth interface. The picture shown in Fig. 10(a) is decomposed into two parts, as depicted in Fig. 10(b) and 10(c). We thus observe the following:

- In layer k, there is an *upgoing* wave that consists of the superposition of the reflection of a downgoing wave incident on the kth interface (i.e., boundary) and the transmission of an upgoing (incident) wave from layer $k + 1$.
- In layer $k + 1$, there is a *downgoing* wave that consists of the superposition of the transmission of a downgoing (incident) wave from layer k and the reflection of an upgoing wave incident on the kth interface.

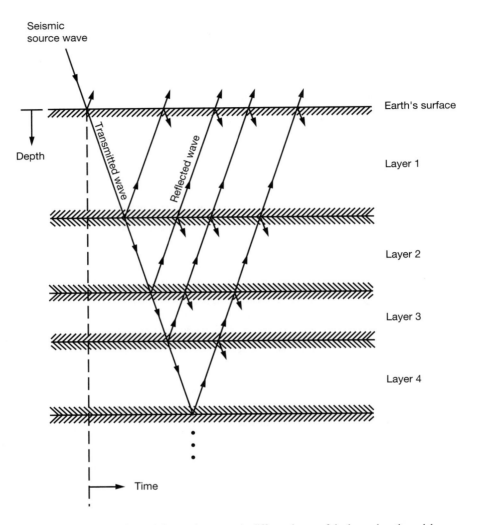

Figure 9 Upgoing and downgoing waves in different layers of the layered earth model.
Note: The layers are unevenly spaced to add a sense of realism.

Lattice Model. Let c_k denote the upward *reflection coefficient* of the kth interface [see Fig. 10(b)]. Let $d_k(n)$ and $u_k(n)$ denote the downgoing and upgoing waves, respectively, at the *top* of layer k, and let $d_k'(n)$ and $u_k'(n)$ denote the downgoing and upgoing waves, respectively, at the *bottom* of layer k, as depicted in Fig. 10(a). The index n denotes discrete time. Ideally, the waves propagate through the medium without distortion, or absorption. Accordingly, we have from Fig. 10(a),

$$d_k'(n) = d_k(n - \tfrac{1}{2}) \tag{9}$$

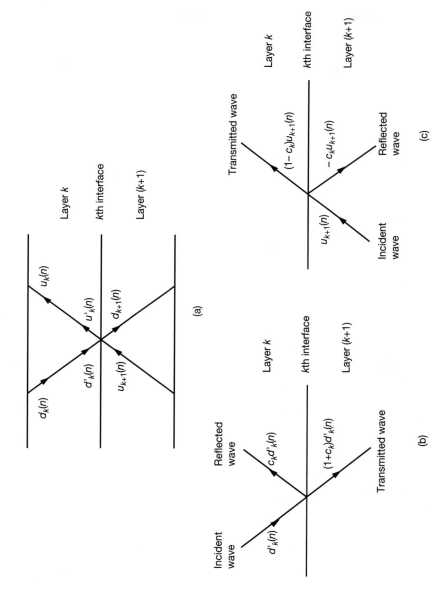

Figure 10 (a) Propagation of seismic waves through a pair of adjacent layers; (b) effects of downgoing incident wave; (c) effects of upgoing incident wave.

and

$$u'_k(n) = u_k(n + \tfrac{1}{2}) \tag{10}$$

where the travel time from the top of a layer to its bottom (or vice versa) is assumed to be one-half of a time unit. The superposition of the pictures depicted in parts (b) and (c) of Fig. 10 and comparison with that of part (a) yields the following interactions between the downgoing and upgoing waves:

$$d_{k+1}(n) = -c_k u_{k+1}(n) + (1 + c_k)d'_k(n) \tag{11}$$

and

$$u'_k(n) = c_k d'_k(n) + (1 - c_k)u_{k+1}(n) \tag{12}$$

The upward *transmission coefficient*[8] of the kth interface is defined by [see Fig. 10(c)]

$$\tau'_k = 1 - c_k \tag{13}$$

Thus, using this definition in Eq. (12), and also using this equation to eliminate $u_{k+1}(n)$ from Eq. (11), we obtain

$$u'_k(n) = c_k d'_k(n) + \tau'_k u_{k+1}(n) \tag{14}$$

and

$$d_{k+1}(n) = \frac{1}{\tau'_k} d'_k(n) - \frac{c_k}{\tau'_k} u'_k(n) \tag{15}$$

Using this pair of equations, we may construct a *lattice model* for layer k, as shown in Fig. 11(a) (Robinson and Durrani, 1986). Moreover, we may extend this idea to develop a *multistage lattice model*, shown in block diagram form in Fig. 11(b), which depicts the propagation of waves through several layers of the medium. The lattice model for each layer has the details given in Fig. 11(a). The combined use of these two figures provides a great deal of physical insight into the interaction of downgoing and upgoing waves as they propagate from one layer to the next.

Examination of Eq. (14) reveals that the evaluation of $u'_k(n)$ at the bottom of layer k requires knowledge of $u_{k+1}(n)$ at the top of layer $k + 1$. But $u_{k+1}(n)$ is not available until the layer $k + 1$ has been dealt with. The lattice model of Fig. 11 is therefore of limited practical use. To overcome this limitation, we may use the z-transform to modify this model. Specifically, applying the z-transform to Eqs. (9), (10), (14), and (15) and manipulating them into matrix form, we get the so-called *scattering equation*:

$$\begin{bmatrix} D_{k+1}(z) \\ U_{k+1}(z) \end{bmatrix} = \frac{z^{1/2}}{\tau'_k} \begin{bmatrix} z^{-1} & -c_k \\ -c_k z^{-1} & 1 \end{bmatrix} \begin{bmatrix} D_k(z) \\ U_k(z) \end{bmatrix} \tag{16}$$

---

[8]The prime in the upward transmission coefficient τ'_k is used to distinguish it from the *downward* transmission coefficient [see Fig. 11(a)], given by

$$\tau_k = 1 + c_k$$

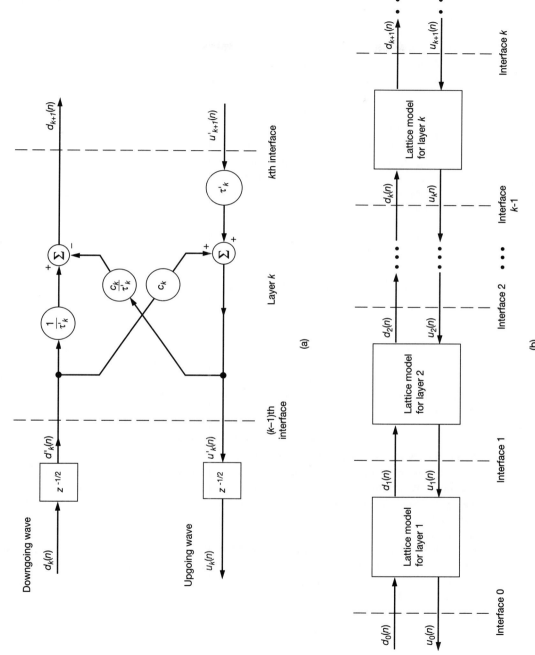

Figure 11 (a) Lattice model for layer k; (b) multistage lattice model of the layered earth model, with each lattice configuration having the form given in part (a).

where z^{-1} is the unit-delay operator. The 2-by-2 matrix on the right-hand side of Eq. (16) is called the *scattering matrix*. Thus, on the basis of Eq. (16), we may construct the *modified lattice model* for layer k, shown in Fig. 12(a) (Robinson and Durrani, 1986). Correspondingly, the multistage version of the modified lattice model is as shown in Fig. 12(b).

The following points are noteworthy in the context of the modified lattice model of Fig. 12 for the propagation of compressional seismic waves in the subsurface of the earth:

1. The lattice structure of the model has *physical* significance, since it follows naturally from the notion of a layered earth.
2. The structure for each layer (state) of the model is *symmetric*.
3. The reciprocal of the transmission coefficient for each layer merely plays the role of a *scaling factor* insofar as input–output relations are concerned. Specifically, for layer k, we may remove $1/\tau_k'$ from the top path of the model in Fig. 12(a) simply by absorbing it in $D_{k+1}(z)$. Similarly, we may remove $1/\tau_k'$ from the bottom path by absorbing it in $U_{k+1}(z)$. Moreover, the values of the transmission coefficients $\tau_1', \tau_2', \ldots, \tau_k', \ldots$ are determined from the respective values of the reflection coefficients $c_1, c_2, \ldots, c_k, \ldots$ by using Eq. (13).
4. The overall model for layers $1, 2, \ldots, k, \ldots$ is *uniquely determined by the sequence of reflection coefficients $c_1, c_2, \ldots, c_k, \ldots$*.
   A case of special interest arises when

$$u_{k+1}(n) = 0, \qquad k \text{ is the deepest layer} \tag{17}$$

This case corresponds to the case when the *final* interface [i.e., the $(k + 1)$th interface] acts as a *perfect absorber*. In other words, there is no outgoing wave from the deepest layer, so Eq. (17) follows. This equation thus represents the *boundary condition* on the lattice model of Fig. 11. The corresponding boundary condition for the modified lattice model of Fig. 12 is

$$U_{k+1}(z) = 0, \quad k \text{ is the deepest layer} \tag{18}$$

Given this boundary condition and the sequence of reflection coefficients $c_1, c_2, \ldots, c_k, \ldots$, we may then use the modified lattice model of Fig. 12(b) (in a stage-by-stage fashion) to determine $U_0(z)$, the z-transform of the output (outgoing) seismic wave $u_0(n)$ at the earth's surface, in terms of $D_0(z)$, the z-transform of the input (downgoing) seismic wave $d_0(n)$.

Tapped-Delay-Line (Transversal Model).　Figure 13 depicts a *tapped-delay-line model* for a layered earth. It provides a local parameterization of the propagation (scattering) phenomenon in the earth's subsurface. According to the alternative model, the input (downgoing) seismic wave $d_0(n)$ and the output (upgoing) seismic wave $u_0(n)$ are, in general, linearly related by the *infinite convolution sum*

$$u_0(n) = \sum_{k=0}^{\infty} w_k d_0(n - k) \tag{19}$$

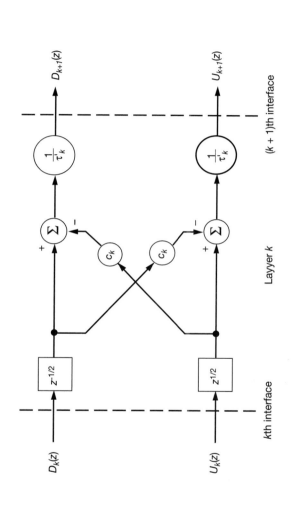

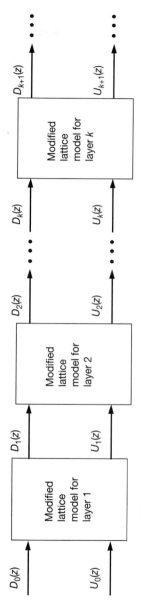

Figure 12 (a) Modified lattice model for layer k; (b) multistage version of modified lattice model, with each stage having the representation shown in part (a).

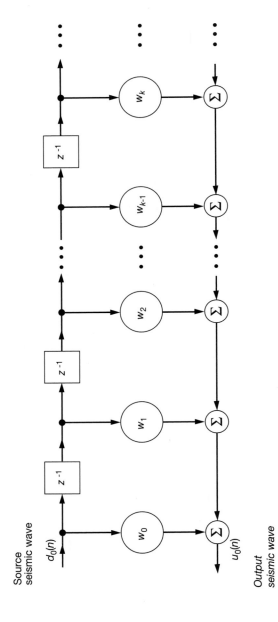

Figure 13 Tapped-delay-line model of layered earth.

where the infinite sequence of tap weights w_n represents the *spatial mapping* of the medium's weighting or the *impulse response* of the medium. Equation (19) states that the output $u_0(n)$ is an infinite series of time-delayed and scaled replicas of the input $d_0(n)$.

There is a *one-to-one correspondence* between the impulse response w_n that characterizes the tapped-delay-line model of Fig. 13 and the sequence of reflection coefficients c_n that characterizes the lattice model of Fig. 11:

$$\{w_n\} \; \rightleftharpoons \; \{c_n\} \tag{20}$$

In other words, given the c_n we may uniquely determine the w_n, and vice versa.

In reflection seismology, the model of Fig. 13 is referred to as the *convolutional model*, in view of the convolution of the impulse response of the medium with the input. This model is the starting point of seismic deconvolution (described in the next application).

Parameter Estimation.[9]   The seismic wave $d_0(n)$ generated by the source of energy acts as a "probing" wave that is transmitted into the earth. Correspondingly, the seismic wave $u_0(n)$ is the output evoked by the propagation of $d_0(n)$ in the earth's subsurface. A *recorded* trace of the output $u_0(n)$ for varying time n is called a *seismogram*. Thus, given digital recordings of the probing wave $d_0(n)$ and the resulting seismogram $u_0(n)$, we may apply an adaptive filtering algorithm to estimate the impulse response w_n of the layered earth. This computation is performed *off-line*, with the probing wave $d_0(n)$ used as input to the adaptive filtering algorithm and the seismogram $u_0(n)$ serving the role of desired response for the algorithm.

Predictive Deconvolution

Convolution is fundamental to the analysis of linear time-invariant systems. Specifically, the output of a linear time-invariant system is the convolution of the input with the impulse response of the system. Convolution is *commutative*. We may therefore also say that the output of the system is the convolution of the impulse response of the system with the input. Moreover, convolution is a linear operation; it therefore holds regardless of the type of signal used as the system input.

Consider the convolutional model for reflection seismology depicted in Fig. 13. We may express the input–output relation of this model simply as

$$u_0(n) = w_n * d_0(n) \tag{21}$$

---

[9]For a survey of different parameter estimation procedures applicable to reflection seismology, see Mendel (1986). This paper also discusses other related issues, namely, *representation* (i.e., how something should be modeled), *measurement* (which physical parameters should be measured and how they should be measured), and *validation* (i.e., demonstrating confidence in the model). For a deterministic approach applicable to reflection seismology, see Bruckstein and Kailath (1987). The approach taken here is based on an *inverse scattering* framework for determining the parameters of a layered wave propagation medium from measurements taken at the boundary.

where $d_0(n)$ is the input, w_n is the impulse response, and $u_0(n)$ is the output. The symbol $*$ is shorthand for convolution. The important point to note here is that given the values of w_n and $d_0(n)$ for varying n, we may determine the corresponding values of $u_0(n)$.

Deconvolution is a linear operation that *removes* the effect of some previous convolution performed on a given data record (time series). Suppose that we are given the input $d_0(n)$ and the output $u_0(n)$. We may then use deconvolution to determine the impulse response w_n. In symbolic form we may thus write

$$w_n = u_0(n) * d_0^{-1}(n) \qquad (22)$$

where $d_0^{-1}(n)$ denotes the *inverse* of $d_0(n)$. Note, however, that $d_0^{-1}(n)$ is *not* the reciprocal of $d_0(n)$; rather, the use of the superscript -1 is merely a flag indicating "inverse."

In *seismic deconvolution*, we are given the seismogram $u_0(n)$ and the requirement is to unravel it so as to obtain an estimate of the impulse response w_n of a layered earth model. The problem, however, is complicated by the fact that in the general case of reflection seismology we do not have an estimate of the input seismic wave (also referred to as the seismic wavelet) $d_0(n)$. To overcome this practical uncertainty, we may use an elegant statistical procedure known as *predictive deconvolution* (Robinson, 1954; Robinson and Durrani, 1986). The term "predictive" arises from the fact that the procedure relies on the use of linear prediction. The derivation of predictive deconvolution rests on two simplifying hypotheses for seismic wave propagation with normal incidence:

1. The input wave $d_0(n)$, generated by the source of seismic energy, is the *impulse response of an all-pole feedback system*, and is thus minimum phase.
2. The impulse response w_n of the layered earth model has the properties of a *white-noise process*.

Condition 1 is referred to as the *feedback hypothesis*, and condition 2 is referred to as the *random hypothesis*. Geophysical experience over three decades has shown that it is indeed possible to satisfy these two hypotheses (Robinson, 1984). As a result, predictive deconvolution is used routinely on all seismic records in every exploration program.

The implication of the feedback hypothesis is that we may express the present value $d_0(n)$ of the input wave as a *linear combination of the past values*, as shown by

$$d_0(n) = -\sum_{k=1}^{M} a_k d_0(n-k) \qquad (23)$$

where the a_k are the *feedback coefficients*, and M is the *order* of the all-pole feedback system. The order M may be fixed in advance; alternatively, it may be determined by a mean-square-error criterion.

According to the random hypothesis, the impulse response w_n has the properties of a white-noise process. We therefore expect the estimate $\hat{w}_n$ produced by the deconvolution filter in Fig. 14 to have similar properties. In other words, the deconvolution filter acts as a *whitening filter*. Furthermore, the deconvolution filter is an *all-zero* filter with a transfer

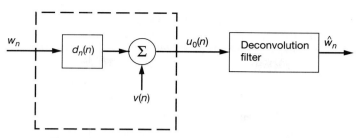

Figure 14 Block diagram illustrating seismic deconvolution.

function equal to the reciprocal of the transfer function of the all-pole feedback system used to model $d_0(n)$. This means that if we express the transfer function of the feedback system [i.e., the z-transform of $d_0(n)$] as:

$$D_0(z) = \frac{1}{1 + a_1 z^{-1} + a_2 z^{-2} + \cdots + a_M z^{-M}} \tag{24}$$

where $a_1, a_2, \ldots, a_M$ are the *feedback coefficients*, and ignore the additive noise $v(n)$ in the model of Fig. 14, then the transfer function of the deconvolution filter is

$$\begin{aligned} A(z) &= \frac{1}{D_0(z)} \\ &= 1 + a_1 z^{-1} + a_2 z^{-2} + \cdots + a_M z^{-M} \end{aligned} \tag{25}$$

To evaluate $A(z)$, we may use a block processing approached based on the *augmented matrix form of the Wiener–Hopf equations for linear preduction*. This relation consists of a system of $(M + 1)$ simultaneous equations that involve the following:

1. A set of $(M + 1)$ known quantities represented by the *estimates* $\hat{r}(0)$, $\hat{r}(1) \ldots$, $\hat{r}(M)$ of the autocorrelation function of the seismogram $u_0(n)$ for varying lags 0, 1, $\ldots$, M, respectively. To get these values, we may use the formula for a *biased estimate* of the autocorrelation function:

$$\hat{r}(l) = \frac{1}{N} \sum_{n=l+1}^{N} d_0(n) d_0(n-l), \qquad l = 0, 1, \ldots, M \tag{26}$$

where N is the *record length* of the seismogram. Typically, N is very large compared to M.

2. A set of $(M + 1)$ unknowns, made up of the feedback coefficients $a_1, a_2, \ldots, a_M$ and the variance σ^2 of the white-noise process assumed to model w_n.

Given the seismogram $u_0(n)$, we may therefore uniquely determine the feedback coefficients $a_1, a_2, \ldots, a_M$ and the variance σ^2 by solving this system of equations.

From Eq. (25), we see that the impulse response of the deconvolution filter consists of the sequence a_k, $k = 1, 2, \ldots, M$. Accordingly, the convolution of this impulse response with $u_0(n)$ yields the desired estimate $\hat{w}_n$, as shown by (see Fig. 14)

$$\hat{w}_n = \sum_{k=0}^{M} a_k u_0(n-k) \tag{27}$$

where $a_0 = 1$. Equation (27) is a description of the deconvolution process. Note, however, the wave $d_0(n)$ generated by the source of seismic energy does not enter this description directly as in the idealized representation of Eq. (23). Rather, the physical nature of $d_0(n)$ influences the deconvolution process by modeling $d_0(n)$ as the impulse response of an all-pole feedback system.

An alternative procedure for constructing the deconvolution filter is to use an adaptive filtering algorithm, as illustrated in Fig. 15. In this application, the present value $u_0(n)$ of the seismic output serves the purpose of a desired response for the algorithm, and the past values $u_0(n-1), u_0(n-2), \ldots, u_0(n-M)$ are used as elements of the input vector. The prediction error controls the adaptation of the M tap weights of the transversal filter component of the algorithm. When the algorithm has converged, the tap weights of the transversal filter provide estimates of the feedback coefficients $a_1, a_2, \ldots, a_M$.

Adaptive Equalization

In digital communications a considerable effort has been devoted to the study of data-transmission systems that utilize the available channel bandwidth efficiently. The objective here is to design a system that accommodates the highest possible rate of data transmission, subject to a specified reliability that is usually measured in terms of the error rate or average probability of symbol error. The transmission of digital data through a linear communication channel is limited by two factors:

1. *Intersymbol interference (ISI).* This is caused by dispersion in the transmit filter, the transmission medium, and the receive filter.
2. *Thermal noise.* This is generated by the receiver at its front end.

For bandwidth-limited channels (e.g., voice-grade telephone channels), we usually find that intersymbol interference is the chief determining factor in the design of high-data-rate transmission systems.

Figure 16 shows the equivalent baseband model of a binary *pulse-amplitude modulation (PAM) system.* The signal applied to the input of the transmitter part of the system consists of a *binary data sequence* b_k, in which each symbol consists of 1 or 0. This sequence is applied to a pulse generator, the output of which is filtered first in the transmitter, then by the medium, and finally in the receiver. Let $u(k)$ denote the sampled output of the receive filter in Fig. 16; the sampling is performed in synchronism with the pulse generator in the transmitter. This output is compared to a *threshold* by means of a *decision device.* If the threshold is exceeded, the receiver makes a decision in favor of symbol 1.

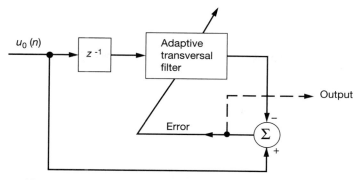

Figure 15 Adaptive filtering scheme for estimating the impulse response of the deconvolution filter.

Otherwise, it decides in favor of symbol 0.

Let a scaling factor a_k be defined by

$$a_k = \begin{cases} +1 & \text{if the input bit } b_k \text{ consists of symbol 1} \\ -1 & \text{if the input bit } b_k \text{ consists of symbol 0} \end{cases} \qquad (28)$$

Then, in the absence of thermal noise, we may express $u(k)$ as

$$u(k) = \sum_n a_n p(k-n)$$

$$= a_k p(0) + \sum_{\substack{n \\ n \neq k}} a_n p(k-n) \qquad (29)$$

where $p(n)$ is the sampled version of the impulse response of the cascade connection of the transmit filter, the transmission medium, and the receive filter. The first term on the right-hand side of Eq. (29) defines the desired symbol, whereas the remaining series represents the intersymbol interference caused by the *channel* (i.e., the combination of the transmit filter, the medium, and the receive filter). This intersymbol interference, if left unchecked, can result in erroneous decisions when the sampled signal at the channel output is compared with some preassigned threshold by means of a decision device.

To overcome the intersymbol interference problem, control of the time-sampled function $p(n)$ is required. In principle, if the characteristics of the transmission medium are known precisely, then it is virtually always possible to design a pair of transmit and receive filters that will make the effect of intersymbol interference (at sampling times) arbitrarily small. This is achieved by proper shaping of the overall response of the channel in accordance with Nyquist's classic work on telegraph transmission theory. The overall frequency response consists of a *flat portion* and a *roll-off portion* that has a cosine form (Haykin, 1994). Correspondingly, the overall impulse response attains its maximum value at time $n = 0$ and is zero at all other sampling instants; the intersymbol interference is therefore zero. In practice we find that the channel is *time varying*, due to variations in the transmission medium, which makes the received signal *nonstationary*. Accordingly, the

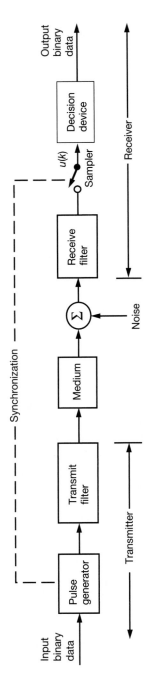

Figure 16 Block diagram of a baseband data transmission system (without equalization).

Introduction

use of a fixed pair of transmit and receive filters, designed on the basis of average channel characteristics, may not adequately reduce intersymbol interference. This suggests the need for an *adaptive equalizer* that provides precise control over the time response of the channel (Lucky, 1965, 1966; Lucky et al., 1968; Proakis, 1975; Quereshi, 1985).

Among the basic philosophies for equalization of datatransmission systems are pre-equalization at the transmitter and postequalization at the receiver. Since the former technique requires the use of a feedback path, we will only consider equalization at the receiver, where the adaptive equalizer is placed after the receive filter–sampler combination in Fig. 16. In theory, the effect of intersymbol interference may be made arbitrarily small by making the number of adjustable coefficients (tap weights) in the adaptive equalizer infinitely large.

An adaptive filtering algorithm requires knowledge of the "desired" response so as to form the error signal needed for the adaptive process to function. In theory, the transmitted sequence (originating at the transmitter output) is the "desired" response for adaptive equalization. In practice, however, with the adaptive equalizer located in the receiver, the equalizer is physically separated from the origin of its ideal desired response. There are two methods in which a *replica (facsimile)* of the desired response may be generated locally in the receiver:

1. *Training method.* In the first method, a replica of the desired response is *stored* in the receiver. Naturally, the generator of this stored reference has to be electronically *synchronized* with the known transmitted sequence. A widely used *test (probing) signal* consists of a *pseudonoise (PN) sequence* (also known as a *maximal-length sequence*) with a broad and even power spectrum. The PN sequence has noiselike properties. Yet it has a deterministic waveform that repeats periodically. For the generation of a PN sequence, we may use a *linear feedback shift register* that consists of a number of consecutive two-state memory stages (flip-flops) regulated by a single timing clock (Golomb, 1964). A *feedback* term, consisting of the modulo-2 sum of the outputs of various memory stages, is applied to the first memory stage of the shift register and thereby prevents it from emptying.

2. *Decision-directed method.* Under normal operating conditions, a good facsimile of the transmitted sequence is being produced at the output of the *decision device* in the receiver. Accordingly, if this output were the correct transmitted sequence, it may be used as the "desired" response for the purpose of adaptive equalization. Such a method of learning is said to be *decision directed*, because the receiver attempts to learn by employing its own decisions (Lucky et al., 1968). If the average probability of symbol error is small (less than 10 percent, say), the decisions made by the receiver are correct enough for the *estimates of the error signal* (used in the adaptive process) to be *accurate most of the time*. This means that, in general, the adaptive equalizer is able to improve the tap-weight settings by virtue of the correlation procedure built into its feedback control loop. The improved tap-weight settings will, in turn, result in a lower average probability of

symbol error and therefore more accurate estimates of the error signal for adaptation, and so it goes on. However, it is also possible for the reverse effect to occur, in which case the tap-weight settings of the equalizer lose acquisition of the channel.

With a known training sequence, as in the first method, the adaptive filtering algorithm used to adjust the equalizer coefficients corresponds mathematically to searching for the unique minimum of a quadratic error-performance surface. The *unimodal* nature of this surface assures convergence of the algorithm. In the decision-directed method, on the other hand, the use of estimated and unreliable data modifies the error performance into a *multimodal* one, in which case complex behavior may result (Mazo, 1980). Specifically, the error performance surface now exhibits two types of local minima:

1. *Desired local minima*, whose positions correspond to coefficient (tap-weight) settings that yield the same performance as that obtained with a known training sequence
2. *Undesired (extraneous) local minima*, whose positions correspond to coefficient settings that yield inferior equalizer performance.

A poor choice of the initial coefficient settings may cause the adaptive equalizer to converge to an undesirable local minimum and stay there. The most significant point to note from this discussion is that, in general, a *linear* adaptive equalizer must be trained before it is switched to the decision-directed mode of operation if we are to be sure of delivering high performance.

A final comment pertaining to performance evaluation is in order. A popular experimental technique for assessing the performance of a data transmission system involves the use of an *eye pattern*. This pattern is obtained by applying (1) the received wave to the vertical deflection plates of an oscilloscope, and (2) a sawtooth wave at the transmitted symbol rate to the horizontal deflection plates. The resulting display is called an *eye pattern* because of its resemblance to the human eye for binary data. Thus, in a system using adaptive equalization, the equalizer attempts to correct for intersymbol interference in the system and thereby open the eye pattern as far as possible.

Thus far we have only discussed adaptive equalizers for baseband PAM systems. However, voice-band data transmission systems employ modulation–demodulation schemes that are commonly known as *modems*. Depending on the speed of operation, we may categorize modems as follows (Qureshi, 1985):

1. *Low-speed* (2400 to 4800 b/s) modems that use *phase-shift keying (PSK)*; PSK is a digital modulation scheme in which the phase of a sinusoidal carrier wave is shifted by $2\pi k/M$ radians in accordance with the input data, where M is the number of phase levels used and $k = 0, 1, \ldots, M - 1$. Specific values of M used in practice are $M = 2$ and 4, representing *binary phase-shift keying (BPSK)* and *quadriphase-shift keying (QPSK)*, respectively.

2. *High-speed* (4800 to 16,800 b/s or possibly even higher) modems that use combined *amplitude and phase modulation* or, equivalently, *quadrature amplitude modulation (QAM)*.

The important point to note is that the baseband model for BPSK is real, whereas the baseband models for QPSK and QAM are complex, involving both in-phase and quadrature channels. Hence, the baseband adaptive equalizer for data transmission systems using BPSK (or its variation) is *real*, whereas the baseband adaptive equalizers for QPSK and QAM are *complex* (i.e., the tap weights of the transversal filter are complex). Note also that a real equalizer processes real inputs to produce a real equalized output, whereas a complex equalizer processes complex inputs to produce complex equalized outputs.

Blind Equalization

In the case of a highly nonstationary communications environment (e.g., digital mobile communications), it is impractical to consider the use of a training sequence. In such a situation, the adaptive filter has to equalize the communication channel in a self-organized (unsupervised) manner, and the resulting operation is referred to as *blind equalization*. Clearly, the design of a blind equalizer is a more challenging task than a conventional adaptive equalizer, because it has to make up for the absence of a training sequence by some practical means. Whereas a conventional adaptive equalizer relies on second-order statistics of the input data, a blind equalizer relies on additional information about the environment.

This additional information may take one of two basic forms:

- *Higher-order statistics (HOS)*, the extraction of which is implicitly or explicitly built into the design of the blind equalizer. For this to be possible, the input data must be non-Gaussian, and the equalizer must include some form of nonlinearity.
- *Cyclostationarity*, which arises when the amplitude, phase, or frequency of a sinusoidal carrier is varied in accordance with an information-bearing signal. In this case, design of the blind equalizer is based on second-order cyclostationary statistics of the input data, and the use of nonlinearity is no longer a requirement.

An advantage of the latter type of blind equalizer is that it exhibits better convergence properties than an HOS-based blind equalizer.

Linear Predictive Coding

The coders used for the digital representation of speech signals fall into two broad classes: *source coders* and *waveform coders*. Source coders are *model dependent*, in that they use *a priori* knowledge about how the speech signal is generated at the source. Source coders for speech are generally referred to as *vocoders* (a contraction of voice coders). They can operate at low coding rates; however, they provide a synthetic quality, with the speech signal having lost substantial naturalness. Waveform coders, on the other hand, essentially

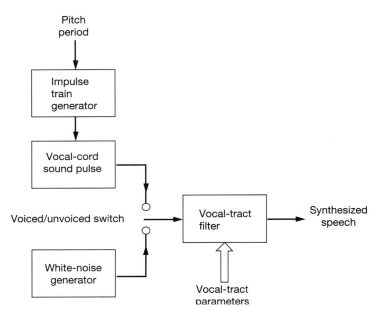

Figure 17   Block diagram of simplified model for the speech production process.

strive for facsimile reproduction of the speech waveform. In principle, these coders are *signal independent*. They may be designed to provide telephone-toll quality for speech at relatively high coding rates. In this subsection we describe a special form of source coder known as a linear predictive coder. Waveform coders are considered in the next subsection.

In the context of speech, *linear predictive coding* (LPC) strives to produce digitized voice data at low bit rates (as low as 2.4 kb/s), with two important motivations in mind. First, the use of linear predictive coding permits the transmission of digitized voice over a *narrow-band channel* (having a bandwidth of approximately 3 kHz). Second, the realization of a low-bit rate makes the *encryption* of voice signals easier and more reliable than would be the case otherwise; encryption is an essential requirement for *secure communications* (as in a military environment). Note that a bit rate of 2.4 kb/s is less than 5 percent of the 64 kb/s used typically for the standard pulse-code modulation (PCM); see the next subsection.

Linear predictive coding achieves a low bit rate for the digital representation of speech by exploiting the special properties of a classical model of the speech production process, which is described next.

Figure 17 shows a simplified block diagram of the classical model for the speech production process. It assumes that the sound-generating mechanism (i.e., the source of excitation) is linearly separable from the intelligence-modulating vocal-tract filter. The precise form of the excitation depends on whether the speech sound is voiced or unvoiced:

1. A *voiced* speech sound (such as[10]/i/ in e̲ve) is generated from quasi-periodic vocal-cord sound. In the model of Fig. 17 the impulse-train generator produces a sequence of impulses (i.e., very short pulses), which are spaced by a fundamental period equal to the *pitch period*. This signal, in turn, excites a linear filter whose impulse response equals the vocal-cord sound pulse.

2. An *unvoiced* speech sound (such as /f/ in fi̲sh) is generated from random sound produced by turbulent airflow. In this case the excitation consists simply of a *white* (i.e., broad spectrum) noise source. The probability distribution of the noise samples does not appear to be critical.

The frequency response of the vocal-tract filter for unvoiced speech or that of the vocal tract multiplied by the spectrum of the vocal-cord sound pulses determines the short-time spectral envelope of the speech signal.

At first sight, it may appear that the speech production model falls under class I of adaptive filtering application (i.e., identification). In reality, however, this is not so. As may be seen in Fig. 17, there is *no* access to the input signal of the vocal tract.

The method of *linear predictive coding (LPC)* is an example of source coding. This method is important, because it provides not only a powerful technique for the digital transmission of speech at low bit rates but also accurate estimates of basic speech parameters.

The development of LPC relies on the model of Fig. 17 for the speech-production process. The frequency response of the vocal tract for unvoiced speech or that of the vocal tract multiplied by the spectrum of the vocal sound pulse for voiced speech is described by the *transfer function*

$$H(z) = \frac{G}{1 + \sum_{k=1}^{M} a_k z^{-k}} \tag{30}$$

where G is a gain parameter and z^{-1} is the unit-delay operator. The form of excitation applied to this filter is changed by switching between voiced and unvoiced sounds. Thus, the filter with transfer function $H(z)$ is excited by a sequence of impulses to generate voiced sounds or a white-noise sequence to generate unvoiced sounds. In this application, the input data are real valued; hence the filter coefficients, a_k, are likewise real valued.

In linear predictive coding, as the name implies, linear prediction is used to estimate the speech parameters. Given a set of past samples of a speech signal, $u(n-1)$, $u(n-2)$, . . . , $u(n-M)$, a linear prediction of $u(n)$, the present sample value of the signal, is defined by

$$\hat{u}(n) = \sum_{k=1}^{M} \hat{w}_k u(n-k) \tag{31}$$

---

[10]The symbol / / is used to denote the *phenome*, a basic linguistic unit.

The predictor coefficients, $\hat{w}_1$, $\hat{w}_2$, . . . , $\hat{w}_M$, are optimized by minimizing the mean-square value of the prediction error, $e(n)$, defined as the difference between $u(n)$ and $\hat{u}(n)$. The use of the minimum-mean-squared-error criterion for optimizing the predictor may be justified for two basic reasons:

1. If the speech signal satisfies the model described by Eq. (30) and if the mean-square value of the error signal $e(n)$ is minimized, then we find that $e(n)$ equals the excitation $u(n)$ multiplied by the gain parameter G in the model of Fig. 18 and $a_k = -\hat{w}_k$, $k = 1, 2, . . . , M$. Thus, the estimation error $e(n)$ consists of quasi-periodic pulses in the case of voiced sounds or a white-noise sequence in the case of unvoiced sounds. In either case, the estimation error $e(n)$ would be small most of the time.

2. The use of the minimum-mean-squared-error criterion leads to tractable mathematics.

Figure 18 shows the block diagram of an LPC vocoder. It consists of a transmitter and a receiver. The transmitter first applies a *window* (typically 10 to 30 ms long) to the input speech signal, thereby identifying a block of speech samples for processing. This window is short enough for the vocal-tract shape to be nearly stationary, so the parameters of the speech-production model in Fig. 18 may be treated as essentially constant for the duration of the window. The transmitter then analyzes the input speech signal in an adaptive manner, block by block, by performing a linear prediction and pitch detection. Finally, it codes the parameters made up of (1) the set of predictor coefficients, (2) the pitch period, (3) the gain parameter, and (4) the voiced–unvoiced parameter, for transmission over the channel. The receiver performs the inverse operations, by first decoding the incoming parameters. In particular, it computes the values of the predictor coefficients, the pitch period, and the gain parameter, and determines whether the segment of interest represents voiced or unvoiced sound. Finally, the receiver uses these parameters to synthesize the speech signal by utilizing the model of Fig. 17.

Adaptive Differential Pulse-Code Modulation

In *pulse-code modulation*, which is the standard technique for waveform coding, three basic operation are performed on the speech signal. The three operations are *sampling* (time discretization), *quantization* (amplitude discretization), and *coding* (digital representation of discrete amplitudes). The operations of sampling and quantization are designed to preserve the shape of the speech signal. As for coding, it is merely a method of translating a discrete sequence of sample values into a more appropriate form of signal representation.

The rationale for sampling follows from a basic property of all speech signals: they are bandlimited. This means that a speech signal can be sampled in time at a finite rate in accordance with the sampling theorem. For example, commercial telephone networks designed to transmit speech signals occupy a bandwidth from 200 to 3200 Hz. To satisfy

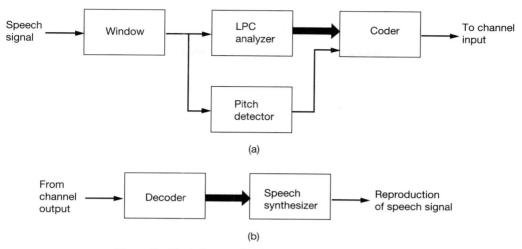

Figure 18   Block diagram of LPC vocoder: (a) transmitter, (b) receiver.

the sampling theorem, a conservative sampling rate of 8 kHz is commonly used in practice.

Quantization is justified on the following grounds. Although a speech signal has a continuous range of amplitudes (and therefore its samples also have a continuous amplitude range), it is not necessary to transmit the exact amplitudes of the samples. Basically, the human ear (as ultimate receiver) can only detect finite amplitude differences.

In PCM, as used in telephony, the speech signal (after low-pass filtering) is sampled at the rate of 8 kHz, nonlinearly (e.g., logarithmically) quantized, and then coded into 8-bit words; see Fig. 19(a). The result is a good signal-to-quantization-noise ratio over a wide dynamic range of input signal levels. This method requires a bit rate of 64 kb/s.

Differential pulse-code modulation (DPCM), another example of waveform coding, involves the use of a predictor as in Fig. 19(b). The predictor is designed to exploit the correlation that exists between adjacent samples of the speech signal, in order to realize a reduction in the number of bits required for the transmission of each sample of the speech signal and yet maintain a prescribed quality of performance. This is achieved by quantizing and then coding the prediction error that results from the subtraction of the predictor output from the input signal. If the prediction is optimized, the variance of the prediction error will be significantly smaller than that of the input signal, so a quantizer with a given number of levels can be adjusted to produce a quantizing error with a smaller variance than would be possible if the input signal were quantized directly as in a standard PCM system. Equivalently, for a quantizing error of prescribed variance, DPCM requires a smaller number of quantizing levels (and therefore a smaller bit rate) than PCM. Differential pulse-code modulation uses a fixed quantizer and a fixed predictor. A further reduction in the transmission rate can be achieved by using an adaptive quantizer together with an adaptive predictor of sufficiently high order, as in Fig. 19(c). This type of waveform coding is called *adaptive differential pulse-code modulation (ADPCM)*, where A denotes

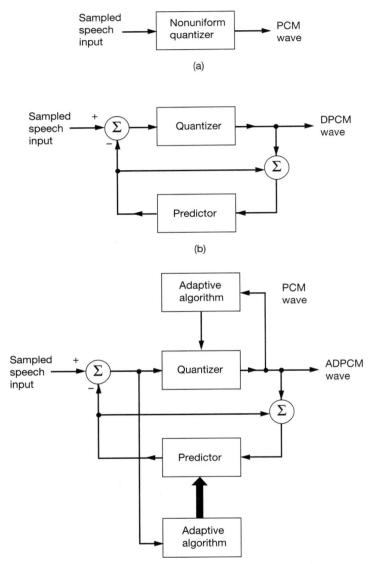

Figure 19 Waveform coders: (a) PCM, (b) DPCM, (c) ADPCM.

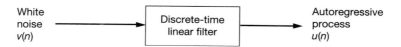

Figure 20   Black box representation of a stochastic model.

adaptation of both quantizer and predictor algorithms. An adaptive predictor is used in order to account for the nonstationary nature of speech signals. ADPCM can digitize speech with toll quality (8-bit PCM quality) at 32 kb/s. It can realize this level of quality with a 4-bit quantizer.[11]

Adaptive Spectrum Estimation

The *power spectrum* provides a quantitative measure of the second-order statistics of a discrete-time stochastic process as a function of frequency. In *parametric spectrum analysis*, we evaluate the power spectrum of the process by assuming a *model* for the process. In particular, the process is modeled as the output of a linear filter that is excited by a *white-noise process*, as in Fig. 20. By definition, a white-noise process has a constant power spectrum. A model that is of practical utility is the *autoregressive (AR) model*, in which the transfer function of the filter is assumed to consist of poles only. Let this transfer function be denoted by

$$H(e^{j\omega}) = \frac{1}{1 + a_1 e^{-j\omega} + \cdots + a_M e^{-jM\omega}}$$

$$= \frac{1}{1 + \sum_{k=1}^{M} a_k e^{-jk\omega}} \tag{32}$$

where the a_k are called the *autoregressive (AR) parameters*, and M is the *model order*. Let σ_v^2 denote the constant power spectrum of the white-noise process $v(n)$ applied to the filter input. Accordingly, the power spectrum of the filter output $u(n)$ equals

$$S_{AR}(\omega) = \sigma_v^2 |H(e^{j\omega})|^2 \tag{33}$$

We refer to $S_{AR}(\omega)$ as the *autoregressive (AR) power spectrum*. Equation (32) assumes that the AR process $u(n)$ is real, in which case the AR parameters themselves assume real values.

---

[11]The International Telephone and Telegraph Consultative Committee (CCITT) has adopted the 32-kb/s ADPCM as an international standard. The adaptive predictor used herein has a transfer function consisting of two poles and six zeros. A two-pole configuration was chosen, because it permits control of decoder stability in the presence of transmission errors. Six zeros were combined with the two poles in order to improve performance. The eight coefficients of the predictor are adapted by using a simplified version of the LMS algorithm; for details, see Benvenuto et al. (1986) and Nishitani et al. (1987).

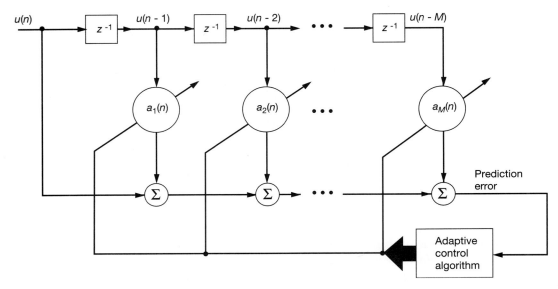

Figure 21   Adaptive prediction-error filter for real-valued data.

When the AR model is time varying, the model parameters become time dependent, as shown by $a_1(n)$, $a_2(n)$, . . . , $a_M(n)$. In this case, we express the power spectrum of the time-varying AR process as

$$S_{AR}(\omega,n) = \frac{\sigma_v^2}{\left|1 + \displaystyle\sum_{k=1}^{M} a_k(n)e^{-jk\omega}\right|^2} \tag{34}$$

We may determine the AR parameters of the time-varying model by applying $u(n)$ to an *adaptive prediction-error filter*, as indicated in Fig. 21. The filter consists of a transversal filter with adjustable tap weights. In the adaptive scheme of Fig. 21, the prediction error produced at the output of the filter is used to control the adjustments applied to the tap weights of the filter.

The *adaptive AR model* provides a practical means for measuring the *instantaneous frequency* of a frequency-modulated process. In particular, we may do this by measuring the frequency at which the AR power spectrum $S_{AR}(\omega, n)$ attains its peak value for varying time n.

Signal Detection

The *detection problem*, that is, the problem of detecting an information-bearing signal in noise, may be viewed as one of *hypothesis testing* with deep roots in *statistical decision theory* (Van Trees, 1968). In the statistical formulation of hypothesis testing, there are two criteria of most interest: the *Bayes criterion* and the *Neyman–Pearson criterion*. In the Bayes test, we minimize the *average cost* or *risk* of the experiment of interest, which incorporates two sets of parameters: (1) *a priori probabilities* that represent the observer's information about the source of information before the experiment is conducted, and (2) a set of *costs* assigned to the various possible courses of action. As such, the Bayes criterion is directly applicable to digital communications. In the Neyman–Pearson test, on the other hand, we maximize the *probability of detection* subject to the constraint that the *probability of false alarm* does *not* exceed some preassigned value. Accordingly, the Neyman–Pearson criterion is directly applicable to radar or sonar. An idea of fundamental importance that emerges in hypothesis testing is that, for a Bayes criterion or Neyman–Pearson criterion, the optimum test consists of two distinct operations: (1) processing the observed data to compute a test statistic called the *likelihood ratio*, and (2) computing the likelihood ratio with a *threshold* to make a *decision* in favor of one of the two hypotheses. The choice of one criterion or the other merely affects the value assigned to the threshold. Let H_1 denote the hypothesis that the observed data consist of noise alone, and H_2 denote the hypothesis that the data consist of signal plus noise. The likelihood ratio is defined as the ratio of two maximum likelihood functions, the numerator assuming that hypothesis H_2 is true and the denominator assuming that hypothesis H_1 is true. If the likelihood ratio exceeds the threshold, the decision is made in favor of hypothesis H_2; otherwise, the decision is made in favor of hypothesis H_1.

In simple binary hypothesis testing, it is assumed that the signal is known, and the noise is both white and Gaussian. In this case, the likelihood ratio test yields a *matched filter* (matched in the sense that its impulse response equals the time-reversed version of the known signal). When the additive noise is a *colored Gaussian noise* of known mean and correlation matrix, the likelihood ratio test yields a filter that consists of two sections: a *whitening filter* that transforms the colored noise component at the input into a white Gaussian noise process, and a *matched filter* that is matched to the new version of the known signal as modified by the whitening filter.

However, in some important operational environments such as *communications, radar*, and *active sonar*, there may be inadequate information on the signal and noise statistics to design a fixed optimum detector. For example, in a sonar environment it may be difficult to develop a precise *model* for the received sonar signal, one that would account for the following factors completely:

- Loss in the signal strength of a *target echo* from an object of interest (e.g., enemy vessel), due to oceanic propagation effects and reflection loss at the target
- Statistical variations in the additive *reverberation* component, produced by reflections of the transmitted signal from scatterers such as the ocean surface, ocean floor, biologies, and inhomogeneities within the ocean volume
- Potential sources of *noise* such as biological, shipping, oil drilling, seismic, and oceanographic phenomena.

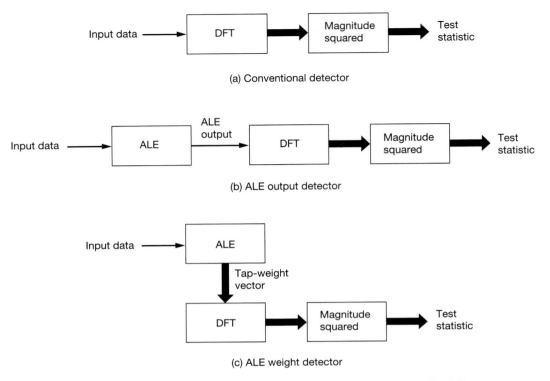

Figure 22 Fixed and adaptive detection schemes: (a) conventional detector. (b) ALE output detector. (c) ALE weight detector.

In situations of this kind, the use of adaptivity offers an attractive approach to solve the target (signal) detection problem. Typically, the design of an *adaptive detector* proceeds by exploiting some knowledge of general characteristics of the signal and noise, and designing the detector in such a way that its internal structure is adjustable in response to changes in the received signal. In general, the incorporation of this adjustment makes the performance analysis of an adaptive detector much more difficult to undertake than that of a fixed detector.

Fixed and adaptive detectors.   Figure 22(a) shows the block diagram of a conventional detector based on the *discrete Fourier transform (DFT)* for the detection of narrow-band signals in white Gaussian noise (Williams and Ricker, 1972). The DFT may be viewed as a bank of nonoverlapping narrow-band filters whose passbands span the frequency range of interest. In the detector of Fig. 22(a) the magnitude of each complex output of the DFT is squared to form a *sufficient statistic*. This statistic is optimum (in the Neyman–Pearson sense) for detecting a sinusoid of known frequency (centered in the pertinent passband of the DFT) but unknown phase, and in the presence of white Gaussian noise. The detector output is compared to a threshold. If the threshold is exceeded, the

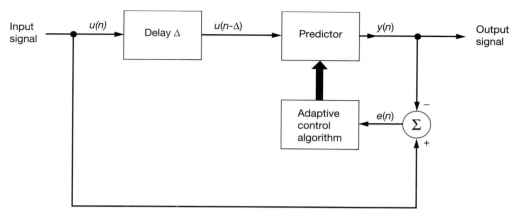

Figure 23   Adaptive line enhancer.

detector decides in favor of the narrow-band signal; otherwise, the detector declares the signal to be absent.

The performance of this conventional noncoherent detector may be improved by using an *adaptive line enhancer (ALE)* as a *prefilter* (preprocessor) to the detector (Widrow et al., 1975b). The ALE is a special form of adaptive noise canceler that is designed to suppress the wide-band noise component of the input, while passing the narrow-band signal component with little attenuation. Figure 23 depicts the block diagram of an ALE. It consists of the interconnection of a delay element and a linear predictor. The predictor output $y(n)$ is subtracted from the input signal $u(n)$ to produce the estimation error $e(n)$. This estimation error is, in turn, used to adaptively control the tap weights of the predictor. The predictor input equals $u(n - \Delta)$, where the delay Δ is equal to or greater than the sampling period. The main function of the *prediction depth* Δ is to remove the correlation between the noise component in the original input signal $u(n)$ and the delayed predictor input $u(n - \Delta)$. It is for this reason that the delay Δ is also called the *decorrelation parameter* of the ALE.

Two types of ALE detection structures have been proposed in the literature (Zeidler, 1990):

1. *ALE output detector.* In this adaptive detector shown in Fig. 22(b), the output of an ALE is applied to a DFT. The magnitude of the resulting DFT output is squared to produce the sufficient statistic for the detector.

2. *ALE weight detector.* In this second adaptive detector, shown in Fig. 22(c), the tap-weight vector of an ALE is applied to a DFT. The magnitude of the DFT output is squared as before to produce the sufficient statistic.

In both cases, the ALE processes N input data points, with the ALE length small compared to N. The real benefit of the ALE is realized in a nonstationary noise background (Zeidler, 1990).

The practical value of an ALE as a preprocessor to a conventional matched filter has been demonstrated by Nielson and Thomas (1988) as a means of improving the performance of the detector in the presence of Arctic ocean noise. This type of noise is known to have highly non-Gaussian and nonstationary characteristics; hence the benefit to be gained from the use of an ALE.

Adaptive Noise Canceling

As the name implies, adaptive noise canceling relies on the use of *noise canceling* by subtracting noise from a received signal, an operation controlled in an *adaptive* manner for the purpose of improved signal-to-noise ratio. Ordinarily, it is inadvisable to subtract noise from a received signal, because such an operation could produce disastrous results by causing an increase in the average power of the output noise. However, when proper provisions are made, and filtering and subtraction are controlled by an adaptive process, it is possible to achieve a superior system performance compared to direct filtering of the received signal (Widrow et al., 1975b; Widrow and Stearns, 1985).

Basically, an adaptive noise canceler is a *dual-input, closed-loop adaptive feedback system* as illustrated in Fig. 24. The two inputs of the system are derived from a pair of sensors: a *primary sensor* and a *reference (auxiliary) sensor*. Specifically, we have the following:

1. The primary sensor receives an *information-bearing signal* $s(n)$ corrupted by *additive noise* $v_0(n)$, as shown by

$$d(n) = s(n) + v_0(n) \qquad (35)$$

The signal $s(n)$ and the noise $v_0(n)$ are uncorrelated with each other; that is,

$$E[s(n)v_0(n-k)] = 0 \qquad \text{for all } k \qquad (36)$$

where $s(n)$ and $v_0(n)$ are assumed to be real valued.

2. The reference sensor receives a noise $v_1(n)$ that is *uncorrelated* with the signal $s(n)$ but *correlated* with the noise $v_0(n)$ in the primary sensor output in an *unknown* way; that is,

$$E[s(n)v_1(n-k)] = 0 \qquad \text{for all } k \qquad (37)$$

and

$$E[v_0(n)v_1(n-k)] = p(k) \qquad (38)$$

where, as before, the signals are real valued and $p(k)$ is an unknown cross-correlation for lag k.

The reference signal $v_1(n)$ is processed by an adaptive filter to produce the output signal:

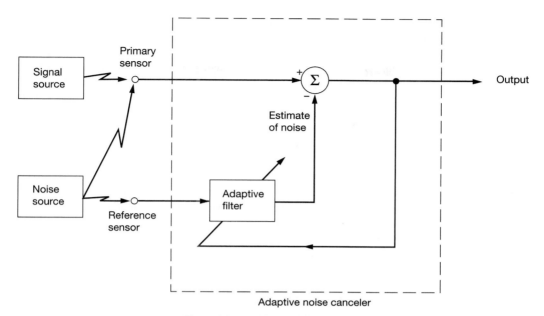

Figure 24 Adaptive noise cancelation.

$$y(n) = \sum_{k=0}^{M-1} \hat{w}_k(n) v_1(n-k) \tag{39}$$

where the $\hat{w}_k(n)$ are the adjustable (real) tap weights of the adaptive filter. The filter output $y(n)$ is subtracted from the primary signal $d(n)$, serving as the "desired" response for the adaptive filter. The error signal is defined by

$$e(n) = d(n) - y(n) \tag{40}$$

Thus, substituting Eq. (35) in (40), we get

$$e(n) = s(n) + v_0(n) - y(n) \tag{41}$$

The error signal is, in turn, used to adjust the tap weights of the adaptive filter, and the control loop around the operations of filtering and subtraction is thereby closed. Note that the information-bearing signal $s(n)$ is indeed part of the error signal $e(n)$, as indicated in Eq. (41).

The error signal $e(n)$ constitutes the overall *system output*. From Eq. (41) we see that the noise component in the system output is $v_0(n) - y(n)$. Now, the adaptive filter attempts to minimize the mean-square value (i.e., average power) of the error signal $e(n)$. The information-bearing signal $s(n)$ is essentially unaffected by the adaptive noise canceler.

Hence, minimizing the mean-square value of the error signal $e(n)$ is equivalent to minimizing the mean-square value of the output noise $v_0(n) - y(n)$. With the signal $s(n)$ remaining essentially constant, it follows that *the minimization of the mean-square value of the error signal is indeed the same as the maximization of the output signal-to-noise ratio of the system.*

The signal-processing operation described herein has two limiting cases that are noteworthy:

1. The adaptive filtering operation is *perfect* in the sense that

$$y(n) = v_0(n)$$

In this case, the system output is *noise free* and the noise cancelation is perfect. Correspondingly, the output signal-to-noise ratio is infinitely large.

2. The reference signal $v_1(n)$ is *completely uncorrelated* with both the signal and noise components of the primary signal $d(n)$; that is,

$$E[d(n)v_1(n - k)] = 0 \qquad \text{for all } k$$

In this case, the adaptive filter "switches itself off," resulting in a zero value for the output $y(n)$. Hence, the adaptive noise canceler has *no* effect on the primary signal $d(n)$, and the output signal-to-noise ratio remains unaltered.

The effective use of adaptive noise canceling therefore requires that we place the reference sensor in the noise field of the primary sensor with two specific objectives in mind. First, the information-bearing signal component of the primary sensor output is *undetectable* in the reference sensor output. Second, the reference sensor output is *highly correlated* with the noise component of the primary sensor output. Moreover, the adaptation of the adjustable filter coefficients must be near optimum.

In the remainder of this subsection, we describe three useful applications of the adaptive noise-canceling operation:

1. *Canceling 60-Hz interference in electrocardiography.* In *electrocardiography (ECG)*, commonly used to monitor heart patients, an *electrical discharge* radiates energy through a human *tissue* and the resulting output is received by an *electrode*. The electrode is usually positioned in such a way that the received energy is maximized. Typically, however, the electrical discharge involves very low potentials. Correspondingly, the received energy is very small. Hence extra care has to be exercised in minimizing signal degradation due to external *interference*. By far, the strongest form of interference is that of a 60-Hz periodic waveform picked up by the receiving electrode (acting like an antenna) from nearby electrical equipment (Huhta and Webster, 1973). Needless to say, this interference has undesirable effects in the interpretation of electrocardiograms. Widrow et al. ((1975b) have demonstrated the use of adaptive noise canceling (based on the LMS algorithm) as a method for reducing this form of interference. Specifically, the primary signal is taken from the ECG preamplifier, and the reference signal is taken from a wall outlet with proper attenuation. Figure 25 shows a block diagram of the adaptive noise canceler used

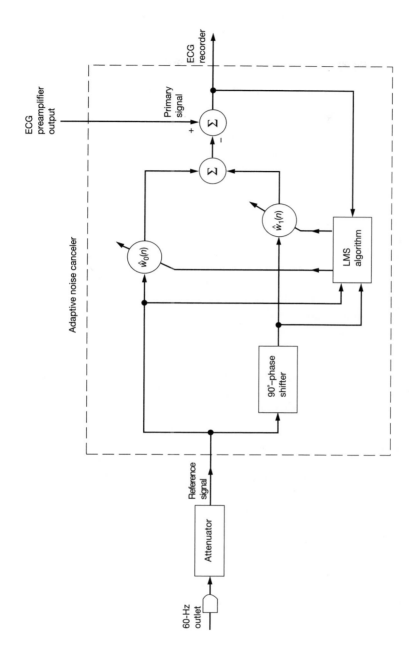

Figure 25 Adaptive noise canceler for suppressing 60-Hz interference in electrocardiography (After Widrow et al., 1975b).

by Widrow et al. (1975b). The adaptive filter has two adjustable weights, $\hat{w}_0(n)$ and $\hat{w}_1(n)$. One weight, $\hat{w}_0(n)$, is fed directly from the reference point. The second weight, $\hat{w}_1(n)$, is fed from a 90°-phase-shifted version of the reference input. The sum of the two weighted versions of the reference signal is then subtracted from the ECG output to produce an error signal. This error signal together with the weighted inputs are applied to the LMS algorithm, which, in turn, controls the adjustments applied to the two weights. In this application, the adaptive noise canceler acts as a variable "notch filter." The frequency of the sinusoidal interference in the ECG output is presumably the same as that of the sinusoidal reference signal. However, the amplitude and phase of the sinusoidal interference in the ECG output are unknown. The two weights $\hat{w}_0(n)$ and $\hat{w}_1(n)$ provide the two *degrees of freedom* required to control the amplitude and phase of the sinusoidal reference signal so as to cancel the 60-Hz interference contained in the ECG output.

2. *Reduction of acoustic noise in speech.* At a noisy site (e.g., the cockpit of a military aircraft), voice communication is affected by the presence of *acoustic noise*. This effect is particularly serious when linear predictive coding (LPC) is used for the digital representation of voice signals at low bit rates; LPC was discussed earlier. To be specific, high-frequency acoustic noise severely affects the estimated LPC spectrum in both the low- and high-frequency regions. Consequently, the intelligibility of digitized speech using LPC often falls below the minimum acceptable level. Kang and Fransen (1987) describe the use of an adaptive noise canceler, based on the LMS algorithm, for reducing acoustic noise in speech. The noise-corrupted speech is used as the primary signal. To provide the reference signal (noise only), a reference microphone is placed in a location where there is sufficient isolation from the source of speech (i.e., the known location of the speaker's mouth). In the experiments described by Kang and Fransen, a reduction of 10 to 15 dB in the acoustic noise floor is achieved, without degrading voice quality. Such a level of noise reduction is significant in improving voice quality, which may be unacceptable otherwise.

3. *Adaptive speech enhancement.* Consider the situation depicted in Fig. 26. The requirement is to listen to the voice of the desired speaker in the presence of background noise, which may be satisfied through the use of adaptive noise canceling. Specifically, *reference microphones* are added at locations far enough away from the desired speaker such that their outputs contain *only* noise. As indicated in Fig. 26, a weighted sum of the auxiliary microphone outputs is subtracted from the output of the desired speech-containing microphone, and an adaptive filtering algorithm (e.g., the LMS algorithm) is used to adjust the weights so as to minimize the average output power. A useful application of the idea described herein is in the adaptive noise cancelation for hearing aids[12] (Chazan et al., 1988). The so-called "cocktail party effect" severely limits the usefulness of hearing aids. The cocktail party phenomenon refers to the ability of a person with normal hearing to focus on a conversation taking place at a distant location in a crowded room. This ability

---

[12]This idea is similar to that of adaptive spatial filtering in the context of antennas, which is considered later in this section.

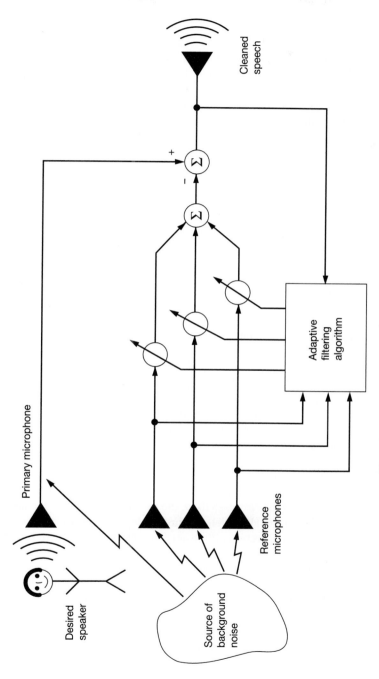

Figure 26   Block diagram of an adaptive noise canceler for speech.

is lacking in a person who wears hearing aids, because of extreme sensitivity to the presence of *background noise*. This sensitivity is attributed to two factors: (a) the loss of directional cues, and (b) the limited channel capacity of the ear caused by the reduction in both dynamic range and frequency response. Chazan et al. (1988) describe an adaptive noise canceling technique aimed at overcoming this problem. The technique involves the use of an *array of microphones* that exploit the difference in spatial characteristics between the desired signal and the noise in a crowded room. The approach taken by Chazan et al. is based on the fact that each microphone output may be viewed as the sum of the signals produced by the individual speakers engaged in conversations in the room. Each signal contribution in a particular microphone output is essentially the result of a speaker's speech signal having passed through the *room filter*. In other words, each speaker (including the desired speaker) produces a signal at the microphone output that is the sum of the direct transmission of his or her speech signal and its reflections from the walls of the room. The requirement is to reconstruct the desired speaker signal, including its room reverberations, while canceling out the source of noise. In general, the transformation undergone by the speech signal from the desired speaker is not known. Also, the characteristics of the background noise are variable. We thus have a signal-processing problem for which adaptive noise canceling offers a feasible solution.

Echo Cancelation

Almost all conversations are conducted in the presence of *echoes*. An echo may be nonnoticeable or distinct, depending on the time delay involved. If the delay between the speech and the echo is short, the echo is not noticeable but perceived as a form of spectral distortion or reverberation. If, on the other hand, the delay exceeds a few tens of milliseconds, the echo is distinctly noticeable. Distinct echoes are annoying.

Echoes may also be experienced on a telephone circuit (Sondhi and Berkley, 1980). When a speech signal encounters an *impedance mismatch* at any point on a telephone circuit, a portion of that signal is reflected (returned) as an echo. An echo represents an *impairment* that can be annoying subjectively as the more obvious impairments of low volume and noise.

To see how echoes occur, consider a long-distance telephone circuit depicted in Fig. 27. Every telephone set in a given geographical area is connected to a central office by a *two-wire line* called the *customer loop*; the two-wire line serves the need for communications in either direction. However, for circuits longer than about 35 miles, a separate path is necessary for each direction of transmission. Accordingly, there has to be provision for connecting the two-wire circuit to the four-wire circuit. This connection is accomplished by means of a *hybrid transformer*, commonly referred to as a *hybrid*. Basically, a hybrid is a bridge circuit with three ports (terminal pairs), as depicted in Fig. 28. If the bridge is *not* perfectly balanced, the "in" port of the hybrid becomes coupled to the "out" port, thereby giving rise to an echo.

Echoes are noticeable when a long-distance call is made on a telephone circuit, particularly one that includes a *geostationary satellite*. Due to the high altitude of such a satellite, there is a one-way travel time of about 300 ms between a ground station and the

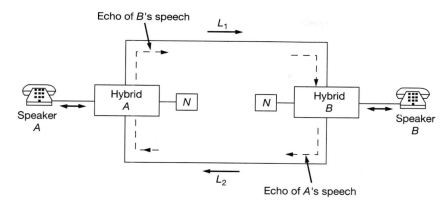

Figure 27 Long-distance telephone circuit; the boxes marked N are balancing impedances.

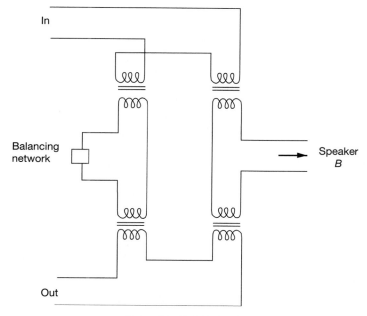

Figure 28 Hybrid circuit.

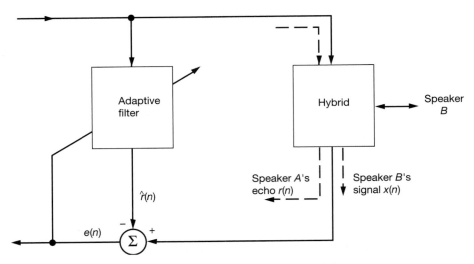

Figure 29   Signal definitions for echo cancelation.

satellite. Thus, the *round-trip delay* in a satellite link (including telephone circuits) can be as long as 600 ms. Generally speaking, the longer the echo delay, the more it must be attenuated before it becomes noticeable.

The question to be answered is: How do we exercise echo control? It appears that the idea with the greatest potential for echo control is that of *adaptive echo cancelation* (Sondhi and Prasti, 1966; Sondhi, 1967; Sondhi and Berkley, 1980; Messerschmitt, 1984; Murano et al., 1990). The basic principle of echo cancelation is to *synthesize a replica of the echo and subtract it from the returned signal.* This principle is illustrated in Fig. 29 for only one direction of transmission (from speaker A on the *far* left of the hybrid to speaker B on the right). The adaptive canceler is placed in the four-wire path *near* the origin of the echo. The synthetic echo, denoted by $\hat{r}(n)$, is generated by passing the speech signal from speaker A (i.e., the "reference" signal for the adaptive canceler) through an adaptive filter that ideally matches the transfer function of the echo path. The reference signal, passing through the hybrid, results in the echo signal $r(n)$. This echo, together with a near-end talker signal $x(n)$ (i.e., the speech signal from speaker B) constitutes the "desired" response for the adaptive canceler. The synthetic echo $\hat{r}(n)$ is subtracted from the desired response $r(n) + x(n)$ to yield the canceler error signal

$$e(n) = r(n) - \hat{r}(n) + x(n) \tag{42}$$

Note that the error signal $e(n)$ also contains the near-end talker signal $x(n)$. In any event, the error signal $e(n)$ is used to control the adjustments made in the coefficients (tap weights) of the adaptive filter. In practice, the echo path is highly variable, depending on the distance to the hybrid, the characteristics of the two-wire circuit, and so on. These variations are taken care of by the adaptive control loop built into the canceler. The control loop continuously adapts the filter coefficients to take care of fluctuations in the echo path.

For the adaptive echo cancelation circuit to operate satisfactorily, the impulse response of the adaptive filter should have a length greater than the longest echo path that needs to be accommodated. Let T_s be the sampling period of the digitized speech signal, M be the number of adjustable coefficients (tap weights) in the adaptive filter, and τ be the longest echo delay to be accommodated. We must then choose

$$MT_s > \tau \tag{43}$$

As mentioned previously (when discussing adaptive differential pulse-code modulation), the sampling rate for speech signals on the telephone network is conservatively chosen as 8 kHz, that is,

$$T_s = 125 \ \mu s$$

Suppose, for example, that the echo delay $\tau = 30$ ms. Then we must choose

$$M > 240 \text{ taps}$$

Thus, the use of an echo canceler with $M = 256$ taps, say, is satisfactory for this situation.

Adaptive Beamforming

For our last application, we describe a *spatial* form of adaptive signal processing that finds practical use in radar, sonar, communications, geophysical exploration, astrophysical exploration, and biomedical signal processing.

In the particular type of spatial filtering of interest to us in this book, a number of independent *sensors* are placed at different points in space to "listen" to the received signal. In effect, the sensors provide a means of *sampling* the received signal *in space*. The set of sensor outputs collected at a particular instant of time constitutes a *snapshot*. Thus, a snapshot of data in spatial filtering (for the case when the sensors lie uniformly on a straight line) plays a role analogous to that of a set of consecutive tap inputs that exist in a transversal filter at a particular instant of time.[13]

In radar, the sensors consist of antenna elements (e.g., dipoles, horns, slotted waveguides) that respond to incident electromagnetic waves. In sonar, the sensors consist of hydrophones designed to respond to acoustic waves. In any event, spatial filtering, known as *beamforming*, is used in these systems to distinguish between the spatial properties of signal and noise. The device used to do the beamforming is called a *beamformer*. The term "beamformer" is derived from the fact that the early forms of antennas (spatial filters) were designed to form *pencil beams*, so as to receive a signal radiating from a specific direction and attenuate signals radiating from other directions of no interest (Van Veen and Buckley, 1988). Note that the beamforming applies to the radiation (transmission) or reception of energy.

---

[13]For a discussion of the analogies between time- and space-domain forms of signal processing, see Bracewell (1986) and Van Veen and Buckley (1988).

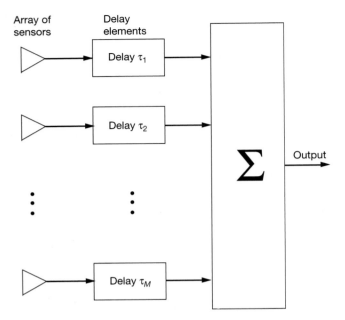

Figure 30   Delay-and-sum beamformer.

In a primitive type of spatial filtering, known as the *delay-and-sum-beamformer*, the various sensor outputs are delayed (by appropriate amounts to align signal components coming from the direction of a target) and then summed, as in Fig. 30. Thus, for a single target, the average power at the output of the delay-and-sum beamformer is maximized when it is steered toward the target. A major limitation of the delay-and-sum beamformer, however, is that it has no provisions for dealing with sources of *interference*.

In order to enable a beamformer to respond to an unknown interference environment, it has to be made *adaptive* in such a way that it places *nulls* in the direction(s) of the source(s) of interference automatically and in real time. By so doing, the output signal-to-noise ratio of the system is increased, and the *directional response* of the system is thereby improved. Below, we consider two examples of *adaptive beamformers* that are well suited for use with narrow-band signals in radar and sonar systems.

Adaptive beamformer with minimum-variance distortionless response.   Consider an adaptive beamformer that uses a linear array of M identical sensors, as in Fig. 31. The individual sensor outputs, assumed to be in *baseband* form, are weighted and then summed. The beamformer has to satisfy two requirements: (1) a *steering* capability whereby the target signal is always protected, and (2) the effects of sources of interference are minimized. One method of providing for these two requirements is to

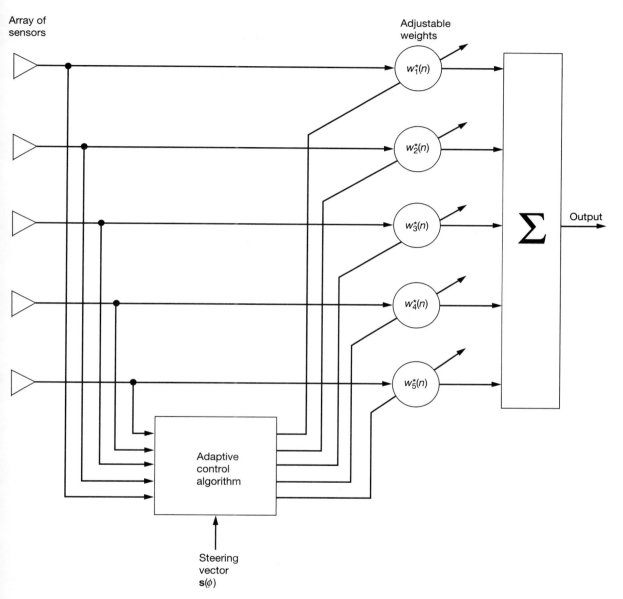

Figure 31 Adaptive beamformer for an array of 5 sensors. The sensor outputs (in baseband form) are complex valued; hence the weights are complex valued.

minimize the variance (i.e., average power) of the beamformer output, subject to the *constraint* that, during the process of adaptation, the weights satisfy the condition:

$$\mathbf{w}^H(n)\mathbf{s}(\phi) = 1 \qquad \text{for all } n, \text{ and } \phi = \phi_t \tag{44}$$

where $\mathbf{w}(n)$ is the M-by-1 weight vector and $\mathbf{s}(\phi)$ is an M-by-1 *steering vector*. The superscript H denotes Hermitian transposition (i.e., transposition combined with complex conjugation). In this application, the baseband data are complex valued; hence the need for complex conjugation. The value of *electrical angle* $\phi = \phi_t$ is determined by the direction of the target. The angle ϕ is itself measured with sensor 1 (at the top end of the array) treated as the point of reference.

The dependence of vector $\mathbf{s}(\phi)$ on the angle ϕ is defined by

$$\mathbf{s}(\phi) = [1, e^{-j\phi}, \ldots, e^{-j(M-1)\phi}]^T$$

The angle ϕ is itself related to incidence angle θ of a plane wave, measured with respect to the normal to the linear array, as follows[14]

$$\phi = \frac{2\pi d}{\lambda}\sin\theta \tag{45}$$

where d is the spacing between adjacent sensors of the array, and λ is the wavelength (see Fig. 32). The incidence angle θ lies inside the range $-\pi/2$ to $\pi/2$. The permissible values that the angle ϕ may assume lie inside the range $-\pi$ to π. This means that we must choose the spacing $d < \lambda/2$, so that there is a one-to-one correspondence between the values of θ and ϕ without ambiguity. The condition $d < \lambda/2$ may be viewed as the spatial analog of the sampling theorem.

The imposition of the *signal-protection constraint* in Eq. (44) ensures that, for a prescribed look direction, the response of the array is maintained constant (i.e., equal to 1), no matter what values are assigned to the weights. An algorithm that minimizes the variance of the beamformer output, subject to this constraint, is therefore referred to as the *minimum-variance distortionless response (MVDR) beamforming algorithm* (Capon, 1969; Owsley, 1985). The imposition of the constraint described in Eq. (44) reduces the number of "degrees of freedom" available to the MVDR algorithm to $M - 2$, where M is the number of sensors in the array. This means that the number of independent nulls produced by the MVDR algorithm (i.e., the number of independent interferences that can be canceled) is $M - 2$.

The MVDR beamforming is a special case of *linearly constrained minimum variance (LCMV) beamforming*. In the latter case, we minimize the variance of the beamformer output, subject to the constraint

$$\mathbf{w}^H(n)\mathbf{s}(\phi) = g \qquad \text{for all } n, \text{ and } \phi = \phi_t \tag{46}$$

---

[14]When a plane wave impinges on a linear array as in Fig. 32 there is a spatial delay of $d \sin\theta$ between the signals received at any pair of adjacent sensors. With a wavelength of λ, this spatial delay is translated into an electrical angular difference defined by $\phi = 2\pi(d \sin\theta/\lambda)$.

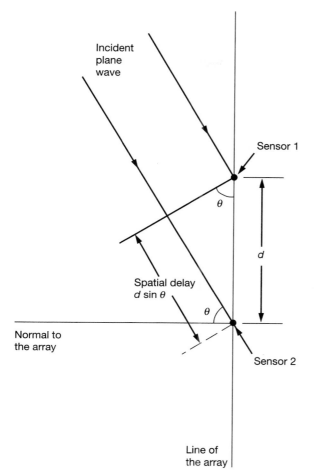

Incident plane wave

Sensor 1

θ

d

Spatial delay $d \sin \theta$

θ

Normal to the array

Sensor 2

Line of the array

Figure 32 Spatial delay incurred when a plane wave impinges on a linear array.

where g is a complex constant. The LCMV beamformer linearly constrains the weights, such that any signal coming from electrical angle ϕ_t is passed to the output with response (gain) g. Comparing the constraint of Eq. (44) with that of Eq. (46), we see that the MVDR beamformer is indeed a special case of the LCMV beamformer for $g = 1$.

Adaptation in beam space. The MVDR beamformer performs adaptation directly in the *data space*. The adaptation process for interference cancelation may also be performed in *beam space*. To do so, the input data (received by the array of sensors) are transformed into the beam space by means of an *orthogonal multiple-beamforming network*, as illustrated in the block diagram of Fig. 33. The resulting output is processed by a *multiple sidelobe canceler* so as to cancel interference(s) from unknown directions.

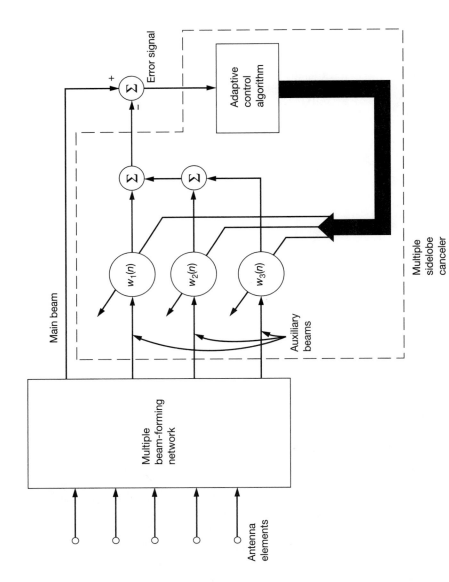

Figure 33 Block diagram of adaptive combiner with fixed beams; owing to the symmetric nature of the multiple beamforming network, final values of the weights are real valued.

The beamforming network is designed to generate a set of *orthogonal* beams. The multiple outputs of the beamforming network are referred to as *beam ports*. Assume that the sensor outputs are equally weighted and have a *uniform* phase. Under this condition, the response of the array produced by an incident plane wave arriving at the array along direction θ, measured with respect to the normal to the array, is given by

$$A(\phi, \alpha) = \sum_{n=-N}^{N} e^{jn\phi} e^{-jn\alpha} \tag{47}$$

where $M = (2N + 1)$ is the total number of sensors in the array, with the sensor at the midpoint of the array treated as the point of reference. The electrical angle ϕ is related to θ by Eq. (45), and α is a constant called the *uniform phase factor*. The quantity $A(\phi, \alpha)$ is called the *array pattern*. For $d = \lambda/2$, we find from Eq. (45) that

$$\phi = \pi \sin \theta$$

Summing the geometric series in Eq. (47), we may express the array pattern as

$$A(\phi, \alpha) = \frac{\sin[\frac{1}{2}(2N + 1)(\phi - \alpha)]}{\sin[\frac{1}{2}(\phi - \alpha)]} \tag{48}$$

By assigning different values to α, the main beam of the antenna is thus scanned across the range $-\pi < \phi \leq \pi$. To generate an orthogonal set of beams, equal to $2N$ in number, we assign the following discrete values to the uniform phase factor

$$\alpha = \frac{\pi}{2N + 1} k, \qquad k = \pm 1, \pm 3, \ldots, \pm 2N - 1 \tag{49}$$

Figure 34 illustrates the variations of the magnitude of the array pattern $A(\phi, \alpha)$ with ϕ for the case of $2N + 1 = 5$ elements and $\alpha = \pm\pi/5, \pm 3\pi/5$. Note that owing to the symmetric nature of the beamformer, the final values of the weights are real valued.

The orthogonal beams generated by the beamforming network represent $2N$ independent *look directions*, one per beam. Depending on the target direction of interest, a particular beam in the set is identified as the *main beam* and the remainder are viewed as *auxiliary beams*. We note from Fig. 34 that each of the auxiliary beams has a *null in the look direction of the main beam*. The auxiliary beams are adaptively weighted by the multiple sidelobe canceler so as to form a cancelation beam that is subtracted from the main beam. The resulting estimation error is fed back to the multiple sidelobe canceler so as to control the corrections applied to its adjustable weights.

Since all the auxiliary beams have nulls in the look direction of the main beam, and the main beam is excluded from the multiple sidelobe canceler, the overall output of the adaptive beamformer is constrained to have a constant response in the look direction of the main beam (i.e., along the direction of the target). Moreover, with $(2N - 1)$ degrees of freedom (i.e., the number of available auxiliary beams) the system is capable of placing up to $(2N - 1)$ nulls along the (unknown) directions of independent interferences.

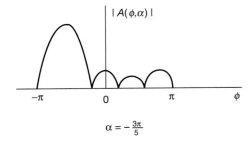

$$\alpha = -\frac{3\pi}{5}$$

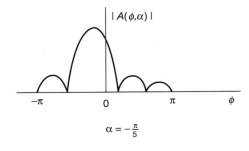

$$\alpha = -\frac{\pi}{5}$$

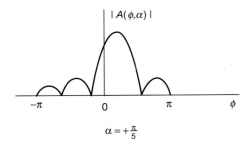

$$\alpha = +\frac{\pi}{5}$$

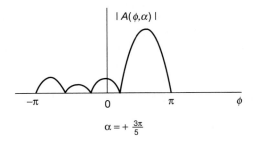

$$\alpha = +\frac{3\pi}{5}$$

Figure 34 Variations of the magnitude of the array pattern $A(\phi, \alpha)$ with ϕ and α.

Introduction

Note that with an array of $(2N + 1)$ sensors, we may produce a beamforming network with $(2N + 1)$ orthogonal beam ports by assigning the uniform phase factor the following set of values:

$$\alpha = \frac{k\pi}{2N - 1}, \qquad k = 0, \pm 2, \ldots, \pm 2N \qquad (50)$$

In this case, a small fraction of the main lobe of the beam port at either end lies in the non-visible region. Nevertheless, with one of the beam ports providing the main beam and the remaining $2N$ ports providing the auxiliary beams, the adaptive beamformer is now capable of producing up to $2N$ independent nulls.

8. SOME HISTORICAL NOTES

> To understand a science it is necessary to know its history.
> Auguste Comte (1798—1857)

We complete this introductory chapter by presenting a brief historical review of developments in four areas that are closely related insofar as the subject matter of this book is concerned. The areas are: linear estimation theory, linear adaptive filters, neural networks, and adaptive signal-processing applications.

Linear Estimation Theory

The earliest stimulus for the development of linear estimation theory[15] was apparently provided by astonomical studies in which the motion of planets and comets was studied using telescopic measurement data. The beginnings of a "theory" of estimation in which attempts are made to minimize various functions of errors can be attributed to Galileo Galilei in 1632. However, the origin of linear estimation theory is credited to Gauss who, at the age of 18 in 1795, invented the *method of least squares* to study the motion of heavenly bodies (Gauss, 1809). Nevertheless, in the early nineteenth century, there was considerable controversy regarding the actual inventor of the method of least squares. The controversy arose because Gauss did not publish his discovery in 1795. Rather, it was first published by Legendre in 1805, who independently invented the method (Legendre, 1810).

The first studies of minimum mean-square estimation in stochastic processes were made by Kolmogorov, Krein, and Wiener during the late 1930s and early 1940s (Kolmogorov, 1939; Krein, 1945; Wiener, 1949). The works of Kolmogorov and Krein were independent of Wiener's, and while there was some overlap in the results, their aims were rather different. There were many conceptual differences (as one would expect after 140 years) between Gauss's problem and the problem treated by Kolmogorov, Krein, and Wiener.

---

[15]The notes presented on linear estimation are influenced by the following review papers: Sorenson (1970), Kailath (1974), and Makhoul (1975).

Kolmogorov, inspired by some early work of Wold on discrete-time stationary processes (Wold, 1938), developed a comprehensive treatment of the linear prediction problem for discrete-time stochastic processes. Krein noted the relationship of Kolmogorov's results to some early work by Szegö on orthogonal polynomials (Szegö, 1939; Grenander and Szegö, 1958) and extended the results to continuous time by clever use of a bilinear transformation.

Wiener, independently, formulated the continuous-time linear prediction problem and derived an explicit formula for the optimum predictor. Wiener also considered the "filtering" problem of estimating a process corrupted by an additive "noise" process. The explicit formula for the optimum estimate required the solution of an integral equation known as the *Wiener–Hopf equation* (Wiener and Hopf, 1931).

In 1947, Levinson formulated the Wiener filtering problem in discrete time. In the case of discrete-time signals, the Wiener–Hopf equation takes on a matrix form described by[16]

$$\mathbf{R}\mathbf{w}_0 = \mathbf{p} \tag{51}$$

where $\mathbf{w}_0$ is the tap-weight vector of the optimum Wiener filter structured in the form of a transversal filter, $\mathbf{R}$ is the correlation matrix of the tap inputs, and $\mathbf{p}$ is the cross-correlation vector between the tap inputs and the desired response. For stationary inputs, the correlation matrix $\mathbf{R}$ assumes a special structure known as *Toeplitz*, so named after the mathematician O. Toeplitz. By exploiting the properties of a Toeplitz matrix, Levinson derived an elegant recursive procedure for solving the matrix form of the Wiener–Hopf equation (Levinson, 1947). In 1960, Durbin rediscovered Levinson's recursive procedure as a scheme for recursive fitting of autoregressive models to scalar time-series data (Durbin, 1960). The problem considered by Durbin is a special case of Eq. (51) in that the column vector $\mathbf{p}$ comprises the same elements found in the correlation matrix $\mathbf{R}$. In 1963, Whittle showed there is a close relationship between the Levinson–Durbin recursion and that for Szegö's orthogonal polynomials, and also derived a multivariate generalization of the Levinson–Durbin recursion (Whittle, 1963).

Wiener and Kolmogorov assumed an infinite amount of data and assumed the stochastic processes to be stationary. During the 1950s, some generalizations of the Wiener–Kolmogorov filter theory were made by various authors to cover the estimation of stationary processes given only for a finite observation interval and to cover the estimation of nonstationary processes. However, there were dissatisfactions with the most significant of the results of this period because they were rather complicated, difficult to update with increases in the observations interval, and difficult to modify for the vector case. These last two difficulties became particularly evident in the late 1950s in the problem of determining satellite orbits. In this application, there were generally vector observations of

---

[16]The Wiener–Hopf equation, originally formulated as an integral equation, specifies the optimum solution of a continuous-time linear filter subject to the constraint of causality. This is a difficult-to-solve equation that has resulted in the development of a considerable amount of theory, including spectral factorization. For a tutorial treatment of this subject, see Gardner (1990).

some combinations of position and velocity, and there were also large amounts of data sequentially accumulated with each pass of the satellite over a tracking station. Swerling was one of the first to tackle this problem by presenting some useful recursive algorithms (Swerling, 1958). For different reasons, Kalman independently developed a somewhat more restricted algorithm than Swerling's, but it was an algorithm that seemed particularly matched to the dynamical estimation problems that were brought by the advent of the space age (Kalman, 1960). After Kalman had published his paper and it had attained considerable fame, Swerling wrote a letter claiming priority for the Kalman filter equations (Swerling, 1963). However, history shows that Swerling's plea has fallen on deaf ears. It is ironic that orbit determination problems provided the stimulus for both Gauss's method of least squares and the Kalman filter, and that there were squabbles concerning their inventors. Kalman's original formulation of the linear filtering problem was derived for discrete-time processes. The continuous-time filter was derived by Kalman in his subsequent collaboration with Bucy; this latter solution is sometimes referred to as the *Kalman–Bucy filter* (Kalman and Bucy, 1961).

In a series of stimulating papers, Kailath reformulated the solution to the linear filtering problem by using the *innovations* approach (Kailath, 1968, 1970; Kailath and Frost, 1968; Kailath and Geesey, 1973). In this approach, a stochastic process $u(n)$ is represented as the output of a causal and causally invertible filter driven by a white-noise process $v(n)$. The white noise process $v(n)$ is called the *innovations process*, with the term "innovation" denoting "newness." The reason for this terminology is that each sample of the process $v(n)$ provides entirely new information, in the sense that it is statistically independent of all past samples of the original process $u(n)$, assuming Gaussianity; otherwise, it is only uncorrelated with all past samples of $u(n)$. The idea of innovations approach was introduced by Kolmogorov (1941).

Linear Adaptive Filters

Stochastic gradient algorithms. The earliest work on adaptive filters may be traced back to the late 1950s, during which time a number of researchers were working independently on different applications of adaptive filters. From this early work, the *least-mean-square (LMS) algorithm* emerged as a simple and yet effective algorithm for the operation of adaptive transversal filters. The LMS algorithm was devised by Widrow and Hoff in 1959 in their study of a pattern recognition scheme known as the *adaptive linear* (threshold logic) *element*, commonly referred to in the literature as the *Adaline* (Widrow and Hoff, 1960; Widrow, 1970). The LMS algorithm is a stochastic gradient algorithm in that it iterates each tap weight of a transversal filter in the direction of the gradient of the squared magnitude of an error signal with respect to the tap weight. As such, the LMS algorithm is closely related to the concept of *stochastic approximation* developed by Robbins and Monro (1951) in statistics for solving certain sequential parameter estimation problems. The primary difference between them is that the LMS algorithm uses a fixed step-size parameter to control the correction applied to each tap weight from one iteration to the next, whereas in stochastic approximation methods the step-size parameter is made inversely proportional to time n or to a power of n. Another stochastic gradient algorithm, closely related to the LMS algorithm, is the *gradient adaptive lattice (GAL) algorithm* (Griffiths, 1977, 1978); the difference between them is structural in that the GAL algorithm is lattice-based, whereas the LMS algorithm uses a transversal filter.

In 1981, Zames introduced the H^{∞} *norm* (or *minimax criterion*) as a robust index of performance for solving problems in estimation and control, and with it the field of robust control took on a new research direction. In this context, it is particularly noteworthy that Hassibi et al. (1996) have shown that the LMS algorithm is indeed optimal under the H^{∞} criterion. Thus, for the first time, theoretical evidence was presented for the robust performance of the LMS algorithm. It is also of interest to note that the zero-forcing algorithm, which represents an alternative to the LMS algorithm for the adaptive equalization of communication channels, also uses a minimax type of performance criterion (Lucky, 1965).

Recursive least-squares algorithms.   Turning next to the recursive least-squares (RLS) family of adaptive filtering algorithms, the original paper on the *standard RLS algorithm* appears to be that of Plackett (1950), though it must be said that many other investigators have derived and rederived the RLS algorithm. In 1974, Godard used Kalman filter theory to derive a variant of the RLS algorithm, which is sometimes referred to in the literature as the *Godard algorithm*. Although prior to this date, several investigators had applied Kalman filter theory to solve the adaptive filtering problem, Godard's approach was widely accepted as the most successful application of Kalman filter theory for a span of two decades. Then, Sayed and Kailath (1994) published an expository paper, in which the *exact* relationship between the RLS algorithm and Kalman filter theory was delineated for the first time, thereby laying the groundwork for how to exploit the vast literature on Kalman filters for solving linear adaptive filtering problems.

In 1981, Gentleman and Kung introduced a numerically robust method, based on the *QR-decomposition* of matrix algebra, for solving the recursive least-squares problem. The resulting adaptive filter structure, sometimes referred to as the *Gentleman–Kung (systolic) array*, was subsequently refined and extended in various ways by many other investigators.

In the 1970s and during subsequent years, a great deal of research effort was expended on the development of numerically stable *fast RLS algorithms*, with the aim of reducing computational complexity to a level comparable to that of the LMS algorithm. In one form or another, the development of these algorithms can be traced back to results derived by Morf in 1974 for solving the deterministic counterpart of the stochastic filtering problem solved efficiently by the Levinson–Durbin algorithm for stationary inputs.

Returning to the paper by Sayed and Kailath (1994), the one-to-one correspondences between RLS and Kalman variables was exploited in that paper to show that QR-decomposition-based RLS algorithms and fast RLS algorithms are all in fact special cases of the Kalman filter, thereby providing a unified treatment of the RLS family of linear adaptive filters in a rather elegant and compact fashion.

Neural Networks[17]

Research interest in neural networks began with the pioneering work of McCulloch and Pitts (1943), who described a logical calculus for neural networks. Then, in 1958, Rosenblatt introduced a new approach to the pattern-classification problem using a neural network known as the *perceptron*. Out of this early work on neural networks, the LMS algorithm was pioneered by Widrow and Hoff in 1959, which, as mentioned previously, was used to formulate the Adaline. In the 1960s, it seemed as if neural networks could solve any problem. But then came the book by Minsky and Papert (1969), who used elegant mathematics to demonstrate that there are fundamental limits on what single-layer perceptrons can compute, and with it interest in neural networks took a sharp downturn.

In 1986, successful development of the *back-propagation algorithm* was reported by Rumelhart, Hinton, and Williams as a device for the training of multilayer perceptrons; the back-propagation algorithm is a generalization of the LMS algorithm. In that same year, the two-volume seminal book, *Parallel Distributed Processing: Explorations in the Microstructures of Cognition*, with Rumelhart and McClelland as editors, was published. This book has been a major influence in reviving interest in the use of neural networks. After the publication of this book, however, it became known that the back-propagation algorithm had actually been described earlier by Werbos in his Ph.D. thesis at Harvard University in 1974.

The multilayer perceptron represents one important type of feedforward layered network that is well suited for adaptive signal processing. Another equally important feedforward layered network is the *radial-basis function (RBF) network*, which was described by Broomhead and Lowe in 1988. However, the basic idea of RBF networks may be traced back to earlier work by Bashkirov, Braverman, and Muchnick in 1964 on the method of potential functions.

The field of neural networks encompasses many other types of network structures and learning algorithms. Indeed, they have been established as an interdisciplinary subject with deep roots in the neurosciences, psychology, mathematics, the physical sciences, and engineering. Needless to say, they have a major impact on adaptive signal processing, particularly in those applications that require the use of nonlinearity.

Adaptive Signal-Processing Applications

Adaptive Equalization. Until the early 1960s, the equalization of telephone channels to combat the degrading effects of intersymbol interference on data transmission was performed by using either fixed equalizers (resulting in a performance loss) or equalizers whose parameters were adjusted manually (a rather cumbersome procedure). In 1965, Lucky made a major breakthrough in the equalization problem by proposing a *zero-forcing algorithm* for automatically adjusting the tap weights of a transversal equalizer. A distinguishing feature of the work by Lucky was the use of a *minimax* type of performance

---

[17]For a more complete historical account of neural networks, see Cowan (1990) and Haykin (1994).

criterion. In particular, he used a performance index called *peak distortion*, which is directly related to the maximum value of intersymbol interference that can occur. The tap weights in the equalizer are adjusted to minimize the peak distortion. This has the effect of *forcing* the intersymbol interference due to those adjacent pulses that are contained in the transversal equalizer to become *zero*; hence the name of the algorithm. A sufficient, but not necessary, condition for optimality of the zero-forcing algorithm is that the *initial distortion* (the distortion that exists at the equalizer input) be less than unity. In a subsequent paper published in 1966, Lucky extended the use of the zero-forcing algorithm to the tracking mode of operation. In 1965, DiToro independently used adaptive equalization for combatting the effect of intersymbol interference on data transmitted over high-frequency links.

The pioneering work by Lucky inspired many other significant contributions to different aspects of the adaptive equalization problem in one way or another. Gersho (1969) and Proakis and Miller (1969) independently reformulated the adaptive equalization problem using a mean-square-error criterion. In 1972, Ungerboeck presented a detailed mathematical analysis of the convergence properties of an adaptive transversal equalizer using the LMS algorithm. In 1974, as mentioned previously, Godard used Kalman filter theory to derive a powerful algorithm for adjusting the tap weights of a transversal equalizer. In 1978, Falconer and Ljung presented a modification of this algorithm that simplified its computational complexity to a level comparable to that of the simple LMS algorithm. Satorius and Alexander (1979) and Satorius and Pack (1981) demonstrated the usefulness of lattice-based algorithms for adaptive equalization of dispersive channels.

This brief historical review pertains to the use of adaptive equalizers for *linear synchronous receivers*; by "synchronous" we mean that the equalizer in the receiver has its taps spaced at the reciprocal of the symbol rate. Even though our interest in adaptive equalizers is largely restricted to this class of receivers, nevertheless, such a historical review would be incomplete without some mention of fractionally spaced equalizers and decision-feedback equalizers.

In a *fractionally spaced equalizer (FSE)*, the equalizer taps are spaced closer than the reciprocal of the symbol rate. An FSE has the capability of compensating for delay distortion much more effectively than a conventional synchronous equalizer. Another advantage of the FSE is the fact that data transmission may begin with an arbitrary sampling phase. However, mathematical analysis of the FSE is much more complicated than for a conventional synchronous equalizer. It appears that early work on the FSE was initiated by Brady (1970). Other contributions to the subject include subsequent work by Ungerboeck (1976) and Gitlin and Weinstein (1981).

A *decision-feedback equalizer* consists of a feedforward section and a feedback section connected as shown in Fig. 35. The feedforward section itself consists of a transversal filter whose taps are spaced at the reciprocal of the symbol rate. The data sequence to be equalized is applied to the input of this section. The feedback section consists of another transversal filter whose taps are also spaced at the reciprocal of the symbol rate. The input applied to the feedback section is made up of decisions on previously detected symbols. The function of the feedback section is to subtract out that portion of intersymbol interference produced by previously detected symbols from the estimates of future symbols. This

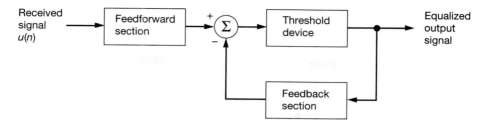

Figure 35   Block diagram of decision-feedback equalizer.

cancelation is an old idea known as the *bootstrap technique*. A decision-feedback equalizer yields good performance in the presence of severe intersymbol interference as experienced in fading radio channels, for example. The first report on decision-feedback equalization was published by Austin (1967), and the optimization of the decision-feedback receiver for minimum mean-squared error was first accomplished by Monsen (1971).

Coding of speech.   In 1966, Saito and Itakura used a *maximum likelihood* approach for the application of prediction to speech. A standard assumption in the application of the maximum likelihood principle is that the input process is Gaussian. Under this condition, the exact application of the maximum likelihood principle yields a set of nonlinear equations for the parameters of the predictor. To overcome this difficulty, Itakura and Saito utilized approximations based on the assumption that the number of available data points greatly exceeds the prediction order. The use of this assumption makes the result obtained from the maximum likelihood principle assume an approximate form that is the same as the *autocorrelation method* of linear prediction. The application of the maximum likelihood principle is justified on the assumption that speech is a stationary Gaussian process, which seems reasonable in the case of unvoiced sounds.

In 1970, Atal presented the first use of the term "linear prediction" for speech analysis. Details of this new approach, linear predictive coding (LPC), to speech analysis and synthesis were published by Atal and Hanauer in 1971, in which the speech waveform is represented directly in terms of time-varying parameters related to the transfer function of the vocal tract and the characteristics of the excitation. The predictor coefficients are determined by minimizing the mean-squared error, with the error defined as the difference between the actual and predicted values of the speech samples. In the work by Atal and Hanauer, the speech wave was sampled at 10 kHz and then analyzed by predicting the present speech sample as a linear combination of the 12 previous samples. Thus 15 parameters [the 12 parameters of the predictor, the pitch period, a binary parameter indicating whether the speech is voiced or unvoiced, and the root-mean-square (rms) value of the speech samples] were used to describe the speech analyzer. For the speech synthesizer, an all-pole filter was used, with a sequence of quasi-periodic pulses or a white-noise source providing the excitation.

Another significant contribution to the linear prediction of speech was made in 1972 by Itakura and Saito; they used partial correlation techniques to develop a new structure, the lattice, for formulating the linear prediction problem.[18] The parameters that characterize the lattice predictor are called *reflection coefficients* or *partial correlation (PARCOR) coefficients*, depending on the algebraic sign used in the definition. Although by that time the essence of the lattice structure had been considered by several other investigators, the invention of the lattice predictor is credited to Saito and Itakura. In 1973, Wakita showed that the filtering actions of the lattice predictor model and an acoustic tube model of speech are identical, with the reflection coefficients in the acoustic tube model as common factors. This discovery made possible the extraction of the reflection coefficients by the use of a lattice predictor.

Early designs of a lattice predictor were based on a *block processing* approach (Burg, 1967). In 1981, Makhoul and Cossell used an *adaptive* approach for designing the lattice predictor for applications in speech analysis and synthesis. They showed that the convergence of the adaptive lattice predictor is fast enough for its performance to equal that of the optimal (but more expensive) adaptive autocorrelation method.

This historical review on speech coding relates to LPC vocoders. We next present a historical review of the adaptive predictive coding of speech, starting with ordinary pulse-code modulation (PCM).

PCM was invented in 1937 by Reeves (1975). This was followed by the invention of differential pulse-code modulation (DPCM) by Cutler (1952). The early use of DPCM for the predictive coding of speech signals was limited to linear predictors with *fixed* parameters (McDonald, 1966). However, due to the nonstationary nature of speech signals, a fixed predictor cannot predict the signal values efficiently at all times. In order to respond to the nonstationary characteristics of speech signals, the predictor has to be adaptive (Atal and Schroeder, 1967). In 1970, Atal and Schroeder described a sophisticated scheme for adaptive predictive coding of speech. The scheme recognizes that there are two main causes of redundancy in speech (Schroeder, 1966): (1) quasi-periodicity during voiced segments, and (2) lack of flatness of the short-time spectral envelope. Thus, the predictor is designed to remove signal redundancy in two stages. The first stage of the predictor removes the quasi-periodic nature of the signal. The second stage removes formant information from the spectral envelope. The scheme achieves dramatic reductions in bit rate at the expense of a significant increase in circuit complexity. Atal and Schroeder (1970) report that the scheme can transmit speech at 10 kb/s, which is several times less than the bit rate required for logarithmic-PCM encoding with comparable speech quality.

Spectrum analysis.   At the turn of the twentieth century, Schuster introduced the *periodogram* for analyzing the power spectrum[19] of a time series (Schuster, 1898). The periodogram is defined as the squared amplitude of the discrete Fourier transform of the time series. The periodogram was originally used by Schuster to detect and estimate the amplitude of a sine wave of known frequency that is buried in noise. Until the work of

---

[18]According to Markel and Gray (1976), the work of Itakura and Saito in Japan on the PARCOR formulation of linear prediction had been presented in 1969.

[19]For a fascinating historical account of the concept of power spectrum, its origin and its estimation, see Robinson (1982).

Yule in 1927, the periodogram was the only numerical method available for spectrum analysis. However, the periodogram suffers from the limitation that when it is applied to empirical time series observed in nature the results obtained are very erratic. This led Yule to introduce a new approach based on the concept of a *finite parameter model* for a stationary stochastic process in his investigation of the periodicities in time series with special reference to Wolfer's sunspot number (Yule, 1927). Yule, in effect, created a stochastic feedback model in which the present sample value of the time series is assumed to consist of a linear combination of past sample values plus an error term. This model is called an autoregressive model in that a sample of the time series regresses on its own past values, and the method of spectrum analysis based on such a model is accordingly called autoregressive spectrum analysis. The name "autoregressive" was coined by Wold in his doctoral thesis (Wold, 1938).

Interest in the autoregressive method was reinitiated by Burg (1967, 1975). Burg introduced the term *maximum-entropy method* to describe an algorithmic approach for estimating the power spectrum directly from the available time series. The idea behind the maximum-entropy method is to extrapolate the autocorrelation function of the time series in such a way that the *entropy* of the corresponding probability density function is maximized at each step of the extrapolation. In 1971, Van den Bos showed that the maximum-entropy method is equivalent to least-squares fitting of an autoregressive model to the known autocorrelation sequence.

Another important contribution made to the literature on spectrum analysis is that by Thomson (1982). His *method of multiple windows*, based on the prolate spheroidal wave functions, represents a nonparametric method for spectrum estimation that overcomes many of the limitations of the above-mentioned techniques.

Adaptive Noise Cancelation.   The initial work on adaptive echo cancelers started around 1965. It appears that Kelly of Bell Telephone Laboratories was the first to propose the use of an adaptive filter for echo cancelation, with the speech signal itself utilized in performing the adaptation; Kelly's contribution is recognized in the paper by Sondhi (1967). This invention and its refinement are described in the patents by Kelly and Logan (1970) and Sondhi (1970).

The adaptive line enhancer was originated by Widrow and his co-workers at Stanford University. An early version of this device was built in 1965 to cancel 60-Hz interference at the output of an electrocardiographic amplifier and recorder. This work is described in the paper by Widrow et al. (1975b). The adaptive line enhancer and its application as an adaptive detector are patented by McCool et al. (1980).

The adaptive echo canceler and the adaptive linear enhancer, although intended for different applications, may be viewed as examples of the *adaptive noise canceler* discussed by Widrow et al. (1975). This scheme operates on the outputs of two sensors: a *primary sensor* that supplies a desired signal of interest buried in noise, and a *reference sensor* that supplies noise alone, as illustrated in Fig. 24. It is assumed that (1) the signal and noise at the output of the primary sensor are uncorrelated, and (2) the noise at the output of the reference sensor is correlated with the noise component of the primary sensor output.

The adaptive noise canceler consists of an adaptive filter that operates on the reference sensor output to produce an *estimate* of the noise, which is subtracted from the primary sensor output. The overall output of the canceler is used to control the adjustments applied to the tap weights in the adaptive filter. The adaptive canceler tends to minimize the mean-square value of the overall output, thereby causing the output to be the best estimate of the desired signal in the minimum-mean-square error sense.

Adaptive beamforming.   The development of adaptive beamforming technology may be traced back to the invention of the *intermediate frequency (IF) sidelobe canceler* by Howells in the late 1950s. In a paper published in the 1976 Special Issue of the IEEE Transactions on Antennas and Propagation, Howells describes his personal observations on early work on adaptive antennas at the General Electric and Syracuse University Research Corporation (Howells, 1976). According to this historic report, Howells had developed by mid-1957 a sidelobe canceler capable of automatically nulling out the effect of one jammer. The sidelobe canceler uses a *primary* (high-gain) antenna and a *reference omni-directional* (low-gain) antenna to form a two-element array with one degree of freedom that makes it possible to steer a deep null anywhere in the sidelobe region of the combined antenna pattern. In particular, a null is placed in the direction of the jammer, with only a minor perturbation of the main lobe. Subsequently, Howells (1965) patented the sidelobe canceler.

The second major contribution to adaptive array antennas was made by Applebaum in 1966. In a classic report, he derived the *control law* governing the operation of an adaptive array antenna, with a control loop for each element of the array (Applebaum, 1966). The algorithm derived by Applebaum was based on maximizing the signal-to-noise ratio (SNR) at the array antenna output for any type of noise environment. Applebaum's theory included the sidelobe canceler as a special case. His 1966 classic report was reprinted in the 1976 Special Issue of IEEE Transactions on Antennas and Propagation.

Another algorithm for the weight adjustment in adaptive array antennas was advanced independently in 1967 by Widrow and his co-workers at Stanford University. They based their theory on the simple and yet effective LMS algorithm. The 1967 paper by Widrow et al. was not only the first publication in the open literature on adaptive array antenna systems, but also it is considered to be another classic of that era.

It is noteworthy that the maximum SNR algorithm (used by Applebaum) and the LMS algorithm (used by Widrow and his co-workers) for adaptive array antennas are rather similar. Both algorithms derive the control law for adaptive adjustment of the weights in the array antenna by sensing the correlation between element signals. Indeed, they both converge toward the optimum Wiener solution for stationary inputs (Gabriel, 1976).

A different method for solving the adaptive beamforming problem was proposed by Capon (1969). Capon realized that the poor performance of the delay-and-sum beamformer is due to the fact that its response along a direction of interest depends not only on the power of the incoming target signal but also undesirable contributions received from other sources of interference. To overcome this limitation of the delay-and-sum beamformer, Capon proposed a new beamformer in which the weight vector $\mathbf{w}(n)$ is chosen so as to *minimize the variance* (i.e., average power) of the beamformer output, subject to the constraint $\mathbf{w}^H(n)\mathbf{s}(\phi) = 1$ for all n, where $\mathbf{s}(\phi)$ is a prescribed *steering vector*. This constrained minimization yields an adaptive beamformer with *minimum-variance distortionless response (MVDR)*.

In 1983, McWhirter proposed a simplification of the Gentleman–Kung (systolic) array for recursive least-squares estimation. The resulting filtering structure, often referred to as the *McWhirter (systolic) array*, is particularly well suited for adaptive beamforming applications.

The historical notes presented in this last section of the chapter on adaptive filter theory and applications are not claimed to be complete. Rather, they are intended to highlight many of the significant contributions made to this important part of the ever-expanding field of signal processing. Above all, it is hoped that they provide a source of inspiration to the reader.

PART 1
Background Material

Part I consists of Chapters 1 through 4. In this part of the book we present background material on discrete-time signals and systems and thereby lay a foundation for the rest of the book, as summarized here:

- Chapter 1 reviews fundamentals of discrete-time signal processing, with emphasis on the *z*-transform, the discrete Fourier transform, and the discrete cosine transform.
- Chapter 2 covers the time-domain characteristics of discrete-time stochastic processes.
- Chapter 3 covers the frequency-domain characteristics of discrete-time stochastic processes, with particular emphasis on the notion of a power spectrum or power spectral density. Higher-order statistics and cyclostationary properties of stochastic processes are also discussed here.
- In Chapter 4 we study the eigenvalue problem, which is central to a detailed mathematical description of discrete-time wide-sense stationary processes.

CHAPTER

1

Discrete-time Signal Processing

Typically, a signal of interest is described as a function of time. The transformation of a signal from the time domain into the frequency domain plays a key role in the study of signal processing. The particular transformation used in practice depends on the type of signal being considered. Given the pervasive nature of digital processing and the benefits (flexibility and accuracy of computation) offered by its use, our interest in this book is confined to discrete-time signals. Specifically, the signal is described as a *time series*, consisting of a sequence of uniformly spaced samples whose varying amplitudes carry the useful information content of the signal. In such a situation, the transforms that immediately come to mind are two closely related transforms, namely, the z-transform and the Fourier transform. The Fourier transform is defined in terms of a real variable (frequency), whereas the z-transform is defined in terms of a complex variable.

In this chapter we present a brief review of discrete-time signal processing,[1] beginning with a definition of the z-transform and its properties.

1.1 *z*-TRANSFORM

Consider a time series (sequence) denoted by the samples $u(n)$, $u(n-1)$, $u(n-2)$, . . . , where n denotes *discrete time*. For convenience of presentation, it is assumed that the

[1]For a detailed treatment of the many facets of discrete-time signal processing, see Oppenheim and Schafer (1989).

spacing between adjacent samples of the sequence is unity. The sequence is written as $\{u(n)\}$ or simply $u(n)$. The two-sided *z-transform* of $u(n)$ is defined as

$$U(z) = z[u(n)]$$

$$= \sum_{n=-\infty}^{\infty} u(n) z^{-n} \tag{1.1}$$

where z is a *complex variable*. The first line of Eq. (1.1) describes the z-transform as an "operator," and the second line defines it as an infinite power series in z. The sequence $u(n)$ and its z-transform form a *z-transform pair*, described by

$$u(n) \rightleftharpoons U(z) \tag{1.2}$$

The power series defined in Eq. (1.1) is a *Laurent series*, which features prominently in the functional theory of complex variables; a brief review of complex variable theory is presented in Appendix A. The important point to note here is that for the z-transform $U(z)$ to be meaningful, the power series defined in Eq. (1.1) must be absolutely summable; that is $U(z)$ is uniformly convergent. For any given time series $u(n)$, the set of values of the complex variable z for which the z-transform $U(z)$ is uniformly convergent is referred to as the *region of convergence* (ROC).

Let the region of convergence of the z-transform $U(z)$ be denoted by the annular domain $R_1 < |z| < R_2$. Let $\mathscr{C}$ be a closed contour that encloses the origin and is contained in this region of convergence. Then, given the z-transform $U(z)$, the original time series $u(n)$ may be uniquely recovered using the *z-transform inversion integral formula* (see Appendix A)

$$u(n) = \frac{1}{2\pi j} \oint_{\mathscr{C}} U(z) z^n \frac{dz}{z} \tag{1.3}$$

where the contour integration is performed by transversing the contour $\mathscr{C}$ in the counterclockwise direction.

Properties of the *z*-Transform

The z-transform is a *linear transform* in that it satisfies the principle of superposition. Specifically, given two sequences $u_1(n)$ and $u_2(n)$ whose z-transforms are denoted by $U_1(z)$ and $U_2(z)$, respectively, we may write

$$a\,u_1(n) + b\,u_2(n) \rightleftharpoons a\,U_1(z) + b\,U_2(z) \tag{1.4}$$

where a and b are scaling factors. The region of convergence, for which Eq. (1.4) holds, contains the intersection of the regions of convergence of $U_1(z)$ and $U_2(z)$.

Another important property of the z-transform is the *time-shifting property*. Let $U(z)$ denote the z-transform of the sequence $u(n)$. The z-transform of $u(n - n_0)$ is described by the relation

$$u(n - n_0) \rightleftharpoons z^{-n_0} U(z) \tag{1.5}$$

where n_0 is an integer. Equation (1.5) holds for the same region of convergence as the original time series $u(n)$, except for a possible addition or deletion of $z = 0$ or $z = \infty$. For the special case of $n_0 = 1$, we see that such a time shift has the effect of multiplying the z-transform $U(z)$ by the factor z^{-1}. It is for this reason that z^{-1} is commonly referred to as a *unit-delay element*.

One other property of the z-transform of particular interest to us is the *convolution theorem*. Let $U_1(z)$ and $U_2(z)$ denote the z-transforms of the time series $u_1(n)$ and $u_2(n)$, respectively. According to the convolution theorem, we have

$$\sum_{i=-\infty}^{\infty} u_1(i) u_2(n-i) \rightleftharpoons U_1(z) U_2(z) \tag{1.6}$$

where the region of convergence includes the intersection of the regions of convergence of $U_1(z)$ and $U_2(z)$. The proof of Eq. (1.6) follows directly from the defining equation (1.1). In other words, convolution of two sequences in the time domain is transformed into multiplication of their z-transforms in the frequency domain.

1.2 LINEAR TIME-INVARIANT FILTERS

The z-transform plays a key role in the study of a particular class of filters known as *linear time-invariant filters*, which are characterized by the following two properties: linearity and time invariance. The *linearity* property means that the filter satisfies the principle of superposition. Specifically, if $v_1(n)$ and $v_2(n)$ are two different *excitations* applied to the filter and $u_1(n)$ and $u_2(n)$ are the *responses* produced by the filter, respectively, then the response of the filter to the composite excitation $a\, v_1(n) + b\, v_2(n)$ is equal to $a\, u_1(n) + b\, u_2(n)$, where a and b are arbitrary constants. The *time-invariance* property means that if $u(n)$ is the response of the filter due to the excitation $v(n)$, then the response of the filter to the new excitation $v(n - k)$ is equal to $u(n - k)$, where k is an arbitrary time shift.

One useful way of describing a linear time-invariant filter is in terms of its *impulse response*, defined as the response of the filter to a unit impulse or delta function applied to the filter at zero time. Let $h(n)$ denote the impulse response of the filter. The response $u(n)$ of the filter produced by an arbitrary excitation $v(n)$ is defined by the *convolution sum*

$$u(n) = \sum_{i=-\infty}^{\infty} h(i) v(n-i) \tag{1.7}$$

Applying the z-transform to both sides of Eq. (1.7) and invoking the convolution theorem, we may write

$$U(z) = H(z)\, V(z) \tag{1.8}$$

where $U(z)$, $V(z)$, and $H(z)$ are the z-transforms of $u(n)$, $v(n)$, and $h(n)$, respectively.

The z-transform $H(z)$ [i.e., the z-transform of the impulse response $h(n)$] is called the *transfer function* of the filter; it provides the basis of another way of describing a linear time-invariant filter. According to Eq. (1.8), we have

$$H(z) = \frac{U(z)}{V(z)} \tag{1.9}$$

Thus, the transfer function $H(z)$ is equal to the ratio of the z-transform of the filter's response to the z-transform of the excitation applied to the filter.

In an important subclass of linear time-invariant filters, the input sequence (excitation) $v(n)$ and the output sequence (response) $u(n)$ are related by a difference equation of order N as follows:

$$\sum_{j=0}^{N} a_j u(n - j) = \sum_{j=0}^{N} b_j v(n - j) \tag{1.10}$$

where the a_j and the b_j are constant coefficients. Applying the z-transform to both sides of Eq. (1.10) and using the time-shifting property of the z-transform, we may readily express the transfer function of the filter as

$$H(z) = \frac{U(z)}{V(z)}$$

$$= \frac{\displaystyle\sum_{j=0}^{N} a_j z^{-j}}{\displaystyle\sum_{j=0}^{N} b_j z^{-j}} \tag{1.11}$$

Equivalently, we may express the rational transfer function of Eq. (1.11) in the factored form

$$H(z) = \frac{a_0}{b_0} \frac{\displaystyle\prod_{k=1}^{N} (1 - c_k z^{-1})}{\displaystyle\prod_{k=1}^{N} (1 - d_k z^{-1})} \tag{1.12}$$

Each factor $(1 - c_k z^{-1})$ in the numerator on the right-hand side of Eq. (1.12) contributes a zero at $z = c_k$ and a pole at $z = 0$, whereas each factor $(1 - d_k z^{-1})$ in the denominator contributes a pole at $z = d_k$ and a zero at $z = 0$. Thus, except for the scaling factor a_0/b_0, the transfer function $H(z)$ of the filter is uniquely defined in terms of its poles and zeros. Note that with the time-domain behavior of the filter defined by a constant-coefficient difference equation of the form given in Eq. (1.10), the poles and zeros of the transfer function $H(z)$ are real or else appear in complex-conjugate pairs.

Based on the representation given in Eq. (1.12), we may distinguish between two distinct types of linear time-invariant filters:

1. *Finite-duration impulse response (FIR) filters.* For this type of filter, d_k is zero for all k, which means that the filter is an *all-zero filter* in that the poles of its transfer function $H(z)$ are all confined to $z = 0$. Correspondingly, the impulse response $h(n)$ of the filter has a finite duration; hence the descriptor "finite-duration impulse response."

2. *Infinite-duration impulse response (IIR) filters.* In this second type of filter, the transfer function $H(z)$ has at least one nonzero pole that is not canceled by a zero. Correspondingly, the impulse response $h(n)$ of the filter has an infinite duration; hence the descriptor "infinite-duration impulse response." When c_k is zero for all k, the IIR filter is said to be an *all pole filter*, in that the zeros of its transfer function $H(z)$ are all confined to $z = 0$.

Figures 1.1(a) and 1.1(b) show examples of FIR and IIR filters, respectively. The boxes labeled z^{-1} represent unit-delay elements, and the circles labeled $a_1, a_2, \ldots, a_N$ represent filter coefficients. Note that the FIR filter of Fig. 1.1(a) involves feedforward paths only, whereas the IIR filter of Fig. 1.1(b) involves both feedforward and feedback paths. In both cases, the basic functional blocks needed to build the filters consist of unit-delay elements, multipliers, and adders.

Causality and Stability

A linear time-invariant filter is said to be *causal* if its impulse response $h(n)$ is zero for negative time, as shown by

$$h(n) = 0 \qquad \text{for } n < 0 \tag{1.13}$$

Clearly, for a filter to operate in real time, it would have to be causal. However, causality is not a necessary requirement for physical realizability. There are many applications in which the signal to be processed is available in stored form; in these situations, the filter can be noncausal and yet physically realizable.

The filter is said to be *stable* if the output sequence (response) of the filter is bounded for all bounded input sequences (excitations). This requirement is called the *bounded input–bounded output (BIBO) stability criterion*, the application of which is well suited for linear time-invariant filters. From Eq. (1.7) we readily see that the necessary and sufficient condition for BIBO stability is

$$\sum_{k=-\infty}^{\infty} |h(k)| < \infty \tag{1.14}$$

That is, the impulse response of the filter must be absolutely summable.

Causality and stability are not necessarily compatible requirements. For a linear time-invariant filter defined by the difference equation (1.10) to be both causal and stable,

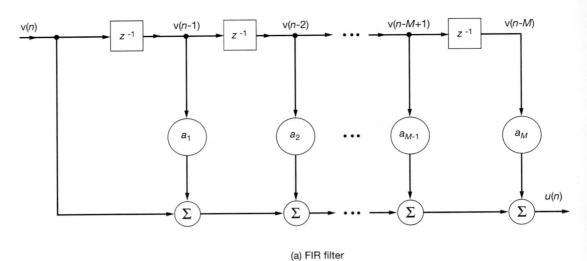

(a) FIR filter

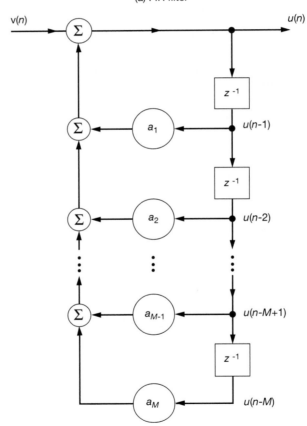

(b) IIR filter

Figure 1.1 Two basic types of filters.

the region of convergence of the filter's transfer function $H(z)$ must satisfy two requirements (Oppenheim and Schafer, 1989):

1. It must lie outside the outermost poles of $H(z)$.
2. It must include the unit circle in the z-plane.

Clearly, these requirements can only be satisfied if all the poles of $H(z)$ lie inside the unit circle, as indicated in Fig. 1.2. We may thus make the following important statement on the issue of stability: *A causal, linear time-invariant filter is stable if and only if all of the*

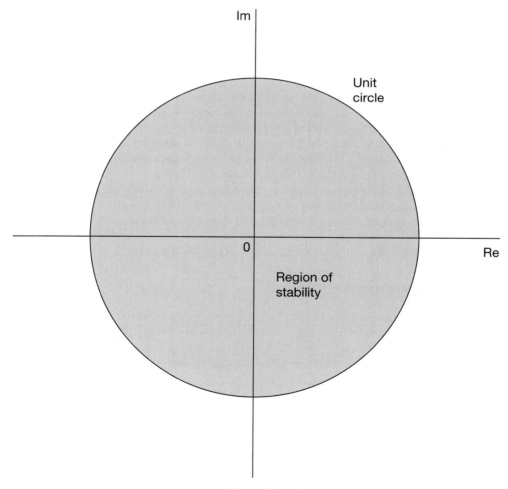

Figure 1.2   *z*-plane.

poles of the filter's transfer function lie inside the unit circle in the z-plane. Note that this statement says nothing about the zeros of the filter's transfer function $H(z)$. Insofar as causality and stability are concerned, the zeros of $H(z)$ can indeed lie anywhere in the z-plane.

1.3 MINIMUM-PHASE FILTERS

The unit circle plays a critical role not only in the stability criterion of a causal filter, but also in the evaluation of its frequency response. Specifically, setting

$$z = e^{j2\pi f}$$

in the expression for the transfer function $H(z)$, we get the filter's *frequency response* denoted by $H(e^{j2\pi f})$, where f denotes the frequency in Hertz. Expressing $H(e^{j2\pi f})$ in its polar form, we may define the frequency response of the filter in terms of two components:

- The *magnitude (amplitude) response*, denoted by $|H(e^{j2\pi f})|$
- The *phase response*, denoted by ang $(H(e^{j2\pi f}))$

In the case of a special class of filters known as *minimum-phase filters*, the magnitude response and phase response of the filter are uniquely related to each other, in that if we are given one of them, we can compute the other component uniquely (Oppenheim and Schafer, 1989). A minimum-phase filter derives its name from the fact that, for a specified magnitude response, it has the minimum phase response possible for all values of z on the unit circle.

The minimum-phase property of a linear time-invariant filter places restrictions of its own on possible locations of the zeros of the filter's transfer function $H(z)$. Specifically, the zeros of $H(z)$ must satisfy the following requirements:

- The zeros of $H(z)$ may lie anywhere inside the unit circle in the z-plane.
- Zeros are permitted to lie on the unit circle, provided that they are *simple* (i.e., they are of order one).

A minimum-phase filter has the following interesting property: given a minimum-phase filter of transfer function $H(z)$, we may define an *inverse filter* with transfer function $1/H(z)$ that is both causal and stable, provided that $H(z)$ does not have zeros on the unit circle. The cascade connection of such a pair of filters has a transfer function equal to unity.

Finally, we note that a *nonminimum-phase filter*, whose transfer function $H(z)$ has zeros outside the unit circle, can always be treated as the cascade connection of a minimum-phase filter and an all-pass filter. An *all-pass filter* is defined as a filter whose transfer function has poles and zeros that are the reciprocals of each other with respect to the unit circle; naturally, the poles are confined to the interior of the unit circle, in which case all the zeros are confined to the exterior of the unit circle. Consequently, the magnitude response of an all-pass filter is equal to unity, which means that it passes all the frequency components of the input signal with no change in amplitude. When the nonminimum-phase filter has all of its zeros located outside the unit circle, it is said to be a *maximum-phase filter*.

1.4 DISCRETE FOURIER TRANSFORM

The *Fourier transform* of a sequence is readily obtained from its z-transform simply by setting the complex variable z equal to $\exp(j2\pi f)$, where f is the real frequency variable. When the sequence of interest has a finite duration, we may go one step further and develop a Fourier representation for it by defining the *discrete Fourier transform (DFT)*. The DFT is itself made up of a sequence of samples, uniformly spaced in frequency. The DFT has established itself as a powerful tool in digital signal processing by virtue of the fact that there exist efficient algorithms for its numerical computation; these algorithms are known collectively as *fast Fourier transform (FFT) algorithms* (Oppenheim and Schafer, 1989).

Consider a finite-duration sequence $u(n)$, assumed to be of length N. The DFT of $u(n)$ is defined by

$$U(k) = \sum_{n=0}^{N-1} u(n) \exp\left(-\frac{j2\pi kn}{N}\right), \quad k = 0, \ldots, N-1 \tag{1.15}$$

The *inverse discrete Fourier transform* (IDFT) of $U(k)$ is defined by

$$u(n) = \frac{1}{N} \sum_{k=0}^{N-1} U(k) \exp\left(\frac{j2\pi kn}{N}\right), \quad n = 0, 1, \ldots, N-1 \tag{1.16}$$

Note that both the original sequence $u(n)$ and its DFT $U(k)$ are of the same length, N. We thus speak of the discrete Fourier transform as an "N-point DFT."

The discrete Fourier transform has an interesting interpretation in terms of the z-transform, as described here: the DFT of a finite-duration sequence may be obtained by evaluating the z-transform of that same sequence at N points uniformly spaced on the unit circle in the z-plane. This "sampling" process is illustrated in Fig. 1.3 for $N = 8$.

Though the sequence $u(n)$ and its DFT $U(k)$ are defined as "finite-length" sequences, in reality they both represent a single period of their respective periodic sequences. This double periodicity is the direct consequence of sampling a continuous-time signal as well as its continuous Fourier transform.

1.5 IMPLEMENTING CONVOLUTIONS USING THE DFT

The underlying "double-periodic" nature of the discrete Fourier transform just mentioned imparts to it certain properties that distinguish it from the continuous Fourier transform. In particular, the *linear convolution* of two sequences, $h(n)$ and $v(n)$, say, involves multiplying one sequence by a time-reversed and linearly shifted version of the other sequence and

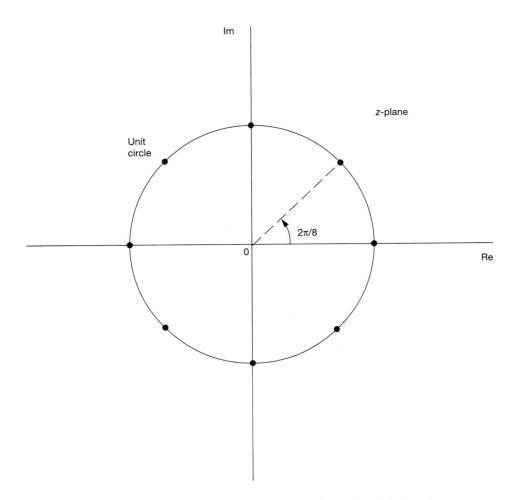

Figure 1.3   A set of $N\ (=8)$ uniformly spaced points on the unit circle in the z-plane.

then summing the product $h(i)v(n-i)$ over all i, as described in Eq. (1.7). In contrast, in the case of DFT we have a *circular convolution* in which the second sequence is circularly time-reversed and circularly shifted with respect to the first sequence. In other words, in circular convolution both sequences have length N (or less) and the sequences are shifted modulo N. It is only when convolution is defined in this way that the convolution of two sequences in the time domain is transformed into the product of their DFTs in the frequency domain (Oppenheim and Schafer, 1989). Stating this property in another way, if we multiply the DFTs of two finite-duration sequences and then evaluate the IDFT of the product, the result so obtained is equivalent to a circular convolution of the original sequences.

With circular convolution being markedly different from linear convolution, the key issue is how to use the DFT to perform linear convolution. To illustrate how we may do this, consider two sequences $v(n)$ and $h(n)$, assuming that they are of lengths L and P, respectively. The linear convolution of these two sequences is a finite-duration sequence of length $L + P - 1$. Recognizing that the convolution of two periodic sequences is another periodic sequence of the same period, we may proceed as follows:

- Append an appropriate number of zero-valued samples to $v(n)$ and $h(n)$ to make them both N-point sequences, where $N = L + P - 1$; this process is referred to as *zero padding*.
- Compute the N-point DFTs of the appended versions of the sequences $v(n)$ and $h(n)$, multiply the DFTs, and then compute the IDFT of the product.
- Use one period of the circular convolution so computed as the linear convolution of the original sequences $v(n)$ and $h(n)$.

The procedure described here works perfectly well for finite-duration sequences. But, what about linear filtering applications where the input signal is, for all practical purposes, of infinite duration? In situations of this kind, we may use two widely used techniques known as the *overlap-add* and *overlap-save* sectioning methods, which are described next.

Overlap-Add Method

The best way to explain the overlap-add method is by way of an example. Consider the sequences $v(n)$ and $h(n)$ shown in Fig. 1.4; it is assumed that the sequence $v(n)$ is effectively of "infinite" length, and the sequence $h(n)$ is of some finite length P. The sequence $v(n)$ is first sectioned into nonoverlapping blocks, each of length $Q = N - P$ for some predetermined N, as illustrated in Fig. 1.5(a). It may therefore be represented as the sum of shifted finite-duration sequences, as shown by

$$v(n) = \sum_{r=0}^{\infty} v_r(n) \tag{1.17}$$

where

$$v_r(n) = \begin{cases} v(n + rQ), & n = 0, 1, \ldots, Q - 1 \\ 0, & \text{otherwise} \end{cases} \tag{1.18}$$

Next, each section is padded with $P - 1$ zero-valued samples to form one period of a periodic sequence, as illustrated in Fig. 1.5(a). We may thus describe the first section by writing

$$v_0(n) = \begin{cases} v(n), & n = 0, 1, \ldots, N - P \\ 0, & n = N - P + 1, \ldots, N - 1 \end{cases} \tag{1.19}$$

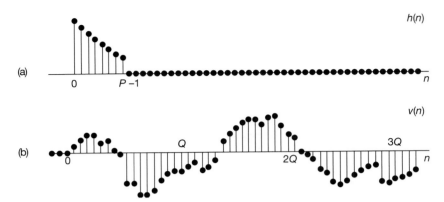

Figure 1.4 Finite-length impulse response $h(n)$ and indefinite-length signal $v(n)$ to be filtered by $h(\text{n})$ (reproduced, with permission, from Oppenheim and Schafer, 1989).

The circular convolution of $v_0(n)$ with $h(n)$ yields the output sequence $u_0(n)$ shown in the first trace of Fig. 1.5(b).

The second section $v_1(n)$ and all other sections of the "infinitely" long sequence $v(n)$ are treated in a similar manner. The resulting output sequences $u_1(n)$, and $u_2(n)$ are also illustrated in Fig. 1.5(b) for the input sections $v_1(n)$ and $v_2(n)$, respectively. Finally, the output sequences $u_0(n)$, $u_1(n)$, $u_2(n)$, . . . are combined to yield the overall output sequence $u(n)$. Note that $u_1(n)$, $u_2(n)$, . . . are shifted by the appropriate values, namely, N, $2N$, . . . , before they are added to $u_0(n)$. The sectioned convolution technique described here is called the *overlap-add method* for two reasons: the output sequences tend to overlap each other, and they are added together to produce the correct result.

Overlap-Save Method

The overlap-save method differs from the overlap-add method in that it involves overlapping input sections rather than output sections. Specifically, the "infinitely" long sequence is sectioned into N-point blocks that overlap by $P - 1$ samples, where P is the length of the "short" sequence $h(n)$, as illustrated in Fig. 1.6(a). The N-point circular convolution of $h(n)$ and $v_r(n)$ is computed for $r = 0,1,2, ...$ The resulting output sequences $u_0(n)$, $u_1(n)$, and $u_2(n)$ for the sections $v_0(n)$, $v_1(n)$, and $v_2(n)$ are illustrated in Fig. 1.6(b). The first $P - 1$ samples of each output sequence $u_r(n)$, $r = 0,1,2, . . .$ are ignored, because they are due to the wraparound (end) effect of the circular convolution. Finally, the remaining samples of the output sequences $u_0(n)$, $u_1(n)$, $u_2(n)$, . . . are added after they have been shifted by appropriate values, yielding the correct output sequence $u(n)$. For obvious reasons, this second sectioning technique is referred to as the *overlap-save method*.

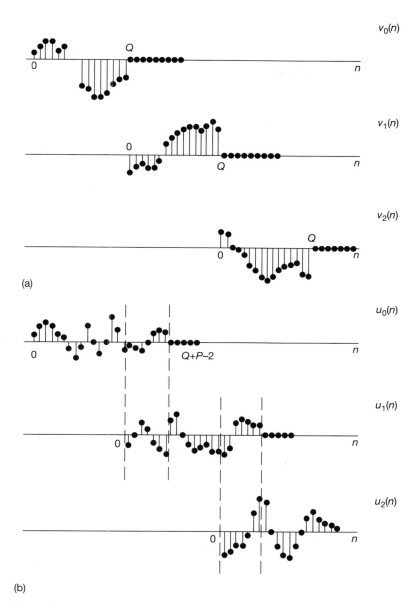

Figure 1.5   (a) Decomposition of the input signal $v(n)$ in Fig. 1.4 into nonoverlapping sections, each of length Q. (b) Result of convolving each such section with $h(n)$ (reproduced, with permission, from Oppenheim and Schafer, 1989).

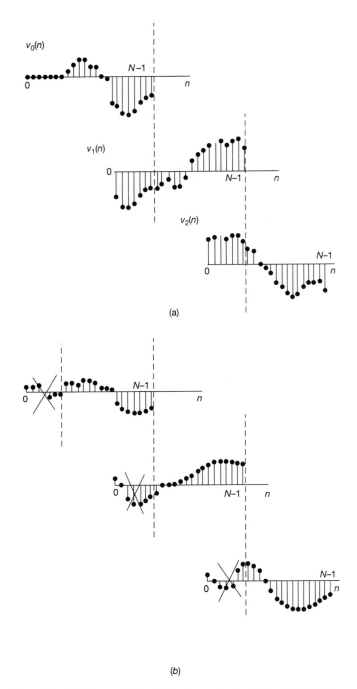

Figure 1.6 (a) Decomposition of the input signal $v(n)$ in Fig. 1.4 into overlapping sections, each of length N. (b) Result of convolving each section with $h(n)$; the portions of each filtered section to be discarded in forming the linear convolution are indicated (reproduced, with permission, from Oppenheim and Schafer, 1989).

Thus, we may use the overlap-add method or the overlap-save method to compute the linear convolution of a short sequence $h(n)$ with a much longer sequence $v(n)$ by first sectioning the latter sequence into small blocks, then indirectly computing the circular convolution of each such block with the short sequence $h(n)$ via the DFT, and finally piecing the individual results together in an appropriate fashion. The utility of the overlap-add and overlap-save methods is made a practical reality by virtue of the availability of highly efficient algorithms (i.e., FFT algorithms) for computing the DFT. The indirect computation of convolution using the overlap-add method or overlap-save method via the FFT is referred to as *fast convolution*, as it is faster than its direct computation.

1.6 DISCRETE COSINE TRANSFORM

Another transform that features in certain applications of digital signal processing is the *discrete cosine transform* (DCT). Unlike the DFT, the DCT may be defined in several different ways (Rao and Yip, 1990). For the purpose of our present discussion, the DCT of an N-point sequence $u(n)$ is defined by

$$U(m) = k_m \sum_{n=0}^{N-1} u(n) \cos\left(\frac{(2n+1)m\pi}{2N}\right), \qquad m = 0, 1, \ldots, N-1 \quad (1.20)$$

and the *inverse discrete cosine transform* (IDCT) of $U(m)$ is defined by

$$u(n) = \frac{2}{N} \sum_{m=0}^{N-1} k_m U(m) \cos\left(\frac{(2n+1)m\pi}{2N}\right), \qquad n = 0, 1, \ldots, N-1 \quad (1.21)$$

The constant k_m in Eqs. (1.20) and (1.21) is itself defined by

$$k_m = \begin{cases} 1/\sqrt{2}, & m = 0 \\ 1, & m = 1, \ldots, N-1 \end{cases} \quad (1.22)$$

The DCT is related to the DFT, as one would expect. Specifically, we first construct a $2N$-point sequence $\tilde{u}(n)$ related to the original sequence $u(n)$ as follows:

$$\tilde{u}(n) = \begin{cases} u(n), & n = 0, 1, \ldots, N-1 \\ u(2N-n-1), & n = N, N+1, \ldots, 2N-1 \end{cases} \quad (1.23)$$

Thus, the $\tilde{u}(n)$ is an even extension of $u(n)$. The $2N$-point DFT of the sequence $\tilde{u}(n)$ is given by

$$\begin{aligned} \tilde{U}(m) &= \sum_{n=0}^{2N-1} \tilde{u}(n) \exp\left(-\frac{j2\pi mn}{2N}\right) \\ &= \sum_{n=0}^{N-1} \tilde{u}(n) \exp\left(-\frac{j2\pi mn}{2N}\right) + \sum_{n=N}^{2N-1} \tilde{u}(n) \exp\left(-\frac{j2\pi mn}{2N}\right) \end{aligned} \quad (1.24)$$

Substituting Eq. (1.23) into (1.24), we get

$$\tilde{U}(m) = \sum_{n=0}^{N-1} u(n)\exp\left(-\frac{j2\pi mn}{2N}\right) + \sum_{n=N}^{2N-1} u(2N-n-1)\exp\left(-\frac{j2\pi mn}{2N}\right)$$

$$= \sum_{n=0}^{N-1} u(n)\left[\exp\left(-\frac{j2\pi mn}{2N}\right) + \exp\left(\frac{2j\pi m(n+1)}{2N}\right)\right]$$

(1.25)

Introducing the phase shift $m\pi/2N$ and the weighting factor $k_m/2$ into Eq. (1.25), we may write

$$\frac{1}{2}k_m\exp\left(-\frac{jm\pi}{2N}\right)\tilde{U}(m) = \frac{1}{2}k_m\sum_{n=0}^{N-1} u(n)\left[\exp\left(-\frac{j(2n+1)m\pi}{2N}\right) + \exp\left(\frac{j(2n+1)m\pi}{2N}\right)\right]$$

$$= k_m\sum_{n=0}^{N-1} u(n)\cos\left(\frac{(2n+1)m\pi}{2N}\right)$$

(1.26)

The right-hand side of Eq. (1.26) is recognized as the definition for the DCT of the original sequence $u(n)$. It follows, therefore, that the discrete cosine transform $U(m)$ of the sequence $u(n)$ and the discrete Fourier transform $\tilde{U}(m)$ of its extended version $\tilde{u}(n)$ are related as follows:

$$U(m) = \frac{1}{2}k_m\exp\left(-\frac{jm\pi}{2N}\right)\tilde{U}(m), \qquad m = 0, 1, \ldots, N-1 \qquad (1.27)$$

This relation shows that, whereas the DFT is periodic with period N, the DCT is periodic with period $2N$.

1.7 SUMMARY AND DISCUSSION

In this chapter we reviewed the z-transform, the discrete Fourier transform, and the discrete cosine transform; these transforms are all related to each other. The discrete Fourier transform represents an important example of a general class of finite-length orthogonal transforms, which may be defined by the following pair of relations:

$$U(k) = \sum_{n=0}^{N-1} u(n)\varphi_k^*(n), \qquad k = 0, 1, \ldots, N-1$$

$$u(n) = \frac{1}{N}\sum_{k=0}^{N-1} U(k)\varphi_k(n), \qquad n = 0, 1, \ldots, N-1$$

where $u(n)$ is the given sequence and $U(k)$ is its discrete transform. The sequences $\varphi_k(n)$ for different k constitute an *orthogonal set*, as shown by

$$\sum_{n=0}^{N-1} \varphi_k(n) \varphi_l^*(n) = \begin{cases} N, & l = k \\ 0, & \text{otherwise} \end{cases}$$

Our interest in the z-transform is motivated by the fact it provides a basic tool for the characterization of linear time-invariant systems, which constitute an important class of systems of particular interest in the study of linear adaptive filtering. As for the discrete Fourier transform and the discrete cosine transform, they provide the necessary tools for the implementation of adaptive filtering in the frequency domain, which is sometimes found to be preferable to adaptive filtering in the time domain, an issue that is discussed later in this book.

PROBLEMS

1. An all-pole filter is characterized by the second-order difference equation:

    $$u(n) - 0.1\, u(n-1) - 0.8\, u(n-2) = v(n)$$

    (a) Determine the transfer function $H(z)$ of the filter.
    (b) Plot the pole-zero map of $H(z)$.
    (c) Find the impulse response of the filter.

2. The inverse of the filter described in Problem 1 consists of an all-zero filter.
    (a) Plot the pole-zero map of the transfer function for this inverse filter.
    (b) Find the difference equation that describes the time-domain behavior of the inverse filter.
    (c) Find the impulse response of the inverse filter.

3. A second-order nonminimum phase system has the transfer function

    $$H(z) = \frac{2(1 + z^{-1} - 2z^{-2})}{1 - 0.2828\, z^{-1} + z^{-2}}$$

    (a) Plot a pole-zero map for $H(z)$.
    (b) The system described here may be considered to be equivalent to the cascade connection of a minimum phase system and an all-pass system. Determine the transfer functions of these two systems, and plot their individual pole-zero maps.

4. When considering the inverse of a minimum-phase system characterized by the transfer function $H(z)$, it is not permissible for $H(z)$ to have any zero on the unit circle in the z-plane. Why?

5. Convolution, be it linear or circular, is a commutative operation. Demonstrate this property.

6. If $U(e^{j2\pi f})$ is the Fourier transform of a finite-duration sequence $u(n)$, the Fourier transform of the time-shifted sequence $u(n-m)$ is $e^{-j2\pi mf}\, U(e^{j2\pi f})$. How is the corresponding time-shifting property of the discrete Fourier transform for the sequence $u(n)$ described?

CHAPTER

2

Stationary Processes and Models

The term *stochastic process* or *random process* is used to describe the time evolution of a statistical phenomenon according to probabilistic laws. The time evolution of the phenomenon means that the stochastic process is a function of time, defined on some observation interval. The statistical nature of the phenomenon means that, before conducting an experiment, it is not possible to define exactly the way it evolves in time. Examples of a stochastic process include speech signals, television signals, radar signals, digital computer data, the output of a communication channel, seismological data, and noise.

The form of a stochastic process that is of interest to us is one that is defined at *discrete and uniformly spaced instants of time* (Box and Jenkins, 1976; Priestley, 1981). Such a restriction may arise naturally in practice, as in the case of radar signals or digital computer data. Alternatively, the stochastic process may be defined originally for a continuous range of real values of time; however, before processing, it is *sampled uniformly* in time, with the sampling rate chosen to be greater than twice the highest frequency component of the process (Haykin, 1994).

A stochastic process is *not* just a single function of time; rather, it represents, in theory, an infinite number of *different* realizations of the process. One particular realization of a discrete-time stochastic process is called a *discrete-time series* or simply *time series*. For convenience of notation, *we normalize time with respect to the sampling period*. For example, the sequence $u(n), u(n-1), \ldots, u(n-M)$ represents a time series that consists of the *present* observation $u(n)$ made at time n and M past observations of the process made at times $n-1, \ldots, n-M$.

We say that a stochastic process is *strictly stationary* if its statistical properties are *invariant* to a shift of time. Specifically, for a discrete-time stochastic process represented by the time series $u(n), u(n-1) \ldots, u(n-M)$ to be strictly stationary, the *joint probability density function* of these observations made at times $n, n-1, \ldots, n-M$ must remain the same no matter what values we assign to n for fixed M.

2.1 PARTIAL CHARACTERIZATION OF A DISCRETE-TIME STOCHASTIC PROCESS

In practice, we usually find that it is not possible to determine (by means of suitable measurements) the joint probability density function for an arbitrary set of observations made on a stochastic process. Accordingly, we must content ourselves with a partial characterization of the process by specifying its first and second moments.

Consider a discrete-time stochastic process represented by the time series $u(n)$, $u(n-1), \ldots, u(n-M)$, which may be complex valued. We define the *mean-value function* of the process as

$$\mu(n) = E[u(n)] \tag{2.1}$$

where E denotes the *statistical expectation operator*. We define the *autocorrelation function* of the process as

$$r(n, n-k) = E[u(n)u^*(n-k)], \quad k = 0, \pm1, \pm2, \ldots, \tag{2.2}$$

where the asterisk denotes *complex conjugation*. We define the *autocovariance function* of the process as

$$c(n, n-k) = E[(u(n) - \mu(n))(u(n-k) - \mu(n-k))^*], \quad k = 0, \pm1, \pm2, \ldots \tag{2.3}$$

From Eqs. (2.1) to (2.3), we see that the mean-value, autocorrelation and autocovariance functions of the process are related by

$$c(n, n-k) = r(n, n-k) - \mu(n)\mu^*(n-k) \tag{2.4}$$

For a partial characterization of the process, we therefore need to specify (1) the mean-value function $\mu(n)$ and (2) the autocorrelation function $r(n, n-k)$ or the autocovariance function $c(n, n-k)$ for various values of n and k that are of interest. Note also the autocorrelation and autocovariance functions have the same value when the mean $\mu(n)$ is zero for all n.

This form of partial characterization offers two important advantages:

1. It lends itself to practical measurements.
2. It is well suited to *linear* operations on stochastic processes.

For a discrete-time stochastic process that is strictly stationary, all three quantities defined in Eqs. (2.1) to (2.3) assume simpler forms. In particular, we find that the mean-value function of the process is a constant μ (say), so we may write

$$\mu(n) = \mu \qquad \text{for all } n \tag{2.5}$$

We also find that both the autocorrelation and autocovariance functions depend only on the *difference* between the observation times n and $n - k$, that is, k, as shown by

$$r(n, n - k) = r(k) \tag{2.6}$$

and

$$c(n, n - k) = c(k) \tag{2.7}$$

Note that when $k = 0$, corresponding to a time difference or *lag* of zero, $r(0)$ equals the *mean-square value* of $u(n)$:

$$r(0) = E[|u(n)|^2] \tag{2.8}$$

and $c(0)$ equals the *variance* of $u(n)$:

$$c(0) = \sigma_u^2 \tag{2.9}$$

The condition of Eqs. (2.5) to (2.7) are *not* sufficient to guarantee that the discrete-time stochastic process is strictly stationary. However, a discrete-time stochastic process that is not strictly stationary, but for which these conditions hold, is said to be *wide-sense stationary*, or *stationary to the second order*. A strictly stationary process $\{u(n)\}$, or $u(n)$ for short, is stationary in the wide sense if and only if (Doob, 1953)

$$E[|u(n)|^2] < \infty \qquad \text{for all } n$$

This condition is ordinarily satisfied by stochastic processes encountered in the physical sciences and engineering.

2.2 MEAN ERGODIC THEOREM

The *expectations* or *ensemble averages* of a stochastic process are averages "across the process." Clearly, we may also define *long-term sample averages* or *time averages* that are averages "along the process." Indeed, time averages may be used to build a *stochastic model* of a physical process by *estimating* unknown parameters of the model. For such an approach to be rigorous, however, we have to show that time averages converge to corresponding ensemble averages of the process in some statistical sense. A popular criterion for convergence is that of mean square-error, as described next.

To be specific, consider a discrete-time stochastic process $u(n)$ that is wide-sense stationary. Let a constant μ denote the mean of the process, and $c(k)$ denote its autocovariance function for lag k. For an estimate of the mean μ, we may use the time average

$$\hat{\mu}(N) = \frac{1}{N} \sum_{n=0}^{N-1} u(n) \tag{2.10}$$

where N is the total number of samples used in the estimation. Note that the estimate $\hat{\mu}(N)$ is a random variable with a mean and variance of its own. In particular, we readily find from Eq. (2.10) that the mean (expectation) of $\hat{\mu}(N)$ is

$$E[\hat{\mu}(N)] = \mu \qquad \text{for all } N \tag{2.11}$$

It is in the sense of Eq. (2.11) that we say the time average $\hat{\mu}(N)$ is an *unbiased* estimator of the ensemble average (mean) of the process.

Moreover, we say that the process $u(n)$ is *mean ergodic in the mean-square error sense* if the mean-square value of the error between the ensemble average μ and the time average $\hat{\mu}(N)$ approaches zero as the number of samples N approaches infinity; that is,

$$\lim_{N \to \infty} [(\mu - \hat{\mu}(N))^2] = 0$$

Using the time average formula of Eq. (2.10), we may write

$$
\begin{aligned}
E[|\mu - \hat{\mu}(N)|^2] &= E\left[\left|\mu - \frac{1}{N}\sum_{n=0}^{N-1} u(n)\right|^2\right] \\
&= \frac{1}{N^2}E\left[\left|\sum_{n=0}^{N-1}(u(n) - \mu)\right|^2\right] \\
&= \frac{1}{N^2}E\left[\sum_{n=0}^{N-1}\sum_{k=0}^{N-1}(u(n) - \mu)(u(k) - \mu)^*\right] \\
&= \frac{1}{N^2}\sum_{n=0}^{N-1}\sum_{k=0}^{N-1}E[(u(n) - \mu)(u(k) - \mu)^*] \\
&= \frac{1}{N^2}\sum_{n=0}^{N-1}\sum_{k=0}^{N-1}c(n - k)
\end{aligned}
\tag{2.12}
$$

Let $l = n - k$. We may then simplify the double summation in Eq. (2.12) as follows:

$$E[|\mu - \hat{\mu}(N)|^2] = \frac{1}{N}\sum_{l=-N+1}^{N-1}\left(1 - \frac{|l|}{N}\right)c(l)$$

Accordingly, we may state that the necessary and sufficient condition for the process $u(n)$ to be mean ergodic in the mean-square error sense is that

$$\lim_{N \to \infty}\frac{1}{N}\sum_{l=-N+1}^{N-1}\left(1 - \frac{|l|}{N}\right)c(l) = 0 \tag{2.13}$$

In other words, if the process $u(n)$ is asymptotically uncorrelated in the sense of Eq. (2.13), then the time average $\hat{\mu}(N)$ of the process converges to the ensemble average μ in the mean-square error sense. This is the statement of a particular form of the *mean ergodic theorem* (Gray and Davisson, 1986).

The use of the mean ergodic theorem may be extended to other time averages of the process. Consider, for example, the following time average used to estimate the autocorrelation function of a wide-sense stationary process:

$$\hat{r}(k, N) = \frac{1}{N} \sum_{n=0}^{N-1} u(n)u(n-k), \qquad 0 \le k \le N-1 \tag{2.14}$$

The process $u(n)$ is said to be *correlation ergodic* in the mean-square error sense if the mean-square value of the difference between the true value $r(k)$ and the estimate $\hat{r}(k, N)$ approaches zero as the number of samples N approaches infinity. Let $z(n, k)$ denote a new discrete-time stochastic process related to the original process $u(n)$ as follows:

$$z(n, k) = u(n)u(n-k) \tag{2.15}$$

Hence, by substituting $z(n, k)$ for $u(n)$, we may use the mean ergodic theorem to establish the conditions for $z(n, k)$ to be mean ergodic or, equivalently, for $u(n)$ to be correlation ergodic.

2.3 CORRELATION MATRIX

Let the M-by-1 *observation vector* $\mathbf{u}(n)$ represent the elements of the time series $u(n)$, $u(n-1), \ldots, u(n-M+1)$. To show the composition of the vector $\mathbf{u}(n)$ explicitly, we write

$$\mathbf{u}(n) = [u(n), u(n-1), \ldots, u(n-M+1)]^T \tag{2.16}$$

where the superscript T denotes *transposition*. We define the *correlation matrix* of a stationary discrete-time stochastic process represented by this time series as *the expectation of the outer product of the observation vector* $\mathbf{u}(n)$ *with itself.* Let $\mathbf{R}$ denote the M-by-M correlation matrix defined in this way. We thus write

$$\mathbf{R} = E[\mathbf{u}(n)\,\mathbf{u}^H(n)] \tag{2.17}$$

where the superscript H denotes *Hermitian transposition* (i.e., the operation of transposition combined with complex conjugation). By substituting Eq. (2.16) in (2.17) and using the condition of wide-sense stationarity, we may express the correlation matrix $\mathbf{R}$ in the expanded form:

$$\mathbf{R} = \begin{bmatrix} r(0) & r(1) & \cdots & r(M-1) \\ r(-1) & r(0) & \cdots & r(M-2) \\ \cdot & \cdot & \cdot & \cdot \\ \cdot & \cdot & \cdot & \cdot \\ \cdot & \cdot & \cdot & \cdot \\ r(-M+1) & r(-M+2) & \cdots & r(0) \end{bmatrix} \tag{2.18}$$

The element $r(0)$ on the main diagonal is always real valued. For complex-valued data, the remaining elements of **R** assume complex values.

Properties of the Correlation Matrix

The correlation matrix **R** plays a key role in the statistical analysis and design of discrete-time filters. It is therefore important that we understand its various properties and their implications. In particular, using the definition of Eq. (2.17), we find that the correlation matrix of a stationary discrete-time stochastic process has the following properties.

Property 1. *The correlation matrix of a stationary discrete-time stochastic process is Hermitian.*

We say that a *complex-valued* matrix is *Hermitian* if it is equal to its *conjugate transpose*. We may thus express the Hermitian property of the correlation matrix **R** by writing

$$\mathbf{R}^H = \mathbf{R} \tag{2.19}$$

This property follows directly from the definition of Eq. (2.17).

Another way of stating the Hermitian property of the correlation matrix **R** is to write

$$r(-k) = r^*(k) \tag{2.20}$$

where $r(k)$ is the autocorrelation function of the stochastic process $u(n)$ for a lag of k. Accordingly, for a wide-sense stationary process we only need M values of the autocorrelation function $r(k)$ for $k = 0, 1, \ldots, M - 1$ in order to completely define the correlation matrix **R**. We may thus rewrite Eq. (2.18) as follows:

$$\mathbf{R} = \begin{bmatrix} r(0) & r(1) & \cdots & r(M-1) \\ r^*(1) & r(0) & \cdots & r(M-2) \\ \cdot & \cdot & \cdot & \cdot \\ \cdot & \cdot & \cdot & \cdot \\ \cdot & \cdot & \cdot & \cdot \\ r^*(M-1) & r^*(M-2) & \cdots & r(0) \end{bmatrix} \tag{2.21}$$

From here on, we will use this representation for the expanded matrix form of the correlation matrix of a wide-sense stationary discrete-time stochastic process. Note that for the special case of *real-valued data*, the autocorrelation function $r(k)$ is real for all k, and the correlation matrix **R** is *symmetric*.

Property 2. *The correlation matrix of a stationary discrete-time stochastic process is Toeplitz.*

We say that a square matrix is *Toeplitz* if all the elements on its main diagonal are equal, and if the elements on any other diagonal parallel to the main diagonal are also equal. From the expanded form of the correlation matrix **R** given in Eq. (2.21), we see that all the elements on the main diagonal are equal to $r(0)$, all the elements on the first diagonal above the main diagonal are equal to $r(1)$, all the elements along the first diagonal below the main diagonal are equal to $r^*(1)$, and so on for the other diagonals. We conclude therefore that the correlation matrix $\mathbf{R}$ is Toeplitz.

It is important to recognize, however, that the Toeplitz property of the correlation matrix $\mathbf{R}$ is a direct consequence of the assumption that the discrete-time stochastic process represented by the observation vector $\mathbf{u}(n)$ is wide-sense stationary. Indeed, we may state that if the discrete-time stochastic process is wide-sense stationary, then its correlation matrix $\mathbf{R}$ must be Toeplitz; and, conversely, if the correlation matrix $\mathbf{R}$ is Toeplitz, then the discrete-time stochastic process must be wide-sense stationary.

Property 3.   *The correlation matrix of a discrete-time stochastic process is always nonnegative definite and almost always positive definite.*

Let $\mathbf{x}$ be an arbitrary (nonzero) M-by-1 complex-valued vector. Define the scalar random variable y as the *inner product* of $\mathbf{x}$ and the observation vector $\mathbf{u}(n)$, as shown by

$$y = \mathbf{x}^H \mathbf{u}(n)$$

Taking the Hermitian transpose of both sides and recognizing that y is a scalar, we get

$$y^* = \mathbf{u}^H(n)\mathbf{x}$$

where the asterisk denotes *complex conjugation*. The mean-square value of the random variable y equals

$$
\begin{aligned}
E[|y|^2] &= E[yy^*] \\
&= E[\mathbf{x}^H \mathbf{u}(n)\mathbf{u}^H(n)\mathbf{x}] \\
&= \mathbf{x}^H E[\mathbf{u}(n)\mathbf{u}^H(n)]\mathbf{x} \\
&= \mathbf{x}^H \mathbf{R}\mathbf{x}
\end{aligned}
$$

where $\mathbf{R}$ is the correlation matrix defined in Eq. (2.17). The expression $\mathbf{x}^H \mathbf{R}\mathbf{x}$ is called a Hermitian form. Since

$$E[|y|^2] \geq 0$$

it follows that

$$\mathbf{x}^H \mathbf{R}\mathbf{x} \geq 0 \tag{2.22}$$

A Hermitian form that satisfies this condition for every nonzero $\mathbf{x}$ is said to be *nonnegative definite* or *positive semidefinite*. Accordingly, we may state that the correlation matrix of a wide-sense stationary process is always nonnegative definite.

If the Hermitian form $\mathbf{x}^H \mathbf{R}\mathbf{x}$ satisfies the condition

$$\mathbf{x}^H \mathbf{R}\mathbf{x} > 0$$

for every nonzero $\mathbf{x}$, we say that the correlation matrix $\mathbf{R}$ is *positive definite*. This condition is satisfied for a wide-sense stationary process unless there are linear dependencies between the random variables that constitute the M elements of the observation vector

$\mathbf{u}(n)$. Such a situation arises essentially only when the process $u(n)$ consists of the sum of K sinusoids with $K \leq M$; see Section 2.4 for more details. In practice, we find that this idealized situation is so rare in occurrence that the correlation matrix $\mathbf{R}$ is almost always positive definite.

The positive definiteness of a correlation matrix implies that its determinant and all principal minors are greater than zero. For example, for $M = 2$, we must have

$$\begin{vmatrix} r(0) & r(1) \\ r^*(1) & r(0) \end{vmatrix} > 0$$

Similarly, for $M = 3$, we must have

$$\begin{vmatrix} r(0) & r(1) \\ r^*(1) & r(0) \end{vmatrix} > 0$$

$$\begin{vmatrix} r(0) & r(2) \\ r^*(2) & r(0) \end{vmatrix} > 0$$

$$\begin{vmatrix} r(0) & r(1) & r(2) \\ r^*(1) & r(0) & r(1) \\ r^*(2) & r^*(1) & r(0) \end{vmatrix} > 0$$

and so on for higher values of M. These conditions, in turn, imply that the correlation matrix is nonsingular. We say that a matrix is *nonsingular* if its inverse exists; otherwise, it is singular. Accordingly, we may state that a correlation matrix is almost always nonsingular.

Property 4.   *When the elements that constitute the observation vector of a stationary discrete-time stochastic process are rearranged backward, the effect is equivalent to the transposition of the correlation matrix of the process.*

Let $\mathbf{u}^B(n)$ denote the M-by-1 vector obtained by rearranging the elements that constitute the observation vector $\mathbf{u}(n)$ *backward*. We illustrate this operation by writing

$$\mathbf{u}^{BT}(n) = [u(n - M + 1), u(n - M + 2), \ldots, u(n)] \tag{2.23}$$

where the superscript B denotes the backward rearrangement of a vector. The correlation matrix of the vector $\mathbf{u}^B(n)$ equals, by definition,

$$E[\mathbf{u}^B(n)\mathbf{u}^{BH}(n)] = \begin{bmatrix} r(0) & r^*(1) & \cdots & r^*(M-1) \\ r(1) & r(0) & \cdots & r^*(M-2) \\ \vdots & \vdots & & \vdots \\ r(M-1) & r(M-2) & \cdots & r(0) \end{bmatrix} \tag{2.24}$$

Hence, comparing the expanded correlation matrix of Eq. (2.24) with that of Eq. (2.21), we see that

$$E\,[\mathbf{u}^B(n)\,\mathbf{u}^{BH}(n)] = \mathbf{R}^T \tag{2.25}$$

which is the desired result.

Property 5.   *The correlation matrices $\mathbf{R}_M$ and $\mathbf{R}_{M+1}$ of a stationary discrete-time stochastic process, pertaining to M and M + 1 observations of the process, respectively, are related by*

$$\mathbf{R}_{M+1} = \left[\begin{array}{c:c} r(0) & \mathbf{r}^H \\ \hdashline \mathbf{r} & \mathbf{R}_M \end{array} \right] \tag{2.26}$$

or equivalently,

$$\mathbf{R}_{M+1} = \left[\begin{array}{c:c} \mathbf{R}_M & \mathbf{r}^{B*} \\ \hdashline \mathbf{r}^{BT} & r(0) \end{array} \right] \tag{2.27}$$

where $r(0)$ is the autocorrelation of the process for a lag of zero, and

$$\mathbf{r}^H = [r(1), r(2), \ldots, r(M)] \tag{2.28}$$

and

$$\mathbf{r}^{BT} = [r(-M), r(-M+1), \ldots, r(-1)] \tag{2.29}$$

Note that in describing Property 5 we have added a subscript, M or $M + 1$, to the symbol for the correlation matrix in order to display dependence on the number of observations used to define this matrix. We follow such a practice (in the context of the correlation matrix and other vector quantities) *only* when the issue at hand involves dependence on the number of observations or dimensions of the matrix.

To prove the relation of Eq. (2.26), we express the correlation matrix $\mathbf{R}_{M+1}$ in its expanded form, partitioned as follows:

$$\mathbf{R}_{M+1} = \left[\begin{array}{c:cccc} r(0) & r(1) & r(2) & \cdots & r(M) \\ \hdashline r^*(1) & r(0) & r(1) & \cdots & r(M-1) \\ r^*(2) & r^*(1) & r(0) & \cdots & r(M-2) \\ \vdots & \vdots & \vdots & \ddots & \vdots \\ r^*(M) & r^*(M-1) & r^*(M-2) & \cdots & r(0) \end{array} \right] \tag{2.30}$$

Using Eqs. (2.18), (2.20), and (2.28) in (2.30), we get the result given in Eq. (2.26). Note that according to this relation, the observation vector $\mathbf{u}_{M+1}(n)$ is *partitioned* in the form

$$\mathbf{u}_{M+1}(n) = \begin{bmatrix} u(n) \\ \text{-------} \\ u(n-1) \\ u(n-2) \\ \bullet \\ \bullet \\ \bullet \\ u(n-M) \end{bmatrix}$$

$$= \begin{bmatrix} u(n) \\ \text{-------} \\ \mathbf{u}_M(n-1) \end{bmatrix} \tag{2.31}$$

where the subscript $M + 1$ is intended to denote the fact that the vector $\mathbf{u}_{M+1}(n)$ has $M + 1$ elements, and likewise for $\mathbf{u}_M(n)$.

To prove the relation of Eq. (2.27), we express the correlation matrix $\mathbf{R}_{M+1}$ in its expanded form, partitioned in the alternative form

$$\mathbf{R}_{M+1} = \begin{bmatrix} r(0) & r(1) & \cdots & r(M-1) & r(M) \\ r^*(1) & r(0) & \cdots & r(M-2) & r(M-1) \\ \bullet & \bullet & \bullet & \bullet & \\ \bullet & \bullet & \bullet & \bullet & \\ r^*(M-1) & r^*(M-2) & \cdots & r(0) & r(1) \\ \text{-----} & \text{-----} & \text{-----} & \text{-----} & \text{-----} \\ r^*(M) & r^*(M-1) & \cdots & r^*(1) & r(0) \end{bmatrix} \tag{2.32}$$

Here again, using Eqs. (2.18), (2.20), and (2.29) in (2.32), we get the result given in Eq. (2.27). Note that according to this second relation the observation vector $\mathbf{u}_{M+1}(n)$ is partitioned in the alternative form

$$\mathbf{u}_{M+1}(n) = \begin{bmatrix} u(n) \\ u(n-1) \\ \bullet \\ \bullet \\ \bullet \\ u(n-M+1) \\ \text{--------} \\ u(n-M) \end{bmatrix}$$

$$= \begin{bmatrix} \mathbf{u}_M(n) \\ \text{-------} \\ u(n-M) \end{bmatrix} \tag{2.33}$$

2.4 CORRELATION MATRIX OF SINE WAVE PLUS NOISE

A time series of special interest is one that consists of a *complex sinusoid corrupted by additive noise*. Such a time series is representative of several important signal-processing applications. In the *temporal context*, for example, this time series represents the composite signal at the input of a receiver, with the complex sinusoid representing a *target signal* and the noise representing thermal noise generated at the front end of the receiver. In the *spatial context*, it represents the received signal in a linear array of sensors, with the complex sinusoid representing a *plane wave* produced by a remote source (emitter) and the noise representing *sensor noise*.

Let α denote the amplitude of the complex sinusoid, and ω denote its angular frequency. Let $v(n)$ denote a sample of the noise, assumed to have zero mean. We may then write a corresponding sample of the time series that consists of the complex sinusoid plus noise as follows:

$$u(n) = \alpha \exp(j\omega n) + v(n), \qquad n = 0, 1, \ldots, N - 1 \qquad (2.34)$$

The sources of the complex sinusoid and the noise are independent of each other. Since the noise component $v(n)$ has zero mean, by assumption, we see from Eq. (2.34) that the mean of $u(n)$ is equal to $\alpha \exp(j\omega n)$.

To calculate the autocorrelation function of the process $u(n)$, we clearly need to know the autocorrelation function of the noise process $v(n)$. To proceed then, we assume a special form of noise characterized by the autocorrelation function

$$E[v(n)v^*(n - k)] = \begin{cases} \sigma_v^2, & k = 0 \\ \\ 0, & k \neq 0 \end{cases} \qquad (2.35)$$

Such a form of noise is commonly referred to as *white noise*; more will be said about it in Chapter 3. Since the sources responsible for the generation of the complex sinusoid and the noise are independent and, therefore, uncorrelated, it follows that the autocorrelation function of the process $u(n)$ equals the sum of the autocorrelation functions of its two individual components. Accordingly, using Eqs. (2.34) and (2.35), we find that the autocorrelation function of the process $u(n)$ for a lag k is given by

$$r(k) = E[u(n)u^*(n - k)]$$

$$= \begin{cases} |\alpha|^2 + \sigma_v^2, & k = 0 \\ \\ |\alpha|^2 \exp(j\omega k), & k \neq 0 \end{cases} \qquad (2.36)$$

where $|\alpha|$ is the magnitude of the complex amplitude α. Note that for a lag $k \neq 0$, the autocorrelation function $r(k)$ varies with k in the same sinusoidal fashion as the sample $u(n)$ varies with n, except for a change in amplitude. Given the series of samples $u(n)$, $u(n - 1)$, $\ldots$, $u(n - M + 1)$, we may thus express the correlation matrix of $u(n)$ as

$$\mathbf{R} = |\alpha|^2 \begin{bmatrix} 1 + \dfrac{1}{\rho} & \exp(j\omega) & \cdots & \exp(j\omega(M-1)) \\ \exp(-j\omega) & 1 + \dfrac{1}{\rho} & \cdots & \exp(j\omega(M-2)) \\ \cdot & \cdot & \cdot & \cdot \\ \cdot & \cdot & \cdot & \cdot \\ \cdot & \cdot & \cdot & \cdot \\ \exp(-j\omega(M-1)) & \exp(-j\omega(M-2)) & \cdots & 1 + \dfrac{1}{\rho} \end{bmatrix} \quad (2.37)$$

where ρ is the *signal-to-noise ratio*, defined by

$$\rho = \frac{|\alpha|^2}{\sigma_v^2} \quad (2.38)$$

The correlation matrix $\mathbf{R}$ of Eq. (2.37) has all of the properties described in Section 2.3; the reader is invited to verify them.

Equation (2.36) provides the mathematical basis of a two-step practical procedure for estimating the parameters of a complex sinusoid in the presence of additive noise:

1. Measure the mean-square value $r(0)$ of the process $u(n)$. Hence, given the noise variance σ_v^2, determine the magnitude $|\alpha|$.
2. Measure the autocorrelation function $r(k)$ of the process $u(n)$ for a lag $k \neq 0$. Hence, given $|\alpha|^2$ from step 1, determine the angular frequency ω.

Note that this estimation procedure is *invariant to the phase of* α, which is a direct consequence of the definition of the autocorrelation function $r(k)$.

Example 1

Consider the idealized case of a noiseless sinusoid of angular frequency ω. For the purpose of illustration, we assume that the time series of interest consists of three uniformly spaced samples drawn from this sinusoid. Hence, setting the signal-to-noise ratio $\rho = \infty$ and the number of samples $M = 3$, we find from Eq. (2.37) that the correlation matrix of the time series so obtained has the following value:

$$\mathbf{R} = |\alpha|^2 \begin{bmatrix} 1 & \exp(j\omega) & \exp(j2\omega) \\ \exp(-j\omega) & 1 & \exp(j\omega) \\ \exp(-j2\omega) & \exp(-j\omega) & 1 \end{bmatrix}$$

From this expression we readily see that the determinant of $\mathbf{R}$ and all principal minors are identically zero. Hence, this correlation matrix is singular.

We may generalize the result of this example by stating that when a process $u(n)$ consists of M samples drawn from the sum of K sinusoids with $K < M$ and there is *no* additive noise, then the correlation matrix of that process is singular.

2.5 STOCHASTIC MODELS

The term *model* is used for any hypothesis that may be applied to explain or describe the hidden laws that are supposed to govern or constrain the generation of physical data of interest. The representation of a stochastic process by a model dates back to an idea by Yule (1927). The idea is that a time series $u(n)$, consisting of highly correlated observations, may be generated by applying a series of statistically independent "shocks" to a linear filter, as in Fig. 2.1. The shocks are random variables drawn from a fixed distribution that is usually assumed to be *Gaussian* with zero mean and constant variance. Such a series of random variables constitutes a purely random process, commonly referred to as *white Gaussian noise*. Specifically, we may describe the input $v(n)$ in Figure 2.1 in statistical terms as follows:

$$E[v(n)] = 0 \qquad \text{for all } n \tag{2.39}$$

and

$$E[v(n)v^*(k)] = \begin{cases} \sigma_v^2, & k = n \\ 0, & \text{otherwise} \end{cases} \tag{2.40}$$

where σ_v^2 is the noise variance. Equation (2.39) follows from the zero-mean assumption, and Eq. (2.40) follows from the white assumption. The implication of the Gaussian assumption is discussed in Section 2.11.

In general, the time-domain description of the input–output relation for the stochastic model of Fig. 2.1 may be described as follows:

$$\begin{pmatrix} \text{present value} \\ \text{of model output} \end{pmatrix} + \begin{pmatrix} \text{linear combination} \\ \text{of past values} \\ \text{of model output} \end{pmatrix} = \begin{pmatrix} \text{linear combination of} \\ \text{present and past values} \\ \text{of model input} \end{pmatrix} \tag{2.41}$$

A stochastic process so described is referred to as a *linear process*.

The structure of the linear filter in Fig. 2.1 is determined by the manner in which the two linear combinations indicated in Eq. (2.41) are formulated. We may thus identify three popular types of linear stochastic models:

1. Autoregressive models, in which no past values of the model input are used.
2. Moving average models, in which no past values of the model output are used.
3. Mixed autoregressive-moving average models, in which the description of Eq. (2.41) applies in its entire form. Hence, this class of stochastic models includes autoregressive and moving average models as special cases.

Figure 2.1 Stochastic model.

These models are described next, in that order.

Autoregressive Models

We say that the time series $u(n)$, $u(n-1)$, ..., $u(n-M)$ represents the realization of an *autoregressive process (AR) of order M* if it satisfies the difference equation

$$u(n) + a_1^* u(n-1) + \cdots + a_M^* u(n-M) = v(n) \qquad (2.42)$$

where $a_1, a_2, \ldots, a_M$ are constants called the AR *parameters*, and $v(n)$ is a white-noise process. The term $a_k^* u(n-k)$ is the scalar version of *inner product* of a_k and $u(n-k)$, where $k = 1, \ldots, M$.

To explain the reason for the term "autoregressive," we rewrite Eq. (2.42) in the form

$$u(n) = w_1^* u(n-1) + w_2^* u(n-2) + \cdots + w_M^* u(n-M) + v(n) \qquad (2.43)$$

where $w_k = -a_k$. We thus see that the present value of the process, that is, $u(n)$, equals a *finite linear combination of past values* of the process, $u(n-1), \ldots, u(n-M)$, plus an *error term* $v(n)$. We now see the reason for the term "autoregressive." Specifically, a linear model

$$y = \sum_{k=1}^{M} w_k^* x_k + v$$

relating a *dependent* variable y to a set of *independent* variables $x_1, x_2, \ldots, x_M$ plus an error term v is often referred to as a *regression model*, and y is said to be "regressed" on $x_1, x_2, \ldots, x_M$. In Eq. (2.43), the variable $u(n)$ is *regressed* on previous values of *itself*; hence the term "autoregressive."

The left-hand side of Eq. (2.42) represents the *convolution* of the input sequence $u(n)$ and the sequence of parameters a_n^*. To highlight this point, we rewrite Eq. (2.42) in the form of a convolution sum:

$$\sum_{k=0}^{M} a_k^* u(n-k) = v(n) \qquad (2.44)$$

where $a_0 = 1$. By taking the *z-transform* of both sides of Eq. (2.44), we transform the convolution sum on the left-hand side of the equation into a multiplication of the z-transforms of the two sequences $u(n)$ and a_n^*. Let $H_A(z)$ denote the z-transform of the sequence a_n^*:

$$H_A(z) = \sum_{n=0}^{M} a_n^* z^{-n} \qquad (2.45)$$

Let $U(z)$ denote the z-transform of the input sequence $u(n)$:

$$U(z) = \sum_{n=0}^{\infty} u(n) z^{-n} \qquad (2.46)$$

where z is a *complex variable*. We may thus transform the difference equation (2.42) into the equivalent form

$$H_A(z)U(z) = V(z) \tag{2.47}$$

where

$$V(z) = \sum_{n=0}^{\infty} v(n)z^{-n} \tag{2.48}$$

The z-transform of Eq. (2.47) offers two interpretations, depending on whether the AR process $u(n)$ is viewed as the input or output of interest:

1. Given the AR process $u(n)$, we may use the filter shown in Fig. 2.2(a) to produce the white noise process $v(n)$ as output. The parameters of this filter bear a one-to-one correspondence with those of the AR process $u(n)$. Accordingly, this filter represents a *process analyzer* with discrete transfer function $H_A(z) = V(z)/U(z)$. The impulse response of the AR process analyzer, that is, the inverse z-transform of $H_A(z)$, has *finite duration*.

2. With the white noise $v(n)$ acting as input, we may use the filter shown in Fig. 2.2(b) to produce the AR process $u(n)$ as output. Accordingly, this second filter represents a *process generator*, whose transfer function equals

$$
\begin{aligned}
H_G(z) &= \frac{U(z)}{V(z)} \\[2mm]
&= \frac{1}{H_A(z)} \\[2mm]
&= \frac{1}{\displaystyle\sum_{n=0}^{M} a_n^* z^{-n}}
\end{aligned} \tag{2.49}
$$

The impulse response of the AR process generator, that is, the inverse z-transform of $H_G(z)$, has *infinite duration*.

The AR process analyzer of Fig. 2.2(a) is an *all-zero filter*. It is so called because its transfer function $H_A(z)$ is completely defined by specifying the locations of its *zeros*. This filter is inherently stable.

The AR process generator of Fig. 2.2(b) is an *all-pole filter*. It is so called because its transfer function $H_G(z)$ is completely defined by specifying the locations of its *poles*, as shown by

$$H_G(z) = \frac{1}{(1 - p_1 z^{-1})(1 - p_2 z^{-1})\cdots(1 - p_M z^{-1})} \tag{2.50}$$

The parameters $p_1, p_2, \ldots, p_M$ are *poles* of $H_G(z)$; they are defined by the roots of the *characteristic equation*

$$1 + a_1^* z^{-1} + a_2^* z^{-2} + \cdots + a_M^* z^{-M} = 0 \tag{2.51}$$

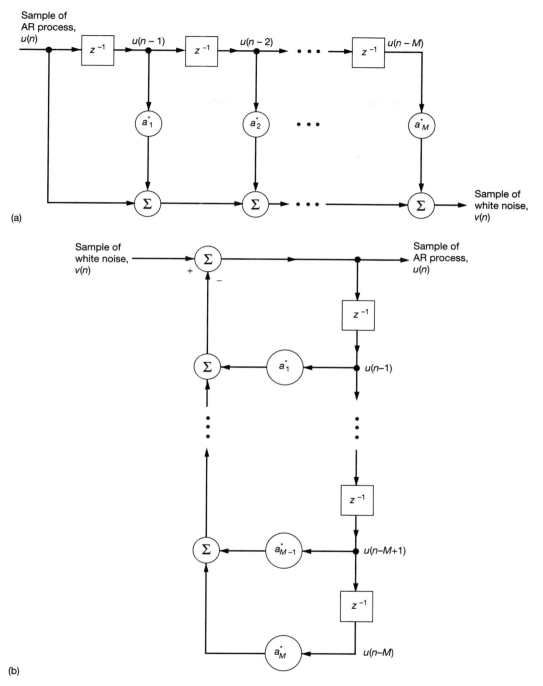

Figure 2.2   (a) AR process analyzer; (b) AR process generator.

For the all-pole AR process generator of Fig. 2.2(b) to be stable, the roots of the characteristic equation (2.51) must all lie inside the unit circle in the z-plane. This is also a necessary and sufficient condition for wide-sense stationarity of the AR process produced by the model of Fig. 2.2(b). We have more to say on the issue of stationarity in Section 2.7.

Moving Average Models

In a *moving average (MA) model*, the discrete-time linear filter of Fig. 2.1 consists of an *all-zero filter* driven by white noise. The resulting process $u(n)$, produced at the filter output, is described by the difference equation:

$$u(n) = v(n) + b_1^* v(n - 1) + \cdots + b_K^* v(n - K) \tag{2.52}$$

where $b_1, \ldots, b_K$ are constants called the *MA parameters*, and $v(n)$ is a white-noise process of zero mean and variance σ_v^2. Except for $v(n)$, each term on the right-hand side of Eq. (2.52) represents the scalar version of an inner product. The *order* of the MA process equals K. The term moving average is a rather quaint one; nevertheless, its use is firmly established in the literature. Its usage arose in the following way: If we are given a complete temporal realization of the white-noise process $v(n)$, we may compute $u(n)$ by constructing a *weighted average* of the sample values $v(n), v(n - 1), \ldots, v(n - K)$.

From Eq. (2.52), we readily obtain the MA model (i.e., process-generator) depicted in Fig. 2.3. Specifically, we start with a white-noise process $v(n)$ at the model input and generate an MA process $u(n)$ of order K at the model output. To proceed in the reverse manner, that is, to produce the white-noise process $v(n)$, given the MA process $u(n)$, we require the use of an *all-pole filter*. In other words, the filters used in the generation and analysis of an MA process are the *opposite* of those used in the case of an AR process.

Autoregressive–Moving Average Models

To generate a mixed *autoregressive–moving average (ARMA) process* $u(n)$, we use a discrete-time linear filter in Fig. 2.1 with a transfer function that contains *both poles and zeros*. Accordingly, given a white-noise process $v(n)$ as the filter input, the ARMA process $u(n)$ produced at the filter output is described by the difference equation

$$u(n) + a_1^* u(n - 1) + \cdots + a_M^* u(n - M) = v(n) + b_1^* v(n - 1) + \cdots + b_K^* v(n - K) \tag{2.53}$$

where $a_1, \ldots, a_M$ and $b_1, \ldots, b_K$ are called the *ARMA parameters*. Except for $u(n)$ on the left-hand side and $v(n)$ on the right-hand side of Eq. (2.53), all of the terms represent scalar versions of inner products. The *order* of the ARMA process equals (M, K).

From Eq. (2.53), we readily deduce the ARMA model (i.e., process generator) depicted in Fig. 2.4. Comparing this figure with Figs. 2.2(b) and 2.3, we clearly see that AR and MA models are indeed special cases of an ARMA model.

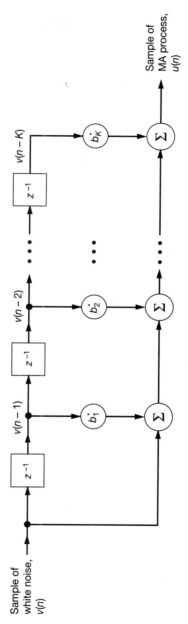

Figure 2.3   Moving average model (process generator).

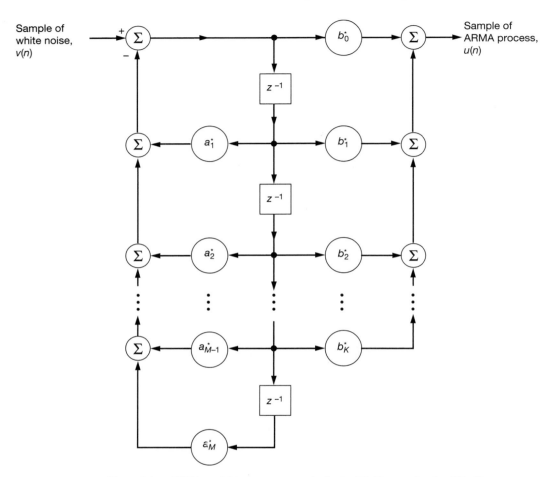

Figure 2.4   ARMA model (process generator) of order (M, K), assuming that $M > K$.

The transfer function of the ARMA process generator in Fig. 2.4 has both poles and zeros. Similarly, the ARMA analyzer used to generate a white-noise process $v(n)$, given an ARMA process $u(n)$, is characterized by a transfer function containing both poles and zeros.

From a computational viewpoint, the AR model has an advantage over the MA and ARMA models. Specifically, the computation of the AR coefficients in the model of Fig. 2.2(a) involves a system of *linear equations* known as the Yule–Walker equations, details of which are given in Section 2.8. On the other hand, the computation of the MA coefficients in the model of Fig. 2.3 and the computation of the ARMA coefficients in the model of Fig. 2.4 are much more complicated. Both of these computations require solving systems of *nonlinear equations*. It is for this reason that, in practice, we find that the use of AR models is more popular than MA and ARMA models. The wide application of AR models may also be justified by virtue of a fundamental theorem of time series analysis, which is discussed next.

2.6 WOLD DECOMPOSITION

Wold (1938) proved a fundamental theorem, which states that any stationary discrete-time stochastic process may be decomposed into the sum of a *general linear process* and a *predictable process*, with these two processes being uncorrelated with each other. More precisely, Wold proved the following result:

Any stationary discrete-time stochastic process $x(n)$ may be expressed in the form

$$x(n) = u(n) + s(n) \tag{2.54}$$

where

1. $u(n)$ and $s(n)$ are uncorrelated processes,

2. $u(n)$ is a general linear process represented by the MA model:

$$u(n) = \sum_{k=0}^{\infty} b_k^* v(n-k) \tag{2.55}$$

with $b_0 = 1$, and

$$\sum_{k=0}^{\infty} |b_k|^2 < \infty,$$

and where $v(n)$ is a white-noise process uncorrelated with $s(n)$; that is,

$$E[v(n)s^*(k)] = 0 \qquad \text{for all } (n, k)$$

3. $s(n)$ is a predictable process; that is, the process can be predicted from its own past with zero prediction variance.

This result is known as *Wold's decomposition theorem*. A proof of this theorem is given in Priestley (1981).

According to Eq. (2.55), the general linear process $u(n)$ may be generated by feeding an *all-zero filter* with the white-noise process $v(n)$ as in Fig. 2.5(a). The zeros of the transfer function of this filter equal the roots of the equation:

$$B(z) = \sum_{n=0}^{\infty} b_n^* z^{-n} = 0$$

A solution of particular interest is an all-zero filter that is *minimum phase*, which means that all the zeros of the polynomial $B(z)$ lie inside the unit circle. In such a case, we may replace the all-zero filter with an *equivalent* all-pole filter that has the same impulse response $h_n = b_n^*$, as in Fig. 2.5(b). This means that except for a predictable component, a stationary discrete-time stochastic process may also be represented as an AR process of the appropriate order, subject to the above-mentioned restriction on $B(z)$. The basic difference between the MA and AR models is that $B(z)$ operates on the input $v(n)$ in the MA model, whereas the inverse $B^{-1}(z)$ operates on the output $u(n)$ in the AR model.

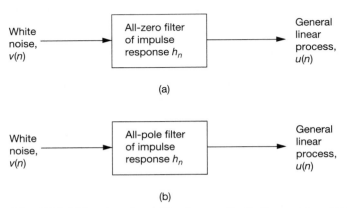

Figure 2.5   (a) Model, based on all-zero filter, for generating the linear process $u(n)$; (b) model, based on all-pole filter, for generating the general linear process $u(n)$. Both filters have exactly the same impulse response.

2.7 ASYMPTOTIC STATIONARITY OF AN AUTOREGRESSIVE PROCESS

Equation (2.42) represents a *linear, constant coefficient, difference equation of order M*, in which $v(n)$ plays the role of *input* or *driving function* and $u(n)$ that of *output* or *solution*. By using the *classical method*[1] for solving such an equation, we may formally express the solution $u(n)$ as the sum of a *complementary function*, $u_c(n)$, and a *particular solution*, $u_p(n)$, as follows:

$$u(n) = u_c(n) + u_p(n) \tag{2.56}$$

The evaluation of the solution $u(n)$ may thus proceed in two stages:

1. The complementary function $u_c(n)$ is the solution of the *homogeneous equation*

$$u(n) + a_1^* u(n-1) + a_2^* u(n-2) + \cdots + a_M^* u(n-M) = 0$$

In general, the complementary function $u_c(n)$ will therefore be of the form

$$u_c(n) = B_1 p_1^n + B_2 p_2^n + \cdots + B_M p_M^n \tag{2.57}$$

where $B_1, B_2, \ldots, B_M$ are arbitrary constants, and $p_1, p_2, \ldots, p_M$ are roots of the characteristic equation (2.51).

2. The particular solution $u_p(n)$ is defined by

$$u_p(n) = H_G(D)[v(n)] \tag{2.58}$$

---

[1]We may also use the z-transform method to solve the difference equation (2.42). However, for the discussion presented here, we find it more informative to use the classical method.

where D is the *unit-delay operator*, and the operator $H_G(D)$ is obtained by substituting D for z^{-1} in the discrete-transfer function of Eq. (2.49). The unit-delay operator D has the property

$$D^k[u(n)] = u(n - k), \qquad k = 0, 1, 2, \ldots \tag{2.59}$$

The constants $B_1, B_2, \ldots, B_M$ are determined by the choice of *initial conditions* that equal M in number. It is customary to set

$$
\begin{aligned}
u(0) &= 0 \\
u(-1) &= 0 \\
&\;\;\vdots \\
u(-M + 1) &= 0
\end{aligned}
\tag{2.60}
$$

This is equivalent to setting the output of the model in Fig. 2.2(b) as well as the succeeding $(M - 1)$ tap inputs equal to zero at time $n = 0$. Thus, by substituting these initial conditions into Eqs. (2.56) – (2.58), we obtain a set of M simultaneous equations that can be solved for the constants $B_1, B_2, \ldots, B_M$.

The result of imposing the initial conditions of Eq. (2.60) on the solution $u(n)$ is to make the discrete-time stochastic process represented by this solution nonstationary. On reflection, it is clear that this must be so, since we have given a "special status" to the time point $n = 0$, and the property of *invariance under a shift of time origin* cannot hold, even for second-order moments. If, however, the solution $u(n)$ is able to "forget" its initial conditions, the resulting process is asymptotically stationary in the sense that it settles down to a stationary behavior as n approaches infinity (Priestley, 1981). This requirement may be achieved by choosing the parameters of the AR model in Fig. 2.2(b) such that the complementary function $u_c(n)$ decays to zero as n approaches infinity. From Eq. (2.57) we see that, for arbitrary constants in the equation, this requirement can be met if and only if

$$|p_k| < 1 \qquad \text{for all } k$$

Hence, *for asymptotic stationarity of the discrete-time stochastic process represented by the solution $u(n)$, we require that all the poles of the filter in the AR model lie inside the unit circle in the z-plane.* This is intuitively satisfying.

Correlation Function of an Asymptotically Stationary AR Process

Assuming that the condition for asymptotic stationarity is satisfied, we may derive an important recursive relation for the autocorrelation function of the resulting AR process $u(n)$ as follows. We first multiply both sides of Eq. (2.42) by $u^*(n - l)$ and then apply the expectation operator, thereby obtaining

$$E\left[\sum_{k=0}^{M} a_k^* u(n - k) u^*(n - l) \right] = E[v(n) u^*(n - l)] \tag{2.61}$$

Next, we simplify the left-hand side of Eq. (2.61) by interchanging the expectation and summation, and recognizing that the expectation $E[u(n - k)u^*(n - l)]$ equals the autocorrelation function of the AR process for a lag of $l - k$. We simplify the right-hand side by observing that the expectation $E[v(n)u^*(n - l)]$ is zero for $l > 0$, since $u(n - l)$ only involves samples of the white-noise process at the filter input in Fig. 2.2(b) up to time $n - l$, which are uncorrelated with the white-noise sample $v(n)$. Accordingly, we simplify Eq. (2.61) as follows:

$$\sum_{k=0}^{M} a_k^* r(l - k) = 0, \qquad l > 0 \tag{2.62}$$

where $a_0 = 1$. We thus see that the autocorrelation function of the AR process satisfies the difference equation

$$r(l) = w_1^* r(l - 1) + w_2^* r(l - 2) + \cdots + w_M^* r(l - M), \qquad l > 0 \tag{2.63}$$

where $w_k = -a_k$, $k = 1, 2, \ldots, M$. Note that Eq. (2.63) is analogous to the difference equation satisfied by the AR process $u(n)$ itself.

We may express the general solution of Eq. (2.63) as follows:

$$r(m) = \sum_{k=1}^{M} C_k p_k^m \tag{2.64}$$

where $C_1, C_2, \ldots, C_M$ are constants, and $p_1, p_2, \ldots, p_M$ are roots of the characteristic equation (2.51). Note that when the AR model of Fig. 2.2(b) satisfies the condition for asymptotic stationarity, $|p_k| < 1$ for all k, in which case the autocorrelation function $r(m)$ approaches zero as the lag m approaches infinity.

The exact form of the contribution made by a pole p_k in Eq. (2.64) depends on whether the pole is real or complex. When p_k is real, the corresponding contribution decays geometrically to zero as the lag m increases. We refer to such a contribution as a *damped exponential*. On the other hand, complex poles occur in conjugate pairs, and the contribution of a complex-conjugate pair of poles is in the form of *a damped sine wave*. We thus find that, in general, the autocorrelation function of an asymptotically stationary AR process consists of a mixture of damped exponentials and damped sine waves.

2.8 YULE–WALKER EQUATIONS

In order to uniquely define the AR model of order M, depicted in Fig. 2.2(b), we need to specify two sets of model parameters:

1. The AR coefficients $a_1, a_2, \ldots, a_M$
2. The variance σ_v^2 of the white noise $v(n)$ used as excitation.

We now address these two issues in turn.

First, writing Eq. (2.63) for $l = 1, 2, \ldots, M$, we get a set of M simultaneous equations with the values $r(0), r(1), \ldots, r(M)$ of the autocorrelation function of the AR process as the known quantities and the AR parameters $a_1, a_2, \ldots, a_M$ as the unknowns. This set of equations may be expressed in the expanded matrix form

$$
\begin{bmatrix}
r(0) & r(1) & \cdots & r(M-1) \\
r^*(1) & r(0) & \cdots & r(M-1) \\
\vdots & \vdots & \ddots & \vdots \\
r^*(M-1) & r^*(M-2) & \cdots & r(0)
\end{bmatrix}
\begin{bmatrix}
w_1 \\
w_2 \\
\vdots \\
w_M
\end{bmatrix}
=
\begin{bmatrix}
r^*(1) \\
r^*(2) \\
\vdots \\
r^*(M)
\end{bmatrix}
\tag{2.65}
$$

where we have $w_k = -a_k$. The set of equations (2.65) is called the *Yule–Walker equations* (Yule, 1927; Walker, 1931).

We may express the Yule–Walker equations in the compact matrix form

$$
\mathbf{Rw} = \mathbf{r} \tag{2.66}
$$

and its solution as (assuming that the correlation matrix $\mathbf{R}$ is nonsingular)

$$
\mathbf{w} = \mathbf{R}^{-1}\mathbf{r} \tag{2.67}
$$

where $\mathbf{R}^{-1}$ is the inverse of matrix $\mathbf{R}$, and the vector $\mathbf{w}$ is defined by

$$
\mathbf{w} = [w_1, w_2, \ldots, w_M]^T
$$

The correlation matrix $\mathbf{R}$ is defined by Eq. (2.21), and vector $\mathbf{r}$ is defined by Eq. (2.28). From these two equations, we see that we may uniquely determine both the matrix $\mathbf{R}$ and the vector $\mathbf{r}$, given the autocorrelation sequence $r(0), r(1), \ldots, r(M)$. Hence, using Eq. (2.67) we may compute the coefficient vector $\mathbf{w}$ and, therefore, the AR coefficients $a_k = -w_k, k = 1, 2, \ldots, M$. In other words, there is a unique relationship between the coefficients $a_1, a_2, \ldots, a_M$ of the AR model and the *normalized* correlation coefficients $\rho_1, \rho_2, \ldots, \rho_M$ of the AR process $u(n)$, as shown by

$$
\{a_1, a_2, \ldots, a_M\} \rightleftharpoons \{\rho_1, \rho_2, \ldots, \rho_M\} \tag{2.68}
$$

where the *correlation coefficient* ρ_k is defined by

$$
\rho_k = \frac{r(k)}{r(0)}, \qquad k = 1, 2, \ldots, M \tag{2.69}
$$

Variance of the White Noise

For $l = 0$, we find that the expectation on the right-hand side of Eq. (2.61) assumes the special form

$$
E[v(n)u^*(n)] = E[v(n)v^*(n)]
$$

$$
= \sigma_v^2 \tag{2.70}
$$

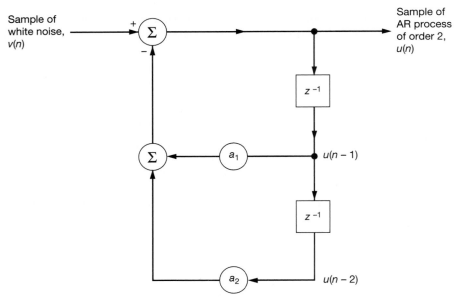

Figure 2.6   Model of (real-valued) autoregressive process of order 2.

where σ_v^2 is the variance of the zero-mean white noise $v(n)$. Accordingly, setting $l = 0$ in Eq. (2.61) and complex-conjugating both sides, we get the following formula for the variance of the white-noise process:

$$\sigma_v^2 = \sum_{k=0}^{M} a_k r(k) \tag{2.71}$$

where $a_0 = 1$. Hence, given the autocorrelations $r(0), r(1), \ldots, r(M)$, we may determine the white-noise variance σ_v^2.

2.9 COMPUTER EXPERIMENT: AUTOREGRESSIVE PROCESS OF ORDER 2

To illustrate the theory developed above for the modeling of an AR process, we consider the example of a second-order AR process that is real valued.[2] Figure 2.6 shows the block diagram of the model used to generate this process. Its time-domain description is governed by the second-order difference equation

$$u(n) + a_1 u(n - 1) + a_2 u(n - 2) = v(n) \tag{2.72}$$

---

[2]In this example, we follow the approach described by Box and Jenkins (1976).

where $v(n)$ is drawn from a white-noise process of zero mean and variance σ_v^2. Figure 2.7(a) shows one realization of this white-noise process. The variance σ_v^2 is chosen to make the variance of $u(n)$ equal unity.

Conditions for Asymptotic Stationarity

The second-order AR process $u(n)$ has the characteristic equation

$$1 + a_1 z^{-1} + a_2 z^{-2} = 0 \tag{2.73}$$

Let p_1 and p_2 denote the two roots of this equation:

$$p_1, \ p_2 = \tfrac{1}{2}(-a_1 \pm \sqrt{a_1^2 - 4a_2}) \tag{2.74}$$

To ensure the asymptotic stationarity of the AR process $u(n)$, we require that these two roots lie inside the unit circle in the z-plane. That is, both p_1 and p_2 must have a magnitude less than 1. This, in turn, requires that the AR parameters a_1 and a_2 lie in the triangular region defined by

$$-1 \leq a_2 + a_1$$
$$-1 \leq a_2 - a_1 \tag{2.75}$$
$$-1 \leq a_2 \leq 1$$

as shown in Fig. 2.8.

Autocorrelation Function

The autocorrelation function $r(m)$ of an asymptotically stationary AR process for lag m satisfies the difference equation (2.62). Hence, using this equation, we obtain the following second-order difference equation for the autocorrelation function of a second-order AR process:

$$r(m) + a_1 r(m-1) + a_2 r(m-2) = 0, \quad m > 0 \tag{2.76}$$

For the initial values, we have (as will be explained later)

$$r(0) = \sigma_u^2$$
$$r(1) = \frac{-a_1}{1 + a_2}\sigma_u^2 \tag{2.77}$$

Thus, solving Eq. (2.76) for $r(m)$, we get (for $m > 0$)

$$r(m) = \sigma_u^2 \left[\frac{p_1(p_2^2 - 1)}{(p_2 - p_1)(p_1 p_2 + 1)}p_1^m - \frac{p_2(p_1^2 - 1)}{(p_2 - p_1)(p_1 p_2 + 1)}p_2^m \right] \tag{2.78}$$

where p_1 and p_2 are defined by Eq. (2.74).

There are two specific cases to be considered, depending on whether the roots p_1 and p_2 are real or complex valued, as described next.

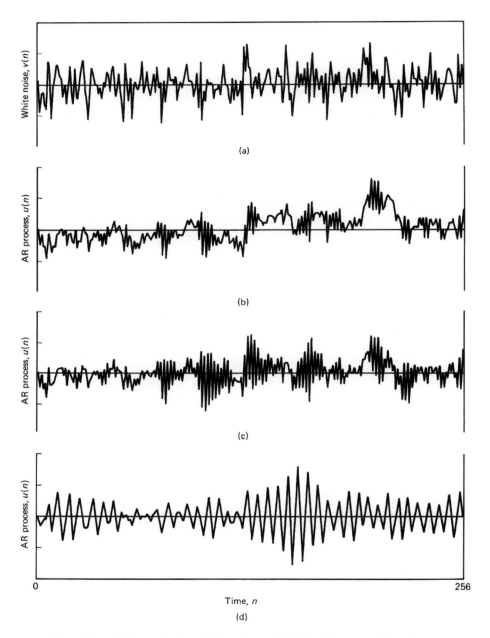

Figure 2.7   (a) One realization of white-noise input; (b), (c), (d) corresponding outputs of AR model of order 2 for parameters of Eqs. (2.79), (2.80), and (2.81), respectively.

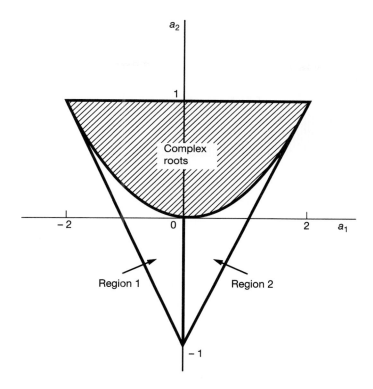

Figure 2.8   Permissible region for the AR parameters a_1 and a_2.

Case 1:   *Real Roots.* This case occurs when

$$a_1^2 - 4a_2 > 0$$

which corresponds to regions 1 and 2 below the parabolic boundary in Fig. 2.8. In region 1, the autocorrelation function remains positive as it damps out, corresponding to a positive dominant root. This situation is illustrated in Fig. 2.9(a) for the AR parameters

$$a_1 = -0.10$$
$$a_2 = -0.8$$

(2.79)

In Fig. 2.7(b), we show the time variation of the output of the model in Fig. 2.6 [with a_1 and a_2 assigned the values given in Eq. (2.79)]. This output is produced by the white-noise input shown in Fig. 2.7(a).

In region 2 of Fig. 2.8, the autocorrelation function alternates in sign as it damps out, corresponding to a negative dominant root. This situation is illustrated in Fig. 2.9(b) for the AR parameters

$$a_1 = 0.1$$
$$a_2 = -0.8$$

(2.80)

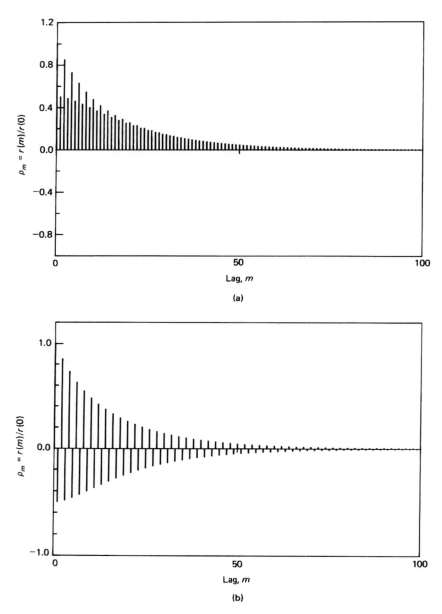

Figure 2.9   Plots of normalized autocorrelation function of real-valued AR(2) process;
(a) $r(1) > 0$; (b) $r(1) < 0$; (c) conjugate roots.

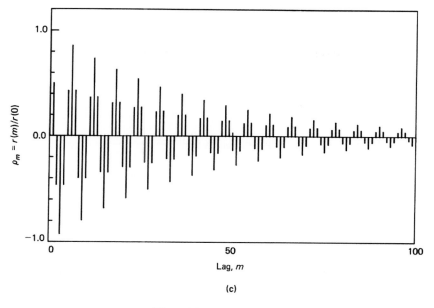

(c)

Figure 2.9   (*continued*)

In Fig. 2.7(c) we show the time variation of the output of the model in Fig. 2.6 [with a_1 and a_2 assigned the values given in Eq. (2.80)]. This output is also produced by the white-noise input shown in Fig. 2.7(a).

Case 2:   *Complex-Conjugate Roots.* This occurs when

$$a_1^2 - 4a_2 < 0$$

which corresponds to the shaded region shown in Fig. 2.8 above the parabolic boundary. In this case, the autocorrelation function displays a pseudoperiodic behavior, as illustrated in Fig. 2.9(c) for the AR parameters

$$a_1 = -0.975$$
$$a_2 = 0.95$$

(2.81)

In Fig. 2.7(d) we show the time variation of the output of the model in Fig. 2.6 [with a_1 and a_2 assigned the values given in Eq. (2.81)], which is produced by the white-noise input shown in Fig. 2.7(a).

Yule–Walker Equations

Substituting the value $M = 2$ for the AR model order in Eq. (2.65), we get the following Yule–Walker equations for the second-order AR process:

$$\begin{bmatrix} r(0) & r(1) \\ r(1) & r(0) \end{bmatrix} \begin{bmatrix} w_1 \\ w_2 \end{bmatrix} = \begin{bmatrix} r(1) \\ r(2) \end{bmatrix} \tag{2.82}$$

where we have used the fact that $r(-1) = r(1)$ for a real-valued process. Solving Eq. (2.82) for w_1 and w_2, we get

$$w_1 = -a_1 = \frac{r(1)\,[r(0) - r(2)\,]}{r^2(0) - r^2(1)}$$

$$w_2 = -a_2 = \frac{r(0)\,r(2) - r^2(1)}{r^2(0) - r^2(1)} \tag{2.83}$$

We may also use Eq. (2.82) to express $r(1)$ and $r(2)$ in terms of the AR parameters a_1 and a_2 as follows:

$$r(1) = \frac{-a_1}{1 + a_2}\,\sigma_u^2$$

$$r(2) = \left(-a_2 + \frac{a_1^2}{1 + a_2}\right)\sigma_u^2 \tag{2.84}$$

where $\sigma_u^2 = r(0)$. This solution explains the initial values for $r(0)$ and $r(1)$ that were quoted in Eq. (2.77).

The conditions for asymptotic stationarity of the second-order AR process are given in terms of the AR parameters a_1 and a_2 in Eq. (2.75). Using the expressions for $r(1)$ and $r(2)$ in terms of a_1 and a_2, given in Eq. (2.84), we may reformulate the conditions for asymptotic stationarity as follows:

$$-1 < \rho_1 < 1$$

$$-1 < \rho_2 < 1$$

$$\rho_1^2 < \frac{1}{2}(1 + \rho_2) \tag{2.85}$$

where ρ_1 and ρ_2 are the normalized *correlation coefficients* defined by

$$\rho_1 = \frac{r(1)}{r(0)}$$

and

$$\rho_2 = \frac{r(2)}{r(0)} \tag{2.86}$$

Figure 2.10 shows the admissible region for ρ_1 and ρ_2.

Variance of the White-Noise Process

Putting $M = 2$ in Eq. (2.71), we may express the variance of the white-noise process $v(n)$ as

$$\sigma_v^2 = r(0) + a_1 r(1) + a_2 r(2) \tag{2.87}$$

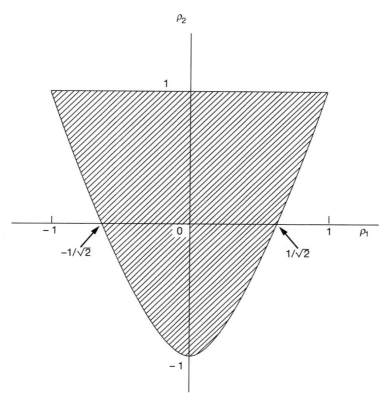

Figure 2.10   Permissible region for parameters of second-order AR process in terms of the normalized correlation coefficients ρ_1 and ρ_2.

Next, substituting Eq. (2.84) in (2.87), and solving for $\sigma_u^2 = r(0)$, we get

$$\sigma_u^2 = \left(\frac{1+a_2}{1-a_2}\right)\frac{\sigma_v^2}{[(1+a_2)^2 - a_1^2]} \tag{2.88}$$

For the three sets of AR parameters considered previously, we thus find that the variance of the white noise $v(n)$ has the values given in Table 2.1, assuming that $\sigma_u^2 = 1$.

TABLE 2.1   AR PARAMETERS AND NOISE VARIANCE

a_1	a_2	σ_v^2
-0.10	-0.8	0.27
0.1	-0.8	0.27
-0.975	0.95	0.0731

2.10 SELECTING THE MODEL ORDER

The representation of a stochastic process by a linear model may be used for synthesis or analysis. In *synthesis*, we generate a desired time series by assigning a prescribed set of values to the parameters of the model and feeding it with white noise of zero mean and prescribed variance. In *analysis*, on the other hand, we *estimate* the parameters of the model by processing a given time series of finite length. Insofar as the estimation is statistical, we need an appropriate measure of the fit between the model and the observed data. This implies that unless we have some prior information, the estimation procedure should include a criterion for selecting the *model order* (i.e., the number of independently adjusted parameters in the model). In the case of an AR process defined by Eq. (2.42), the model order equals M. In the case of an MA process defined by Eq. (2.52), the model order equals K. In the case of an ARMA process defined by Eq. (2.53), the model order equals (M, K). Various criteria for model-order selection are described in the literature (Priestley, 1981; Kay, 1988). In this section we describe two important criteria for selecting the order of the model, one of which was pioneered by Akaike (1973, 1974) and the other by Rissanen (1978) and Schwartz (1978); both criteria result from the use of information-theoretic arguments, but in entirely different ways.

An Information-Theoretic Criterion

Let $u_i = u(i)$, $i = 1, 2, \ldots, N$, denote the data obtained by N independent observations of a stationary discrete-time stochastic process, and $g(u_i)$ denote the probability density function of u_i. Let $f_U(u_i|\hat{\boldsymbol{\theta}}_m)$ denote the conditional probability density function of u_i, given $\hat{\boldsymbol{\theta}}_m$, where $\hat{\boldsymbol{\theta}}_m$ is the *estimated* vector of parameters that model the process. Let m be the model order, so that we may write

$$\hat{\boldsymbol{\theta}}_m = \begin{bmatrix} \hat{\theta}_{1m} \\ \hat{\theta}_{2m} \\ \cdot \\ \cdot \\ \cdot \\ \hat{\theta}_{mm} \end{bmatrix} \tag{2.89}$$

We thus have several models that compete with each other to represent the process of interest. An *information-theoretic criterion* (AIC) proposed by Akaike selects the model for which the quantity

$$\text{AIC}(m) = -2L(\hat{\boldsymbol{\theta}}_m) + 2m \tag{2.90}$$

is a minimum. The function $L(\hat{\boldsymbol{\theta}}_m)$ is defined by

$$L(\hat{\boldsymbol{\theta}}_m) = \max \sum_{i=1}^{N} \ln f_U(u_i|\hat{\boldsymbol{\theta}}_m) \tag{2.91}$$

where ln denotes the natural logarithm. The criterion of Eq. (2.91) is derived by minimizing the *Kullback–Leibler mean information*,[3] which is used to provide a measure of the separation or distance between the "unknown" true probability density function $g(u)$ and the conditional probability density function $f_U(u_i|\hat{\boldsymbol{\theta}}_m)$ given by the model in light of the observed data.

The function $L(\hat{\boldsymbol{\theta}}_m)$, constituting the first term on the right-hand side of Eq. (2.90), except for a scalar, is recognized as a *logarithm* of the *maximum-likelihood estimates* of the parameters in the model; for a discussion of the method of maximum likelihood, see Appendix D. The second term, $2m$, represents a *model complexity penalty* that makes AIC(m) an estimate of the Kullback–Leibler mean information.

The first term of Eq. (2.90) tends to decrease rapidly with model order m. On the other hand, the second term increases linearly with m. The result is that if we plot AIC(m) versus model order m, the graph will, in general, show a definite minimum value, and the *optimum order* of the model is determined by that value of m at which AIC(m) attains its minimum value. The minimum value of AIC is called MAIC(minimum AIC).

Minimum Description Length Criterion

Rissanen (1978, 1989) has used an entirely different approach to solve the statistical model identification problem. Specifically, he starts with the notion that a model may be viewed as a device for describing the regular features of a set of observed data, with the objective being that of searching for a model that best captures the regular features or constraints that give the data their special structure. Recognizing that the presence of constraints reduces uncertainty about the data, the objective may equally be that of *encoding* the data in the shortest or least redundant manner; the term "encoding" used here refers to an exact description of the observed data. Accordingly, the number of binary digits needed to encode both the observed data, when advantage is taken of the constraints offered by a model, and the model itself may be used as a criterion for *measuring the amount of the same constraints* and therefore the goodness of the model.

We may thus formally state Rissanen's *minimum description length (MDL) criterion*[4] as follows: Given a data set of interest and a family of competing statistical

---

[3]In Akaike (1973, 1974, 1977) and Ulrych and Ooe (1983), the criterion of Eq. (2.90) is derived from the principle of minimizing the expectation $E[I(g;f(\bullet|\hat{\boldsymbol{\theta}}_m)]$, where

$$I(g;f(\bullet|\hat{\boldsymbol{\theta}}_m)) = \int_{-\infty}^{\infty} g(u) \ln g(u)\, du - \int_{-\infty}^{\infty} g(u) \ln f_U(u|\hat{\boldsymbol{\theta}}_m)\, du$$

We refer to $I(g;f(\bullet|\hat{\boldsymbol{\theta}}))$ as the *Kullback–Leibler mean information* for discrimination between $g(u)$ and $f_U(u|\hat{\boldsymbol{\theta}}_m)$ (Kullback and Leibler, 1951). The idea is to minimize the information added to the time series by modeling it as an AR, MA, or ARMA process of finite order, since any information added is virtually false information in a real-world situation. Since $g(u)$ is fixed and unknown, the problem reduces to one of maximizing the second term that makes up $I(g;f(\bullet|\hat{\boldsymbol{\theta}}_m))$.

[4]The idea of *minimum description length* of individual recursively definable objects may be traced to Kolmogorov (1968).

models, the best model is the particular one that provides the shortest description length for the data. In mathematical terms, it is defined by[5] (Rissanen, 1978, 1989; Wax, 1995)

$$\text{MDL}(m) = -L(\hat{\boldsymbol{\theta}}_m) + \frac{1}{2} m \ln N \tag{2.92}$$

where m is the number of independently adjusted parameters in the model, and N is the sample size (i.e., the number of observations). As with Akaike's information–theoretic criterion, $L(\hat{\boldsymbol{\theta}}_m)$ is the logarithm of the maximum likelihood estimates of the model parameters. In comparing Eqs. (2.90) and (2.92), we see that the principal difference between the AIC and MDL criterion lies in the structure-dependent term.

According to Rissanen (1989), the MDL criterion offers the following attributes:

- The model permits the shortest encoding of the observed data and captures all the *learnable* properties of the observed data in the best possible manner.
- The MDL criterion is a *consistent* model-order estimator in the sense that it converges to the true model order as the sample size increases.
- The model is optimal in the context of linear regression problems as well as ARMA models.

Perhaps the most significant point to note is the fact that in all of the applications involving the MDL criterion, there has been no anomalous result or a model with undesirable properties reported in the literature.

2.11 COMPLEX GAUSSIAN PROCESSES

Gaussian stochastic processes, or simply *Gaussian processes*, are frequently encountered in both theoretical and applied analysis. In this section we present a summary of some important properties of Gaussian processes that are *complex valued*.[6]

Let $u(n)$ denote a complex Gaussian process consisting of N samples. For the first- and second-order statistics of this process, we assume the following:

1. A *mean* of zero as shown by

$$\mu = E[u(n)] = 0 \qquad \text{for } n = 1, 2, \ldots, N \tag{2.93}$$

---

[5]Schwartz (1989) has derived a similar result, using a Bayesian approach. In particular, he considers the asymptotic behavior of Bayes estimators under a special class of *priors*. These priors put positive probability on the subspaces that correspond to the competing models. The decision is made by selecting the model that yields the maximum *a posteriori probability*.

It turns out that, in the large sample limit, the two approaches taken by Schwartz and Rissanen yield essentially the same result. However, Rissanen's approach is much more general, whereas Schwartz's approach is restricted to the case that the observations are independent and come from an exponential distribution.

[6]For a detailed treatment of complex Gaussian processes, see the book by Miller (1974). Properties of complex Gaussian processes are also discussed in Kelly et al. (1960), Reed (1962), and McGee (1971).

2. An *autocorrelation function* denoted by

$$r(k) = E[u(n)u^*(n-k)], \quad k = 0, 1, \ldots, N-1 \tag{2.94}$$

The set of autocorrelation functions $\{r(k), k = 0, 1, \ldots, N-1\}$ defines the correlation matrix $\mathbf{R}$ of the Gaussian process $u(n)$.

The shorthand notation $\mathcal{N}(\mathbf{0}, \mathbf{R})$ is commonly used to refer to a Gaussian process with a mean vector of zero and correlation matrix $\mathbf{R}$.

Equations (2.93) and (2.94) imply wide-sense stationarity of the process. Knowledge of the mean μ and the autocorrelation function $r(k)$ for varying values of lag k is indeed sufficient for the complete characterization of the complex Gaussian process $u(n)$. In particular, it may be shown that the *joint probability density function* of N samples of the process so described is as follows (Kelly et al., 1960):

$$\mathbf{f}_{\mathbf{U}}(\mathbf{u}) = \frac{1}{(2\pi)^N \det{(\Lambda)}} \exp\left(-\frac{1}{2}\mathbf{u}^H \Lambda^{-1} \mathbf{u}\right) \tag{2.95}$$

where $\mathbf{u}$ is the N-by-1 data vector; that is,

$$\mathbf{u} = [u(1), u(2), \ldots, u(N)]^T \tag{2.96}$$

and Λ is the N-by-N Hermitian-symmetric *moment matrix* of the process, defined in terms of the correlation matrix $\mathbf{R} = \{r(k)\}$ as

$$\Lambda = \tfrac{1}{2} E[\mathbf{u}\mathbf{u}^H] \tag{2.97}$$

$$= \tfrac{1}{2}\mathbf{R}$$

Note that the joint probability density function $f_{\mathbf{U}}(\mathbf{u})$ is $2N$-dimensional, where the factor 2 accounts for the fact that each of the N samples of the process has a real and an imaginary part. Note also that the probability density function of a single sample $u(n)$ of the process, which is a special case of Eq. (2.95), is given by

$$f_U(u) = \frac{1}{\pi\sigma^2}\exp\left(-\frac{|u|^2}{\sigma^2}\right) \tag{2.98}$$

where $|u|$ is the magnitude of the sample $u(n)$ and σ^2 is its variance.

Based on the representation described herein, we may now summarize some important properties of a *zero-mean complex Gaussian process $u(n)$ that is wide-sense stationary* as follows:

1. The process $u(n)$ is *stationary in the strict sense.*

2. The process $u(n)$ is *circularly complex* in the sense that any two different samples $u(n)$ and $u(k)$ of the process satisfy the condition

$$E[u(n)u(k)] = 0 \qquad \text{for } n \neq k \tag{2.99}$$

It is for this reason that the process $u(n)$ is often referred to as a *circularly complex Gaussian process.*

3. Suppose that $u_n = u(n)$, for $n = 1, 2, \ldots, N$, are samples picked from a zero-mean, complex Gaussian process $u(n)$. We may thus state Property 3 in two parts (Reed, 1962):

   (a) If $k \neq l$, then

$$E[u_{s_1}^* u_{s_2}^* \cdots u_{s_k}^* u_{t_1} u_{t_2} \cdots u_{t_l}] = 0 \qquad (2.100)$$

   where s_i and t_j are integers selected from the available set $\{1, 2, \ldots, N\}$.

   (b) If $k = l$, then

$$E[u_{s_1}^* u_{s_2}^* \cdots u_{s_l}^* u_{t_1} u_{t_2} \cdots u_{t_l}] = E[u_{s_{\pi(1)}}^* u_{t_1}] E[u_{s_{\pi(2)}}^* u_{t_2}] \cdots E[u_{s_{\pi(l)}}^* u_{t_1}] \qquad (2.101)$$

   where π is a permutation of the set of integers $\{1, 2, \ldots, l\}$, and $\pi(j)$ is the jth element of that permutation. For the set of integers $\{1, 2, \ldots, l\}$ we have a total of $l!$ possible permutations. This means that the right-hand side of Eq. (2.101) consists of the product of $l!$ expectation product terms. Equation (2.101) is called the *Gaussian moment factoring theorem*.

Example 2

Consider the case of $N = 4$, for which the complex Gaussian process $u(n)$ consists of the four samples u_1, u_2, u_3, and u_4. Hence, the use of the Gaussian moment factoring theorem given in Eq. (2.101) yields the following useful identity:

$$E[u_1^* u_2^* u_3 u_4] = E[u_1^* u_3] E[u_2^* u_4] + E[u_2^* u_3] E[u_1^* u_4] \qquad (2.102)$$

For other useful identities derived from the Gaussian moment factoring theorem, see Problem 11.

2.12 SUMMARY AND DISCUSSION

In this chapter we studied the partial characterization of a stationary discrete-time stochastic process. Such a characterization is uniquely described in terms of two statistical parameters:

1. The mean, which is a constant
2. The autocorrelation function, which depends only on the time difference between any two samples of the process

The mean of the process may naturally be zero, or it can always be subtracted from the process to yield a new process of zero mean. For this reason, in much of the discussion in subsequent chapters of this book, the mean of the process is assumed to be zero. Thus, given an M-by-1 observation vector $\mathbf{u}(n)$ known to belong to a complex, stationary, discrete-time stochastic process of zero mean, we may partially describe it by defining an M-by-M correlation matrix $\mathbf{R}$ as the statistical expectation of the outer product of $\mathbf{u}(n)$ with itself. The matrix $\mathbf{R}$ is Hermitian, Toeplitz, and almost always positive definite; the latter property means that $\mathbf{R}$ is almost always nonsingular, and therefore the inverse matrix $\mathbf{R}^{-1}$ exists.

Another topic discussed in the chapter is the notion of a stochastic model, the need for which arises when we are given a set of experimental data known to be of a statistical nature, and the requirement is to analyze the data. In this context, we may mention two general requirements for a suitable model:

1. An *adequate number of adjustable parameters* for the model to capture the essential information content of the input data
2. *Mathematical tractability* of the model

The first requirement, in effect, means that the complexity of the model should closely match the complexity of the underlying physical mechanism responsible for generating the input data; in so doing, problems associated with underfitting or overfitting the input data are avoided. The second requirement is usually satisfied by the choice of a linear model.

Within the family of linear stochastic models, the autoregressive (AR) model is often preferred over the moving average (MA) model and the autoregressive-moving average (ARMA) model for an important reason: unlike an MA or ARMA model, computation of the AR coefficients is governed by a system of linear equations, namely, the Yule–Walker equations. Moreover, except for a predictable component, we may approximate a stationary discrete-time stochastic process by an AR model of sufficiently high order, subject to certain restrictions. To select a suitable value for the model order, we may use an information-theoretic criterion (AIC) according to Akaike or the minimum-description length (MDL) criterion according to Rissanen. A useful feature of the MDL criterion is that it is a consistent model-order estimator.

PROBLEMS

1. The sequences $y(n)$ and $u(n)$ are related by the difference equation

$$y(n) = u(n + a) - u(n - a)$$

where a is a constant. Evaluate the autocorrelation function of $y(n)$ in terms of that of $u(n)$.

2. Consider a correlation matrix $\mathbf{R}$ for which the inverse matrix $\mathbf{R}^{-1}$ exists. Show that $\mathbf{R}^{-1}$ is Hermitian.

3. (a) Equation (2.26) relates the $(M + 1)$-by-$(M + 1)$ correlation matrix $\mathbf{R}_{M+1}$, pertaining to the observation vector $\mathbf{u}_{M+1}(n)$ taken from a stationary stochastic process, to the M-by-M correlation matrix $\mathbf{R}_M$ of the observation vector $\mathbf{u}_M(n)$ taken from the same process. Evaluate the inverse of the correlation matrix $\mathbf{R}_{M+1}$ in terms of the inverse of the correlation matrix $\mathbf{R}_M$.

   (b) Repeat your evaluation using Eq. (2.27).

4. A first-order autoregressive (AR) process $u(n)$, which is real-valued, satisfies the real-valued difference equation

$$u(n) + a_1 u(n - 1) = v(n)$$

where a_1 is a constant, and $v(n)$ is a white-noise process of variance σ_ν^2.

(a) Show that if $v(n)$ has a nonzero mean, the AR process $u(n)$ is nonstationary.

(b) For the case when $v(n)$ has zero mean, and the constant a_1 satisfies the condition $|a_1| < 1$, show that the variance of $u(n)$ equals

$$\mathrm{var}[u(n)] = \frac{\sigma_\nu^2}{1 - a_1^2}$$

(c) For the conditions specified in part (b), find the autocorrelation function of the AR process $u(n)$. Sketch this autocorrelation function for the two cases $0 < a_1 < 1$ and $-1 < a_1 < 0$.

5. Consider an autoregressive process $u(n)$ of order 2, described by the difference equation

$$u(n) = u(n - 1) - 0.5u(n - 2) + v(n)$$

where $v(n)$ is a white-noise process of zero mean and variance 0.5.

(a) Write the Yule–Walker equations for the process.

(b) Solve these two equations for the autocorrelation function values $r(1)$ and $r(2)$.

(c) Find the variance of $u(n)$.

6. Consider a wide-sense stationary process that is modeled as an AR process $u(n)$ of order M. The set of parameters made up of the average power P_0 and the AR coefficients $a_1, a_2, \ldots, a_M$ bear a one-to-one correspondence with the autocorrelation sequence $r(0), r(1), r(2), \ldots, r(M)$, as shown by

$$\{r(0), r(1), r(2), \ldots, r(M)\} \; \rightleftharpoons \; \{P_0, a_1, a_2, \ldots, a_M\}$$

Justify the validity of this statement.

7. Evaluate the transfer functions of the following two stochastic models:

(a) The MA model of Fig. 2.3

(b) The ARMA model of Fig. 2.4.

(c) Specify the conditions for which the transfer function of the ARMA model of Fig. 2.4 reduces (1) to that of an AR model, and (2) to that of an MA model.

8. Consider an MA process $x(n)$ of order 2 described by the difference equation

$$x(n) = v(n) + 0.75v(n - 1) + 0.25v(n - 2)$$

where $v(n)$ is a zero-mean white-noise process of unit variance. The requirement is to approximate this process by an AR process $u(n)$ of order M. Do this approximation for the following orders:

(a) $M = 2$

(b) $M = 5$

(c) $M = 10$

Comment on your results. How big would the order M of the AR process $u(n)$ have to be for it to be equivalent to the MA process $x(n)$ exactly?

9. A time series $u(n)$ obtained from a wide-sense stationary stochastic process of zero mean and correlation matrix $\mathbf{R}$ is applied to an FIR filter of impulse response w_n. This impulse response defines the coefficient vector $\mathbf{w}$.

(a) Show that the average power of the filter output is equal to $\mathbf{w}^H \mathbf{R} \mathbf{w}$.

(b) How is the result in part (a) modified if the stochastic process at the filter input is a white noise of variance σ^2?

10. A general linear complex-valued process $u(n)$ is described by

$$u(n) = \sum_{k=0}^{\infty} b_k^* v(n-k)$$

where $v(n)$ is a white noise process, and b_k is a complex coefficient. Justify the following statements:

   (a) If the process $v(n)$ is Gaussian, then the original process $u(n)$ is also Gaussian.

   (b) Conversely, a Gaussian process $u(n)$ implies that the process $v(n)$ is Gaussian.

11. Consider a complex Gaussian process $u(n)$. Let $u(n) = u_n$. Using the Gaussian moment factoring theorem, demonstrate the following identities:

   (a) $E[(u_1^* u_2)^k] = k! \, (E[u_1^* u_2])^k$

   (b) $E[|u|^{2k}] = k! \, (E[|u|^2])^k$

CHAPTER

3

Spectrum Analysis

The autocorrelation function is a *time-domain description* of the second-order statistics of a stochastic process. The *frequency-domain description* of the second-order statistics of such a process is the *power spectral density*, which is also commonly referred to as the *power spectrum* or simply *spectrum*. Indeed, the power spectral density of a stochastic process is firmly established as the most useful description of the time series commonly encountered in engineering and physical sciences.

This chapter is devoted in part to the definition of the power spectral density of a wide-sense stationary discrete-time stochastic process, the properties of power spectral density, and methods for its estimation. We begin the discussion by establishing a mathematical definition of the power spectral density of a stationary process in terms of the Fourier transform of a single realization of the process.

3.1 POWER SPECTRAL DENSITY

Consider an infinitely long time series $u(n)$, $n = 0, \pm1, \pm2, \ldots$, that represents a *single realization* of a wide-sense stationary discrete-time stochastic process. For convenience of presentation, we assume that the process has zero mean. Initially, we focus our attention on a *windowed* portion of this time series, written as

$$u_N(n) = \begin{cases} u(n), & n = 0, 1, \ldots, N-1 \\ 0, & n > N, n < 0 \end{cases} \tag{3.1}$$

where N is the *total length (duration) of the window*. By definition, the *discrete-time Fourier transform* of the windowed time series $u_N(n)$ is given by

$$U_N(\omega) = \sum_{n=0}^{N-1} u_N(n) e^{-j\omega n} \tag{3.2}$$

where ω is the *angular frequency*, lying in the interval $(-\pi, \pi]$. In general, $U_N(\omega)$ is complex valued; specifically, its *complex conjugate* is given by

$$U_N^*(\omega) = \sum_{k=0}^{N-1} u_N^*(k) e^{j\omega k} \tag{3.3}$$

where the asterisk denotes complex conjugation. In Eq. (3.3) we have used the variable k to denote discrete time for reasons that will become apparent immediately. In particular, we may multiply Eq. (3.2) by (3.3) to express the squared magnitude of $U_N(n)$ as follows:

$$|U_N(\omega)|^2 = \sum_{n=0}^{N-1} \sum_{k=0}^{N-1} u_N(n) u_N^*(k) e^{-j\omega(n-k)} \tag{3.4}$$

Each realization $U_N(n)$ produces such a result. The *expected* result is obtained by taking the statistical expectation of both sides of Eq. (3.4), and interchanging the order of expectation and double summation:

$$E[|U_N(\omega)|^2] = \sum_{n=0}^{N-1} \sum_{k=0}^{N-1} E[u_N(n) u_N^*(k)] e^{-j\omega(n-k)} \tag{3.5}$$

We now recognize that for the wide-sense stationary discrete-time stochastic process under discussion, the autocorrelation function of $u_N(n)$ for lag $n - k$ is

$$r_N(n - k) = E[u_N(n) u_N^*(k)] \tag{3.6}$$

which may be rewritten as follows, in light of the defining equation (3.1):

$$r_N(n - k) = \begin{cases} E[u(n)u^*(k)] = r(n - k) & \text{for } 0 \leq (n,k) \leq N - 1 \\ 0 & \text{otherwise} \end{cases} \tag{3.7}$$

Accordingly, Eq. (3.6) takes on the form

$$E[|U_N(\omega)|^2] = \sum_{n=0}^{N-1} \sum_{k=0}^{N-1} r(n - k) e^{-j\omega(n-k)} \tag{3.8}$$

Let $l = n - k$, and so rewrite Eq. (3.8) as follows:

$$\frac{1}{N} E[|U_N(\omega)|^2] = \sum_{l=-N+1}^{N-1} \left(1 - \frac{|l|}{N}\right) r(l) e^{-j\omega l} \tag{3.9}$$

Equation (3.9) may be interpreted as the discrete-time Fourier transform of the product of two time functions: the autocorrelation function $r_N(l)$ for lag l, and a triangular window

$w_B(l)$ known as the *Barlett window*. The latter function is defined by

$$w_B(l) = \begin{cases} 1 - \dfrac{|l|}{N}, & |l| \leq N-1 \\ 0, & |l| \geq N \end{cases} \tag{3.10}$$

As N approaches infinity, the Barlett window $w_B(l)$ approaches unity for all l. Correspondingly, we may write

$$\lim_{N \to \infty} \frac{1}{N} E[|U_N(\omega)|^2] = \sum_{l=-\infty}^{\infty} r(l)e^{-j\omega l} \tag{3.11}$$

where $r(l)$ is the autocorrelation function of the original time series $u(n)$, assumed to have infinite length. The quantity $U_N(\omega)$ is the discrete-time Fourier transform of a rectangular *windowed* portion of this time series that has length N.

Equation (3.11) leads us to define the quantity

$$S(\omega) = \lim_{N \to \infty} \frac{1}{N} E[|U_N(\omega)|^2] \tag{3.12}$$

where the quantity $|U_N(\omega)|^2/N$ is called the *periodogram* of the windowed time series $u_N(n)$. Note that the order of expectation and limiting operations indicated in Eq. (3.12) cannot be changed. Note also that the periodogram converges to $S(\omega)$ only in the mean value, but *not* in mean square or any other meaningful way.

When the limit in Eq. (3.12) exists, the quantity $S(\omega)$ has the following interpretation (Priestley, 1981):

$$S(\omega)\, d\omega =\ \text{average of the contribution to the total power from components} \tag{3.13}$$
of a wide-sense stationary stochastic process with angular
frequencies located between ω and $\omega + d\omega$; the average is
taken over all possible realizations of the process

Accordingly, the quantity $S(\omega)$ is the "spectral density of expected power," which is abbreviated as the *power spectral density* of the process. Thus, equipped with the definition of power spectral density given in Eq. (3.12), we may now rewrite Eq. (3.11) as

$$S(\omega) = \sum_{l=-\infty}^{\infty} r(l)e^{-j\omega l} \tag{3.14}$$

In summary, Eq. (3.12) gives a basic definition of the power spectral density of a wide-sense stationary stochastic process, and Eq. (3.14) defines the mathematical relationship between the autocorrelation function and the power spectral density of such a process.

3.2 PROPERTIES OF POWER SPECTRAL DENSITY

Property 1.   *The autocorrelation function and power spectral density of a wide-sense stationary stochastic process form a Fourier transform pair.*

Consider a wide-sense stationary stochastic process represented by the time series $u(n)$, assumed to be of infinite length. Let $r(l)$ denote the autocorrelation function of such a process for lag l, and let $S(\omega)$ denote its power spectral density. According to Property 1, these two quantities are related by the pair of relations:

$$S(\omega) = \sum_{l=-\infty}^{\infty} r(l)e^{-j\omega l}, \qquad -\pi < \omega \le \pi \tag{3.15}$$

and

$$r(l) = \frac{1}{2\pi}\int_{-\pi}^{\pi} S(\omega)e^{j\omega l}d\omega, \qquad l = 0, \pm 1, \pm 2, \ldots \tag{3.16}$$

Equation (3.15) states that *the power spectral density is the discrete-time Fourier transform of the autocorrelation function.* On the other hand, Eq. (3.16) states that *the autocorrelation function is the inverse discrete-time Fourier transform of the power spectral density.* This fundamental pair of equations constitutes the *Einstein–Wiener–Khintchine* relations.

In a way, we already have a proof of this property. Specifically, Eq. (3.15) is merely a restatement of Eq. (3.14), previously established in Section 3.1. Equation (3.16) follows directly from this result by invoking the formula for the inverse discrete-time Fourier transform.

Property 2.   *The frequency support of the power spectral density $S(\omega)$ is the Nyquist interval $-\pi < \omega \le \pi$.*

Outside this interval, $S(\omega)$ is periodic as shown by

$$S(\omega + 2k\pi) = S(\omega) \quad \text{for integer } k \tag{3.17}$$

Property 3.   *The power spectral density of a stationary discrete-time stochastic process is real.*

To prove this property, we rewrite Eq. (3.15) as

$$S(\omega) = r(0) + \sum_{k=1}^{\infty} r(k)e^{-j\omega k} + \sum_{k=-\infty}^{-1} r(k)e^{-j\omega k}$$

Replacing k with $-k$ in the third term on the right-hand side of this equation, and recognizing that $r(-k) = r^*(k)$, we get

$$S(\omega) = r(0) + \sum_{k=1}^{\infty} [r(k)e^{-j\omega k} + r^*(k)e^{j\omega k}]$$

$$= r(0) + 2\sum_{k=1}^{\infty} \text{Re}[r(k)e^{-j\omega k}] \tag{3.18}$$

where Re denotes the *real part operator.* Equation (3.18) shows that the power spectral density $S(\omega)$ is a real-valued function of ω. It is because of this property that we have used the notation $S(\omega)$ rather than $S(e^{j\omega})$ for the power spectral density.

Property 4. *The power spectral density of a real-valued stationary discrete-time stochastic process is even (i.e., symmetric); if the process is complex-valued, its power spectral density is not necessarily even.*

For a real-valued stochastic process, we find that $S(-\omega) = S(\omega)$, indicating that $S(\omega)$ is an even function of ω; that is, it is symmetric about the origin. If, however, the process is complex-valued, then $r(-k) = r^*(k)$, in which case we find that $S(-\omega) \neq S(\omega)$, and $S(\omega)$ is *not* an even function of ω.

Property 5. *The mean-square value of a stationary discrete-time stochastic process equals, except for the scaling factor $1/2\pi$, the area under the power spectral density curve for $-\pi < \omega \leq \pi$.*

This property follows directly from Eq. (3.16), evaluated for $l = 0$. For this condition, we may thus write

$$r(0) = \frac{1}{2\pi} \int_{-\pi}^{\pi} S(\omega) \, d\omega \qquad (3.19)$$

Since $r(0)$ equals the mean-square value of the process, we see that Eq. (3.19) is a mathematical description of Property 5. The mean-square value of a process is equal to the *expected power* of the process developed across a load resistor of 1 ohm. On this basis, the terms "expected power" and "mean-square value" are used interchangeably in what follows.

Property 6. *The power spectral density of a stationary discrete-time stochastic process is nonnegative.*

That is,

$$S(\omega) \geq 0 \qquad \text{for all } \omega \qquad (3.20)$$

This property follows directly from the basic formula of Eq. (3.12), reproduced here for convenience of presentation:

$$S(\omega) = \lim_{N \to \infty} \frac{1}{N} E[|U_N(\omega)|^2]$$

We first note that $|U_N(\omega)|^2$, representing the squared magnitude of the discrete-time Fourier transform of a windowed portion of the time series $u(n)$, is nonnegative for all ω. The expectation $E[|U_N(\omega)|^2]$ is also nonnegative for all ω. Thus, using the basic definition of $S(\omega)$ in terms of $U_N(\omega)$, the property described by Eq. (3.20) follows immediately.

3.3 TRANSMISSION OF A STATIONARY PROCESS THROUGH A LINEAR FILTER

Consider a discrete-time filter that is *linear, time invariant,* and *stable*. Let the filter be characterized by the *discrete transfer function H(z),* defined as the *ratio of the z-transform of the filter output to the z-transform of the filter input.* Suppose that we feed the filter with

Figure 3.1   Transmission of stationary process through a discrete-time linear filter.

a stationary discrete-time stochastic process of power spectral density $S(\omega)$, as in Fig. 3.1. Let $S_o(\omega)$ denote the power spectral density of the filter output. We may then write

$$S_o(\omega) = |H(e^{j\omega})|^2 S(\omega) \qquad (3.21)$$

where $H(e^{j\omega})$ is the *frequency response* of the filter. The frequency response $H(e^{j\omega})$ equals the discrete transfer function $H(z)$ evaluated on the unit circle in the z-plane. The important feature of this result is that the value of the output spectral density at angular frequency ω depends purely on the squared *amplitude response* of the filter and the input power spectral density at the same angular frequency ω.

Equation (3.21) is a fundamental relation in stochastic process theory. To prove it, we may proceed as follows. Let the time series $y(n)$ denote the filter output in Fig. 3.1, produced in response to the time series $u(n)$ applied to the filter input. Assuming that $u(n)$ represents a single realization of a wide-sense stationary discrete-time stochastic process, we find that $y(n)$ also represents a single realization of a wide-sense stationary discrete-time stochastic process modified by the filtering operation. Thus, given that the autocorrelation function of the filter input $u(n)$ is written as

$$r_u(l) = E[u(n)u^*(n - l)]$$

we may express the autocorrelation function of the filter output $y(n)$ in a corresponding way as

$$r_y(l) = E[y(n)y^*(n - l)] \qquad (3.22)$$

where $y(n)$ is related to $u(n)$ by the convolution sum

$$y(n) = \sum_{i=-\infty}^{\infty} h(i)u(n - i) \qquad (3.23)$$

Similarly, we may write

$$y^*(n - l) = \sum_{k=-\infty}^{\infty} h^*(k)u^*(n - l - k) \qquad (3.24)$$

Substituting Eqs. (3.23) and (3.24) in (3.22), and interchanging the orders of expectation and summations, we find that the autocorrelation functions $r_y(l)$ and $r_u(l)$, for lag l, are related as follows:

$$r_y(l) = \sum_{i=-\infty}^{\infty} \sum_{k=-\infty}^{\infty} h(i)h^*(k)r_u(k - i + l) \qquad (3.25)$$

Finally, taking the discrete-time Fourier transforms of both sides of Eq. (3.25), and invoking Property 1 of the power spectral density and the fact that the transfer function of a linear filter is equal to the Fourier transform of its impulse response, we get the result described in Eq. (3.21).

Power Spectrum Analyzer

Suppose that the discrete-time linear filter in Fig. 3.1 is designed to have a bandpass characteristic. That is, the amplitude response of the filter is defined by

$$|H(e^{j\omega})| = \begin{cases} 1, & |\omega - \omega_c| \leq \Delta\omega \\ 0, & \text{remainder of the interval } -\pi < \omega \leq \pi \end{cases} \tag{3.26}$$

This amplitude response is depicted in Fig. 3.2. We assume that the *angular bandwidth* of the filter, $2\Delta\omega$, is small enough for the spectrum inside this bandwidth to be essentially constant. Then using Eq. (3.21) we may write

$$S_o(\omega) = \begin{cases} S(\omega_c), & |\omega - \omega_c| \leq \Delta\omega \\ 0, & \text{remainder of the interval } -\pi < \omega \leq \pi \end{cases} \tag{3.27}$$

Next, using Properties 4 and 5 of the power spectral density, we may express the mean-square value of the filter output resulting from a real-valued stochastic input as

$$P_o = \frac{1}{2\pi} \int_{-\pi}^{\pi} S_o(\omega) \, d\omega$$

$$= \frac{2\Delta\omega}{2\pi} S(\omega_c) + \frac{2\Delta\omega}{2\pi} S(-\omega_c)$$

$$= 2\frac{\Delta\omega}{\pi} S(\omega_c) \qquad \text{for real data}$$

Equivalently, we may write

$$S(\omega_c) = \frac{\pi P_o}{2\Delta\omega} \tag{3.28}$$

where $\Delta\omega/\pi$ is that fraction of the Nyquist interval that corresponds to the passband of the filter. Equation (3.28) states that the value of the power spectral density of the filter input $u(n)$, measured at the center frequency ω_c of the filter, is equal to the mean-square value P_o of the filter output, scaled by a constant factor. We may thus use Eq. (3.28) as the mathematical basis for building a *power spectrum analyzer*, as depicted in Fig. 3.3. Ideally, the discrete-time bandpass filter employed here should satisfy two requirements: *fixed bandwidth* and *adjustable center frequency*. Clearly, in a practical filter design, we can only approximate these two ideal requirements. Note also that the reading of the *average power meter* at the output end of Fig. 3.3 approximates (for finite averaging time) the expected power of an ergodic process $y(n)$.

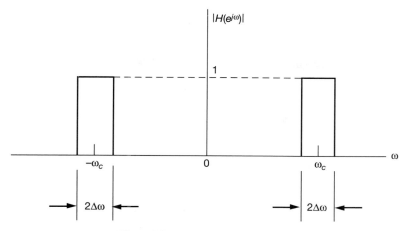

Figure 3.2   Ideal bandpass characteristic.

Example 1: White Noise

A stochastic process of zero mean is said to be *white* if its power spectral density $S(\omega)$ is constant for all frequencies, as shown by

$$S(\omega) = \sigma^2 \quad \text{for } -\pi < \omega \leq \pi$$

where σ^2 is the variance of a sample taken from the process. Suppose that this process is passed through a discrete-time bandpass filter, characterized as in Fig. 3.2. Hence, from Eq. (3.28), we find that the mean-square value of the filter output is

$$P_o = \frac{2\sigma^2 \Delta \omega}{\pi}$$

White noise has the property that any two of its samples are uncorrelated, as shown by the autocorrelation function

$$r(\tau) = \sigma^2 \delta_{\tau,0}$$

where $\delta_{\tau,0}$ is the Kronecker delta:

$$\delta_{\tau,0} = \begin{cases} 1, & \tau = 0 \\ 0, & \text{otherwise} \end{cases}$$

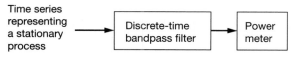

Figure 3.3   Power spectrum analyzer.

3.4 CRAMÉR SPECTRAL REPRESENTATION FOR A STATIONARY PROCESS

Equation (3.12) provides one way of defining the power spectral density of a wide-sense stationary process. Another way of defining the power spectral density is to use the *Cramér spectral representation for a stationary process*. According to this representation, a sample $u(n)$ of a discrete-time stochastic process is written as an inverse Fourier transform (Thomson, 1982, 1988):

$$u(n) = \frac{1}{2\pi} \int_{-\pi}^{\pi} e^{j\omega n} dZ(\omega) \qquad (3.29)$$

If the process represented by the time series $u(n)$ is wide-sense stationary with no periodic components, then the *increment process* $dZ(\omega)$ has the following three properties:

1. The mean of the increment process $dZ(\omega)$ is zero; that is,

$$E[dZ(\omega)] = 0 \qquad \text{for all } \omega \qquad (3.30)$$

2. The energy of the increment process $dZ(\omega)$ at different frequencies is uncorrelated; that is,

$$E[dZ(\omega)dZ^*(v)] = 0 \qquad \text{for } v \neq \omega \qquad (3.31)$$

3. The expected value of $|dZ(\omega)|^2$ defines the spectrum $S(\omega)\,d\omega$; that is,

$$E[|dZ(\omega)|^2] = S(\omega)\,d\omega \qquad (3.32)$$

In other words, for a wide-sense stationary discrete-time stochastic process represented by the time series $u(n)$, the increment process $dZ(\omega)$ defined by Eq. (3.29) is a *zero-mean orthogonal process*. More precisely, $dZ(\omega)$ may be viewed as a "white process" described in the frequency domain in a manner similar to the time-domain description of ordinary white noise.

Equation (3.32), in conjunction with Eq. (3.31), provides another basic definition for the power spectral density $S(\omega)$. Complex-conjugating both sides of Eq. (3.29) and using v in place of ω, we get

$$u^*(n) = \frac{1}{2\pi} \int_{-\pi}^{\pi} e^{-jvn} dZ^*(v) \qquad (3.33)$$

Hence, multiplying Eq. (3.29) by (3.33), we may express the squared magnitude of $u(n)$ as

$$|u(n)|^2 = \frac{1}{(2\pi)^2} \int_{-\pi}^{\pi} \int_{-\pi}^{\pi} e^{jn(\omega-v)} dZ(\omega)\,dZ^*(v) \qquad (3.34)$$

Next, taking the statistical expectation of Eq. (3.34), and interchanging the order of expectation and double integration, we get

$$E[|u(n)|]^2 = \frac{1}{(2\pi)^2} \int_{-\pi}^{\pi} \int_{-\pi}^{\pi} e^{jn(\omega-v)} E[dZ(\omega)\,dZ^*(v)] \qquad (3.35)$$

If we now use the two basic properties of the increment process $dZ(\omega)$ described by Eqs. (3.31) and (3.32), we may simplify Eq. (3.35) into the form

$$E[|u(n)|^2] = \frac{1}{2\pi}\int_{-\pi}^{\pi} S(\omega)\, d\omega \tag{3.36}$$

The expectation $E[|u(n)|^2]$ on the left-hand side of Eq. (3.36) is recognized as the mean-square value of the complex sample $u(n)$. The right-hand side of this equation equals the total area under the curve of the power spectral density $S(\omega)$, scaled by the factor $1/2\pi$. Accordingly, Eq. (3.36) is merely a restatement of Property 5 of the power spectral density $S(\omega)$, described by Eq. (3.19).

The Fundamental Equation

Consider the time series $u(0), u(1), \ldots, u(N-1)$, consisting of N observations (samples) of a wide-sense stationary stochastic process. The discrete-time Fourier transform of this time series is given by

$$U_N(\omega) = \sum_{n=0}^{N-1} u(n)e^{-j\omega n} \tag{3.37}$$

According to the Cramér spectral representation of the process, the observation $u(n)$ is given by Eq. (3.29). Hence, using the dummy variable v in place of ω in Eq. (3.29), and then substituting the result in Eq. (3.37), we get

$$U_N(\omega) = \frac{1}{2\pi}\int_{-\pi}^{\pi} \sum_{n=0}^{N-1} (e^{-j(\omega-v)n})\, dZ(v) \tag{3.38}$$

where we have interchanged the order of summation and integration. Define

$$K_N(\omega) = \sum_{n=0}^{N-1} e^{-j\omega n} \tag{3.39}$$

which is known as the *Dirichlet kernel*. The kernel $K_N(\omega)$ represents a geometric series with a first term of unity, a common ratio of $e^{-j\omega}$, and a total number of terms equal to N. Summing this series, we may redefine the kernel $K_N(\omega)$ as follows:

$$\begin{aligned} K_N(\omega) &= \frac{1-e^{-j\omega N}}{1-e^{-j\omega}} \\ &= \frac{\sin(N\omega/2)}{\sin(\omega/2)}\exp\left[-\frac{1}{2}j\omega(N-1)\right] \end{aligned} \tag{3.40}$$

Note that $K_N(0) = N$. Returning to Eq. (3.38), we may use the definition of the Dirichlet kernel $K_N(\omega)$ given in Eq. (3.39) to rewrite $U_N(\omega)$ as follows:

$$U_N(\omega) = \frac{1}{2\pi}\int_{-\pi}^{\pi} K_N(\omega-v)\, dZ(v) \tag{3.41}$$

The integral equation (3.41) is a *linear* relation, referred to as the *fundamental equation* of power spectrum analysis.

An integral equation is one that involves an *unknown* function under the integral sign. In the context of power spectrum analysis as described by Eq. (3.41), the increment variable $dZ(\omega)$ is the unknown function, and $U_N(\omega)$ is known. Accordingly, Eq. (3.41) may be viewed as an example of a *Fredholm integral equation of the first kind* (Morse and Feshbach, 1953; Whittaker and Watson, 1965).

Note that $U_N(\omega)$ may be inverse Fourier transformed to recover the original data. It follows therefore that $U_N(\omega)$ is a *sufficient statistic* of the power spectral density. This property makes the use of Eq. (3.41) for spectrum analysis all the more important.

3.5 POWER SPECTRUM ESTIMATION

An issue of practical importance is how to *estimate* the power spectral density of a wide-sense stationary process. Unfortunately, this issue is complicated by the fact that there is a bewildering array of power spectrum estimation procedures, with each procedure purported to have or to show some optimum property. The situation is made worse by the fact that unless care is taken in the selection of the right method, we may end up with misleading conclusions.

Two philosophically different families of power spectrum estimation methods may be identified in the literature: *parametric methods* and *nonparametric methods*. The basic ideas behind these methods are discussed in the sequel.

Parametric Methods

In parametric methods of spectrum estimation we begin by postulating a *stochastic model* for the situation at hand. Depending on the specific form of stochastic model adopted, we may identify three different parametric approaches for spectrum estimation.

1. *Model identification procedures.* In this class of parametric methods, a rational function or a polynomial in $e^{-j\omega}$ is assumed for the transfer function of the model, and a white-noise source is used to drive the model, as depicted in Fig. 3.4. The power spectrum of the resulting model output provides the desired spectrum estimate. Depending on the application of interest, we may adopt one of the following models (Kay and Marple, 1981; Marple, 1987; Kay, 1988):

   (i) *Autoregressive (AR) model* with an all-pole transfer function.

   (ii) *Moving average (MA) model* with an all-zero transfer function.

   (iii) *Autoregressive–moving average (ARMA) model* with pole-zero transfer function.

   The resulting power spectra measured at the outputs of these models are referred to as AR, MA, and ARMA spectra, respectively. With reference to the input–output relation of Eq. (3.21), let the power spectrum $S(\omega)$ of the model input be put

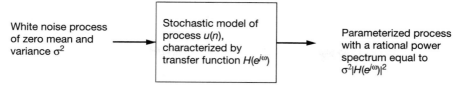

Figure 3.4   Rationale of model identification procedure for power spectrum estimation.

equal to the white noise variance σ^2. We then find that the power spectrum $S_o(\omega)$ of the model output is equal to the squared amplitude response $|H(e^{j\omega})|^2$ of the model, multiplied by σ^2. The problem thus becomes one of estimating the model parameters [i.e., parametrizing the transfer function $H(e^{j\omega})$] such that the process produced at the model output provides an acceptable representation (in some statistical sense) of the stochastic process under study. Such an approach to power spectrum estimation may indeed be viewed as a problem in *model (system) identification*.

Among the model-dependent spectra defined herein, the AR spectrum is by far the most popular. The reason for this popularity is twofold: (1) the *linear* form of the system of simultaneous equations involving the unknown AR model parameters, and (2) the availability of efficient algorithms for computing the solution.

2. *Minimum variance distortionless response method.* To describe this second parametric approach for power spectrum estimation, consider the situation depicted in Fig. 3.5. The process $u(n)$ is applied to a transversal filter (i.e., discrete-time filter with an all-zero transfer function). In the *minimum variance distortionless response (MVDR) method*, the filter coefficients are chosen so as to minimize the variance (which is the same as expected power for a zero-mean process) of the filter output, subject to the constraint that the frequency response of the filter is equal to unity at some angular frequency ω_0. Under this constraint, the process $u(n)$ is passed through the filter with *no distortion* at the angular frequency ω_0. Moreover, signals at angular frequencies other than ω_0 tend to be attenuated.

3. *Eigendecomposition-based methods.* In this final class of parametric spectrum estimation methods, the eigendecomposition of the ensemble-averaged correla-

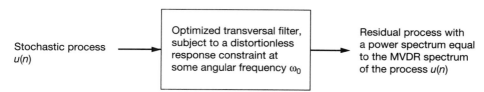

Figure 3.5   Rationale of MVDR procedure for power spectrum estimation.

tion matrix **R** of the process $u(n)$ is used to define two disjoint subspaces: *signal subspace* and *noise subspace*. This form of partitioning is then exploited to derive an appropriate algorithm for estimating the power spectrum (Schmidt, 1979, 1981). Eigenanalysis and the notion of subspace decomposition are discussed in the next chapter.

Nonparametric Methods

In nonparametric methods of power spectrum estimation, on the other hand, no assumptions are made with respect to the stochastic process under study. The starting point in the discussion is the fundamental equation (3.41). Depending on the way in which this equation is interpreted, we may distinguish two different nonparametric approaches:

1. *Periodogram-based methods.* Traditionally, the fundamental equation (3.41) is treated as a *convolution* of two frequency functions. One frequency function, $U(\omega)$, represents the discrete-time Fourier transform of an *infinitely long* time series, $u(n)$; this function arises from the definition of the increment variable $dZ(\omega)$ as the product of $U(\omega)$ and the frequency increment $d\omega$. The other frequency function is the kernel $K_N(\omega)$, defined by Eq. (3.40). This approach leads us to consider Eq. (3.12) as the basic definition of the power spectral density $S(\omega)$, and therefore the *periodogram* $|U_N(\omega)|^2/N$ as the starting point for the data analysis. However, the periodogram suffers from a *serious limitation in the sense that it is not a sufficient statistic for the power spectral density*. This implies that the phase information ignored in the use of the periodogram is essential. Consequently, the statistical insufficiency of the periodogram is inherited by any estimate that is based on or equivalent to the periodogram.

2. *Multiple-window method.* A more constructive nonparametric approach is to treat the fundamental equation (3.41) as a *Fredholm integral equation of the first kind* for the increment variable $dZ(\omega)$; the goal here is to obtain an *approximate solution* for the equation with statistical properties that are close to those of $dZ(\omega)$ in some sense (Thomson, 1982). The key to the attainment of this important goal is the use of windows defined by a set of special sequences known as *Slepian sequences*[1] or *discrete prolate spheroidal sequences*, which are fundamental to the study of time- and frequency-limited systems. The remarkable property of this family of windows is that their energy distributions add up in a very special way that collectively defines an ideal (ideal in the sense of the total in-bin versus out-of-bin energy concentration) rectangular frequency bin. This property, in turn, allows us to trade spectral resolution for improved spectral properties (i.e., reduced variance of the spectral estimate).

---

[1]Detailed information on Slepian sequences is given in Slepian (1978). A method for computing them, for large data length, is given in the appendix of the paper by Thomson (1982). For additional information, see the references listed in Thomson's paper. Mullis and Scharf (1991) also present an informative discussion of the role of Slepian sequences in spectrum analysis.

In general, a discrete-time stochastic process $u(n)$ has a *mixed spectrum*, in that its power spectrum contains two components: a deterministic component and a continuous component. The *deterministic component* represents the *first moment* of the increment process $dZ(\omega)$; it is explicitly given by

$$E[dZ(\omega)] = \sum_k a_k \delta(\omega - \omega_k)\, d\omega \qquad (3.42)$$

where $\delta(\omega)$ is the *Dirac delta function* defined in the frequency domain. The ω_k are the angular frequencies of *periodic* or *line components* contained in the process $u(n)$, and the a_k are their amplitudes. The continuous component, on the other hand, represents the *second central moment* of the increment process $dZ(\omega)$, as shown by

$$E[|dZ(\omega) - E[dZ(\omega)]|^2] = S(\omega)\, d\omega \qquad (3.43)$$

It is important that the distinction between the first and second moments is carefully noted.

Spectra computed using the parametric methods tend to have sharper peaks and higher resolution than those obtained from the nonparametric (classical) methods. The application of these parametric methods is therefore well suited for estimating the deterministic component and, in particular, for locating the frequencies of periodic components in additive white noise when the signal-to-noise ratio is high. Another well-proven technique for estimating the deterministic component is the classical method of maximum likelihood, which is discussed in Appendix D. Of course, if the physical laws governing the generation of a process match a stochastic model (e.g., AR model) in an exact manner or approximately in some statistical sense, then the parametric method corresponding to that model may be used to estimate the power spectrum of the process. If, however, the stochastic process of interest has a purely continuous power spectrum, and the underlying physical mechanism responsible for the generation of the process is unknown, then the recommended procedure is the non-parametric method of multiple windows.

In this book, we confine our attention to classes 1 and 2 of parametric methods of spectrum estimation, as their theory fits naturally under the umbrella of adaptive filters. For a comprehensive discussion of the other methods of spectrum analysis, the reader is referred to the books by Gardner (1987), Marple (1987), and Kay (1988), the paper by Thomson (1982), and a chapter contribution by Mullis and Scharf (1991).

3.6 OTHER STATISTICAL CHARACTERISTICS OF A STOCHASTIC PROCESS

In the material presented in the previous chapter and up to this point in the present chapter, we have focused our attention on a partial characterization of a discrete-time stochastic process. According to this particular characterization, we only need to specify the mean as the first moment of the process and its autocorrelation function as the second moment. Since the autocorrelation function and power spectral density form a Fourier-transform pair, we may equally well specify the power spectral density in place of the autocorrelation function. The use of second-order statistics as described herein is adequate for the study of linear adaptive filters. However, when we move on later in the book to consider difficult applications (e.g., blind deconvolution) that are beyond the reach of linear adaptive filters, we will have to resort to the use of other statistical properties of a stochastic process.

Two particular statistical properties that bring in additional information about a stochastic process, which can prove useful in practice, are as follows:

1. *High-order statistics.* An obvious way of expanding the characterization of a stationary stochastic process is to include *higher-order statistics* (HOS) of the process. This is done by invoking the use of *cumulants* and their Fourier transforms, known as *polyspectra.* Indeed, cumulants and polyspectra of a zero-mean stochastic process may be viewed as generalizations of the autocorrelation function and power spectral density, respectively. It is important to note that higher-order statistics are only meaningful in the context of *non-Gaussian processes.* Furthermore, to exploit them, we need to use some form of nonlinear filtering.

2. *Cyclostationarity.* In an important class of stochastic processes commonly encountered in practice, the mean and autocorrelation function of the process exhibit periodicity, as in

$$\mu(t_1 + T) = \mu(t_1) \tag{3.44}$$

$$r(t_1 + T, t_2 + T) = r(t_1, t_2) \tag{3.45}$$

for all t_1 and t_2. Both t_1 and t_2 represent values of the continuous-time variable t, and T denotes period. A stochastic process satisfying Eqs. (3.44) and (3.45) is said to be *cyclostationary* in the wide sense (Franks, 1969; Gardner and Franks, 1975; Gardner, 1994). Modeling a stochastic process as cyclostationary adds a new dimension, namely, the period T, to the partial description of the process. Examples of cyclostationary processes include a modulated process obtained by varying the amplitude, phase, or frequency of a sinusoidal carrier. Note that, unlike higher-order statistics, cyclostationarity can be exploited by means of linear filtering.

In the sequel, we will discuss these two specific aspects of stochastic processes under the section headings "polyspectra" and "spectral-correlation density." As already mentioned, polyspectra provide a frequency-domain description of the higher-order statistics of a stationary stochastic process. By the same token, spectral-correlation density provides a frequency-domain description of a cyclostationary stochastic process.

3.7 POLYSPECTRA

Consider a stationary stochastic process $u(n)$ with zero mean; that is,

$$E[u(n)] = 0 \qquad \text{for all } n$$

Let $u(n)$, $u(n + \tau_1)$, . . . , $u(n + \tau_{k-1})$ denote the random variables obtained by observing this stochastic process at times n, $n + \tau_1$, . . . , $n + \tau_{k-1}$, respectively. These random variables form the k-by-1 vector:

$$\mathbf{u} = [u(n), u(n + \tau_1), \ldots, u(n + \tau_{k-1})]^T \tag{3.46}$$

Correspondingly, define a k-by-1 vector:

$$\mathbf{z} = [z_1, z_2, \ldots, z_k]^T \tag{3.47}$$

We may then define the kth-*order cumulant* of the stochastic process $u(n)$, denoted by $c_k(t_1, t_2, \ldots, \tau_{k-1})$, as the coefficient of the vector $\mathbf{z}$ in the Taylor expansion of the *cumulant-generating function* (Priestley, 1981; Swami and Mendel, 1990; Gardner, 1994):

$$K(\mathbf{z}) = \text{In } E[\exp(\mathbf{z}^T \mathbf{u})] \tag{3.48}$$

The kth-order cumulant of the process $u(n)$ is thus defined in terms of its joint moments of orders up to k; to simplify the presentation in this section, we assume that $u(n)$ is real valued. Specifically, the second-, third-, and fourth-order cumulants are given, respectively, by

$$c_2(\tau) = E[u(n)u(n + \tau)] \tag{3.49}$$

$$c_3(\tau_1, \tau_2) = E[u(n)u(n + \tau_1)u(n + \tau_2)] \tag{3.50}$$

and

$$\begin{aligned} c_4(\tau_1, \tau_2, \tau_3) = {} & E[u(n)u(n + \tau_1)u(n + \tau_2)u(n + \tau_3)] \\ & - E[u(n)u(n + \tau_1)]E[u(n + \tau_2)u(n + \tau_3)] \\ & - E[u(n)u(n + \tau_2)]E[u(n + \tau_3)u(n + \tau_1)] \\ & - E[u(n)u(n + \tau_3)]E[u(n + \tau_1)u(n + \tau_2)] \end{aligned} \tag{3.51}$$

From the definitions given in Eqs. (3.49) to (3.51), we note the following:

1. The second-order cumulant $c_2(\tau)$ is the same as the autocorrelation function $r(t)$.
2. The third-order cumulant $c_3(\tau_1, \tau_2)$ is the same as the third-order moment $E[u(n)u(n + \tau_1)u(n + \tau_2)]$.
3. The fourth-order cumulant $c_4(\tau_1, \tau_2, \tau_3)$ is *different* from the fourth-order moment $E[u(n)u(n + \tau_1)u(n + \tau_2)u(n + \tau_3)]$. In order to generate the fourth-order cumulant, we need to know the fourth-order moment and six different values of the autocorrelation function.

Note that the kth-order cumulant $c(\tau_1, \tau_2, \ldots, \tau_{k-1})$ does not depend on time n. For this to be valid, however, the process $u(n)$ has to be stationary up to order k. A process $u(n)$ is said to be *stationary up to order k* if, for any admissible $\{n_1, n_2, \ldots, n_p\}$ all the joint moments up to order k of $\{u(n_1), u(n_2), \ldots, u(n_p)\}$ exist and equal the corresponding joint moments up to order k of $\{u(n_1 + \tau), u(n_2 + \tau), \ldots, u(n_p + \tau)\}$ where $\{n_1 + \tau, n_2 + \tau, \ldots, n_p + \tau\}$ is an admissible set too (Priestley, 1981).

Consider next a linear time-invariant system, characterized by the impulse response h_n. Let the system be excited by a process $x(n)$ consisting of independent and identically distributed (iid) random variables. Let $u(n)$ denote the resulting system output. The kth-order cumulant of $u(n)$ is given by

$$c_k(\tau_1, \tau_2, \cdots, \tau_{k-1}) = \gamma_k \sum_{i=-\infty}^{\infty} h_i h_{i+\tau_1} \cdots h_{i+\tau_{k-1}} \qquad (3.52)$$

where γ_k is the kth-order cumulant of the input process $x(n)$. Note that the summation term on the right-hand side of Eq. (3.52) has a form similar to that of a kth-order moment, except that the expectation operator has been replaced by a summation.

The kth-*order polyspectrum* (or kth-*order cumulant spectrum*) is defined by (Priestley, 1981; Nikias and Raghuveer, 1987):

$$C_k(\omega_1, \omega_2, \ldots, \omega_{k-1}) = \sum_{\tau_1=-\infty}^{\infty} \cdots \sum_{\tau_{k-1}=-\infty}^{\infty} c_k(\tau_1, \tau_2, \ldots, \tau_{k-1})$$
$$\cdot \exp[-j(\omega_1\tau_1 + \omega_2\tau_2 + \ldots + \omega_{k-1}\tau_{k-1}) \qquad (3.53)$$

A sufficient condition for the existence of the polyspectrum $C_k(\omega_1, \omega_2, \ldots, \omega_{k-1})$ is that the associated kth-order cumulant $c_k(\tau_1, \tau_2, \ldots, \tau_{k-1})$ be absolutely summable, as shown by

$$\sum_{\tau_1=-\infty}^{\infty} \cdots \sum_{\tau_{k-1}=-\infty}^{\infty} |c_k(\tau_1, \tau_2, \ldots, \tau_{k-1})| < \infty \qquad (3.54)$$

The *power spectrum, bispectrum,* and *trispectrum* are special cases of the kth-order polyspectrum defined in Eq. (3.53). Specifically, we may state the following:

1. For $k = 2$, we have the ordinary power spectrum:

$$C_2(\omega_1) = \sum_{\tau_1=-\infty}^{\infty} c_2(\tau_1) \exp(-j\omega_1\tau_1) \qquad (3.55)$$

which is a restatement of the Einstein–Wiener–Khintchine relation, namely, Eq. (3.15).

2. For $k = 3$, we have the *bispectrum*, defined by

$$C_3(\omega_1, \omega_2) = \sum_{\tau_1=-\infty}^{\infty} \sum_{\tau_2=-\infty}^{\infty} c_3(\tau_1, \tau_2) \exp[-j(\omega_1\tau_1 + \omega_2\tau_2)] \qquad (3.56)$$

3. For $k = 4$, we have the *trispectrum*, defined by

$$C_4(\omega_1, \omega_2, \omega_3) = \sum_{\tau_1=-\infty}^{\infty} \sum_{\tau_2=-\infty}^{\infty} \sum_{\tau_3=-\infty}^{\infty} c_4(\tau_1, \tau_2, \tau_3) \exp[-j(\omega_1\tau_1 + \omega_2\tau_2 + \omega_3\tau_3)]$$
$$(3.57)$$

An outstanding property of polyspectrum is that all polyspectra of order higher than the second vanish when the process $u(n)$ is Gaussian. This property is a direct consequence of the fact that all joint cumulants of order higher than the second are zero for multivariate Gaussian distributions. Accordingly, the bispectrum, trispectrum, and all higher-order polyspectra are identically zero if the process $u(n)$ is Gaussian. Thus, higher-order spectra provide measures of the *departure of a stochastic process from Gaussianity* (Priestley, 1981; Nikias and Raghuveer, 1987).

The kth-order cumulant $c_k(\tau_1, \tau_2, \ldots, \tau_{k-1})$ and the kth-order polyspectrum $C_k(\omega_1, \omega_2, \ldots, \omega_{k-1})$ form a pair of multidimensional Fourier transforms. Specifically, the polyspectrum $C_k(\omega_1, \omega_2, \ldots, \omega_{k-1})$, is the *multidimensional discrete-time Fourier transform* of $c_k(\tau_1, \tau_2, \ldots, \tau_{k-1})$, and $c_k(\tau_1, \tau_2, \ldots, \tau_{k-1})$ is the *inverse multidimensional discrete-time Fourier transform* of $C_k(\omega_1, \omega_2, \ldots, \omega_{k-1})$.

For example, given the bispectrum $C_3(\omega_1, \omega_2)$, we may determine the third-order cumulant $c_3(\tau_1, \tau_2)$ by using the inverse two-dimensional discrete-time Fourier transform:

$$c_3(\tau_1, \tau_2) = \left(\frac{1}{2\pi}\right)^2 \int_{-\pi}^{\pi} \int_{-\pi}^{\pi} C_3(\omega_1, \omega_2) \exp\left[j(\omega_1\tau_1 + \omega_2\tau_2)\right] d\omega_1 d\omega_2 \qquad (3.58)$$

We may use this relation to develop an alternative definition of the bispectrum as follows. According to the *Cramér spectral representation*, we have

$$u(n) = \frac{1}{2\pi} \int_{-\pi}^{\pi} e^{j\omega n} dZ(\omega) \qquad \text{for all } n \qquad (3.59)$$

Hence, using Eq. (3.59) in (3.50), we get

$$c_3(\tau_1, \tau_2) = \left(\frac{1}{2\pi}\right)^3 \int_{-\pi}^{\pi} \int_{-\pi}^{\pi} \int_{-\pi}^{\pi} \exp\left[jn(\omega_1 + \omega_2 + \omega_3)\right]$$

$$\cdot \exp[j(\omega_1\tau_1 + \omega_2\tau_2)]E[dZ(\omega_1)\, dZ(\omega_2)\, dZ(\omega_3)] \qquad (3.60)$$

Comparing the right-hand sides of Eqs. (3.58) and (3.60), we deduce the following result:

$$E[dZ(\omega_1)\, dZ(\omega_2)\, dZ(\omega_3)] = \begin{cases} C_3(\omega_1, \omega_2)\, d\omega_1\, d\omega_2, & \omega_1 + \omega_2 + \omega_3 = 0 \\ 0, & \text{otherwise} \end{cases} \qquad (3.61)$$

It is apparent from Eq. (3.61) that the bispectrum $C_3(\omega_1, \omega_2)$ represents the contribution to the mean product of three Fourier components whose *individual frequencies add up to zero*. This is an extension of the interpretation developed for the ordinary power spectrum in Section 3.3. In a similar manner we may develop an interpretation of the trispectrum.

In general, the polyspectrum $C_k(\omega_1, \omega_2, \ldots, \omega_{k-1})$ is *complex for order k higher than two*, as shown by

$$C_k(\omega_1, \omega_2, \ldots, \omega_{k-1}) = \left|C_k(\omega_1, \omega_2, \ldots, \omega_{k-1})\right| \exp[j\phi_k(\omega_1, \omega_2, \ldots, \omega_{k-1})] \qquad (3.62)$$

where we note that $\left|C_k(\omega_1, \omega_2, \ldots, \omega_{k-1})\right|$ is the *magnitude* of the polyspectrum, and $\phi_k(\omega_1, \omega_2, \ldots, \omega_{k-1})$ is the *phase*. Moreover, the polyspectrum is a *periodic* function with period 2π; that is,

$$C_k(\omega_1, \omega_2, \ldots, \omega_{k-1}) = C_k(\omega_1 + 2\pi, \omega_2 + 2\pi, \ldots, \omega_{k-1} + 2\pi) \qquad (3.63)$$

Whereas the power spectral density of a stationary stochastic process is *phase blind*, the polyspectra of the process are *phase sensitive*. More specifically, the power spectral density is real-valued; referring to the input–output relation of Eq. (3.21), we clearly see that in passing a stationary stochastic process through a linear system, information about the phase response of the system is completely destroyed in the power spectrum of the output. In contrast, the polyspectrum is complex-valued, with the result that in a similar situation the polyspectrum of the output signal preserves information about the phase response of the system. It is for this reason that polyspectra provide a useful tool for the "blind" identification of an unknown system, where we only have access to the output signal and some additional information in the form of a probabilistic model of the input signal. We will have more to say on this issue in Chapter 18.

3.8 SPECTRAL-CORRELATION DENSITY

Polyspectra preserve phase information about a stochastic process by invoking higher-order statistics of the process, which is feasible only if the process is non-Gaussian. The preservation of phase information is also possible if the process is *cyclostationary* in the wide sense, as defined in Eqs. (3.44) and (3.45). This latter approach has two important advantages over the higher-order statistics approach:

- The phase information is contained in second-order cyclostationary statistics of the process; hence, the phase information can be exploited in a computationally efficient manner that avoids the use of higher-order statistics.
- Preservation of the phase information holds, irrespective of Gaussianity.

Consider then a discrete-time stochastic process $u(n)$ that is cyclostationary in the wide sense. Without loss of generality, the process is assumed to have zero mean. The ensemble-average autocorrelation function of the process $u(n)$ is defined in the usual way by Eq. (2.6), reproduced here for convenience of presentation:

$$r(n, n - k) = E[u(n)u^*(n - k)] \qquad (3.64)$$

Under the condition of cyclostationarity, the autocorrelation function $r(n, n - k)$ is periodic in n for every k. Keeping in mind the discrete-time nature of the process $u(n)$, we may expand the autocorrelation function $r(n, n - k)$ into a complex Fourier series as follows (Gardner, 1994):

$$r(n, n - k) = \sum_{\{\alpha\}} r^\alpha(k) e^{j2\pi\alpha n - j\pi\alpha k} \qquad (3.65)$$

where both n and k take on only integer values, and the set $\{\alpha\}$ includes all values of α for which the corresponding Fourier coefficient $r^\alpha(k)$ is not zero. The Fourier coefficient $r^\alpha(k)$

is itself defined by

$$r^{\alpha}(k) = \frac{1}{N} \sum_{n=0}^{N-1} r(n, n-k) e^{-j2\pi\alpha n + j\pi\alpha k} \tag{3.66}$$

where the number of samples N denotes the period. Equivalently, in light of Eq. (3.64), we may define $r^{\alpha}(k)$ as

$$r^{\alpha}(k) = \frac{1}{N} \left\{ \sum_{n=0}^{N-1} E[u(n)u*(n-k)e^{-j2\pi\alpha n}] \right\} e^{j\pi\alpha k} \tag{3.67}$$

The quantity $r^{\alpha}(k)$ is called the *cyclic autocorrelation function*, which has the following properties:

1. The cyclic autocorrelation function $r^{\alpha}(k)$ is periodic in α with period two.

2. For any α, we have from Eq. (3.67):

$$r^{\alpha+1}(k) = (-1)^k \, r^{\alpha}(k) \tag{3.68}$$

3. For the special case of $\alpha = 0$, Eq. (3.67) reduces to

$$r^0(k) = r(k) \tag{3.69}$$

where $r(k)$ is the ordinary autocorrelation function of a stationary process.

According to the Einstein–Wiener–Khintchine relations of Eqs. (3.15) and (3.16), the ordinary versions of the autocorrelation function and power spectral density of a wide-sense stationary stochastic process form a Fourier-transform pair. In a corresponding way, we may define the discrete-time Fourier transform of the cyclic autocorrelation function $r^{\alpha}(k)$ as follows (Gardner, 1994):

$$S^{\alpha}(\omega) = \sum_{k=-\infty}^{\infty} r^{\alpha}(k) e^{-j\omega k}, \quad -\pi < \omega \leq \pi \tag{3.70}$$

The new quantity $S^{\alpha}(\omega)$ is called the *spectral-correlation density*, which is complex valued for $\alpha \neq 0$. Note that for the special case of $\alpha = 0$, Eq. (3.70) reduces to

$$S^0(\omega) = S(\omega) \tag{3.71}$$

where $S(\omega)$ is the ordinary power spectral density.

In light of the defining equations (3.67) and (3.70), we may set up the block diagram of Fig. 3.6 for measuring the spectral-correlation density $S^{\alpha}(\omega)$. For this measurement, it is assumed that the process $u(n)$ is *cycloergodic* (Gardner, 1994), which means that time averages may be substituted for ensemble averages "with samples taken once per period." According to the instrumentation described here, $S^{\alpha}(\omega)$ is the bandwidth-normalized version of the cross-correlation narrow-band spectral components contained in the time series $u(n)$ at the angular frequencies $\omega + \alpha\pi$ and $\omega - \alpha\pi$, in the limit as the bandwidth of these spectral components is permitted to approach zero (Gardner, 1994). Note that the two narrow-band filters in Fig. 3.6 are identical, both having a mid-band (angular) frequency ω and a bandwidth $\Delta\omega$ that is small compared to ω, but large compared to the reciprocal of

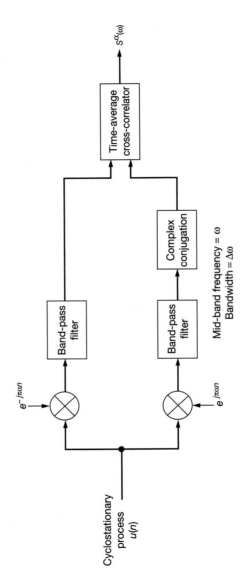

Figure 3.6 Scheme for measuring the spectral-correlation density of a cyclostationary process.

the averaging time used in the cross-correlator at the output end in Fig. 3.6. In one channel of this scheme the input $u(n)$ is multiplied by $\exp(-j\pi\alpha n)$, and in the other channel it is multiplied by $\exp(j\pi\alpha n)$; the resulting filtered signals are then applied to the cross-correlator. It is these two multiplications (prior to correlation) that provide the spectral-correlation density $S^\alpha(\omega)$ with a phase-preserving property for nonzero values of α.

3.9 SUMMARY AND DISCUSSION

In this chapter we discussed various aspects of spectrum analysis pertaining to a discrete-time stochastic process. In particular, we identified three distinct spectral parameters, depending on the statistical characterization of the process, as summarized here:

1. *Power spectral density, $S(\omega)$,* defined as the discrete-time Fourier transform of the ordinary autocorrelation function of a wide-sense stationary process. For such a process, the autocorrelation function is Hermitian, which makes the power spectral density $S(\omega)$ a real-valued quantity. Accordingly, $S(\omega)$ destroys phase information about the process. Despite this limitation, the power spectral density is commonly accepted as a useful parameter for displaying the correlation properties of a wide-sense stationary process.

2. *Polyspectra, $C_k(\omega_1, \omega_2, \dots, \omega_{k-1})$,* defined as the multidimensional Fourier transform of the cumulants of a stationary process. For second-order statistics, $k = 2$, and $C_2(\omega_1)$ reduces to the ordinary power spectral density $S(\omega)$. For higher-order statistics, $k > 2$, and the polyspectra $C_k(\omega_1, \omega_1, \dots, \omega_{k-1})$ take on complex forms. It is this property of polyspectra that makes them a useful tool for dealing with situations where knowledge of phase is a necessary requirement. However, for polyspectra to be meaningful, the process has to be non-Gaussian, and the exploitation of phase information contained in polyspectra requires the use of nonlinear filtering.

3. *Spectral-correlation density, $S^\alpha(\omega)$,* defined as the discrete-time Fourier transform of the cyclic autocorrelation function of a process that is cyclostationary in the wide sense. For $\alpha \neq 0$, $S^\alpha(\omega)$ is complex valued; for $\alpha = 0$, it reduces to $S(\omega)$. The useful feature of $S^\alpha(\omega)$ is that it preserves phase information, which can be exploited by means of linear filtering.

The different properties of the ordinary power spectral density, polyspectra, and spectral-correlation density give these statistical parameters their own individual areas of application.

One last comment is in order. The theories of second-order cyclostationary processes and conventional polyspectra have been brought together under the umbrella of *cyclic polyspectra.* Simply stated, cyclic polyspectra are spectral cumulants, in which the individual frequencies involved can add up to any cycle frequency α, whereas they must add up to zero for conventional polyspectra. For a detailed treatment of cyclic polyspectra, the interested reader is referred to (Gardner and Spooner, 1994; Spooner and Gardner, 1994).

PROBLEMS

1. Consider the definition given in Eq. (3.12) for the power spectral density. Is it permissible to interchange the operation of taking the limit and that of the expectation in this equation? Justify your answer.

2. In deriving Eq. (3.25), we invoked the notion that if a wide-sense stationary process is applied to a linear, time-invariant, and stable filter, the stochastic process produced at the filter output is wide-sense stationary too. Show that, in general,

$$r_y(n, m) = \sum_{i=-\infty}^{\infty} \sum_{k=-\infty}^{\infty} h(i) h^*(k) r_u(n - i, m - k)$$

which includes the result of Eq. (3.25) as a special case.

3. The mean-square value of the filter output in Eq. (3.28) assumes that the bandwidth of the filter is small compared to its midband frequency. Is this assumption necessary for the corresponding result obtained in Example 1 for a white-noise process? Justify your answer.

4. A white-noise process with a variance of 0.1 V squared is applied to a low-pass discrete-time filter whose bandwidth is 1 Hz. The process is real.
   (a) Calculate the variance of the filter output.
   (b) Assuming that the input is Gaussian, determine the probability density function of the filter output.

5. Justify the fact that the expectation of $|dZ(\omega)|^2$ has the physical significance of power.

6. Show that the third- and higher-order cumulants of a Gaussian process are all identically zero.

7. Develop a physical interpretation of the trispectrum $C_4(\omega_1, \omega_2, \omega_3)$ of a stationary stochastic process $u(n)$; assume that $u(n)$ is real valued.

8. Consider a linear time-invariant system whose transfer function is $H(z)$. The system is excited by a real-valued sequence $x(n)$ of independently and identically distributed (iid) random variables with zero mean and unit variance. The probability distribution of $x(n)$ is nonsymmetric.
   (a) Evaluate the third-order cumulant and bispectrum of the system output $u(n)$.
   (b) Show that the phase component of the bispectrum of $u(n)$ is related to the phase response of the system transfer function $H(z)$ as follows:

$$\arg[C_3(\omega_1, \omega_2)] = \arg[H(e^{j\omega_1})] + \arg\{H(e^{j\omega_2})] - \arg[H(e^{j(\omega_1 + \omega_2)})]$$

9. Equation (3.52) gives the kth-order cumulant of the output of a linear time-invariant system of impulse response h_n that is driven by a sequence $x(n)$ of independent and identically distributed random variables. Prove the validity of this equation.

10. Show that for a stochastic process $u(n)$ that is cyclostationary in the wide sense, the cyclic autocorrelation function $r^{\alpha}(k)$ satisfies the property

$$r^{\alpha}(-k) = r^{\alpha}*(k)$$

where the asterisk denotes complex conjugation.

11. Figure 3.6 describes a method for measuring the spectral-correlation density of a time series $u(n)$ that is representative of a cyclostationary process in the wide sense. For $\alpha = 0$, show that Fig. 3.6 reduces to the simpler form shown in Fig. 3.3.

CHAPTER
4

Eigenanalysis

In this chapter we continue the statistical characterization of a discrete-time stochastic process that is stationary in the wide sense. From Chapter 2 we recall that the ensemble-averaged correlation matrix of such a process is Hermitian. An important aspect of a Hermitian matrix is that it permits a useful decomposition of the matrix in terms of its eigenvalues and associated eigenvectors. This form of representation is commonly referred to as *eigenanalysis*, which is basic to the study of digital signal processing.

We begin the discussion of eigenanalysis by outlining the eigenvalue problem in the context of the correlation matrix. We then study the properties of eigenvalues and eigenvectors of the correlation matrix and a related optimum filtering problem. We finish the discussion by briefly describing canned routines for eigenvalue computations and some related issues.

4.1 THE EIGENVALUE PROBLEM

Let the Hermitian matrix $\mathbf{R}$ denote the M-by-M correlation matrix of a wide-sense stationary discrete-time stochastic process represented by the M-by-1 observation vector $\mathbf{u}(n)$. In general, this matrix may contain complex elements. We wish to find an M-by-1 vector $\mathbf{q}$ that satisfies the condition

$$\mathbf{Rq} = \lambda \mathbf{q} \tag{4.1}$$

for some constant λ. This condition states that the vector $\mathbf{q}$ is linearly transformed to the vector $\lambda\mathbf{q}$ by the Hermitian matrix $\mathbf{R}$. Since λ is a constant, the vector $\mathbf{q}$ therefore has special significance in that it is left *invariant in direction* (in the M-dimensional space) by a linear transformation. For a typical M-by-M matrix $\mathbf{R}$, there will be M such vectors. To show this, we first rewrite Eq. (4.1) in the form

$$(\mathbf{R} - \lambda\mathbf{I})\mathbf{q} = \mathbf{0} \qquad\qquad (4.2)$$

where $\mathbf{I}$ is the M-by-M identity matrix, and $\mathbf{0}$ is the M-by-1 null vector. The matrix $(\mathbf{R} - \lambda\mathbf{I})$ has to be singular. Hence, Eq. (4.2) has a nonzero solution in the vector $\mathbf{q}$ if and only if the determinant of the matrix $(\mathbf{R} - \lambda\mathbf{I})$ equals zero; that is,

$$\det(\mathbf{R} - \lambda\mathbf{I}) = 0 \qquad\qquad (4.3)$$

This determinant, when expanded, is clearly a polynomial in λ of degree M. We thus find that, in general, Eq. (4.3) has M distinct roots. Correspondingly, Eq. (4.3) has M solutions in the vector $\mathbf{q}$.

Equation (4.3) is called the *characteristic equation* of the matrix $\mathbf{R}$. Let $\lambda_1, \lambda_2, \ldots,$ λ_M denote the M roots of this equation. These roots are called the *eigenvalues* of the matrix $\mathbf{R}$. Although the M-by-M matrix $\mathbf{R}$ has M eigenvalues, they need not be distinct. When the characteristic equation (4.3) has multiple roots, the matrix $\mathbf{R}$ is said to have *degenerate* eigenvalues. Note that, in general, the use of root finding in the characteristic equation (4.3) is a poor method for computing the eigenvalues of the matrix $\mathbf{R}$; the issue of eigenvalue computations is considered later in Section 4.5.

Let λ_i denote the ith eigenvalue of the matrix $\mathbf{R}$. Also, let $\mathbf{q}_i$ be a nonzero vector such that

$$\mathbf{R}\mathbf{q}_i = \lambda_i\mathbf{q}_i \qquad\qquad (4.4)$$

The vector $\mathbf{q}_i$ is called the *eigenvector* associated with λ_i. An eigenvector can correspond to only one eigenvalue. However, an eigenvalue may have many eigenvectors. For example, if $\mathbf{q}_i$ is an eigenvector associated with eigenvalue λ_i, then so is $a\mathbf{q}_i$ for any scalar $a \neq 0$.

Example 1: White Noise

Consider the M-by-M correlation matrix of a white-noise process that is described by the diagonal matrix

$$\mathbf{R} = \text{diag}(\sigma^2, \sigma^2, \ldots, \sigma^2)$$

where σ^2 is the variance of a sample of the process. *This correlation matrix $\mathbf{R}$ has a single degenerate eigenvalue equal to the variance σ^2 with multiplicity M. Any M-by-1 random vector qualifies as the associated eigenvector*, which shows that (for white noise) one eigenvalue σ^2 has M eigenvectors.

Example 2: Complex Sinusoid

Consider next the M-by-M correlation matrix of a time series whose elements are samples of a complex sinusoid with random phase and unit power. This correlation matrix may be written as

$$\mathbf{R} = \begin{bmatrix} 1 & e^{j\omega} & \cdots & e^{j(M-1)\omega} \\ e^{-j\omega} & 1 & \cdots & e^{j(M-2)\omega} \\ \bullet & \bullet & \bullet & \bullet \\ \bullet & \bullet & \bullet & \bullet \\ \bullet & \bullet & \bullet & \bullet \\ e^{-j(M-1)\omega} & e^{-j(M-2)\omega} & \cdots & 1 \end{bmatrix}$$

where ω is the angular frequency of the complex sinusoid. The M-by-1 vector

$$\mathbf{q} = [1, e^{-j\omega}, \ldots, e^{-j(M-1)\omega}]^T$$

is an eigenvector of the correlation matrix $\mathbf{R}$, and the corresponding eigenvalue is M (i.e., the dimension of the matrix $\mathbf{R}$). In other words, a complex sinusoidal vector represents an eigenvector of its own correlation matrix, except for the trivial operation of complex conjugation.

Note that the correlation matrix $\mathbf{R}$ has *rank* 1, which means that any column of $\mathbf{R}$ may be expressed as a linear combination of the remaining columns (i.e., the matrix $\mathbf{R}$ has only one independent column). It also means that the other eigenvalue is zero with multiplicity $M - 1$, and this eigenvalue has $M - 1$ eigenvectors.

4.2 PROPERTIES OF EIGENVALUES AND EIGENVECTORS

In this section we discuss the various properties of the eigenvalues and eigenvectors of the correlation matrix $\mathbf{R}$ of a stationary discrete-time stochastic process. Some of the properties derived here are direct consequences of the Hermitian property and the nonnegative definiteness of the correlation matrix $\mathbf{R}$, which were established in Section 2.3.

Property 1. *If $\lambda_1, \lambda_2, \ldots, \lambda_M$ denote the eigenvalues of the correlation matrix $\mathbf{R}$, then the eigenvalues of the matrix $\mathbf{R}^k$ equal $\lambda^k_1, \lambda^k_2, \ldots, \lambda^k_M$ for any integer $k > 0$.*
Repeated premultiplication of both sides of Eq. (4.1) by the matrix $\mathbf{R}$ yields

$$\mathbf{R}^k\mathbf{q} = \lambda^k\mathbf{q} \tag{4.5}$$

This shows that (1) if λ is an eigenvalue of $\mathbf{R}$, then λ^k is an eigenvalue of $\mathbf{R}^k$, which is the desired result, and (2) every eigenvector of $\mathbf{R}$ is also an eigenvector of $\mathbf{R}^k$.

Property 2. *Let $\mathbf{q}_1, \mathbf{q}_2, \ldots, \mathbf{q}_M$ be the eigenvectors corresponding to the distinct eigenvalues $\lambda_1, \lambda_2, \ldots, \lambda_M$ of the M-by-M correlation matrix $\mathbf{R}$, respectively. Then the eigenvectors $\mathbf{q}_1, \mathbf{q}_2, \ldots, \mathbf{q}_M$ are linearly independent.*
We say that the eigenvectors $\mathbf{q}_1, \mathbf{q}_2, \ldots, \mathbf{q}_M$ are *linearly dependent* if there are scalars $v_1, v_2, \ldots, v_M$, not all zero, such that

$$\sum_{i=1}^{M} v_i\mathbf{q}_i = \mathbf{0} \tag{4.6}$$

If no such scalars exist, we say that the eigenvectors are *linearly independent*.

We will prove the validity of Property 2 by contradiction. Suppose that Eq. (4.6) holds for certain not all zero scalars v_i. Repeated multiplication of Eq. (4.6) by matrix $\mathbf{R}$ and the use of Eq. (4.5) yield the following set of M equations:

$$\sum_{i-1}^{M} v_i \lambda_i^{k-1} \mathbf{q}_i = \mathbf{0}, \qquad k = 1, 2, \ldots, M \tag{4.7}$$

This set of equations may be written in the form of a single matrix equation:

$$[v_1 \mathbf{q}_1, v_2 \mathbf{q}_2, \ldots, v_M \mathbf{q}_M] \mathbf{S} = \mathbf{0} \tag{4.8}$$

where

$$\mathbf{S} = \begin{bmatrix} 1 & \lambda_1 & \lambda_1^2 & \cdots & \lambda_1^{M-1} \\ 1 & \lambda_2 & \lambda_2^2 & \cdots & \lambda_2^{M-1} \\ \cdot & \cdot & \cdot & \cdot & \cdot \\ \cdot & \cdot & \cdot & \cdot & \cdot \\ \cdot & \cdot & \cdot\cdot & & \cdot \\ 1 & \lambda_M & \lambda_M^2 & \cdots & \lambda_M^{M-1} \end{bmatrix} \tag{4.9}$$

The matrix $\mathbf{S}$ is called a *Vandermonde matrix* (Strang, 1980). When the λ_i are distinct, the Vandermonde matrix $\mathbf{S}$ is nonsingular. Therefore, we may postmultiply Eq. (4.8) by the inverse matrix $\mathbf{S}^{-1}$, obtaining

$$[v_1 \mathbf{q}_1, v_2 \mathbf{q}_2, \ldots, v_M \mathbf{q}_M] = \mathbf{O}$$

Hence, each column $v_i \mathbf{q}_i = \mathbf{0}$. Since the eigenvectors $\mathbf{q}_i$ are not zero, this condition can be satisfied if and only if the v_i are all zero. This contradicts the assumption that the scalars v_i are not all zero. In other words, the eigenvectors are linearly independent.

We may put this property to an important use by having the linearly independent eigenvectors $\mathbf{q}_1, \mathbf{q}_2, \ldots, \mathbf{q}_M$ serve as a *basis* for the representation of an arbitrary vector $\mathbf{w}$ with the same dimension as the eigenvectors themselves. In particular, we may express the arbitrary vector $\mathbf{w}$ as a linear combination of the eigenvectors $\mathbf{q}_1, \mathbf{q}_2, \ldots, \mathbf{q}_M$ as follows:

$$\mathbf{w} = \sum_{i=1}^{M} v_i \mathbf{q}_i \tag{4.10}$$

where $v_1, v_2, \ldots, v_M$ are constants. Suppose now we apply a linear transformation to the vector $\mathbf{w}$ by premultiplying it by the matrix $\mathbf{R}$, obtaining

$$\mathbf{Rw} = \sum_{i=1}^{M} v_i \mathbf{Rq}_i \tag{4.11}$$

By definition, we have $\mathbf{Rq}_i = \lambda_i \mathbf{q}_i$. Therefore, we may express the result of this linear transformation in the equivalent form

$$\mathbf{Rw} = \sum_{i=1}^{M} v_i \lambda_i \mathbf{q}_i \tag{4.12}$$

We thus see that when a linear transformation is applied to an arbitrary vector $\mathbf{w}$ defined in Eq. (4.10), the eigenvectors remain independent of each other, and the effect of the transformation is simply to multiply each eigenvector by its respective eigenvalue.

Property 3.   *Let $\lambda_1, \lambda_2, \ldots, \lambda_M$ be the eigenvalues of the M-by-M correlation matrix $\mathbf{R}$. Then all these eigenvalues are real and nonnegative.*

To prove this property, we first use Eq. (4.1) to express the condition on the ith eigenvalue λ_i as

$$\mathbf{R}\mathbf{q}_i = \lambda_i\mathbf{q}_i, \qquad i = 1, 2, \ldots, M \tag{4.13}$$

Premultiplying both sides of this equation by $\mathbf{q}_i^H$, the Hermitian transpose of eigenvector $\mathbf{q}_i$, we get

$$\mathbf{q}_i^H \mathbf{R}\mathbf{q}_i = \lambda_i\mathbf{q}_i^H\mathbf{q}_i, \qquad i = 1, 2, \ldots, M \tag{4.14}$$

The inner product $\mathbf{q}_i^H\mathbf{q}_i$ is a positive scalar, representing the squared Euclidean length of the eigenvector $\mathbf{q}_i$; that is, $\mathbf{q}_i^H\mathbf{q}_i > 0$. We may therefore divide both sides of Eq. (4.14) by $\mathbf{q}_i^H\mathbf{q}_i$ and so express the ith eigenvalue λ_i as the ratio

$$\lambda_i = \frac{\mathbf{q}_i^H\mathbf{R}\mathbf{q}_i}{\mathbf{q}_i^H\mathbf{q}_i}, \qquad i = 1, 2, \ldots, M \tag{4.15}$$

Since the correlation matrix $\mathbf{R}$ is always nonnegative definite, the Hermitian form $\mathbf{q}_i^H\mathbf{R}\mathbf{q}_i$ in the numerator of this ratio is always real and nonnegative; that is $\mathbf{q}_i^H \mathbf{R}\mathbf{q}_i \geq 0$. Therefore, it follows from Eq. (4.15) that $\lambda_i \geq 0$ for all i. That is, all the eigenvalues of the correlation matrix $\mathbf{R}$ are always real and nonnegative.

The correlation matrix $\mathbf{R}$ is positive definite, except in noise-free sinusoidal and noise-free array signal-processing problems; and so we have $\mathbf{q}_i^H\mathbf{R}\mathbf{q}_i > 0$ and, correspondingly, $\lambda_i > 0$ for all i. That is, the eigenvalues of the correlation matrix $\mathbf{R}$ are almost always real and positive.

The ratio of the Hermitian form $\mathbf{q}_i^H\mathbf{R}\mathbf{q}_i$ to the inner product $\mathbf{q}_i^H\mathbf{q}_i$ on the right-hand side of Eq. (4.15) is called the *Rayleigh quotient* of the vector $\mathbf{q}_i$. We may thus state that an eigenvalue of the correlation matrix equals the Rayleigh quotient of the corresponding eigenvector.

Property 4.   *Let $\mathbf{q}_1, \mathbf{q}_2, \ldots, \mathbf{q}_M$ be the eigenvectors corresponding to the distinct eigenvalues $\lambda_1, \lambda_2, \ldots, \lambda_M$ of the M-by-M correlation matrix $\mathbf{R}$, respectively. Then the eigenvectors $\mathbf{q}_1, \mathbf{q}_2, \ldots, \mathbf{q}_M$ are orthogonal to each other.*

Let $\mathbf{q}_i$ and $\mathbf{q}_j$ denote any two eigenvectors of the correlation matrix $\mathbf{R}$. We say that these two eigenvectors are *orthogonal* to each other if

$$\mathbf{q}_i^H\mathbf{q}_j = 0, \qquad i \neq j \tag{4.16}$$

Using Eq. (4.1) we may express the conditions on the eigenvectors $\mathbf{q}_i$ and $\mathbf{q}_j$ as follows, respectively,

$$\mathbf{R}\mathbf{q}_i = \lambda_i\mathbf{q}_i \tag{4.17}$$

and

$$\mathbf{Rq}_j = \lambda_j \mathbf{q}_j \tag{4.18}$$

Premultiplying both sides of Eq. (4.17) by the Hermitian-transposed vector $\mathbf{q}_j^H$, we get

$$\mathbf{q}_j^H \mathbf{Rq}_i = \lambda_i \mathbf{q}_j^H \mathbf{q}_i \tag{4.19}$$

Since the correlation matrix $\mathbf{R}$ is Hermitian, we have $\mathbf{R}^H = \mathbf{R}$. Also, from Property 3 we know that the eigenvalue λ_j is real for all j. Hence, taking the Hermitian transpose of both sides of Eq. (4.18) and using these two properties, we get

$$\mathbf{q}_j^H \mathbf{R} = \lambda_j \mathbf{q}_j^H \tag{4.20}$$

Postmultiplying both sides of Eq. (4.20) by the vector $\mathbf{q}_i$,

$$\mathbf{q}_j^H \mathbf{Rq}_i = \lambda_j \mathbf{q}_j^H \mathbf{q}_i \tag{4.21}$$

Subtracting Eq.(4.21) from (4.19), we get

$$(\lambda_i - \lambda_j)\mathbf{q}_j^H \mathbf{q}_i = 0 \tag{4.22}$$

Since the eigenvalues of the correlation matrix $\mathbf{R}$ are assumed to be distinct, we have $\lambda_i \neq \lambda_j$. Accordingly, the condition of Eq. (4.22) holds if and only if

$$\mathbf{q}_j^H \mathbf{q}_i = 0, \qquad i \neq j \tag{4.23}$$

which is the desired result. That is, the eigenvectors $\mathbf{q}_i$ and $\mathbf{q}_j$ are *orthogonal* to each other for $i \neq j$.

Property 5: Unitary Similarity Transformation.   *Let* $\mathbf{q}_1, \mathbf{q}_2, \ldots, \mathbf{q}_M$ *be the eigenvectors corresponding to the distinct eigenvalues* $\lambda_1, \lambda_2, \ldots, \lambda_M$ *of the M-by-M correlation matrix* $\mathbf{R}$, *respectively. Define the M-by-M matrix*

$$\mathbf{Q} = [\mathbf{q}_1, \mathbf{q}_2, \ldots, \mathbf{q}_M]$$

where

$$\mathbf{q}_i^H \mathbf{q}_j = \begin{cases} 1, & i = j \\ 0, & i \neq j \end{cases}$$

Define the M-by-M diagonal matrix

$$\mathbf{\Lambda} = \mathrm{diag}(\lambda_1, \lambda_2, \ldots, \lambda_M)$$

Then the original matrix $\mathbf{R}$ *may be diagonalized as follows:*

$$\mathbf{Q}^H \mathbf{R} \mathbf{Q} = \mathbf{\Lambda}$$

The condition that $\mathbf{q}_i^H \mathbf{q}_i = 1$ for $i = 1, 2, \ldots, M$ requires that each eigenvector be *normalized* to have a *length* of 1. The *squared length* or *squared norm* of a vector $\mathbf{q}_i$ is defined as the inner product $\mathbf{q}_i^H \mathbf{q}_i$. The orthogonality condition that $\mathbf{q}_i^H \mathbf{q}_j = 0$, for $i \neq j$, follows from Property 4. When both of these conditions are simultaneously satisfied; that is,

$$\mathbf{q}_i^H \mathbf{q}_j = \begin{cases} 1, & i = j \\ 0, & i \neq j \end{cases} \tag{4.24}$$

we say the eigenvectors $\mathbf{q}_1, \mathbf{q}_2, \ldots \mathbf{q}_M$ form an *orthonormal* set. By definition, the eigenvectors $\mathbf{q}_1, \mathbf{q}_2, \ldots, \mathbf{q}_M$ satisfy the equations [see Eq. (4.1)]

$$\mathbf{R}\mathbf{q}_i = \lambda_i \mathbf{q}_i, \quad i = 1, 2, \ldots, M \tag{4.25}$$

The M-by-M matrix $\mathbf{Q}$ has as its columns the orthonormal set of eigenvectors $\mathbf{q}_1, \mathbf{q}_2, \ldots, \mathbf{q}_M$; that is,

$$\mathbf{Q} = [\mathbf{q}_1, \mathbf{q}_2, \ldots, \mathbf{q}_M] \tag{4.26}$$

The M-by-M diagonal matrix $\mathbf{\Lambda}$ has the eigenvalues $\lambda_1, \lambda_2, \ldots, \lambda_M$ for the elements of its main diagonal:

$$\mathbf{\Lambda} = \text{diag} \, (\lambda_1, \lambda_2, \ldots, \lambda_M) \tag{4.27}$$

Accordingly, we may rewrite the set of M equations (4.25) as a single matrix equation:

$$\mathbf{R}\mathbf{Q} = \mathbf{Q}\mathbf{\Lambda} \tag{4.28}$$

Owing to the orthonormal nature of the eigenvectors, as defined in Eq. (4.24), we find that

$$\mathbf{Q}^H \mathbf{Q} = \mathbf{I}$$

Equivalently, we may write

$$\mathbf{Q}^{-1} = \mathbf{Q}^H \tag{4.29}$$

That is, the matrix $\mathbf{Q}$ is nonsingular with an inverse $\mathbf{Q}^{-1}$ equal to the Hermitian transpose of $\mathbf{Q}$. A matrix that has this property is called a *unitary matrix*.

Thus, premultiplying both sides of Eq. (4.28) by the Hermitian-transposed matrix $\mathbf{Q}^H$ and using the property of Eq. (4.29), we get the desired result:

$$\mathbf{Q}^H \mathbf{R} \mathbf{Q} = \mathbf{\Lambda} \tag{4.30}$$

This transformation is called the *unitary similarity transformation*.

We have thus proved an important result. The correlation matrix $\mathbf{R}$ may be *diagonalized* by a unitary similarity transformation. Furthermore, the matrix $\mathbf{Q}$ that is used to diagonalize $\mathbf{R}$ has as its columns an orthonormal set of eigenvectors for $\mathbf{R}$. The resulting diagonal matrix $\mathbf{\Lambda}$ has as its diagonal elements the eigenvalues of $\mathbf{R}$.

By postmultiplying both sides of Eq. (4.28) by the inverse matrix $\mathbf{Q}^{-1}$ and then using the property of Eq. (4.29), we may also write

$$\mathbf{R} = \mathbf{Q}\mathbf{\Lambda}\mathbf{Q}^H$$

$$= \sum_{i=1}^{M} \lambda_i \mathbf{q}_i \mathbf{q}_i^H \tag{4.31}$$

where M is the dimension of matrix $\mathbf{R}$. Let the projection $\mathbf{P}_i$ denote the outer product $\mathbf{q}_i\mathbf{q}_i^H$. Then, it is a straightforward matter to show that

$$\mathbf{P}_i = \mathbf{P}_i^2 = \mathbf{P}_i^H$$

which, in effect, means that $\mathbf{P}_i = \mathbf{q}_i\mathbf{q}_i^H$ is a *rank-one projection*. Thus, Eq. (4.31) states that the correlation matrix of a wide-sense stationary process equals the linear combination of all such rank-one projections, with each projection being weighted by the respective eigenvalue. This result is known as *Mercer's theorem*. It is also referred to as the *spectral theorem*.

Property 6.   *Let $\lambda_1, \lambda_2, \ldots, \lambda_M$ be the eigenvalues of the M-by-M correlation matrix $\mathbf{R}$. Then the sum of these eigenvalues equals the trace of matrix $\mathbf{R}$.*

The *trace* of a square matrix is defined as the sum of the diagonal elements of the matrix. Taking the trace of both sides of Eq. (4.30), we may write

$$\text{tr}[\mathbf{Q}^H\mathbf{R}\mathbf{Q}] = \text{tr}[\mathbf{\Lambda}] \tag{4.32}$$

The diagonal matrix $\mathbf{\Lambda}$ has as its diagonal elements the eigenvalues of $\mathbf{R}$. Hence, we have

$$\text{tr}[\mathbf{\Lambda}] = \sum_{i=1}^{M} \lambda_i \tag{4.33}$$

Using a rule in matrix algebra, we may write[1]

$$\text{tr}[\mathbf{Q}^H\mathbf{R}\mathbf{Q}] = \text{tr}[\mathbf{R}\mathbf{Q}\mathbf{Q}^H]$$

However, $\mathbf{Q}\mathbf{Q}^H$ equals the identity matrix $\mathbf{I}$. Hence we have

$$\text{tr}[\mathbf{Q}^H\mathbf{R}\mathbf{Q}] = \text{tr}[\mathbf{R}]$$

Accordingly, we may rewrite Eq. (4.32) as

$$\text{tr}[\mathbf{R}] = \sum_{i=1}^{M} \lambda_i \tag{4.34}$$

We have thus shown that the trace of the correlation matrix $\mathbf{R}$ equals the sum of its eigenvalues. Although in proving this result we used a property that requires the matrix $\mathbf{R}$ to be Hermitian with distinct eigenvalues, nevertheless, the result applies to any square matrix.

Property 7.   *The correlation matrix $\mathbf{R}$ is ill conditioned if the ratio of the largest eigenvalue to the smallest eigenvalue of $\mathbf{R}$ is large.*

To appreciate the impact of Property 7, it is important that we recognize the fact that the development of an algorithm for the effective solution of a signal processing problem

---

[1] This result follows from the following rule in matrix algebra. Let $\mathbf{A}$ be an M-by-N matrix and $\mathbf{B}$ be an N-by-M matrix. The trace of the matrix product $\mathbf{A}\mathbf{B}$ equals the trace of $\mathbf{B}\mathbf{A}$.

and the understanding of associated *perturbation theory* go hand-in-hand (Van Loan, 1989). We may illustrate the synergism between these two fields by considering the following linear system of equations:

$$\mathbf{Aw} = \mathbf{d}$$

where the matrix $\mathbf{A}$ and the vector $\mathbf{d}$ are data-dependent quantities, and $\mathbf{w}$ is a coefficient vector characterizing a linear FIR filter of interest. An elementary formulation of perturbation theory tells us that if the matrix $\mathbf{A}$ and vector $\mathbf{d}$ are perturbed by small amounts $\delta\mathbf{A}$ and $\delta\mathbf{d}$, respectively, and if $\|\delta\mathbf{A}\|/\|\mathbf{A}\|$ and $\|\delta\mathbf{d}\|/\|\mathbf{d}\|$ are both on the order of some ϵ with $\epsilon \ll 1$, we then have (Golub and Van Loan, 1989)

$$\frac{\|\delta\mathbf{w}\|}{\|\mathbf{w}\|} \leq \epsilon\,\chi(\mathbf{A})$$

where $\delta\mathbf{w}$ is the change produced in $\mathbf{w}$, and $\chi(\mathbf{A})$ is the *condition number* of matrix $\mathbf{A}$ with respect to inversion. The condition number is so called because it describes the ill condition or bad behavior of matrix $\mathbf{A}$ quantitatively. In particular, it is defined as follows (Wilkinson, 1963; Strang, 1980; Golub and Van Loan, 1989):

$$\chi(\mathbf{A}) = \|\mathbf{A}\|\,\|\mathbf{A}^{-1}\| \tag{4.35}$$

where $\|\mathbf{A}\|$ is a *norm* of matrix $\mathbf{A}$, and $\|\mathbf{A}^{-1}\|$ is the corresponding norm of the inverse matrix $\mathbf{A}^{-1}$. The norm of a matrix is a number assigned to the matrix that is in some sense a measure of the magnitude of the matrix. We find it natural to require that the norm of a matrix satisfy the following conditions:

1. $\|\mathbf{A}\| \geq 0$, $\|\mathbf{A}\| = 0$ if and only if $\mathbf{A} = \mathbf{O}$
2. $\|c\mathbf{A}\| = |c|\,\|\mathbf{A}\|$, where c is any real number and $|c|$ is its magnitude
3. $\|\mathbf{A} + \mathbf{B}\| \leq \|\mathbf{A}\| + \|\mathbf{B}\|$
4. $\|\mathbf{AB}\| \leq \|\mathbf{A}\|\,\|\mathbf{B}\|$

Condition 3 is the *triangle inequality*, and condition 4 is the *mutual consistency*. There are several ways of defining the norm $\|\mathbf{A}\|$, which satisfy the preceding conditions (Ralston, 1965). For our present discussion, however, we find it convenient to use the *spectral norm*[2] defined as the square root of the largest eigenvalue of the matrix product $\mathbf{A}^H\mathbf{A}$, where $\mathbf{A}^H$ is the Hermitian transpose of $\mathbf{A}$; that is,

$$\|\mathbf{A}\|_s = (\text{largest eigenvalue of } \mathbf{A}^H\mathbf{A})^{1/2} \tag{4.36}$$

---

[2] Another matrix norm of interest is the *Frobenius norm*, $\|\mathbf{A}\|_F$, defined by (Stewart, 1973):

$$\|\mathbf{A}\|_F = \sqrt{\sum_{j=1}^{M}\sum_{j=1}^{N} |a_{ij}|^2}$$

where M and N are the dimensions of matrix $\mathbf{A}$, and a_{ij} is its ijth element.

Since for any matrix $\mathbf{A}$ the product $\mathbf{A}^H\mathbf{A}$ is always Hermitian and nonnegative definite, it follows that the eigenvalues of $\mathbf{A}^H\mathbf{A}$ are all real and nonnegative, as required. Moreover, from Eq. (4.15) we note that an eigenvalue of $\mathbf{A}^H\mathbf{A}$ equals the Rayleigh coefficient of the corresponding eigenvector. Squaring both sides of Eq. (4.36) and using this property, we may therefore write[3]

$$\|\mathbf{A}\|_s^2 = \max \frac{\mathbf{x}^H\mathbf{A}^H\mathbf{A}\mathbf{x}}{\mathbf{x}^H\mathbf{x}}$$

$$= \max \frac{\|\mathbf{A}\mathbf{x}\|^2}{\|\mathbf{x}\|^2}$$

where $\|\mathbf{x}\|^2$ is the inner product of vector $\mathbf{x}$ with itself, and likewise for $\|\mathbf{A}\mathbf{x}\|^2$. We refer to $\|\mathbf{x}\|$ as the *Euclidean norm* or *length* of vector $\mathbf{x}$. We may thus express the spectral norm of matrix $\mathbf{A}$ in the equivalent form

$$\|\mathbf{A}\|_s = \max \frac{\|\mathbf{A}\mathbf{x}\|}{\|\mathbf{x}\|} \tag{4.37}$$

According to this relation, the spectral norm of $\mathbf{A}$ measures the largest amount by which any vector (eigenvector or not) is amplified by matrix multiplication, and the vector that is amplified the most is the eigenvector that corresponds to the largest eigenvalue of $\mathbf{A}^H\mathbf{A}$ (Strang, 1980).

Consider now the application of the definition in Eq. (4.36) to the correlation matrix $\mathbf{R}$. Since $\mathbf{R}$ is Hermitian, we have $\mathbf{R}^H = \mathbf{R}$. Hence, from Property 1 we deduce that if λ_{max} is the largest eigenvalue of $\mathbf{R}$, the largest eigenvalue of $\mathbf{R}^H\mathbf{R}$ equals λ_{max}^2. Accordingly, the spectral norm of the correlation matrix $\mathbf{R}$ is

$$\|\mathbf{R}\|_s = \lambda_{max} \tag{4.38}$$

Similarly, we may show that the spectral norm of $\mathbf{R}^{-1}$, the inverse of the correlation matrix, is

$$\|\mathbf{R}^{-1}\|_s = \frac{1}{\lambda_{min}} \tag{4.39}$$

where λ_{min} is the smallest eigenvalue of $\mathbf{R}$. Thus, by adopting the spectral norm as the basis of the condition number, we have shown that the condition number of the correlation matrix $\mathbf{R}$ equals

$$\chi(\mathbf{R}) = \frac{\lambda_{max}}{\lambda_{min}} \tag{4.40}$$

This ratio is commonly referred to as the *eigenvalue spread* or the *eigenvalue ratio* of the correlation matrix. Note that we always have $\chi(\mathbf{R}) \geq 1$.

---

[3] Note that the vector $\mathbf{x}$ is one of the eigenvectors. Hence, at this stage, we can only say that $\|\mathbf{A}\|_s^2$ is the *maximum Rayleigh quotient* of the eigenvectors. However, this may be extended to any vector after the minimax theorem is proved; see Property 9.

Suppose that the correlation matrix $\mathbf{R}$ is *normalized* so that the magnitude of the largest element, $r(0)$, equals 1. Then, if the condition number or eigenvalue spread of the correlation matrix $\mathbf{R}$ is large, we find that the inverse matrix $\mathbf{R}^{-1}$ contains some very large elements. This behavior may cause trouble in solving a system of equations involving $\mathbf{R}^{-1}$. In such a case, we say that the correlation matrix $\mathbf{R}$ is *ill conditioned*, hence the justification of Property 7.

Property 8.   *The eigenvalues of the correlation matrix of a discrete-time stochastic process are bounded by the minimum and maximum values of the power spectral density of the process.*

Let λ_i and $\mathbf{q}_i$, $i = 1, 2, \ldots, M$, denote the eigenvalues of the M-by-M correlation matrix $\mathbf{R}$ of a discrete-time stochastic process $u(n)$ and their associated eigenvectors, respectively. From Eq. (4.15), we have

$$\lambda_i = \frac{\mathbf{q}_i^H \mathbf{R} \mathbf{q}_i}{\mathbf{q}_i^H \mathbf{q}_i}, \qquad i = 1, 2, \ldots, M \tag{4.41}$$

The Hermitian form in the numerator may be expressed in its expanded form as follows

$$\mathbf{q}_i^H \mathbf{R} \mathbf{q}_i = \sum_{k=1}^{M} \sum_{l=1}^{M} q_{ik}^* \, r(l - k) q_{il} \tag{4.42}$$

where q_{ik}^* is the kth element of the row vector $\mathbf{q}_i^H$, $r(l - k)$ is the klth element of the matrix $\mathbf{R}$, and q_{il} is the lth element of the column vector $\mathbf{q}_i$. Using the Einstein–Wiener–Khintchine relation of Eq. (3.16), we may write

$$r(l - k) = \frac{1}{2\pi} \int_{-\pi}^{\pi} S(\omega) e^{j\omega(l-k)} \, d\omega \tag{4.43}$$

where $S(\omega)$ is the power spectral density of the process $u(n)$. Hence, we may rewrite Eq. (4.42) as

$$\begin{aligned}
\mathbf{q}_i^H \mathbf{R} \mathbf{q}_i &= \frac{1}{2\pi} \sum_{k=1}^{M} \sum_{l=1}^{M} q_{ik}^* q_{il} \int_{-\pi}^{\pi} S(\omega) e^{j\omega(l-k)} \, d\omega \\
&= \frac{1}{2\pi} \int_{-\pi}^{\pi} d\omega S(\omega) \sum_{k=1}^{M} q_{ik}^* e^{-j\omega k} \sum_{l=1}^{M} q_{il} e^{j\omega l}
\end{aligned} \tag{4.44}$$

Let the discrete-time Fourier transform of the sequence $q_{i1}^*, q_{i2}^*, \ldots, q_{iM}^*$ be denoted by

$$Q_i'(e^{j\omega}) = \sum_{k=1}^{M} q_{ik}^* e^{-j\omega k} \tag{4.45}$$

Therefore, using Eq. (4.45) in (4.44), we get

$$\mathbf{q}_i^H \mathbf{R} \mathbf{q}_i = \frac{1}{2\pi} \int_{-\pi}^{\pi} |Q_i'(e^{j\omega})|^2 S(\omega) d\omega \tag{4.46}$$

Similarly, we may show that

$$\mathbf{q}_i^H \mathbf{q}_i = \frac{1}{2\pi} \int_{-\pi}^{\pi} |Q_i'(e^{j\omega})|^2 d\omega \qquad (4.47)$$

Accordingly, we may use Eq. (4.15) to redefine the eigenvalue λ_i of the correlation matrix $\mathbf{R}$ in terms of the associated power spectral density as

$$\lambda_i = \frac{\displaystyle\int_{-\pi}^{\pi} |Q_i'(e^{j\omega})|^2 S(\omega) d\omega}{\displaystyle\int_{-\pi}^{\pi} |Q_i'(e^{j\omega})|^2 d\omega} \qquad (4.48)$$

Let $S_{\min}$ and $S_{\max}$ denote the absolute minimum and maximum values of the power spectral density $S(\omega)$, respectively. Then it follows that

$$\int_{-\pi}^{\pi} |Q_i'(e^{j\omega})|^2 S(\omega) d\omega \geq S_{\min} \int_{-\pi}^{\pi} |Q_i'(e^{j\omega})|^2 d\omega \qquad (4.49)$$

and

$$\int_{-\pi}^{\pi} |Q_i'(e^{j\omega})|^2 S(\omega) d\omega \leq S_{\max} \int_{-\pi}^{\pi} |Q_i'(e^{j\omega})|^2 d\omega \qquad (4.50)$$

Hence, we deduce that the eigenvalues λ_i are bounded by the maximum and minimum values of the associated power spectral density as follows:

$$S_{\min} \leq \lambda_i \leq S_{\max}, \qquad i = 1, 2, \ldots, M \qquad (4.51)$$

Correspondingly, the eigenvalue spread $\chi(\mathbf{R})$ is bounded as

$$\chi(\mathbf{R}) = \frac{\lambda_{\max}}{\lambda_{\min}} \leq \frac{S_{\max}}{S_{\min}} \qquad (4.52)$$

It is of interest to note that as the dimension M of the correlation matrix approaches infinity, the maximum eigenvalue $\lambda_{\max}$ approaches $S_{\max}$, and the minimum eigenvalue $\lambda_{\min}$ approaches $S_{\min}$. Accordingly, the eigenvalue spread $\chi(\mathbf{R})$ of the correlation matrix $\mathbf{R}$ approaches the ratio $S_{\max}/S_{\min}$ as the dimension M of the matrix $\mathbf{R}$ approaches infinity.

Property 9. Minimax Theorem. *Let the M-by-M correlation matrix $\mathbf{R}$ have eigenvalues $\lambda_1, \lambda_2, \ldots, \lambda_M$ that are arranged in decreasing order as follows:*

$$\lambda_1 \leq \lambda_2 \leq \ldots \leq \lambda_M$$

The minimax theorem states that

$$\lambda_k = \min_{\dim(\mathscr{S})=k} \max_{\substack{\mathbf{x}\in\mathscr{S} \\ \mathbf{x}\neq\mathbf{0}}} \frac{\mathbf{x}^H \mathbf{R} \mathbf{x}}{\mathbf{x}^H \mathbf{x}}, \qquad k = 1, 2, \ldots M \qquad (4.53)$$

where $\mathcal{S}$ is a subspace of the vector space of all M-by-1 complex vectors, dim ($\mathcal{S}$) denotes the dimension of subspace $\mathcal{S}$, and $\mathbf{x} \in \mathcal{S}$ signifies that the vector $\mathbf{x}$ (assumed to be nonzero) varies over the subspace $\mathcal{S}$.

Let $\mathbb{C}^M$ denote a complex vector space of dimension M. For the purpose of our present discussion, we define the *complex (linear) vector space $\mathbb{C}^M$* as the set of all complex vectors that can be expressed as a linear combination of M *basis vectors*. Specifically, we may write

$$\mathbb{C}^M = \{\mathbf{y}\} \tag{4.54}$$

where $\mathbf{y}$ is any complex vector defined by

$$\mathbf{y} = \sum_{i=1}^{M} a_i \mathbf{q}_i \tag{4.55}$$

The $\mathbf{q}_i$ are the basis vectors, and the a_i are scalars. For the basis vectors we may use any *orthonormal set* of vectors $\mathbf{q}_1, \mathbf{q}_2, \ldots, \mathbf{q}_M$ that satisfy two requirements:

$$\mathbf{q}_i^H \mathbf{q}_j = \begin{cases} 1, & i = j \\ 0, & i \neq j \end{cases} \tag{4.56}$$

In other words, each basis vector is *normalized* to have a *Euclidean length* or *norm* of unity, and it is *orthogonal* to every other basis vector in the set. The *dimension M* of the complex vector space $\mathbb{C}^M$ is the minimum number of basis vectors required to span the entire space.

The basis functions define the "coordinates" of a complex vector space. Any complex vector of compatible dimension may then be represented simply as a "point" in that space. Indeed, the idea of a complex vector space is a natural generalization of Euclidean geometry. Central to this idea is that of a *subspace*. We say that $\mathcal{S}$ is a subspace of the complex vector space $\mathbb{C}^M$ if it involves a *subset* of the M basis vectors that define $\mathbb{C}^M$. In other words, a subspace of dimension k is defined as the set of complex vectors that can be written as a linear combination of the basis vectors $\mathbf{q}_1, \mathbf{q}_2, \ldots, \mathbf{q}_k$, as shown by

$$\mathbf{x} = \sum_{i=1}^{k} a_i \mathbf{q}_i \tag{4.57}$$

Obviously, we have $k \leq M$. Note, however, that the dimension of the vector $\mathbf{x}$ is M.

These ideas are illustrated in the three-dimensional (real) vector space depicted in Fig. 4.1. The $\mathbf{q}_1, \mathbf{q}_2$-plane represents a subspace $\mathcal{S}$ of dimension 2. The representation of vector $\mathbf{y}$ and that of vector $\mathbf{x}$ (i.e., the part of $\mathbf{y}$ lying in subspace $\mathcal{S}$) are indicated in Fig. 4.1.

Returning to the issue at hand, namely, a proof of the *minimax theorem* described in Eq. (4.53), we may proceed as follows. We first use the spectral theorem of Eq. (4.31) to decompose the M-by-M correlation matrix $\mathbf{R}$ as

$$\mathbf{R} = \sum_{i=1}^{M} \lambda_i \mathbf{q}_i \mathbf{q}_i^H$$

where the λ_i are the eigenvalues of the correlation matrix $\mathbf{R}$ and the $\mathbf{q}_i$ are the associated eigenvectors. In view of the orthonormality conditions of Eq. (4.24) satisfied by the eigen-

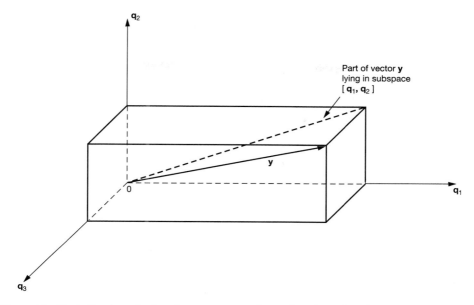

Figure 4.1   Illustrating the projection of a vector onto a subspace for a three-dimensional (real) vector space.

vectors $\mathbf{q}_1, \mathbf{q}_2, \ldots, \mathbf{q}_M$, we may adopt them as the M basis vectors of the complex vector space $\mathbb{C}^M$. Let an M-by-1 vector $\mathbf{x}$ be constrained to lie in a subspace $\mathcal{S}$ of dimension k, as defined in Eq. (4.57). Then, using Eq. (4.31), we may express the Rayleigh quotient of the vector $\mathbf{x}$ as

$$\frac{\mathbf{x}^H \mathbf{R} \mathbf{x}}{\mathbf{x}^H \mathbf{x}} = \frac{\displaystyle\sum_{i=1}^{k} a_i^2 \lambda_i}{\displaystyle\sum_{i=1}^{k} a_i^2} \tag{4.58}$$

Equation (4.58) states that the Rayleigh quotient of a vector $\mathbf{x}$ lying in the subspace $\mathcal{S}$ of dimension k (i.e., the subspace spanned by the eigenvectors $\mathbf{q}_1, \mathbf{q}_2, \ldots, \mathbf{q}_k$) is a *weighted mean* of the eigenvalues $\lambda_1, \lambda_2, \ldots, \lambda_k$. Since, by assumption, we have $\lambda_1 \leq \lambda_2 \leq \ldots \leq \lambda_k$, it follows that for any subspace $\mathcal{S}$ of dimension k,

$$\max_{\substack{\mathbf{x} \in \mathcal{S} \\ \mathbf{x} \neq 0}} \frac{\mathbf{x}^H \mathbf{R} \mathbf{x}}{\mathbf{x}^H \mathbf{x}} \leq \lambda_k$$

This result implies that

$$\min_{\dim(\mathcal{S})=k} \max_{\substack{\mathbf{x} \in \mathcal{S} \\ \mathbf{x} \neq 0}} \frac{\mathbf{x}^H \mathbf{R} \mathbf{x}}{\mathbf{x}^H \mathbf{x}} \leq \lambda_k \tag{4.59}$$

We next prove that for any subspace $\mathcal{S}$ of dimension k, spanned by the eigenvectors $\mathbf{q}_{i_1}, \mathbf{q}_{i_2}, \ldots, \mathbf{q}_{i_k}$ where $\{i_1, i_2, \ldots, i_k\}$ is a subset of $\{1, 2, \ldots, M\}$, there exists at least one nonzero vector $\mathbf{x}$ common to $\mathcal{S}$ and the subspace $\mathcal{S}'$ spanned by the eigenvectors $\mathbf{q}_k$, $\mathbf{q}_{k+1}, \ldots, \mathbf{q}_M$. To do so, we consider the system of M homogeneous equations:

$$\sum_{j=1}^{k} a_j \mathbf{q}_{i_j} = \sum_{i=k}^{M} b_i \mathbf{q}_i \tag{4.60}$$

where the $(M + 1)$ unknowns are made up as follows:

1. A total of k scalars, namely $a_1, a_2, \ldots, a_k$, on the left-hand side.
2. A total of $M - k + 1$ scalars, namely $b_k, b_{k+1}, \ldots, b_M$ on the right-hand side.

Hence the system of equations (4.60) will always have a nontrivial solution. Moreover, we know from Property 2 that the eigenvectors $\mathbf{q}_{i_1}, \mathbf{q}_{i_2}, \ldots, \mathbf{q}_{i_k}$ are *linearly independent*, as are the eigenvectors $\mathbf{q}_k, \mathbf{q}_{k+1}, \ldots, \mathbf{q}_M$. It follows therefore that there is at least one *nonzero vector* $\mathbf{x} = \sum_{j=1}^{k} a_j \mathbf{q}_{i_j}$ that is *common* to the space of $\mathbf{q}_{i_1}, \mathbf{q}_{i_2}, \ldots, \mathbf{q}_{i_k}$ and the space of $\mathbf{q}_k$, $\mathbf{q}_{k+1}, \ldots, \mathbf{q}_M$. Thus, using Eqs. (4.60), (4.57), and (4.31), we may also express the Rayleigh quotient of the vector $\mathbf{x}$ as a weighted mean of the eigenvalues $\lambda_k, \lambda_{k+1}, \ldots, \lambda_M$, as shown by

$$\frac{\mathbf{x}^H \mathbf{R} \mathbf{x}}{\mathbf{x}^H \mathbf{x}} = \frac{\sum_{i=k}^{M} b_i^2 \lambda_i}{\sum_{i=k}^{M} b_i^2} \tag{4.61}$$

Since, by assumption, we have $\lambda_k \leq \lambda_{k+1} \leq \ldots \leq \lambda_M$, and since $\mathbf{x}$ is also a vector in the subspace $\mathcal{S}$, we may write

$$\max_{\substack{\mathbf{x} \in \mathcal{S} \\ \mathbf{x} \neq \mathbf{0}}} \frac{\mathbf{x}^H \mathbf{R} \mathbf{x}}{\mathbf{x}^H \mathbf{x}} \geq \lambda_k$$

Therefore,

$$\min_{\dim(\mathcal{S})=k} \max_{\substack{\mathbf{x} \in \mathcal{S} \\ \mathbf{x} \neq \mathbf{0}}} \frac{\mathbf{x}^H \mathbf{R} \mathbf{x}}{\mathbf{x}^H \mathbf{x}} \geq \lambda_k \tag{4.62}$$

because $\mathcal{S}$ is an arbitrary subspace of dimension k.

All that remains for us to do is to combine the results of Eqs. (4.59) and (4.62), and the minimax theorem of Eq. (4.53) describing Property 9 follows immediately.

From Property 9, we may make two important observations:

1. The minimax theorem as stated in Eq. (4.53) does not require any special knowledge of the eigenstructure (i.e., eigenvalues and eigenvectors) of the correlation matrix $\mathbf{R}$. Indeed, it may be adopted as the basis for defining the eigenvalues λ_k for $k = 1, 2, \ldots, M$.
2. The minimax theorem points to a unique two-fold feature of the eigenstructure of the correlation matrix: (a) the eigenvectors represent the particular basis for an M-dimensional space that is most efficient in the energy sense, and (b) the eigen-

values are certain energies of the M-by-1 input (observation) vector $\mathbf{u}(n)$. This issue is pursued in greater depth under Property 10.

Another noteworthy point is that Eq. (4.53) may also be formulated in the following alternative but equivalent form:

$$\lambda_k = \max_{\dim(\mathcal{S}')=M-k+1} \min_{\substack{\mathbf{x}\in\mathcal{S}' \\ \mathbf{x}\neq\mathbf{0}}} \frac{\mathbf{x}^H\mathbf{R}\mathbf{x}}{\mathbf{x}^H\mathbf{x}} \tag{4.63}$$

Equation (4.63) is referred to as the *maximin theorem*.

From Eqs. (4.53) and (4.63) we may readily deduce the following two special cases:

1. For $k = M$, the subspace $\mathcal{S}$ occupies the complex vector space $\mathbb{C}^M$ entirely. Under this condition, Eq. (4.53) reduces to

$$\lambda_M = \max_{\substack{\mathbf{x}\in\mathbb{C}^M \\ \mathbf{x}\neq\mathbf{0}}} \frac{\mathbf{x}^H\mathbf{R}\mathbf{x}}{\mathbf{x}^H\mathbf{x}} \tag{4.64}$$

where λ_M is the *largest eigenvalue* of the correlation matrix $\mathbf{R}$.

2. For $k = 1$, the subspace $\mathcal{S}'$ occupies the complex vector space $\mathbb{C}^M$ entirely. Under this condition, Eq. (4.63) reduces to

$$\lambda_1 = \min_{\substack{\mathbf{x}\in\mathbb{C}^M \\ \mathbf{x}\neq\mathbf{0}}} \frac{\mathbf{x}^H\mathbf{R}\mathbf{x}}{\mathbf{x}^H\mathbf{x}} \tag{4.65}$$

where λ_1 is the *smallest eigenvalue* of the correlation matrix $\mathbf{R}$.

Property 10. Karhunen–Loève expansion.   *Let the M-by-1 vector $\mathbf{u}(n)$ denote a data sequence drawn from a wide-sense stationary process of zero mean and correlation matrix $\mathbf{R}$. Let $\mathbf{q}_1, \mathbf{q}_2, \ldots, \mathbf{q}_M$ be eigenvectors associated with the M eigenvalues of the matrix $\mathbf{R}$. The vector $\mathbf{u}(n)$ may be expanded as a linear combination of these eigenvectors as follows:*

$$\mathbf{u}(n) = \sum_{i=1}^{M} c_i(n)\mathbf{q}_i \tag{4.66}$$

The coefficients of the expansion are zero-mean, uncorrelated random variables defined by the inner product

$$c_i(n) = \mathbf{q}_i^H\mathbf{u}(n), \qquad i = 1, 2, \ldots, M \tag{4.67}$$

The representation of the random vector $\mathbf{u}(n)$ described by Eqs. (4.66) and (4.67) is the discrete-time version of the *Karhunen–Loève expansion*. In particular, Eq. (4.67) is the "analysis" part of the expansion in that it defines the $c_i(n)$ in terms of the input vector $\mathbf{u}(n)$. On the other hand, Eq. (4.66) is the "synthesis" part of the expansion in that it reconstructs the original input vector $\mathbf{u}(n)$ from the $c_i(n)$. Given the expansion of Eq. (4.66), the definition of $c_i(n)$ in Eq. (4.67) follows directly from the fact that the eigenvectors $\mathbf{q}_1, \mathbf{q}_2, \ldots,$ $\mathbf{q}_M$ form an orthonormal set, assuming they are all normalized to have unit length. Conversely, this same property may be used to derive Eq. (4.66), given (4.67).

The coefficients of the expansion are random variables characterized as follows:

$$E[c_i(n)] = 0, \qquad i = 1, 2, \ldots, M \tag{4.68}$$

and

$$E[c_i(n)c_j^*(n)] = \begin{cases} \lambda_i & i = j \\ 0, & i \neq j \end{cases} \tag{4.69}$$

Equation (4.68) states that all the coefficients of the expansion have zero mean; this follows directly from (Eq. 4.67) and the fact the random vector $\mathbf{u}(n)$ is itself assumed to have zero mean. Equation (4.69) states that the coefficients of the expansion are uncorrelated, and that each one of them has a mean-square value equal to the respective eigenvalue. This second equation is readily obtained by using the expansion of Eq. (4.66) in the definition of the correlation matrix $\mathbf{R}$ as the expectation of the outer product $\mathbf{u}(n)\mathbf{u}^H(n)$, and then invoking the unitary similarity transformation (i.e., Property 5).

For a physical interpretation of the Karhunen–Loève expansion, we may view the eigenvectors $\mathbf{q}_1, \mathbf{q}_2, \ldots, \mathbf{q}_M$ as the coordinates of an M-dimensional space, and thus represent the random vector $\mathbf{u}(n)$ by the set of its projections $c_1(n), c_2(n), \ldots, c_M(n)$ onto these axes, respectively. Moreover, we deduce from Eq. (4.66) that

$$\sum_{i=1}^{M} |c_i(n)|^2 = \|\mathbf{u}(n)\|^2 \tag{4.70}$$

where $\|\mathbf{u}(n)\|$ is the Euclidean norm of $\mathbf{u}(n)$. That is to say, the coefficient $c_i(n)$ has an energy equal to that of the observation vector $\mathbf{u}(n)$ measured along the ith coordinate. Naturally, this energy is a random variable whose mean value equals the ith eigenvalue, as shown by

$$E[|c_i(n)|^2] = \lambda_i, \qquad i = 1, 2, \ldots, M \tag{4.71}$$

This result follows directly from Eqs. (4.67) and (4.69).

4.3 LOW-RANK MODELING

A key problem in statistical signal processing is that of *feature selection*, which refers to a process whereby a *data space* is transformed into a *feature space* that, in theory, has exactly the same dimension as the original data space. However, it would be desirable to design the transformation in such a way that the data vector can be represented by a reduced number of "effective" features and yet retain most of the intrinsic information content of the input data. In other words, the data vector undergoes a *dimensionality reduction*.

To be specific, suppose we have an M-dimensional data vector $\mathbf{u}(n)$ representing a particular realization of a wide-sense stationary process. We would like to transmit this vector over a noisy channel using a new set of p distinct numbers, where $p < M$. Basically, this is a feature-selection problem, which may be solved using the Karhunen–Loève expansion, as described next.

According to Eq. (4.66), the data vector $\mathbf{u}(n)$ may be expanded as a linear combination of the eigenvectors $\mathbf{q}_1, \mathbf{q}_2, \ldots, \mathbf{q}_M$ associated with the respective eigenvalues λ_1, $\lambda_2, \ldots, \lambda_M$ of the correlation matrix $\mathbf{R}$ of $\mathbf{u}(n)$. It is assumed that the eigenvalues are all distinct and arranged in decreasing order, as shown by

$$\lambda_1 > \lambda_2 > \ldots > \lambda_i > \ldots > \lambda_M \tag{4.72}$$

The data representation described in Eq. (4.66) using all the eigenvalues of matrix $\mathbf{R}$ is *exact* in the sense that it involves *no* loss of information. Suppose, however, that we have *prior knowledge* that the $(M - p)$ eigenvalues $\lambda_{p+1}, \ldots, \lambda_M$ at the tail end of Eq. (4.72) are all very small. We may take advantage of this prior knowledge by retaining the p largest eigenvalues of matrix $\mathbf{R}$ and thereby truncating the Karhunen–Loève expansion of Eq. (4.66) at the term $i = p$. Accordingly, we may define an *approximate reconstruction* of the data vector $\mathbf{u}(n)$ as follows:

$$\hat{\mathbf{u}}(n) = \sum_{i=1}^{p} c_i(n)\, \mathbf{q}_i, \qquad p < M \tag{4.73}$$

The vector $\hat{\mathbf{u}}(n)$ has rank p, which is lower than the rank M of the original data vector $\mathbf{u}(n)$. For this reason, the data model defined by Eq. (4.73) is referred to as a *low-rank model*. The important point to note here is that we may reconstruct the approximation $\hat{\mathbf{u}}(n)$ by using the set of p numbers: $\{c_i(n); i = 1, 2, \ldots, p\}$. The $c_i(n)$ are themselves defined in terms of the data vector $\mathbf{u}(n)$ by Eq. (4.67). In other words, the new vector $\mathbf{c}(n)$, having $c_1(n), c_2(n), \ldots, c_p(n)$ as elements, may be viewed as the reduced-rank *representation* for the original data vector $\mathbf{u}(n)$.

Figure 4.2 depicts the essence of the feature selection procedure described above. We start with an M-dimensional *data space*, in which a particular point defines the location of the data vector $\mathbf{u}(n)$. This point is transformed, via Eq. (4.67), into a new point in a *feature space* of dimension p that is lower than M. The transformation described here is sometimes referred to as a *subspace decomposition*.

Clearly, in using Eq. (4.73) to reconstruct the data vector $\mathbf{u}(n)$, an error is incurred due to the fact that $\hat{\mathbf{u}}(n)$ is of lower rank than $\mathbf{u}(n)$. The *reconstruction error vector* is defined by

$$\mathbf{e}(n) = \mathbf{u}(n) - \hat{\mathbf{u}}(n) \tag{4.74}$$

Hence, using Eqs. (4.66) and (4.73) in Eq. (4.74) yields

$$\mathbf{e}(n) = \sum_{i=p+1}^{M} c_i(n)\mathbf{q}_i \tag{4.75}$$

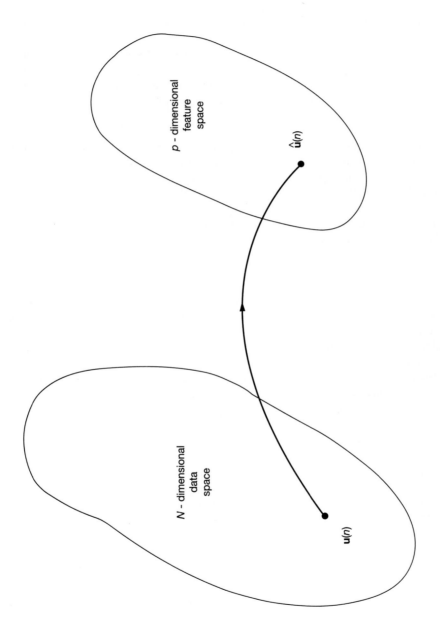

Figure 4.2   Illustrating the transformation involved in subspace decomposition.

The mean-square error is therefore

$$\epsilon = E[\|\mathbf{e}(n)\|^2]$$
$$= E[\mathbf{e}^H(n)\mathbf{e}(n)]$$
$$= E\left[\sum_{i=p+1}^{M} \sum_{j=p+1}^{M} c_i^*(n)c_j(n)\mathbf{q}_i^H\mathbf{q}_j\right] \qquad (4.76)$$
$$= \sum_{i=p+1}^{M} \sum_{j=p+1}^{M} E[c_i^*(n)c_j(n)]\mathbf{q}_i^H\mathbf{q}_j$$
$$= \sum_{i=p+1}^{M} \lambda_i$$

which confirms that the data reconstruction defined by Eq. (4.73) is a good one, provided that the eigenvalues $\lambda_{p+1}, \ldots, \lambda_M$ are all very small.

An Application of Low-rank Modeling

To appreciate the practical value of the low-rank model based on Eq. (4.73), consider the transmission of data vector $\mathbf{u}(n)$ over a *noisy communication channel*. In particular, the received signal is corrupted by *channel noise* $\mathbf{v}(n)$, which is modeled as additive white noise of zero mean.

Specifically, we have

$$E[\mathbf{u}(n)\mathbf{v}^H(n)] = \mathbf{O} \qquad (4.77)$$

and

$$E[\mathbf{v}(n)\mathbf{v}^H(n)] = \sigma^2\mathbf{I} \qquad (4.78)$$

Equation (4.77) says that the noise vector $\mathbf{v}(n)$ is uncorrelated with the data vector $\mathbf{u}(n)$. Equation (4.78), with $\mathbf{I}$ denoting the identity matrix, says that the elements of the noise vector are uncorrelated with each other and that each element has a variance of σ^2.

In Fig. 4.3 we describe two methods for accomplishing the data transmission over the channel. One method is *direct*, and the other is *indirect*, as described next.

In the direct method depicted in Fig. 4.3(a), the received signal vector is given by

$$\mathbf{y}_{\text{direct}}(n) = \mathbf{u}(n) + \mathbf{v}(n) \qquad (4.79)$$

The mean-square value of the *transmission error* is therefore

$$\epsilon_{\text{direct}} = E[\|\mathbf{y}_{\text{direct}}(n) - \mathbf{u}(n)\|^2]$$
$$= E[\|\mathbf{v}(n)\|^2]$$
$$= E[\mathbf{v}^H(n)\mathbf{v}(n)]$$

From Eq. (4.78) we see that each element $v_i(n)$, say, of the noise vector $\mathbf{v}(n)$ has variance σ^2. We may therefore express ϵ_{direct} simply as

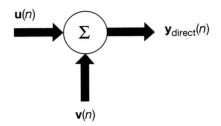

(a)

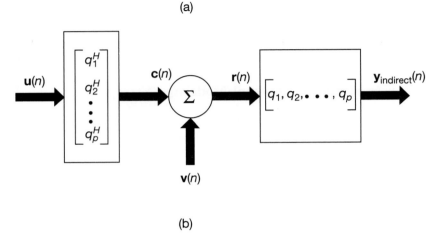

(b)

Figure 4.3 Data transmission using (a) direct method, and (b) indirect method inspired by low-rank modeling.

$$\epsilon_{\text{direct}} = \sum_{i=1}^{M} E[|v_i(n)|^2]$$
$$= M \sigma^2 \tag{4.80}$$

where M is the size of the noise vector $\mathbf{v}(n)$.

Consider next the indirect method depicted in Fig. 4.3(b), where the input vector $\mathbf{u}(n)$ is first applied to a *transmit filter bank*, whose individual tap-weight vectors are set equal to the Hermitian transpose of the eigenvectors $\mathbf{q}_1, \mathbf{q}_2, \ldots, \mathbf{q}_p$ associated with the p largest eigenvalues $\lambda_1, \lambda_2, \ldots, \lambda_p$ of the correlation matrix $\mathbf{R}$ of the input vector $\mathbf{u}(n)$. The resulting p-by-1 vector $\mathbf{c}(n)$, whose elements are made up of the inner products of $\mathbf{u}(n)$ with $\mathbf{q}_1, \mathbf{q}_2, \ldots, \mathbf{q}_p$ in accordance with Eq. (4.67), constitutes the transmitted signal vector $\mathbf{c}(n)$ as shown by

$$\mathbf{c}(n) = [\mathbf{q}_1, \mathbf{q}_2, \ldots, \mathbf{q}_p]^H \mathbf{u}(n) \tag{4.81}$$

Correspondingly, the received signal vector is defined by

$$\mathbf{r}(n) = \mathbf{c}(n) + \mathbf{v}(n) \tag{4.82}$$

where the channel noise vector $\mathbf{v}(n)$ is now of size p to be compatible with that of $\mathbf{c}(n)$. To reconstruct the original data vector, the received signal vector $\mathbf{r}(n)$ is applied to a *receive filter bank*, whose individual tap-weight vectors are defined by the eigenvectors $\mathbf{q}_1, \mathbf{q}_2, \ldots, \mathbf{q}_p$. The resulting output vector of the receiver in Fig. 4.3(b) is given by

$$
\begin{aligned}
\mathbf{y}_{\text{indirect}}(n) &= [\mathbf{q}_1, \mathbf{q}_2, \ldots, \mathbf{q}_p]\mathbf{r}(n) \\
&= [\mathbf{q}_1, \mathbf{q}_2, \ldots, \mathbf{q}_p]\mathbf{c}(n) + [\mathbf{q}_1, \mathbf{q}_2, \ldots, \mathbf{q}_p]\mathbf{v}(n)
\end{aligned}
\tag{4.83}
$$

Hence, evaluating the mean-square value of the overall reconstruction error for the indirect method, we get (see Problem 20)

$$
\begin{aligned}
\epsilon_{\text{indirect}} &= E[\|\mathbf{y}_{\text{indirect}}(n) - \mathbf{u}(n)\|^2] \\
&= \sum_{i=p+1}^{M} \lambda_i + p\sigma^2
\end{aligned}
\tag{4.84}
$$

The first term of Eq. (4.84) is due to the low-rank modeling of the data vector $\mathbf{u}(n)$ prior to transmission over the channel. The second term is due to the effect of channel noise.

Comparing Eq. (4.84) for the indirect method with Eq. (4.81) for the direct method, we readily see that the use of low-rank modeling offers an advantage, provided that we have

$$
\sum_{i=p+1}^{M} \lambda_i < (M - p)\sigma^2
\tag{4.85}
$$

This is an interesting result (Scharf and Tufts, 1987). It states that if the tail-end eigenvalues $\lambda_{p+1}, \ldots, \lambda_M$ of the correlation matrix of the data vector $\mathbf{u}(n)$ are all very small, the mean-square error produced by transmitting a low-rank approximation to the original data vector [as in Fig. 4.3(b)] is less than the mean-square error produced by transmitting the original data vector without any approximation [as in Fig. 4.3(a)].

The result described in Eq. (4.84) is particularly important in that it highlights the essence of what is commonly referred to as the "bias–variance tradeoff." Specifically, a low-rank model is used for representing the data vector $\mathbf{u}(n)$, thereby incurring a *bias*. Interestingly enough, this is done knowingly, in return for a reduction in *variance*, namely, the part of the mean-square error due to the additive noise vector $v(n)$. Indeed, the example described herein clearly illustrates the motivation for using a parsimonious (i.e., simpler) model that may not exactly match the underlying physics responsible for generating the data vector $\mathbf{u}(n)$, hence the bias; but the model is less susceptible to noise, hence a reduction in variance.

4.4 EIGENFILTERS

A fundamental issue in communication theory is that of determining an optimum finite (duration) impulse response (FIR) filter, with the optimization criterion being that of maximizing the output signal-to-noise ratio. In this section we show that this filter optimization is linked to an eigenvalue problem.

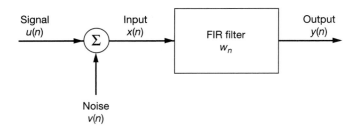

Figure 4.4   Linear filtering.

Consider a linear FIR filter whose impulse response is denoted by the sequence w_n. The sequence $x(n)$ applied to the filter input consists of a useful *signal* component $u(n)$ plus an additive *noise* component $v(n)$. The signal $u(n)$ is drawn from a wide-sense stationary stochastic process of zero mean and correlation matrix $\mathbf{R}$. The zero-mean noise $v(n)$ is white with a constant power spectral density determined by the variance σ^2. It is assumed that the signal $u(n)$ and the noise $v(n)$ are uncorrelated; that is,

$$E[u(n)v^*(m)] = 0 \qquad \text{for all } (n, m)$$

The filter output is denoted by $y(n)$. The situation described herein is depicted in Fig. 4.4.

Since the filter is linear, the principle of superposition applies. We may therefore consider the effects of signal and noise separately. Let P_o denote the average power of the signal component of the filter output $y(n)$. We may therefore show that (see Problem 9 of Chapter 2):

$$P_o = \mathbf{w}^H \mathbf{R} \mathbf{w} \tag{4.86}$$

where the elements of the vector $\mathbf{w}$ are the filter coefficients, and $\mathbf{R}$ is the correlation matrix of the signal component $u(n)$ in the filter input.

Consider next the effect of noise acting alone. Let N_o denote the average power of the noise component in the filter output $y(n)$. This is a special case of Eq. (4.86), as shown by

$$N_o = \sigma^2 \mathbf{w}^H \mathbf{w} \tag{4.87}$$

where σ^2 is the variance of the white noise in the filter input.

Let $(\text{SNR})_o$ denote the *output signal-to-noise ratio*. Dividing Eq. (4.86) by (4.87), we may thus write

$$(\text{SNR})_o = \frac{P_o}{N_o}$$

$$= \frac{\mathbf{w}^H \mathbf{R} \mathbf{w}}{\sigma^2 \mathbf{w}^H \mathbf{w}} \tag{4.88}$$

The optimum problem may now be stated as follows:

Determine the coefficient vector $\mathbf{w}$ of an FIR filter so as to maximize the output signal-to-noise ratio $(SNR)_o$ subject to the constraint

$$\mathbf{w}^H \mathbf{w} = 1$$

Equation (4.88) shows that except for the scaling factor $1/\sigma^2$, the output signal-to-noise ratio $(SNR)_o$ is equal to the Rayleigh quotient of the coefficient vector $\mathbf{w}$ of the FIR filter. We see therefore that the optimum filtering problem, as stated herein, may be viewed as an eigenvalue problem. Indeed, the solution to the problem follows directly from the minimax theorem. Specifically, using the special form of the minimax theorem given in Eq. (4.64), we may state the following:

- The maximum value of the output signal-to-noise ratio is given by

$$(SNR)_{o,max} = \frac{\lambda_{max}}{\sigma^2} \qquad (4.89)$$

  where λ_{max} is the largest eigenvalue of the correlation matrix $\mathbf{R}$. Note that λ_{max} and σ^2 have the same units but different physical interpretations.
- The coefficient vector of the optimum FIR filter that yields the maximum output signal-to-noise ratio of Eq. (4.89) is defined by

$$\mathbf{w}_o = \mathbf{q}_{max} \qquad (4.90)$$

  where $\mathbf{q}_{max}$ is the eigenvector associated with the largest eigenvalue λ_{max} of the correlation matrix $\mathbf{R}$. The correlation matrix $\mathbf{R}$ belongs to the signal component $u(n)$ in the filter input.

An FIR filter whose impulse response has coefficients equal to the elements of an eigenvector is called an *eigenfilter* (Makhoul, 1981). Accordingly, we may state that *the maximum eigenfilter (i.e., the eigenfilter associated with the largest eigenvalue of the correlation matrix of the signal component in the filter input) is the optimum filter.* It is important to note that the optimum filter described in this way is uniquely characterized by an eigendecomposition of the correlation matrix of the signal component in the filter input. The power spectrum of the white noise at the filter input merely affects the maximum value of the output signal-to-noise ratio. In particular, we may proceed as follows:

1. An eigendecomposition of the correlation matrix $\mathbf{R}$ is performed.
2. *Only the largest eigenvalue λ_{max} and the associated eigenvector $\mathbf{q}_{max}$ are* retained.
3. The eigenvector $\mathbf{q}_{max}$ defines the impulse response of the optimum filter. The eigenvalue λ_{max}, divided by the noise variance σ^2, defines the maximum value of the output signal-to-noise ratio.

The optimum filter so characterized may be viewed as the "stochastic" counterpart of a matched filter. The optimum filter described herein maximizes the output signal-to-noise ratio for a *random signal* (i.e., a sample function of a discrete-time wide-sense stationary stochastic process) in additive white noise. A *matched filter*, on the other hand, maximizes the output signal-to-noise ratio for a *known signal* in additive white noise (North, 1963; Haykin, 1994).

4.5 EIGENVALUE COMPUTATIONS

The computation of the eigenvalues of a square matrix can, in general, be a complicated issue. Special and cooperative efforts by a group of experts between 1958 and 1970 resulted in the development of several *canned routines* that are widely available for matrix eigenvalue computations (Parlett, 1985). Special mention should be made of the following program libraries:

- MATLAB: a matrix-based numerical system for interactive computation, visualization, modeling, and algorithm development (Riddle, 1994)
- MATHEMATICA: an integrated mathematical system for numerical, symbolic, and graphical computation and visualization (Riddle, 1994)
- LINPACK: standard subroutine packages for computational linear algebra (Dongarra et al., 1979)
- LAPACK: a linear algebra library for single-address space machines
- ScaLAPACK: a linear algebra library for multiple-address space machines (Demmel, 1994)

The canned *eigen-routines* in these libraries are well documented and well tested.

The origin of almost all canned eigen-routines may be traced back to routines published in Volume II, *Linear Algebra*, of the *Handbook for Automatic Computation* co-edited by Wilkinson and Reinsch (1971). This reference is the bible of eigenvalue computations.

Another useful source of routines, written in the *C programming language*, is the book by Press et al. (1988); a companion book by the same authors, with routines written in FORTRAN and Pascal, is also available. The eigen-routines written in C can only handle real matrices. It is, however, a straightforward matter to extend the use of these eigen-routines to deal with Hermitian matrices, as shown next.

Let $\mathbf{A}$ denote an M-by-M Hermitian matrix, written in terms of its real and imaginary parts as follows:

$$\mathbf{A} = \mathbf{A}_r + j\mathbf{A}_i \tag{4.91}$$

Correspondingly, let an associated M-by-1 eigenvector $\mathbf{q}$ be written as

$$\mathbf{q} = \mathbf{q}_r + j\mathbf{q}_i \tag{4.92}$$

The M-by-M complex eigenvalue problem

$$(\mathbf{A}_r + j\mathbf{A}_i)(\mathbf{q}_r + j\mathbf{q}_i) = \lambda(\mathbf{q}_r + j\mathbf{q}_i) \tag{4.93}$$

may then be reformulated as the $2M$-by-$2M$ real eigenvalue problem:

$$\begin{bmatrix} \mathbf{A}_r & -\mathbf{A}_i \\ \mathbf{A}_i & \mathbf{A}_r \end{bmatrix} \begin{bmatrix} \mathbf{q}_r \\ \mathbf{q}_i \end{bmatrix} = \lambda \begin{bmatrix} \mathbf{q}_r \\ \mathbf{q}_i \end{bmatrix} \tag{4.94}$$

where the eigenvalue λ is a real number. The Hermitian property

$$\mathbf{A}^H = \mathbf{A}$$

is equivalent to $\mathbf{A}^T_r = \mathbf{A}_r$ and $\mathbf{A}^T_i = -\mathbf{A}_i$. Accordingly, the $2M$-by-$2M$ matrix in Eq. (4.94) is not only real but also symmetric. Note, however, that for a given eigenvalue λ, the vector

$$\begin{bmatrix} -\mathbf{q}_i \\ \mathbf{q}_r \end{bmatrix}$$

is also an eigenvector. This means that if $\lambda_1, \lambda_2, \ldots, \lambda_M$ are the eigenvalues of the M-by-M Hermitian matrix $\mathbf{A}$, then the eigenvalues of the $2M$-by-$2M$ symmetric matrix of Eq. (4.94) are $\lambda_1, \lambda_1, \lambda_2, \lambda_2, \ldots, \lambda_M, \lambda_M$. We may therefore make two observations.

1. Each eigenvalue of the matrix in Eq. (4.94) has a multiplicity of 2.
2. The associated eigenvectors consist of pairs, each of the form $\mathbf{q}_r + j\mathbf{q}_i$ and $j(\mathbf{q}_r + j\mathbf{q}_i)$, differing merely by a rotation through $90°$.

Thus, to solve the M-by-M complex eigenvalue problem of Eq. (4.93) with the aid of real eigen-routines, we choose one eigenvalue and eigenvector from each pair associated with the augmented $2M$-by-$2M$ real eigenvalue problem of Eq. (4.94).

Strategies for Matrix Eigenvalue Computations

There are two different "strategies" behind practically all modern eigen-routines: *diagonalization* and *triangularization*. Since not all matrices can be diagonalized through a sequence of unitary similarity transformations, the diagonalization strategy applies only to Hermitian matrices such as a correlation matrix. On the other hand, the triangularization strategy is general in that it applies to any square matrix. These two strategies are described in the following sections.

Diagonalization. The idea behind this strategy is to nudge a Hermitian matrix $\mathbf{A}$ toward a diagonal form by the repeated application of unitary similarity transformations, as described here:

$$\begin{aligned} \mathbf{A} &\rightarrow \mathbf{Q}^H_1 \mathbf{A} \mathbf{Q}_1 \\ &\rightarrow \mathbf{Q}^H_2 \mathbf{Q}^H_1 \mathbf{A} \mathbf{Q}_1 \mathbf{Q}_2 \\ &\rightarrow \mathbf{Q}^H_3 \mathbf{Q}^H_2 \mathbf{Q}^H_1 \mathbf{A} \mathbf{Q}_1 \mathbf{Q}_2 \mathbf{Q}_3 \end{aligned} \tag{4.95}$$

and so on. This sequence of unitary similarity transformation is, in theory, infinitely long. In practice, however, it is continued until we are close to a diagonal matrix. The elements of the diagonal matrix so obtained define the eigenvalues of the original Hermitian matrix **A**. The associated eigenvectors are the column vectors of the accumulated sequence of transformations, as shown by

$$\mathbf{Q} = \mathbf{Q}_1 \mathbf{Q}_2 \mathbf{Q}_3 \cdots \tag{4.96}$$

One method for implementing the diagonalization strategy of Eq. (4.95) is to use *Givens rotations*. This method is discussed in Chapter 12.

Triangularization. The idea behind this second strategy is to reduce a Hermitian matrix **A** to a triangular form by a sequence of unitary similarity transformations. The resulting iterative procedure is called the *QL algorithm*.[4] Suppose that we are given an *M*-by-*M* Hermitian matrix $\mathbf{A}_n$, where the subscript n refers to a particular step in the iterative procedure. Let the matrix $\mathbf{A}_n$ be *factored* in the form

$$\mathbf{A}_n = \mathbf{Q}_n \mathbf{L}_n \tag{4.97}$$

where $\mathbf{Q}_n$ is a *unitary matrix* and $\mathbf{L}_n$ is a *lower triangular matrix* (i.e., the elements of the matrix $\mathbf{L}_n$ located above the main diagonal are all zero). At step $n + 1$ in the iterative procedure, we use the known matrices $\mathbf{Q}_n$ and $\mathbf{L}_n$ to compute a new *M*-by-*M* matrix

$$\mathbf{A}_{n+1} = \mathbf{L}_n \mathbf{Q}_n \tag{4.98}$$

Note that the factorization in Eq. (4.98) is written in the opposite order to that in Eq. (4.97). Since $\mathbf{Q}_n$ is a unitary matrix, we have $\mathbf{Q}_n^{-1} = \mathbf{Q}_n^H$, so we may rewrite Eq. (4.97) as

$$\mathbf{L}_n = \mathbf{Q}_n^{-1} \mathbf{A}_n$$
$$= \mathbf{Q}_n^H \mathbf{A}_n \tag{4.99}$$

Therefore, substituting Eq. (4.99) into (4.98), we get

$$\mathbf{A}_{n+1} = \mathbf{Q}_n^H \mathbf{A}_n \mathbf{Q}_n \tag{4.100}$$

Equation (4.100) shows that the Hermitian matrix $\mathbf{A}_{n+1}$ at iteration $n + 1$ is indeed unitarily related to the Hermitian matrix $\mathbf{A}_n$ at iteration n.

The QL algorithm thus consists of a sequence of unitary similarity transformations, summarized by writing

$$\mathbf{A}_n = \mathbf{Q}_n \mathbf{L}_n$$
$$\mathbf{A}_{n+1} = \mathbf{L}_n \mathbf{Q}_n$$

---

[4] The QL algorithm uses a lower triangular matrix. There is a companion algorithm, called the QR algorithm, which uses an upper triangular matrix. The QR algorithm is not to be confused with the QR-decomposition; the latter is discussed in Chapter 14.

where $n = 0, 1, 2, \ldots$. The algorithm is *initialized* by setting

$$\mathbf{A}_0 = \mathbf{A}$$

where $\mathbf{A}$ is the given M-by-M Hermitian matrix.

For general matrix $\mathbf{A}$, the following theorem is the basis of the QL algorithm[5]:

If matrix $\mathbf{A}$ has eigenvalues of different absolute values, then the matrix $\mathbf{A}_n$ approaches a lower triangular form as the number of iterations n approaches infinity.

The eigenvalues of the original matrix $\mathbf{A}$ appear on the main diagonal of the lower triangular matrix resulting from the QL algorithm in increasing order of absolute value.

To implement the factorization in Eq. (4.98), we may use Givens rotations. Here again, however, we defer a discussion of the Givens rotation to Chapter 12. In that chapter we discuss computations for the singular value decomposition of a general matrix, which includes eigendecomposition as a special case.

4.6 SUMMARY AND DISCUSSION

In this chapter we studied the decomposition of the ensemble-averaged correlation matrix of a discrete-time wide-sense stationary stochastic process in terms of its eigenvalues and associated eigenvectors. Eigendecomposition provides an invaluable bridge between matrix algebra and stochastic processes, thereby placing it at the forefront of discrete-time linear filter theory.

Building on the properties of a discrete-time wide-sense stationary stochastic process described in Chapters 2 and 3, we established the following properties:

- The eigenvalues of the correlation matrix of the process are always nonnegative and bounded by the maximum and minimum values of the power spectral density of the process.
- The associated eigenvectors form an orthonormal set.

Another important result that we established is the Karhunen–Loève expansion, according to which a data vector (drawn from a wide-sense stationary stochastic process) may be expanded as a linear combination of the eigenvectors pertaining to the correlation matrix of the process. This important result provides the theoretical basis for the design of a low-rank model of the data vector, which means that the dimensionality of the data vector may be reduced without sacrificing the intrinsic information content of the data.

---

[5]For a proof of this theorem, see Stoer and Bulirsch (1980). See also Stewart (1973), Golub and Van Loan (1989), and Press et al. (1992) for an improved version of the QL algorithm.

The final result established in the chapter is the notion of a maximum eigenfilter, defined by the eigenvector associated with the largest eigenvalue of the correlation matrix of a wide-sense stationary stochastic process. This filter optimizes the detection of a random signal, representing a particular realization of the process, embedded in a white-noise background.

PROBLEMS

1. The correlation matrix $\mathbf{R}$ of a wide-sense stationary process $u(n)$ has the following values for its two eigenvalues:

$$\lambda_1 = 0.5$$

$$\lambda_2 = 1.5$$

   (a) Find the trace of matrix $\mathbf{R}$.
   (b) Write an expression for the decomposition of matrix $\mathbf{R}$ in terms of its two eigenvalues and associated eigenvectors. Comment on the uniqueness of this decomposition.

2. Show that the eigenvalues of a triangular matrix equal the diagonal elements of the matrix.

3. Consider the $2M$-by-$2M$ real eigenvalue problem described in Eq. (4.94). Show that if $\mathbf{q}_r + j\mathbf{q}_i$ is an eigenvector of the matrix described herein, so is $\mathbf{q}_i - j\mathbf{q}_r$, with both eigenvectors being associated with the same eigenvalue.

4. Let $\lambda_1, \lambda_2, \ldots, \lambda_M$ denote the eigenvalues of the correlation matrix of an observation vector $\mathbf{u}(n)$ taken from a stationary process of zero mean and variance σ_u^2. Show that

$$\sum_{i=1}^{M} \lambda_i = M\sigma_u^2$$

5. An M-by-M correlation matrix $\mathbf{R}$ is represented in terms of its eigenvalues $\lambda_1, \lambda_2, \ldots, \lambda_M$ and their associated eigenvectors $\mathbf{q}_1, \mathbf{q}_2, \ldots, \mathbf{q}_M$ as follows:

$$\mathbf{R} = \sum_{i=1}^{M} \lambda_i \mathbf{q}_i \mathbf{q}_i^H$$

   (a) Show that the corresponding representation for the *square root* of matrix $\mathbf{R}$ is

$$\mathbf{R}^{1/2} = \sum_{i=1}^{M} \lambda_i^{1/2} \mathbf{q}_i \mathbf{q}_i^H$$

   (b) By definition, we have $\mathbf{R} = \mathbf{R}^{1/2} \mathbf{R}^{1/2}$. Using this result, describe a procedure for computing the square root of a square matrix.

6. Consider a stationary process $u(n)$ whose M-by-M correlation matrix equals $\mathbf{R}$. Show that the determinant of the correlation matrix $\mathbf{R}$ equals

$$\det(\mathbf{R}) = \prod_{i=1}^{M} \lambda_i$$

7. (a) Show that the product of two unitary matrices is also a unitary matrix.
   (b) Show that the inverse of a unitary matrix is also a unitary matrix.

8. Let $\mathbf{A}$ be an M-by-M matrix. The *Schur decomposition theorem* states there exists a unitary matrix $\mathbf{Z}$ such that

$$\mathbf{Z}^H\mathbf{A}\mathbf{Z} = \mathbf{T}$$

where $\mathbf{T}$ is an upper triangular matrix. The theorem also states that:
   (i) The diagonal of matrix $\mathbf{T}$ is made up of the eigenvalues of the matrix $\mathbf{A}$.
   (ii) If $\mathbf{Z} = [\mathbf{z}_1, \mathbf{z}_2, \ldots, \mathbf{z}_M]$, then span $(\mathbf{z}_1, \mathbf{z}_2, \ldots, \mathbf{z}_k)$ is an invariant subspace associated with the eigenvalues $t_{11}, t_{22}, \ldots, t_{kk}$ where $k \leq M$.
   (a) Apply the Schur decomposition to the correlation matrix $\mathbf{R}$ of a wide-stationary stochastic process. Hence, show that in this case the matrix $\mathbf{T}$ is a diagonal matrix.
   (b) What is the implication of the statement under (ii) in the context of the correlation matrix $\mathbf{R}$?

9. Consider the factorization

$$\mathbf{A}_n - k_n\mathbf{I} = \mathbf{Q}_n\mathbf{L}_n$$

where $\mathbf{A}_n$ is an M-by-M Hermitian matrix, $\mathbf{I}$ is the M-by-M identity matrix, $\mathbf{Q}_n$ is an M-by-M unitary matrix, $\mathbf{L}_n$ is an M-by-M lower triangular matrix, and k_n is a scalar. Define the matrix

$$\mathbf{A}_{n+1} = \mathbf{L}_n\mathbf{Q}_n + k_n\mathbf{I}$$

Hence, show that

$$\mathbf{A}_{n+1} = \mathbf{Q}_n^H\mathbf{A}_n\mathbf{Q}_n$$

10. In this problem we consider a *Fourier analyzer* for a single channel. The *Fourier basis* is described by

$$\mathbf{v}_i = \frac{1}{\sqrt{M}} \, [1, e^{j2\pi i/M}, e^{j4\pi i/M}, \ldots, e^{j(M-1)2\pi i/M}]^T$$

where $i = 0, 1, \ldots, M - 1$. Let an arbitrary M-by-1 vector $\mathbf{u}(n)$ be expanded in terms of this orthonormal set as follows:

$$\mathbf{u}(n) = \sum_{i=0}^{M-1} c_i(n)\mathbf{v}_i$$

   (a) Evaluate the *Fourier coefficients* $c_0(n), c_1(n), \ldots, c_{M-1}(n)$ in terms of the vector $\mathbf{u}(n)$.
   (b) Are the Fourier coefficients correlated? Justify your answer.
   (c) What does the expectation of $|c_i(n)|^2$ approximate?

11. Show that the condition number of matrix $\mathbf{A}$ is unchanged when this matrix is multiplied by a unitary matrix of compatible dimensions.

12. Consider an L-by-M matrix $\mathbf{A}$. Show that the M-by-M matrix $\mathbf{A}^H\mathbf{A}$ and the L-by-L matrix $\mathbf{A}\mathbf{A}^H$ have the same nonzero eigenvalues.

13. A stochastic process $v(n)$ with a wide-band power spectrum is applied to a discrete-time linear filter whose amplitude response $|H(e^{j\omega})|$ is nonuniform. The maximum and minimum values of this response are denoted by $H_{\max}$ and $H_{\min}$, respectively. Let $\chi(\mathbf{R})$ denote the eigenvalue spread of the correlation matrix $\mathbf{R}$ of the stochastic process $u(n)$ produced at the output of the filter. Show that

$$\chi(\mathbf{R}) \simeq \left(\frac{H_{\max}}{H_{\min}}\right)^2$$

14. *Szegö's theorem* states that if $g(\cdot)$ is a continuous function, then

$$\lim_{M \to \infty} \frac{g(\lambda_1) + g(\lambda_2) + \ldots + g(\lambda_M)}{M} = \frac{1}{2\pi} \int_{-\pi}^{\pi} g[S(\omega)]d\omega$$

where $S(\omega)$ is the power spectral density of a stationary discrete-time stochastic process $u(n)$, and $\lambda_1, \lambda_2, \ldots, \lambda_M$ are the eigenvalues of the associated correlation matrix $\mathbf{R}$. It is assumed that the process $u(n)$ is limited to the interval $-\pi < \omega \leq \pi$. Using this theorem, show that

$$\lim_{M \to \infty} [\det(\mathbf{R})]^{1/M} = \exp\left(\frac{1}{2\pi} \int_{-\pi}^{\pi} \ln [S(\omega)]d\omega\right)$$

15. Consider a linear system of equations described by

$$\mathbf{R}\mathbf{w}_o = \mathbf{p}$$

where $\mathbf{R}$ is an M-by-M matrix, and $\mathbf{w}_o$ and $\mathbf{p}$ are M-by-1 vectors. The vector $\mathbf{w}_o$ represents the set of unknown parameters. Due to a combination of factors (e.g., measurement inaccuracies, computational errors), the matrix $\mathbf{R}$ is perturbed by a small amount $\delta\mathbf{R}$, producing a corresponding change $\delta\mathbf{w}$ in the vector of unknowns.

(a) Show that

$$\frac{\|\delta\mathbf{w}\|}{\|\mathbf{w}_o\|} \leq \chi(\mathbf{R}) \frac{\|\delta\mathbf{R}\|}{\|\mathbf{R}\|}$$

where $\chi(\mathbf{R})$ is the condition number of $\mathbf{R}$, and $\|\cdot\|$ denotes the norm of the quantity enclosed within.

(b) Develop the corresponding formula for a small change in the vector $\mathbf{p}$. *Hint:* Use the inequality

$$\|\mathbf{A}\mathbf{x}\| \leq \|\mathbf{A}\| \|\mathbf{x}\|$$

16. Consider the three-dimensional vector space of Fig. 4.1. Let the subspace $\mathscr{S}$ denote the $\mathbf{q}_{i1}$, $\mathbf{q}_{i2}$-plane where i_1, i_2 is a subset of $\{1, 2, 3\}$. Let the subspace $\mathscr{S}'$ denote the $\mathbf{q}_2$, $\mathbf{q}_3$-plane.

(a) Specify a vector $\mathbf{x}$ of unit length that is common to the subspaces $\mathscr{S}$ and $\mathscr{S}'$.

(b) What is the Rayleigh coefficient of the vector $\mathbf{x}$ specified in part (a)? Justify your answer in light of the minimax theorem.

17. Consider an M-by-M doubly symmetric matrix $\mathbf{R}$ that is symmetric about both the main diagonal and the secondary diagonal. Let $\mathbf{J}$ denote an M-by-M matrix that consists of 1's along the secondary diagonal and zeros everywhere else. The matrix $\mathbf{J}$ is called a *reverse operator* or *exchange matrix* because $\mathbf{J}\mathbf{R}$ reverses the rows of matrix $\mathbf{R}$, $\mathbf{R}\mathbf{J}$ reverses the columns of $\mathbf{R}$, and $\mathbf{J}\mathbf{R}\mathbf{J}$ reverses both the rows and columns of $\mathbf{R}$.

(a) Show that for the matrix $\mathbf{R}$ to be doubly symmetric, a necessary and sufficient condition is

$$\mathbf{J}\mathbf{R}\mathbf{J} = \mathbf{R}$$

Noting that $\mathbf{J}^{-1} = \mathbf{J}$, show that the inverse of matrix $\mathbf{R}$ is also doubly symmetric, as shown by

$$\mathbf{J}\mathbf{R}^{-1}\mathbf{J} = \mathbf{R}^{-1}$$

(b) Assume that the doubly symmetric matrix **R** has distinct eigenvalues. Hence show that the matrix **R** has $\lfloor (M + 1)/2 \rfloor$ symmetric eigenvectors and $\lfloor M/2 \rfloor$ skew-symmetric eigenvectors, where $\lfloor X \rfloor$ denotes the largest integer less than or equal to X. An eigenvector **q** is said to be *symmetric* if

$$\mathbf{Jq} = \mathbf{q}$$

and *skew symmetric* if

$$\mathbf{Jq} = -\mathbf{q}$$

where **J** is the reverse operator.

(c) Let $A(z)$ denote the transfer function of an eigenfilter of the doubly symmetric matrix **R**. Show that if $A(z)$ is associated with a symmetric eigenvector or skew-symmetric eigenvector of **R** and if z_i is a zero of $A(z)$, then so is $1/z_i$.

[The properties described in this problem and the next three are taken from Makhoul (1981) and Reddi (1984).]

18. Let **R** denote an M-by-M nonsingular symmetric Toeplitz matrix. Naturally, the properties described in Problem 17 apply to the matrix **R**. Note, however, that the inverse matrix $\mathbf{R}^{-1}$ is not Toeplitz, in general. But owing to the special structure of a Toeplitz matrix, the matrix **R** has two additional properties:

(a) Let $\lambda_{\max}$ denote the largest eigenvalue of the matrix **R**, which is assumed to be distinct. Show that the discrete transfer function of the eigenfilter associated with $\lambda_{\max}$ has all of its zeros located on the unit circle in the z-plane.

(b) Let $\lambda_{\min}$ denote the smallest eigenvalue of the matrix **R**, which is assumed to be distinct. Show that the discrete transfer function of the eigenfilter associated with $\lambda_{\min}$ has all of its zeros located on the unit circle in the z-plane.

19. Consider the *normalized* 3-by-3 correlation matrix

$$\mathbf{R} = \begin{bmatrix} 1 & \rho_1 & \rho_2 \\ \rho_1 & 1 & \rho_1 \\ \rho_2 & \rho_1 & 1 \end{bmatrix}$$

where

$$\rho_i = \frac{r(i)}{r(0)}, \qquad i = 1, 2$$

(a) Using properties (b) and (c) of Problem 17, demonstrate the following results:

(1) The matrix **R** has a single skew-symmetric eigenvector of the form

$$\mathbf{q}_1 = \frac{1}{\sqrt{2}} [1, 0, -1]^T$$

that is associated with the eigenvalue

$$\lambda_1 = 1 - \rho_2$$

(2) The matrix **R** has two symmetric eigenvectors of the form

$$\mathbf{q}_i = \frac{1}{\sqrt{1 + c_i^2}} [1, c_i, 1]^T, \qquad i = 2, 3$$

where c_i is related to the corresponding eigenvalue λ_i by

$$c_i = \frac{2\rho_1}{\lambda_i - 1} = \frac{\lambda_i - 1 - \rho_2}{\rho_1}, \qquad i = 2, 3$$

Hence, complete the specification of the eigenvalues and the eigenvectors of the matrix $\mathbf{R}$.

(b) Given that the eigenvalues of matrix $\mathbf{R}$ are distinct and ordered as $\lambda_1 > \lambda_2 > \lambda_3$, and given that the eigenfilters associated with λ_1 and λ_3 have their zeros on the unit circle in accordance with properties (a) and (b) of Problem 18, respectively, find the condition that the coefficient c_2 must satisfy for the following two situations to occur:

(1) The eigenfilter associated with eigenvalue λ_2 will *also* have its zeros on the unit circle.

(2) The eigenfilter associated with eigenvalue λ_2 will *not* have its zeros on the unit circle.

Illustrate both of these situations with selected values for the correlation coefficients ρ_2 and ρ_3.

20. In this problem, we wish to establish the result given in Eq. (4.84) for the mean-square error produced by the transmission system of Fig. 4.3(b), using a low-rank model of the input data $\mathbf{u}(n)$.

The transmitted signal vector $\mathbf{c}(n)$ and the reconstructed signal vector $\mathbf{y}_{\text{direct}}(n)$ at the receiver output are defined by Eqs. (4.81) and (4.83), respectively. Derive Eq. (4.84) using the following properties:

(a) The eigenvectors $\mathbf{q}_1, \mathbf{q}_2, \ldots, \mathbf{q}_p$ associated with the eigenvalues $\lambda_1, \lambda_2, \ldots, \lambda_p$ of the correlation matrix $\mathbf{R}$ of the input vector $\mathbf{u}(n)$ form an orthonormal set.

(b) The data vector $\mathbf{u}(n)$ and the noise vector $\mathbf{v}(n)$ are uncorrelated.

(c) The elements of the noise vector $\mathbf{v}(n)$ are drawn from a white noise process of zero mean and variance σ^2.

21. To solve the optimum filtering problem described in Section 4.5, we selected an eigenfilter associated with the largest eigenvalue of the correlation matrix of the signal component at the filter input. What would be the result of selecting an eigenfilter associated with the smallest eigenvalue of this correlation matrix? Justify your answer.

PART 2
Linear Optimum Filtering

Part II of the book consists of Chapters 5 through 7. It is devoted to a detailed treatment of linear optimum filter theory for discrete-time wide-sense stationary stochastic processes. Adaptive filters are derived from this theory. Chapter 5 covers the classical Wiener filter. Chapter 6 builds on the Wiener filter theory to solve the linear prediction problem. Chapter 7 covers the classical Kalman filter for solving the optimum filtering problem (formulated in terms of a state vector) in a recursive manner.

CHAPTER

5

Wiener Filters

This chapter deals with a class of *linear optimum discrete-time filters* known collectively as *Wiener filters*. The theory for a Wiener filter is formulated for the general case of *complex-valued* time series with the filter specified in terms of its impulse response. The reason for using complex-valued time series is that in many practical situations (e.g., communications, radar, sonar) the *baseband* signal of interest appears in complex form; the term *baseband* is used to designate a band of frequencies representing the original signal as delivered by a source of information. The case of real-valued time series may of course be considered as a special case of this theory.

We begin our study of Wiener filters by outlining the linear optimum filtering problem and setting the stage for the rest of the chapter.

5.1 LINEAR OPTIMUM FILTERING: PROBLEM STATEMENT

Consider the block diagram of Fig. 5.1 built around a *linear discrete-time filter*. The filter *input* consists of a *time series* $u(0)$, $u(1)$, $u(2)$, . . . , and the filter is itself characterized by the *impulse response* w_0, w_1, w_2, ... At some *discrete time n*, the filter produces an *output* denoted by $y(n)$. This output is used to provide an *estimate* of a *desired response* denoted by $d(n)$. With the filter input and the desired response representing single realizations of respective stochastic processes, the estimation is accompanied by an error with

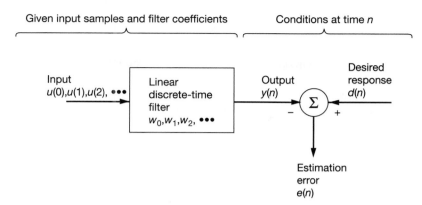

Figure 5.1   Block diagram representation of the statistical filtering problem.

statistical characteristics of its own. In particular, the *estimation error*, denoted by $e(n)$, is defined as the difference between the desired response $d(n)$ and the filter output $y(n)$. The requirement is to make the estimation error $e(n)$ "as small as possible" in some statistical sense.

Two restrictions have so far been placed on the filter:

1. The filter is *linear*, which makes the mathematical analysis easy to handle.
2. The filter operates in *discrete time*, which makes it possible for the filter to be implemented using digital hardware/software.

The final details of the filter specification, however, depend on two other choices that have to be made:

1. Whether the impulse response of the filter has *finite* or *infinite* duration.
2. The type of *statistical criterion* used for the optimization.

The choice of a *finite-duration impulse response* (FIR) or an *infinite-duration impulse response* (IIR) for the filter is dictated by *practical considerations*. The choice of a statistical criterion for optimizing the filter design is influenced by *mathematical tractability*. These two issues are considered in turn.

For the initial development of the Wiener filter theory, we will assume an IIR filter; the theory so developed includes that for FIR filters as a special case. However, for much of the material presented in this chapter, and indeed in the rest of the book, we will confine our attention to the use of FIR filters. We do so for the following reason. An FIR filter is *inherently stable*, because its structure involves the use of *forward paths only*. In other words, the only mechanism for input–output interaction in the filter is via forward paths from the filter input to its output. Indeed, it is this form of signal transmission through the filter that limits its impulse response to a finite duration. On the other hand, an IIR filter involves *both feedforward and feedback*. The presence of feedback means that portions of the filter output and possibly other *internal* variables in the filter are fed back to the input. Consequently, unless it is properly designed, feedback in the filter can indeed make it *unstable* with the result that the filter *oscillates*; this kind of operation is clearly unacceptable when the requirement is that of filtering for which stability is a "must." By itself, the stability problem in IIR filters is manageable in both theoretical and practical terms. However, when the filter is required to be *adaptive*, bringing with it stability problems of its own, the inclusion of adaptivity combined with feedback that is inherently present in an IIR filter makes a difficult problem that much more difficult to handle. It is for this reason that we find that in the majority of applications requiring the use of adaptivity, the use of an FIR filter is preferred over an IIR filter even though the latter is less demanding in computational requirements.

Turning next to the issue of what criterion to choose for statistical optimization, there are indeed several criteria that suggest themselves. Specifically, we may consider optimizing the filter design by *minimizing a cost function*, or *index of performance*, selected from the following short list of possibilities:

1. Mean-square value of the estimation error
2. Expectation of the absolute value of the estimation error
3. Expectation of third or higher powers of the absolute value of the estimation error

Option 1 has a clear advantage over the other two, because it leads to tractable mathematics. In particular, the choice of *the mean-square error criterion results in a second-order dependence for the cost function on the unknown coefficients in the impulse response of the filter. Moreover, the cost function has a distinct minimum that uniquely defines the optimum statistical design of the filter.*

We may now summarize the essence of the filtering problem by making the following statement:

Design a linear discrete-time filter whose output $y(n)$ provides an estimate of a desired response $d(n)$, given a set of input samples $u(0)$, $u(1)$, $u(2)$, . . . , such that the mean-square value of the estimation error $e(n)$, defined as the difference between the desired response $d(n)$ and the actual response $y(n)$, is minimized.

We may develop the mathematical solution to this statistical optimization problem by following two entirely different approaches that are complementary. One approach leads to the development of an important theorem commonly known as the principle of orthogonality. The other approach highlights the error-performance surface that describes the second-order dependence of the cost function on the filter coefficients. We will proceed by deriving the principle of orthogonality first, because the derivation is relatively simple and because the principle of orthogonality is highly insightful.

5.2 PRINCIPLE OF ORTHOGONALITY

Consider again the statistical filtering problem described in Fig. 5.1. The filter input is denoted by the time series $u(0)$, $u(1)$, $u(2)$, . . . , and the impulse response of the filter is denoted by w_0, w_1, w_2, . . . , both of which are assumed to have *complex values* and *infinite duration*. The filter output $y(n)$ at discrete time n is defined by the *linear convolution sum*:

$$y(n) = \sum_{k=0}^{\infty} w_k^* u(n - k), \qquad n = 0, 1, 2, \ldots \tag{5.1}$$

where the asterisk denotes *complex conjugation*. Note that in complex terminology, the term $w_k^* u(n - k)$ represents the scalar version of an *inner product of the filter coefficient w_k and the filter input $u(n - k)$*. Figure 5.2 illustrates the steps involved in computing the linear discrete-time form of convolution described in Eq. (5.1) for real data .

The purpose of the filter in Fig. 5.1 is to produce an estimate of the desired response $d(n)$. We assume that the filter input and the desired response are single realizations of *jointly wide-sense stationary stochastic processes*, both with zero mean. Accordingly, the estimation of $d(n)$ is accompanied by an error, defined by the difference

$$e(n) = d(n) - y(n) \tag{5.2}$$

The estimation error $e(n)$ is the sample value of a random variable. *To optimize the filter design, we choose to minimize the mean-square value of the estimation error $e(n)$.* We may thus define the cost function as the *mean-squared error*

$$J = E[e(n)e^*(n)] \tag{5.3}$$

$$= E[|e(n)|^2]$$

where E denotes the *statistical expectation operator*. The problem is therefore to determine the operating conditions for which J attains its minimum value.

For complex input data, the filter coefficients are in general complex, too. Let the kth filter coefficient w_k be denoted in terms of its real and imaginary parts as follows:

$$w_k = a_k + jb_k, \qquad k = 0, 1, 2, \ldots \tag{5.4}$$

Correspondingly, we may define a *gradient operator* ∇, the kth element of which is written in terms of first-order partial derivatives with respect to the real part a_k and the imaginary part b_k, for the kth filter coefficient, as

$$\nabla_k = \frac{\partial}{\partial a_k} + j\frac{\partial}{\partial b_k}, \qquad k = 0, 1, 2, \ldots \tag{5.5}$$

Thus, for the situation at hand, applying the operator ∇ to the cost function J, we obtain a multidimensional complex *gradient vector* ∇J, the kth element of which is

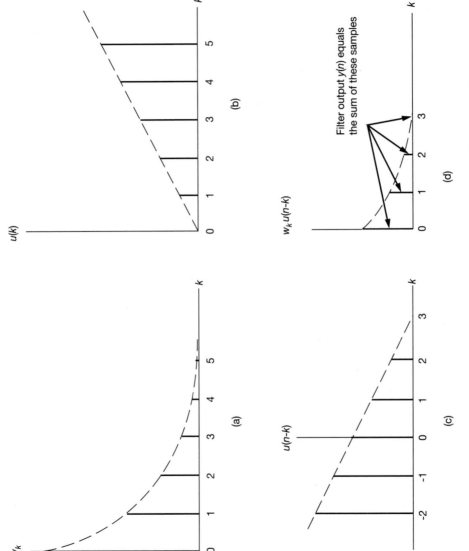

Figure 5.2 Linear convolution: (a) impulse response; (b) filter input; (c) time-reversed and shifted version of filter input; (d) calculation of filter output at time $n = 3$.

$$\nabla_k J = \frac{\partial J}{\partial a_k} + j \frac{\partial J}{\partial b_k}, \qquad k = 0, 1, 2, \ldots \tag{5.6}$$

Equation (5.6) represents a natural extension of the customary definition of a gradient for a function of real coefficients to the more general case of a function of complex coefficients.[1] Note that for the definition of the complex gradient given in Eq. (5.6) to be valid, it is essential that J be *real*. The gradient operator is always used in the context of finding the *stationary points* of a function of interest. This means that a complex constraint must be converted to a pair of *real* constraints. In Eq. (5.6), the pair of real constraints are obtained by setting both the real and imaginary parts of $\nabla_k J$ equal to zero.

For the cost function J to attain its minimum value, all the elements of the gradient vector ∇J must be simultaneously equal to zero, as shown by

$$\nabla_k J = 0, \qquad k = 0, 1, 2, \ldots \tag{5.7}$$

Under this set of conditions, the filter is said to be *optimum in the mean-squared-error sense*.[2]

According to Eq. (5.3), the cost function J is a scalar independent of time n. Hence, substituting the first line of Eq. (5.3) in (5.6), we get

$$\nabla_k J = E\left[\frac{\partial e(n)}{\partial a_k} e^*(n) + \frac{\partial e^*(n)}{\partial a_k} e(n) + \frac{\partial e(n)}{\partial b_k} je^*(n) + \frac{\partial e^*(n)}{\partial b_k} je(n) \right] \tag{5.8}$$

Using Eqs. (5.2) and (5.4), we get the following partial derivatives:

$$\frac{\partial e(n)}{\partial a_k} = -u(n - k)$$

$$\frac{\partial e(n)}{\partial b_k} = ju(n - k)$$

$$\frac{\partial e^*(n)}{\partial a_k} = -u^*(n - k) \tag{5.9}$$

$$\frac{\partial e^*(n)}{\partial b_k} = -ju^*(n - k)$$

---

[1] The concept of a gradient is commonly discussed in books on optimization (see, e.g., Dorny, 1975). For the complex case, it is discussed in Widrow et al. (1975a) and Monzingo and Miller (1980).

Note that the cost function J, for the general case of complex data, is *not* an analytic function (see Problem 1). Hence the definition of the derivative of the cost function J with respect to a filter coefficient w_k, say, requires particular attention. This issue is discussed in Appendix B. In this appendix we also discuss the relationship between the concepts of a gradient and a derivative for the case of complex coefficients.

[2] Note that in Eq. (5.7), we have presumed optimality at a stationary point. In the linear filtering problem, finding a stationary point assures global optimization of the filter by virtue of the quadratic nature of the error-performance surface; see Section 5.5.

Thus, substituting these partial derivatives in Eq. (5.8) and then canceling common terms, we finally get the result

$$\nabla_k J = -2E[u(n - k)e^*(n)] \tag{5.10}$$

We are now ready to specify the operating conditions required for minimizing the cost function J. Let e_o *denote the special value of the estimation error that results when the filter operates in its optimum condition.* We then find that the conditions specified in Eq. (5.7) are indeed equivalent to

$$E[u(n - k)e_o^*(n)] = 0, \qquad k = 0, 1, 2, \dots \tag{5.11}$$

In words, Eq. (5.11) states the following:

The necessary and sufficient condition for the cost function J to attain its minimum value is that the corresponding value of the estimation error $e_o(n)$ is orthogonal to each input sample that enters into the estimation of the desired response at time n.

Indeed, this statement constitutes the *principle of orthogonality*; it represents one of the most elegant theorems in the subject of linear optimum filtering. It also provides the mathematical basis of a procedure for testing that the linear filter is operating in its optimum condition.

Corollary to the Principle of Orthogonality

There is a corollary to the principle of orthogonality that we may derive by examining the correlation between the filter output $y(n)$ and the estimation error $e(n)$. Using Eq. (5.1), we may express this correlation as follows:

$$
\begin{aligned}
E[y(n)e^*(n)] &= E\left[\sum_{k=0}^{\infty} w_k^* u(n - k)e^*(n)\right] \\
&= \sum_{k=0}^{\infty} w_k^* E[u(n - k)e^*(n)]
\end{aligned}
\tag{5.12}
$$

Let y_o *denote the output produced by the filter optimized in the mean-squared-error sense, with $e_o(n)$ denoting the corresponding estimation error.* Hence, using the principle of orthogonality described by Eq. (5.11), we get the desired result:

$$E[y_o(n)e_o^*(n)] = 0 \tag{5.13}$$

We may thus state the corollary to the principle of orthogonality as follows:

When the filter operates in its optimum condition, the estimate of the desired response defined by the filter output, $y_o(n)$, and the corresponding estimation error, $e_o(n)$, are orthogonal to each other.

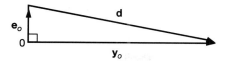

Figure 5.3 Geometric interpretation of the relationship between the desired response, the estimate at the filter output, and the estimation error.

Let $\hat{d}(n|\mathcal{U}_n)$ *denote the estimate of the desired response that is optimized in the mean-squared-error sense, given the input data that span the space* $\mathcal{U}_n$ *up to and including time* n.[3] We may then write

$$\hat{d}(n|\mathcal{U}_n) = y_o(n) \tag{5.14}$$

Note that the estimate $\hat{d}(n|\mathcal{U}_n)$ has zero mean, because the tap inputs are assumed to have zero mean. This condition matches the assumed zero mean of the desired response $d(n)$.

Geometric Interpretation of the Corollary to the Principle of Orthogonality

Equation (5.13) offers an interesting geometric interpretation of the conditions that exist at the output of the optimum filter, as illustrated in Fig. 5.3. In this figure, the desired response, the filter output, and the corresponding estimation error are represented by vectors labeled **d**, $\mathbf{y}_o$, and $\mathbf{e}_o$, respectively; the subscript o in $\mathbf{y}_o$ and $\mathbf{e}_o$ refers to the optimum condition. We see that for the optimum filter the vector representing the estimation error is *normal* (i.e., perpendicular) to the vector representing the filter output. It should, however, be emphasized that the situation depicted in Fig. 5.3 is merely an *analogy*, where random variables and expectations are replaced with vectors and vector inner products, respectively. Also, for obvious reasons the geometry depicted in this figure may be viewed as a *Statistician's Pythagorean theorem* (Scharf and Thomas, 1995).

5.3 MINIMUM MEAN-SQUARED ERROR

When the linear discrete-time filter in Fig. 5.1 operates in its optimum condition, Eq. (5.2) takes on the following special form

$$\begin{aligned}
e_o(n) &= d(n) - y_o(n) \\
&= d(n) - \hat{d}(n|\mathcal{U}_n)
\end{aligned} \tag{5.15}$$

---

[3] If a space $\mathcal{U}_n$ consists of all linear combinations of random variables, $u_1, u_2, \ldots, u_n$, then these random variables are said to *span* that particular space. In other words, every random variable in $\mathcal{U}_n$ can be expressed as some combination of the u's, as shown by

$$u = w_1^* u_1 + \ldots + w_n^* u_n$$

for some coefficients w_n. This assumes that the space $\mathcal{U}_n$ has a finite dimension.

where, in the second line, we have made use of Eq. (5.14). Rearranging the terms in Eq. (5.15), we have

$$d(n) = \hat{d}(n|\mathcal{U}_n) + e_o(n) \tag{5.16}$$

Let J_{min} denote the minimum mean-squared error, defined by

$$J_{min} = E[|e_o(n)|^2] \tag{5.17}$$

Hence, evaluating the mean-square values of both sides of Eq. (5.16), and applying to it the corollary to the principle of orthogonality described by Eqs. (5.13) and (5.14), we get

$$\sigma_d^2 = \sigma_{\hat{d}}^2 + J_{min} \tag{5.18}$$

where σ_d^2 is the variance of the desired response, and $\sigma_{\hat{d}}^2$ is the variance of the estimate $\hat{d}(n|\mathcal{U}_n)$; both of these random variables are assumed to be of zero mean. Solving Eq. (5.18) for the minimum mean-squared error, we get

$$J_{min} = \sigma_d^2 - \sigma_{\hat{d}}^2 \tag{5.19}$$

This relation shows that for the optimum filter, the minimum mean-squared error equals the difference between the variance of the desired response and the variance of the estimate that the filter produces at its output.

It is convenient to normalize the expression in Eq. (5.19) in such a way that the minimum value of the mean-squared error always lies between zero and one. We may do this by dividing both sides of Eq. (5.19) by σ_d^2, obtaining

$$\frac{J_{min}}{\sigma_d^2} = 1 - \frac{\sigma_{\hat{d}}^2}{\sigma_d^2} \tag{5.20}$$

Clearly, this is possible because σ_d^2 is never zero, except in the trivial case of a desired response $d(n)$ that is zero for all n. Let

$$\epsilon = \frac{J_{min}}{\sigma_d^2} \tag{5.21}$$

The quantity ϵ is called the *normalized mean-squared error*, in terms of which we may rewrite Eq. (5.20) in the form

$$\epsilon = 1 - \frac{\sigma_{\hat{d}}^2}{\sigma_d^2} \tag{5.22}$$

We note that (1) the ratio ϵ can never be negative, and (2) the ratio $\sigma_{\hat{d}}^2/\sigma_d^2$ is always positive. We therefore have

$$0 \leq \epsilon \leq 1 \tag{5.23}$$

If ϵ is zero, the optimum filter operates perfectly in the sense that there is complete agreement between the estimate $\hat{d}(n|\mathcal{U}_n)$ at the filter output and the desired response $d(n)$. On the other hand, if ϵ is unity, there is no agreement whatsoever between these two quantities; this corresponds to the worst possible situation.

5.4 WIENER–HOPF EQUATIONS

The principle of orthogonality, described in Eq. (5.11), specifies the necessary and sufficient condition for the optimum operation of the filter. We may reformulate the necessary and sufficient condition for optimality by substituting Eqs. (5.1) and (5.2) in (5.11). In particular, we may write

$$E\left[u(n-k)\left(d^*(n) - \sum_{i=0}^{\infty} w_{oi}u^*(n-i)\right)\right] = 0, \qquad k = 0, 1, 2, \ldots$$

where w_{oi} is the ith coefficient in the impulse response of the optimum filter. Expanding this equation and rearranging terms, we get

$$\sum_{i=0}^{\infty} w_{oi}E[u(n-k)u^*(n-i)] = E[u(n-k)d^*(n)], \qquad k = 0, 1, 2, \ldots \qquad (5.24)$$

The two expectations in Eq. (5.24) may be interpreted as follows:

1. The expectation $E[u(n-k)u^*(n-i)]$ is equal to the *autocorrelation function of the filter input* for a lag of $i - k$. We may thus express this expectation as

$$r(i-k) = E[u(n-k)u^*(n-i)] \qquad (5.25)$$

2. The expectation $E[u(n-k)d^*(n)]$ is equal to the cross-correlation between the filter input $u(n-k)$ and the desired response $d(n)$ for a lag of $-k$. We may thus express this second expectation as

$$p(-k) = E[u(n-k)d^*(n)] \qquad (5.26)$$

Accordingly, using the definitions of Eqs. (5.25) and (5.26) in (5.24), we get an infinitely large system of equations as the necessary and sufficient condition for the optimality of the filter:

$$\sum_{i=0}^{\infty} w_{oi}r(i-k) = p(-k), \qquad k = 0, 1, 2, \ldots \qquad (5.27)$$

The system of equations (5.27) defines the optimum filter coefficients, in the most general setting, in terms of two correlation functions: the autocorrelation function of the filter input, and the cross-correlation between the filter input and the desired response. These equations are called the *Wiener–Hopf equations*.[4]

---

[4] In order to solve the Wiener–Hopf equations for the optimum filter coefficients, we need to use a special technique known as *spectral factorization*. For a description of this technique and its use in solving the Wiener–Hopf equations (5.27), the interested reader is referred to Haykin (1989).

It should also be noted that the defining equation for a linear optimum filter was formulated originally by Wiener and Hopf (1931) for the case of a continuous-time filter, whereas, of course the system of equations (5.27) is formulated for a discrete-time filter.

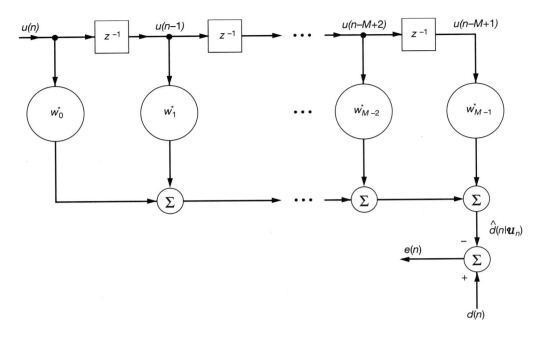

Figure 5.4   Transversal filter.

Solution of the Wiener–Hopf Equations for Linear Transversal Filters

The solution of the set of Wiener–Hopf equations is greatly simplified for the special case when a *linear transversal filter*, or FIR filter, is used to perform the estimation of desired response $d(n)$ in Fig. 5.1. Consider then the structure shown in Fig. 5.4. The transversal filter involves a combination of three basic operations: *storage*, *multiplication*, and *addition*, as described here:

1. The storage is represented by a cascade of $M - 1$ *one-sample delays*, with the block for each such unit labeled as z^{-1}. We refer to the various points at which the one-sample delays are accessed as *tap points*. The tap inputs are denoted by $u(n)$, $u(n-1)$, ..., $u(n-M+1)$. Thus, with $u(n)$ viewed as the *current* value of the filter input, the remaining $M - 1$ tap inputs, $u(n-1)$, ..., $u(n-M+1)$, represent *past values of the input*.

2. The scalar *inner products* of tap inputs $u(n)$, $u(n-1)$, ..., $u(n-M+1)$ and *tap weights* w_0, w_1, ..., w_{M-1}, respectively, are formed by using a correspond-

ing set of multipliers. In particular, the multiplication involved in forming the scalar inner product of $u(n)$ and w_0 is represented by a block labeled w_0^*, and so on for the other inner products.

3. The function of the adders is to sum the multiplier outputs to produce an overall output for the filter.

The impulse response of the transversal filter in Fig. 5.4 is defined by the finite set of tap weights $w_0, w_1, \ldots, w_{M-1}$. Accordingly, the Wiener–Hopf equations (5.27) reduce to a *system of M simultaneous equations*, as shown by

$$\sum_{i=0}^{M-1} w_{oi} r(i - k) = p(-k), \qquad k = 0, 1, \ldots, M - 1 \tag{5.28}$$

where $w_{o0}, w_{o1}, \ldots, w_{o,M-1}$ are the optimum values of the tap weights of the filter.

Matrix Formulation of the Wiener–Hopf Equations

Let $\mathbf{R}$ denote the *M-by-M correlation matrix* of the tap inputs $u(n)$, $u(n - 1)$, $\ldots$, $u(n - M + 1)$ in the transversal filter of Fig. 5.4:

$$\mathbf{R} = E[\mathbf{u}(n)\mathbf{u}^H(n)] \tag{5.29}$$

where $\mathbf{u}(n)$ is the *M-by-*1 tap-input vector:

$$\mathbf{u}(n) = [u(n), u(n - 1), \ldots, u(n - M + 1)]^T \tag{5.30}$$

In expanded form, we have

$$\mathbf{R} = \begin{bmatrix} r(0) & r(1) & \cdots & r(M - 1) \\ r^*(1) & r(0) & \cdots & r(M - 2) \\ \cdot & \cdot & \cdot & \cdot \\ \cdot & \cdot & \cdot & \cdot \\ \cdot & \cdot & \cdot & \cdot \\ r^*(M - 1) & r^*(M - 2) & \cdots & r(0) \end{bmatrix} \tag{5.31}$$

Correspondingly, let $\mathbf{p}$ denote the *M-by-*1 *cross-correlation vector* between the tap inputs of the filter and the desired response $d(n)$:

$$\mathbf{p} = E[\mathbf{u}(n)d^*(n)] \tag{5.32}$$

In expanded form, we have

$$\mathbf{p} = [p(0), p(-1), \ldots, p(1 - M)]^T \tag{5.33}$$

Note that the lags used in the definition of $\mathbf{p}$ are either zero or else negative. We may thus rewrite the Wiener–Hopf equations (5.28) in the compact matrix form:

$$\mathbf{R}\mathbf{w}_o = \mathbf{p} \tag{5.34}$$

where $\mathbf{w}_o$ denotes the M-by-1 *optimum tap-weight vector* of the transversal filter; that is,

$$\mathbf{w}_o = [\mathbf{w}_{o0}, \mathbf{w}_{o1}, \ldots, \mathbf{w}_{o,M-1}]^T \tag{5.35}$$

To solve the Wiener–Hopf equations (5.34) for $\mathbf{w}_o$, we assume that the correlation matrix $\mathbf{R}$ is nonsingular. We may then premultiply both sides of Eq. (5.34) by $\mathbf{R}^{-1}$, the *inverse* of the correlation matrix, obtaining

$$\mathbf{w}_o = \mathbf{R}^{-1}\mathbf{p} \tag{5.36}$$

The computation of the optimum tap-weight vector $\mathbf{w}_o$ requires knowledge of two quantities: (1) the correlation matrix $\mathbf{R}$ of the tap-input vector $\mathbf{u}(n)$, and (2) the cross-correlation vector $\mathbf{p}$ between the tap-input vector $\mathbf{u}(n)$ and the desired response $d(n)$.

5.5 ERROR-PERFORMANCE SURFACE

The Wiener–Hopf equations (5.34), as derived above, are traceable to the principle of orthogonality, which itself was derived in Section 5.2. We may also derive the Wiener–Hopf equations by examining the dependence of the cost function J on the tap weights of the transversal filter in Fig. 5.4. First, we write the estimation error $e(n)$ as follows:

$$e(n) = d(n) - \sum_{k=0}^{M-1} w_k^* u(n - k) \tag{5.37}$$

where $d(n)$ is the desired response; $w_0, w_1, \ldots, w_{M-1}$ are the tap weights of the filter; and $u(n), u(n-1), \ldots, u(n-M+1)$ are the corresponding tap inputs. Accordingly, we may define the cost function for the transversal filter structure of Fig. 5.4 as

$$\begin{aligned} J &= E[e(n)e^*(n)] \\ &= E[|d(n)|^2] - \sum_{k=0}^{M-1} w_k^* E[u(n-k)d^*(n)] - \sum_{k=0}^{M-1} w_k E[u^*(n-k)d(n)] \\ &\quad + \sum_{k=0}^{M-1}\sum_{i=0}^{M-1} w_k^* w_i E[u(n-k)u^*(n-i)] \end{aligned} \tag{5.38}$$

We may now recognize the four expectations on the right-hand side of the second line in Eq. (5.38), as follows:

1. For the first expectation, we have

$$\sigma_d^2 = E[|d(n)|^2] \tag{5.39}$$

where σ_d^2 is the *variance* of the desired response $d(n)$, assumed to be of zero mean.

2. For the second and third expectations, we have, respectively,

$$p(-k) = E[u(n - k)d^*(n)] \qquad (5.40)$$

and

$$p^*(-k) = E[u^*(n - k)d(n)] \qquad (5.41)$$

where $p(-k)$ is the cross-correlation between the tap input $u(n - k)$ and the desired response $d(n)$.

3. Finally, for the fourth expectation, we have

$$r(i - k) = E[u(n - k)u^*(n - i)] \qquad (5.42)$$

where $r(i - k)$ is the autocorrelation function of the tap inputs for lag $i - k$.

We may thus rewrite Eq. (5.38) in the form

$$J = \sigma_d^2 - \sum_{k=0}^{M-1} w_k^* p(-k) + \sum_{k=0}^{M-1} w_k^* p^*(-k) + \sum_{k=0}^{M-1}\sum_{i=0}^{M-1} w_k^* w_i r(i - k) \qquad (5.43)$$

   Equation (5.43) states that for the case when the tap inputs of the transversal filter and the desired response are jointly stationary, the cost function, or mean-squared error, J is precisely a *second-order function of the tap weights in the filter*. Consequently, we may visualize the dependence of the cost function J on the tap weights $w_0, w_1, \ldots, w_{M-1}$ as a *bowl-shaped* $(M + 1)$-*dimensional surface with M degrees of freedom represented by the tap weights of the filter*. This surface is characterized by a unique minimum. We refer to the surface so described as the *error-performance surface* of the transversal filter in Fig. 5.4.

   At the *bottom* or *minimum point* of the error-performance surface, the cost function J attains its *minimum value* denoted by $J_{\min}$. At this point, the *gradient vector* ∇J is identically zero. In other words,

$$\nabla_k J = 0, \qquad k = 0, 1, \ldots, M - 1 \qquad (5.44)$$

where $\nabla_k J$ is the kth element of the gradient vector. As before, we write the kth tap weight w_k as

$$w_k = a_k + jb_k$$

Hence, using Eq. (5.43), we may express $\nabla_k J$ as

$$\nabla_k J = \frac{\partial J}{\partial a_k} + j\frac{\partial J}{\partial b_k}$$

$$= -2p(-k) + 2\sum_{i=0}^{M-1} w_i r(i - k) \qquad (5.45)$$

Applying the necessary and sufficient condition of Eq. (5.44) for optimality to Eq. (5.45), we find that the optimum tap weights $w_{o0}, w_{o1}, \ldots, w_{o,M-1}$ for the transversal filter in Fig. 5.4 are defined by the system of equations:

$$\sum_{i=0}^{M-1} w_{oi} r(i - k) = p(-\mathrm{k}), \qquad k = 0, 1, \ldots, M - 1$$

This system of equations is identical to the Wiener–Hopf equations (5.28) derived in Section 5.4.

Minimum Mean-squared Error

Let $\hat{d}(n|\mathcal{U}_n)$ denote the estimate of the desired response $d(n)$, produced at the output of the transversal filter in Fig. 5.4 that is optimized in the mean-squared-error sense, given the tap inputs $u(n), u(n - 1), \ldots, u(n - M + 1)$ that span the space $\mathcal{U}_n$. Then from Fig. 5.4 we deduce that

$$\hat{d}(n|\mathcal{U}_n) = \sum_{k=0}^{M-1} w_{ok}^* u(n - k)$$

$$= \mathbf{w}_o^H \mathbf{u}(n) \tag{5.46}$$

where $\mathbf{w}_o$ is the tap-weight vector of the optimum filter with elements $w_{o0}, w_{o1}, \ldots, w_{o,M-1}$, and $\mathbf{u}(n)$ is the tap-input vector defined in Eq. (5.30). Note that $\mathbf{w}_o^H \mathbf{u}(n)$ denotes an inner product of the optimum tap-weight vector $\mathbf{w}_o$ and the tap-input vector $\mathbf{u}(n)$. We assume that $\mathbf{u}(n)$ has zero mean, making the estimate $\hat{d}(n|\mathcal{U}_n)$ have zero mean too. Hence, we may use Eq. (5.46) to evaluate the variance of $\hat{d}(n|\mathcal{U}_n)$, obtaining

$$\sigma_{\hat{d}}^2 = E[\mathbf{w}_o^H \mathbf{u}(n)\mathbf{u}^H(n)\mathbf{w}_o]$$

$$= \mathbf{w}_o^H E[\mathbf{u}(n)\mathbf{u}^H(n)]\mathbf{w}_o \tag{5.47}$$

$$= \mathbf{w}_o^H \mathbf{R} \mathbf{w}_o$$

where $\mathbf{R}$ is the correlation matrix of the tap-weight vector $\mathbf{u}(n)$, as defined in Eq. (5.29). We may eliminate the dependence of the variance $\sigma_{\hat{d}}^2$ on the optimum tap-weight vector $\mathbf{w}_o$ by using Eq. (5.34). In particular, we may rewrite Eq. (5.47) as

$$\sigma_{\hat{d}}^2 = \mathbf{w}_o^H \mathbf{p}$$

$$= \mathbf{p}^H \mathbf{w}_o \tag{5.48}$$

To evaluate the minimum mean-squared error produced by the transversal filter in Fig. 5.4, we may use Eq. (5.48) in (5.19), obtaining

$$J_{\min} = \sigma_{\hat{d}}^2 - \mathbf{p}^H \mathbf{w}_o$$

$$= \sigma_{\hat{d}}^2 - \mathbf{p}^H \mathbf{R}^{-1} \mathbf{p} \tag{5.49}$$

which is the desired result.

Canonical Form of the Error-Performance Surface

Equation (5.43) defines the expanded form of the mean-squared error J produced by the transversal filter in Fig. 5.4. We may rewrite this equation in matrix form, by using the definitions for the correlation matrix $\mathbf{R}$ and the cross-correlation vector $\mathbf{p}$ given in Eqs. (5.29) and (5.33), respectively, as shown by

$$J(\mathbf{w}) = \sigma_d^2 - \mathbf{w}^H\mathbf{p} - \mathbf{p}^H\mathbf{w} + \mathbf{w}^H\mathbf{R}\mathbf{w} \tag{5.50}$$

where the mean-squared error is written as $J(\mathbf{w})$ to emphasize its dependence on the tap-weight vector $\mathbf{w}$. As mentioned in Chapter 2 the correlation matrix $\mathbf{R}$ is almost always positive definite, so that the inverse matrix $\mathbf{R}^{-1}$ exists. Accordingly, expressing $J(\mathbf{w})$ as a "perfect square" in $\mathbf{w}$, we may rewrite Eq. (5.50) in the form

$$J(\mathbf{w}) = \sigma_d^2 - \mathbf{p}^H\mathbf{R}^{-1}\mathbf{p} + (\mathbf{w} - \mathbf{R}^{-1}\mathbf{p})^H\mathbf{R}(\mathbf{w} - \mathbf{R}^{-1}\mathbf{p}) \tag{5.51}$$

From Eq. (5.51), we now immediately see that

$$\min_{\mathbf{w}} J(\mathbf{w}) = \sigma_d^2 - \mathbf{p}^H\mathbf{R}^{-1}\mathbf{p}$$

for

$$\mathbf{w}_o = \mathbf{R}^{-1}\mathbf{p}$$

In effect, starting from Eq. (5.50), we have rederived the Wiener filter in a rather simple way. Moreover, we may use the defining equations for the Wiener filter to explicitly show the unique optimality of the minimizing tap-weight vector $\mathbf{w}_o$ by writing

$$J(\mathbf{w}) = J_{\min} + (\mathbf{w} - \mathbf{w}_o)^H\mathbf{R}(\mathbf{w} - \mathbf{w}_o) \tag{5.52}$$

This equation shows explicitly the unique optimality of the minimizing tap-weight vector $\mathbf{w}_o$.

Although the quadratic form on the right-hand side of Eq. (5.52) is quite informative, nevertheless, it is desirable to change the basis on which it is defined so that the representation of the error-performance surface is considerably simplified. To do this, we recall from Chapter 4 that the correlation matrix $\mathbf{R}$ of the tap-input vector may be expressed in terms of eigenvalues and eigenvectors as follows:

$$\mathbf{R} = \mathbf{Q}\mathbf{\Lambda}\mathbf{Q}^H \tag{5.53}$$

where $\mathbf{\Lambda}$ is a diagonal matrix consisting of the eigenvalues $\lambda_1, \lambda_2, \ldots, \lambda_M$ of the correlation matrix, and the matrix $\mathbf{Q}$ has for its columns the eigenvectors $\mathbf{q}_1, \mathbf{q}_2, \ldots, \mathbf{q}_M$ associated with these eigenvalues, respectively. Hence, substituting Eq. (5.53) into (5.52), we get

$$J = J_{\min} + (\mathbf{w} - \mathbf{w}_o)^H\mathbf{Q}\mathbf{\Lambda}\mathbf{Q}^H(\mathbf{w} - \mathbf{w}_o) \tag{5.54}$$

Define a *transformed* version of the difference between the tap-weight vector $\mathbf{w}$ and the optimum solution $\mathbf{w}_o$ as

$$\mathbf{v} = \mathbf{Q}^H(\mathbf{w} - \mathbf{w}_o) \tag{5.55}$$

Then we may put the quadratic form of Eq. (5.54) into its *canonical form* defined by

$$J = J_{\min} + \mathbf{v}^H \mathbf{\Lambda} \mathbf{v} \tag{5.56}$$

This new formulation of the mean-squared error contains no cross-product terms, as shown by

$$J = J_{\min} + \sum_{k=1}^{M} \lambda_k v_k v_k^*$$

$$= J_{\min} + \sum_{k=1}^{M} \lambda_k |v_k|^2 \tag{5.57}$$

where v_k is the kth component of the vector $\mathbf{v}$. The feature that makes the canonical form of Eq. (5.57) a rather useful representation of the error-performance surface is the fact that the components of the transformed coefficient vector $\mathbf{v}$ constitute the *principal axes* of the error-performance surface. The practical significance of this result will become apparent in later chapters.

5.6 NUMERICAL EXAMPLE

To illustrate the filtering theory developed above, we consider the example depicted in Fig. 5.5. The desired response $d(n)$ is modeled as an AR process of order 1; that is, it may be produced by applying a white-noise process $v_1(n)$ of zero mean and variance $\sigma_1^2 = 0.27$ to the input of an all-pole filter of order 1, whose transfer function equals [see Fig. 5.5(a)]

$$H_1(z) = \frac{1}{1 + 0.8458z^{-1}}$$

The process $d(n)$ is applied to a communication channel modeled by the all-pole transfer function

$$H_2(z) = \frac{1}{1 - 0.9458z^{-1}}$$

The channel output $x(n)$ is corrupted by an additive white-noise process $v_2(n)$ of zero mean and variance $\sigma_2^2 = 0.1$, so a sample of the received signal $u(n)$ equals [see Fig. 5.5(b)]

$$u(n) = x(n) + v_2(n)$$

The white-noise processes $v_1(n)$ and $v_2(n)$ are uncorrelated. It is also assumed that $d(n)$ and $u(n)$, and therefore $v_1(n)$ and $v_2(n)$, are all real valued.

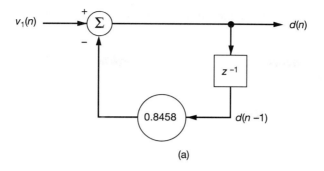

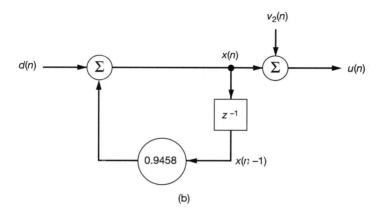

Figure 5.5   (a) Autoregressive model of desired response $d(n)$; (b) model of noisy communication channel.

The requirement is to specify a Wiener filter consisting of a transversal filter with two taps, which operates on the received signal $u(n)$ so as to produce an estimate of the desired response that is optimum in the mean-square sense.

Statistical Characterization of the Desired Response $d(n)$ and the Received Signal $u(n)$

We begin the analysis by considering the difference equations that characterize the various processes described by the models of Fig. 5.5. First, the generation of the desired response $d(n)$ is governed by the first-order difference equation

$$d(n) + a_1 d(n - 1) = v_1(n) \tag{5.58}$$

where $a = 0.8458$. The variance of the process $d(n)$ equals (see Problem 4 of Chapter 2)

$$\sigma_d^2 = \frac{\sigma_1^2}{1 - a_1^2}$$

$$= \frac{0.27}{1 - (0.8458)^2} \tag{5.59}$$

$$= 0.9486$$

The process $d(n)$ acts as input to the channel. Hence, from Fig. 5.5(b), we find that the channel output $x(n)$ is related to the channel input $d(n)$ by the first-order difference equation

$$x(n) + b_1 x(n - 1) = d(n) \tag{5.60}$$

where $b_1 = -0.9458$. We also observe from the two parts of Fig. 5.5 that the channel output $x(n)$ may be generated by applying the white-noise process $v_1(n)$ to a second-order all-pole filter whose transfer function equals

$$H(z) = H_1(z)H_2(z) \tag{5.61}$$

$$= \frac{1}{(1 + 0.8458z^{-1})(1 - 0.9458z^{-1})}$$

Accordingly, $x(n)$ is a second-order AR process described by the difference equation

$$x(n) + a_1 x(n - 1) + a_2 x(n - 2) = v(n) \tag{5.62}$$

where $a_1 = -0.1$ and $a_2 = -0.8$. Note that both AR processes $d(n)$ and $x(n)$ are wide-sense stationary.

To characterize the Wiener filter, we need to solve the Wiener–Hopf equations (5.34). This set of equations requires knowledge of two quantities: (1) the correlation matrix $\mathbf{R}$ pertaining to the received signal $u(n)$, and (2) the cross-correlation vector $\mathbf{p}$ between $u(n)$ and the desired response $d(n)$. In our example, $\mathbf{R}$ is a 2-by-2 matrix and $\mathbf{p}$ is a 2-by-1 vector, since the transversal filter used to implement the Wiener filter is assumed to have two taps.

The received signal $u(n)$ consists of the channel output $x(n)$ plus the additive white noise $v_2(n)$. Since the process $x(n)$ and $v_2(n)$ are uncorrelated, it follows that the correlation matrix $\mathbf{R}$ equals the correlation matrix of $x(n)$ plus the correlation matrix of $v_2(n)$. That is,

$$\mathbf{R} = \mathbf{R}_x + \mathbf{R}_2 \tag{5.63}$$

For the correlation matrix $\mathbf{R}_x$, we write [since the process $x(n)$ is real valued]

$$\mathbf{R}_x = \begin{bmatrix} r_x(0) & r_x(1) \\ r_x(1) & r_x(0) \end{bmatrix}$$

where $r_x(0)$ and $r_x(1)$ are the autocorrelation functions of the received signal $x(n)$ for lags of 0 and 1, respectively. From Section 2.9 we have

$$r_x(0) = \sigma_x^2$$

$$= \left(\frac{1 + a_2}{1 - a_2}\right) \frac{\sigma_1^2}{[(1 + a_2)^2 - a_1^2]}$$

$$= \left(\frac{1 - 0.8}{1 + 0.8}\right) \frac{0.27}{[(1 - 0.8)^2 - (0.1)^2]}$$

$$= 1$$

$$r_x(1) = \frac{-a_1}{1 + a_2}$$

$$= \frac{0.1}{1 - 0.8}$$

$$= 0.5$$

Hence,

$$\mathbf{R}_x = \begin{bmatrix} 1 & 0.5 \\ 0.5 & 1 \end{bmatrix} \tag{5.64}$$

Next we observe that since $v_2(n)$ is a white-noise process of zero mean and variance $\sigma_2^2 = 0.1$, the 2-by-2 correlation matrix $\mathbf{R}_2$ of this process equals

$$\mathbf{R}_2 = \begin{bmatrix} 0.1 & 0 \\ 0 & 0.1 \end{bmatrix} \tag{5.65}$$

Thus, substituting Eqs. (5.64) and (5.65) in Eq. (5.63), we find that the 2-by-2 correlation matrix of the received signal $x(n)$ equals

$$\mathbf{R} = \begin{bmatrix} 1.1 & 0.5 \\ 0.5 & 1.1 \end{bmatrix} \tag{5.66}$$

For the 2-by-1 cross-correlation vector $\mathbf{p}$, we write

$$\mathbf{p} = \begin{bmatrix} p(0) \\ p(-1) \end{bmatrix}$$

where $p(0)$ and $p(-1)$ are the cross-correlation functions between $d(n)$ and $u(n)$ for lags of 0 and -1, respectively. Since these two processes are real valued, we have

$$p(k) = p(-k) = E[u(n - k)d(n)], \qquad k = 0, 1 \tag{5.67}$$

Substituting Eqs. (5.57) and (5.60) into Eq. (5.67), and recognizing that the channel output $x(n)$ is uncorrelated with the white-noise process $v_2(n)$, we get

$$p(k) = r_x(k) + b_1 r_x(k - 1), \quad k = 0, 1$$

Putting $b_1 = -0.9458$ and using the element values for the correlation matrix $\mathbf{R}_x$ given in Eq. (5.64), we obtain

$$\begin{aligned} p(0) &= r_x(0) + b_1 r_x(-1) \\ &= 1 - 0.9458 \times 0.5 \\ &= 0.5272 \\ p(1) &= r_x(1) + b_1 r_x(0) \\ &= 0.5 - 0.9458 \times 1 \\ &= -0.4458 \end{aligned}$$

Hence,

$$\mathbf{p} = \begin{bmatrix} 0.5272 \\ -0.4458 \end{bmatrix} \tag{5.68}$$

Error-Performance Surface

The dependence of the mean-squared error on the 2-by-1 tap-weight vector $\mathbf{w}$ is defined by Eq. (5.50). Hence, substituting Eqs. (5.59), (5.66), and (5.68) into (5.50), we get

$$\begin{aligned} J(w_0, w_1) &= 0.9486 - 2[0.5272, -0.4458] \begin{bmatrix} w_0 \\ w_1 \end{bmatrix} + [w_0, w_1] \begin{bmatrix} 1.1 & 0.5 \\ 0.5 & 1.1 \end{bmatrix} \begin{bmatrix} w_0 \\ w_1 \end{bmatrix} \\ &= 0.9486 - 1.0544 w_0 + 0.8916 w_1 + w_0 w_1 + 1.1(w_0^2 + w_1^2) \end{aligned}$$

Using a three-dimensional computer plot, the mean-squared error $J(w_0, w_1)$ is plotted versus the tap weights w_0 and w_1. The result is shown in Fig. 5.6.

Figure 5.7 shows contour plots of the tap weight w_1 versus w_0 for varying values of the mean-squared error J. We see that the locus of w_1 versus w_0 for a fixed J is in the form of an ellipse. The elliptical locus shrinks in size as the mean-squared error J approaches the minimum value $J_{\min}$. For $J = J_{\min}$, the locus reduces to a point with coordinates w_{o0} and w_{o1}.

Wiener Filter

The 2-by-1 optimum tap-weight vector $\mathbf{w}_o$ of the Wiener filter is defined by Eq. (5.36). In particular, it consists of the inverse matrix $\mathbf{R}^{-1}$ multiplied by the cross-correlation vector $\mathbf{p}$. Inverting the correlation matrix $\mathbf{R}$ of Eq. (5.66), we get

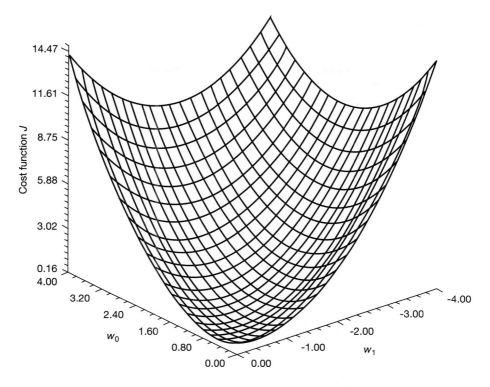

Figure 5.6   Error-performance surface of the two-tap transversal filter described in the numerical example.

$$\mathbf{R}^{-1} = \begin{bmatrix} r(0) & r(1) \\ r(1) & r(0) \end{bmatrix}^{-1}$$

$$= \frac{1}{r^2(0) - r^2(1)} \begin{bmatrix} r(0) & -r(1) \\ -r(1) & r(0) \end{bmatrix} \qquad (5.69)$$

$$= \begin{bmatrix} 1.1456 & -0.5208 \\ -0.5208 & 1.1456 \end{bmatrix}$$

Hence, substituting Eqs. (5.68) and (5.69) into Eq. (5.36), we get the desired result:

$$\mathbf{w}_o = \begin{bmatrix} 1.1456 & -0.5208 \\ -0.5208 & 1.1456 \end{bmatrix} \begin{bmatrix} 0.5272 \\ -0.4458 \end{bmatrix}$$

$$= \begin{bmatrix} 0.8360 \\ -0.7853 \end{bmatrix} \qquad (5.70)$$

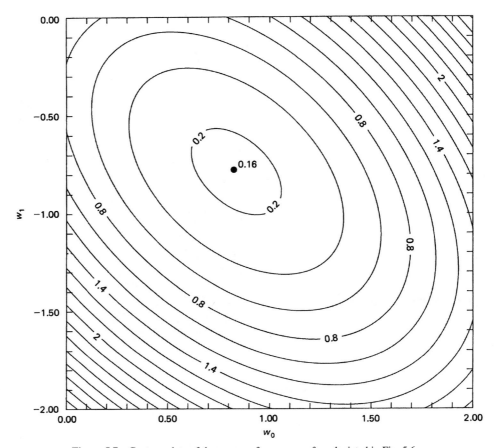

Figure 5.7   Contour plots of the error-performance surface depicted in Fig. 5.6.

Minimum Mean-Squared Error

To evaluate the minimum value of the mean-squared error, $J_{\min}$, which results from the use of the optimum tap-weight vector $\mathbf{w}_o$, we use Eq. (5.49). Hence, substituting Eqs. (5.59), (5.68), and (5.70) into Eq. (5.49), we get

$$J_{\min} = 0.9486 - [0.5272, \ -0.4458] \begin{bmatrix} 0.8360 \\ -0.7853 \end{bmatrix}$$

$$= 0.1579 \tag{5.71}$$

The point represented jointly by the optimum tap-weight vector $\mathbf{w}_o$ of Eq. (5.70) and the minimum mean-squared error of Eq. (5.71) defines the bottom of the error-performance surface in Fig. 5.6, or the center of the contour plots in Fig. 5.7.

Canonical Error-Performance Surface

The characteristic equation of the 2-by-2 correlation matrix $\mathbf{R}$ of Eq. (5.66) is

$$(1.1 - \lambda)^2 - (0.5)^2 = 0$$

The two eigenvalues of the correlation matrix $\mathbf{R}$ are therefore

$$\lambda_1 = 1.6$$

$$\lambda_2 = 0.6$$

The canonical error-performance surface is therefore defined by [see Eq. (5.57)]

$$J(v_1, v_2) = J_{\min} + 1.6v_1^2 + 0.6v_2^2 \tag{5.72}$$

The locus of v_2 versus v_1, as defined in Eq. (5.72), traces an *ellipse* for a fixed value of $J - J_{\min}$. In particular, the ellipse has a minor axis of $[(J - J_{\min})/\lambda_1]^{1/2}$ along the v_1-coordinate and a major axis of $[(J - J_{\min}/\lambda_2)]^{1/2}$ along the v_2-coordinate; this assumes that $\lambda_1 > \lambda_2$, which is how they are related.

5.7 CHANNEL EQUALIZATION

We turn next to some applications of Wiener filter theory. In this section we consider a temporal signal-processing problem, namely, that of channel equalization. This is followed by a spatial signal processing problem, namely, that of beamforming, which is presented in the next two sections.

A communication channel well suited for the transmission of digital data (e.g., computer data) is the *telephone channel*, which is characterized by a high signal-to-noise ratio. However, a practical shortcoming of the telephone channel is the fact that it is *bandwidth-limited*. Consequently, when data are transmitted over the channel by means of discrete pulse-amplitude modulation combined with a linear modulation scheme (e.g., quadriphase-shift keying), the number of detectable levels that the telephone channel can support is essentially limited by *intersymbol interference* rather than by additive noise. In what follows, we are therefore justified in ignoring channel noise. Intersymbol interference (ISI) arises because of the "spreading" of a transmitted pulse due to the dispersive nature of the channel, which results in an overlap of adjacent pulses. If ISI is left unchecked, it can produce errors in the reconstructed data stream at the receiver output. An effective method for combatting the system degradation due to ISI is to connect an *equalizer* in cascade with the channel, as in Fig. 5.8(a). A structure well suited for this application is the tapped-delay-line filter shown in Fig. 5.8(b). For equalizer *symmetry*, the total number of taps in the equalizer is chosen to be $(2N + 1)$, with the tap-weights themselves denoted by $h_{-N}, \ldots, h_{-1}, h_0, h_1, \ldots, h_N$. The impulse response of the equalizer is, therefore,

$$h(n) = \sum_{k=-N}^{N} h_k \, \delta(n - k) \tag{5.73}$$

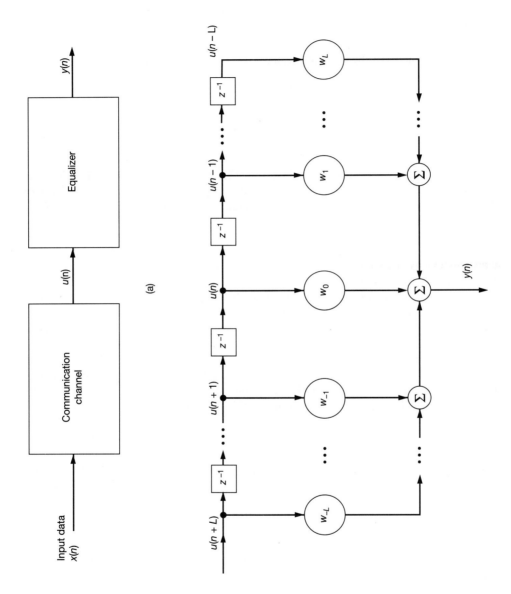

Figure 5.8 (a) Block diagram of equalized channel. (b) Symmetric tapped-delay-line filter implementation of the equalizer.

where $\delta(n)$ is the Dirac delta function. Similarly, we may express the impulse response of the channel as

$$c(n) = \sum_{k} c_k \, \delta(n - k) \tag{5.74}$$

In light of what we said earlier, we may ignore the effect of channel noise. Thus, the cascade connection of the channel and the equalizer is equivalent to a single tapped-delay-line filter. Let the impulse response of this equivalent filter be defined by

$$w(n) = \sum_{k=-N}^{N} w_k \, \delta(n - k) \tag{5.75}$$

where the sequence w_n is equal to the *convolution* of the sequences c_n and h_n. That is, we have

$$w_l = \sum_{k=-N}^{N} h_k c_{l-k}, \qquad l = 0, \pm 1, \ldots, \pm N \tag{5.76}$$

Let the data sequence $u(n)$ applied to the channel input consist of a white-noise sequence of zero mean and unit variance. In practice, such a sequence may be closely approximated by a pseudo-inverse sequence generated using a feedback shift register. Accordingly, we may express the elements of the correlation matrix $\mathbf{R}$ of the channel input as follows:

$$r(l) = \begin{cases} 1, & l = 0 \\ 0, & l \neq 0 \end{cases} \tag{5.77}$$

For the desired response $d(n)$ supplied to the equalizer, we assume the availability of a delayed "replica" of the transmitted sequence. This desired response may be generated by using another feedback shifter of identical design to that used to supply the original data sequence $u(n)$. The two feedback shift registers are synchronized with each other, such that we may set

$$d(n) = u(n)$$

where time n is measured with respect to the center tap of the equalizer. Thus, the cross-correlation between the transmitted sequence $u(n)$ and the desired response $d(n)$ is defined by

$$p(l) = \begin{cases} 1, & l = 0 \\ 0, & l = \pm 1, \pm 2, \ldots, \pm N \end{cases} \tag{5,78}$$

The stage is now set for the application of the Wiener–Hopf equations (5.28). Specifically, in light of Eqs. (5.77) and (5.78), we may write

$$w_l = \begin{cases} 1, & l = 0 \\ 0, & l = \pm 1, \pm 2, \ldots, \pm N \end{cases} \tag{5.79}$$

Equivalently, invoking the convolution sum of Eq. (5.76), we have

$$\sum_{k=-N}^{N} h_k c_{l-k} = \begin{cases} 1, & l = 0 \\ 0, & l = \pm 1, \pm 2, \ldots, \pm N \end{cases} \tag{5.80}$$

This system of simultaneous equations may be rewritten in the expanded matrix form:

$$\begin{bmatrix} c_0 & \cdots & c_{-N+1} & c_{-N} & c_{-N-1} & \cdots & c_{-2N} \\ \cdot & & \cdot & \cdot & \cdot & & \cdot \\ \cdot & & \cdot & \cdot & \cdot & & \cdot \\ \cdot & & \cdot & \cdot & \cdot & & \cdot \\ c_{N-1} & \cdots & c_0 & c_{-1} & c_{-2} & \cdots & c_{-N-1} \\ c_N & \cdots & c_1 & c_0 & c_{-1} & \cdots & c_{-N} \\ c_{N+1} & \cdots & c_2 & c_1 & c_0 & \cdots & c_{-N+1} \\ \cdot & & \cdot & \cdot & \cdot & & \cdot \\ \cdot & & \cdot & \cdot & \cdot & & \cdot \\ \cdot & & \cdot & \cdot & \cdot & & \cdot \\ c_{2N} & \cdots & c_{N+1} & c_N & c_{N-1} & \cdots & c_0 \end{bmatrix} \begin{bmatrix} h_{-N} \\ \cdot \\ \cdot \\ \cdot \\ h_{-1} \\ h_0 \\ h_1 \\ \cdot \\ \cdot \\ \cdot \\ h_N \end{bmatrix} = \begin{bmatrix} 0 \\ \cdot \\ \cdot \\ \cdot \\ 0 \\ 1 \\ 0 \\ \cdot \\ \cdot \\ \cdot \\ 0 \end{bmatrix} \tag{5.81}$$

Thus, given the impulse response of the channel characterized by the coefficients c_{-N}, . . . , $c_{-1}, c_0, c_1, \ldots, c_N$, we may use Eq. (5.81) to solve for the unknown tap-weights h_{-N}, . . . , $h_{-1}, h_0, h_1, \ldots, h_N$ of the equalizer.

In the literature on digital communications, an equalizer designed in accordance with Eq. (5.81) is referred to as a *zero-forcing equalizer* (Lucky et al., 1968). The equalizer is so called because, with a single pulse transmitted over the channel, it "forces" the receiver output to be zero at all the sampling instances, except for the time instant that corresponds to the transmitted pulse.

5.8 LINEARLY CONSTRAINED MINIMUM VARIANCE FILTER

The essence of a Wiener filter is that it minimizes the mean-square value of an estimation error, defined as the difference between a desired response and the actual filter output. In solving this optimization (minimization) problem, there are *no* constraints imposed on the solution. In some filtering applications, however, it may be desirable (or even mandatory) to design a filter that minimizes a mean-square criterion, subject to a specific *constraint*. For example, the requirement may be that of minimizing the average output power of a linear filter while the response of the filter measured at some specific frequency of interest is constrained to remain constant. In this section, we consider one such solution.

Consider a linear transversal filter, as in Fig. 5.9. The filter output, in response to the tap inputs $u(n), u(n-1), \ldots, u(n-M+1)$, is given by

$$y(n) = \sum_{k=0}^{M-1} w_k^* u(n-k) \tag{5.82}$$

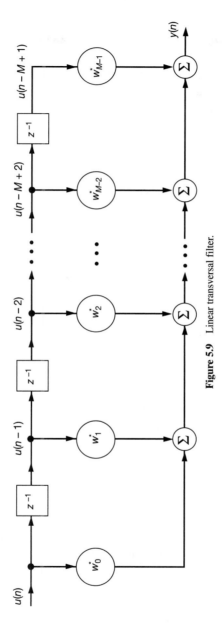

Figure 5.9 Linear transversal filter.

For the special case of a *sinusoidal excitation*

$$u(n) = e^{j\omega n} \tag{5.83}$$

we may rewrite Eq. (5.82) as

$$y(n) = e^{j\omega n} \sum_{k=0}^{M-1} w_k^* e^{-j\omega k} \tag{5.84}$$

where ω is the angular frequency of the excitation, which is normalized with respect to the sampling rate.

The *constrained optimization problem* we wish to solve may now be stated as follows:

Find the optimum set of filter coefficients $w_{o0}, w_{o1}, \ldots, w_{o,M-1}$ that minimizes the mean-square value of the filter output $y(n)$, subject to the linear constraint

$$\sum_{k=0}^{M-1} w_k^* e^{-j\omega_0 k} = g \tag{5.85}$$

where ω_0 is a prescribed value of the normalized angular frequency ω, lying inside the range $-\pi < \omega \leq \pi$, and g is a complex-valued gain.

The constrained optimization filtering problem described by Eqs. (5.82) and (5.85) is *temporal* in nature. We may formulate the *spatial* version of this constrained optimization problem by considering the *beamformer* depicted in Fig. 5.10, which consists of a linear array of uniformly spaced antenna elements. The array is illuminated by an isotropic source located in the *far field*, such that, at time n, a *plane wave* impinges on the array along a direction specified by the angle θ_0 with respect to the perpendicular to the array. It is also assumed that the interelement spacing of the array is less than $\lambda/2$, where λ is the wavelength of the transmitted signal so as to avoid the appearance of grating lobes (Skolnik, 1980). The resulting beamformer output is given by

$$y(n) = u_0(n) \sum_{k=0}^{M-1} w_k^* e^{-jk\phi_0} \tag{5.86}$$

where the *direction of arrival* is defined by the electrical angle ϕ_0 that is related to the angle of incidence θ_0 by Eq. (45) of the introductory chapter, $u_0(n)$ is the electrical signal picked up by the antenna element labeled 0 in Fig. 5.10 that is treated as the point of reference, and the w_k denote the *elemental weights* of the beamformer. The spatial version of the constrained optimization problem may thus be stated as follows:

Find the optimum set of elemental weights $w_{o0}, w_{o1}, \ldots, w_{o,M-1}$ that minimizes the mean-square value of the beamformer output, subject to the linear constraint:

$$\sum_{k=0}^{M-1} w_k^* e^{-jk\phi_0} = g \tag{5.87}$$

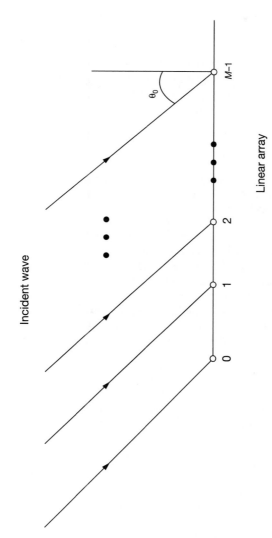

Figure 5.10   Illustrating a plane wave incident on a linear array antenna.

where ϕ_0 is a prescribed value of the electrical angle ϕ, lying inside the range $-\pi < \phi \leq \pi$, and g is a complex-valued gain. The beamformer is narrowband in the sense that its response needs to be constrained only at a single frequency.

Comparing the transversal filter and beamformer described in Figs. 5.9 and 5.10, respectively, we see that although they address entirely different physical situations, their formulations are equivalent in mathematical terms. Indeed, in both cases we have exactly the same constrained optimization problem on our hands.

To solve this constrained optimization problem, we use the *method of Lagrange multipliers*.[5] We begin by defining a *real-valued* cost function J that combines the two parts of the constrained optimization problem. Specifically, we write

$$J = \underbrace{\sum_{k=0}^{M-1} \sum_{i=0}^{M-1} w_k^* \, w_i r(i-k)}_{\text{output power}} + \underbrace{\text{Re}\left[\lambda^*\left(\sum_{k=0}^{M-1} w_k^* e^{-j\phi_0 k} - g\right)\right]}_{\text{linear constraint}} \tag{5.88}$$

where λ is a *complex Lagrange multiplier*. Note that there is no desired response in the definition of the cost function J; rather, it includes a linear constraint that has to be satisfied for the prescribed electrical angle ϕ_0 in the context of beamforming, or equivalently the angular frequency ω_0 in transversal filtering. In any event, imposition of the linear constraint preserves the signal of interest, and minimization of the cost function J attenuates interference or noise that can be troublesome if left unchecked.

We wish to solve for the optimum values of the elemental weights of the beamformer that minimize J defined in Eq. (5.88). To do so, we may determine the gradient vector ∇J, and then set it equal to zero. Thus, proceeding in a manner similar to that described in Section 5.2, we find that the kth element of the gradient vector ∇J is

$$\nabla_k J = 2 \sum_{i=0}^{M-1} w_i r\,(i-k) + \lambda^* e^{-j\phi_0 k} \tag{5.89}$$

Let w_{oi} be the ith element of the optimum weight vector $\mathbf{w}_o$. Then the condition for optimality of the beamformer is described by

$$\sum_{i=0}^{M-1} w_{oi} r\,(i-k) = -\frac{\lambda^*}{2}\,e^{-j\phi_0 k}, \qquad k = 0, 1, \ldots, M-1 \tag{5.90}$$

This system of M simultaneous equations defines the optimum values of the beamformer's elemental weights. It has a form somewhat similar to that of the Wiener–Hopf equations (5.28).

At this point in the analysis, we find it convenient to switch to matrix notation. In particular, we may rewrite the system of M simultaneous equations given in (5.90) simply as

$$\mathbf{R}\mathbf{w}_o = -\frac{\lambda^*}{2}\,\mathbf{s}(\phi_0) \tag{5.91}$$

---

[5] The method of Lagrange multipliers is described in Appendix C.

where $\mathbf{R}$ is the M-by-M correlation matrix, and $\mathbf{w}_o$ is the M-by-1 optimum weight vector of the constrained beamformer. The M-by-1 *steering* vector $\mathbf{s}(\phi_0)$ is defined by

$$\mathbf{s}(\phi_0) = [1, e^{-j\phi_0}, \ldots, e^{-j(M-1)\phi_0}]^T \tag{5.92}$$

Solving Eq. (5.91) for $\mathbf{w}_o$, we thus have

$$\mathbf{w}_o = -\frac{\lambda^*}{2}\mathbf{R}^{-1}\mathbf{s}(\phi_0) \tag{5.93}$$

where $\mathbf{R}^{-1}$ is the inverse of the correlation matrix $\mathbf{R}$, assuming that $\mathbf{R}$ is nonsingular. This assumption is perfectly justified in practice by virtue of the fact that, in the context of a beamformer, the received signal at the output of each antenna element of the beamformer includes a white (thermal) noise component.

The solution for the optimum weight vector $\mathbf{w}_o$ given in Eq. (5.93) is not quite complete, as it involves the unknown Lagrange multiplier λ (or its complex conjugate to be precise). To eliminate λ^* from this expression, we first use the linear constraint of Eq. (5.87) to write

$$\mathbf{w}_o^H\mathbf{s}(\phi_0) = g \tag{5.94}$$

Hence, taking the Hermitian transpose of both sides of Eq. (5.93), postmultiplying by $\mathbf{s}(\phi_0)$, and then using the linear constraint of Eq. (5.94), we get

$$\lambda = -\frac{2g}{\mathbf{s}^H(\phi_0)\mathbf{R}^{-1}\mathbf{s}(\phi_0)} \tag{5.95}$$

where we have used the fact that $\mathbf{R}^{-H} = \mathbf{R}^{-1}$. The quadratic form $\mathbf{s}^H(\phi_0)\mathbf{R}^{-1}\mathbf{s}(\phi_0)$ is real valued. Hence, substituting Eq. (5.95) in (5.93), we get the desired formula for the optimum weight vector

$$\mathbf{w}_o = \frac{g^*\mathbf{R}^{-1}\mathbf{s}(\phi_0)}{\mathbf{s}^H(\phi_0)\mathbf{R}^{-1}\mathbf{s}(\phi_0)} \tag{5.96}$$

Note that by minimizing the output power, subject to the linear constraint of Eq. (5.87), signals incident on the array along directions different from the prescribed value ϕ_0 tend to be attenuated.

For obvious reasons, a beamformer characterized by the weight vector $\mathbf{w}_o$ is referred to as a *linearly constrained minimum variance* (LCMV) *beamformer*. For a zero-mean input and therefore zero-mean output, "minimum variance" and "minimum mean-square value" are indeed synonymous. Also, in light of what we said previously, the solution defined by Eq. (5.96) with ω_0 substituted for ϕ_0 may be referred to as an *LCMV filter*.

Minimum Variance Distortionless Response Beamformer

The complex constant g defines the response of an LCMV beamformer at the electrical angle ϕ_0. For the special case of $g = 1$, the optimum solution given in Eq. (5.96) reduces to

$$\mathbf{w}_o = \frac{\mathbf{R}^{-1}\mathbf{s}(\phi_0)}{\mathbf{s}^H(\phi_0)\mathbf{R}^{-1}\mathbf{s}(\phi_0)} \tag{5.97}$$

The response of the beamformer defined by Eq. (5.97) is constrained to equal unity at the electrical angle ϕ_0. In other words, this beamformer is constrained to produce a distortionless response along the look direction corresponding to ϕ_0.

Now, the minimum mean-square value (average power) of the optimum beamformer output may be expressed as the quadratic form

$$J_{\min} = \mathbf{w}_o^H \mathbf{R} \mathbf{w}_o \tag{5.98}$$

Hence, substituting Eq. (5.97) in (5.98) and simplifying, we get the result

$$J_{\min} = \frac{1}{\mathbf{s}^H(\phi_0)\mathbf{R}^{-1}\mathbf{s}(\phi_0)} \tag{5.99}$$

The optimum beamformer is constrained to pass the target signal with unit response, while at the same time minimizing the total output variance. This variance minimization process attenuates interference and noise not originating at the electrical angle ϕ_0. Hence, $J_{\min}$ represents an estimate of the variance of the signal impinging on the array along the direction corresponding to ϕ_0. We may generalize this result and obtain an estimate of variance as a function of direction by formulating $J_{\min}$ as a function of ϕ. In so doing, we obtain the MVDR (spatial) power spectrum defined as

$$S_{\text{MVDR}}(\phi) = \frac{1}{\mathbf{s}^H(\phi)\mathbf{R}^{-1}\mathbf{s}(\phi)} \tag{5.100}$$

where

$$\mathbf{s}(\phi) = [1, e^{-j\phi}, \ldots, e^{-j\phi(M-1)}]^T \tag{5.101}$$

The M-by-1 vector $\mathbf{s}(\phi)$ is called a *spatial scanning vector* in the context of the beamformer of Fig. 5.10, and a *frequency scanning vector* with ω in place of ϕ for the transversal filter of Fig. 5.9. By definition, $S_{\text{MVDR}}(\phi)$ or $S_{\text{MVDR}}(\omega)$ has the dimension of power. Its dependence on the electrical angle ϕ at the beamformer input or the angular frequency ω at the transversal filter input therefore justifies referring to it as a power spectrum estimate. Indeed, it is commonly referred to as the *minimum variance distortionless response* (MVDR) *spectrum*.[6] Note that at any ω in the temporal context, power due to other angular frequencies is minimized. Accordingly, the MVDR spectrum tends to have sharper peaks and higher resolution, compared to nonparametric (classical) methods based on the definition of power spectrum discussed in Chapter 3.

In Appendix F we present a fast algorithm for computing the MVDR spectrum when the correlation matrix $\mathbf{R}$ is known. Also, it is noteworthy that the MVDR beamformer/spectrum analyzer is an important member of the family of *superresolution algo-*

---

[6] The formula given in Eq. (5.100) is credited to Capon (1969). It is also referred to in the literature as the *maximum-likelihood method (MLM)*. In reality, however, this formula has no bearing on the classical principle of maximum likelihood. The use of the terminology MLM for this formula is therefore not recommended.

rithms. The term "super-resolution" or "high-resolution" refers to the fact that a frequency estimation or angle-of-arrival estimation algorithm so termed has, under carefully controlled conditions, the ability to surpass the limiting behavior of classical Fourier-based methods, with the limitation being imposed by the finite length of the transversal filter or finite aperture of the linear array.

5.9 GENERALIZED SIDELOBE CANCELERS

Continuing with the discussion of the LCMV narrow-band beamformer defined by the linear constraint of Eq. (5.87), we note that this constraint represents the inner product

$$\mathbf{w}^H \mathbf{s}(\phi_0) = g$$

in which $\mathbf{w}$ is the weight vector and $\mathbf{s}(\phi_0)$ is the steering vector pointing along the electrical angle ϕ_0. The steering vector is an M-by-1 vector, where M is the number of antenna elements in the beamformer. We may generalize the notion of a linear constraint by introducing *multiple linear constraints* defined by

$$\mathbf{C}^H \mathbf{w} = \mathbf{g} \tag{5.102}$$

The matrix $\mathbf{C}$ is termed the *constraint matrix*; and the vector $\mathbf{g}$, termed the *gain vector*, has constant elements. Assuming that there are L linear constraints, the matrix $\mathbf{C}$ is an M-by-L matrix, and $\mathbf{g}$ is an L-by-1 vector; each column of the matrix $\mathbf{C}$ represents a single linear constraint. Furthermore, it is assumed that the constraint matrix $\mathbf{C}$ has linearly independent columns. For example, with

$$[s(\phi_0), s(\phi_1)]^H \, \mathbf{w} = \begin{bmatrix} 1 \\ 0 \end{bmatrix}$$

the narrow-band beamformer is constrained to preserve a signal of interest impinging on the array along the electrical angle ϕ_0 and, at the same time, to suppress an interference known to originate along the electrical angle ϕ_1.

   Let the columns of an M-by-$(M-L)$ matrix $\mathbf{C}_a$ be defined as a basis for the *orthogonal complement* of the space spanned by the columns of matrix $\mathbf{C}$. Using the definition of an orthogonal complement, we may thus write

$$\mathbf{C}^H \mathbf{C}_a = \mathbf{O} \tag{5.103}$$

or, just as well, write

$$\mathbf{C}_a^H \, \mathbf{C} = \mathbf{O} \tag{5.104}$$

The null matrix $\mathbf{O}$ in Eq. (5.103) is L-by-$(M-L)$, whereas in Eq. (5.104) it is $(M-L)$-by-L; we naturally have $M > L$. Define the M-by-M partitioned matrix

$$\mathbf{U} = [\mathbf{C} : \mathbf{C}_a] \tag{5.105}$$

whose columns span the entire M-dimensional signal space. The inverse matrix $\mathbf{U}^{-1}$ exists by virtue of the fact that the determinant of matrix $\mathbf{U}$ is nonzero.

Next, let the M-by-1 weight vector of the beamformer be written in terms of the matrix $\mathbf{U}$ as

$$\mathbf{w} = \mathbf{U}\mathbf{q} \tag{5.106}$$

Equivalently, the M-by-1 vector $\mathbf{q}$ is defined by

$$\mathbf{q} = \mathbf{U}^{-1}\mathbf{w} \tag{5.107}$$

Let the vector $\mathbf{q}$ be partitioned in a compatible way to that in Eq. (5.105), as shown by

$$\mathbf{q} = \left[\begin{array}{c} \mathbf{v} \\ \hline -\mathbf{w}_a \end{array}\right] \tag{5.108}$$

where $\mathbf{v}$ is an L-by-1 vector, and the $(M-L)$-by-1 vector $\mathbf{w}_a$ is that portion of the weight vector $\mathbf{w}$ that is not affected by the constraints. We may then use the definitions of Eqs. (5.105) and (5.108) in Eq. (5.106) to write

$$\mathbf{w} = [\mathbf{C} \vdots \mathbf{C}_a] \left[\begin{array}{c} \mathbf{v} \\ \hline -\mathbf{w}_a \end{array}\right] \tag{5.109}$$

$$= \mathbf{C}\mathbf{v} - \mathbf{C}_a\mathbf{w}_a$$

We may now apply the multiple linear constraints of Eq. (5.102), obtaining

$$\mathbf{C}^H\mathbf{C}\mathbf{v} - \mathbf{C}^H\mathbf{C}_a\mathbf{w}_a = \mathbf{g} \tag{5.110}$$

But, from Eq. (5.103) we know that $\mathbf{C}^H\mathbf{C}_a$ is zero; hence, Eq. (5.110) reduces to

$$\mathbf{C}^H\mathbf{C}\mathbf{v} = \mathbf{g} \tag{5.111}$$

Solving for the vector $\mathbf{v}$, we thus get

$$\mathbf{v} = (\mathbf{C}^H\mathbf{C})^{-1}\mathbf{g} \tag{5.112}$$

which shows that the multiple linear constraints do not affect $\mathbf{w}_a$.

Define a nonadaptive beamformer component represented by

$$\mathbf{w}_q = \mathbf{C}\mathbf{v} = \mathbf{C}(\mathbf{C}^H\mathbf{C})^{-1}\mathbf{g} \tag{5.113}$$

which is orthogonal to the columns of matrix $\mathbf{C}_a$ by virtue of the property described in Eq. (5.104); the rationale for using the subscript q in $\mathbf{w}_q$ will become apparent later. Using this definition, we may use Eq. (5.109) to express the overall weight vector of the beamformer as

$$\mathbf{w} = \mathbf{w}_q - \mathbf{C}_a\mathbf{w}_a \tag{5.114}$$

Substituting Eq. (5.114) in (5.102) yields

$$\mathbf{C}^H\mathbf{w}_q - \mathbf{C}^H\mathbf{C}_a\mathbf{w}_a = \mathbf{g}$$

which, by virtue of Eq. (5.103), reduces to

$$\mathbf{C}^H\mathbf{w}_q = \mathbf{g} \tag{5.115}$$

Equation (5.115) shows that the weight vector $\mathbf{w}_q$ is that part of the weight vector $\mathbf{w}$ that satisfies the constraints. In contrast, the vector $\mathbf{w}_a$ is unaffected by the constraints; it therefore provides the degrees of freedom built into the design of the beamformer. Thus, in light of Eq. (5.114), the beamformer may be represented by the block diagram shown in Fig. 5.11(a). The beamformer described herein is referred to as a *generalized sidelobe canceler* (GSC).[7]

In light of Eq. (5.115), we may now perform an unconstrained minimization of the mean-square value of the beamformer output $y(n)$ with respect to the adjustable weight vector $\mathbf{w}_a$. According to Eq. (5.86), the beamformer output is defined by the inner product

$$y(n) = \mathbf{w}^H \mathbf{u}(n) \tag{5.116}$$

where $\mathbf{u}(n)$ is the input signal vector:

$$\mathbf{u}(n) = u_0(n) \, [1, e^{-j\phi_0}, \ldots, e^{-j(M-1)\phi_0}]^T \tag{5.117}$$

where the electrical angle ϕ_0 is defined by the direction of arrival of the incoming plane wave, and $u_0(n)$ is the electrical signal picked up by antenna element 0 of the linear array in Fig. 5.10 at time n. Hence, substituting Eq. (5.114) in (5.116) yields

$$y(n) = \mathbf{w}_q^H \mathbf{u}(n) - \mathbf{w}_a^H \mathbf{C}_a^H \mathbf{u}(n) \tag{5.118}$$

Define

$$\mathbf{w}_q^H \mathbf{u}(n) = d(n) \tag{5.119}$$

$$\mathbf{C}_a^H \mathbf{u}(n) = \mathbf{x}(n) \tag{5.120}$$

We may then rewrite Eq. (5.118) in a form that resembles the standard Wiener filter exactly, as shown by

$$y(n) = d(n) - \mathbf{w}_a^H \mathbf{x}(n) \tag{5.121}$$

where $d(n)$ plays the role of a "desired response" for the GSC and $\mathbf{x}(n)$ plays the role of input vector, as depicted in Fig. 5.11(b). We thus see that the combined use of vector $\mathbf{w}_q$ and matrix $\mathbf{C}_a$ has converted the linearly constrained optimization problem into a standard optimum filtering problem. In particular, we now have an unconstrained optimization problem involving the adjustable portion $\mathbf{w}_a$ of the weight vector, which may be formally written as

$$\min_{\mathbf{w}_a} E[\,|y(n)|^2] = \min_{\mathbf{w}_a} (\sigma_d^2 - \mathbf{p}_x^H \mathbf{w}_a - \mathbf{w}_a^H \mathbf{p}_x + \mathbf{w}_a^H \mathbf{R}_x \mathbf{w}_a) \tag{5.122}$$

where the $(M-L)$-by-1 vector $\mathbf{p}_x$ is defined by

$$\mathbf{p}_x = E[\mathbf{x}(n)d^*(n)] \tag{5.123}$$

---

[7] The essence of the generalized sidelobe canceler may be traced back to a method for solving linearly constrained quadratic minimization problems originally proposed by Hanson and Lawson (1969). The term "generalized sidelobe canceler" was coined by Griffiths and Jim (1982). For a discussion of the generalized sidelobe canceler, see Van Veen and Buckley (1988) and Van Veen (1992).

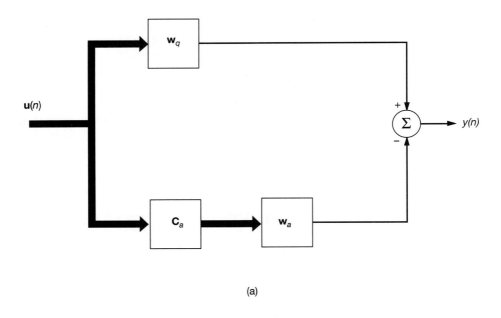

(a)

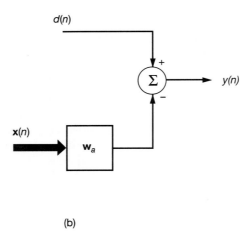

(b)

Figure 5.11   (a) Block diagram of generalized sidelobe canceler. (b) Reformulation of the
generalized sidelobe canceling problem as a standard optimum filtering problem.

and the $(M-L)$-by-$(M-L)$ matrix $\mathbf{R}_x$ is defined by

$$\mathbf{R}_x = E[\mathbf{x}(n)\mathbf{x}^H(n)] \tag{5.124}$$

The cost function of Eq. (5.122) is a quadratic in the unknown vector $\mathbf{w}_a$, which, as previously stated, embodies the available degrees of freedom in the GSC. Most importantly, this cost function has exactly the same mathematical form as that of the standard Wiener filter defined in Eq. (5.50). Accordingly, we may readily use our previous results to obtain the optimum value of $\mathbf{w}_a$ as

$$\mathbf{w}_{ao} = \mathbf{R}_x^{-1}\mathbf{p}_x \tag{5.125}$$

Using the definitions of Eqs. (5.119) and (5.120) in Eq. (5.123), we may express the vector $\mathbf{p}_x$ as

$$\begin{aligned}
\mathbf{p}_x &= E[\mathbf{C}_a^H\mathbf{u}(n)\mathbf{u}^H(n)\mathbf{w}_q] \\
&= \mathbf{C}_a^H E[\mathbf{u}(n)\mathbf{u}^H(n)]\mathbf{w}_q \\
&= \mathbf{C}_a^H\mathbf{R}\mathbf{w}_q
\end{aligned} \tag{5.126}$$

where $\mathbf{R}$ is the correlation matrix of the incoming data vector $\mathbf{u}(n)$. Similarly, using the definition of Eq. (5.120) in (5.124), we may express the matrix $\mathbf{R}_x$ as

$$\begin{aligned}
\mathbf{R}_x &= E[\mathbf{C}_a^H\mathbf{u}(n)\mathbf{u}^H(n)\mathbf{C}_a] \\
&= \mathbf{C}_a^H\mathbf{R}\mathbf{C}_a
\end{aligned} \tag{5.127}$$

The matrix $\mathbf{C}_a$ has full rank, and the correlation matrix $\mathbf{R}$ is positive definite since the incoming data always contain some form of additive receiver noise. Accordingly, we may rewrite the optimum solution $\mathbf{w}_{ao}$ of Eq. (5.125) as

$$\mathbf{w}_{ao} = (\mathbf{C}_a^H\mathbf{R}\mathbf{C}_a)^{-1}\mathbf{C}_a^H\mathbf{R}\mathbf{w}_q \tag{5.128}$$

Let P_o denote the minimum output power of the GSC attained by using the optimum solution $\mathbf{w}_{ao}$. Then, adapting the previous result derived in Eq. (5.49) for the standard Wiener filter and proceeding in a manner similar to that described above, we may express P_o as follows:

$$\begin{aligned}
P_o &= \sigma_d^2 - \mathbf{p}_x^H\mathbf{R}_x^{-1}\mathbf{p}_x \\
&= \mathbf{w}_q^H\mathbf{R}\mathbf{w}_q - \mathbf{w}_q^H\mathbf{R}\mathbf{C}_a(\mathbf{C}_a^H\mathbf{R}\mathbf{C}_a)^{-1}\mathbf{C}_a^H\mathbf{R}\mathbf{w}_q
\end{aligned} \tag{5.129}$$

Consider the special case of a *quiet environment*, for which the received signal consists of white noise acting alone. Let the corresponding value of the correlation matrix $\mathbf{R}$ be written as

$$\mathbf{R} = \sigma^2\mathbf{I} \tag{5.130}$$

where $\mathbf{I}$ is the M-by-M identity matrix, and σ^2 is the noise variance. Under this condition, we readily find from Eq. (5.128) that

$$\mathbf{w}_{ao} = (\mathbf{C}_a^H\mathbf{C}_a)^{-1}\mathbf{C}_a^H\mathbf{w}_q$$

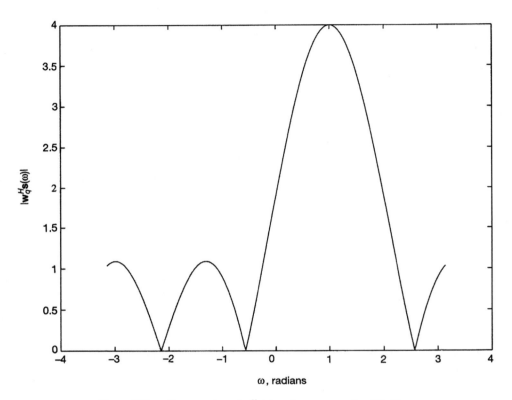

Figure 5.12a   Interpretation of $\mathbf{w}_q^H \mathbf{s}(\omega)$ as the response of an FIR filter.

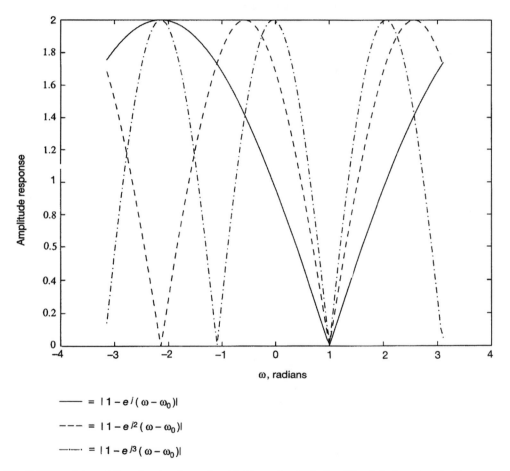

$$\text{——} \quad = |1 - e^{j}(\omega - \omega_0)|$$

$$\text{- - -} \quad = |1 - e^{j2}(\omega - \omega_0)|$$

$$\text{—·—·} \quad = |1 - e^{j3}(\omega - \omega_0)|$$

Figure 5.12b Interpretation of each column of matrix $\mathbf{C}_a$ as a band-rejection filter. In both parts of the figure, part a on the previous page and part b on this page, it is assumed that $\omega_0 = 1$.

By definition, the weight vector $\mathbf{w}_q$ is orthogonal to the columns of matrix $\mathbf{C}_a$. It follows therefore that the optimum weight vector $\mathbf{w}_{ao}$ is identically zero for a quiet environment described by Eq. (5.130). Thus, with $\mathbf{w}_{ao}$ equal to zero, we find from Eq. (5.114) that $\mathbf{w} = \mathbf{w}_q$. It is for this reason that $\mathbf{w}_q$ is often referred to as the *quiescent weight vector*, hence the use of subscript q for denoting it.

Filtering Interpretations of $\mathbf{w}_q$ and $\mathbf{C}_a$

The quiescent weight vector $\mathbf{w}_q$ and matrix $\mathbf{C}_a$ play critical roles of their own in the operation of the GSC. To develop physical interpretations of them, consider an MVDR spectrum estimator (formulated in temporal terms) for which we have

$$\mathbf{C} = \mathbf{s}(\omega_0)$$
$$= [1, e^{-j\omega_0}, \ldots, e^{-j(M-1)\omega_0}]^T \tag{5.131}$$

and

$$\mathbf{g} = 1$$

Hence, the use of these values in Eq. (5.113) yields the corresponding value of the quiescent weight vector to be

$$\mathbf{w}_q = \mathbf{C}(\mathbf{C}^H\mathbf{C})^{-1}\mathbf{g}$$
$$= \frac{1}{M}[1, e^{-j\omega_0}, \ldots, e^{-j(M-1)\omega_0}]^T \tag{5.132}$$

which represents an FIR filter of length M. The frequency response of this filter is given by

$$\mathbf{w}_q^H\mathbf{s}(\omega) = \frac{1}{M}\sum_{k=0}^{M-1} e^{jk(\omega_0-\omega)}$$
$$= \frac{1 - e^{jM(\omega_0-\omega)}}{1 - e^{j(\omega_0-\omega)}} \tag{5.133}$$
$$= \left(\frac{\sin\left(\dfrac{M}{2}(\omega_0 - \omega)\right)}{\sin\left(\dfrac{1}{2}(\omega_0 - \omega)\right)}\right) \exp\left(\frac{j(M-1)}{2}(\omega_0 - \omega)\right)$$

Figure 5.12(a) shows the amplitude response of this filter for $M = 4$ and $\omega_0 = 1$. From this figure we clearly see that the FIR filter representing the quiescent weight vector $\mathbf{w}_q$ acts like a bandpass filter tuned to the angular frequency ω_0, for which the MVDR spectrum estimator is constrained to produce a distortionless response.

Consider next a physical interpretation of the matrix $\mathbf{C}_a$. The use of Eq. (5.131) in (5.103) yields

$$\mathbf{s}^H(\omega_0)\mathbf{C}_a = \mathbf{0} \tag{5.134}$$

According to Eq. (5.134), each of the $(M-L)$ columns of matrix $\mathbf{C}_a$ represents an FIR filter with an amplitude response that is zero at ω_0 as illustrated in Fig. 5.12b for $\omega_0 = 1$, $M = 4$, $L = 1$, and

$$\mathbf{C}_a = \begin{bmatrix} -1 & -1 & -1 \\ e^{-j\omega_0} & 0 & 0 \\ 0 & e^{-j2\omega_0} & 0 \\ 0 & 0 & e^{-j3\omega_0} \end{bmatrix}$$

In other words, the matrix $\mathbf{C}_a$ is represented by a bank of band-rejection filters, each of which is tuned to ω_0. Thus, the matrix $\mathbf{C}_a$ is referred to as a *signal-blocking matrix*, since it blocks (rejects) the received signal at the angular frequency ω_0. The function of the matrix $\mathbf{C}_a$ is to cancel interference that leaks through the sidelobes of the band-pass filter representing the quiescent weight vector $\mathbf{w}_q$.

5.10 SUMMARY AND DISCUSSION

The discrete-time version of the Wiener filter theory, as described in this chapter, has evolved from the pioneering work of Norbert Wiener on linear optimum filters for continuous-time signals. The importance of the Wiener filter lies in the fact that it provides a frame of reference for the linear filtering of stochastic signals, assuming wide-sense stationarity.

The filtering structures that fall under the umbrella of Wiener filter theory are of two different physical types:

- Transversal filters, which are characterized by an impulse response of finite duration
- Narrow-band beamformers, which consist of a set of uniformly spaced antenna elements with adjustable weights.

These two structures share a common feature: they are both examples of a linear device whose output is defined by the inner product of its weight vector and the input vector. The optimum filter involving such a structure is embodied in the Wiener–Hopf equations, the solution of which involves two ensemble-averaged parameters:

- The correlation matrix of the input vector
- The cross-correlation vector between the input vector and desired response

The standard formulation of Wiener filtering requires the availability of a desired response. There are, however, applications where it is not feasible to provide a desired response. In such situations, we may use a class of linear optimum filters known as linearly constrained minimum variance (LCMV) filters or LCMV beamformers, depending on whether the application is temporal or spatial in nature. The essence of the LCMV approach is that it minimizes the average output power, subject to a set of linear constraints on the weight vector. The constraints are imposed so as to prevent the weight vector from canceling the signal of interest. In a special form of LCMV beamformers known as generalized sidelobe cancelers, the weight vector is separated into two components:

- A quiescent weight vector, which satisfies the prescribed constraints
- An unconstrained weight vector, the optimization of which in accordance with the Wiener filter theory minimizes the effects of receiver noise and interfering signals

PROBLEMS

1. A necessary condition for a function $f(z)$ of the complex variable $z = x + jy$ to be an analytic function is that its real part $u(x, y)$ and imaginary part $v(x, y)$ must satisfy the Cauchy–Riemann equations. For a discussion of complex variables, see Appendix A. Demonstrate that the cost function J defined in Eq. (5.43) is *not* an analytic function of the weights by doing the following:
   (a) Show that the product $p^*(-k)w_k$ is an analytic function of the complex tap-weight (filter coefficient) w_k.
   (b) Show that the second product $w_k^* p(-k)$ is *not* an analytic function.

2. Consider a Wiener filtering problem characterized as follows: The correlation matrix $\mathbf{R}$ of the tap-input vector $\mathbf{u}(n)$ is

$$\mathbf{R} = \begin{bmatrix} 1 & 0.5 \\ 0.5 & 1 \end{bmatrix}$$

The cross-correlation vector $\mathbf{p}$ between the tap-input vector $\mathbf{u}(n)$ and the desired response $d(n)$ is

$$\mathbf{p} = \begin{bmatrix} 0.5 \\ 0.25 \end{bmatrix}$$

   (a) Evaluate the tap weights of the Wiener filter.
   (b) What is the minimum mean-squared error produced by this Wiener filter?
   (c) Formulate a representation of the Wiener filter in terms of the eigenvalues of matrix $\mathbf{R}$ and associated eigenvectors.

3. The tap-weight vector of a transversal filter is defined by

$$\mathbf{u}(n) = \alpha(n)\mathbf{s}(n) + \mathbf{v}(n)$$

   where

$$\mathbf{s}(\omega) = [1, e^{-j\omega}, \ldots, e^{-j\omega(M-1)}]^T$$

   and

$$\mathbf{v}(n) = [v(n), v(n-1), \ldots, v(n-M+1)]^T$$

The complex amplitude of the sinusoidal vector $\mathbf{s}(\omega)$ is a random variable with zero mean and variance $\sigma_\alpha^2 = E[|\alpha(n)|^2]$.

(a) Determine the correlation matrix of the tap-input vector $\mathbf{u}(n)$.
(b) Suppose that the desired response $d(n)$ is uncorrelated with $\mathbf{u}(n)$. What is the value of the tap-weight vector of the corresponding Wiener filter?
(c) Suppose that the variance σ_α^2 is zero, and the desired response is defined by

$$d(n) = v(n - k)$$

where $0 \leq k \leq M - 1$. What is the new value of the tap-weight vector of the Wiener filter?
(d) Determine the tap-weight vector of the Wiener filter for a desired response defined by

$$d(n) = \alpha(n)e^{-j\omega\tau}$$

where τ is a prescribed delay.

4. Show that the Wiener–Hopf equations (5.34), defining the tap-weight vector $\mathbf{w}_o$ of the Wiener filter, and Eq. (5.49), defining the minimum mean-squared error $J_{\min}$, may be combined into a single matrix relation

$$\mathbf{A}\begin{bmatrix} 1 \\ -\mathbf{w}_o \end{bmatrix} = \begin{bmatrix} J_{\min} \\ \mathbf{0} \end{bmatrix}$$

The matrix $\mathbf{A}$ is the correlation matrix of the augmented vector

$$\begin{bmatrix} d(n) \\ \mathbf{u}(n) \end{bmatrix}$$

where $d(n)$ is the desired response and $\mathbf{u}(n)$ is the tap-input vector of the Wiener filter.

5. The minimum mean-squared error $J_{\min}$ is defined by [see Eq. (5.49)]

$$J_{\min} = \sigma_d^2 - \mathbf{p}^H\mathbf{R}^{-1}\mathbf{p}$$

where σ_d^2 is the variance of the desired response $d(n)$, $\mathbf{R}$ is the correlation matrix of the tap-input vector $\mathbf{u}(n)$, and $\mathbf{p}$ is the cross-correlation vector between $\mathbf{u}(n)$ and $d(n)$. By applying the unitary similarity transformation to the inverse of the correlation matrix, that is, $\mathbf{R}^{-1}$, show that

$$J_{\min} = \sigma_d^2 - \sum_{k=1}^{M} \frac{|\mathbf{q}_k^H\mathbf{p}|^2}{\lambda_k}$$

where λ_k is the kth eigenvalue of the correlation matrix $\mathbf{R}$, and $\mathbf{q}_k$ is the corresponding eigenvector. Note that $\mathbf{q}_k^H\mathbf{p}$ is a scalar.

6. In this problem, we explore the extent of the improvement that may result from using a more complex Wiener filter for the environment described in Section 5.6. To be specific, the new formulation of the Wiener filter has three taps.
(a) Find the 3-by-3 correlation matrix of the tap inputs of this filter and the 3-by-1 cross-correlation vector between the desired response and the tap inputs.
(b) Compute the 3-by-1 tap-weight vector of the Wiener filter, and also compute the new value for the minimum mean-squared error.

7. In this problem we explore an application of Wiener filtering to *radar*. The sampled form of the transmitted radar signal is $A_0e^{j\omega_0 n}$ where ω_0 is the transmitted angular frequency, and A_0 is the transmitted complex amplitude. The received signal is

$$\mathbf{u}(n) = A_1e^{-j\omega_1 n} + v(n)$$

where $|A_1| < |A_0|$ and ω_1 differs from ω_0 by virtue of the *Doppler* shift produced by the motion of a target of interest, and $v(n)$ is a sample of white noise.

(a) Show that the correlation matrix of the time series $u(n)$, made up of M elements, may be written as

$$\mathbf{R} = \sigma_v^2 \mathbf{I} + \sigma_1^2 \mathbf{s}(\omega_1) \mathbf{s}^H(\omega_1)$$

where σ_v^2 is the variance of the zero-mean white noise $v(n)$, and

$$\sigma_1^2 = E[|A_1|^2]$$

and

$$\mathbf{s}(\omega_1) = [1, e^{-j\omega_1}, \ldots, e^{-j\omega_1(M-1)}]^T$$

(b) The time series $u(n)$ is applied to an M-tap Wiener filter with the cross-correlation vector $\mathbf{p}$ between $u(n)$ and the desired response $d(n)$ preset to

$$\mathbf{p} = \sigma_0^2 \mathbf{s}(\omega_0)$$

where

$$\sigma_0^2 = E[|A_0|^2]$$

and

$$\mathbf{s}(\omega_0) = [1, e^{-j\omega_0}, \ldots, e^{-j\omega_0(M-1)}]^T$$

Derive an expression for the tap-weight vector of the Wiener filter.

8. An array processor consists of a primary sensor and a reference sensor interconnected with each other. The output of the reference sensor is weighted by w and then subtracted from the output of the primary sensor. Show that the mean-square value of the output of the array processor is minimized when the weight w attains the optimum value

$$w_o = \frac{E[u_1(n)u_2^*(n)]}{E[|u_2(n)|^2]}$$

where $u_1(n)$ and $u_2(n)$ are the primary- and reference-sensor outputs at time n, respectively.

9. Consider a discrete-time stochastic process $u(n)$ that consists of K (uncorrelated) complex sinusoids plus additive white noise of zero mean and variance σ^2. That is,

$$u(n) = \sum_{k=1}^{K} A_k e^{j\omega_k n} + v(n)$$

where the terms $A_k \exp(j\omega_k n)$ and $v(n)$ refer to the kth sinusoid and noise, respectively. The process $u(n)$ is applied to a transversal filter with M taps, producing the output

$$e(n) = \mathbf{w}^H \mathbf{u}(n)$$

Assume that $M > K$. The requirement is to choose the tap-weight vector $\mathbf{w}$ so as to minimize the mean-square value of $e(n)$, subject to the multiple signal-protection constraint

$$\mathbf{S}^H \mathbf{w} = \mathbf{D}^{1/2} \mathbf{1}$$

where $\mathbf{S}$ is the M-by-K signal matrix whose kth column has $1, \exp(j\omega_k), \ldots, \exp[j\omega_k(M-1)]$ for its elements, $\mathbf{D}$ is the K-by-K diagonal matrix whose nonzero elements equal the average powers of the individual sinusoids, and the K-by-1 vector $\mathbf{1}$ has 1's for all its K elements. Using the method of Lagrange multipliers, show that the value of the optimum weight vector that results from this constrained optimization equals

$$\mathbf{w}_o = \mathbf{R}^{-1}\mathbf{S}(\mathbf{S}^H\mathbf{R}^{-1}\mathbf{S})^{-1}\mathbf{D}^{1/2}\mathbf{1}$$

where $\mathbf{R}$ is the correlation matrix of the M-by-1 tap-input vector $\mathbf{u}(n)$. This formula represents a temporal generalization of the MVDR formula.

10. The weight vector $\mathbf{w}$ of the LCMV beamformer is defined by Eq. (5.96). In general, the LCMV beamformer so defined does *not* maximize the output signal-to-noise ratio. To be specific, let the input vector $\mathbf{u}(n)$ be written as

$$\mathbf{u}(n) = \mathbf{s}(n) + \mathbf{v}(n)$$

where the vector $\mathbf{s}(n)$ represents the signal component, and the vector $\mathbf{v}(n)$ represents the additive noise component. Show that the weight vector $\mathbf{w}$ does *not* satisfy the condition

$$\max_{\mathbf{w}} \frac{\mathbf{w}^H\mathbf{R}_s\mathbf{w}}{\mathbf{w}^H\mathbf{R}_v\mathbf{w}}$$

where $\mathbf{R}_s$ is the correlation matrix of $\mathbf{s}(n)$, and $\mathbf{R}_v$ is the correlation matrix of $\mathbf{v}(n)$.

11. In this problem, we explore the design of constraints for a beamformer using a *nonuniformly spaced array* of antenna elements. Let t_i denote the propagation delay to the ith element for a plane wave impinging on the array from look direction θ; the delay t_i is measured with respect to the zero-time reference.
    (a) Find the response of the beamformer with elemental weight vector $\mathbf{w}$ to a signal of angular frequency ω that originates from the look direction θ.
    (b) Hence, specify the linear constraint imposed on the array to produce a response equal to g along the direction θ.

12. Consider the problem of detecting a known signal in the presence of additive noise. The noise is assumed to be Gaussian, to be independent of the signal, and to have zero mean and a positive definite correlation matrix $\mathbf{R}_v$. The aim of the problem is to show that under these conditions the three criteria: minimum mean-squared error, maximum signal-to-noise ratio, and the likelihood ratio test yield identical designs for the transversal filter.

    Let $u(n)$, $n = 1, 2, \ldots, M$, denote a set of M complex-valued data samples. Let $v(n)$, $n = 1, 2, \ldots, M$, denote a set of samples taken from a Gaussian noise process of zero mean. Finally, let $s(n)$, $n = 1, 2, \ldots, M$, denote samples of the signal. The detection problem is to determine whether the input consists of signal plus noise or noise alone. That is, the two hypotheses to be tested for are

$$\text{hypothesis } H_2\text{:} \quad u(n) = s(n) + v(n), \qquad n = 1, 2, \ldots, M$$
$$\text{hypothesis } H_1\text{:} \quad u(n) = v(n), \qquad\qquad n = 1, 2, \ldots, M$$

    (a) The *Wiener filter* minimizes the mean-squared error. Show that this criterion yields an optimum tap-weight vector for estimating s_k, the kth component of signal vector $\mathbf{s}$, that equals

$$\mathbf{w}_o = \frac{s_k}{1 + \mathbf{s}^H\mathbf{R}_v^{-1}\mathbf{s}}\mathbf{R}_v^{-1}\mathbf{s}$$

    Hint: To evaluate the inverse of the correlation matrix of $u(n)$ under hypothesis H_2, you may use the matrix inversion lemma. Let

$$\mathbf{A} = \mathbf{B}^{-1} + \mathbf{C}\mathbf{D}^{-1}\mathbf{C}^H$$

where $\mathbf{A}$, $\mathbf{B}$ and $\mathbf{D}$ are positive-definite matrices. Then

$$\mathbf{A}^{-1} = \mathbf{B} - \mathbf{BC}(\mathbf{D} + \mathbf{C}^H \mathbf{BC})^{-1} \mathbf{C}^H \mathbf{B}$$

(b) The *maximum signal-to-noise ratio filter* maximizes the ratio

$$\rho = \frac{\text{average power of filter output due to signal}}{\text{average power of filter output due to noise}}$$

$$= \frac{E[(\mathbf{w}^H \mathbf{s})^2]}{E[(\mathbf{w}^H \mathbf{v})^2]}$$

Show that the tap-weight vector for which the output signal-to-noise ratio ρ is at maximum equals

$$\mathbf{w}_{SN} = \mathbf{R}_v^{-1} \mathbf{s}$$

Hint: Since $\mathbf{R}_v$ is positive definite, you may use $\mathbf{R}_v = \mathbf{R}_v^{1/2} \mathbf{R}_v^{1/2}$.

(c) The *likelihood ratio processor* computes the log-likelihood ratio and compares it to a threshold. If the threshold is exceeded, it decides in favor of hypothesis H_2; otherwise, it decides in favor of hypothesis H_1. The likehood ratio is defined by

$$\Lambda = \frac{\mathbf{f}_U(\mathbf{u}|H_2)}{\mathbf{f}_U(\mathbf{u}|H_1)}$$

where $\mathbf{f}_U(\mathbf{u}|H_i)$ is the conditional joint probability density function of the observation vector $\mathbf{u}$, given that hypothesis H_i is true, where $i = 1, 2$. Show that the likelihood ratio test is equivalent to the test

$$\mathbf{w}_{ml}^H \mathbf{u} \underset{H_1}{\overset{H_2}{\gtrless}} \eta$$

where η is the threshold and

$$\mathbf{w}_{ml} = \mathbf{R}_v^{-1} \mathbf{s}$$

Hint: Refer to Section 2.11 for the joint probability function of the M-by-1 Gaussian noise vector $\mathbf{v}$ with zero mean and correlation matrix $\mathbf{R}_v$.

CHAPTER

6

Linear Prediction

One of the most celebrated problems in time-series analysis is that of *predicting* a future value of a stationary discrete-time stochastic process, given a set of past samples of the process. To be specific, consider the time series $u(n)$, $u(n-1)$, ..., $u(n-M)$, representing $(M+1)$ samples of such a process up to and including time n. The operation of prediction may, for example, involve using the samples $u(n-1)$, $u(n-2)$, ..., $u(n-M)$ to make an estimate of $u(n)$. Let $\mathcal{U}_{n-1}$ denote the M-dimensional space spanned by the samples $u(n-1)$, $u(n-2)$, ..., $u(n-M)$, and use $\hat{u}(n|\mathcal{U}_{n-1})$ to denote the *predicted value* of $u(n)$ given this set of samples. In *linear prediction*, we express this predicted value as a linear combination of the samples $u(n-1)$, $u(n-2)$, ..., $u(n-M)$. This operation corresponds to one-step prediction into the future, measured with respect to time $n-1$. Accordingly, we refer to this form of prediction as *one-step linear prediction in the forward direction* or simply *forward linear prediction*. In another form of prediction, we use the samples $u(n)$, $u(n-1)$, ..., $u(n-M+1)$ to make a prediction of the past sample $u(n-M)$. We refer to this second form of prediction as *backward linear prediction*.[1]

In this chapter, we study forward linear prediction (FLP) as well as backward linear prediction (BLP). In particular, we use the Wiener filter theory of Chapter 5 to optimize

---

[1] The term "backward prediction" is somewhat of a misnomer. A more appropriate description for this operation is "hindsight." Correspondingly, the use of "forward" in the associated operation of forward prediction is superfluous. Nevertheless, the terms "forward prediction" and "backward prediction" have become deeply embedded in the literature on linear prediction.

the design of a forward or backward *predictor* in the mean-square sense for the case of a wide-sense stationary discrete-time stochastic process. As explained in Chapter 2, the correlation matrix of such a process has a Toeplitz structure. We will put this Toeplitz structure to good use in developing algorithms that are computationally efficient.

6.1 FORWARD LINEAR PREDICTION

Figure 6.1(a) shows a *forward predictor* that consists of a linear transversal filter with M tap weights $w_{f,1}, w_{f,2}, \ldots, w_{f,M}$ and tap inputs $u(n-1), u(n-2), \ldots, u(n-M)$, respectively. We assume that these tap inputs are drawn from a wide-sense stationary stochastic process of zero mean. We further assume that the tap weights are optimized in the mean-square sense in accordance with the Wiener filter theory. The predicted value $\hat{u}(n|\mathcal{U}_{n-1})$ is defined by

$$\hat{u}(n|\mathcal{U}_{n-1}) = \sum_{k=1}^{M} w_{f,k}^* u(n-k) \tag{6.1}$$

For the situation described herein, the desired response $d(n)$ equals $u(n)$, representing the actual sample of the input process at time n. We may thus write

$$d(n) = u(n) \tag{6.2}$$

The *forward prediction error* equals the difference between the input sample $u(n)$ and its predicted value $\hat{u}(n|\mathcal{U}_{n-1})$. We denote the forward prediction error by $f_M(n)$ and thus write

$$f_M(n) = u(n) - \hat{u}(n|\mathcal{U}_{n-1}) \tag{6.3}$$

The subscript M in the symbol for the forward prediction error signifies *order* of the predictor, defined as *the number of unit-delay elements needed to store the given set of samples used to make the prediction*. The reason for using the subscript will become apparent later in the chapter.

Let P_M denote the *minimum mean-squared prediction error*:

$$P_M = E[|f_M(n)|^2] \tag{6.4}$$

With the tap inputs assumed to have zero mean, the forward prediction error $f_M(n)$ will likewise have zero mean. Under this condition, P_M will also equal the variance of the forward prediction error. Yet another interpretation for P_M is that it may be viewed as the ensemble-averaged *forward prediction error power*, assuming that $f_M(n)$ is developed across a 1-Ω load. We will use the latter description to refer to P_M.

Let $\mathbf{w}_f$ denote the M-by-1 optimum tap-weight vector of the forward predictor in Fig. 6.1(a). We write it in expanded form as

$$\mathbf{w}_f = [w_{f,1}, w_{f,2}, \ldots, w_{f,M}]^T \tag{6.5}$$

To solve the Wiener–Hopf equations for the weight vector $\mathbf{w}_f$, we require knowledge of two quantities: (1) the M-by-M correlation matrix of the tap inputs $u(n-1), u(n-2),$

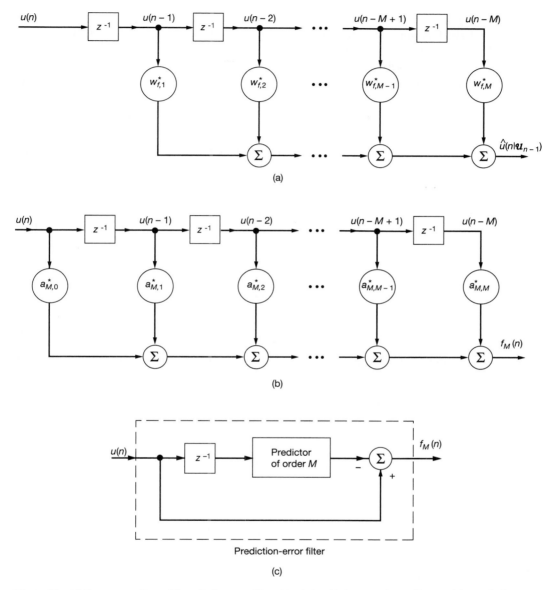

Figure 6.1 (a) One-step predictor; (b) prediction-error filter; (c) relationship between the predictor and the prediction-error filter.

. . . , $u(n - M)$, and (2) the M-by-1 cross-correlation vector between these tap inputs and the desired response $u(n)$. To evaluate P_M, we require a third quantity, the variance of $u(n)$. We now consider these three quantities, one by one:

1. The tap inputs $u(n - 1)$, $u(n - 2)$, . . . , $u(n - M)$ define the M-by-1 tap-input vector, $\mathbf{u}(n - 1)$, as shown by

$$\mathbf{u}(n - 1) = [u(n - 1), u(n - 2), \ldots, u(n - M)]^T \tag{6.6}$$

Hence, the correlation matrix of the tap inputs equals

$$\mathbf{R} = E[\mathbf{u}(n - 1)\mathbf{u}^H(n - 1)]$$

$$= \begin{bmatrix} r(0) & r(1) & \cdots & r(M - 1) \\ r*(1) & r(0) & \cdots & r(M - 2) \\ \cdot & \cdot & \cdot & \cdot \\ \cdot & \cdot & \cdot & \cdot \\ \cdot & \cdot & \cdot & \cdot \\ r*(M - 1) & r*(M - 2) & \cdots & r(0) \end{bmatrix} \tag{6.7}$$

where $r(k)$ is the autocorrelation function of the input process for lag k, where $k = 0, 1, \ldots, M - 1$. Note that the symbol used for the correlation matrix of the tap inputs in Fig. 6.1(a) is the same as that for the correlation matrix of the tap inputs in the transversal filter of Fig. 5.4. We are justified to do this since the input process in both cases is assumed to be wide-sense stationary, so the correlation matrix of the process is invariant to a time shift.

2. The cross-correlation vector between the tap inputs $u(n - 1), \ldots, u(n - M)$ and the desired response $u(n)$ equals

$$\mathbf{r} = E[\mathbf{u}(n - 1)u*(n)]$$

$$= \begin{bmatrix} r*(1) \\ r*(2) \\ \cdot \\ \cdot \\ \cdot \\ r*(M) \end{bmatrix} = \begin{bmatrix} r(-1) \\ r(-2) \\ \cdot \\ \cdot \\ \cdot \\ r(-M) \end{bmatrix} \tag{6.8}$$

3. The variance of $u(n)$ equals $r(0)$, since $u(n)$ has zero mean.

In Table 6.1, we summarize the various quantities pertaining to the Wiener filter of Fig. 5.4 and the corresponding quantities pertaining to the forward predictor of Fig. 6.1(a). The last column of this table pertains to the backward predictor, on which more will be said later.

TABLE 6.1 SUMMARY OF WIENER FILTER VARIABLES

Quantity	Wiener filter of Fig. 5.4	Forward predictor of Fig. 6.1(a)	Backward predictor of Fig. 6.2(a)
Tap-input vector	$\mathbf{u}(n)$	$\mathbf{u}(n-1)$	$\mathbf{u}(n)$
Desired response	$d(n)$	$u(n)$	$u(n-M)$
Tap-weight vector	$\mathbf{w}_o$	$\mathbf{w}_f$	$\mathbf{w}_b$
Estimation error	$e(n)$	$f_M(n)$	$b_M(n)$
Correlation matrix of tap inputs	$\mathbf{R}$	$\mathbf{R}$	$\mathbf{R}$
Cross-correlation vector between tap inputs and desired response	$\mathbf{p}$	$\mathbf{r}$	$\mathbf{r}^{B*}$
Minimum mean-squared error	$J_{\min}$	P_M	P_M

Thus, using the correspondences of this table, we may adapt the Wiener–Hopf equations (5.45) to solve the forward linear prediction (FLP) problem for stationary inputs and so write

$$\mathbf{R}\mathbf{w}_f = \mathbf{r} \tag{6.9}$$

Similarly, the use of Eq. (5.49), together with Eq. (6.8), yields the following expression for the forward prediction-error power:

$$P_M = r(0) - \mathbf{r}^H \mathbf{w}_f \tag{6.10}$$

From Eqs. (6.8) and (6.9), we see that the M-by-1 tap-weight vector of the forward predictor and the forward prediction-error power are determined solely by the set of $(M+1)$ autocorrelation function values of the input process for lags $0, 1, \ldots, M$.

Relation between Linear Prediction and Autoregressive Modeling

It is highly informative to compare the Wiener-Hopf equations (6.9) for linear prediction with the Yule–Walker equations (2.66) for an autoregressive (AR) model. We see that these two systems of simultaneous equations are of exactly the same mathematical form. Furthermore, Eq. (6.10) defining the average power (i.e., variance) of the forward prediction error is also of the same mathematical form as Eq. (2.71) defining the variance of the white-noise process used to excite the autoregressive model. For the case of an AR process for which we know the model order M, we may thus state that when a forward predictor is optimized in the mean-square sense, in theory, its tap weights take on the same values as the corresponding parameters of the process. This relationship should not be surprising since the equation defining the forward prediction error and the difference equation defining the autoregressive model have the same mathematical form. When the process is not autoregressive, however, the use of a predictor provides an approximation to the process.

Forward Prediction-Error Filter

The forward predictor of Fig. 6.1(a) consists of M unit-delay elements and M tap weights $w_{f,1}, w_{f,2}, \ldots, w_{f,M}$ that are fed with the respective samples $u(n-1), u(n-2), \ldots,$ $u(n-M)$ as inputs. The resultant output is the predicted value of $u(n)$, which is defined by Eq. (6.1). Hence, substituting Eq. (6.1) in (6.3), we may express the forward prediction error as

$$f_M(n) = u(n) - \sum_{k=1}^{M} w_{f,k}^* u(n-k) \qquad (6.11)$$

Let $a_{M,k}, k = 0, 1, \ldots, M,$ denote the tap weights of a new transversal filter, which are related to the tap weights of the forward predictor as follows:

$$a_{M,k} = \begin{cases} 1, & k = 0 \\ -w_{f,k}, & k = 1, 2, \ldots, M \end{cases} \qquad (6.12)$$

Then we may combine the two terms on the right-hand side of Eq. (6.11) into a single summation as follows:

$$f_M(n) = \sum_{k=0}^{M} a_{M,k}^* u(n-k) \qquad (6.13)$$

This input–output relation is represented by the transversal filter shown in Fig. 6.1(b). A filter that operates on the set of samples $u(n), u(n-1), \ldots, u(n-M)$ to produce the forward prediction error $f_M(n)$ at its output is called a forward *prediction-error filter* (PEF).

The relationship between the forward prediction-error filter and the forward predictor is illustrated in block diagram form in Fig. 6.1(c). Note that the length of the prediction-error filter exceeds the length of the one-step prediction filter by 1. However, both filters have the same order, M, as they both involve the same number of delay elements for the storage of past data.

Augmented Wiener–Hopf Equations for Forward Prediction

The Wiener–Hopf equations (6.9) define the tap-weight vector of the forward predictor, while Eq. (6.10) defines the resulting forward prediction-error power P_M. We may combine these two equations into a single matrix relation as follows:

$$\begin{bmatrix} r(0) & \mathbf{r}^H \\ \mathbf{r} & \mathbf{R} \end{bmatrix} \begin{bmatrix} 1 \\ -\mathbf{w}_f \end{bmatrix} = \begin{bmatrix} P_M \\ \mathbf{0} \end{bmatrix} \qquad (6.14)$$

where $\mathbf{0}$ is the M-by-1 null vector. The M-by-M correlation matrix $\mathbf{R}$ is defined in Eq. (6.7), and the M-by-1 correlation vector $\mathbf{r}$ is defined in Eq. (6.8). The partitioning of the $(M + 1)$-by-$(M + 1)$ correlation matrix on the left-hand side of Eq. (6.14) into the form shown therein was discussed in Section 2.3. Note that this $(M + 1)$-by-$(M + 1)$ matrix equals the correlation matrix of the tap inputs $u(n), u(n-1), \ldots, u(n-M)$ in the prediction-error filter of Fig. 6.1(b). Moreover, the $(M + 1)$-by-1 coefficient vector on the left-hand side of Eq. (6.14) equals the *forward prediction-error filter vector*:

$$\mathbf{a}_M = \begin{bmatrix} 1 \\ -\mathbf{w}_f \end{bmatrix} \tag{6.15}$$

We may also express the matrix relation of Eq. (6.14) as a system of $(M + 1)$ simultaneous equations as follows:

$$\sum_{l=0}^{M} a_{M,l} r(l - i) = \begin{cases} P_M, & i = 0 \\ 0, & i = 1, 2, \ldots, M \end{cases} \tag{6.16}$$

We refer to Eq. (6.14) or (6.16) as the *augmented Wiener–Hopf equations* of a forward prediction-error filter of order M.

Example 1

For the case of a prediction-error filter of order $M = 1$, Eq. (6.14) yields a pair of simultaneous equations described by

$$\begin{bmatrix} r(0) & r(1) \\ r^*(1) & r(0) \end{bmatrix} \begin{bmatrix} a_{1,0} \\ a_{1,1} \end{bmatrix} = \begin{bmatrix} P_1 \\ 0 \end{bmatrix}$$

Solving for $a_{1,0}$, and $a_{1,1}$, we get

$$a_{1,0} = \frac{P_1}{\Delta_r} r(0)$$

$$a_{1,1} = -\frac{P_1}{\Delta_r} r^*(1)$$

where Δ_r is the determinant of the correlation matrix; thus

$$\Delta_r = \begin{vmatrix} r(0) & r(1) \\ r^*(1) & r(0) \end{vmatrix}$$

$$= r^2(0) - |r(1)|^2$$

But $a_{1,0}$ equals 1. Hence,

$$P_1 = \frac{\Delta_r}{r(0)}$$

$$a_{1,1} = -\frac{r^*(1)}{r(0)}$$

Consider next the case of a prediction-error filter of order $M = 2$. Equation (6.14) yields a system of three simultaneous equations, as shown by

$$\begin{bmatrix} r(0) & r(1) & r(2) \\ r^*(1) & r(0) & r(1) \\ r^*(2) & r^*(1) & r(0) \end{bmatrix} \begin{bmatrix} a_{2,0} \\ a_{2,1} \\ a_{2,2} \end{bmatrix} = \begin{bmatrix} P_2 \\ 0 \\ 0 \end{bmatrix}$$

Solving for $a_{2,0}$, $a_{2,1}$, and $a_{2,2}$, we get

$$a_{2,0} = \frac{P_2}{\Delta_r}[r^2(0) - |r(1)|^2]$$

$$a_{2,1} = -\frac{P_2}{\Delta_r}[r^*(1)r(0) - r(1)r^*(2)]$$

$$a_{2,2} = \frac{P_2}{\Delta_r}[(r^*(1))^2 - r(0)r^*(2)]$$

where Δ_r is the determinant of the correlatin matrix:

$$\Delta_r = \begin{vmatrix} r(0) & r(1) & r(2) \\ r^*(1) & r(0) & r(1) \\ r^*(2) & r^*(1) & r(0) \end{vmatrix}$$

The coefficient $a_{2,0}$ equals 1. Accordingly, we may express the prediction-error power P_2 as

$$P_2 = \frac{\Delta_r}{r^2(0) - |r(1)|^2}$$

and the prediction-error filter coefficients $a_{2,1}$ and $a_{2,2}$ as

$$a_{2,1} = -\frac{r^*(1)(r(0) - r(1)r^*(2))}{r^2(0) - |r(1)|^2}$$

$$a_{2,2} = \frac{(r^*(1))^2 - r(0)r^*(2)}{r^2(0) - |r(1)|^2}$$

6.2 BACKWARD LINEAR PREDICTION

The form of linear prediction considered in Section 6.1 is said to be in the *forward* direction. That is, given the time series $u(n)$, $u(n - 1)$, . . . , $u(n - M)$, we use the subset of M samples $u(n - 1)$, $u(n - 2)$, . . . , $u(n - M)$ to make a prediction of the sample $u(n)$. This operation corresponds to *one-step linear prediction into the future*, measured with respect to time $n - 1$. Naturally, we may also operate on this time series in the *backward* direction. That is, we may use the subset of M samples $u(n)$, $u(n - 1)$, . . . , $u(n - M + 1)$ to make a prediction of the sample $u(n - M)$. This second operation corresponds to *backward linear prediction by one step*, measured with respect to time $n - M + 1$.

Let $\mathcal{U}_n$ denote the M-dimensional space spanned by $u(n)$, $u(n - 1)$, . . . , $u(n - M + 1)$ that are used in making the backward prediction. Then, using this set of samples as tap inputs, we make a linear prediction of the sample $u(n - M)$, as shown by

$$\hat{u}(n - M|\mathcal{U}_n) = \sum_{k=1}^{M} w_{b,k}^* u(n - k + 1) \tag{6.17}$$

where $w_{b,1}$, $w_{b,2}$, . . . , $w_{b,M}$ are the tap weights. Figure 6.2(a) shows a representation of the

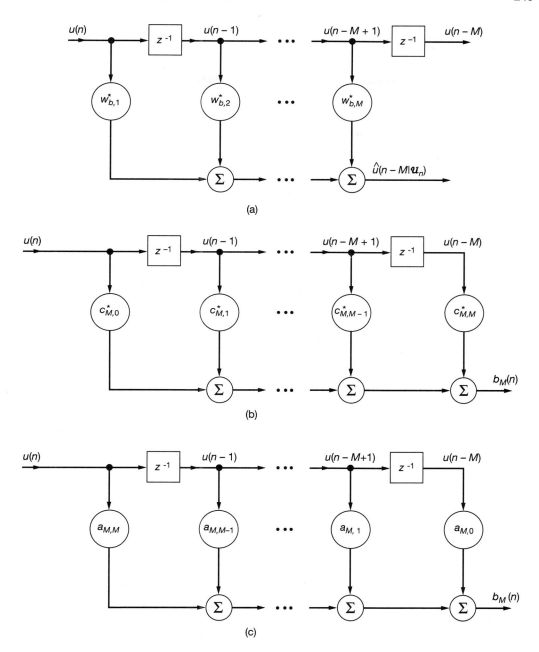

Figure 6.2   (a) Backward one-step predictor; (b) backward prediction-error filter; (c) backward prediction-error filter defined in terms of the tap weights of the corresponding forward prediction-error filter.

backward predictor as described by Eq. (6.17). We assume that these tap weights are optimized in the mean-square sense in accordance with the Wiener filter theory.

In the case of backward prediction, the desired response equals

$$d(n) = u(n - M) \tag{6.18}$$

The *backward prediction error* equals the difference between the actual sample value $u(n - M)$ and its predicted value $\hat{u}(n - M|\mathcal{U}_n)$. We denote the backward prediction error by $b_M(n)$ and thus write

$$b_M(n) = u(n - M) - \hat{u}(n - M|\mathcal{U}_n) \tag{6.19}$$

Here, again, the subscript M in the symbol for the backward prediction error $b_M(n)$ signifies the number of unit-delay elements needed to store the given set of samples used to make the prediction; that is, M is the order of the predictor.

Let P_M denote the *minimum mean-squared prediction error*

$$P_M = E[|b_M(n)|^2] \tag{6.20}$$

We may also view P_M as the ensemble-averaged *backward prediction-error power*, assuming that $b_M(n)$ is developed across a 1-Ω load.

Let $\mathbf{w}_b$ denote the M-by-1 optimum tap-weight vector of the backward predictor in Fig. 6.2(a). We express it in the expanded form

$$\mathbf{w}_b = [w_{b,1}, w_{b,2}, \ldots, w_{b,M}]^T \tag{6.21}$$

To solve the Wiener–Hopf equations for the weight vector $\mathbf{w}_b$, we require knowledge of two quantities: (1) the M-by-M correlation matrix of the tap inputs $u(n)$, $u(n - 1)$, ..., $u(n - M + 1)$, and (2) the M-by-1 cross-correlation vector between the desired response $u(n - M)$ and these tap inputs. To evaluate P_M, we need a third quantity, the variance of $u(n - M)$. We consider these three quantities in turn:

1. Let $\mathbf{u}(n)$ denote the M-by-1 tap-input vector in the backward predictor of Fig. 6.2(a). We write it in expanded form as

$$\mathbf{u}(n) = [u(n), u(n - 1), \ldots, u(n - M + 1)]^T \tag{6.22}$$

The M-by-M correlation matrix of the tap inputs in Fig. 6.2(a) thus equals

$$\mathbf{R} = E[\mathbf{u}(n)\mathbf{u}^H(n)]$$

The expanded form of the correlation matrix $\mathbf{R}$ is given in Eq. (6.7).

2. The M-by-1 cross-correlation vector between the tap inputs $u(n)$, $u(n - 1)$, ..., $u(n - M + 1)$ and the desired response $u(n - M)$ equals

$$\mathbf{r}^{B*} = E[\mathbf{u}(n)u^*(n - M)]$$

$$= \begin{bmatrix} r(M) \\ r(M - 1) \\ \cdot \\ \cdot \\ \cdot \\ r(1) \end{bmatrix} \tag{6.23}$$

The expanded form of the correlation vector $\mathbf{r}$ is given in Eq. (6.8). As usual, the superscript B denotes backward arrangement and the asterisk denotes complex conjugation.

3. The variance of the desired response $u(n - M)$ equals $r(0)$.

In the last column of Table 6.1, we summarize the various quantities pertaining to the backward predictor of Fig. 6.2(a).

Accordingly, using the correspondences of Table 6.1, we may adapt the Wiener–Hopf equations (5.34) to solve the backward linear prediction (BLP) problem for stationary inputs and so write

$$\mathbf{R}\mathbf{w}_b = \mathbf{r}^{B*} \tag{6.24}$$

Similarly, the use of Eq. (5.49), together with Eq. (6.24), yields the following expression for the backward prediction-error power:

$$P_M = r(0) - \mathbf{r}^{BT}\mathbf{w}_b \tag{6.25}$$

Here again we see that the M-by-1 tap-weight vector $\mathbf{w}_b$ of a backward predictor and the backward prediction-error power P_M are uniquely defined by knowledge of the set of auto-correlation function values of the process for lags $0, 1, \ldots, M$.

Relations between Backward and Forward Predictors

In comparing the two sets of Wiener–Hopf equations (6.9) and (6.24), pertaining to forward prediction and backward prediction, respectively, we see that the vector on the right-hand side of Eq. (6.24) differs from that of Eq. (6.9) in two respects: (1) its elements are arranged backward, and (2) they are complex conjugated. To correct for the first difference, we reverse the order in which the elements of the vector on the right-hand side of Eq. (6.24) are arranged. This operation has the effect of replacing the left-hand side of Eq. (6.24) by $\mathbf{R}^T\mathbf{w}_b^B$, where $\mathbf{R}^T$ is the transpose of the correlation matrix $\mathbf{R}$ and $\mathbf{w}_b^B$ is the backward version of the tap-weight vector $\mathbf{w}_b$ (see Problem 3). We may thus write

$$\mathbf{R}^T\mathbf{w}_b^B = \mathbf{r}^* \tag{6.26}$$

To correct for the remaining difference, we complex-conjugate both sides of Eq. (6.26), obtaining

$$\mathbf{R}^H\mathbf{w}_b^{B*} = \mathbf{r}$$

Since the correlation matrix $\mathbf{R}$ is Hermitian, that is, $\mathbf{R}^H = \mathbf{R}$, we may thus reformulate the Wiener–Hopf equations for backward prediction as

$$\mathbf{R}\mathbf{w}_b^{B*} = \mathbf{r} \tag{6.27}$$

Now we may compare Eq. (6.27) with Eq. (6.9) and thus deduce the following fundamental relationship between the tap-weight vectors of a backward predictor and the corresponding forward predictor:

$$\mathbf{w}_b^{B*} = \mathbf{w}_f \tag{6.28}$$

Equation (6.28) states that *we may modify a backward predictor into a forward predictor by reversing the sequence in which its tap weights are positioned and also complex-conjugating them.*

Next we wish to show that the ensemble-averaged error powers for backward prediction and forward prediction have exactly the same value. To do this, we first observe that the product $\mathbf{r}^{BT}\mathbf{w}_b$ equals $\mathbf{r}^T\mathbf{w}_b^B$, so we may rewrite Eq. (6.25) as

$$P_M = r(0) - \mathbf{r}^T\mathbf{w}_b^B \tag{6.29}$$

Taking the complex conjugate of both sides of Eq. (6.29), and recognizing that both P_M and $r(0)$ are unaffected by this operation since they are real-valued scalars, we get

$$P_M = r(0) - \mathbf{r}^H\mathbf{w}_b^B{}^* \tag{6.30}$$

Comparing this result with Eq. (6.10) and using the equivalence of Eq. (6.28), we find that the backward prediction-error power has exactly the same value as the forward prediction-error power. Indeed, it is in anticipation of this equality that we have used the same symbol P_M to denote both the forward prediction-error power and the backward prediction-error power.

Backward Prediction-Error Filter

The backward prediction error $b_M(n)$ equals the difference between the desired response $u(n - M)$ and the linear prediction of it, given the samples $u(n)$, $u(n - 1)$, . . . , $u(n - M + 1)$. This prediction is defined by Eq. (6.17). Therefore, substituting Eq. (6.17) in (6.19), we get

$$b_M(n) = u(n - M) - \sum_{k=1}^{M} w_{bk}^* u(n - k + 1) \tag{6.31}$$

Define the tap weights of the backward prediction-error filter in terms of the corresponding backward predictor as follows:

$$c_{M,k} = \begin{cases} -w_{b,k+1} & k = 0, 1, \ldots, M - 1 \\ 1, & k = M \end{cases} \tag{6.32}$$

Hence, we may rewrite Eq. (6.31) as [see Fig. 6.2(b)]

$$b_M(n) = \sum_{k=0}^{M} c_{M,k}^* u(n - k) \tag{6.33}$$

Equation (6.28) defines the tap-weight vector of the backward predictor in terms of that of the forward predictor. We may express the scalar version of this relation as

$$w_{b,M-k+1}^* = w_{f,k}, \qquad k = 1, 2, \ldots, M$$

or, equivalently

$$w_{b,k} = w_{f,M-k+1}^* \qquad k = 1, 2, \ldots, M \tag{6.34}$$

Hence, substituting Eq. (6.34) in (6.32), we get

$$c_{M,k} = \begin{cases} -w_{f,M-k}^*, & k = 0, 1, \ldots, M - 1 \\ 1, & k = M \end{cases} \tag{6.35}$$

Thus, using the relationship between the tap weights of the forward prediction-error filter and those of the forward predictor as given in Eq. (6.12), we may write

$$c_{M,k} = a_{M,M-k}^*, \qquad k = 0, 1, \ldots, M \tag{6.36}$$

Accordingly, we may express the input–output relation of the backward prediction-error filter in the equivalent form

$$b_M(n) = \sum_{k=0}^{M} a_{M,M-k} u(n - k) \tag{6.37}$$

The input–output relation of Eq. (6.37) is depicted in Fig. 6.2(c). Comparison of this representation for a backward prediction-error filter with that of Fig. 6.1(b) for the corresponding forward prediction-error filter reveals that these two forms of a prediction-error filter for *stationary inputs* are uniquely related to each other. In particular, *we may modify a forward prediction-error filter into the corresponding backward prediction-error filter by reversing the sequence in which the tap weights are positioned and complex-conjugating them*. Note that in both figures, the respective tap inputs have the same values.

Augmented Wiener–Hopf Equations for Backward Prediction

The set of Wiener–Hopf equations for backward prediction is defined by Eq. (6.24), and the resultant backward prediction-error power is defined by Eq. (6.25). We may combine these two equations into a single relation as follows:

$$\begin{bmatrix} \mathbf{R} & \mathbf{r}^{B*} \\ \mathbf{r}^{BT} & r(0) \end{bmatrix} \begin{bmatrix} -\mathbf{w}_b \\ 1 \end{bmatrix} = \begin{bmatrix} \mathbf{0} \\ P_M \end{bmatrix} \tag{6.38}$$

where $\mathbf{0}$ is the M-by-1 null vector. The M-by-M matrix $\mathbf{R}$ is the correlation matrix of the M-by-1 tap-input vector $\mathbf{u}(n)$; it has the expanded form shown in the second line of Eq. (6.7) by virtue of the assumed wide-sense stationarity of the input process. The M-by-1 vector $\mathbf{r}^{B*}$ is the cross-correlation vector between the input vector $\mathbf{u}(n)$ and the desired response $u(n - M)$; here again, the assumed wide-sense stationarity of the input process means that the vector $\mathbf{r}$ has the expanded form shown in the second line of Eq. (6.8). The $(M + 1)$-by-$(M + 1)$ matrix on the left-hand side of Eq. (6.38) equals the correlation matrix of the tap inputs in the backward prediction-error filter of Fig. 6.2(c). The partitioning of this $(M + 1)$-by-$(M + 1)$ into the form shown in Eq. (6.38) was discussed in Section 2.3.

We may also express the matrix relation of Eq. (6.38) as a system of $(M + 1)$ simultaneous equations:

$$\sum_{l=0}^{M} a_{M,M-l}^* r(l - i) = \begin{cases} 0, & i = 0, \ldots, M - 1 \\ P_M, & i = M \end{cases} \tag{6.39}$$

We refer to Eq. (6.38) or (6.39) as the *augmented Wiener–Hopf equations* of a backward prediction-error filter of order M.

Note that in the matrix form of the augmented Wiener–Hopf equations for backward prediction defined by Eq. (6.38), the correlation matrix of the tap inputs is equivalent to that in the corresponding equation (6.14). This is merely a restatement of the fact that the tap inputs in the backward prediction-error filter of Fig. 6.2(c) are exactly the same as those in the forward prediction-error filter of Fig. 6.1(b).

6.3 LEVINSON–DURBIN ALGORITHM

We now describe a direct method for computing the prediction-error filter coefficients and prediction-error power by solving the augmented Wiener–Hopf equations. The method is recursive in nature and makes particular use of the Toeplitz structure of the correlation matrix of the tap inputs of the filter. It is known as the *Levinson–Durbin algorithm*, so named in recognition of its use first by Levinson (1947) and then its independent refor-mulation at a later date by Durbin (1960). Basically, the procedure utilizes the solution of the augmented Wiener–Hopf equations for a prediction-error filter of order $m - 1$ to com-pute the corresponding solution for a prediction-error filter of order m (i.e., one order higher). The order $m = 1, 2, \ldots, M$, where M is the *final* order of the filter. The impor-tant virtue of the Levinson–Durbin algorithm is its computational efficiency, in that its use results in a big saving in the number of operations (multiplications or divisions) and stor-age locations compared to standard methods such as the Gauss elimination method (Makhoul, 1975). To derive the Levinson–Durbin recursive procedure, we will use the matrix formulation of both forward and backward predictions in an elegant way (Burg, 1968, 1975).

Let the $(m + 1)$-by-1 vector $\mathbf{a}_m$ denote the tap-weight vector of a forward prediction-error filter of order m. The $(m + 1)$-by-1 tap-weight vector of the corresponding backward prediction-error filter is obtained by backward rearrangement of the elements of vector $\mathbf{a}_m$ and their complex conjugation. We denote the combined effect of these two operations by $\mathbf{a}_m^{B*}$. Let the m-by-1 vectors $\mathbf{a}_{m-1}$ and $\mathbf{a}_{m-1}^{B*}$ denote the tap-weight vectors of the corre-sponding forward and backward prediction-error filters of order $m - 1$, respectively. The Levinson–Durbin recursion may be stated in one of two equivalent ways:

1. The tap-weight vector of a *forward* prediction-error filter may be order-updated as follows:

$$\mathbf{a}_m = \begin{bmatrix} \mathbf{a}_{m-1} \\ 0 \end{bmatrix} + \kappa_m \begin{bmatrix} 0 \\ \mathbf{a}_{m-1}^{B*} \end{bmatrix} \tag{6.40}$$

where κ_m is a constant. The scalar version of this *order update* is

$$a_{m,l} = a_{m-1,l} + \kappa_m a_{m-1,m-l}^*, \qquad l = 0, 1, \ldots, m \tag{6.41}$$

where $a_{m,l}$ is the lth tap weight of a backward prediction-error filter of order m, and likewise for $a_{m-1,l}^*$. The element $a_{m-1,m-l}^*$ is the lth tap weight of a backward prediction-error filter of order $m - 1$. In Eq. (6.41), note that $a_{m-1,0} = 1$ and $a_{m-1,m} = 0$.

2. The tap-weight vector of a *backward* prediction-error filter may be order-updated as follows:

$$\mathbf{a}_m^{B*} = \begin{bmatrix} 0 \\ \mathbf{a}_{m-1}^{B*} \end{bmatrix} + \kappa_m^* \begin{bmatrix} \mathbf{a}_{m-1} \\ 0 \end{bmatrix} \tag{6.42}$$

The scalar version of this order update is

$$a_{m,m-l}^* = a_{m-1,m-l}^* + \kappa_m^* a_{m-1,l}, \qquad l = 0, 1, \ldots, m \tag{6.43}$$

where $a_{m,m-l}^*$ is the lth tap weight of the backward prediction-error filter of order m, and the other elements are as defined previously.

The Levinson–Durbin recursion is usually formulated in the context of forward prediction, in vector form as in Eq. (6.40) or scalar form as in Eq. (6.41). The formulation of the recursion in the context of backward prediction, in vector form as in Eq. (6.42) or scalar form as in Eq. (6.43), follows directly from that of Eq. (6.40) or (6.41), respectively, through a combination of backward rearrangement and complex conjugation (see Problem 8).

To establish the condition that the constant κ_m has to satisfy in order to justify the validity of the Levinson–Durbin algorithm, we proceed in four stages as follows:

1. We premultiply both sides of Eq. (6.40) by $\mathbf{R}_{m+1}$, the $(m + 1)$-by-$(m + 1)$ correlation matrix of the tap inputs $u(n), u(n - 1), \ldots, u(n - m)$ in the forward prediction-error filter of order m. For the left-hand side of Eq. (6.40), we thus get [see Eq. (6.14)]

$$\mathbf{R}_{m+1}\mathbf{a}_m = \begin{bmatrix} P_m \\ \mathbf{0}_m \end{bmatrix} \tag{6.44}$$

where P_m is the forward prediction-error power, and $\mathbf{0}_m$ is the m-by-1 null vector. The subscripts in the matrix $\mathbf{R}_{m+1}$ and the vector $\mathbf{0}_m$ refer to their dimensions, whereas the subscripts in the vector $\mathbf{a}_m$ and the scalar P_m refer to prediction order.

2. For the first term of the right-hand side of Eq. (6.40), we use the following partitioned form of the correlation matrix $\mathbf{R}_{m+1}$ (see Section 2.3).

$$\mathbf{R}_{m+1} = \begin{bmatrix} \mathbf{R}_m & \mathbf{r}_m^{B*} \\ \mathbf{r}_m^{BT} & r(0) \end{bmatrix}$$

where $\mathbf{R}_m$ is the m-by-m correlation matrix of the tap inputs $u(n), u(n - 1), \ldots, u(n - m + 1)$, and $\mathbf{r}_m^{B*}$ is the cross-correlation vector between these tap inputs and $u(n - m)$. We may thus write

$$\mathbf{R}_{m+1} \begin{bmatrix} \mathbf{a}_{m-1} \\ 0 \end{bmatrix} = \begin{bmatrix} \mathbf{R}_m & \mathbf{r}_m^{B*} \\ \mathbf{r}_m^{BT} & r(0) \end{bmatrix} \begin{bmatrix} \mathbf{a}_{m-1} \\ 0 \end{bmatrix} \tag{6.45}$$

$$= \begin{bmatrix} \mathbf{R}_m \mathbf{a}_{m-1} \\ \mathbf{r}_m^{BT} \mathbf{a}_{m-1} \end{bmatrix}$$

The set of augmented Wiener–Hopf equations for the forward prediction-error filter of order $m - 1$ is

$$\mathbf{R}_m \mathbf{a}_{m-1} = \begin{bmatrix} P_{m-1} \\ \mathbf{0}_{m-1} \end{bmatrix} \tag{6.46}$$

where P_{m-1} is the prediction-error power for this filter, and $\mathbf{0}_{m-1}$ is the $(m - 1)$-by-1 null vector. Define the scalar

$$\Delta_{m-1} = r_m^{BT} \mathbf{a}_{m-1}$$

$$= \sum_{l=0}^{m-1} r(l - m) a_{m-1,l} \tag{6.47}$$

Substituting Eqs. (6.46) and (6.47) in Eq. (6.45), we may therefore write

$$\mathbf{R}_{m+1} \begin{bmatrix} \mathbf{a}_{m-1} \\ 0 \end{bmatrix} = \begin{bmatrix} P_{m-1} \\ \mathbf{0}_{m-1} \\ \Delta_{m-1} \end{bmatrix} \tag{6.48}$$

3. For the second term on the right-hand side of Eq. (6.40), we use the following partitioned form of the correlation matrix $\mathbf{R}_{m+1}$ (see Section 2.3):

$$\mathbf{R}_{m+1} = \begin{bmatrix} r(0) & \mathbf{r}_m^H \\ \mathbf{r}_m & \mathbf{R}_m \end{bmatrix}$$

where $\mathbf{R}_m$ is the m-by-m correlation matrix of the tap inputs $u(n - 1)$, $u(n - 2)$, . . . , $u(n - m)$, and $\mathbf{r}_m$ is the m-by-1 cross-correlation vector between these tap inputs and $u(n)$. We may thus write

$$\mathbf{R}_{m+1} \begin{bmatrix} 0 \\ \mathbf{a}_{m-1}^{B*} \end{bmatrix} = \begin{bmatrix} r(0) & \mathbf{r}_m^H \\ \mathbf{r}_m & \mathbf{R}_m \end{bmatrix} \begin{bmatrix} 0 \\ \mathbf{a}_{m-1}^{B*} \end{bmatrix}$$

$$= \begin{bmatrix} \mathbf{r}_m^H \mathbf{a}_{m-1}^{B*} \\ \mathbf{R}_m \mathbf{a}_{m-1}^{B*} \end{bmatrix} \tag{6.49}$$

The scalar $\mathbf{r}_m^H \mathbf{a}_{m-1}^{B*}$ equals

$$\mathbf{r}_m^H \mathbf{a}_{m-1}^{B*} = \sum_{k=1}^{m} r*(-k) a_{m-1,m-k}^{*}$$

$$= \sum_{l=0}^{m-1} r*(l - m) a_{m-1,l}^{*} \tag{6.50}$$

$$= \Delta_{m-1}^{*}$$

Also, the set of augmented Wiener–Hopf equations for the backward prediction-error filter of order $m - 1$ is

$$\mathbf{R}_m \mathbf{a}_{m-1}^{B*} = \begin{bmatrix} \mathbf{0}_{m-1} \\ P_{m-1} \end{bmatrix} \tag{6.51}$$

Substituting Eqs. (6.50) and (6.51) in (6.49), we may therefore write

$$\mathbf{R}_{m+1} \begin{bmatrix} 0 \\ \mathbf{a}_{m-1}^{B*} \end{bmatrix} = \begin{bmatrix} \Delta_{m-1}^* \\ \mathbf{0}_{m-1} \\ P_{m-1} \end{bmatrix} \tag{6.52}$$

4. Summarizing the results obtained in stages 1, 2, and 3 and, in particular, using Eqs. (6.44), (6.48), and (6.52), we may now state that the premultiplication of both sides of Eq. (6.40) by the correlation matrix $\mathbf{R}_{m+1}$ yields

$$\begin{bmatrix} P_m \\ \mathbf{0}_m \end{bmatrix} = \begin{bmatrix} P_{m-1} \\ \mathbf{0}_{m-1} \\ \Delta_{m-1} \end{bmatrix} + \kappa_m \begin{bmatrix} \Delta_{m-1}^* \\ \mathbf{0}_{m-1} \\ P_{m-1} \end{bmatrix} \tag{6.53}$$

We conclude therefore that, if the order-update recursion of Eq. (6.40) holds, the results described by Eq. (6.53) are direct consequences of this recursion. Conversely, we may state that, if the conditions described by Eq. (6.53) apply, the tap-weight vector of a forward prediction-error filter may be order-updated as in Eq. (6.40).

From Eq. (6.53), we may make two important deductions:

1. By considering the first elements of the vectors on the left-hand and right-hand sides of Eq. (6.53), we have

$$P_m = P_{m-1} + \kappa_m \Delta_{m-1}^* \tag{6.54}$$

2. By considering the last elements of the vectors on the left-hand and right-hand sides of Eq. (6.53), we have

$$0 = \Delta_{m-1} + \kappa_m P_{m-1} \tag{6.55}$$

From Eq. (6.55), we see that the constant κ_m has the value

$$\kappa_m = -\frac{\Delta_{m-1}}{P_{m-1}} \tag{6.56}$$

where Δ_{m-1} is itself defined by Eq. (6.47). Furthermore, eliminating Δ_{m-1} between Eqs. (6.54) and (6.55), we get the following relation for the order update of the prediction-error power:

$$P_m = P_{m-1} (1 - |\kappa_m|^2) \tag{6.57}$$

As the order m of the prediction-error filter increases, the corresponding value of the prediction-error power P_m normally decreases or else remains the same. Of course, P_m can never be negative. Hence, we must always have

$$0 \le P_m \le P_{m-1}, \qquad m \ge 1 \tag{6.58}$$

For the elementary case of a prediction-error filter of order zero, we naturally have

$$P_0 = r(0)$$

where $r(0)$ is the autocorrelation function of the input process for zero lag.

Starting with $m = 0$, and increasing the filter order by 1 at a time, we find that through the repeated application of Eq. (6.57) the prediction-error power for a prediction-error filter of *final* order M equals

$$P_M = P_0 \prod_{m=1}^{M} (1 - |\kappa_m|^2) \tag{6.59}$$

Interpretations of the Parameters κ_m and Δ_{m-1}

The parameters κ_m, $1 \leq m \leq M$, resulting from the application of the Levinson–Durbin recursion to a prediction-error filter of final order M, are called *reflection coefficients*. The use of this term comes from the analogy of Eq. (6.57) with transmission line theory, where (in the latter context) κ_m may be considered as the reflection coefficient at the boundary between two sections with different characteristic impedances. Note that the condition on the reflection coefficient corresponding to that of Eq. (6.58) is

$$|\kappa_m| \leq 1 \qquad \text{for all } m \tag{6.60}$$

From Eq. (6.41), we see that for a prediction-error filter of order m, the reflection coefficient κ_m equals the *last* tap-weight $a_{m,m}$ of the filter. That is,

$$\kappa_m = a_{m,m}$$

As for the parameter Δ_{m-1}, it may be interpreted as a cross-correlation between the forward prediction error $f_{m-1}(n)$ and the delayed backward prediction error $b_{m-1}(n - 1)$. Specifically, we may write (see Problem 9)

$$\Delta_{m-1} = E[b_{m-1}(n - 1)f_{m-1}^*(n)] \tag{6.61}$$

where $f_{m-1}(n)$ is produced at the output of a forward prediction-error filter of order $m - 1$ in response to the tap inputs $u(n), u(n - 1) \ldots , u(n - m + 1)$, and $b_{m-1}(n - 1)$ is produced at the output of a backward prediction-error filter of order $m - 1$ in response to the tap inputs $u(n - 1), u(n - 2), \ldots , u(n - m)$.

Note that

$$f_0(n) = b_0(n) = u(n)$$

where $u(n)$ is the prediction-error filter input at time n. Accordingly, from Eq. (6.61) we find that this cross-correlation parameter has the *zero-order value*

$$\Delta_0 = E[b_0(n - 1)f_0^*(n)]$$

$$= E[u(n - 1)u^*(n)]$$

$$= r^*(1)$$

where $r(1)$ is the autocorrelation function of the input for a lag of 1.

We may also use Eqs. (6.56) and (6.61) to develop a second interpretation for the parameter κ_m. In particular, since P_{m-1} may be viewed as the mean-square value of the forward prediction error $f_{m-1}(n)$, we may write

$$\kappa_m = \frac{E[b_{m-1}(n-1)f^*_{m-1}(n)]}{E[|f_{m-1}(n)|^2]} \tag{6.62}$$

The right-hand side of Eq. (6.62), except for the minus sign, is referred to as a *partial correlation (PARCOR) coefficient*. This terminology is widely used in the statistics literature (Box and Jenkins, 1976). Hence, the reflection coefficient, as defined here, is the negative of the PARCOR coefficient.

Application of the Levinson–Durbin Algorithm

There are two possible ways of applying the Levinson–Durbin algorithm to compute the prediction-error filter coefficients $a_{M,k}$, $k = 0, 1, \ldots, M$, and the prediction-error power P_M for a final prediction order M:

1. We have explicit knowledge of the autocorrelation function of the input process; in particular, we have $r(0), r(1), \ldots, r(M)$, denoting the values of the autocorrelation function for lags $0, 1, \ldots, M$, respectively. For example, we may compute *biased* estimates of these parameters by means of the *time-average formula*

$$\hat{r}(k) = \frac{1}{N} \sum_{n=1+k}^{N} u(n)u^*(n-k), \qquad k = 0, 1, \ldots, M \tag{6.63}$$

   where N is the total length of the input time series, with $N >> M$. There are, of course, other estimators that we may use.[2] In any event, given $r(0), r(1), \ldots, r(M)$, the computation proceeds by using Eq. (6.47) for Δ_{m-1} and Eq. (6.57) for P_m. The recursion is initiated with $m = 0$, for which we have $P_0 = r(0)$ and $\Delta_0 = r^*(1)$. Note also that $a_{m,0}$ equals 1 for all m, and $a_{m,k}$ is zero for all $k > m$. The computation is terminated when $m = M$. The resulting estimates of the prediction-error filter coefficients and prediction error power obtained by using this procedure are known as the *Yule–Walker estimates*.

2. We have explicit knowledge of the reflection coefficients $\kappa_1, \kappa_2, \ldots, \kappa_M$ and the autocorrelation function $r(0)$ for a lag of zero. Later in the chapter we describe a procedure for estimating these reflection coefficients directly from the given data. In this second application of the Levinson–Durbin recursion, we only need the pair of relations:

---

[2] In practice, the biased estimate of Eq. (6.63) is preferred over an unbiased estimate because it yields a much lower variance for the estimate $\hat{r}(k)$ for values of the lag k close to the data length N; for more details, see Box and Jenkins (1976). For a more refined estimate of the autocorrelation function $r(k)$, we may use the multiple-window method described in McWhorter and Scharf (1995). This method uses a multiplicity of special windows, resulting in the most general Hermitian, nonnegative-definite, and modulation-invariant estimate. The Hermitian and nonnegative-definite properties were described in Chapter 2. To define the modulation-invariant property, let $\hat{\mathbf{R}}$ denote an estimate of the correlation matrix, given the input vector $\mathbf{u}$. The estimate is said to be *modulation-invariant* if $\mathbf{D}(e^{j\phi})\mathbf{u}$ has a correlation matrix equal to $\mathbf{D}(e^{j\phi})\hat{\mathbf{R}}\mathbf{D}(e^{-j\phi})$, where $\mathbf{D}(e^{j\phi}) = \text{diag}(\boldsymbol{\psi}(e^{j\phi}))$ is a modulation matrix, and $\boldsymbol{\psi}(e^{j\phi}) = [1, e^{j\phi}, \ldots, e^{j\phi(M-1)}]^T$.

$$a_{m,k} = a_{m-1,k} + \kappa_m a^*_{m-1,m-k}, \qquad k = 0, 1, \ldots, m$$

$$P_m = P_{m-1}(1 - |\kappa_m|^2)$$

Here, again, the recursion is initiated with $m = 0$ and stopped when the order m reaches the final value M.

Example 2

To illustrate the second method for the application of the Levinson–Durbin recursion, suppose we are given the reflection coefficients κ_1, κ_2, κ_3 and average power P_0. The problem we wish to solve is to use these parameters to determine the corresponding tap weights $a_{3,1}$, $a_{3,2}$, $a_{3,3}$ and the prediction-error power P_3 for a prediction-error filter of order 3. The application of the Levinson–Durbin recursion, described by Eqs. (6.41) and (6.57), yields the following results for $m = 1, 2, 3$:

1. Prediction-error filter order $m = 1$:

$$a_{1,0} = 1$$

$$a_{1,1} = \kappa_1$$

$$P_1 = P_0(1 - |\kappa_1|^2)$$

2. Prediction-error filter order $m = 2$:

$$a_{2,0} = 1$$

$$a_{2,1} = \kappa_1 + \kappa_2 \kappa_1^*$$

$$a_{2,2} = \kappa_2$$

$$P_2 = P_1(1 - |\kappa_2|^2)$$

where P_1 is as defined above.

3. Prediction-error filter order $m = 3$:

$$a_{3,0} = 1$$

$$a_{3,1} = a_{2,1} + \kappa_3 \kappa_2^*$$

$$a_{3,2} = \kappa_2 + \kappa_3 a_{2,1}^*$$

$$a_{3,3} = \kappa_3$$

$$P_3 = P_2(1 - |\kappa_3|^2)$$

where $a_{2,1}$ and P_2 are as defined above.

The interesting point to observe from this example is that the Levinson–Durbin recursion yields not only the values of the tap weights and prediction-error power for the prediction-error filter of final order M, but also the corresponding values of these parameters for the prediction-error filters of intermediate orders $M - 1, \ldots, 1$.

Inverse Levinson–Durbin Algorithm

In the normal application of the Levinson–Durbin recursion, as illustrated in Example 2, we are given the set of reflection coefficients $\kappa_1, \kappa_2, \ldots \kappa_M$ and the requirement is to compute the corresponding set of tap weights $a_{M,1}, a_{M,2}, \ldots, a_{M,M}$ for a prediction-error filter of final order M. Of course, the remaining coefficient of the filter, $a_{M,0} = 1$. Frequently, however, the need arises to solve the following *inverse* problem: Given the set of tap weights $a_{M,1}, a_{M,2}, \ldots, a_{M,M}$, solve for the corresponding set of reflection coefficients κ_1, $\kappa_2, \ldots, \kappa_M$. We may solve this problem by applying the inverse form of Levinson–Durbin recursion, which we refer to simply as the *inverse recursion*.

To derive the inverse recursion, we first combine Eqs. (6.41) and (6.43), representing the scalar versions of the Levinson–Durbin recursion for forward and backward prediction-error filters, respectively, in matrix form as follows:

$$\begin{bmatrix} a_{m,k} \\ a^*_{m,m-k} \end{bmatrix} = \begin{bmatrix} 1 & \kappa_m \\ \kappa^*_m & 1 \end{bmatrix} \begin{bmatrix} a_{m-1,k} \\ a^*_{m-1,m-k} \end{bmatrix}, \qquad k = 0, 1, \ldots, m \qquad (6.64)$$

where the order $m = 1, 2, \ldots, M$. Then, assuming that $|\kappa_m| < 1$ and solving Eq. (6.64) for the tap weight $a_{m-1,k}$, we get

$$a_{m-1,k} = \frac{a_{m,k} - a_{m,m}a^*_{m,m-k}}{1 - |a_{m,m}|^2}, \qquad k = 0, 1, \ldots, m \qquad (6.65)$$

where we have used the fact that $\kappa_m = a_{m,m}$. We may now describe the procedure. Starting with the set of tap weights $\{a_{M,k}\}$ for which the prediction-error filter order equals M, we use the inverse recursion, Eq. (6.65), with decreasing filter order $m = M, M - 1, \ldots,$ 2 to compute the tap weights of the corresponding prediction-error filters of order $M - 1$, $M - 2, \ldots, 1$, respectively. Finally, knowing the tap weights of all the prediction-error filters of interest (whose order ranges all the way from M down to 1), we use the fact that

$$\kappa_m = a_{m,m}, \qquad m = M, M - 1, \ldots, 1$$

to determine the desired set of reflection coefficients $\kappa_M, \kappa_{M-1}, \ldots, \kappa_1$. Example 3 illustrates the application of the inverse recursion.

Example 3

Suppose we are given the tap weights $a_{3,1}, a_{3,2}, a_{3,3}$ of a prediction-error filter of order 3, and the requirement is to determine the corresponding reflection coefficients $\kappa_1, \kappa_2, \kappa_3$. Application of the inverse recursion, described by Eq. (6.65), for filter order $m = 3, 2$ yields the following set of tap weights:

1. Prediction-error filter of order 2 [corresponding to $m = 3$ in Eq. (6.65)]:

$$a_{2,1} = \frac{a_{3,1} - a_{3,3}a^*_{3,2}}{1 - |a_{3,3}|^2}$$

$$a_{2,2} = \frac{a_{3,2} - a_{3,3}a^*_{3,1}}{1 - |a_{3,3}|^2}$$

2. Prediction-error filter of order 1 [corresponding to $m = 2$ in Eq. (6.65)]:

$$a_{1,1} = \frac{a_{2,1} - a_{2,2}a_{2,1}^*}{1 - |a_{2,2}|^2}$$

where $a_{2,1}$ and $a_{2,2}$ are as defined above. Thus, the required reflection coefficients are given by

$$\kappa_3 = a_{3,3}$$

$$\kappa_2 = a_{2,2}$$

$$\kappa_1 = a_{1,1}$$

where $a_{3,3}$ is given, and $a_{2,2}$ and $a_{1,1}$ are computed as shown above.

6.4 PROPERTIES OF PREDICTION-ERROR FILTERS

Property 1. *Relations between the autocorrelation function and the reflection coefficients*   It is customary to represent the second-order statistics of a stationary time series in terms of its autocorrelation function or, equivalently, the power spectrum. The autocorrelation function and power spectrum form a discrete-time Fourier-transform pair (see Chapter 3). Another way of describing the second-order statistics of a stationary time series is to use the set of numbers $P_0, \kappa_1, \kappa_2, \ldots, \kappa_M$, where $P_0 = r(0)$ is the value of the autocorrelation function of the process for a lag of zero, and $\kappa_1, \kappa_2, \ldots, \kappa_M$ are the reflection coefficients for a prediction-error filter of final order M. This is a consequence of the fact that the set of numbers $P_0, \kappa_1, \kappa_2, \ldots, \kappa_M$ uniquely determines the corresponding set of autocorrelation function values $r(0), r(1), \ldots, r(M)$, and vice versa.

To prove this relationship, we first eliminate Δ_{m-1} between Eqs. (6.47) and (6.55), obtaining

$$\sum_{k=0}^{m-1} a_{m-1,k} r(k - m) = -\kappa_m P_{m-1} \tag{6.66}$$

Solving Eq. (6.66) for $r(m) = r^*(-m)$ and recognizing that $a_{m-1,0}$ equals 1, we get

$$r(m) = -\kappa_m^* P_{m-1} - \sum_{k=1}^{m-1} a_{m-1,k}^* \, r(m - k) \tag{6.67}$$

This is the desired recursive relation. If we are given the set of numbers $r(0), \kappa_1, \kappa_2, \ldots, \kappa_M$, then by using Eq. (6.67), together with the Levinson–Durbin recursive equations (6.41) and (6.57), we may recursively generate the corresponding set of numbers $r(0), r(1), \ldots, r(M)$.

For $|\kappa_m| \leq 1$, we find from Eq. (6.67) that the permissible region for $r(m)$, the value of the autocorrelation function of the input signal for a lag of m, is the interior (including circumference) of a circle of radius P_{m-1} and center at the complex-valued quantity

$$-\sum_{k=1}^{m-1} a_{m-1,k}^* r(m - k)$$

This is illustrated in Fig. 6.3.

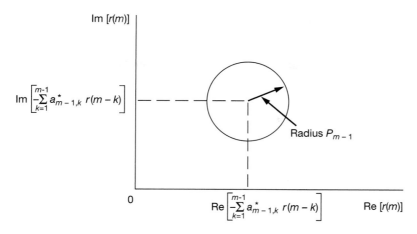

Figure 6.3   Permissible region for $r(m)$ for the case when $|\kappa_m| \leq 1$.

Suppose now that we are given the set of autocorrelation function values $r(1), \ldots,$ $r(M)$. Then we may recursively generate the corresponding set of numbers $\kappa_1, \kappa_2, \ldots, \kappa_M$ by using the relation:

$$\kappa_m = -\frac{1}{P_{m-1}} \sum_{k=0}^{m-1} a_{m-1,k} r(k - m) \tag{6.68}$$

which is obtained by solving Eq. (6.66) for κ_m. In Eq. (6.68), it is assumed that P_{m-1} is nonzero. If P_{m-1} is zero, this would have been the result of $|\kappa_{m-1}| = 1$, and the sequence of reflection coefficients $\kappa_1, \kappa_2, \ldots, \kappa_{m-1}$ is terminated.

We may therefore make the statement: *There is a one-to-one correspondence between the two sets of quantities* $\{P_0, \kappa_1, \kappa_2, \ldots, \kappa_M\}$ *and* $\{r(0), r(1), r(2), \ldots, r(M)\}$, *in that if we are given the one we may uniquely determine the other in a recursive manner.*

Example 4

Suppose that we are given P_0, and $\kappa_1, \kappa_2, \kappa_3$ and the requirement is to compute $r(0)$, $r(1)$, $r(2)$, and $r(3)$. We start with $m = 1$, for which Eq. (6.67) yields

$$r(1) = -P_0 \kappa_1^*$$

where

$$P_0 = r(0)$$

For $m = 2$, the use of Eq. (6.67) yields

$$r(2) = -P_1 \kappa_2^* - r(1) \kappa_1^*$$

where

$$P_1 = P_0(1 - |\kappa_1|^2)$$

Finally, for $m = 3$, the use of Eq. (6.67) yields

$$r(3) = -P_2\kappa_3^* - [a_{2,1}^* r(2) + \kappa_2^* r(1)]$$

where

$$P_2 = P_1(1 - |\kappa_2|^2)$$

$$a_{2,1} = \kappa_1 + \kappa_2\kappa_1^*$$

Property 2. *Transfer function of a forward prediction-rrror filter*   Let $H_{f,m}(z)$ denote the *transfer function* of a forward prediction-error filter of order m, and whose impulse response is defined by the sequence of numbers $a_{m,k}^*$, $k = 0, 1, \ldots, m$, as illustrated in Fig. 6.1(b) for $m = M$. From Chapter 1 we recall that the transfer function of a discrete-time filter equals the z-transform of its impulse response. We may therefore write

$$H_{f,m}(z) = \sum_{k=0}^{m} a_{m,k}^* z^{-k} \tag{6.69}$$

where z is a complex variable. Based on the Levinson–Durbin recursion, in particular Eq. (6.41), we may relate the coefficients of this filter of order m to those of a corresponding prediction-error filter of order $m - 1$ (i.e., one order smaller). In particular, substituting Eq. (6.41) into (6.69), we get

$$H_{f,m}(z) = \sum_{k=0}^{m} a_{m-1,k}^* z^{-k} + \kappa_m^* \sum_{k=0}^{m} a_{m-1,m-k} z^{-k}$$

$$= \sum_{k=0}^{m-1} a_{m-1,k}^* z^{-k} + \kappa_m^* z^{-1} \sum_{k=0}^{m-1} a_{m-1,m-1-k} z^{-k} \tag{6.70}$$

where, in the second line, we have used the fact that $a_{m-1,m} = 0$. The sequence of numbers $a_{m-1,k}^*$, $k = 0, 1, \ldots, m - 1$, defines the impulse response of a forward prediction-error filter of order $m - 1$. Hence, we may write

$$H_{f,m-1}(z) = \sum_{k=0}^{m-1} a_{m-1,k}^* z^{-k} \tag{6.71}$$

The sequence of numbers $a_{m-1,m-1-k}$, $k = 0, 1, \ldots, m - 1$, defines the impulse response of a backward prediction-error filter of order $m - 1$; this is illustrated in Fig. 6.2(c) for the case of prediction order $m = M$. Hence, the second summation on the right-hand side of Eq. (6.70) represents the transfer function of this backward prediction-error filter. Let $H_{b,m-1}(z)$ denote this transfer function, as shown by

$$H_{b,m-1}(z) = \sum_{k=0}^{m-1} a_{m-1,m-1-k} z^{-k} \tag{6.72}$$

Hence, substituting Eqs. (6.71) and (6.72) in (6.70), we may write

$$H_{f,m}(z) = H_{f,m-1}(z) + \kappa_m^* z^{-1} H_{b,m-1}(z) \tag{6.73}$$

On the basis of the order update recursion of Eq. (6.73), we may now state: *Given the reflection coefficient κ_m and the transfer functions of the forward and backward predic-*

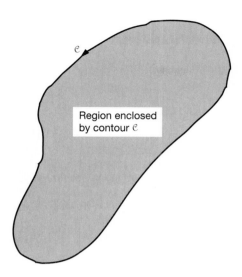

Figure 6.4 Contour $\mathscr{C}$ (traversed in the counterclockwise direction) in the z-plane and the region enclosed by it.

tion-error filters of order $m - 1$, the transfer function of the corresponding forward prediction-error filter of order m *is uniquely determined.*

    Property 3. *A forward prediction-error filter is minimum phase*   On the unit circle in the z-plane (i.e., for $|z| = 1$), we find that

$$|H_{f,m-1}(z)| = |H_{b,m-1}(z)|, \qquad |z| = 1$$

This is readily proved by substituting $z = \exp(j\omega)$, $-\pi < \omega \leq \pi$, in Eqs. (6.71) and (6.72). Suppose that the reflection coefficient κ_m satisfies the requirement $|\kappa_m| < 1$, for all m. Then we find that on the unit circle in the z-plane the magnitude of the second term in the right-hand side of Eq. (6.73) satisfies the conditions

$$|\kappa_m^* z^{-1} H_{b,m-1}(z)| < |H_{b,m-1}(z)| = |H_{f,m-1}(z)|, \qquad |z| = 1 \qquad (6.74)$$

At this stage in our discussion, it is useful to recall *Rouché's theorem* from the theory of complex variables.[3] Rouché's theorem states:

If a function $F(z)$ is analytic upon a contour $\mathscr{C}$ in the z-plane and within the region enclosed by this contour, and if a second function $G(z)$, in addition to satisfying the same analyticity conditions, also fulfills the condition $|G(z)| < F(z)$ on the contour $\mathscr{C}$, then the function $F(z) + G(z)$ has the same number of zeros within the region enclosed by the contour $\mathscr{C}$ as does the function $F(z)$.

    Ordinarily, the enclosed contour $\mathscr{C}$ is transversed in the *counterclockwise* direction, and the region enclosed by the contour lies to the *left* of it, as illustrated in Fig. 6.4. We

---

[3] For a review of complex variable theory, including Rouché's theorem, see Appendix A.

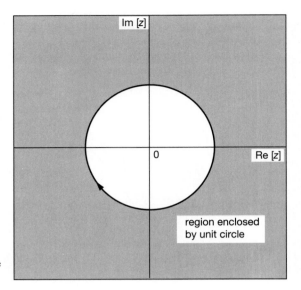

region enclosed
by unit circle

Figure 6.5 Unit circle (traversed in the clockwise direction) used as contour $\mathscr{C}$.

say that a function is analytic upon the contour $\mathscr{C}$ and within the region enclosed by it if the function has a continuous derivative everywhere upon the contour $\mathscr{C}$ and within the region enclosed by this contour. For this requirement to be satisfied, the function must have no poles upon the contour $\mathscr{C}$ or inside the region enclosed by the contour.

Let the contour $\mathscr{C}$ be the unit circle in the z-plane, which is traversed in the *clockwise* direction, as in Fig. 6.5. According to the convention just described, this assumption implies that the region enclosed by the contour $\mathscr{C}$ is represented by the entire part of the z-plane that lies *outside* the unit circle.

Let the functions $F(z)$ and $G(z)$ be identified with the two terms in the right-hand side of Eq. (6.73), as shown by

$$F(z) = H_{f,m-1}(z) \tag{6.75}$$

$$G(z) = \kappa_m^* z^{-1} H_{b,m-1}(z) \tag{6.76}$$

We observe that:

- The functions $F(z)$ and $G(z)$ have no poles inside the contour $\mathscr{C}$ defined in Fig. 6.5. Indeed, their derivatives are continuous throughout the region enclosed by this contour. Therefore, they are both analytic everywhere upon the unit circle and the region outside it.
- In view of Eq. (6.74), we have $|G(z)| < |F(z)|$ on the unit circle.

Accordingly, the functions $F(z)$ and $G(z)$ defined by Eqs. (6.75) and (6.76), respectively, satisfy all the conditions required by Rouché's theorem with respect to the contour $\mathscr{C}$ defined as the unit circle in Fig. 6.5.

Suppose that $H_{f,m-1}(z)$ and therefore $F(z)$ are known to have no zeros outside the unit circle in the z-plane. Then, by applying Rouché's theorem, we find that $F(z) + G(z)$, or, equivalently, $H_{f,m}(z)$ also has no zeros on or outside the unit circle in the z-plane.

In particular, for $m = 0$, the transfer function $H_{f,0}(z)$ is a constant equal to 1; therefore, it has no zeros at all. Using the result just derived, we may state that since $H_{f,0}(z)$ has no zeros outside the unit circle, then $H_{f,1}(z)$ will also have no zeros in this region of the z-plane, provided that $|\kappa_1| < 1$. Indeed, we can easily prove this latter result by noting that

$$H_{f,1}(z) = a_{1,0}^* + a_{1,1}^* z^{-1}$$
$$= 1 + \kappa_1^* z^{-1}$$

Hence, $H_{f,1}(z)$ has a single zero located at $z = -\kappa_1^*$ and a pole at $z = 0$. With the reflection coefficient κ_1 constrained by the condition $|\kappa_1| < 1$, it follows that this zero must lie inside the unit circle. In other words, $H_{f,1}(z)$ has no zeros on or outside the unit circle. If $H_{f,1}(z)$ has no zeros on or outside the unit circle, then $H_{f,2}(z)$ will also have no zeros on or outside the unit circle provided that $|\kappa_2| < 1$, and so on.

We may thus state that the transfer function $H_{f,m}(z)$ of a forward prediction-error filter of order m has no zeros on or outside the unit circle in the z-plane for all values of m, if and only if the reflection coefficients satisfy the condition $|\kappa_m| < 1$ for all m. Such a filter is said to be minimum phase in the sense that, for a specified amplitude response, it has the minimum phase response possible for all values of z on the unit circle (see Chapter 1). Moreover, the amplitude response and phase response of the filter are uniquely related to each other. Based on these findings we may now make the statement: The transfer function $H_{f,m}(z)$ of a forward prediction-error filter of order m has no zeros on or outside the unit circle in the z-plane for all values of m, if and only if the reflection coefficients satisfy the condition $|\kappa_m| < 1$ for all m. In other words, *a forward prediction-error filter is minimum phase in the sense that, for a specified amplitude response, it has the minimum phase response possible for all values of* z *on the unit circle.*

Property 4. *A backward prediction-error filter is maximum phase*   The transfer functions of backward and forward prediction-error filters of the same order are related, in that if we are given one we may uniquely determine the order. To find this relationship, we first evaluate $H_{f,m}*(z)$, the complex conjugate of the transfer function of a forward prediction-error filter of order m, and so write [see Eq. 6.69)]

$$H_{f,m}^*(z) = \sum_{k=0}^{m} a_{m,k}(z^*)^{-k} \qquad (6.77)$$

Replacing z by the reciprocal of its complex conjugate z^*, we may rewrite Eq. (6.77) as

$$H_{f,m}^*\left(\frac{1}{z^*}\right) = \sum_{k=0}^{m} a_{m,k} z^k$$

Next, replacing k by $m - k$, we get

$$H_{f,m}^*\left(\frac{1}{z^*}\right) = z^m \sum_{k=0}^{m} a_{m,m-k} z^{-k} \qquad (6.78)$$

The summation on the right-hand side of Eq. (6.78) constitutes the transfer function of a backward prediction-error filter of order m, as shown by

$$H_{b,m}(z) = \sum_{k=0}^{m} a_{m,m-k} z^{-k} \tag{6.79}$$

We thus find that $H_{b,m}(z)$ and $H_{f,m}(z)$ are related as follows:

$$H_{b,m}(z) = z^{-m} H_{f,m}^{*}\left(\frac{1}{z^{*}}\right) \tag{6.80}$$

where $H_{f,m}^{*}(1/z^{*})$ is obtained by complex-conjugating $H_{f,m}(z)$, the transfer function of a forward prediction-error filter of order m, and replacing z by the reciprocal of z^{*}. Equation (6.80) states that multiplication of the new function obtained in this way by z^{-m} yields $H_{b,m}(z)$, the transfer function of the corresponding backward prediction-error filter.

Let the transfer function $H_{f,m}(z)$ be expressed in its factored form as follows:

$$H_{f,m}(z) = \prod_{i=1}^{m} (1 - z_i z^{-1}) \tag{6.81}$$

where z_i, $i = 1, 2, \ldots, m$, denote the zeros of the forward prediction-error filter. Hence, substituting Eq. (6.81) into (6.80), we may express the transfer function of the corresponding backward prediction-error filter in the factored form

$$\begin{aligned} H_{b,m}(z) &= z^{-m} \prod_{i=1}^{m} (1 - z_i^{*} z) \\ &= \prod_{i=1}^{m} (z^{-1} - z_i^{*}) \end{aligned} \tag{6.82}$$

The zeros of this transfer function are located at $1/z_i^{*}$, $i = 1, 2, \ldots m$. That is, the zeros of the backward and forward prediction-error filters are the *inverse* of each other with respect to the unit circle in the z-plane. The geometric nature of this relationship is illustrated for $m = 1$ in Fig. 6.6. The forward prediction-error filter has a zero at $z = -\kappa_1^{*}$, as in Fig. 6.6(a), and the backward prediction-error filter has a zero at $z = -1/\kappa_1$, as in Fig. 6.6(b). In this figure, it is assumed that the reflection coefficient κ_1 has a complex value. Consequently, a backward prediction-error filter has all of its zeros located outside the unit circle in the z-plane, because $|\kappa_m| < 1$ for all m.

We may therefore formally make the statement: *A backward prediction-error filter is maximum-phase in the sense that, for a specified amplitude response, it has the maximum phase responsible possible for all values of z on the unit circle.*

Property 5. *A forward prediction-error filter is a whitening filter*   By definition, a *white-noise* process consists of a sequence of uncorrelated random variables. Thus, assuming that such a process, denoted by $v(n)$, has zero mean and variance σ_v^2, we may write (see Section 2.5)

$$E[v(k)v^{*}(n)] = \begin{cases} \sigma_v^2, & k = n \\ 0, & k \neq n \end{cases} \tag{6.83}$$

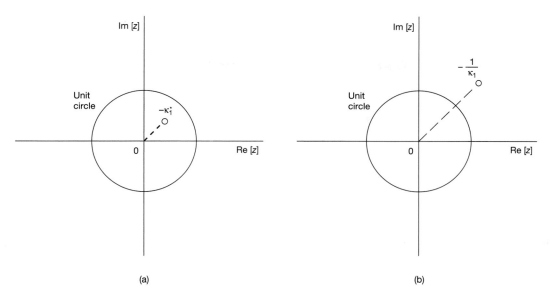

Figure 6.6   (a) Zero of forward prediction-error filter at $z = -\kappa_1^*$; (b) corresponding zero of backward prediction-error filter at $z = -1/\kappa_1$.

Accordingly, we say that white noise is purely unpredictable in the sense that the value of the process at time n is uncorrelated with all past values of the process up to and including time $n - 1$ (and, indeed, with all future values of the process too).

We may now state another important property of prediction-error filters: *A prediction-error filter is capable of whitening a stationary discrete-time stochastic process applied to its input, provided that the order of the filter is high enough.* Basically, prediction relies on the presence of correlation between adjacent samples of the input process. The implication of this is that, as we increase the order of the prediction-error filter, we successively reduce the correlation between adjacent samples of the input process, until ultimately we reach a point at which the filter has a high enough order to produce an output process that consists of a sequence of uncorrelated samples. The whitening of the original process applied to the filter input will have thereby been accomplished.

Property 6. *Eigenvector representation of forward prediction-error filters*   The representation of a forward prediction-error filter is naturally related to eigenvalues (and associated eigenvectors) of the correlation matrix of the tap inputs in the filter. To develop such a representation, we first rewrite the augmented Wiener–Hopf equations (6.14), pertaining to a forward prediction-error filter of order M, in the compact matrix form

$$\mathbf{R}_{M+1}\mathbf{a}_M = P_M \mathbf{i}_{M+1} \tag{6.84}$$

where $\mathbf{R}_{M+1}$ is the $(M + 1)$-by-$(M + 1)$ correlation matrix of the tap inputs $u(n)$, $u(n - 1), \ldots, u(n - M)$ in the filter of Fig. 6.1(b), $\mathbf{a}_M$ is the $(M + 1)$-by-1 tap-weight vector of the filter, and the scalar P_M is the prediction error power. The $(M + 1)$-by-1 vector $\mathbf{i}_{M+1}$ is called the *first coordinate vector*; it has unity for its first element and zero for all the others. We illustrate this by writing

$$\mathbf{i}_{M+1} = [1, 0, \ldots, 0]^T \tag{6.85}$$

Solving Eq. (6.84) for $\mathbf{a}_M$, we get

$$\mathbf{a}_M = P_M \mathbf{R}_{M+1}^{-1} \mathbf{i}_{M+1} \tag{6.86}$$

where $\mathbf{R}_{M+1}^{-1}$ is the inverse of the correlation matrix $\mathbf{R}_{M+1}$; it is assumed that the matrix $\mathbf{R}_{M+1}$ is nonsingular, so that its inverse exists. Using the eigenvalue–eigenvector representation of a correlation matrix, which was discussed in Section 4.2, we may express the inverse matrix $\mathbf{R}_{M+1}^{-1}$ as follows:

$$\mathbf{R}_{M+1}^{-1} = \mathbf{Q}\boldsymbol{\Lambda}^{-1}\mathbf{Q}^H \tag{6.87}$$

where $\boldsymbol{\Lambda}$ is an $(M + 1)$-by-$(M + 1)$ diagonal matrix consisting of the eigenvalues of the correlation matrix $\mathbf{R}_{M+1}$, and $\mathbf{Q}$ is an $(M + 1)$-by-$(M + 1)$ matrix whose columns are the associated eigenvectors. That is,

$$\boldsymbol{\Lambda} = \text{diag}[\lambda_0, \lambda_1, \ldots, \lambda_M] \tag{6.88}$$

and

$$\mathbf{Q} = [\mathbf{q}_0, \mathbf{q}_1, \ldots, \mathbf{q}_M] \tag{6.89}$$

where $\lambda_0, \lambda_1, \ldots, \lambda_M$ are the real-valued eigenvalues of the correlation matrix $\mathbf{R}_{M+1}$, and $\mathbf{q}_0, \mathbf{q}_1, \ldots, \mathbf{q}_M$ are the respective eigenvectors. Thus, substituting Eqs. (6.87), (6.88), and (6.89) in Eq. (6.86), we get

$$
\begin{aligned}
\mathbf{a}_M &= P_M \mathbf{Q}\boldsymbol{\Lambda}^{-1}\mathbf{Q}^H \mathbf{i}_{M+1} \\[2mm]
&= P_M[\mathbf{q}_0, \mathbf{q}_1, \ldots, \mathbf{q}_M]\text{diag}[\lambda_0^{-1}, \lambda_1^{-1}, \ldots, \lambda_M^{-1}]
\begin{bmatrix} \mathbf{q}_0^H \\ \mathbf{q}_1^H \\ \cdot \\ \cdot \\ \cdot \\ \mathbf{q}_M^H \end{bmatrix}
\begin{bmatrix} 1 \\ 0 \\ \cdot \\ \cdot \\ \cdot \\ 0 \end{bmatrix} \\[2mm]
&= P_M \sum_{k=0}^{M} \left(\frac{q_{k0}^*}{\lambda_k} \right) \mathbf{q}_k
\end{aligned}
\tag{6.90}
$$

where q_{k0} is the first element of the kth eigenvector of the correlation matrix $\mathbf{R}_{M+1}$. We note that the first element of the forward prediction-error filter vector $\mathbf{a}_M$ equals 1. Therefore, using this fact, we find from Eq. (6.90) that the prediction-error power equals

$$P_M = \frac{1}{\displaystyle\sum_{k=0}^{M} |q_{k0}|^2 \lambda_k^{-1}} \tag{6.91}$$

Thus, on the basis of Eqs. (6.90) and (6.91) we may make the statement: *The tap-weight vector of a forward prediction-error filter of order M and the resultant prediction-error power are uniquely defined by specifying the (M + 1) eigenvalues and the corresponding (M + 1) eigenvectors of the correlation matrix of the tap inputs of the filter.*

6.5 SCHUR–COHN TEST

The test described under Property 3 in Section 6.4 for the minimum-phase condition of a forward prediction-error of order M is relatively simple to apply if we know the associated set of reflection coefficients $\kappa_1, \kappa_2, \ldots, \kappa_M$. For the filter to be minimum phase [i.e., for all the zeros of the transfer function of the filter, $H_{f,m}(z)$, to lie inside the unit circle], we simply require that $|\kappa_m| < 1$ for all m. Suppose, however, that instead of these reflection coefficients we are given the tap weights of the filter, $a_{M,1}, a_{M,2}, \ldots, a_{M,M}$. In this case we first apply the inverse recursion [described by Eq. (6.65)] to compute the corresponding set of reflection coefficients $\kappa_1, \kappa_2, \ldots, \kappa_M$. Then, as before, we check whether or not $|\kappa_m| < 1$ for all m.

The method just described for determining whether or not $H_{f,m}(z)$ has zeros inside the unit circle, given the coefficients $a_{M,1}, a_{M,2}, \ldots, a_{M,M}$, is essentially the same as the *Schur–Cohn test.*[4]

To formulate the Schur–Cohn test, let

$$x(z) = a_{M,M}\, z^M + a_{M,M-1}z^{M-1} + \cdots + a_{M,0} \tag{6.92}$$

which is a polynomial in z, with $x(0) = a_{M,0} = 1$. Define

$$x'(z) = z^M x^*(1/z^*)$$
$$= a_{M,M}^* + a_{M,M-1}^* z + \cdots + a_{M,0}^* z^M \tag{6.93}$$

which is the *reciprocal polynomial* associated with $x(z)$. The polynomial $x'(z)$ is so called since its zeros are the reciprocals of the zeros of $x(z)$. For $z = 0$, we have $x'(0) = a_{M,M}^*$. Next, define the linear combination

$$T[x(z)] = a_{M,0}^* x(z) - a_{M,M} x'(z) \tag{6.94}$$

so that, in particular, the value

$$T[x(0)] = a_{M,0}^* x(0) - a_{M,M} x'(0)$$
$$= 1 - |a_{M,M}|^2 \tag{6.95}$$

---

[4] The classical Schur–Cohn test is discussed in Marden (1949) and Tretter (1976). The origin of the test can be traced back to Schur (1917) and Cohn (1922), hence the name. The test is also referred to as the Lehmer–Schur method (Ralston, 1965); this is in recognition of the application of Schur's theorem by Lehmer (1961).

is real, as it should be. Note also that $T[x(z)]$ has no term in z^M. Repeat this operation as far as possible, so that if we define

$$T^i[x,(z)] = T\{T^{i-1}[x(z)]\} \tag{6.96}$$

we generate a finite sequence of polynomials in z of *decreasing* order. The coefficient $a_{M,0}$ is equal to 1. Let it also be assumed that:

- The polynomial $x(z)$ has no zeros on the unit circle.
- The integer m is the smallest for which

$$T^m[x(z)] = 0, \text{ where } m \leq M + 1$$

Then, we may state the Schur–Cohn theorem as follows (Lehmer, 1961):

If for some i such that $1 \leq i \leq m$, we have $T^i[x(0)] < 0$, then $x(z)$ has at least one zero inside the unit circle. If, on the other hand, $T^i[x(0)] > 0$ for $1 \leq i < m$, and $T^{m-1}[x(z)]$ is a constant, then no zero of $x(z)$ lies inside the unit circle.

To apply this theorem to determine whether or not the polynomial $x(z)$ of Eq. (6.92), with $a_{M,0} \neq 0$, has a zero inside the unit circle, we proceed as follows (Ralston, 1965):

1. Calculate $T[x(z)]$. Is $T[x(0)]$ negative? If so, there is a zero inside the unit circle; if not, proceed to step 2.
2. Calculate $T^i[x(z)]$, $i = 1, 2, \ldots$, until $T^i[x(0)] < 0$ for $i < m$, or $T^i[x(0)] > 0$ for $i < m$. If the former occurs, there is a zero inside the unit circle. If the latter occurs, and if $T^{m-1}[x(z)]$ is a constant, then there is no zero inside the unit circle.

Note that when the polynomial $x(z)$ has zeros inside the unit circle, this algorithm does not tell us how many; rather, it only confirms their existence.

The connection between the Schur–Cohn method and the inverse recursion is readily established by observing that (see Problem 10):

1. The polynomial $x(z)$ is related to the transfer function of a backward prediction-error filter of order M as follows:

$$x(z) = z^M H_{b,M}(z) \tag{6.97}$$

Accordingly, if the Schur–Cohn test indicates that $x(z)$ has zero(s) inside the unit circle, we may conclude that the transfer function $H_{b,M}(z)$ is *not* maximum phase.

2. The reciprocal polynomial $x'(z)$ is related to the transfer function of the corresponding forward prediction-error filter of order M as follows:

$$x'(z) = z^M H_{f,M}(z) \tag{6.98}$$

Accordingly, if the Schur–Cohn test indicates that the original polynomial $x(z)$, with which $x'(z)$ is associated, has no zero(s) inside the unit circle, we may then conclude that the transfer function $H_{f,M}(z)$ is *not* minimum phase.

3. In general, we have

$$T^i[x(0)] = \prod_{j=0}^{i-1} (1 - |a_{M-j,M-j}|^2), \qquad 1 \le i \le M \qquad (6.99)$$

and

$$H_{b,M-i}(z) = \frac{z^{i-M} T^i[x(z)]}{T^i[x(0)]} \qquad (6.100)$$

where $H_{b,M-i}(z)$ is the transfer function of the backward prediction-error filter of order $M - i$.

6.6 AUTOREGRESSIVE MODELING OF A STATIONARY STOCHASTIC PROCESS

The whitening property of a forward prediction-error filter, operating on a stationary discrete-time stochastic process, is intimately related to the *autoregressive modeling* of the process. Indeed, we may view these two operations as *complementary*, as illustrated in Fig. 6.7. Part (a) of the figure depicts a forward prediction-error filter of order M, whereas part (b) depicts the corresponding autoregressive model. We may make the following observations:

1. We may view the operation of prediction-error filtering applied to a stationary process $u(n)$ as one of *analysis*. In particular, we may use such an operation to whiten the process $u(n)$ by choosing the prediction-error filter order M sufficiently large, in which case the prediction error process $f_M(n)$ at the filter output consists of uncorrelated samples. When this unique condition has been established, the original stochastic process $u(n)$ is represented by the set of tap weights of the filter, $\{a_{M,k}\}$, and the prediction error power, P_M.

2. We may view the autoregressive (AR) modeling of the stationary process $u(n)$ as one of *synthesis*. In particular, we may generate the AR process $u(n)$ by applying a white-noise process $v(n)$ of zero mean and variance σ_v^2 to the input of an *inverse* filter whose parameters are set equal to the AR parameters $w_{ok}, k = 1, 2, \ldots, M$.

The two filter structures of Fig. 6.7 constitute a *matched pair*, with their parameters related as follows:

$$a_{M,k} = -w_{ok}, \qquad k = 1, 2, \ldots, M$$

and

$$P_M = \sigma_v^2$$

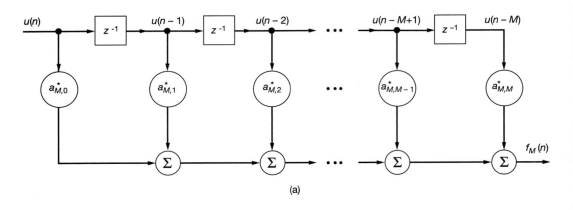

(a)

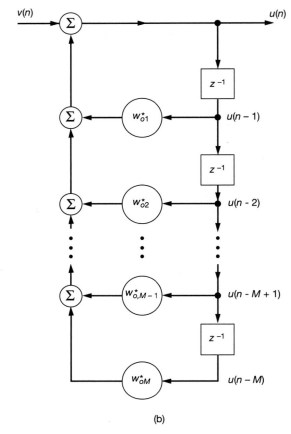

(b)

Figure 6.7 (a) Prediction-error (all-zero) filter; (b) autoregressive (all-pole) model with $w_{ok} = -a_{M,K}$, where $k = 1$, $2, \ldots, M$; the input $v(n)$ is white noise.

The prediction-error filter of Fig. 6.7(a) is an *all-zero filter with an impulse response of finite duration*. On the other hand, the inverse filter in the AR model of Fig. 6.7(b) is an *all-pole filter with an impulse response of infinite duration*. The prediction-error filter is minimum phase, with the zeros of its transfer function located at exactly the same positions (inside the units circle in the z-plane) as the poles of the transfer function of the inverse filter in part (b) of the figure. This assures the stability of the inverse filter in the bounded input-bounded output sense or, equivalently, the asymptotic stationarity of the AR process generated at the output of this filter.

Equivalence of the Autoregressive and Maximum Entropy Power Spectra

Consider the AR model of Fig. 6.7(b). The process $v(n)$ applied to the model input consists of a white noise of zero mean and variance σ_v^2. To find the power spectral density of the AR process $u(n)$ produced at the model output, we multiply the power spectral density of the model input $v(n)$ by the squared amplitude response of the model (see Chapter 2). Let $S_{AR}(\omega)$ denote the power spectral density of the AR process $u(n)$. We may therefore write (Priestley, 1981)

$$S_{AR}(\omega) = \frac{\sigma_v^2}{\left| 1 - \sum_{k=1}^{M} w_{ok}^* e^{-j\omega k} \right|^2} \tag{6.101}$$

The formula of Eq. (6.101) is called the *autoregressive power spectrum* or simply the *AR spectrum*.

A power spectrum of closely related interest is obtained using the *maximum entropy method (MEM)*. Suppose that we are given $2M + 1$ values of the autocorrelation function of a wide-sense stationary process $u(n)$. The essence of the maximum entropy method is to determine the particular power spectrum of the process that corresponds to the most random time series whose autocorrelation function is consistent with the set of $2M + 1$ known values (Burg, 1968, 1975). The result so obtained is referred to as the *maximum entropy power spectrum* or simply the *MEM spectrum*; see Appendix E. Let $S_{MEM}(\omega)$ denote the MEM spectrum. The determination of $S_{MEM}(\omega)$ is linked with the characterization of a prediction-error filter of order M, as shown in Appendix E. There it is shown that

$$S_{MEM}(\omega) = \frac{P_M}{\left| 1 + \sum_{k=1}^{M} a_{M,k}^* e^{-jk\omega} \right|^2} \tag{6.102}$$

where the $a_{M,k}$ are the prediction-error filter coefficients, and P_M is the prediction-error power, all of which correspond to a prediction order M.

In view of the one-to-one correspondence that exists between the prediction-error filter of Fig. 6.7(a) and the AR model of Fig. 6.7(b), we have

$$a_{M,k} = -w_{ok}, \qquad k = 1, 2, \ldots, M \tag{6.103}$$

and

$$P_M = \sigma_v^2 \tag{6.104}$$

Accordingly, the formulas given in Eqs. (6.101) and (6.102) are one and the same. In other words, for the case of a wide-sense stationary process, the AR spectrum (for model order M) and the MEM spectrum (for prediction order M) are indeed *equivalent* (Van den Bos, 1971).

6.7 CHOLESKY FACTORIZATION

Consider a stack of backward prediction-error filters of orders 0 to M that are connected in parallel as in Fig. 6.8. The filters are all fed at time n, with the same input denoted by $u(n)$. Note that for the case of zero prediction order, we simply have a direct connection as shown at the top end of Fig. 6.8. Let $b_0(n), b_1(n), \ldots, b_M(n)$ denote the sequence of backward prediction errors produced by these filters. We may express these errors in terms of the respective filter inputs and filter coefficients as follows [see Fig. 6.2(c)]:

$$
\begin{aligned}
b_0(n) &= u(n) \\
b_1(n) &= a_{1,1}u(n) + a_{1,0}u(n-1) \\
b_2(n) &= a_{2,2}u(n) + a_{2,1}u(n-1) + a_{2,0}u(n-2)
\end{aligned}
$$

$$\bullet$$
$$\bullet$$
$$\bullet$$

$$b_M(n) = a_{M,M}u(n) + a_{M,M-1}u(n-1) + \ldots + a_{M,0}u(n-M)$$

Combining this system of $(M+1)$ simultaneous linear equations into a compact matrix form, we have

$$\mathbf{b}(n) = \mathbf{L}\mathbf{u}(n) \tag{6.105}$$

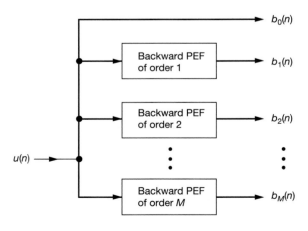

Figure 6.8 Parallel connection of a stack of backward prediction-error filters of orders 0 to M. *Note*: The direct connection represents a PEF with a single coefficient equal to unity.

where $\mathbf{u}(n)$ is the $(M + 1)$-by-1 input vector:

$$\mathbf{u}(n) = [u(n), u(n - 1), \ldots, u(n - M)]^T$$

and $\mathbf{b}(n)$ is the $(M + 1)$-by-1 output vector of backward prediction errors:

$$\mathbf{b}(n) = [b_0(n), b_1(n), \ldots, b_M(n)]^T$$

The $(M + 1)$-by-$(M + 1)$ coefficient matrix on the right-hand side of Eq. (6.105) is defined in terms of the backward prediction-error filter coefficients of orders 0 to M as follows

$$\mathbf{L} = \begin{bmatrix} 1 & 0 & \cdots & 0 \\ a_{1,1} & 1 & \cdots & 0 \\ \vdots & \vdots & \ddots & \vdots \\ a_{M,M} & a_{M,M-1} & \cdots & 1 \end{bmatrix} \tag{6.106}$$

The matrix $\mathbf{L}$ has three useful properties:

1. The matrix $\mathbf{L}$ is a *lower triangular matrix* with 1's along its main diagonal; all of its elements above the main diagonal are zero.
2. The determinant of matrix $\mathbf{L}$ is unity; hence, it is nonsingular (i.e., its inverse exists).
3. The nonzero elements of each row of the matrix $\mathbf{L}$, except for complex conjugation, equal the weights of a backward prediction-error filter whose order corresponds to the position of that row in the matrix.

The transformation of Eq. (6.105) is known as the *Gram–Schmidt orthogonalization algorithm*.[5] According to this algorithm, there is a *one-to-one correspondence between* the input vector $\mathbf{u}(n)$ and the backward prediction-error vector $\mathbf{b}(n)$. In particular, given $\mathbf{u}(n)$ we may obtain $\mathbf{b}(n)$ by using Eq. (6.105). Conversely, given $\mathbf{b}(n)$, we may obtain the corresponding vector $\mathbf{u}(n)$ by using the inverse of Eq. (6.105), as shown by

$$\mathbf{u}(n) = \mathbf{L}^{-1}\mathbf{b}(n) \tag{6.107}$$

where $\mathbf{L}^{-1}$ is the inverse of the matrix $\mathbf{L}$.

Orthogonality of Backward Prediction Errors

The backward prediction errors $b_0(n), b_1(n), \ldots, b_M(n)$, constituting the elements of the vector $\mathbf{b}(n)$, have an important property: *They are all orthogonal to each other, as shown by*

$$E[b_m(n)b_i^*(n)] = \begin{cases} P_m, & i = m \\ 0, & i \neq m \end{cases} \tag{6.108}$$

---

[5] For a full discussion of the Gram–Schmidt algorithm and the various methods for its implementation, see Haykin (1989).

To prove this property, we may proceed as follows. First of all, without loss of generality, we may assume that $m \geq i$. To prove Eq. (6.108), we express the backward prediction error $b_i(n)$ in terms of the input $u(n)$ as the linear convolution sum

$$b_i(n) = \sum_{k=0}^{i} a_{i,i-k} u(n - k) \tag{6.109}$$

which is a rewrite of Eq. (6.37) with prediction order i used in place of M. Hence, using this relation to evaluate the expectation of $b_m(n)b_i^*(n)$, we get

$$E[b_m(n)b_i^*(n)] = E[b_m(n) \sum_{k=0}^{i} a_{i,i-k}^* u^*(n - k)] \tag{6.110}$$

From the principle of orthogonality, we have

$$E[b_m(n)u^*(n - k)] = 0, \quad 0 \leq k \leq m - 1 \tag{6.111}$$

For $m > i$ and $0 \leq k \leq i$, we therefore find that all the expectation terms inside the summation on the right-hand side of Eq. (6.110) are zero. Correspondingly,

$$E[b_m(n)b_i^*(n)] = 0, \qquad m \neq i$$

When $m = i$, Eq. (6.110) reduces to

$$E[b_m(n)b_i^*(n)] = E[b_m(n)b_m^*(n)]$$
$$= P_m, \qquad m = i$$

This completes the proof of Eq. (6.108). It is important, however, to note that *this property holds only for wide-sense stationary input data.*

We thus see that the Gram–Schmidt orthogonalization algorithm of Eq. (6.105) transforms the input vector $\mathbf{u}(n)$ consisting of correlated samples into an equivalent vector $\mathbf{b}(n)$ of uncorrelated backward prediction errors.[6]

Factorization of the Inverse of Correlation Matrix R

Having equipped ourselves with the property that backward prediction errors are indeed orthogonal to each other, we may return to the transformation described by the Gram–Schmidt algorithm in Eq. (6.105). Specifically, using this transformation, we may

---

[6] Two random variables X and Y are said to be *orthogonal* to each other if

$$E[XY^*] = 0$$

They are said to be *uncorrelated* with each other if

$$E[X - E[X])(Y - E[Y])^*] = 0$$

If one or both of X and Y have zero mean, then these two conditions become one and the same. For the discussion presented herein, the input data, and therefore the backward predicted errors, are assumed to have zero mean. Under this assumption, we may interchange orthogonality with uncorrelatedness when we refer to backward prediction errors.

express the correlation matrix of the backward prediction-error vector $\mathbf{b}(n)$ in terms of the correlation matrix of the input vector $\mathbf{u}(n)$ as follows:

$$E[\mathbf{b}(n)\mathbf{b}^H(n)] = E[\mathbf{L}\mathbf{u}(n)\mathbf{u}^H(n)\mathbf{L}^H]$$
$$= \mathbf{L}E[\mathbf{u}(n)\mathbf{u}^H(n)]\mathbf{L}^H \tag{6.112}$$

Let $\mathbf{D}$ denote the correlation matrix of the backward prediction-error vector $\mathbf{b}(n)$, as shown by

$$\mathbf{D} = E[\mathbf{b}(n)\mathbf{b}^H(n)] \tag{6.113}$$

As before, the correlation matrix of the input vector $\mathbf{u}(n)$ is denoted by $\mathbf{R}$. We may therefore rewrite Eq. (6.112) as

$$\mathbf{D} = \mathbf{L}\mathbf{R}\mathbf{L}^H \tag{6.114}$$

We now make two observations:

1. When the correlation matrix $\mathbf{R}$ of the input vector $\mathbf{u}(n)$ is positive definite and therefore its inverse exists, the correlation matrix $\mathbf{D}$ of the backward prediction-error vector $\mathbf{b}(n)$ is also positive definite and likewise its inverse exists.
2. The correlation matrix $\mathbf{D}$ is a diagonal matrix, because $\mathbf{b}(n)$ consists of elements that are all orthogonal to each other. In particular, we may express the correlation matrix $\mathbf{D}$ in the expanded form:

$$\mathbf{D} = \text{diag}(P_0, P_1, \ldots, P_M) \tag{6.115}$$

where P_i is the average power of the ith backward prediction error $b_i(n)$; that is,

$$P_i = E[|b_i(n)|^2], \qquad i = 0, 1, \ldots, M \tag{6.116}$$

The inverse of matrix $\mathbf{D}$ is also a diagonal matrix, as shown by

$$\mathbf{D}^{-1} = \text{diag}(P_0^{-1}, P_1^{-1}, \ldots, P_M^{-1}) \tag{6.117}$$

Accordingly, we may use Eq. (6.114) to express the inverse of the correlation matrix $\mathbf{R}$ as follows:

$$\mathbf{R}^{-1} = \mathbf{L}^H\mathbf{D}^{-1}\mathbf{L}$$
$$= (\mathbf{D}^{-1/2}\mathbf{L})^H(\mathbf{D}^{-1/2}\mathbf{L}) \tag{6.118}$$

which is the desired result.

The inverse matrix $\mathbf{D}^{-1}$, in the first line of Eq. (6.118), is a diagonal matrix defined by Eq. (6.117). The matrix $\mathbf{D}^{-1/2}$, the *square root* of $\mathbf{D}^{-1}$, in the second line of Eq. (6.118) is also a diagonal matrix defined by

$$\mathbf{D}^{-1/2} = \text{diag}(P_0^{-1/2}, P_1^{-1/2}, \ldots, P_M^{-1/2})$$

The transformation of Eq. (6.118) is called the *Cholesky factorization of the inverse matrix* $\mathbf{R}^{-1}$ (Stewart, 1973). Note that the matrix $\mathbf{D}^{-1/2}\mathbf{L}$ is a lower triangular matrix; however, it differs from the lower triangular matrix $\mathbf{L}$ of Eq. (6.106) in that its diagonal elements are different from 1. Note also that the Hermitian-transposed matrix product $(\mathbf{D}^{-1/2}\mathbf{L})^H$ is an upper triangular matrix whose diagonal elements are different from 1. Thus, *according to the Cholesky factorization, the inverse correlation matrix $\mathbf{R}^{-1}$ may be factored into the product of an upper triangular matrix and a lower triangular matrix that are the Hermitian transpose of each other.*

6.8 LATTICE PREDICTORS

To implement the Gram–Schmidt algorithm of Eq. (6.105) for transforming an input vector $\mathbf{u}(n)$ consisting of correlated samples into an equivalent vector $\mathbf{b}(n)$ consisting of uncorrelated backward prediction errors, we may use the parallel connection of a direct path and an appropriate number of backward prediction-error filters, as illustrated in Fig. 6.8. The vectors $\mathbf{b}(n)$ and $\mathbf{u}(n)$ are said to be "equivalent" in the sense that they contain the same amount of information (see Problem 21). A much more efficient method of implementing the Gram–Schmidt orthogonalization algorithm, however, is to use an order-recursive structure in the form of a ladder, known as a *lattice predictor*. This device combines several forward and backward prediction-error filtering operations into a single structure. Specifically, a lattice predictor consists of a cascade connection of elementary units (stages), all of which have a structure similar to that of a lattice, hence the name. The number of stages in a lattice predictor equals the prediction order. Thus, for a prediction-error filter of order m, there are m stages in the lattice realization of the filter.

Order-update Recursions for the Prediction Errors

The input–output relations that characterize a lattice predictor may be derived in various ways, depending on the particular form in which the Levinson–Durbin algorithm is utilized. For the derivation presented here, we start with the matrix formulations of this algorithm given by Eqs. (6.40) and (6.42) that pertain to the forward and backward operations of a prediction-error filter, respectively. For convenience of presentation, we reproduce these two relations here:

$$\mathbf{a}_m = \begin{bmatrix} \mathbf{a}_{m-1} \\ 0 \end{bmatrix} + \kappa_m \begin{bmatrix} 0 \\ \mathbf{a}^{B*}_{m-1} \end{bmatrix} \tag{6.119}$$

$$\mathbf{a}^{B*}_m = \begin{bmatrix} 0 \\ \mathbf{a}^{B*}_{m-1} \end{bmatrix} + \kappa^*_m \begin{bmatrix} \mathbf{a}_{m-1} \\ 0 \end{bmatrix} \tag{6.120}$$

The $(m + 1)$-by-1 vector $\mathbf{a}_m$ and the m-by-1 vector $\mathbf{a}_{m-1}$ refer to forward prediction-error filters of order m and $m - 1$, respectively. The $(m + 1)$-by-1 vector $\mathbf{a}^{B*}_m$ and the m-by-1 vector $\mathbf{a}^{B*}_{m-1}$ refer to the corresponding backward prediction-error filters of order m and $m - 1$, respectively. The scalar κ_m is the associated reflection coefficient.

Consider first the forward prediction-error filter of order m, with its tap inputs denoted by $u(n), u(n-1), \ldots, u(n-m)$. We may partition $\mathbf{u}_{m+1}(n)$, the $(m+1)$-by-1 tap-input vector of this filter, in the form

$$\mathbf{u}_{m+1}(n) = \left[\begin{array}{c} \mathbf{u}_m(n) \\ \hline u(n-m) \end{array} \right] \tag{6.121}$$

or in the equivalent form

$$\mathbf{u}_{m+1}(n) = \left[\begin{array}{c} u(n) \\ \hline \mathbf{u}_m(n-1) \end{array} \right] \tag{6.122}$$

Next, we form the inner product of the $(m+1)$-by-1 vectors $\mathbf{a}_m$ and $\mathbf{u}_{m+1}(n)$. This is done by premultiplying $\mathbf{u}_{m+1}(n)$ by the Hermitian transpose of $\mathbf{a}_m$. Thus, using Eq. (6.119) for $\mathbf{a}_m$, we may treat the terms resulting from this multiplication as follows:

1. For the left-hand side of Eq. (6.119), premultiplication of $\mathbf{u}_{m+1}(n)$ by $\mathbf{a}_m^H$ yields

    $$f_m(n) = \mathbf{a}_m^H \mathbf{u}_{m+1}(n) \tag{6.123}$$

    where $f_m(n)$ is the forward prediction error produced at the output of the forward prediction-error filter of order m.

2. For the first term on the right-hand side of Eq. (6.119), we use the partitioned form of $\mathbf{u}_{m+1}(n)$ given in Eq. (6.121). We may therefore write

    $$[\mathbf{a}_{m-1}^H \,\vdots\, 0]\mathbf{u}_{m+1}(n) = [\mathbf{a}_{m-1}^H \,\vdots\, 0] \left[\begin{array}{c} \mathbf{u}_m(n) \\ \hline u(n-m) \end{array} \right]$$

    $$= \mathbf{a}_{m-1}^H \mathbf{u}_m(n) \tag{6.124}$$

    $$= f_{m-1}(n)$$

    where $f_{m-1}(n)$ is the forward prediction error produced at the output of the forward prediction-error filter of order $m-1$.

3. For the second matrix term on the right-hand side of Eq. (6.119), we use the partitioned form of $\mathbf{u}_{m+1}(n)$ given in Eq. (6.122). We may therefore write

    $$[0 \,\vdots\, \mathbf{a}_{m-1}^{BT}] \,\mathbf{u}_{m+1}(n) = [0 \,\vdots\, \mathbf{a}_{m-1}^{BT}] \left[\begin{array}{c} u(n) \\ \hline \mathbf{u}_m(n-1) \end{array} \right]$$

    $$= \mathbf{a}_{m-1}^{BT} \mathbf{u}_m(n-1) \tag{6.125}$$

    $$= b_{m-1}(n-1)$$

    where $b_{m-1}(n-1)$ is the *delayed* backward prediction error produced at the output of the backward prediction-error filter of order $m-1$.

Combining the results of the multiplications, described by Eqs. (6.123), (6.124), and (6.125), we may thus write

$$f_m(n) = f_{m-1}(n) + \kappa_m^* b_{m-1}(n-1) \tag{6.126}$$

Consider next the backward prediction-error filter of order m, with its tap inputs denoted by $u(n)$, $u(n-1)$, ..., $u(n-m)$. Here gain we may express $\mathbf{u}_{m+1}(n)$, the $(m+1)$-by-1 tap-input vector of this filter, in the partitioned form of Eq. (6.121) or that of Eq. (6.122). In this case, the terms resulting from the formation of the inner product of the vectors $\mathbf{a}_m^{B*}$ and $\mathbf{u}_{m+1}(n)$ are treated as follows:

1. For the left-hand side of Eq. (6.120), premultiplication of $\mathbf{u}_{m+1}(n)$ by the Hermitian transpose of $\mathbf{a}_m^{B*}$ yields

$$b_m(n) = \mathbf{a}_m^{BT}\,\mathbf{u}_{m+1}(n) \tag{6.127}$$

where $b_m(n)$ is the backward prediction error produced at the output of the backward prediction-error filter order m.

2. For the first term on the right-hand side of Eq. (6.120), we use the partitioned form of the tap-input vector $\mathbf{u}_{m+1}(n)$ given in Eq. (6.122). Thus, multiplying the Hermitian transpose of this first term by $\mathbf{u}_{m+1}(n)$, we get

$$[0 \mathrel{\vdots} \mathbf{a}_{m-1}^{BT}]\mathbf{u}_{m+1}(n) = [0 \mathrel{\vdots} \mathbf{a}_{m-1}^{BT}]\left[\begin{array}{c} u(n) \\ \hline \mathbf{u}_m(n-1) \end{array}\right]$$

$$= \mathbf{a}_{m-1}^{BT}\mathbf{u}_m(n-1) \tag{6.128}$$

$$= b_{m-1}(n-1)$$

3. For the second matrix term on the right-hand side of Eq. (6.120), we use the partitioned form of the tap-input vector $\mathbf{u}_{m+1}(n)$ given in Eq. (6.121). Thus, multiplying the Hermitian transpose of this second term by $\mathbf{u}_{m+1}(n)$, we get

$$[\mathbf{a}_{m-1}^{H} \mathrel{\vdots} 0]\mathbf{u}_{m+1}(n) = [\mathbf{a}_{m-1}^{H} \mathrel{\vdots} 0]\left[\begin{array}{c} u(n) \\ \hline \mathbf{u}_m(n-m) \end{array}\right]$$

$$= \mathbf{a}_{m-1}^{H}\mathbf{u}_m(n) \tag{6.129}$$

$$= f_{m-1}(n)$$

Combining the results of Eqs. (6.127), (6.128), and (6.129), we thus find that the inner product of $\mathbf{a}_m^{B*}$ and $\mathbf{u}_{m+1}(n)$ yields

$$b_m(n) = b_{m-1}(n-1) + \kappa_m f_{m-1}(n) \tag{6.130}$$

Equations (6.126) and (6.130) are the sought-after pair of *order-update recursions* that characterize stage m of the lattice predictor. They are reproduced here in matrix form as follows:

$$\begin{bmatrix} f_m(n) \\ b_m(n) \end{bmatrix} = \begin{bmatrix} 1 & \kappa_m^* \\ \kappa_m & 1 \end{bmatrix}\begin{bmatrix} f_{m-1}(n) \\ b_{m-1}(n-1) \end{bmatrix}, \qquad m = 1, 2, \ldots, M \tag{6.131}$$

We may view $b_{m-1}(n-1)$ as the result of applying the unit-delay operator z^{-1} to the backward prediction error $b_{m-1}(n)$; that is,

$$b_{m-1}(n-1) = z^{-1}[b_{m-1}(n)] \tag{6.132}$$

Thus, using Eqs. (6.131) and (6.132), we may represent stage m of the lattice predictor by the signal-flow graph shown in Fig. 6.9(a). Except for the branch pertaining to the block labeled z^{-1}, this signal-flow graph has the appearance of a lattice, hence the name "lattice predictor."[7] Note also that the parameterization of stage m of the lattice predictor is uniquely defined by the reflection coefficient κ_m.

For the elementary case of $m = 0$, we get the *initial conditions*:

$$f_0(n) = b_0(n) = u(n) \tag{6.133}$$

where $u(n)$ is the input signal at time n. Therefore, starting with $m = 0$, and progressively increasing the order of the filter by 1, we obtain the *lattice equivalent model* shown in Fig. 6.9(b) for a prediction-error filter of final order M. In this figure we merely require knowledge of the complete set of reflection coefficients $\kappa_1, \kappa_2, \ldots, \kappa_M$, one for each stage of the filter.

The lattice filter, depicted in Fig. 6.9(b), offers the following attractive features:

1. A lattice filter is a highly efficient structure for generating the sequence of forward prediction errors and the corresponding sequence of backward prediction errors simultaneously.

2. The various stages of a lattice predictor are "decoupled" from each other. This decoupling property was indeed derived in Section 6.7, where it was shown that the backward prediction errors produced by the various stages of a lattice predictor are "orthogonal" to each other for wide-sense stationary input data.

3. The lattice filter is *modular* in structure; hence, if the requirement calls for increasing the order of the predictor, we simply add one or more stages (as desired) without affecting earlier computations.

4. All the stages of a lattice predictor have a similar structure; hence, it lends itself to the use of *very large scale integration* (VLSI) technology if the use of this technology is considered beneficial to the application of interest.

Inverse Filtering Using the Lattice Structure

The multistage lattice predictor of Fig. 6.9(b) may be viewed as an *analyzer*. That is, it enables us to represent an autoregressive (AR) process $u(n)$ by a corresponding sequence of reflection coefficients $\{\kappa_m\}$. By rewiring this multistage lattice predictor in the manner depicted in Fig. 6.10, we may use this new structure as a *synthesizer* or *inverse filter*. That is, given the sequence of reflection coefficients $\{\kappa_m\}$, we may reproduce the original AR process by applying a white-noise process $v(n)$ to the input of the structure in Fig. 6.10.

---

[7] The first application of lattice filters in on-line adaptive signal processing was apparently made by Itakura and Saito (1971) in the field of speech analysis. Equivalent lattice-filter models, however, were familiar in geophysical signal processing as "layered earth models" (Robinson, 1967; Burg, 1968). It is also of interest to note that such lattice filters have been well studied in network theory, especially in the cascade synthesis of multiport networks (Dewilde, 1969).

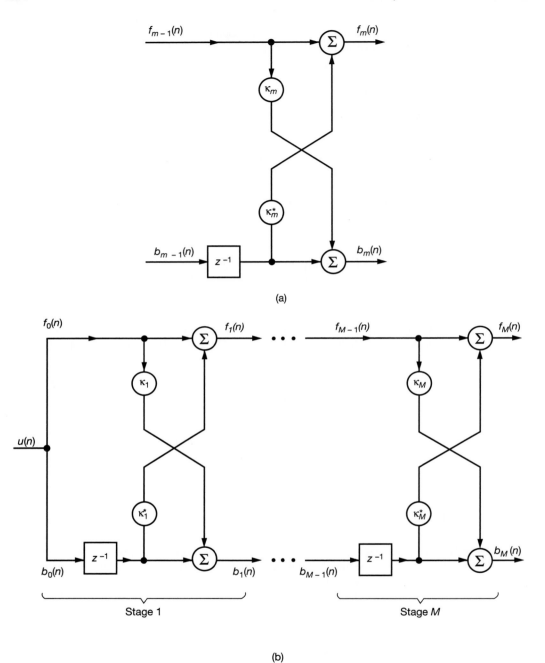

(a)

(b)

Figure 6.9 Signal-flow graph for stage m of a lattice predictor; (b) lattice equivalent model of prediction-error filter of order M.

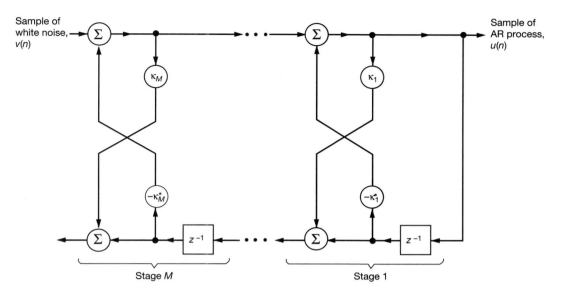

Figure 6.10   Lattice inverse filter of order M.

We illustrate the operation of the lattice inverse filter with an example.

Example 5

Consider the two-stage lattice inverse filter of Fig. 6.11(a). There are four possible paths that can contribute to the makeup of the sample $u(n)$ at the output, as illustrated in Fig. 6.11(b). In particular, we have

$$u(n) = v(n) - \kappa_1^* u(n-1) - \kappa_1 \kappa_2^* u(n-1) - \kappa_2^* u(n-2)$$

$$= v(n) - (\kappa_1^* + \kappa_1 \kappa_2^*) u(n-1) - \kappa_2^* u(n-2)$$

From Example 2, we recall that

$$a_{2,1} = \kappa_1 + \kappa_1^* \kappa_2$$
$$a_{2,2} = \kappa_2$$

We may therefore express the mechanism governing the generation of process $u(n)$ as follows:

$$u(n) + a_{2,1}^* u(n-1) + a_{2,2}^* u(n-2) = v(n)$$

which is recognized as the difference equation of a second-order AR process.

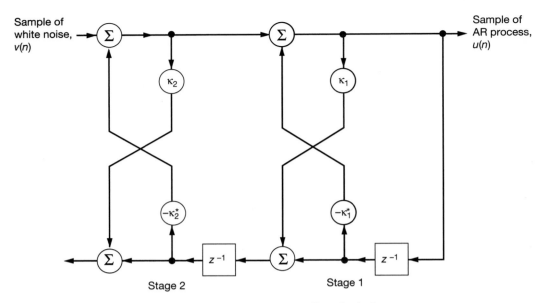

Figure 6.11   (a) Lattice inverse filter of order 2.

6.9 JOINT-PROCESS ESTIMATION

In this section, we use the lattice predictor as a subsystem to solve a *joint-process estimation problem* that is optimal in the mean-square sense (Griffiths, 1978; Makhoul, 1978). In particular, we consider the minimum mean-square estimation of a process $d(n)$, termed the desired response, by using a set of *observables* derived from a related process $u(n)$. We assume that the processes $d(n)$ and $u(n)$ are jointly stationary. This estimation problem is similar to that considered in Chapter 5, with one basic difference. In Chapter 5 we used samples of the process $u(n)$ as the observables directly. Our approach here is different in that for the observables we use the set of backward prediction errors obtained by feeding the input of a multistage lattice predictor with samples of the process $u(n)$. The fact that the backward prediction errors are orthogonal to each other simplifies the solution to the problem significantly.

     The structure of the *joint-process estimator* is shown in Fig. 6.12. This device performs two optimum estimations jointly:

     1. The *lattice predictor*, consisting of a cascade of M stages, characterized individually by the reflection coefficients $\kappa_1, \kappa_2, \ldots, \kappa_M$, performs predictions (of varying orders) on the input. In particular, it transforms the sequence of (correlated) input samples $u(n), u(n - 1), \ldots, u(n - M)$ into a corresponding sequence of (uncorrelated) backward prediction errors $b_0(n), b_1(n), \ldots, b_M(n)$.

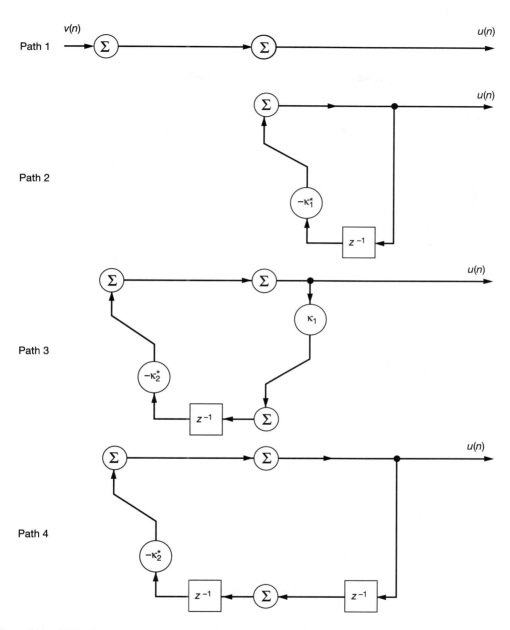

Figure 6.11 (b) The four possible paths that contribute to the makeup of the output $u(n)$ in the lattice inverse filter of Fig. 6.11(a).

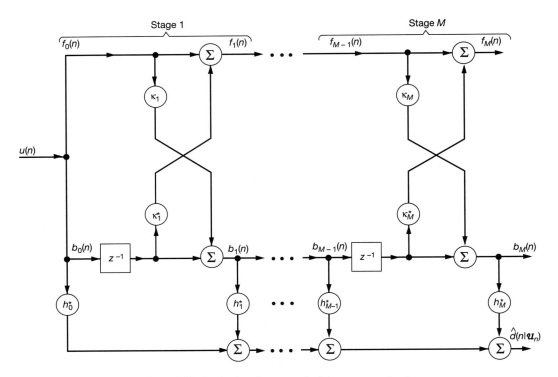

Figure 6.12   Lattice-based structure for joint-process estimation.

2. The multiple regression filter, characterized by the set of weights $h_0, h_1, \ldots, h_M$, operates on the sequence of backward prediction errors $b_0(n), b_1(n), \ldots, b_M(n)$ as inputs, respectively, to produce an estimate of the desired response $d(n)$. The resulting estimate is defined as the sum of the respective scalar inner products of these two sets of quantities, as shown by

$$\hat{d}(n|\mathcal{U}_n) = \sum_{i=0}^{M} h_i^* b_i(n) \tag{6.134}$$

where $\mathcal{U}_n$ is the space spanned by the inputs $u(n), u(n-1), \ldots, u(n-M)$. We may rewrite Eq. (6.134) in matrix form as follows:

$$\hat{d}(n|\mathcal{U}_n) = \mathbf{h}^H \mathbf{b}(n) \tag{6.135}$$

where $\mathbf{h}$ is an $(M+1)$-by-1 vector defined by

$$\mathbf{h} = [h_0, h_1, \ldots, h_M]^T \tag{6.136}$$

We refer to $h_0, h_1, \ldots, h_M$ as the regression coefficients of the estimator, and to **h** as the regression-coefficient vector.

Let **D** denote the $(M + 1)$-by-$(M + 1)$ correlation matrix of **b**(n), the $(M + 1)$-by-1 vector of backward prediction errors $b_0(n), b_1(n), \ldots, b_M(n)$. Let **z** denote the $(M + 1)$-by-1 cross-correlation vector between the backward prediction errors and the desired response as shown by

$$\mathbf{z} = E[\mathbf{b}(n)d^*(n)] \tag{6.137}$$

Therefore, applying the Wiener–Hopf equations to our present situation, we find that the optimum regression-coefficient vector $\mathbf{h}_o$ is defined by

$$\mathbf{D}\mathbf{h}_o = \mathbf{z} \tag{6.138}$$

Solving for $\mathbf{h}_o$, we get

$$\mathbf{h}_o = \mathbf{D}^{-1}\mathbf{z} \tag{6.139}$$

where the inverse matrix $\mathbf{D}^{-1}$ is a diagonal matrix, defined in terms of various prediction-error powers as in Eq. (6.117). Note that, unlike the ordinary transversal filter realization of the Wiener filter, the computation of $\mathbf{h}_o$ in the joint-process estimator of Fig. 6.12 is relatively simple to accomplish.

Relationship between the Optimum Regression-Coefficient Vector and the Wiener Solution

From the Cholesky factorization given in Eq. (6.118), we deduce that

$$\mathbf{D}^{-1} = \mathbf{L}^{-H}\mathbf{R}^{-1}\mathbf{L}^{-1} \tag{6.140}$$

Hence, substituting Eq. (6.140) in (6.139) yields

$$\mathbf{h}_o = \mathbf{L}^{-H}\mathbf{R}^{-1}\mathbf{L}^{-1}\mathbf{z} \tag{6.141}$$

Moreover, from Eq. (6.105) we note that

$$\mathbf{b}(n) = \mathbf{L}\mathbf{u}(n) \tag{6.142}$$

Therefore, substituting Eq. (6.142) in (6.137) yields

$$\mathbf{z} = \mathbf{L}E[\mathbf{u}(n)d^*(n)] \tag{6.143}$$
$$= \mathbf{L}\mathbf{p}$$

where **p** is the cross-correlation vector between the tap-input vector $\mathbf{u}(n)$ and the desired response $d(n)$. Thus, using Eq. (6.143) in (6.141), we finally obtain

$$\mathbf{h}_o = \mathbf{L}^{-H}\mathbf{R}^{-1}\mathbf{L}^{-1}\mathbf{L}\mathbf{p}$$
$$= \mathbf{L}^{-H}\mathbf{R}^{-1}\mathbf{p} \tag{6.144}$$
$$= \mathbf{L}^{-H}\mathbf{w}_o$$

where $\mathbf{L}$ is a lower triangular matrix defined in terms of the equivalent forward prediction-error filter coefficients, as in Eq. (6.106). Equation (6.144) is the sought-after relationship between the optimum regression-coefficient vector $\mathbf{h}_o$ and the Wiener solution $\mathbf{w}_o = \mathbf{R}^{-1}\mathbf{p}$.

6.10 BLOCK ESTIMATION

In this chapter, we have discussed two basic structures for building a *linear predictor* or its natural extension, a *prediction-error filter*; the two structures are a *transversal filter* and a *lattice filter*.[8] The transversal filter is characterized by a set of *tap weights*, whereas the lattice filter is characterized by a corresponding set of *reflection coefficients*. In both cases, the filter coefficients provide the designer with *degrees of freedom*, the number of which equals the *prediction order*. The mathematical link between the tap weights of a transversal predictor and the reflection coefficients of a lattice predictor is provided by the Levinson–Durbin algorithm.

Regardless of the particular structure chosen, we clearly need a procedure for *estimating* the filter coefficients. To carry out this estimation we have two approaches to consider:

- Block estimation
- Adaptive estimation

In *block estimation*, the available data are divided into individual *blocks*, each of length N, say. The *block length* N is usually chosen short enough to ensure pseudostationarity of the input data over the length N. Under this assumption, the filter coefficients of interest are then computed on a *block-by-block* basis. Typically, the filter coefficients vary from one block of data to another. The block estimation algorithms may be categorized as follows:

1. *Indirect methods.* For each block of data, estimates of the autocorrelation function of the input are computed for different lags. The Levinson–Durbin algorithm is then used to compute the corresponding set of tap weights for a transversal predictor, or the corresponding set of reflection coefficients for a lattice predictor, depending on the application of interest.

---

[8] In actual fact, there is a third structure for building a linear predictor that is based on the *Schur algorithm* (Schur, 1917). Like the Levinson–Durbin algorithm, the Schur algorithm provides a procedure for computing the reflection coefficients from a known autocorrelation sequence. The Schur algorithm, however, lends itself to parallel implementation, with the result that it achieves a throughput rate higher than that obtained using the Levinson–Durbin algorithm. For a discussion of the Schur algorithm, including mathematical details and implementational considerations, see Haykin (1989).

2. *Direct methods*. For each block of data, estimates of the reflection coefficients for the different stages of a lattice predictor are computed directly from the data. The reflection coefficients lend themselves to direct computation because of the decoupling property of a multistage lattice predictor for wide-sense stationary inputs.

From a computational viewpoint, direct methods are more efficient than indirect methods if the application of interest requires knowledge of the reflection coefficients. If, on the other hand, the application is that of system identification in terms of the tap weights of a transversal filter, as in autoregressive modeling, for example, the computational advantage of direct methods over indirect methods may well disappear. In this section, we develop a popular algorithm known as the *Burg algorithm*, which may be classified as a direct block estimation method.[9]

Consider a lattice predictor consisting of M stages connected in cascade as in Fig. 6.9(b). For wide-sense stationary input data, the M stages of this lattice predictor are decoupled from each other by virtue of the orthogonality of its backward prediction-error outputs; see Section 6.7. This decoupling property makes it possible for us to accomplish the *global optimization of a multistage lattice predictor as a sequence of local optimizations, one at each stage of the structure*. Moreover, it is a straightforward matter to increase the order of the predictor by simply adding one or more stages, as required, without affecting the earlier design computations. For example, suppose that we have optimized the design of a lattice predictor consisting of M stages. To increase the order of the predictor by 1, we simply add a new stage that is locally optimized, leaving the optimum design of the earlier stages unchanged.

Consider stage m of the lattice predictor shown in Fig. 6.9(a), for which the input–output relations are given in matrix form in Eq. (6.131). For convenience of presentation, these relations are reproduced here in the expanded form:

$$f_m(n) = f_{m-1}(n) + \kappa_m^* b_{m-1}(n-1) \tag{6.145}$$

$$b_m(n) = b_{m-1}(n-1) + \kappa_m f_{m-1}(n) \tag{6.146}$$

where $m = 1, 2, \ldots, M$, and M is the final order of the predictor. Several criteria may be used to optimize the design of this stage (Makhoul, 1977). However, one particular criterion yields a design with interesting properties that conform to the lattice predictor theory. Specifically, the reflection coefficient κ_m is chosen so as to minimize the sum of the mean-squared values of the forward and backward prediction errors. Let the cost function J_m denote this sum at the output of stage m of the lattice predictor:

$$J_m = E[|f_m(n)|^2] + E[|b_m(n)|^2] \tag{6.147}$$

---

[9] For a more complete discussion of block estimation, including both indirect and direct methods, see Marple (1987), Kay (1988), and Haykin (1989).

Substituting Eqs. (6.145) and (6.146) in (6.147), we get

$$J_m = \{E[|f_{m-1}(n)|^2] + E[|b_{m-1}(n-1)|^2]\}[1 + |\kappa_m|^2]$$
$$+ 2\kappa_m E[f_{m-1}(n)b_{m-1}^*(n-1)] \qquad (6.148)$$
$$+ 2\kappa_m^* E[b_{m-1}(n-1)f_{m-1}^*(n)]$$

In general, the reflection coefficient κ_m is complex valued, as shown by

$$\kappa_m = \alpha_m + j\beta_m \qquad (6.149)$$

Therefore, differentiating the cost function J_m with respect to both the real and imaginary parts of κ_m, we get the complex-valued *gradient*

$$\nabla J_m = \frac{\partial J_m}{\partial \alpha_m} + j\frac{\partial J_m}{\partial \beta_m}$$
$$= 2\kappa_m\{E[|f_{m-1}(n)|^2] + E[|b_{m-1}(n-1)|^2]\} \qquad (6.150)$$
$$+ 4E[b_{m-1}(n-1)f_{m-1}^*(n)]$$

Putting this gradient to zero, we find that the optimum value of the reflection coefficient, for which the cost function J_m is minimum, equals

$$\kappa_{m,o} = -\frac{2E[b_{m-1}(n-1)f_{m-1}^*(n)]}{E[|f_{m-1}(n)|^2 + |b_{m-1}(n-1)|^2]} \quad , \qquad m = 1, 2, \ldots, M \qquad (6.151)$$

Equation (6.151) for the reflection coefficient is known as the *Burg formula* (Burg, 1968).[10] Its use offers two interesting properties (see Problem 23):

1. The reflection coefficient $\kappa_{m,o}$ satisfies the condition

$$|\kappa_{m,o}| \leq 1 \quad \text{for all } m \qquad (6.152)$$

   In other words, the Burg formula always yields a minimum-phase design for the lattice predictor.

2. The mean-square values of the forward and backward prediction errors at the output of stage m are related to those at its own input as follows, respectively:

$$E[|f_m(n)|^2] = (1 - |\kappa_{m,o}|^2)E[|f_{m-1}(n)|^2] \qquad (6.153)$$

and

$$E[|b_m(n)|^2] = (1 - |\kappa_{m,o}|^2)E[|b_{m-1}(n-1)|^2] \qquad (6.154)$$

The Burg formula, as described in Eq. (6.151), involves the use of ensemble averaging. Assuming that the input $u(n)$ is *ergodic*, we may substitute time averages for the

---

[10] The 1968 paper by Burg is reproduced in the book by Childers (1978).

expectations in the dominator and denominator of this equation. We thus get the *Burg estimate*[11] for the reflection coefficient of stage m in the lattice predictor

$$\hat{\kappa}_m = - \frac{2\displaystyle\sum_{n=m+1}^{N} b_{m-1}(n-1)f^*_{m-1}(n)}{\displaystyle\sum_{n=m+1}^{N} [|f_{m-1}(n)|^2 + |b_{m-1}(n-1)|^2]}, \qquad m = 1, 2, \ldots M \qquad (6.155)$$

where N is the length of a block of input data, and $f_0(n) = b_0(n) = u(n)$.

   With a lattice predictor of m stages, each of which contains a single unit-delay element, and with the input $u(n)$ zero for $n \le 0$, we find that *all* the samples in the input data contribute to the outputs of stage m in the predictor for the first time at $n = m + 1$, hence the use of this value for the lower limits of the summation terms in Eq. (6.155). Note also that the estimate $\hat{\kappa}_m$ for the mth reflection coefficient is dependent on data length N. The choice of N is usually dictated by two conflicting factors. First, it should be large enough to smooth out the effects of noise in computing the time averages in the numerator and denominator of Eq. (6.155). Second, as mentioned previously, it should be small enough to ensure quasi-statistical stationarity of the input data during the computations, and thereby justify the application of Burg's formula.

   The block-estimation approach usually requires a large amount of computation, as well as a large amount of storage. Furthermore, in this approach, we find that for any stage of the predictor the estimate of the reflection coefficient at time $n + 1$ does not depend in a simple way on its previous estimate at time n. This behavior is to be contrasted with the adaptive estimation procedure described in subsequent chapters of the book.

6.11 SUMMARY AND DISCUSSION

In this chapter we presented a detailed study of the linear prediction problem pertaining to wide-sense stationary stochastic processes. In particular, we used the Wiener filter theory of Chapter 5 to develop optimum solutions for the two basic forms of linear prediction:

- Forward linear prediction, in which case we are given the input sequence $u(n-1)$, $u(n-2)$, . . . , $u(n-M)$ and the requirement is to make an optimum prediction of the next sample value $u(n)$.

---

[11] For some practical applications of the Burg estimator given in Eq. (6.155), see Haykin et al. (1982) and Swingler and Walker (1989). The first of these two papers describes a (temporal) procedure based on this estimator for classifying the different forms of radar clutter (e.g., radar returns from different targets) as encountered in an air traffic control environment. The second paper presents a demonstration of a linear array beamformer based on a spatial interpretation of the Burg estimator for sonar environment. The studies reported in both of these papers employ real-life data.

- Backward linear prediction, in which case we are given the input sequence $u(n), u(n - 1), \ldots, u(n - M + 1)$ and the requirement is to make an optimum prediction of the old sample value $u(n - M)$.

In both cases, the desired response is derived from the time series itself. In forward linear prediction $u(n)$ acts as the desired response, whereas in backward linear prediction $u(n - M)$ acts as the desired response.

The prediction process may be described in terms of a predictor or, equivalently, a prediction-error filter. These two linear devices differ from each other in their respective outputs. The output of a forward predictor is a one-step prediction of the input. On the other hand, the output of a forward prediction-error filter is the prediction error. In a similar way, we may distinguish between a backward predictor and a backward prediction-error filter.

The two structures most commonly used for building a prediction-error filter are:

- Transversal filter, where the issue of concern is the determination of the tap weights
- Lattice filter, where the issue of concern is the determination of the reflection coefficients

These two sets of parameters are in fact uniquely related to each other via the Levinson–Durbin recursion.

The important properties of prediction-error filters may be summarized as follows:

- The forward prediction-error filter is minimum phase, which means that all the zeros of its transfer function lie inside the unit circle in the z-plane. The corresponding inverse filter, representing an autoregressive model of the input process, is therefore stable.
- The backward prediction-error filter is maximum phase, which means that all the zeros of its transfer function lie outside the unit circle in the z-plane. In this case, the inverse filter is unstable and therefore of no practical value.
- The forward prediction-error filter is a whitening filter, whereas the backward prediction-error filter is an anticausal whitening filter (see Problem 14).

The lattice predictor offers some highly desirable properties:

- *Order-recursive structure*, which means that the prediction order may be increased by adding one or more stages to the structure without destroying the previous calculations.
- *Modularity*, which is exemplified by the fact that all the stages of the lattice predictor have exactly the same physical structure.
- *Simultaneous computation of forward and backward prediction errors*, which provides for computational efficiency.

- *Statistical decoupling of the individual stages*, which is another way of saying that the backward prediction errors of varying orders produced by the different stages of the lattice predictor are uncorrelated with each other. This property, embodied in the Cholesky factorization, is exploited in the joint-estimation process, where the backward prediction errors are used to provide an estimate of some desired response.

PROBLEMS

1. The augmented Wiener–Hopf equations (6.14) of a forward prediction-error filter were derived by first optimizing the linear prediction filter in the mean-square sense and then combining the two resultants: the Wiener–Hopf equations for the tap-weight vector and the minimum mean-squared prediction error. This problem addresses the issue of deriving Eq. (6.14) directly by proceeding as follows:
   (a) Formulate the expression for the mean-square value of the forward prediction error as a function of the tap-weight vector of the forward prediction-error filter.
   (b) Minimize this mean-squared prediction error, subject to the constraint that the leading element of the tap-weight vector of the forward prediction-error filter equals 1.
   Hint: Use the method of Lagrange multipliers to solve the constrained optimization problem. For details of this method, see Appendix C. This hint also applies to part (b) of Problem 2.

2. The augmented Wiener–Hopf equations (6.38) of a backward prediction-error filter were derived indirectly in Section 6.2. This problem addresses the issue of deriving Eq. (6.38) directly by proceeding as follows:
   (a) Formulate the expression for the mean-square value of the backward prediction error in terms of the tap-weight vector of the backward prediction-error filter.
   (b) Minimize this mean-squared prediction error, subject to the constraint that the last element of the tap-weight vector of the backward prediction-error filter equals 1.

3. (a) Equation (6.24) defines the Wiener–Hopf equations for backward linear prediction. This system of equations is reproduced here for convenience:

$$\mathbf{R}\mathbf{w}_b = \mathbf{r}^{B*}$$

where $\mathbf{w}_b$ is the tap-weight vector of the predictor, $\mathbf{R}$ is the correlation matrix of the tap inputs $u(n), u(n-1), \ldots, u(n-M+1)$, and $\mathbf{r}^{B*}$ is the cross-correlation vector between these tap inputs and the desired response $u(n-M)$. Show that if the elements of the column vector $\mathbf{r}^{B*}$ are rearranged in reverse order, the effect of this reversal is to modify the Wiener–Hopf equations as

$$\mathbf{R}^T\mathbf{w}_b^B = \mathbf{r}^*$$

   (b) Show that the inner products $\mathbf{r}^{BT}\mathbf{w}_b$ and $\mathbf{r}^T\mathbf{w}_b^B$ are equal.

4. Consider a wide-sense stationary process $u(n)$ whose autocorrelation function has the following values for different lags:

$$r(0) = 1$$
$$r(1) = 0.8$$
$$r(2) = 0.6$$
$$r(3) = 0.4$$

(a) Use the Levinson–Durbin recursion to evaluate the reflection coefficients κ_1, κ_2, and κ_3.

(b) Set up a three-stage lattice predictor for this process using the values for the reflection coefficients found in part (a).

(c) Evaluate the average power of the prediction error produced at the output of each of the three stages in this lattice predictor. Hence, make a plot of prediction-error power versus prediction order. Comment on your results.

5. Consider the filtering structure described in Fig. P6.1, where the delay $\triangle$ is an integer greater than one. The requirement is to choose the weight vector **w** so as to minimize the mean-square value of the estimation error $e(n)$. Find the optimum value of $\mathbf{w}(n)$.

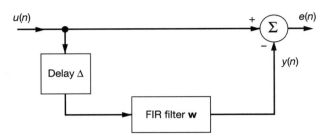

Figure P6.1

6. Consider the linear prediction of a stationary autoregressive process $u(n)$, generated from the first-order difference equation

$$u(n) = 0.9u(n - 1) + v(n)$$

where $v(n)$ is a white noise process of zero mean and unit variance. The prediction order is two.

(a) Determine the tap weights $a_{2,1}$ and $a_{2,2}$ of the forward prediction-error filter.

(b) Determine the reflection coefficients κ_1 and κ_2 of the corresponding lattice predictor. Comment on your results in parts (a) and (b).

7. (a) A process $u_1(n)$ consists of a single sinusoidal process of complex amplitude α and angular frequency ω in additive white noise of zero mean and variance σ_v^2, as shown by

$$u_1(n) = \alpha e^{j\omega n} + v(n)$$

where

$$E[|\alpha|^2] = \sigma_\alpha^2$$

and

$$E[|v(n)|^2] = \sigma_v^2$$

The process $u_1(n)$ is applied to a linear predictor of order M, optimized in the mean-squared error sense. Do the following:

(i) Determine the tap weights of the prediction-error filter of order M, and the final value of the prediction error power P_M.

(ii) Determine the reflection coefficients κ_1, κ_2, . . . , κ_M of the corresponding lattice predictor.

(iii) How are the results in part (i) and (ii) modified when we let the noise variance σ_v^2 approach zero?

(b) Consider next an AR process $u_2(n)$ described by

$$u_2(n) = -\alpha e^{j\omega} u_2(n-1) + v(n)$$

where, as before, $v(n)$ is an additive white noise process of zero mean and variance σ_v^2. Assume that $0 < |\alpha| < 1$ but very close to 1. The process $u_2(n)$ is also applied to a linear predictor of order M, optimized in the mean-squared error sense.

 (i) Determine the tap weights of the new prediction-error filter of order M.

 (ii) Determine the reflection coefficients $\kappa_1, \kappa_2, \ldots, \kappa_M$ of the corresponding lattice predictor.

(c) Use your results in parts (a) and (b) to compare the similarities and differences between the linear prediction of the processes $u_1(n)$ and $u_2(n)$.

8. Equation (6.40) defines the Levinson–Durbin recursion for forward linear prediction. By rearranging the elements of the tap-weight vector $\mathbf{a}_m$ backward and then complex-conjugating them, reformulate the Levinson–Durbin recursion for backward linear prediction as in Eq. (6.42).

9. Starting with the definition of Eq. (6.47) for Δ_{m-1}, show that Δ_{m-1} equals the cross-correlation between the delayed backward prediction error $b_{m-1}(n-1)$ and the forward prediction error $f_{m-1}(n)$.

10. Develop in detail the relationship between the Schur–Cohn method and the inverse recursion as outlined by Eqs. (6.97) through (6.100).

11. Consider an autoregressive process $u(n)$ of order 2, described by the difference equation

$$u(n) = u(n-1) - 0.5u(n-2) + v(n)$$

where $v(n)$ is a white-noise process of zero mean and variance 0.5.

(a) Find the average power of $u(n)$.

(b) Find the reflection coefficients κ_1 and κ_2.

(c) Find the average prediction-error powers P_1 and P_2.

12. Using the one-to-one correspondence between the two sequences of numbers $\{P_0, \kappa_1, \kappa_2\}$ and $\{r(0), r(1), r(2)\}$, compute the autocorrelation function values $r(1)$ and $r(2)$ that correspond to the reflection coefficients κ_1, and κ_2 found in Problem 11 for the second-order autoregressive process $u(n)$.

13. In Section 6.4, we presented a proof of the minimum-phase property of a prediction-error filter by using Rouché's theorem. In this problem, we explore another proof of this property by contradiction. Consider Fig. P6.2, which shows the prediction-error filter (of order M) represented as the cascade of two functional blocks, one with transfer function $C_i(z)$ and the other with its transfer function equal to the zero factor $(1 - z_i z^{-1})$. Let $S(\omega)$ denote the power spectral density of the process $u(n)$ applied to the input of the prediction-error filter.

(a) Show that the mean-square value of the forward prediction error $f_M(n)$ equals

$$\epsilon = \int_{-\pi}^{\pi} S(\omega)|C_i(e^{j\omega})|^2[1 - 2\rho_i \cos(\omega - \omega_i) + \rho_i^2]d\omega$$

 where

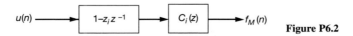

Figure P6.2

(b) Suppose that $\rho_i > 1$ so that the complex zero lies outside the unit circle. Hence, show that under this condition $\partial\epsilon/\partial\rho_i > 0$. Is such a condition possible and at the same time the filter operates at its optimum condition? What conclusion can you draw from your answers?

14. When an autoregressive process of order M is applied to a forward prediction-error filter of order M, the output consists of white noise. Show that when such a process is applied to a backward prediction-error filter of order M, the output consists of an anticausal realization of white noise.

15. Consider a forward prediction-error filter characterized by a real-valued set of coefficients $a_{m,1}$, $a_{m,2}, \ldots, a_{m,m}$. Define a polynomial $\phi_m(z)$ as follows:

$$\sqrt{P_m}\,\phi_m(z) = z^m + a_{m,1}z^{m-1} + \cdots + a_{m,m}$$

where P_m is the average prediction-error power of order m, and z^{-1} is the unit-delay operator. [Note the difference between the definition of $\phi_m(z)$ and that of the corresponding transfer function $H_{f,m}(z)$ of the filter.] The filter coefficients bear a one-to-one correspondence with the sequence of autocorrelations $r(0), r(1), \ldots, r(m)$. Define.

$$S(z) = \sum_{i=-m}^{m} r(i)z^{-i}$$

Show that

$$\frac{1}{2\pi j}\oint_{\mathscr{C}} \phi_m(z)\phi_k(z^{-1})S(z)\,dz = \delta_{mk}$$

where δ_{mk} is the Kronecker delta:

$$\delta_{mk} = \begin{cases} 1, & k = m \\ 0, & k \neq m \end{cases}$$

and the contour $\mathscr{C}$ is the unit circle. The polynomial $\phi_m(z)$ is referred to as a *Szegö polynomial*.

16. **(a)** Construct the two-stage lattice predictor for the second-order autoregressive process $u(n)$ considered in Problem 11.

   (b) Given a white-noise process $v(n)$, construct the two-stage lattice synthesizer for generating the autoregressive process $u(n)$. Check your answer against the second-order difference equation for the process $u(n)$ that was considered in Problem 11.

17. In a *normalized* lattice predictor, the forward and backward prediction errors at the various stages of the predictor are all normalized to have *unit variance*. Such an operation makes it possible to utilize the full dynamic range of multipliers used in the hardware implementation of a lattice predictor. For stage m of the normalized lattice predictor, the normalized forward and backward prediction errors are defined as follows, respectively:

$$\bar{f}_m(n) = \frac{f_m(n)}{P_m^{1/2}}$$

and

$$\bar{b}_m(n) = \frac{b_m(n)}{P_m^{1/2}}$$

where P_m is the average power (or variance) of the forward prediction error $f_m(n)$ or that of the backward prediction error $b_m(n)$. Show that the structure of stage m of the normalized lattice predictor is as shown in Fig. P6.3 for real-valued data.

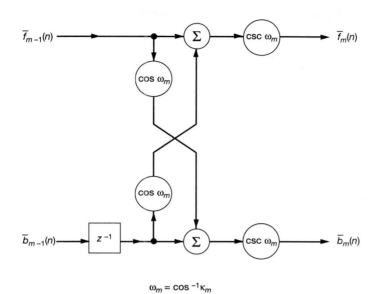

$$\omega_m = \cos^{-1} \kappa_m$$

Figure P6.3

18. **(a)** Consider the matrix product **LR** that appears in the decomposition of Eq. (6.114), where the $(M + 1)$-by-$(M + 1)$ lower triangular matrix **L** is defined in Eq. (6.106) and **R** is the $(M + 1)$-by-$(M + 1)$ correlation matrix. Let **Y** denote this matrix product, and let y_{mk} denote the mkth element of **Y**. Hence, show that

$$y_{mm} = P_m, \qquad m = 0, 1, \ldots, M$$

where P_m is the prediction-error power for order m.

(b) Show that the mth column of matrix **Y** is obtained by passing the autocorrelation sequence $\{r(0), r(1), \ldots, r(m)\}$ through a corresponding sequence of backward prediction-error filters represented by the transfer functions $H_{b,0}(z), H_{b,1}(z), \ldots, H_{b,m}(z)$.

(c) Suppose that we apply the autocorrelation sequence $\{r(0), r(1), \ldots, r(m)\}$ to the input of a lattice predictor of order m. Show that the variables appearing at the various points on the lower line of the predictor at time m equal the elements of the mth column of matrix **Y**.

(d) For the situation described in part (c), show that the lower output of stage m in the predictor at time m equals P_m, and that the upper output of this same stage at time $m + 1$ equals Δ_m^*. How is the ratio of these two outputs related to the reflection coefficient of stage $m + 1$?

(e) Use the results of part (d) to develop a recursive procedure for computing the sequence of reflection coefficients from the autocorrelation sequence.

19. In Section 6.8, we considered the use of an inverse lattice filter as the generator of an autoregressive process. The lattice inverse filter may also be used to efficiently compute the autocor-

relation sequence $r(1), r(2), \ldots, r(m)$ normalized with respect to $r(0)$. The procedure involves initializing the states (i.e., unit-delay elements) of the lattice inverse filter to $1, 0, \ldots, 0$ and then allowing the filter to operate with zero input. In effect, this procedure provides a lattice interpretation of Eq. (6.67) that relates the autocorrelation sequence $\{r(0), r(1), \ldots, r(M)\}$ and the augmented sequence of reflection coefficients $\{P_0, \kappa_1, \ldots, \kappa_M\}$. Demonstrate the validity of this procedure for the following values of final order M:

(a) $M = 1$

(b) $M = 2$

(c) $M = 3$

20. Prove the following correlation properties of lattice filters:

(a) $E[f_m(n)u^*(n - k)] = 0, \;\; 1 \le k \le m$

   $E[b_m(n)u^*(n - k)] = 0, \;\; 0 \le k \le m - 1$

(b) $E[f_m(n)u^*(n)] = E[b_m(n)u^*(n - m)] = P_m$

(c) $E[b_m(n)b_i^*(n)] = \begin{cases} P_m, & m = i \\ 0, & m \ne i \end{cases}$

(d) $E[f_m(n)f_i^*(n - l)] = E[f_m(n + l)f_i^*(n)] = 0, \qquad \begin{array}{l} 1 \le l \le m - i \\ m > i \end{array}$

   $E[b_m(n)b_i^*(n - l)] = E[b_m(n + l)b_i^*(n)] = 0, \qquad \begin{array}{l} 0 \le l \le m - i - 1 \\ m > i \end{array}$

(e) $E[f_m(n + m)f_i^*(n + i)] = \begin{cases} P_m, & m = i \\ 0, & m \ne i \end{cases}$

   $E[b_m(n + m)b_i^*(n + i)] = P_m, \qquad m \ge i$

(f) $E[f_m(n)b_i^*(n)] = \begin{cases} \kappa_i^* P_m, & m \ge i \\ 0, & m < i \end{cases}$

21. The *entropy* of a random input vector $\mathbf{u}(n)$ of joint probability density function $f_U(\mathbf{u})$ is defined by the multiple integral

$$H_u = -\int_{-\infty}^{\infty} f_U(\mathbf{u}) \ln [f_U(\mathbf{u})] \, d\mathbf{u}$$

Given that the backward prediction-error vector $\mathbf{b}(n)$ is related to $\mathbf{u}(n)$ by the Gram–Schmidt algorithm of Eq. (6.105), show that the vectors $\mathbf{b}(n)$ and $\mathbf{u}(n)$ are equivalent in the sense that they have the same entropy, and therefore convey the same amount of information.

22. Consider the problem of optimizing stage m of the lattice predictor. The cost function to be used in the optimization is described by

$$J_m(\kappa_m) = aE[|f_m(n)|^2] + (1 - a)E[|b_m(n)|^2]$$

where a is a constant that lies between zero and one; $f_m(n)$ and $b_m(n)$ denote the forward and backward prediction errors at the output of stage m, respectively.

(a) Show that the optimum value of the reflection coefficient κ_m for which J_m is at minimum equals

$$\kappa_{m,o}(a) = -\frac{E[b_{m-1}(n - 1)f_{m-1}^*(n)]}{(1 - a)E[|f_{m-1}(n)|^2] + aE[|b_{m-1}(n - 1)|^2]}$$

(b) Evaluate $\kappa_{m,o}(a)$ for each of the following three special conditions:

$$(1)\ \ a = 1$$

$$(2)\ \ a = 2$$

$$(3)\ \ a = \tfrac{1}{2}$$

Notes: When the parameter $a = 1$, the cost function reduces to

$$J_m(\kappa_m) = E[|f_m(n)|^2]$$

We refer to this criterion as the *forward method*.

When the parameter $a = 0$, the cost function reduces to

$$J_m(\kappa_m) = E[|b_m(n)|^2]$$

We refer to this criterion as the *backward method*.

When the parameter $a = \tfrac{1}{2}$, the formula for $\kappa_{m,o}(a)$ reduces to the Burg formula.

23. Let $\kappa_{m,o}(1)$ and $\kappa_{m,o}(0)$ denote the optimum values of the reflection coefficient κ_m for stage m of the lattice predictor using the forward method and backward method, respectively, as determined in Problem 22.

(a) Show that the optimum value of $\kappa_{m,o}$ obtained from the Burg formula equals the harmonic mean of the two values $\kappa_{m,o}(1)$ and $\kappa_{m,o}(0)$, as shown by

$$\frac{2}{\kappa_{m,o}} = \frac{1}{\kappa_{m,o}(1)} + \frac{1}{\kappa_{m,o}(0)}$$

(b) Using the result of part (a), show that

$$|\kappa_{m,o}| \leq 1 \qquad \text{for all } m$$

(c) For the case of a lattice predictor using the Burg formula, show that the mean-square values of the forward and backward prediction errors at the output of stage m are related to those at the input as follows, respectively:

$$E[|f_m(n)|^2] = (1 - |\kappa_{m,o}|^2)E[|f_{m-1}(n)|^2]$$

and

$$E[|b_m(n)|^2] = (1 - |\kappa_{m,o}|^2)E[|b_{m-1}(n-1)|^2]$$

CHAPTER

7

Kalman Filters

In this chapter we complete our study of linear optimum filters by developing the basic ideas of *Kalman filtering*. A distinctive feature of a Kalman filter is that its mathematical formulation is described in terms of *state-space concepts*. Another novel feature of a Kalman filter is that its solution is computed *recursively*. In particular, each updated estimate of the state is computed from the previous estimate and the new input data, so only the previous estimate requires storage. In addition to eliminating the need for storing the entire past observed data, a Kalman filter is computationally more efficient than computing the estimate directly from the entire past observed data at each step of the filtering process. The Kalman filter is thus ideally suited for implementation on a digital computer. Most importantly, it has been applied successfully to many practical problems in diverse fields, particularly in aerospace and aeronautical applications.

Our interest in the Kalman filter is motivated by the fact that it provides a unifying framework for the derivation of an important family of adaptive filters known as recursive least-squares filters, as demonstrated in subsequent chapters of the book. To pave the way for the development of the Kalman filter, we begin by solving the *recursive minimum mean-squared estimation problem* for the simple case of scalar random variables. For this solution, we use the *innovations approach* that exploits the correlation properties of a special stochastic process known as the *innovations process* (Kailath, 1968, 1970).

7.1 RECURSIVE MINIMUM MEAN-SQUARE ESTIMATION FOR SCALAR RANDOM VARIABLES

Let us assume that, based on a complete set of observed random variables $y(1)$, $y(2)$, ..., $y(n - 1)$, starting with the first observation at time 1 and extending up to and including time $n - 1$, we have found the minimum mean-square estimate $\hat{x}(n - 1|\mathcal{Y}_{n-1})$ of a related zero-mean random variable $x(n - 1)$. We are assuming that the observation at (or before) $n = 0$ is zero. The space spanned by the observations $y(1)$, ..., $y(n - 1)$ is denoted by $\mathcal{Y}_{n-1}$. Suppose that we now have an additional observation $y(n)$ at time n, and the requirement is to compute an *updated* estimate $\hat{x}(n|\mathcal{Y}_n)$ of the related random variable $x(n)$, where $\mathcal{Y}_n$ denotes the space spanned by $y(1)$, ..., $y(n)$. We may do this computation by storing the *past* observations, $y(1)$, $y(2)$, ..., $y(n - 1)$, and then redoing the whole problem with the available data $y(1)$, $y(2)$, ..., $y(n - 1)$, $y(n)$, including the new observation. Computationally, however, it is much more efficient to use a *recursive estimation procedure*. In this procedure we *store* the previous estimate $\hat{x}(n - 1|\mathcal{Y}_{n-1})$ and exploit it to compute the updated estimate $\hat{x}(n|\mathcal{Y}_n)$ in light of the new observation $y(n)$. There are several ways of developing the algorithm to do this recursive estimation. We will use the notion of *innovations* (Kailath, 1968, 1970), the origin of which may be traced back to Kolmogorov (1941).

Define the forward prediction error

$$f_{n-1}(n) = y(n) - \hat{y}(n|\mathcal{Y}_{n-1}), \quad n = 1, 2, \ldots \tag{7.1}$$

where $\hat{y}(n|\mathcal{Y}_{n-1})$ is the *one-step prediction* of the observed random variable $y(n)$ at time n, using *all* past observations available up to and including time $n - 1$. The past observations used in this estimation are $y(1)$, $y(2)$, ..., $y(n - 1)$, so the order of the prediction equals $n - 1$. We may view $f_{n-1}(n)$ as the output of a forward prediction-error filter of order $n - 1$, and with the filter input fed by the time series $y(1)$, $y(2)$, ..., $y(n)$. Note that the *prediction order $n - 1$ increases linearly with n*. According to the principle of orthogonality, the prediction error $f_{n-1}(n)$ is orthogonal to all past observations $y(1)$, $y(2)$, ..., $y(n - 1)$ and may therefore be regarded as a *measure* of the new information in the random variable $y(n)$ observed at time n, hence the name "innovation." The fact is that the observation $y(n)$ does not itself convey completely new information, since the predictable part, $\hat{y}(n|\mathcal{Y}_{n-1})$, is already completely determined by the past observations $y(1)$, $y(2)$, ..., $y(n - 1)$. Rather, the part of the observation $y(n)$ that is really new is contained in the forward prediction error $f_{n-1}(n)$. We may therefore refer to this prediction error as the *innovation*, and for simplicity of notation write

$$\alpha(n) = f_{n-1}(n), \quad n = 1, 2, \ldots \tag{7.2}$$

The innovation $\alpha(n)$ has several important properties, as described here.

Property 1. *The innovation $\alpha(n)$, associated with the observed random variable $y(n)$, is orthogonal to the past observations $y(1)$, $y(2)$, ..., $y(n - 1)$, as shown by*

$$E[\alpha(n)y^*(k)] = 0, \quad 1 \leq k \leq n - 1 \tag{7.3}$$

This is simply a restatement of the principle of orthogonality.

Property 2. *The innovations* $\alpha(1)$, $\alpha(2)$, . . . , $\alpha(n)$ *are orthogonal to each other, as shown by*

$$E[\alpha(n)\alpha^*(k)] = 0, \quad 1 \le k \le n - 1 \tag{7.4}$$

This is a restatement of the fact that [see part (e) of Problem 20, Chapter 6]:

$$E[f_{n-1}(n)f^*_{k-1}(k)] = 0, \quad 1 \le k \le n - 1$$

Equation (7.4), in effect, states that the innovation process $\alpha(n)$, described by Eqs. (7.1) and (7.2), is *white*.

Property 3. *There is a one-to-one correspondence between the observed data* $\{y(1), y(2), . . . , y(n)\}$ *and the innovations* $\{\alpha(1), \alpha(2), . . . , \alpha(n)\}$, *in that the one sequence may be obtained from the other by means of a causal and causally invertible filter without any loss of information. We may thus write*

$$\{y(1), y(2), . . . , y(n)\} \rightleftharpoons \{\alpha(1), \alpha(2), . . . , \alpha(n)\} \tag{7.5}$$

To prove this property, we use a form of the Gram–Schmidt orthogonalization procedure (described in Chapter 6). The procedure assumes that the observations $y(1), y(2), . . . , y(n)$ are linearly independent in an algebraic sense. We first put

$$\alpha(1) = y(1) \tag{7.6}$$

where it is assumed that $\hat{y}(1|\mathcal{Y}_0)$ is zero. Next we put

$$\alpha(2) = y(2) + a_{1,1}y(1) \tag{7.7}$$

The coefficient $a_{1,1}$ is chosen such that the innovations $\alpha(1)$ and $\alpha(2)$ are orthogonal, as shown by

$$E[\alpha(2)\alpha^*(1)] = 0 \tag{7.8}$$

This requirement is satisfied by choosing

$$a_{1,1} = -\frac{E[y(2)y^*(1)]}{E[y(1)y^*(1)]} \tag{7.9}$$

Except for the minus sign, $a_{1,1}$ is a partial correlation coefficient in that it equals the cross-correlation between the observations $y(2)$ and $y(1)$, normalized with respect to the mean-square value of $y(1)$.

Next, we put

$$\alpha(3) = y(3) + a_{2,1}y(2) + a_{2,2}y(1) \tag{7.10}$$

where the coefficients $a_{2,1}$ and $a_{2,2}$ are chosen such that $\alpha(3)$ is orthogonal to both $\alpha(1)$ and $\alpha(2)$, and so on. Thus, in general, we may express the transformation of the observed data $y(1), y(2), . . . , y(n)$ into the innovations $\alpha(1), \alpha(2), . . . , \alpha(n)$ by writing

$$
\begin{bmatrix} \alpha(1) \\ \alpha(2) \\ \bullet \\ \bullet \\ \bullet \\ \alpha(n) \end{bmatrix} = \begin{bmatrix} 1 & 0 & \bullet\bullet\bullet & 0 \\ a_{1,1} & 1 & \bullet\bullet\bullet & 0 \\ \bullet & \bullet & \bullet & \bullet \\ \bullet & \bullet & \bullet & \bullet \\ \bullet & \bullet & \bullet & \bullet \\ a_{n-1,n-1} & a_{n-1,n-2} & \bullet\bullet\bullet & 1 \end{bmatrix} \begin{bmatrix} y(1) \\ y(2) \\ \bullet \\ \bullet \\ \bullet \\ y(n) \end{bmatrix}
\tag{7.11}
$$

The nonzero elements of row k of the *lower triangular transformation matrix* on the right-hand side of Eq. (7.11) are deliberately denoted as $a_{k-1,k-1}, a_{k-1,k-2} \cdots, 1$, where $k = 1$, $2, \ldots, n$. These elements represent the coefficients of a *forward prediction-error filter* of order $k - 1$. Note that $a_{k,0} = 1$ for all k. Accordingly, given the observed data $y(1), y(2),$ $\ldots, y(n)$, we may compute the innovations $\alpha(1), \alpha(2), \ldots, \alpha(n)$. There is no loss of information in the course of this transformation, since we may recover the original observed data $y(1), y(2), \ldots, y(n)$ from the innovations $\alpha(1), \alpha(2), \ldots, \alpha(n)$. This we do by premultiplying both sides of Eq. (7.11) by the inverse of the lower triangular transformation matrix. This matrix is nonsingular, since its determinant equals 1 for all n. The transformation is therefore reversible.

Using Eq. (7.5), we may thus write

$$\hat{x}(n|\mathcal{Y}_n) = \text{minimum mean-square estimate of } x(n)$$
$$\text{given the observed data } y(1), y(2), \ldots, y(n)$$

or, equivalently,

$$\hat{x}(n|\mathcal{Y}_n) = \text{minimum mean-square estimate of } x(n)$$
$$\text{given the innovations } \alpha(1), \alpha(2), \ldots, \alpha(n)$$

Define the estimate $\hat{x}(n|\mathcal{Y}_n)$ as a linear combination of the innovations $\alpha(1), \alpha(2), \ldots,$ $\alpha(n)$:

$$
\hat{x}(n|\mathcal{Y}_n) = \sum_{k=1}^{n} b_k \alpha(k)
\tag{7.12}
$$

where the b_k are to be determined. With the innovations $\alpha(1), \alpha(2), \ldots, \alpha(n)$ orthogonal to each other, and the b_k chosen to minimize the mean-square value of the estimation error $x(n) - \hat{x}(n|\mathcal{Y}_n)$, we find that

$$
b_k = \frac{E[x(n)\alpha^*(k)]}{E[\alpha(k)\alpha^*(k)]}, \quad 1 \le k \le n
\tag{7.13}
$$

We rewrite Eq. (7.12) in the form

$$
\hat{x}(n|\mathcal{Y}_n) = \sum_{k=0}^{n-1} b_k \alpha(k) + b_n \alpha(n)
\tag{7.14}
$$

where

$$
b_n = \frac{E[x(n)\alpha^*(n)]}{E[\alpha(n)\alpha^*(n)]}
\tag{7.15}
$$

However, by definition, the summation term on the right-hand side of Eq. (7.14) equals the previous estimate $\hat{x}(n - 1|\mathcal{Y}_{n-1})$. We may thus express the recursive estimation algorithm that we are seeking as

$$\hat{x}(n|\mathcal{Y}_n) = \hat{x}(n - 1|\mathcal{Y}_{n-1}) + b_n\alpha(n) \tag{7.16}$$

where b_n is defined by Eq. (7.15). Thus, by adding a *correction term* $b_n\alpha(n)$ to the previous estimate $\hat{x}(n - 1|\mathcal{Y}_{n-1})$, with the correction being proportional to the innovation $\alpha(n)$, we get the updated estimate $\hat{x}(n|\mathcal{Y}_n)$.

The simple formulas of Eq. (7.15) and (7.16) are the basis of all recursive linear estimation schemes. Equipped with these simple and yet powerful ideas, we are now ready to study the more general Kalman filtering problem.

7.2 STATEMENT OF THE KALMAN FILTERING PROBLEM

Consider a *linear, discrete-time dynamical system* described by the signal-flow graph shown in Fig. 7.1. The time-domain description of the system presented here offers the following advantages (Gelb, 1974):

- Mathematical and notational convenience
- Close relationship to physical reality
- Useful basis for accounting for statistical behavior of the system

The notation of *state* plays a key role in this formulation. The *state vector*, denoted by $\mathbf{x}(n)$ in Fig. 7.1, is defined as any set of quantities that would be sufficient to uniquely describe the unforced dynamical behavior of the system. Typically, the state vector $\mathbf{x}(n)$, assumed to be of dimension M, is unknown. To estimate it, we use a set of observed data, denoted by the vector $\mathbf{y}(n)$ in Fig. 7.1. The *observation vector* $\mathbf{y}(n)$ is assumed to be of dimension N.

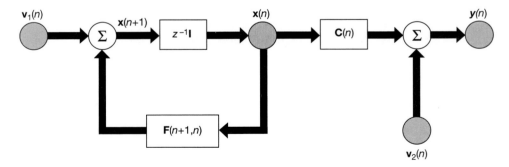

Figure 7.1   Signal-flow graph representation of a linear, discrete-time dynamical system.

In mathematical terms, the signal-flow graph of Fig. 7.1 embodies the following pair of equations:

1. A *process equation*

$$\mathbf{x}(n + 1) = \mathbf{F}(n + 1, n)\mathbf{x}(n) + \mathbf{v}_1(n) \tag{7.17}$$

where $\mathbf{F}(n + 1, n)$ is a known *M-by-M state transition matrix* relating the state of the system at times $n + 1$ and n. The M-by-1 vector $\mathbf{v}_1(n)$ represents *process noise*. The vector $\mathbf{v}_1(n)$ is modeled as a zero-mean, white-noise process whose correlation matrix is defined by

$$E[\mathbf{v}_1(n)\mathbf{v}_1^H(k)] = \begin{cases} \mathbf{Q}_1(n), & n = k \\ \mathbf{O}, & n \neq k \end{cases} \tag{7.18}$$

2. A *measurement equation*, describing the observation vector as

$$\mathbf{y}(n) = \mathbf{C}(n)\mathbf{x}(n) + \mathbf{v}_2(n) \tag{7.19}$$

where $\mathbf{C}(n)$ is a known *N-by-M measurement matrix*. The N-by-1 vector $\mathbf{v}_2(n)$ is called *measurement noise*. It is modeled as a zero-mean, white-noise process whose correlation matrix is

$$E[\mathbf{v}_2(n)\mathbf{v}_2^H(k)] = \begin{cases} \mathbf{Q}_2(n), & n = k \\ \mathbf{O}, & n \neq k \end{cases} \tag{7.20}$$

It is assumed that $\mathbf{x}(0)$, the initial value of the state, is uncorrelated with both $\mathbf{v}_1(n)$ and $\mathbf{v}_2(n)$ for $n \geq 0$. The noise vectors $\mathbf{v}_1(n)$ and $\mathbf{v}_2(n)$ are statistically independent, so we may write

$$E[\mathbf{v}_1(n)\, \mathbf{v}_2^H(k)] = \mathbf{O} \qquad \text{for all } n \text{ and } k \tag{7.21}$$

The Kalman filtering problem may now be formally stated as follows: Use the entire observed data, consisting of the vectors $\mathbf{y}(1), \mathbf{y}(2), \ldots, \mathbf{y}(n)$, to find for each $n \geq 1$ the minimum mean-square estimates of the components of the state $\mathbf{x}(i)$. The problem is called the *filtering* problem if $i = n$, the *prediction* problem if $i > n$, and the *smoothing* problem if $1 \leq i < n$. In this chapter we will only be concerned with the filtering and prediction problems, which are closely related. As remarked earlier in the introduction, we will solve the Kalman filtering problem by using the innovations approach (Kailath, 1968, 1970, 1981; Tretter, 1976).

7.3 THE INNOVATIONS PROCESS

Let the vector $\hat{\mathbf{y}}(n|\mathcal{Y}_{n-1})$ denote the minimum mean-square estimate of the observed data $\mathbf{y}(n)$ at time n, given all the past values of the observed data starting at time $n = 1$ and extending up to and including time $n - 1$. These past values are represented by the vectors $\mathbf{y}(1), \mathbf{y}(n), \ldots, \mathbf{y}(n-1)$, which span the vector space $\mathcal{Y}_{n-1}$. We define the *innovations process* associated with $\mathbf{y}(n)$ as

$$\boldsymbol{\alpha}(n) = \mathbf{y}(n) - \hat{\mathbf{y}}(n|\mathcal{Y}_{n-1}), \quad n = 1, 2, \ldots \tag{7.22}$$

The M-by-1 vector $\boldsymbol{\alpha}(n)$ represents the new information in the observed data $\mathbf{y}(n)$.

Generalizing the results of Eqs. (7.3), (7.4) and (7.5), we find that the innovations process $\boldsymbol{\alpha}(n)$ has the following properties:

1. The innovations process $\boldsymbol{\alpha}(n)$, associated with the observed data $\mathbf{y}(n)$ at time n, is orthogonal to all past observations $\mathbf{y}(1), \mathbf{y}(2), \ldots, \mathbf{y}(n-1)$ as shown by

$$E[\boldsymbol{\alpha}(n)\mathbf{y}^H(k)] = \mathbf{O}, \quad 1 \leq k \leq n - 1 \tag{7.23}$$

2. The innovations process consists of a sequence of vector random variables that are orthogonal to each other, as shown by

$$E[\boldsymbol{\alpha}(n)\boldsymbol{\alpha}^H(k)] = \mathbf{O}, \quad 1 \leq k \leq n - 1 \tag{7.24}$$

3. There is a one-to-one correspondence between the sequence of vector random variables $\{\mathbf{y}(1), \mathbf{y}(2), \ldots, \mathbf{y}(n)\}$ representing the observed data and the sequence of vector random variables $\{\boldsymbol{\alpha}(1), \boldsymbol{\alpha}(2), \ldots, \boldsymbol{\alpha}(n)\}$ representing the innovations process, in that the one sequence may be obtained from the other by means of linear stable operators without loss of information. Thus, we may state that

$$\{\mathbf{y}(1), \mathbf{y}(2), \ldots, \mathbf{y}(n)\} \rightleftharpoons \{\boldsymbol{\alpha}(1), \boldsymbol{\alpha}(2), \ldots, \boldsymbol{\alpha}(n)\} \tag{7.25}$$

To form the sequence of vector random variables defining the innovations process, we may use a Gram–Schmidt orthogonalization procedure similar to that described in Section 7.1, except that the procedure is now formulated in terms of vectors and matrices (see Problem 1).

Correlation Matrix of the Innovations Process

To determine the correlation matrix of the innovations process $\boldsymbol{\alpha}(n)$, we first solve the state equation (7.17) recursively to obtain

$$\mathbf{x}(k) = \mathbf{F}(k, 0)\mathbf{x}(0) + \sum_{i=1}^{k-1} \mathbf{F}(k, i+1)\mathbf{v}_1(i) \tag{7.26}$$

where we have made use of the following assumptions and properties:

1. The initial value of the state vector $\mathbf{x}(0)$.
2. As previously assumed, the observed data [and therefore the noise vector $\mathbf{v}_1(n)$] are zero for $n \leq 0$.

3. The state transition matrix has the properties

$$\mathbf{F}(k, k - 1)\mathbf{F}(k - 1, k - 2) \ldots \mathbf{F}(i + 1, i) = \mathbf{F}(k, i)$$

and

$$\mathbf{F}(k, k) = \mathbf{I}$$

where $\mathbf{I}$ is the identity matrix. Note that for a time-invariant system we have

$$\mathbf{F}(n + 1, n) = \mathbf{F}(n + 1 - n) = \mathbf{F}(1) = \text{constant}.$$

Equation (7.26) shows that $\mathbf{x}(k)$ is a linear combination of $\mathbf{x}(0)$ and $\mathbf{v}_1(1)$, $\mathbf{v}_1(2)$, . . . , $\mathbf{v}_1(k - 1)$.

By hypothesis, the measurement noise vector $\mathbf{v}_2(n)$ is uncorrelated with both the initial state vector $\mathbf{x}(0)$ and the process noise vector $\mathbf{v}_1(n)$. Accordingly, premultiplying both sides of Eq. (7.26) by $\mathbf{v}_2^H(n)$, and taking expectations, we deduce that

$$E[\mathbf{x}(k)\mathbf{v}_2^H(n)] = \mathbf{O}, \quad k, n \leq 0 \tag{7.27}$$

Correspondingly, we deduce from the measurement equation (7.19) that

$$E[\mathbf{y}(k)\mathbf{v}_2^H(n)] = \mathbf{O}, \quad 0 \leq k \leq n - 1 \tag{7.28}$$

Moreover, we may write

$$E[\mathbf{y}(k)\mathbf{v}_1^H(n)] = \mathbf{O}, \quad 0 \leq k \leq n \tag{7.29}$$

Given the past observations $\mathbf{y}(1)$, . . . , $\mathbf{y}(n - 1)$ that span the space $\mathcal{Y}_{n-1}$, we also find from the measurement equation (7.19) that the minimum mean-square estimate of the present value $\mathbf{y}(n)$ of the observation vector equals

$$\hat{\mathbf{y}}(n|\mathcal{Y}_{n-1}) = \mathbf{C}(n)\hat{\mathbf{x}}(n|\mathcal{Y}_{n-1}) + \hat{\mathbf{v}}_2(n|\mathcal{Y}_{n-1})$$

However, the estimate $\hat{\mathbf{v}}_2(n|\mathcal{Y}_{n-1})$ of the measurement noise vector is zero since $\mathbf{v}_2(n)$ is orthogonal to the past observations $\mathbf{y}(1)$, . . . , $\mathbf{y}(n - 1)$; see Eq. (7.28). Hence, we may simply write

$$\hat{\mathbf{y}}(n|\mathcal{Y}_{n-1}) = \mathbf{C}(n)\hat{\mathbf{x}}(n|\mathcal{Y}_{n-1}) \tag{7.30}$$

Therefore, using Eqs. (7.22) and (7.30), we may express the innovations process in the form

$$\alpha(n) = \mathbf{y}(n) - \mathbf{C}(n)\hat{\mathbf{x}}(n\,|\mathcal{Y}_{n-1}) \tag{7.31}$$

Substituting the measurement equation (7.19) in (7.31), we get

$$\alpha(n) = \mathbf{C}(n)\epsilon(n, n - 1) + \mathbf{v}_2(n) \tag{7.32}$$

where $\epsilon(n, n - 1)$ is the *predicted state-error vector* at time n, using data up to time

$n - 1$. That is, $\boldsymbol{\epsilon}(n, n - 1)$ is the difference between the state vector $\mathbf{x}(n)$ and the one-step prediction vector $\hat{\mathbf{x}}(n \mid \mathcal{Y}_{n-1})$, as shown by

$$\boldsymbol{\epsilon}(n, n - 1) = \mathbf{x}(n) - \hat{\mathbf{x}}(n \mid \mathcal{Y}_{n-1}) \tag{7.33}$$

Note that the predicted state-error vector is orthogonal to both the process noise vector $\mathbf{v}_1(n)$ and the measurement noise vector $\mathbf{v}_2(n)$; see Problem 2.

The correlation matrix of the innovations process $\boldsymbol{\alpha}(n)$ is defined by

$$\mathbf{R}(n) = E[\boldsymbol{\alpha}(n)\boldsymbol{\alpha}^H(n)] \tag{7.34}$$

Therefore, substituting Eq. (7.32) in (7.34), expanding the pertinent terms, and then using the fact that the vectors $\boldsymbol{\epsilon}(n, n - 1)$ and $\mathbf{v}_2(n)$ are orthogonal, we obtain the result:

$$\mathbf{R}(n) = \mathbf{C}(n)\mathbf{K}(n, n - 1)\mathbf{C}^H(n) + \mathbf{Q}_2(n) \tag{7.35}$$

where $\mathbf{Q}_2(n)$ is the correlation matrix of the noise vector $\mathbf{v}_2(n)$. The M-by-M matrix $\mathbf{K}(n, n - 1)$ is called the *predicted state-error correlation matrix*; it is defined by

$$\mathbf{K}(n, n - 1) = E[\boldsymbol{\epsilon}(n, n - 1)\boldsymbol{\epsilon}^H(n, n - 1)] \tag{7.36}$$

where $\boldsymbol{\epsilon}(n, n - 1)$ is the predicted state-error vector. The matrix $\mathbf{K}(n, n - 1)$ is used as the statistical description of the error in the predicted estimate $\hat{\mathbf{x}}(n \mid \mathcal{Y}_{n-1})$.

7.4 ESTIMATION OF THE STATE USING THE INNOVATIONS PROCESS

Consider next the problem of deriving the minimum mean-square estimate of the state $\mathbf{x}(i)$ from the innovations process. From the discussion presented in Section 7.1, we deduce that this estimate may be expressed as a linear combination of the sequence of innovations processes $\boldsymbol{\alpha}(1), \boldsymbol{\alpha}(2), \ldots, \boldsymbol{\alpha}(n)$ [see Eq. (7.12) for comparison]:

$$\hat{\mathbf{x}}(i \mid \mathcal{Y}_n) = \sum_{k=1}^{n} \mathbf{B}_i(k)\boldsymbol{\alpha}(k) \tag{7.37}$$

where $\mathbf{B}_i(k)$, $k = 1, 2, \ldots, n$, is a set of M-by-N matrices to be determined. According to the principle of orthogonality, the predicted state-error vector is orthogonal to the innovation process, as shown by

$$E[\boldsymbol{\epsilon}(i, n)\boldsymbol{\alpha}^H(m)] = E\{[\mathbf{x}(i) - \hat{\mathbf{x}}(i \mid \mathcal{Y}_n)]\boldsymbol{\alpha}^H(m)\}$$
$$= \mathbf{O}, \qquad m = 1, 2, \ldots, n \tag{7.38}$$

Substituting Eq. (7.37) in (7.38) and using the orthogonality property of the innovations process, namely, Eq. (7.24), we get

$$E[\mathbf{x}(i)\boldsymbol{\alpha}^H(m)] = \mathbf{B}_i(m)E[\boldsymbol{\alpha}(m)\boldsymbol{\alpha}^H(m)]$$
$$= \mathbf{B}_i(m)\mathbf{R}(m) \tag{7.39}$$

Hence, postmultiplying both sides of Eq. (7.39) by the inverse matrix $\mathbf{R}^{-1}(m)$, we find that $\mathbf{B}_i(m)$ is given by

$$\mathbf{B}_i(m) = E[\mathbf{x}(i)\boldsymbol{\alpha}^H(m)] \, \mathbf{R}^{-1}(m) \tag{7.40}$$

Finally, substituting Eq. (7.40) in (7.37), we get the minimum mean-square estimate

$$\hat{\mathbf{x}}(i|\mathcal{Y}_n) = \sum_{k=1}^{n} E[\mathbf{x}(i)\boldsymbol{\alpha}^H(k)] \, \mathbf{R}^{-1}(k)\boldsymbol{\alpha}(k)$$

$$= \sum_{k=1}^{n-1} E[\mathbf{x}(i)\boldsymbol{\alpha}^H(k)] \, \mathbf{R}^{-1}(k) \, \boldsymbol{\alpha}(k)$$

$$+ E[\mathbf{x}(i)\boldsymbol{\alpha}^H(n)] \, \mathbf{R}^{-1}(n)\boldsymbol{\alpha}(n)$$

For $i = n + 1$, we may therefore write

$$\hat{\mathbf{x}}(n + 1|\mathcal{Y}_n) = \sum_{k=1}^{n-1} E[\mathbf{x}(n + 1)\boldsymbol{\alpha}^H(k)] \, \mathbf{R}^{-1}(k)\boldsymbol{\alpha}(k)$$

$$+ E[\mathbf{x}(n + 1)\boldsymbol{\alpha}^H(n)] \, \mathbf{R}^{-1}(n)\boldsymbol{\alpha}(n) \tag{7.41}$$

However, the state $\mathbf{x}(n + 1)$ at time $n + 1$ is related to the state $\mathbf{x}(n)$ at time n by Eq. (7.17). Therefore, using this relation, we may write for $0 \leq k \leq n$:

$$E[\mathbf{x}(n + 1)\boldsymbol{\alpha}^H(k)] = E\{[\mathbf{F}(n + 1, n)\mathbf{x}(n) + \mathbf{v}_1(n)]\boldsymbol{\alpha}^H(k)\}$$

$$= \mathbf{F}(n + 1, n)E[\mathbf{x}(n)\boldsymbol{\alpha}^H(k)] \tag{7.42}$$

where we have made use of the fact that $\boldsymbol{\alpha}(k)$ depends only on the observed data $\mathbf{y}(1), \ldots, \mathbf{y}(k)$, and therefore from Eq. (7.29) we see that $\mathbf{y}(n)$ and $\boldsymbol{\alpha}(k)$ are orthogonal for $0 \leq k \leq n$. We may thus rewrite the summation term on the right-hand side of Eq. (7.41) as follows:

$$\sum_{k=1}^{n-1} E[\mathbf{x}(n + 1)\boldsymbol{\alpha}^H(k)] \, \mathbf{R}^{-1}(k)\boldsymbol{\alpha}(k) = \mathbf{F}(n + 1, n) \sum_{k=1}^{n-1} E[\mathbf{x}(n)\boldsymbol{\alpha}^H(k)] \, \mathbf{R}^{-1}(k)\boldsymbol{\alpha}(k)$$

$$= \mathbf{F}(n + 1, n)\hat{\mathbf{x}}(n|\mathcal{Y}_{n-1}) \tag{7.43}$$

To proceed further, we introduce some basic definitions, as described next.

Kalman Gain

Define the M-by-N matrix:

$$\mathbf{G}(n) = E[\mathbf{x}(n + 1)\boldsymbol{\alpha}^H(n)] \, \mathbf{R}^{-1}(n) \tag{7.44}$$

Then, using this definition and the result of Eq. (7.43), we may rewrite Eq. (7.41) as follows:

$$\hat{\mathbf{x}}(n + 1|\mathcal{Y}_n) = \mathbf{F}(n + 1, n)\hat{\mathbf{x}}(n|\mathcal{Y}_{n-1}) + \mathbf{G}(n)\boldsymbol{\alpha}(n) \tag{7.45}$$

Equation (7.45) is of fundamental significance. It shows that we may compute the minimum mean-square estimate $\hat{\mathbf{x}}(n + 1|\mathcal{Y}_n)$ of the state of a linear dynamical system by adding to the previous estimate $\hat{\mathbf{x}}(n|\mathcal{Y}_{n-1})$, which is premultiplied by the state transition matrix $\mathbf{F}(n + 1, n)$, a correction term equal to $\mathbf{G}(n)\boldsymbol{\alpha}(n)$. The correction term equals the innovations process $\boldsymbol{\alpha}(n)$ premultiplied by the matrix $\mathbf{G}(n)$. Accordingly, and in recognition of the pioneering work by Kalman, the matrix $\mathbf{G}(n)$ is called the *Kalman gain*.

There now remains only the problem of expressing the Kalman gain $\mathbf{G}(n)$ in a form convenient for computation. To do this, we first use Eqs. (7.32) and (7.42) to express the expectation of the product of $\mathbf{x}(n + 1)$ and $\boldsymbol{\alpha}^H(n)$ as follows:

$$
\begin{aligned}
E[\mathbf{x}(n + 1)\boldsymbol{\alpha}^H(n)] &= \mathbf{F}(n + 1, n)E[\mathbf{x}(n)\boldsymbol{\alpha}^H(n)] \\
&= \mathbf{F}(n + 1, n)E[\mathbf{x}(n)(\mathbf{C}(n)\boldsymbol{\epsilon}(n, n - 1) + \mathbf{v}_2(n))^H] \\
&= \mathbf{F}(n + 1, n)E[\mathbf{x}(n)\boldsymbol{\epsilon}^H(n, n - 1)]\mathbf{C}^H(n)
\end{aligned}
\tag{7.46}
$$

where we have used the fact that the state $\mathbf{x}(n)$ and noise vector $\mathbf{v}_2(n)$ are uncorrelated [see Eq. (7.27)]. We further note that the predicted state-error vector $\boldsymbol{\epsilon}(n, n - 1)$ is orthogonal to the estimate $\hat{\mathbf{x}}(n|\mathcal{Y}_{n-1})$. Therefore, the expectation of the product of $\mathbf{x}(n|\mathcal{Y}_{n-1})$ and $\boldsymbol{\epsilon}^H(n, n - 1)$ is zero, and so we may rewrite Eq. (7.46) by replacing the multiplying factor $\mathbf{x}(n)$ by the predicted state-error vector $\boldsymbol{\epsilon}(n, n - 1)$ as follows:

$$
E[\mathbf{x}(n + 1)\boldsymbol{\alpha}^H(n)] = \mathbf{F}(n + 1, n)E[\boldsymbol{\epsilon}(n, n - 1)\boldsymbol{\epsilon}^H(n, n - 1)]\mathbf{C}^H(n)
\tag{7.47}
$$

From Eq. (7.36), we see that the expectation on the right-hand side of Eq. (7.47) equals the predicted state-error correlation matrix. Hence, we may rewrite Eq. (7.47) as follows:

$$
E[\mathbf{x}(n + 1)\boldsymbol{\alpha}^H(n)] = \mathbf{F}(n + 1, n)\mathbf{K}(n, n - 1)\mathbf{C}^H(n)
\tag{7.48}
$$

We may now redefine the Kalman gain. In particular, substituting Eq. (7.48) in (7.44), we get

$$
\mathbf{G}(n) = \mathbf{F}(n + 1, n)\mathbf{K}(n, n - 1)\mathbf{C}^H(n)\mathbf{R}^{-1}(n)
\tag{7.49}
$$

where the correlation matrix $\mathbf{R}(n)$ is itself defined in Eq. (7.35).

The block diagram of Fig. 7.2 shows the signal-flow graph representation of Eq. (7.49) for computing the Kalman gain $\mathbf{G}(n)$. Having computed the Kalman gain $\mathbf{G}(n)$, we may then use Eq. (7.45) to update the one-step prediction, that is, to compute $\hat{\mathbf{x}}(n + 1|\mathcal{Y}_n)$ given its old value $\hat{\mathbf{x}}(n|\mathcal{Y}_{n-1})$, as illustrated in Fig. 7.3. In this figure we have also used Eq. (7.31) for the innovations process $\boldsymbol{\alpha}(n)$.

Riccati Equation

As it stands, the formula of Eq. (7.49) is not particularly useful for computing the Kalman gain $\mathbf{G}(n)$, since it requires that the predicted state-error correlation matrix $\mathbf{K}(n, n - 1)$ be known. To overcome this difficulty, we derive a formula for the recursive computation of $\mathbf{K}(n, n - 1)$.

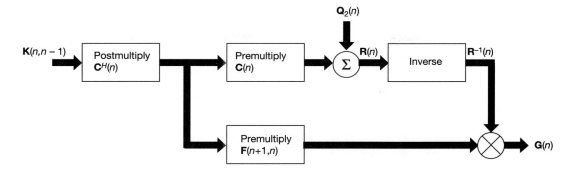

Figure 7.2   Kalman gain computer.

The predicted state-error vector $\boldsymbol{\epsilon}(n + 1, n)$ equals the difference between the state $\mathbf{x}(n + 1)$ and the one-step prediction $\hat{\mathbf{x}}(n + 1|\mathcal{Y}_n)$[see Eq. (7.33)]:

$$\boldsymbol{\epsilon}(n + 1, n) = \mathbf{x}(n + 1) - \hat{\mathbf{x}}(n + 1|\mathcal{Y}_n) \tag{7.50}$$

Substituting Eqs. (7.17) and (7.45) in (7.50), and using Eq. (7.31) for the innovations process $\boldsymbol{\alpha}(n)$, we get

$$\begin{aligned}\boldsymbol{\epsilon}(n + 1, n) = \mathbf{F}(n + 1, n)[\mathbf{x}(n) - \hat{\mathbf{x}}(n|\mathcal{Y}_{n-1})] \\ - \mathbf{G}(n)\,[\mathbf{y}(n) - \mathbf{C}(n)\hat{\mathbf{x}}(n|\mathcal{Y}_{n-1})] + \mathbf{v}_1(n)\end{aligned} \tag{7.51}$$

Next, using the measurement equation (7.19) to eliminate $\mathbf{y}(n)$ in Eq. (7.51), we get the following difference equation for recursive computation of the predicted state-error vector:

$$\begin{aligned}\boldsymbol{\epsilon}(n + 1, n) = [\mathbf{F}(n + 1, n) - \mathbf{G}(n)\mathbf{C}(n)]\,\boldsymbol{\epsilon}(n, n - 1) \\ + \mathbf{v}_1(n) - \mathbf{G}(n)\,\mathbf{v}_2(n)\end{aligned} \tag{7.52}$$

The correlation matrix of the predicted state-error vector $\boldsymbol{\epsilon}(n + 1, n)$ equals [see Eq. (7.36)]

$$\mathbf{K}(n + 1, n) = E[\boldsymbol{\epsilon}(n + 1, n)\boldsymbol{\epsilon}^H(n + 1, n)] \tag{7.53}$$

Substituting Eq. (7.52) in (7.53), and recognizing that the error vector $\boldsymbol{\epsilon}(n, n - 1)$ and the noise vectors $\mathbf{v}_1(n)$ and $\mathbf{v}_2(n)$ are mutually uncorrelated, we may express the predicted state-error correlation matrix as follows:

$$\begin{aligned}\mathbf{K}(n + 1, n) = [\mathbf{F}(n + 1, n) - \mathbf{G}(n)\mathbf{C}(n)]\mathbf{K}(n, n - 1)\,[\mathbf{F}(n + 1, n) - \mathbf{G}(n)\mathbf{C}(n)]^H \\ + \mathbf{Q}_1(n) + \mathbf{G}(n)\mathbf{Q}_2(n)\mathbf{G}^H(n)\end{aligned} \tag{7.54}$$

where $\mathbf{Q}_1(n)$ and $\mathbf{Q}_2(n)$ are the correlation matrices of $\mathbf{v}_1(n)$ and $\mathbf{v}_2(n)$, respectively. By expanding the right-hand side of Eq. (7.54), and then using Eqs. (7.49) and (7.35) for the

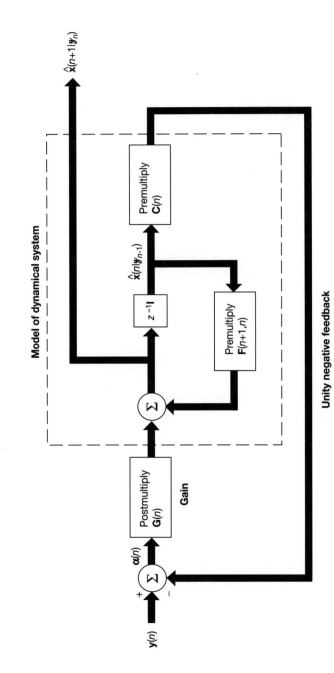

Figure 7.3 One-step predictor.

Kalman gain, we get the *Riccati difference equation*[1] for the recursive computation of the predicted state-error correlation matrix:

$$\mathbf{K}(n + 1, n) = \mathbf{F}(n + 1, n)\mathbf{K}(n)\mathbf{F}^H(n + 1, n) + \mathbf{Q}_1(n) \tag{7.55}$$

The M-by-M matrix $\mathbf{K}(n)$ is described by the recursion:

$$\mathbf{K}(n) = \mathbf{K}(n, n - 1) - \mathbf{F}(n, n + 1)\mathbf{G}(n)\mathbf{C}(n)\mathbf{K}(n, n - 1) \tag{7.56}$$

Here we have used the property

$$\mathbf{F}(n + 1, n)\,\mathbf{F}(n, n + 1) = \mathbf{I} \tag{7.57}$$

where $\mathbf{I}$ is the identity matrix. This property follows from the definition of the transition matrix. The mathematical significance of the matrix $\mathbf{K}(n)$ in Eq. (7.56) will be explained later in Section 7.5.

Figure 7.4 is a signal-flow graph representation of Eqs. (7.56) and (7.55), in that order. This diagram may be viewed as a representation of the *Riccati equation solver* in that, given $\mathbf{K}(n, n - 1)$, it computes the updated value $\mathbf{K}(n + 1, n)$.

Equations (7.49), (7.35), (7.31), (7.45), (7.56), and (7.55), in that order, define Kalman's one-step prediction algorithm.

Comments

The process applied to the input of the Kalman filter consists of the observed data $\mathbf{y}(1)$, $\mathbf{y}(2), \ldots, \mathbf{y}(n)$ that span the space $\mathcal{Y}_n$. The resulting filter output equals the predicted state vector $\hat{\mathbf{x}}(n + 1|\mathcal{Y}_n)$. Given that the matrices $\mathbf{F}(n + 1, n)$, $\mathbf{C}(n)$, $\mathbf{Q}_1(n)$, and $\mathbf{Q}_2(n)$ are all known quantities, we find from Eqs. (7.44), (7.55), and (7.56) that the predicted state-error correlation matrix $\mathbf{K}(n + 1, n)$ is actually independent of the input $\mathbf{y}(n)$, which it has to be. The Kalman gain $\mathbf{G}(n)$ is also independent of the input $\mathbf{y}(n)$. Consequently, the predicted state-error correlation matrix $\mathbf{K}(n + 1, n)$ and the Kalman gain $\mathbf{G}(n)$ may be computed before the Kalman filter is actually put into operation. With the correlation matrix $\mathbf{K}(n + 1, n)$ providing a statistical description of the error in the predicted state vector $\hat{\mathbf{x}}(n + 1|\mathcal{Y}(n))$, we may examine this matrix before actually using the Kalman filter to produce a realization of a physical system of interest; in this way, we may determine whether the solution supplied by the Kalman filter is indeed satisfactory.

As already mentioned, the Kalman filter theory assumes knowledge of the matrices $\mathbf{F}(n + 1, n)$, $\mathbf{C}(n)$, $\mathbf{Q}_1(n)$ and $\mathbf{Q}_2(n)$. However, the theory may be *generalized* to include a situation where one or more of these matrices may assume values that depend on the input $\mathbf{y}(n)$. In such a situation, we find that although $\hat{\mathbf{x}}(n + 1|\mathcal{Y}_n)$ and $\mathbf{K}(n + 1, n)$ are still given by Eqs. (7.45) and (7.55), respectively, the Kalman gain $\mathbf{G}(n)$ and the predicted state-error correlation matrix $\mathbf{K}(n + 1, n)$ are *not* precomputable (Anderson and Moore, 1979).

---

[1] The Riccati difference equation is named in honor of Count Jacopo Francisco Riccati. This equation has become of particular importance in control theory.

316

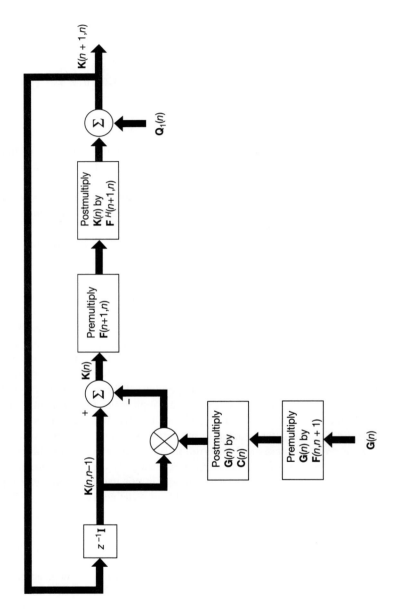

Figure 7.4   Riccati equation solver.

Rather, they both now depend on the input $\mathbf{y}(n)$. This means that $\mathbf{K}(n + 1, n)$ is a *conditional* error-correlation matrix, conditional on the input $\mathbf{y}(n)$.

7.5 FILTERING

The next signal-processing operation we wish to consider is that of filtering. In particular, we wish to compute the *filtered estimate* $\hat{\mathbf{x}}(n|\mathcal{Y}_n)$ by using the one-step prediction algorithm described previously.

We first note that the state $\mathbf{x}(n)$ and the noise vector $\mathbf{v}_1(n)$ are independent of each other. Hence, from the state equation (7.17) we find that the minimum mean-square estimate of the state $\mathbf{x}(n + 1)$ at time $n + 1$, given the observed data up to and including time n [i.e., given $\mathbf{y}(1), \ldots, \mathbf{y}(n)$], equals

$$\hat{\mathbf{x}}(n + 1|\mathcal{Y}_n) = \mathbf{F}(n + 1, n)\hat{\mathbf{x}}(n|\mathcal{Y}_n) + \hat{\mathbf{v}}_1(n|\mathcal{Y}_n) \tag{7.58}$$

Since the noise vector $\mathbf{v}_1(n)$ is independent of the observed data $\mathbf{y}(1), \ldots, \mathbf{y}(n)$, it follows that the corresponding minimum mean-square estimate $\hat{\mathbf{v}}_1(n|\mathcal{Y}_n)$ is zero. Accordingly, Eq. (7.58) simplifies to

$$\hat{\mathbf{x}}(n + 1|\mathcal{Y}_n) = \mathbf{F}(n + 1, n) \, \hat{\mathbf{x}}(n|\mathcal{Y}_n) \tag{7.59}$$

To find the filtered estimate $\hat{\mathbf{x}}(n|\mathcal{Y}_n)$, we premultiply both sides of Eq. (7.59) by the inverse of the transition matrix $\mathbf{F}(n + 1, n)$, and thus write

$$\hat{\mathbf{x}}(n|\mathcal{Y}_n) = \mathbf{F}^{-1}(n + 1, n)\hat{\mathbf{x}}(n + 1|\mathcal{Y}_n) \tag{7.60}$$

Using the property of the state transition matrix given in Eq. (7.57), we have

$$\mathbf{F}^{-1}(n + 1, n) = \mathbf{F}(n, n + 1) \tag{7.61}$$

We may therefore rewrite Eq. (7.60) in the equivalent form:

$$\hat{\mathbf{x}}(n|\mathcal{Y}_n) = \mathbf{F}(n, n + 1) \, \hat{\mathbf{x}}(n + 1|\mathcal{Y}_n) \tag{7.62}$$

This shows that knowing the solution to the one-step prediction problem, that is, the minimum mean-square estimate $\hat{\mathbf{x}}(n + 1|\mathcal{Y}_n)$, we may determine the corresponding filtered estimate $\hat{\mathbf{x}}(n|\mathcal{Y}_n)$ simply by multiplying $\hat{\mathbf{x}}(n + 1|\mathcal{Y}_n)$ by the state transition matrix $\mathbf{F}(n, n + 1)$.

Filtered Estimation Error and Conversion Factor

In a filtering framework, it is natural that we define a *filtered estimation error vector* in terms of the filtered estimate of the state as follows:

$$\mathbf{e}(n) = \mathbf{y}(n) - \mathbf{C}(n)\hat{\mathbf{x}}(n \, |\mathcal{Y}_n) \tag{7.63}$$

This definition is similar to that of Eq. (7.31) for the innovations vector $\boldsymbol{\alpha}(n)$, except that we have substituted the filtered estimate $\hat{\mathbf{x}}(n|\mathcal{Y}_n)$ for the predicted estimate $\hat{\mathbf{x}}(n|\mathcal{Y}_{n-1})$. Using Eqs. (7.45) and (7.62) in (7.63), we get

$$
\begin{aligned}
\mathbf{e}(n) &= \mathbf{y}(n) - \mathbf{C}(n)\hat{\mathbf{x}}(n\,|\mathcal{Y}_{n-1}) - \mathbf{C}(n)\mathbf{F}(n,\,n+1)\mathbf{G}(n)\boldsymbol{\alpha}(n) \\
&= \boldsymbol{\alpha}(n) - \mathbf{C}(n)\mathbf{F}(n,\,n+1)\,\mathbf{G}(n)\boldsymbol{\alpha}(n) \hspace{2cm} (7.64) \\
&= [\mathbf{I} - \mathbf{C}(n)\mathbf{F}(n,\,n+1)\mathbf{G}(n)]\boldsymbol{\alpha}(n)
\end{aligned}
$$

The matrix-valued quantity inside the square brackets in Eq. (7.64) is called the *conversion factor*, which provides a formula for converting the innovations vector $\boldsymbol{\alpha}(n)$ into the filtered estimation error vector $\mathbf{e}(n)$. Using Eq. (7.49) to eliminate the Kalman gain $\mathbf{G}(n)$ from this definition and canceling common terms, we may rewrite Eq. (7.64) in the equivalent form:

$$
\mathbf{e}(n) = \mathbf{Q}_2(n)\,\mathbf{R}^{-1}(n)\boldsymbol{\alpha}(n) \hspace{2cm} (7.65)
$$

where $\mathbf{Q}_2(n)$ is the correlation matrix of the measurement noise process $\mathbf{v}_2(n)$, and the matrix $\mathbf{R}(n)$ is itself defined in Eq. (7.35) as the correlation matrix of the innovations process $\boldsymbol{\alpha}(n)$. Thus, except for a premultiplication by $\mathbf{Q}_2(n)$, Eq. (7.65) shows that the inverse matrix $\mathbf{R}^{-1}(n)$ plays the role of a conversion factor in the Kalman filter theory. Indeed, for the special case of $\mathbf{Q}_2(n)$ equal to the identity matrix, the inverse matrix $\mathbf{R}^{-1}$ is exactly the conversion factor defined herein.

Filtered State-Error Correlation Matrix

Earlier we introduced the M-by-M matrix $\mathbf{K}(n)$ in the formulation of the Riccati difference equation (7.55). We conclude our present discussion of the standard Kalman filter theory by showing that this matrix equals the correlation matrix of the error inherent in the filtered estimate $\hat{\mathbf{x}}(n|\mathcal{Y}_n)$.

Define the *filtered state-error vector* $\boldsymbol{\epsilon}(n)$ as the difference between the state $\mathbf{x}(n)$ and the filtered estimate $\hat{\mathbf{x}}(n|\mathcal{Y}_n)$, as shown by

$$
\boldsymbol{\epsilon}(n) = \mathbf{x}(n) - \hat{\mathbf{x}}(n|\mathcal{Y}_n) \hspace{2cm} (7.66)
$$

Substituting Eqs. (7.45) and (7.62) in (7.66), and recognizing that the product of $\mathbf{F}(n,\,n+1)$ and $\mathbf{F}(n+1,\,n)$ equals the identity matrix, we get

$$
\begin{aligned}
\boldsymbol{\epsilon}(n) &= \mathbf{x}(n) - \hat{\mathbf{x}}(n|\mathcal{Y}_{n-1}) - \mathbf{F}(n,\,n+1)\,\mathbf{G}(n)\,\boldsymbol{\alpha}(n) \\
&= \boldsymbol{\epsilon}(n,\,n-1) - \mathbf{F}(n,\,n+1)\mathbf{G}(n)\,\boldsymbol{\alpha}(n)
\end{aligned} \hspace{1.5cm} (7.67)
$$

where $\boldsymbol{\epsilon}(n,\,n,\,-1)$ is the predicted state-error vector at time n, using data up to time $n-1$, and $\boldsymbol{\alpha}(n)$ is the innovations process.

By definition, the correlation matrix of the filtered state-error vector $\boldsymbol{\epsilon}(n)$ equals the expectation $E[\boldsymbol{\epsilon}(n)\boldsymbol{\epsilon}^H(n)]$. Hence, using Eq. (7.67), we may express this expectation as follows:

$$E[\boldsymbol{\epsilon}(n)\boldsymbol{\epsilon}^H(n)] = E[\boldsymbol{\epsilon}(n, n-1)\,\boldsymbol{\epsilon}^H(n, n-1)]$$
$$+ \mathbf{F}(n, n+1)\,\mathbf{G}(n)\,E[\boldsymbol{\alpha}(n)\boldsymbol{\alpha}^H(n)]\,\mathbf{G}^H(n)\,\mathbf{F}^H(n, n+1) \qquad (7.68)$$
$$- 2E[\boldsymbol{\epsilon}(n, n-1)\boldsymbol{\alpha}^H(n)]\,\mathbf{G}^H(n)\,\mathbf{F}^H(n, n+1)$$

Examining the right-hand side of Eq. (7.68), we find that the three expectations contained in it may be interpreted individually as follows:

1. The first expectation equals the predicted state-error correlation matrix:

$$\mathbf{K}(n, n-1) = E[\boldsymbol{\epsilon}(n, n-1)\,\boldsymbol{\epsilon}^H(n, n-1)]$$

2. The expectation in the second term equals the correlation matrix of the innovations process $\boldsymbol{\alpha}(n)$:

$$\mathbf{R}(n) = E[\boldsymbol{\alpha}(n)\boldsymbol{\alpha}^H(n)]$$

3. The expectation in the third term may be expressed as follows:

$$E[\boldsymbol{\epsilon}(n, n-1)\,\boldsymbol{\alpha}^H(n)] = E[(\mathbf{x}(n) - \hat{\mathbf{x}}(n|\mathcal{Y}_{n-1}))\boldsymbol{\alpha}^H(n)]$$
$$= E[\mathbf{x}(n)\,\boldsymbol{\alpha}^H(n)]$$

where, in the last line, we have used the fact that the estimate $\hat{\mathbf{x}}(n|\mathcal{Y}_{n-1})$ is orthogonal to the innovations process $\boldsymbol{\alpha}(n)$ acting as input. Next, from Eq. (7.42) we see, by putting $k = n$ and then premultiplying both sides by the inverse matrix $\mathbf{F}^{-1}(n+1, n) = \mathbf{F}(n, n+1)$, that

$$E[\mathbf{x}(n)\,\boldsymbol{\alpha}^H(n)] = \mathbf{F}(n, n+1)\,E[\mathbf{x}(n+1)\boldsymbol{\alpha}^H(n)]$$
$$= \mathbf{F}(n, n+1)\mathbf{G}(n)\,\mathbf{R}(n)$$

where, in the last line, we have made use of Eq. (7.44). Hence,

$$E[\boldsymbol{\epsilon}(n, n-1)\,\boldsymbol{\alpha}^H(n)] = \mathbf{F}(n, n+1)\,\mathbf{G}(n)\,\mathbf{R}(n)$$

We may now use these results in Eq. (7.68), and so obtain

$$E[\boldsymbol{\epsilon}(n)\boldsymbol{\epsilon}^H(n)] = \mathbf{K}(n, n-1) - \mathbf{F}(n, n+1)\,\mathbf{G}(n)\,\mathbf{R}(n)\mathbf{G}^H(n)\,\mathbf{F}^H(n, n+1) \qquad (7.69)$$

We may further simplify this result by noting that [see Eq. (7.49)]

$$\mathbf{G}(n)\,\mathbf{R}(n) = \mathbf{F}(n+1, n)\,\mathbf{K}(n, n-1)\,\mathbf{C}^H(n) \qquad (7.70)$$

Accordingly, using Eqs. (7.69) and (7.70), and recognizing that the product of $\mathbf{F}(n, n+1)$ and $\mathbf{F}(n+1, n)$ equals the identity matrix, we get the desired result for the filtered state-error correlation matrix:

$$E[\boldsymbol{\epsilon}(n)\boldsymbol{\epsilon}^H(n)] = \mathbf{K}(n, n-1) - \mathbf{K}(n, n-1)\mathbf{C}^H(n)\,\mathbf{G}^H(n)\mathbf{F}^H(n, n+1)$$

Equivalently, using the Hermitian property of $E[\boldsymbol{\epsilon}(n)\boldsymbol{\epsilon}^H(n)]$ and that of $\mathbf{K}(n, n-1)$, we may write

$$E[\boldsymbol{\epsilon}(n)\boldsymbol{\epsilon}^H(n)] = \mathbf{K}(n, n-1) - \mathbf{F}(n, n+1)\mathbf{G}(n)\, \mathbf{C}(n)\, \mathbf{K}(n, n-1) \tag{7.71}$$

Comparing Eq. (7.71) with (7.56), we readily see that

$$E[\boldsymbol{\epsilon}(n)\, \boldsymbol{\epsilon}^H(n)] = \mathbf{K}(n)$$

This shows that the matrix $\mathbf{K}(n)$ used in the Riccati difference equation (7.55) is in fact the *filtered state-error correlation matrix*. The matrix $\mathbf{K}(n)$ is used as the statistical description of the error in the filtered estimate $\hat{\mathbf{x}}(n|\mathcal{Y}_n)$.

7.6 INITIAL CONDITIONS

To operate the one-step prediction and filtering algorithms described in Sections 7.4 and 7.5, we obviously need to specify the *initial conditions*. We now address this issue.

The initial state of the process equation (7.17) is not known precisely. Rather, it is usually described by its mean and correlation matrix. In the absence of any observed data at time $n = 0$, we may choose the *initial predicted estimate* as

$$\hat{\mathbf{x}}(1|\mathcal{Y}_0) = E[\mathbf{x}(1)] \tag{7.72}$$

and its *correlation matrix*

$$\begin{aligned}\mathbf{K}(1,0) &= E[(\mathbf{x}(1) - E[\mathbf{x}(1)])(\mathbf{x}(1) - E[\mathbf{x}(1)])^H] \\ &= \Pi_0\end{aligned} \tag{7.73}$$

This choice for the initial conditions is not only intuitively satisfying but also has the advantage of yielding a filtered estimate of the state $\hat{\mathbf{x}}(n|\mathcal{Y}_n)$ that is *unbiased* (see Problem 10). Assuming that the state vector $\mathbf{x}(n)$ has *zero mean*, we may simplify Eqs. (7.72) and (7.73) by setting

$$\hat{\mathbf{x}}(1|\mathcal{Y}_0) = \mathbf{0}$$

and

$$\mathbf{K}(1,0) = E[\mathbf{x}(1)\, \mathbf{x}^H(1)] = \Pi_0$$

7.7 SUMMARY OF THE KALMAN FILTER

Table 7.1 presents a summary of the variables used to formulate the solution to the Kalman filtering problem. The input of the filter is the vector process $\mathbf{y}(n)$, represented by the vector space $\mathcal{Y}_n$, and the output is the filtered estimate $\hat{\mathbf{x}}(n|\mathcal{Y}_n)$ of the state vector. In Table 7.2, we present a summary of the Kalman filter (including initial conditions) based on the one-step prediction algorithm.

TABLE 7.1   SUMMARY OF THE KALMAN VARIABLES

Variable	Definition	Dimension
$\mathbf{x}(n)$	State vector at time n	M-by-1
$\mathbf{y}(n)$	Observation vector at time n	N-by-1
$\mathbf{F}(n + 1, n)$	State transition matrix from time n to $n + 1$	M-by-M
$\mathbf{C}(n)$	Measurement matrix at time n	N-by-M
$\mathbf{Q}_1(n)$	Correlation matrix of process noise vector $\mathbf{v}_1(n)$	M-by-M
$\mathbf{Q}_2(n)$	Correlation matrix of measurement noise vector $\mathbf{v}_2(n)$	N-by-N
$\hat{\mathbf{x}}(n + 1 \vert \mathcal{Y}_n)$	Predicted estimate of the state vector at time $n + 1$, given the observation vectors $\mathbf{y}(1)$, $\mathbf{y}(2)$, . . . , $\mathbf{y}(n)$	M-by-1
$\hat{\mathbf{x}}(n \vert \mathcal{Y}_n)$	Filtered estimate of the state vector at time n, given the observation vectors $\mathbf{y}(1)$, $\mathbf{y}(2)$, . . . , $\mathbf{y}(n)$	M-by-1
$\mathbf{G}(n)$	Kalman gain at time n	M-by-N
$\boldsymbol{\alpha}(n)$	Innovations vector at time n	N-by-1
$\mathbf{R}(n)$	Correlation matrix of the innovations vector $\boldsymbol{\alpha}(n)$	N-by-N
$\mathbf{K}(n + 1, n)$	Correlation matrix of the error in $\hat{\mathbf{x}}(n + 1 \vert \mathcal{Y}_n)$	M-by-M
$\mathbf{K}(n)$	Correlation matrix of the error in $\hat{\mathbf{x}}(n \vert \mathcal{Y}_n)$	M-by-M

TABLE 7.2   SUMMARY OF THE KALMAN FILTER BASED ON ONE-STEP PREDICTION

Input vector process
    Observations = $\{\mathbf{y}(1), \mathbf{y}(2), . . . , \mathbf{y}(n)\}$
Known parameters
    State transition matrix = $\mathbf{F}(n + 1, n)$
    Measurement matrix = $\mathbf{C}(n)$
    Correlation matrix of process noise vector = $\mathbf{Q}_1(n)$
    Correlation matrix of measurement noise vector = $\mathbf{Q}_2(n)$
Computation: $n = 1, 2, 3, . . .$
    $\mathbf{G}(n) = \mathbf{F}(n + 1, n)\mathbf{K}(n, n - 1)\mathbf{C}^H(n)[\mathbf{C}(n)\mathbf{K}(n, n - 1)\mathbf{C}^H(n) + \mathbf{Q}_2(n)]^{-1}$
    $\boldsymbol{\alpha}(n) = \mathbf{y}(n) - \mathbf{C}(n)\hat{\mathbf{x}}(n \vert \mathcal{Y}_{n-1})$
    $\hat{\mathbf{x}}(n + 1 \vert \mathcal{Y}_n) = \mathbf{F}(n + 1, n)\hat{\mathbf{x}}(n \vert \mathcal{Y}_{n-1}) + \mathbf{G}(n)\boldsymbol{\alpha}(n)$
    $\mathbf{K}(n) = \mathbf{K}(n, n - 1) - \mathbf{F}(n, n + 1)\mathbf{G}(n)\mathbf{C}(n)\mathbf{K}(n, n - 1)$
    $\mathbf{K}(n + 1, n) = \mathbf{F}(n + 1, n)\mathbf{K}(n)\mathbf{F}^H(n + 1, n) + \mathbf{Q}_1(n)$
Initial conditions:
    $\hat{\mathbf{x}}(1 \vert \mathcal{Y}_0) = E[\mathbf{x}(1)]$
    $\mathbf{K}(1, 0) = E[(\boldsymbol{x}(1) - E[\boldsymbol{x}(1)])(\boldsymbol{x}(1) - E[\boldsymbol{x}(1)])^H] = \boldsymbol{\Pi}_0$

A block diagram representation of the Kalman filter is given in Fig. 7.5, which is based on three functional blocks:

- Kalman gain computer, described in Fig. 7.2
- One-step predictor, described in Fig. 7.3.
- Riccati equation solver, described in Fig. 7.4

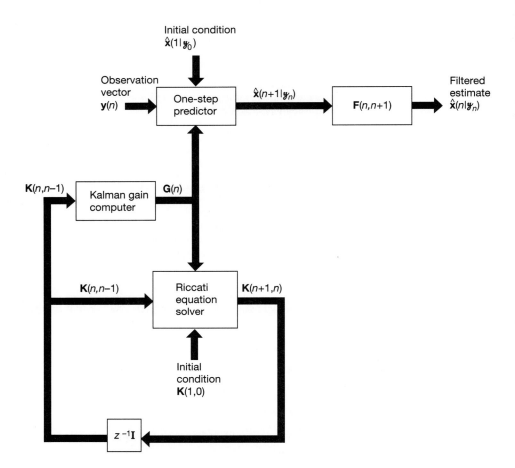

Figure 7.5   Black diagram of the Kalman filter based on one-step prediction.

7.8 VARIANTS OF THE KALMAN FILTER

As mentioned in the introductory remarks to this chapter, the main reason for our interest
in Kalman filter theory in this book is that it provides a general framework for the derivation of certain adaptive filtering algorithms known collectively as the family of recursive
lest-squares (RLS) algorithms.

The application of Kalman filter theory to adaptive filtering was apparently first
reported in the literature by Lawrence and Kaufman (1971); see Problem 8. This was followed by Godard (1974), who used an approach different from that of Lawrence and Kaufman. In particular, Godard formulated the adaptive filtering problem (using a tap-
delay-line structure) as the estimation of a state vector in Gaussian noise, which represents

a classical Kalman filtering problem. Godard's paper prompted many other investigators to explore the application of Kalman filter theory to adaptive filtering problems.

However, we had to await the paper by Sayed and Kailath (1994) to discover how indeed the Riccati-based Kalman filtering algorithm and its variants can be correctly framed into one-to-one correspondences with all the known algorithms in the RLS family. We will take up the details of this unifying framework later in the book. For now, we will focus our attention on a special dynamical model that befits our future needs.

Special Case: Unforced Dynamics

Consider a linear dynamical system whose state-space model is described by the following pair of state equations (Sayed and Kailath, 1994):

$$\mathbf{x}(n + 1) = \lambda^{-1/2}\mathbf{x}(n) \tag{7.74}$$

$$y(n) = \mathbf{u}^H(n)\mathbf{x}(n) + v(n) \tag{7.75}$$

where λ is a positive real scalar. According to this model, the process noise is zero, and the measurement noise, denoted by the scalar $v(n)$, is a zero-mean white noise process with unit variance, as shown by

$$E[v(n)v^*(k)] = \begin{cases} 1, & n = k \\ 0, & n \neq k \end{cases} \tag{7.76}$$

Thus, comparing this model with the general model described by Eqs. (7.17) to (7.21), we note the following:

$$\mathbf{F}(n + 1, n) = \lambda^{-1/2}\mathbf{I} \tag{7.77}$$

$$\mathbf{Q}_1(n) = \mathbf{O} \tag{7.78}$$

$$\mathbf{C}(n) = \mathbf{u}^H(n) \tag{7.79}$$

$$\mathbf{Q}_2(n) = 1 \tag{7.80}$$

The state-space model described by Eqs. (7.74) to (7.76) is referred to as an *unforced dynamical model* by virtue of the fact that the process equation (7.74) is free of an external force. Most importantly, the state transition matrix of the model is equal to the identity matrix $\mathbf{I}$ scaled by the constant $\lambda^{-1/2}$. Consequently, the predicted state-error correlation matrix $\mathbf{K}(n + 1, n)$ and the filtered state-error correlation matrix $\mathbf{K}(n)$ assume a common value; see Problem 9.

This special unforced dynamical model holds the key to the formulation of a general framework for deriving the RLS family of adaptive filtering algorithms. As we shall see later in the book, the constant λ has a significant role in the operation of these algorithms. For now we content ourselves by considering variants of the Kalman filtering algorithm based on this model.

TABLE 7.3 SUMMARY OF THE COVARIANCE (KALMAN) FILTERING ALGORITHM FOR THE SPECIAL UNFORCED DYNAMICAL MODEL

Input scalar process:
  Observations: $y(1), y(2), \ldots y(n)$
Known parameters:
  state transition matrix $= \lambda^{-1/2}\mathbf{I}, \quad \mathbf{I} = $ identity matrix
  measurement matrix $= \mathbf{u}^H(n)$
  variance of measurement noise $v(n) = 1$
Initial conditions:
  $\hat{\mathbf{x}}(1 \,|\, \mathcal{Y}_0) = E[\mathbf{x}(1)]$
  $\mathbf{K}(1,0) = E[(\mathbf{x}(1) - E[\mathbf{x}(1)])(\mathbf{x}(1) - E[\mathbf{x}(1)])^H] = \Pi_0$
Computation: $n = 1, 2, 3, \ldots$

$$\mathbf{g}(n) = \frac{\lambda^{-1/2}\mathbf{K}(n-1)\mathbf{u}(n)}{\mathbf{u}^H(n)\mathbf{K}(n-1)\mathbf{u}(n) + 1}$$

$$\alpha(n) = y(n) - \mathbf{u}^H(n)\hat{\mathbf{x}}(n \,|\, \mathcal{Y}_{n-1})$$

$$\hat{\mathbf{x}}(n+1 \,|\, \mathcal{Y}_n) = \lambda^{-1/2}\hat{\mathbf{x}}(n \,|\, \mathcal{Y}_{n-1}) + \mathbf{g}(n)\alpha(n)$$

$$\mathbf{K}(n) = \lambda^{-1}\mathbf{K}(n-1) - \lambda^{-1/2}\mathbf{g}(n)\mathbf{u}^H(n)\mathbf{K}(n-1)$$

Covariance (Kalman) Filtering Algorithm

The Kalman filtering algorithm summarized in Table 7.2 is designed to propagate the correlation (covariance) matrix $\mathbf{K}(n+1, n)$ that refers to the error in the state's estimate $\hat{\mathbf{x}}(n+1|\mathcal{Y}_n)$. This algorithm is therefore commonly referred to as the *covariance (Kalman) filtering algorithm*. For the unforced dynamical model at hand, we find that substituting Eqs. (7.77) to (7.80) in Table 7.2 yields the simplified covariance filtering algorithm summarized in Table 7.3. In this table we have used $\mathbf{g}(n)$ to denote the Kalman gain, as it takes the form of a vector here.

Information Filtering Algorithm

The Kalman filter may also be implemented by propagating the inverse matrix $\mathbf{K}^{-1}(n)$ which accentuates the recursive least-squares nature of the filtering process. The *inverse state-error correlation matrix*, $\mathbf{K}^{-1}(n)$, is related to *Fisher's information matrix*[2], which permits an interpretation of filter performance in information-theoretic terms. For this reason, an implementation of the Kalman filtering algorithm based on $\mathbf{K}^{-1}(n)$ is termed the *information filtering algorithm* (Fraser, 1967).

For the derivation of the information filtering algorithm, we may proceed in the manner described next.

Step 1. We start with the Riccati difference equation which, for the special unforced dynamical model, has the form (see the last line of the algorithm in Table 7.3):

---

[2]Fisher's information matrix is discussed in Appendix D.

$$\mathbf{K}(n) = \lambda^{-1}\mathbf{K}(n-1) - \lambda^{-1/2}\mathbf{g}(n)\mathbf{u}^H(n)\mathbf{K}(n-1) \tag{7.81}$$

Solving this equation for the matrix product $\mathbf{g}(n)\mathbf{u}^H(n)\mathbf{K}(n-1)$, we get

$$\mathbf{g}(n)\mathbf{u}^H(n)\mathbf{K}(n-1) = \lambda^{-1/2}\mathbf{K}(n-1) - \lambda^{1/2}\mathbf{K}(n) \tag{7.82}$$

Next, from the first line of the algorithm in Table 7.3, the Kalman gain for the unforced dynamical model of interest is defined by

$$\mathbf{g}(n) = \frac{\lambda^{-1/2}\mathbf{K}(n-1)\mathbf{u}(n)}{\mathbf{u}^H(n)\mathbf{K}(n-1)\mathbf{u}(n) + 1} \tag{7.83}$$

Cross-multiplying and rearranging terms, we may rewrite Eq. (7.83) as

$$\mathbf{g}(n) = \lambda^{-1/2}\mathbf{K}(n-1)\mathbf{u}(n) - (\mathbf{g}(n)\mathbf{u}^H(n)\mathbf{K}(n-1))\mathbf{u}(n) \tag{7.84}$$

Substituting Eq. (7.82) in (7.84), and then canceling common terms, we get a new definition for the Kalman gain:

$$\mathbf{g}(n) = \lambda^{1/2}\mathbf{K}(n)\mathbf{u}(n) \tag{7.85}$$

Next, eliminating $\mathbf{g}(n)$ between Eqs. (7.82) and (7.85), and multiplying the result by $\lambda^{1/2}$, we get

$$\mathbf{K}(n-1) = \lambda\mathbf{K}(n)\mathbf{u}(n)\mathbf{u}^H(n)\mathbf{K}(n-1) + \lambda\mathbf{K}(n) \tag{7.86}$$

Premultiplying Eq. (7.86) by the inverse matrix $\mathbf{K}^{-1}(n)$ and postmultiplying it by $\mathbf{K}^{-1}(n-1)$, we get the first recursion of the information filtering algorithm:

$$\mathbf{K}^{-1}(n) = \lambda\mathbf{K}^{-1}(n-1) + \lambda\mathbf{u}(n)\mathbf{u}^H(n) \tag{7.87}$$

Step 2. From the second and third lines of the algorithm summarized in Table 7.3, we have, respectively,

$$\alpha(n) = y(n) - \mathbf{u}^H(n)\hat{\mathbf{x}}(n \mid \mathcal{Y}_{n-1}) \tag{7.88}$$

and

$$\hat{\mathbf{x}}(n+1 \mid \mathcal{Y}_n) = \lambda^{-1/2}\hat{\mathbf{x}}(n \mid \mathcal{Y}_{n-1}) + \mathbf{g}(n)\alpha(n) \tag{7.89}$$

Therefore, substituting Eq. (7.85) in (7.89), we get

$$\hat{\mathbf{x}}(n+1 \mid \mathcal{Y}_n) = \lambda^{-1/2}\hat{\mathbf{x}}(n \mid \mathcal{Y}_{n-1}) + \lambda^{1/2}\mathbf{K}(n)\mathbf{u}(n)\alpha(n) \tag{7.90}$$

Next, eliminating $\alpha(n)$ between Eqs. (7.88) and (7.90) yields

$$\hat{\mathbf{x}}(n+1 \mid \mathcal{Y}_n) = [\lambda^{-1/2}\mathbf{I} - \lambda^{1/2}\mathbf{K}(n)\mathbf{u}(n)\mathbf{u}^H(n)]\hat{\mathbf{x}}(n \mid \mathcal{Y}_{n-1}) + \lambda^{1/2}\mathbf{K}(n)\mathbf{u}(n)y(n) \tag{7.91}$$

But, from Eq. (7.86), we readily deduce the following relation:

$$\lambda^{-1/2}\mathbf{I} - \lambda^{1/2}\mathbf{K}(n)\mathbf{u}(n)\mathbf{u}^H(n) = \lambda^{1/2}\mathbf{K}(n)\mathbf{K}^{-1}(n-1) \tag{7.92}$$

Accordingly, we may simplify Eq. (7.91) as follows:

$$\hat{\mathbf{x}}(n+1 \mid \mathcal{Y}_n) = \lambda^{1/2}\mathbf{K}(n)\mathbf{K}^{-1}(n-1)\hat{\mathbf{x}}(n \mid \mathcal{Y}_{n-1}) + \lambda^{1/2}\mathbf{K}(n)\mathbf{u}(n)y(n)$$

Premultiplying this equation by the inverse matrix $\mathbf{K}^{-1}(n)$, we get the second recursion of the information filtering algorithm:

$$\mathbf{K}^{-1}(n)\hat{\mathbf{x}}(n+1 \mid \mathcal{Y}_n) = \lambda^{1/2}[\mathbf{K}^{-1}(n-1)\hat{\mathbf{x}}(n \mid \mathcal{Y}_{n-1}) + \mathbf{u}(n)y(n)] \qquad (7.93)$$

Note that in Eq. (7.93) the algorithm propagates the product $\mathbf{K}^{-1}(n-1)\hat{\mathbf{x}}(n|\mathcal{Y}_{n-1})$ rather than the estimate $\hat{\mathbf{x}}(n \mid \mathcal{Y}_{n-1})$ by itself.

Step 3. Finally, the updated value of the state's estimate is computed by combining the results of steps 1 and 2 as follows:

$$\hat{\mathbf{x}}(n + 1|\mathcal{Y}_n) = \mathbf{K}(n) \, (\mathbf{K}^{-1}(n)\hat{\mathbf{x}}(n + 1|\mathcal{Y}_n)) \qquad (7.94)$$

$$= [\mathbf{K}^{-1}(n)]^{-1}(\mathbf{K}^{-1}(n)\hat{\mathbf{x}}(n + 1|\mathcal{Y}_n))$$

Equations (7.87), (7.93), and (7.94), in that order, constitute the information-filtering algorithm for the unforced dynamical model of Eqs. (7.74) to (7.76). A summary of the algorithm is presented in Table 7.4.

Although the covariance and information implementations of the Kalman filter, as described herein, are algebraically equivalent, the numerical properties of these two algorithms may differ substantially from each other (Kaminski et al., 1971). However, both algorithms require the same number of algebraic operations (i.e., multiplications and additions), which, for the special model at hand, is $O(M^2)$, where M is the state dimension.

Square-root Filtering

The covariance implementation of the Kalman filter, summarized in Table 7.2, is the optimal solution to the linear filtering problem posed in Section 7.2. However, this algorithm is prone to serious numerical difficulties that are well documented in the literature (Kamin-

---

TABLE 7.4 SUMMARY OF THE INFORMATION-FILTERING ALGORITHM FOR THE SPECIAL UNFORCED DYNAMICAL MODEL

Input scalar process:
    observations $= y(1), y(2), \ldots, y(n)$
Known parameters:
    state transition matrix $= \lambda^{-1/2}\mathbf{I}$, $\mathbf{I} =$ identity matrix
    measurement matrix $= \mathbf{u}^H(n)$
    variance of measurement noise $v(n) = 1$
Initial conditions:

$$\hat{\mathbf{x}}(1 \mid \mathcal{Y}_0) = E[\mathbf{x}(1)]$$

$$\mathbf{K}(1,0) = E[(\mathbf{x}(1) - E[\mathbf{x}(1)])(\mathbf{x}(1) - E[\mathbf{x}(1)])^H] = \Pi_0$$

Computation: $n = 1, 2, 3 \ldots$

$$\mathbf{K}^{-1}(n) = \lambda[\mathbf{K}^{-1}(n-1) + \mathbf{u}(n)\mathbf{u}^H(n)]$$

$$\mathbf{K}^{-1}(n)\hat{\mathbf{x}}(n + 1 \mid \mathcal{Y}_n) = \lambda^{1/2}[\mathbf{K}^{-1}(n-1)\hat{\mathbf{x}}(n|\mathcal{Y}_{n-1}) + \mathbf{u}(n)y(n)]$$

$$\hat{\mathbf{x}}(n + 1 \mid \mathcal{Y}_n) = [\mathbf{K}^{-1}(n)]^{-1}\mathbf{K}^{-1}(n)\hat{\mathbf{x}}(n + 1|\mathcal{Y}_n)$$

---

ski et al., 1971; Bierman and Thornton, 1977). For example, according to Eq. (7.56) the matrix $\mathbf{K}(n)$ is defined as the difference between two nonnegative definite matrices; hence, unless the numerical accuracy employed at every iteration of the algorithm is high enough, the matrix $\mathbf{K}(n)$ resulting from this computation may *not* be nonnegative definite. Such a situation is clearly unacceptable, because $\mathbf{K}(n)$ represents a correlation matrix. The unstable behavior of the Kalman filter, which results from numerical inaccuracies due to the use of finite wordlength arithmetic, is called the *divergence phenomenon*.

This problem may be overcome by using numerically stable unitary transformations at every iteration of the Kalman filtering algorithm (Potter, 1963; Kaminski et al., 1971; Morf and Kailath, 1975). In particular, the matrix $\mathbf{K}(n)$ is propagated in a square-root form by using the *Cholesky factorization*[3]:

$$\mathbf{K}(n) = \mathbf{K}^{1/2}(n)\mathbf{K}^{H/2}(n) \tag{7.95}$$

where $\mathbf{K}^{1/2}(n)$ is reserved for a lower triangular matrix, and $\mathbf{K}^{H/2}$ is its Hermitian transpose. In linear algebra, the Cholesky factor $\mathbf{K}^{1/2}(n)$ is commonly referred to as the *square root* of the matrix $\mathbf{K}(n)$. Accordingly, any variant of the Kalman filtering algorithm based on the Cholesky factorization is referred to as *square-root filtering*. The important point to note here is that the matrix product $\mathbf{K}^{1/2}(n)\mathbf{K}^{H/2}(n)$ is much less likely to become indefinite, because the product of any square matrix and its Hermitian transpose is always positive definite. Indeed, even in the presence of roundoff errors, the numerical conditioning of the Cholesky factor $\mathbf{K}^{1/2}(n)$ is generally much better than that of $\mathbf{K}(n)$ itself; see Problem 12.

The information filtering algorithm may also be implemented in a square-root form of its own by propagating the square root $\mathbf{K}^{-1/2}(n)$ rather than the inverse matrix $\mathbf{K}^{-1}(n)$ itself (Kaminski et al., 1971; Bierman, 1977). In this variant of the Kalman filter, the Cholesky factorization is used to express the inverse matrix $\mathbf{K}^{-1}(n)$ as follows:

$$\mathbf{K}^{-1}(n) = \mathbf{K}^{-H/2}(n)\mathbf{K}^{-1/2}(n) \tag{7.96}$$

where $\mathbf{K}^{-1/2}(n)$ is a lower triangular matrix, and $\mathbf{K}^{-H/2}$ is its Hermitian transpose.

UD-factorization

The square-root implementation of a Kalman filter requires more computation than the conventional Kalman filter. This problem of computational efficiency led to the development of a modified version of the square-root filtering algorithm known as the *UD-factorization algorithm* (Bierman, 1977). In this second approach, the filtered state-error correlation matrix $\mathbf{K}(n)$ is factored into an upper triangular matrix $\mathbf{U}(n)$ with 1's along its main diagonal and a real diagonal matrix $\mathbf{D}(n)$, as shown by

$$\mathbf{K}(n) = \mathbf{U}(n)\mathbf{D}(n)\mathbf{U}^H(n) \tag{7.97}$$

---

[3]The Cholesky factorization was also discussed in Section 6.7 in the context of linear prediction.

Equivalently, the factorization may be written as

$$\mathbf{K}(n) = (\mathbf{U}(n)\mathbf{D}^{1/2}(n)) \, (\mathbf{U}(n)\mathbf{D}^{1/2}(n))^H \qquad (7.98)$$

where $\mathbf{D}^{1/2}(n)$ is the *square-root* of $\mathbf{D}(n)$. The nonnegative definiteness of the computed matrix $\mathbf{K}(n)$ is guaranteed by updating the factors $\mathbf{U}(n)$ and $\mathbf{D}(n)$ instead of $\mathbf{K}(n)$ itself. However, a Kalman filter based on the UD-factorization does *not* possess the numerical advantage of a standard square-root Kalman filter. Moreover, a Kalman filter using UD-factorization may suffer from serious overflow/underflow problems (Stewart and Chapman, 1990). When an arithmetic operation produces a resultant number with too large or too small a characteristic, it is said to suffer from *overflow* or *underflow*, respectively.

One final comment is in order. With the ever-increasing improvements in digital technology, the old argument that square roots are expensive and awkward to calculate is no longer as compelling as it used to be. Accordingly, to avoid the divergence of a Kalman filter, we will only pursue a detailed discussion of square-root filtering in this book. This we do in Chapter 14, after equipping ourselves with certain unitary transformations in Chapter 12.

7.9 THE EXTENDED KALMAN FILTER

The Kalman filtering problem considered up to this point in the discussion has addressed the estimation of a state vector in a linear model of a dynamical system. If, however, the model is *nonlinear*, we may extend the use of Kalman filtering through a linearization procedure. The resulting filter is naturally referred to as the *extended Kalman filter* (EKF). Such an extension is feasible by virtue of the fact that the Kalman filter is described in terms of differential equations (in the case of continuous-time systems) or difference equations (in the case of discrete-time systems). This is in contrast to the Wiener filter that is limited to linear systems, since the notion of an impulse response (on which the Wiener filter is based) is meaningful only in the context of linear systems. Here is another important advantage of the Kalman filter over the Wiener filter.

To set the stage for a development of the extended Kalman filter in the discrete-time domain, consider first the standard linear state-space model that we studied in the earlier part of this chapter [Eqs. (7.17) and (7.19)], reproduced here for convenience of presentation:

$$\mathbf{x}(n + 1) = \mathbf{F}(n + 1, n)\mathbf{x}(n) + \mathbf{v}_1(n) \qquad (7.99)$$

$$\mathbf{y}(n) = \mathbf{C}(n)\mathbf{x}(n) + \mathbf{v}_2(n) \qquad (7.100)$$

where $\mathbf{v}_1(n)$ and $\mathbf{v}_2(n)$ are uncorrelated zero-mean white-noise processes with correlation matrices $\mathbf{Q}_1(n)$ and $\mathbf{Q}_2(n)$, respectively, as defined in Eqs. (7.18), (7.20), and (7.21). The corresponding Kalman filter equations are summarized in Table 7.2. In this section, however, we will rewrite these equations in a slightly modified form that is more convenient for our present discussion. Specifically, the update of the state estimate is performed in two steps. The first step updates $\hat{\mathbf{x}}(n|\mathcal{Y}_n)$ to $\hat{\mathbf{x}}(n + 1|\mathcal{Y}_n)$; this update equation is simply (7.59). The second step updates $\hat{\mathbf{x}}(n|\mathcal{Y}_{n-1})$ to $\hat{\mathbf{x}}(n|\mathcal{Y}_n)$ and is obtained by substituting Eq. (7.45) into Eq. (7.60), and by defining a new gain matrix:

$$\mathbf{G}_f(n) = \mathbf{F}^{-1}(n + 1, n)\mathbf{G}(n): \tag{7.101}$$

We may thus write

$$\hat{\mathbf{x}}(n + 1|\mathcal{Y}_n) = \mathbf{F}(n + 1, n)\hat{\mathbf{x}}(n|\mathcal{Y}_n) \tag{7.102}$$

$$\hat{\mathbf{x}}(n|\mathcal{Y}_n) = \hat{\mathbf{x}}(n|\mathcal{Y}_{n-1}) + \mathbf{G}_f(n)\boldsymbol{\alpha}(n) \tag{7.103}$$

$$\boldsymbol{\alpha}(n) = \mathbf{y}(n) - \mathbf{C}(n)\hat{\mathbf{x}}(n|\mathcal{Y}_{n-1}) \tag{7.104}$$

$$\mathbf{G}_f(n) = \mathbf{K}(n, n - 1)\mathbf{C}^H(n)[\mathbf{C}(n)\mathbf{K}(n, n - 1)\mathbf{C}^H(n) + \mathbf{Q}_2(n)]^{-1} \tag{7.105}$$

$$\mathbf{K}(n + 1, n) = \mathbf{F}(n + 1, n)\mathbf{K}(n)\mathbf{F}^H(n + 1, n) + \mathbf{Q}_1(n) \tag{7.106}$$

$$\mathbf{K}(n) = [\mathbf{I} - \mathbf{G}_f(n)\mathbf{C}(n)]\mathbf{K}(n, n - 1) \tag{7.107}$$

We next make the following observation. Suppose that instead of the state equations (7.99) and (7.100), we are given the alternative state-space model

$$\mathbf{x}(n + 1) = \mathbf{F}(n + 1, n)\mathbf{x}(n) + \mathbf{v}_1(n) + \mathbf{d}(n) \tag{7.108}$$

$$\mathbf{y}(n) = \mathbf{C}(n)\mathbf{x}(n) + \mathbf{v}_2(n) \tag{7.109}$$

where $\mathbf{d}(n)$ is a known (i.e., nonrandom) vector. In this case, it is easily verified that the same Kalman equations (7.103) through (7.107) apply except for a modification in the first equation (7.102), which now reads as follows:

$$\hat{\mathbf{x}}(n + 1|\mathcal{Y}_n) = \mathbf{F}(n + 1, n)\hat{\mathbf{x}}(n|\mathcal{Y}_n) + \mathbf{d}(n) \tag{7.110}$$

This modification arises in the derivation of the extended Kalman filter, as discussed in the sequel.

As mentioned previously, the extended Kalman filter (EKF) is an *approximate* solution that allows us to extend the Kalman filtering idea to *nonlinear* state-space models (Jazwinski, 1970; Maybeck, 1982; Ljung and Söderstrom, 1983). In particular, the nonlinear model considered here has the following form:

$$\mathbf{x}(n + 1) = \mathbf{F}(n, \mathbf{x}(n)) + \mathbf{v}_1(n) \tag{7.111}$$

$$\mathbf{y}(n) = \mathbf{C}(n, \mathbf{x}(n)) + \mathbf{v}_2(n) \tag{7.112}$$

where, as before, $\mathbf{v}_1(n)$ and $\mathbf{v}_2(n)$ are uncorrelated zero-mean white-noise processes with correlation matrices $\mathbf{Q}_1(n)$ and $\mathbf{Q}_2(n)$, respectively. Here, however, the functional

$\mathbf{F}(n,\mathbf{x}(n))$ denotes a nonlinear transition matrix function that is possibly time-variant. In the linear case, we simply have

$$\mathbf{F}(n,\mathbf{x}(n)) = \mathbf{F}(n+1,n)\mathbf{x}(n)$$

But in a general nonlinear setting, the entries of the state vector $\mathbf{x}(n)$ may be combined nonlinearly by the action of the functional $\mathbf{F}(n,\mathbf{x}(n))$. Moreover, this nonlinear operation may vary with time. Likewise, the functional $\mathbf{C}(n,\mathbf{x}(n))$ denotes a *nonlinear measurement matrix* that may be time-variant too.

As an example, consider the following two-dimensional nonlinear state-space model:

$$\begin{bmatrix} x_1(n+1) \\ x_2(n+1) \end{bmatrix} = \begin{bmatrix} x_1(n) + x_2^2(n) \\ nx_1(n) - x_1(n)x_2(n) \end{bmatrix} + \begin{bmatrix} v_{1,1}(n) \\ v_{1,2}(n) \end{bmatrix}$$

$$y(n) = x_1(n)x_2^2(n) + v_2(n)$$

In this example, we have

$$\mathbf{F}(n,\mathbf{x}(n)) = \begin{bmatrix} x_1(n) + x_2^2(n) \\ nx_1(n) - x_1(n)x_2(n) \end{bmatrix}$$

and

$$\mathbf{C}(n,\mathbf{x}(n)) = x_1(n)x_2^2(n)$$

The basic idea of the extended Kalman filter is to *linearize* the state-space model of Eqs. (7.111) and (7.112) at each time instant around the most recent state estimate, which is taken to be either $\hat{\mathbf{x}}(n|\mathcal{Y}_n)$ or $\hat{\mathbf{x}}(n|\mathcal{Y}_{n-1})$, depending on which particular functional is being considered. Once a linear model is obtained, the standard Kalman filter equations are applied.

More explicitly, the approximation proceeds in two stages.

Stage 1. The following two matrices are constructed

$$\mathbf{F}(n+1,n) = \left.\frac{\partial \mathbf{F}(n,\mathbf{x})}{\partial \mathbf{x}}\right|_{\mathbf{x}=\hat{\mathbf{x}}(n|\mathcal{Y}_n)} \tag{7.113}$$

and

$$\mathbf{C}(n) = \left.\frac{\partial \mathbf{C}(n,\mathbf{x})}{\partial \mathbf{x}}\right|_{\mathbf{x}=\hat{\mathbf{x}}(n|\mathcal{Y}_{n-1})} \tag{7.114}$$

That is, the *ij*th entry of $\mathbf{F}(n+1,n)$ is equal to the partial derivative of the *i*th component of $\mathbf{F}(n,\mathbf{x})$ *with respect to the j*th component of $\mathbf{x}$. Likewise, the *ij*th entry of $\mathbf{C}(n)$ is equal to the partial derivative of the *i*th component of $\mathbf{C}(n,\mathbf{x})$ with respect to the *j*th component of $\mathbf{x}$. In the former case, the derivatives are evaluated at $\hat{\mathbf{x}}(n|\mathcal{Y}_n)$, while in the latter case the derivatives are evaluated at $\hat{\mathbf{x}}(n|\mathcal{Y}_{n-1})$. The entries of the matrices $\mathbf{F}(n+1,n)$ and $\mathbf{C}(n)$ are all known (i.e., computable), since $\hat{\mathbf{x}}(n|\mathcal{Y}_n)$ and $\hat{\mathbf{x}}(n|\mathcal{Y}_{n-1})$ are made available as described later.

Applying the definitions of Eqs. (7.113) and (7.114) to the previous example, we get

$$\frac{\partial \mathbf{F}(n, \mathbf{x})}{\partial \mathbf{x}} = \begin{bmatrix} 1 & 2x_2 \\ n - x_2 & -x_1 \end{bmatrix}$$

$$\frac{\partial \mathbf{C}(n, \mathbf{x})}{\partial \mathbf{x}} = [x_2^2 \quad 2x_1 x_2]$$

which leads to

$$\mathbf{F}(n+1, n) = \begin{bmatrix} 1 & 2\hat{x}_2(n|\mathcal{Y}_n) \\ n - \hat{x}_2(n|\mathcal{Y}_n) & -\hat{x}_1(n|\mathcal{Y}_n) \end{bmatrix}$$

and

$$\mathbf{C}(n) = [\hat{x}_2^2(n|\mathcal{Y}_{n-1}) \quad 2\hat{x}_1(n|\mathcal{Y}_{n-1})\, \hat{x}_2(n|\mathcal{Y}_{n-1})]$$

Stage 2. Once the matrices $\mathbf{F}(n+1,n)$ and $\mathbf{C}(n)$ are evaluated, they are then employed in a *first-order Taylor approximation* of the nonlinear functionals $\mathbf{F}(n,\mathbf{x}(n))$ and $\mathbf{C}(n,\mathbf{x}(n))$ around $\hat{\mathbf{x}}(n|\mathcal{Y}_n)$ and $\hat{\mathbf{x}}(n|\mathcal{Y}_{n-1})$, respectively. Specifically, $\mathbf{F}(n,\mathbf{x}(n))$ and $\mathbf{C}(n,\mathbf{x}(n))$ are approximated as follows, respectively:

$$\mathbf{F}(n,\mathbf{x}(n)) \simeq \mathbf{F}(n, \hat{\mathbf{x}}(n|\mathcal{Y}_n)) + \mathbf{F}(n+1,n)\,[\mathbf{x}(n) - \hat{\mathbf{x}}(n|\mathcal{Y}_n)] \tag{7.115}$$

$$\mathbf{C}(n, \mathbf{x}(n)) \simeq \mathbf{C}(n, \hat{\mathbf{x}}(n|\mathcal{Y}_{n-1})) + \mathbf{C}(n)[\mathbf{x}(n) - \hat{\mathbf{x}}(n|\mathcal{Y}_{n-1})] \tag{7.116}$$

With the above approximate expressions at hand, we may now proceed to approximate the nonlinear state-equations (7.111) and (7.112) as shown by, respectively,

$$\mathbf{x}(n+1) \simeq \mathbf{F}(n+1, n)\mathbf{x}(n) + \mathbf{v}_1(n) + \mathbf{d}(n) \tag{7.117}$$

$$\bar{\mathbf{y}}(n) \simeq \mathbf{C}(n)\mathbf{x}(n + \mathbf{v}_2(n) \tag{7.118}$$

where we have introduced two new quantities:

$$\bar{\mathbf{y}}(n) = \mathbf{y}(n) - [\mathbf{C}(n,\hat{\mathbf{x}}(n|\mathcal{Y}_{n-1})) - \mathbf{C}(n)\hat{\mathbf{x}}(n|\mathcal{Y}_{n-1})] \tag{7.119}$$

and

$$\mathbf{d}(n) = \mathbf{F}(n,\hat{\mathbf{x}}(n|\mathcal{Y}_n)) - \mathbf{F}(n+1,n)\hat{\mathbf{x}}(n|\mathcal{Y}_n) \tag{7.120}$$

The entries in the term $\bar{\mathbf{y}}(n)$ are all known at time n, and, therefore, $\bar{\mathbf{y}}(n)$ can be regarded as an observation vector at time n. Likewise, the entries in the term $\mathbf{d}(n)$ are all known at time n.

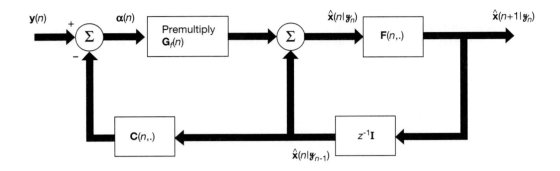

Figure 7.6   One-step predictor for the extended Kalman filter.

The approximate state-space model of Eqs. (7.117) and (7.118) is a linear model of the same mathematical form as that described in Eqs. (7.108) and (7.109); indeed, it is with this objective in mind that earlier on we formulated the state-space model of Eqs. (7.108) and (7.109). The extended Kalman filter equations simply correspond to applying the standard Kalman equations (7.103) through (7.109) and (7.110) to the above linear model. This leads to the following set of equations:

$$\hat{\mathbf{x}}(n+1|\mathcal{Y}_n) = \mathbf{F}(n+1,n)\hat{\mathbf{x}}(n|\mathcal{Y}_n) + \mathbf{d}(n)$$

$$= \mathbf{F}(n+1,n)\hat{\mathbf{x}}(n|\mathcal{Y}_n) + [\mathbf{F}(n,\hat{\mathbf{x}}(n|\mathcal{Y}_n)) - \mathbf{F}(n+1,n)\hat{\mathbf{x}}(n|\mathcal{Y}_n)]$$

$$= \mathbf{F}(n,\hat{\mathbf{x}}(n|\mathcal{Y}_n)) \tag{7.121}$$

$$\hat{\mathbf{x}}(n|\mathcal{Y}_n) = \hat{\mathbf{x}}(n|\mathcal{Y}_{n-1}) + \mathbf{G}_f(n)\boldsymbol{\alpha}(n)$$

$$\boldsymbol{\alpha}(n) = \bar{\mathbf{y}}(n) - \mathbf{C}(n)\hat{\mathbf{x}}(n|\mathcal{Y}_{n-1})$$

$$= \mathbf{y}(n) - \mathbf{C}(n,\hat{\mathbf{x}}(n|\mathcal{Y}_{n-1})) + \mathbf{C}(n)\hat{\mathbf{x}}(n|\mathcal{Y}_{n-1}) - \mathbf{C}(n)\hat{\mathbf{x}}(n|\mathcal{Y}_{n-1})$$

$$= \mathbf{y}(n) - \mathbf{C}(n,\hat{\mathbf{x}}(n|\mathcal{Y}_{n-1})) \tag{7.122}$$

On the basis of Eqs. (7.121) and (7.122), we may formulate the signal-flow graph of Fig. 7.6 for updating the one-step prediction in the extended Kalman filter.

In Table 7.5 we present a summary of the extended Kalman filtering algorithm, where the linearized matrices $\mathbf{F}(n+1, n)$ and $\mathbf{C}(n)$ are computed from their respective nonlinear counterparts using Eqs. (7.113) and (7.114). Given a nonlinear state-space model of the form described in Eqs. (7.111) and (7.112), we may thus use this algorithm to compute state estimates recursively. Comparing the equations of the extended Kalman filter

TABLE 7.5   SUMMARY OF THE EXTENDED KALMAN FILTER

Input vector process
   Observations $= \{\mathbf{y}(1), \mathbf{y}(2), \ldots, \mathbf{y}(n)\}$
Known parameters
   Nonlinear state transition matrix $= \mathbf{F}(n, \mathbf{x}(n))$
   Nonlinear measurement matrix $= \mathbf{C}(n, \mathbf{x}(n))$
   Correlation matrix of process noise vector $= \mathbf{Q}_1(n)$
   Correlation matrix of measurement noise vector $= \mathbf{Q}_2(n)$
Computation: $n = 1, 2, 3, \ldots$.

$$\mathbf{G}_f(n) = \mathbf{K}(n, n-1)\mathbf{C}^H(n)[\mathbf{C}(n)\mathbf{K}(n, n-1)\mathbf{C}^H(n) + \mathbf{Q}_2(n)]^{-1}$$

$$\boldsymbol{\alpha}(n) = \mathbf{y}(n) - \mathbf{C}(n,\hat{\mathbf{x}}(n|\mathcal{Y}_{n-1}))$$

$$\hat{\mathbf{x}}(n|\mathcal{Y}_n) = \hat{\mathbf{x}}(n|\mathcal{Y}_{n-1}) + \mathbf{G}_f(n)\boldsymbol{\alpha}(n)$$

$$\hat{\mathbf{x}}(n+1|\mathcal{Y}_n) = \mathbf{F}(n,\hat{\mathbf{x}}(n|\mathcal{Y}_n))$$

$$\mathbf{K}(n) = [\mathbf{I} - \mathbf{G}_f(n)\mathbf{C}(n)]\mathbf{K}(n, n-1)$$

$$\mathbf{K}(n+1,n) = \mathbf{F}(n+1,n)\mathbf{K}(n)\mathbf{F}^H(n+1,n) + \mathbf{Q}_1(n)$$

Note: The linearized matrices $\mathbf{F}(n+1,n)$ and $\mathbf{C}(n)$ are computed from their nonlinear counterparts $\mathbf{F}(n,\mathbf{x}(n))$ and $\mathbf{C}(n,\mathbf{x}(n))$ using Eqs. (7.113) and (7.114), respectively.
Initial conditions
   $$\hat{\mathbf{x}}(1|\mathcal{Y}_0) = E[\mathbf{x}(1)]$$
   $$\mathbf{K}(1,0) = E[(\mathbf{x}(1) - E[\mathbf{x}(1)])(\mathbf{x}(1) - E[\mathbf{x}(1)])^H] = \Pi_0$$

summarized herein with those of the standard Kalman filter given in Eqs. (7.102) through (7.107), we see that the only differences between them arise in the computations of the innovations vector $\boldsymbol{\alpha}(n)$ and the updated estimate $\hat{\mathbf{x}}(n+1|\mathcal{Y}_n)$. Specifically, the linear terms $\mathbf{F}(n+1,n)\hat{\mathbf{x}}(n|\mathcal{Y}_n)$ and $\mathbf{C}(n)\hat{\mathbf{x}}(n|\mathcal{Y}_{n-1})$ in the standard Kalman filter are replaced by the approximate terms $\mathbf{F}(n,\hat{\mathbf{x}}(n|\mathcal{Y}_n))$ and $\mathbf{C}(n,\hat{\mathbf{x}}(n|\mathcal{Y}_{n-1}))$, respectively, in the extended Kalman filter. These differences also show up in comparing the signal-flow graph of Fig. 7.3 for one-step prediction in the standard Kalman filter with that of Fig. 7.6 for one-step prediction in the extended Kalman filter.

7.10 SUMMARY AND DISCUSSION

The Kalman filter is a linear, discrete-time, finite-dimensional system, the implementation of which is well suited for a digital computer. A key property of the Kalman filter is that it leads to minimization of the trace of the filtered state error correlation matrix $\mathbf{K}(n)$. This, in turn, means that the Kalman filter is the *linear minimum variance estimator* of the state vector $\mathbf{x}(n)$ (Anderson and Moore, 1979; Goodwin and Sin, 1984).

The Kalman filter has been successfully applied to solve many real-world problems as can be seen in the literature on control systems (Sorenson, 1985). Moreover, the Kalman filter provides the general framework for deriving *all* of the known algorithms that constitute the recursive least-squares family of adaptive filters (Sayed and Kailath, 1994). In the intervening two decades between the paper by Sayed and Kailath and the seminal paper by Godard in 1974, many attempts were made to incorporate this important family of adaptive filtering algorithms into the framework of Kalman filter theory. However, some annoying discrepancies always remained, thereby hindering the full application of the extensive control literature on Kalman filters to adaptive filtering problems. For the first time, the paper by Sayed and Kailath has shown us how to devise a state-space model for the adaptive filtering problem that is a perfect match for the application of Kalman filter theory. It has been said by many that many of the problems encountered in signal processing and control theory are mathematically equivalent. The link between Kalman filter theory and adaptive filter theory demonstrated in the paper by Sayed and Kailath is further testimony to the validity of this mathematical equivalence.

PROBLEMS

1. The Gram–Schmidt orthogonalization procedure enables the set of observation vectors $\mathbf{y}(1)$, $\mathbf{y}(2), \ldots, \mathbf{y}(n)$ to be transformed into the set of innovations processes $\boldsymbol{\alpha}(1), \boldsymbol{\alpha}(2), \ldots, \boldsymbol{\alpha}(n)$ without loss of information, and vice versa. Illustrate this procedure for $n = 2$, and comment on the procedure for $n > 2$.

2. The predicted state-error vector is defined by

$$\boldsymbol{\epsilon}(n, n - 1) = \mathbf{x}(n) - \hat{\mathbf{x}}(n|\mathcal{Y}_{n-1})$$

where $\hat{\mathbf{x}}(n|\mathcal{Y}_{n-1})$ is the minimum mean-square estimate of the state $\mathbf{x}(n)$, given the space $\mathcal{Y}_{n-1}$ that is spanned by the observed data $\mathbf{y}(1), \ldots, \mathbf{y}(n - 1)$. Let $\mathbf{v}_1(n)$ and $\mathbf{v}_2(n)$ denote the process noise and measurement noise vectors, respectively. Show that $\boldsymbol{\epsilon}(n, n - 1)$ is orthogonal to both $\mathbf{v}_1(n)$ and $\mathbf{v}_2(n)$; that is,

$$E[\boldsymbol{\epsilon}(n, n - 1)\mathbf{v}_1^H(n)] = \mathbf{O}$$

and

$$E[\boldsymbol{\epsilon}(n, n - 1)\mathbf{v}_2^H(n)] = \mathbf{O}$$

3. Consider a set of scalar observations $y(n)$ of zero mean, which is transformed into the corresponding set of innovations $\alpha(n)$ of zero mean and variance $\sigma_\alpha^2(n)$. Let the estimate of the state vector $\mathbf{x}(i)$, given this set of data, be expressed as

$$\hat{\mathbf{x}}(i|\mathcal{Y}_n) = \sum_{k=1}^{n} \mathbf{b}_i(k)\alpha(k)$$

where $\mathcal{Y}_n$ is the space spanned by $y(1), \ldots, y(n)$, and $\mathbf{b}_i(k)$, $k = 1, 2, \ldots, n$, is a set of vectors to be determined. The requirement is to choose the $\mathbf{b}_i(k)$ so as to minimize the expected value of the squared norm of the estimated state-error vector

$$\boldsymbol{\epsilon}(i|\mathcal{Y}_n) = \mathbf{x}(i) - \hat{\mathbf{x}}(i|\mathcal{Y}_n)$$

Show that this minimization yields the result

$$\hat{\mathbf{x}}(i|\mathcal{Y}_n) = \sum_{k=1}^{n} E[\mathbf{x}(i)\phi^*(k)]\phi(k)$$

where $\phi(k)$ is the normalized innovation

$$\phi(k) = \frac{\alpha(k)}{\sigma_\alpha(k)}$$

This result may be viewed as a special case of Eqs. (7.37) and (7.40).

4. The Kalman gain $\mathbf{G}(n)$ defined in Eq. (7.49) involves the inverse matrix $\mathbf{R}^{-1}(n)$. The matrix $\mathbf{R}(n)$ is itself defined in Eq. (7.35), reproduced here for convenience:

$$\mathbf{R}(n) = \mathbf{C}(n)\mathbf{K}(n, n-1)\mathbf{C}^H(n) + \mathbf{Q}_2(n)$$

The matrix $\mathbf{C}(n)$ is nonnegative definite but not necessarily nonsingular.
(a) Why is $\mathbf{R}(n)$ nonnegative definite?
(b) What prior condition would you impose on the matrix $\mathbf{Q}_2(n)$ to ensure that the inverse matrix $\mathbf{R}^{-1}(n)$ exists?

5. In many cases the predicted state-error correlation matrix $\mathbf{K}(n+1, n)$ converges to the steady-state value $\mathbf{K}$ as the number of iterations n approaches infinity. Show that the limiting value $\mathbf{K}$ satisfies the *algebraic Riccati equation*

$$\mathbf{KC}^H(\mathbf{CKC}^H + \mathbf{Q}_2)^{-1}\mathbf{CK} - \mathbf{Q}_1 = \mathbf{O}$$

where it is assumed that the state transition matrix equals the identity matrix, and the matrices $\mathbf{C}$, $\mathbf{Q}_1$, and $\mathbf{Q}_2$ are the limiting values of $\mathbf{C}(n)$, $\mathbf{Q}_1(n)$, and $\mathbf{Q}_2(n)$, respectively.

6. Consider a stochastic process $y(n)$ represented by an autoregressive-moving average (ARMA) model of order (1, 1):

$$y(n) + ay(n-1) = v(n) + bv(n-1)$$

where a and b are the ARMA parameters and $v(n)$ is a zero-mean white-noise process of variance σ^2.
(a) Show that a state-space representation of this model is

$$\mathbf{x}(n+1) = \begin{bmatrix} -a & 1 \\ 0 & 0 \end{bmatrix}\mathbf{x}(n) + \begin{bmatrix} 1 \\ b \end{bmatrix}v(n+1)$$

$$y(n) = [1 \quad 0]\mathbf{x}(n)$$

where $\mathbf{x}(n)$ is a 2-by-1 state vector.
(b) Assume the applicability of the algebraic Riccati equation described in Problem 5. Hence, show that the solution of this equation is

$$\mathbf{K} = \sigma^2 \begin{bmatrix} 1+c & b \\ b & b^2 \end{bmatrix}$$

where c is a scalar that satisfies the second-order equation

$$c = (b-a)^2 + a^2c - \frac{(b-a-ac)^2}{1+c}$$

What are the two values of c that satisfy this equation? Determine the corresponding values of the matrix $\mathbf{K}$.

(c) Show that the Kalman gain is

$$\mathbf{G} = \frac{b - a - ac}{1 + c} \begin{bmatrix} 1 \\ 0 \end{bmatrix}$$

Determine the values of $\mathbf{G}$ that correspond to the solutions for the scalar c found in part (b).

7. In this problem we consider the general case of *time-varying real-valued ARMA process* $y(n)$ described by the difference equation:

$$y(n) + \sum_{k=1}^{M} a_k(n)y(n - k) = \sum_{k=1}^{N} a_{M+k}(n)v(n - k) + v(n)$$

where $a_1(n), a_2(n), \ldots, a_M(n), a_{M+1}(n), a_{M+2}(n), \ldots, a_{M+N}(n)$ are the ARMA coefficients, the process $v(n)$ is the input, and process $y(n)$ is the output. The process $v(n)$ is a white Gaussian noise process of zero mean and variance σ^2. The ARMA coefficients are subject to random fluctuations, as shown in the model

$$a_k(n + 1) = a_k(n) + w_k(n), \quad k = 1, \ldots, M + N$$

where $w_k(n)$ is a zero-mean, white Gaussian noise process that is independent of $w_j(n)$ for $j \neq k$, and also independent of $v(n)$. The issue of interest is to provide a technique based on the Kalman filter for identifying the coefficients of the ARMA process. To do this, we define an $(M + N)$-dimensional state vector:

$$\mathbf{x}(n) = [a_1(n), \ldots, a_M(n), \ldots, a_{M+N}(n)]^T$$

We also define the measurement matrix (actually, a row vector):

$$\mathbf{C}(n) = [-y(n - 1), \ldots, -y(n - M), v(n - 1), \ldots, v(n - N)]$$

On this basis, do the following:

(a) Formulate the state-space equations for the ARMA process.
(b) Find an algorithm for computing the predicted value of the state vector $\mathbf{x}(n + 1)$, given the observation $y(n)$.
(c) How would you initialize the algorithm in part (b)?

8. Consider a communication channel modeled as an FIR filter of known impulse response. The channel output $y(n)$ is defined by

$$y(n) = \mathbf{h}^T\mathbf{x}(n) + w(n)$$

where $\mathbf{h}$ is an M-by-1 vector representing the channel impulse response, $\mathbf{x}(n)$ is an M-by-1 vector representing the present value $u(n)$ of the channel input and $(M - 1)$ previous transmissions, and $w(n)$ is a white Gaussian noise process of zero mean and variance σ_w^2. At time n, the channel input $u(n)$ consists of a coded binary sequence of zeros and ones, statistically independent of $w(n)$. This model suggests that we may view $\mathbf{x}(n)$ as a state vector, in which case the state equation is written as[4]

$$\mathbf{x}(n + 1) = \mathbf{A}\mathbf{x}(n) + \mathbf{b}v(n)$$

---

[4]This problem is adapted from Lawrence and Kaufman (1971).

where $v(n)$ is a white Gaussian noise process of zero mean and variance σ_v^2, which is indepen-dent of $w(n)$. The matrix $\mathbf{A}$ is an M-by-M matrix whose ijth element is defined by

$$a_{ij} = \begin{cases} 1, & i = j + 1 \\ 0, & \text{otherwise} \end{cases}$$

The vector $\mathbf{b}$ is an M-by-1 vector whose ith element is defined by

$$b_i = \begin{cases} 1, & i = 1 \\ 0, & i = 2, \ldots, M \end{cases}$$

We may now state the problem: Given the foregoing channel model and a sequence $y(n)$ of noisy measurements made at the channel output, use the Kalman filter to construct an equalizer that yields a good estimate of the channel input $u(n)$ at some delayed time $(n + D)$, where $0 \le D \le M - 1$. Show that the equalizer so constructed is an IIR filter where coefficients are determined by two distinct sets of parameters: (a) the M-by-1 channel impulse response vector, and (b) the Kalman gain, which (in this problem) is an M-by-1 vector.

9. For the case when the transition matrix $\mathbf{F}(n + 1, n)$ is the identity matrix and the state noise vec-tor is zero, show that the predicted state-error correlation matrix $\mathbf{K}(n + 1, n)$ and the filtered state-error correlation matrix $\mathbf{K}(n)$ are equal.

10. Using the initial conditions described in Eqs. (7.72) and (7.73), show that the resulting filtered estimate $\hat{\mathbf{x}}(n \,|\, \mathcal{Y}_n)$ produced by the Kalman filter is unbiased; that is,

$$E[\hat{\mathbf{x}}(n) \,|\, \mathcal{Y}(n)] = \mathbf{x}(n)$$

11. In the UD-factorization algorithm, the filtered state-error correlation matrix $\mathbf{K}(n)$ is expressed as follows

$$\mathbf{K}(n) = \mathbf{U}(n)\mathbf{D}(n)\mathbf{U}^H(n)$$

where $\mathbf{U}(n)$ is an upper triangular matrix with 1's along its main diagonal, and $\mathbf{D}(n)$ is a real diagonal matrix. Let $\lambda_{\max}$ and $\lambda_{\min}$ denote the maximum and minimum eigenvalues of the matrix $\mathbf{K}(n)$. Show that the condition number of the diagonal matrix $\mathbf{D}(n)$ is governed by

$$\chi(\mathbf{D}) \ge \frac{\lambda_{\max}}{\lambda_{\min}} = \chi(\mathbf{K})$$

12. Let $\chi(\mathbf{K})$ denote the condition number of the filtered state-error correlation matrix $\mathbf{K}(n)$, defined as the ratio of the largest eigenvalue $\lambda_{\max}$ to the smallest eigenvalue $\lambda_{\min}$. Show that

$$\chi(\mathbf{K}) = (\chi(\mathbf{K}^{1/2}))^2$$

where $\mathbf{K}^{1/2}(n)$ is the square-root of $\mathbf{K}(n)$. What is the computational implication of this relation?

13. Consider the state-space model described in Eqs. (7.108) and (7.109). Show that the one-step prediction $\mathbf{x}(n + 1 \,|\, \mathcal{Y}_n)$ of the state vector in this model is given by Eq. (7.110).

14. (a) Figures 7.3 and 7.6 are signal-flow graph representations of the one-step predictor for the standard Kalman filter and extended Kalman filter, respectively. Show that for a linear model of a dynamical system, these two representations are equivalent.

(b) Figure 7.5 shows a block diagram representation of the standard Kalman filter. How is this block diagram modified for representation of the extended Kalman filter?

PART 3
Linear Adaptive Filtering

Part III, by far the largest portion of the book, consists of Chapters 8 through 17. It is devoted to a detailed treatment of linear finite-duration impulse response (FIR) adaptive filters.

In Chapter 8, we develop the method of steepest descent for computing the tap-weight vector of the Wiener filter in a recursive fashion. In Chapter 9, we use the method of steepest descent to derive the least-mean-square (LMS) algorithm and study its important characteristics. Chapter 10 discusses frequency-domain adaptive filters, designed to extend the utility of the LMS algorithm.

Chapter 11 covers the fundamentals of linear least-squares estimation. In this chapter we also develop the singular value decomposition (SVD), which provides a powerful tool for solving the linear least-square estimation problem and related ones. Chapter 12 discusses the unitary rotations that are basic to the design of square-root Kalman and least-squares filters.

In Chapter 13, we derive the standard recursive least squares (RLS) algorithm, which may be viewed as a special case of the Kalman filter. In Chapter 14, we study square-root variants of the RLS algorithm. Chapter 15 discusses order-recursive least-squares filters, built around the lattice structure.

In Chapter 16 we study the tracking performance of linear adaptive filters. Part III finishes with Chapter 17 on finite precision (numerical) effects that arise when linear adaptive filters are implemented on a general-purpose or special-purpose digital machine.

CHAPTER

8

Method of Steepest Descent

In this chapter we begin our study of gradient-based adaptation by describing an old optimization technique known as the *method of steepest descent*. This method is basic to the understanding of the various ways in which gradient-based adaptation is implemented in practice. The method of steepest descent is *recursive* in the sense that starting from some initial (arbitrary) value for the tap-weight vector, it improves with the increased number of iterations. The final value so computed for the tap-weight vector converges to the Wiener solution. The important point to note is that the method of steepest descent is descriptive of a *deterministic feedback system* that finds the minimum point of the ensemble-averaged error-performance surface without knowledge of the surface itself. Accordingly, it provides some heuristics for writing the recursions that describe the least-mean-square (LMS) algorithm, an issue that is taken up in the next chapter.

8.1 SOME PRELIMINARIES

Consider a transversal filter with tap inputs $u(n)$, $u(n-1)$, ..., $u(n-M+1)$ and a corresponding set of *tap weights* $w_0(n)$, $w_1(n)$, ..., $w_{M-1}(n)$. The tap inputs represent samples drawn from a wide-sense stationary stochastic process of zero mean and correlation matrix **R**. In addition to these inputs, the filter is supplied with a *desired response $d(n)$* that provides a frame of reference for the optimum filtering action. Figure 8.1 depicts the filtering action described herein.

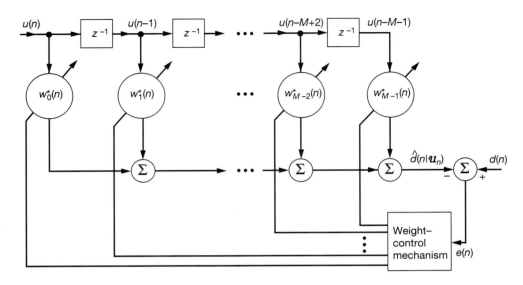

Figure 8.1   Structure of adaptive transversal filter.

The vector of tap inputs at time n is denoted by $\mathbf{u}(n)$, and the corresponding *estimate* of the desired response at the filter output is denoted by $\hat{d}(n|\mathcal{U}_n)$, where $\mathcal{U}_n$ is the space spanned by the tap inputs $u(n), u(n-1), \ldots, u(n-M+1)$. By comparing this estimate with the desired response $d(n)$, we produce an *estimation error* denoted by $e(n)$. We may thus write

$$e(n) = d(n) - \hat{d}(n|\mathcal{U}_n)$$

$$= d(n) - \mathbf{w}^H(n)\mathbf{u}(n) \tag{8.1}$$

where the term $\mathbf{w}^H(n)\mathbf{u}(n)$ is the inner product of the tap-weight vector $\mathbf{w}(n)$ and the tap-input vector $\mathbf{u}(n)$. The expanded form of the tap-weight vector is described by

$$\mathbf{w}(n) = [w_0(n), w_1(n), \ldots, w_{M-1}(n)]^T \tag{8.2}$$

and that of the tap-input vector is described by

$$\mathbf{u}(n) = [u(n), u(n-1), \ldots, (n-M+1)]^T \tag{8.3}$$

If the tap-input vector $\mathbf{u}(n)$ and the desired $d(n)$ are jointly stationary, then the *mean-squared error* or *cost function* $J(n)$ at time n is a quadratic function of the tap-weight vector, so we may write [see Eq. (5.50)]

$$J(n) = \sigma_d^2 - \mathbf{w}^H(n)\mathbf{p} - \mathbf{p}^H\mathbf{w}(n) + \mathbf{w}^H(n)\mathbf{R}\mathbf{w}(n) \tag{8.4}$$

where σ_d^2 = variance of the desired response $d(n)$

$\mathbf{p}$ = cross-correlation vector between the tap-input vector $\mathbf{u}(n)$ and the desired response $d(n)$

$\mathbf{R}$ = correlation matrix of the tap-input vector $\mathbf{u}(n)$

Equation (8.4) defines the mean-squared error that would result if the tap-weight vector in the transversal filter were *fixed* at the value $\mathbf{w}(n)$. Since $\mathbf{w}(n)$ varies with time n, it is only natural that the mean-squared error varies with time n in a corresponding fashion, hence the use of $J(n)$ for the mean-squared error in Eq. (8.4). The variation of the mean-squared error $J(n)$ with time n signifies the fact that the estimation error process $e(n)$ is nonstationary.

We may visualize the dependence of the mean-squared error $J(n)$ on the elements of the tap-weight vector $\mathbf{w}(n)$ as a *bowl-shaped surface* with a unique minimum. We refer to this surface as the *error-performance surface* of the adaptive filter. The adaptive process has the task of continually seeking the *bottom* or *minimum point* of this surface. At the minimum point of the error-performance surface, the tap-weight vector takes on the optimum value $\mathbf{w}_o$, which is defined by the Wiener–Hopf equations (5.34), reproduced here for convenience,

$$\mathbf{R}\mathbf{w}_o = \mathbf{p} \tag{8.5}$$

The minimum mean-square error equals [see Eq. (5.49)]

$$J_{\min} = \sigma_d^2 - \mathbf{p}^H\mathbf{w}_o \tag{8.6}$$

8.2 STEEPEST-DESCENT ALGORITHM

The requirement that an adaptive transversal filter has to satisfy is to find a solution for its tap-weight vector that satisfies the Wiener–Hopf equations (8.5). One way of doing this would be to solve this system of equations by some analytical means. In general, this procedure is quite straightforward. However, it presents serious computational difficulties, especially when the filter contains a large number of tap weights and when the input data rate is high.

An alternative procedure is to use the *method of steepest descent*, which is one of the oldest methods of optimization.[1] To find the minimum value of the mean-squared error, $J_{\min}$, by the steepest-descent algorithm, we proceed as follows:

1. We begin with an initial value $\mathbf{w}(0)$ for the tap-weight vector, which provides an initial guess as to where the minimum point of the error-performance surface may be located. Unless some prior knowledge is available, $\mathbf{w}(0)$ is usually set equal to the null vector.

2. Using this initial or present guess, we compute the *gradient vector*, the real and imaginary parts of which are defined as the derivative of the mean-squared error

---

[1] The steepest-descent algorithm belongs to a family of *iterative methods of optimization* (Luenberger, 1969); it provides a method of searching a multidimensional performance surface. Another method in this family that may be used for this purpose is *Newton's algorithm*, which is primarily a method for finding the zeros of a function. In the context of adaptive filtering, the use of the method of steepest descent results in an algorithm that is much simpler, but slower, than Newton's method (Widrow and Stearns, 1985).

$J(n)$, evaluated with respect to the real and imaginary parts of the tap-weight vector $\mathbf{w}(n)$ at time n (i.e., the nth iteration).

3. We compute the next guess at the tap-weight vector by making a change in the initial or present guess in a direction opposite to that of the gradient vector.

4. We go back to step 2 and repeat the process.

It is intuitively reasonable that successive corrections to the tap-weight vector in the direction of the negative of the gradient vector (i.e., in the direction of the steepest descent of the error-performance surface) should eventually lead to the minimum mean-squared error $J_{\min}$, at which point the tap-weight vector assumes its optimum value $\mathbf{w}_o$.

Let $\boldsymbol{\nabla}J(n)$ denote the value of the *gradient vector* at time n. Let $\mathbf{w}(n)$ denote the value of the tap-weight vector at time n. According to the method of steepest descent, the updated value of the tap-weight vector at time $n + 1$ is computed by using the simple recursive relation

$$\mathbf{w}(n + 1) = \mathbf{w}(n) + \tfrac{1}{2}\mu[-\boldsymbol{\nabla}J(n)] \tag{8.7}$$

where μ is a positive real-valued constant. The factor $\tfrac{1}{2}$ is used merely for the purpose of canceling a factor 2 that appears in the formula for $\boldsymbol{\nabla}J(n)$; see Eq. (8.8).

From Chapter 4 we find that the gradient vector $\boldsymbol{\nabla}J(n)$ is given by

$$\boldsymbol{\nabla}J(n) = \begin{bmatrix} \dfrac{\partial J(n)}{\partial a_0(n)} + j\,\dfrac{\partial J(n)}{\partial b_0(n)} \\[2mm] \dfrac{\partial J(n)}{\partial a_1(n)} + j\,\dfrac{\partial J(n)}{\partial b_1(n)} \\[2mm] \cdot \\ \cdot \\ \cdot \\[2mm] \dfrac{\partial J(n)}{\partial a_{M-1}(n)} + j\,\dfrac{\partial J(n)}{\partial b_{M-1}(n)} \end{bmatrix} \tag{8.8}$$

$$= -2\mathbf{p} + 2\mathbf{R}\mathbf{w}(n)$$

where, in the expanded column vector, $\partial J(n)/\partial a_k(n)$ and $\partial J(n)/\partial b_k(n)$ are the partial derivatives of the cost function $J(n)$ with respect to the real part $a_k(n)$ and the imaginary part $b_k(n)$ of the kth tap weight $w_k(n)$, respectively, with $k = 1, 2, \ldots, M - 1$. For the application of the steepest-descent algorithm, we assume that in Eq. (8.8) the correlation matrix $\mathbf{R}$ and the cross-correlation vector $\mathbf{p}$ are known so that we may compute the gradient vector $\boldsymbol{\nabla}J(n)$ for a given value of the tap-weight vector $\mathbf{w}(n)$. Thus, substituting Eq. (8.8) in (8.7), we may compute the updated value of the tap-weight vector $\mathbf{w}(n + 1)$ by using the simple recursive relation

$$\mathbf{w}(n + 1) = \mathbf{w}(n) + \mu[\mathbf{p} - \mathbf{R}\mathbf{w}(n)], \quad n = 0, 1, 2, \ldots \tag{8.9}$$

We observe that the parameter μ controls the size of the incremental correction applied to the tap-weight vector as we proceed from one iteration cycle to the next. We therefore refer to μ as the *step-size parameter*. Equation (8.9) describes the mathematical formulation of the steepest-descent algorithm.

According to Eq. (8.9), the correction $\delta\mathbf{w}(n)$ applied to the tap-weight vector at time $n + 1$ is equal to $\mu[\mathbf{p} - \mathbf{R}\mathbf{w}(n)]$. This correction may also be expressed as μ times the expectation of the inner product of the tap-input vector $\mathbf{u}(n)$ and the estimation error $e(n)$; see Problem 4. This suggests that we may use a bank of cross-correlators to compute the correction $\delta\mathbf{w}(n)$ applied to the tap-weight vector $\mathbf{w}(n)$ as indicated in Fig. 8.2. In this figure, the elements of the correction vector $\delta\mathbf{w}(n)$ are denoted by $\delta w_0(n)$, $\delta w_1(n)$, . . . , $\delta w_{M-1}(n)$.

Another point of interest is that we may view the steepest-descent algorithm of Eq. (8.9) as a *feedback model*, as illustrated by the *signal-flow graph* shown in Fig. 8.3. This model is multidimensional in the sense that the "signals" at the *nodes* of the graph consist of vectors and that the *transmittance* of each branch of the graph is a scalar or a square matrix. For each branch of the graph, the signal vector flowing out equals the signal vector flowing in multiplied by the transmittance matrix of the branch. For two branches connected in parallel, the overall transmittance matrix equals the sum of the transmittance matrices of the individual branches. For two branches connected in cascade, the overall transmittance matrix equals the product of the individual transmittance matrices arranged in the same order as the pertinent branches. Finally, the symbol z^{-1} is the unit-delay operator, and $z^{-1}\mathbf{I}$ is the transmittance matrix of a unit-delay branch representing a delay of one iteration cycle.

8.3 STABILITY OF THE STEEPEST-DESCENT ALGORITHM

Since the steepest-descent algorithm involves the presence of *feedback*, as exemplified by the model of Fig. 8.3, the algorithm is subject to the possibility of becoming *unstable*. From the feedback model of Fig. 8.3, we observe that the *stability performance* of the steepest-descent algorithm is determined by two factors: (1) the step-size parameter μ, and (2) the correlation matrix $\mathbf{R}$ of the tap-input vector $\mathbf{u}(n)$, as these two parameters completely control the transfer function of the *feedback loop*. To determine *the condition for the stability* of the steepest-descent algorithm, we examine the *natural modes* of the algorithm (Widrow, 1970). In particular, we use the representation of the correlation matrix $\mathbf{R}$ in terms of its eigenvalues and eigenvectors to define a transformed version of the tap-weight vector.

We begin the analysis by defining a *weight-error vector* at time n as

$$\mathbf{c}(n) = \mathbf{w}(n) - \mathbf{w}_o \tag{8.10}$$

where $\mathbf{w}_o$ is the optimum value of the tap-weight vector, as defined by the Wiener–Hopf equations (8.5). Then, eliminating the cross-correlation vector $\mathbf{p}$ between Eqs. (8.5) and (8.9), and rewriting the result in terms of the weight-error vector $\mathbf{c}(n)$, we get

$$\mathbf{c}(n + 1) = (\mathbf{I} - \mu\mathbf{R})\mathbf{c}(n) \tag{8.11}$$

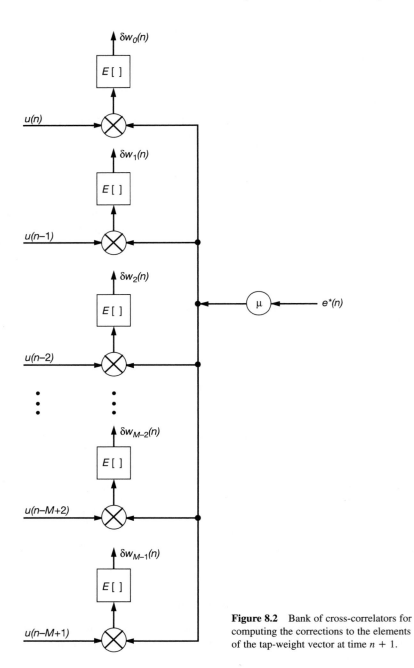

Figure 8.2 Bank of cross-correlators for computing the corrections to the elements of the tap-weight vector at time $n + 1$.

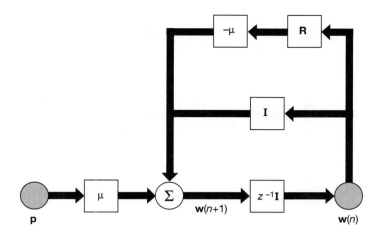

Figure 8.3   Signal-flow-graph representation of the steepest-descent algorithm.

where $\mathbf{I}$ is the identity matrix. Equation (8.11) is represented by the feedback model shown in Fig. 8.4. This diagram further emphasizes the fact that the stability performance of the steepest-descent algorithm is dependent exclusively on μ and $\mathbf{R}$.

Using the unitary similarity transformation, we may express the correlation matrix $\mathbf{R}$ as follows (see Chapter 4):

$$\mathbf{R} = \mathbf{Q}\mathbf{\Lambda}\mathbf{Q}^H \qquad (8.12)$$

The matrix $\mathbf{Q}$ has as its columns an orthogonal set of *eigenvectors* associated with the eigenvalues of the matrix $\mathbf{R}$. The matrix $\mathbf{Q}$ is called the *unitary matrix* of the transformation. The matrix $\mathbf{\Lambda}$ is a *diagonal* matrix and has as its diagonal elements the eigenvalues of the correlation matrix $\mathbf{R}$. These eigenvalues, denoted by $\lambda_1, \lambda_2, \ldots, \lambda_M$, are all positive and real. Each eigenvalue is associated with a corresponding eigenvector or column of matrix $\mathbf{Q}$. Substituting Eq. (8.12) in (8.11), we get

$$\mathbf{c}(n + 1) = (\mathbf{I} - \mu\mathbf{Q}\mathbf{\Lambda}\mathbf{Q}^H)\mathbf{c}(n) \qquad (8.13)$$

Premultiplying both sides of this equation by $\mathbf{Q}^H$ and using the property of the unitary matrix $\mathbf{Q}$ that $\mathbf{Q}^H$ equals the inverse $\mathbf{Q}^{-1}$ (see Chapter 4), we get

$$\mathbf{Q}^H\mathbf{c}(n + 1) = (\mathbf{I} - \mu\mathbf{\Lambda})\mathbf{Q}^H\mathbf{c}(n) \qquad (8.14)$$

We now define a new set of coordinates as follows:

$$\begin{aligned}
\boldsymbol{v}(n) &= \mathbf{Q}^H\mathbf{c}(n) \\
&= \mathbf{Q}^H[\mathbf{w}(n) - \mathbf{w}_o]
\end{aligned} \qquad (8.15)$$

Accordingly, we may rewrite Eq. (8.13) in the transformed form:

$$\boldsymbol{v}(n + 1) = (\mathbf{I} - \mu\mathbf{\Lambda})\boldsymbol{v}(n) \qquad (8.16)$$

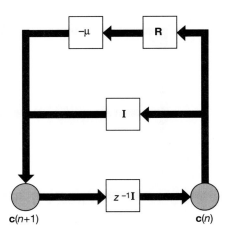

Figure 8.4   Signal-flow-graph representation of the steepest-descent algorithm based on the weight-error vector.

The initial value of $\boldsymbol{v}(n)$ equals

$$\boldsymbol{v}(0) = \mathbf{Q}^H[\mathbf{w}(0) - \mathbf{w}_o] \tag{8.17}$$

Assuming that the initial tap-weight vector is zero, Eq. (8.17) reduces to

$$\boldsymbol{v}(0) = -\mathbf{Q}^H \mathbf{w}_o \tag{8.18}$$

For the kth *natural mode* of the steepest descent algorithm, we thus have

$$v_k(n + 1) = (1 - \mu\lambda_k)v_k(n), \quad k = 1, 2, \ldots, M \tag{8.19}$$

where λ_k is the kth eignevalue of the correlation matrix $\mathbf{R}$. This equation is represented by the scalar-valued feedback model of Fig. 8.5, where z^{-1} is the unit-delay operator. Clearly, the structure of this model is much simpler than that of the original matrix-valued feedback model of Fig. 8.3. These two models represent different and yet equivalent ways of viewing the steepest-descent algorithm.

Equation (8.19) is a homogeneous *difference equation of the first order*. Assuming that $v_k(n)$ has the initial value $v_k(0)$, we readily obtain the solution

$$v_k(n) = (1 - \mu\lambda_k)^n v_k(0), \quad k = 1, 2, \ldots, M \tag{8.20}$$

Since all eignevalues of the correlation matrix $\mathbf{R}$ are positive and real, the response $v_k(n)$ will exhibit no oscillations. Furthermore, as illustrated in Fig. 8.6, the numbers generated by Eq. (8.20) represent a *geometric series* with a geometric ratio equal to $1 - \mu\lambda_k$. For *stability* or *convergence* of the steepest-descent algorithm, the magnitude of this geometric ratio must be less than 1 for all k. That is, provided we have

$$-1 < 1 - \mu\lambda_k < 1 \quad \text{for all } k$$

then as the number of iterations, n, approaches infinity, all the natural modes of the steepest-descent algorithm die out, irrespective of the initial conditions. This is equivalent to saying that the tap-weight vector $\mathbf{w}(n)$ approaches the optimum solution $\mathbf{w}_o$ as n approaches infinity.

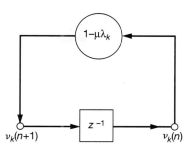

Figure 8.5 Signal-flow-graph representation of the kth natural mode of the steepest-descent algorithm.

Since the eigenvalues of the correlation matrix **R** are all real and positive, it therefore follows that the necessary and sufficient condition for the convergence or stability of the steepest-descent algorithm is to require the step-size parameter μ satisfy the following condition:

$$0 < \mu < \frac{2}{\lambda_{\max}} \tag{8.21}$$

where $\lambda_{\max}$ is the largest eigenvalue of the correlation matrix **R**.

Referring to Fig. 8.6, we see that an exponential envelope of *time constant* τ_k can be fitted to the geometric series by assuming the unit of time to be the duration of one iteration cycle and by choosing the time constant τ_k such that

$$1 - \mu\lambda_k = \exp\left(-\frac{1}{\tau_k}\right)$$

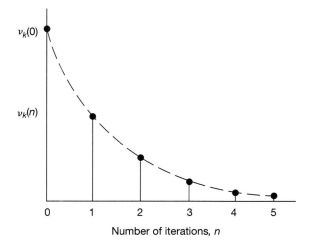

Figure 8.6 Variation of the kth natural mode of the steepest-descent algorithm with time, assuming that the magnitude of $1 - \mu\lambda_k$ is less than 1.

Hence, the kth time constant τ_k can be expressed in terms of the step-size parameter μ and the kth eigenvalue as follows:

$$\tau_k = \frac{-1}{\ln(1 - \mu\lambda_k)} \tag{8.22}$$

The time constant τ_k defines the number of iterations required for the amplitude of the kth natural mode $v_k(n)$ to decay to $1/e$ of its initial value $v_k(0)$, where e is the base of the natural logarithm. For the special case of slow adaptation, for which the step-size parameter μ is small, we may approximate the time constant τ_κ as

$$\tau_k \simeq \frac{1}{\mu\lambda_k}, \qquad \mu << 1 \tag{8.23}$$

We may now formulate the transient behavior of the original tap-weight vector $\mathbf{w}(n)$. In particular, premultiplying both sides of Eq. (8.15) by $\mathbf{Q}$, using the fact that $\mathbf{QQ}^H = \mathbf{I}$, and solving for $\mathbf{w}(n)$, we get

$$\mathbf{w}(n) = \mathbf{w}_o + \mathbf{Q}\mathbf{v}(n)$$

$$= \mathbf{w}_o + [\mathbf{q}_1, \mathbf{q}_2, \ldots, \mathbf{q}_M] \begin{bmatrix} v_1(n) \\ v_2(n) \\ \cdot \\ \cdot \\ \cdot \\ v_M(n) \end{bmatrix} \tag{8.24}$$

$$= \mathbf{w}_o + \sum_{k=1}^{M} \mathbf{q}_k v_k(n)$$

where $\mathbf{q}_1, \mathbf{q}_2, \ldots, \mathbf{q}_M$ are the eigenvectors associated with the eigenvalues $\lambda_1, \lambda_2, \ldots, \lambda_M$ of the correlation matrix $\mathbf{R}$, respectively, and the kth natural mode $v_k(n)$ is defined by Eq. (8.20). Thus, substituting Eq. (8.20) in (8.24), we find that the transient behavior of the ith tap weight is described by (Griffiths, 1975)

$$w_i(n) = w_{oi} + \sum_{k=1}^{N} q_{ki}v_k(0)(1 - \mu\lambda_k)^n, \qquad i = 1, 2, \ldots, M \tag{8.25}$$

where w_{oi} is the optimum value of the ith tap weight, and q_{ki} is the ith element of the kth eigenvector $\mathbf{q}_k$.

Equation (8.25) shows that each tap weight in the steepest-descent algorithm converges as the weighted sum of exponentials of the form $(1 - \mu\lambda_k)^n$. The time τ_k required for each term to reach $1/e$ of its initial value is given by Eq. (8.22). However, the *overall time constant*, τ_a, defined as the time required for the summation term in Eq. (8.25) to decay to $1/e$ of its initial value, cannot be expressed in a simple closed form similar to Eq. (8.22). Nevertheless, the *slowest rate of convergence* is attained when $q_{ki}v_k(0)$ is zero for all k except for that mode corresponding to the smallest eigenvalue $\lambda_{\min}$ of matrix $\mathbf{R}$, so the upper bound on τ_a is defined by $-1/\ln(1 - \mu\lambda_{\min})$. The *fastest rate of convergence* is attained when all the $q_{ki}v_k(0)$ are zero except for that mode corresponding to the largest eigenvalue $\lambda_{\max}$, and so the lower bound on τ_a is defined by $-1/\ln(1 - \mu\lambda_{\max})$. Accordingly, the overall time constant τ_a for any tap weight of the steepest-descent algorithm is bounded as follows (Griffiths, 1975):

$$\frac{-1}{\ln(1 - \mu\lambda_{\max})} \leq \tau_a \leq \frac{-1}{\ln(1 - \mu\lambda_{\min})} \tag{8.26}$$

We see therefore that, when the eigenvalues of the correlation matrix $\mathbf{R}$ are widely spread (i.e., the correlation matrix of the tap inputs is ill conditioned), the settling time of the steepest-descent algorithm is limited by the smallest eigenvalues or the slowest modes.

Transient Behavior of the Mean-Squared Error

We may develop further insight into the operation of the steepest-descent algorithm by examining the transient behavior of the mean-squared error $J(n)$. From Eq. (5.56) we have

$$J(n) = J_{\min} + \sum_{k=1}^{M} \lambda_k |v_k(n)|^2 \tag{8.27}$$

where $J_{\min}$ is the minimum mean-squared error. The transient behavior of the kth natural mode, $v_k(n)$, is defined by Eq. (8.20). Hence substituting Eq. (8.20) into (8.27), we get

$$J(n) = J_{\min} + \sum_{k=1}^{M} \lambda_k (1 - \mu\lambda_k)^{2n} |v_k(0)|^2 \tag{8.28}$$

where $v_k(0)$ is the initial value of $v_k(n)$. When the steepest-descent algorithm is convergent, that is, the step-size parameter μ is chosen within the bounds defined by Eq. (8.21), we see that, irrespective of the initial conditions,

$$\lim_{n \to \infty} J(n) = J_{\min} \tag{8.29}$$

The curve obtained by plotting the mean-squared error $J(n)$ versus the number of iterations, n, is called a *learning curve*. From Eq. (8.28), we see that *the learning curve of the steepest-descent algorithm consists of a sum of exponentials, each of which corresponds to a natural mode of the algorithm.* In general, the number of natural modes equals the number of tap weights. In going from the initial value $J(0)$ to the final value $J_{\min}$, the exponential decay for the kth natural mode has a time constant equal to

$$\tau_{\kappa,\text{mse}} \simeq \frac{-1}{2\ln(1 - \mu\lambda_k)} \tag{8.30}$$

For small values of the step-size parameter μ, we may approximate this time constant as

$$\tau_{\kappa,\text{mse}} \simeq \frac{1}{2\mu\lambda_k} \tag{8.31}$$

Equation (8.31) shows that the smaller the step-size parameter μ, the slower will be the rate of decay of each natural mode of the LMS algorithm.

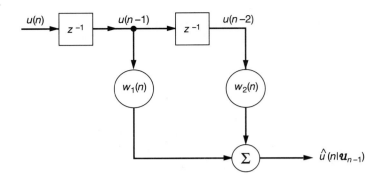

Figure 8.7   Two-tap predictor for real-valued input.

8.4 EXAMPLE

In this example, we examine the transient behavior of the steepest-descent algorithm applied to a predictor that operates on a real-valued autoregressive (AR) process. Figure 8.7 shows the structure of the predictor, assumed to contain two tap weights that are denoted by $w_1(n)$ and $w_2(n)$; the dependence of these tap weights on the number of iterations n emphasizes the transient condition of the predictor. The AR process $u(n)$ is described by the second-order difference equation

$$u(n) + a_1u(n-1) + a_2u(n-2) = v(n) \tag{8.32}$$

where the sample $v(n)$ is drawn from a white-noise process of zero mean and variance σ_v^2. The AR parameters a_1 and a_2 are chosen so that the roots of the characteristic equation

$$1 + a_1z^{-1} + a_2z^{-2} = 0$$

are complex; that is, $a_1^2 < 4a_2$. The particular values assigned to a_1 and a_2 are determined by the desired eigenvalue spread $\chi(\mathbf{R})$. For specified values of a_1 and a_2, the variance σ_v^2 of the white-noise process is chosen to make the process $u(n)$ have variance $\sigma_u^2 = 1$.

The requirement is to evaluate the transient behavior of the steepest-descent algorithm for the following conditions:

- Varying eigenvalue spread $\chi(\mathbf{R})$ and fixed step-size parameter μ.
- Varying step-size parameter μ and fixed eigenvalue spread $\chi(\mathbf{R})$.

Characterization of the AR Process

Since the predictor of Fig. 8.7 has two tap weights and the AR process $u(n)$ is real valued, it follows that the correlation matrix $\mathbf{R}$ of the tap inputs is a 2-by-2 symmetric matrix:

$$\mathbf{R} = \begin{bmatrix} r(0) & r(1) \\ r(1) & r(0) \end{bmatrix}$$

where (see Chapter 2)

$$r(0) = \sigma_u^2$$

$$r(1) = -\frac{a_1}{1 + a_2} \sigma_u^2$$

$$\sigma_u^2 = \left(\frac{1 + a_2}{1 - a_2}\right) \frac{\sigma_v^2}{(1 + a_2)^2 - a_1^2}$$

The two eigenvalues of $\mathbf{R}$ are

$$\lambda_1 = \left(1 - \frac{a_1}{1 + a_2}\right) \sigma_u^2$$

$$\lambda_2 = \left(1 + \frac{a_1}{1 + a_2}\right) \sigma_u^2$$

Hence, the eigenvalue spread equals (assuming that a_1 is negative)

$$\chi(\mathbf{R}) = \frac{\lambda_1}{\lambda_2} = \frac{1 - a_1 + a_2}{1 + a_1 + a_2}$$

The eigenvectors $\mathbf{q}_1$ and $\mathbf{q}_2$ associated with the respective eigenvalues λ_1 and λ_2 are

$$\mathbf{q}_1 = \frac{1}{\sqrt{2}} \begin{bmatrix} 1 \\ 1 \end{bmatrix}$$

$$\mathbf{q}_2 = \frac{1}{\sqrt{2}} \begin{bmatrix} 1 \\ -1 \end{bmatrix}$$

both of which are normalized to unit length.

Experiment 1: Varying Eigenvalue Spread

In this experiment, the step-size parameter μ is fixed at 0.3, and the evaluations are made for the four sets of AR parameters given in Table 8.1.

TABLE 8.1 SUMMARY OF PARAMETER VALUES CHARACTERIZING THE SECOND-ORDER AR MODELING PROBLEM

Case	AR parameters		Eigenvalues		Eigenvalue spread, $\chi = \lambda_1/\lambda_2$	Minimum mean squared error, $J_{min} = \sigma_v^2$
	a_1	a_2	λ_1	λ_2		
1	−0.1950	0.95	1.1	0.9	1.22	0.0965
2	−0.9750	0.95	1.5	0.5	3	0.0731
3	−1.5955	0.95	1.818	0.182	10	0.0322
4	−1.9114	0.95	1.957	0.0198	100	0.0038

For a given set of parameters, we use a two-dimensional plot of the transformed tap-weight error $v_1(n)$ versus $v_2(n)$ to display the transient behavior of the steepest-descent algorithm. In particular, the use of Eq. (8.20) yields

$$\boldsymbol{v}(n) = \begin{bmatrix} v_1(n) \\ v_2(n) \end{bmatrix} \tag{8.33}$$

$$= \begin{bmatrix} (1 - \mu\lambda_1)^n v_1(0) \\ (1 - \mu\lambda_2)^n v_2(0) \end{bmatrix}, \qquad n = 1, 2, \ldots$$

To calculate the initial value $\boldsymbol{v}(0)$, we use Eq. (8.18), assuming that the initial value $\mathbf{w}(0)$ of the tap-weight vector $\mathbf{w}(n)$ is zero. This equation requires knowledge of the optimum tap-weight vector $\mathbf{w}_o$. Now when the two-tap predictor of Fig. 8.7 is optimized, with the second-order AR process of Eq. (8.32) supplying the tap inputs, we find that the optimum tap-weight vector equals

$$\mathbf{w}_o = \begin{bmatrix} -a_1 \\ -a_2 \end{bmatrix}$$

and the minimum mean-squared error equals

$$J_{\min} = \sigma_v^2$$

Accordingly, the use of Eq. (8.18) yields the initial value:

$$\boldsymbol{v}(0) = \begin{bmatrix} v_1(0) \\ v_2(0) \end{bmatrix}$$

$$= \frac{-1}{\sqrt{2}} \begin{bmatrix} 1 & 1 \\ 1 & -1 \end{bmatrix} \begin{bmatrix} -a_1 \\ -a_2 \end{bmatrix} \tag{8.34}$$

$$= \frac{1}{\sqrt{2}} \begin{bmatrix} a_1 + a_2 \\ a_1 - a_2 \end{bmatrix}$$

Thus, for specified parameters, we use Eq. (8.34) to compute the initial value $\boldsymbol{v}(0)$, and then use Eq. (8.33) to compute $\boldsymbol{v}(1), \boldsymbol{v}(2), \ldots$. By joining the points defined by these values of $\boldsymbol{v}(n)$ for varying n, we obtain a *trajectory* that describes the transient behavior of the steepest-descent algorithm for the particular set of parameters.

It is informative to include in the two-dimensional plot of $v_1(n)$ versus $v_2(n)$ loci representing Eq. (8.27) for fixed values of n. For our example, Eq. (8.27) yields

$$J(n) - J_{\min} = \lambda_1 v_1^2(n) + \lambda_2 v_2^2(n) \tag{8.35}$$

When $\lambda_1 = \lambda_2$ and n is fixed, Eq. (8.35) represents a circle with center at the origin and radius equal to the square root of $[J(n) - J_{\min}]/\lambda$, where λ is the common value of the two eigenvalues. When, on the other hand, $\lambda_1 \neq \lambda_2$, Eq. (8.35) represents (for fixed n) an ellipse with major axis equal to the square root of $[J(n) - J_{\min}]/\lambda_2$ and minor axis equal to the square root of $[J(n) - J_{\min}]/\lambda_1$.

Case 1: Eigenvalue Spread $\chi(\mathbf{R}) = 1.22$. For the parameter values given for Case 1 in Table 8.1, the eigenvalue spread $\chi(\mathbf{R})$ equals 1.22; that is, the eigenvalues λ_1 and λ_2 are approximately equal. The use of these parameter values in Eqs. (8.33) and (8.34) yields the trajectory of $[v_1(n), v_2(n)]$ shown in Fig. 8.8(a), with n as running parameter, and their use in Eq. (8.35) yields the (approximately) circular loci shown for fixed values of $J(n)$, corresponding to $n = 0, 1, 2, 3, 4, 5$.

We may also display the transient behavior of the steepest-descent algorithm by plotting the tap weight $w_1(n)$ versus $w_2(n)$. In particular, for our example the use of Eq. (8.24) yields the tap-weight vector

$$\mathbf{w}(n) = \begin{bmatrix} w_1(n) \\ w_2(n) \end{bmatrix}$$

$$= \begin{bmatrix} -a_1 + (v_1(n) + v_2(n))/\sqrt{2} \\ -a_2 + (v_1(n) - v_2(n))/\sqrt{2} \end{bmatrix} \tag{8.36}$$

The corresponding trajectory of $[w_1(n), w_2(n)]$, with n as a running parameter, obtained by using Eq. (8.36), is shown plotted in Fig. 8.9(a). Here again we have included the loci of $[w_1(n), w_2(n)]$ for fixed values of $J(n)$ corresponding to $n = 0, 1, 2, 3, 4, 5$. Note that these loci, unlike Fig. 8.8(a), are ellipsoidal.

Case 2: Eigenvalue Spread $\chi(\mathbf{R}) = 3$. The use of the parameter values for Case 2 in Eqs. (8.33) and (8.34) yields the trajectory of $[v_1(n), v_2(n)]$ shown in Fig. 8.8(b), with n as running parameter, and their use in Eq. (8.35) yields the ellipsoidal loci shown for the fixed values of $J(n)$ for $n = 0, 1, 2, 3, 4, 5$. Note that for this set of parameter values the initial value $v_2(0)$ is approximately zero, so the initial value $\mathbf{v}(0)$ lies practically on the v_1-axis.

The corresponding trajectory of $[w_1(n), w_2(n)]$, with n as running parameter, is shown in Fig. 8.9(b).

Case 3: Eigenvalue Spread $\chi(\mathbf{R}) = 10$. For this case, the application of Eqs. (8.33) and (8.34) yields the trajectory of $[v_1(n), v_2(n)]$ shown in Fig. 8.8(c), with n as running parameter, and the application of Eq. (8.35) yields the ellipsoidal loci included in this figure for fixed values of $J(n)$ for $n = 0, 1, 2, 3, 4, 5$. The corresponding trajectory of $[w_1(n), w_2(n)]$, with n as running parameter, is shown in Fig. 8.9(c).

Case 4: Eigenvalue Spread $\chi(\mathbf{R}) = 100$. For this case, application of the preceding equations yields the results shown in Fig. 8.8(d) for the trajectory of $[v_1(n), v_2(n)]$ and the ellipsoidal loci for fixed values of $J(n)$. The corresponding trajectory of $[w_1(n), w_2(n)]$ is shown in Fig. 8.9(d).

In Fig. 8.10 we have plotted the mean-squared error $J(n)$ versus n for the four eigenvalue spreads 1.22, 3, 10, and 100. We see that as the eigenvalue spread increases (and the input process becomes more correlated), the minimum mean-squared error $J_{\min}$ decreases.

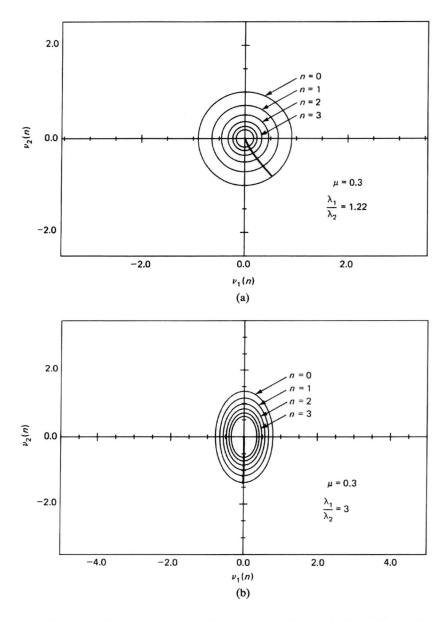

Figure 8.8 Loci of $v_1(n)$ versus $v_2(n)$ for the steepest-descent algorithm with step-size parameter $\mu = 0.3$ and varying eigenvalue spread: (a) $\chi(\mathbf{R}) = 1.22$; (b) $\chi(\mathbf{R}) = 3$; (c) $\chi(\mathbf{R})$ = 10; (d) $\chi(\mathbf{R}) = 100$.

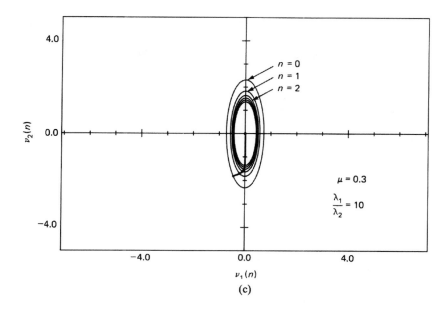

(c)

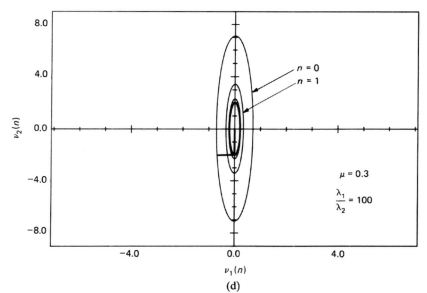

(d)

Figure 8.8   *(Cont.)*

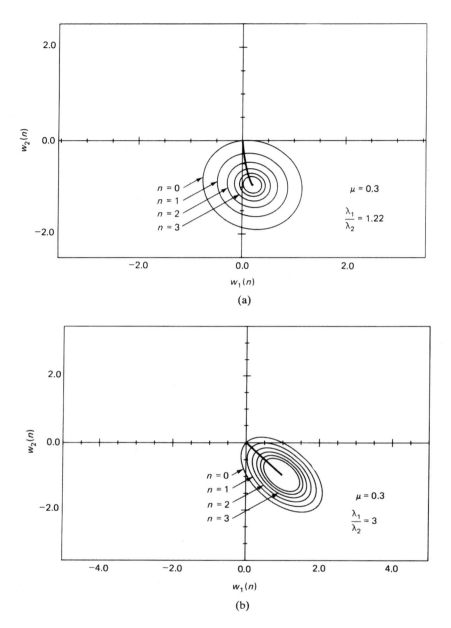

Figure 8.9 Loci of $w_1(n)$ versus $w_2(n)$ for the steepest-descent algorithm with step-size parameter $\mu = 0.3$ and varying eigenvalue spread: (a) $\chi(\mathbf{R}) = 1.22$; (b) $\chi(\mathbf{R}) = 3$; (c) $\chi(\mathbf{R}) = 10$; (d) $\chi(\mathbf{R}) = 100$.

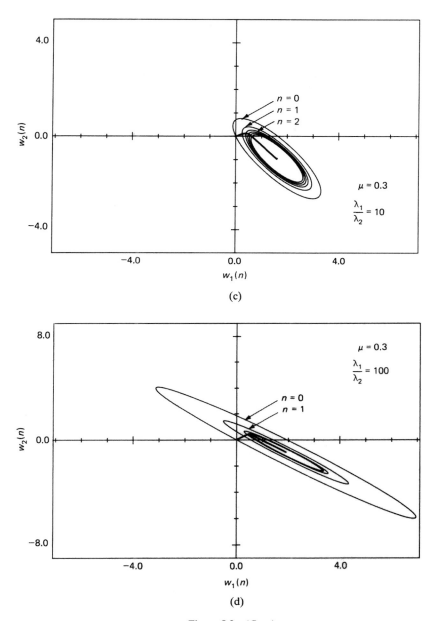

(c)

(d)

Figure 8.9   (*Cont.*)

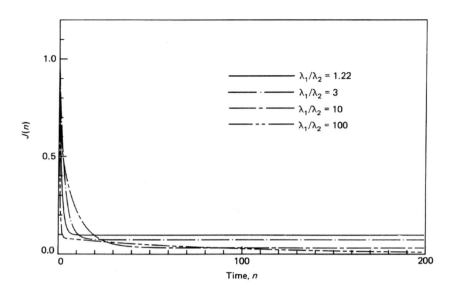

Figure 8.10 Learning curves of steepest-descent algorithm with step-size parameter $\mu = 0.3$ and varying eigenvalue spread.

This makes intuitive sense: The predictor should do a better job tracking a highly correlated input process than a weakly correlated one.

Experiment 2: Varying Step-size Parameter

In this experiment the eigenvalue spread is fixed at $\chi(\mathbf{R}) = 10$, and the step-size parameter μ is varied. In particular, we examine the transient behavior of the steepest-descent algorithm for $\mu = 0.3$ and 1.0. The corresponding results in terms of the transformed tap-weight errors $v_1(n)$ and $v_2(n)$ are shown in parts (a) and (b) of Fig. 8.11, respectively. The results included in part (a) of this figure are the same as those in Fig. 8.8(c). Note also that in accordance with Eq. (8.21), the critical value of the step-size parameter equals $\mu_{max} = 2/\lambda_{max} = 1.1$, which is slightly in excess of the actual value $\mu = 1$ used in Fig. 8.11(b).

The results for $\mu = 0.3$ and 1.0 in terms of the tap weights $w_1(n)$ and $w_2(n)$ are shown in parts (a) and (b) of Fig. 8.12, respectively. Here again, the results included in part (a) of the figure are the same as those in Fig. 8.9(c).

Observations

Based on the results presented for Experiments 1 and 2, we may make the following observations:

1. The trajectory of $[v_1(n), v_2(n)]$, with the number of iterations n as running parameter, is normal to the locus of $[v_1(n), v_2(n)]$ for fixed $J(n)$. This also applies to the trajectory of $[w_1(n), w_2(n)]$ for fixed $J(n)$.

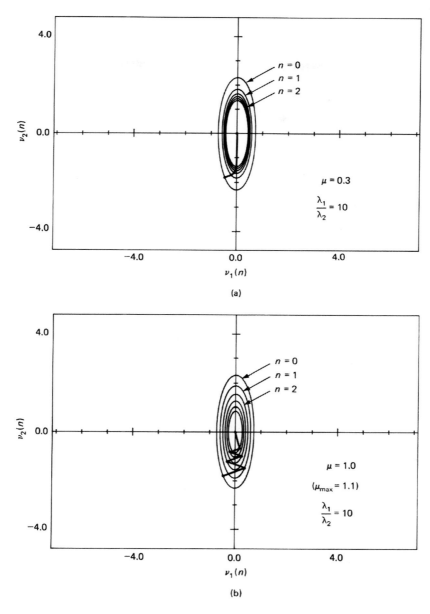

Figure 8.11   Loci of $v_1(n)$ versus $v_2(n)$ for the steepest-descent algorithm with eigenvalue spread $\chi(\mathbf{R}) = 10$ and varying step-size parameters: (a) overdamped, $\mu = 0.3$; (b) under-damped, $\mu = 1.0$.

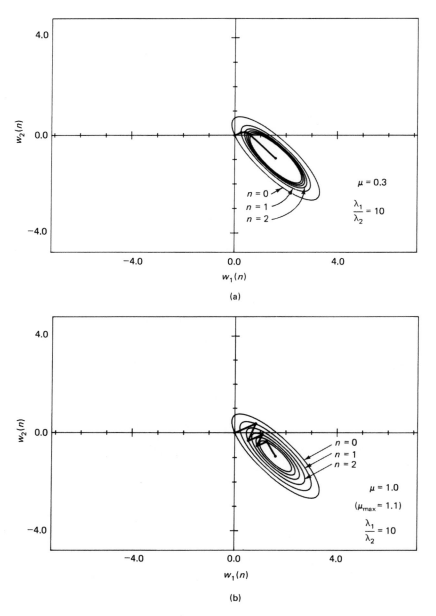

Figure 8.12   Loci of $w_1(n)$ versus $w_2(n)$ for the steepest-descent algorithm with eigenvalue spread $\chi(\mathbf{R}) = 10$ and varying step-size parameters: (a) overdamped, $\mu = 0.3$; underdamped, $\mu = 1.0$.

2. When the eigenvalues λ_1 and λ_2 are equal, the trajectory of $[v_1(n), v_2(n)]$ or that of $[w_1(n), w_2(n)]$, with n as running parameter, is a straight line. This is illustrated in Fig. 8.8(a) or 8.9(a) for which the eigenvalues λ_1 and λ_2 are approximately equal.

3. When the conditions are right for the initial value $\boldsymbol{v}(0)$ of the transformed tap-weight error vector $\boldsymbol{v}(n)$ to lie on the v_1-axis or v_2-axis, the trajectory of $[v_1(n), v_2(n)]$, with n as running parameter, is a straight line. This is illustrated in Fig. 8.8(b), where $v_1(0)$ is approximately zero. Correspondingly, the trajectory of $[w_1(n), w_2(n)]$, with n as running parameter, is also a straight line, as illustrated in Fig. 8.9(b).

4. Except for two special cases—(1) equal eigenvalues, and (2) the right choice of initial conditions—the trajectory of $[v_1(n), v_2(n)]$, with n as running parameter, follows a curved path, as illustrated in Fig. 8.8(c). Correspondingly, the trajectory of $[w_1(n), w_2(n)]$, with n as running parameter, also follows a curved path as illustrated in Fig. 8.9(c). When the eigenvalue spread is very high (i.e., the input data are very highly correlated), two things happen:

   - The error performance surface assumes the shape of a deep valley.
   - The trajectories of $[v_1(n), v_2(n)]$ and $[w_1(n), w_2(n)]$ develop distinct bends. Both of these points are well illustrated in Figs. 8.8(d) and 8.9(d), respectively, for the case of $\chi(\mathbf{R}) = 100$.

5. The steepest-descent algorithm converges fastest when the eigenvalues λ_1 and λ_2 are equal or the starting point of the algorithm is chosen right, for which cases the trajectory formed by joining the points $v(0), v(1), v(2), \ldots$ is a straight line, the shortest possible path.

6. For fixed step-size parameter μ, as the eigenvalue spread $\chi(\mathbf{R})$ increases (i.e., the correlation matrix $\mathbf{R}$ of the tap inputs becomes more ill conditioned), the ellipsoidal loci of $[v_1(n), v_2(n)]$ for fixed values of $J(n)$, for $n = 0, 1, 2, \ldots$, become increasingly narrower (i.e., the minor axis becomes smaller) and more crowded.

7. When the step-size parameter μ is small, the transient behavior of the steepest-descent algorithm is *overdamped,* in that the trajectory formed by joining the points $\boldsymbol{v}(0), \boldsymbol{v}(1), \boldsymbol{v}(2), \ldots$ follows a continuous path. When, on the other hand μ approaches the maximum allowable value, $\mu_{max} = 2/\lambda_{max}$, the transient behavior of the steepest-descent algorithm is *underdamped*, in that this trajectory exhibits oscillations. These two different forms of transient behavior are illustrated in parts (a) and (b) of Fig. 8.11 in terms of $v_1(n)$ and $v_2(n)$. The corresponding results in terms of $w_1(n)$ and $w_2(n)$ are presented in parts (a) and (b) of Fig. 8.12.

The conclusion to be drawn from these observations is that the transient behavior of the steepest-descent algorithm is highly sensitive to variations in both the step-size parameter μ and the eigenvalue spread of the correlation matrix of the tap inputs.

8.5 SUMMARY AND DISCUSSION

The method of steepest descent provides a simple procedure for computing the tap-weight vector of a Wiener filter, given knowledge of two ensemble-averaged quantities:

- The correlation matrix of the tap-input vector
- The cross-correlation vector between the tap-input vector and the desired response.

A critical feature of the method of steepest descent is the presence of feedback, which is another way of saying that the underlying algorithm is recursive in nature. As such, we have to pay particular attention to the issue of stability, which is governed by two parameters in the feedback loop of the algorithm:

- The step-size parameter μ
- The correlation matrix $\mathbf{R}$ of the tap-input vector.

Specifically, the necessary and sufficient condition for stability of the algorithm is embodied in the following:

$$0 < \mu < \frac{2}{\lambda_{max}}$$

where λ_{max} is the largest eigenvalue of the correlation matrix $\mathbf{R}$.

Moreover, depending on the value assigned to the step-size parameter μ, the transient response of the steepest descent algorithm may exhibit one of three forms of behavior:

- *Underdamped response*, in which case the trajectory followed by the tap-weight vector toward the optimum Wiener solution exhibits oscillations; this response arises when the step-size parameter μ is large.
- *Overdamped response*, which is a nonoscillatory behavior that arises when μ is small.
- *Critically damped response*, which is the fine dividing line between the underdamped and overdamped conditions.

Unfortunately, in general, these conditions do not lend themselves to an exact mathematical analysis; they are usually evaluated by experimentation.

PROBLEMS

1. Consider a Wiener filtering problem characterized by the following values for the correlation matrix $\mathbf{R}$ of the tap-input vector $\mathbf{u}(n)$ and the cross-correlation vector $\mathbf{p}$ between $\mathbf{u}(n)$ and the desired response $d(n)$:

$$\mathbf{R} = \begin{bmatrix} 1 & 0.5 \\ 0.5 & 1 \end{bmatrix}$$

$$\mathbf{p} = \begin{bmatrix} 0.5 \\ 0.25 \end{bmatrix}$$

(a) Suggest a suitable value for the step-size parameter μ that would ensure convergence of the method of steepest descent, based on the given value for matrix $\mathbf{R}$.

(b) Using the value proposed in part (a), determine the recursions for computing the elements $w_1(n)$ and $w_2(n)$ of the tap-weight vector $\mathbf{w}(n)$. For this computation, you may assume the initial values

$$w_1(0) = w_2(0) = 0$$

(c) Investigate the effect of varying the step-size parameter μ on the trajectory of the tap-weight vector $\mathbf{w}(n)$ as n varies from zero to infinity.

2. Start with the formula for the estimation error:

$$e(n) = d(n) - \mathbf{w}^H(n)\mathbf{u}(n)$$

where $d(n)$ is the desired response, $\mathbf{u}(n)$ is the tap-input vector, and $\mathbf{w}(n)$ is the tap-weight vector in the transversal filter. Hence, show that the gradient of the instantaneous squared error $|e(n)|^2$ equals

$$\hat{\nabla} J(n) = -2\mathbf{u}(n)d^*(n) + 2\mathbf{u}(n)\mathbf{u}^H(n)\mathbf{w}(n)$$

3. In this problem we explore another way of deriving the steepest-descent algorithm of Eq. (8.9) used to adjust the tap-weight vector in a transversal filter. The inverse of a positive-definite matrix may be expanded in a series as follows:

$$\mathbf{R}^{-1} = \mu \sum_{k=0}^{\infty} (\mathbf{I} - \mu\mathbf{R})^k$$

where $\mathbf{I}$ is the identity matrix, and μ is a positive constant. To ensure convergence of the series, the constant μ must lie inside the range

$$0 < \mu < \frac{2}{\lambda_{\max}}$$

where $\lambda_{\max}$ is the largest eigenvalue of the matrix $\mathbf{R}$. By using this series expansion for the inverse of the correlation matrix in the Wiener–Hopf equations, develop the recursion

$$\mathbf{w}(n + 1) = \mathbf{w}(n) + \mu[\mathbf{p} - \mathbf{R}\mathbf{w}(n)]$$

where $\mathbf{w}(n)$ is the approximation to the Wiener solution for the tap-weight vector:

$$\mathbf{w}(n) = \mu \sum_{k=0}^{n-1} (\mathbf{I} - \mu\mathbf{R})^k \mathbf{p}$$

4. In the method of steepest descent, show that the correction applied to the tap-weight vector after $n + 1$ iterations may be expressed as follows:

$$\delta\mathbf{w}(n + 1) = \mu E[\mathbf{u}(n)e^*(n)]$$

where $\mathbf{u}(n)$ is the tap-input vector and $e(n)$ is the estimation error. What happens to this correction at the minimum point of the error performance surface? Discuss your answer in light of the principle of orthogonality.

5. Consider the method of steepest descent involving a single weight $w(n)$. Do the following:
   (a) Determine the mean-squared error $J(n)$ as a function of $w(n)$.
   (b) Find the Wiener solution w_o, and the minimum mean-squared error J_{min}.
   (c) Sketch a plot of $J(n)$ versus $w(n)$.

6. Equation (8.28) defines the transient behavior of the mean-squared error $J(n)$ for varying n that is produced by the steepest-descent algorithm. Let $J(0)$ and $J(\infty)$ denote the initial and final values of $J(n)$. Suppose that we approximate this transient behavior with a single exponential, as follows:

$$J_{approx}(n) = [J(0) - J(\infty)]e^{-n/\tau} + J(\infty)$$

where τ is termed the *effective time constant*. Let τ be chosen such that

$$J_{approx}(1) = J(1)$$

Hence, show that *the initial rate of convergence* of the steepest-descent algorithm, defined as the inverse of τ, is given by

$$\frac{1}{\tau} = \ln\left[\frac{J(0) - J(\infty)}{J(1) - J(\infty)}\right]$$

Using Eq. (8.28), find the value of $1/\tau$. Assume that the initial value $\mathbf{w}(0)$ is zero, and that the step-size parameter μ is small.

7. Consider an autoregressive (AR) process $u(n)$ of order 1, described by the difference equation

$$u(n) = -au(n-1) + v(n)$$

where a is the AR parameter of the process, and $v(n)$ is a zero-mean white-noise process of variance σ_v^2.
   (a) Set up a linear predictor of order 1 to compute the parameter a. To be specific, use the method of steepest descent for the recursive computation of the Wiener solution for the parameter a.
   (b) Plot the error-performance curve for this problem, identifying the minimum point of the curve in terms of known parameters.
   (c) What is the condition on the step-size parameter μ to ensure stability? Justify your answer.

CHAPTER

9

Least-Mean-Square Algorithm

In this chapter we develop the theory of a widely used algorithm named the *least-mean-square (LMS) algorithm* by its originators, Widrow and Hoff (1960). The LMS algorithm is an important member of the family of *stochastic gradient algorithms*. The term "stochastic gradient" is intended to distinguish the LMS algorithm from the method of steepest descent that uses a deterministic gradient in a recursive computation of the Wiener filter for stochastic inputs. A significant feature of the LMS algorithm is its *simplicity*. Moreover, it does not require measurements of the pertinent correlation functions, nor does it require matrix inversion. Indeed, it is the simplicity of the LMS algorithm that has made it the *standard* against which other adaptive filtering algorithms are benchmarked.

The material presented in this chapter also includes a detailed analysis of the convergence behavior of the LMS algorithm, supported by computer experiments. We begin our study of the LMS algorithm by presenting an overview of its structure and operation.

9.1 OVERVIEW OF THE STRUCTURE AND OPERATION OF THE LEAST-MEAN-SQUARE ALGORITHM

The least-mean-square (LMS) algorithm is a linear adaptive filtering algorithm that consists of two basic processes:

1. A *filtering process*, which involves (a) computing the output of a transversal filter produced by a set of tap inputs, and (b) generating an estimation error by comparing this output to a desired response.

     2. An *adaptive process*, which involves the automatic adjustment of the tap weights of the filter in accordance with the estimation error.

Thus, the combination of these two processes working together constitutes a *feedback loop* around the LMS algorithm, as illustrated in the block diagram of Fig. 9.1(a). First, we have a transversal filter, around which the LMS algorithm is built; this component is responsible for performing the filtering process. Second, we have a mechanism for performing the adaptive control process on the tap weights of the transversal filter, hence the designation "adaptive weight-control mechanism" in Fig. 9.1(a).

     Details of the *transversal filter* component are presented in Fig. 9.1(b). The tap inputs $u(n)$, $u(n - 1)$, . . ., $u(n - M + 1)$ form the elements of the M-by-1 *tap-input vector* $\mathbf{u}(n)$, where $M - 1$ is the number of delay elements; these tap inputs span a multidimensional space denoted by $\mathcal{U}_n$. Correspondingly, the tap weights $\hat{w}_0(n)$, $\hat{w}_1(n)$, . . . , $\hat{w}_{M-1}(n)$ form the elements of the M-by-1 tap-weight vector $\hat{\mathbf{w}}(n)$. The value computed for the tap-weight vector $\hat{\mathbf{w}}(n)$ using the LMS algorithm represents an estimate whose expected value approaches the Wiener solution $\mathbf{w}_o$ (for a wide-sense stationary environment) as the number of iterations n approaches infinity.

     During the filtering process the *desired response* $d(n)$ is supplied for processing, alongside the tap-input vector $\mathbf{u}(n)$. Given this input, the transversal filter produces an output $\hat{d}(n|\mathcal{U}_n)$ used as an *estimate* of the desired response $d(n)$. Accordingly, we may define an *estimation error* $e(n)$ as the difference between the desired response and the actual filter output, as indicated in the output end of Fig. 9.1(b). The estimation error $e(n)$ and the tap-input vector $\mathbf{u}(n)$ are applied to the control mechanism, and the feedback loop around the tap weights is thereby closed.

     Figure 9.1(c) presents details of the *adaptive weight-control mechanism*. Specifically, a scalar version of the inner product of the estimation error $e(n)$ and the tap input $u(n - k)$ is computed for $k = 0, 1, 2, . . ., M - 2, M - 1$. The result so obtained defines the *correction* $\delta\hat{w}_k(n)$ applied to the tap weight $\hat{w}_k(n)$ at iteration $n + 1$. The scaling factor used in this computation is denoted by μ in Fig. 9.1(c). It is called the *step-size parameter*.

     Comparing the control mechanism of Fig. 9.1(c) for the LMS algorithm with that of Fig. 8.2 for the method of steepest descent, we see that the LMS algorithm uses the product $u(n - k)e^*(k)$ as an estimate of element k in the gradient vector $\nabla J(n)$ that characterizes the method of steepest descent. In other words, the expectation operator is missed out from all the paths in Fig. 9.1(c). Accordingly, the recursive computation of each tap weight in the LMS algorithm suffers from a *gradient noise*.

     Throughout this chapter, it is assumed that the tap-input vector $\mathbf{u}(n)$ and the desired response $d(n)$ are drawn from a jointly wide-sense stationary environment. For such an environment, we know from Chapter 8 that the method of steepest descent computes a tap-weight vector $\mathbf{w}(n)$ that moves down the ensemble-averaged error-performance surface along a deterministic trajectory, which terminates on the Wiener solution, $\mathbf{w}_o$. The LMS algorithm, on the other hand, behaves differently because of the presence of gradient noise.

Rather than terminating on the Wiener solution, the tap-weight vector $\hat{\mathbf{w}}(n)$ [different from $\mathbf{w}(n)$] computed by the LMS algorithm executes a *random motion* around the minimum point of the error-performance surface.

Earlier we pointed out that the LMS algorithm involves feedback in its operation, which therefore raises the related issue of *stability*. In this context, a meaningful criterion is to require that

$$J(n) \rightarrow J(\infty) \qquad \text{as } n \rightarrow \infty$$

where $J(n)$ is the mean-squared error produced by the LMS algorithm at time n, and its final value $J(\infty)$ is a constant. An algorithm that satisfies this requirement is said to be *convergent in the mean square*. For the LMS algorithm to satisfy this criterion, the step-size parameter μ has to satisfy a certain condition related to the eigenstructure of the correlation matrix of the tap inputs.

The difference between the final value $J(\infty)$ and the minimum value $J_{\min}$ attained by the Wiener solution is called the *excess mean-squared error $J_{\text{ex}}(\infty)$*. This difference represents the price paid for using the adaptive (stochastic) mechanism to control the tap weights in the LMS algorithm in place of a deterministic approach as in the method of steepest descent. The ratio of $J_{\text{ex}}(\infty)$ to $J_{\min}$ is called the *misadjustment*, which is a measure of how far the steady-state solution computed by the LMS algorithm is away from the Wiener solution. It is important to realize, however, that the misadjustment $\mathcal{M}$ is under the designer's control. In particular, the feedback loop acting around the tap weights behaves like a *low-pass filter*, whose "average" time constant is *inversely* proportional to the step-size parameter μ. Hence, by assigning a small value to μ the adaptive process is made to progress slowly, and the effects of gradient noise on the tap weights are largely filtered out. This, in turn, has the effect of reducing the misadjustment.

We may therefore justifiably say that the LMS algorithm is simple in implementation, yet capable of delivering high performance by adapting to its external environment. To do so, however, we have to pay particular attention to the choice of a suitable value for the step-size parameter μ.

A closely related issue is that of specifying a suitable value for the filter order $M-1$; for this specification, we may use the AIC or MDL criterion for the model-order selection that was discussed in Chapter 2.

The factors influencing the choice of μ are covered in Sections 9.4 and 9.7. First and foremost, however, we wish to derive the LMS algorithm. This we do in the next section by building on our previous knowledge of the method of steepest descent.

9.2 LEAST-MEAN-SQUARE ADAPTATION ALGORITHM

If it were possible to make exact measurements of the gradient vector $\nabla J(n)$ at each iteration n, and if the step-size parameter μ is suitably chosen, then the tap-weight vector computed by using the steepest-descent algorithm would indeed converge to the optimum

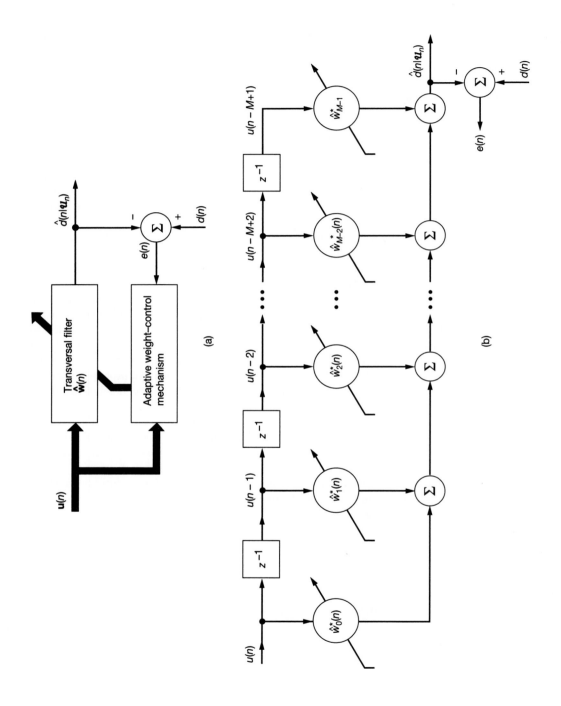

(a)

(b)

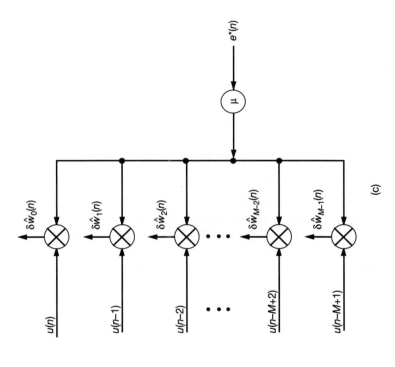

(c)

Figure 9.1 (a) Block diagram of adaptive transversal filter. (b) Detailed structure of the transversal filter component. (c) Detailed structure of the adaptive weight-control mechanism.

Wiener solution. In reality, however, exact measurements of the gradient vector are not possible since this would require prior knowledge of both the correlation matrix $\mathbf{R}$ of the tap inputs and the cross-correlation vector $\mathbf{p}$ between the tap inputs and the desired response [see Eq. (8.8)]. Consequently, the gradient vector must be *estimated* from the available data

To develop an estimate of the gradient vector $\nabla J(n)$, the most obvious strategy is to substitute estimates of the correlation matrix $\mathbf{R}$ and the cross-correlation vector $\mathbf{p}$ in the formula of Eq. (8.8), which is reproduced here for convenience:

$$\nabla J(n) = -2\mathbf{p} + 2\mathbf{R}\mathbf{w}(n) \tag{9.1}$$

The *simplest* choice of estimators for $\mathbf{R}$ and $\mathbf{p}$ is to use *instantaneous estimates* that are based on sample values of the tap-input vector and desired response, as defined by, respectively,

$$\hat{\mathbf{R}}(n) = \mathbf{u}(n)\mathbf{u}^H(n) \tag{9.2}$$

and

$$\hat{\mathbf{p}}(n) = \mathbf{u}(n)d^*(n) \tag{9.3}$$

Correspondingly, the instantaneous estimate of the gradient vector is

$$\hat{\nabla} J(n) = -2\mathbf{u}(n)d^*(n) + 2\mathbf{u}(n)\mathbf{u}^H(n)\hat{\mathbf{w}}(n) \tag{9.4}$$

Generally speaking, this estimate is *biased* because the tap-weight estimate vector $\hat{\mathbf{w}}(n)$ is a random vector that depends on the tap-input vector $\mathbf{u}(n)$. Note that the estimate $\hat{\nabla} J(n)$ may also be viewed as the gradient operator ∇ applied to the instantaneous squared error $|e(n)|^2$; see Problem 2 of Chapter 8.

Substituting the estimate of Eq. (9.4) for the gradient vector $\nabla J(n)$ in the steepest-descent algorithm described in Eq. (8.7), we get a new recursive relation for updating the tap-weight vector:

$$\hat{\mathbf{w}}(n + 1) = \hat{\mathbf{w}}(n) + \mu \mathbf{u}(n)[d^*(n) - \mathbf{u}^H(n)\hat{\mathbf{w}}(n)] \tag{9.5}$$

Here we have used a hat over the symbol for the tap-weight vector to distinguish it from the value obtained by using the steepest-descent algorithm. Equivalently, we may write the result in the form of three basic relations as follows:

1. *Filter output:*

$$y(n) = \hat{\mathbf{w}}^H(n)\mathbf{u}(n) \tag{9.6}$$

2. *Estimation error:*

$$e(n) = d(n) - y(n) \tag{9.7}$$

3. *Tap-weight adaptation:*

$$\hat{\mathbf{w}}(n + 1) = \hat{\mathbf{w}}(n) + \mu \mathbf{u}(n)e^*(n) \tag{9.8}$$

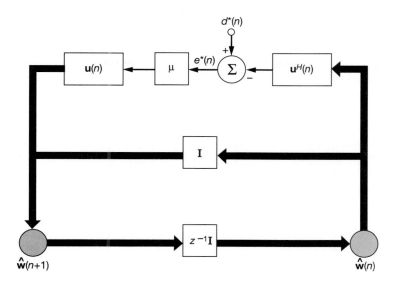

Figure 9.2   Signal-flow graph representation of the LMS algorithm.

Equations (9.6) and (9.7) define the estimation error $e(n)$, the computation of which is based on the *current estimate* of the tap-weight vector, $\hat{\mathbf{w}}(n)$. Note also that the second term, $\mu\mathbf{u}(n)e^*(n)$, on the right-hand side of Eq. (9.8) represents the *correction* that is applied to the current estimate of the tap-weight vector, $\hat{\mathbf{w}}(n)$. The iterative procedure is started with an initial guess $\hat{\mathbf{w}}(0)$.

The algorithm described by Eqs. (9.6) to (9.8) is the complex form of the adaptive *least-mean-square (LMS) algorithm.*[1] At each iteration or time update, it also requires knowledge of the most recent values: $\mathbf{u}(n)$, $d(n)$, and $\hat{\mathbf{w}}(n)$. The LMS algorithm is a member of the family of *stochastic gradient algorithms*. In particular, when the LMS algorithm operates on stochastic inputs, the allowed set of directions along which we "step" from one iteration cycle to the next is quite random and cannot therefore be thought of as being true gradient directions.

Figure 9.2 shows a signal-flow graph representation of the LMS algorithm in the form of a feedback model. This model bears a close resemblance to the feedback model of

---

[1]The complex form of the LMS algorithm, as originally proposed by Widrow et al. (1975b), differs slightly from that described in Eqs. (9.6) to (9.8). Widrow and other authors have based their derivation on the definition $\mathbf{R} = E[\mathbf{u}^*(n)\mathbf{u}^T(n)]$ for the correlation matrix. On the other hand, the LMS algorithm described by Eqs. (9.6) to (9.8) is based on the definition $\mathbf{R} = E[\mathbf{u}(n)\mathbf{u}^H(n)]$ for the correlation matrix. The adoption of the latter definition for the correlation matrix of complex-valued data is the natural extension of the definition for real-valued data.

Fig. 8.3 describing the steepest-descent algorithm. The signal-flow graph of Fig. 9.2 clearly illustrates the simplicity of the LMS algorithm. In particular, we see from this figure that the LMS algorithm requires only $2M + 1$ *complex multiplications and $2M$ complex additions per iteration,* where M is the number of tap weights used in the adaptive transversal filter. In other words, the computational complexity of the LMS algorithm is $O(M)$.

The instantaneous estimates of $\mathbf{R}$ and $\mathbf{p}$ given in Eqs. (9.2) and (9.3), respectively, have relatively large variances. At first sight, it may therefore seem that the LMS algorithm is incapable of good performance since the algorithm uses these instantaneous estimates. However, we must remember that the LMS algorithm is recursive in nature, with the result that the algorithm itself effectively averages these estimates, in some sense, during the course of adaptation.

9.3 EXAMPLES

Before proceeding further with a convergence analysis of the LMS algorithm, it is instructive to develop an appreciation for the versatility of this important algorithm. We do this by presenting six examples that relate to widely different applications of the LMS algorithm.

Example 1: Canonical Model of the Complex LMS Algorithm

The LMS algorithm described in Eqs. (9.6) to (9.8) is *complex* in the sense that the input and output data as well as the tap weights are all complex valued. To emphasize the complex nature of the algorithm, we use complex notation to express the data and tap weights as follows:

Tap-input vector:
$$\mathbf{u}(n) = \mathbf{u}_I(n) + j\mathbf{u}_Q(n) \tag{9.9}$$

Desired response:
$$d(n) = d_I(n) + jd_Q(n) \tag{9.10}$$

Tap-weight vector:
$$\hat{\mathbf{w}}(n) = \hat{\mathbf{w}}_I(n) + j\hat{\mathbf{w}}_Q(n) \tag{9.11}$$

Transversal filter output:
$$y(n) = y_I(n) + jy_Q(n) \tag{9.12}$$

Estimation error:
$$e(n) = e_I + je_Q(n) \tag{9.13}$$

The subscripts I and Q denote "in-phase" and "quadrature" components, respectively; that is, the real and imaginary parts, respectively. Using these definitions in Eqs. (9.6) to (9.8), expanding terms, and then equating real and imaginary parts, we get

$$y_I(n) = \hat{\mathbf{w}}_I^T(n)\mathbf{u}_I(n) - \hat{\mathbf{w}}_Q^T(n)\mathbf{u}_Q(n) \tag{9.14}$$

$$y_Q(n) = \hat{\mathbf{w}}_I^T(n)\mathbf{u}_Q(n) + \hat{\mathbf{w}}_Q^T(n)\mathbf{u}_I(n) \tag{9.15}$$

$$e_I(n) = d_I(n) - y_I(n) \tag{9.16}$$

$$e_Q(n) = d_Q(n) - y_Q(n) \tag{9.17}$$

$$\hat{\mathbf{w}}_I(n + 1) = \hat{\mathbf{w}}_I(n) + \mu[e_I(n)\mathbf{u}_I(n) - e_Q(n)\mathbf{u}_Q(n)] \tag{9.18}$$

$$\hat{\mathbf{w}}_Q(n + 1) = \hat{\mathbf{w}}_Q(n) + \mu[e_I(n)\mathbf{u}_Q(n) + e_Q(n)\mathbf{u}_I(n)] \tag{9.19}$$

Equations (9.14) to (9.17), defining the error and output signals, are represented by the cross-coupled signal-flow graph shown in Fig. 9.3(a). The update equations (9.18) and (9.19) are likewise represented by the cross-coupled signal-flow graph shown in Fig. 9.3(b). The combination of this pair of signal-flow graphs constitutes the *canonical model* of the complex LMS algorithm. This canonical model clearly illustrates that a complex LMS algorithm is equivalent to a set of four real LMS algorithms with *cross-coupling* between them. Its use may arise, for example, in the adaptive equalization of a communication channel for the transmission of binary data by means of a multiphase modulation scheme such as quadriphase-shift keying (QPSK).

Example 2: Adaptive Deconvolution for Processing of Time-Varying Seismic Data

In Section 7 of the introductory chapter, we described the idea of *predictive deconvolution* as a method for processing reflection seismograms. In this example we discuss the use of the LMS algorithm for the adaptive implementation of predictive deconvolution. It is assumed that the seismic data are stationary over the design gate used to generate the deconvolution operation.

Let the set of *real* data points in the seismogram to be processed be denoted by $u(n)$, $n = 1, 2, \ldots, N$, where N is the data length. The real LMS-based adaptive deconvolution method, which is appropriate for this example, proceeds as follows (Griffiths et al., 1977):

- An M-dimensional operator $\hat{\mathbf{w}}(n)$ is used to generate a predicted trace from the data; that is,

$$\hat{\mathbf{u}}(n + \Delta) = \hat{\mathbf{w}}^T(n)\mathbf{u}(n) \tag{9.20}$$

where

$$\hat{\mathbf{w}}(n) = [\hat{w}_0(n), \hat{w}_1(n), \ldots, \hat{w}_{M-1}(n)]^T$$

$$\mathbf{u}(n) = [u(n), u(n - 1), \ldots, u(n - M + 1)]^T$$

and $\Delta \geq 1$ is the *prediction depth*.

- The deconvolved trace $y(n)$ defining the difference between the input and predicted traces is evaluated:

$$y(n) = u(n) - \hat{u}(n)$$

- The operator $\hat{\mathbf{w}}(n)$ is updated:

$$\hat{\mathbf{w}}(n + 1) = \hat{\mathbf{w}}(n) + \mu[u(n + \Delta) - \hat{u}(n + \Delta)]\mathbf{u}(n) \tag{9.21}$$

Equations (9.20) and (9.21) constitute the LMS-based adaptive seismic deconvolution algorithm. The adaptation is begun with an initial guess $\hat{\mathbf{w}}(0)$.

Example 3: Instantaneous Frequency Measurement

In this example we study the use of the LMS algorithm as the basis for estimating the frequency content of a *narrow-band signal* characterized by a rapidly varying power spectrum (Griffiths, 1975). In so doing we illustrate the linkage between three basic ideas: an autoregressive (AR) model for describing a stochastic process (studied in Chapter 2), a linear

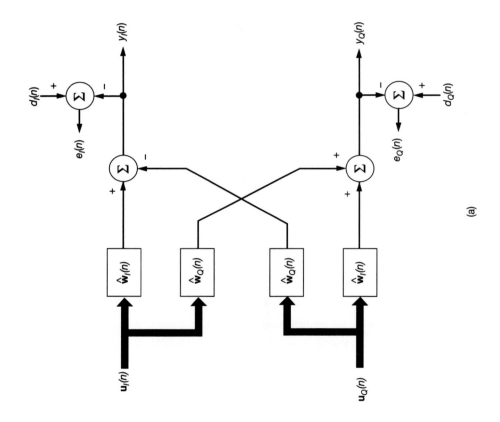

(a)

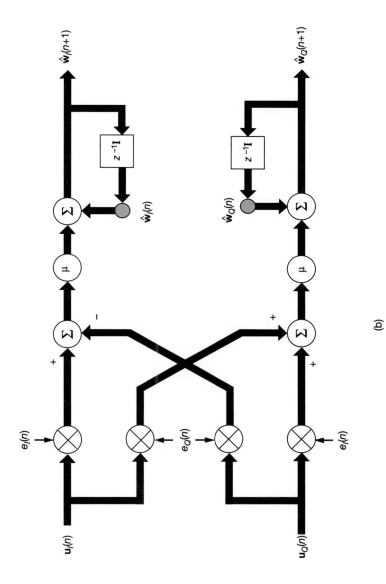

Figure 9.3 Canonical signal-flow graph representation of the complex LMS algorithm. (a) Error and output signals. (b) Tap-weight update equation.

(b)

predictor for analyzing the process (studied in Chapter 6), and the LMS algorithm for estimating the AR parameters.

By a "narrow-band signal" we mean a signal whose bandwidth Ω is small compared to the midband angular frequency ω_c, as illustrated in Fig. 9.4. A *frequency-modulated (FM) signal* is an example of a narrow-band signal, provided that the carrier frequency is high enough. The *instantaneous frequency* (defined as the derivative of phase with respect to time) of an FM signal varies linearly with the modulating signal. Consider then a narrow-band process $u(n)$ generated by a *time varying* AR model of order M, as shown by the difference equation (assuming *real* data):

$$u(n) = -\sum_{k=1}^{M} a_k(n)u(n - k) + v(n) \tag{9.22}$$

where the $a_k(n)$ are the time-varying model parameters, and $v(n)$ is a zero-mean white-noise process of time-varying variance $\sigma_v^2(n)$. The *time-varying AR (power) spectrum* of the narrow-band process $u(n)$ is given by

$$S_{\text{AR}}(\omega; n) = \frac{\sigma_v^2(n)}{\left| 1 + \displaystyle\sum_{k=1}^{M} a_k(n)e^{-j\omega k} \right|^2}, \qquad -\pi < \omega \leq \pi \tag{9.23}$$

Note that an AR process whose poles are near the unit circle in the z-plane has the characteristics of a narrow-band process.

To estimate the model parameters, we use an adaptive transversal filter employed as a linear predictor of order M. Let the tap weights of the predictor be denoted by $\hat{w}_k(n)$, $k = 1, 2, \ldots, M$. The tap weights are adapted continuously as the input signal $u(n)$ is received. In particular, we use the LMS algorithm for adapting the tap weights, as shown by

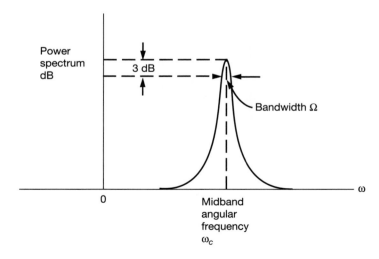

Figure 9.4   Definition of a narrow-band signal in terms of its spectrum.

$$\hat{w}_k(n + 1) = \hat{w}_k(n) + \mu u(n - k) f_M(n), \qquad k = 1, 2, \ldots, M \qquad (9.24)$$

where $f_M(n)$ is the *prediction error*:

$$f_M(n) = u(n) - \sum_{k=1}^{M} \hat{w}_k(n) u(n - k) \qquad (9.25)$$

The tap weights of the adaptive predictor are related to the AR model parameters as follows:

$$-\hat{w}_k(n) = \text{estimate of } a_k(n) \text{ at time } n \qquad \text{for } k = 1, 2, \ldots, M$$

Moreover, the average power of the prediction error $f_M(n)$ provides an estimate of the noise variance $\sigma_v^2(n)$. Our interest is in locating the frequency of a narrow-band signal. Accordingly, in the sequel, we ignore the estimation of $\sigma_v^2(n)$. Specifically, we only use the tap weights of the adaptive predictor to define the *time-varying frequency function*:

$$F(\omega; n) = \frac{1}{\left| 1 - \sum_{k=1}^{M} \hat{w}_k(n) e^{-j\omega k} \right|^2} \qquad (9.26)$$

Given the relationship between $\hat{w}_k(n)$ and $a_k(n)$, we see that the essential difference be-tween the frequency function $F(\omega; n)$ in Eq. (9.26) and the AR power spectrum $S_{AR}(\omega; n)$ in Eq. (9.23) lies in their numerator scale factors. The numerator of $F(\omega; n)$ is a constant equal to 1, whereas that of $S_{AR}(\omega; n)$ is a time-varying constant equal to $\sigma_v^2(n)$. The advantages of the frequency function $F(\omega; n)$ over the AR spectrum $S_{AR}(\omega; n)$ are twofold. First, the 0/0 indeterminacy inherent in the narrow-band spectrum of Eq. (9.23) is replaced by a "computationally tractable" limit of 1/0 in Eq. (9.26). Second, the frequency function $F(\omega; n)$ is not affected by amplitude scale changes in the input signal $u(n)$, with the result that the peak value of $F(\omega; n)$ is related directly to the spectral width of the input signal.

We may use the function $F(\omega; n)$ to measure the instantaneous frequency of a frequency-modulated signal $u(n)$, provided that the following assumptions are justified (Griffiths, 1975):

- The adaptive predictor has been in operation sufficiently long, so as to ensure that any transients caused by the initialization of the tap weights have died out.
- The step-size parameter μ is chosen correctly for the adaptive predictor to track well; that is to say, the prediction error $f_M(n)$ is small for all n.
- The modulating signal is essentially constant over the sampling range of the adaptive predictor, which extends from time $(n - M)$ to time $(n - 1)$.

Given the validity of these assumptions, we find that the frequency function $F(\omega; n)$ has a peak at the instantaneous frequency of the input signal $u(n)$. Moreover, the LMS algorithm will track the time variation of the instantaneous frequency.

Example 4: Adaptive Noise Canceling Applied to a Sinusoidal Interference

The traditional method of suppressing a *sinusoidal interference* corrupting an information-bearing signal is to use a fixed *notch filter* tuned to the frequency of the interference. To design the filter, we naturally need to know the precise frequency of the interference. But what if the notch is required to be very sharp and the interfering sinusoid is known to *drift* slowly? Clearly, then, we have a problem that calls for an *adaptive* solution. One such solution is provided by the use of adaptive noise canceling.

Figure 9.5 shows the block diagram of a dual-input *adaptive noise canceler*. The primary input supplies an information-bearing signal and a sinusoidal interference that are uncorrelated with each other. The reference input supplies a correlated version of the sinusoidal interference. For the adaptive filter, we may use a transversal filter whose tap weights are adapted by means of the LMS algorithm. The filter uses the reference input to provide (at its output) an estimate of the sinusoidal interfering signal contained in the primary input. Thus, by subtracting the adaptive filter output from the primary input, the effect of the sinusoidal interference is diminished. In particular, an adaptive noise canceler using the LMS algorithm has two important characteristics (Widrow et al., 1976; Glover, 1977):

1. The canceler behaves as an adaptive notch filter whose null point is determined by the angular frequency ω_0 of the sinusoidal interference. Hence, it is tunable, and the tuning frequency moves with ω_0.
2. The notch in the frequency response of the canceler can be made very sharp at precisely the frequency of the sinusoidal interference by choosing a small enough value for the step-size parameter μ.

In the example considered here, the input data are assumed to be real valued:

- *Primary input:*

$$d(n) = s(n) + A_0 \cos(\omega_0 n + \phi_0) \qquad (9.27)$$

where $s(n)$ is an information-bearing signal; A_0 is the amplitude of the sinusoidal interference, ω_0 is the *normalized* angular frequency, and ϕ_0 is the phase.

- *Reference input:*

$$u(n) = A \cos(\omega_0 n + \phi) \qquad (9.28)$$

where the amplitude A and the phase ϕ are different from those in the primary input, but the angular frequency ω_0 is the same.

Using the real, expanded form of the LMS algorithm, the tap-weight update is described by the following equations:

$$y(n) = \sum_{i=0}^{M-1} \hat{w}_i(n)u(n-i) \qquad (9.29)$$

$$e(n) = d(n) - y(n) \qquad (9.30)$$

$$\hat{w}_i(n+1) = \hat{w}_i(n) + \mu u(n-i)e(n), \qquad i = 0, 1, \ldots, M-1 \qquad (9.31)$$

where M is the total number of tap weights in the transversal filter, and the constant μ is the step-size parameter. Note that the sampling period in the input data and all other signals in the LMS algorithm is assumed to be unity for convenience of presentation; as mentioned previously this practice is indeed followed throughout the book.

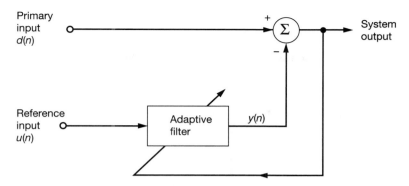

Figure 9.5   Block diagram of adaptive noise canceler.

With a sinusoidal excitation as the input of interest, we restructure the block diagram of the adaptive noise canceler as in Fig. 9.6(a). According to this new representation, we may lump the sinusoidal input $u(n)$, the transversal filter, and the weight-update equation of the LMS algorithm into a single (open-loop) system defined by a transfer function $G(z)$, as in the equivalent model of Fig. 9.6(b). The transfer function $G(z)$ is

$$G(z) = \frac{Y(z)}{E(z)}$$

where $Y(z)$ and $E(z)$ are the z-transforms of the reference input $u(n)$ and the estimation error $e(n)$, respectively. Given $E(z)$, our task is to find $Y(z)$, and therefore $G(z)$.

To do so, we use the detailed signal-flow-graph representation of the LMS algorithm depicted in Fig. 9.7 (Glover, 1977). In this diagram, we have singled out the ith tap weight for specific attention. The corresponding value of the tap input is

$$u(n - i) = A \cos[\omega_0(n - i) + \phi] \tag{9.32}$$

$$= \frac{A}{2} [e^{j(\omega_0 n + \phi_i)} + e^{-j(\omega_0 n + \phi_i)}]$$

where

$$\phi_i = \phi - \omega_0 i$$

In Fig. 9.7, the input $u(n - i)$ is multiplied by the estimation error $e(n)$. Hence, taking the z-transform of the product $u(n - i)e(n)$ and using $z[\bullet]$ to denote this operation, we obtain

$$z[u(n - i)e(n)] = \frac{A}{2} e^{j\phi_i} E(ze^{-j\omega_0}) + \frac{A}{2} e^{-j\phi_i} E(ze^{j\omega_0}) \tag{9.33}$$

where $E(ze^{-j\omega_0})$ is the z-transform $E(z)$ rotated counterclockwise around the unit circle through the angle ω_0. Similarly, $E(ze^{j\omega_0})$ represents a clockwise rotation through ω_0.

Next, taking the z-transform of Eq. (9.31), we get

$$z\hat{W}_i(z) = \hat{W}_i(z) + \mu\, z[u(n - i)e(n)] \tag{9.34}$$

where $\hat{W}_i(z)$ is the z-transform of $\hat{w}_i(n)$. Solving Eq. (9.34) for $\hat{W}_i(z)$ and using the z-transform

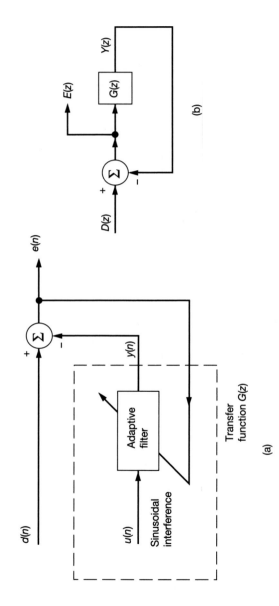

Figure 9.6 (a) New representation of adaptive noise canceler, (b) Equivalent model in the z-domain.

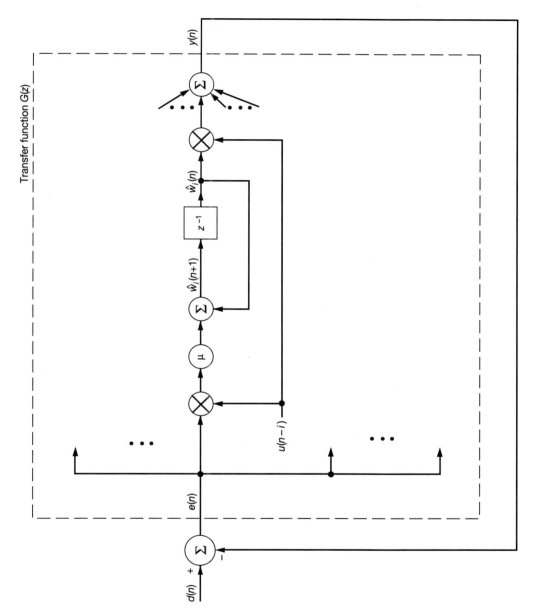

Figure 9.7 Signal-flow graph representation of adaptive noise canceler, singling out the ith tap weight for detailed attention.

given in Eq. (9.33), we get

$$\hat{W}_i(z) = \frac{\mu A}{2} \frac{1}{z-1} [e^{j\phi_i} E(ze^{-j\omega_0}) + e^{-j\phi_i} E(ze^{j\omega_0})] \tag{9.35}$$

We turn next to Eq. (9.29) that defines the adaptive filter output $y(n)$. Substituting Eq. (9.32) in (9.29), we get

$$y(n) = \frac{A}{2} \sum_{i=0}^{M-1} \hat{w}_i(n)[e^{j(\omega_0 n + \phi_i)} + e^{-j(\omega_0 n + \phi_i)}]$$

Hence, evaluating the z-transform of $y(n)$, we obtain

$$Y(z) = \frac{A}{2} \sum_{i=0}^{M-1} [e^{j\phi_i} \hat{W}_i(ze^{-j\omega_0}) + e^{-j\phi_i} \hat{W}_i(ze^{j\omega_0})] \tag{9.36}$$

Thus, using Eq. (9.35) in (9.36), we obtain an expression for $Y(z)$ that consists of the sum of two components (Glover, 1977):

1. A *time-invariant component* defined by

$$\frac{\mu M A^2}{4} \left(\frac{1}{ze^{-j\omega_0} - 1} + \frac{1}{ze^{j\omega_0} - 1} \right)$$

   which is independent of the phase ϕ_i and, therefore, the time index i.

2. A *time-varying component* that is dependent on the phase ϕ_i, hence the variation with time i. This second component is scaled in amplitude by the factor

$$\beta(\omega_0, M) = \frac{\sin(M\omega_0)}{\sin \omega_0}$$

For a given angular frequency ω_0, we assume that the total number of tap weights M in the transversal filter is large enough to satisfy the following approximation:

$$\frac{\beta(\omega_0, M)}{M} = \frac{\sin(M\omega_0)}{M \sin \omega_0} \approx 0 \tag{9.37}$$

Accordingly, we may justifiably ignore the time-varying component of the z-transform $Y(z)$, and so approximate $Y(z)$ by retaining the time-invariant component only:

$$Y(z) \approx \frac{\mu M A^2}{4} E(z) \left(\frac{1}{ze^{-j\omega_0} - 1} + \frac{1}{ze^{j\omega_0} - 1} \right) \tag{9.38}$$

The open-loop transfer function $G(z)$ is therefore

$$G(z) = \frac{Y(z)}{E(z)}$$

$$\approx \frac{\mu M A^2}{4} \left(\frac{1}{ze^{-j\omega_0} - 1} + \frac{1}{ze^{j\omega_0} - 1} \right) \tag{9.39}$$

$$\approx \frac{\mu M A^2}{2} \left(\frac{z \cos \omega_0 - 1}{z^2 - 2z \cos \omega_0 + 1} \right)$$

The transfer function $G(z)$ has two complex-conjugate poles on the unit circle at $z = e^{\pm j\omega_0}$ and a real zero at $z = 1/\cos \omega_0$, as illustrated in Fig. 9.8(a). In other words, the adaptive noise canceler has a null point determined by the angular frequency ω_0 of the sinusoidal interference, as stated previously (see characteristic 1). Indeed, according to Eq. (9.39), we may view $G(z)$ as a pair of *integrators* that have been rotated by $\pm\omega_0$. In actual fact, we see from Fig. 9.7 that it is the input that is first shifted in frequency by an amount $\pm\omega_0$ due to the first multiplication by the reference sinusoid $u(n)$, digitally integrated at zero frequency, and then shifted back again by the second multiplication. This overall operation is similar to a well-known technique in communications for obtaining a resonant filter that involves the combined use of two low-pass filters and heterodyning with sine and cosine at the resonant frequency (Wozencraft and Jacobs, 1965; Glover, 1977).

The model of Fig. 9.6 is recognized as a *closed-loop feedback system*, whose transfer function $H(z)$ is related to the open-loop transfer function $G(z)$ as follows:

$$H(z) = \frac{E(z)}{D(z)} \tag{9.40}$$

$$= \frac{1}{1 + G(z)}$$

where $E(z)$ is the z-transform of the system output $e(n)$, and $D(z)$ is the z-transform of the system input $d(n)$. Accordingly, substituting Eq. (9.39) in (9.40), we get the approximate result

$$H(z) \approx \frac{z^2 - 2z \cos \omega_0 + 1}{z^2 - 2(1 - \mu MA^2/4)z \cos \omega_0 + (1 - \mu MA^2/2)} \tag{9.41}$$

Equation (9.41) is the transfer function of a *second-order digital notch filter* with a notch at the normalized angular frequency ω_0. The zeros of $H(z)$ are at the poles of $G(z)$; that is, they are located on the unit circle at $z = e^{\pm j\omega_0}$. For a small value of the step-size parameter μ (i.e., a slow adaptation rate), such that

$$\frac{\mu MA^2}{4} \ll 1$$

we find that the poles of $H(z)$ are approximately located at

$$z \approx \left(1 - \frac{\mu MA^2}{4}\right) e^{\pm j\omega_0} \tag{9.42}$$

In other words, the two poles of $H(z)$ lie inside the unit circle, a radial distance approximately equal to $\mu MA^2/4$ behind the zeros, as indicated in Fig. 9.8(b). The fact that the poles of $H(z)$ lie inside the unit circle means that the adaptive noise canceler is stable, as it should be for practical use in real time.

Figure 9.8(b) also includes the *half-power points* of $H(z)$. Since the zeros of $H(z)$ lie on the unit circle, the adaptive noise canceler has (in theory) a notch of infinite depth (in dB) at $\omega = \omega_0$. The *sharpness* of the notch is determined by the closeness of the poles of $H(z)$ to its zeros. The *3-dB bandwidth, B,* is determined by locating the two half-power points on the unit circle that are $\sqrt{2}$ times as far from the poles as they are from the zeros. Using this geometric approach, we find that the 3-dB bandwidth of the adaptive noise canceler is approximately

$$B \approx \frac{\mu MA^2}{2} \text{ radians} \tag{9.43}$$

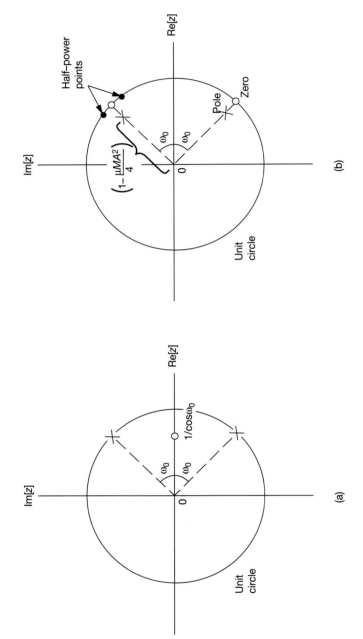

Figure 9.8 Approximate pole-zero patterns of (a) the open-loop transfer function $G(z)$. and (b) the closed-loop transfer function $H(z)$.

The smaller we therefore make μ, the smaller the bandwidth B is, and therefore the sharper the notch is. This confirms characteristic 2 of the adaptive noise canceler that was mentioned previously. Its analysis is thereby completed.

Example 5: Adaptive Line Enhancement

The *adaptive line enhancer (ALE)*, illustrated in Fig. 9.9, is a device that may be used to detect a periodic signal buried in a broad-band noise background.[2] This figure shows that the ALE is in fact a degenerate form of the adaptive noise canceler in that its reference signal, instead of being derived separately, consists of a delayed version of the primary (input) signal. The delay, denoted by Δ in Fig. 9.9, is called the *prediction depth* or *decorrelation delay* of the ALE, measured in units of the sampling period. The reference signal $u(n - \Delta)$ is processed by a transversal filter to produce an error signal $e(n)$, defined as the difference between the actual input $u(n)$ and the ALE's output $y(n) = u(n)$. The error signal $e(n)$ is, in turn, used to actuate the LMS algorithm for adjusting the M tap weights of the transversal filter.

Consider an input signal $u(n)$ that consists of a sinusoidal component $A\sin(\omega_0 n + \phi_0)$ buried in wide-band noise $v(n)$, as shown by

$$u(n) = A \sin(\omega_0 n + \phi_0) + v(n) \tag{9.44}$$

where ϕ_0 is an arbitrary phase shift, and the noise $v(n)$ is assumed to have zero mean and variance σ_v^2. The ALE acts as a signal detector by virtue of two actions (Treichler, 1979):

- The prediction depth Δ is assigned a value large enough to *remove* the correlation between the noise $v(n)$ in the original input signal and the noise $v(n - \Delta)$ in the reference signal, while a simple phase shift equal to $\omega_0\Delta$ is introduced between the sinusoidal components in these two inputs.

- The tap weights of the transversal filter are adjusted by the LMS algorithm so as to minimize the mean-square value of the error signal and thereby compensate for the phase shift $\omega_0\Delta$.

The net result of these two actions is the production of an output signal $y(n)$ that consists of a scaled sinusoid in noise of zero mean. In particular, when ω_0 is several multiples of π/M away from zero or π, Rickard and Zeidler (1979) have shown that

$$y(n) = aA \sin(\omega_0 n + \phi) + v_{\text{out}}(n) \tag{9.45}$$

where ϕ denotes a phase shift, and $v_{\text{out}}(n)$ denotes the output noise. The scaling factor a is defined by

$$a = \frac{(M/2)\,\text{SNR}}{1 + (M/2)\,\text{SNR}} \tag{9.46}$$

---

[2]The ALE owes its origin to Widrow et al. (1975). For a statistical analysis of its performance for the detection of sinusoidal signals in wide-band noise, see Zeidler et al. (1978), Treichler (1979), and Rickard and Zeidler (1979). For a tutorial treatment of the ALE, see Zeidler (1990); the effects of signal bandwidth, input SNR, noise correlation, and noise nonstationarity are explicitly considered in Zeidler's paper.

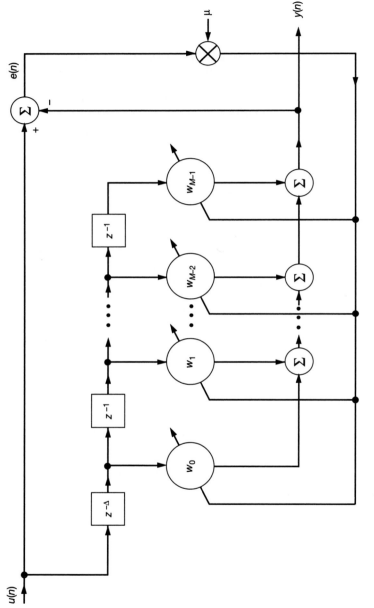

Figure 9.9 Adaptive line enhancer.

where M is the length of the transversal filter, and SNR denotes the *signal-to-noise ratio* at the input of the ALE:

$$\text{SNR} = \frac{A^2}{2\sigma_v^2} \tag{9.47}$$

According to Eq. (9.45), the ALE acts as a *self-tuning filter* whose frequency response exhibits a peak at the angular frequency ω_0 of the incoming sinusoid, hence the name "spectral line enhancer" or simply "line enhancer."

Rickard and Zeidler (1979) have also shown that the power spectral density of the ALE output $y(n)$ may be expressed as

$$S(\omega) = \frac{\pi A^2}{2}\,(a^2 + \mu\sigma_v^2 M)\,[\delta(\omega - \omega_0) + \delta(\omega + \omega_0)] + \mu\sigma_v^4 M$$

$$+ \frac{a^2\sigma_v^2}{M^2}\left[\frac{1 - \cos M(\omega - \omega_0)}{1 - \cos(\omega - \omega_0)} + \frac{1 + \cos M(\omega - \omega_0)}{1 + \cos(\omega - \omega_0)}\right], \quad -\pi < \omega \le \pi \tag{9.48}$$

where $\delta(\cdot)$ denotes a Dirac delta function. To understand the composition of Eq. (9.48), we first recall from the overview of the LMS algorithm presented in Section 9.1 that, in a stationary environment, the mean of the weight vector $\hat{\mathbf{w}}(n)$ converges to the Wiener solution $\mathbf{w}_o(n)$. A formal analysis of this behavior is presented in the next section. For now it suffices to say that the steady-state model of the converged weight vector consists of the Wiener solution $\mathbf{w}_o$ acting in parallel with a slowly fluctuating, zero-mean random component $\boldsymbol{\epsilon}(n)$ due to gradient noise. The ALE may thus be modeled as shown in Fig. 9.10.

Recognizing that the ALE input itself consists of two components, a sinusoid of angular frequency ω_0 and a wide-band noise $v(n)$ of zero mean and variance σ_v^2, we may distinguish four components in the power spectrum of Eq. (9.48) as described here (Zeidler, 1990):

- A sinusoidal component of angular frequency ω_0 and average power $\pi a^2 A^2/2$, which is the result of processing the input sinusoid by the Wiener filter represented by the weight vector $\mathbf{w}_o$
- A sinusoidal component of angular frequency ω_0 and average power $\pi\mu A^2\sigma_v^2 M/2$, which is due to the stochastic filter represented by the weight vector $\boldsymbol{\epsilon}(n)$ acting on the input sinusoid
- A wide-band noise component of variance $\mu\sigma_v^4 M$, which is due to the action of the stochastic filter on the noise $v(n)$
- A narrow-band filtered noise component centered on ω_0, which results from processing the noise $v(n)$ by the Wiener filter

These four components are depicted in Fig. 9.11, respectively. Thus, the power spectrum of the ALE output consists of a sinusoid at the center of a pedestal of narrow-band filtered noise, the combination of which is embedded in a wide-band noise background. Most importantly, when an adequate SNR exists at the ALE input, the ALE output is, on the average, approximately equal to the sinusoidal component present at the input, thereby providing a simple adaptive device for the detection of a sinusoidal in wide-band noise.

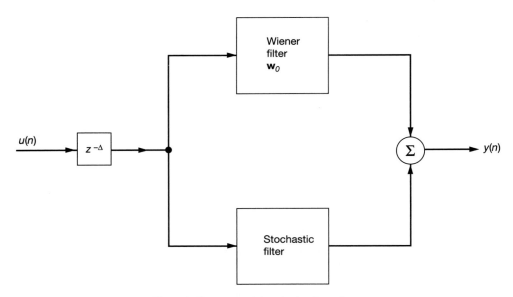

Figure 9.10   Model of the adaptive line enhancer.

Example 6: Adaptive Beamforming

In this final example, we consider a spatial application of the LMS algorithm, namely, that of adaptive beamforming. In particular, we revisit the *generalized sidelobe canceler* (GSC) that was studied under the umbrella of Wiener filter theory in Chapter 5.

Figure 9.12 shows a block diagram of the GSC, the operation of which hinges on the combination of two actions:

- The imposition of linear multiple constraints, designed to preserve an incident signal along a direction of interest
- The adjustment of some weights, in accordance with the LMS algorithm, so as to minimize the effects of interference and noise at the beamformer output

The multiple linear constraints are described by an M-by-L matrix $\mathbf{C}$, on the basis of which a signal blocking matrix $\mathbf{C}_a$ of size M-by-$(M - L)$ is defined by

$$\mathbf{C}_a^H \mathbf{C} = \mathbf{O} \tag{9.49}$$

In the GSC, the vector of weights assigned to the linear array of antenna elements is represented by

$$\mathbf{w}(n) = \mathbf{w}_q - \mathbf{C}_a \mathbf{w}_a(n) \tag{9.50}$$

where $\mathbf{w}_a(n)$ is the *adjustable-weight vector* and $\mathbf{w}_q$ is the *quiescent-weight vector*. The latter component is defined in terms of the constraint matrix $\mathbf{C}$ by

$$\mathbf{w}_q = \mathbf{C}(\mathbf{C}^H \mathbf{C})^{-1} \mathbf{g} \tag{9.51}$$

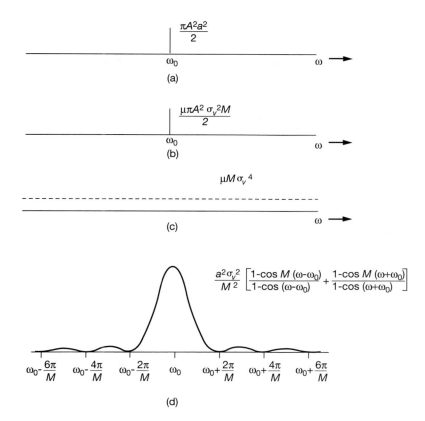

Figure 9.11 The four primary spectral components of the power spectral density at the ALE output. (a) Component due to Wiener filter acting on the input sinusoid. (b) Component due to stochastic filter acting on the input sinusoid. (c) Wide-band due to the stochastic filter acting on the noise $v(n)$. (d) Narrow-band noise due to the Wiener filter acting on $v(n)$.

where **g** is a prescribed gain vector.

The beamformer output is

$$e(n) = \mathbf{w}^H(n)\mathbf{u}(n)$$

$$= (\mathbf{w}_q - \mathbf{C}_a\mathbf{w}_a(n))^H\mathbf{u}(n) \tag{9.52}$$

$$= \mathbf{w}_q^H\mathbf{u}(n) - \mathbf{w}_a^H(n)\mathbf{C}_a^H\mathbf{u}(n)$$

In words, the quiescent weight vector $\mathbf{w}_q$ influences that part of the input vector $\mathbf{u}(n)$ that lies in the subspace spanned by the columns of constraint matrix $\mathbf{C}$, whereas the adjustable weight vector $\mathbf{w}_a(n)$ influences the remaining part of the input vector $\mathbf{u}(n)$ that lies in the complementary subspace spanned by the columns of signal blocking matrix $\mathbf{C}_a$. Note also the $e(n)$ in Eq. (9.52) is the same as $y(n)$ in Eq. (5.118).

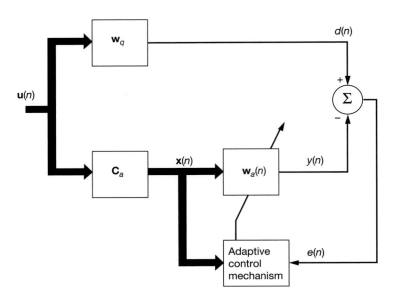

Figure 9.12   Block diagram of the generalized sidelobe canceler.

According to Eq. (9.52), the inner product $\mathbf{w}_q^H\mathbf{u}(n)$ plays the role of desired response:

$$d(n) = \mathbf{w}_q^H\mathbf{u}(n)$$

By the same token, the matrix product $\mathbf{C}_a^H\mathbf{u}(u)$ plays the role of input vector for the adjustable weight vector $\mathbf{w}_a(n)$; to emphasize this point, we let

$$\mathbf{x}(n) = \mathbf{C}_a^H\mathbf{u}(n)$$

We are now ready to formulate the LMS algorithm for the adaptation of weight vector $\mathbf{w}_a(n)$ in the GSC. Specifically, we may write

$$
\begin{aligned}
\mathbf{w}_a(n+1) &= \mathbf{w}_a(n) + \mu\mathbf{x}(n)e^*(n) \\
&= \mathbf{w}_a(n) + \mu\mathbf{C}_a^H\mathbf{u}(n)\,(\mathbf{w}_q^H\mathbf{u}(n) - \mathbf{w}_a^H(n)\mathbf{C}_a^H\mathbf{u}(n))^* \qquad (9.53) \\
&= \mathbf{w}_a(n) + \mu\mathbf{C}_a^H\mathbf{u}(n)\mathbf{u}^H(n)\,(\mathbf{w}_q - \mathbf{C}_a\mathbf{w}_a(n))
\end{aligned}
$$

where μ is the step-size parameter, and all of the remaining quantities are displayed in the block diagram of Fig. 9.12.

9.4 STABILITY AND PERFORMANCE ANALYSIS OF THE LMS ALGORITHM

In this section we present a stability and performance analysis of the LMS algorithm that is largely based on the mean-squared value of the estimation error $e(n)$. For this analysis, we find it convenient to work with the weight-error vector rather than the tap-weight vector itself. To avoid confusion with the notation used in the study of the method of steepest descent in Chapter 8, we denote the *weight-error vector* in the LMS algorithm by

$$\boldsymbol{\epsilon}(n) = \hat{\mathbf{w}}(n) - \mathbf{w}_o \tag{9.54}$$

where, as before, $\mathbf{w}_o$ denotes the optimum Wiener solution for the tap-weight vector, and $\hat{\mathbf{w}}(n)$ is the estimate produced by the LMS algorithm at iteration n. Subtracting the optimum tap-weight vector $\mathbf{w}_o$ from both sides of Eq. (9.5), and using the definition of Eq. (9.54) to eliminate $\hat{\mathbf{w}}(n)$ from the correction term on the right-hand side of Eq. (9.5), we may rewrite the LMS algorithm in terms of the weight-error vector $\boldsymbol{\epsilon}(n)$ as follows:

$$\boldsymbol{\epsilon}(n+1) = [\mathbf{I} - \mu\mathbf{u}(n)\mathbf{u}^H(n)]\boldsymbol{\epsilon}(n) + \mu\mathbf{u}(n)e_o^*(n) \tag{9.55}$$

where $\mathbf{I}$ is the identity matrix, and $e_o(n)$ is the *estimation error produced in the optimum Wiener solution*:

$$e_o(n) = d(n) - \mathbf{w}_o^H\mathbf{u}(n)$$

Direct-Averaging Method

Equation (9.55) is a *stochastic difference equation* in the weight-error vector $\boldsymbol{\epsilon}(n)$ with the following characteristic feature:

- A system matrix equal to $[\mathbf{I} - \mu\mathbf{u}(n)\mathbf{u}^H(n)]$, which is approximately equal to the identity matrix $\mathbf{I}$ for all n, provided that the step-size parameter μ is sufficiently small.

To study the convergence behavior of such a stochastic algorithm in an average sense, we may invoke the *direct-averaging method* described in Kushner (1984). According to this method, the solution of the stochastic difference equation (9.55), operating under the assumption of a *small* step-size parameter μ, is close to the solution of another stochastic difference equation whose system matrix is equal to the ensemble average:

$$E[\mathbf{I} - \mu\mathbf{u}(n)\mathbf{u}^H(n)] = \mathbf{I} - \mu\mathbf{R}$$

where $\mathbf{R}$ is the correlation matrix of the tap-input vector $\mathbf{u}(n)$. More specifically, we may replace the stochastic difference equation (9.55) with another stochastic difference equation described by

$$\boldsymbol{\epsilon}(n+1) = (\mathbf{I} - \mu\mathbf{R})\,\boldsymbol{\epsilon}(n) + \mu\mathbf{u}(n)e_o^*(n) \tag{9.56}$$

To be exact, the notation used in the new stochastic difference equation (9.56) should be different from that in the original stochastic difference equation (9.55). We have chosen *not* to do so here merely for convenience of presentation.

The direct averaging method is a reasonable heuristic, since it is based on the idea that, with small step sizes, the randomness of $\boldsymbol{\epsilon}(n)$ will tend to average out. The method applies rigorously, under appropriate conditions, to general stochastic approximation problems with small step sizes.

Independence Theory

In what follows, we will restrict ourselves to a statistical analysis of the LMS algorithm under the *independence assumption*, consisting of four points:

1. The tap-input vectors $\mathbf{u}(1)$, $\mathbf{u}(2)$, . . . $\mathbf{u}(n)$ constitute a sequence of statistically independent vectors.
2. At time n, the tap-input vector $\mathbf{u}(n)$ is statistically independent of all previous samples of the desired response, namely, $d(1)$, $d(2)$, . . . , $d(n-1)$.
3. At time n, the desired response $d(n)$ is dependent on the corresponding tap-input vector $\mathbf{u}(n)$, but statistically independent of all previous samples of the desired response.
4. The tap-input vector $\mathbf{u}(n)$ and the desired response $d(n)$ consist of *mutually Gaussian-distributed* random variables for all n.

The statistical analysis of the LMS algorithm so based is called the *independence theory*.[3]

From Eq. (9.5), we observe that the tap-weight vector $\hat{\mathbf{w}}(n+1)$ at time $n+1$ depends *only* on three inputs:

1. The previous sample vectors of the input process, $\mathbf{u}(n)$, $\mathbf{u}(n-1)$, . . . , $\mathbf{u}(1)$.
2. The previous samples of the desired response, $d(n)$, $d(n-1)$, . . . , $d(1)$.
3. The initial value of the tap-weight vector, $\hat{\mathbf{w}}(0)$.

Accordingly, in view of points 1 and 2 in the independence assumption, we find that the tap-weight vector $\hat{\mathbf{w}}(n+1)$, and therefore the weight-error vector $\boldsymbol{\epsilon}(n+1)$, is independent of both $\mathbf{u}(n+1)$ and $d(n+1)$. This is a very useful observation and one that will be used repeatedly in the sequel. The significance of the other two points in the independence assumption will become apparent as we proceed with the analysis.

---

[3]The independence theory of the LMS algorithm may be traced back to Widrow et al. (1976) and Mazo (1979).

Convergence analysis of the LMS algorithm remains an active area of research, which is motivated by a desire to relax the independence assumption. For a detailed discussion of a realistic model to prove convergence of the LMS algorithm, see Chapter 4 of the book by Macchi (1995); the model described therein assumes *no* independence between successive input vectors.

The independence assumption may be justified in certain applications such as adaptive beamforming, where it is possible for successive snapshots of data (i.e., input vectors) received by an array of antenna elements from the surrounding environment to be independent from each other. However, in adaptive filtering applications in communications (e.g., signal prediction, channel equalization, and echo cancelation), the sequence of input vectors that direct the "hunting" of the weight vector toward the optimum Wiener solution are in fact statistically dependent. This dependence arises because of the *shifting property* of the input data. Specifically, the tap-input vector at time n is

$$\mathbf{u}(n) = [u(n), u(n-1), \cdots, u(n-M+1)]^T$$

At time $n + 1$, it takes on the new value

$$\mathbf{u}(n+1) = [u(n+1), u(n), \cdots, u(n-M)]^T$$

Thus, with the arrival of the new sample $u(n+1)$, the oldest sample $u(n-M+1)$ is discarded from $\mathbf{u}(n)$, and the remaining samples $u(n), u(n-1) \cdots, u(n-M+2)$ are shifted back in time by one time unit to make room for the new sample. We see therefore that in a temporal setting the tap-input vectors, and correspondingly the gradient directions computed by the LMS algorithm, are indeed *statistically dependent*.

The independence theory *ignores* the statistical dependence among successive tap-input vectors at certain points in the development of the theory. For example, to evaluate the expectation of the term $\mathbf{u}(n)\mathbf{u}^H(n)\boldsymbol{\epsilon}(n)\boldsymbol{\epsilon}^H(n)$, it is assumed that the weight-error vector $\boldsymbol{\epsilon}(n)$ and the tap-input vector $\mathbf{u}(n)$ are statistically independent (even though, in reality, they may not be), and so this expectation is treated as the product of two expectations:

$$E[\mathbf{u}(n)\mathbf{u}^H(n)\boldsymbol{\epsilon}(n)\boldsymbol{\epsilon}^H(n)] = E[\mathbf{u}(n)\mathbf{u}^H(n)] \, E[\boldsymbol{\epsilon}(n)\boldsymbol{\epsilon}^H(n)]$$

Ironically, when it comes to evaluating the expectation $E[\mathbf{u}(n)\mathbf{u}^H(n)]$, the correlation structure of the source producing $\mathbf{u}(n)$ is fully preserved, likewise for $E[\boldsymbol{\epsilon}(n)\boldsymbol{\epsilon}^H(n)]$. In so doing, the independence theory retains sufficient information about the structure of the adaptive process for the results of the theory to serve as reliable design guidelines.

Convergence Criteria

On the basis of Eq. (9.56), we could go on to establish the condition necessary for the *convergence of the mean*; that is,

$$E[\boldsymbol{\epsilon}(n)] \to \mathbf{0} \qquad \text{as } n \to \infty$$

or, equivalently,

$$E[\hat{\mathbf{w}}(n)] \to \mathbf{w}_o \qquad \text{as } n \to \infty \qquad (9.57)$$

where $\mathbf{w}_o$ is the Wiener solution (see Problem 4 for details). Unfortunately, such a convergence criterion is of little practical value, since a sequence of zero-mean, but otherwise arbitrary, random variables converges in this sense. A stronger criterion is *convergence in the mean*, which is described by

$$E[\|\boldsymbol{\epsilon}(n)\|] \to 0 \qquad \text{as } n \to \infty$$

where $\|\boldsymbol{\epsilon}(n)\|$ is the Euclidean norm of the weight-error vector $\boldsymbol{\epsilon}(n)$. However, a proof of convergence in the mean is rather tedious, because $\|\boldsymbol{\epsilon}(n)\|$ is singular at the origin (Macchi, 1995).

To avoid the shortcomings of the two criteria, convergence of the mean and convergence in the mean, we may consider *convergence in the mean square*. Specifically, we say that the LMS algorithm is convergent in the mean square if

$$\mathcal{D}(n) = E[\|\boldsymbol{\epsilon}(n)\|^2] \to \text{constant} \qquad \text{as } n \to \infty \tag{9.58}$$

where the scalar quantity $\mathcal{D}(n)$ is called the *squared error deviation*. Another way of describing convergence of the LMS algorithm in the mean square is to require that

$$J(n) = E[|e(n)|^2] \to \text{constant} \qquad \text{as } n \to \infty \tag{9.59}$$

where $e(n)$ is the estimation error and $J(n)$ is the mean-squared error. Later in the section, we will show that

$$\lambda_{\min} \mathcal{D}(n) \le J_{\text{ex}}(n) \le \lambda_{\max} \mathcal{D}(n) \qquad \text{for all } n \tag{9.60}$$

where $\lambda_{\min}$ and $\lambda_{\max}$ are, respectively, the minimum and maximum eigenvalues of the correlation matrix $\mathbf{R}$; and $J_{\text{ex}}(n)$ is the excess mean-squared error, that is, the difference between $J(n)$ and the minimum mean-squared error $J_{\min}$ produced by the optimum Wiener filter. Hence, in light of this two-fold inequality, we may state that the decays of $J_{\text{ex}}(n)$ and $\mathcal{D}(n)$ for increasing n are mathematically equivalent (Macchi, 1995). Therefore, it suffices for us to focus our attention on the convergence criterion described in (9.59).

With this convergence criterion for the LMS algorithm in mind, we propose to proceed as follows:

1. We use the stochastic difference equation (9.56) to derive a recursive equation for computing the correlation matrix of the weight-error vector $\boldsymbol{\epsilon}(n)$.

2. We then go on to derive an expression for the excess mean-squared error $J_{\text{ex}}(n)$.

These derivations are presented in the next two subsections, respectively.

Weight-Error Correlation Matrix

By definition, the correlation matrix of the weight-error vector $\boldsymbol{\epsilon}(n)$ is

$$\mathbf{K}(n) = E[\boldsymbol{\epsilon}(n)\boldsymbol{\epsilon}^H(n)] \tag{9.61}$$

Hence, applying this definition to the stochastic difference equation (9.56) and then invoking the independence assumption, we get

$$\mathbf{K}(n + 1) = (\mathbf{I} - \mu\mathbf{R})\mathbf{K}(n)(\mathbf{I} - \mu\mathbf{R}) + \mu^2 J_{\min}\mathbf{R} \tag{9.62}$$

Equation (9.62) may be justified as follows:

- The first term, $(\mathbf{I} - \mu\mathbf{R})\mathbf{K}(n) (\mathbf{I} - \mu\mathbf{R})$, is the result of evaluating the expectation of the outer product of $(\mathbf{I} - \mu\mathbf{R})\boldsymbol{\epsilon}(n)$ with itself.
- The expectation of the cross-product term, $\mu e_o(n)(\mathbf{I} - \mu\mathbf{R})\boldsymbol{\epsilon}(n)\mathbf{u}^H(n)$, is zero by virtue of the implied independence of $\boldsymbol{\epsilon}(n)$ and $\mathbf{u}(n)$.
- The last term, $\mu^2 J_{\min}\mathbf{R}$, is obtained by applying the Gaussian factorization theorem to the product term $\mu^2 e_o{}^*(n)\mathbf{u}(n)\mathbf{u}^H(n)e_o(n)$; for details, see Problem 5.

The last term, $\mu^2 J_{\min}\mathbf{R}$, on the right-hand side of Eq. (9.62) prevents $\mathbf{K}(n) = \mathbf{O}$ from being a solution to this equation. Accordingly, the correlation matrix $\mathbf{K}(n)$ is prevented from going to zero by this small forcing term. In particular, the weight-error vector $\boldsymbol{\epsilon}(n)$ only approaches zero, but then executes small fluctuations about zero.[4] This formally confirms the point we made earlier under Example 5 that the convergent weight vector of the LMS algorithm may be modeled as shown in Fig. 9.10.

Since the correlation matrix $\mathbf{R}$ is positive definite, and μ is small, it follows that the first term of the expression on the right-hand side of Eq. (9.62) is also positive definite, provided that $\mathbf{K}(n)$ is positive definite. Clearly, the last term of this expression is always positive definite. It follows therefore that $\mathbf{K}(n + 1)$ is positive definite, provided that $\mathbf{K}(n)$ is. The proof by induction is completed by noting that $\mathbf{K}(0)$ is positive definite, where

$$\mathbf{K}(0) = \boldsymbol{\epsilon}(0)\boldsymbol{\epsilon}^H(0)$$

$$= [\hat{\mathbf{w}}(0) - \mathbf{w}_o][\hat{\mathbf{w}}^H(0) - w_o^H]$$

In summary, Eq. (9.62) represents a recursive relationship for updating the weight-error correlation matrix $\mathbf{K}(n)$, starting with $n = 0$, for which we have $\mathbf{K}(0)$. Furthermore, after each iteration it does yield a positive-definite answer for the updated value of the weight-error correlation matrix.

Excess Mean-Squared Error

The matrix difference equation(9.62) provides a useful tool for determining the transient behavior of the mean-square error of the LMS algorithm, based on the independence

---

[4]In Bucklew et al., (1993), it is shown that the fluctuations of the weight-error vector $\boldsymbol{\epsilon}(n)$ about zero are asymptotically Gaussian in distribution. This asymptotic distribution is the result of two notions:
- a form of the central limit theorem
- assumed ergodicity of the input and disturbances in the LMS algorithm.

assumption. Specifically, we may proceed as follows. First, we use Eqs. (9.6) and (9.7) to express the estimation error, $e(n)$, produced by the LMS algorithm as

$$e(n) = d(n) - \hat{\mathbf{w}}^H(n)\mathbf{u}(n)$$
$$= d(n) - \mathbf{w}_o^H\mathbf{u}(n) - \boldsymbol{\epsilon}^H(n)\mathbf{u}(n) \qquad (9.63)$$
$$= e_o(n) - \boldsymbol{\epsilon}^H(n)\mathbf{u}(n)$$

where $e_o(n)$ is the estimation error in the optimum Wiener solution, and $\boldsymbol{\epsilon}(n)$ is the tap-weight error vector. Let $J(n)$ denote the mean-squared error due to the LMS algorithm at iteration n. Hence, using Eq. (9.63) to evaluate $J(n)$ and then invoking the independence assumption, we get

$$J(n) = E[|e(n)|^2]$$
$$= E[(e_o(n) - \boldsymbol{\epsilon}^H(n)\mathbf{u}(n))(e_o^*(n) - \mathbf{u}^H(n)\boldsymbol{\epsilon}(n))] \qquad (9.64)$$
$$= J_{\min} + E[\boldsymbol{\epsilon}^H(n)\mathbf{u}(n)\mathbf{u}^H(n)\boldsymbol{\epsilon}(n)]$$

where $J_{\min}$ is the minimum mean-squared error produced by the optimum Wiener filter.

Our next task is to evaluate the expectation term in the final line of Eq. (9.64). Here we note that this term is the expected value of a scalar random variable represented by a triple vector product; and the trace of a scalar is the scalar itself. We may therefore rewrite it as[5]

$$E[\boldsymbol{\epsilon}^H(n)\mathbf{u}(n)\mathbf{u}^H(n)\boldsymbol{\epsilon}(n)] = E[\text{tr}\{\boldsymbol{\epsilon}^H(n)\mathbf{u}(n)\mathbf{u}^H(n)\boldsymbol{\epsilon}(n)\}]$$
$$= E[\text{tr}\{\mathbf{u}(n)\mathbf{u}^H(n)\boldsymbol{\epsilon}(n)\boldsymbol{\epsilon}^H(n)\}]$$
$$= \text{tr}\{E[\mathbf{u}(n)\mathbf{u}^H(n)\boldsymbol{\epsilon}(n)\boldsymbol{\epsilon}^H(n)]\}$$

Invoking the independence assumption again, we may reduce this expectation to

$$E[\boldsymbol{\epsilon}^H(n)\mathbf{u}(n)\mathbf{u}^H(n)\boldsymbol{\epsilon}(n)] = \text{tr}\{E[\mathbf{u}(n)\mathbf{u}^H(n)]E[\boldsymbol{\epsilon}(n)\boldsymbol{\epsilon}^H(n)]\} \qquad (9.65)$$
$$= \text{tr}[\mathbf{R}\mathbf{K}(n)]$$

where $\mathbf{R}$ is the correlation matrix of the tap inputs and $\mathbf{K}(n)$ is the weight-error correlation matrix.

---

[5]Let $\mathbf{A}$ and $\mathbf{B}$ denote a pair of matrices of compatible dimensions. The trace of matrix product $\mathbf{AB}$ equals the trace of matrix product $\mathbf{BA}$; that is,

$$\text{tr}[\mathbf{AB}] = \text{tr}[\mathbf{BA}]$$

For the problem at hand, we have

$$\mathbf{A} = \boldsymbol{\epsilon}^H(n)$$
$$\mathbf{B} = \mathbf{u}(n)\mathbf{u}^H(n)\boldsymbol{\epsilon}(n)$$

Therefore,

$$\text{tr}[\boldsymbol{\epsilon}^H(n)\mathbf{u}(n)\mathbf{u}^H(n)\boldsymbol{\epsilon}(n)] = \text{tr}[\mathbf{u}(n)\mathbf{u}^H(n)\boldsymbol{\epsilon}(n)\boldsymbol{\epsilon}^H(n)]$$

Accordingly, using Eq. (9.65) in Eq. (9.64), we may rewrite the expression for the mean-squared error in the LMS algorithm simply as

$$J(n) = J_{\min} + \text{tr}[\mathbf{R}\mathbf{K}(n)] \tag{9.66}$$

Equation (9.66) indicates that for all n, the mean-square value of the estimation error in the LMS algorithm consists of two components: the minimum mean-squared error $J_{\min}$, and a component depending on the transient behavior of the weight-error correlation matrix $\mathbf{K}(n)$. Since the latter component is positive definite for all n, *the LMS algorithm always produces a mean-squared error J(n) that is in excess of the minimum mean-squared error $J_{\min}$.*

We now formally define the *excess mean-squared error* as the difference between the mean-squared error, $J(n)$, produced by the adaptive algorithm at time n and the minimum value, $J_{\min}$, pertaining to the optimum Wiener solution. Denoting the excess mean-squared error by $J_{ex}(n)$, we have

$$J_{ex}(n) = J(n) - J_{\min}$$

$$= \text{tr}[\mathbf{R}\mathbf{K}(n)] \tag{9.67}$$

For $\mathbf{K}(n)$ we use the recursive relation of Eq. (9.62). However, when the mean-squared error is of primary interest, another form of this equation obtained by a simple rotation of coordinates is more useful. The particular rotation of coordinates we have in mind is described by the unitary similarity transformation of Eq. (4.30), reproduced here for convenience:

$$\mathbf{Q}^H \mathbf{R} \mathbf{Q} = \boldsymbol{\Lambda} \tag{9.68}$$

where $\boldsymbol{\Lambda}$ is a diagonal matrix consisting of the eigenvalues of the correlation matrix $\mathbf{R}$, and $\mathbf{Q}$ is the unitary matrix consisting of the eigenvectors associated with these eigenvalues. Note that the matrix $\boldsymbol{\Lambda}$ is real valued. Furthermore, let

$$\mathbf{Q}^H \mathbf{K}(n) \mathbf{Q} = \mathbf{X}(n) \tag{9.69}$$

In general, $\mathbf{X}(n)$ is not a diagonal matrix. Using Eqs. (9.68) and (9.69), we get

$$\text{tr}[\mathbf{R}\mathbf{K}(n)] = \text{tr}[\mathbf{Q}\boldsymbol{\Lambda}\mathbf{Q}^H\mathbf{Q}\mathbf{X}(n)\mathbf{Q}^H]$$

$$= \text{tr}[\mathbf{Q}\boldsymbol{\Lambda}\mathbf{X}(n)\mathbf{Q}^H]$$

$$= \text{tr}[\mathbf{Q}^H\mathbf{Q}\boldsymbol{\Lambda}\mathbf{X}(n)]$$

$$= \text{tr}[\boldsymbol{\Lambda}\mathbf{X}(n)] \tag{9.70}$$

where, in the third line, we used the matrix property described in footnote 5, and in the second and last lines we used the property $\mathbf{Q}^H\mathbf{Q} = \mathbf{I}$. Accordingly, we have

$$J_{ex}(n) = \text{tr}[\boldsymbol{\Lambda}\mathbf{X}(n)] \tag{9.71}$$

Since $\mathbf{\Lambda}$ is a diagonal matrix, we may also write[6]

$$J_{\text{ex}}(n) = \sum_{i=1}^{M} \lambda_i x_i(n) \tag{9.72}$$

where the $x_i(n)$, $i = 1, 2, \ldots, M$, are the diagonal elements of the matrix $\mathbf{X}(n)$, and λ_i are the eigenvalues of the correlation matrix $\mathbf{R}$.

Next, using the transformations described by Eqs. (9.68) and (9.69), we may rewrite the recursive equation (9.62) in terms of $\mathbf{X}(n)$ and $\mathbf{\Lambda}$ as follows:

$$\mathbf{X}(n + 1) = (\mathbf{I} - \mu\mathbf{\Lambda})\mathbf{X}(n)(\mathbf{I} - \mu\mathbf{\Lambda}) + \mu^2 J_{\min}\mathbf{\Lambda} \tag{9.73}$$

We observe from Eq. (9.72) that $J_{\text{ex}}(n)$ depends on the $x_i(n)$. This suggests that we need only look at the diagonal terms of the recursive equation (9.73). Because of the form of this equation, the x_i decouple from the off-diagonal terms, and so we have

$$x_i(n) = (1 - \mu\lambda_i)^2 x_i(n) + \mu^2 J_{\min}\lambda_i, \qquad i = 1, 2, \ldots, M \tag{9.74}$$

Define the M-by-1 vectors $\mathbf{x}(n)$ and $\mathbf{\lambda}$ as follows, respectively:

$$\mathbf{x}(n) = [x_1(n), x_2(n), \ldots, x_M(n)]^T$$

$$\mathbf{\lambda} = [\lambda_1, \lambda_2, \ldots, \lambda_M]^T$$

Then we may rewrite Eq. (9.74) in matrix form as

$$\mathbf{x}(n + 1) = \mathbf{B}\mathbf{x}(n) + \mu^2 J_{\min}\mathbf{\lambda} \tag{9.75}$$

where $\mathbf{B}$ is an M-by-M matrix with elements

$$b_{ij} = \begin{cases} (1 - \mu\lambda_i)^2, & i = j \\ \mu^2\lambda_i\lambda_j, & i \neq j \end{cases} \tag{9.76}$$

From Eq. (9.76), we readily see that the matrix $\mathbf{B}$ is *real, positive,* and *symmetric.*

---

[6]The two-fold inequality (9.60), referred to earlier under the subsection on convergence criteria, may be derived from Eq. (9.72) as follows. Starting with the definition of matrix $\mathbf{X}(n)$ given in Eq. (9.69), we note that

$$\mathbf{X}(n) = \mathbf{Q}^H E[\mathbf{\epsilon}(n)\mathbf{\epsilon}^H(n)]\mathbf{Q} = E[\mathbf{Q}^H\mathbf{\epsilon}(n)\mathbf{\epsilon}^H(n)\mathbf{Q}]$$

Let $\mathbf{\omega}(n) = \mathbf{Q}^H\mathbf{\epsilon}(n)$. Then, noting that $\mathbf{Q}$ is a unitary matrix, we may express the *i*th diagonal element of matrix $\mathbf{X}(n)$ as

$$x_i(n) = E[|\omega_i(n)|^2] = E[|\epsilon_i(n)|^2] \qquad \text{for all } i$$

where $\omega_i(n)$ and $\epsilon_i(n)$ are the *i*th elements of $\mathbf{\omega}(n)$ and $\mathbf{\epsilon}(n)$, respectively. Accordingly, using Eq. (9.72), we may write

$$J_{\text{ex}}(n) = \sum_{i=1}^{M} \lambda_i E[|\epsilon_i(n)|^2]$$

Let $\lambda_1 = \lambda_{\max}$ and $\lambda_M = \lambda_{\min}$. Since $\|\mathbf{\epsilon}(n)\|^2 = \sum_{i=1}^{M} |\epsilon_i(n)|^2$, we may bound $J_{\text{ex}}(n)$ as shown by

$$\lambda_{\min} E[\|\mathbf{\epsilon}(n)\|^2] \leq J_{\text{ex}}(n) \leq \lambda_{\max} E[\|\mathbf{\epsilon}(n)\|^2]$$

from which (9.60) follows immediately(Macchi, 1995).

In Appendix H it is shown that the solution to the difference equation (9.75) is given by

$$\mathbf{x}(n) = \sum_{i=1}^{M} c_i^n \mathbf{g}_i \mathbf{g}_i^T [\mathbf{x}(0) - \mathbf{x}(\infty)] + \mathbf{x}(\infty) \tag{9.77}$$

where the various terms are defined as follows:

- The coefficient c_i is the ith eigenvalue of matrix $\mathbf{B}$, and $\mathbf{g}_i$ is the associated eigenvector; that is,

$$\mathbf{G}^T \mathbf{B} \mathbf{G} = \mathbf{C} \tag{9.78}$$

  where

$$\mathbf{C} = \text{diag}[c_1, c_2, \ldots, c_M]$$
$$\mathbf{G} = [\mathbf{g}_1, \mathbf{g}_2, \ldots, \mathbf{g}_M]$$

- The vector $\mathbf{x}(0)$ is the initial value of $\mathbf{x}(n)$, and $\mathbf{x}(\infty)$ is its final value.

The excess mean-squared error equals [see Eq. (9.72)]

$$
\begin{aligned}
J_{\text{ex}}(n) &= \boldsymbol{\lambda}^T \mathbf{x}(n) \\
&= \sum_{i=1}^{M} c_i^n \boldsymbol{\lambda}^T \mathbf{g}_i \mathbf{g}_i^T [\mathbf{x}(0) - \mathbf{x}(\infty)] + \boldsymbol{\lambda}^T \mathbf{x}(\infty) \\
&= \sum_{i=1}^{M} c_i^n \boldsymbol{\lambda}^T \mathbf{g}_i \mathbf{g}_i^T [\mathbf{x}(0) - \mathbf{x}(\infty)] + J_{\text{ex}}(\infty)
\end{aligned} \tag{9.79}
$$

where

$$
\begin{aligned}
J_{\text{ex}}(\infty) &= \boldsymbol{\lambda}^T \mathbf{x}(\infty) \\
&= \sum_{j=1}^{M} \lambda_j x_j(\infty)
\end{aligned} \tag{9.80}
$$

In Eq. (9.79), the first term on the right-hand side describes the transient behavior of the mean-squared error, whereas the second term represents the final value of the excess mean-squared error after adaptation is completed (i.e., its steady-state value).

Transient Behavior of the Mean-squared Error

Using Eq. (9.79) and the first line of Eq. (9.67), we may express the time evolution of the mean-squared error for the LMS algorithm by the equation

$$J(n) = \sum_{i=1}^{M} \gamma_i c_i^n + J_{\text{min}} + J_{\text{ex}}(\infty) \tag{9.81}$$

where c_i is the ith eigenvalue of matrix $\mathbf{B}$, and γ_i is defined by

$$\gamma_i = \boldsymbol{\lambda}^T \mathbf{g}_i \mathbf{g}_i^T [\mathbf{x}(0) - \mathbf{x}(\infty)], \qquad i = 1, 2, \ldots, M \tag{9.82}$$

Equation (9.81) provides the basis for a deeper understanding of the operation of the LMS algorithm in a wide-sense stationary environment, as described next in the form of four properties.

Property 1. *The transient component of the mean-squared error, J (n), does not exhibit oscillations.*

The transient component of $J(n)$ equals

$$\sum_{i=1}^{M} \gamma_i c_i^n$$

where the γ_i are constant coefficients and the c_i, $i = 1, 2, \ldots, M$, are the eigenvalues of matrix **B**. These eigenvalues are all real positive numbers, since **B** is a real symmetric positive-definite matrix [see Eq. (9.76)]. Hence, the ensemble-averaged learning curve, that is, a plot of the mean-squared error $J(n)$ versus the number of iterations, n, consists only of exponentials.

Note, however, that the learning curve represented by a plot of the squared error $|e(n)|^2$, without ensemble averaging, versus n consists of *noisy* exponentials. The amplitude of the noise becomes smaller as the step-size parameter μ is reduced.

Property 2. *The transient component of the mean-squared error J (n) dies out; that is, the LMS algorithm is convergent in the mean square if and only if the step-size parameter μ satisfies the condition*

$$0 < \mu < \frac{2}{\lambda_{\max}} \tag{9.83}$$

where $\lambda_{\max}$ is the largest eigenvalue of the correlation matrix **R**.

For this property to hold, all the eigenvalues of matrix **B** must be less than 1 in magnitude. Let **g** be an eigenvector of matrix **B**, associated with eigenvalue c. Then, by definition, we have

$$\mathbf{Bg} = c\mathbf{g}$$

Equivalently, we may write

$$\sum_{j=1}^{M} b_{ij} g_j = c g_i, \quad i = 1, 2, \ldots, M \tag{9.84}$$

where the g_i are the elements of the eigenvector **g**. Using Eq. (9.76) for the elements of matrix **B** in Eq. (9.84), we get

$$(1 - \mu\lambda_i)^2 \, g_i + \mu^2 \lambda_i \sum_{\substack{j=1 \\ j \neq i}}^{M} \lambda_j g_j = c g_i, \quad i = 1, 2, \ldots, M \tag{9.85}$$

Solving Eq. (9.85) for g_i, we may thus write

$$g_i = \frac{\mu^2 \lambda_i}{c - (1 - \mu\lambda_i)^2} \sum_{\substack{j=1 \\ j \neq i}}^{M} \lambda_j g_j, \quad i = 1, 2, \ldots, M \tag{9.86}$$

Next, we acknowledge the fact that the square matrix **B** is a *positive matrix* since all of its elements are positive. This means that we may use *Perron's theorem*,[7] which applies to a positive square matrix. Perron's theorem states that (Bellman, 1960):

If **B** is a positive square matrix, there is a unique eigenvalue of **B**, which has the largest magnitude. This eigenvalue is positive and simple (i.e., of multiplicity 1), and its associated eigenvector consists entirely of positive elements.

Accordingly, we may associate a *positive eigenvector* (i.e., a vector consisting entirely of positive elements) with the special eigenvalue of matrix **B** that has the largest magnitude. Thus setting the eigenvalue c equal to 1 in Eq. (9.86), and then simplifying, we get

$$g_i = \frac{\mu}{2 - \mu\lambda_i} \sum_{\substack{j=1 \\ j \neq i}}^{M} \lambda_j g_j, \quad i = 1, 2, \ldots, M \tag{9.87}$$

from which we readily see that for g_i to be positive for all i, the step-size parameter μ must be upper bounded as in (9.83).

Property 3. *The final value of the excess mean-squared error is less than the minimum mean-squared error if the step-size parameter μ satisfies the condition*

$$\sum_{i=1}^{M} \frac{2\lambda_i}{2 - \mu\lambda_i} < 1 \tag{9.88}$$

where the λ_i, $i = 1, 2, \ldots, M$, are the eigenvalues of the correlation matrix **R**.

Given that the LMS algorithm is convergent in the mean square and therefore Property 2 holds, we find that as the number of iterations approaches infinity:

$$J_{ex}(\infty) = J(\infty) - J_{min}$$

To find the final value of the excess mean-squared error, $J_{ex}(\infty)$, we may go back to Eq. (9.74). In particular, setting $n = \infty$ and then solving the resulting equation for $x_i(\infty)$, we get

$$x_i(\infty) = \frac{\mu J_{min}}{2 - \mu\lambda_i}, \quad i = 1, 2, \ldots, M \tag{9.89}$$

---

[7]Perron's theorem is also known as the *Perron–Frobenius theorem*.

Hence, evaluating Eq. (9.72) for $n = \infty$ and then substituting Eq. (9.89) in the resultant, we get

$$J_{\text{ex}}(\infty) = \sum_{i=1}^{M} \lambda_i x_i(\infty)$$

$$= J_{\min} \sum_{i=1}^{M} \frac{\mu\lambda_i}{2 - \mu\lambda_i} \tag{9.90}$$

From this equation we readily see that $J_{\text{ex}}(\infty)$ is indeed less than $J_{\min}$, provided that the step-size parameter μ satisfies the condition described in Eq. (9.88).

Property 4. *The misadjustment, defined as the ratio of the steady-state value $J_{\text{ex}}(\infty)$ of the excess mean-squared error to the minimum mean-squared error $J_{\min}$, equals*

$$\mathcal{M} = \frac{J_{\text{ex}}(\infty)}{J_{\min}}$$

$$= \sum_{i=1}^{M} \frac{\mu\lambda_i}{2 - \mu\lambda_i} \tag{9.91}$$

which is less than unity if the step-size parameter μ satisfies the condition of Eq. (9.88).

The formula for $\mathcal{M}$ given in Eq. (9.91) follows immediately from (9.90).

The misadjustment $\mathcal{M}$ is a dimensionless quantity, providing a measure of how close the LMS algorithm is to optimality in the mean-square error sense. The smaller $\mathcal{M}$ is compared to unity, the more *accurate* is the adaptive filtering action being performed by the LMS algorithm. It is customary to express the misadjustment $\mathcal{M}$ as a percentage. Thus, for example, a misadjustment of 10 percent means that the LMS algorithm produces a mean-squared error (after adaptation is completed) that is 10 percent greater than the minimum mean-squared error $J_{\min}$. Such performance is ordinarily considered to be satisfactory in practice.

Simple Working Rules

In this rather long section we have presented a theoretical (albeit approximate) analysis of the stability and mean-squared error performance of the LMS algorithm when operating in a wide-sense stationary environment. The analysis has been based on (1) the direct-averaging method, assuming that the step-size parameter μ is assigned a small value and (2) the independence assumption. Notwithstanding these assumptions, it appears that in practice the theory presented herein holds for a reasonably wide range of values of μ (Widrow et al., 1976). In any event, given the lengthy material presented in this section, it is befitting that we finish the discussion with some helpful rules for the LMS design of adaptive filters.

The condition for the LMS algorithm to be convergent in the mean square, described in Eq. (9.83), requires knowledge of the largest eigenvalue $\lambda_{\max}$ of the correlation matrix **R**. In typical applications of the LMS algorithm, knowledge of $\lambda_{\max}$ is not available. To overcome this practical difficulty, the trace of $\mathbf{R}$ may be taken as a conservative estimate for $\lambda_{\max}$, in which case the condition of (9.83) may be reformulated as

$$0 < \mu < \frac{2}{\text{tr}[\mathbf{R}]} \tag{9.92}$$

where $\text{tr}[\mathbf{R}]$ denotes the trace of matrix $\mathbf{R}$. We may go one step further by noting that the correlation matrix $\mathbf{R}$ is not only positive definite but also Toeplitz with all of the elements on its main diagonal equal to $r(0)$. Since $r(0)$ is itself equal to the mean-square value of the input at each of the M taps in the transversal filter, we have

$$\text{tr}[\mathbf{R}] = M\,r(0)$$
$$= \sum_{k=0}^{M-1} E[\,|u(n-k)|^2\,] \tag{9.93}$$

Thus, using the term "tap-input power" to refer to the sum of the mean-square values of tap inputs $u(n), u(n-1), \ldots, u(n-M+1)$ in Fig. 9.1(b), we may restate the condition of Eq. (9.92) for convergence of the LMS algorithm in the mean square as

$$0 < \mu < \frac{2}{\text{tap-input power}} \tag{9.94}$$

Another formula that needs to be revisited is that of Eq. (9.91), pertaining to the misadjustment $\mathcal{M}$. In its present form, this formula is impractical for it requires knowledge of all the eigenvalues of the correlation matrix $\mathbf{R}$. However, assuming that the step-size parameter μ is small compared to the largest eigenvalue $\lambda_{\max}$, we may approximate Eq. (9.91) as follows:

$$\mathcal{M} \approx \frac{\mu}{2} \sum_{i=1}^{M} \lambda_i$$
$$= \frac{\mu}{2}\,(\text{tap-input power}) \tag{9.95}$$

Thus, the practical condition of (9.94) imposed on the step-size parameter μ not only assures the convergence of the LMS algorithm in the mean square, but also results in a misadjustment $\mathcal{M}$ that is less than unity, both of which are desirable goals in their own individual ways.

Define an *average eigenvalue* for the underlying correlation matrix $\mathbf{R}$ of the tap inputs as

$$\lambda_{\text{av}} = \frac{1}{M} \sum_{i=1}^{M} \lambda_i \tag{9.96}$$

Suppose also that the ensemble-averaged learning curve of the LMS algorithm is approximated by a single exponential with time constant $(\tau)_{\text{mse,av}}$. We may then use Eq. (8.31), developed for the method of steepest descent, to define the following *average time constant* for the LMS algorithm:

$$(\tau)_{\text{mse,av}} \approx \frac{1}{2\mu\lambda_{\text{av}}} \tag{9.97}$$

Hence, using the average values of Eqs. (9.96) and (9.97) in (9.95), we may redefine the misadjustment approximately as follows (Widrow and Stearns, 1985):

$$\mathcal{M} \approx \frac{\mu M \lambda_{av}}{2}$$

$$\simeq \frac{M}{4\tau_{mse,av}} \tag{9.98}$$

On the basis of this formula, we may now make the following observations:

1. The misadjustment $\mathcal{M}$ increases linearly with the filter length (number of taps) denoted by M, for a fixed $\tau_{mse,av}$.

2. The *settling time* of the LMS algorithm (i.e., the time taken for the transients to die out) is proportional to the average time constant $\tau_{mse,av}$. It follows therefore that the misadjustment $\mathcal{M}$ is inversely proportional to the settling time.

3. The misadjustment $\mathcal{M}$ is directly proportional to the step-size parameter μ, whereas the average time constant $\tau_{mse,av}$ is inversely proportional to μ. We therefore have conflicting requirements in that if μ is reduced so as to reduce the misadjustment, then the settling time of the LMS algorithm is increased. Conversely, if μ is increased so as to reduce the settling time, then the misadjustment is increased. Careful attention has therefore to be given to the choice of μ. (In Chapter 16 we will evaluate the additional constraints on the choice of μ when the environment in which the adaptive filter operates is time varying.)

Comparison of the LMS Algorithm with the Method of Steepest Descent

At this point in the discussion, it is informative for us to look at the LMS algorithm for stochastic inputs in light of what we know about the steepest-descent algorithm that we studied in Chapter 8.

Ideally, the minimum mean-squared error J_{min} is realized when the coefficient vector $\hat{\mathbf{w}}(n)$ of the transversal filter approaches the optimum value $\mathbf{w}_o$, defined by the Wiener–Hopf equations. Indeed, as shown in Section 8.5, the steepest-descent algorithm does realize this idealized condition as the number of iterations, n, approaches infinity. The steepest-descent algorithm has the capability to do this, because it uses *exact* measurements of the gradient vector at each iteration of the algorithm. On the other hand, the LMS algorithm relies on a *noisy* estimate for the gradient vector, with the result that the tap-weight vector estimate $\hat{\mathbf{w}}(n)$ only approaches the optimum value $\mathbf{w}_o$. Consequently, use of the LMS algorithm, after a large number of iterations, results in a mean-squared error $J(\infty)$ that is greater than the minimum mean-squared error J_{min}. The amount by which the actual value of $J(\infty)$ is greater than J_{min} is the excess mean-squared error.

There is another basic difference between the steepest-descent algorithm and the LMS algorithm. In Section 8.5, we showed that the steepest-descent algorithm has a well-defined learning curve, obtained by plotting the mean-squared error versus the number of iterations. For this algorithm, the learning curve consists of a sum of decaying exponentials, the number of which equals (in general) the number of tap coefficients. On the other hand, in individual applications of the LMS algorithm, we find that the learning curve consists of *noisy, decaying exponentials*. The amplitude of the noise usually becomes smaller as the step-size parameter μ is reduced.

Imagine now an *ensemble of adaptive transversal filters*. Each filter is assumed to use the LMS algorithm with the same step-size parameter μ and the same initial tap-weight vector $\hat{\mathbf{w}}(0)$. Also, each adaptive filter has individual stationary ergodic inputs that are selected at random from the same statistical population. We compute the *noisy learning curves* for this ensemble of adaptive filters by plotting the squared magnitude of the estimation error $e(n)$ versus the number of iterations n. To compute the *ensemble-averaged learning curve* of the LMS algorithm, that is, the plot of the mean-squared error $J(n)$ versus n, we take the average of these noisy learning curves over the ensemble of adaptive filters.

Thus two entirely different ensemble-averaging operations are used in the steepest-descent and LMS algorithms for determining their learning curves (i.e., plots of their mean-squared errors versus the learning period). In the steepest-descent algorithm, the correlation matrix $\mathbf{R}$ and the cross-correlation vector $\mathbf{p}$ are first computed through the use of ensemble-averaging operations applied to statistical populations of the tap inputs and the desired response; these values are then used to compute the learning curve of the algorithm. In the LMS algorithm, on the other hand, noisy learning curves are computed for an ensemble of adaptive LMS filters with identical parameters; the learning curve is then smoothed by averaging over the ensemble of noisy learning curves.

9.5 SUMMARY OF THE LMS ALGORITHM

Parameters: M = number of taps
$\qquad\qquad\ \ \mu$ = step-size parameter

$$0 < \mu < \frac{2}{\text{tap-input power}}$$

$$\text{tap-input power} = \sum_{k=0}^{M-1} E[|u(n-k)|^2]$$

Initialization: If prior knowledge on the tap-weight vector $\hat{\mathbf{w}}(n)$ is available, use it to select an appropriate value for $\hat{\mathbf{w}}(0)$. Otherwise, set $\hat{\mathbf{w}}(0) = \mathbf{0}$.
Data:
• Given: $\mathbf{u}(n)$ = M-by-1 tap-input vector at time n
$\qquad\qquad d(n)$ = desired response at time n
• To be computed:
$\qquad\ \ \hat{\mathbf{w}}(n+1)$ = estimate of tap-weight vector at time $n+1$
Computation: For $n = 0, 1, 2, \ldots$, compute
$$e(n) = d(n) - \hat{\mathbf{w}}^H(n)\mathbf{u}(n)$$
$$\hat{\mathbf{w}}(n+1) = \hat{\mathbf{w}}(n) + \mu\mathbf{u}(n)e^*(n)$$

9.6 COMPUTER EXPERIMENT ON ADAPTIVE PREDICTION

For our first computer experiment involving the LMS algorithm, we use a first-order, autoregressive (AR) process to study the effects of ensemble averaging on the transient characteristics of the LMS algorithm for *real data*.

Consider then an AR process $u(n)$ of order 1, described by the difference equation

$$u(n) = -au(n - 1) + v(n) \tag{9.99}$$

where a is the (one and only) parameter of the process, and $v(n)$ is a zero-mean white-noise process of variance σ_v^2. To estimate the parameter a, we may use an adaptive predictor of order 1, as depicted in Fig. 9.13. The *real* LMS algorithm for the adaptation of the (one and only) tap weight of the predictor is written as

$$\hat{w}(n + 1) = \hat{w}(n) + \mu u(n - 1)f(n) \tag{9.100}$$

where $f(n)$ is the prediction error, defined by

$$f(n) = u(n) - \hat{w}(n)u(n - 1) \tag{9.101}$$

Figure 9.14 shows plots of $\hat{w}(n)$ versus the number of iterations n for a single trial of the experiment, and the following two sets of conditions:

1. AR parameter: $a = -0.99$
   Variance of AR process $u(n)$: $\sigma_u^2 = 0.93627$
2. AR parameter: $a = +0.99$
   Variance of AR process $u(n)$: $\sigma_u^2 = 0.995$

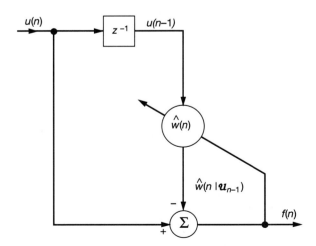

Figure 9.13   Adaptive first-order predictor.

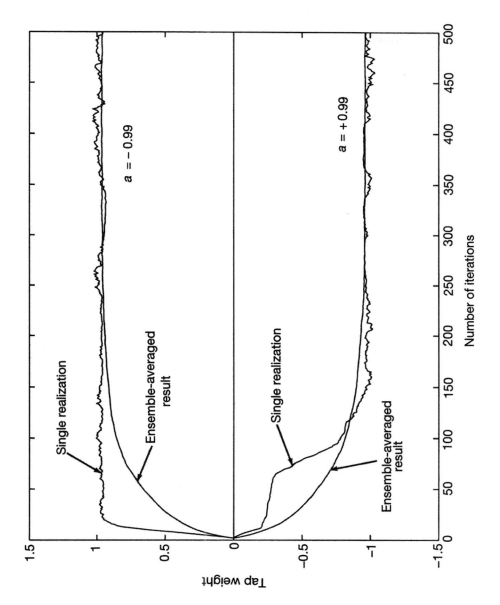

Figure 9.14 Transient behavior of weight $\hat{w}(n)$ of adaptive first-order predictor.

In both cases, the step-size parameter $\mu = 0.05$, and the initial condition is $\hat{w}(0) = 0$. We see that the transient behavior of $\hat{w}(n)$ follows a *noisy exponential curve*. Figure 9.14 also includes the corresponding plot of $E[\hat{w}(n)]$ obtained by ensemble averaging over 100 *independent trials* of the experiment. For each trial, a different computer realization of the AR process $u(n)$ is used. We see that the ensemble averaging has the effect of smoothing out the effects of gradient noise.

Figure 9.15 shows a plot of the squared prediction error $f^2(n)$ versus the number of iterations n for the set of AR parameters listed under (1) specified above; the step-size parameter $\mu = 0.05$, as before. We see that the learning curve for a single realization of the LMS algorithm exhibits a very noise form. Figure 9.15 also includes the corresponding plot of $E[f^2(n)]$ obtained by ensemble averaging over 100 independent trials of the experiment. The smoothing effect of the ensemble averaging operation on the learning curve of the LMS algorithm is again visible in the figure.

Figure 9.16 shows experimental plots of the learning curves of the LMS algorithm [i.e., the mean-squared error $J(n)$ versus the number of iterations n] for the set of AR parameters listed under (1) specified above and varying step-size parameter μ. Specifically, the values used for μ are 0.01, 0.05, and 0.1. The ensemble averaging was performed over 100 independent trials of the experiment. From Fig. 9.16, we observe the following:

- As the step-size parameter μ is reduced, the rate of convergence of the LMS algorithm is correspondingly decreased.
- A reduction in the step-size parameter μ also has the effect of reducing the variation in the experimentally computed learning curve.

Comparison of Experimental Results with Theory

In Fig. 9.17, we have plotted two pairs of curves for $\hat{w}(n)$ versus n, corresponding to the two different sets of parameter values listed under (1) and (2) above, and step-size parameter $\mu = 0.05$. One pair of curves corresponds to experimentally derived results obtained by ensemble-averaging over 100 independent trials of the experiment; these curves are labeled "Experiment" in Fig. 9.17. The other pair of curves is computed from theory. In particular, for the situation at hand, we have:

1. The autocorrelation function of the AR process $u(n)$ for zero lag is

$$r(0) = \sigma_u^2$$

2. The Wiener solution for the tap weight of the order-1 predictor is

$$w_o = -a$$

Taking the expectation of Eq. (9.100) and invoking the use of Eqs. (9.99) and (9.101), we find that in light of the independence assumption:

$$E[\hat{w}(n+1)] = (1 - \mu\sigma_u^2)E[\hat{w}(n)] - \mu a\sigma_u^2 \qquad (9.102)$$

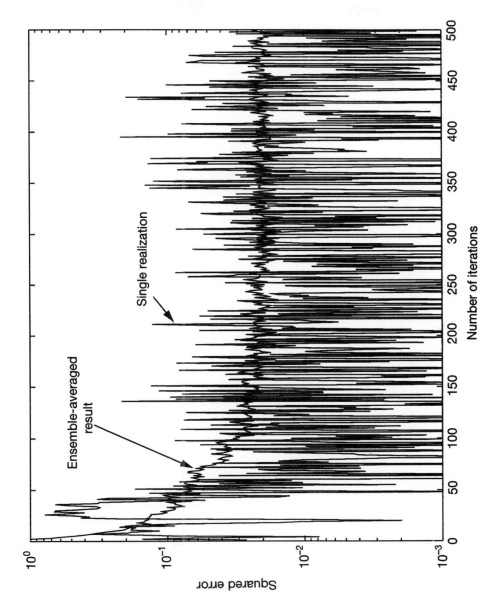

Figure 9.15 Transient behavior of squared prediction error in adaptive first-order for $\mu = 0.05$.

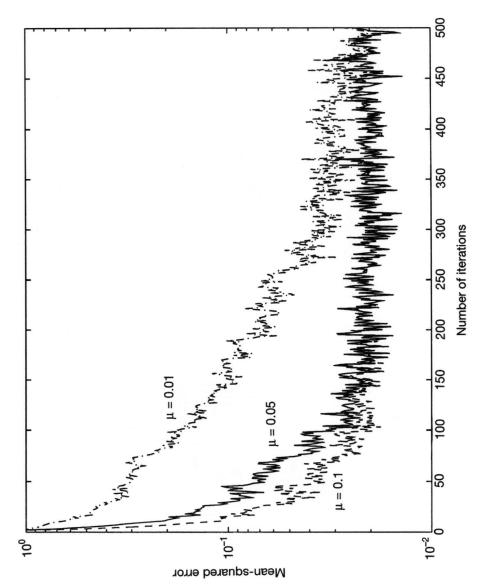

Figure 9.16 Experimental learning curves of adaptive first-order predictor for varying step-size parameter μ.

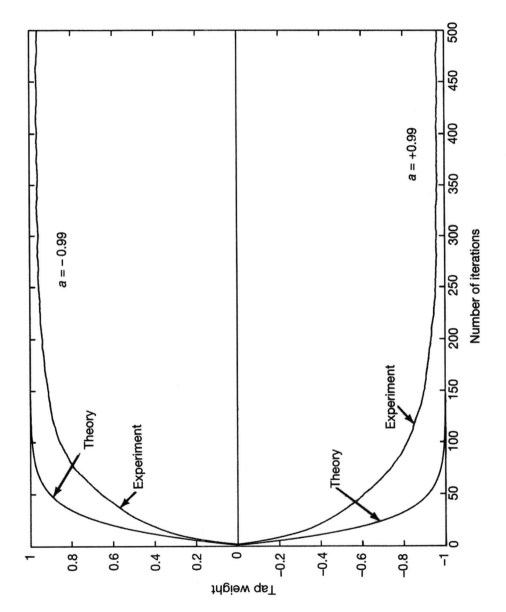

Figure 9.17 Comparison of experimental results with theory, based on $\hat{w}(n)$.

The use of Eq. (9.102), for the two sets of AR parameters listed under (1) and (2) and $\mu = 0.05$, yields the two curves labeled "theoretical" in Fig. 9.17. This figure demonstrates a reasonably good agreement between theory and experiment, which improves with increasing number of iterations.

In Fig. 9.18, we have plotted two learning curves for the LMS algorithm, one obtained experimentally and the other computed from theory. The experimental curve, labeled "Experiment," was obtained by ensemble-averaging the squared value of the prediction error $f(n)$ over 100 independent trials and for varying n. The theoretical curve, labeled "Theory" in Fig. 9.18, was obtained from the following equation:

$$J(n) = \left(\sigma_u^2 - \sigma_v^2 (1 + \frac{\mu}{2}\sigma_u^2) \right)(1 - \mu\sigma_u^2)^{2n} + \sigma_v^2\left(1 + \frac{\mu}{2}\,\sigma_u^2\right) \tag{9.103}$$

where

$$\sigma_v^2 = (1 - a^2)\sigma_u^2 \tag{9.104}$$

Equation (9.103) follows from the simple working rules presented at the end of Section 9.4, as explained here for the problem at hand:

- The initial value of the mean-squared error $J(n)$ is

$$J(0) = \sigma_u^2$$

- The final value of $J(n)$ is

$$J(\infty) = \sigma_v^2\left(1 + \frac{\mu}{2}\,\sigma_u^2\right)$$

  which represents the sum of the minimum mean-squared error

$$J_{\min} = \sigma_v^2$$

  and the excess mean-squared error

$$J_{ex}(\infty) \simeq \frac{\mu}{2}\,\sigma_v^2\lambda_1 = \frac{\mu}{2}\,\sigma_u^2\,\sigma_v^2$$

- The average time constant (for small μ) is

$$(\tau)_{\text{mse,av}} = -\frac{1}{2\ln(1 - \mu\lambda_1)} = -\frac{1}{2\ln(1 - \mu\sigma_u^2)} \simeq \frac{1}{2\mu\sigma_u^2}$$

Here also we observe reasonably good agreement between theory and experiment, with the agreement improving as the number of iterations is increased.

9.7 COMPUTER EXPERIMENT ON ADAPTIVE EQUALIZATION

In this second computer experiment we study the use of the LMS algorithm for *adaptive equalization* of a linear dispersive channel that produces (unknown) distortion. Here again we assume that the data are all *real valued*. Figure 9.19 shows the block diagram of the

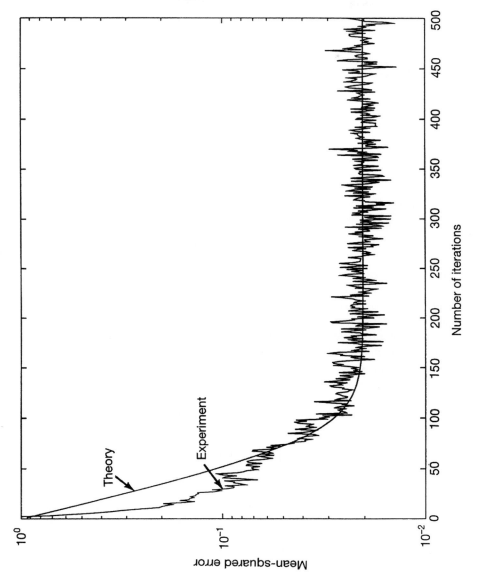

Figure 9.18 Comparison of experimental results with theory for the adaptive predictor, based on the mean-squared error.

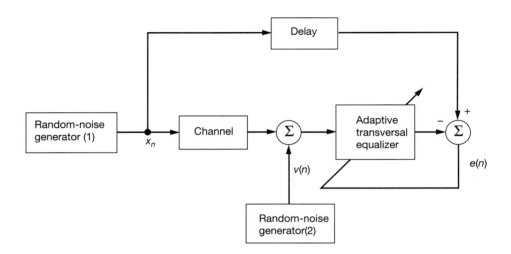

Figure 9.19   Block diagram of adaptive equalizer experiment.

system used to carry out the study. Random number generator 1 provides the test signal x_n, used for probing the channel, whereas random-number generator 2 serves as the source of additive white noise $v(n)$ that corrupts the channel output. These two random-number generators are independent of each other. The adaptive equalizer has the task of correcting for the distortion produced by the channel in the presence of the additive white noise. Random-number generator 1, after suitable delay, also supplies the desired response applied to the adaptive equalizer in the form of a training sequence.

The random sequence $\{x_n\}$ applied to the channel input consists of a *Bernoulli sequence*, with $x_n = \pm 1$ and the random variable x_n having zero mean and unit variance. The impulse response of the channel is described by the raised cosine:[8]

$$h_n = \begin{cases} \dfrac{1}{2}\left[1 + \cos\left(\dfrac{2\pi}{W}(n-2)\right)\right], & n = 1, 2, 3 \\ 0, & \text{otherwise} \end{cases} \tag{9.105}$$

where the parameter W controls the amount of amplitude distortion produced by the channel, with the distortion increasing with W.

Equivalently, the parameter W controls the eigenvalue spread $\chi(\mathbf{R})$ of the correlation matrix of the tap inputs in the equalizer, with the eigenvalue spread increasing with W. The

---

[8]The parameters specified in this experiment closely follow the paper by Satorius and Alexander (1979).

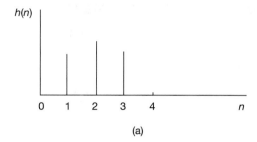

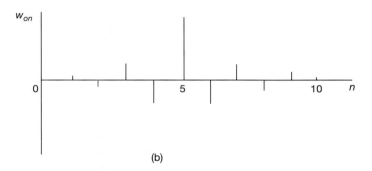

Figure 9.20   (a) Impulse response of channel; (b) impulse response of optimum transversal equalizer.

sequence $v(n)$, produced by the second random generator, has zero mean and variance $\sigma_v^2 = 0.001$.

The equalizer has $M = 11$ taps. Since the channel has an impulse response h_n that is symmetric about time $n = 2$, as depicted in Fig. 9.20(a), it follows that the optimum tap weights w_{on} of the equalizer are likewise symmetric about time $n = 5$, as depicted in Fig. 9.20(b). Accordingly, the channel input x_n is delayed by $\delta = 2 + 5 = 7$ samples to provide the desired response for the equalizer. By selecting the delay δ to match the midpoint of the transversal equalizer, the LMS algorithm is enabled to provide an approximate inversion of both the minimum-phase and nonminimum-phase components of the channel response.

The experiment is in two parts that are intended to evaluate the response of the adaptive equalizer using the LMS algorithm to changes in the eigenvalue spread $\chi(\mathbf{R})$ and step-size parameter μ. Before proceeding to describe the results of the experiment, however, we first compute the eigenvalues of the correlation matrix $\mathbf{R}$ of the 11 tap inputs in the equalizer.

Correlation Matrix of the Equalizer Input

The first tap input of the equalizer at time n equals

$$u(n) = \sum_{k=1}^{3} h_k a(n-k) + v(n) \tag{9.106}$$

where all the parameters are real valued. Hence, the correlation matrix $\mathbf{R}$ of the 11 tap inputs of the equalizer, $u(n), u(n-1), \ldots, u(n-10)$, is a symmetric 11-by-11 matrix. Also, since the impulse response h_n has nonzero values only for $n = 1, 2, 3$, and the noise process $v(n)$ is white with zero mean and variance σ_v^2, the correlation matrix $\mathbf{R}$ is *quintdiagonal*. That is, the only nonzero elements of $\mathbf{R}$ are on the main diagonal and the four diagonals directly above and below it, two on either side, as shown by the special structure:

$$\mathbf{R} = \begin{bmatrix} r(0) & r(1) & r(2) & 0 & \cdots & 0 \\ r(1) & r(0) & r(1) & r(2) & \cdots & 0 \\ r(2) & r(1) & r(0) & r(1) & \cdots & 0 \\ 0 & r(2) & r(1) & r(0) & \cdots & 0 \\ \cdot & \cdot & \cdot & \cdot & \cdot & \cdot \\ \cdot & \cdot & \cdot & \cdot & \cdot & \cdot \\ \cdot & \cdot & \cdot & \cdot & \cdot & \cdot \\ 0 & 0 & 0 & 0 & \cdots & r(0) \end{bmatrix} \tag{9.107}$$

where

$$r(0) = h_1^2 + h_2^2 + h_3^2 + \sigma_v^2$$
$$r(1) = h_1 h_2 + h_2 h_3$$
$$r(2) = h_1 h_3$$

The variance $\sigma_v^2 = 0.001$; hence, h_1, h_2, h_3 are determined by the value assigned to parameter W in Eq. (9.105).

In Table 9.1, we have listed (1) values of the autocorrelation function $r(l)$ for lag $l = 0, 1, 2$, and (2) the smallest eigenvalue, $\lambda_{\min}$, the largest eigenvalue, $\lambda_{\max}$, and the

TABLE 9.1 SUMMARY OF PARAMETERS FOR THE EXPERIMENT ON ADAPTIVE EQUALIZATION

W	2.9	3.1	3.3	3.5
$r(0)$	1.0963	1.1568	1.2264	1.3022
$r(1)$	0.4388	0.5596	0.6729	0.7774
$r(2)$	0.0481	0.0783	0.1132	0.1511
$\lambda_{\min}$	0.3339	0.2136	0.1256	0.0656
$\lambda_{\max}$	2.0295	2.3761	2.7263	3.0707
$\chi(\mathbf{R}) = \lambda_{\max}/\lambda_{\min}$	6.0782	11.1238	21.7132	46.8216

eigenvalue spread $\chi(\mathbf{R}) = \lambda_{\max}/\lambda_{\min}$. We thus see that the eigenvalue spread ranges from 6.0782 (for $W = 2.9$) to 46.8216 (for $W = 3.5$).

Experiment 1: Effect of Eigenvalue Spread.   For the first part of the experiment, the step-size parameter was held fixed at $\mu = 0.075$. This is in accordance with the condition of Eq. (9.94) for convergence in the mean square for the worst eigenvalue spread of 46.8216 (corresponding to $W = 3.5$):

$$\mu_{\text{crit}} = \frac{2}{\text{tap-input power}}$$

$$= \frac{2}{Mr(0)}$$

$$= 0.14$$

The choice of $\mu = 0.075$ therefore assures the convergence of the adaptive equalizer in the mean square for all the conditions listed in Table 9.1.

For each eigenvalue spread, an approximation to the ensemble-averaged learning curve of the adaptive equalizer is obtained by averaging the instantaneous squared error "$e^2(n)$ versus n" curve over 200 independent trials of the computer experiment. The results of this computation are shown in Fig. 9.21.

We thus see from Fig. 9.21 that increasing the eigenvalue spread $\chi(\mathbf{R})$ has the effect of slowing down the rate of convergence of the adaptive equalizer and also increasing the steady-state value of the average squared error. For example, when $\chi(\mathbf{R}) = 6.0782$, approximately 80 iterations are required for the adaptive equalizer to converge in the mean square, and the average squared error (after 500 iterations) approximately equals 0.003. On the other hand, when $\chi(\mathbf{R}) = 46.8216$ (i.e., the equalizer input is ill conditioned), the equalizer requires approximately 200 iterations to converge in the mean square, and the resulting average squared error (after 500 iterations) approximately equals 0.03.

In Fig. 9.22, we have plotted the ensemble-averaged impulse response of the adaptive equalizer after 1000 iterations for each of the four eigenvalue spreads of interest. As before, the ensemble averaging was carried out over 200 independent trials of the experiment. We see that in each case the ensemble-averaged impulse response of the adaptive equalizer is very close to being symmetric with respect to the center tap, as expected. The variation in the impulse response from one eigenvalue spread to another merely reflects the effect of a corresponding change in the impulse response of the channel.

Experiment 2: Effect of Step-Size Parameter.   For the second part of the experiment, the parameter W in Eq. (9.105) was fixed at 3.1, yielding an eigenvalue spread of 11.1238 for the correlation matrix of the tap inputs in the equalizer. The step-size parameter μ was this time assigned one of three values: 0.075, 0.025, 0.0075.

Figure 9.23 shows the results of this computation. As before, each learning curve is the result of ensemble averaging the instantaneous squared error "$e^2(n)$ versus n" curve over 200 independent trials of the computer experiment.

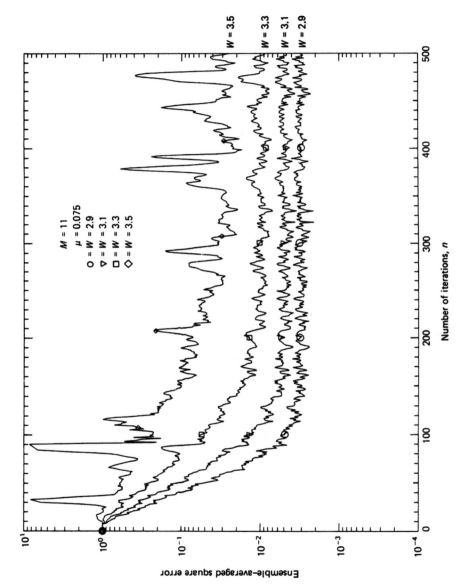

Figure 9.21 Learning curves of the LMS algorithm for adaptive equalizer with number of taps $M = 11$, step-size parameter $\mu = 0.075$, and varying eigenvalue spread $\chi(\mathbf{R})$.

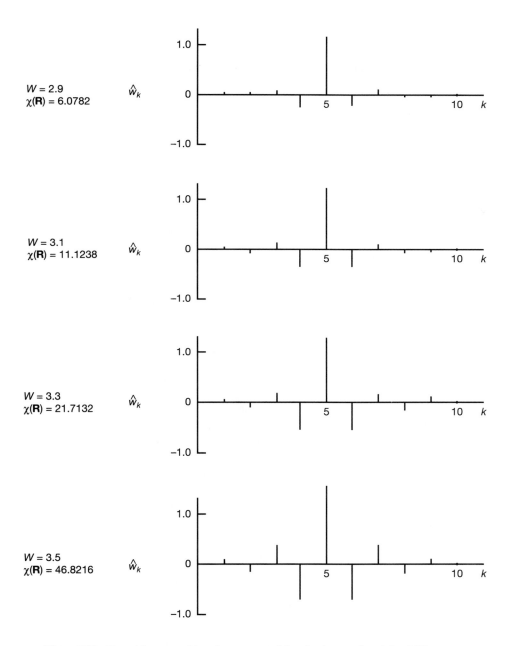

Figure 9.22 Ensemble-averaged impulse response of the adaptive equalizer (after 1000 iterations) for each of four different eigenvalue spreads.

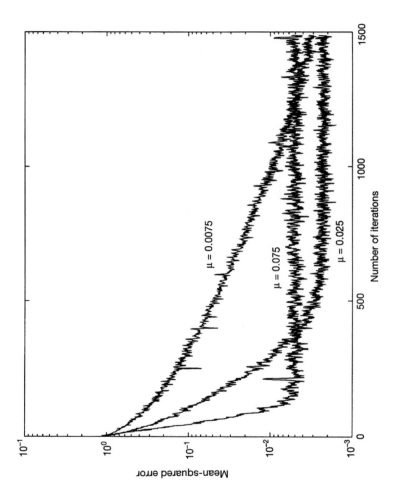

Figure 9.23 Learning curves of the LMS algorithm for adaptive equalizer with the number of taps $M = 11$, fixed eigenvalue spread, and varying step-size parameter μ.

The results confirm that the rate of convergence of the adaptive equalizer is highly dependent on the step-size parameter μ. For large step-size parameter ($\mu = 0.075$), the equalizer converged to steady-state conditions in approximately 120 iterations. On the other hand, when μ is small (equal to 0.0075), the rate of convergence slowed down by more than an order of magnitude. The results also show that the steady-state value of the average squared error (and hence the misadjustment) increases with increasing μ.

9.8 COMPUTER EXPERIMENT ON MINIMUM-VARIANCE DISTORTIONLESS RESPONSE BEAMFORMER

For our final experiment we consider the LMS algorithm applied to an adaptive minimum-variance distortionless response (MVDR) beamformer consisting of a linear array of five uniformly spaced sensors (e.g., antenna elements), as depicted in Fig. 9.24. The spacing d between adjacent elements of the array equals one half of the received wavelength so as to avoid the appearance of grating lobes. The beamformer operates in an environment that consists of two components: a target signal impinging on the array along a direction of interest, and a single source of interference originating from an unknown direction. It is assumed that these two components originate from independent sources, and that the received signal includes additive white Gaussian noise at the output of each sensor.

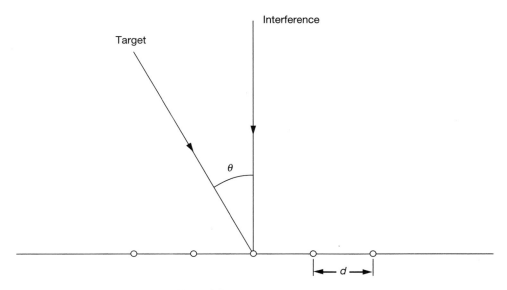

Figure 9.24 Linear array antenna.

The aims of the experiment are twofold:

- To examine the evolution of the adapted spatial response (pattern) of the MVDR beamformer with time for a prescribed target signal-to-interference ratio
- To evaluate the effect of varying the target-to-interference ratio on the interference-nulling performance of the beamformer

The angles of incidence of the target and interfering signals, measured in radians with respect to the normal to the line of the array, are as follows:

- *Target signal*:

$$\theta_{\text{target}} = \sin^{-1}(-0.2)$$

- *Interference*:

$$\theta_{\text{interf}} = \sin^{-1}(0)$$

The design of the LMS algorithm for adjusting the weight vector of the adaptive MVDR beamformer follows the theory presented in Section 5.8. For the application at hand, the gain vector $\mathbf{g} = 1$.

Figure 9.25 shows the adapted spatial response of the MVDR beamformer for signal-to-noise ratio = 10 dB, varying interference-to-noise ratio (INR) and varying number

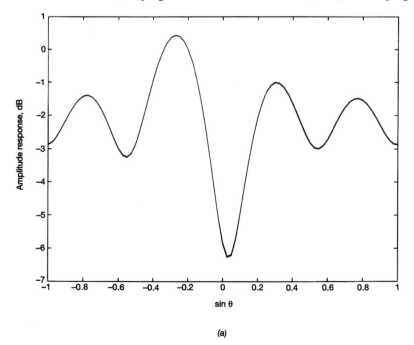

(a)

Figure 9.25   Adapted spatial response of MVDR beamformer for varying interference-to-noise ratio, and varying number of iterations. (a) $n = 20$. (b) $n = 100$. (c) $n = 200$. In each part of the figure, the interference-to-noise ratio assumes one of three values; in part (a) the number of iterations is too small for these variations to have a noticeable effect. Parts (b) and (c) are shown on the next page.

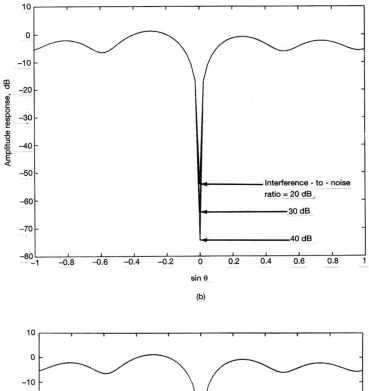

(b)

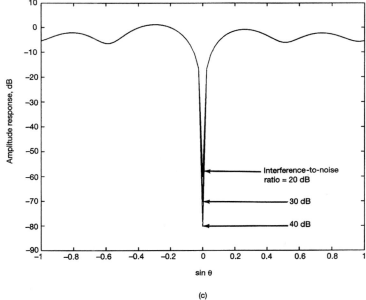

(c)

Figure 9.25   (*Contd.*)

of iterations. The spatial response is defined by $20 \log_{10}|\hat{\mathbf{w}}^H(n)\mathbf{s}(\phi)|^2$, where $\mathbf{s}(\phi)$ is the steering vector:

$$\mathbf{s}(\phi) = [1, e^{-j\phi}, e^{-j2\phi}, e^{-j3\phi}, e^{-j4\phi}]^T$$

The electrical angle ϕ, measured in radians, is related to the angle of incidence θ as follows:

$$\phi = \pi\sin\theta$$

The weight vector $\hat{\mathbf{w}}(n)$ of the beamformer is computed using the LMS algorithm with step-size parameter $\mu = 10^{-8}$, 10^{-9}, and 10^{-10} for INR = 20, 30, and 40 dB, respectively. The reason for varying μ is to ensure convergence for a prescribed interference-to-noise ratio, as the largest eigenvalue λ_{max} of the correlation matrix of the input data depends on the interference-to-noise ratio.

Figure 9.26 shows the adapted spatial response of the MVDR beamformer after 20, 25, and 30 iterations. The three curves of the figure pertain to INR = 20 dB and a fixed target signal-to-noise ratio = 10 dB.

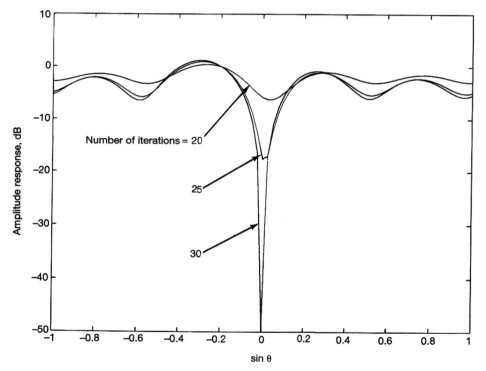

Figures 9.26 Adapted spatial response of MVDR beamformer for signal-to-noise ratio = 10dB, interference-to-noise ratio = 20dB, step-size parameter = 10^{-8}, and varying number of iterations.

On the basis of the results presented in Figs. 9.25 and 9.26, we may make the following observations:

- The response of the MVDR beamformer is always held fixed at the value of unity along the prescribed angle of incidence $\theta_{target} = \sin^{-1}(-0.2)$, as required.
- The interference-nulling capability of the beamformer improves with (a) increasing number of iterations (snapshots of data), and (b) increasing interference-to-target signal ratio.

9.9 DIRECTIONALITY OF CONVERGENCE OF THE LMS ALGORITHM FOR NON-WHITE INPUTS

The eigenstructure of the correlation matrix $\mathbf{R}$ of a transversal filter's tap inputs has a profound impact on the convergence behavior of the LMS algorithm used to adapt the filter's tap weights. When the tap inputs are drawn from a white noise process, the tap inputs are uncorrelated and the eigenvalue spread $\chi(\mathbf{R})$ of the correlation matrix $\mathbf{R}$ is unity, with the result that the LMS algorithm enjoys a non-directional convergence. At the other extreme, when the tap inputs are highly correlated and the eigenvalue spread $\chi(\mathbf{R})$ is large, which does happen under non-white inputs, convergence of the LMS algorithm (like the steepest descent algorithm from which it is derived) takes on a directional nature. Moreover, in such an environment, initialization may play a significant role in determining the rate of convergence. Indeed, it is the combination of these factors that is responsible for the relatively slow rate of convergence observed in the results on adaptive equalization presented in Fig. 9.21.

The *directionality of convergence*[9], exhibited by the LMS algorithm under non-white inputs, manifests itself in two ways:

1. The speed of convergence of the LMS algorithm is faster in some directions in the algorithm's weight space than in some other directions.
2. Depending on the direction along which the convergence of the LMS algorithm takes place, it is possible for the convergence to be accelerated by an increase in the eigenvalue spread $\chi(\mathbf{R})$.

These two aspects of the directionality of convergence are illustrated in the following example.

---

[9]The material presented in this section, including Example 7, is based on: LeBlanc, J.P., private communication, 1995.

Example 7: Sinusoidal Process

Consider the simple example of a two-tap transversal filter, whose true tap-weight vector (i.e., the Wiener solution) is denoted by $\mathbf{w}_o$. The tap inputs $u(n)$ and $u(n-1)$ are drawn from a deterministic process consisting of two sinusoids, as shown by

$$u(n) = A_1 \cos(\omega_1 n) + A_2 \cos(\omega_2 n)$$

where ω_1 and ω_2 are the angular frequencies of the two sinusoids, and A_1 and A_2 are their respective amplitudes. The correlation matrix of the tap inputs is

$$\mathbf{R} = \begin{bmatrix} E[u^2(n)] & E[u(n-1)u(n)] \\ E[u(n)u(n-1)] & E[u^2(n-1)] \end{bmatrix}$$

$$= \frac{1}{2} \begin{bmatrix} A_1^2 + A_2^2 & A_1^2 \cos \omega_1 + A_2^2 \cos \omega_2 \\ A_1^2 \cos \omega_1 + A_2^2 \cos \omega_2 & A_1^2 + A_2^2 \end{bmatrix}$$

This two-by-two matrix is doubly symmetric, which means that its two eigenvalues and associated eigenvectors are as follows (see Problem 17 of Chapter 4):

$$\lambda_1 = \frac{1}{2} A_1^2 (1 + \cos \omega_1) + \frac{1}{2} A_2^2 (1 + \cos \omega_2); \quad \mathbf{q}_1 = [1, 1]^T$$

$$\lambda_2 = \frac{1}{2} A_1^2 (1 - \cos \omega_1) + \frac{1}{2} A_2^2 (1 - \cos \omega_2); \quad \mathbf{q}_2 = [-1, 1]^T$$

In the sequel, we study the convergence behavior of the LMS algorithm with the following specifications:

> step-size parameter, $\mu = 0.01$
> initial condition, $\hat{\mathbf{w}}(0) = [0, 0]^T$
> total number of iterations, $n = 200$

In particular, we consider two different filters and two different inputs. One filter is the *minimum eigenfilter*, whose tap-weight vector is defined by the eigenvector $\mathbf{q}_2$ associated with the smallest eigenvalue (i.e., λ_2) of the correlation matrix $\mathbf{R}$, and the other filter is the *maximum eigenfilter* defined by the eigenvector $\mathbf{q}_1$ associated with the largest eigenvalue (i.e., λ_1). The two inputs are

$$u_a(n) = \cos(1.2n) + 0.5 \cos(0.1n)$$

$$u_b(n) = \cos(0.6n) + 0.5 \cos(0.23n)$$

The first input has an eigenvalue spread $\chi(\mathbf{R}) = 2.9$, and the second input has an eigenvalue spread $\chi(\mathbf{R}) = 12.9$. Thus, there are four distinct combinations to be considered, which we do under the following two cases.

Case 1. Minimum eigenfilter.   For this case, the true tap-weight vector of the transversal filter is

$$\mathbf{w}_o = \mathbf{q}_2 = [-1, 1]^T$$

Here under input $u_a(n)$, the convergence of the LMS algorithm is along a "slow" trajectory, and transverses about halfway to the true parameterization of the filter in 200 iterations of the algorithm starting from $\mathbf{w}(0) = [0, 0]^T$, as shown in Fig. 9.27(a).

Next, the input signal is chosen as $u_b(n)$, for which the eigenvalue spared $\chi(\mathbf{R})$ is 12.9, compared to 2.9 for $u_a(n)$. The increased eigenvalue spread is evidenced by an increased eccentricity of the error surface contours, as portrayed in Fig. 9.27(b). Comparing Fig. 9.27(b) with 9.27(a), we see that in Case 1 the convergence of the LMS algorithm has been *decelerated* by the increase in the eigenvalue spread $\chi(\mathbf{R})$.

Case 2. Maximum eigenfilter.   For this second case, the true tap-weight vector of the transversal filter is

$$\mathbf{w}_o = \mathbf{q}_1 = [1, 1]^T$$

Reverting to the input $u_a(n)$, we now find that the convergence of the LMS algorithm is along a "fast" trajectory, and traverses the error surface contours, as shown in Fig. 9.28(a). Moreover, when the input signal $u_b(n)$ is used, the convergence of the LMS algorithm is *accelerated* by the increase in the eigenvalue spread $\chi(\mathbf{R})$, as shown in Fig. 9.28(b).

In the example described here, the initial condition $\hat{\mathbf{w}}(0)$ is fixed, but the true tap-weight vector $\mathbf{w}_o$ is varied from case 1 to case 2. In the usual application of the LMS algorithm to stationary inputs, the true tap-weight vector $\mathbf{w}_o$ is fixed but unknown. For some fixed $\mathbf{w}_o$, we may equivalently specify the initial condition for this example as follows:

$$\hat{\mathbf{w}}(0) = \begin{cases} \mathbf{w}_o - \mathbf{q}_2 & \text{for case 1 (minimum eigenfilter)} \\ \mathbf{w}_o - \mathbf{q}_1 & \text{for case 2 (maximum eigenfilter)} \end{cases}$$

Thus, in light of the results presented in this example, the directionality of convergence of the LMS algorithm may be exploited by choosing a suitable value for the initial condition $\hat{\mathbf{w}}(0)$, such that the algorithm is guided along a fast trajectory. This, of course, assumes the availability of *prior knowledge* about the environment in which the LMS algorithm is operating. In such a scenario, the LMS algorithm performs essentially the role of "tuning" the tap-weights of the transversal filter.

9.10 ROBUSTNESS OF THE LMS ALGORITHM

The development of the LMS algorithm presented in Section 9.2 was carried out in a heuristic manner, starting from the method of steepest descent as the basis for computing the Wiener solution of an adaptive transversal filter. However, once instantaneous estimates of the correlation matrix $\mathbf{R}$ and cross-correlation vector $\mathbf{p}$ are invoked in this development, links with the least-mean-square estimate implicit in the Wiener solution are destroyed. If then a "single" realization of the LMS algorithm is not optimum in the least-mean-square sense, what is the actual criterion on the basis of which it is optimum? The

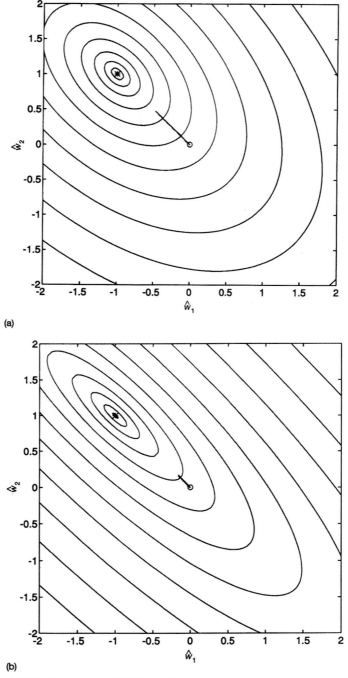

(a)

(b)

Figure 9.27 Convergence of the LMS algorithm, for a deterministic sinusoidal process, along "slow" eigenvector (i.e., minimum eigenfilter) for (a) input $u_a(n)$, and (b) input $u_b(n)$.

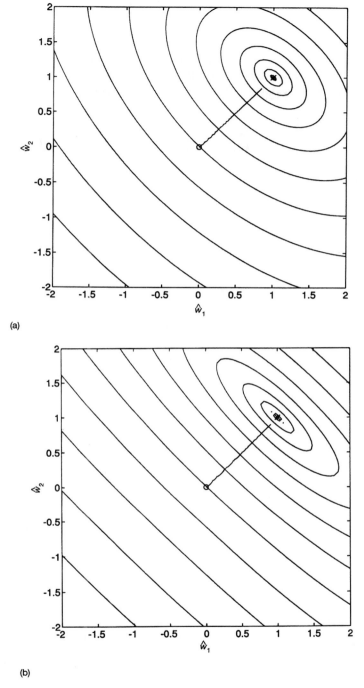

(a)

(b)

Figure 9.28 Convergence of the LMS algorithm, for deterministic sinusoidal process, along "fast" eigenvector (i.e., maximum eigenfilter) for (a) input $u_a(n)$, and (b) input $u_b(n)$.

answer to this fundamental question lies in the so-called H^{∞} *(or minimax) criterion,*[10] on which extensive studies were carried out in the field of robust control during the 1980s.

To proceed, suppose that we have a set of noisy measurements that fit into a multiple regression model of order M as follows:

$$d(i) = \mathbf{w}_o^H \mathbf{u}(i) + v(i), \quad i = 0, 1, \ldots, n \tag{9.108}$$

The issue of concern is to formulate a recursive estimate of the unknown weight vector $\mathbf{w}_o$, given input-output pairs $\{\mathbf{u}(i), d(i) | i = 0, 1, \ldots, n\}$ that are corrupted by the additive white noise $v(i)$. The estimate, denoted by $\hat{\mathbf{w}}(i)$, is required to be optimum in a certain sense, yet to be defined. Suppose, further, we do two other things:

- A positive real number μ is chosen so as to satisfy the condition

$$0 < \mu \leq \frac{1}{\|\mathbf{u}(i)\|^2} \quad \text{for } 0 \leq i \leq n \tag{9.109}$$

- Subsequently, an estimate $\hat{\mathbf{w}}(i)$ is chosen *at will* for the unknown weight vector $\mathbf{w}_o$ at iteration i. This estimate always satisfies the condition

$$\frac{|\hat{\mathbf{w}}^H(i)\mathbf{u}(i) - \mathbf{w}_o^H \mathbf{u}(i)|^2}{\mu^{-1}\|\hat{\mathbf{w}}(i) - \mathbf{w}_o\|^2} \leq 1 \quad \text{for } 0 \leq i \leq n \tag{9.110}$$

  where the numerator of the left-hand side is the squared error between the estimate $\hat{\mathbf{w}}^H(i)\mathbf{u}(i)$ and the true inner product $\mathbf{w}_o^H \mathbf{u}(i)$ in Eq. (9.108), and the denominator (except for the scaling factor μ^{-1}) is the squared Euclidean distance between the estimate $\hat{\mathbf{w}}(i)$ and the true weight vector $\mathbf{w}_o$.

Note that the condition described in (9.110) follows from that of (9.109). First, we write

$$|\hat{\mathbf{w}}^H(i)\mathbf{u}(i) - \mathbf{w}_o^H\mathbf{u}(i)|^2 = |(\hat{\mathbf{w}}(i) - \mathbf{w}_o)^H\mathbf{u}(i)|^2$$

$$\leq \|\hat{\mathbf{w}}(i) - \mathbf{w}_o\|^2\|\mathbf{u}(i)\|^2 \tag{9.111}$$

where, in the last line, we have invoked the *Cauchy–Schwarz inequality.*[11] Hence, in light of the inequality in (9.109), we may readily recast that of Eq. (9.111) in the form previously specified in (9.110).

---

[10]The H^{∞} optimality of the LMS algorithm is discussed in detail in (Hassibi et al., (1996); see also Sayed and Kailath (1994). The H^{∞} criterion is due to Zames (1981), and it is developed in Zames and Francis (1983) and Kimura (1984). The criterion is discussed in Doyle et al. (1989), Khargonekar and Nagpal (1991), and Green and Limebeer (1995). The paper by Sayed and Rupp (1994) also deals with time-variant step-sizes $\mu(i)$, and their results are therefore applicable to other forms of the LMS algorithm, e.g., normalized LMS; the latter algorithm is discussed in the next section.

[11]Consider the inner product $\mathbf{a}^H\mathbf{b}$ of two vectors $\mathbf{a}$ and $\mathbf{b}$ of compatible dimensions. The Cauchy–Schwarz inequality states that

$$|\mathbf{a}^H\mathbf{b}|^2 \leq \|\mathbf{a}\|^2 \|\mathbf{b}\|^2$$

where $\|\cdot\|$ denotes Euclidean norm of the enclosed vector:

Given that the inequality of (9.110) holds for arbitrary $\hat{\mathbf{w}}(i)$, then it will still hold if, say, the denominator is increased by the squared amplitude of the noise, namely, $|v(i)|^2$, as shown by

$$\frac{|\hat{\mathbf{w}}^H(i)\mathbf{u}(i) - \mathbf{w}_o^H\mathbf{u}(i)|^2}{\mu^{-1}\|\hat{\mathbf{w}}(i) - \mathbf{w}_o\|^2 + |v(i)|^2} \leq 1, \qquad 0 \leq i \leq n \qquad (9.112)$$

At this point in the discussion, we may ask: In what particular way would this inequality be modified if $\hat{\mathbf{w}}(i)$ is chosen according to the weight update computed by the LMS algorithm? It turns out that, in fact, the inequality is further tightened by such an update, in that we may also write

$$\frac{|\hat{\mathbf{w}}^H(i)\mathbf{u}(i) - \mathbf{w}_o^H\mathbf{u}(i)|^2}{\mu^{-1}\|\hat{\mathbf{w}}(i) - \mathbf{w}_o\|^2 - \mu^{-1}\|\hat{\mathbf{w}}(i+1) - \mathbf{w}_o\|^2 + |v(i)|^2} \leq 1 \qquad (9.113)$$

Comparing this inequality with that of (9.112), we see that the denominator has been reduced by subtracting the new term $\mu^{-1}\|\hat{\mathbf{w}}(i+1) - \mathbf{w}_o\|^2$. To verify the validity of the inequality of (9.113), we first rewrite it in the equivalent form

$$\mu^{-1}\|\hat{\mathbf{w}}(i) - \mathbf{w}_o\|^2 - \mu^{-1}\|\hat{\mathbf{w}}(i+1) - \mathbf{w}_o\|^2 + |v(i)|^2 - |\hat{\mathbf{w}}^H(i)\mathbf{u}(i) - \mathbf{w}_o^H\mathbf{u}(i)|^2 \geq 0 \qquad (9.114)$$

Next, using the update recursion for the LMS algorithm, namely,

$$\hat{\mathbf{w}}(i+1) = \hat{\mathbf{w}}(i) + \mu\,\mathbf{u}(i)\,(d(i) - \hat{\mathbf{w}}^H(i)\mathbf{u}(i))^* \qquad (9.115)$$

it is a straightforward matter to show that the left-hand side of the inequality of (9.114) may be simplified, as shown here (see Problem 11):

$$(1 - \mu\|\mathbf{u}(i)\|^2)|d(i) - \hat{\mathbf{w}}^H(i)\mathbf{u}(i)|^2 \geq 0 \qquad (9.116)$$

Finally, we see that this inequality does indeed hold, provided that in the first place the step-size parameter μ of the LMS algorithm is chosen to satisfy the condition of (9.109).

Suppose, now, the LMS algorithm is computed up to some iteration n, such that we satisfy the requirement[12]

$$0 < \mu < \min_{0 \leq i \leq n} \frac{1}{\|\mathbf{u}(i)\|^2} \qquad (9.117)$$

Then, by virtue of the inequality described in (9.114), we have for each iteration i of the LMS algorithm:

$$|\hat{\mathbf{w}}^H(i)\mathbf{u}(u) - \mathbf{w}_o^H\mathbf{u}(i)|^2 \leq \mu^{-1}\|\hat{\mathbf{w}}(i) - \mathbf{w}_o\|^2 - \mu^{-1}\|\hat{\mathbf{w}}(i+1) - \mathbf{w}_o\|^2 + |v(i)|^2 \qquad (9.118)$$

Hence, summing both sides of this inequality over $0 \leq i \leq n$, we get

$$\sum_{i=0}^{n} |\hat{\mathbf{w}}^H(i)\mathbf{u}(i) - \mathbf{w}_o^H\mathbf{u}(i)|^2 \leq \mu^{-1}\|\hat{\mathbf{w}}(0) - \mathbf{w}_o\|^2 - \mu^{-1}\|\hat{\mathbf{w}}(n+1) - \mathbf{w}_o\|^2 + \sum_{i=0}^{n} |v(i)|^2$$

---

[12]Note that the restriction imposed on the step-size parameter μ in Eq. (9.117) is somewhat different from the condition imposed on μ for convergence of the LMS algorithm in the mean square.

If this condition holds, then we certainly have

$$\sum_{i=0}^{n} |\hat{\mathbf{w}}^H(i)\mathbf{u}(i) - \mathbf{w}_o^H\mathbf{u}(i)|^2 \leq \mu^{-1}\|\hat{\mathbf{w}}(0) - \mathbf{w}_o\|^2 + \sum_{i=0}^{n} |v(i)|^2 \tag{9.119}$$

On the basis of this last inequality, we may equivalently write

$$\frac{\displaystyle\sum_{i=0}^{n} |\hat{\mathbf{w}}^H(i)\mathbf{u}(i) - \mathbf{w}_o^H\mathbf{u}(i)|^2}{\mu^{-1}\|\hat{\mathbf{w}}(0) - \mathbf{w}_o\|^2 + \displaystyle\sum_{i=0}^{n} |v(i)|^2} \leq 1 \tag{9.120}$$

which is the fundamental result that we have been seeking. The numerator and denominator of the left-hand side of this inequality may be interpreted as follows:

- The difference term $\hat{\mathbf{w}}^H(i)\mathbf{u}(i) - \mathbf{w}_o^H\mathbf{u}(i)$ represents the error produced in using the inner product $\hat{\mathbf{w}}^H(i)\mathbf{u}(i)$ as an estimate of the true quantity $\mathbf{w}_o^H\mathbf{u}(i)$. Accordingly, the numerator may be viewed as the *sum of squared errors* so incurred over the entire computation interval $0 \leq i \leq n$.

- The denominator term consists of the sum of two terms: a scaled version of the squared Euclidean distance between the initial weight vector $\hat{\mathbf{w}}(0)$ and the true weight vector $\mathbf{w}_o$, and the sum of squared noise $v(i)$ over the same interval $0 \leq i \leq n$.

Let T denote the operator that maps the set of disturbances $\{v(i)|i = 0, 1, \ldots, n\}$ and the initial weight uncertainty $\mu^{-1/2}(\hat{\mathbf{w}}(0) - \mathbf{w}_o)$ to the corresponding set of errors $\{\hat{\mathbf{w}}^H(i)\mathbf{u}(i) - \mathbf{w}_o^H\mathbf{u}(i) \mid i = 0, 1, \ldots, n\}$. Then, the inequality described in (9.120) states that *the Euclidean norm induced by the operator T is always bounded by one* (Sayed and Kailath, 1994). Stated in another way, the sum of squared errors is always upper bounded by the combined effects of the initial weight uncertainty $(\hat{\mathbf{w}}(0) - \mathbf{w}_o)$ and the noise $v(i)$, which explains the "robust" behavior of the LMS algorithm in its endeavour to estimate the uncorrupted term $\mathbf{w}_o^H \mathbf{u}(i)$ defined in (9.108).

In conclusion, we may say that a single realization of the LMS algorithm (despite its name) is not optimal in the least-mean-square sense, but is optimal in the H^∞ sense.

9.11 NORMALIZED LMS ALGORITHM

In the standard form of LMS algorithm, the correction $\mu\mathbf{u}(n)e^*(n)$ applied to the tap-weight vector $\hat{\mathbf{w}}(n)$ at iteration $n + 1$ is directly proportional to the tap-input vector $\mathbf{u}(n)$.

Therefore, when $\mathbf{u}(n)$ is large, the LMS algorithm experiences a *gradient noise amplification* problem. To overcome this difficulty, we may use the *normalized LMS algorithm*,[13] which is the companion to the ordinary LMS algorithm. In particular, the correction applied to the tap-weight vector $\hat{\mathbf{w}}(n)$ at iteration $n + 1$ is "normalized" with respect to the squared Euclidean norm of the tap-input vector $\mathbf{u}(n)$ at iteration n, hence the term "normalized."

We may formulate the normalized LMS algorithm as a natural modification of the ordinary LMS algorithm. Alternatively, we may derive the normalized LMS algorithm in its own rightful manner; we follow the latter procedure here as it provides insight into its operation.

Normalized LMS Algorithm as the Solution to a Constrained Optimization Problem

The normalized LMS algorithm may be viewed as the solution to a constrained optimization (minimization) problem (Goodwin and Sin, 1984). Specifically, the problem of interest may be stated as follows:

> Given the tap-input vector $\mathbf{u}(n)$ and the desired response $d(n)$, determine the tap-weight vector $\hat{\mathbf{w}}(n + 1)$ so as to minimize the squared Euclidean norm of the change

$$\delta\hat{\mathbf{w}}(n + 1) = \hat{\mathbf{w}}(n + 1) - \hat{\mathbf{w}}(n) \tag{9.121}$$

in the tap-weight vector $\hat{\mathbf{w}}(n + 1)$ with respect to its old value $\hat{\mathbf{w}}(n)$, subject to the constraint

$$\hat{\mathbf{w}}^H(n + 1)\mathbf{u}(n) = d(n) \tag{9.122}$$

To solve this constrained optimization problem, we may use the *method of Lagrange multipliers*.[14]

The squared norm of the change $\delta\hat{\mathbf{w}}(n + 1)$ in the tap-weight vector $\hat{\mathbf{w}}(n + 1)$ may be expressed as

$$
\begin{aligned}
\|\delta\hat{\mathbf{w}}(n + 1)\|^2 &= \delta\hat{\mathbf{w}}^H(n + 1)\delta\hat{\mathbf{w}}(n + 1) \\
&= [\hat{\mathbf{w}}(n + 1) - \hat{\mathbf{w}}(n)]^H[\hat{\mathbf{w}}(n + 1) - \hat{\mathbf{w}}(n)] \\
&= \sum_{k=0}^{M-1} |\hat{w}_k(n + 1) - \hat{w}_k(n)|^2
\end{aligned}
\tag{9.123}
$$

---

[13]The stochastic gradient algorithm known as the normalized LMS algorithm was suggested independently by Nagumo and Noda (1967) and Albert and Gardner (1967). Nagumo and Noda did not use any special name for the algorithm, whereas Albert and Gardner referred to it as a "quick and dirty regression" scheme. It appears that Bitmead and Anderson (1980) coined the name "normalized LMS algorithm."

[14]For a discussion of the method of Lagrange multipliers, see Appendix C.

Define the tap weight $\hat{w}_k(n)$ for $k = 0, 1, \ldots, M - 1$ in terms of its real and imaginary parts by writing

$$\hat{w}_k(n) = a_k(n) + jb_k(n), \qquad k = 0, 1, \ldots, M - 1 \tag{9.124}$$

We then have

$$\|\delta\hat{\mathbf{w}}(n + 1)\|^2 = \sum_{k=0}^{M-1} ([a_k(n + 1) - a_k(n)]^2 + [b_k(n + 1) - b_k(n)]^2) \tag{9.125}$$

Let the tap input $u(n - k)$ and the desired response $d(n)$ be defined in terms of their respective real and imaginary parts as follows:

$$d(n) = d_1(n) + jd_2(n) \tag{9.126}$$

$$u(n - k) = u_1(n - k) + ju_2(n - k) \tag{9.127}$$

Accordingly, we may rewrite the complex constraint of Eq. (9.122) as an equivalent pair of real constraints:

$$\sum_{k=0}^{M-1} (a_k(n + 1)u_1(n - k) + b_k(n + 1)u_2(n - k)) = d_1(n) \tag{9.128}$$

and

$$\sum_{k=0}^{M-1} (a_k(n + 1)u_2(n - k) - b_k(n + 1)u_1(n - k)) = d_2(n) \tag{9.129}$$

We are now ready to formulate a real-valued cost function $J(n)$ for the constrained optimization problem at hand. In particular, we combine Eqs. (9.125), (9.128), and (9.129) into a single relation:

$$
\begin{aligned}
J(n) = &\sum_{k=0}^{M-1} ([a_k(n + 1) - a_k(n)]^2 + [b_k(n + 1) - b_k(n)]^2) \\
&+ \lambda_1 \left[d_1(n) - \sum_{k=0}^{M-1} (a_k(n + 1)u_1(n - k) + b_k(n + 1)u_2(n - k)) \right] \\
&+ \lambda_2 \left[d_2(n) - \sum_{k=0}^{M-1} (a_k(n + 1)u_2(n - k) - b_k(n + 1)u_1(n - k)) \right]
\end{aligned}
\tag{9.130}
$$

where λ_1 and λ_2 are *Lagrange multipliers*. To find the optimum values of $a_k(n + 1)$ and $b_k(n + 1)$, we differentiate the cost function $J(n)$ with respect to these two parameters and then set the results equal to zero. Hence, the use of Eq. (9.130) in the equation

$$\frac{\partial J(n)}{\partial a_k(n + 1)} = 0$$

yields the result

$$2[a_k(n + 1) - a_k(n)] - \lambda_1 u_1(n - k) - \lambda_2 u_2(n - k) = 0 \tag{9.131}$$

Similarly, the use of Eq. (9.130) in the complementary equation

$$\frac{\partial J(n)}{\partial b_k(n + 1)} = 0$$

yields the complementary result

$$2[b_k(n + 1) - b_k(n)] - \lambda_1 u_2(n - k) + \lambda_2 u_1(n - k) = 0 \qquad (9.132)$$

Next, we use the definitions of Eqs. (9.124) and (9.127) to combine these two real results into a single complex one, as shown by

$$2[\hat{w}_k(n + 1) - \hat{w}_k(n)] = \lambda^* u(n - k), \qquad k = 0, 1, \ldots, M - 1 \quad (9.133)$$

where λ *is a* complex Lagrange multiplier:

$$\lambda = \lambda_1 + j\lambda_2 \qquad (9.134)$$

To solve for the unknown λ^*, we multiply both sides of Eq. (9.133) by $u^*(n - k)$ and then sum over all possible integer values of k for 0 to $M - 1$. We thus get

$$\lambda^* = \frac{2}{\displaystyle\sum_{k=0}^{M-1} |u(n - k)|^2} \left[\sum_{k=0}^{M-1} \hat{w}_k(n + 1)u^*(n - k) - \sum_{k=0}^{M-1} \hat{w}_k(n)u^*(n - k) \right]$$

$$(9.135)$$

$$= \frac{2}{\|\mathbf{u}(n)\|^2} [\hat{\mathbf{w}}^T(n + 1)\mathbf{u}^*(n) - \hat{\mathbf{w}}^T(n)\mathbf{u}^*(n)]$$

where $\|\mathbf{u}(n)\|$ is the Euclidean norm of the tap-input vector $\mathbf{u}(n)$. Next, we use the complex constraint of Eq. (9.122) in (9.135) and thus formulate λ^* as follows:

$$\lambda^* = \frac{2}{\|\mathbf{u}(n)\|^2} [d^*(n) - \hat{\mathbf{w}}^T(n)\mathbf{u}^*(n)] \qquad (9.136)$$

However, from the definition of the estimation error $e(n)$, we have

$$e(n) = d(n) - \hat{\mathbf{w}}^H(n)\mathbf{u}(n)$$

Accordingly, we may further simplify the expression given in Eq. (9.136) and thus write

$$\lambda^* = \frac{2}{\|\mathbf{u}(n)\|^2} e^*(n) \qquad (9.137)$$

Finally, we substitute Eq. (9.137) into (9.133), obtaining

$$\delta\hat{w}_k(n + 1) = \hat{w}_k(n + 1) - \hat{w}_k(n)$$

$$= \frac{1}{\|\mathbf{u}(n)\|^2} u(n - k)e^*(n), \qquad k = 0, 1, \ldots, M - 1 \quad (9.138)$$

In vector form, we may equivalently write

$$\delta \hat{\mathbf{w}}(n + 1) = \hat{\mathbf{w}}(n + 1) - \hat{\mathbf{w}}(n)$$

$$= \frac{1}{\|\mathbf{u}(n)\|^2} \, \mathbf{u}(n)e^*(n) \tag{9.139}$$

In order to exercise control over the change in the tap-weight vector from one iteration to the next without changing its direction, we introduce a positive real scaling factor denoted by $\tilde{\mu}$. That is, we redefine the change $\delta \hat{\mathbf{w}}(n + 1)$ simply as

$$\delta \hat{\mathbf{w}}(n + 1) = \hat{\mathbf{w}}(n + 1) - \hat{\mathbf{w}}(n)$$

$$= \frac{\tilde{\mu}}{\|\mathbf{u}(n)\|^2} \, \mathbf{u}(n)e^*(n) \tag{9.140}$$

Equivalently, we may write

$$\hat{\mathbf{w}}(n + 1) = \hat{\mathbf{w}}(n) + \frac{\tilde{\mu}}{\|\mathbf{u}(n)\|^2} \, \mathbf{u}(n)e^*(n) \tag{9.141}$$

Indeed, this is the desired recursion for computing the M-by-1 tap-weight vector in the normalized LMS algorithm.

Equation (9.141) clearly shows the reason for using the term "normalized." In particular, we see that the product vector $\mathbf{u}(n)e^*(n)$ is normalized with respect to the squared Euclidean norm of the tap-input vector $\mathbf{u}(n)$.

The important point to note from the analysis presented above is that given new input data (at time n) represented by the tap-input vector $\mathbf{u}(n)$ and desired response $d(n)$, the normalized LMS algorithm updates the tap-weight vector in such a way that the value $\hat{\mathbf{w}}(n + 1)$ computed at time $n + 1$ exhibits the *minimum change* (in a Euclidean norm sense) with respect to the known value $\hat{\mathbf{w}}(n)$ at time n; for example, *no* charge may represent minimum change. Hence, the normalized LMS algorithm (and for that matter the conventional LMS algorithm) is a manifestation of the *principle of minimal disturbance* (Widrow and Lehr, 1990). The principle of minimal disturbance states that, *in the light of new input data, the parameters of an adaptive system should only be disturbed in a minimal fashion.*

Moreover, comparing the recursion of Eq. (9.141) for the normalized LMS algorithm with that of Eq. (9.8) for the conventional LMS algorithm, we may make the following observations:

- The adaptation constant $\tilde{\mu}$ for the normalized LMS algorithm is *dimensionless*, whereas the adaptation constant μ for the LMS algorithm has the dimensions of *inverse power*.
- Setting

$$\mu(n) = \frac{\tilde{\mu}}{\|\mathbf{u}(n)\|^2} \tag{9.142}$$

TABLE 9.2 SUMMARY OF THE NORMALIZED LMS ALGORITHM

Parameters: M = number of taps
$\tilde{\mu}$ = adaptation constant
$0 < \tilde{\mu} < 2$
a = positive constant

Initialization. If prior knowledge on the tap-weight vector $\hat{\mathbf{w}}(n)$ is available, use it to select an appropriate value for $\hat{\mathbf{w}}(0)$. Otherwise, set $\hat{\mathbf{w}}(0) = \mathbf{0}$.

Data
(a) Given: $\mathbf{u}(n)$: M-by-1 tap input vector at time n
$d(n)$: desired response at time n

(b) To be computed: $\hat{\mathbf{w}}(n + 1)$ = estimate of tap-weight vector at time $n + 1$
Computation: $n = 0, 1, 2, \ldots$

$$e(n) = d(n) - \hat{\mathbf{w}}^H(n)\mathbf{u}(n)$$

$$\hat{\mathbf{w}}(n + 1) = \hat{\mathbf{w}}(n) + \frac{\tilde{\mu}}{a + \|\mathbf{u}(n)\|^2}\,\mathbf{u}(n)e^*(n)$$

---

we may view the normalized LMS algorithm as an LMS algorithm with a *time-varying step-size parameter*.

• The normalized LMS algorithm is *convergent in the mean square* if the adaptation constant $\tilde{\mu}$ satisfies the following condition (Weiss and Mitra,1979; Hsia, 1983):

$$0 < \tilde{\mu} < 2 \tag{9.143}$$

Most importantly, the normalized LMS algorithm exhibits a rate of convergence that is potentially faster than that of the standard LMS algorithm for both uncorrelated and correlated input data (Nagumo and Noda, 1967; Douglas and Meng, 1994). Another point of interest is that in overcoming the gradient noise amplification problem associated with the LMS algorithm, the normalized LMS algorithm introduces a problem of its own. Specifically, when the tap-input vector $\mathbf{u}(n)$ is small, numerical difficulties may arise because then we have to divide by a small value for the squared norm $\|\mathbf{u}(n)\|^2$. To overcome this problem, we slightly modify the recursion of Eq. (9.141) as follows:

$$\hat{\mathbf{w}}(n + 1) = \hat{\mathbf{w}}(n) + \frac{\tilde{\mu}}{a + \|\mathbf{u}(n)\|^2}\,\mathbf{u}(n)e^*(n) \tag{9.144}$$

where $a > 0$, and as before $0 < \tilde{\mu} < 2$. For $a = 0$, Eq. (9.144) reduces to the previous form given in Eq. (9.141). The normalized LMS algorithm is summarized in Table 9.2.

9.12 SUMMARY AND DISCUSSION

In this rather long chapter, we have presented a detailed study of the least-mean-square (LMS) algorithm, which represents the workhorse of linear adaptive filtering. The practical importance of the LMS algorithm is largely due to two unique attributes:

- Simplicity of implementation
- Model-independent and therefore robust performance

The main limitation of the LMS algorithm is its relatively slow rate of convergence.

Two principal factors affect the convergence behavior of the LMS algorithm: the step-size parameter μ, and the eigenvalues of the correlation matrix $\mathbf{R}$ of the tap-input vector. In light of the analysis of the LMS algorithm, using the independence theory, their individual effects may be summarized as follows:

1. Convergence of the LMS algorithm in the mean square is assured by choosing the step-size parameter μ in accordance with the practical condition:

$$0 < \mu < \frac{2}{\text{tap-input power}}$$

   where the tap-input power is the sum of the mean-square values of all the tap inputs in the transversal filter.

2. When a small value is assigned to μ, the adaptation is slow, which is equivalent to the LMS algorithm having a long "memory." Correspondingly, the excess mean-squared error after adaptation is small, on the average, because of the large amount of data used by the algorithm to estimate the gradient vector. On the other hand, when μ is large, the adaptation is relatively fast, but at the expense of an increase in the average excess mean-squared error after adaptation. In this case, less data enter the estimation, hence a degraded estimation error performance. Thus, the reciprocal of the parameter μ may be viewed as the *memory* of the LMS algorithm.

3. When the eigenvalues of the correlation matrix $\mathbf{R}$ are widely spread, the excess mean-squared error produced by the LMS algorithm is primarily determined by the largest eigenvalues, and the time taken by the average tap-weight vector $E[\hat{\mathbf{w}}(n)]$ to converge is limited by the smallest eigenvalues. However, the speed of convergence of the mean-squared error, $J(n)$, is affected by a spread of the eigenvalues of $\mathbf{R}$ to a lesser extent than the convergence of $E[\hat{\mathbf{w}}(n)]$. When the eigenvalue spread is large (i.e., when the correlation matrix of the tap inputs is ill conditioned), the convergence of the LMS algorithm may slow down. However, this need not always be so, as the convergence behavior of the LMS algorithm takes on a directional nature under non-white inputs. This property may indeed be exploited in initializing the LMS algorithm, thereby improving the convergence process; for this to be possible, prior knowledge is required.

A basic limitation of the independence theory is the fact that it ignores the statistical dependence between the "gradient" directions as the algorithm proceeds from one iteration to the next. Several worthwhile results have been obtained in the literature on the practical case of *statistically dependent inputs*. In this regard, special mention should be made of the papers by Mazo (1979), Farden (1981), Jones et al. (1982), Macchi and Eweda (1984), and Gardner (1984), and the book by Macchi (1995).

Sethares (1993) presents a detailed discussion of the convergence behavior of the LMS algorithm using two other approaches: the stochastic approximation approach and the deterministic approach. In the *stochastic approximation approach*, developed independently by Ljung (1977) and Kushner and Clark (1978), the discrete-time evolution of the parameter estimation errors of the LMS algorithm is related to the behavior of an unforced deterministic ordinary differential equation; in particular, it is shown that stability of the ordinary differential equation so derived implies convergence of the LMS algorithm. In the *deterministic approach*, the basic update equation of the LMS algorithm [i.e., Eq. (9.8)] is interpreted as the state equation of a nonlinear, time-varying system; the system is then linearized and averaged to derive the operating conditions for which the LMS algorithm may be expected to succeed in its linear adaptive filtering task.

In yet another approach described in Butterweck (1994), a steady-state analysis of the LMS algorithm is presented, which relies on the use of a power series solution for the weight-error vector $\boldsymbol{\epsilon}(n)$ without invoking the independence assumption. The essence of this latter approach is described in Appendix I.

One last comment is in order. In the study of digital filters, frequency-domain performance measures play a central role alongside their time-domain counterparts. Yet the convergence analysis of the LMS algorithm presented in this chapter (and for that matter, in much of the literature on adaptive filters) has been confined to the time domain. The paper by Johnson et al. (1994) attempts to redress this imbalance by presenting a fundamental re-evaluation of adaptive filter performance in frequency-domain terms, and uses an adaptive equalizer as an illustrative example. An interesting point that emerges from the study presented therein is that there is a correspondence between (a) the rates at which the different bands in the equalizer's frequency response adapt toward their steady-state values, and (b) the way in which the eigenfilters are grouped according to the eigenvalues of the correlation matrix of the channel output (i.e., the equalizer's input). Another noteworthy paper on the frequency-domain analysis of linear adaptive filters is that of Gunnarsson and Ljung (1989). The focal point of this latter paper is the formulation of a performance measure in terms of the mean-square error between the true (momentary) transfer function and the one being estimated by the adaptive filter. The evaluation is done in the context of tracking a linear time-variant system, which is the subject matter of Chapter 16.

PROBLEMS

1. The LMS algorithm is used to implement a dual-input, single-weight adaptive noise canceler. Set up the equations that define the operation of this algorithm.

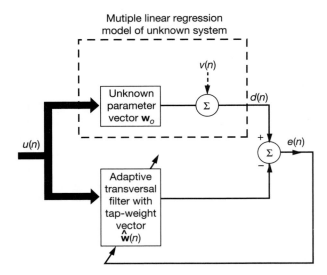

Figure P9.1

2. The LMS-based adaptive deconvolution procedure discussed in Example 2 of Section 9.3 applies to forward-time adaptation (i.e., forward prediction). Reformulate this procedure for reverse-time adaptation (i.e., backward prediction).

3. The zero-mean output $d(n)$ of an unknown real-valued system is represented by the *multiple linear regression model*

$$d(n) = \mathbf{w}_o^T \mathbf{u}(n) + v(n)$$

where $\mathbf{w}_o$ is the (unknown) parameter vector of the model, $\mathbf{u}(n)$ is the input vector (regressor), and $v(n)$ is the sample value of an immeasurable white-noise process of zero mean and variance σ_v^2. The block diagram of Fig. P9.1 shows the adaptive modeling of the unknown system, in which the adaptive transversal filter is controlled by a *modified* version of the LMS algorithm. In particular, the tap-weight vector $\mathbf{w}(n)$ of the transversal filter is chosen to minimize the index of performance

$$J(\mathbf{w}, K) = E[e^{2K}(n)]$$

for $K = 1, 2, 3, \ldots$

(a) By using the instantaneous gradient vector, show that the new adaptation rule for the corresponding estimate of the tap-weight vector is

$$\hat{\mathbf{w}}(n + 1) = \hat{\mathbf{w}}(n) + \mu K \mathbf{u}(n) e^{2K-1}(n)$$

where μ is the step-size parameter, and $e(n)$ is the estimation error

$$e(n) = d(n) - \mathbf{w}^T(n)\mathbf{u}(n)$$

(b) Assume that the weight-error vector

$$\boldsymbol{\epsilon}(n) = \hat{\mathbf{w}}(n) - \mathbf{w}_o$$

is close to zero, and that $v(n)$ is independent of $\mathbf{u}(n)$. Hence, show that

$$E[\boldsymbol{\epsilon}(n+1)] = (\mathbf{I} - \mu K(2K-1)E[v^{2K-2}(n)]\mathbf{R}E[\boldsymbol{\epsilon}(n)]$$

where $\mathbf{R}$ is the correlation matrix of the input vector $\mathbf{u}(n)$.

(c) Show that the modified LMS algorithm described in part (a) converges in the mean value if the step-size parameter μ satisfies the condition

$$0 < \mu < \frac{2}{K(2K-1)E[v^{2(K-1)}(n)]\lambda_{\max}}$$

where $\lambda_{\max}$ is the largest eigenvalue of matrix $\mathbf{R}$.

(d) For $K = 1$, show that the results given in parts (a), (b), and (c) reduce to those in the conventional LMS algorithm.

4. (a) Let $\mathbf{m}(n)$ denote the *mean weight vector* in the LMS algorithm at iteration n; that is

$$\mathbf{m}(n) = E[\hat{\mathbf{w}}(n)]$$

Using the independence assumption of Section 9.4, show that

$$\mathbf{m}(n) = (\mathbf{I} - \mu\mathbf{R})^n[\mathbf{m}(0) - \mathbf{m}(\infty)] + \mathbf{m}(\infty)$$

where μ is the step-size parameter, $\mathbf{R}$ is the correlation matrix of the input vector, and $\mathbf{m}(0)$ and $\mathbf{m}(\infty)$ are the initial and final values of the mean weight vector, respectively.

(b) Hence, show that for convergence of the mean value $\mathbf{m}(n)$, the step-size parameter μ must satisfy the condition

$$0 < \mu < \frac{2}{\lambda_{\max}}$$

where $\lambda_{\max}$ is the largest eigenvalue of the correlation matrix $\mathbf{R}$.

5. Consider the product $x_1 x_2^* x_3 x_4^*$, where x_1, x_2, x_3, and x_4 are complex Gaussian random variables. The Gaussian moment factoring theorem states that (see Chapter 2)

$$E[x_1 x_2^* x_3 x_4^*] = E[x_1 x_2^*]E[x_3 x_4^*] + E[x_1 x_4^*]E[x_3 x_2^*]$$

Using this theorem, show that

$$E[e_o^*(n)\mathbf{u}(n)\mathbf{u}^H(n)e_o(n)] = J_{\min}\mathbf{R}$$

6. Consider the use of a white noise sequence of zero mean and variance σ^2 as the input to the LMS algorithm. Evaluate the following:

(a) Condition for convergence of the algorithm in the mean square.

(b) The excess mean-squared error.

7. *The leaky LMS algorithm.* Consider the time-varying cost function

$$J(n) = |e(n)|^2 + \alpha\|\mathbf{w}(n)\|^2$$

where $\mathbf{w}(n)$ is the tap-weight vector of a transversal filter, $e(n)$ is the estimation error, and α is a constant. As usual, $e(n)$ is defined by

$$e(n) = d(n) - \mathbf{w}^H(n)\mathbf{u}(n)$$

where $d(n)$ is the desired response, and $\mathbf{u}(n)$ is the tap-input vector. In the *leaky LMS algorithm*, the cost function $J(n)$ is minimized with respect to the weight vector $\mathbf{w}(n)$.

(a) Show that the time update for the tap-weight vector $\hat{\mathbf{w}}(n)$ is defined by

$$\hat{\mathbf{w}}(n + 1) = (1 - \mu\alpha)\hat{\mathbf{w}}(n) + \mu\mathbf{u}(n)e^*(n)$$

(b) Using the independence theory, show that

$$\lim_{n \to \infty} E[\hat{\mathbf{w}}(n)] = (\mathbf{R} + \alpha\mathbf{I})^{-1}\mathbf{p}$$

where $\mathbf{R}$ is the correlation matrix of the tap inputs and $\mathbf{p}$ is the cross-correlation vector between the tap inputs and the desired response. What is the condition for the algorithm to converge in the mean value?

(c) How would you modify the tap-input vector in the conventional LMS algorithm to get the equivalent result described in part (a)?

8. Consider the operation of an adaptive line enhancer using the LMS algorithm under a low signal-to-noise ratio condition. The correlation matrix of the input vector is defined by

$$\mathbf{R} = \sigma_v^2\mathbf{I}$$

where $\mathbf{I}$ is the identity matrix. Show that the steady-state value of the weight-error correlation matrix $\mathbf{K}(n)$ is given by

$$\mathbf{K}(\infty) \simeq \frac{\mu}{2}J_{\min}\,\mathbf{I}$$

where μ is the step-size parameter, and $J_{\min}$ is the minimum mean-squared error. You may assume that the number of taps in the adaptive transversal filter is large.

9. (a) The LMS algorithm is usually referred to as a stochastic gradient algorithm. Yet, in examining Figs. 9.27 and 9.28 of Example 7 involving a purely sinusoidal process, the trajectories displayed therein are all well defined (i.e., the parameter estimates produced by the LMS algorithm are deterministic). Both of these statements are valid in their own ways; how do you reconcile them?

(b) Suppose a white noise process of zero mean and various σ_v^2 is added to the sinusoidal process considered in Example 7. How would the results of that example be modified?

10. The convergence ratio, $\mathscr{C}(n)$, of an adaptive algorithm is defined by

$$\mathscr{C}(n) = \frac{E[\|\boldsymbol{\epsilon}(n + 1)\|^2]}{E[\|\boldsymbol{\epsilon}(n)\|^2]}$$

Show that, for small n, the convergence ratio of the LMS algorithm for stationary inputs approximately equals

$$\mathscr{C}(n) \simeq (1 - \mu\sigma_u^2)^2 \qquad n \text{ small}$$

Here it is assumed that the correlation matrix of the tap-input vector $\mathbf{u}(n)$ is approximately equal to $\sigma_u^2\,\mathbf{I}$.

11. Starting from Eq. (9.114) and using the update recursion of Eq. (9.115) for the LMS algorithm, verify the validity of the inequality described in Eq. (9.116).

12. Expanding on the result described in (9.120), show that the LMS algorithm also satisfies the bound (Sayed and Rupp, 1994)

$$\frac{\mu^{-1}\|\hat{\mathbf{w}}(n+1) - \mathbf{w}_o\|^2 + \sum_{i=0}^{n} |\hat{\mathbf{w}}^H(i)\mathbf{u}(i) - \mathbf{w}_o^H\mathbf{u}(i)|^2}{\mu^{-1}\|\hat{\mathbf{w}}(0) - \mathbf{w}_o\|^2 + \sum_{i=1}^{n} |v(i)|^2} \leq 1$$

In light of this result, what can you say about the operator that maps $\{\mu^{-1/2}(\hat{\mathbf{w}}(0) - \mathbf{w}_o), v(0), v(1), \ldots, v(n)\}$ to the sequence of errors $\{\mu^{-1/2}(\hat{\mathbf{w}}(n+1) - \mathbf{w}_o), \hat{\mathbf{w}}^H(i)\mathbf{u}(i) - \mathbf{w}_o^H\mathbf{u}(i)|i = 0, 1, \ldots, n\}$?

13. The *normalized LMS algorithm* is described by the following recursion for the tap-weight vector:

$$\hat{\mathbf{w}}(n+1) = \hat{\mathbf{w}}(n) + \frac{\tilde{\mu}}{\|\mathbf{u}(n)\|^2}\mathbf{u}(n)e^*(n)$$

where $\tilde{\mu}$ is a positive constant, and $\|\mathbf{u}(n)\|$ is the *norm* of the tap-input vector. The estimation error $e(n)$ is defined by

$$e(n) = d(n) - \mathbf{w}^H(n)\mathbf{u}(n)$$

where $d(n)$ is the desired response.

Using the independence theory, show that the necessary and sufficient condition for the normalized LMS algorithm to be convergent in the mean square is $0 < \tilde{\mu} < 2$. *Hint*: You may use the following approximation:

$$E\left[\frac{\mathbf{u}(n)\mathbf{u}^H(n)}{\|\mathbf{u}(n)\|^2}\right] \simeq \frac{E[\mathbf{u}(n)\mathbf{u}^H(n)]}{E[\|\mathbf{u}(n)\|^2]}$$

14. In Section 9.11 we presented a derivation of the normalized LMS algorithm in its own right. In this problem we explore another derivation of this algorithm by modifying the method of steepest descent that led to the development of the conventional LMS algorithm. The modification involves writing the tap-weight vector update in the method of steepest descent as follows:

$$\mathbf{w}(n+1) = \mathbf{w}(n) - \frac{1}{2}\mu(n)\nabla(n)$$

where $\mu(n)$ is a *time-varying step-size parameter*, and $\nabla(n)$ is the gradient vector defined by

$$\nabla(n) = 2[\mathbf{R}\mathbf{w}(n) - \mathbf{p}]$$

where $\mathbf{R}$ is the correlation matrix of the tap-input vector $\mathbf{u}(n)$, and $\mathbf{p}$ is the cross-correlation vector between the tap-input vector $\mathbf{u}(n)$ and the desired response $d(n)$.

(a) At time $n + 1$, the mean-squared error is defined by

$$J(n+1) = E[|e(n+1)|^2]$$

where

$$e(n+1) = d(n+1) - \mathbf{w}^H(n+1)\mathbf{u}(n+1)$$

Determine the value of the step-size parameter $\mu_o(n)$ that minimizes $J(n + 1)$ as a function of $\mathbf{R}$ and $\mathbf{\nabla}(n)$.

(b) Using instantaneous estimates for $\mathbf{R}$ and $\mathbf{\nabla}(n)$ in the expression for $\mu_o(n)$ derived in part (a), determine the corresponding instantaneous estimate for $\mu_o(n)$. Hence, formulate the update equation for the tap-weight vector $\hat{\mathbf{w}}(n)$, and compare your result with that obtained for the normalized LMS algorithm.

15. When conducting a computer experiment that involves the generation of an AR process, sometimes not enough time is allowed for the transients to die out. The purpose of this experiment is to evaluate the effects of such transients on the operation of the LMS algorithm. Consider then the AR process $u(n)$ of order 1 described in Section 9.6. The parameters of this process are as follows:

AR parameter: $a = 0.99$
AR process variance: $\sigma_u^2 = 1.00$
Noise variance: $\sigma_v^2 = 0.02$

Generate the process $u(n)$ so described for $1 \leq n \leq 100$, assuming zero initial conditions. Use the process $u(n)$ as the input of a linear adaptive predictor that is based on the LMS algorithm using a step-size parameter $\mu = 0.05$. In particular, plot the learning curve of the predictor by ensemble averaging over 100 independent realizations of the squared value of its output versus time n for $1 \leq n \leq 100$. Unlike the normal operation of the LMS algorithm, the learning curve so computed should start at the origin, rise to a peak, and then decay toward a steady-state value. Explain the reasons for this phenomenon.

10

Frequency-Domain Adaptive Filters

In the case of the LMS algorithm described in the previous chapter, adaptation of the tap weights (free parameters) of a finite-duration impulse response (FIR) filter is performed in the time domain. Recognizing that the *Fourier transform* maps time-domain signals into the frequency domain and that the inverse Fourier transform provides the inverse mapping that takes us back into the time domain, it is equally feasible to perform the adaptation of filter parameters in the frequency domain. In such a case we speak of *frequency-domain adaptive filtering (FDAF)*, the origin of which may be traced back to an early paper by Walzman and Schwartz (1973).

There are two main reasons for seeking the use of frequency-domain adaptive filtering in one form or another:

1. In certain applications, such as acoustic echo cancelation in teleconferencing, for example, the adaptive filter is required to have a long impulse response (i.e., long memory) to cope with an equally long echo duration (Murano et al., 1990). When the LMS algorithm is adapted in the time domain, we find that the requirement of a long memory results in a significant increase in the computational complexity of the algorithm. How then do we deal with this problem? There are two options available to us. We may choose an infinite-duration impulse response (IIR) filter and adapt it in the time domain (Shynk, 1989; Regalia, 1994); the difficulty with this approach is that we inherit a new problem, namely, that of filter

instability. Alternatively, we may use a particular type of frequency-domain adaptive filtering that combines two complementary methods widely used in digital signal processing (Ferrara, 1980, 1985; Clark et al., 1981, 1983; Shynk, 1992):

- *Block implementation* of an FIR filter, which allows the efficient use of parallel processing and thereby results in a gain in computational speed
- *Fast Fourier transform (FFT)* algorithms for performing fast convolution (filtering), which permits adaptation of filter parameters in the frequency domain in a computationally efficient manner

This approach to frequency-domain adaptive filtering builds on the so-called *block LMS algorithm* that includes the standard LMS algorithm as a special case. The principal virtue of this approach is that it makes it feasible to apply adaptive FIR filtering with long memory in a computationally efficient manner.

2. Frequency-domain adaptive filtering, mechanized in a different way from that described under point 1, is used to improve the convergence performance of the standard LMS algorithm. In this second situation a more uniform convergence rate is attained by exploiting the *orthogonality properties* of the discrete Fourier transform (DFT) and related discrete transforms (Narayan et al., 1981, 1983; Widrow et al., 1987, 1994; Shynk, 1992).

Both of these approaches to frequency-domain adaptive filtering are discussed in this chapter. We begin the discussion by considering the idea of block adaptive filtering that paves the way for the implementation of FFT-based frequency-domain adaptive filtering.

To simplify the presentation, we will confine our attention in this chapter to the case of real-valued data.

10.1 BLOCK ADAPTIVE FILTERS

In a *block adaptive filter*, depicted in Fig. 10.1, the incoming data sequence $u(n)$ is sectioned into L-point blocks by means of a serial-to-parallel converter, and the blocks of input data so produced are applied to an FIR filter of length M, one block at a time. The tap weights of the filter are held *fixed* over each block of data, so that adaptation of the filter proceeds on a block-by-block basis rather than on a sample-by-sample basis as in the standard LMS algorithm (Clark et al., 1981; Shynk, 1992). Let k refer to *block time* and $\hat{\mathbf{w}}(k)$ denote the tap-weight vector of the filter for the kth block, as shown by

$$\hat{\mathbf{w}}(k) = [\hat{w}_0(k), \hat{w}_1(k), \ldots, \hat{w}_{M-1}(k)]^T, \qquad k = 0, 1, \ldots \tag{10.1}$$

The index n is reserved for the original *sample time*, written in terms of the block time as follows:

$$n = kL + i, \qquad i = 0, 1, \ldots, M - 1, \qquad k = 0, 1, \ldots, \tag{10.2}$$

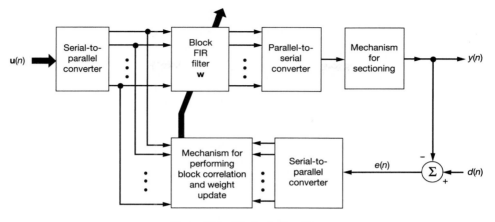

Figure 10.1   Block-adaptive filter.

Let the input signal vector $\mathbf{u}(n)$ at time n be written as

$$\mathbf{u}(n) = [u(n), u(n-1), \ldots, u(n-M+1)]^T \tag{10.3}$$

Accordingly, at time n the output $y(n)$ produced by the filter in response to the input signal vector $\mathbf{u}(n)$ is defined by the inner product

$$y(n) = \hat{\mathbf{w}}^T(k)\mathbf{u}(n) \tag{10.4}$$

Equivalently, in light of Eq. (10.2) we may write

$$
\begin{aligned}
y(kL+i) &= \hat{\mathbf{w}}^T(k)\mathbf{u}(kL+i) \\
&= \sum_{l=0}^{M-1} \hat{w}_l(k)u(kL+i-l), \qquad i = 0, 1, \ldots, M-1
\end{aligned}
\tag{10.5}
$$

Let

$$d(n) = d(kL+i) \tag{10.6}$$

denote the corresponding value of the desired response. An error signal $e(n)$ is produced by comparing the filter output $y(n)$ against the desired response $d(n)$, as shown in Fig. 10.1; the error signal is defined by

$$e(n) = d(n) - y(n) \tag{10.7}$$

or equivalently

$$e(kL+i) = d(kL+i) - y(kL+i) \tag{10.8}$$

Thus, the error signal is permitted to vary at the sampling rate as in the standard LMS algorithm. The error sequence $e(n)$ is sectioned into L-point blocks in a synchronous manner with that at the input end of the block adaptive filter and then used to compute the correction to be applied to the tap weights of the filter, as depicted in Fig. 10.1.

Example 1

To illustrate the operation of the block adaptive filter, consider the example of a filter for which the filter length M and block size L are both equal to 3. We may then express the output sequence computed by the filter for three consecutive blocks, $k - 1$, k, and $k + 1$, as follows:

$$(k - 1)\text{th block} \begin{cases} \begin{bmatrix} u(3k-3) & u(3k-4) & u(3k-5) \\ u(3k-2) & u(3k-3) & u(3k-4) \\ u(3k-1) & u(3k-2) & u(3k-3) \end{bmatrix} \begin{bmatrix} w_0(k - 1) \\ w_1(k - 1) \\ w_2(k - 1) \end{bmatrix} = \begin{bmatrix} y(3k - 3) \\ y(3k - 2) \\ y(3k - 1) \end{bmatrix} \end{cases}$$

$$k\text{th block} \begin{cases} \begin{bmatrix} u(3k) & u(3k-1) & u(3k-2) \\ u(3k+1) & u(3k) & u(3k-1) \\ u(3k+2) & u(3k+1) & u(3k) \end{bmatrix} \begin{bmatrix} w_0(k) \\ w_1(k) \\ w_2(k) \end{bmatrix} = \begin{bmatrix} y(3k) \\ y(3k + 1) \\ y(3k + 2) \end{bmatrix} \end{cases}$$

$$(k + 1)\text{th block} \begin{cases} \begin{bmatrix} u(3k+3) & u(3k+2) & u(3k+1) \\ u(3k+4) & u(3k+3) & u(3k+2) \\ u(3k+5) & u(3k+4) & u(3k+3) \end{bmatrix} \begin{bmatrix} w_0(k + 1) \\ w_1(k + 1) \\ w_2(k + 1) \end{bmatrix} = \begin{bmatrix} y(3k + 3) \\ y(3k + 4) \\ y(3k + 5) \end{bmatrix} \end{cases}$$

Note that the data matrix defined here is a Toeplitz matrix by virtue of the fact that the elements on any principal diagonal of the matrix are all the same.

Block LMS Algorithm

From the development of the LMS algorithm presented in the previous chapter, we recall the following formula for the "correction" applied to the tap-weight vector from one iteration of the algorithm to the next (assuming real-valued data here):

$$\begin{pmatrix} \text{Correction to the} \\ \text{weight vector} \end{pmatrix} = \begin{pmatrix} \text{Step-size} \\ \text{parameter} \end{pmatrix} \begin{pmatrix} \text{tap-input} \\ \text{vector} \end{pmatrix} (\text{error signal})$$

Recognizing that in the block LMS algorithm the error signal is allowed to vary at the sampling rate, it follows that for each block of data we have different values of the error signal for use in the adaptive process. Accordingly, for the kth block, we may sum the product $\mathbf{u}(kL + i)e(kL + i)$ over all possible values of i, and so define the following update equation for the tap-weight vector of the block LMS algorithm operating on real-valued data:

$$\hat{\mathbf{w}}(k + 1) = \hat{\mathbf{w}}(k) + \mu \sum_{i=0}^{L-1} \mathbf{u}(kM + i)e(kM + i) \tag{10.9}$$

where μ is the step-size parameter. For convenience of presentation (which will become apparent in the next section), we rewrite Eq. (10.9) in the form

$$\hat{\mathbf{w}}(k + 1) = \hat{\mathbf{w}}(k) + \mu\boldsymbol{\phi}(k) \tag{10.10}$$

where the M-by-1 vector $\boldsymbol{\phi}(k)$ is a cross-correlation defined by

$$\boldsymbol{\phi}(k) = \sum_{i=0}^{L-1} \mathbf{u}(kM + i)e(kM + i) \tag{10.11}$$

The jth element of the vector $\boldsymbol{\phi}(k)$ is defined by

$$\varphi_j(k) = \sum_{i=0}^{L-1} u(kM + i - j)e(kM + i), \qquad j = 0, 1, \ldots, M - 1 \tag{10.12}$$

A distinctive feature of the block LMS algorithm described herein is that its design incorporates an *averaged estimate* of the gradient vector, as shown by

$$\hat{\nabla}(k) = -\frac{2}{L} \sum_{i=0}^{L-1} \mathbf{u}(kL + i)e(kL + i) \tag{10.13}$$

where the factor 2 is included to be consistent with the definition of the gradient vector used in Chapters 8 and 9, and the factor $1/L$ is included for $\hat{\nabla}(k)$ to be an unbiased time average. Then, in terms of $\hat{\nabla}(k)$ we may reformulate the block LMS algorithm as follows:

$$\hat{\mathbf{w}}(k + 1) = \hat{\mathbf{w}}(k) - \frac{1}{2} \mu_B \hat{\nabla}(k) \tag{10.14}$$

where μ_B may be viewed as the "effective" step-size parameter of the block LMS algorithm; it is defined by

$$\mu_B = L\mu \tag{10.15}$$

Convergence Properties of the Block LMS Algorithm

The block LMS algorithm has properties similar to those of the standard LMS algorithm in that they both attempt to minimize the same mean-square error function

$$J = \frac{1}{2} E[e^2(n)] \tag{10.16}$$

where E is the statistical expectation operator. The fundamental difference between these two algorithms lies in the estimates of the gradient vector used in their respective implementations. Comparing the estimate of Eq. (10.13) for the block LMS algorithm with that of Eq. (9.4) from the previous chapter for the conventional LMS algorithm, we see that the block LMS algorithm uses a more accurate estimate of the gradient vector because of the time averaging, with the estimation accuracy increasing as the block size L is increased. However, this improvement does not imply faster adaptation, a fact that is revealed by examining the convergence properties of the block LMS algorithm.

We may proceed through a convergence analysis of the block LMS algorithm in a manner similar to that described in Chapter 9 for the conventional LMS algorithm. Indeed, such an analysis follows the same steps as those described there. There is only a minor modification to be considered, namely, the summation of certain expectations over the index $i = 0, 1, \ldots, L - 1$, which is related to the sample time n as in Eq. (10.2) and the use of which arises by virtue of Eq. (10.8). We may thus summarize the convergence properties of the block LMS algorithm as follows:

1. *Condition for convergence.* The mean of the tap-weight vector $\hat{\mathbf{w}}(k)$ computed by using the block LMS algorithm converges to the optimum Wiener solution $\mathbf{w}_o$ as the number of block iterations k approaches infinity, as shown by

$$\lim_{k \to \infty} E[\hat{\mathbf{w}}(k)] = \mathbf{R}^{-1}\mathbf{p} = \mathbf{w}_o \tag{10.17}$$

where

$$\mathbf{R} = E[\mathbf{u}(n)\mathbf{u}^T(n)] \tag{10.18}$$

$$\mathbf{p} = E[\mathbf{u}(n)d(n)] \tag{10.19}$$

The condition that has to be satisfied by the step-size parameter μ for convergence of the block LMS algorithm in the mean value is described by (Clark et al., 1981)

$$0 < \mu < \frac{2}{\lambda_{\max}} \tag{10.20}$$

where L is the block size, and $\lambda_{\max}$ is the largest eigenvalue of the correlation matrix $\mathbf{R}$ of the input signal vector $\mathbf{u}(n)$.

2. *Misadjustment.* Invoking the definitions of the excess mean-squared error $J_{\text{ex}}(k)$ and the minimum mean-squared $J_{\min}$ that were given in Chapter 9, we note that for the $J_{\text{ex}}(k)$ computed by the block LMS algorithm to converge to a constant value $J_{\text{ex}}(\infty) < J_{\min}$ as the number of block iterations k approaches infinity, the step-size parameter μ has to satisfy the more stringent condition

$$0 < \mu < \frac{2}{L \sum_{i=1}^{M} \lambda_i} \tag{10.21}$$

The corresponding value of the misadjustment is

$$\mathcal{M} = \frac{\mu}{2} \sum_{i=1}^{M} \lambda_i \tag{10.22}$$

Comparing the results described here for the block LMS algorithm with the corresponding results derived in Chapter 9 for the standard LMS algorithm, we may make the following observations when operating in a wide-sense stationary environment:

- The converged mean weight vector and misadjustment of the block LMS algorithm are identical to those of the standard LMS algorithm. The same holds for the average time constant.
- For an input signal vector $\mathbf{u}(n)$ whose correlation matrix $\mathbf{R}$ has a prescribed eigenstructure, the condition imposed on the block LMS for convergence in the mean square is more restrictive than the corresponding condition for the standard LMS algorithm. This is readily confirmed by comparing Eqs. (10.20) and (10.21) for the

block LMS algorithm with Eqs. (9.57) and (9.72), respectively, for the standard LMS algorithm. In particular, the tighter bound on the step-size parameter μ may cause the block LMS algorithm to converge more slowly than the standard LMS algorithm, particularly when the eigenvalue spread of the correlation matrix $\mathbf{R}$ is large. More seriously, we may be confronted with a situation that requires fast adaptation and therefore a large μ, but the required block size L is so large that the conditions for convergence are not satisfied, making it impractical to use the block LMS algorithm.

Choice of Block Size

An important issue that needs to be considered in the design of a block adaptive filter is how to choose the block size L. From Eq. (10.9) we observe that the operation of the block LMS algorithm holds true for any integer value of L equal to or greater than unity. Nevertheless, the option of choosing the block size L equal to the filter length M is preferred in most applications of block adaptive filtering. This choice may be justified on the following grounds (Clark et al., 1981):

- When $L > M$, redundant operations are involved in the adaptive process, because then the estimation of the gradient vector (computed over L points) uses more input information than the filter itself.
- When $L < M$, some of the tap weights in the filter are wasted, because the sequence of tap inputs is not long enough to feed the whole filter.

It thus appears that the most practical choice is $L = M$.

10.2 FAST LMS ALGORITHM

Given an adaptive signal-processing application for which the block LMS algorithm is a satisfactory solution, the key question to be addressed is how to implement it in a computationally efficient manner. Referring to Eqs. (10.5) and (10.12), where the computational burden of the block LMS algorithm lies, we observe the following:

- Equation (10.5) defines a *linear convolution* of the tap inputs and tap weights of the filter.
- Equation (10.12) defines a *linear correlation* between the taps inputs of the filter and the error signal.

Now, from the material presented in Chapter 1, we know that the *fast Fourier transform (FFT) algorithm* provides a powerful tool for performing *fast convolution* and *fast correlation*. These observations point to a frequency-domain method for efficient implementation of the block LMS algorithm. Specifically, rather than performing the adaptation in the time domain as described in the previous section, the adaptation of filter parameters is actually performed in the frequency domain by using the FFT algorithm. The block LMS algorithm so implemented is referred to as the *fast LMS algorithm*, which was developed independently by Clark et al. (1980, 1982) and Ferrara (1980).

From Chapter 2 we recall that fast convolution may be performed using the overlap-save method or, alternatively, the overlap-add method. However, in implementing the fast LMS algorithm, the overlap-add method[1] results in more computations than that needed in the overlap-save method (Clark et al., 1983). According to Clark et al. (1981) the most efficient implementation of the overlap-save method is obtained by using 50 percent overlap. Hence, the description of the fast LMS algorithm presented here uses the overlap-save method with 50 percent overlap.

According to this method, the M tap weights of the filter are padded with an equal number of zeros, and an N-point FFT is used for the computation, where

$$N = 2M \tag{10.23}$$

Thus let the N-by-1 vector $\hat{\mathbf{W}}(k)$ denote the FFT coefficients of the zero-padded, tap-weight vector $\hat{\mathbf{w}}(k)$, as follows:

$$\hat{\mathbf{W}}(k) = \mathrm{FFT}\begin{bmatrix} \hat{\mathbf{w}}(k) \\ \mathbf{0} \end{bmatrix} \tag{10.24}$$

where $\mathbf{0}$ is the M-by-1 null vector and FFT [ ] denotes fast Fourier transformation. Note that the frequency-domain weight vector $\hat{\mathbf{W}}(k)$ is *twice* as long as the time-domain weight vector $\hat{\mathbf{w}}(k)$. Correspondingly, let $\mathbf{U}(k)$ denote an N-by-N diagonal matrix derived from the input data as follows:

$$\mathbf{U}(k) = \mathrm{diag}\{\mathrm{FFT}[\underbrace{u(kM - M), \ldots, u(kM - 1)}_{(k-1)\text{th block}}, \underbrace{u(kM), \ldots, u(kM + M - 1)}_{k\text{th block}}]\} \tag{10.25}$$

We could use a vector to define the transformed version of the input signal vector $\mathbf{u}(M)$; however, for our present needs the matrix notation of Eq. (10.25) is considered to be more appropriate. Hence, applying the overlap-save method to the linear convolution of Eq. (10.5) yields the M-by-1 vector

$$\begin{aligned} \mathbf{y}^T(k) &= [y(kM), y(kM + 1), \ldots, y(kM + M - 1)] \\ &= \text{last } M \text{ elements of } \mathrm{IFFT}[\mathbf{U}(k)\hat{\mathbf{W}}(k)] \end{aligned} \tag{10.26}$$

where IFFT[ ] denotes inverse fast Fourier transformation. Only the last M elements in Eq. (10.26) are retained, because the first M elements correspond to a circular convolution.

Consider next the linear correlation of Eq. (10.12). For the kth block, define the M-by-1 desired response vector

$$\mathbf{d}(k) = [d(kM), d(kM + 1), \ldots, d(kM + M - 1)]^T \tag{10.27}$$

---

[1] In Sommen and Jayasinghe (1988), a simplified form of the overlap-add method is described, saving two inverse DFTs.

and the corresponding M-by-1 error signal vector

$$\mathbf{e}(k) = [e(kM), e(kM + 1), \ldots, e(kM + M - 1)]^T \tag{10.28}$$
$$= \mathbf{d}(k) - \mathbf{y}(k)$$

Noting that in implementing the linear convolution described in Eq. (10.26) the first M elements are discarded from the output, we may transform the error signal vector $\mathbf{e}(k)$ into the frequency domain as follows:

$$\mathbf{E}(k) = \text{FFT}\begin{bmatrix} \mathbf{0} \\ \mathbf{e}(k) \end{bmatrix} \tag{10.29}$$

Next, recognizing that a linear correlation is basically a "reversed" form of linear convolution, we find that applying the overlap-save method to the linear correlation of Eq. (10.12) yields

$$\boldsymbol{\phi}(k) = \text{first } M \text{ elements of IFFT}[\mathbf{U}^H(k)\mathbf{E}(k)] \tag{10.30}$$

Note that whereas in the case of linear convolution considered in Eq. (10.26) the first M elements are discarded, in the case of Eq. (10.30) the last M elements are discarded.

Finally, consider Eq. (10.10) for updating the tap-weight vector of the filter. Noting that in the definition of the frequency-domain weight vector $\hat{\mathbf{W}}(k)$ of Eq. (10.24) the time-domain weight vector $\hat{\mathbf{w}}(k)$ is followed by M zeros, we may correspondingly transform Eq. (10.10) into the frequency domain as follows:

$$\hat{\mathbf{W}}(k + 1) = \hat{\mathbf{W}}(k) + \mu\, \text{FFT}\begin{bmatrix} \boldsymbol{\phi}(k) \\ \mathbf{0} \end{bmatrix} \tag{10.31}$$

Equations (10.24) to (10.31), in that order, define the fast LMS algorithm. Figure 10.2 shows a signal-flow graph representation of the fast LMS algorithm (Shynk, 1992). This algorithm represents a precise frequency-domain implementation of the block LMS algorithm. As such, its convergence properties are identical to those of the block LMS algorithm discussed in Section 10.1.

Computational Complexity

The computational complexity of the fast LMS algorithm operating in the frequency domain is now compared with that of the standard LMS algorithm operating in the time domain. The comparison is based on a count of the total number of multiplications involved in each of these two implementations for a block size M. Although in an actual implementation, there are other factors to be considered (e.g., the number of additions, storage requirements), the use of multiplications provides a reasonably accurate basis for comparing the computational complexity of these two algorithms (Shynk, 1992).

Consider first the standard LMS algorithm with M tap weights operating on real data. In this case, M multiplications are performed to compute the output and a further M multiplications are performed to update the tap weights, making for a total of $2M$ multi-

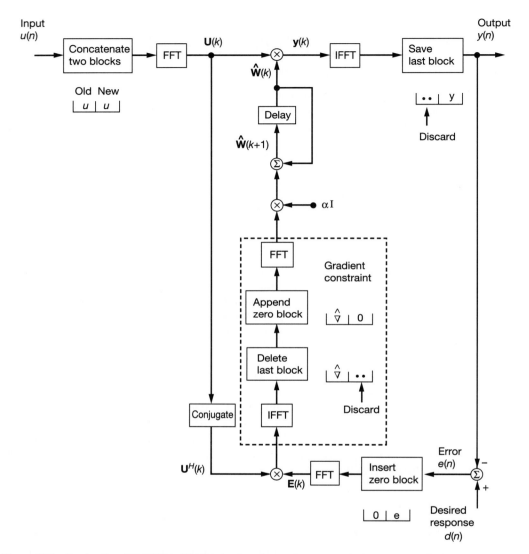

Figure 10.2   Overlap-Save FDAF. This FDAF is basd on the overlap-save sectioning procedure for implementing linear convolutions and linear correlations. (Taken from *IEEE SP Magazine* with permission of the IEEE)

plications per iteration. Hence, for a block of M output samples, the total number of multiplications is $2M^2$.

Consider next the fast LMS algorithm. Each N-point FFT (and IFFT) requires approximately $N \log_2 N$ real multiplications (Oppenheim and Schafer, 1989), where $N = 2M$. According to the structure of the fast LMS algorithm shown in Fig. 10.2, there are five frequency transformations performed, which therefore account for $5N\log_2 N$ multiplications. In addition, the computation of the frequency-domain output vector requires $4N$ multiplications, and so does the computation of the cross-correlations relating to the gradient vector estimation. Hence, the total corresponding number of multiplications performed in the fast LMS algorithm is

$$5N\log_2 N + 8N = 10M\log_2(2M) + 16M$$

$$= 10M\log_2 M + 26M$$

The *complexity ratio* for the fast LMS to the standard LMS algorithm is therefore (Shynk, 1992)

$$
\begin{aligned}
\text{Complexity ratio} &= \frac{10M\log_2 M + 26M}{2M^2} \\
&= \frac{5\log_2 M + 13}{M}
\end{aligned}
\tag{10.32}
$$

For example, for $M = 1024$, the use of Eq. (10.32) shows that the fast LMS algorithm is roughly 16 times faster than the standard LMS algorithm in computational terms.

Convergence Rate Improvement

Even though the block LMS algorithm, from which the fast LMS algorithm is derived, uses a more accurate estimate of the gradient vector than the standard LMS algorithm, the weights must be updated in increments small enough to ensure stability of the algorithm. We say this in light of the condition imposed on the step-size parameter μ as described in Eq. (10.21), which ensures a midadjustment less than 100 percent and covers for convergence of the LMS algorithm in the mean square. Stability of the algorithm is particularly serious when operating in an environment for which the eigenvalues of the correlation matrix of the input signal vector are highly disparate.

We may improve the convergence performance of the fast LMS algorithm by making the following observations (Ferrara, 1985):

- The weights are adapted independently from each other, which means that each weight is associated with one mode of the adaptive process. Since the modes are easily accessible, their individual convergence rates may be varied in a straightforward manner. Thus, whereas in the standard LMS algorithm each weight is responsible for a *mixture* of modes, in the fast LMS algorithm it is responsible for *one* specific mode and its rate of convergence may therefore be optimized for that mode.

- Assuming wide-sense stationary inputs, the convergence time for the ith mode is inversely proportional to $\mu\lambda_i$, where λ_i is the ith eigenvalue of the correlation matrix $\mathbf{R}$ of the input signal vector $\mathbf{u}(n)$; the eigenvalue λ_i is a measure of the average input power in the ith frequency bin.

Accordingly, we may make all the modes of the adaptive process converge essentially at the same rate by assigning to each weight an individual step-size parameter of its own, defined by (Ferrara, 1985)

$$\mu_i = \frac{\alpha}{P_i}, \qquad i = 0, 1, \ldots, M - 1 \tag{10.33}$$

where α is a constant and P_i is an estimate of the average power in the ith bin. Under this condition, the weights tend to converge at the same rate with a time constant defined by

$$\tau = \frac{2M}{\alpha} \text{ samples} \tag{10.34}$$

The conditions described in Eqs. (10.33) and (10.34) apply when the environment in which the fast LMS algorithm operates is wide-sense stationary. When, however, the environment is nonstationary, or if an estimate of the average input power in each bin is not available, then we may use the following simple recursion (based on the idea of convex combination) for its estimation (Griffiths, 1978; Ferrara, 1985; Shynk, 1992):

$$P_i(k) = \gamma P_i(k - 1) + (1 - \gamma) |U_i(k)|^2, \qquad i = 0, 1, \ldots, 2M - 1 \tag{10.35}$$

where $U_i(k)$ is the input applied to the ith weight in the fast LMS algorithm at time k, and γ is a constant chosen in the range $0 < \gamma < 1$. The parameter γ is a *forgetting factor* that controls the effective "memory" of the iterative process described in Eq. (10.35). In particular, we may express the input power $P_i(k)$ as an exponentially weighted sum of the magnitude squared of the input values, as shown by

$$P_i(k) = (1 - \gamma) \sum_{l=0}^{\infty} \gamma^l |U_i(k - l)|^2 \tag{10.36}$$

Thus, given the estimate $P_i(k)$ for the average signal power in the ith bin, the step-size parameter μ is replaced by an M-by-M diagonal matrix in accordance with Eq. (10.33) as follows:

$$\mu(k) = \alpha \mathbf{D}(k) \tag{10.37}$$

where

$$\mathbf{D}(k) = \text{diag}[P_0^{-1}(k), P_1^{-1}(k), \ldots, P_{M-1}^{-1}(k)] \tag{10.38}$$

Correspondingly, the fast LMS algorithm is modified as follows (Ferrara, 1985; Shynk, 1992):

1. In Eq. (10.30) involving the computation of the cross-correlation vector $\boldsymbol{\phi}(k)$, the product term $\mathbf{U}^H(k)\mathbf{E}(k)$ is replaced by $\mathbf{D}(k)\mathbf{U}^H(k)\mathbf{E}(k)$, as shown by

$$\boldsymbol{\phi}(k) = \text{first } M \text{ elements of IFFT } [\mathbf{D}(k)\mathbf{U}^H(k)\mathbf{E}(k)] \tag{10.39}$$

   where the inverse Fourier transformation now includes the the inverse powers in the individual bins.

2. In Eq. (10.31), μ is replaced by the constant α; otherwise, the computation of the frequency-domain weight vector $\hat{\mathbf{w}}(k)$ is the same as before.

TABLE 10.1 SUMMARY OF THE FAST LMS ALGORITHM BASED ON OVERLAP-SAVE
SECTIONING (ASSUMING REAL-VALUED DATA)

Initialization:
$$\hat{\mathbf{W}}(0) = 2M\text{-by-1 null vector}$$
$$P_i(0) = \delta_i, \qquad i = 0, \ldots, 2M - 1$$

Notations:
0   = M-by-1 null vector
FFT = fast Fourier transformation
IFFT = inverse fast Fourier transformation
α = adaptation constant

Computation: For each new block of M input samples, compute
$$\mathbf{U}(k) = \text{diag}\{\text{FFT}[u(kM - M), \ldots, u(kM - 1), u(kM), \ldots, u(kM + M - 1)]^T\}$$
$$\mathbf{y}(k) = \text{last } M \text{ elements of IFFT}[\mathbf{U}(k)\hat{\mathbf{W}}(k)]$$
$$\mathbf{e}(k) = \mathbf{d}(k) - \mathbf{y}(k)$$
$$\mathbf{E}(k) = \text{FFT}\begin{bmatrix} \mathbf{0} \\ \mathbf{e}(k) \end{bmatrix}$$
$$P_i(k) = \gamma P_i(k - 1) + (1 - \gamma) |U_i(k)|^2, \qquad i = 0, 1, \ldots, 2M - 1$$
$$\mathbf{D}(k) = \text{diag}[P_0^{-1}(k), P_1^{-1}(k), \ldots, P_{2M-1}^{-1}(k)]$$
$$\boldsymbol{\phi}(k) = \text{first } M \text{ elements of IFFT}[\mathbf{D}(k)\mathbf{U}^H(k)\mathbf{E}(k)]$$
$$\hat{\mathbf{W}}(k + 1) = \hat{\mathbf{W}}(k) + \alpha \, \text{FFT}\begin{bmatrix} \boldsymbol{\phi}(k) \\ \mathbf{0} \end{bmatrix}$$

Table 10.1 presents a summary of the fast LMS algorithm, incorporating the modifications described herein (Shynk, 1992).

10.3 UNCONSTRAINED FREQUENCY-DOMAIN ADAPTIVE FILTERING

The fast LMS algorithm described by the signal-flow graph of Fig. 10.2 may be viewed as a constrained form of frequency-domain adaptive filtering. Specifically, two of the five FFTs involved in its operation are needed to impose a *time-domain constraint* for the purpose of performing a linear correlation as specified in Eq. (10.11). The time-domain constraint consists of the following operations:

- Discarding the last M elements of the inverse FFT of $\mathbf{U}^H(k)\mathbf{E}(k)$, as described in Eq. (10.30)
- Replacing the elements so discarded by a block of M zeros before reapplying the FFT, as described in Eq. (10.31)

The combination of four operations described herein is contained inside the dashed rectangle of Fig. 10.2; this combination is referred to as a *gradient constraint* in recognition of the fact that it is involved in computing an estimate of the gradient vector. Note that the gradient constraint is actually a time-domain constraint. Basically, it ensures that the $2M$ frequency-domain weights correspond to only M time-domain weights. This is the reason why a zero block is appended in the gradient constraint in Fig. 10.2.

In the *unconstrained frequency-domain adaptive filter* (Mansour and Gray, 1982), the gradient constraint is removed completely from the signal-flow graph of Fig. 10.2. The net result is a simpler implementation that involves only three FFTs. Thus, the combination of Eqs. (10.30) and (10.31) in the fast LMS algorithm is now replaced by the much simpler algorithm

$$\hat{\mathbf{W}}(k + 1) = \hat{\mathbf{W}}(k) + \mu\, \mathbf{U}^H(k)\mathbf{E}(k) \tag{10.40}$$

It is important to note, however, that the estimate of the gradient vector computed here no longer corresponds to a linear correlation as specified in Eq. (10.13); rather, we now have a circular correlation.

Consequently, we find that in general the unconstrained frequency-domain adaptive filtering algorithm of Eq. (10.40) deviates from the fast LMS algorithm, in that the tap-weight vector no longer converges to the Wiener solution as the number of block iterations approaches infinity (Sommen et al., 1987; Lee and Un, 1989; Shynk, 1992). Another point to note is that although the convergence rate of the unconstrained frequency-domain adaptive filtering algorithm is increased with time-varying step sizes, the improvement is offset by a worsening of the misadjustment. Indeed, according to Lee and Un (1989), the unconstrained algorithm requires twice as many iterations as the constrained algorithm to produce the same level of misadjustment.

10.4 SELF-ORTHOGONALIZING ADAPTIVE FILTERS

In the previous sections we addressed the issue of how to use frequency-domain techniques to improve the computational efficiency of the LMS algorithm when the application of interest requires a long filter memory. In this section we consider another important adaptive filtering issue, namely, that of improving the convergence properties of the LMS algorithm. This improvement is, however, attained at the cost of an increase in computational complexity.

To motivate the discussion, consider an input signal vector $\mathbf{u}(n)$ characterized by the correlation matrix $\mathbf{R}$. The *self-orthogonalizing adaptive filtering algorithm* for such a wide-sense stationary environment is described by (Chang, 1971; Cowan, 1987)

$$\hat{\mathbf{w}}(n + 1) = \hat{\mathbf{w}}(n) + \alpha\mathbf{R}^{-1}\mathbf{u}(n)e(n) \tag{10.41}$$

where $\mathbf{R}^{-1}$ is the inverse of the correlation matrix $\mathbf{R}$, and $e(n)$ is the error signal defined in the usual way. The constant α lies in the range $0 < \alpha < 1$; according to Cowan (1987), it may be set at the value

$$\alpha = \frac{1}{2M} \tag{10.42}$$

where M is the filter length. An important property of the self-organizing filtering algorithm of Eq. (10.41) is that, in theory, it guarantees a constant rate of convergence, irrespective of the input statistics.

To prove this useful property, define the weight-error vector

$$\boldsymbol{\epsilon}(n) = \hat{\mathbf{w}}(n) - \mathbf{w}_o \qquad (10.43)$$

where the weight vector $\mathbf{w}_o$ is the Wiener solution. Hence, we may rewrite the algorithm of Eq. (10.41) in terms of $\boldsymbol{\epsilon}(n)$ as follows:

$$\boldsymbol{\epsilon}(n + 1) = (\mathbf{I} - \alpha\mathbf{R}^{-1}\mathbf{u}(n)\mathbf{u}^T(n))\boldsymbol{\epsilon}(n) + \alpha\mathbf{R}^{-1}\mathbf{u}(n)e_o(n) \qquad (10.44)$$

where $\mathbf{I}$ is the identity matrix, and $e_o(n)$ is the optimum value of the error signal that is produced by the Wiener solution. Applying the statistical expectation operator to both sides of Eq. (10.44), and invoking the independence assumption [i.e., the tap-weight vector $\hat{\mathbf{w}}(n)$ is independent of the input vector $\mathbf{u}(n)$], we obtain the following result:

$$E[\boldsymbol{\epsilon}(n + 1)] = (\mathbf{I} - \alpha\,\mathbf{R}^{-1})E[\mathbf{u}(n)\mathbf{u}^T(n)]E[\boldsymbol{\epsilon}(n)] + \alpha\mathbf{R}^{-1}E[\mathbf{u}(n)e_o(n)] \qquad (10.45)$$

We now recognize the following points (for real-valued data):

- From the definition of a correlation matrix for a wide-sense stationary input, we have (see Eq. (10.18))

$$E[\mathbf{u}(n)\mathbf{u}^T(n)] = \mathbf{R}$$

- From the principle of orthogonality, we have (see Section 5.2)

$$E[\mathbf{u}(n)e_o(n)] = \mathbf{0}$$

Accordingly, we may simplify Eq. (10.45) as follows:

$$\begin{aligned} E[\boldsymbol{\epsilon}(n + 1) &= (\mathbf{I} - \alpha\mathbf{R}^{-1}\mathbf{R})E[\boldsymbol{\epsilon}(n)] \\ &= (1 - \alpha)E[\boldsymbol{\epsilon}(n)] \end{aligned} \qquad (10.46)$$

Equation (10.46) represents a first-order difference equation, the solution of which is

$$E[\boldsymbol{\epsilon}(n)] = (1 - \alpha)^n\, E[\boldsymbol{\epsilon}(0)] \qquad (10.47)$$

where $\boldsymbol{\epsilon}(0)$ is the initial value of the weight-error vector. Hence, with the value of α lying in the range $0 < \alpha < 1$, we may write

$$\lim_{n \to \infty} E[\boldsymbol{\epsilon}(n)] = \mathbf{0} \qquad (10.48)$$

or, equivalently,

$$\lim_{n \to \infty} E[\hat{\mathbf{w}}(n)] = \mathbf{w}_o \qquad (10.49)$$

Most importantly, we note from Eq. (10.47) that the rate of convergence is completely independent of the input statistics, as stated previously.

Example 2: White Gaussian Noise Input

To illustrate the convergence properties of the self-organizing adaptive filtering algorithm, consider the case of a white Gaussian noise input process, whose correlation matrix is defined by

$$\mathbf{R} = \sigma^2 \mathbf{I} \tag{10.50}$$

where σ^2 is the noise variance, and $\mathbf{I}$ is the identity matrix. For this input, the use of Eq. (10.41) yields (with $\alpha = 1/2M$)

$$\hat{\mathbf{w}}(n + 1) = \hat{\mathbf{w}}(n) + \frac{1}{2M\sigma^2} \mathbf{u}(n)e(n) \tag{10.51}$$

This algorithm is recognized as the standard LMS algorithm with a step-size parameter defined by

$$\mu = \frac{1}{2M\sigma^2} \tag{10.52}$$

In other words, for the special case of a white Gaussian noise sequence characterized by an eigenvalue spread of unity, the standard LMS algorithm behaves in the same way as the self-orthogonalizing adaptive filtering algorithm.

Two-Stage Adaptive Filter

This last example suggests that we may mechanize a self-orthogonalizing-adaptive filter for an arbitrary environment by proceeding in two stages [Narayan et al., 1983; Cowan and Grant, 1985]:

1. The input vector $\mathbf{u}(n)$ is transformed into a corresponding vector of uncorrelated variables.
2. The transformed vector is used as the input to an LMS algorithm.

From the discussion presented in Section 4.3, we recall that, in theory, the first objective may be realized by using the *Karhunen–Loève transform (KLT)*. Specifically, given an input vector $\mathbf{u}(n)$ of zero mean, drawn from a wide-sense stationary environment, the ith output of the KLT is defined by (for real-valued data)

$$v_i(n) = \mathbf{q}_i^T \mathbf{u}(n), \qquad i = 0, 1, \ldots, M - 1 \tag{10.53}$$

where $\mathbf{q}_i$ is the eigenvector associated with the ith eigenvalue λ_i belonging to the correlation matrix $\mathbf{R}$ of the input vector $\mathbf{u}(n)$. The individual outputs of the KLT are zero-mean, uncorrelated variables as shown by

$$E[v_i(n)v_j(n)] = \begin{cases} \lambda_i, & j = i \\ 0, & j \neq i \end{cases} \tag{10.54}$$

Accordingly, we may express the correlation matrix of the M-by-1 vector $\mathbf{v}(n)$ produced by the KLT as the diagonal matrix:

$$\begin{aligned}\mathbf{\Lambda} &= E[\mathbf{v}(n)\mathbf{v}^T(n)] \\ &= \text{diag}[\lambda_0, \lambda_1, \ldots, \lambda_{M-1}]\end{aligned} \tag{10.55}$$

The inverse of $\mathbf{\Lambda}$ is also a diagonal matrix, as shown by

$$\mathbf{\Lambda}^{-1} = \text{diag}[\lambda_0^{-1}, \lambda_1^{-1}, \ldots, \lambda_{M-1}^{-1}] \tag{10.56}$$

Consider now the self-orthogonalizing adaptive filtering algorithm of Eq. (10.41) with the transformed vector $\mathbf{v}(n)$ and its inverse correlation matrix $\mathbf{\Lambda}^{-1}$ used in place of $\mathbf{u}(n)$ and $\mathbf{R}^{-1}$, respectively. Under these new circumstances, Eq. (10.41) takes the form

$$\hat{\mathbf{w}}(n + 1) = \hat{\mathbf{w}}(n) + \alpha\mathbf{\Lambda}^{-1}\mathbf{v}(n)e(n) \tag{10.57}$$

the ith element of which may written as

$$\hat{w}_i(n + 1) = \hat{w}_i(n) + \frac{\alpha}{\lambda_i} v_i(n)e(n), \qquad i = 0, 1, \ldots, M - 1 \tag{10.58}$$

Equation (10.58) is immediately recognized as a normalized form of the LMS algorithm. Normalization here means that each tap-weight is assigned its own step-size parameter that is related to the corresponding eigenvalue of the correlation matrix of the original input vector $\mathbf{u}(n)$. Thus, Eq. (10.58) takes care of the second point mentioned above. Note, however, that the algorithm described herein is different from the traditional normalized LMS algorithm discussed in Section 9.11.

The KLT is a signal-dependent transformation, the implementation of which requires the estimation of the correlation matrix of the input vector, the diagonalization of this matrix, and the construction of the required basis vectors. These computations make the KLT impractical for real-time applications. Fortunately, the *discrete cosine transform (DCT)*, discussed in Chapter 1, provides a predetermined set of basis vectors that are good approximation to the KLT. Indeed, for a stationary zero-mean, first-order Markov process that is deemed to be sufficiently general in signal-processing studies, the DCT is asymptotically equivalent to the KLT,[2] with this asymptotic equivalence being demonstrated both as the sequence length increases and also as the adjacent correlation coefficient tends to 1 (Rao and Yip, 1990); the *adjacent correlation coefficient* of a stochastic process is defined as the autocorrelation function of the process for a unit lag, divided by the autocorrelation function of the process for zero lag (i.e., the mean-square value). Whereas the KLT is signal dependent, the DCT is signal independent and can therefore be implemented in a computationally efficient manner.

---

[2] Interestingly enough, the DFT is also asymptotically equivalent to the KLT (Grenander and Szegö, 1958; Gray, 1972). However, the asymptotic eigenvalue spread for the DFT with first-order Markov inputs is much worse than that for the DCT, which makes the DCT the preferred approximant to the KLT. Specifically, for such inputs, the asymptotic eigenvalue spread is equal to $[(1 + \rho)/(1 - \rho)]$ for the DFT versus $(1 + \rho)$ for the DCT, where ρ is the adjacent correlation coefficient of the input (Beaufays, 1995a).

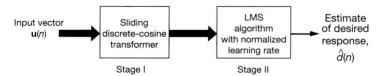

Figure 10.3   Block diagram of the DCT–LMS algorithm.

We are now equipped with the tools we need to formulate a practical approximation to the self-orthogonalizing adaptive filter that combines the desirable properties of the DCT with those of the LMS algorithm. Figure 10.3 shows a block diagram of the filter. It consists of two stages, with stage I providing the implementation of a *sliding* DCT algorithm and stage II implementing a *normalized* version of the LMS algorithm (Beaufays and Widrow, 1994; Beaufays, 1995a). In effect, stage I acts as a preprocessor that performs the "orthogonalization" of the input vector, albeit in an approximate manner.

Sliding DCT

The DCT we have in mind for our present application uses a sliding window, with the computation being performed for *each* new input sample. This, in turn, enables the LMS algorithm (following the DCT) to operate at the incoming data rate as in its conventional form. Thus, unlike the fast LMS algorithm, the frequency-domain adaptive filtering algorithm described here is a nonblock algorithm, and therefore not as computationally efficient.

From the discussion presented in Chapter 1, we recall that the discrete Fourier transform of an even function results in the discrete cosine transform. We may exploit this simple property to develop an efficient algorithm for computing the sliding DCT. To proceed, consider a sequence of M samples denoted by $u(n), u(n-1), \ldots, u(n-M+1)$. We may then construct an extended sequence $a(n)$ that is symmetric about the point $n - M + 1/2$ as follows (see Fig. 10.4):

$$a(i) = \begin{cases} u(i), & i = n, n-1, \ldots, n-M+1 \\ u(-i+2n-2M+1), & i = n-M, n-M-1, \ldots, n-2M+1 \end{cases} \tag{10.59}$$

For convenience of presentation, define

$$W_{2M} = \exp\left(-\frac{j2\pi}{2M}\right) \tag{10.60}$$

The mth element of the $2M$-point DFT of the extended sequence in Eq. (10.59) at time n is defined by

$$A_m(n) = \sum_{i=n-2M+1}^{n} a(i) W_{2M}^{m(n-i)} \tag{10.61}$$

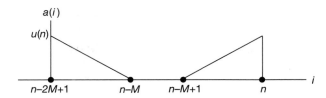

Figure 10.4   Illustrating the construction of the extended sequence $a(n)$.

Using Eq. (10.60) in (10.61), we may write

$$A_m(n) = \sum_{i=n-M+1}^{n} a(i)W_{2M}^{m(n-i)} + \sum_{i=n-2M+1}^{n-M} a(i)W_{2M}^{m(n-i)}$$

$$= \sum_{i=n-M+1}^{n} u(i)W_{2M}^{m(n-i)} + \sum_{i=n-2M+1}^{n-M} u(-i + 2n - 2M + 1)W_{2M}^{m(n-i)} \qquad (10.62)$$

$$= \sum_{i=n-M+1}^{n} u(i)W_{2m}^{m(n-i)} + \sum_{i=n-M+1}^{n} u(i)W_{2M}^{m(i-n+2M-1)}$$

Factoring out the term $W_{2M}^{m(M-1/2)}$ and combining the two summations together, we may redefine $A_m(n)$ as

$$A_m(n) = W_{2M}^{m(M-1/2)} \sum_{i=n-M+1}^{n} u(i)(W_{2M}^{-m(i-n+M-1/2)} + W_{2M}^{m(i-n+M-1/2)})$$

$$= 2(-1)^m\, W_{2M}^{-m/2} \sum_{i=n-M+1}^{n} u(i)\cos\left(\frac{m(i - n + M - 1/2)\pi}{M} \right) \qquad (10.63)$$

where, in the last line, we have used the definition of W_{2M} and Euler's formula for a cosine function. Except for a scaling factor, the summation in Eq. (10.63) is recognized as the DCT of the sequence $u(n)$ at time n; specifically, we have

$$C_m(n) = k_m \sum_{i=n-M+1}^{n} u(i)\cos\left(\frac{m(i - n + M - 1/2)\pi}{M} \right) \qquad (10.64)$$

where the constant k_m is defined by

$$k_m = \begin{cases} 1/\sqrt{2}, & m = 0 \\ 1, & \text{otherwise} \end{cases} \qquad (10.65)$$

Accordingly, in light of Eqs. (10.63) and (10.64) the DCT of the sequence $u(n)$ is related to the DFT of the extended sequence $a(n)$ as follows:

$$C_m(n) = \frac{1}{2} k_m(-1)^m\, W_{2M}^{m/2}\, A_m(n), \qquad m = 0, 1, \ldots, M - 1 \qquad (10.66)$$

The DFT of the extended sequence $a(n)$ given in Eq. (10.62) may be viewed as the sum of two complementary DFTs as follows:

$$A_m(n) = A_m^{(1)}(n) + A_m^{(2)}(n) \tag{10.67}$$

where

$$A_m^{(1)}(n) = \sum_{i=n-M+1}^{n} u(i)W_{2M}^{m(n-i)} \tag{10.68}$$

$$A_m^{(2)}(n) = \sum_{i=n-M+1}^{n} u(i)W_{2M}^{m(i-n+2M-1)} \tag{10.69}$$

Consider first the DFT denoted by $A_m^{(1)}(n)$. Separating out the sample $u(n)$, we may rewrite this DFT (computed at time n) as

$$A_m^{(1)}(n) = u(n) + \sum_{i=n-M+1}^{n-1} u(i)W_{2M}^{m(n-i)} \tag{10.70}$$

Next, we note from Eq. (10.68) that the previous value of this DFT, computed at time $n - 1$, is given by

$$\begin{aligned}
A_m^{(1)}(n-1) &= \sum_{i=n-M}^{n-1} u(i)W_{2M}^{m(n-1-i)} \\
&= (-1)^m W_{2m}^{-m} u(n-M) + W_{2M}^{-m} \sum_{i=n-M+1}^{n} u(i)W_{2M}^{m(n-i)}
\end{aligned} \tag{10.71}$$

where, in the first term of the last line, we have used the fact that

$$W_{2M}^{mM} = e^{-jm\pi} = (-1)^m$$

Hence, multiplying Eq. (10.71) by the factor W_{2M}^m and subtracting the result from Eq. (10.70), we get (after a rearrangement of terms)

$$A_m^{(1)}(n) = W_{2M}^m A_m^{(1)}(n-1) + u(n) - (-1)^m u(n-M), \qquad m = 0, 1, \ldots, M-1 \tag{10.72}$$

Equation (10.72) represents a first-order difference equation, which may be used to update the computation of $A_m^{(1)}(n)$, given its previous value $A_m^{(1)}(n-1)$, the new sample $u(n)$, and the very old sample $u(n-M)$.

Consider next the recursive computation of the second DFT $A_m^{(2)}(n)$ defined in Eq. (10.69). We recognize that $W_{2M}^{2mM} = 1$ for all integer m. Hence, separating out the term that involves the sample $u(n)$, we may express this DFT in the following form:

$$A_m^{(2)}(n) = W_{2M}^{-m} u(n) + W_{2M}^{-m} \sum_{i=n-M+1}^{n-1} u(i)W_{2M}^{m(i-n)} \tag{10.73}$$

Next, using Eq. (10.69) to evaluate this second DFT at time $n - 1$, and then proceeding to separate out the term involving the sample $u(n-M)$, we may write

$$
\begin{aligned}
A_m^{(2)}(n-1) &= \sum_{i=n-M}^{n-1} u(i)W_{2M}^{m(i-n)} \\
&= W_{2M}^{mM}\, u(n-M) + \sum_{i=n-M+1}^{n-1} u(i)W_{2M}^{m(i-n)} \\
&= (-1)^m u(n-M) + \sum_{i=n-M+1}^{n-1} u(i)\, W_{2M}^{m(i-n)}
\end{aligned}
\tag{10.74}
$$

Hence, multiplying Eq. (10.74) by the factor W_{2M}^{-m} and then subtracting the result from Eq. (10.73), we get (after a rearrangement of terms)

$$
A_m^{(2)}(n) = W_{2M}^{-m} A_m^{(2)}(n-1) + W_{2M}^{-m}\left(u(n) - (-1)^m u(n-M)\right)
\tag{10.75}
$$

Finally, using Eqs. (10.66), (10.67), (10.72), and (10.75), we may construct the block diagram shown in Fig. 10.5 for the recursive computation of the discrete cosine transform $C_m(n)$ of the sequence $u(n)$. The construction has been simplified by noting the following points:

- The operations involving the present sample $u(n)$ and the old sample $u(n-M)$ are common to the computations of both discrete Fourier transforms: $A_m^{(1)}(n)$ and $A_m^{(2)}(n)$, hence the common front end of Fig. 10.5.
- The operator z^{-M} in the forward path and the operators z^{-1} inside the two feedback loops in Fig. 10.5 are each multiplied by a new parameter β; the reason for its inclusion is explained below.

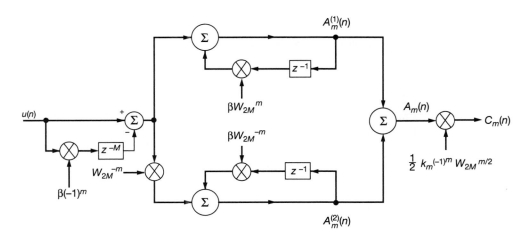

Figure 10.5   Indirect computation of the sliding discrete cosine transform.

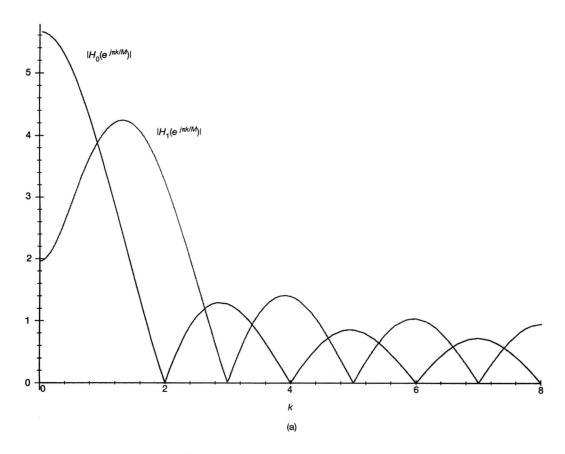

Figure 10.6a   Amplitude responses of frequency-sampling filters for $m = 0$ and $m = 1$.

The discrete-time network of Fig. 10.5 is called a *frequency-sampling filter*. It exhibits a form of structural symmetry that is inherited from the mathematical symmetry built into the definition of the discrete cosine transform.

The transfer function of the filter shown in Fig. 10.5 from the input $u(n)$ to the mth DCT output $C_m(n)$, is given by (with $\beta = 1$)

$$H_m(z) = \frac{1}{2} k_m \left(\exp\left(-\frac{jm\pi}{2M}\right) \frac{(-1)^m - z^{-M}}{1 - \exp\left(\frac{jm\pi}{M}\right) z^{-1}} + \exp\left(\frac{jm\pi}{2M}\right) \frac{(-1)^m - z^{-M}}{1 - \exp\left(-\frac{jm\pi}{M}\right) z^{-1}} \right)$$

$$(10.76)$$

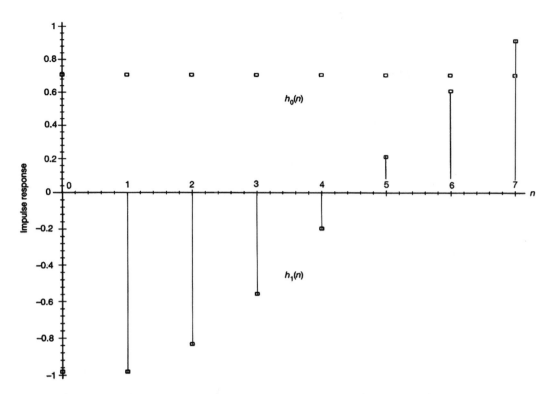

Figure 10.6b   Corresponding impulse responses $h_0(n)$ and $h_1(n)$ of the two frequency-sampling filters.

The common numerator of Eq. (10.76), namely, the factor $((-1)^m - z^{-M})$ represents a set of zeros that are uniformly spaced around the unit circle in the z-plane. These zeros are given by

$$z_m = \exp\left(\frac{j\pi m}{M}\right), \qquad m = 0, \pm 1, \dots, \pm (M-1) \qquad (10.77)$$

The denominator of the first partial fraction in Eq. (10.76) has a single pole at $z = \exp(jm\pi/M)$, whereas the denominator of the second partial fraction has a single pole at $z_m = \exp(-jm\pi/M)$. Accordingly, each of these poles exactly cancels a particular zero of the numerator term. The net result is that the filter structure of Fig. 10.5 is equivalent to two banks of narrow-band all-zero filters operating in parallel; each filter bank corresponds to the M bins of the DCT. Figure 10.6(a) shows the frequency responses of the frequency-sampling filters pertaining to two adjacent bins of the DCT represented by the coefficients $m = 0$ and $m = 1$, and Fig. 10.6(b) shows the corresponding impulse responses of the filters, for $M = 8$ bins.

With $\beta = 1$, the frequency-sampling filters described herein are "marginally" stable, because for each bin of the DCT the poles of the two feedback paths in Fig. 10.5 lie exactly on the unit circle, and round-off errors (however small) may give rise to instability by pushing one or the other (or both) of these poles outside the unit circle. This problem may be alleviated by shifting the zeros of the forward path and the poles of the feedback paths slightly inside the unit circle (Shynk and Widrow, 1986), hence the inclusion of parameter β in Fig. 10.5 with $0 < \beta < 1$. For example, with $\beta = 0.99$, all of the poles and zeros of the two terms in the partial fraction expansion of the transfer function $H(z)$ of a frequency-sampling filter as is Eq. (10.69) are now made to lie on a circle with radius $\beta = 0.99$; the stability of the frequency-sampling filters is thereby ensured even if exact pole-zero cancelations are not realized (Shynk, 1992).

Eigenvalue Estimation

The only issue that remains to be considered in the design of the DCT–LMS algorithm is how to estimate the eigenvalues of the correlation matrix $\mathbf{R}$ of the input vector $\mathbf{u}(n)$, which define the step-sizes used to adapt the individual weights in the LMS algorithm of Eq. (10.58). Assuming that the stochastic process responsible for generating the input vector $\mathbf{u}(n)$ is ergodic, we may define an estimate of its correlation matrix $\mathbf{R}$ (for real-valued data) as follows:

$$\hat{\mathbf{R}}(n) = \frac{1}{n} \sum_{i=1}^{n} \mathbf{u}(i)\mathbf{u}^{T}(i) \tag{10.78}$$

which is known as the *sample correlation matrix*. The coefficients of the DCT provide an approximation to the M-by-M matrix $\mathbf{Q}$ whose columns represent the eigenvectors associated with the eigenvalues of the correlation matrix $\mathbf{R}$. Let $\hat{\mathbf{Q}}$ denote this approximating matrix. The vector of outputs $C_0(n), C_1(n), \ldots, C_{M-1}(n)$ produced by the DCT in response to the input vector $\mathbf{u}(n)$ may thus be expressed as follows:

$$\hat{\mathbf{v}}(n) = [C_0(n), C_1(n), \ldots, C_{M-1}(n)]^{T}$$
$$= \hat{\mathbf{Q}}\mathbf{u}(n) \tag{10.79}$$

Furthermore, the approximation to the orthogonal transformation realized by the DCT may be written [in light of Eqs. (10.78) and (10.79)] in the form

$$\hat{\mathbf{\Lambda}}(n) = \hat{\mathbf{Q}}\,\hat{\mathbf{R}}(n)\hat{\mathbf{Q}}^{T}$$

$$= \frac{1}{n} \sum_{i=1}^{n} \hat{\mathbf{Q}}\mathbf{u}(i)\,\mathbf{u}^{T}(i)\,\mathbf{Q}^{T} \tag{10.80}$$

$$= \frac{1}{n} \sum_{i=1}^{n} \hat{\mathbf{v}}(i)\,\hat{\mathbf{v}}^{T}(i)$$

Equivalently, we have

$$\hat{\lambda}_m(n) = \frac{1}{n} \sum_{i=1}^{n} C_m^2(i), \qquad m = 0, 1, \ldots, M - 1 \tag{10.81}$$

Equation (10.81) may be cast into a recursive form by writing

$$
\begin{aligned}
\hat{\lambda}_m(n) &= \frac{1}{n} C_m^2(n) + \frac{1}{n} \sum_{i=1}^{n-1} C_m^2(i) \\
&= \frac{1}{n} C_m^2(n) + \frac{n-1}{n} \cdot \frac{1}{n-1} \sum_{i=1}^{n-1} C_m^2(i)
\end{aligned} \tag{10.82}
$$

From the defining equation (10.81), we note that

$$
\hat{\lambda}_m(n-1) = \frac{1}{n-1} \sum_{i=1}^{n-1} C_m^2(i)
$$

Accordingly, we may rewrite Eq. (10.82) in the recursive form

$$
\hat{\lambda}_m(n) = \hat{\lambda}_m(n-1) + \frac{1}{n}(C_m^2(n) - \hat{\lambda}_m(n-1)) \tag{10.83}
$$

Equation (10.83) applies to a wide-sense stationary environment. To account for adaptive filtering operation in a nonstationary environment, we may modify the recursive equation (10.83) as follows (Chao et al., 1990):

$$
\hat{\lambda}_m(n) = \gamma\hat{\lambda}_m(n-1) + \frac{1}{n}(C_m^2(n) - \gamma\hat{\lambda}_m(n-1)), \qquad m = 0, 1, \ldots, M-1 \tag{10.84}
$$

where γ is a *forgetting factor* that lies in the range $0 < \gamma < 1$. Equation (10.84) is the desired formula for recursive computation of the eigenvalues of the correlation matrix of the input vector $\mathbf{u}(n)$.

Summary of the DCT–LMS Algorithm

We are now ready to summarize the steps involved in computing the DCT–LMS algorithm. This summary is presented in Table 10.2, which follows from Figure 10.5, Eqs. (10.58) and (10.84), and Eqs. (10.72) and (10.75).

10.5 COMPUTER EXPERIMENT ON ADAPTIVE EQUALIZATION

In this computer experiment we revisit the adaptive channel equalization discussed in Section 9.7, where the standard LMS algorithm was used to perform the adaptation. This time, however, we use the DCT–LMS algorithm derived in the previous section. For details of the channel impulse response and the random sequence applied to the channel input, the reader is referred to Section 9.7.

The experiment is in three parts, as described here:

- In part 1, we highlight some numerical anomalies that have to be cared for in computing the DCT–LMS algorithm.
- In part 2, we study the transient behavior of the DCT–LMS algorithm for different values of the eigenvalue spread of the correlation matrix of the equalizer input.

TABLE 10.2 SUMMARY OF THE DCT–LMS ALGORITHM:

Initialization:
For $m = 0, 1, \ldots, M - 1$, set

$$A_m^{(1)}(0) = A_m^{(2)}(0) = 0$$

$$\hat{\lambda}_m(0) = 0$$

$$\hat{w}_m(0) = 0$$

$$k_m = \begin{cases} 1/\sqrt{2}, & m = 0 \\ 1, & \text{otherwise} \end{cases}$$

Selection of parameters:

$$\alpha = \frac{1}{2M}$$
$$\beta = 0.99$$
$$0 < \gamma < 1$$

Sliding DCT:
For $m = 0, 1, \ldots, M-1$ and $n = 1, 2, \cdots$, compute

$$A_m^{(1)}(n) = \beta W_{2M}^m A_m^{(1)}(n - 1) + u(n) - \beta(-1)^m u(n - M)$$

$$A_m^{(2)}(n) = \beta W_{2M}^{-m} A_m^{(2)}(n - 1) + W_{2M}^{-m}(u(n) - \beta(-1)^m u(n - M))$$

$$A_m(n) = A_m^{(1)}(n) + A_m^{(2)}(n)$$

$$C_m(n) = \frac{1}{2} k_m (-1)^m W_{2M}^{m/2} A_m(n)$$

where W_{2M} is defined by

$$W_{2M} = \exp\left(\frac{-j2\pi}{2M}\right)$$

LMS algorithm:

$$y(n) = \sum_{m=0}^{M-1} C_m(n)\hat{w}_m(n)$$

$$e(n) = d(n) - y(n)$$

$$\hat{\lambda}_m(n) = \gamma\hat{\lambda}_m(n - 1) + \frac{1}{n}(C_m^2(n) - \gamma\hat{\lambda}_m(n - 1))$$

$$\hat{w}_m(n + 1) = \hat{w}_m(n) + \frac{\alpha}{\hat{\lambda}_m(n)} C_m(n)e(n)$$

Note. In computing the updated weight $\hat{w}_m(n + 1)$, care should be taken to prevent instability of the LMS algorithm, which can arise if some of the eigenvalue estimates are close to zero. Adding a small constant δ to $\hat{\lambda}_m(n)$ could do the trick, but it appears that a better strategy is to condition the correlation matrix of the input signal vector by adding a small amount of white noise (F. Beaufays, private communication, 1995).

- In part 3, we compare the transient behavior of the DCT–LMS algorithm to that of the standard LMS algorithm.

Throughout these experiments, the signal-to-noise ratio is maintained at a high value of 30 dB.

Experiment 1: Some Numerical Considerations.   The sliding DCT may be computed indirectly, as summarized in Table 10.2. In other words, we first compute two sliding DFTs, namely, $A_m^{(1)}(n)$ and $A_m^{(2)}(n)$, and then use the results so obtained to compute the DCT, as summarized in Table 10.2. Alternatively, the DCT may be computed directly in a recursive manner, as outlined in Problem 6. Although, in theory, these two procedures are equivalent, it appears however that for finite-precision arithmetic, the indirect procedure is numerically more robust than the direct procedure. This is illustrated in Fig. 10.7, where the ensemble-averaged learning curve of the DCT–LMS algorithm is plotted versus the number of iterations, n, for channel parameter $W = 2.9$, which corresponds to the

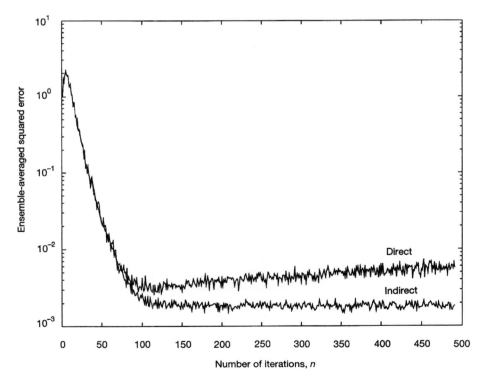

Figure 10.7   Learning curves of the DCT–LMS algorithm using (a) indirect, and (b) direct computation of the DCT for eigenvalue spread $\chi(\mathbf{R}) = 6.0782$.

eigenvalue spread $\chi(\mathbf{R}) = 6.0782$; for the definition of W, see Eq. (9.105). Figure 10.7 clearly shows that within the finite precision available to the software, the DCT–LMS algorithm computed using the direct procedure shows a tendency to diverge with increasing n. On the other hand, the algorithm behaves in its normal fashion when the indirect procedure is used to compute the DCT.

Another noteworthy observation is that in performing a large number of trials, say 400, we may find that a few sample paths exhibit very large fluctuations. This behavior is illustrated in Fig. 10.8. In the ensemble-averaged learning curve labeled 1, all 400 sample paths were included. In the learning curve labeled 2, the five worst-case sample paths were removed from the computation. In the learning curve labeled 3, the eight worst-case sample paths were removed from the computation. Finally, in the learning curve labeled 4, the 66 worst-case sample paths were removed from the computation. The worst-case sample paths were determined on the basis of the highest error contribution for iteration n from 1

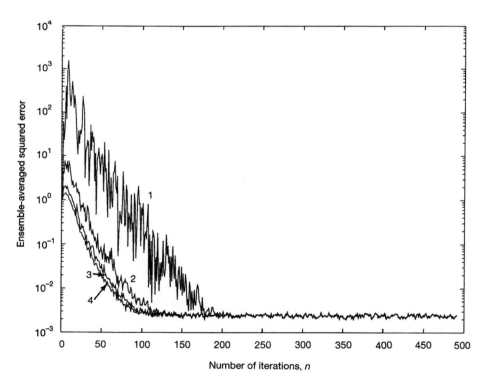

Figure 10.8   Learning curves of the DCT–LMS algorithm. Curve 1: All 400 sample realizations of the algorithm included. Curve 2: The 5 worst realizations of the algorithm excluded. Curve 3: The 8 worst realizations of the algorithm excluded. Curve 4: The 66 worst realizations of the algorithm excluded.

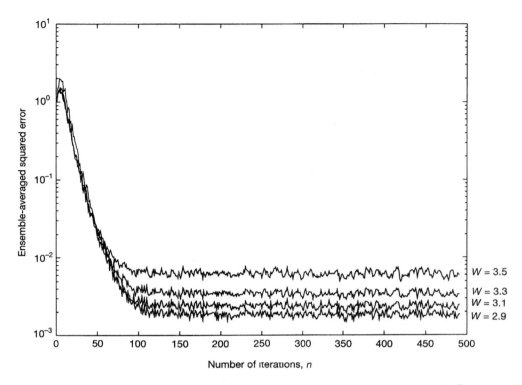

Figure 10.9   Learning curves of the DCT–LMS algorithm for varying eigenvalue spread $\chi(\mathbf{R})$.

to the number at which steady-state error performance is achieved. The results presented in Fig. 10.8 reveal the following:

- A substantial reduction in error is achieved by removing as few as five worst-case sample paths from computation of the ensemble-averaged learning curve.
- The (final) steady-state ensemble-averaged squared error is identical for all the four situations described in Fig. 10.8. This indicates that the worst-case sample paths affect only the transient behavior of the DCT–LMS algorithm.

Experiment 2: Transient Behavior of the DCT–LMS Algorithm.   In Fig. 10.9 the ensemble-averaged learning curve of the DCT–LMS algorithm is plotted for varying channel parameter W. Specifically, we have $W = 2.9, 3.1, 3.3,$ and 3.5, which corresponds to eigenvalue spread $\chi(\mathbf{R}) = 6.0782, 11.1238, 21.7132,$ and 46.8216, respectively; see Table 9.2. The results presented in Fig. 10.9 clearly show that, unlike the standard LMS algorithm, the ensemble-averaged transient behavior of the DCT–LMS algorithm is less sensitive to variations in the eigenvalue spread of the correlation matrix **R** of the input vector $\mathbf{u}(n)$ applied to the channel equalizer. This desirable property is due to the "orthogonalizing" action of the DCT as a preprocessor to the LMS algorithm.

Experiment 3: Comparison of the DCT–LMS Algorithm with Other Adaptive Filtering Algorithms.   Figures 10.10(a) to 10.10(d) present a comparison of the ensemble-averaged error performance of the DCT–LMS algorithm to two other algorithms, the standard LMS algorithm and the recursive least-squares (RLS) algorithm for four different values of channel parameter W. The operation of the standard LMS algorithm follows the theory presented in Chapter 9. The theory of the RLS algorithm is presented in Chapter 13; we have included it here as another interesting frame of reference. On the basis of the

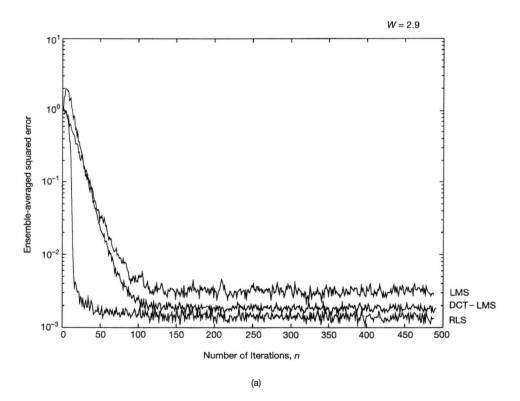

(a)

Figure 10.10   Comparison of the learning curves of the standard LMS, DCT–LMS, and RLS algorithms: (a) $\chi(\mathbf{R}) = 6.0782$. This figure is continued on the next two pages.

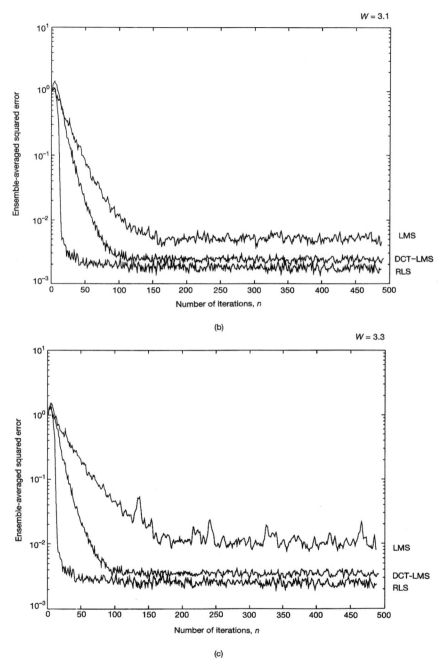

Figure 10.10   (b) $\chi(\mathbf{R}) = 11.1238$. (c) $\chi(\mathbf{R}) = 21.7132$.

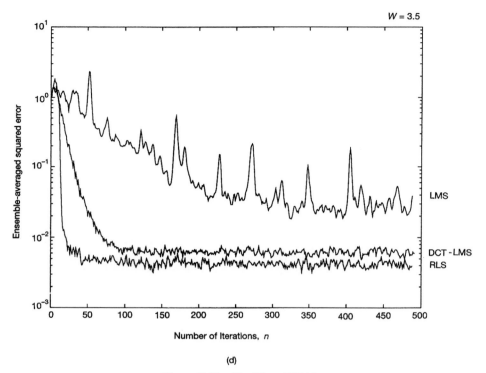

(d)

Figure 10.10   (d) $\chi(\mathbf{R}) = 46.8216$.

results presented in Fig. 10.10, we may make the following observations on the transient performance of the three adaptive filtering algorithms considered here:

- The standard LMS algorithm consistently behaves worst, in that it exhibits the slowest rate of convergence, the greatest sensitivity to variations in the parameter W [and therefore the eigenvalue spread $\chi(\mathbf{R})$], and the largest excess mean-squared error.

- The RLS algorithm consistently achieves the fastest rate of convergence and the smallest excess mean-squared error, with the least sensitivity to variations in the eigenvalue spread $\chi(\mathbf{R})$.

- For a prescribed eigenvalue spread $\chi(\mathbf{R})$, the transient behavior of the DCT–LMS algorithm lies between those of the standard LMS and RLS algorithms. Most importantly, however, we note the following:
  - The rate of convergence of the DCT–LMS algorithm is relatively insensitive to variations in the eigenvalue spread $\chi(\mathbf{R})$, as already noted under experiment 2. Most importantly, the rate of convergence is predictable.

TABLE 10.3 CLASSIFICATION OF LINEAR ADAPTIVE FILTERING ALGORITHMS

Class	Sample Update	Block Update
Stochastic gradient • self-orthogonalizing	LMS DCT–LMS; GAL	block LMS SOBAF
Least squares	RLS	block LS

- The excess mean-squared error produced by the DCT–LMS algorithm is smaller than that of the standard LMS algorithm.

In other words, the ensemble-averaged squared-error performance of the DCT–LMS algorithm is closer to that of the RLS algorithm than that of the standard LMS algorithm.

10.6 CLASSIFICATION OF ADAPTIVE FILTERING ALGORITHMS

In light of the material covered in this chapter and the previous one and, in a certain sense, anticipating the material to be covered in subsequent chapters of the book, we may classify linear adaptive filtering algorithms as shown in Table 10.3.[3] Here, we have identified two main classes of adaptive filtering algorithms:

- stochastic gradient algorithms
- exact least-squares algorithms

Stochastic gradient algorithms include self-orthogonalizing algorithms as a subclass. In each case, there are two basic ways in which the free parameters of the algorithm are updated:

- sample-by-sample basis
- block-by-block basis

The standard LMS algorithm (with sample update) and the block LMS algorithm (with block update) are both stochastic gradient algorithms. The use of "Fourier-for-fast convolution" provides an efficient method for computing the block LMS algorithm. This is one way in which Fourier transformation may be used to advantage in performing linear adaptive filtering.

---

[3] The material covered in this section, including Table 10.3, is based on B. Mulgrew, private communication, 1995.

Another way in which Fourier transformation plays a useful role is in "Fourier-for-orthogonality," as exemplified in the DCT–LMS algorithm. Specifically, the discrete cosine transform (DCT) provides a frequency-domain method for approximating an orthogonal set of samples. The DCT–LMS algorithm uses a sample update. In contrast, in the *self-orthogonalizing block adaptive filter* (SOBAF) described in (Panda et al., 1986) the single-sample gradient estimate employed in Eq. (10.41) is replaced with the block estimate of Eq. (10.13). Indeed, the algorithm of Table 10.1 may be viewed as a form of approximation to SOBAF; the sequence P_i is an estimate of the power spectral density of the input signal vector $\mathbf{u}(n)$.

In the DCT–LMS algorithm, self-orthogonalization of the input data is approximated in the frequency domain via eigenvalue decomposition. On the other hand, in the *gradient adaptive lattice (GAL) algorithm* due to Griffiths (1977, 1978), approximate self-orthogonalization of the input data is performed in the time domain via Cholesky factorization as described in Sections 6.7 and 6.8; a derivation of the GAL algorithm is presented in Appendix G. Although the DCT–LMS and GAL algorithms approach self-orthogonalization of the input data in entirely different ways, they are, however, similar in that they both impose a Toeplitz structure (implicitly or explicitly) on an estimate of the correlation matrix $\mathbf{R}$ of the input signal vector $\mathbf{u}(n)$. Stated in another way, the derivations of both the DCT–LMS and GAL algorithms are rooted in wide-sense stationary process theory since, as explained in Chapter 2, wide-sense stationarity of a stochastic process and the Toeplitz property of the correlation matrix $\mathbf{R}$ are indeed synonymous. The Toeplitz assumption also applies to SOBAF.

In the *least-squares* family of linear adaptive filtering algorithms, exemplified by the *recursive least-squares (RLS) algorithm* with sample update (to be described in Chapter 13) and the *block least-squares algorithm* (to be described in Chapter 11), *exact* orthogonalization of the input data is performed in the time domain. In other words, no approximations are made in the derivations of these algorithms, hence the rapid rate of convergence and other important properties that characterize the least-squares family of adaptive filters.

10.7 SUMMARY AND DISCUSSION

Summarizing the material discussed in this chapter, frequency-domain adaptive filtering techniques provide an alternative route to LMS adaptation in the time domain. The fast LMS algorithm, based on the idea of block adaptation filtering, provides a computationally efficient algorithm for building an adaptive FIR filter with long memory. This algorithm exploits the computational advantage offered by a fast convolution technique known as the overlap-save method that relies on the fast Fourier transform algorithm for its implementation. The fast LMS algorithm exhibits convergence properties that are similar to those of the standard LMS algorithm. In particular, the converged weight vector, misadjustment, and average time constant of the fast LMS algorithm are exactly the same as those of the standard LMS algorithm. The main differences between the two algorithms are (1) the fast LMS algorithm has a tighter stability bound than the standard LMS algorithm, and (2) the fast LMS algorithm provides a more accurate estimate of the gradient vector than the standard LMS algorithm, with the estimation accuracy increasing with block size. Unfortunately, this improvement does not imply a faster convergence behavior, because the eigenvalue spread of the correlation matrix of the input vector (which determines the convergence behavior of the algorithm) is independent of the block size.

The other frequency-domain adaptive filtering technique discussed in the chapter exploits the asymptotic equivalence of the discrete cosine transform to the statistically optimum Karhunen–Loève transform. The algorithm, termed the DCT–LMS algorithm, provides a close approximation to the method of self-orthogonalizing adaptive filtering. Unlike the fast LMS algorithm, the DCT–LMS algorithm is a nonblock algorithm that operates at the incoming data rate, and it is therefore not as computationally efficient as the fast LMS algorithm. The DCT part of the algorithm uses a sliding window, with the computation being performed recursively in $O(M)$ operations, where M is the filter memory. For the recursive computation, we may use a bank of frequency-sampling filters, each of which consists of a forward path and a pair of feedback paths, with the latter ones operating in parallel. Beaufays and Widrow (1994) describe an alternative procedure, based on the idea of LMS spectrum analyzers, for computing the sliding DCT in $O(M)$ operations; this adaptive implementation of the DCT is claimed to be both stable and exact. In any event, the DCT–LMS algorithm achieves a significant improvement in convergence behavior at the expense of an increase in computational complexity.

The fast LMS algorithm and the DCT–LMS algorithm share a common feature: they are both convolution-based frequency-domain adaptive filtering algorithms. As an alternative, we may use *adaptive filtering in subbands*. One motivation for such an approach is mainly to achieve computational efficiency by decimating the signals before performing the adaptive process (Gilloire and Vetterli, 1988, 1992; Petraglia and Mitra, 1993). *Decimation* refers to the process of digitally converting the sampling rate of a signal of interest from a given rate to a lower rate. The use of this approach makes it possible to implement an adaptive FIR filter of long memory, which is computationally efficient. Specifically, the task of designing a single long filter is replaced by one of designing a bank of smaller filters that operate in parallel and at a lower rate. However, when critical subsampling is used, aliased versions of the original signal are generated in the subbands, thereby causing a degradation in the adaptive performance of the algorithm. According to de Courville and Duhamel (1995), a possible explanation of the problem encountered with this approach is that the use of subbands in a "fast" convolution algorithm can only be done in an approximate fashion. To avoid this problem, de Courville and Duhamel propose an algorithm that updates each portion of the frequency response of the adaptive filter in accordance with the error signal measured in the same subband. Thus, the convergence improvement is separated from the reduction of computational complexity.

PROBLEMS

1. Using the average time constant of the LMS algorithm given in Eq. (9.97) as a guide, propose a formula for the average time constant of the block LMS algorithm. Hence, make a comparison between these two time constants.

2. The purpose of this problem is to develop a matrix formulation of the fast LMS algorithm described by the signal-flow graph of Fig. 10.2.

   (a) To define one time-domain constraint built into the operation of this algorithm, let

   $$\mathbf{G}_1 = \begin{bmatrix} \mathbf{I} & \mathbf{O} \\ \mathbf{O} & \mathbf{O} \end{bmatrix}$$

   where **I** is the M-by-M identity matrix and **O** is the M-by-M null matrix. Show that the weight-update equation (10.31) may be rewritten in the following compact form (Shynk, 1992):

   $$\hat{\mathbf{W}}(k + 1) = \hat{\mathbf{W}}(k) + \mu\mathbf{G}\mathbf{U}^H(k)\mathbf{W}(k)$$

   where the matrix **G** represents a constraint imposed on the computation of the gradient vector; it is defined in terms of $\mathbf{G}_1$ by

   $$\mathbf{G} = \mathbf{F}\mathbf{G}_1\mathbf{F}^{-1}$$

   where the matrix operator **F** signifies discrete Fourier transformation and $\mathbf{F}^{-1}$ signifies inverse discrete Fourier transformation.

   (b) To define the other time-domain constraint built into the operation of the fast LMS algorithm, let

   $$\mathbf{G}_2 = [\mathbf{O}, \mathbf{I}]$$

   where, as before, **I** and **O** denote the identity and null matrices, respectively. Hence, show that Eq. (10.29) may be redefined in the compact form (Shynk, 1992)

   $$\mathbf{E}(k) = \mathbf{F}\mathbf{G}_2^T\mathbf{e}(k)$$

   (c) Using the time-domain constraints represented by the matrices $\mathbf{G}_1$ and $\mathbf{G}_2$, formulate the corresponding matrix representations of the steps involved in the fast LMS algorithm.

   (d) What is the value of matrix **G** for which the fast LMS algorithm reduces to the unconstrained frequency-domain adaptive filtering algorithm of Section 10.3?

3. The unconstrained frequency-domain adaptive filtering algorithm of Section 10.3 has a limited range of applications. Identify and discuss at least one adaptive filtering application that is unaffected by ignoring the gradient constraint in Fig. 10.2.

4. The definition of the discrete Fourier transform described in Eq. (10.61) is different from that introduced in Chapter 1. Justify the validity of the definition given in Eq. (10.61).

5. Figure P10.1 shows the block diagram of a *transform-domain LMS filter* (Narayan et al., 1983). The tap-input vector $\mathbf{u}(n)$ is first applied to a bank of bandpass digital filters, implemented by means of the *discrete-Fourier transform* (DFT). Let $\mathbf{x}(n)$ denote the transformed vector produced at the DFT output. In particular, element k of the vector $\mathbf{x}(n)$ is given by

$$x_k(n) = \sum_{i=0}^{M-1} u(n-i)e^{-j(2\pi/M)ik}, \qquad k = 0, 1, \dots, M - 1$$

where $u(n - i)$ is element i of the tap-input vector $\mathbf{u}(n)$. Each $x_k(n)$ is normalized with respect to an estimate of its average power. The inner product of the vector $\mathbf{x}(n)$ and a frequency-domain weight vector $\mathbf{h}(n)$ is formed, obtaining the filter output

$$y(n) = \mathbf{h}^H(n)\mathbf{x}(n)$$

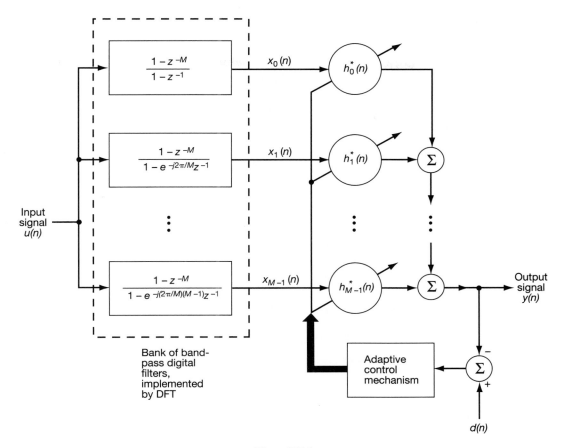

Figure P10.1

The weight vector update equation is

$$\mathbf{h}(n + 1) = \mathbf{h}(n) + \mu \mathbf{D}^{-1}(n)\mathbf{x}(n)e^*(n).$$

where $\mathbf{D}(n) = M$-by-M diagonal matrix whose kth element denotes the average power estimate of the DFT output $x_k(n)$ for $k = 0,1, \ldots, M - 1$

μ = adaptation constant

As usual, the estimation error $e(n)$ is defined by

$$e(n) = d(n) - y(n)$$

where $d(n)$ is the desired response.

(a) Show that the DFT output $x_k(n)$ may be computed recursively, using the relation

$$x_k(n) = e^{j(2\pi/M)k}x_k(n - 1) + u(n) - u(n - M), \qquad k = 0,1, \ldots, M - 1$$

(b) Assuming that μ is chosen properly, show that the weight vector $\mathbf{h}(n)$ converges to the frequency-domain optimum solution:

$$\mathbf{h}_o = \mathbf{Q}\mathbf{w}_o$$

where $\mathbf{w}_o$ is the (time-domain) Wiener solution, and $\mathbf{Q}$ is a unitary matrix defined by the DFT. Determine the components of the unitary matrix $\mathbf{Q}$.

(c) The use of the matrix $\mathbf{D}^{-1}$ in controlling the correction applied to the frequency-domain weight vector, in conjunction with the DFT, has the approximate effect of prewhitening the tap-input vector $\mathbf{u}(n)$. Do the following:

   (i) Demonstrate the prewhitening effect.

   (ii) Discuss how this effect compresses the eigenvalue spread of the DFT output vector $\mathbf{x}(n)$.

   (iii) The transform-domain LMS algorithm has a faster rate of convergence than the conventional LMS algorithm. Why?

6. The discrete cosine transform $C_m(n)$ of the sequence $u(n)$ may be decomposed as

$$C_m(n) = \frac{1}{2}k_m[C_m^{(1)}(n) + C_m^{(2)}(n)]$$

where k_m is defined by Eq. (10.65).

(a) Show that $C_m^{(1)}(n)$ and $C_m^{(2)}(n)$ may be computed recursively as follows:

$$C_m^{(1)}(n) = W_{2M}^{m/2}[W_{2M}^{m/2}C_m^{(1)}(n-1) + (-1)^m u(n) - u(n-M)]$$

$$C_m^{(2)}(n) = W_{2M}^{-m/2}[W_{2M}^{-m/2}C_m^{(2)}(n-1) + (-1)^m u(n) - u(n-M)]$$

where W_{2M} is defined by

$$W_{2M} = \exp\left(-\frac{j2\pi}{2M}\right)$$

(b) How is the computation of $C_m(n)$ modified in light of the operator z^{-M} in the forward path and the operators z^{-1} in the feedback paths of Fig. 10.5, each being multiplied by the factor β, where $0 < \beta < 1$?

CHAPTER

11

Method of Least Squares

In this chapter, we use a model-dependent procedure known as the *method of least squares* to solve the linear filtering problem, without invoking assumptions on the statistics of the inputs applied to the filter. To illustrate the basic idea of least squares, suppose we have a set of real-valued measurements $u(1)$, $u(2)$, . . . , $u(N)$, made at times t_1, t_2, . . . , t_N, respectively, and the requirement is to construct a curve that is used to *fit* these points in some optimum fashion. Let the time dependence of this curve be denoted by $f(t_i)$. According to the method of least squares, the "best" fit is obtained by *minimizing the sum of squares of difference* between $f(t_i)$ and $u(i)$ for $i = 1, 2, . . . , N$, hence the name of the method.

The method of least squares may be viewed as an alternative to Wiener filter theory. Basically, Wiener filters are derived from *ensemble averages* with the result that one filter (optimum in a probabilistic sense) is obtained for all realizations of the operational environment, assumed to be wide-sense stationary. On the other hand, the method of least squares is *deterministic* in approach. Specifically, it involves the use of time averages, with the result that the filter depends on the number of samples used in the computation. We begin our study in the next section by outlining the essence of the linear least-squares estimation problem.

11.1 STATEMENT OF THE LINEAR LEAST-SQUARES ESTIMATION PROBLEM

Consider a physical phenomenon that is characterized by two sets of variables, $d(i)$ and $u(i)$. The variable $d(i)$ is observed at time i in *response* to the subset of variables $u(i)$, $u(i - 1)$, . . . , $u(i - M + 1)$ applied as *inputs*. That is, $d(i)$ is a function of the inputs $u(i)$,

$u(i - 1), \ldots, u(i - M + 1)$. This functional relationship is hypothesized to be *linear*. In particular, the response $d(i)$ is modeled as

$$d(i) = \sum_{k=0}^{M-1} w_{ok}^* u(i - k) + e_o(i) \tag{11.1}$$

where the w_{ok} are *unknown parameters* of the *model*, and $e_o(i)$ represents the *measurement error* to which the statistical nature of the phenomenon is ascribed; each term in the summation in Eq. (11.1) represents a scalar inner product. In effect, the model of Eq. (11.1) says that the variable $d(i)$ may be determined as a linear combination of the input variables $u(i), u(i - 1), \ldots, u(i - M + 1)$, except for the error $e_o(i)$. This model, represented by the signal-flow graph shown in Fig. 11.1, is called a *multiple linear regression model*.

The *measurement error* $e_o(i)$ is an *unobservable* random variable that is introduced into the model to account for its inaccuracy. It is customary to assume that the measurement error process $e_o(i)$ is white with zero mean and variance σ^2. That is,

$$E[e_o(i)] = 0 \qquad \text{for all } i$$

and

$$E[e_o(i)e_o^*(k)] = \begin{cases} \sigma^2, & i = k \\ 0, & i \neq k \end{cases}$$

The implication of this assumption is that we may rewrite Eq. (11.1) in the ensemble-averaged form

$$E[d(i)] = \sum_{k=0}^{M-1} w_{ok}^* u(i - k)$$

where the values of $u(i), u(i - 1), \ldots, u(i - M + 1)$ are known. Hence, the mean of the response $d(i)$, in theory, is uniquely determined by the model.

The problem we have to solve is to *estimate* the unknown parameters of the multiple linear regression model of Fig. 11.1, the w_{ok}, given the two *observable* sets of variables: $u(i)$ and $d(i)$, $i = 1, 2, \ldots, N$. To do this, we postulate the linear transversal filter of Fig. 11.2 as the model of interest. By forming inner scalar products of the *tap inputs* $u(i)$, $u(i - 1), \ldots, u(i - M + 1)$ and the corresponding *tap weights* $w_0, w_1, \ldots, w_{M-1}$, and by utilizing $d(i)$ as the *desired response*, we define the *estimation error* or *residual* $e(i)$ as the difference between the desired response $d(i)$ and the *filter output* $y(i)$, as shown by

$$e(i) = d(i) - y(i) \tag{11.2}$$

where

$$y(i) = \sum_{k=0}^{M-1} w_k^* u(i - k) \tag{11.3}$$

That is,

$$e(i) = d(i) - \sum_{k=0}^{M-1} w_k^* u(i - k) \tag{11.4}$$

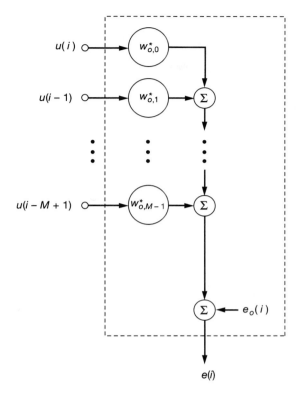

$e(i)$

Figure 11.1 Multiple linear regression model.

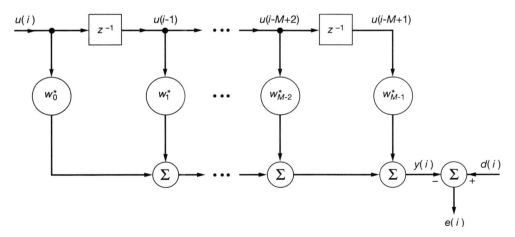

Figure 11.2 Linear transversal filter model.

In the method of least squares, we choose the tap weights of the transversal filter, the w_k, so as to minimize a cost function that consists of the *sum of error squares*:

$$\mathcal{E}(w_0, \ldots, w_{M-1}) = \sum_{i=i_1}^{i_2} |e(i)|^2 \tag{11.5}$$

where i_1 and i_2 define the index limits at which the error minimization occurs; this sum may also be viewed as an *error energy*. The values assigned to these limits depend on the type of *data windowing* employed, as discussed in Section 11.2. Basically, the problem we have to solve is to substitute Eq. (11.4) into (11.5) and then minimize the cost function $\mathcal{E}(w_0, \ldots, w_{M-1})$ with respect to the tap weights of the transversal filter in Fig. 11.2. For this minimization, the tap weights of the filter $w_0, w_1, \ldots, w_{M-1}$ are held *constant* during the interval $i_1 \leq i \leq i_2$. The filter resulting from the minimization is termed a *linear least-squares filter*.

11.2 DATA WINDOWING

Given M as the number of tap weights used in the transversal filter model of Fig. 11.2, the rectangular matrix constructed from the input data, $u(1), u(2), \ldots, u(N)$, may assume different forms, depending on the values assigned to the limits i_1 and i_2 in Eq. (11.5). In particular, we may distinguish four different methods of *windowing* the input data:

1. *Covariance method*, which makes no assumptions about the data outside the interval $[1, N]$. Thus, by defining the limits of interest as $i_1 = M$ and $i_2 = N$, the input data may be arranged in the matrix form

$$\begin{bmatrix} u(M) & u(M+1) & \cdots & u(N) \\ u(M-1) & u(M) & \cdots & u(N-1) \\ \cdot & \cdot & \cdot & \cdot \\ \cdot & \cdot & \cdot & \cdot \\ \cdot & \cdot & \cdot & \cdot \\ u(1) & u(2) & \cdots & u(N-M+1) \end{bmatrix}$$

2. *Autocorrelation method*, which makes the assumption that the data prior to time $i = 1$ and the data after $i = N$ are zero. Thus, by using $i_1 = 1$ and $i_2 = N + M - 1$, the matrix of input data takes on the form

$$\begin{bmatrix} u(1) & u(2) & \cdots & u(M) & u(M+1) & \cdots & u(N) & 0 & \cdots & 0 \\ 0 & u(1) & \cdots & u(M-1) & u(M) & \cdots & u(N-1) & u(N) & \cdots & 0 \\ \cdot & \cdot & \cdot & \cdot & \cdot & \cdot & \cdot & \cdot & \cdot & \cdot \\ \cdot & \cdot & \cdot & \cdot & \cdot & \cdot & \cdot & \cdot & \cdot & \cdot \\ \cdot & \cdot & \cdot & \cdot & \cdot & \cdot & \cdot & \cdot & \cdot & \cdot \\ 0 & 0 & \cdots & u(1) & u(2) & \cdots & u(N-M+1) & u(N-M) & \cdots & u(N) \end{bmatrix}$$

3. *Prewindowing method*, which makes the assumption that the input data prior to $i = 1$ are zero, but makes no assumption about the data after $i = N$. Thus, by using $i_1 = 1$ and $i_2 = N$, the matrix of the input data assumes the form

$$
\begin{bmatrix}
u(1) & u(2) & \cdots & u(M) & u(M+1) & \cdots & u(N) \\
0 & u(1) & \cdots & u(M-1) & u(M) & \cdots & u(N-1) \\
\cdot & \cdot & \cdot & \cdot & \cdot & \cdot & \cdot \\
\cdot & \cdot & \cdot & \cdot & \cdot & \cdot & \cdot \\
\cdot & \cdot & \cdot & \cdot & \cdot & \cdot & \cdot \\
0 & 0 & \cdots & u(1) & u(2) & \cdots & u(N-M+1)
\end{bmatrix}
$$

4. *Postwindowing method*, which makes no assumption about the data prior to time $i = 1$, but makes the assumption that the data after $i = N$ are zero. Thus, by using $i_1 = M$ and $i_2 = N + M - 1$, the matrix of input data takes on the form

$$
\begin{bmatrix}
u(M) & u(M+1) & \cdots & u(N) & 0 & \cdots & 0 \\
u(M-1) & u(M) & \cdots & u(N-1) & u(N) & \cdots & 0 \\
\cdot & \cdot & \cdot & \cdot & \cdot & \cdot & \cdot \\
\cdot & \cdot & \cdot & \cdot & \cdot & \cdot & \cdot \\
\cdot & \cdot & \cdot & \cdot & \cdot & \cdot & \cdot \\
u(1) & u(2) & \cdots & u(N-M+1) & u(N-M) & \cdots & u(N)
\end{bmatrix}
$$

The terms "covariance method" and "autocorrelation method" are commonly used in speech-processing literature (Makhoul, 1975; Markel and Gray, 1976). It should, however, be emphasized that the use of these two terms is *not* based on the standard definition of the covariance function as the correlation function with the means removed. Rather, these two terms derive their names from the way we interpret the meaning of the *known parameters* contained in the system of equations that result from minimizing the index of performance of Eq. (11.5). The covariance method derives its name from control theory literature where, with zero-mean tap inputs, these known parameters represent the elements of a *covariance matrix,* hence the name of the method. The autocorrelation method, on the other hand, derives its name from the fact that, for the conditions stated, these known parameters represent the *short-term autocorrelation function* of the tap inputs, hence the name of the second method. It is of interest to note that, among the four windowing methods described above, the autocorrelation method is the only one that yields a *Toeplitz* correlation matrix for the input data.

In the remainder of this chapter, except for Problem 4, which deals with the autocorrelation method, we will be exclusively concerned with the covariance method. The prewindowing method is considered in subsequent chapters.

11.3 PRINCIPLE OF ORTHOGONALITY (REVISITED)

When we developed the Wiener filter theory in Chapter 5, we proceeded by first deriving the principle of orthogonality (in the ensemble sense) for wide-sense stationary discrete-time stochastic processes, which were then used to derive the Wiener–Hopf equations that provide the mathematical basis of Wiener filters. In this chapter we proceed in a similar fashion by first deriving the principle of orthogonality based on time averages, and then use it to derive a system of equations known as the normal equations that provides the mathematical basis of linear least-squares filters. The development of this theory will be done for the covariance method.

The cost function or the sum of the error squares in the covariance method is defined by

$$\mathscr{E}(w_o, \ldots, w_{M-1}) = \sum_{i=M}^{N} |e(i)|^2 \tag{11.6}$$

By choosing the limits on the time index i in this way, in effect, we make sure that for each value of i, all the M tap inputs of the transversal filter in Fig. 11.2 have nonzero values. As mentioned previously, the problem we have to solve is to determine the tap weights of the transversal filter of Fig. 11.2 for which the sum of error squares is minimum.

We first rewrite Eq. (11.6) as

$$\mathscr{E}(w_0, \ldots, w_{M-1}) = \sum_{i=M}^{N} e(i)e^*(i) \tag{11.7}$$

where the estimation error $e(i)$ is defined in Eq. (11.4). Let the kth tap-weight w_k be expressed in terms of its real and imaginary parts as follows:

$$w_k = a_k + jb_k, \qquad k = 0, 1, \ldots, M-1 \tag{11.8}$$

Thus, substituting Eq. (11.8) in (11.4), we get

$$e(i) = d(i) - \sum_{k=0}^{M-1} (a_k - jb_k)u(i-k) \tag{11.9}$$

We define the kth component of the gradient vector $\nabla\mathscr{E}$ as the derivative of the cost function $\mathscr{E}(w_0, \ldots, w_{M-1})$ with respect to the real and imaginary parts of tap-weight w_k, as shown by

$$\nabla_k\mathscr{E} = \frac{\partial\mathscr{E}}{\partial a_k} + j\frac{\partial\mathscr{E}}{\partial b_k} \tag{11.10}$$

Hence, substituting Eq. (11.7) in (11.10), and recognizing that the estimation error $e(i)$ is complex valued, in general, we get

$$\nabla_k\mathscr{E} = -\sum_{i=M}^{N} \left[e(i)\frac{\partial e^*(i)}{\partial a_k} + e^*(i)\frac{\partial e(i)}{\partial a_k} + je(i)\frac{\partial e^*(i)}{\partial b_k} + je(i)\frac{\partial e(i)}{\partial b_k} \right] \tag{11.11}$$

Next, differentiating $e(i)$ in Eq. (11.9) with respect to the real and imaginary parts of w_k, we get the following four partial derivatives:

$$\frac{\partial e(i)}{\partial a_k} = -u(i - k)$$

$$\frac{\partial e^*(i)}{\partial a_k} = -u^*(i - k)$$

$$\frac{\partial e(i)}{\partial b_k} = ju(i - k) \tag{11.12}$$

$$\frac{\partial e^*(i)}{\partial b_k} = -ju^*(i - k)$$

Thus, the substitution of these four partial derivatives in Eq. (11.11) yields the result:

$$\nabla_k \mathcal{E} = -2 \sum_{i=M}^{N} u(i - k)e^*(i) \tag{11.13}$$

For the minimization of the cost function $\mathcal{E}(w_0, \ldots, w_{M-1})$ with respect to the tap weights $w_0, \ldots, w_{M-1}$ of the transversal filter in Fig. 11.2, we require that the following conditions be satisfied simultaneously:

$$\nabla_k \mathcal{E} = 0, \qquad k = 0, 1, \ldots, M - 1 \tag{11.14}$$

Let $e_{\min}(i)$ denote the special value of the estimation error $e(i)$ that results when the cost function $\mathcal{E}(w_0, \ldots, W_{M-1})$ is minimized (i.e., the transversal filter is optimized) in accordance with Eq. (11.14). From Eq. (11.13) we then readily see that the set of conditions (11.14) is equivalent to the following:

$$\sum_{i=M}^{N} u(i - k)e_{\min}^*(i) = 0, \qquad k = 0, 1, \ldots, M - 1 \tag{11.15}$$

Equation (11.15) is the mathematical description of the temporal version of the *principle of orthogonality*. The *time average*[1] on the left-hand side of Eq. (11.15) represents the cross-correlation between the tap input $u(i - k)$ and the minimum estimation error $e_{\min}(i)$ over the values of time i in the interval $[M, N]$, for a fixed value of k. Accordingly, we may state the *principle of orthogonality* as follows:

The minimum error time series $e_{\min}(i)$ is orthogonal to the time series $u(i - k)$ applied to tap k of a transversal filter of length M for $k = 0, 1, \ldots, M - 1$, when the filter is operating in its least-squares condition.

This principle provides the basis of a simple *test* that we can carry out in practice to check whether or not the transversal filter is operating in its *least-square condition*. We

---

[1]To be precise in the use of the term "time average," we should divide the sum on the left-hand side of Eq. (11.15) by the number of terms $(N - M + 1)$ used in the summation. Clearly, such an operation has no effect on Eq. (11.15). We have chosen to ignore the inclusion of this scaling factor merely for convenience of presentation.

merely have to determine the time-averaged cross-correlation between the estimation error and the time series applied to *each* tap input of the filter. It is *only* when *all* these M cross-correlation functions are identically zero that we find the cost function $\mathcal{E}(w_0, \ldots, w_{M-1})$ is minimum.

Corollary

Let $\hat{w}_0, \hat{w}_1, \ldots, \hat{w}_{M-1}$ denote the special values of the tap weights $w_0, w_1, \ldots, w_{M-1}$ that result when the transversal filter of Fig. 11.2 is optimized to operate in its least-squares condition. The filter output, denoted by $y_{\min}(i)$, is obtained from Eq. (11.3) to be

$$y_{\min}(i) = \sum_{k=0}^{M-1} \hat{w}_k^* u\,(i - k) \tag{11.16}$$

This filter output provides a *least-squares estimate* of the desired response $d(i)$; the estimate is said to be *linear* because it is a linear combination of the tap inputs $u(i)$, $u(i - 1)$, $\ldots$, $u(i - M + 1)$. Let $\mathcal{U}_i$ denote the space spanned by the tap inputs $u(i), \ldots, u(i - M + 1)$. Let $\hat{d}(i|\mathcal{U}_i)$ denote the least-squares estimate of the desired response $d(i)$, given the tap inputs spanned by the space $\mathcal{U}_i$. We may thus write

$$\hat{d}(i|\mathcal{U}_i) = y_{\min}(i) \tag{11.17}$$

or, equivalently,

$$\hat{d}(i|\mathcal{U}_i) = \sum_{k=0}^{M-1} \hat{w}_k^* u\,(i - k) \tag{11.18}$$

Returning to Eq. (11.15), suppose we multiply both sides of this equation by $\hat{w}_k^*$ and then sum the result over the values of k in the interval $[0, M - 1]$. We then get (after interchanging the order of summation):

$$\sum_{i=M}^{N} \left[\sum_{k=0}^{M-1} \hat{w}_k^* u(i - k) \right] e_{\min}^*(i) = 0 \tag{11.19}$$

The summation term inside the parentheses on the left-hand side of Eq. (11.19) is recognized to be the least-squares estimate $\hat{d}(i|\mathcal{U}_i)$ of Eq. (11.18). Accordingly, we may simplify Eq. (11.19) to

$$\sum_{i=M}^{N} \hat{d}(i|\mathcal{U}_i) e_{\min}^*(i) = 0 \tag{11.20}$$

Equation (11.20) is a mathematical description of the *corollary to the principle of orthogonality*. We recognize the time average on the left-hand side of Eq. (11.20) is the cross-correlation of the two time series $\hat{d}(i|\mathcal{U}_i)$ and $e_{\min}(i)$. Accordingly, we may state the corollary to the principle of orthogonality as follows:

When a transversal filter operates in its least-squares condition, the least-squares estimate of the desired response, produced at the filter output and represented by the time series

$\hat{d}(i|\mathcal{U}_i)$, and the minimum estimation error time series $e_{\min}(i)$ are orthogonal to each other over time i.

A geometric illustration of this corollary to the principle of orthogonality is deferred to Section 11.6.

11.4 MINIMUM SUM OF ERROR SQUARES

The principle of orthogonality, given in Eq. (11.15), describes the least-squares condition of the transversal filter in Fig. 11.2 when the cost function $\mathscr{E}(w_0, \ldots, w_{M-1})$ is minimized with respect to the tap weights $w_0, \ldots, w_{M-1}$ in the filter. To find the minimum value of this cost function, that is, the *minimum sum of error squares* $\mathscr{E}_{\min}$, it is obvious that we may write

$$\underbrace{d(i)}_{\substack{\text{desired} \\ \text{response}}} = \underbrace{\hat{d}(i|\mathcal{U}_i)}_{\substack{\text{estimate of} \\ \text{desired} \\ \text{response}}} + \underbrace{e_{\min}(i)}_{\substack{\text{estimation} \\ \text{error}}} \tag{11.21}$$

Hence, evaluating the energy of the time series $d(i)$ for values of time i in the interval $[M, N]$, and using the corollary to the principle of orthogonality [i.e., Eq. (11.20)], we get the simple result

$$\mathscr{E}_d = \mathscr{E}_{\text{est}} + \mathscr{E}_{\min} \tag{11.22}$$

where

$$\mathscr{E}_d = \sum_{i=M}^{N} |d(i)|^2 \tag{11.23}$$

$$\mathscr{E}_{\text{est}} = \sum_{i=M}^{N} |\hat{d}(i|\mathcal{U}_i)|^2 \tag{11.24}$$

$$\mathscr{E}_{\text{est}} = \sum_{i=M}^{N} |e_{\min}(i)|^2 \tag{11.25}$$

Rearranging Eq. (11.22), we may express the minimum sum of error squares $\mathscr{E}_{\min}$ in terms of the energy $\mathscr{E}_d$ and the energy $\mathscr{E}_{\text{est}}$, contained in the time series $d(i)$ and $d(i|\mathcal{U}_i)$, respectively, as follows:

$$\mathscr{E}_{\min} = \mathscr{E}_d - \mathscr{E}_{\text{est}} \tag{11.26}$$

Clearly, given the specification of the desired response $d(i)$ for varying i, we may use Eq. (11.23) to evaluate the energy $\mathscr{E}_d$. As for the energy $\mathscr{E}_{\text{est}}$ contained in the time series $\hat{d}(i|\mathcal{U}_i)$ representing the estimate of the desired response, we are going to defer its evaluation to the next section.

Since $\mathscr{E}_{\min}$ is nonnegative, it follows that the second term on the right-hand side of Eq. (11.26) can never exceed $\mathscr{E}_d$. Indeed, it reaches the value of $\mathscr{E}_d$ when the measurement error $e_o(i)$ in the multiple linear regression model of Fig. 11.1 is zero for all i, which is a practical impossibility.

Another case for which $\mathscr{E}_{\min}$ equals $\mathscr{E}_d$ occurs when the least-squares problem is *underdetermined*. Such a situation arises when there are fewer data points than parameters, in which case the estimation error and therefore $\mathscr{E}_{\text{est}}$ is zero. Note, however, that when the least-squares problem is underdetermined, there is no unique solution to the problem. Discussion of this issue is deferred to the latter part of the chapter.

11.5 NORMAL EQUATIONS AND LINEAR LEAST-SQUARES FILTERS

There are two different, and yet basically equivalent, methods of describing the least-squares condition of the linear transversal filter in Fig. 11.1. The principle of orthogonality, described in Eq. (11.15), represents one method. The system of *normal equations* represents the other method; interestingly enough, the system of normal equations derives its name from the corollary to the principle of orthogonality. Naturally, we may derive this system of equations in its own independent way by formulating the gradient vector $\nabla\mathscr{E}$ in terms of the tap weights of the filter, and then solving for the tap-weight vector $\hat{\mathbf{w}}$ for which $\nabla\mathscr{E}$ is zero. Alternatively, we may derive the system of normal equations from the principle of orthogonality. We are going to pursue the latter (indirect) approach in this section, and leave the former (direct) approach to the interested reader as Problem 7.

The principle of orthogonality in Eq. (11.15) is formulated in terms of a set of tap inputs and the minimum estimation error $e_{\min}(i)$. Setting the tap weights in Eq. (11.4) to their least-squares values, we get

$$e_{\min}(i) = d(i) - \sum_{t=0}^{M-1} \hat{w}_t^* u(i - t) \tag{11.27}$$

where on the right-hand side we have purposely used t as the dummy summation index. Hence, substituting Eq. (11.27) in (11.15), and then rearranging terms, we get a system of M simultaneous equations:

$$\sum_{t=0}^{M-1} \hat{w}_t \sum_{i=M}^{N} u(i - k)u^*(i - t) = \sum_{i=M}^{N} u(i - k)d^*(i), \qquad k = 0, \ldots, M-1 \tag{11.28}$$

The two summations in Eq. (11.28) involving the index i represent time-averages, except for a scaling factor. They have the following interpretations:

1. The time average (over i) on the left-hand side of Eq. (11.28) represents the *time averaged autocorrelation function* of the tap inputs in the transversal filter of Fig. 11.2. In particular, we may write

$$\phi(t, k) = \sum_{i=M}^{N} u(i - k)u^*(i - t), \qquad 0 \le (t, k) \le M - 1 \tag{11.29}$$

2. The time average (also over i) on the right-hand side of Eq. (11.28) represents the *cross-correlation* between the tap inputs and the desired response. In particular, we may write

$$z(-k) = \sum_{i=M}^{N} u(i - k)d^*(i), \qquad 0 \leq k \leq M - 1 \tag{11.30}$$

Accordingly, we may rewrite the system of simultaneous equations (11.28) as follows:

$$\sum_{t=0}^{M-1} \hat{w}_t \phi(t, k) = z(-k), \qquad k = 0, 1, \ldots, M - 1 \tag{11.31}$$

The system of equations (11.31) represents *the expanded system of the normal equations* for a linear least-squares filter.

Matrix Formulation of the Normal Equations

We may recast this system of equations in matrix form by first introducing the following definitions:

1. The M-by-M *time-averaged correlation matrix* of the tap inputs $u(i)$, $u(i - 1)$, ..., $u(i - M + 1)$:

$$\mathbf{\Phi} = \begin{bmatrix} \phi(0, 0) & \phi(1, 0) & \cdots & \phi(M - 1, 0) \\ \phi(0, 1) & \phi(1, 1) & \cdots & \phi(M - 1, 1) \\ \cdot & \cdot & \cdot & \cdot \\ \cdot & \cdot & \cdot & \cdot \\ \cdot & \cdot & \cdot & \cdot \\ \phi(0, M - 1) & \phi(1, M - 1) & \cdots & \phi(M - 1, M - 1) \end{bmatrix} \tag{11.32}$$

2. The M-by-1 *time-averaged cross-correlation vector* between the tap inputs $u(i)$, $u(i - 1), \ldots, u(i - M + 1)$ and the desired response $d(i)$:

$$\mathbf{z} = [z(0), z(-1), \ldots, z(-M + 1)]^T \tag{11.33}$$

3. The M-by-1 tap-weight vector of the least-squares filter:

$$\hat{\mathbf{w}} = [\hat{w}_0, \hat{w}_1, \ldots, \hat{w}_{M-1}]^T \tag{11.34}$$

Hence, in terms of these matrix definitions, we may now rewrite the system of M simultaneous equations (11.31) simply as

$$\mathbf{\Phi}\hat{\mathbf{w}} = \mathbf{z} \tag{11.35}$$

Equation (11.35) is *the matrix form of the normal equations for linear least-squares filters*.

Assuming that $\mathbf{\Phi}$ is nonsingular and therefore the inverse matrix $\mathbf{\Phi}^{-1}$ exists, we may solve Eq. (11.35) for the tap-weight vector of the linear least-squares filter:

$$\hat{\mathbf{w}} = \mathbf{\Phi}^{-1}\mathbf{z} \tag{11.36}$$

The condition for the existence of the inverse matrix $\boldsymbol{\Phi}^{-1}$ is discussed in Section 11.6.

Equation (11.36) is a very important result. In particular, it is the linear least-squares counterpart to the solution of the matrix form of the Wiener–Hopf equations (5.36). Basically, Eq. (11.36) states that the tap-weight vector $\hat{\mathbf{w}}$ of a linear least-squares filter is uniquely defined by the product of the inverse of the time-averaged correlation matrix $\boldsymbol{\Phi}$ of the tap inputs of the filter and the time-averaged cross-correlation vector $\mathbf{z}$ between the tap inputs and the desired response. Indeed, this equation is fundamental to the development of all recursive formulations of the linear least-squares filter, as pursued in subsequent chapters of the book.

Minimum Sum of Error Squares

Equation (11.26), derived in the preceding section, defines the minimum sum of error squares $\mathscr{E}_{\min}$. We now complete the evaluation of $\mathscr{E}_{\min}$, expressed as the difference between the energy $\mathscr{E}_d$ of the desired response and the energy $\mathscr{E}_{\text{est}}$ of its estimate. Usually, $\mathscr{E}_d$ is determined from the time series representing the desired response. To evaluate $\mathscr{E}_{\text{est}}$, we write

$$\mathscr{E}_{\text{est}} = \sum_{i=M}^{N} |\hat{d}(i|\mathcal{U}_i)|^2$$

$$= \sum_{i=M}^{N} \sum_{t=0}^{M-1} \sum_{k=0}^{M-1} \hat{w}_t \hat{w}_k^* u(i-k) u^*(i-t) \qquad (11.37)$$

$$= \sum_{t=0}^{M-1} \sum_{k=0}^{M-1} \hat{w}_t \hat{w}_k^* \sum_{i=M}^{N} u(i-k) u^*(i-t)$$

where, in the second line, we have made use of Eq. (11.18). The inner summation over time i in the final line of Eq. (11.37) represents the time-averaged autocorrelation function $\phi(t, k)$ [see Eq. (11.29)]. Hence, we may rewrite Eq. (11.37) as

$$\mathscr{E}_{\text{est}} = \sum_{t=0}^{M-1} \sum_{k=0}^{M-1} \hat{w}_k^* \phi(t, k) \hat{w}_t \qquad (11.38)$$

$$= \hat{\mathbf{w}}^H \boldsymbol{\Phi} \hat{\mathbf{w}}$$

where $\hat{\mathbf{w}}$ is the least-squares tap-weight vector and $\boldsymbol{\Phi}$ is the time-averaged correlation matrix of the tap inputs. We may further simplify the formula for $\mathscr{E}_{\text{est}}$ by noting that from the normal equations (11.35), the matrix product $\boldsymbol{\Phi}\hat{\mathbf{w}}$ equals the cross-correlation vector $\mathbf{z}$. Accordingly, we have

$$\mathscr{E}_{\text{est}} = \hat{\mathbf{w}}^H \mathbf{z}$$

$$= \mathbf{z}^H \hat{\mathbf{w}} \qquad (11.39)$$

Finally, substituting Eq. (11.39) in (11.26), and then using Eq. (11.36) for $\hat{\mathbf{w}}$, we get

$$\mathscr{E}_{\min} = \mathscr{E}_d - \mathbf{z}^H \hat{\mathbf{w}} \qquad (11.40)$$

$$= \mathscr{E}_d - \mathbf{z}^H \Phi^{-1} \mathbf{z}$$

Equations (11.40) is the formula for the minimum sum of error squares, expressed in terms of three known quantities: the energy $\mathscr{E}_d$ of the desired response, the time-averaged correlation matrix Φ of the tap inputs, and the time-averaged cross-correlation vector $\mathbf{z}$ between the tap inputs and the desired response.

11.6 TIME-AVERAGED CORRELATION MATRIX Φ

The time-averaged correlation matrix or simply the correlation matrix Φ of the tap inputs is shown in its expanded form in Eq. (11.32), with the element $\phi(t, k)$ defined in Eq. (11.29). The index k in $\phi(t, k)$ refers to the row number in the matrix Φ, and t refers to the column number. Let the M-by-1 tap-input vector $\mathbf{u}(i)$ be defined by

$$\mathbf{u}(i) = [u(i), u(i - 1), \ldots, u(i - M + 1)]^T \qquad (11.41)$$

Hence, we may use Eqs. (11.29) and (11.41) to redefine the correlation matrix Φ as the time average of the outer product $\mathbf{u}(i)\mathbf{u}^H(i)$ over i as follows:

$$\Phi = \sum_{i=M}^{N} \mathbf{u}(i)\mathbf{u}^H(i) \qquad (11.42)$$

To restate what we said earlier under footnote 1, the summation in Eq. (11.42) should be divided by the scaling factor $(N - M + 1)$ for the correlation matrix Φ to be a time average in precise terms. In the statistics literature, this scaled form of Φ is referred to as the *sample correlation matrix*. In any event, on the basis of the definition given in Eq. (11.42), we may readily establish the following properties of the correlation matrix:

Property 1.   *The correlation matrix Φ is Hermitian; that is*

$$\Phi^H = \Phi$$

This property follows directly from Eq. (11.42).

Property 2.   *The correlation matrix Φ is nonnegative definite; that is,*

$$\mathbf{x}^H \Phi \mathbf{x} \geq 0$$

for any M-by-1 vector $\mathbf{x}$.

Using the definition of Eq. (11.42), we may write

$$\mathbf{x}^H \mathbf{\Phi} \mathbf{x} = \sum_{i=M}^{N} \mathbf{x}^H \mathbf{u}(i) \mathbf{u}^H(i) \mathbf{x}$$

$$= \sum_{i=M}^{N} [\mathbf{x}^H \mathbf{u}(i)][\mathbf{x}^H \mathbf{u}(i)]^H$$

$$= \sum_{i=M}^{N} |\mathbf{x}^H \mathbf{u}(i)|^2 \geq 0$$

which proves Property 2. The fact that the correlation matrix $\mathbf{\Phi}$ is nonnegative definite means that its determinant and all principal minors are nonnegative. When the above condition is satisfied with the inequality sign, the determinant of $\mathbf{\Phi}$ and its principal minors are likewise nonzero. In the latter case, $\mathbf{\Phi}$ is nonsingular and the inverse $\mathbf{\Phi}^{-1}$ exists.

Property 3.   *The eigenvalues of the correlation matrix $\mathbf{\Phi}$ are all real and non-negative.*

The real requirement on the eigenvalues of $\mathbf{\Phi}$ follows from Property 1. The fact that all these eigenvalues are also nonnegative follows from Property 2.

Property 4.   *The correlation matrix is the product of two rectangular Toeplitz matrices that are the Hermitian transpose of each other.*

The correlation matrix $\mathbf{\Phi}$ is, in general, non-Toeplitz, which is clearly seen by examining the expanded form of the correlation matrix given in Eq. (11.32). The elements on the main diagonal, $\phi(0, 0)$, $\phi(1, 1)$, . . . , $\phi(M - 1, M - 1)$, have different values; this also applies to secondary diagonal above or below the main diagonal. However, the matrix $\mathbf{\Phi}$ has a special structure in the sense that it is the product of two Toeplitz rectangular matrices. To prove this property, we first use Eq. (11.42) to express the matrix $\mathbf{\Phi}$ as follows:

$$\mathbf{\Phi} = [\mathbf{u}(M), \mathbf{u}(M + 1), \ldots, \mathbf{u}(N)] \begin{bmatrix} \mathbf{u}^H(M) \\ \mathbf{u}^H(M - 1) \\ \cdot \\ \cdot \\ \cdot \\ \mathbf{u}^H(N) \end{bmatrix} \tag{11.43}$$

Next, for convenience of presentation, we introduce a *data matrix* **A**, whose Hermitian transpose is defined by

$$\mathbf{A}^H = \quad [\mathbf{u}(M), \qquad \mathbf{u}(M+1), \quad \cdots, \mathbf{u}(N)]$$

$$= \begin{bmatrix} u(M) & u(M+1) & \cdots & u(N) \\ u(M-1) & u(M) & \cdots & u(N-1) \\ \cdot & \cdot & & \cdot \\ \cdot & \cdot & & \cdot \\ \cdot & \cdot & & \cdot \\ u(1) & u(2) & \cdots & u(N-M+1) \end{bmatrix} \qquad (11.44)$$

The expanded matrix on the right-hand side of Eq. (11.44) is recognized to be the matrix of input data for the covariance method of data windowing (see point 1 of Section 11.2). Thus, using the definition of Eq. (11.44), we may rewrite Eq. (11.43) in the compact form

$$\mathbf{\Phi} = \mathbf{A}^H \mathbf{A} \qquad (11.45)$$

From the expanded form of the matrix given in the second line of Eq. (11.44), we see that $\mathbf{A}^H$ consists of an M- by-$(N-M+1)$ *rectangular Toeplitz matrix*. The data matrix **A** itself is likewise an $(N-M+1)$-by-M rectangular Toeplitz matrix. According to Eq. (11.45), therefore, the correlation matrix $\mathbf{\Phi}$ is the product of two rectangular Toeplitz matrices that are the Hermitian transpose of each other: this completes the proof of Property 4.

11.7 REFORMULATION OF THE NORMAL EQUATIONS IN TERMS OF DATA MATRICES

The system of normal equations for a least-squares transversal filter is given by Eq. (11.35) in terms of the correlation matrix $\mathbf{\Phi}$ and the cross-correlation vector **z.** We may reformulate the normal equations in terms of data matrices by using Eq. (11.45) for the correlation matrix $\mathbf{\Phi}$ of the tap inputs, and a corresponding relation for the cross-correlation vector **z** between the tap inputs and the desired response. To do this, we introduce a *desired data vector* **d**, consisting of the *desired response* $d(i)$ for values of i in the interval $[M, N]$; in particular, we define

$$\mathbf{d}^H = [d(M), d(M+1), \ldots, d(N)] \qquad (11.46)$$

Note that we have purposely used Hermitian transposition rather than ordinary transposition in the definition of vector **d** to be consistent with the definition of the data matrix **A** in Eq. (11.44). With the definitions of Eqs. (11.44) and (11.46) at hand, we may now use Eqs. (11.30) and (11.33) to express the cross-correlation vector **z** as

$$\mathbf{z} = \mathbf{A}^H \mathbf{d} \qquad (11.47)$$

Furthermore, we may use Eqs. (11.45) and (11.47) in (11.35), and so express the system of normal equations in terms of the data matrix $\mathbf{A}$ and the desired data vector $\mathbf{d}$ as

$$\mathbf{A}^H\mathbf{A}\mathbf{w} = \mathbf{A}^H\mathbf{d}$$

Hence, the system of equations used in the minimization of the cost function $\mathscr{E}$ may be represented by $\mathbf{A}\mathbf{w} = \mathbf{d}$. Furthermore, assuming that the inverse matrix $(\mathbf{A}^H\mathbf{A})^{-1}$ exists, we may solve this system of equations by expressing the tap-weight vector $\hat{\mathbf{w}}$ as

$$\hat{\mathbf{w}} = (\mathbf{A}^H\mathbf{A})^{-1}\mathbf{A}^H\mathbf{d} \tag{11.48}$$

We may complete the reformulation of our results for the linear least-squares problem in terms of the data matrices $\mathbf{A}$ and $\mathbf{d}$ by using (1) the definitions of Eqs. (11.45) and (11.47) in (11.40), and (2) the definitions of Eq. (11.46) in (11.23). By so doing, we may rewrite the formula for the minimum sum of error squares as

$$\mathscr{E}_{\min} = \mathbf{d}^H\mathbf{d} - \mathbf{d}^H\mathbf{A}(\mathbf{A}^H\mathbf{A})^{-1}\mathbf{A}^H\mathbf{d} \tag{11.49}$$

Although this formula looks somewhat cumbersome, its nice feature is that it is expressed explicitly in terms of the data matrix $\mathbf{A}$ and the desired data vector $\mathbf{d}$.

Projection Operator

Equation (11.48) defines the least-squares tap-weight vector $\hat{\mathbf{w}}$ in terms of the data matrix $\mathbf{A}$ and the desired data vector $\mathbf{d}$. The least-squares estimate of $\mathbf{d}$ is therefore given by

$$\hat{\mathbf{d}} = \mathbf{A}\hat{\mathbf{w}}$$

$$= \mathbf{A}(\mathbf{A}^H\mathbf{A})^{-1}\mathbf{A}^H\mathbf{d} \tag{11.50}$$

Accordingly, we may view the multiple matrix product $\mathbf{A}(\mathbf{A}^H\mathbf{A})^{-1}\mathbf{A}^H$ as a *projection operator* onto the linear space spanned by the columns of the data matrix $\mathbf{A}$, which is the same space $\mathcal{U}_i$ mentioned previously for $i = N$. Denoting this projection operator by $\mathbf{P}$, we may thus write

$$\mathbf{P} = \mathbf{A}(\mathbf{A}^H\mathbf{A})^{-1}\mathbf{A}^H \tag{11.51}$$

The matrix difference

$$\mathbf{I} - \mathbf{A}(\mathbf{A}^H\mathbf{A})^{-1}\mathbf{A}^H = \mathbf{I} - \mathbf{P}$$

is the *orthogonal complement projector.* Note that both the projection operator and its complement are uniquely determined by the data matrix A. The projection operator, $\mathbf{P}$, applied to the desired data vector $\mathbf{d}$, yields the corresponding estimate $\hat{\mathbf{d}}$. On the other hand, the orthogonal complement projector, $\mathbf{I} - \mathbf{P}$, applied to the desired data vector $\mathbf{d}$, yields the estimation error vector $\mathbf{e}_{\min} = \mathbf{d} - \hat{\mathbf{d}}$. Figure 11.3 illustrates the functions of the projection operator and the orthogonal complement projector as described herein.

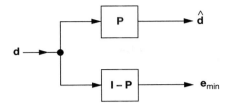

Figure 11.3 Projection operator **P** and orthogonal complement projector **I** − **P**.

Example 1

Consider the example of a linear least-squares filter with two taps (i.e., $M = 2$) and a *real-valued* input time series consisting of four samples (i.e., $N = 4$), hence $N - M + 1 = 3$. The input data matrix **A** and the desired data vector **d** have the following values:

$$\mathbf{A} = \begin{bmatrix} u(2) & u(1) \\ u(3) & u(2) \\ u(4) & u(3) \end{bmatrix}$$

$$= \begin{bmatrix} 2 & 3 \\ 1 & 2 \\ -1 & 1 \end{bmatrix}$$

$$\mathbf{d} = \begin{bmatrix} d(2) \\ d(3) \\ d(4) \end{bmatrix}$$

$$= \begin{bmatrix} 2 \\ 1 \\ 1/34 \end{bmatrix}$$

The purpose of this example is to evaluate the projection operator and the orthogonal complement projector, and use them to illustrate the principle of orthogonality.

The use of Eq. (11.51), reformulated for real data, yields the value of the projection operator **P** as

$$\mathbf{P} = \mathbf{A}(\mathbf{A}^T\mathbf{A})^{-1}\mathbf{A}^T$$

$$= \frac{1}{35} \begin{bmatrix} 26 & 15 & -2 \\ 15 & 10 & 5 \\ -3 & 5 & 34 \end{bmatrix}$$

The corresponding value of the orthogonal complement projector is

$$\mathbf{I} - \mathbf{P} = \frac{1}{35} \begin{bmatrix} 9 & -15 & 3 \\ -15 & 25 & -5 \\ -3 & -5 & 1 \end{bmatrix}$$

Accordingly, the estimate of the desired data vector and the estimation error vector have the following values, respectively:

$$\hat{\mathbf{d}} = \mathbf{Pd}$$

$$= \begin{bmatrix} 1.91 \\ 1.15 \\ 0 \end{bmatrix}$$

$$\mathbf{e}_{min} = (\mathbf{I} - \mathbf{P})\mathbf{d}$$

$$= \begin{bmatrix} 0.09 \\ -0.15 \\ 0.03 \end{bmatrix}$$

Figure 11.4 depicts three-dimensional geometric representations of the vectors $\hat{\mathbf{d}}$ and $\mathbf{e}_{min}$. This figure clearly shows that these two vectors are *normal* (i.e., *perpendicular*) to each other in accordance with the corollary to the principle of orthogonality, hence the terminology "normal" equations. This condition is the geometric portrayal of the fact that in a linear least-squares filter the inner product $\mathbf{e}_{min}^{H}\hat{\mathbf{d}}$ is zero. Figure 11.4 also depicts the desired data vector $\mathbf{d}$ as the "vector sum" of the estimate $\hat{\mathbf{d}}$ and the error $\mathbf{e}_{min}$. Note also that the vector $\mathbf{e}_{min}$ is orthogonal to span($\mathbf{A}$), defined as the set of all linear combinations of the column vectors of the data matrix $\mathbf{A}$. The estimate $\hat{\mathbf{d}}$ is just one vector in span($\mathbf{A}$).

Uniqueness Theorem

The linear least-squares problem of minimizing the sum of error squares, $\mathscr{E}(n)$, always has a solution. That is, for given values of the data matrix $\mathbf{A}$ and the desired data vector $\hat{\mathbf{d}}$, we can always find a vector $\hat{\mathbf{w}}$ that satisfies the normal equations. It is therefore important that we know if and when the solution is *unique*. This requirement is covered by the following *uniqueness theorem* (Stewart, 1973):

The least-squares estimate $\hat{\mathbf{w}}$ is unique if and only if the nullity of the data matrix A equals zero.

Let $\mathbf{A}$ be a K-by-M matrix; in the case of the data matrix $\mathbf{A}$ defined in Eq. (11.44), we have $K = N - M + 1$. We define the *null space* of matrix $\mathbf{A}$, written as $\mathcal{N}(\mathbf{A})$, as the space of all vectors $\mathbf{x}$ such that $\mathbf{Ax} = \mathbf{0}$. We define the *nullity* of matrix $\mathbf{A}$, written as null($\mathbf{A}$), as the dimension of the null space $\mathcal{N}(\mathbf{A})$. In general, we find that

$$\text{null}(\mathbf{A}) \neq \text{null}(\mathbf{A}^{H}).$$

In light of the uniqueness theorem, which is intuitively satisfying, we may expect a unique solution to the linear least-squares problem *only* when the data matrix $\mathbf{A}$ has *linearly independent columns;* that is, when the data matrix $\mathbf{A}$ is of *full column rank*. This implies that the matrix $\mathbf{A}$ has at least as many rows as columns; that is, $(N - M + 1) \geq M$. This latter condition means that the system of equations represented by $\mathbf{A}\hat{\mathbf{w}} = \mathbf{d}$ used in the minimization is *overdetermined*, in that it has more equations than unknowns. Thus,

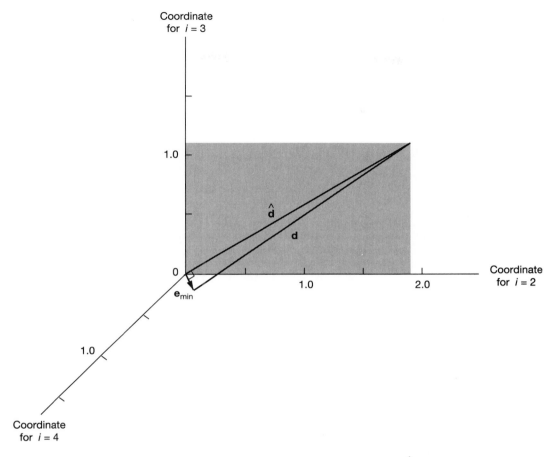

Figure 11.4   Three-dimensional geometric interpretations of vectors $\mathbf{d}$, $\hat{\mathbf{d}}$, and $\mathbf{e}_{\min}$.

provided that the data matrix $\mathbf{A}$ is of full column rank, the M-by-M matrix $\mathbf{A}^H\mathbf{A}$ is *non-singular*, and the least-squares estimate has the unique value given in Eq. (11.48).

When, however, the matrix $\mathbf{A}$ has *linearly dependent columns*, that is, it is *rank deficient*, the nullity of the matrix $\mathbf{A}$ is nonzero, and the result is that an infinite number of solutions can be found for minimizing the sum of error squares. In such a situation, the linear least-squares problem becomes quite involved, in that we now have the new problem of deciding which particular solution to adopt. We defer discussion of this issue to the latter part of the chapter. In the meantime, we assume that the data matrix $\mathbf{A}$ is of full column rank, so that the least-squares estimate $\hat{\mathbf{w}}$ has the unique value defined by Eq. (11.48).

11.8 PROPERTIES OF LEAST-SQUARES ESTIMATES

The method of least squares has a strong intuitive feel that is reinforced by several out-standing properties of the method, assuming that the data matrix $\mathbf{A}$ is known with *no* uncertainty. These properties, four in number, are described next (Miller, 1974; Goodwin and Payne, 1977).

Property 1. *The least-squares estimate $\hat{\mathbf{w}}$ is unbiased, provided that the measure-*ment error process $e_o(i)$ *has zero mean.*

From the multiple linear regression model of Fig. 11.1, we have [using the defini-tions of Eqs. (11.44) and (11.46)]

$$\mathbf{d} = \mathbf{A}\mathbf{w}_o + \boldsymbol{\epsilon}_o \tag{11.52}$$

Hence, substituting Eq. (11.52) into (11.48), we may express the least-squares estimate $\hat{\mathbf{w}}$ as

$$\begin{aligned}
\hat{\mathbf{w}} &= (\mathbf{A}^H\mathbf{A})^{-1}\mathbf{A}^H\mathbf{A}\mathbf{w}_o + (\mathbf{A}^H\mathbf{A})^{-1}\mathbf{A}^H\boldsymbol{\epsilon}_o \\
&= \mathbf{w}_o + (\mathbf{A}^H\mathbf{A})^{-1}\mathbf{A}^H\boldsymbol{\epsilon}_o
\end{aligned} \tag{11.53}$$

The matrix product $(\mathbf{A}^H\mathbf{A})^{-1}\mathbf{A}^H$ is a known quantity, since the data matrix $\mathbf{A}$ is completely defined by the set of given observations $u(1), u(2), \ldots, u(N)$; see Eq. (11.44). Hence, if the measurement error process $e_o(i)$ or, equivalently, the measurement error vector $\boldsymbol{\epsilon}_o$ has zero mean, we find by taking the expectation of both sides of Eq. (11.53) that the estimate $\hat{\mathbf{w}}$ is *unbiased*; that is,

$$E[\hat{\mathbf{w}}] = \mathbf{w}_o \tag{11.54}$$

Property 2. *When the measurement error process $e_o(i)$ is white with zero mean and* variance σ^2, *the covariance matrix of the least-squares estimate $\hat{\mathbf{w}}$ equals $\sigma^2\boldsymbol{\Phi}^{-1}$.*

Using the relation of Eq. (11.53), we find that the covariance matrix of the least-squares estimate $\hat{\mathbf{w}}$ equals

$$\begin{aligned}
\text{cov}[\hat{\mathbf{w}}] &= E[(\hat{\mathbf{w}} - \mathbf{w}_o)(\hat{\mathbf{w}} - \mathbf{w}_o)^H] \\
&= E[(\mathbf{A}^H\mathbf{A})^{-1}\mathbf{A}^H\boldsymbol{\epsilon}_o\boldsymbol{\epsilon}_o^H\mathbf{A}(\mathbf{A}^H\mathbf{A})^{-1}] \\
&= (\mathbf{A}^H\mathbf{A})^{-1}\mathbf{A}^H E[\boldsymbol{\epsilon}_o\boldsymbol{\epsilon}_o^H]\mathbf{A}(\mathbf{A}^H\mathbf{A})^{-1}
\end{aligned} \tag{11.55}$$

With the measurement error process $e_o(i)$ assumed to be white with zero mean and vari-ance σ^2, we have

$$E[\boldsymbol{\epsilon}_o\boldsymbol{\epsilon}_o^H] = \sigma^2\mathbf{I} \tag{11.56}$$

where $\mathbf{I}$ is the identity matrix. Hence, Eq. (11.55) reduces to

$$\text{cov}[\hat{\mathbf{w}}] = \sigma^2(\mathbf{A}^H\mathbf{A})^{-1}\mathbf{A}^H\mathbf{A}(\mathbf{A}^H\mathbf{A})^{-1}$$
$$= \sigma^2(\mathbf{A}^H\mathbf{A})^{-1} \tag{11.57}$$
$$= \sigma^2\mathbf{\Phi}^{-1}$$

which proves Property 2.

Property 3. *When the measurement error process $e_o(i)$ is white with zero mean, the least-squares estimate $\hat{\mathbf{w}}$ is the best linear unbiased estimate.*

Consider any linear unbiased estimator $\tilde{\mathbf{w}}$ that is defined by

$$\tilde{\mathbf{w}} = \mathbf{Bd} \tag{11.58}$$

where $\mathbf{B}$ is an M-by-$(N - M + 1)$ matrix. Substituting Eq. (11.52) into (11.58), we get

$$\tilde{\mathbf{w}} = \mathbf{BAw}_o + \mathbf{B\epsilon}_o \tag{11.59}$$

With the measurement error vector $\mathbf{\epsilon}_o$ having zero mean in accordance with Property 1, we find that the expected value of $\tilde{\mathbf{w}}$ equals

$$E[\tilde{\mathbf{w}}] = \mathbf{BAw}_o$$

For the linear estimator $\tilde{\mathbf{w}}$ to be unbiased, we therefore require that the matrix $\mathbf{B}$ satisfy the condition

$$\mathbf{BA} = \mathbf{I}$$

Accordingly, we may rewrite Eq. (11.59) as follows:

$$\tilde{\mathbf{w}} = \mathbf{w}_o + \mathbf{B\epsilon}_o$$

The covariance matrix of $\tilde{\mathbf{w}}$ equals

$$\text{cov}[\tilde{\mathbf{w}}] = E[(\tilde{\mathbf{w}} - \mathbf{w}_o)(\tilde{\mathbf{w}} - \mathbf{w}_o)^H]$$
$$= E[\mathbf{B\epsilon}_o\mathbf{\epsilon}_o^H\mathbf{B}^H]$$
$$= \sigma^2\mathbf{BB}^H \tag{11.60}$$

Here, we have made use of Eq. (11.56), which describes the assumption that the elements of the measurement error vector $\mathbf{\epsilon}_o$ are uncorrelated and have a common variance σ^2; that is, the measurement error process $e_o(i)$ is white. We next define a new matrix $\mathbf{\Psi}$ in terms of $\mathbf{B}$ as

$$\mathbf{\Psi} = \mathbf{B} - (\mathbf{A}^H\mathbf{A})^{-1}\mathbf{A}^H \tag{11.61}$$

Now we form the matrix product $\boldsymbol{\Psi}\boldsymbol{\Psi}^H$ and note that $\mathbf{BA} = \mathbf{I}$:

$$
\begin{aligned}
\boldsymbol{\Psi}\boldsymbol{\Psi}^H &= [\mathbf{B} - (\mathbf{A}^H\mathbf{A})^{-1}\mathbf{A}^H][\mathbf{B}^H - \mathbf{A}(\mathbf{A}^H\mathbf{A})^{-1}] \\
&= \mathbf{BB}^H - \mathbf{BA}(\mathbf{A}^H\mathbf{A})^{-1} - (\mathbf{A}^H\mathbf{A})^{-1}\mathbf{A}^H\mathbf{B}^H + (\mathbf{A}^H\mathbf{A})^{-1} \\
&= \mathbf{BB}^H - (\mathbf{A}^H\mathbf{A})^{-1}
\end{aligned}
$$

Since the diagonal elements of $\boldsymbol{\Psi}\boldsymbol{\Psi}^H$ are always nonnegative, we may use this relation to write

$$
\sigma^2 \operatorname{diag}[\mathbf{BB}^H] \geq \sigma^2 \operatorname{diag}[(\mathbf{A}^H\mathbf{A})^{-1}] \tag{11.62}
$$

The term $\sigma^2\mathbf{BB}^H$ equals the covariance matrix of the linear estimate $\tilde{\mathbf{w}}$, as in Eq. (11.60). From Property 2, we also know that the term $\sigma^2(\mathbf{A}^H\mathbf{A})^{-1}$ equals the covariance matrix of the least-squares estimate $\hat{\mathbf{w}}$. Thus, Eq. (11.62) shows that within the class of linear unbiased estimates the least-squares estimate $\hat{\mathbf{w}}$ is the "best" estimate of the unknown parameter vector $\mathbf{w}_o$ of the multiple linear regression model, in the sense that each element of $\hat{\mathbf{w}}$ has the smallest possible variance. Accordingly, when the measurement error process e_o contained in this model is white with zero mean, the least-squares estimate $\hat{\mathbf{w}}$ is the *best linear unbiased estimate* (BLUE).

Thus far we have not made any assumption about the statistical distribution of the measurement error process $e_o(i)$ other than that it is a zero mean white-noise process. By making the further assumption that the process $e_o(i)$ is *Gaussian* distributed, we obtain a stronger result on the optimality of the linear least-squares estimate, as discussed next.

Property 4. *When the measurement error process $e_o(i)$ is white and Gaussian, with zero mean, the least-squares estimate $\hat{\mathbf{w}}$ achieves the Cramér–Rao lower bound for unbiased estimates.*

Let $f_E(\boldsymbol{\epsilon}_o)$ denote the joint probability density function of the measurement error vector $\boldsymbol{\epsilon}_o$. Let $\hat{\mathbf{w}}$ denote any unbiased estimate of the unknown parameter vector $\hat{\mathbf{w}}_o$ of the multiple linear regression model. Then the covariance matrix of $\hat{\mathbf{w}}$ satisfies the inequality

$$
\operatorname{cov}[\hat{\mathbf{w}}] \geq \mathbf{J}^{-1} \tag{11.63}
$$

where

$$
\operatorname{cov}[\hat{\mathbf{w}}] = E[(\hat{\mathbf{w}} - \mathbf{w}_o)(\hat{\mathbf{w}} - \mathbf{w}_o)^H] \tag{11.64}
$$

The matrix $\mathbf{J}$ is called *Fisher's information matrix*; it is defined by[2]

$$
\mathbf{J} = E\!\left[\left(\frac{\partial l}{\partial \mathbf{w}_o^*}\right)\!\left(\frac{\partial l}{\partial \mathbf{w}_o^T}\right)\right] \tag{11.65}
$$

where l is the *log-likelihood function*, that is, the natural logarithm of the joint probability density of $\boldsymbol{\epsilon}_o$, as shown by

$$
l = \ln f_E(\boldsymbol{\epsilon}_o) \tag{11.66}
$$

---

[2]Fisher's information matrix is discussed in Appendix D for the case of real parameters.

Since the measurement error process $e_o(n)$ is white, the elements of the vector $\boldsymbol{\epsilon}_o$ are uncorrelated. Furthermore, since the process $e_o(n)$ is Gaussian, the elements of $\boldsymbol{\epsilon}_o$ are statistically independent. With $e_o(i)$ assumed to be complex with a mean of zero and variance σ^2, we have (see Section 2.11)

$$f_E(\boldsymbol{\epsilon}_o) = \frac{1}{(\pi\sigma^2)^{(N-M+1)}} \exp\left[-\frac{1}{\sigma^2} \sum_{i=M}^{N} |e_o(i)|^2\right] \tag{11.67}$$

The log-likelihood function is therefore

$$l = F - \frac{1}{\sigma^2} \sum_{i=M}^{N} |e_o(i)|^2$$

$$= F - \frac{1}{\sigma^2} \boldsymbol{\epsilon}_o^H \boldsymbol{\epsilon}_o \tag{11.68}$$

where F is a constant defined by

$$F = -(N - M + 1) \ln(\pi\sigma^2)$$

From Eq. (11.52), we have

$$\boldsymbol{\epsilon}_o = \mathbf{d} - \mathbf{A}\mathbf{w}_o$$

Using this relation in Eq. (11.68), we may rewrite l in terms of $\mathbf{w}_o$ as

$$l = F - \frac{1}{\sigma^2} \mathbf{d}^H\mathbf{d} + \frac{1}{\sigma^2} \mathbf{w}_o^H\mathbf{A}^H\mathbf{d} + \frac{1}{\sigma^2} \mathbf{d}^H\mathbf{A}\mathbf{w}_o - \frac{1}{\sigma^2} \mathbf{w}_o^H \mathbf{A}^H\mathbf{A}\mathbf{w}_o \tag{11.69}$$

Differentiating the real-valued log-likelihood function l with respect to the complex-valued unknown parameter vector $\mathbf{w}_o$ in accordance with the notation described in Appendix B, we get

$$\frac{\partial l}{\partial \mathbf{w}_o^*} = \frac{1}{\sigma^2} \mathbf{A}^H(\mathbf{d} - \mathbf{A}\mathbf{w}_o)$$

$$= \frac{1}{\sigma^2} \mathbf{A}^H \boldsymbol{\epsilon}_o \tag{11.70}$$

Thus, substituting Eq. (11.70) into (11.65) yields Fisher's information matrix for the problem at hand as

$$\mathbf{J} = \frac{1}{\sigma^4} E[\mathbf{A}^H \boldsymbol{\epsilon}_o \boldsymbol{\epsilon}_o^H \mathbf{A}]$$

$$= \frac{1}{\sigma^4} \mathbf{A}^H E[\boldsymbol{\epsilon}_o \boldsymbol{\epsilon}_o^H]\mathbf{A} \tag{11.71}$$

$$= \frac{1}{\sigma^2} \mathbf{A}^H\mathbf{A}$$

$$= \frac{1}{\sigma^2} \boldsymbol{\Phi}$$

where, in the third line, we have made use of Eq. (11.56) describing the assumption that the measurement error process $e_o(i)$ is white with zero mean and variance σ^2. Accordingly, the use of Eq. (11.63) shows that the covariance matrix of the unbiased estimate $\tilde{\mathbf{w}}$ satisfies the inequality

$$\text{cov}[\tilde{\mathbf{w}}] \geq \sigma^2 \mathbf{\Phi}^{-1} \tag{11.72}$$

However, from Property 2, we know that $\sigma^2 \mathbf{\Phi}^{-1}$ equals the covariance matrix of the least-squares estimate $\hat{\mathbf{w}}$. Accordingly, $\hat{\mathbf{w}}$ achieves the Cramér–Rao lower bound. Moreover, using Property 1, we conclude that when the measurement error process $e_o(i)$ is a zero-mean white Gaussian noise process, the least-squares estimate $\hat{\mathbf{w}}$ is a *minimum variance unbiased estimate* (MVUE).

11.9 PARAMETRIC SPECTRUM ESTIMATION

The method of least squares is particularly well suited for solving *parametric spectrum estimation* problems. In this section we study this important application of the method of least squares. We first consider the case of *autoregressive (AR) spectrum estimation*, assuming the use of an AR model of *known order*. From the discussion of linear prediction presented in Chapter 6, we know that there is a one-to-one correspondence between the coefficients of a prediction-error filter and those of an AR model of similar order. Next, we consider the case of *minimum variance distortionless response (MVDR) spectrum estimation*. In this second case, we have a constrained optimization problem to solve.

AR Spectrum Estimation

The specific estimation procedure described herein relies on the combined use of *forward and backward linear prediction (FBLP)*.[3] Since the method of least squares is basically a *block* estimation procedure, we may therefore view the FBLP algorithm as an alternative to the Burg algorithm (described in Section 6.15) for solving AR modeling problems. There are, however, three basic differences between the FBLP and the Burg algorithms:

1. The FBLP algorithm estimates the coefficients of a *transversal-equivalent model* for the input data, whereas the Burg algorithm estimates the reflection coefficients of a *lattice-equivalent model*.

2. In the method of least squares, and therefore the FBLP algorithm, no assumptions are made concerning the statistics of the input data. The Burg algorithm, on the other hand, exploits the decoupling property of a multistage lattice predictor, which, in turn, assumes wide-sense stationarity of the input data. Accordingly, the

---

[3]The first application of the FBLP method to the design of a linear predictor that has a transversal filter structure, in accordance with the method of least squares, was developed independently by Ulrych and Clayton (1976) and Nuttall (1976).

FBLP algorithm does not suffer from some of the anomalies that are known to arise in the application of the Burg algorithm.[4]

3. The Burg algorithm yields a minimum-phase solution in the sense that the reflection coefficients of the equivalent lattice predictor have a magnitude less than or equal to unity. The FBLP algorithm, on the other hand, does *not* guarantee such a solution. In spectrum estimation, however, the lack of a minimum-phase solution is of no particular concern.

Consider then the *forward linear predictor*, shown in Fig. 11.5(a). The tap weights of the predictor are denoted by $\hat{w}_1, \hat{w}_2, \ldots, \hat{w}_M$ and the tap inputs by $u(i-1), u(i-2), \ldots, u(i-M)$, respectively. The forward prediction error, denoted by $f_M(i)$, equals

$$f_M(i) = u(i) - \sum_{k=1}^{M} \hat{w}_k^* u(i-k) \tag{11.73}$$

The first term, $u(i)$, represents the desired response. The convolution sum, constituting the second term, represents the predictor output; it consists of the sum of scalar inner products. Using matrix notation, we may also express the forward prediction error as

$$f_M(i) = u(i) - \mathbf{w}^H \mathbf{u}(i-1) \tag{11.74}$$

where $\hat{\mathbf{w}}$ is the M-by-1 tap-weight vector of the predictor:

$$\hat{\mathbf{w}} = [\hat{w}_1, \hat{w}_2, \ldots, \hat{w}_M]^T$$

and $\mathbf{u}(i-1)$ is the corresponding tap-input vector:

$$\mathbf{u}(i-1) = [u(i-1), u(i-2), \ldots, u(i-M)]^T$$

Consider next Fig. 11.5(b), which depicts the reconfiguration of the predictor so that it performs backward linear prediction. We have *purposely* retained $\hat{w}_1, \hat{w}_2, \ldots, \hat{w}_M$ as the tap weights of the predictor. The change in the format of the tap inputs is inspired by the discussion presented in Section 6.2 on backward linear prediction and its relation to forward linear prediction for the case of wide-sense stationary inputs. In particular, the tap inputs in the predictor of Fig. 11.5(b) differ from those of the forward linear predictor of Fig. 11.5(a) in two respects:

1. The tap inputs in Fig. 11.5(b) are *time reversed*, in that they appear from right to left whereas in Fig. 11.5(a) they appear from left to right.

---

[4]For example, when the Burg algorithm is used to estimate the frequency of an unknown sine wave in additive noise, under certain conditions a phenomenon commonly referred to as *spectral line splitting* may occur. This phenomenon refers to the occurrence of two (or more) closely spaced spectral peaks where there should only be a single peak; for a discussion of spectral line splitting, see Marple (1987), Kay (1988), and Haykin (1989); the original reference is Fougere et al. (1976). This anomaly, however, does not arise in the application of the FBLP algorithm.

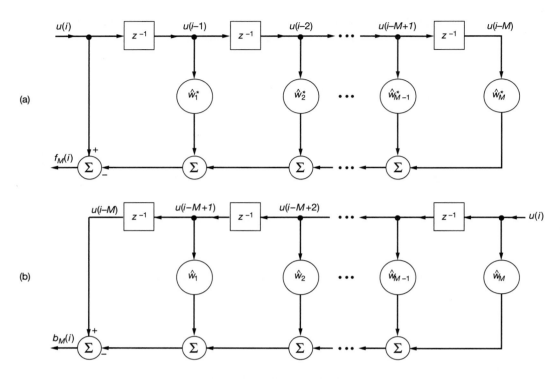

Figure 11.5 (a) Forward linear predictor; (b) reconfiguration of the predictor so as to perform backward linear prediction.

2. With $u(i), u(i-1), \ldots, u(i-M+1)$ used as tap inputs, the structure of Fig. 11.5(b) produces a linear prediction of $u(i-M)$. In other words, it performs backward linear prediction. Denoting the backward prediction error by $b_M(i)$, we may thus express it as

$$b_M(i) = u(i-M) - \sum_{k=1}^{M} \hat{w}_k u(i-M+k) \tag{11.75}$$

where the first term represents the desired response and the second term is the predictor output. Equivalently, in terms of matrix notation, we may write

$$b_M(i) = u(i-M) - \mathbf{u}^{BT}(i)\hat{\mathbf{w}} \tag{11.76}$$

where $\mathbf{u}^B(i)$ is the *time-reversed tap-input vector*:

$$\mathbf{u}^{BT}(i) = [u(i-M+1), \ldots, u(i-1), u(i)]$$

Let $\mathcal{E}_M$ denote the *minimum value of the forward-backward prediction-error energy.* In accordance with the method of least squares, we may therefore write

$$\mathcal{E}_M = \sum_{i=M+1}^{N} [|f_M(i)|^2 + |b_M(i)|^2] \tag{11.77}$$

where the subscript M signifies the order of the predictor or that of the AR model. The lower limit on the time index i equals $M + 1$ so as to ensure that the forward and backward prediction errors are formed only when all the tap inputs of interest assume nonzero values. In particular, we may make two observations:

1. The variable $u(i - M)$, representing the last tap input in the forward prediction of Fig. 11.5(a), assumes a nonzero value for the first time when $i = M + 1$.
2. The variable $u(i - M)$, playing the role of desired response in the backward predictor of Fig. 11.5(b), also assumes a nonzero value for the first time when $i = M + 1$.

Thus, by choosing $(M + 1)$ as the lower limit on i and N as the upper limit, as in Eq. (11.77), we make no assumptions about the data outside the interval $[1, N]$, as required by the covariance method.

Let $\mathbf{A}$ denote the $2(N - M)$-by-M *data matrix*, whose Hermitian transpose is defined by

$$\mathbf{A}^H = \begin{bmatrix} u(M) & \cdots & u(N-1) & u^*(2) & \cdots & u^*(N-M+1) \\ u(M-1) & \cdots & u(N-2) & u^*(3) & \cdots & u^*(N-M+2) \\ \cdot & \cdot & \cdot & \cdot & \cdot & \cdot \\ \cdot & \cdot & \cdot & \cdot & \cdot & \cdot \\ \cdot & \cdot & \cdot & \cdot & \cdot & \cdot \\ u(1) & \cdots & u(N-M) & u^*(M+1) & \cdots & u^*(N) \end{bmatrix}$$

$$\underbrace{\qquad\qquad\qquad}_{\text{forward half}} \qquad \underbrace{\qquad\qquad\qquad}_{\text{backward half}} \tag{11.78}$$

The elements constituting the left half of matrix $\mathbf{A}^H$ represent the various sets of tap inputs used to make a total of $(N - M)$ *forward* linear predictions. The complex-conjugated elements constituting the right half of matrix $\mathbf{A}^H$ represent the corresponding sets of tap inputs used to make a total of $(N - M)$ *backward* linear predictions. Note that as we move from one column to the next in the forward or backward half in Eq. (11.78), we drop a sample, add a new one, and reorder the samples.

Let $\mathbf{d}$ denote the $2(N - M)$-by-1 *desired data vector,* defined in a manner corresponding to that shown in Eq. (11.78):

$$\mathbf{d}^H = [\underbrace{u(M + 1), \ldots, u(N)}_{\text{forward half}}, \underbrace{u^*(1), \ldots, u^*(N - M)}_{\text{backward half}}] \tag{11.79}$$

Each element in the left half of the vector $\mathbf{d}^H$ represents a desired response for forward linear prediction. Each complex-conjugated element in the right half represents a desired response for backward linear prediction.

The FBLP method is a product of the method of least squares; it is therefore described by the system of normal equations [see Eq. (11.48)]

$$\mathbf{A}^H \mathbf{A} \hat{\mathbf{w}} = \mathbf{A}^H \mathbf{d} \tag{11.80}$$

The resulting minimum value of the forward-backward prediction error energy equals [see Eq. (11.49)]

$$\mathscr{E}_{\min} = \mathbf{d}^H \mathbf{d} - \mathbf{d}^H \mathbf{A} (\mathbf{A}^H \mathbf{A})^{-1} \mathbf{A}^H \mathbf{d} \tag{11.81}$$

The data matrix $\mathbf{A}$ and the desired data vector $\mathbf{d}$ are defined by Eqs. (11.78) and (11.79), respectively.

We may combine Eqs. (11.80) and (11.81) into a single matrix relation, as shown by

$$\begin{bmatrix} \mathbf{d}^H \mathbf{d} & \mathbf{d}^H \mathbf{A} \\ \mathbf{A}^H \mathbf{d} & \mathbf{A}^H \mathbf{A} \end{bmatrix} \begin{bmatrix} 1 \\ -\hat{\mathbf{w}} \end{bmatrix} = \begin{bmatrix} \mathscr{E}_{\min} \\ \mathbf{0} \end{bmatrix} \tag{11.82}$$

where $\mathbf{0}$ is the M-by-1 null vector. Equation (11.82) is the matrix form of the *augmented normal equations for FBLP*. Define the $(M + 1)$-by-$(M + 1)$ *augmented correlation matrix*:

$$\mathbf{\Phi} = \begin{bmatrix} \mathbf{d}^H \mathbf{d} & \mathbf{d}^H \mathbf{A} \\ \mathbf{A}^H \mathbf{d} & \mathbf{A}^H \mathbf{A} \end{bmatrix} \tag{11.83}$$

The $\mathbf{\Phi}$ in Eq. (11.83) is an $(M + 1)$-by-$(M + 1)$ matrix; it is *not* to be confused with the $\mathbf{\Phi}$ in Eq. (11.45) that is an M-by-M matrix. Define the $(M + 1)$-by-1 tap-weight vector of the *prediction-error filter of order M*:

$$\hat{\mathbf{a}} = \begin{bmatrix} 1 \\ -\hat{\mathbf{w}} \end{bmatrix} \tag{11.84}$$

Figure 11.6 shows the transversal structure of the prediction-error filter, where a_0, a_1, . . . , a_M denote the tap weights[5] and $a_0 = 1$. Then

$$\mathbf{\Phi} \hat{\mathbf{a}} = \begin{bmatrix} \mathscr{E}_{\min} \\ \mathbf{0} \end{bmatrix} \tag{11.85}$$

The augmented correlation matrix $\mathbf{\Phi}$ is *Hermitian persymmetric*; that is, the individual elements of the matrix $\mathbf{\Phi}$ satisfy two conditions:

$$\phi(k, t) = \phi^*(t, k) \qquad 0 \le (t, k) \le M \tag{11.86}$$

$$\phi(M - k, M - t) = \phi^*(k, t), \qquad 0 \le (t, k) \le M \tag{11.87}$$

---

[5]The subscripts assigned to the tap weights in the prediction-error filter of Fig. 11.6 do not include a direct reference to the prediction order M, unlike the terminology used in Chapter 6. The reason for this simplification is that, in the material presented here, there is no order update to be considered.

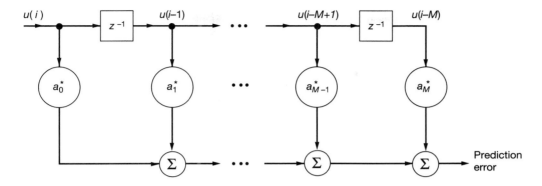

Figure 11.6   Forward prediction-error filter.

The property described in Eq. (11.87) is unique to a correlation matrix that is obtained by time averaging of the input data in the forward as well as backward direction; see the data matrix **A** and the desired data vector **d** defined in Eqs. (11.78) and (11.79), respectively. The matrix $\mathbf{\Phi}$ has another property: it is composed of the sum of two Toeplitz matrix products. The special Toeplitz structure of the matrix $\mathbf{\Phi}$ has been exploited in the development of fast recursive algorithms[6] for the efficient solution of the augmented normal equations (11.85).

Starting with the time series $u(i)$, $1 \leq i \leq N$, the FBLP algorithm is used to compute the tap-weight vector $\hat{\mathbf{w}}$ of a forward linear predictor or, equivalently, the tap-weight vector $\hat{\mathbf{a}}$ of the corresponding prediction-error filter. The vector $\hat{\mathbf{a}}$ represents an estimate of the coefficient vector of an *autoregressive (AR) model* used to fit the time series $u(i)$. Similarly, the minimum mean-squared error $\mathscr{E}_{\min}$, except for a scaling factor, represents an estimate of the white-noise variance σ^2 in the AR model. We may thus use Eq. (6.101) to formulate an *estimate of the AR spectrum* as follows:

$$\hat{S}_{\text{AR}}(\omega) = \frac{\mathscr{E}_{\min}}{\left| 1 + \displaystyle\sum_{k=1}^{M} \hat{a}_k^* e^{-j\omega k} \right|^2} \tag{11.88}$$

where the $\hat{a}_k$ are the elements of the vector $\hat{\mathbf{a}}$; the leading element $\hat{a}_0$ of the vector $\hat{\mathbf{a}}$ is equal to unity, by definition. We may also express $\hat{S}_{\text{AR}}(\omega)$ as

$$\hat{S}_{\text{AR}}(\omega) = \frac{\mathscr{E}_{\min}}{|\hat{\mathbf{a}}^H \mathbf{s}(\omega)|^2} \tag{11.89}$$

---

[6]The correlation matrix $\mathbf{\Phi}$ of Eq. (11.83) does *not* possess a Toeplitz structure. Accordingly, we cannot use the Levinson recursion to develop a fast solution of the augmented normal equations (11.85), as was the case with the augmented Wiener–Hopf equations for stationary inputs. However, Marple (1980, 1981) describes fast recursive algorithms for the efficient solution of the augmented normal equations (11.85). Marple exploits the special Toeplitz structure of the correlation matrix $\mathbf{\Phi}$. The computational complexity of Marple's fast algorithm is proportional to M^2. When the predictor order M is large, the use of Marple's algorithm results in significant savings in computation.

where $\mathbf{s}(\omega)$ is a *variable-frequency vector* or *frequency scanning vector*:

$$\mathbf{s}(\omega) = [1, e^{-j\omega}, \ldots, e^{-j\omega M}]^T, \qquad -\pi < \omega \leq \pi \tag{11.90}$$

Intuitively, the model order M should be as large as possible in order to have a large aperture for the predictor. However, in applying the FBLP algorithm the use of large values of M gives rise to spurious spectral peaks in the AR spectrum. For best performance of the FBLP algorithm, Lang and McClellan (1980) suggest the value

$$M \approx \frac{N}{3} \tag{11.91}$$

where N is the data length.

MVDR Spectrum Estimation

In the method of least squares, as described up to this point in our discussion, there are no *constraints* imposed on the solution. In certain applications, however, the use of such an approach may be unsatisfactory, in which case we may resort to a *constrained* version of the method of least squares. For example, in *adaptive beamforming* that involves spatial processing, we may wish to *minimize the variance (i.e., average power) of the beamformer output while a distortionless response is maintained along the direction of a target signal of interest*. Correspondingly, in the temporal counterpart to this problem, we may be required to *minimize the average power of the spectrum estimator, while a distortionless response is maintained at a particular frequency*. In such applications, the resulting solution is referred to as a *minimum-variance distortionless response (MVDR) estimator* for obvious reasons. To be consistent with the material presented heretofore, we will formulate the temporal version of the MVDR algorithm.

Consider then a linear transversal filter, as depicted in Fig. 11.7. Let the filter output be denoted by $y(i)$. This output is in response to the tap inputs $u(i)$, $u(i-1)$, . . . , $u(i-M)$. Specifically, we have

$$y(i) = \sum_{t=0}^{M} a_t^* u(i-t) \tag{11.92}$$

where $a_0, a_1, \ldots, a_m$ are the transversal filter coefficients. Note, however, that unlike the prediction-error filter of Fig. 11.6, there is no restriction on the filter coefficient a_0; the only reason for using the same terminology as in Fig. 11.6 is because of a desire to be consistent. The requirement is to minimize the *output energy* (assuming the use of the covariance method of data windowing):

$$\mathscr{E}_{\text{out}} = \sum_{i=M+1}^{N} |y(i)|^2$$

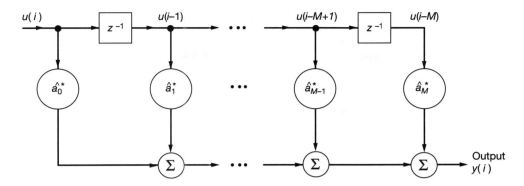

Figure 11.7   Transversal filter.

subject to the *constraint*

$$\sum_{k=0}^{M} a_k^* e^{-jk\omega_0} = 1 \tag{11.93}$$

where ω_0 is an angular frequency of special interest. As in the conventional method of least squares, the filter coefficients $a_0, a_1, \ldots, a_M$ are held constant for the observation interval $1 \le i \le N$, where N is the total data length.

To solve this *constrained minimization* problem, we use the *method of Lagrange multipliers.*[7] Specifically, we define the *constrained cost function*

$$\mathscr{E} = \underbrace{\sum_{i=M+1}^{N} |y(i)|^2}_{\text{output energy}} + \underbrace{\lambda\left(\sum_{k=0}^{M} a_k^* e^{-jk\omega_0} - 1\right)}_{\text{linear constraints}} \tag{11.94}$$

output energy   linear constraints

where λ is a *complex Lagrange multiplier*. Note that in the constrained approach described herein, there is *no* desired response; in place of it, however, we have a set of linear constraints. Note also that in the absence of a desired response and therefore no frame of reference, the principle of orthogonality loses its meaning in this new setting.

To solve for the optimum values of the filter coefficients, we first determine the gradient vector $\nabla\mathscr{E}$ and then set it equal to zero. Thus, proceeding in a manner similar to that described in Section 11.3, we find that the kth element of the gradient vector for the constrained cost function of Eq. (11.94) is

$$\nabla_k \mathscr{E} = 2\sum_{i=M+1}^{N} u(i-k)y^*(i) + \lambda^* e^{-jk\omega_0} \tag{11.95}$$

---

[7]The method of Lagrange multipliers is described in Appendix C.

Next, substituting Eq. (11.92) in (11.95), and rearranging terms, we get

$$\nabla_k \mathcal{E} = 2 \sum_{t=0}^{M} a_t \sum_{i=M+1}^{N} u(i-k)u^*(i-t) + \lambda^* e^{-jk\omega_0}$$

$$= 2 \sum_{t=0}^{M} a_t \phi(t, k) + \lambda^* e^{-jk\omega_0} \tag{11.96}$$

where, in the first term of the second line, we have made use of the definition of Eq. (11.29) for the time-averaged autocorrelation function $\phi(t, k)$ of the tap inputs. To minimize the constrained cost function $\mathcal{E}$, we set

$$\nabla_k \mathcal{E} = 0, \qquad k = 0, 1, \ldots, M \tag{11.97}$$

Accordingly, we find from Eq. (11.96) that the tap-weights of the *optimized* transversal filter satisfy the following system of $M + 1$ simultaneous equations:

$$\sum_{t=0}^{M} \hat{a}_t \phi(t, k) = -\frac{1}{2} \lambda^* e^{-jk\omega_0}, \qquad k = 0, 1, \ldots, M \tag{11.98}$$

Using matrix notation, we may rewrite this system of equations in the compact form

$$\mathbf{\Phi}\hat{\mathbf{a}} = -\frac{1}{2} \lambda^* \mathbf{s}(\omega_0) \tag{11.99}$$

where $\mathbf{\Phi}$ is the $(M + 1)$-by-$(M + 1)$ time-averaged correlation matrix of the tap inputs; $\hat{\mathbf{a}}$ is the $(M\ 1)$-by-1 vector of optimum tap weights; and $\mathbf{s}(\omega_0)$ is the $(M + 1)$-by-1 *fixed frequency vector*:

$$\mathbf{s}(\omega_0) = [1, e^{-j\omega_0}, \ldots, e^{-jM\omega_0}]^T \tag{11.100}$$

Assuming $\mathbf{\Phi}$ is nonsingular and therefore its inverse $\mathbf{\Phi}^{-1}$ exists, we may solve Eq. (11.99) for the optimum tap-weight vector:

$$\hat{\mathbf{a}} = -\frac{1}{2} \lambda^* \mathbf{\Phi}^{-1} \mathbf{s}(\omega_0) \tag{11.101}$$

There only remains the problem of evaluating the Lagrange multiplier λ. To solve for λ, we use the linear constraint in Eq. (11.93) for the optimized transversal filter, written in matrix form as

$$\hat{\mathbf{a}}^H \mathbf{s}(\omega_0) = 1 \tag{11.102}$$

Hence, evaluating the *inner product* of the vector $\mathbf{s}_0$ and the vector $\hat{\mathbf{a}}$ in Eq. (11.101), setting the result equal to 1 and solving for λ, we get

$$\lambda^* = -\frac{2}{\mathbf{s}^H(\omega_0) \mathbf{\Phi}^{-1} \mathbf{s}(\omega_0)} \tag{11.103}$$

Finally, substituting this value of λ in Eq. (11.101), we get the MVDR solution:[8]

$$\hat{\mathbf{a}} = \frac{\boldsymbol{\Phi}^{-1}\mathbf{s}(\omega_0)}{\mathbf{s}^H(\omega_0)\boldsymbol{\Phi}^{-1}\mathbf{s}(\omega_0)} \tag{11.104}$$

Thus, given the time-averaged correlation matrix $\boldsymbol{\Phi}$ of the tap inputs and the frequency vector $\mathbf{s}(\omega_0)$, we may use the *MVDR formula* of (11.104) to compute the optimum tap-weight vector $\hat{\mathbf{a}}$ of the transversal filter in Fig. 11.7.

Let $\hat{S}_{\mathrm{MVDR}}(\omega_0)$ denote the minimum value of the output energy $\mathscr{E}_{\mathrm{out}}$, which results when the MVDR solution $\hat{\mathbf{a}}$ of Eq. (11.104) is used for the tap-weight vector under the condition that the response is tuned to the angular frequency ω_0. We may then write

$$\hat{S}_{\mathrm{MVDR}}(\omega_0) = \hat{\mathbf{a}}^H \boldsymbol{\Phi}\, \hat{\mathbf{a}} \tag{11.105}$$

Substituting Eq. (11.104) in (11.105), and then simplifying the result, we finally get

$$\hat{S}_{\mathrm{MVDR}}(\omega_0) = \frac{1}{\mathbf{s}^H(\omega_0)\boldsymbol{\Phi}^{-1}\mathbf{s}(\omega_0)} \tag{11.106}$$

Equation (11.106) may be given a more general interpretation. Suppose that we define a frequency-scanning vector $\mathbf{s}(\omega)$ as in Eq. (11.90), where the angular frequency ω is now variable in the interval $(-\pi, \pi]$. For each ω, let the tap-weight vector of the transversal filter be assigned a corresponding MVDR estimate. The output energy of the optimized filter then becomes a function of ω. Let $\hat{S}_{\mathrm{MVDR}}(\omega)$ describe this functional dependence, and so we may write[9]

$$\hat{S}_{\mathrm{MVDR}}(\omega) = \frac{1}{\mathbf{s}^H(\omega)\boldsymbol{\Phi}^{-1}\mathbf{s}(\omega)} \tag{11.107}$$

We refer to Eq. (11.107) as the *MVDR spectrum estimate*, and the solution given in Eq. (11.104) as the *MVDR estimate* of the tap-weight vector. Note that at any ω, power due to other frequencies is minimized. Hence, the MVDR spectrum computed in accordance with Eq. (11.107) exhibits relatively sharp peaks.

The MVDR spectrum and AR spectrum are commonly referred to as *super-resolution* or *high-resolution spectra*, in the sense that they both exhibit sub-Rayleigh resolution as power spectrum estimators. For the numerical computation of these spectra, and linear least-squares solutions in general, the recommended procedure is to use a technique known as singular value decomposition, which is considered next.

---

[8]Equation (11.104) is of the same form as that of Eq. (5.97), except for the use of the time-averaged correlation matrix $\boldsymbol{\Phi}$ in place of the ensemble-averaged correlation matrix $\mathbf{R}$, and the use of symbol $\mathbf{a}$ in place of $\mathbf{w}_o$ for the tap-weight vector.

[9]The method for computing the spectrum in Eq. (11.107) is also referred to in the literature as *Capon's method* (Capon, 1969). The term "minimum-variance distortionless response" owes its origin to Owsley (1984).

11.10 SINGULAR VALUE DECOMPOSITION

The analytic power of *singular-value decomposition* lies in the fact that it applies to square as well as rectangular matrices, be they real or complex. As such, it is extremely well suited for the numerical solution of linear least-squares problems in the sense that *it can be applied directly to the data matrix.*

In Sections 11.5 and 11.7 we described two different forms of the normal equations for computing the linear least-squares solution:

1. The form given in Eq. (11.36), namely,

$$\hat{\mathbf{w}} = \mathbf{\Phi}^{-1}\mathbf{z}$$

where $\hat{\mathbf{w}}$ is the least-squares estimate of the tap-weight vector of a transversal filter model, $\mathbf{\Phi}$ is the time-averaged correlation matrix of the tap inputs, and $\mathbf{z}$ is the time-averaged cross-correlation vector between the tap inputs and some desired response.

2. The form given in Eq. (11.48) *directly in terms of data matrices*, namely,

$$\hat{\mathbf{w}} = (\mathbf{A}^H\mathbf{A})^{-1}\mathbf{A}^H\mathbf{d}$$

where $\mathbf{A}$ is the data matrix representing the time evolution of the tap input vectors, and $\mathbf{d}$ is the desired data vector representing the time evolution of the desired response.

These two forms are indeed mathematically equivalent. Yet they point to different computational procedures for evaluating the least-squares solution $\hat{\mathbf{w}}$. Equation (11.36) requires knowledge of the time-averaged correlation matrix $\mathbf{\Phi}$ that involves computing the product of $\mathbf{A}^H$ and $\mathbf{A}$. On the other hand, in Eq. (11.48) the entire term $(\mathbf{A}^H\mathbf{A})^{-1}\mathbf{A}$ can be interpreted, in terms of the singular-value decomposition applied directly to the data matrix $\mathbf{A}$, in such a way that the solution computed for $\hat{\mathbf{w}}$ has *twice the number of correct digits* as the solution computed by means of Eq. (11.36) for the same numerical precision. To be specific, define the matrix

$$\mathbf{A}^+ = (\mathbf{A}^H\mathbf{A})^{-1}\mathbf{A}^H \tag{11.108}$$

Then we may rewrite Eq. (11.36) simply as

$$\hat{\mathbf{w}} = \mathbf{A}^+\mathbf{d} \tag{11.109}$$

The matrix $\mathbf{A}^+$ is called the *pseudoinverse* or the *Moore-Penrose generalized inverse* of the matrix $\mathbf{A}$ (Stewart, 1973; Golub and Van Loan 1989). Equation (11.109) represents a convenient way of saying that "the vector $\hat{\mathbf{w}}$ solves the linear least-squares problem." Indeed, it was with the simple format of Eq. (11.109) in mind and also the desire to be consistent with definitions of the time-averaged correlation matrix $\mathbf{\Phi}$ and the cross-correlation vector $\mathbf{z}$ used in Section 11.5 that we defined the data matrix $\mathbf{A}$ and the desired data vector $\mathbf{d}$ in the manner shown in Eqs. (11.44) and (11.46), respectively.

In practice, we often find that the data matrix **A** contains linearly dependent columns. Consequently, we are faced with a new situation where we now have to decide on which of an infinite number of possible solutions to the least-squares problem to work with. This issue can indeed be resolved by using the singular-value decomposition technique as described in Section 11.12, even when null(**A**) $\neq \emptyset$, where $\emptyset$ denotes the *null set*.

The Singular-Value Decomposition Theorem

The *singular-value decomposition (SVD)* of a matrix is one of the most elegant algorithms in numerical algebra for providing quantitative information about the structure of a system of linear equations (Klema and Laub, 1980). The system of linear equations that is of specific interest to us is described by

$$A\hat{w} = d \tag{11.110}$$

in which **A** is a K-by-M matrix, **d** is a K-by-1 vector, and $\hat{w}$ (representing an estimate of the unknown parameter vector) is an M-by-1 vector. Equation (11.110) represents a simplified matrix form of the normal equations. In particular, premultiplication of both sides of the equation by the vector A^H yields the normal equations for the least-squares weight vector $\hat{w}$.

Given the data matrix **A**, there are two unitary matrices **V** and **U**, such that we may write

$$U^H A V = \begin{bmatrix} \Sigma & 0 \\ 0 & 0 \end{bmatrix} \tag{11.111}$$

where Σ is a diagonal matrix:

$$\Sigma = \text{diag}(\sigma_1, \sigma_2, \ldots, \sigma_W) \tag{11.112}$$

The σ's are ordered as $\sigma_1 \geq \sigma_2 \geq \ldots \geq \sigma_W > 0$. Equation (11.111) is a mathematical statement of the *singular-value decomposition theorem*. This theorem is also referred to as the *Autonne–Eckart–Young theorem* in recognition of its originators.[10]

Figure 11.8 presents a diagrammatic interpretation of the singular value decomposition theorem, as described in Eq. (11.111). In this diagram we have assumed that the number of rows K contained in the data matrix **A** is larger than the number of columns M, and that the number of nonzero singular values W is less than M. We may of course restructure the diagrammatic interpretation of the singular value decomposition theorem by expressing the data matrix in terms of the unitary matrices **U** and **V**, and the diagonal matrix Σ; this is left as an exercise for the reader.

---

[10]According to DeMoor and Golub (1989), the singular-value decomposition was introduced in its general form by Autonne in 1902, and an important characterization of it was described by Eckart and Young (1936). For additional notes on the history of the singular-value decomposition, see Klema and Laub (1980).

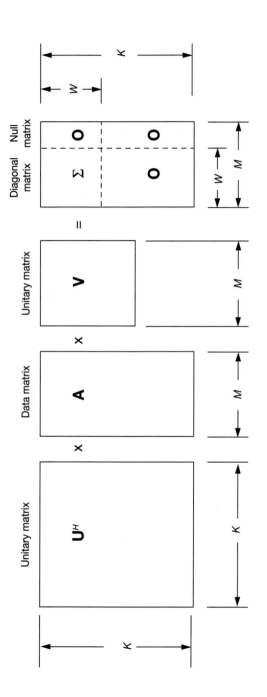

Figure 11.8 Diagrammatic interpretation of the singular value decomposition theorem.

The subscript W in Eq. (11.112) is the *rank* of matrix $\mathbf{A}$, written as rank($\mathbf{A}$); it is defined as the number of linearly independent columns in the matrix $\mathbf{A}$. Note that we always have rank($\mathbf{A}^H$) = rank($\mathbf{A}$).

Since it is possible to have $K > M$ or $K < M$, there are two distinct cases to be considered. We prove the singular-value decomposition theorem by considering both cases, independently of each other. For the case when $K > M$, we have an *overdetermined system* in that we have more equations than unknowns. On the other hand, when $K < M$, we have an *underdetermined system* in that we have more unknowns than equations. In the sequel, we consider these two cases in turn.

Case 1: Overdetermined System.   For the case when $K > M$, we form the M-by-M matrix $\mathbf{A}^H\mathbf{A}$ by premultiplying the matrix $\mathbf{A}$ by its Hermitian transpose $\mathbf{A}^H$. Since the matrix $\mathbf{A}^H\mathbf{A}$ is Hermitian and nonnegative definite, its eigenvalues are all real nonnegative numbers. Let these eigenvalues be denoted by $\sigma_1^2, \sigma_2^2, \ldots, \sigma_M^2$, where $\sigma_1 \geq \sigma_2 \geq \ldots \geq \sigma_W > 0$, and $\sigma_{W+1}, \sigma_{W+2}, \ldots$ are all zero, where $1 \leq W \leq M$. The matrix $\mathbf{A}^H\mathbf{A}$ has the same rank as $\mathbf{A}$; hence, there are W nonzero eigenvalues. Let $\mathbf{v}_1, \mathbf{v}_2, \ldots, \mathbf{v}_M$ denote a set of orthonormal eigenvectors of $\mathbf{A}^H\mathbf{A}$ that are associated with the eigenvalues $\sigma_1^2, \sigma_2^2, \ldots, \sigma_M^2$, respectively. Also, let $\mathbf{V}$ denote the M-by-M unitary matrix whose columns are made up of the eigenvectors $\mathbf{v}_1, \mathbf{v}_2, \ldots, \mathbf{v}_M$. Thus, using the eigendecomposition of the matrix $\mathbf{A}^H\mathbf{A}$, we may write

$$\mathbf{V}^H\mathbf{A}^H\mathbf{A}\mathbf{V} = \begin{bmatrix} \boldsymbol{\Sigma}^2 & \mathbf{0} \\ \mathbf{0} & \mathbf{0} \end{bmatrix} \tag{11.113}$$

Let the unitary matrix $\mathbf{V}$ be partitioned as

$$\mathbf{V} = [\mathbf{V}_1, \mathbf{V}_2] \tag{11.114}$$

where $\mathbf{V}_1$ is an M-by-W matrix,

$$\mathbf{V}_1 = [\mathbf{v}_1, \mathbf{v}_2, \ldots \mathbf{v}_W] \tag{11.115}$$

and $\mathbf{V}_2$ is an M-by-$(M - W)$ matrix,

$$\mathbf{V}_2 = [\mathbf{v}_{W+1}, \mathbf{v}_{W+2}, \ldots, \mathbf{v}_M] \tag{11.116}$$

with

$$\mathbf{V}_1^H\mathbf{V}_2 = \mathbf{0} \tag{11.117}$$

We may therefore make two deductions from Eq. (11.113):

1. For matrix $\mathbf{V}_1$, we have

$$\mathbf{V}_1^H\mathbf{A}^H\mathbf{A}\mathbf{V}_1 = \boldsymbol{\Sigma}^2$$

Consequently,

$$\boldsymbol{\Sigma}^{-1}\mathbf{V}_1^H\mathbf{A}^H\mathbf{A}\mathbf{V}_1\boldsymbol{\Sigma}^{-1} = \mathbf{I} \tag{11.118}$$

2. For matrix $\mathbf{V}_2$, we have

$$\mathbf{V}_2^H \mathbf{A}^H \mathbf{A} \mathbf{V}_2 = \mathbf{0}$$

Consequently,

$$\mathbf{A}\mathbf{V}_2 = \mathbf{0} \tag{11.119}$$

We now define a new K-by-W matrix

$$\mathbf{U}_1 = \mathbf{A}\mathbf{V}_1 \mathbf{\Sigma}^{-1} \tag{11.120}$$

Then, from Eq. (11.118) it follows that

$$\mathbf{U}_1^H \mathbf{U}_1 = \mathbf{I} \tag{11.121}$$

which means that the columns of the matrix $\mathbf{U}_1$ are orthonormal with respect to each other. Next, we choose another K-by-$(K - W)$ matrix $\mathbf{U}_2$ such that the K-by-K matrix formed from $\mathbf{U}_1$ and $\mathbf{U}_2$, namely,

$$\mathbf{U} = [\mathbf{U}_1, \mathbf{U}_2] \tag{11.122}$$

is a unitary matrix. This means that

$$\mathbf{U}_1^H \mathbf{U}_2 = \mathbf{0} \tag{11.123}$$

Accordingly, we may use Eqs. (11.114), (11.122), (11.119), (11.120), and (11.123), in that order, and so write

$$\mathbf{U}^H \mathbf{A} \mathbf{V} = \begin{bmatrix} \mathbf{U}_1^H \\ \mathbf{U}_2^H \end{bmatrix} \mathbf{A} [\mathbf{V}_1, \mathbf{V}_2]$$

$$= \begin{bmatrix} \mathbf{U}_1^H \mathbf{A}\mathbf{V}_1 & \mathbf{U}_1^H \mathbf{A}\mathbf{V}_2 \\ \mathbf{U}_2^H \mathbf{A}\mathbf{V}_1 & \mathbf{U}_2^H \mathbf{A}\mathbf{V}_2 \end{bmatrix}$$

$$= \begin{bmatrix} (\mathbf{\Sigma}^{-1}\mathbf{V}_1^H \mathbf{A}^H)\mathbf{A}\mathbf{V}_1 & \mathbf{U}_1^H(\mathbf{0}) \\ \mathbf{U}_2^H(\mathbf{U}_1 \mathbf{\Sigma}) & \end{bmatrix}$$

$$= \begin{bmatrix} \mathbf{\Sigma} & \mathbf{0} \\ \mathbf{0} & \mathbf{0} \end{bmatrix}$$

which proves Eq. (11.111) for the overdetermined case.

Case 2: Underdetermined System.   Consider next the case when $K < M$. This time we form the K-by-K matrix $\mathbf{A}\mathbf{A}^H$ by postmultiplying the matrix $\mathbf{A}$ by its Hermitian transpose $\mathbf{A}^H$. The matrix $\mathbf{A}\mathbf{A}^H$ is also Hermitian and nonnegative definite, so its eigenvalues are likewise real nonnegative numbers. The nonzero eigenvalues of $\mathbf{A}\mathbf{A}^H$ are the *same* as those of $\mathbf{A}^H\mathbf{A}$. We may therefore denote the eigenvalues of $\mathbf{A}\mathbf{A}^H$ as σ_1^2, $\sigma_2^2, \ldots, \sigma_K^2$, where $\sigma_1 \geq \sigma_2 \geq \ldots \geq \sigma_W > 0$, and σ_{W+1}, $\sigma_{W+2}, \ldots$ are all zero, where

$1 \leq W \leq K$. Let $\mathbf{u}_1, \mathbf{u}_2, \ldots, \mathbf{u}_K$ denote a set of orthonormal eigenvctors of the matrix $\mathbf{AA}^H$ that are associated with the eigenvalues $\sigma_1^2, \sigma_2^2, \ldots, \sigma_K^2$, respectively. Also, let $\mathbf{U}$ denote the unitary matrix whose columns are made up of the eigenvectors $\mathbf{u}_1, \mathbf{u}_2, \ldots, \mathbf{u}_K$. Thus, using the eigendecomposition of $\mathbf{AA}^H$, we may write

$$\mathbf{U}^H \mathbf{AA}^H \mathbf{U} = \begin{bmatrix} \mathbf{\Sigma}^2 & \mathbf{0} \\ \mathbf{0} & \mathbf{0} \end{bmatrix} \tag{11.124}$$

Let the unitary matrix $\mathbf{U}$ be partitioned as

$$\mathbf{U} = [\mathbf{U}_1, \mathbf{U}_2] \tag{11.125}$$

where

$$\mathbf{U}_1 = [\mathbf{u}_1, \mathbf{u}_2, \ldots, \mathbf{u}_W] \tag{11.126}$$

$$\mathbf{U}_2 = [\mathbf{u}_{W+1}, \mathbf{u}_{W+2}, \ldots, \mathbf{u}_K] \tag{11.127}$$

and

$$\mathbf{U}_1^H \mathbf{U}_2 = \mathbf{0} \tag{11.128}$$

We may therefore make two deductions from Eq. (11.124):

1. For matrix $\mathbf{U}_1$, we have

$$\mathbf{U}_1^H \mathbf{AA}^H \mathbf{U}_1 = \mathbf{\Sigma}^2$$

Consequently,

$$\mathbf{\Sigma}^{-1} \mathbf{U}_1^H \mathbf{AA}^H \mathbf{U}_1 \mathbf{\Sigma}^{-1} = \mathbf{I} \tag{11.129}$$

2. For matrix $\mathbf{U}_2$, we have

$$\mathbf{U}_2^H \mathbf{AA}^H \mathbf{U}_2 = \mathbf{0}$$

Consequently,

$$\mathbf{A}^H \mathbf{U}_2 = \mathbf{0} \tag{11.130}$$

We now define an M-by-W matrix

$$\mathbf{V}_1 = \mathbf{A}^H \mathbf{U}_1 \mathbf{\Sigma}^{-1} \tag{11.131}$$

Then from Eq. (11.129), it follows that

$$\mathbf{V}_1^H \mathbf{V}_1 = \mathbf{I} \tag{11.132}$$

which means that the columns of the matrix $\mathbf{V}_1$ are orthonormal with respect to each other. Next, we choose another M-by-$(M - W)$ matrix $\mathbf{V}_2$ such that the M-by-M matrix formed from $\mathbf{V}_1$ and $\mathbf{V}_2$, namely,

$$\mathbf{V} = [\mathbf{V}_1, \mathbf{V}_2] \tag{11.133}$$

is a unitary matrix. This means that

$$\mathbf{V}_2^H \mathbf{V}_1 = \mathbf{0} \tag{11.134}$$

Accordingly, we may use Eqs. (11.125), (11.133), (11.130), (11.131), and (11.134), in that order, and so write

$$\mathbf{U}^H \mathbf{A} \mathbf{V} = \begin{bmatrix} \mathbf{U}_1^H \\ \mathbf{U}_2^H \end{bmatrix} \mathbf{A} [\mathbf{V}_1, \mathbf{V}_2]$$

$$= \begin{bmatrix} \mathbf{U}_1^H \mathbf{A} \mathbf{V}_1 & \mathbf{U}_1^H \mathbf{A} \mathbf{V}_2 \\ \mathbf{U}_2^H \mathbf{A} \mathbf{V}_1 & \mathbf{U}_2^H \mathbf{A} \mathbf{V}_2 \end{bmatrix}$$

$$= \begin{bmatrix} \mathbf{U}_1^H \mathbf{A} (\mathbf{A}^H \mathbf{U}_1 \mathbf{\Sigma}^{-1}) & (\mathbf{\Sigma} \mathbf{V}_1^H) \mathbf{V}_2 \end{bmatrix}$$

$$= \begin{bmatrix} \mathbf{\Sigma} & \mathbf{0} \\ \mathbf{0} & \mathbf{0} \end{bmatrix}$$

This proves Eq. (11.111) for the underdetermined case, and with it the proof of the singular-value decomposition (SVD) theorem is completed.

Terminology and Relation to Eigenanalysis

The numbers $\sigma_1, \sigma_2, \ldots, \sigma_W$, constituting the diagonal matrix $\mathbf{\Sigma}$, are called the *singular values* of the matrix $\mathbf{A}$. The columns of the unitary matrix $\mathbf{V}$, that is, $\mathbf{v}_1, \mathbf{v}_2, \ldots, \mathbf{v}_M$, are the *right singular vectors* of $\mathbf{A}$, and the columns of the second unitary matrix $\mathbf{U}$, that is, $\mathbf{u}_1, \mathbf{u}_2, \ldots, \mathbf{u}_K$ are the *left singular vectors* of $\mathbf{A}$. We note from the preceding discussion that the right singular vectors $\mathbf{v}_1, \mathbf{v}_2, \ldots, \mathbf{v}_M$ are eigenvectors of $\mathbf{A}^H \mathbf{A}$, whereas the left singular vectors $\mathbf{u}_1, \mathbf{u}_2, \ldots, \mathbf{u}_K$ are eigenvectors of $\mathbf{A} \mathbf{A}^H$. Note that the number of positive singular values is equal to the rank of the data matrix $\mathbf{A}$. The singular-value decomposition therefore provides the basis of a practical method for determining the rank of a matrix.

Since $\mathbf{U} \mathbf{U}^H$ equals the identity matrix, we find from Eq. (11.111) that

$$\mathbf{A} \mathbf{V} = \mathbf{U} \begin{bmatrix} \mathbf{\Sigma} & \mathbf{0} \\ \mathbf{0} & \mathbf{0} \end{bmatrix}$$

It follows therefore that

$$\begin{aligned} \mathbf{A} \mathbf{v}_i &= \sigma_i \mathbf{u}_i, & i &= 1, 2, \ldots, W \\ \mathbf{A} \mathbf{v}_i &= \mathbf{0}, & i &= W + 1, \ldots, K \end{aligned} \tag{11.135}$$

Correspondingly, we may express the data matrix $\mathbf{A}$ in the expanded form

$$\mathbf{A} = \sum_{i=1}^{W} \sigma_i \mathbf{u}_i \mathbf{v}_i^H \tag{11.136}$$

Since $\mathbf{VV}^H$ equals the identity matrix, we also find from Eq. (11.111) that

$$\mathbf{U}^H\mathbf{A} = \begin{bmatrix} \boldsymbol{\Sigma} & \mathbf{0} \\ \mathbf{0} & \mathbf{0} \end{bmatrix} \mathbf{V}^H$$

or, equivalently,

$$\mathbf{A}^H\mathbf{U} = \mathbf{V}\begin{bmatrix} \boldsymbol{\Sigma} & \mathbf{0} \\ \mathbf{0} & \mathbf{0} \end{bmatrix}$$

It follows therefore that

$$\mathbf{A}^H\mathbf{u}_i = \sigma_i\mathbf{v}_i, \quad i = 1, 2, \ldots, W$$
$$\mathbf{A}^H\mathbf{u}_i = \mathbf{0}, \quad\quad i = W + 1, \ldots, M \tag{11.137}$$

In this case, we may express the Hermitian transpose of the data matrix $\mathbf{A}$ in the expanded form

$$\mathbf{A}^H = \sum_{i=1}^{W} \sigma_i\mathbf{v}_i\mathbf{u}_i^H \tag{11.138}$$

which checks exactly with Eq. (11.136), and so it should.

Example 2

In this example, we use the SVD to deal with the different facets of *matrix rank*. To be specific, let $\mathbf{A}$ be a K-by-M data matrix with rank W. The matrix $\mathbf{A}$ is said to be of *full rank* if

$$W = \min(K, M)$$

Otherwise, the matrix $\mathbf{A}$ is *rank deficient*. As mentioned previously, the rank W is simply the number of nonzero singular values of matrix $\mathbf{A}$.

Consider next a computational environment that yields a numerical value for each element of the matrix $\mathbf{A}$ that is accurate to within $\pm\epsilon$. Let $\mathbf{B}$ denote the approximate value of matrix $\mathbf{A}$ so obtained. We define the ϵ-*rank* of matrix $\mathbf{A}$ as follows (Golub and Van Loan, 1989):

$$\text{rank}(\mathbf{A}, \epsilon) = \min_{\|\mathbf{A}-\mathbf{B}\|<\epsilon} \text{rank}(\mathbf{B}) \tag{11.139}$$

where $\|\mathbf{A} - \mathbf{B}\|$ is the *spectral norm* of the error matrix $\mathbf{A} - \mathbf{B}$ that results from the use of inaccurate computations. Extending the definition of spectral norm of the matrix introduced in Chapter 4 to the situation at hand, the spectral norm $\|\mathbf{A} - \mathbf{B}\|$ equals the largest singular value of the difference $\mathbf{A} - \mathbf{B}$. In any event, the K-by-M matrix $\mathbf{A}$ is said to be *numerically rank deficient* if

$$\text{rank}(\mathbf{A}, \epsilon) < \min(K, M)$$

The SVD provides a sensible method for characterizing the ϵ-rank and the numerical rank deficiency of the matrix, because the singular values resulting from its use indicate how close a given matrix $\mathbf{A}$ is to another matrix $\mathbf{B}$ of lower rank in a simple fashion.

11.11 PSEUDOINVERSE

Our interest in the singular-value decomposition is to formulate a general definition of pseudoinverse. Let $\mathbf{A}$ denote a K-by-M matrix that has the singular-value decomposition described in Eq. (11.111). We define the pseudoinverse of the matrix $\mathbf{A}$ as (Stewart, 1973; Golub and Van Loan, 1989):

$$\mathbf{A}^+ = \mathbf{V}\begin{bmatrix} \boldsymbol{\Sigma}^{-1} & \mathbf{0} \\ \mathbf{0} & \mathbf{0} \end{bmatrix}\mathbf{U}^H \tag{11.140}$$

where

$$\boldsymbol{\Sigma}^{-1} = \mathrm{diag}(\sigma_1^{-1}, \sigma_2^{-1}, \ldots, \sigma_W^{-1})$$

and W is the rank of the data matrix $\mathbf{A}$. The pseudoinverse $\mathbf{A}^+$ may be expressed in the expanded form:

$$\mathbf{A}^+ = \sum_{i=1}^{W} \frac{1}{\sigma_i} \mathbf{v}_i \mathbf{u}_i^H \tag{11.141}$$

We may identify two special cases that can arise as described next.

Case 1: Overdetermined System.   In this case, we have $K > M$, and we assume that the rank W equals M so that the inverse matrix $(\mathbf{A}^H\mathbf{A})^{-1}$ exists. The pseudoinverse of the data matrix $\mathbf{A}$ is defined by

$$\mathbf{A}^+ = (\mathbf{A}^H\mathbf{A})^{-1}\mathbf{A}^H \tag{11.142}$$

To show the validity of this special formula, we note from Eqs. (11.118) and (11.120) that

$$(\mathbf{A}^H\mathbf{A})^{-1} = \mathbf{V}_1 \boldsymbol{\Sigma}^{-2} \mathbf{V}_1^H$$

and

$$\mathbf{A}^H = \mathbf{V}_1 \boldsymbol{\Sigma} \mathbf{U}_1^H$$

Therefore, using this pair of relations, we may express the right-hand side of Eq. (11.142) as follows:

$$\begin{aligned} (\mathbf{A}^H\mathbf{A})^{-1}\mathbf{A}^H &= (\mathbf{V}_1 \boldsymbol{\Sigma}^{-2} \mathbf{V}_1^H)(\mathbf{V}_1 \boldsymbol{\Sigma} \mathbf{U}_1^H) \\ &= \mathbf{V}_1 \boldsymbol{\Sigma}^{-1} \mathbf{U}_1^H \\ &= \mathbf{V}\begin{bmatrix} \boldsymbol{\Sigma}^{-1} & \mathbf{0} \\ \mathbf{0} & \mathbf{0} \end{bmatrix}\mathbf{U}^H \\ &= \mathbf{A}^+ \end{aligned}$$

Case 2: Underdetermined System.   In this second case, we have $M > K$, and we assume that the rank W equals K so that the inverse matrix $(\mathbf{A}\mathbf{A}^H)^{-1}$ exists. The pseudoinverse of the data matrix $\mathbf{A}$ is now defined by

$$\mathbf{A}^+ = \mathbf{A}^H(\mathbf{A}\mathbf{A}^H)^{-1} \tag{11.143}$$

To show the validity of this second special formula, we note from Eqs. (11.129) and (11.131) that

$$(\mathbf{A}\mathbf{A}^H)^{-1} = \mathbf{U}_1\mathbf{\Sigma}^{-2}\mathbf{U}_1^H$$

and

$$\mathbf{A}^H = \mathbf{V}_1\mathbf{\Sigma}\mathbf{U}_1^H$$

Therefore, using this pair of relations in the right-hand side of Eq. (11.143), we get

$$\begin{aligned}
\mathbf{A}^H(\mathbf{A}\mathbf{A}^H)^{-1} &= (\mathbf{V}_1\mathbf{\Sigma}\mathbf{U}_1^H)(\mathbf{U}_1\mathbf{\Sigma}^{-2}\mathbf{U}_1^H) \\
&= \mathbf{V}_1\mathbf{\Sigma}^{-1}\mathbf{U}_1^H \\
&= \mathbf{V}\begin{bmatrix} \mathbf{\Sigma}^{-1} & \mathbf{0} \\ \mathbf{0} & \mathbf{0} \end{bmatrix}\mathbf{U}^H \\
&= \mathbf{A}^+
\end{aligned}$$

Note, however, the pseudoinverse $\mathbf{A}^+$ as described in Eq. (11.140) or equivalently, Eq. (11.141) is of general application, in that it applies whether the data matrix $\mathbf{A}$ refers to an overdetermined or an underdetermined system and regardless of what the rank W is. Most importantly, it is numerically stable.

11.12 INTERPRETATION OF SINGULAR VALUES AND SINGULAR VECTORS

Consider a K-by-M data matrix $\mathbf{A}$, for which the singular-value decomposition is given in Eq. (11.111) and the pseudoinverse is correspondingly given in Eq. (11.140). We assume that the system is overdetermined. Define a K-by-1 vector $\mathbf{y}$ and an M-by-1 vector $\mathbf{x}$ that are related to each other by the transformation matrix $\mathbf{A}$, as shown by

$$\mathbf{y} = \mathbf{A}\mathbf{x} \tag{11.144}$$

The vector $\mathbf{x}$ is constrained to have a Euclidean norm of unity; that is,

$$\|\mathbf{x}\| = 1 \tag{11.145}$$

Given the transformation of Eq. (11.144) and the constraint of Eq. (11.145), we wish to find the resulting locus of the points defined by the vector $\mathbf{y}$ in a K-dimensional space.

Solving Eq. (11.144) for $\mathbf{x}$, we get

$$\mathbf{x} = \mathbf{A}^+\mathbf{y} \tag{11.146}$$

where $\mathbf{A}^+$ is the pseudoinverse of $\mathbf{A}$. Substituting Eq. (11.142) in (11.146), we get

$$\mathbf{x} = \sum_{i=1}^{W} \frac{1}{\sigma_i} \mathbf{v}_i \mathbf{u}_i^H \mathbf{y}$$

$$= \sum_{i=1}^{W} \frac{(\mathbf{u}_i^H \mathbf{y})}{\sigma_i} \mathbf{v}_i \tag{11.147}$$

where W is the rank of matrix $\mathbf{A}$, and the inner product $\mathbf{u}_i^H \mathbf{y}$ is a scalar. Imposing the constraint of Eq. (11.145) on (11.147), and recognizing that the right singular vectors $\mathbf{v}_1$, $\mathbf{v}_2$, ..., $\mathbf{v}_W$ form an orthonormal set, we get

$$\sum_{i=1}^{W} \frac{|\mathbf{y}^H \mathbf{u}_i|^2}{\sigma_i^2} = 1 \tag{11.148}$$

Equation (11.148) defines the locus traced out by the tip of vector $\mathbf{y}$ in a K-dimensional space. Indeed, this is the equation of a *hyperellipsoid* (Golub and Van Loan, 1989).

To see this interpretation in a better way, define the complex scalar

$$\zeta_i = \mathbf{y}^H \mathbf{u}_i \tag{11.149}$$

$$= \sum_{k=1}^{K} y_k^* u_{ik}, \qquad i = 1, \ldots, W$$

In other words, the complex scalar ζ_i is a linear combination of all possible values of the elements of the left singular vector $\mathbf{u}_i$, so ζ_i is referred to as the "span" of $\mathbf{u}_i$. We may thus rewrite Eq. (11.148) as

$$\sum_{i=1}^{W} \frac{|\zeta_i|^2}{\sigma_i^2} = 1 \tag{11.150}$$

This is the equation of a hyperellipsoid with coordinates $|\zeta_1|, \ldots, |\zeta_W|$ and semi-axis whose lengths are the singular values $\sigma_1, \ldots, \sigma_W$, respectively. Figure 11.9 illustrates the locus traced out by Eq. (11.148) for the case of $W = 2$ and $\sigma_1 > \sigma_2$, assuming that the data matrix A is real.

11.13 MINIMUM NORM SOLUTION TO THE LINEAR LEAST-SQUARES PROBLEM

Having equipped ourselves with the general definition of the pseudoinverse of a matrix $\mathbf{A}$ in terms of its singular-value decomposition, we are now ready to tackle the solution to the linear least-squares problem even when null$(\mathbf{A}) \neq \varnothing$. In particular, we define the solution to the least-squares problem as in Eq. (11.109), reproduced here for convenience:

$$\hat{\mathbf{w}} = \mathbf{A}^+ \mathbf{d} \tag{11.151}$$

The pseudoinverse matrix $\mathbf{A}^+$ is itself defined by Eq. (11.140). We thus find that, out of the many vectors that solve the least-squares problem when null $(\mathbf{A}) \neq \varnothing$, the one defined

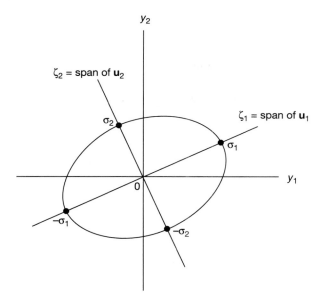

Figure 11.9   Locus of Eq. (11.150) for real data with $W = 2$ and $\sigma_1 > \sigma_2$.

by Eq. (11.151) is *unique* in that it has the *shortest length possible in the Euclidean sense* (Stewart, 1973).

We prove this important result by manipulating the equation that defines the minimum value of the sum of error squares produced in the method of least squares. We note that both matrix products $\mathbf{VV}^H$ and $\mathbf{UU}^H$ equal identity matrices. Hence, we may start with Eq. (11.49) and combine it with Eq. (11.48), and then write

$$
\begin{aligned}
\mathscr{E}_{\min} &= \mathbf{d}^H\mathbf{d} - \mathbf{d}^H\mathbf{A}\hat{\mathbf{w}} \\
&= \mathbf{d}^H(\mathbf{d} - \mathbf{A}\hat{\mathbf{w}}) \\
&= \mathbf{d}^H\mathbf{UU}^H(\mathbf{d} - \mathbf{AVV}^H\hat{\mathbf{w}}) \\
&= \mathbf{d}^H\mathbf{U}(\mathbf{U}^H\mathbf{d} - \mathbf{U}^H\mathbf{AVV}^H\hat{\mathbf{w}})
\end{aligned}
\tag{11.152}
$$

Let

$$
\begin{aligned}
\mathbf{V}^H\hat{\mathbf{w}} &= \mathbf{b} \\
&= \begin{bmatrix} \mathbf{b}_1 \\ \mathbf{b}_2 \end{bmatrix}
\end{aligned}
\tag{11.153}
$$

and

$$
\begin{aligned}
\mathbf{U}^H\mathbf{d} &= \mathbf{c} \\
&= \begin{bmatrix} \mathbf{c}_1 \\ \mathbf{c}_2 \end{bmatrix}
\end{aligned}
\tag{11.154}
$$

where $\mathbf{b}_1$ and $\mathbf{c}_1$ are W-by-1 vectors, and $\mathbf{b}_2$ and $\mathbf{c}_2$ are two other vectors. Thus, substituting Eqs. (11.111), (11.153), and (11.154) in (11.152), we get

$$\mathscr{E}_{\min} = \mathbf{d}^H \mathbf{U}\left(\begin{bmatrix} \mathbf{c}_1 \\ \mathbf{c}_2 \end{bmatrix} - \begin{bmatrix} \mathbf{\Sigma} & \mathbf{0} \\ \mathbf{0} & \mathbf{0} \end{bmatrix}\begin{bmatrix} \mathbf{b}_1 \\ \mathbf{b}_2 \end{bmatrix}\right)$$

$$= \mathbf{d}^H \mathbf{U}\begin{bmatrix} \mathbf{c}_1 - \mathbf{\Sigma}\mathbf{b}_1 \\ \mathbf{c}_2 \end{bmatrix} \qquad (11.155)$$

For $\mathscr{E}_{\min}$ to be minimum, we require that

$$\mathbf{c}_1 = \mathbf{\Sigma}\mathbf{b}_1 \qquad (11.156)$$

or, equivalently,

$$\mathbf{b}_1 = \mathbf{\Sigma}^{-1}\mathbf{c}_1 \qquad (11.57)$$

We observe that $\mathscr{E}_{\min}$ is independent of $\mathbf{b}_2$. Hence, the value of $\mathbf{b}_2$ is arbitrary. However, if we let $\mathbf{b}_2 = \mathbf{0}$, we get the special result

$$\hat{\mathbf{w}} = \mathbf{V}\mathbf{b}$$

$$= \mathbf{V}\begin{bmatrix} \mathbf{\Sigma}^{-1}\mathbf{c}_1 \\ \mathbf{0} \end{bmatrix} \qquad (11.158)$$

We may also express $\hat{\mathbf{w}}$ in the equivalent form:

$$\hat{\mathbf{w}} = \mathbf{V}\begin{bmatrix} \mathbf{\Sigma}^{-1} & \mathbf{0} \\ \mathbf{0} & \mathbf{0} \end{bmatrix}\begin{bmatrix} \mathbf{c}_1 \\ \mathbf{c}_2 \end{bmatrix}$$

$$= \mathbf{V}\begin{bmatrix} \mathbf{\Sigma}^{-1} & \mathbf{0} \\ \mathbf{0} & \mathbf{0} \end{bmatrix}\mathbf{U}^H\mathbf{d}$$

$$= \mathbf{A}^+\mathbf{d}$$

This coincides exactly with the value defined by Eq. (11.151), where the pseudoinverse $\mathbf{A}^+$ is defined by Eq. (11.140). In effect, we have shown that this value of $\hat{\mathbf{w}}$ does indeed solve the linear least-squares problem.

Moreover, the vector $\hat{\mathbf{w}}$ so defined is *unique*, in that it has the minimum Euclidean norm possible. In particular, since $\mathbf{V}\mathbf{V}^H = \mathbf{I}$, we find from Eq. (11.158) that the squared Euclidean norm of $\hat{\mathbf{w}}$ equals

$$\|\hat{\mathbf{w}}\|^2 = \|\mathbf{\Sigma}^{-1}\mathbf{c}_1\|^2$$

Consider now another possible solution to the linear least-squares problem that is defined by

$$\mathbf{w}' = \mathbf{V}\begin{bmatrix} \mathbf{\Sigma}^{-1}\mathbf{c}_1 \\ \mathbf{b}_2 \end{bmatrix}, \qquad \mathbf{b}_2 \neq \mathbf{0}$$

The squared Euclidean norm of $\mathbf{w}'$ equals

$$\|\mathbf{w}'\|^2 = \|\mathbf{\Sigma}^{-1}\mathbf{c}_1\|^2 + \|\mathbf{b}_2\|^2$$

For any $\mathbf{b}_2 \neq \mathbf{0}$, we see therefore that

$$\|\mathbf{w}\| < \|\mathbf{w}'\| \tag{11.159}$$

In summary, the tap-weight $\hat{\mathbf{w}}$ of a linear transversal filter defined in by Eq. (11.151) is a unique solution to the linear least-squares problem, even when null($\mathbf{A}$) $\neq \emptyset$. *The vector $\hat{\mathbf{w}}$ is unique in the sense that it is the only tap-weight vector that simultaneously satisfies two requirements: (1) it produces the minimum sum of error squares, and (2) it has the smallest Euclidean norm possible. This special value of the tap-weight vector $\hat{\mathbf{w}}$ is called the minimum-norm solution.*

Another Formulation of the Minimum-Norm Solution

We may develop an expanded formulation of the minimum-norm solution, depending on whether we are dealing with the overdetermined or underdetermined case. These two cases are considered in turn.

Case 1: Overdetermined. For this case, the number of equations K is greater than the number of unknown parameters M. To proceed then, we substitute Eq. (11.140) in (11.151), and then use the partitioned forms of the unitary matrices $\mathbf{V}$ and $\mathbf{U}$. We may thus write

$$
\begin{aligned}
\hat{\mathbf{w}} &= (\mathbf{V}_1\boldsymbol{\Sigma}^{-1})(\mathbf{A}\mathbf{V}_1\boldsymbol{\Sigma}^{-1})^H\mathbf{d} \\
&= \mathbf{V}_1\boldsymbol{\Sigma}^{-1}\boldsymbol{\Sigma}^{-1}\mathbf{V}_1^H\mathbf{A}^H\mathbf{d} \\
&= \mathbf{V}_1\boldsymbol{\Sigma}^{-2}\mathbf{V}_1\mathbf{A}^H\mathbf{d}
\end{aligned}
\tag{11.160}
$$

Hence, using the definition [see Eq. (11.115)]

$$\mathbf{V}_1 = [\mathbf{v}_1, \mathbf{v}_2, \dots, \mathbf{v}_W]$$

in Eq. (11.160), we get the following expanded formulation for $\hat{\mathbf{w}}$ for the overdetermined case:

$$\hat{\mathbf{w}} = \sum_{i=1}^{W} \frac{\mathbf{v}_i}{\sigma_i^2} \mathbf{v}_i^H\mathbf{A}^H\mathbf{d} \tag{11.161}$$

Case 2: Underdetermined. For this second case, the number of equations K is smaller than the number of unknowns M. This time we find it appropriate to use the representation given in Eq. (11.131) for the submatrix $\mathbf{V}_1$ in terms of the data matrix $\mathbf{A}$. Thus, substituting Eq. (11.131) in (11.151), we get

$$
\begin{aligned}
\hat{\mathbf{w}} &= (\mathbf{A}^H\mathbf{U}_1\boldsymbol{\Sigma}^{-1})(\boldsymbol{\Sigma}^{-1}\mathbf{U}_1\mathbf{d}) \\
&= \mathbf{A}^H\mathbf{U}_1\boldsymbol{\Sigma}^{-2}\mathbf{U}_1^H\mathbf{d}
\end{aligned}
\tag{11.162}
$$

Substituting the definition [see Eq. (11.126)]

$$\mathbf{U}_1 = [\mathbf{u}_1, \mathbf{u}_2, \dots, \mathbf{u}_W]$$

in Eq. (11.162), we get the following expanded formulation for $\hat{\mathbf{w}}$ for the underdetermined case:

$$\hat{\mathbf{w}} = \sum_{i=1}^{W} \frac{(\mathbf{u}_i^H \mathbf{d})}{\sigma_i^2} \mathbf{A}^H \mathbf{u}_i \tag{11.163}$$

which is different from that of Eq. (11.161) for the overdetermined case.

The important point to note is that the expanded solutions of $\hat{\mathbf{w}}$ given in Eqs. (11.161) and (11.163) for the overdetermined and underdetermined systems, respectively, are both contained in the compact formula of Eq. (11.151). Indeed, from a numerical computation point of view, the use of Eq. (11.151) is the preferred method for computing the least-squares estimatee $\hat{\mathbf{w}}$.

11.14 NORMALIZED LMS ALGORITHM VIEWED AS THE MINIMUM-NORM SOLUTION TO AN UNDERDETERMINED LEAST-SQUARES ESTIMATION PROBLEM

In Chapter 9 we derived the normalized least-mean-square (LMS) algorithm as the solution to a constrained minimization problem. In this section we revisit this algorithm in light of the theory developed on singular-value decomposition. In particular, we show that the normalized LMS algorithm is indeed the minimum-norm solution to an underdetermined linear least-squares problem involving a single error equation with M unknowns, where M is the dimension of the tap-weight vector in the algorithm.

Consider the error equation

$$\epsilon(n) = d(n) - \hat{\mathbf{w}}^H(n+1)\mathbf{u}(n) \tag{11.164}$$

where $d(n)$ is a desired response and $\mathbf{u}(n)$ is a tap-input vector, both measured at time n. The requirement is to find the tap-weight vector $\hat{\mathbf{w}}(n+1)$, measured at time $n+1$, such that the change in the tap-weight vector given by

$$\delta\hat{\mathbf{w}}(n+1) = \hat{\mathbf{w}}(n+1) - \hat{\mathbf{w}}(n) \tag{11.165}$$

is minimized, subject to the constraint

$$\epsilon(n) = 0 \tag{11.166}$$

Using Eq. (11.165) in (11.164), we may reformulate the error $\epsilon(n)$ as

$$\epsilon(n) = d(n) - \hat{\mathbf{w}}^H(n)\mathbf{u}(n) - \delta\hat{\mathbf{w}}^H(n+1)\mathbf{u}(n) \tag{11.167}$$

We now recognize the customary definition of the estimation error, namely,

$$e(n) = d(n) - \hat{\mathbf{w}}^H(n)\mathbf{u}(n) \tag{11.168}$$

Hence, we may simplify Eq. (11.168) as

$$\epsilon(n) = e(n) - \delta\hat{\mathbf{w}}^H(n+1)\mathbf{u}(n) \tag{11.169}$$

TABLE 11.1 SUMMARY OF CORRESPONDENCES BETWEEN LINEAR LEAST-SQUARES ESTIMATION AND NORMALIZED LMS ALGORITHM

	Linear least-squares estimation (underdetermined)	Normalized LMS algorithm
Data matrix	$\mathbf{A}$	$\mathbf{u}^H(n)$
Desired data vector	$\mathbf{d}$	$e^*(n)$
Parameter vector	$\hat{\mathbf{w}}$	$\delta\hat{\mathbf{w}}(n + 1)$
Rank	W	1
Eigenvalue	$\sigma_i^2, i = 1, \ldots, W$	$\|\mathbf{u}(n)\|^2$
Eigenvector	$\mathbf{u}_i, i = 1, \ldots, W$	1

Thus, complex conjugating both sides of Eq. (11.169), we note that the constraint of Eq. (11.166) is equivalent to

$$\mathbf{u}^H(n)\delta\hat{\mathbf{w}}(n + 1) = e^*(n) \tag{11.170}$$

Accordingly, we may restate our constrained minimization problem as follows:

Find the minimum-norm solution for the change $\delta\hat{\mathbf{w}}(n + 1)$ in the tap-weight vector at time $n + 1$, which satisfies the constraint

$$\mathbf{u}^H(n)\delta\hat{\mathbf{w}}(n + 1) = e^*(n)$$

This problem is one of linear least-squares estimation that is underdetermined. To solve it, we may use the method of singular-value decomposition described in Eq. (11.163). To help us in the application of this method, we use Eq. (11.170) to make the identifications listed in Table 11.1 between the normalized LMS algorithm and linear least-squares estimation. In particular, we note that the normalized LMS algorithm has only one nonzero singular value equal to the squared norm of the tap-input vector $\mathbf{u}(n)$; that is, the rank $W = 1$. The corresponding left-singular vector is therefore simply equal to one. Hence, with the aid of Table 11.1, the application of Eq. (11.163) yields

$$\delta\hat{\mathbf{w}}(n + 1) = \frac{1}{\|\mathbf{u}(n)\|^2} \mathbf{u}(n)e^*(n) \tag{11.171}$$

This is precisely the result that we derived previously in Chapter 9; see Eq. (9.139).

We may next follow a reasoning similar to that described in Section 9.10 and redefine the change $\delta\hat{\mathbf{w}}(n + 1)$ by introducing a scaling factor $\tilde{\mu}$ as shown by [see Eq. (9.140)]

$$\delta\hat{\mathbf{w}}(n + 1) = \frac{\tilde{\mu}}{\|\mathbf{u}(n)\|^2} \mathbf{u}(n)e^*(n)$$

or, equivalently, we may write

$$\hat{\mathbf{w}}(n + 1) = \hat{\mathbf{w}}(n) + \frac{\tilde{\mu}}{\|\mathbf{u}(n)\|^2} \mathbf{u}(n)e^*(n) \tag{11.172}$$

By so doing, we are able to exercise control over the change in the tap-weight vector from one iteration to the next without changing its direction. Equation (11.172) is the tap-weight vector update for the normalized LMS algorithm.

The important point to note from the discussion presented in this section is that the singular-value decomposition provides an insightful link between the underdetermined form of linear least-squares estimation and LMS theory. In particular, we have shown that the weight update in the normalized LMS algorithm may indeed be viewed as the minimum norm solution to an underdetermined form of the linear least-squares problem. The problem involves a single error equation with a number of unknowns equal to the dimension of the tap-weight vector in the algorithm.

11.15 SUMMARY AND DISCUSSION

In this chapter we presented a detailed discussion of the method of least-squares for solving the linear adaptive filtering problem. The distinguishing features of this approach include the following:

- It is a model-dependent procedure that operates on the input data on a block-by-block basis.
- It yields a solution for the tap-weight vector of an adaptive transversal filter that is the best linear unbiased estimate (BLUE), assuming that the measurement error process in the underlying model is white with zero mean.

The method of least squares is well suited for solving super-resolution spectrum estimation/beamforming problems, such as those based on autoregressive (AR) and minimum-variance distortionless response (MVDR) models. For the efficient computation of these spectra, and linear least-squares solution in general, the recommended procedure is to use singular value decomposition (SVD) that operates on the input data directly. The SVD is defined by the following parameters:

- A set of left singular vectors that form a unitary matrix
- A set of right singular vectors that form another unitary matrix
- A corresponding set of nonzero singular values

The important advantage of using the SVD to solve a linear least-squares problem is that the solution, defined in terms of the pseudoinverse of the input data matrix, is numerically stable. An algorithm is said to be *numerically stable* if it does not introduce any more sensitivity to perturbation than that which is inherently present in the problem under study (Klema and Laub, 1980).

Another useful application of the SVD is in *rank determination*. The *column rank* of a matrix is defined by the number of linearly independent columns of the matrix. Specifically, we say that an M-by-K matrix, with $M \geq K$, has *full* column rank if and only if it has K independent columns. In theory, the issue of full rank determination is a yes–no type of proposition in the sense that either the matrix in question has full rank or it does not. In practice, however, the fuzzy nature of a data matrix and the use of inexact (finite-precision) arithmetic complicate the rank determination problem. The SVD provides a practical method for determining the rank of a matrix, given fuzzy data and roundoff errors due to finite-precision computations.

PROBLEMS

1. Consider a linear array consisting of M uniformly spaced sensors. The output of sensor k observed at time i is denoted by $u(k, i)$ where $k = 1, 2, \ldots, M$ and $i = 1, 2, \ldots, n$. In effect, the observations $u(1, i), u(2, i), \ldots, u(M, i)$ define snapshot i. Let $\mathbf{A}$ denote the n-by-M data matrix, whose Hermitian transpose is defined by

$$\mathbf{A}^H = \begin{bmatrix} u(1, 1) & u(1, 2) & \cdots & u(1, n) \\ u(2, 1) & u(2, 2) & \cdots & u(2, n) \\ \cdot & \cdot & \cdot & \cdot \\ \cdot & \cdot & \cdot & \cdot \\ \cdot & \cdot & \cdot & \cdot \\ u(M, 1) & u(M, 2) & \cdots & u(M, n) \end{bmatrix}$$

where the number of columns equals the number of snapshots, and the number of rows equals the number of sensors in the array. Demonstrate the following interpretations:

   (a) The M-by-M matrix $\mathbf{A}^H\mathbf{A}$ is the *spatial* correlation matrix with temporal averaging. This form of averaging assumes that the environment is temporally stationary.

   (b) The n-by-n matrix $\mathbf{A}\mathbf{A}^H$ is the *temporal* correlation matrix with spatial averaging. This form of averaging assumes that the environment is spatially stationary.

2. We say that the least-squares estimate $\hat{\mathbf{w}}$ is *consistent* if, in the long run, the difference between $\hat{\mathbf{w}}$ and the unknown parameter vector $\mathbf{w}_o$ of the multiple linear regression model becomes negligibly small in the mean-square sense. Hence, show that the least-squares estimate $\hat{\mathbf{w}}$ is consistent if the error vector $\boldsymbol{\epsilon}_o$ has zero mean and its elements are uncorrelated and if the trace of the inverse matrix $\boldsymbol{\Phi}^{-1}$ approaches zero as the number of observations, N, approaches infinity.

3. In Example 1 in Section 11.6, we used a 3-by-2 input data matrix and 3-by-1 desired data vector to illustrate the corollary to the principle of orthogonality. Use the data given in that example to calculate the two tap-weights of the linear least-squares filter.

4. In the autocorrelation method of linear prediction, we choose the tap-weight vector of a transversal predictor to minimize the error energy

$$\mathcal{E}_f = \sum_{n=1}^{\infty} |f(n)|^2$$

where $f(n)$ is the prediction error. Show that the transfer function $H(z)$ of the (forward) prediction-error filter is minimum phase, in that its roots must lie strictly within the unit circle.

Hints: (1) Express the transfer function $H(z)$ of order M (say) as the product of a simple zero factor $(1 - z_i z^{-1})$ and a function $H'(z)$. Hence, minimize the prediction-error energy with respect to the magnitude of zero z_i.

(2) Use the Cauchy–Schwartz inequality:

$$\text{Re}\left[\sum_{n=1}^{\infty} e^{j\theta} g(n-1) g^*(n)\right] \le \left[\sum_{n=1}^{\infty} |g(n)|^2\right]^{1/2} \left[\sum_{n=1}^{\infty} |e^{j\theta} g(n-1)|^2\right]^{1/2}$$

The equality holds if and only if $g(n) = e^{j\theta} g(n-1)$ for $n = 1, 2, \ldots, \infty$.

5. Figure 11.5(a) shows a *forward linear predictor* using a transversal structure, with the tap inputs $u(i-1), u(i-2), \ldots, u(i-M)$ used to make a linear prediction of $u(i)$. The problem is to find the tap-weight vector $\hat{\mathbf{w}}$ that minimizes the sum of forward prediction-error squares:

$$\mathscr{E}_f = \sum_{i=M+1}^{N} |f_M(i)|^2$$

where $f_M(i)$ is the forward prediction error. Find the following parameters:

(a) The M-by-M correlation matrix of the tap inputs of the predictor.

(b) The M-by-1 cross-correlation vector between the tap inputs of the predictor and the desired response $u(i)$.

(c) The minimum value of $\mathscr{E}_f$.

6. Figure 11.5(b) shows a *backward linear predictor* using a transversal structure, with the tap inputs $u(i-M+1), \ldots, u(i-1), u(i)$ used to make a linear prediction of the input $u(i-M)$. The problem is to find the tap-weight vector $\hat{\mathbf{w}}$ that minimizes the sum of backward prediction-error squares

$$\mathscr{E}_b = \sum_{i=M+1}^{N} |b_M(i)|^2$$

where $b_M(i)$ is the backward prediction error. Find the following parameters:

(a) The M-by-M correlation matrix of the tap inputs.

(b) The M-by-1 correlation vector between the tap inputs and the desired response $u(i-M)$.

(c) The minimum value of $\mathscr{E}_b$.

7. Use a direct approach to derive the system of normal equations given in expanded form in Eq. (11.31).

8. Calculate the singular values and singular vectors of the 2-by-2 real matrix:

$$\mathbf{A} = \begin{bmatrix} 1 & -1 \\ 0.5 & 2 \end{bmatrix}$$

Do the calculation using two different methods:

(a) Eigendecomposition of the matrix product $\mathbf{A}^T \mathbf{A}$.

(b) Eigendecomposition of the matrix product $\mathbf{A}\mathbf{A}^T$.

Hence, find the pseudoinverse of matrix $\mathbf{A}$.

9. Consider the 2-by-2 complex matrix

$$\mathbf{A} = \begin{bmatrix} 1 + j & 1 + 0.5j \\ 0.5 - j & 1 - j \end{bmatrix}$$

Calculate the singular values and singular vectors of the matrix **A** by proceeding as follows:

(a) Construct the matrix $\mathbf{A}^H\mathbf{A}$; hence, evaluate the eigenvalues and eigenvectors of $\mathbf{A}^H\mathbf{A}$.

(b) Construct the matrix $\mathbf{A}\mathbf{A}^H$; hence, evaluate the eigenvalues and eigenvectors of $\mathbf{A}\mathbf{A}^H$.

(c) Relate the eigenvalues and eigenvectors calculated in parts (a) and (b) to the singular values and singular vectors of **A**.

10. Refer back to Example 1 in Section 11.7. For the sets of data given in that example, do the following:

(a) Calculate the pseudo-inverse of the 3-by-2 data matrix **A**.

(b) Use this value of the pseudo-inverse $\mathbf{A}^+$ to calculate the two tap weights of the linear least-squares filter.

11. In this problem we explore the derivation of the weight update for the normalized LMS algorithm described in Eq. (9.144) using the idea of singular-value decomposition. This problem may be viewed as an extension of the discussion presented in Section 11.14. Find the minimum norm solution for the coefficient vector

$$\mathbf{c}(n+1) = \begin{bmatrix} \delta\hat{\mathbf{w}}(n+1) \\ 0 \end{bmatrix}$$

that satisfies the equation

$$\mathbf{x}^H(n)\mathbf{c}(n+1) = e^*(n)$$

where

$$\mathbf{x}(n) = \begin{bmatrix} \mathbf{u}(n) \\ \sqrt{a} \end{bmatrix}$$

Hence, show that

$$\hat{\mathbf{w}}(n+1) = \hat{\mathbf{w}}(n) + \frac{\tilde{\mu}}{a + \|\mathbf{u}(n)\|^2}\,\mathbf{u}(n)e^*(n)$$

where $a > 0$, and $0 < \tilde{\mu} < 2$. [This is the weight update described in Eq. (9.144).]

12. You are given a processor that is designed to perform the singular-value composition of a K-by-M data matrix **A**. Using such a processor, develop block diagrams for the following two super-resolution algorithms:

(a) The autoregressive (AR) algorithm

(b) The minimum-variance distortionless response (MVDR) algorithm

CHAPTER

12

Rotations and Reflections

In the previous chapter we emphasized the importance of singular value decomposition (SVD) as a tool for solving the linear least-squares problem. In this chapter we turn our attention to the practical issue of how to compute the SVD of a data matrix. With numerical stability as a primary design objective, the recommended procedure for SVD computation is to work directly with the data matrix. In this context, we may mention two different algorithms for SVD computation:

- *QR algorithm*, which proceeds by using a sequence of planar reflections known as Householder transformations
- *Cyclic Jacobi algorithm*, which employs a sequence of 2-by-2 plane rotations known as Jacobi rotations or Givens rotations

The cyclic Jacobi algorithm and QR algorithm are both *data adaptive* and *block-processing* oriented. They share a common goal, albeit in different ways:

- The diagonalization of the data matrix in a step-by-step fashion, and to within some prescribed numerical precision

It is important to note that plane rotations and reflections are wide ranging in their applications. In particular, they play a key role in the design of square-root Kalman filters and related linear adaptive filters. We therefore have reason in later chapters of the book to refer back to some of the fundamental concepts presented herein. However, the main focus of attention in this chapter is on numerically stable algorithms for SVD computation using rotations and reflections. We begin the discussion by considering plane rotations; planar reflections are considered later in this chapter.

12.1 PLANE ROTATIONS

An algebraic tool that is fundamental to the cyclic Jacobi algorithm is the 2-by-2 *orthogonal matrix*:[1]

$$\Theta = \begin{bmatrix} c & s \\ -s & c \end{bmatrix} \tag{12.1}$$

where c and s are real parameters defined by

$$c = \cos \theta \tag{12.2}$$

and

$$s = \sin \theta \tag{12.3}$$

with the trigonometric *constraint*:

$$c^2 + s^2 = 1 \tag{12.4}$$

We refer to the transformation Θ as a "plane rotation," because multiplication of a 2-by-1 data vector by Θ amounts to a plane rotation of that vector. This property holds whether the data vector is premultiplied or postmultiplied by Θ.

The transformation of Eq. (12.1) is referred to as the *Jacobi rotation* in honor of Jacobi (1846), who proposed a method for reducing a symmetric matrix to diagonal form. It is also referred to as the *Givens rotation*. In this book we will use the latter terminology, or simply plane rotation.

To illustrate the nature of this plane rotation, consider the case of a real 2-by-1 vector:

$$\mathbf{a} = \begin{bmatrix} a_i \\ a_k \end{bmatrix}$$

Then premultiplication of the vector $\mathbf{a}$ by Θ yields

$$\mathbf{x} = \Theta \mathbf{a}$$

$$= \begin{bmatrix} c & s \\ -s & c \end{bmatrix} \begin{bmatrix} a_i \\ a_k \end{bmatrix}$$

$$= \begin{bmatrix} ca_i + sa_k \\ -sa_i + ca_k \end{bmatrix}$$

---

[1]In SVD terminology (and eigenanalysis for that matter), the term "orthogonal matrix" is used in the context of *real* data, whereas the term "unitary matrix" is used for complex data.

We may readily show, in view of the definitions of the rotation parameters c and s, that the vector $\mathbf{x}$ has the same Euclidean length as the vector $\mathbf{a}$. Moreover, given that the angle θ is positive, the transformation Θ *rotates* the vector $\mathbf{a}$ in a clockwise direction into the new position defined by $\mathbf{x}$, as illustrated in Fig. 12.1. Note that the vectors $\mathbf{a}$ and $\mathbf{x}$ *remain in the same (i, k) plane*, hence the name "plane rotation."

12.2 TWO-SIDED JACOBI ALGORITHM

To pave the way for a development of the cyclic Jacobi algorithm, consider the simple case of a *real* 2-by-2 data matrix:

$$\mathbf{A} = \begin{bmatrix} a_{ii} & a_{ik} \\ a_{ki} & a_{kk} \end{bmatrix} \tag{12.5}$$

We assume that $\mathbf{A}$ is nonsymmetric; that is, $a_{ki} \neq a_{ik}$. The requirement is to *diagonalize* this 2-by-2 matrix. We do so by means of two plane rotations Θ_1 and Θ_2, as shown by

$$\underbrace{\begin{bmatrix} c_1 & s_1 \\ -s_1 & c_i \end{bmatrix}^T}_{\Theta_1} \underbrace{\begin{bmatrix} a_{ii} & a_{ik} \\ a_{ki} & a_{kk} \end{bmatrix}}_{\mathbf{A}} \underbrace{\begin{bmatrix} c_2 & s_2 \\ -s_2 & c_2 \end{bmatrix}}_{\Theta_2} = \underbrace{\begin{bmatrix} d_1 & 0 \\ 0 & d_2 \end{bmatrix}}_{\substack{\text{diagonal} \\ \text{matrix}}} \tag{12.6}$$

To design the two plane rotations indicated in Eq. (12.5), we proceed in two stages. Stage I transforms the 2-by-2 data matrix $\mathbf{A}$ into a symmetric matrix; we refer to this stage as "symmetrization." Stage II diagonalizes the symmetric matrix resulting from stage I; we refer to this second stage as "diagonalization." Of course, if the data matrix is symmetric to begin with, we proceed to stage II directly.

Stage I: Symmetrization. To transform the 2-by-2 data matrix $\mathbf{A}$ into a symmetric matrix, we premultiply it by the transpose of a plane rotation Θ and thus write

$$\underbrace{\begin{bmatrix} c & s \\ -s & c \end{bmatrix}^T}_{\Theta^T} \underbrace{\begin{bmatrix} a_{ii} & a_{ik} \\ a_{ki} & a_{kk} \end{bmatrix}}_{\mathbf{A}} = \underbrace{\begin{bmatrix} y_{ii} & y_{ik} \\ y_{ki} & y_{kk} \end{bmatrix}}_{\mathbf{Y}} \tag{12.7}$$

Expanding the left-hand side of Eq. (12.7) and equating terms, we get

$$y_{ii} = ca_{ii} - sa_{ki} \tag{12.8}$$

$$y_{kk} = sa_{ik} + ca_{kk} \tag{12.9}$$

$$y_{ik} = ca_{ik} - sa_{kk} \tag{12.10}$$

$$y_{ki} = sa_{ii} + ca_{ki} \tag{12.11}$$

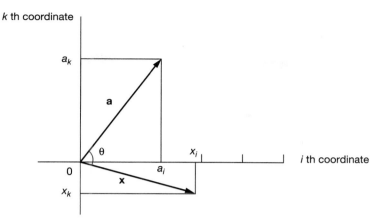

Figure 12.1   Plane rotation of a real 2-by-1 vector.

The purpose of stage I is to compute the cosine–sine pair (c, s) such that the 2-by-2 matrix $\mathbf{Y}$ produced by the plane rotation $\mathbf{\Theta}$ is symmetric. In other words, the elements y_{ik} and y_{ki} are to equal each other.

Define a parameter ρ as the ratio of c to s; that is,

$$\rho = \frac{c}{s} \tag{12.12}$$

We may relate ρ to the elements of the data matrix by setting $y_{ik} = y_{ki}$. Thus, using Eqs. (12.10) and (12.11), we obtain

$$\rho = \frac{a_{ii} + a_{kk}}{a_{ik} - a_{ki}}, \qquad a_{ki} \neq a_{ik} \tag{12.13}$$

Next, we determine the value of s by eliminating c between Eqs. (12.4) and (12.12); hence,

$$s = \frac{\text{sgn}(\rho)}{\sqrt{1 + \rho^2}} \tag{12.14}$$

The computation of c and s thus proceeds as follows:

- Use Eq. (12.13) to evaluate ρ.
- Use Eq. (12.14) to evaluate the sine parameter s.
- Use Eq. (12.12) to evaluate the cosine parameter c.

If $\mathbf{A}$ is symmetric to begin with, then $a_{ki} = a_{ik}$, in which case we have $s = 0$ and $c = 1$; that is, stage I is bypassed.

Stage II: Diagonalization. The purpose of stage II is to diagonalize the symmetric matrix $\mathbf{Y}$ produced in stage I. To do so, we premultiply and postmultiply it by $\boldsymbol{\Theta}_2^T$ and $\boldsymbol{\Theta}_2$, respectively, where $\boldsymbol{\Theta}_2$ is a second plane rotation to be determined. This operation is simply an orthogonal similarity transformation applied to a symmetric matrix. We may thus write

$$
\underbrace{\begin{bmatrix} c_2 & s_2 \\ -s_2 & c_2 \end{bmatrix}^T}_{\boldsymbol{\Theta}_2^T} \underbrace{\begin{bmatrix} y_{ii} & y_{ki} \\ y_{ki} & y_{kk} \end{bmatrix}}_{\mathbf{Y}} \underbrace{\begin{bmatrix} c_2 & s_2 \\ -s_2 & c_2 \end{bmatrix}}_{\boldsymbol{\Theta}_2} = \underbrace{\begin{bmatrix} d_1 & 0 \\ 0 & d_2 \end{bmatrix}}_{\mathbf{D}}
\tag{12.15}
$$

Expanding the left-hand side of Eq. (12.15) and then equating the respective diagonal terms, we get

$$
d_1 = c_2^2 y_{ii} - 2c_2 s_2 y_{ki} + s_2^2 y_{kk}
\tag{12.16}
$$

$$
d_2 = s_2^2 y_{ii} - 2c_2 s_2 y_{ki} + s_2^2 y_{kk}
\tag{12.17}
$$

Let o_1 and o_2 denote the off-diagonal terms of the 2-by-2 matrix formed by carrying out the matrix multiplications indicated on the left-hand side of Eq. (12.15). From symmetry considerations, we have

$$
o_1 = o_2
\tag{12.18}
$$

Evaluating these off-diagonal terms and equating them to zero for diagonalization, we get

$$
0 = (y_{ii} - y_{kk}) - \left(\frac{s_2}{c_2}\right) y_{ki} + \left(\frac{c_2}{s_2}\right) y_{ki}
\tag{12.19}
$$

Equation (12.19) suggests that we introduce the following two definitions:

$$
t = \frac{s_2}{c_2}
\tag{12.20}
$$

and

$$
\zeta = \frac{y_{kk} - y_{ii}}{2y_{ki}}
\tag{12.21}
$$

Hence, we may rewrite Eq. (12.19) as

$$
t^2 + 2\zeta t - 1 = 0
\tag{12.22}
$$

Equation (12.22) is a quadratic in t; it therefore has two possible solutions, yielding the following different plane rotations:

1. *Inner rotation,* for which we have the solution

$$
t = \frac{\text{sign}(\zeta)}{|\zeta| + \sqrt{1 + \zeta^2}}
\tag{12.23}
$$

Having computed t, we may use Eqs. (12.4) and (12.20) to solve for c_2 and s_2, obtaining

$$c_2 = \frac{1}{\sqrt{1 + t^2}} \tag{12.24}$$

and

$$s_2 = tc_2 \tag{12.25}$$

We note from Eqs. (12.2), (12.3), and (12.20) that the rotation angle θ_2 is related to t as follows:

$$\theta_2 = \arctan t \tag{12.26}$$

Hence, adoption of the solution given in Eq. (12.23) produces a plane rotation $\mathbf{\Theta}_2$ for which $|\theta_2|$ lies in the interval $[0, \pi/4]$; this rotation is therefore called an *inner rotation*. The computation thus proceeds as follows:

(a) Use Eq. (12.21) to compute ζ.
(b) Use Eq. (12.23) to compute t.
(c) Use Eqs. (12.24) and (12.25) to compute c_2 and s_2, respectively.

If the original matrix $\mathbf{A}$ is diagonal, then $a_{ik} = a_{ki} = 0$, in which case the angle $\theta_2 = 0$, and so the matrix remains unchanged.

2. *Outer rotation*, for which we have the solution:

$$t = - \,\text{sign}(\zeta)\, (|\zeta| + \sqrt{1 + \zeta^2}) \tag{12.27}$$

Having computed t, we may then evaluate c_2 and s_2 using the formulas (12.24) and (12.25), respectively; we may do so because the derivations of these two equations are independent of the quadratic equation (12.22). In this second case, however, the use of Eq. (12.27) in (12.26) yields a plane rotation for which $|\theta_2|$ lies in the interval $[\pi/4, \pi/2]$. The rotation associated with the second solution is therefore referred to as the *outer rotation*. Note that if the original matrix $\mathbf{A}$ is diagonal, then $a_{ik} = a_{ki} = 0$, in which case $\theta_2 = \pi/2$. In this special case, the diagonal elements of the matrix are interchanged, as shown by

$$\begin{bmatrix} 0 & -1 \\ 1 & 0 \end{bmatrix} \begin{bmatrix} a_{ii} & 0 \\ 0 & a_{kk} \end{bmatrix} \begin{bmatrix} 0 & 1 \\ -1 & 0 \end{bmatrix} = \begin{bmatrix} a_{kk} & 0 \\ 0 & a_{ii} \end{bmatrix} \tag{12.28}$$

Fusion of Rotations $\mathbf{\Theta}$ and $\mathbf{\Theta}_2$

Substituting the matrix $\mathbf{Y}$ of Eq. (12.7) in (12.15) and comparing the resulting equation with (12.6), we deduce the following definition for $\mathbf{\Theta}_1$ in terms of $\mathbf{\Theta}$ (determined in the symmetrization stage) and $\mathbf{\Theta}_2$ (determined in the diagonalization stage):

$$\mathbf{\Theta}_1^T = \mathbf{\Theta}_2^T \mathbf{\Theta}^T$$

or, equivalently,

$$\Theta_1 = \Theta \, \Theta_2 \qquad (12.29)$$

In other words, in terms of the cosine–sine parameters, we have

$$\underbrace{\begin{bmatrix} c_1 & s_1 \\ -s_1 & c_1 \end{bmatrix}}_{\Theta_1} = \underbrace{\begin{bmatrix} c & s \\ -s & c \end{bmatrix}}_{\Theta} \underbrace{\begin{bmatrix} c_2 & s_2 \\ -s_2 & c_2 \end{bmatrix}}_{\Theta_2} \qquad (12.30)$$

Expanding and equating terms, we thus obtain

$$c_1 = cc_2 - ss_2 \qquad (12.31)$$

and

$$s_1 = sc_2 + cs_2 \qquad (12.32)$$

For real data, from Eqs. (12.31) and (12.32) we find that the angles θ and θ_2, associated with the plane rotations Θ and Θ_2, respectively, add to produce the angle θ_1 associated with Θ_1.

Two Special Cases

For reasons that will become apparent later in Section 12.3, the Jacobi algorithm for computing the singular-value decomposition has to be capable of handling two special cases:

Case 1: $a_{kk} = a_{ik} = 0$. In this case we need only to perform the symmetrization of **A**, as shown by

$$\begin{bmatrix} c_1 & s_1 \\ -s_1 & c_1 \end{bmatrix}^T \begin{bmatrix} a_{ii} & 0 \\ a_{ki} & 0 \end{bmatrix} \begin{bmatrix} 1 & 0 \\ 0 & 1 \end{bmatrix} = \begin{bmatrix} d_1 & 0 \\ 0 & 0 \end{bmatrix} \qquad (12.33)$$

Case 2: $a_{kk} = a_{ki} = 0$. In this case, we have

$$\begin{bmatrix} 1 & 0 \\ 0 & 1 \end{bmatrix} \begin{bmatrix} a_{ii} & a_{ik} \\ 0 & 0 \end{bmatrix} \begin{bmatrix} c_2 & s_2 \\ -s_2 & c_2 \end{bmatrix} = \begin{bmatrix} d_1 & 0 \\ 0 & 0 \end{bmatrix} \qquad (12.34)$$

Additional Operations for Complex Data

The plane rotation described in Eq. (12.6) applies to real data only, because (to begin with) the cosine and sine parameters defining the rotation were all chosen to be real. To extend its application to the more general case of complex data, we have to perform additional operations on the data. At first sight, it may appear that we merely have to modify stage I (symmetrization) of the two-sided Jacobi algorithm so as to accommodate a complex 2-by-2 matrix. In reality, however, the issue of dealing with complex data (in the context of the Jacobi algorithm) is not so simple. The approach taken here is first to reduce the complex 2-by-2 matrix of Eq. (12.5) to a real form, and then proceed with the application of the two-sided Jacobi algorithm in the usual way.[2] The *complex-to-real data reduction* is performed by following a two-stage procedure, as described next.

Stage I: Triangularization.   Consider a complex 2-by-2 data matrix **A** having the form given in Eq. (12.5). Without loss of generality, we assume that the leading element a_{ii} is a positive real number. This assumption may be justified (if need be) by factoring out the exponential term $e^{j\theta_{ii}}$, where θ_{ii} is the phase angle of a_{ii}. The factorization has the effect of leaving inside the 2-by-2 matrix a positive real term equal to the magnitude of a_{ii}, and subtracting θ_{ii} from the phase angle of each of the remaining three complex terms in the matrix.

Let the matrix **A** so described be premultiplied by a 2-by-2 plane rotation for the purpose of its triangularization, as shown by

$$\begin{bmatrix} c & s^* \\ -s & c \end{bmatrix} \begin{bmatrix} a_{ii} & a_{ik} \\ a_{ki} & a_{kk} \end{bmatrix} = \begin{bmatrix} \omega_{ii} & \omega_{ik} \\ 0 & \omega_{kk} \end{bmatrix} \tag{12.35}$$

The cosine parameter c is real, but the sine parameter s is now *complex*. To emphasize this point, we write

$$s = |s| e^{j\alpha} \tag{12.36}$$

where $|s|$ is the magnitude of s, and α is its phase angle. In addition, the (c, s) pair is required to satisfy the constraint

$$c^2 + |s|^2 = 1 \tag{12.37}$$

The objective is to choose the (c, s) pair so as to annihilate the kith (off-diagonal) term. To do this, we must satisfy the condition

$$-sa_{ii} + ca_{ki} = 0$$

or, equivalently,

$$s = \frac{a_{ki}}{a_{ii}} c \tag{12.38}$$

Substituting Eq. (12.38) in (12.37), and solving for the cosine parameter, we get

$$c = \frac{|a_{ii}|}{\sqrt{|a_{ii}|^2 + |a_{ki}|^2}} \tag{12.39}$$

Note that, in Eq. (12.39), we have chosen to work with the *positive* real root for the cosine parameter c. Also, if a_{ki} is zero, that is, the data matrix is upper triangular to begin with, then $c = 1$ and $s = 0$, in which case we may bypass stage I. If, by the same token, a_{ik} is zero, we apply transposition and proceed to stage II.

---

[2]F. T. Luk, private communication, 1990.

Having determined the values of c and s needed for the triangularization of the 2-by-2 matrix $\mathbf{A}$, we may now determine the elements of the resulting upper triangular matrix shown on the right-hand side of Eq. (12.35) as follows:

$$\omega_{ii} = ca_{ii} + s^*a_{ki} \tag{12.40}$$

$$\omega_{ik} = ca_{ik} + s^*a_{kk} \tag{12.41}$$

$$\omega_{kk} = -sa_{ik} + ca_{kk} \tag{12.42}$$

Given that a_{ii} is positive real, by assumption, the use of Eqs. (12.38) and (12.39) in (12.40) reveals that the diagonal element ω_{ii} is real and nonnegative; that is,

$$\omega_{ii} \geq 0 \tag{12.43}$$

In general, however, the remaining two elements ω_{ik} and ω_{kk} of the upper triangular matrix on the right-hand side of Eq. (12.35) are complex valued.

Stage II: Phase Cancelation.   As already mentioned, the elements ω_{ik} and ω_{kk} may be complex. To reduce them to real form, we premultiply and postmultiply the upper triangular matrix on the right-hand side of Eq. (12.35) by a pair of *phase-canceling diagonal matrices* as follows:

$$\begin{bmatrix} e^{-j\beta} & 0 \\ 0 & e^{-j\gamma} \end{bmatrix} \begin{bmatrix} \omega_{ii} & \omega_{ik} \\ 0 & \omega_{kk} \end{bmatrix} \begin{bmatrix} e^{j\beta} & 0 \\ 0 & 1 \end{bmatrix} = \begin{bmatrix} \omega_{ii} & |\omega_{ik}| \\ 0 & |\omega_{kk}| \end{bmatrix} \tag{12.44}$$

The *rotation angles* β and γ of the premultiplying matrix are chosen so as to cancel the phase angles of ω_{ik} and ω_{kk}, respectively, as shown by

$$\beta = \arg(\omega_{ik}) \tag{12.45}$$

$$\gamma = \arg(\omega_{kk}) \tag{12.46}$$

The postmultiplying matrix is included so as to *correct* for the phase change in the element ω_{ii} produced by the premultiplying matrix. In other words, the combined process of premultiplication and postmultiplication in Eq. (12.44) leaves the diagonal element ω_{ii} unchanged.

Stage II thus yields an upper triangular matrix whose three nonzero elements are all real and nonnegative. Note that the procedure described herein for reducing the complex 2-by-2 matrix $\mathbf{A}$ to a real upper triangular form requires four degrees of freedom, namely, the (c, s) pair, and the angles β and γ. The way is now paved for us to proceed with the application of the Jacobi method for a real 2-by-2 matrix, as described earlier in the section.

12.3 CYCLIC JACOBI ALGORITHM

We are now ready to describe the *cyclic Jacobi algorithm* or *generalized Jacobi algorithm* for a square data matrix by solving an appropriate sequence of 2-by-2 singular-value decomposition problems. The description will be presented for real data. To deal with complex data, we incorporate the complex-to-real data reduction developed in the preceding section.

Let $\Theta_1(i, k)$ denote a plane rotation in the (i, k) plane, where $k > i$. The matrix $\Theta_1(i, k)$ is the same as the M-by-M identity matrix, except for the four strategic elements located on rows i, k and columns i, k, as shown by

$$
\Theta_1(i, k) =
\begin{bmatrix}
1 & 0 & \cdots & 0 & \cdots & 0 \\
 & \ddots & & & & \\
0 & c_1 & \cdots & s_1 & \cdots & 0 \\
 & & & & & \\
0 & -s_1 & \cdots & c_1 & \cdots & 0 \\
 & & & & & \\
0 & 0 & \cdots & 0 & \cdots & 1
\end{bmatrix}
\begin{array}{l} \\ \\ \longleftarrow \text{row } i \\ \\ \\ \longleftarrow \text{row } k \\ \\ \\ \\ \end{array}
\qquad (12.47)
$$

$$\underset{\text{column } i}{\uparrow} \qquad \underset{\text{column } k}{\uparrow}$$

Let $\Theta_2(i, k)$ denote a second plane rotation in the (i, k) plane that is similarly defined; the dimension of this second transformation is also M. The Jacobi transformation of the data matrix $\mathbf{A}$ is thus described by

$$\mathbf{T}_{ik}: \mathbf{A} \leftarrow \Theta_1^T(i, k)\mathbf{A}\Theta_2(i, k) \qquad (12.48)$$

The Jacobi rotations $\Theta_1(i, k)$ and $\Theta_2(i, k)$ are designed to annihilate the (i, k) and (k, i) elements of $\mathbf{A}$. Accordingly, the transformation $\mathbf{T}_{ik}$ produces a matrix $\mathbf{X}$ (equal to the updated value of $\mathbf{A}$) that is more diagonal than the original $\mathbf{A}$ in the sense that

$$\text{off}(\mathbf{X}) = \text{off}(\mathbf{A}) - a_{ik}^2 - a_{ki}^2 \qquad (12.49)$$

where off($\mathbf{A}$) is the *norm of the off-diagonal elements*:

$$\text{off}(\mathbf{A}) = \sum_{i=1}^{M} \sum_{\substack{k=1 \\ k \neq i}}^{M} a_{ik}^2 \qquad \text{for } \mathbf{A} = \{a_{ik}\} \qquad (12.50)$$

In the cyclic Jacobi algorithm the transformation (12.48) is applied for a total of $m = M(M - 1)/2$ different index pairs ("pivots") that are selected in some fixed order. Such a sequence of m transformations is called a *sweep*. The construction of a sweep may be *cyclic by rows* or *cyclic by columns*, as illustrated in Example 1 below. In either case, we obtain a new matrix $\mathbf{A}$ after each sweep, for which we compute off($\mathbf{A}$). If off($\mathbf{A}$) $\leq \delta$, where δ is some small machine-dependent number, we stop the computation. If on the other hand, off($\mathbf{A}$) $> \delta$, we repeat the computation. For typical values of δ [e.g., $\delta = 10^{-12}$ off($\mathbf{A}_0$), where $\mathbf{A}_0$ is the original matrix], the algorithm converges in about 4 to 10 sweeps for values of M in the range of 4 to 2000.

As far as we know, the row ordering or the column ordering is the only ordering that guarantees convergence of the Jacobi cyclic algorithm.[3] By "convergence" we mean

$$\text{off}(\mathbf{A}^{(k)}) \longrightarrow 0 \qquad \text{as } k \longrightarrow \infty \tag{12.51}$$

where $\mathbf{A}^{(k)}$ is the M-by-M matrix computed after sweep number k.

Example 1

Consider a 4-by-4 *real* matrix $\mathbf{A}$. With the matrix dimension $M = 4$, we have a total of six orderings in each sweep. A sweep of orderings cyclic by rows is represented by

$$\mathbf{T}_R = \mathbf{T}_{34}\mathbf{T}_{24}\mathbf{T}_{23}\mathbf{T}_{14}\mathbf{T}_{13}\mathbf{T}_{12}$$

A sweep of orderings cyclic by columns is represented by

$$\mathbf{T}_C = \mathbf{T}_{34}\mathbf{T}_{24}\mathbf{T}_{14}\mathbf{T}_{23}\mathbf{T}_{13}\mathbf{T}_{12}$$

It is easily checked that the transformation $\mathbf{T}_{ik}$ and $\mathbf{T}_{pq}$ *commute* if two conditions hold:

1. The index i is neither p nor q.
2. The index k is neither p nor q.

Accordingly, we find that the transformations $\mathbf{T}_R$ and $\mathbf{T}_C$ are indeed equivalent, as they should be.

Consider next the application of the transformation $\mathbf{T}_R$ (obtained from the sweep of orderings cyclic by rows) to the data matrix $\mathbf{A}$. In particular, using the rotation of Eq. (12.48), we may write the following transformations:

$$\mathbf{T}_{12} : \mathbf{A} \leftarrow \mathbf{\Theta}_1^T(1, 2)\mathbf{A}\mathbf{\Theta}_2(1, 2)$$

$$\mathbf{T}_{13}\mathbf{T}_{12} : \mathbf{A} \leftarrow \mathbf{\Theta}_3^T(1, 3)\mathbf{\Theta}_1^T(1, 2)\mathbf{A}\mathbf{\Theta}_2(1, 2)\mathbf{\Theta}_4(1, 3)$$

$$\mathbf{T}_{14}\mathbf{T}_{13}\mathbf{T}_{12} : \mathbf{A} \leftarrow \mathbf{\Theta}_5^T(1, 4)\mathbf{\Theta}_3^T(1, 3)\mathbf{\Theta}_1^T(1, 2)\mathbf{A}\mathbf{\Theta}_2(1, 2)\mathbf{\Theta}_4(1, 3)\mathbf{\Theta}_6(1, 4)$$

and so on. The final step in this sequence of transformations may be written as

$$\mathbf{T}_R : \mathbf{A} \longleftarrow \mathbf{U}^T\mathbf{A}\mathbf{V}$$

---

[3]A proof of convergence of the Jacobi cyclic algorithm, based on row ordering or column ordering, is given in Forsythe and Henrici (1960). Subsequently, Luk and Park (1989) proved that many of the orderings used in parallel implementation of the algorithm are equivalent to the row ordering, and thus guarantee convergence as well.

which defines the singular value decomposition of the real data matrix $\mathbf{A}$. The orthogonal matrices $\mathbf{U}$ and $\mathbf{V}$ are respectively defined by

$$\mathbf{U} = \mathbf{\Theta}_1(1, 2)\mathbf{\Theta}_3(1, 3)\mathbf{\Theta}_5(1, 4)\mathbf{\Theta}_7(2, 3)\mathbf{\Theta}_9(2, 4)\mathbf{\Theta}_{11}(3, 4)$$

and

$$\mathbf{V} = \mathbf{\Theta}_2(1, 2)\mathbf{\Theta}_4(1, 3)\mathbf{\Theta}_6(1, 4)\mathbf{\Theta}_8(2, 3)\mathbf{\Theta}_{10}(2, 4)\mathbf{\Theta}_{12}(3, 4)$$

Rectangular Data Matrix

Thus far we have focused attention on the cyclic Jacobi algorithm for computing the singular-value decomposition of a square matrix. To handle the more general case of a rectangular matrix, we may extend the use of this algorithm by proceeding as follows. Consider first the case of a K-by-M real data matrix $\mathbf{A}$, for which K is greater than M. We generate a square matrix by appending $(K - M)$ columns of zeros to $\mathbf{A}$. We may thus write

$$\tilde{\mathbf{A}} = [\mathbf{A}, \mathbf{O}] \tag{12.52}$$

We refer to $\tilde{\mathbf{A}}$ as the *augmented data matrix*. We then proceed as before by applying the cyclic Jacobi algorithm to the K-by-K matrix $\tilde{\mathbf{A}}$. In performing this computation, we require the use of special case 1 described in Eq. (12.33). In any event, we emerge with the factorization

$$\mathbf{U}^T[\mathbf{A}, \mathbf{O}]\begin{bmatrix} \mathbf{V} & \mathbf{O} \\ \mathbf{O} & \mathbf{I} \end{bmatrix} = \text{diag}(\sigma_1, \ldots, \sigma_M, 0, \ldots, 0) \tag{12.53}$$

The desired factorization of the original data matrix $\mathbf{A}$ is obtained by writing

$$\mathbf{U}^T\mathbf{A}\mathbf{V} = \text{diag}(\sigma_1, \ldots, \sigma_M) \tag{12.54}$$

If, on the other hand, the dimension M of matrix $\mathbf{A}$ is greater than K, we augment it by adding $(M - K)$ rows; we may thus write

$$\tilde{\mathbf{A}} = \begin{bmatrix} \mathbf{A} \\ \mathbf{O} \end{bmatrix} \tag{12.55}$$

We then treat the square matrix $\tilde{\mathbf{A}}$ in the same way as before. In this second situation, we require the use of special case 2 described in Eq. (12.34).

In the case of a complex rectangular data matrix $\mathbf{A}$, we may proceed in a fashion similar to that described above, except for a change in the characterization of matrices $\mathbf{U}$ and $\mathbf{V}$. For a real data matrix, the matrices $\mathbf{U}$ and $\mathbf{V}$ are both orthogonal, whereas for a complex data matrix they are both unitary.

The strategy of matrix augmentation described herein represents a straightforward extension of the cyclic Jacobi algorithm for a square matrix. A drawback of this approach, however, is that the algorithm becomes too inefficient if the dimension K of matrix $\mathbf{A}$ is much greater than the dimension M, or vice versa.[4]

12.4 HOUSEHOLDER TRANSFORMATION

We turn next to the *Householder transformation* or the *Householder matrix*, which is so named in recognition of its originator (Householder, 1958 a,b, 1964). To proceed with a discussion of this issue, let $\mathbf{u}$ be an M-by-1 vector whose Euclidean norm is

$$\|\mathbf{u}\| = (\mathbf{u}^H\mathbf{u})^{1/2}$$

Then, the Householder transformation, denoted by an M-by-M matrix $\mathbf{Q}$, is defined by

$$\mathbf{Q} = \mathbf{I} - \frac{2\mathbf{u}\mathbf{u}^H}{\|\mathbf{u}\|^2} \tag{12.56}$$

where $\mathbf{I}$ is the M-by-M identity matrix.

For a geometric interpretation of the Householder transformation, consider an M-by-1 vector $\mathbf{x}$ premultiplied by the matrix $\mathbf{Q}$, as shown by

$$\mathbf{Q}\mathbf{x} = \left(\mathbf{I} - \frac{2\mathbf{u}\mathbf{u}^H}{\|\mathbf{u}\|^2}\right)\mathbf{x}$$

$$= \mathbf{x} - \frac{2\mathbf{u}^H\mathbf{x}}{\|\mathbf{u}\|^2}\mathbf{u} \tag{12.57}$$

By definition, the *projection* of $\mathbf{x}$ onto $\mathbf{u}$ is given by

$$\mathbf{P}_u(\mathbf{x}) = \frac{\mathbf{u}^H\mathbf{x}}{\|\mathbf{u}\|^2}\mathbf{u} \tag{12.58}$$

This projection is illustrated in Fig. 12.2. In this figure we have also included the vector representation of the product $\mathbf{Q}\mathbf{x}$. We thus see that $\mathbf{Q}\mathbf{x}$ is the mirror-image *reflection* of the vector $\mathbf{x}$ with respect to the hyperplane span $\{\mathbf{u}\}^\perp$ which is perpendicular to the vector $\mathbf{u}$. It is for this reason that the Householder transformation is also known as the *Householder reflection.*[5]

---

[4]An alternative approach that overcomes this difficulty is to proceed as follows (Luk, 1986):

1. *Triangularize* the K-by-M data matrix $\mathbf{A}$ by performing a QR-decomposition, defined by

$$\mathbf{A} = \mathbf{Q}\begin{bmatrix}\mathbf{R}\\\mathbf{O}\end{bmatrix}$$

where $\mathbf{Q}$ is a K-by-K orthogonal matrix, and $\mathbf{R}$ is an M-by-M upper triangular matrix.

2. Diagonalize the matrix $\mathbf{R}$ using the cyclic Jacobi algorithm.

3. Combine the results of steps 1 and 2.

[5]For a tutorial review of the Householder transformation and its use in adaptive signal processing, see Steinhardt (1988).

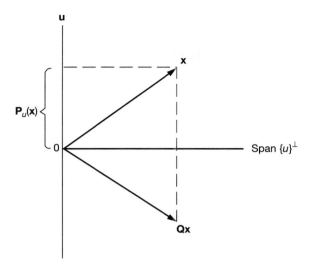

Figure 12.2 Geometric interpretation of the Householder transformation.

The Householder transformation, defined in Eq. (12.56), has the following properties:

Property 1.   *The Householder transformation* $\mathbf{Q}$ *is Hermitian*; that is,

$$\mathbf{Q}^H = \mathbf{Q} \tag{12.59}$$

Property 2.   *The Householder transformation* $\mathbf{Q}$ *is unitary*; that is,

$$\mathbf{Q}^{-1} = \mathbf{Q}^H \tag{12.60}$$

Property 3.   *The Householder transformation is length preserving*; that is,

$$\|\mathbf{Q}\mathbf{x}\| = \|\mathbf{x}\| \tag{12.61}$$

This property is illustrated in Fig. 12.2, where we see that the vector $\mathbf{x}$ and its reflection $\mathbf{Q}\mathbf{x}$ have exactly the same length.

Property 4.   *If two vectors undergo the same Householder transformation, their inner product remains unchanged.*

Consider any three vectors, $\mathbf{x}$, $\mathbf{y}$, and $\mathbf{u}$. Let the Householder matrix $\mathbf{Q}$ be defined in terms of the vector $\mathbf{u}$, as in Eq. (12.56). Let the remaining two vectors $\mathbf{x}$ and $\mathbf{y}$ be transformed by $\mathbf{Q}$, yielding $\mathbf{Q}\mathbf{x}$ and $\mathbf{Q}\mathbf{y}$, respectively. The inner product of these two transformed vectors is

$$(\mathbf{Q}\mathbf{x})^H(\mathbf{Q}\mathbf{y}) = \mathbf{x}^H\mathbf{Q}^H\mathbf{Q}\mathbf{y} \tag{12.62}$$

$$= \mathbf{x}^H\mathbf{y}$$

where we have made use of Property 2. Hence, the transformed vectors $\mathbf{Qx}$ and $\mathbf{Qy}$ have the same inner product as the original vectors $\mathbf{x}$ and $\mathbf{y}$.

Property 4 has important practical implications in the numerical solution of linear least-squares problems. Specifically, Householder transformations are used to reduce the given data matrix to a *sparse* matrix (i.e., one that consists mostly of zeros), but which is "equivalent" to the original data matrix in some mathematical sense. Needless to say, the particular form of matrix sparseness used depends on the application of interest. Whatever the application, however, the data reduction is used in order to simplify the numerical computations involved in solving the problem. In this context, a popular form of data reduction is that of *triangularization*, referring to the reduction of a full data matrix to an upper triangular one. Given this form of data reduction, we may then simply use *Gaussian elimination* to perform the matrix inversion and thereby compute the least-squares solution to the problem.

Properties 1 through 4 apply not only to the Householder transformations but also to the Givens rotations. It is the next two properties that distinguish Householder transformations from Givens rotations.

Property 5. *Given the Householder transformation $\mathbf{Q}$, the transformed vector $\mathbf{Qx}$ is a reflection of $\mathbf{x}$ above the hyperplane perpendicular to the vector $\mathbf{u}$ involved in the definition of $\mathbf{Q}$.*

This property is merely a restatement of Eq. (12.57). The following two limiting cases of Property 5 are especially noteworthy:

- The vector $\mathbf{x}$ is a scalar multiple of $\mathbf{u}$: In this case, Eq. (12.57) simplifies to

$$\mathbf{Qx} = -\mathbf{x}$$

- The vector $\mathbf{x}$ is orthogonal to $\mathbf{u}$; that is, their inner product is zero: In this second case, Eq. (12.57) reduces to

$$\mathbf{Qx} = \mathbf{x}$$

Property 6. *Let $\mathbf{x}$ be any nonzero M-by-1 vector with Euclidean norm $\|\mathbf{x}\|$. Let the M-by-1 vector $\mathbf{1}$ denote the first column of the identity matrix; that is,*

$$\mathbf{1} = [1, 0, \ldots, 0]^T \tag{12.63}$$

Then there exists a Householder transformation $\mathbf{Q}$ defined by the vector

$$\mathbf{u} = \mathbf{x} - \|\mathbf{x}\|\mathbf{1} \tag{12.64}$$

such that the transformed vector $\mathbf{Qx}$ corresponding to $\mathbf{u}$ is a linear multiple of the vector $\mathbf{1}$.

With the vector $\mathbf{u}$ assigned the value in Eq. (12.64), we have

$$
\begin{aligned}
\|\mathbf{u}\|^2 &= \mathbf{u}^H\mathbf{u} \\
&= (\mathbf{x} - \|\mathbf{x}\|\mathbf{1})^H(\mathbf{x} - \|\mathbf{x}\|\mathbf{1}) \\
&= 2\|\mathbf{x}\|^2 - 2\|\mathbf{x}\|x_1 \\
&= 2\|\mathbf{x}\|(\|\mathbf{x}\| - x_1)
\end{aligned}
\tag{12.65}
$$

where x_1 is the first element of the vector $\mathbf{x}$. Similarly, we may write

$$
\begin{aligned}
\mathbf{u}^H\mathbf{x} &= (\mathbf{x} - \|\mathbf{x}\|\mathbf{1})^H\mathbf{x} \\
&= \|\mathbf{x}\|^2 - \|\mathbf{x}\|x_1 \\
&= \|\mathbf{x}\|(\|\mathbf{x}\| - x_1)
\end{aligned} \tag{12.66}
$$

Accordingly, substituting Eqs. (12.65) and (12.66) in Eq. (12.57), we find that the transformed vector $\mathbf{Qx}$ corresponding to the defining vector $\mathbf{u}$ of Eq. (12.64) is given by

$$
\begin{aligned}
\mathbf{Qx} &= \mathbf{x} - \mathbf{u} \\
&= \mathbf{x} - (\mathbf{x} - \|\mathbf{x}\|\mathbf{1}) \\
&= \|\mathbf{x}\|\mathbf{1}
\end{aligned} \tag{12.67}
$$

which proves Property 6.

From Eq. (12.65), we observe that the first element x_1 of the vector $\mathbf{x}$ has to be real, and the Euclidean norm of $\mathbf{x}$ has to satisfy the condition

$$
\|\mathbf{x}\| > |x_1| \tag{12.68}
$$

This condition merely says that not only the first element of $\mathbf{x}$ but also one other element must be nonzero. Then, the vector $\mathbf{u}$ defined by Eq. (12.64) is indeed effective.

Property 6 makes the Householder transformation a very powerful computational tool. Given a vector $\mathbf{x}$, we may use Eq. (12.64) to define the vector $\mathbf{u}$ such that the corresponding Householder transformation $\mathbf{Q}$ annihilates all the M elements of the vector $\mathbf{x}$ except for the first one. This result is equivalent to the application of $(M - 1)$ plane rotations, with a minor difference: The determinant of the Householder matrix $\mathbf{Q}$ defined in Eq. (12.56) is

$$
\begin{aligned}
\det(\mathbf{Q}) &= \det\left(\mathbf{I} - \frac{2\mathbf{u}\mathbf{u}^H}{\|\mathbf{u}\|^2}\right) \\
&= -1
\end{aligned} \tag{12.69}
$$

Hence, the Householder transformation reverses the orientation of the configuration.

Having familiarized ourselves with the Householder transformation, we are ready to resume our discussion of SVD computation by describing the QR algorithm, which we do in the next section.

12.5 THE QR ALGORITHM

The starting point in the development of the *QR algorithm* for SVD computation is that of finding a class of unitary matrices, which preserve the singular values of a data matrix $\mathbf{A}$. In this context, the matrix $\mathbf{A}$ is said to be *unitarily equivalent* to another matrix $\mathbf{B}$ if

$$
\mathbf{B} = \mathbf{PAQ} \tag{12.70}
$$

where $\mathbf{P}$ and $\mathbf{Q}$ are unitary matrices; that is,

$$\mathbf{P}^H\mathbf{P} = \mathbf{I}$$

$$\mathbf{Q}^H\mathbf{Q} = \mathbf{I}$$

Consequently, we have

$$\mathbf{B}^H\mathbf{B} = \mathbf{Q}^H\mathbf{A}^H\mathbf{P}^H\mathbf{P}\mathbf{A}\mathbf{Q} \qquad (12.71)$$

$$= \mathbf{Q}^H\mathbf{A}^H\mathbf{A}\mathbf{Q}$$

Postmultiplying the correlation matrix $\mathbf{A}^H\mathbf{A}$ by a unitary matrix $\mathbf{Q}$ and premultiplying it by the Hermitian transpose of the matrix $\mathbf{Q}$ leaves the eigenvalues of $\mathbf{A}^H\mathbf{A}$ unchanged. Accordingly, the correlation matrices $\mathbf{A}^H\mathbf{A}$ and $\mathbf{B}^H\mathbf{B}$, or more simply, the matrices $\mathbf{A}$ and $\mathbf{B}$ themselves are said to be *eigen-equivalent*.

The purpose of using the transformation defined in Eq. (12.70) is to reduce the data matrix $\mathbf{A}$ to *upper bidiagonal* form, with eigen-equivalence maintained, for which Householder transformations are well suited. The reduced data matrix $\mathbf{B}$ is said to be upper bidiagonal if all of its elements except for those on the main diagonal and the superdiagonal are zero; that is, the ijth element of $\mathbf{B}$ is

$$b_{ij} = 0 \qquad \text{whenever } i > j \text{ or } j > i + 1 \qquad (12.72)$$

Having reduced the data matrix $\mathbf{A}$ to upper bidiagonal form, the next step is the application of the Golub–Kahan SVD algorithm. These two steps, in turn, are considered next.

Householder Bidiagonalization

Consider a K-by-M data matrix $\mathbf{A}$, where $K \geq M$. Let $\mathbf{Q}_1, \mathbf{Q}_2, \ldots, \mathbf{Q}_M$ denote a set of K-by-K Householder matrices, and let $\mathbf{P}_1, \mathbf{P}_2, \ldots, \mathbf{P}_{M-2}$ denote another set of M-by-M Householder matrices. In order to reduce the data matrix $\mathbf{A}$ to upper bidiagonal form, we determine the products of Householder matrices

$$\mathbf{Q}_B = \begin{cases} \mathbf{Q}_1\mathbf{Q}_2 \cdots \mathbf{Q}_{M-1}, & K = M \\ \mathbf{Q}_1\mathbf{Q}_2 \cdots \mathbf{Q}_M, & K > M \end{cases} \qquad (12.73)$$

and

$$\mathbf{P}_B = \mathbf{P}_1\mathbf{P}_2 \cdots \mathbf{P}_{M-2} \qquad (12.74)$$

such that

$$\mathbf{Q}_B^H\mathbf{A}\mathbf{P}_B = \mathbf{B} = \left[\begin{array}{c} \begin{matrix} d_1 & f_2 & & & \mathbf{O} \\ & d_2 & \cdot & \cdot & \\ & & \cdot & \cdot & f_M \\ & & & \cdot & d_M \end{matrix} \\ \hline \mathbf{O} \hspace{3cm} \\ \mathbf{O} \end{array}\right] \left.\begin{array}{c} \\ \\ \end{array}\right\} (K-M)\text{-by-}M \text{ null matrix} \qquad (12.75)$$

For $K > M$, premultiplication of the data matrix $\mathbf{A}$ by the Householder matrices $\mathbf{Q}_1$, $\mathbf{Q}_2$, ..., $\mathbf{Q}_M$ corresponds to reflecting the respective columns of $\mathbf{A}$, whereas postmultiplication by the Householder matrices $\mathbf{P}_1$, $\mathbf{P}_2$, ..., $\mathbf{P}_{M-2}$ corresponds to reflecting the respective rows of $\mathbf{A}$. The desired upper bidiagonal form is attained by "ping-ponging" column and row reflections. Note that for $K > M$ the number of Householder matrices constituting $\mathbf{Q}_B$ is M, whereas those constituting $\mathbf{P}_B$ number $M - 2$. Note also that, by construction, the matrix product $\mathbf{P}_1\mathbf{P}_2 \ldots \mathbf{P}_{M-2}$ does *not* alter the first column of any matrix that it postmultiplies.

We illustrate this data reduction process by way of an example.

Example 2

Consider a 5-by-4 data matrix $\mathbf{A}$ written in expanded form as follows:

$$\mathbf{A} = \begin{bmatrix} x & x & x & x \\ x & x & x & x \\ x & x & x & x \\ x & x & x & x \\ x & x & x & x \end{bmatrix}$$

where the x's denote nonzero matrix entries. The upper bidiagonalization of $\mathbf{A}$ proceeds as follows. First, $\mathbf{Q}_1$ is chosen so that $\mathbf{Q}_1^H \mathbf{A}$ has zeros in the positions distinguished below:

$$\begin{bmatrix} x & x & x & x \\ \otimes & x & x & x \\ \otimes & x & x & x \\ \otimes & x & x & x \\ \otimes & x & x & x \end{bmatrix}$$

Thus, $\mathbf{Q}_1^H \mathbf{A}$ may be written as

$$\mathbf{Q}_1^H \mathbf{A} = \begin{bmatrix} x & x & \otimes & \otimes \\ 0 & x & x & x \\ 0 & x & x & x \\ 0 & x & x & x \\ 0 & x & x & x \end{bmatrix} \tag{12.76}$$

Next, $\mathbf{P}_1$ is chosen so that $\mathbf{Q}_1^H \mathbf{A} \mathbf{P}_1$ has zeros in the positions distinguished in the first row of $\mathbf{Q}_1^H \mathbf{A}$, as in Eq. (12.76). Hence, $\mathbf{Q}_1^H \mathbf{A} \mathbf{P}_1$ has the form

$$\mathbf{Q}_1^H \mathbf{A} \mathbf{P}_1 = \begin{bmatrix} x & x & 0 & 0 \\ 0 & x & x & x \\ 0 & x & x & x \\ 0 & x & x & x \\ 0 & x & x & x \end{bmatrix} \tag{12.77}$$

Note that $\mathbf{P}_1$ does not affect the first column of the matrix.

The data reduction is continued by operating on the trailing 4-by-3 submatrix of $\mathbf{Q}^H \mathbf{A} \mathbf{P}_1$ that has nonzero entries. Specifically, we choose $\mathbf{Q}_2$ and $\mathbf{P}_2$ so that $\mathbf{Q}_2^H \mathbf{Q}_1^H \mathbf{A} \mathbf{P}_1 \mathbf{P}_2$ has the form

$$\mathbf{Q}_2^H \mathbf{Q}_1^H \mathbf{A} \mathbf{P}_1 \mathbf{P}_2 = \begin{bmatrix} x & x & 0 & 0 \\ 0 & x & x & 0 \\ \hline 0 & 0 & x & x \\ 0 & 0 & x & x \\ 0 & 0 & x & x \end{bmatrix} \tag{12.78}$$

Next, we operate on the trailing 3-by-2 submatrix of $\mathbf{Q}_2^H \mathbf{Q}_1^H \mathbf{A} \mathbf{P}_1 \mathbf{P}_2$ that has nonzero entries. Specifically, we choose $\mathbf{Q}_3$ such that $\mathbf{Q}_3^H \mathbf{Q}_2^H \mathbf{Q}_1^H \mathbf{A} \mathbf{P}_1 \mathbf{P}_2$ has the form

$$\mathbf{Q}_3^H \mathbf{Q}_2^H \mathbf{Q}_1^H \mathbf{A} \mathbf{P}_1 \mathbf{P}_2 = \begin{bmatrix} x & x & 0 & 0 \\ 0 & x & x & 0 \\ 0 & 0 & x & x \\ \hline 0 & 0 & 0 & x \\ 0 & 0 & 0 & x \end{bmatrix} \tag{12.79}$$

Finally, we choose $\mathbf{Q}_4$ to operate on the trailing 2-by-1 submatrix of $\mathbf{Q}_3^H \mathbf{Q}_2^H \mathbf{Q}_1^H \mathbf{Q} \mathbf{P}_1 \mathbf{P}_2$, such that we may write

$$\mathbf{B} = \mathbf{Q}_4^H \mathbf{Q}_3^H \mathbf{Q}_2^H \mathbf{Q}_1^H \mathbf{A} \mathbf{P}_1 \mathbf{P}_2 = \begin{bmatrix} x & x & 0 & 0 \\ 0 & x & x & 0 \\ 0 & 0 & x & x \\ 0 & 0 & 0 & x \\ 0 & 0 & 0 & 0 \end{bmatrix} \tag{12.80}$$

This completes the upper bidiagonalization of the data matrix $\mathbf{A}$.

The Golub–Kahan Step

The bidiagonalization of the data matrix $\mathbf{A}$ is followed by an *iterative process* that reduces it further to *diagonal* form. Referring to Eq. (12.75), we see that the matrix $\mathbf{B}$, resulting from the bidiagonalization of $\mathbf{A}$, is zero below the Mth row. Evidently, the last $K - M$ rows of zeros in the matrix $\mathbf{B}$ do *not* contribute to the singular values of the original data matrix $\mathbf{A}$. Accordingly, it is convenient to delete the last $K - M$ rows of matrix $\mathbf{B}$ and thus treat it as a square matrix with dimension M. The basis of the diagonalization of matrix $\mathbf{B}$ is the *Golub–Kahan algorithm* (Golub and Kahan, 1965), which is an adaptation of the QR *algorithm* developed originally for solving the symmetric eigenvalue problem.[6]

---

[6]The explicit form of the QR algorithm is a variant of the QL algorithm discussed in Chapter 4.

Let **B** denote an M-by-M upper bidiagonal matrix having no zeros on its main diagonal or superdiagonal. The first iteration of the Golub–Kahan algorithm proceeds as follows (Golub and Kahan, 1965; Golub and Van Loan, 1989):

1. Identify the trailing 2-by-2 submatrix of the product $\mathbf{T} = \mathbf{B}^H \mathbf{B}$, which has the form

$$
\begin{bmatrix}
|d_{M-1}|^2 + |f_{M-1}|^2 & d_{M-1}^* f_M \\
f_M^* d_{M-1} & |d_M|^2 + |f_M|^2
\end{bmatrix}
\tag{12.81}
$$

where d_{M-1} and d_M are the trailing diagonal elements of matrix **B**, and f_{M-1} and f_M are the trailing superdiagonal elements; see the right-hand side of Eq. (12.75). Compute the eigenvalue λ of this 2-by-2 submatrix, which is closer to $|d_M|^2 + |f_M|^2$; this particular eigenvalue λ is known as the *Wilkinson shift*.

2. Compute the Givens rotation parameters c_1 and s_1 such that

$$
\begin{bmatrix}
c_1 & s_1^* \\
-s_1 & c_1
\end{bmatrix}
\begin{bmatrix}
|d_1|^2 - \lambda \\
f_2^* d_1
\end{bmatrix}
=
\begin{bmatrix}
\star \\
0
\end{bmatrix}
\tag{12.82}
$$

where d_1 and f_2 are the leading main diagonal and superdiagonal elements of matrix **B**, respectively; see the right-hand side of Eq. (12.75). The element marked $\star$ on the right-hand side of Eq. (12.82) indicates a nonzero element. Set

$$
\boldsymbol{\Theta}_1 =
\left[
\begin{array}{cc:c}
c_1 & s_1^* & \\
-s_1 & c_1 & \mathbf{O} \\
\hdashline
& \mathbf{O} & \mathbf{I}
\end{array}
\right]
\tag{12.83}
$$

3. Apply the Givens rotation $\boldsymbol{\Theta}_1$ to matrix **B** directly. Since **B** is upper bidiagonal, and $\boldsymbol{\Theta}_1$ is a rotation in the (2, 1) plane, it follows that the matrix product **B** has the following form (illustrated for the case of $M = 4$):

$$
\mathbf{B}\boldsymbol{\Theta}_1 =
\begin{bmatrix}
\mathrm{x} & \mathrm{x} & 0 & 0 \\
z^{(1)} & \mathrm{x} & \mathrm{x} & 0 \\
0 & 0 & \mathrm{x} & \mathrm{x} \\
0 & 0 & 0 & \mathrm{x}
\end{bmatrix}
$$

where $z^{(1)}$ is a new element produced by the Givens rotation $\boldsymbol{\Theta}_1$.

4. Determine the sequence of Givens rotations $\mathbf{U}_1, \mathbf{V}_2, \mathbf{U}_2, \ldots, \mathbf{V}_{M-1}$, and $\mathbf{U}_{M-1}$ operating on $\mathbf{B}\boldsymbol{\Theta}_1$ in a "ping-pong" fashion so as to chase the unwanted nonzero element $z^{(1)}$ down the bidiagonal. This sequence of operations is illustrated below, again for the case of $M = 4$:

$$
\mathbf{U}_1^H \mathbf{B} \boldsymbol{\Theta}_1 =
\begin{bmatrix}
x & x & z^{(2)} & 0 \\
0 & x & x & 0 \\
0 & 0 & x & x \\
0 & 0 & 0 & x
\end{bmatrix}
$$

$$
\mathbf{U}_1^H \mathbf{V} \mathbf{B} \boldsymbol{\Theta}_1 \mathbf{V}_2 =
\begin{bmatrix}
x & x & 0 & 0 \\
0 & x & x & 0 \\
0 & z^{(3)} & x & x \\
0 & 0 & 0 & x
\end{bmatrix}
$$

$$
\mathbf{U}_2^H \mathbf{U}_1^H \mathbf{B} \boldsymbol{\Theta}_1 \mathbf{V}_2 =
\begin{bmatrix}
x & x & 0 & 0 \\
0 & x & x & z^{(4)} \\
0 & 0 & x & x \\
0 & 0 & 0 & x
\end{bmatrix}
$$

$$
\mathbf{U}_2^H \mathbf{U}_1^H \mathbf{B} \boldsymbol{\Theta}_1 \mathbf{V}_2 \mathbf{V}_3 =
\begin{bmatrix}
x & x & 0 & 0 \\
0 & x & x & 0 \\
0 & 0 & x & x \\
0 & 0 & z^{(5)} & x
\end{bmatrix}
$$

$$
\mathbf{U}_3^H \mathbf{U}_2^H \mathbf{U}_1^H \mathbf{B} \boldsymbol{\Theta}_1 \mathbf{V}_2 \mathbf{V}_3 =
\begin{bmatrix}
x & x & 0 & 0 \\
0 & x & x & 0 \\
0 & 0 & x & x \\
0 & 0 & 0 & x
\end{bmatrix}
$$

The iteration thus terminates with a new bidiagonal matrix $\mathbf{B}$ that is related to the original bidiagonal matrix $\mathbf{B}$ as follows:

$$
\mathbf{B} \leftarrow (\mathbf{U}_{M-1}^H \ldots \mathbf{U}_2^H \mathbf{U}_1^H) \mathbf{B} (\boldsymbol{\Theta}_1 \mathbf{V}_2 \ldots \mathbf{V}_{M-1}) = \mathbf{U}^H \mathbf{B} \mathbf{V} \tag{12.84}
$$

where

$$
\mathbf{U} = \mathbf{U}_1 \mathbf{U}_2 \ldots \mathbf{U}_{M-1} \tag{12.85}
$$

and

$$
\mathbf{V} = \boldsymbol{\Theta}_1 \mathbf{V}_2 \ldots \mathbf{V}_{M-1} \tag{12.86}
$$

Steps 1 through 4 constitute one iteration of the Golub–Kahan algorithm. Typically, after a few iterations of this algorithm, the superdiagonal entry f_M becomes negligible. When f_M

becomes sufficiently small, we can *deflate* the matrix and apply the algorithm to the smaller matrix. The criterion for the smallness of f_M is usually of the following form:

$$|f_M| \leq \epsilon \, (|d_{M-1}| + |d_M|) \text{ where } \epsilon \text{ is the machine precision} \qquad (12.87)$$

The description just presented leaves much unsaid about the Golub–Kahan algorithm for the diagonaliztion of a square data matrix. For a more detailed treatment of the algorithm, the reader is referred to the original paper of Golub and Kahan (1965) or the book by Golub and Van Loan (1989).

Recently, there has been a significant improvement in the Golub–Kahan algorithm. The Golub–Kahan algorithm has the property that it computes every singular value of a bidiagonal matrix **B** with an absolute error bound of about $\epsilon\|\mathbf{B}\|$, where ϵ is the machine precision. Thus large singular values (those near $\|\mathbf{B}\|$) are computed with high relative accuracy, but small ones (those near $\epsilon\|\mathbf{B}\|$ or smaller) may have no relative accuracy at all. The new algorithm computes every singular value to high relative accuracy independent of its size. It also computes the singular vectors much more accurately. It is also approximately as fast as the old algorithm (and occasionally much faster). The new algorithm is a hybrid of the Golub–Kahan algorithm and a simplified version that corresponds to taking $\lambda = 0$ in Eq. (12.82). When $\lambda = 0$, the remainder of the algorithm can be stabilized so as to compute every matrix entry to high relative accuracy, whence the final accuracy of the singular values. The analysis of this algorithm can be found in Demmel and Kahan (1990) and Deift et al. (1989).

Summary of the QR Algorithm

The QR algorithm is not only mathematically elegant, but also a computationally powerful and highly versatile algorithm for SVD computation. Given a K-by-M data matrix **A**, the QR algorithm used to compute its SVD proceeds as follows:

1. Compute a sequence of Householder transformations that reduce the matrix **B** to upper bidiagonal form.
2. Apply the Golub–Kahan algorithm to the M-by-M nonzero submatrix resulting from step 1, and iterate this application until the superdiagonal elements become negligible in accordance with the criterion defined in Eq. (12.87).
3. The SVD of the data matrix **A** is determined as follows:
   - The diagonal elements of the matrix resulting from step 2 are the singular values of matrix **A**.
   - The product of the Householder transformations of step 1 and the Givens rotations of step 2 involved in premultiplication defines the left singular vectors of **A**. The product of the Householder transformations and Givens rotations involved in postmultiplication define the right-singular vectors of **A**.

TABLE 12.1   ILLUSTRATING THE FIRST TWO
ITERATIONS OF THE GOLUB–KAHAN
ALGORITHM

Iteration number	Matrix **B**		
0	1.0000	1.0000	0.0000
	0.0000	2.0000	1.0000
	0.0000	0.0000	3.0000
1	0.9155	0.6627	0.0000
	0.0000	2.0024	0.0021
	0.0000	0.0000	3.2731
2	0.8817	0.4323	0.0000
	0.0000	2.0791	0.0000
	0.0000	0.0000	3.2731

Example 3

Consider the real valued 3-by-3 bidiagonal matrix:

$$\mathbf{B} = \begin{bmatrix} 1 & 1 & 0 \\ 0 & 2 & 1 \\ 0 & 0 & 3 \end{bmatrix}$$

The iterative application of the Golub–Kahan algorithm to this matrix yields the sequence of results shown in Table 12.1 for $\epsilon = 10^{-4}$ in the stopping rule defined in Eq. (12.87). After two iterations of the algorithm, matrix **B** becomes small, at which point it is deflated. Specifically, we now work on the 2-by-2 leading principal submatrix:

$$\begin{matrix} 0.8817 & 0.4323 \\ 0.0000 & 2.0791 \end{matrix}$$

This submatrix is finally diagonalized in one step, yielding

$$\begin{matrix} 0.8596 & 0.0000 \\ 0.0000 & 2.1326 \end{matrix}$$

The singular values of the bidiagonal matrix are thus computed to be:

$$\sigma_1 = 0.8596$$
$$\sigma_2 = 2.1326$$
$$\sigma_3 = 3.2731$$

12.6 SUMMARY AND DISCUSSION

The *singular-value decomposition (SVD)* has become a fundamental tool in linear algebra, system theory, and signal processing (Kung et al., 1985; Deprettere, 1988; Van Loan, 1989; Haykin, 1989). Not only does the SVD permit an *elegant problem formulation, it also provides geometrical and algebraic insight together with a numerically robust implementation* (Golub and Van Loan, 1989). It includes the eigenvalue decomposition of a nonnegative-definite matrix (e.g., correlation matrix) as a special case. In the context of our present discussion, the SVD provides a direct and numerically robust solution for the linear least-squares estimation problem, be it overdetermined or underdetermined; by "direct" we mean that the solution is obtained by applying the SVD *directly* to the data matrix.

The basic idea behind an algorithm used to compute the singular-value decomposition is to "nudge" a data matrix toward a diagonal form in a step-by-step fashion. The two most common iterative algorithms used to do this nudging are (1) the *cyclic Jacobi algorithm*, and (2) the *QR algorithm* (not to be confused with the QR-decomposition). The QR algorithm, in general, is computationally more *efficient* (i.e., requires less operations) than the cyclic Jacobi algorithm. On the other hand, the cyclic Jacobi algorithm is the preferred method when accuracy demands are extraordinary.

In Mathias (1995), building on and greatly simplifying the previous work by Demmel and Veselic (1989), it is shown that Jacobi's method (involving a sequence of elementary orthogonal matrices) is guaranteed to compute the eigenvalues and eigenvectors of a real-valued positive definite matrix more accurately than the QR algorithm. It is also shown that in the case of an M-by-N matrix with size M much bigger than N, Jacobi's method computes the singular values of the matrix essentially as quickly as the QR algorithm, but potentially much more accurately. With regard to the latter point, Jacobi's method and the QR algorithm start by reducing the matrix to an N-by-N matrix, which requires the same amount of work in both methods. Then the QR algorithm requires less work than Jacobi's method, but this exra work is just on N-by-N matrices and so it is negligible compared to the work required to reduce the M-by-N matrix to N-by-N.

There are two types of new algorithms for SVD computation which deserve special mention:

1. For singular values, the algorithm described in Fernando and Parlett, (1994) is several times faster, and more accurate, than its predecessor in LAPACK; for a short note on LAPACK; see Section 4.5. This new algorithm can be implemented in either parallel or pipelined form, with each iteration (performed on an M-by-M symmetric positive-definite matrix) nominally taking $O(\log_2 M)$ operations. The interesting point to note is that the development of the algorithm by Fernando and Parlett breaks away from the traditional *orthogonal paradigm* that has dominated the field of matrix computations since the 1960's. Specifically, the QR algorithm is abandoned in favor of the *Cholesky LR algorithm* that consists of successive applications of the Cholesky factorization. Given a symmetric positive-definite matrix $\mathbf{A}$, its Cholesky factorization may be written as

$$\mathbf{A} = \mathbf{L}^H \mathbf{L}$$

where $\mathbf{L}$ is a lower triangular matrix and $\mathbf{L}^H$ is its Hermitian transpose.

2. For singular vectors, the algorithm described in Gu and Eisenstat (1994) and Gu et al. (1994) is faster, and more accurate, than the QR algorithm. This second algorithm is based on a *divide-and-conquer* strategy that involves removing a whole column/row of the bidiagonal matrix (resulting from the Householder transformation of the data matrix) one at a time.

The two new algorithms mentioned here point to the fact that SVD computation is indeed an active area of research.

PROBLEMS

1. Repeat the calculation of the singular values and singular vectors of the matrix **A** given in Problem 9 of Chapter 11 by using the two-sided Jacobi algorithm.
2. Demonstrate that the sweep of orderings by rows is equivalent to the sweep of orderings by columns described in Example 1.
3. The transformation of a 2-by-2 complex matrix into a real one involves a plane rotation, followed by certain forms of premultiplication and postmultiplication.
   (a) Show that the combined effect of the plane rotation in Eq. (12.35) and the premultiplication in Eq. (12.44) is equivalent to a unitary matrix.
   (b) Show that the postmultiplying matrix on the left-hand side of Eq. (12.44) is also a unitary matrix.
4. Consider an M-by-M matrix **A** that is triangularized by the use of Householder transformations. After $M - 1$ steps, at most, the matrix **A** is triangularized as follows:

$$\mathbf{QA} = \mathbf{R}$$

where **R** is an upper triangular matrix, and

$$\mathbf{Q} = \mathbf{Q}_{M-1}\mathbf{Q}_{M-2}\cdots\mathbf{Q}_1$$

   (a) Show that

$$\det(\mathbf{A}) = (-1)^{m-1}\det(\mathbf{R})$$

   (b) Using the fact that the Euclidean norm of a matrix is preserved under multiplication by a unitary matrix, and that each diagonal element of the triangular matrix **R** is the norm of the projection of that column on a certain subspace, show that

$$|\det(\mathbf{A})| \leq \prod_{i=1}^{M}\|\mathbf{a}_i\|$$

   where $\mathbf{a}_i$ is the ith column of matrix **A**. (This result is known as the *Hadamard theorem*.)
5. Consider the 4-by-3 data matrix

$$\mathbf{A} = \begin{bmatrix} 1 & 1 & 1 \\ 2 & 1.5 & 2 \\ 3 & 3 & 4 \\ 4 & 4.5 & 8 \end{bmatrix}$$

Using a sequence of Householder transformations, reduce the matrix to upper bidiagonal form.

6. Consider the data matrix

$$A = \begin{bmatrix} 1 & 1 & 1 \\ 2 & 1.5 & 2 \\ 3 & 3 & 4 \end{bmatrix}$$

Reduce the matrix A to upper triangular form, using:

(a) Householder transformations.

(b) Givens rotations.

7. Use the Golub–Kahan algorithm to compute the singular-value decomposition of the bidiagonal matrix:

$$A = \begin{bmatrix} 1.5 & 1.5 & 0 & 0 \\ 0 & 3 & 1.5 & 0 \\ 0 & 0 & 4.5 & 1.5 \\ 0 & 0 & 0 & 6 \end{bmatrix}$$

CHAPTER

13

Recursive Least-Squares Algorithm

In this chapter we extend the use of the method of least squares to develop a recursive algorithm for the design of adaptive transversal filters such that, given the least-squares estimate of the tap-weight vector of the filter at iteration $n - 1$, we may compute the updated estimate of this vector at iteration n upon the arrival of new data. We refer to the resulting algorithm as the *recursive least-squares (RLS) algorithm*.

The RLS algorithm may be viewed as a special case of the Kalman filter. Indeed, this special relationship between the RLS algorithm and the Kalman filter is considered later in the chapter. Our main mission in this chapter, however, is to develop the basic theory of the RLS algorithm as an important tool for linear adaptive filtering in its own right.

We begin the development of the RLS algorithm by reviewing some basic relations that pertain to the method of least squares. Then, by exploiting a relation in matrix algebra known as the *matrix inversion lemma*, we develop the RLS algorithm. An important feature of the RLS algorithm is that it utilizes information contained in the input data, extending back to the instant of time when the algorithm is initiated. The resulting rate of convergence is therefore typically an order of magnitude faster than the simple LMS algorithm. This improvement in performance, however, is achieved at the expense of a large increase in computational complexity.

13.1 SOME PRELIMINARIES

In *recursive* implementations of the method of least squares, we start the computation with *known initial conditions* and use the information contained in new data samples to *update* the old estimates. We therefore find that the length of observable data is variable. Accordingly, we express the *cost function* to be minimized as $\mathscr{E}(n)$, where n is the variable length of the observable data. Also, it is customary to introduce a *weighting factor* into the definition of the cost function $\mathscr{E}(n)$. We thus write

$$\mathscr{E}(n) = \sum_{i=1}^{n} \beta(n, i)|e(i)|^2 \tag{13.1}$$

where $e(i)$ is the difference between the *desired response* $d(i)$ and the *output* $y(i)$ produced by a transversal filter whose tap inputs (at time i) equal $u(i), u(i-1), \ldots, u(i-M+1)$, as in Fig. 13.1. That is, $e(i)$ is defined by

$$e(i) = d(i) - y(i) \tag{13.2}$$

$$= d(i) - \mathbf{w}^H(n)\mathbf{u}(i)$$

where $\mathbf{u}(i)$ is the *tap-input vector at time i*, defined by

$$\mathbf{u}(i) = [u(i), u(i-1), \ldots, u(i-M+1)]^T \tag{13.3}$$

and $\mathbf{w}(n)$ is the *tap-weight vector at time n*, defined by

$$\mathbf{w}(n) = [w_0(n), w_1(n), \ldots, w_{M-1}(n)]^T \tag{13.4}$$

Note that the tap weights of the transversal filter remain *fixed* during the observation interval $1 \leq i \leq n$ for which the cost function $\mathscr{E}(n)$ is defined.

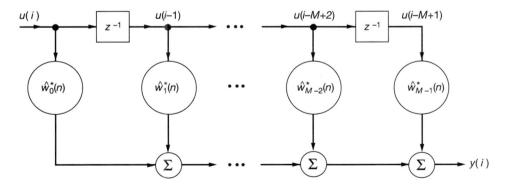

Figure 13.1   Transversal filter.

The weighting factor $\beta(n, i)$, in Eq. (13.1) has the property that

$$0 < \beta(n, i) \le 1, \qquad i = 1, 2, \ldots, n \tag{13.5}$$

The use of the weighting factor $\beta(n, i)$, in general, is intended to ensure that data in the distant past are "forgotten" in order to afford the possibility of following the statistical variations of the observable data when the filter operates in a nonstationary environment. A special form of weighting that is commonly used is the *exponential weighting factor* or *forgetting factor* defined by

$$\beta(n, i) = \lambda^{n-i}, \qquad i = 1, 2, \ldots, n \tag{13.6}$$

where λ is a positive constant close to, but less than, 1. When λ equals 1, we have the ordinary method of least squares. The inverse of $1 - \lambda$ is, roughly speaking, a measure of the *memory* of the algorithm. The special case $\lambda = 1$ corresponds to *infinite memory*. Thus, in the *method of exponentially weighted least squares*, we minimize the cost function

$$\mathscr{E}(n) = \sum_{i=1}^{n} \lambda^{n-i} |e(i)|^2 \tag{13.7}$$

The optimum value of the tap-weight vector, $\hat{\mathbf{w}}(n)$, for which the cost function $\mathscr{E}(n)$ of Eq. (13.7) attains its minimum value is defined by the *normal equations* written in matrix form:

$$\mathbf{\Phi}(n)\hat{\mathbf{w}}(n) = \mathbf{z}(n) \tag{13.8}$$

The M-by-M correlation matrix $\mathbf{\Phi}(n)$ is now defined by

$$\mathbf{\Phi}(n) = \sum_{i=1}^{n} \lambda^{n-i} \mathbf{u}(i)\mathbf{u}^H(i) \tag{13.9}$$

The M-by-1 cross-correlation vector $\mathbf{z}(n)$ between the tap inputs of the transversal filter and the desired response is correspondingly defined by

$$\mathbf{z}(n) = \sum_{i=1}^{n} \lambda^{n-i} \mathbf{u}(i)d^*(i) \tag{13.10}$$

where the asterisk denotes complex conjugation.

The correlation matrix $\mathbf{\Phi}(n)$ of Eq. (13.9) differs from the time-averaged version of Eq. (11.45) in two respects.

1. The matrix product $\mathbf{u}(i)\mathbf{u}^H(i)$ inside the summation on the right-hand side of Eq. (11.45) is weighted by the exponential factor λ^{n-i}, which arises naturally from the adoption of Eq. (13.7) as the cost function.

2. The use of *prewindowing* is assumed, according to which the input data prior to time $i = 1$ are equal to zero, hence the use of $i = 1$ as the lower limit of the summation.

Similar remarks apply to the cross-correlation vector $\mathbf{z}(n)$ compared to its time-averaged counterpart of Chapter 11.

Isolating the term corresponding to $i = n$ from the rest of the summation on the right-hand side of Eq. (13.9), we may write

$$\mathbf{\Phi}(n) = \lambda \left[\sum_{i=1}^{n-1} \lambda^{n-1-i}\, \mathbf{u}(i)\mathbf{u}^H(i) \right] + \mathbf{u}(n)\mathbf{u}^H(n) \tag{13.11}$$

However, by definition, the expression inside the square brackets on the right-hand side of Eq. (13.11) equals the correlation matrix $\mathbf{\Phi}(n - 1)$. Hence, we have the following recursion for updating the value of the correlation matrix of the tap inputs:

$$\mathbf{\Phi}(n) = \lambda\mathbf{\Phi}(n - 1) + \mathbf{u}(n)\mathbf{u}^H(n) \tag{13.12}$$

where $\mathbf{\Phi}(n - 1)$ is the "old" value of the correlation matrix, and the matrix product $\mathbf{u}(n)\mathbf{u}^H(n)$ plays the role of a "correction" term in the updating operation.

Similarly, we may use Eq. (13.10) to derive the following recursion for updating the cross-correlation vector between the tap inputs and the desired response:

$$\mathbf{z}(n) = \lambda\mathbf{z}(n - 1) + \mathbf{u}(n)d^*(n) \tag{13.13}$$

To compute the least-square estimate $\hat{\mathbf{w}}(n)$ for the tap-weight vector in accordance with Eq. (13.8), we have to determine the inverse of the correlation matrix $\mathbf{\Phi}(n)$. In practice, however, we usually try to avoid performing such an operation as it can be very time consuming, particularly if the number of tap weights, M, is high. Also, we would like to be able to compute the least-squares estimate $\hat{\mathbf{w}}(n)$ for the tap-weight vector recursively for $n = 1, 2, \ldots, \infty$. We can realize both of these objectives by using a basic result in matrix algebra known as the *matrix inversion lemma*. We assume that the initial conditions have been chosen to ensure the nonsingularity of the correlation matrix $\mathbf{\Phi}(n)$; this issue is discussed later in Section 13.3.

13.2 THE MATRIX INVERSION LEMMA

Let $\mathbf{A}$ and $\mathbf{B}$ be two positive-definite M-by-M matrices related by

$$\mathbf{A} = \mathbf{B}^{-1} + \mathbf{C}\mathbf{D}^{-1}\mathbf{C}^H \tag{13.14}$$

where $\mathbf{D}$ is another positive-definite N-by-M matrix, and $\mathbf{C}$ is an M-by-N matrix. According to the *matrix inversion lemma*, we may express the inverse of the matrix $\mathbf{A}$ as follows:

$$\mathbf{A}^{-1} = \mathbf{B} - \mathbf{B}\mathbf{C}(\mathbf{D} + \mathbf{C}^H\mathbf{B}\mathbf{C})^{-1}\mathbf{C}^H\mathbf{B} \tag{13.15}$$

The proof of this lemma is established by multiplying Eq. (13.14) by (13.15) and recognizing that the product of a square matrix and its inverse is equal to the identity matrix (see Problem 2). The matrix inversion lemma states that if we are given a matrix $\mathbf{A}$ as defined in Eq. (13.14), we can determine its inverse $\mathbf{A}^{-1}$ by using the relation of Eq. (13.15). In effect, the lemma is described by this pair of equations. The matrix inversion lemma is also referred to in the literature as *Woodbury's identity*.[1]

In the next section we show how the matrix inversion lemma can be applied to obtain a recursive equation for computing the least-squares solution $\hat{\mathbf{w}}(n)$ for the tap-weight vector.

13.3 THE EXPONENTIALLY WEIGHTED RECURSIVE LEAST-SQUARES ALGORITHM

With the correlation matrix $\mathbf{\Phi}(n)$ assumed to be positive definite and therefore nonsingular, we may apply the matrix inversion lemma to the recursive equation (13.12). We first make the following identifications:

$$\mathbf{A} = \mathbf{\Phi}(n)$$

$$\mathbf{B}^{-1} = \lambda\mathbf{\Phi}(n-1)$$

$$\mathbf{C} = \mathbf{u}(n)$$

$$\mathbf{D} = 1$$

Then, substituting these definitions in the matrix inversion lemma of Eq. (13.15), we obtain the following recursive equation for the inverse of the correlation matrix:

$$\mathbf{\Phi}^{-1}(n) = \lambda^{-1}\mathbf{\Phi}^{-1}(n-1) - \frac{\lambda^{-2}\mathbf{\Phi}^{-1}(n-1)\mathbf{u}(n)\mathbf{u}^{H}(n)\mathbf{\Phi}^{-1}(n-1)}{1 + \lambda^{-1}\mathbf{u}^{H}(n)\mathbf{\Phi}^{-1}(n-1)\mathbf{u}(n)} \qquad (13.16)$$

For convenience of computation, let

$$\mathbf{P}(n) = \mathbf{\Phi}^{-1}(n) \qquad (13.17)$$

and

$$\mathbf{k}(n) = \frac{\lambda^{-1}\mathbf{P}(n-1)\mathbf{u}(n)}{1 + \lambda^{-1}\mathbf{u}^{H}(n)\mathbf{P}(n-1)\mathbf{u}(n)} \qquad (13.18)$$

Using these definitions, we may rewrite Eq. (13.16) as follows:

$$\mathbf{P}(n) = \lambda^{-1}\mathbf{P}(n-1) - \lambda^{-1}\mathbf{k}(n)\mathbf{u}^{H}(n)\mathbf{P}(n-1) \qquad (13.19)$$

---

[1] The exact origin of the matrix inversion lemma is not known. Householder (1964) attributes it to Woodbury (1950). Nevertheless, application of the matrix inversion lemma in the filtering literature was first made by Kailath, who used a form of this lemma to prove the equivalence of the Wiener filter and the maximum-likelihood procedure for estimating the output of a random linear time-invariant channel that is corrupted by additive white Gaussian noise (Kailath, 1960). Early use of the matrix inversion lemma was also made by Ho (1963). Another interesting application of the matrix inversion lemma was made by Brooks and Reed, who used it to prove the equivalence of the Wiener filter, the maximum signal-to-noise ratio filter, and the likelihood ratio processor for detecting a signal in additive white Gaussian noise (Brooks and Reed, 1972).

The M-by-M matrix $\mathbf{P}(n)$ is referred to as the *inverse correlation matrix*. The M-by-1 vector $\mathbf{k}(n)$ is referred to as the *gain vector* for reasons that will become apparent later in the section. Equation (13.19) is the *Riccati equation* for the RLS algorithm.

By rearranging Eq. (13.18), we have

$$
\begin{aligned}
\mathbf{k}(n) &= \lambda^{-1}\mathbf{P}(n-1)\mathbf{u}(n) - \lambda^{-1}\mathbf{k}(n)\mathbf{u}^H(n)\mathbf{P}(n-1)\mathbf{u}(n) \\
&= [\lambda^{-1}\mathbf{P}(n-1) - \lambda^{-1}\mathbf{k}(n)\mathbf{u}^H(n)\mathbf{P}(n-1)]\mathbf{u}(n)
\end{aligned}
\tag{13.20}
$$

We see from Eq. (13.19) that the expression inside the brackets on the right-hand side of Eq. (13.20) equals $\mathbf{P}(n)$. Hence. we may simplify Eq. (13.20) to

$$
\mathbf{k}(n) = \mathbf{P}(n)\mathbf{u}(n)
\tag{13.21}
$$

This result, together with $\mathbf{P}(n) = \boldsymbol{\Phi}^{-1}(n)$, may be used as the definition for the gain vector:

$$
\mathbf{k}(n) = \boldsymbol{\Phi}^{-1}(n)\mathbf{u}(n)
\tag{13.22}
$$

In other words, the gain vector $\mathbf{k}(n)$ is defined as the tap-input vector $\mathbf{u}(n)$ transformed by the inverse of the correlation matrix $\boldsymbol{\Phi}(n)$.

Time Update for the Tap-Weight Vector

Next, we wish to develop a recursive equation for updating the least-squares estimate $\hat{\mathbf{w}}(n)$ for the tap-weight vector. To do this, we use Eqs. (13.8), (13.13), and (13.17) to express the least-squares estimate $\hat{\mathbf{w}}(n)$ for the tap-weight vector at iteration n as follows:

$$
\begin{aligned}
\hat{\mathbf{w}}(n) &= \boldsymbol{\Phi}^{-1}(n)\mathbf{z}(n) \\
&= \mathbf{P}(n)\mathbf{z}(n) \\
&= \lambda\mathbf{P}(n)\mathbf{z}(n-1) + \mathbf{P}(n)\mathbf{u}(n)d^*(n)
\end{aligned}
\tag{13.23}
$$

Substituting Eq. (13.19) for $\mathbf{P}(n)$ in the first term only in the right-hand side of Eq. (13.23), we get

$$
\begin{aligned}
\hat{\mathbf{w}}(n) &= \mathbf{P}(n-1)\mathbf{z}(n-1) - \mathbf{k}(n)\mathbf{u}^H(n)\mathbf{P}(n-1)\mathbf{z}(n-1) \\
&\quad + \mathbf{P}(n)\mathbf{u}(n)d^*(n) \\
&= \boldsymbol{\Phi}^{-1}(n-1)\mathbf{z}(n-1) - \mathbf{k}(n)\mathbf{u}^H(n)\boldsymbol{\Phi}^{-1}(n-1)\mathbf{z}(n-1) \\
&\quad + \mathbf{P}(n)\mathbf{u}(n)d^*(n) \\
&= \hat{\mathbf{w}}(n-1) - \mathbf{k}(n)\mathbf{u}^H(n)\hat{\mathbf{w}}(n-1) + \mathbf{P}(n)\mathbf{u}(n)d^*(n)
\end{aligned}
\tag{13.24}
$$

Finally, using the fact that $\mathbf{P}(n)\mathbf{u}(n)$ equals the gain vector $\mathbf{k}(n)$, as in Eq. (13.21), we get the desired recursive equation for updating the tap-weight vector:

$$
\begin{aligned}
\hat{\mathbf{w}}(n) &= \hat{\mathbf{w}}(n-1) + \mathbf{k}(n)[d^*(n) - \mathbf{u}^H(n)\hat{\mathbf{w}}(n-1)] \\
&= \hat{\mathbf{w}}(n-1) + \mathbf{k}(n)\xi^*(n)
\end{aligned}
\tag{13.25}
$$

where $\xi(n)$ is the *a priori estimation error* defined by

$$\xi(n) = d(n) - \mathbf{u}^T(n)\hat{\mathbf{w}}^*(n - 1) \qquad (13.26)$$

$$= d(n) - \hat{\mathbf{w}}^H(n - 1)\mathbf{u}(n)$$

The inner product $\hat{\mathbf{w}}^H(n - 1)\mathbf{u}(n)$ represents an estimate of the desired response $d(n)$, based on the *old* least-squares estimate of the tap-weight vector that was made at time $n - 1$.

Equation (13.25) for the adjustment of the tap-weight vector and Eq. (13.26) for the *a priori* estimation error suggest the block-diagram representation depicted in Fig. 13.2(a) for the *recursive least-squares RLS algorithm.*

The *a priori* estimation error $\xi(n)$ is, in general, different from the *a posteriori esti-mation error*

$$e(n) = d(n) - \hat{\mathbf{w}}^H(n)\mathbf{u}(n) \qquad (13.27)$$

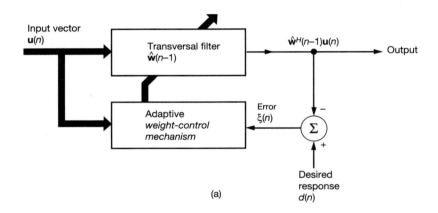

(a)

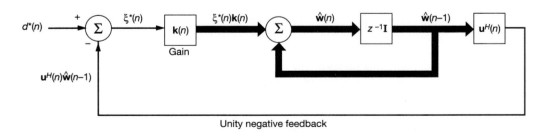

(b)

Figure 13.2 Representations of the RLS algorithm: (1) block diagram; (b) signal-flow graph.

TABLE 13.1 SUMMARY OF THE RLS ALGORITHM

Initialize the algorithm by setting
$\quad\mathbf{P}(0) = \delta^{-1}\mathbf{I}, \qquad \delta = $ small positive constant
$\quad\hat{\mathbf{w}}(0) = \mathbf{0}$
For each instant of time, $n = 1, 2, \ldots$, compute

$$k(n) = \frac{\lambda^{-1}\mathbf{P}(n-1)\mathbf{u}(n)}{1 + \lambda^{-1}\mathbf{u}^H(n)\mathbf{P}(n-1)\mathbf{u}(n)}$$

$$\xi(n) = d(n) - \hat{\mathbf{w}}^H(n-1)\mathbf{u}(n)$$

$$\hat{\mathbf{w}}(n) = \hat{\mathbf{w}}(n-1) + \mathbf{k}(n)\xi^*(n)$$

$$\mathbf{P}(n) = \lambda^{-1}\mathbf{P}(n-1) - \lambda^{-1}\mathbf{k}(n)\mathbf{u}^H(n)\mathbf{P}(n-1)$$

the computation of which involves the *current* least-squares estimate of the tap-weight vector available at time n. Indeed, we may view $\xi(n)$ as a "tentative" value of $e(n)$ before updating the tap-weight vector. Note, however, in the least-squares optimization that led to the recursive algorithm of Eq. (13.25) for the tap-weight vector, we actually minimized a cost function based on $e(n)$ and *not* $\xi(n)$.

Summary of the RLS Algorithm

Equations (13.18), (13.26), (13.25), and (13.19), collectively and in that order, constitute the *RLS algorithm*, as summarized in Table 13.1. We note that, in particular, Eq. (13.26) describes the filtering operation of the algorithm, whereby the transversal filter is excited to compute the *a priori* estimation error $\xi(n)$. Equation (13.25) describes the adaptive operation of the algorithm, whereby the tap-weight vector is updated by incrementing its old value by an amount equal to the complex conjugate of the *a priori* estimation error $\xi(n)$ times the time-varying gain vector $\mathbf{k}(n)$, hence the name "gain vector." Equations (13.18) and (13.19) enable us to update the value of the gain vector itself. An important feature of the RLS algorithm described by these equations is that the inversion of the correlation matrix $\boldsymbol{\Phi}(n)$ is replaced at each step by a simple scalar division. Figure 13.2(b) depicts a signal-flow-graph representation of the RLS algorithm that complements the block diagram of Fig. 13.2(a).

Initialization of the RLS Algorithm

The applicability of the RLS algorithm requires that we initialize the recursion of Eq. (13.19) by choosing a starting value $\mathbf{P}(0)$ that assures the nonsingularity of the correlation matrix $\boldsymbol{\Phi}(n)$. We may do this by evaluating the inverse

$$\left[\sum_{i=-n_0}^{0} \lambda^{-i}\mathbf{u}(i)\mathbf{u}^H(i) \right]^{-1}$$

where the tap-weight vector $\mathbf{u}(i)$ is obtained from an initial block of data for $-n_0 \le i \le 0$.

A simpler procedure, however, is to modify the expression slightly for the correlation matrix $\boldsymbol{\Phi}(n)$ by writing

$$\boldsymbol{\Phi}(n) = \sum_{i=1}^{n} \lambda^{n-i} \mathbf{u}(i)\mathbf{u}^H(i) + \delta\lambda^n \mathbf{I} \tag{13.28}$$

where $\mathbf{I}$ is the M-by-M identity matrix, and δ is a small positive constant. Thus putting $n = 0$ in Eq. (13.28), we have

$$\boldsymbol{\Phi}(0) = \delta\mathbf{I}$$

Correspondingly, for the initial value of $\mathbf{P}(n)$ equal to the inverse of the correlation matrix $\boldsymbol{\Phi}(n)$, we set

$$\mathbf{P}(0) = \delta^{-1}\mathbf{I} \tag{13.29}$$

The initialization described in Eq. (13.29) is equivalent to forcing the unknown data sample $u(-M + 1)$ equal to the value $\lambda^{(-M+1)/2}\delta^{1/2}$ instead of zero. In other words, during the initialization period we modify the prewindowing method by writing

$$u(n) = \begin{cases} \lambda^{(-M+1)/2}\delta^{1/2}, & n = -M + 1 \\ 0, & n < 0, n \ne -M + 1 \end{cases} \tag{13.30}$$

Note that for a transversal filter with M taps, the index $n = M + 1$ refers to the *last* tap in the filter. When the first nonzero data sample $u(i)$ enters the filter, the initializing tap input $u(-M + 1)$ leaves the filter and from then on the RLS algorithm takes over.

It only remains for us to choose an initial value for the tap-weight vector. It is customary to set

$$\hat{\mathbf{w}}(0) = \mathbf{0} \tag{13.31}$$

where $\mathbf{0}$ is the M-by-1 null vector.

The initialization procedure incorporating Eqs. (13.29) and (13.31) is referred to as a *soft-constrained initialization*. The positive constant δ is the only parameter required for this initialization. The recommended choice of δ is that it should be small compared to $0.01\sigma_u^2$, where σ_u^2 is the variance of a data sample $u(n)$. Such a choice is based on practical experience with the RLS algorithm, supported by a statistical analysis of the soft-constrained initialization of the algorithm (Hubing and Alexander, 1990). For large data lengths, the exact value of the initializing constant δ has an insignificant effect.

It is important to note that by using the initialization procedure defined by Eqs. (13.29) and (13.31), we are no longer computing a solution that minimizes the cost function $\xi(n)$ of Eq. (13.7) exactly. Instead, we are computing the solution that minimizes the modified cost function:

$$\mathscr{E}(n) = \delta\lambda^n\|\mathbf{w}(n)\|^2 + \sum_{i=1}^{n} \lambda^{n-i} |e(i)|^2$$

In other words, the RLS algorithm summarized in Table 13.1 yields the exact recursive solution to the following optimization problem (Sayed and Kailath, 1994):

$$\min_{\mathbf{w}(n)} [\delta\lambda^n \|\mathbf{w}(n)\|^2 + \sum_{i=1}^{n} \lambda^{n-i}|e(i)|^2]$$

where $e(i)$ is defined by Eq. (13.2).

13.4 UPDATE RECURSION FOR THE SUM OF WEIGHTED ERROR SQUARES

The minimum value of the sum of weighted error squares, $\mathscr{E}_{\min}(n)$, results when the tap-weight vector is set equal to the least-squares estimate $\hat{\mathbf{w}}(n)$. To compute $\mathscr{E}_{\min}(n)$, we may therefore use the relation [see first line of Eq. (10.40)]:

$$\mathscr{E}_{\min}(n) = \mathscr{E}_d(n) - \mathbf{z}^H(n)\hat{\mathbf{w}}(n) \tag{13.32}$$

where $\mathscr{E}_d(n)$ is defined by (using the notation of this chapter)

$$\mathscr{E}_d(n) = \sum_{i=1}^{n} \lambda^{n-i}|d(i)|^2$$
$$= \lambda \mathscr{E}_d(n-1) + |d(n)|^2 \tag{13.33}$$

Therefore, substituting Eqs. (13.13), (13.25), and (13.33) in (13.32), we get

$$\mathscr{E}_{\min}(n) = \lambda[\mathscr{E}_d(n-1) - \mathbf{z}^H(n-1)\hat{\mathbf{w}}(n-1)]$$
$$+ d(n)[d^*(n) - \mathbf{u}^H(n)\hat{\mathbf{w}}(n-1)] \tag{13.34}$$
$$- \mathbf{z}^H(n)\mathbf{k}(n)\xi^*(n)$$

where, in the last term, we have restored $\mathbf{z}(n)$ to its original form. By definition, the expression inside the first set of brackets on the right-hand side of Eq. (13.34) equals $\mathscr{E}_{\min}(n-1)$. Also, by definition, the expression inside the second set of brackets equals the complex conjugate of the *a priori* estimation error $\xi(n)$. For the last term, we use the definition of the gain vector $\mathbf{k}(n)$ to express the inner product $\mathbf{z}^H(n)\mathbf{k}(n)$ as

$$\mathbf{z}^H(n)\mathbf{k}(n) = \mathbf{z}^H(n)\mathbf{\Phi}^{-1}(n)\mathbf{u}(n)$$
$$= [\mathbf{\Phi}^{-1}(n)\mathbf{z}(n)]^H\mathbf{u}(n)$$
$$= \hat{\mathbf{w}}^H(n)\mathbf{u}(n)$$

where (in the second line) we have used the Hermitian property of the correlation matrix $\mathbf{\Phi}(n)$, and (in the third line) we have used the fact that $\mathbf{\Phi}^{-1}(n)\mathbf{z}(n)$ equals the least-squares estimate $\hat{\mathbf{w}}(n)$. Accordingly, we may simplify Eq. (13.34) to

$$\mathscr{E}_{\min}(n) = \lambda\mathscr{E}_{\min}(n-1) + d(n)\xi^*(n) - \hat{\mathbf{w}}^H(n)\mathbf{u}(n)\xi^*(n)$$
$$= \lambda\mathscr{E}_{\min}(n-1) + \xi^*(n)[d(n) - \hat{\mathbf{w}}^H(n)\mathbf{u}(n)] \tag{13.35}$$
$$= \lambda\mathscr{E}_{\min}(n-1) + \xi^*(n)e(n)$$

where $e(n)$ is the *a posteriori* estimation error. Equation (13.35) is the recursion for updating the sum of weighted error squares. We thus see that the product of the complex conjugate of $\xi(n)$ and $e(n)$ represents the correction term in this update. Note that this product is real valued, which implies that we always have

$$\xi(n)e^*(n) = \xi^*(n)e(n) \tag{13.36}$$

Conversion Factor

The formula of Eq. (13.35) involves two different estimation errors: the *a priori* estimation error $\xi(n)$ and the *a posteriori* estimation error $e(n)$, which are naturally related. To establish the relationship between these two estimation errors, we start with the defining equation (13.27) and substitute the update equation (13.25), obtaining

$$\begin{aligned}
e(n) &= d(n) - [\hat{\mathbf{w}}(n-1) + \mathbf{k}(n)\xi^*(n)]^H\mathbf{u}(n) \\
&= d(n) - \hat{\mathbf{w}}^H(n-1)\mathbf{u}(n) - \mathbf{k}^H(n)\mathbf{u}(n)\xi(n) \\
&= (1 - \mathbf{k}^H(n)\mathbf{u}(n))\xi(n)
\end{aligned} \tag{13.37}$$

where, in the last line, we have made use of the definition given in Eq. (13.26). The ratio of the *a posteriori* estimation $e(n)$ to the *a priori* estimation $\xi(n)$ is called the *conversion factor*, denoted by $\gamma(n)$. We may thus write

$$\begin{aligned}
\gamma(n) &= \frac{e(n)}{\xi(n)} \\
&= 1 - \mathbf{k}^H(n)\mathbf{u}(n)
\end{aligned} \tag{13.38}$$

the value of which is uniquely determined by the gain vector $\mathbf{k}(n)$ and the tap-input vector $\mathbf{u}(n)$.

13.5 EXAMPLE: SINGLE-WEIGHT ADAPTIVE NOISE CANCELER

Consider the *single-weight, dual-input adaptive noise canceler* depicted in Fig. 13.3. The two inputs are represented by the *primary signal* $d(n)$ and the *reference signal* $u(n)$ that are characterized as follows. First, the primary signal consists of an *information-bearing signal* component and an additive *interference*. Second, the reference signal $u(n)$ is correlated with the interference and has no detectable contribution to the information-bearing signal. The requirement is to exploit the properties of the reference signal in relation to the primary signal to suppress the interference at the adaptive noise canceler output.

Application of the RLS algorithm yields the following set of equations for this canceler (after reorganization of terms):

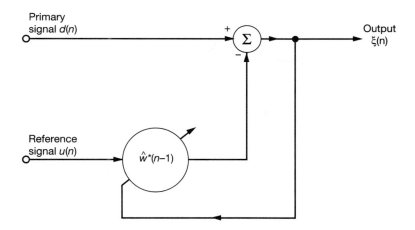

Figure 13.3   Single-weight adaptive noise canceler.

$$k(n) = \left[\frac{1}{\lambda\hat{\sigma}^2(n-1) + |u(n)|^2} \right] u(n) \tag{13.39}$$

$$\xi(n) = d(n) - \hat{w}^*(n-1)u(n) \tag{13.40}$$

$$\hat{w}(n) = \hat{w}(n-1) + k(n)\xi^*(n) \tag{13.41}$$

$$\hat{\sigma}^2(n) = \lambda\hat{\sigma}^2(n-1) + |u(n)|^2 \tag{13.42}$$

where $\hat{\sigma}^2(n)$ is an estimate of the error variance. It is the inverse of $P(n)$, the scalar version of the matrix $\mathbf{P}(n)$ in the RLS algorithm, as shown by

$$\hat{\sigma}^2(n) = P^{-1}(n) \tag{13.43}$$

It is informative to compare the algorithm described in Eqs. (13.39) to (13.42) with its counterpart obtained using the normalized LMS algorithm; the version of the normalized LMS algorithm of particular interest in the context of our present situation is that given in Eq. (9.144). The *major difference* between these two algorithms is that the constant a in the normalized LMS algorithm is replaced by the time-varying term $\lambda\hat{\sigma}^2(n-1)$ in the denominator of the gain factor $k(n)$ that controls the correction applied to the tap weight in Eq. (13.41).

13.6 CONVERGENCE ANALYSIS OF THE RLS ALGORITHM

In this section we demonstrate the convergence of the RLS algorithm operating in a stationary environment. The treatment presented here is rigorous, within the confines of the independence theory, the elements of which were described in Section 9.4. We say "rigor-

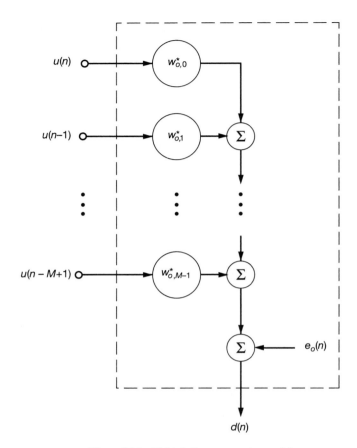

Figure 13.4 Multiple linear regression model.

ous" in the sense that we will *not* invoke the direct-averaging method, as was the case with the LMS algorithm.

To proceed with the analysis, we assume that the desired response $d(n)$ and the tap-input vector $\mathbf{u}(n)$ are related by the *multiple linear regression model* of Fig. 13.4. In particular, we may write

$$d(n) = e_o(n) + \mathbf{w}_o^H \mathbf{u}(n) \tag{13.44}$$

where the M-by-1 vector $\mathbf{w}_o$ denotes the *regression parameter vector* of the model, and $e_o(n)$ is the *measurement error*. The measurement error process $e_o(n)$ is white with zero mean and variance σ^2. The parameter vector $\mathbf{w}_o$ is constant. The latter assumption is equivalent to saying that the adaptive transversal filter operates in a stationary environment, with $\lambda = 1$.

Convergence of the RLS Algorithm in the Mean Value

Starting with $\boldsymbol{\Phi}(0) = \mathbf{0}$ that corresponds to $\mathbf{u}(0) = \mathbf{0}$ (i.e., prewindowing of the input data), we find that with the soft-constrained initialization procedure described in Section 13.3, the weight vector $\hat{\mathbf{w}}(n)$ computed by the RLS algorithm is almost exactly the same as that computed by the method of least-squares for $n \geq M$, where M is the number of taps in the adaptive transversal filter. Accordingly, we may use the normal equations to express $\hat{\mathbf{w}}(n)$ by the formula

$$\hat{\mathbf{w}}(n) = \boldsymbol{\Phi}^{-1}(n)\mathbf{z}(n), \qquad n \geq M \tag{13.45}$$

where, for $\lambda = 1$,

$$\boldsymbol{\Phi}(n) = \sum_{i=1}^{n} \mathbf{u}(i)\mathbf{u}^{H}(i) \tag{13.46}$$

and

$$\mathbf{z}(n) = \sum_{i=1}^{n} \mathbf{u}(i)d^*(i) \tag{13.47}$$

Substituting Eq. (13.44) in (13.47) yields

$$\mathbf{z}(n) = \sum_{i=1}^{n} \mathbf{u}(i)\mathbf{u}^{H}(i)\mathbf{w}_o + \sum_{i=1}^{n} \mathbf{u}(i)e_o^*(i)$$

$$= \boldsymbol{\Phi}(n)\mathbf{w}_0 + \sum_{i=1}^{n} \mathbf{u}(i)e_o^*(i) \tag{13.48}$$

This, in turn, means that we may rewrite Eq. (13.45) as

$$\hat{\mathbf{w}}(n) = \boldsymbol{\Phi}^{-1}(n)\boldsymbol{\Phi}(n)\mathbf{w}_o + \boldsymbol{\Phi}^{-1}(n)\sum_{i=1}^{n} \mathbf{u}(i)e_o^*(i)$$

$$= \mathbf{w}_o + \boldsymbol{\Phi}^{-1}(n)\sum_{i=1}^{n} \mathbf{u}(i)e_o^*(i) \tag{13.49}$$

Next, we invoke the expectation property of a random variable:

$$E[x] = E[E[x|y]] \tag{13.50}$$

where $E[x|y]$ is the conditional expectation of a random variable x, given another random variable y; the remaining expectation on the right-hand side of Eq. (13.50) is with respect to y. Hence, in light of this property, we may use Eq. (13.49) to express the expected value of $\hat{\mathbf{w}}(n)$ as follows:

$$E[\hat{\mathbf{w}}(n)] = \mathbf{w}_o + E[\boldsymbol{\Phi}^{-1}(n)\sum_{i=1}^{n} \mathbf{u}(i)e_o^*(i)]$$

$$= \mathbf{w}_o + E\left[E[\boldsymbol{\Phi}^{-1}(n)\sum_{i=1}^{n} \mathbf{u}(i)e_o^*(i) \,|\mathbf{u}(i), i = 1, 2, \ldots, n]\right] \tag{13.51}$$

Recognizing that

- The time-averaged correlation matrix $\mathbf{\Phi}(n)$ is uniquely defined by the sequence of input vectors $\mathbf{u}(1), \mathbf{u}(2), \ldots, \mathbf{u}(n)$
- The measurement error $e_o(i)$ is independent of the input vector $\mathbf{u}(i)$ for all i
- The measurement error $e(i)$ has zero mean

we may reduce Eq. (13.51) to

$$E[\hat{\mathbf{w}}(n)] = \mathbf{w}_o, \qquad n \geq M \tag{13.52}$$

Equation (13.52) states that the *RLS algorithm is convergent in the mean value* for $n \geq M$, where M is the number of taps in the adaptive transversal filter. Note that, unlike the LMS algorithm, the RLS algorithm does not have to wait for $n \rightarrow \infty$ for convergence in the mean value to be attained.

The Mean-squared Error in the Weight Vector $\hat{\mathbf{w}}(n)$

Consider next the convergence of the mean-squared error in the weight vector $\hat{\mathbf{w}}(n)$. To demonstrate this convergence, we first use Eq. (13.49) to express the *weight-error vector* as

$$\boldsymbol{\epsilon}(n) = \hat{\mathbf{w}}(n) - \mathbf{w}_o \tag{13.53}$$

$$= \mathbf{\Phi}^{-1}(n) \sum_{i=1}^{n} \mathbf{u}(i) e_o^*(i)$$

Next, using the definition of the *weight-error correlation matrix*

$$\mathbf{K}(n) = E[\boldsymbol{\epsilon}(n)\boldsymbol{\epsilon}^H(n)] \tag{13.54}$$

we have

$$\mathbf{K}(n) = E\left[\mathbf{\Phi}^{-1}(n) \sum_{i=1}^{n} \sum_{j=1}^{n} \mathbf{u}(i) e_o^*(i) e_o(j) \mathbf{u}^H(j) \mathbf{\Phi}^{-1}(n) \right] \tag{13.55}$$

Again, invoking the expectation property described in Eq. (13.50), we may rewrite Eq. (13.55) as

$$\mathbf{K}(n) = E\left[\mathbf{\Phi}^{-1}(n) \sum_{i=1}^{n} \sum_{j=1}^{n} \mathbf{u}(i) E[e_o^*(i) e_0(j)] \mathbf{u}^H(n) \mathbf{\Phi}^{-1}(n)] \right] \tag{13.56}$$

Since the measurement error $e_o(i)$ is assumed to be drawn from a white-noise process of variance σ^2, we have

$$E[e_o(i) e_o^*(j)] = \begin{cases} \sigma^2, & j = i \\ 0, & j \neq i \end{cases} \tag{13.57}$$

It follows therefore that we may simplify Eq. (13.56) to

$$\mathbf{K}(n) = \sigma^2 E\left[\boldsymbol{\Phi}^{-1}(n) \sum_{i=1}^{n} \mathbf{u}(i)\mathbf{u}^H(i)\boldsymbol{\Phi}^{-1}(n)\right]$$

$$= \sigma^2 E[\boldsymbol{\Phi}^{-1}(n)\boldsymbol{\Phi}(n)\boldsymbol{\Phi}^{-1}(n)] \qquad (13.58)$$

$$= \sigma^2 E[\boldsymbol{\Phi}^{-1}(n)]$$

Next, we invoke two elements of the *independence assumption* described in Section 9.4:

1. The input vectors $\mathbf{u}(1)$, $\mathbf{u}(2)$, ..., $\mathbf{u}(n)$ are *independently and identically distributed* (iid).
2. The input vectors $\mathbf{u}(1)$, $\mathbf{u}(2)$, ..., $\mathbf{u}(n)$ are drawn from a stochastic process with a *multivariate Gaussian distribution* of zero mean and ensemble-averaged correlation matrix $\mathbf{R}$.

Then, in light of the material presented in Appendix J, the correlation matrix $\boldsymbol{\Phi}(n)$ is described by a *complex Wishart distribution*, which is so named in honor of Wishart (1982). In particular, in that appendix it is shown that the expectation of the inverse correlation matrix $\boldsymbol{\Phi}^{-1}(n)$ is exactly

$$E[\boldsymbol{\Phi}^{-1}(n)] = \frac{1}{n-M-1}\mathbf{R}^{-1}, \qquad n > M + 1 \qquad (13.59)$$

Substituting Eq. (13.59) in (13.58), we may therefore express the weight-error correlation matrix $\mathbf{K}(n)$ as

$$\mathbf{K}(n) = \frac{\sigma^2}{n-M-1}\mathbf{R}^{-1}, \qquad n > M + 1 \qquad (13.60)$$

from which we readily deduce that

$$E[\boldsymbol{\epsilon}^H(n)\boldsymbol{\epsilon}(n)] = \text{tr}[\mathbf{K}(n)]$$

$$= \frac{\sigma^2}{n-M-1}\text{tr}[\mathbf{R}^{-1}] \qquad (13.61)$$

$$= \frac{\sigma^2}{n - M - 1}\sum_{i=1}^{M}\frac{1}{\lambda_i}, \qquad n > M + 1$$

where the λ_i are the eigenvalues of the ensemble-averaged correlation matrix $\mathbf{R}$.

On the basis of Eq. (13.61), we may now make the following two important observations for $n > M + 1$:

1. The mean-squared error in the weight vector $\hat{\mathbf{w}}(n)$ is *magnified by the inverse of the smallest eigenvalue* $\lambda_{\min}$. Hence, to a first order of approximation, the sensi-

tivity of the RLS algorithm to eigenvalue spread is determined initially in proportion to the inverse of the smallest eigenvalue. Therefore, ill-conditioned least-squares problems may lead to *bad* convergence properties.

2. The mean-squared error in the weight-vector $\hat{\mathbf{w}}(n)$ decays almost linearly with the number of iterations, n. Hence, the estimate $\hat{\mathbf{w}}(n)$ produced by the RLS algorithm for the tap-weight vector converges in the norm (i.e., mean square) to the parameter vector $\mathbf{w}_o$ of the multiple linear regression model almost *linearly with time*.

Learning Curve of the RLS Algorithm

In the RLS algorithm there are two errors, the *a priori* estimation error $\xi(n)$ and the *a posteriori* estimation error $e(n)$, to be considered. Given the initial conditions of Section 13.3 we find that the mean-square values of these two errors vary differently with time n. At time $n = 1$, the mean-square value of $\xi(n)$ attains a *large* value, equal to the mean-square value of the desired response $d(n)$, and then *decays* with increasing n. The mean-square value of $e(n)$, on the other hand, attains a *small* value at $n = 1$, and then *rises* with increasing n. Accordingly, the choice of $\xi(n)$ as the error of interest yields a learning curve for the RLS algorithm that has the same general shape as that for the LMS algorithm. By so doing, we can then make a direct graphical comparison between the learning curves of the RLS and LMS algorithms. We will therefore base computation of the ensemble-averaged learning curve of the RLS algorithm on the *a priori* estimation error $\xi(n)$.

Eliminating the desired response $d(n)$ between Eqs. (13.26) and (13.44), we may express the *a priori* estimation error $\xi(n)$ as

$$\xi(n) = e_o(n) - [\hat{\mathbf{w}}(n-1) - \mathbf{w}_o]^H \mathbf{u}(n)$$
$$= e_o(n) - \boldsymbol{\epsilon}^H(n-1)\mathbf{u}(n) \tag{13.62}$$

where $\boldsymbol{\epsilon}(n-1)$ is the weight-error vector at time $n-1$. As an *index of statistical performance* for the RLS algorithm, it is convenient to use the *a priori* estimation error $\xi(n)$ to define the *mean-squared error*:

$$J'(n) = E[|\xi(n)|^2] \tag{13.63}$$

The prime in the symbol $J'(n)$ is intended to distinguish the mean-square value of $\xi(n)$ from that of $e(n)$. Substituting Eq. (13.62) in (13.63), and then expanding terms, we get

$$J'(n) = E[|e_o(n)|^2] + E[\mathbf{u}^H(n)\boldsymbol{\epsilon}(n-1)\boldsymbol{\epsilon}^H(n-1)\mathbf{u}(n)]$$
$$- E[\boldsymbol{\epsilon}^H(n-1)\mathbf{u}(n)e_o^*(n)] - E[e_o(n)\mathbf{u}^H(n)\boldsymbol{\epsilon}(n-1)] \tag{13.64}$$

With the measurement $e_o(n)$ assumed to be of zero mean, the first expectation on the right-hand side of Eq. (13.64) is simply the variance of $e_o(n)$, which is denoted by σ^2. As for the remaining three expectations, we may make the following observations in light of the independence assumption described previously:

1. The estimate $\hat{\mathbf{w}}(n - 1)$, and therefore the weight-error vector $\boldsymbol{\epsilon}(n - 1)$, is independent of the tap-input vector $\mathbf{u}(n)$; the latter is assumed to be drawn from a wide-sense stationary process of zero mean. Hence, we may use this statistical independence together with well-known results from matrix algebra to express the second expectation on the right-hand side of Eq. (13.64) as follows:

$$
\begin{aligned}
E[\mathbf{u}^H(n)\boldsymbol{\epsilon}(n - 1)\boldsymbol{\epsilon}^H(n - 1)\mathbf{u}(n)] &= E[\text{tr}\{\mathbf{u}^H(n)\boldsymbol{\epsilon}(n - 1)\boldsymbol{\epsilon}^H(n - 1)\mathbf{u}(n)\}] \\
&= E[\text{tr}\{\mathbf{u}(n)\mathbf{u}^H(n)\boldsymbol{\epsilon}(n - 1)\boldsymbol{\epsilon}^H(n - 1)\}] \\
&= \text{tr}\{E[\mathbf{u}(n)\mathbf{u}^H(n)\boldsymbol{\epsilon}(n - 1)\boldsymbol{\epsilon}^H(n - 1)]\} \\
&= \text{tr}\{E[\mathbf{u}(n)\mathbf{u}^H(n)]E[\boldsymbol{\epsilon}(n - 1)\boldsymbol{\epsilon}^H(n - 1)]\} \\
&= \text{tr}\{\mathbf{R}\mathbf{K}(n - 1)\} \qquad (13.65)
\end{aligned}
$$

where, in the last line, we have made use of the definitions of the ensemble-averaged correlation matrix $\mathbf{R}$ and weight-error correlation matrix $\mathbf{K}(n - 1)$.

2. The measurement error $e_o(n)$ depends on the tap-input vector $\mathbf{u}(n)$; this follows from a simple rearrangement of Eq. (13.44). The weight-error vector $\boldsymbol{\epsilon}(n - 1)$ is therefore independent of both $\mathbf{u}(n)$ and $e_o(n)$. Accordingly, we may show that the third expectation on the right-hand side of Eq. (13.64) is zero by first reformulating it as follows:

$$
E[\boldsymbol{\epsilon}^H(n - 1)\mathbf{u}(n)e_o^*(n)] = E[\boldsymbol{\epsilon}^H(n - 1)]E[\mathbf{u}(n)e_o^*(n)]
$$

We now recognize from the principle of orthogonality that all the elements of the tap-input vector $\mathbf{u}(n)$ are orthogonal to the measurement error $e_o(n)$. We therefore have

$$
E[\boldsymbol{\epsilon}^H(n - 1)\mathbf{u}(n)e_o^*(n)] = 0 \qquad (13.66)
$$

3. The fourth and final expectation on the right-hand side of Eq. (13.64) has the same mathematical form as that just considered in point 2, except for a trivial complex conjugation. We may therefore set this expectation equal to zero, too:

$$
E[e_o(n)\mathbf{u}^H(n)\boldsymbol{\epsilon}(n - 1)] = 0 \qquad (13.67)
$$

Thus, recognizing that $E[|e_o(n)|^2] = \sigma^2$, and using the results of Eqs. (13.65) to (13.67) in (13.64), we get the following simple formula for the mean-squared error in the RLS algorithm.

$$
J'(n) = \sigma^2 + \text{tr}[\mathbf{R}\mathbf{K}(n - 1)] \qquad (13.68)
$$

Next, substituting Eq. (13.60) in (13.68), we get (for $\lambda = 1$)

$$
J'(n) = \sigma^2 + \frac{M\sigma^2}{n - M - 1}, \qquad n > M + 1 \qquad (13.69)
$$

Based on this result, we may make the following deductions:

1. The ensemble-averaged learning curve of the RLS algorithm converges in about $2M$ iterations, where M is the number of taps in the transversal filter. This means that the rate of convergence of the RLS algorithm is typically an order of magnitude *faster* than that of the LMS algorithm.

2. As the number of iterations, n, approaches infinity, the mean-squared error $J'(n)$ approaches a final value equal to the variance σ^2 of the measurement error $e_o(n)$. In other words the RLS algorithm, in theory, produces zero excess mean-squared error (or, equivalently, zero misadjustment) when operating in a stationary environment.

3. Convergence of the RLS algorithm in the mean square is independent of the eigenvalues of the ensemble-averaged correlation matrix $\mathbf{R}$ of the input vector $\mathbf{u}(n)$.

It should be emphasized that the above-mentioned improvement in the rate of convergence of the RLS algorithm over the LMS algorithm holds only when the measurement error $e_o(n)$ is small compared to the desired response $d(n)$, that is, when the signal-to-noise ratio is high. Also, the zero misadjustment property of the RLS algorithm assumes that the exponential weighting factor λ equals unity; that is, the algorithm operates with infinite memory.

13.7 COMPUTER EXPERIMENT ON ADAPTIVE EQUALIZATION

For our computer experiment, we use the RLS algorithm with the exponential weighting factor $\lambda = 1$, for the adaptive equalization of a linear dispersive communication channel. The LMS version of this study was presented in Section 9.7. The block diagram of the system used in the study is depicted in Fig. 13.5. Two independent random-number generators are used, one, denoted by x_n, for probing the channel, and the other, denoted by $v(n)$, for simulating the effect of additive white noise at the receiver input. The sequence x_n is a Bernoulli sequence with $x_n = \pm 1$; the random variable x_n has zero mean and unit variance. The second sequence $v(n)$ has no zero mean; its variance σ_v^2 is determined by the desired signal-to-noise ratio. The equalizer has 11 taps. The impulse response of the channel is defined by

$$h_n = \begin{cases} \dfrac{1}{2}\left[1 + \cos\left(\dfrac{2\pi}{W}(n-2)\right)\right], & n = 1, 2, 3 \\ 0 & \text{otherwise} \end{cases}$$

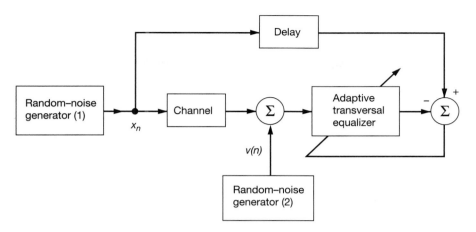

Figure 13.5   Block diagram of adaptive equalizer for computer experiment.

where W controls the amount of amplitude distortion and therefore the eigenvalue spread produced by the channel. The channel input x_n, after a delay of seven samples, provides the desired response for the equalizer (see Section 9.13 for details).

The experiment is in two parts: In part 1 the signal-to-noise ratio is high, and in part 2 it is low. In both parts of the experiment, the constant $\delta = 0.004$.

1. Signal-to-Noise Ratio = 30 dB.   The results of the experiment for a fixed signal-to-noise ratio of 30 dB (equivalently, variance $\sigma_\nu^2 = 0.001$) and varying W or eigenvalue spread $\chi(\mathbf{R})$ were presented previously in Chapter 10; see Fig. 10.10. The four parts of that figure correspond to the parameter $W = 2.9, 3.1, 3.3,$ and 3.5, or equivalently $\chi(\mathbf{R}) = 6.0782, 11.1238, 21.7132,$ and 46.8216, respectively (see Table 9.2 for details). Each part of the figure includes learning curves for the LMS, DCT-LMS, and RLS algorithms. The present discussion pertains to the RLS and LMS algorithms. The learning curves of the RLS algorithm were obtained by ensemble-averaging the squared value of the *a priori* estimation error $\xi(n)$ for each iteration n, and those for the LMS algorithm were obtained by ensemble-averaging the squared value of the *a posteriori* estimation error $e(n)$. The ensemble-averaging was performed over 200 independent trials of the experiment. For the LMS algorithm, the step-size parameter $\mu = 0.075$ was used. Based on the results shown in Fig. 10.10, we may make the following observations:

- Convergence of the RLS algorithm is attained in about 20 iterations, approximately twice the number of taps in the transversal equalizer.
- Rate of convergence of the RLS algorithm is relatively insensitive to variations in the eigenvalue spread $\chi(\mathbf{R})$ compared to the LMS algorithm. This property is

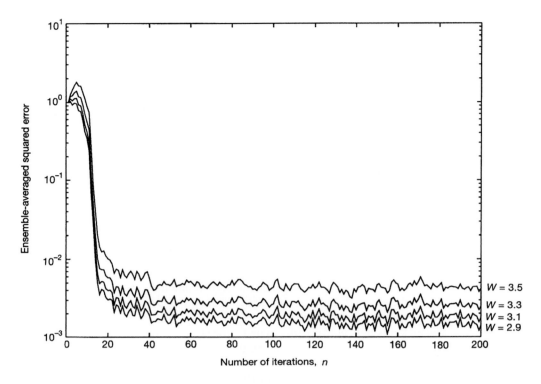

Figure 13.6 Learning curves for the RLS algorithm with four different eigenvalue spreads, and $\delta = 0.004$, $\lambda = 1.0$.

clearly illustrated in Fig. 13.6, where we have reproduced the learning curves of the RLS algorithm, corresponding to the four different values of the eigenvalue spread.

- The RLS algorithm converges much faster than the LMS algorithm.
- The steady-state value of the averaged squared error produced by the RLS algorithm is much smaller than in the case of the LMS algorithm, confirming what we said earlier: The RLS algorithm produces zero misadjustment, in theory.

The results presented in Figs. 10.10 and 13.6 clearly show the superior rate of convergence of the RLS over the LMS algorithm; for it to be realized, however, the signal-to-noise ratio has to be high. This advantage is lost when the signal-to-noise ratio is not high, as demonstrated next.

2. Signal-to-Noise Ratio = 10 dB.   Figure 13.7 shows the learning curves for the RLS algorithm and the LMS algorithm (with the step-size parameter $\mu = 0.075$) for

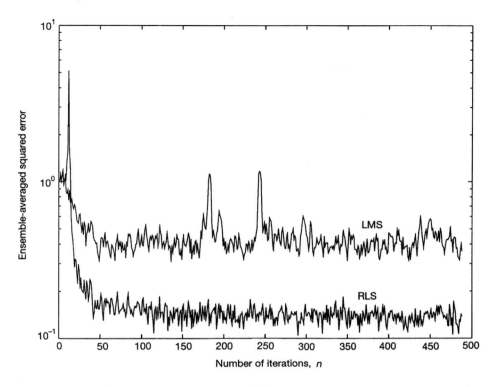

Figure 13.7   Learning curves for the RLS and LMS algorithms for $W = 3.1$ (i.e., eigen-value spread $\chi(\mathbf{R}) = 11.1238$), and SNR = 10 dB. RLS: $\delta = 0.004$ and $\lambda = 1.0$. LMS: Step size parameter $\mu = 0.075$.

$W = 3.1$ and signal-to-noise ratio of 10 dB. Insofar as the rate of convergence is concerned, we now see that the RLS and LMS algorithms perform in roughly the same manner, both requiring about 40 iterations to converge.

13.8 STATE-SPACE FORMULATION OF THE RLS PROBLEM

The exponentially weighted RLS algorithm was derived from first principles in Section 13.3 using the matrix inversion lemma. The underlying mathematical model used in this derivation is a *deterministic* one, in that the only source of uncertainty in the model resides in the measurement error $e_o(n)$. The RLS algorithm may also be *deduced in its exact form* directly from the covariance Kalman filtering algorithm of Chapter 7 by using a state-space model that matches the RLS problem (Sayed and Kailath, 1994). The state-space model used here is naturally stochastic in its formulation. This alternative approach to the solution of the RLS problem is important because it enables us to establish a highly valuable list of one-to-one correspondences between the RLS variables rooted in a linear regression model and the Kalman variables rooted in a state-space model. With such a list at our disposal, we may exploit the vast control literature on Kalman filters to solve the RLS problem in all of its different forms in a *unified manner*, which is precisely our ultimate objective.

A Comparison of Stochastic and Deterministic Models

To proceed, consider the unforced dynamical model described by Eqs. (7.74) to (7.76), reproduced here for convenience of presentation:

$$\mathbf{x}(n + 1) = \lambda^{-1/2}\mathbf{x}(n) \tag{13.70}$$

$$y(n) = \mathbf{u}^H(n)\mathbf{x}(n) + v(n) \tag{13.71}$$

where $\mathbf{x}(n)$ is the state vector of the model, $y(n)$ is a scalar observation or reference signal, $\mathbf{u}^H(n)$ is the measurement matrix, and $v(n)$ is a scalar white noise process of zero mean and unit variance. The model parameter λ is a positive real constant. From Eq. (13.70), we readily see that

$$\mathbf{x}(n) = \lambda^{-n/2}\mathbf{x}(0) \tag{13.72}$$

where $\mathbf{x}(0)$ is the initial value of the state vector. Hence, evaluating Eq. (13.71) for time $n = 0, 1, \ldots$, and then utilizing Eq. (13.72) to express the state vectors at different times in terms of the common value $\mathbf{x}(0)$, we obtain the following stochastic system of linear simultaneous equations:

$$
\begin{aligned}
y(0) &= \mathbf{u}^H(0)\mathbf{x}(0) + v(0) \\
y(1) &= \lambda^{-1/2}\mathbf{u}^H(1)\mathbf{x}(0) + v(1) \\
&\;\bullet \\
&\;\bullet \\
&\;\bullet \\
y(n) &= \lambda^{-n/2}\mathbf{u}^H(n)\mathbf{x}(0) + v(n)
\end{aligned}
\tag{13.73}
$$

Equivalently, we may write

$$
\begin{aligned}
y(0) &= \mathbf{u}^H(0)\mathbf{x}(0) + v(0) \\
\lambda^{1/2}y(1) &= \mathbf{u}^H(1)\mathbf{x}(0) + \lambda^{1/2}v(1) \\
&\;\bullet \\
&\;\bullet \\
&\;\bullet \\
\lambda^{n/2}y(n) &= \mathbf{u}^H(n)\mathbf{x}(0) + \lambda^{n/2}v(n)
\end{aligned}
\tag{13.74}
$$

The system of Eqs. (13.74) represents a stochastic characterization of the unforced dynamical model pertaining to a Kalman filter point of view.

Consider next a deterministic formulation of the problem as seen from the RLS point of view. Adapting the linear regression model of Eq. (13.44) to the problem at hand, we may write the following deterministic system of linear simultaneous equations:

$$d*(0) = \mathbf{u}^H(0)\mathbf{w}_o + e_o^*(0)$$

$$d*(1) = \mathbf{u}^H(1)\mathbf{w}_o + e_o^*(1)$$

$$\cdot$$
$$\cdot \qquad\qquad\qquad\qquad\qquad\qquad (13.75)$$
$$\cdot$$

$$d*(n) = \mathbf{u}^H(n)\mathbf{w}_o + e_o^*(n)$$

where $\mathbf{w}_o$ is the unknown parameter vector of the model, $\mathbf{u}(n)$ is the input vector, $d(n)$ is the reference signal or desired response, and $e_o(n)$ is the measurement error.

We thus have two systems of linear simultaneous equations for solving essentially the same problem. One system of equations is stochastic, rooted in Kalman filter theory; and the other system is deterministic, rooted in least-squares estimation theory. We would intuitively expect that both approaches lead to exactly the same solution for the problem at hand. Moreover, recognizing that these two systems of equations have the same mathematical form, it seems reasonable for us to set

$$\mathbf{x}(0) = \mathbf{w}_o \qquad\qquad\qquad (13.76)$$

On this basis, a comparison between the stochastic equations (13.74) and the deterministic equations (13.75) immediately reveals the following one-to-one correspondences:

$$y(n) = \lambda^{-n/2}d*(n) \qquad\qquad\qquad (13.77)$$

$$v(n) = \lambda^{-n/2}e_o*(n) \qquad\qquad\qquad (13.78)$$

where the asterisk denotes complex conjugation, the variables on the left-hand side refer to the state-space model, and those on the right-hand side refer to the linear regression model.

A Comparison of Covariance Kalman Filtering and RLS Algorithm

We may expand on this list of one-to-one correspondences by comparing the covariance Kalman filtering algorithm summarized in Table 7.3 of Chapter 7 and the RLS algorithm summarized in Table 13.1 of this chapter. Indeed, a comparison of these two algorithms, line by line, immediately leads us to write [assuming $\mathbf{K}(1, 0) = \lambda^{-1}\delta^{-1}\mathbf{I}$ and $\hat{\mathbf{x}}(1|\mathcal{Y}_0) = \mathbf{0}$ in Table 7.3]:

$$\mathbf{K}(n-1) = \lambda^{-1}\mathbf{P}(n-1) \qquad\qquad\qquad (13.79)$$

$$\mathbf{g}(n) = \lambda^{-1/2}\mathbf{k}(n) \qquad\qquad\qquad (13.80)$$

$$\alpha(n) = \lambda^{-n/2}\xi*(n) \qquad\qquad\qquad (13.81)$$

$$\hat{\mathbf{x}}(n+1|\mathcal{Y}_n) = \lambda^{-(n+1)/2}\hat{\mathbf{w}}(n) \qquad\qquad\qquad (13.82)$$

TABLE 13.2 SUMMARY OF CORRESPONDENCES BETWEEN KALMAN VARIABLES AND RLS VARIABLES

Kalman		RLS	
Description	Variable	Variable	Description
Initial value of state vector	$\mathbf{x}(0)$	$\mathbf{w}_o$	Unknown regression-coefficient vector
State vector	$\mathbf{x}(n)$	$\lambda^{-n/2}\mathbf{w}_o$	Exponentially weighted version of unknown coefficient-regression vector
Reference (observation) signal	$y(n)$	$\lambda^{-n/2}d^*(n)$	Desired response
Measurement noise	$v(n)$	$\lambda^{-n/2}e_o^*(n)$	Measurement error
One-step prediction of state vector	$\hat{\mathbf{x}}(n+1\,|\,\mathcal{Y}_n)$	$\lambda^{-(n+1)/2}\hat{\mathbf{w}}(n)$	Estimate of tap-weight vector
Correlation matrix of error in state prediction	$\mathbf{K}(n)$	$\lambda^{-1}\mathbf{P}(n)$	Inverse of correlation matrix of input vector
Kalman gain	$\mathbf{g}(n)$	$\lambda^{-1/2}\mathbf{k}(n)$	Gain vector
Innovation	$\alpha(n)$	$\lambda^{-n/2}\xi^*(n)$	A *priori* estimation error
Conversion factor	$r^{-1}(n)$	$\gamma(n)$	Conversion factor
Initial conditions	$\hat{\mathbf{x}}(1\,|\,\mathcal{Y}_0)=\mathbf{0}$ $\mathbf{K}(0)$	$\hat{\mathbf{w}}(0)=\mathbf{0}$ $\lambda^{-1}\mathbf{P}(0)$	Initial conditions

Moreover, comparing Eqs. (7.65) and (13.38), we see that the conversion factor $r^{-1}(n)$ in the specialized form of the covariance Kalman filtering algorithm[2] and the conversion factor $\gamma(n)$ in the RLS algorithm are exactly the same, as shown by

$$r^{-1}(n) = \gamma(n) \tag{13.83}$$

Thus, collecting the results described by Eqs. (13.77) to (13.83) and other related results under one umbrella, we may set up the one-to-one correspondences listed in Table 13.2 between the Kalman and RLS variables, assuming complex-valued data.[3] The left half of the table pertains to the Kalman variables and their descriptions, whereas the right half per-

---

[2]Adapting Eq. (7.65) to the specialized form of the covariance (Kalman) filtering algorithm summarized in Table 7.3, we have

$$r^{-1}(n) = \frac{e(n)}{\xi(n)}$$
$$= \frac{1}{\mathbf{u}^H(n)\mathbf{K}(n-1)\mathbf{u}(n)+1}$$

[3]The list of correspondences presented in Table 13.2 is the same as that in the paper by Sayed and Kailath (1994); some minor differences are merely the result of differences in notation.

tains to the RLS variables. In making up the descriptions for the latter, we have focused on the variables of interest, ignoring (for the sake of simplicity) references to the operation of complex conjugation and multiplication by powers of the exponential weighting factor λ involved here.

13.9 SUMMARY AND DISCUSSION

In this chapter we first derived the recursive least-squares (RLS) algorithm as a natural extension of the method of least squares. The derivation was based on a lemma in matrix algebra known as the matrix inversion lemma.

The fundamental difference between the RLS algorithm and the LMS algorithm may be stated as follows: The step-size parameter μ in the LMS algorithm is replaced by $\Phi^{-1}(n)$, that is, the inverse of the correlation matrix of the input vector $\mathbf{u}(n)$. This modification has a profound impact on the convergence behavior of the RLS algorithm for a stationary environment, as summarized here:

1. The rate of convergence of the RLS algorithm is typically an order of magnitude faster than that of the LMS algorithm.
2. The rate of convergence of the RLS algorithm is invariant to the eigenvalue spread (i.e., condition number) of the ensemble-averaged correlation matrix $\mathbf{R}$ of the input vector $\mathbf{u}(n)$.
3. The excess mean-squared error $J'_{ex}(n)$ of the RLS algorithm converges to zero as the number of iterations, n, approaches infinity.

The operation of the RLS algorithm described herein refers to a stationary environment with the exponential weighting factor $\lambda = 1$. The case of $\lambda \neq 1$ is considered in Chapter 16, where it is shown that Properties 1 and 2 still hold but the excess mean-squared error $J'_{ex}(n)$ is no longer zero. In any event, computation of the mean-squared error $J'(n)$, produced by the RLS algorithm, is based on the *a priori* estimation error $\xi(n)$.

Another important result that we established in this chapter is that, although the RLS algorithm is deterministic and the Kalman filter is stochastic, there exist one-to-one correspondences between their individual variables. In particular, we may use these correspondences to derive important variants of the RLS algorithm, in a unified manner, from their Kalman filter counterparts, as demonstrated in the next two chapters.

PROBLEMS

1. To permit a recursive implementation of the method of least squares, the window or weighting function $\beta(n, i)$ must have a suitable structure. Assume that $\beta(n, i)$ may be expressed as

$$\beta(n, i) = \lambda(i)\beta(n, i - 1), \qquad i = 1, \ldots, n$$

where $\beta(n, n) = 1$. Hence, show that

$$\beta(n, i) = \prod_{k=i+1}^{n} \lambda^{-1}(k)$$

What is the form of $\lambda(k)$ for which $\beta(n, i) = \lambda^{n-i}$ is obtained?

2. Establish the validity of the matrix inversion lemma.

3. Consider a correlation matrix $\mathbf{\Phi}(n)$ defined by

$$\mathbf{\Phi}(n) = \mathbf{u}(n)\mathbf{u}^H(n) + \delta\mathbf{I}$$

where $\mathbf{u}(n)$ is a tap-input vector and δ is a small positive constant. Use the matrix inversion lemma to evaluate $\mathbf{P}(n) = \mathbf{\Phi}^{-1}(n)$.

4. Consider the modified definition of the correlation matrix $\mathbf{\Phi}(n)$ given in Eq. (13.28), which is reproduced here for convenience.

$$\mathbf{\Phi}(n) = \sum_{i=1}^{n} \lambda^{n-i}\mathbf{u}(i)\mathbf{u}^H(i) + \delta\lambda^n\mathbf{I}$$

where $\mathbf{u}(i)$ is the tap-input vector, λ is the exponential weighting factor, and δ is a small positive constant. Show that the use of this new definition for $\mathbf{\Phi}(n)$ leaves the equations that define the RLS algorithm completely unchanged.

5. Let $\xi(n)$ denote the *a priori* estimation error

$$\xi(n) = d(n) - \hat{\mathbf{w}}^H(n - 1)\mathbf{u}(n)$$

where $d(n)$ is the desired response, $\mathbf{u}(n)$ is the tap-input vector, and $\hat{\mathbf{w}}(n - 1)$ is the old estimate of the tap-weight vector. Let $e(n)$ denote the *a posteriori* estimation error

$$e(n) = d(n) - \hat{\mathbf{w}}^H(n)\mathbf{u}(n)$$

where $\hat{\mathbf{w}}(n)$ is the current estimate of the tap-weight vector. For complex-valued data, both $\xi(n)$ and $e(n)$ are likewise complex valued. Show that the product $\xi(n)e^*(n)$ is always real valued.

6. Given the initial conditions of Section 13.3 for the RLS algorithm, explain the reasons for the fact that the mean-square value of the *a posteriori* estimation error $e(n)$ attains a small value at $n = 1$ and then rises with increasing n.

7. The last two entries in Table 13.2 pertain to the one-to-one correspondences between the initial conditions of the Kalman variables and those of the RLS variables. Justify the validity of these two entries.

CHAPTER
—14—
Square-Root Adaptive Filters

One of the problems encountered in applying the RLS algorithm of Chapter 13 is that of numerical instability, which can arise because of the way in which the Riccati difference equation is formulated. This same problem is also known to arise in the classical Kalman filtering algorithm for exactly the same reason. In Chapter 7 we pointed out that the instability (divergence) problem encountered in a Kalman filter can be ameliorated by using a square-root variant of the filter. At that point in the discussion, we deferred a detailed treatment of square-root Kalman filtering until we are ready for it. In this chapter we will take up a full discussion of this issue, which we do in the next section. The solution to the square-root Kalman filtering problem sets the stage for deriving the corresponding variants of the RLS algorithm in light of the one-to-one correspondences that exist between the Kalman variables and the RLS variables, established in the previous chapter.

14.1 SQUARE-ROOT KALMAN FILTERS

The recursions in a Kalman filter of the covariance type propagate the matrix $\mathbf{K}(n)$, which denotes the correlation matrix of the error in the filtered state estimate; this propagation takes place via the Riccati difference equation. The recursions in a root-square Kalman filter, on the other hand, propagate a lower triangular matrix $\mathbf{K}^{1/2}(n)$, defined as the *square root* of $\mathbf{K}(n)$. The relation between $\mathbf{K}(n)$ and $\mathbf{K}^{1/2}(n)$ is defined by

$$\mathbf{K}(n) = \mathbf{K}^{1/2}(n)\mathbf{K}^{H/2}(2) \tag{14.1}$$

where the upper triangular matrix $\mathbf{K}^{H/2}(n)$ is the Hermitian transpose of $\mathbf{K}^{1/2}(n)$. Unlike the situation that may exist in the covariance Kalman filter, the nonnegative definite character of $\mathbf{K}(n)$ as a correlation matrix is preserved by virtue of the fact that the product of any square matrix and its Hermitian transpose is always a nonnegative definite matrix.

In this section we derive the square-root forms of the covariance and information implementations of the Kalman filter. But, with variants of the RLS algorithm in mind, we will focus our attention on the special unforced dynamical model (Sayed and Kailath, 1994):

$$\mathbf{x}(n + 1) = \lambda^{-1/2}\mathbf{x}(n + 1) \tag{14.2}$$

$$y(n) = \mathbf{u}^H(n)\mathbf{x}(n) + v(n) \tag{14.3}$$

where $\mathbf{x}(n)$ is the state vector, the row vector $\mathbf{u}^H(n)$ is the measurement matrix, the scalar $y(n)$ is an observation or reference signal, and the scalar $v(n)$ is a white noise process of zero mean and unit variance; the positive real scalar λ is a constant of the model. However, before proceeding with the derivations, we digress briefly to consider a lemma in matrix algebra that is pivotal to our present discussion.

Matrix Factorization Lemma

Given any two N-by-M matrices, $\mathbf{A}$ and $\mathbf{B}$, with dimension $N \leq M$, the *matrix factorization lemma* states that (Stewart, 1973; Golub and Van Loan, 1989; Sayed and Kailath, 1994)

$$\mathbf{A}\mathbf{A}^H = \mathbf{B}\mathbf{B}^H \tag{14.4}$$

if, and only if, there exists a unitary matrix $\mathbf{\Theta}$ such that

$$\mathbf{B} = \mathbf{A}\mathbf{\Theta} \tag{14.5}$$

Assuming that the condition (14.5) holds, we readily find that

$$\mathbf{B}\mathbf{B}^H = \mathbf{A}\mathbf{\Theta}\mathbf{\Theta}^H\mathbf{A}^H \tag{14.6}$$

From the definition of a unitary matrix, we have

$$\mathbf{\Theta}\mathbf{\Theta}^H = \mathbf{I} \tag{14.7}$$

where $\mathbf{I}$ is the identity matrix. Hence, Eq. (14.6) reduces immediately to Eq. (14.4).

Conversely, the equality described in Eq. (14.4) implies that the matrices $\mathbf{A}$ and $\mathbf{B}$ must be related. We may prove the converse implication of the matrix factorization lemma by invoking the singular value decomposition theorem. According to this theorem, the matrix $\mathbf{A}$ may be factored as follows (see Section 11.10):

$$\mathbf{A} = \mathbf{U}_A\mathbf{\Sigma}_A\mathbf{V}_A^H \tag{14.8}$$

where $\mathbf{U}_A$ and $\mathbf{V}_A$ are N-by-N and M-by-M unitary matrices, respectively, and $\mathbf{\Sigma}_A$ is an N-by-M matrix defined by the singular values of matrix $\mathbf{A}$. Similarly, the second matrix $\mathbf{B}$ may be factored as follows:

$$\mathbf{B} = \mathbf{U}_B \boldsymbol{\Sigma}_B \mathbf{V}_B^H \tag{14.9}$$

The identity $\mathbf{A}\mathbf{A}^H = \mathbf{B}\mathbf{B}^H$ implies that we have

$$\mathbf{U}_A = \mathbf{U}_B \tag{14.10}$$

and

$$\boldsymbol{\Sigma}_A = \boldsymbol{\Sigma}_B \tag{14.11}$$

Now let

$$\boldsymbol{\Theta} = \mathbf{V}_A \mathbf{V}_B^H \tag{14.12}$$

Then, substituting Eqs. (14.8) and (14.10) in (14.12) in the matrix product $\mathbf{A}\boldsymbol{\Theta}$ yields a result equal to matrix $\mathbf{B}$ by virtue of Eq. (14.9), which is precisely the converse implication of the matrix factorization lemma.

Square-root Covariance Filter

Returning to the issue of square-root Kalman filtering, we note that the Riccati difference equation for the covariance Kalman filter may be expressed as follows (by combining the first and final lines of the algorithm summarized in Table 7.3):

$$\mathbf{K}(n) = \lambda^{-1}\mathbf{K}(n - 1) - \lambda^{-1}\mathbf{K}(n - 1)\mathbf{u}(n)r^{-1}(n)\mathbf{u}^H(n)\mathbf{K}(n - 1) \tag{14.13}$$

The scalar $r(n)$ is the variance of the filtered estimation error; it is defined by

$$r(n) = \mathbf{u}^H(n)\mathbf{K}(n - 1)\mathbf{u}(n) + 1 \tag{14.14}$$

There are four distinct matrix terms that constitute the right-hand side of the Riccati equation (14.13), in light of which we may introduce the following two-by-two block matrix:

$$\mathbf{M}(n) = \begin{bmatrix} \mathbf{u}^H(n)\mathbf{K}(n - 1)\mathbf{u}(n) + 1 & \lambda^{-1/2}\mathbf{u}^H(n)\mathbf{K}(n - 1) \\ \lambda^{-1/2}\mathbf{K}(n - 1)\mathbf{u}(n) & \lambda^{-1}\mathbf{K}(n - 1) \end{bmatrix} \tag{14.15}$$

Expressing the correlation matrix $\mathbf{K}(n-1)$ in its factored form:

$$\mathbf{K}(n - 1) = \mathbf{K}^{1/2}(n - 1)\mathbf{K}^{H/2}(n - 1) \tag{14.16}$$

and recognizing that the matrix $\mathbf{M}(n)$ is a nonnegative-definite matrix, we may use the Cholesky factorization to express Eq. (14.15) as follows:

$$\mathbf{M}(n) = \begin{bmatrix} 1 & \mathbf{u}^H(n)\mathbf{K}^{1/2}(n - 1) \\ \mathbf{0} & \lambda^{-1/2}\mathbf{K}^{1/2}(n - 1) \end{bmatrix} \begin{bmatrix} 1 & \mathbf{0}^T \\ \mathbf{K}^{H/2}(n - 1)\mathbf{u}(n) & \lambda^{-1/2}\mathbf{K}^{H/2}(n - 1) \end{bmatrix} \tag{14.17}$$

where $\mathbf{0}$ is the null vector.

The matrix product on the right-hand side of Eq. (14.17) may be interpreted as the product of matrix $\mathbf{A}$, say, and its Hermitian transpose $\mathbf{A}^H$. The stage is therefore set for invoking the matrix factorization lemma, according to which we may write

$$\begin{bmatrix} 1 & \mathbf{u}^H(n)\mathbf{K}^{1/2}(n - 1) \\ \mathbf{0} & \lambda^{-1/2}\mathbf{K}^{1/2}(n - 1) \end{bmatrix} \boldsymbol{\Theta}(n) = \begin{bmatrix} b_{11}(n) & \mathbf{0}^T \\ \mathbf{b}_{21}(n) & \mathbf{B}_{22}(n) \end{bmatrix} \tag{14.18}$$

where $\boldsymbol{\Theta}(n)$ is a unitary rotation; and the scalar $b_{11}(n)$, the vector $\mathbf{b}_{21}(n)$, and the matrix $\mathbf{B}_{22}(n)$ denote the nonzero block elements of matrix $\mathbf{B}$. In Eq. (14.18), we may distinguish two arrays of numbers:

- A *prearray*, which is operated on by a unitary rotation.
- A *postarray*, which is characterized by a block zero entry resulting from the action of the unitary rotation. The postarray therefore has a "triangular" structure in a block sense.

To evaluate the unknown block elements b_{11}, $\mathbf{b}_{21}$, and $\mathbf{B}_{22}$ of the postarray, we proceed by squaring both sides of Eq. (14.18). Then, recognizing that $\boldsymbol{\Theta}(n)$ is a unitary matrix, and therefore $\boldsymbol{\Theta}(n)\boldsymbol{\Theta}^H(n)$ equals the identity matrix for all n, we may write

$$\underbrace{\begin{bmatrix} 1 & \mathbf{u}^H(n)\mathbf{K}^{1/2}(n-1) \\ \mathbf{0} & \lambda^{-1/2}\mathbf{K}^{1/2}(n-1) \end{bmatrix}}_{\mathbf{A}} \underbrace{\begin{bmatrix} 1 & \mathbf{0}^T \\ \mathbf{K}^{H/2}(n-1)\mathbf{u}(n) & \lambda^{-1/2}\mathbf{K}^{H/2}(n-1) \end{bmatrix}}_{\mathbf{A}^H}$$

$$= \underbrace{\begin{bmatrix} b_{11}(n) & \mathbf{0}^T \\ \mathbf{b}_{21}(n) & \mathbf{B}_{22}(n) \end{bmatrix}}_{\mathbf{B}} \underbrace{\begin{bmatrix} b_{11}^*(n) & \mathbf{b}_{21}^H(n) \\ \mathbf{0} & \mathbf{B}_{22}^H(n) \end{bmatrix}}_{\mathbf{B}^H}$$

(14.19)

Thus, comparing the respective terms on both sides of the equality (14.19), we get the following identities:

$$|b_{11}(n)|^2 = \mathbf{u}^H(n)\mathbf{K}(n-1)\mathbf{u}(n) + 1 = r(n) \tag{14.20}$$

$$\mathbf{b}_{21}(n)b_{11}^*(n) = \lambda^{-1/2}\mathbf{K}(n-1)\mathbf{u}(n) \tag{14.21}$$

$$\mathbf{b}_{21}(n)\mathbf{b}_{21}^H(n) + \mathbf{B}_{22}(n)\mathbf{B}_{22}^H(n) = \lambda^{-1}\mathbf{K}(n-1) \tag{14.22}$$

Equations (14.20) to (14.22) may be satisfied by choosing

$$b_{11}(n) = r^{1/2}(n) \tag{14.23}$$

$$\mathbf{b}_{21}(n) = \lambda^{-1/2}\mathbf{K}(n-1)\mathbf{u}(n)r^{-1/2}(n) = \mathbf{g}(n)r^{1/2}(n) \tag{14.24}$$

$$\mathbf{B}_{22}(n) = \mathbf{K}^{1/2}(n) \tag{14.25}$$

where, in the second line, $\mathbf{g}(n)$ denotes the Kalman gain; see the first computation step of Table 7.3 for a definition of the Kalman gain $\mathbf{g}(n)$.

We may thus rewrite Eq. (14.18) as

$$\begin{bmatrix} 1 & \mathbf{u}^H(n)\mathbf{K}^{1/2}(n-1) \\ \mathbf{0} & \lambda^{-1/2}\mathbf{K}^{1/2}(n-1) \end{bmatrix}\boldsymbol{\Theta}(n) = \begin{bmatrix} r^{1/2}(n) & \mathbf{0}^T \\ \mathbf{g}(n)r^{1/2}(n) & \mathbf{K}^{1/2}(n) \end{bmatrix} \tag{14.26}$$

The block elements of the prearray and postarray in Eq. (14.26) deserve close scrutiny, as they reveal some useful properties of their own:

- The block elements $\{\lambda^{-1/2}\mathbf{K}^{1/2}(n-1), \mathbf{u}^{H}(n)\,\mathbf{K}^{1/2}(n-1)\}$ of the prearray uniquely characterize the constitution of the quantities contained in the right-hand side of the Riccati difference equation (14.13), except for $r(n)$. Correspondingly, the block element $\mathbf{K}^{1/2}(n)$ of the postarray provides the quantity needed to update the prearray and therefore initiate the next iteration of the algorithm.
- Inclusion of the block elements $\{1, \mathbf{0}\}$ in the prearray induces the generation of two block elements in the postarray, namely, $\{r^{1/2}(n), \mathbf{g}(n)r^{1/2}(n)\}$. These elements make it possible to calculate two useful variables: the Kalman gain $\mathbf{g}(n)$ and the variance $r(n)$ of the filtered estimation error. The variance $r(n)$ is obtained simply by squaring the scalar entry $r^{1/2}(n)$. The Kalman gain $\mathbf{g}(n)$ is obtained equally simply by dividing the block entry $\mathbf{g}(n)r^{1/2}(n)$ by $r^{1/2}(n)$.

Building on the latter result, we may now readily update the state estimate as follows:

$$\hat{\mathbf{x}}(n + 1\,|\mathcal{Y}_n) = \lambda^{-1/2}\hat{\mathbf{x}}(n)\,|\mathcal{Y}_{n-1}) + \mathbf{g}(n)\alpha(n) \qquad (14.27)$$

where $\alpha(n)$ is the innovation defined by

$$\alpha(n) = y(n) - \mathbf{u}^{H}(n\hat{\mathbf{x}}(n\,|\mathcal{Y}_{n-1}) \qquad (14.28)$$

Table 14.1 presents a summary of the computations performed in the square-root covariance filtering algorithm (Sayed and Kailath, 1994). The initialization of the algorithm proceeds in exactly the same way as for the conventional covariance Kalman filtering algorithm (see Table 7.3).

Square-root Information Filters

Consider next the square-root implementation of Kalman's filtering algorithm, which propagates the inverse matrix $\mathbf{K}^{-1}(n)$ rather than $\mathbf{K}(n)$ itself. This form of propagation is useful, particularly when there exist large initial uncertainties (i.e., the initial value $\mathbf{\Pi}_0$ of the correlation matrix $\mathbf{K}(0)$ is large). A summary of the information filtering algorithm is presented in Table 7.4. The first two recursions of the algorithm are reproduced here for convenience of presentation:

$$\mathbf{K}^{-1}(n) = \lambda\mathbf{K}^{-1}(n - 1) + \lambda\mathbf{u}(n)\mathbf{u}^{H}(n) \qquad (14.29)$$

$$\mathbf{K}^{-1}(n)\hat{\mathbf{x}}(n + 1\,|\mathcal{Y}_n) = \lambda^{1/2}\mathbf{K}^{-1}(n - 1)\hat{\mathbf{x}}(n\,|\mathcal{Y}_{n-1}) + \lambda^{1/2}\mathbf{u}(n)y(n) \qquad (14.30)$$

Let the inverse matrix $\mathbf{K}^{-1}(n)$ be expressed in its factored form:

$$\mathbf{K}^{-1}(n) = \mathbf{K}^{-H/2}(n)\mathbf{K}^{-1/2}(n) \qquad (14.31)$$

TABLE 14.1 SUMMARY OF THE COMPUTATIONS PERFORMED IN SQUARE-ROOT VARIANTS OF THE KALMAN FILTER

1. *Square-root covariance filter:*

$$\begin{bmatrix} 1 & \mathbf{u}^H(n)\mathbf{K}^{1/2}(n-1) \\ 0 & \lambda^{-1/2}\mathbf{K}^{1/2}(n-1) \end{bmatrix} \Theta(n) = \begin{bmatrix} r^{1/2}(n) & \mathbf{0}^T \\ \mathbf{g}(n)r^{1/2}(n) & \mathbf{K}^{1/2}(n) \end{bmatrix}$$

$$\mathbf{g}(n) = (\mathbf{g}(n)r^{1/2}(n))\,(r^{1/2}(n))^{-1}$$

$$\alpha(n) = y(n) - \mathbf{u}^H(n)\hat{\mathbf{x}}(n \mid \mathcal{Y}_{n-1})$$

$$\hat{\mathbf{x}}(n+1 \mid \mathcal{Y}_n) = \lambda^{-1/2}\hat{\mathbf{x}}(n \mid \mathcal{Y}_{n-1}) + \mathbf{g}(n)\alpha(n)$$

2. *Square-root information filter:*

$$\begin{bmatrix} \lambda^{1/2}\mathbf{K}^{-H/2}(n-1) & \lambda^{1/2}\mathbf{u}(n) \\ \hat{\mathbf{x}}^H(n \mid \mathcal{Y}_{n-1})\mathbf{K}^{-H/2}(n-1) & y^*(n) \\ \mathbf{0}^T & 1 \end{bmatrix} \Theta(n) = \begin{bmatrix} \mathbf{K}^{-H/2}(n) & \mathbf{0} \\ \hat{\mathbf{x}}^H(n+1 \mid \mathcal{Y}_n)\mathbf{K}^{-H/2}(n) & r^{-1/2}(n)\alpha^*(n) \\ \lambda^{1/2}\mathbf{u}^H(n)\mathbf{K}^{1/2}(n) & r^{-1/2}(n) \end{bmatrix}$$

$$\hat{\mathbf{x}}^H(n+1 \mid \mathcal{Y}_n) = (\hat{\mathbf{x}}^H(n+1 \mid \mathcal{Y}_n)\mathbf{K}^{-H/2}(n))(\mathbf{K}^{-H/2}(n))^{-1}$$

3. *Extended square-root information filter:*

$$\begin{bmatrix} \lambda^{1/2}\mathbf{K}^{-H/2}(n-1) & \lambda^{1/2}\mathbf{u}(n) \\ \hat{\mathbf{x}}^H(n \mid \mathcal{Y}_{n-1})\mathbf{K}^{-H/2}(n-1) & y^*(n) \\ \mathbf{0}^T & 1 \\ \lambda^{-1/2}\mathbf{K}^{1/2}(n-1) & \mathbf{0} \end{bmatrix} \Theta(n) = \begin{bmatrix} \mathbf{K}^{-H/2}(n) & \mathbf{0} \\ \hat{\mathbf{x}}^H(n+1 \mid \mathcal{Y}_n)\mathbf{K}^{-H/2}(n) & r^{-1/2}(n)\alpha^*(N) \\ \lambda^{1/2}\mathbf{u}^H(n)\mathbf{K}^{1/2}(n) & r^{-1/2}(n) \\ \mathbf{K}^{1/2}(n) & -\mathbf{g}(n)r^{1/2}(n) \end{bmatrix}$$

$$\hat{\mathbf{x}}(n+1 \mid \mathcal{Y}_n) = \lambda^{-1/2}\hat{\mathbf{x}}(n \mid \mathcal{Y}_{n-1}) + (\mathbf{g}(n)r^{1/2}(n))(r^{-1/2}(n)\alpha^*(n))^*$$

$$= (\mathbf{K}^{1/2}(n))\,(\mathbf{K}^{-1/2}(n)\hat{\mathbf{x}}(n+1 \mid \mathcal{Y}_n))$$

Note: In all three cases, $\Theta(n)$ is a unitary rotation that produces a block zero entry in the top row of the postarray.

For reasons that will become apparent presently, we find it more convenient to express Eqs. (14.29) and (14.30) in their Hermitian transposed forms, in which case we may express the four quantities on the right-hand sides of these equations in their individual factored forms as follows:

$$\lambda\mathbf{K}^{-H}(n-1) = (\lambda^{1/2}\mathbf{K}^{-H/2}(n-1))\,(\lambda^{1/2}\mathbf{K}^{-1/2}(n-1))$$

$$\lambda\,\mathbf{u}(n)\mathbf{u}^H(n) = (\lambda^{1/2}\mathbf{u}(n))\,(\lambda^{1/2}\mathbf{u}^H(n))$$

$$\lambda^{1/2}\hat{\mathbf{x}}^H(n \mid \mathcal{Y}_{n-1})\mathbf{K}^{-H}(n-1) = (\hat{\mathbf{x}}^H(n \mid \mathcal{Y}_{n-1})\mathbf{K}^{-H/2}(n-1))\,(\lambda^{1/2}\mathbf{K}^{-1/2}(n-1))$$

$$\lambda^{1/2}y^*(n)\mathbf{u}^H(n) = (y^*(n))\,(\lambda^{1/2}\mathbf{u}^H(n))$$

We may now identify four distinct factors as the block elements of the prearray, which are paired in the following manner:

- $\lambda^{1/2}\mathbf{K}^{-H/2}(n-1)$ and $\lambda^{1/2}\mathbf{u}(n)$, which are of dimensions M-by-M and M-by-1, respectively.

- $\hat{\mathbf{x}}^H(n \mid \mathcal{Y}_{n-1})\mathbf{K}^{-H/2}(n-1)$ and $y^*(n)$, which are of dimensions 1-by-M and 1-by-1, respectively.

The first pair of factors is naturally compatible by virtue of being made up of a matrix and a vector. The compatibility of the last pair of factors as row vectors is the reason for working with the Hermitian transposed forms of Eqs. (14.29) and (14.30). We may thus construct the following prearray:

$$
\begin{bmatrix}
\lambda^{1/2}\mathbf{K}^{-H/2}(n-1) & \lambda^{1/2}\mathbf{u}(n) \\
\hat{\mathbf{x}}^H(n \mid \mathcal{Y}_{n-1})\,\mathbf{K}^{-H/2}(n-1) & y^*(n) \\
\mathbf{0}^T & 1
\end{bmatrix}
$$

The last row, made up of a block of zeros followed by a unity term, has been added in order to make room for the generation of other Kalman variables in the postarray (Morf and Kailath, 1975; Sayed and Kailath, 1994). Suppose next we choose a unitary rotation $\mathbf{\Theta}(n)$ that transforms this prearray so as to produce a block zero in the second entry of the postarray's top block row, as shown by

$$
\begin{bmatrix}
\lambda^{1/2}\mathbf{K}^{-H/2}(n-1) & \lambda^{1/2}\mathbf{u}(n) \\
\hat{\mathbf{x}}^H(n \mid \mathcal{Y}_{n-1})\mathbf{K}^{-H/2}(n-1) & y^*(n) \\
\mathbf{0}^T & 1
\end{bmatrix}
\mathbf{\Theta}(n) =
\begin{bmatrix}
\mathbf{B}_{11}^H(n) & \mathbf{0} \\
\mathbf{b}_{21}^H(n) & b_{22}^*(n) \\
\mathbf{b}_{31}^H(n) & b_{32}^*(n)
\end{bmatrix}
\quad (14.32)
$$

By proceeding in a manner similar to that described for the square-root covariance filter, that is, by squaring both sides of Eq. (14.32) and then comparing respective terms on both sides of the resulting equality, we may choose the block elements of the postarray as follows (see Problem 1):

$$
\mathbf{B}_{11}^H(n) = \mathbf{K}^{-H/2}(n) \tag{14.33}
$$

$$
\mathbf{b}_{21}^H(n) = \hat{\mathbf{x}}^H(n+1 \mid \mathcal{Y}_n)\mathbf{K}^{-H/2}(n) \tag{14.34}
$$

$$
\mathbf{b}_{31}^H(n) = \lambda^{1/2}\mathbf{u}^H(n)\mathbf{K}^{1/2}(n) \tag{14.35}
$$

$$
b_{22}^*(n) = r^{-1/2}(n)\alpha^*(n) \tag{14.36}
$$

$$
b_{32}^*(n) = r^{-1/2}(n) \tag{14.37}
$$

Accordingly, we may rewrite Eq. (14.32) in the form:

$$
\begin{bmatrix}
\lambda^{1/2}\mathbf{K}^{-H/2}(n-1) & \lambda^{1/2}\mathbf{u}(n) \\
\hat{\mathbf{x}}^H(n \mid \mathcal{Y}_{n-1})\mathbf{K}^{-H/2}(n-1) & y^*(n) \\
\mathbf{0}^T & 1
\end{bmatrix}
\mathbf{\Theta}(n)
$$

$$
=
\begin{bmatrix}
\mathbf{K}^{-H/2}(n) & \mathbf{0} \\
\hat{\mathbf{x}}^H(n+1 \mid \mathcal{Y}_n)\mathbf{K}^{-H/2}(n) & r^{-1/2}(n)\alpha^*(n) \\
\lambda^{1/2}\mathbf{u}^H(n)\mathbf{K}^{1/2}(n) & r^{-1/2}(n)
\end{bmatrix}
\tag{14.38}
$$

The block elements of the postarray provide two sets of useful results:

1. *Updated block elements of the prearray*:
   - The updated square root $\mathbf{K}^{-H/2}(n)$ is given by $\mathbf{B}_{11}^H(n)$.
   - The updated matrix product $\hat{\mathbf{x}}^H(n+1 \mid \mathcal{Y}_n)\mathbf{K}^{-H/2}(n)$ is given by $\mathbf{b}_{21}^H(n)$.
2. *Other Kalman variables*:
   - The conversion factor, $r^{-1}(n)$, is obtained by squaring $b_{32}(n)$, which is real.
   - The innovation, $\alpha(n)$, is obtained by dividing $b_{22}(n)$ by $b_{32}(n)$.

The updated value of the state estimate $\hat{\mathbf{x}}(n+1 \mid \mathcal{Y}_n)$ is computed from the upper triangular system of equation (14.34), where $\mathbf{K}^{-H/2}(n)$ is known by virtue of Eq. (14.33). Specifically, the individual elements of $\hat{\mathbf{x}}(n+1 \mid \mathcal{Y}_n)$ are computed by using the method of *back substitution* that exploits the upper triangular structure of the square root $\mathbf{K}^{-H/2}(n)$.

Table 14.1 presents a summary of the square-root information filtering algorithm; initialization of the algorithm proceeds in the usual way.

Extended Square-root Information Filter

The need for back-substitution, required in the square-root information filtering algorithm for computing the elements of the updated state estimate $\hat{\mathbf{x}}(n+1 \mid \mathcal{Y}_n)$, may be avoided by "expanding" the prearray of Eq. (14.38) as follows (Sayed and Kailath, 1994):

$$
\begin{bmatrix}
\lambda^{1/2}\mathbf{K}^{-H/2}(n-1) & \lambda^{1/2}\mathbf{u}(n) \\
\hat{\mathbf{x}}^H(n \mid \mathcal{Y}_{n-1})\mathbf{K}^{-H/2}(n-1) & y^*(n) \\
\mathbf{0}^T & 1 \\
\lambda^{-1/2}\mathbf{K}^{1/2}(n-1) & \mathbf{0}
\end{bmatrix}
$$

The last block line of this prearray is borrowed from the square-root Kalman filtering algorithm; see the last line of the prearray in Eq. (14.18). Then, following a procedure similar to that described for the conventional square-root information filtering algorithm, we may show that (see Problem 3):

$$
\begin{bmatrix}
\lambda^{1/2}\mathbf{K}^{-H/2}(n-1) & \lambda^{1/2}\mathbf{u}(n) \\
\hat{\mathbf{x}}^H(n \mid \mathcal{Y}_{n-1})\mathbf{K}^{-H/2}(n-1) & y^*(n) \\
\mathbf{0}^T & 1 \\
\lambda^{-1/2}\mathbf{K}^{1/2}(n-1) & \mathbf{0}
\end{bmatrix} \Theta(n)
$$

$$
=
\begin{bmatrix}
\mathbf{K}^{-H/2}(n) & \mathbf{0} \\
\hat{\mathbf{x}}^H(n+1 \mid \mathcal{Y}_n)\mathbf{K}^{-H/2}(n) & r^{-1/2}(n)\alpha^*(n) \\
\lambda^{1/2}\mathbf{u}^H(n)\mathbf{K}^{1/2}(n) & r^{-1/2}(n) \\
\mathbf{K}^{1/2}(n) & -\mathbf{g}(n)r^{1/2}(n)
\end{bmatrix}
\tag{14.39}
$$

where, as before, $\Theta(n)$ is a unitary rotation that produces a block zero entry in the top block row of the postarray.

Now we can see the benefit of using the extended prearray, as described here. Specifically, the updated value of the state estimate may be computed as follows:

$$\hat{x}(n + 1 \mid \mathcal{Y}_n) = \lambda^{-1/2}\hat{x}(n \mid \mathcal{Y}_{n-1}) + \mathbf{g}(n)\alpha(n)$$

$$= \lambda^{-1/2}\hat{x}(n \mid \mathcal{Y}_{n-1}) + (\mathbf{g}(n)r^{1/2}(n))(r^{-1/2}(n)\alpha^*(n))^* \qquad (14.40)$$

$$= (\mathbf{K}^{1/2}(n)) \, (\mathbf{K}^{-1/2}(n)\hat{x}(n + 1 \mid \mathcal{Y}_n))$$

where the quantities $\{\mathbf{g}(n)r^{1/2}(n), \, r^{-1/2}(n)\alpha^*(n)\}$ and $\{\mathbf{K}^{1/2}(n), \, \mathbf{K}^{-1/2}(n)\hat{x}(n + 1 \mid \mathcal{Y}_n)\}$ are read directly from the postarray in Eq. (14.39). The last two lines of Eq. (14.40) provide two different but equivalent methods for computing the updated state estimate. Thus the cumbersome operation of back substitution is replaced by a simple multiplication.

A summary of the *extended square-root information filtering algorithm* is presented in Table 14.1; initialization of the algorithm proceeds in the usual manner.

The square-root covariance filter, the conventional square-root information filter, and the extended square-root information filter share a common feature. The number of operations (multiplications and additions) needed to proceed from one iteration of the algorithm to the next, in all three cases, is $O(M^2)$, where M is the state dimension. This computational complexity is of the same order as that of the conventional Riccati-based Kalman filtering algorithm.

14.2 BUILDING SQUARE-ROOT ADAPTIVE FILTERING ALGORITHMS ON THEIR KALMAN FILTER COUNTERPARTS

The square-root variants of the Kalman filter described in the previous section provide the general framework for the derivation of known square-root adaptive filtering algorithms for exponentially weighted recursive least-squares (RLS) estimation. We say so in light of the one-to-one correspondences that exist between the Kalman variables and RLS variables, as demonstrated in Chapter 13.

The square-root adaptive filtering algorithms for RLS estimation are known as the *QR–RLS algorithm, extended QR–RLS algorithm,* and *inverse QR–RLS algorithm.* The reason for this terminology is that the derivation of these algorithms has traditionally relied, in one form or another, on the use of an orthogonal triangularization process known in matrix algebra as the *QR decomposition.* The motivation for using the QR decomposition in adaptive filtering is to exploit its good numerical properties.

For a matrix $\mathbf{A}(n)$, say, the QR decomposition may be written as follows (Stewart, 1973; Golub and Van Loan, 1989):

$$\mathbf{Q}(n)\mathbf{A}(n) = \begin{bmatrix} \mathbf{R}(n) \\ \mathbf{O} \end{bmatrix} \qquad (14.41)$$

where $\mathbf{Q}(n)$ is a unitary matrix, $\mathbf{R}(n)$ is an upper triangular matrix, and $\mathbf{O}$ is the null matrix. The pervasive use of the symbols $\mathbf{Q}$ and $\mathbf{R}$ in such a transformation has led, in the course of time, to the common use of "QR decomposition." By the same token, adaptive RLS filtering algorithms based on the QR decomposition, in a broad sense, became known as "QR–RLS algorithms." Prior to the 1994 paper by Sayed and Kailath, the QR–RLS algorithms for exponentially weighted RLS estimation were derived starting from the prewindowed version of a data matrix, which was then triangularized by applying the QR decomposition. The paper by Sayed and Kailath revealed for the first time how these different adaptive filtering algorithms can indeed be deduced directly from their square-root Kalman filter counterparts, thereby achieving two highly desirable objectives:

- The unified treatment of QR–RLS adaptive filtering algorithms for exponentially weighted RLS estimation
- Consolidating the bridge between the deterministic RLS estimation theory and the stochastic Kalman filter theory

In the remainder of this chapter, we follow the paper by Sayed and Kailath in deriving the different QR–RLS adaptive filtering algorithms. However, the order in which these algorithms are considered follows the traditional development of RLS estimation theory rather than the order of square-root Kalman filters summarized in Table 14.1.

14.3 QR–RLS ALGORITHM

The *QR–RLS algorithm*, or more precisely, the *QR decomposition-based RLS algorithm*, derives its name from the fact that the computation of the least-squares weight vector in a finite-duration impulse response (FIR) filter implementation of the adaptive filtering algorithm is accomplished by working directly with the incoming data matrix via the QR decomposition rather than working with the (time-averaged) correlation matrix of the input data as in the standard RLS algorithm (Gentleman and Kung, 1981; McWhirter, 1983; Haykin, 1991). Accordingly, the QR–RLS algorithm is numerically more stable than the standard RLS algorithm.

Assuming the use of prewindowing on the input data, the *data matrix* is defined by

$$\mathbf{A}^H(n) = [\mathbf{u}(1), \mathbf{u}(2), \ldots, \mathbf{u}(M), \ldots, \mathbf{u}(n)]$$

$$= \begin{bmatrix} u(1) & u(2) & \cdots & u(M) & \cdots & u(n) \\ 0 & u(1) & \cdots & u(M-1) & \cdots & u(n-1) \\ \cdot & \cdot & & \cdot & & \cdot \\ \cdot & \cdot & & \cdot & & \cdot \\ \cdot & \cdot & & \cdot & & \cdot \\ 0 & 0 & \cdots & u(1) & \cdots & u(n-M+1) \end{bmatrix} \quad (14.42)$$

where M is the number of FIR filter coefficients. Correspondingly, the correlation matrix of the input data is defined by

$$\Phi(n) = \sum_{i=1}^{n} \lambda^{n-i} \mathbf{u}(i)\mathbf{u}^H(i)$$

$$= \mathbf{A}^H(n)\Lambda(n)\mathbf{A}(n) \qquad (14.43)$$

The matrix $\Lambda(n)$ is called the *exponential weighting matrix*, defined by

$$\Lambda(n) = \text{diag}[\lambda^{n-1}, \lambda^{n-2}, \ldots, 1] \qquad (14.44)$$

where λ is the *exponential weighting factor*. Equation (14.43) represents a generalization of Eq. (10.45) used in the method of least squares.

From Chapter 13 we recall that the matrix $\mathbf{P}(n)$, used in deriving the RLS algorithm, is defined as the inverse of the correlation matrix $\Phi(n)$, as shown by [see Eq. (13.17)]

$$\mathbf{P}(n) = \Phi^{-1}(n) \qquad (14.45)$$

From Chapter 13, we also note the following correspondences between the Kalman variables and RLS variables:

KALMAN VARIABLE	RLS VARIABLE	DESCRIPTION
$\mathbf{K}^{-1}(n)$	$\lambda\mathbf{P}^{-1}(n) = \lambda\Phi(n)$	Correlation matrix
$r^{-1}(n)$	$\gamma(n)$	Conversion factor
$\mathbf{g}(n)$	$\lambda^{-1/2}\mathbf{k}(n)$	Gain vector
$\alpha(n)$	$\lambda^{-n/2}\,\xi^*(n)$	*A priori* estimation error
$y(n)$	$\lambda^{-n/2}\,d^*(n)$	Desired response
$\hat{\mathbf{x}}(n\,|\,\mathcal{Y}_{n-1})$	$\lambda^{-n/2}\,\hat{\mathbf{w}}(n-1)$	Estimate of tap-weight vector

From the first line of this table, we immediately see that the QR–RLS algorithm corresponds to the square-root information filtering algorithm (14.38) of Kalman filter theory.

Before proceeding to formulate the QR–RLS algorithm in light of this correspondence, we find it convenient to make a change of notation. According to the normal equations, the least squares estimate of the tap-weight vector, $\hat{\mathbf{w}}(n)$, is defined by [see Eq. (10.35)]

$$\Phi(n)\hat{\mathbf{w}}(n) = \mathbf{z}(n) \qquad (14.46)$$

where $\mathbf{z}(n)$ is the cross-correlation vector between the desired response $d(n)$ and input data vector $\mathbf{u}(n)$. Let $\Phi(n)$ be expressed in its factored form:

$$\Phi(n) = \Phi^{1/2}(n)\,\Phi^{H/2}(n) \qquad (14.47)$$

Then, premultiplying both sides of Eq. (14.46) by the square root $\Phi^{-1/2}(n)$, we may introduce a new vector variable defined by

$$\mathbf{p}(n) = \Phi^{H/2}(n)\,\hat{\mathbf{w}}(n) = \Phi^{-1/2}(n)\,\mathbf{z}(n) \qquad (14.48)$$

We are now ready to formulate the QR–RLS algorithm for linear adaptive filtering. Specifically, we may translate Eq. (14.38) pertaining to the square-root information filtering algorithm into the corresponding prearray-to-postarray transformation for the QR–RLS algorithm as follows (after cancelation of common terms):

$$
\begin{bmatrix}
\lambda^{1/2}\boldsymbol{\Phi}^{1/2}(n-1) & \mathbf{u}(n) \\
\lambda^{1/2}\mathbf{p}^{H}(n-1) & d(n) \\
\mathbf{0}^{T} & 1
\end{bmatrix}
\boldsymbol{\Theta}(n) =
\begin{bmatrix}
\boldsymbol{\Phi}^{1/2}(n) & \mathbf{0} \\
\mathbf{p}^{H}(n) & \xi(n)\,\gamma^{12}(n) \\
\mathbf{u}^{H}(n)\boldsymbol{\Phi}^{-H/2}(n) & \gamma^{1/2}(n)
\end{bmatrix}
\quad (14.49)
$$

Basically $\boldsymbol{\Theta}(n)$ is any unitary rotation that operates on the elements of the input data vector $\mathbf{u}(n)$ in the prearray, *annihilating* them one by one so as to produce a block zero entry in the top block row of the postarray. Naturally, the lower triangular structure of the square root of the correlation matrix, namely, $\boldsymbol{\Phi}^{1/2}$, is preserved in its exact form before and after the transformation. This is indeed the very essence of the QR decomposition for RLS estimation, hence the name "QR–RLS algorithm."

Having computed the updated block values $\boldsymbol{\Phi}^{1/2}(n)$ and $\mathbf{p}^{H}(n)$, we may then solve for the least-squares weight vector $\hat{\mathbf{w}}(n)$ by using the formula [see Eq. (14.48)]

$$
\hat{\mathbf{w}}^{H}(n) = \mathbf{p}^{H}(n)\,\boldsymbol{\Phi}^{-1/2}(n) \quad (14.50)
$$

The computation of this solution is accomplished using the *method of back substitution* that exploits the lower triangular structure of $\boldsymbol{\Phi}^{1/2}(n)$. Note, however, that this computation is feasible only for time $n > M$, for which the data matrix $\mathbf{A}(n)$, and therefore $\boldsymbol{\Phi}^{1/2}(n)$, is of full column rank.

To initialize the QR–RLS algorithm, we may set $\boldsymbol{\Phi}^{1/2}(0) = \mathbf{O}$ and $\mathbf{p}(0) = \mathbf{0}$. The *exact initialization* of the QR–RLS algorithm occupies the period $\mathbf{O} \leq n \leq M$ for which the *a posteriori* estimation error $e(n)$ is zero. At iteration $n = M$, the initialization is completed, whereafter $e(n)$ may assume a nonzero value.

A summary of the QR–RLS algorithm is presented in Table 14.2, including details of the initialization and other matters of interest.

Implementation Considerations

Thus far we have not focused on the particulars of the unitary rotation $\boldsymbol{\Theta}(n)$, other than to require that it be chosen to produce a block zero entry in the top block row of the postarray. A unitary matrix that befits this requirement is the transformation based on the *Givens rotation* discussed in Chapter 12. Through successive applications of the Givens rotation, we may develop a systematic procedure for the efficient annihilation of the block entry $\mathbf{u}(n)$ in the prearray, as prescribed by Eq. (14.49).

Moreover, the use of Givens rotations lends itself to a *parallel implementation* in the form of a *systolic array*, the idea of which was developed originally by Kung and Leiserson (1978). A systolic array consists of an array of individual *processing cells* arranged as a regular structure. Each cell in the array is provided with local memory of its own, and each cell is connected only to its nearest neighbors. The array is designed such that regular streams of data are clocked through it in a highly rhythmic fashion, much like the

TABLE 14.2   SUMMARY OF THE QR–RLS ALGORITHM AND ITS EXTENDED FORM FOR EXPONENTIALLY WEIGHTED RLS ESTIMATION

Inputs:
  input signal vector $= \{\mathbf{u}(1), \mathbf{u}(2), \ldots, \mathbf{u}(n)\}$
  desired response $= \{d(1)\, d(2), \ldots, d(n)\}$

Known parameter:
  exponential weighting factor $= \lambda$

Initial conditions:
  $\mathbf{\Phi}^{1/2}(0) = \mathbf{O}$
  $\mathbf{p}(0) = \mathbf{0}$

1. *QR–RLS algorithm*:
   For $n = 1, 2, \ldots$, compute

$$
\begin{bmatrix}
\lambda^{1/2}\mathbf{\Phi}^{1/2}(n-1) & \mathbf{u}(n) \\
\lambda^{1/2}\mathbf{p}^{H}(n-1) & d(n) \\
\mathbf{0}^{T} & 1
\end{bmatrix}
\mathbf{\Theta}(n) =
\begin{bmatrix}
\mathbf{\Phi}^{1/2}(n) & \mathbf{0} \\
\mathbf{p}^{H}(n) & \xi(n)\gamma^{1/2}(n) \\
\mathbf{u}^{H}(n)\mathbf{\Phi}^{-H/2}(n) & \gamma^{1/2}(n)
\end{bmatrix}
$$

$$\hat{\mathbf{w}}^{H}(n) = \mathbf{p}^{H}(n)\,\mathbf{\Phi}^{-1/2}(n)$$

2. *Extended QR–RLS algorithm*:
   For $n = 1, 2, \ldots$, compute

$$
\begin{bmatrix}
\lambda^{1/2}\mathbf{\Phi}^{1/2}(n-1) & \mathbf{u}(n) \\
\lambda^{1/2}\mathbf{p}^{H}(n-1) & d(n) \\
\mathbf{0}^{T} & 1 \\
\lambda^{-1/2}\mathbf{\Phi}^{-H/2}(n-1) & \mathbf{0}
\end{bmatrix}
\mathbf{\Theta}(n) =
\begin{bmatrix}
\mathbf{\Phi}^{1/2}(n) & \mathbf{0} \\
\mathbf{p}^{H}(n) & \xi(n)\gamma^{1/2}(n) \\
\mathbf{u}^{H}(n)\mathbf{\Phi}^{-H/2}(n) & \gamma^{1/2}(n) \\
\mathbf{\Phi}^{-H/2}(n) & -\mathbf{k}(n)\gamma^{-1/2}(n)
\end{bmatrix}
$$

$$\hat{\mathbf{w}}(n) = \hat{\mathbf{w}}(n-1) + (\mathbf{k}(n)\,\gamma^{-1/2}(n))(\xi(n)\,\gamma^{-1/2}(n))^{*}$$

Note: In both cases, $\mathbf{\Theta}(n)$ is a unitary rotation that operates on the prearray to produce a block zero entry in the top block row of the postarray.

pumping action of the human heart, hence the name "systolic" (Kung, 1982). The important point to note here is that systolic arrays are well suited for implementing complex signal processing algorithms such as the QR–RLS algorithm, particularly when the requirement is to operate in *real time* and at *high data bandwidths*.

Turning to the QR–RLS algorithm as described herein, we may identify two systolic implementations of it, referred to as implementation I and implementation II. Basically, these two implementations differ from each other in their specific computation products.

Systolic Array Implementation I

Figure 14.1 shows a systolic array structure for implementing a simplified form of the QR–RLS algorithm for the case when the weight vector $\hat{\mathbf{w}}(n)$ has three elements (i.e., $M = 3$). The simplification merely involves deleting the last rows of the prearray and the postarray in Eq. (14.49), with a corresponding reduction in the dimensions of the unitary matrix $\mathbf{\Theta}(n)$. Specifically, we now have[1]

$$\begin{bmatrix} \lambda^{1/2}\mathbf{\Phi}^{1/2}(n-1) & \mathbf{u}(n) \\ \lambda^{1/2}\mathbf{p}^{H}(n-1) & d(n) \end{bmatrix} \mathbf{\Theta}(n) = \begin{bmatrix} \mathbf{\Phi}^{1/2}(n) & \mathbf{0} \\ \mathbf{p}^{H}(n) & \xi(n)\gamma^{1/2}(n) \end{bmatrix} \quad (14.51)$$

The systolic structure of Fig. 14.1 is arranged with two specific points in mind. First, data flow through it from left to right, consistent with all other adaptive filters considered in previous chapters. Second, the systolic array operates directly on the input data that are represented by successive values of the input data vector $\mathbf{u}(n)$ and the desired response $d(n)$.

The structure of Fig. 14.1 consists of two distinct sections: a *triangular systolic array* and a *linear systolic array* (Gentleman and Kung, 1981). The entire systolic array is controlled by a single *clock*. Each section of the array consists of two types of processing cells: *internal cells* (squares) and *boundary cells* (circles). The specific arithmetic functions of these cells are defined later. Each cell receives its input data from the directions indicated for one clock cycle, performs the specific arithmetic functions, and then, on the next clock cycle, delivers the resulting output values to neighboring cells as indicated. A distinctive feature of systolic arrays is that each processing cell is always kept active as data flow across the array. The triangular systolic array section implements the Givens rotations part of the QR–RLS algorithm, whereas the linear systolic array section computes the weight vector *at the end of the entire recursion*. If we were to compute the weight vector at each iteration of the QR–RLS algorithm, which we indeed can, the operation of the systolic array processor in Fig. 14.1 would be prohibitively slow, hence the idea of deferring the computation of $\hat{\mathbf{w}}(n)$ to the end of the recursion.

The dashed squares shown in Fig. 14.1 are merely included to represent delays in the transfer of data from the triangular array to the linear section. These delays are needed to ensure that the data transfer takes place at the correct instants of time.

Consider first the operation of the triangular systolic array section labeled ABC in Fig. 14.1. The boundary and internal cells of this section are given in Fig. 14.2. Basically,

---

[1] The $(M+1)$-by-$(M+1)$ unitary matrix $\mathbf{\Theta}$ in Eq. (14.51) for the QR–RLS algorithm is implemented as a sequence of M Givens rotations, each of which is configured to annihilate a particular element of the M-by-1 vector $\mathbf{u}(n)$ in the prearray. We may thus write

$$\mathbf{\Theta} = \prod_{k=1}^{M} \mathbf{\Theta}_k$$

where $\mathbf{\Theta}_k$ consists of a unitary matrix except for four strategic elements located at the points where the pair of rows k and $M+1$ intersects the pair of columns k and $M+1$. These four elements, denoted by θ_{kk}, $\theta_{M+1,k}$, $\theta_{k,M+1}$, and $\theta_{M+1,M+1}$ are defined as follows:

$$\theta_{kk} = \theta_{M+1,M+1} = c_k$$
$$\theta_{M+1,k} = s_k^*$$
$$\theta_{k,M+1} = -s_k$$

where $k = 1, 2, \ldots, M$. The cosine parameter c_k is real, whereas the sine parameter s_k is complex. The remarks made in this footnote also apply to the unitary matrix $\mathbf{\Theta}$ in Eq. (14.51) for the extended QR–RLS algorithm.

The design equations for the cells in the triangular array in Fig. 14.2 for the QR–RLS algorithm and the corresponding ones in Fig. 14.9 for the extended QR–RLS algorithm are based on this particular description of the unitary matrix $\mathbf{\Theta}$.

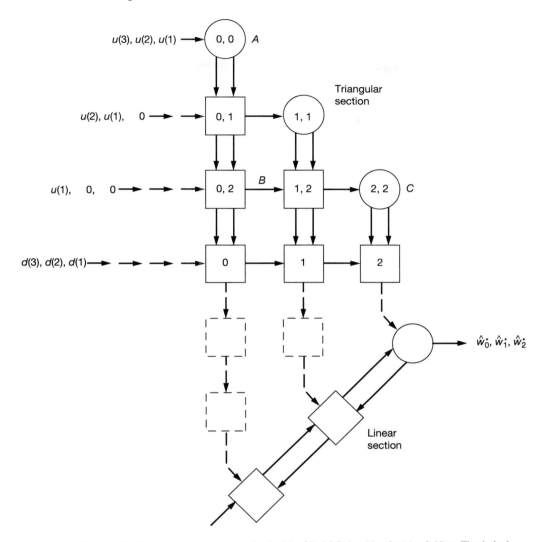

Figure 14.1   Systolic array implementation I of the QR–RLS algorithm for $M = 3$. Note: The dashed squares represent delays in the transfer of computations from the triangular section to the linear section.

the internal cells perform only additions and multiplications, as described in Fig. 14.2(b). The boundary cells, on the other hand, are considerably more complex, in that they compute square roots and reciprocals, as described in Fig. 14.2(a). Each cell of the triangular systolic array section (depending on its location) stores a particular element of the lower triangular matrix $\boldsymbol{\Phi}^{1/2}(n)$, which, at the outset of the least-squares recursion, is initialized to zero and thereafter updated every clock cycle. The function of each column of processing cells in the triangular systolic array section is to *rotate* one column of the stored triangular matrix with a vector of data received from the left in such a way that the leading *ele-*

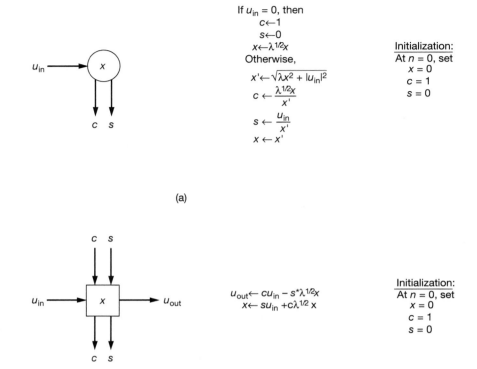

(a)

(b)

Figure 14.2 Cells for systolic array I: (a) boundary cell; (b) internal cell. *Note*: The stored value x is initialized to be zero (i.e., real). For the boundary cell, it always remains *real*; this is consistent with the property that the diagonal elements of the upper triangular matrix $\mathbf{R}$ are all real. Hence, the formulas for the rotation parameters c and s computed by the boundary cell can be simplified considerably, as shown in part (a). Also, note that the values x stored in the array are elements of the lower triangular matrix $\mathbf{R}^H$; hence, we may identify $r^* = x$ for all elements of the triangular array.

ment of the received data vector is annihilated. The reduced data vector is then passed to the right on to the next column of cells. The boundary cell in each column of the section computes the pertinent rotation parameters and then passes them downward on the next clock cycle. The internal cells subsequently apply the same rotation to all other elements in the received data vector. Since a delay of one clock cycle per cell is incurred in passing the rotation parameters downward along a column, it is necessary that the input data vectors enter the triangular systolic array in a *skewed order*, as illustrated in Fig. 14.1 for the case of $M = 3$. This arrangement of the input data ensures that as each column vector $\mathbf{u}(n)$ of the data matrix $\mathbf{A}^H(n)$ propagates through the array, it interacts with the previously stored triangular matrix $\mathbf{\Phi}^{1/2}(n-1)$ and thereby undergoes the sequence of Givens rotations denoted by $\mathbf{\Theta}(n)$, as required. Accordingly, all the elements of the column vector $\mathbf{u}(n)$

are annihilated, one by one, and an updated lower triangular matrix $\mathbf{\Phi}^{1/2}(n)$ is produced and stored in the process.

The systolic array operates in a highly pipelined manner, whereby, as (time-skewed) input data vectors enter the array from the left, we find that in effect each such vector defines a processing *wavefront* that moves across the array. It should therefore be appreciated that, on any particular clock cycle, elements of the pertinent lower triangular matrix $\mathbf{\Phi}^{1/2}(n)$ only exist along the corresponding wave-front. This phenomenon is illustrated later in Example 1.

At the same time that the orthogonal triangularization process is being performed by the triangular systolic array section labeled *ABC* in Fig. 14.1, the row vector $\mathbf{p}^H(n)$ is computed by the appended bottom row of internal cells.

When the entire orthogonal triangularization process is completed, the data flow *stops*, and then the stored data can be *clocked out* for subsequent processing by the linear systolic array section. The dashed lines in Fig. 14.3 depict the clock-out paths for the final values of the elements of both $\mathbf{\Phi}^{1/2}(n)$ and $\mathbf{p}^H(n)$ into the linear section of the systolic processor.

The linear section of the processor computes the Hermitian transposed least-squares weight vector, namely $\hat{\mathbf{w}}^H(n)$. For convenience of presentation, let

$$\mathbf{R}^H(n) = \mathbf{\Phi}^{1/2}(n) \tag{14.52}$$

Then in accordance with Eq. (14.49), the elements of the vector $\hat{\mathbf{w}}^H(n)$ are computed by using the method of back substitution (Kung and Leiserson, 1978). Taking the Hermitian transpose of both sides of Eq. (14.52), we have

$$\mathbf{\Phi}^{H/2}(n) = \mathbf{R}(n) \tag{14.53}$$

We may then write

$$z_i^{(M-1)} = 0$$
$$z_i^{(k-1)} = z_i^{(k)} + r_{ik}^*(n)\hat{w}_k^*(n), \qquad k = M-1, \ldots, i; \quad i = M-1, \ldots, 0 \tag{14.54}$$
$$\hat{w}_i^*(n) = \frac{p_i^*(n) - z_i^{(i)}}{r_{ii}(n)}$$

where the $z_i^{(k)}$ are intermediate variables, the $r_{ik}(n)$ are elements of upper triangular matrix $\mathbf{R}(n)$, the $p_i(n)$ are elements of the vector $\mathbf{p}(n)$, and the $\hat{w}_k(n)$ are elements of the weight vector $\hat{\mathbf{w}}(n)$. The linear systolic array section consists of one boundary cell and $(M-1)$ internal cells that perform the arithmetic functions defined in Fig. 14.4, in accordance with Eq. (14.54). The boundary cell performs subtraction and division, whereas the internal cells perform additions and multiplications. The elements of the complex-conjugated weight vector leave the linear array every second clock cycle with $\hat{w}_{M-1}^*(n)$ leaving first, followed by $\hat{w}_{M-2}^*(n)$, and so on right up to $\hat{w}_0^*(n)$. In effect, the elements of the weight vector $\hat{\mathbf{w}}^H(n)$ are read out *backward*. Thus, by chaining the linear and triangular systolic array sections together in the manner shown in Fig. 14.1, we produce a device capable of solving the exact least-squares problem recursively.

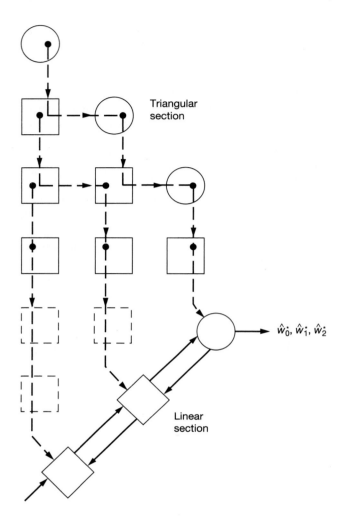

Triangular
section

$\hat{w}_0^*, \hat{w}_1^*, \hat{w}_2^*$

Linear
section

Figure 14.3   Clock-out paths of the triangular section into the linear section of the systolic array. The dots represent cell contents.

The following points are noteworthy in the context of the two-section systolic processor of Fig. 14.1 for the general case of *complex-valued data*:

1. Initially, zeros are stored in all boundary and internal cells. Also, the parameters of the Givens rotation at the output of each boundary cell (and therefore every other cell in the triangular systolic array section) are initially set at the values $c_{\text{out}} = 1$ and $s_{\text{out}} = 0$. The initialization of the complete systolic array shown in Fig. 14.1 occupies a total of $3M$ *clock cycles*, where M is the dimension of the weight vector.

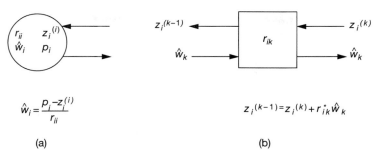

Figure 14.4   Cells for linear systolic array: (a) boundary cell; (b) internal cell.

2. The element $r_{ik}^*(n)$ of the lower triangular matrix $\mathbf{R}^H(n) = \boldsymbol{\Phi}^{1/2}(n)$ is computed in the kth cell on the ith row of the triangular section at time $n + (i - 1) + (k - 1)$, where $i, k = 0, 1, \ldots, M - 1$; note that the diagonal elements of $\mathbf{R}^H(n)$ are all *real valued*. The element $p_k^*(n)$ of the row vector $\mathbf{p}^H(n)$ is computed in the kth cell of the row of cells appended to the triangular section at time $n + M + k$, where $k = 0, 1, \ldots, M - 1$.

3. The systolic processor of Fig. 14.1 experiences *delays* in the various stages of computation. In particular, for the general case of an M-by-1 weight vector, $2M$ clock cycles are required to compute the Givens rotations, and another $2M$ clock cycles are required to clock out and compute the weight vector. Accordingly, a *total delay* or *latency of* $4M$ *clock cycles* is experienced in computing the complete M-by-1 weight vector $\hat{\mathbf{w}}(n)$. For real-time operations requiring the use of a large M, this latency may be too large and therefore unacceptable.

4. The element values of the matrix $\mathbf{R}^H(n)$ do not all propagate from the triangular section straight down to the linear section. Rather, except for $r_{i,M-1}*(n)$, $i = 0, 1, \ldots, M - 1$, the propagation of all the other elements of $\mathbf{R}^H(n)$ follows zig-zag paths, as illustrated in Fig. 14.3 for the case of $M = 3$.

5. The linear systolic array section has a lower data throughput than the triangular systolic array section. Consequently, the elements of the least-squares weight vector $\hat{\mathbf{w}}(n)$ are computed serially only when the entire sequence of computations in the triangular section comes to an end, largely for the sake of ensuring computational efficiency. It is, of course, a straightforward matter to modify these two array structures to make "on-the-fly" computation of $\hat{\mathbf{w}}(n)$ possible by the addition of extra data paths at the cost of additional computations; Varvisiotis et al. (1989) describe a scheme for parallel implementation of the linear section.

Example 1

In this example, we illustrate the operation of the systolic array structure of Fig. 14.1 for the case when the input data are *real valued* and $M = 3$.

TABLE 14.3   INPUTS AND STATES OF THE TRIANGULAR SYSTOLIC ARRAY FOR $M = 3$ AND REAL DATA

$r_{00}(n-1)$			$u(n)$	$r_{00}(n)$		
$r_{01}(n-2)$	$r_{11}(n-3)$		$u(n-2)$	$r_{01}(n-1)$	$r_{11}(n-2)$	
$r_{02}(n-3)$	$r_{12}(n-4)$	$r_{22}(n-5)$	$u(n-4)$	$r_{02}(n-2)$	$r_{12}(n-3)$	$r_{22}(n-4)$
$p_0(n-4)$	$p_1(n-5)$	$p_2(n-6)$	$d(n-3)$	$p_0(n-3)$	$p_1(n-4)$	$p_2(n-5)$

States at time n_- Inputs at   States at time n_+
time n

Note: The initialization procedure can also be represented by this stage graph, provided that we set $u(k) = 0$, $r_{ij}(k) = 0$, and $p_i(k) = 0$, when $k < 1$.

The inputs and states of the triangular systolic array for $M = 3$ are summarized in Table 14.3. In particular, we show the states of the individual cells in this section at time n_-, the external inputs applied to them at time n, and the states of the individual cells at time n_+.

The elements of column 1 of the lower triangular 3-by-3 matrix $\mathbf{\Phi}^{1/2}(n) = \mathbf{R}^T(n)$ for $M = 3$, that is, the elements $r_{00}(n)$, $r_{01}(n)$, and $r_{02}(n)$, are computed in the cells of column 1 of the triangular section at times n, $n + 1$, and $n + 2$, respectively. The elements of column 2 of $\mathbf{R}^T(n)$, that is, the elements $r_{11}(n)$ and $r_{12}(n)$, are computed in the cells of column 2 of the triangular section at times $n + 2$ and $n + 3$, respectively. The remaining element of $\mathbf{R}^T(n)$, that is, $r_{33}(n)$, is computed in the only cell of column 3 of the triangular section at time $n + 4$. The elements of the 1-by-3 vector $\mathbf{p}^T(n)$, that is, $p_0(n)$, $p_1(n)$, and $p_2(n)$, are computed in the elements of the row of cells appended to the triangular section at times $n + 4$, $n + 5$, and $n + 6$, respectively. When the orthogonal triangularization of the weighted data matrix is completed, the data flow is terminated, and the stored contents of the internal and boundary cells are clocked out in the manner described in Fig. 14.3. In particular, the clocked-out data are processed by a linear systolic array. For the example at hand, Fig. 14.5 presents the details of the operation of this section. The clock cycle numbers included in Fig. 14.5 are measured from the instant when the linear section begins its operation.

Systolic Array Implementation II

The triangular systolic array section shown in Fig. 14.1 may be viewed as a *partial* implementation of the transformation described in Eq. (14.49) that constitutes what the QR–RLS algorithm is all about. Figure 14.6 shows a systolic array implementation of this transformation in *full* (McWhirter, 1983). The internal cells of this structure are identical

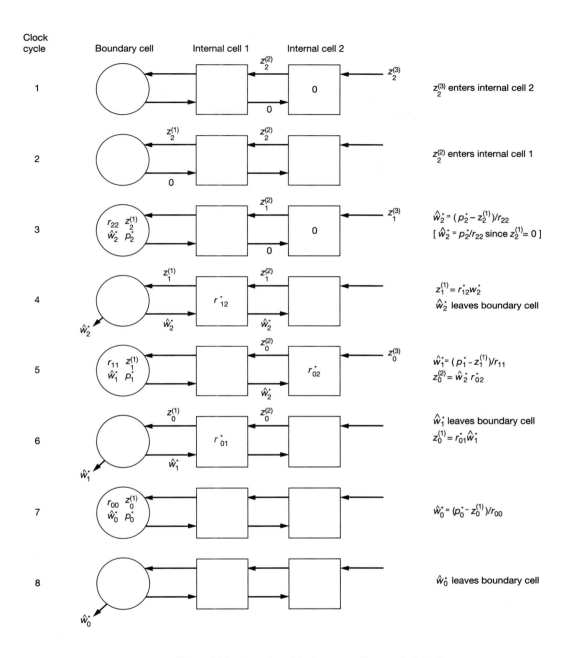

Figure 14.5 Operation of the linear systolic array for $M = 3$.

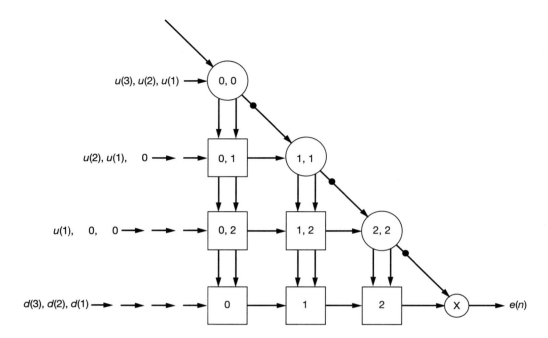

Figure 14.6　Systolic array implementation II of the QR–RLS algorithm. The dots along the diagonal of the array represent storage elements. This processing delay, which is a consequence of the temporal skew imposed on the input data, may be incorporated within the associated boundary cell.

to those in the triangular section of Fig. 14.1, but the boundary cells are now more complex, as shown in Fig. 14.7 to account for the structure of the full postarray in Eq. (14.49).

The computations of $\mathbf{\Phi}^{1/2}(n)$ and $\mathbf{p}^H(n)$ proceed along exactly the same lines as those described for the systolic structure of Fig. 14.1.

From Fig. 14.7(a), we note that the kth boundary cell in the systolic array structure of Fig. 14.6 performs the computation:

$$\gamma_{\text{out},k}^{1/2} = c_k(n)\gamma_{\text{in},k}^{1/2}(n), \qquad k = 1, 2, \ldots, M \tag{14.55}$$

where $c_k(n)$ is the cosine rotation parameter of that cell. Accordingly, with a set of M boundary cells connected together as in Fig. 14.6, the output of the last boundary cell produced in response to a unit input applied to the first boundary cell, may be expressed as follows:

$$\gamma^{1/2}(n) = \gamma_{\text{out},M}^{1/2}(n) \left| \gamma_{\text{in},1}^{1/2}(n) = 1 \right.$$
$$= \prod_{k=1}^{M} c_k(n) \tag{14.56}$$

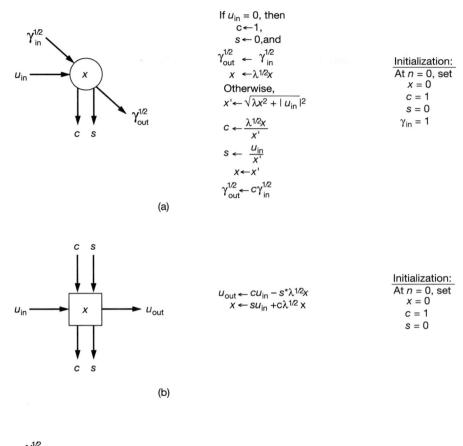

If $u_{in} = 0$, then
$$c \leftarrow 1,$$
$$s \leftarrow 0, \text{and}$$
$$\gamma_{out}^{1/2} \leftarrow \gamma_{in}^{1/2}$$
$$x \leftarrow \lambda^{1/2} x$$
Otherwise,
$$x' \leftarrow \sqrt{\lambda x^2 + |u_{in}|^2}$$
$$c \leftarrow \frac{\lambda^{1/2} x}{x'}$$
$$s \leftarrow \frac{u_{in}}{x'}$$
$$x \leftarrow x'$$
$$\gamma_{out}^{1/2} \leftarrow c \gamma_{in}^{1/2}$$

Initialization:
At $n = 0$, set
$$x = 0$$
$$c = 1$$
$$s = 0$$
$$\gamma_{in} = 1$$

(a)

(b)

$$u_{out} \leftarrow c u_{in} - s^* \lambda^{1/2} x$$
$$x \leftarrow s u_{in} + c \lambda^{1/2} x$$

Initialization:
At $n = 0$, set
$$x = 0$$
$$c = 1$$
$$s = 0$$

(b)

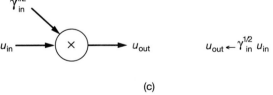

$$u_{out} \leftarrow \gamma_{in}^{1/2} u_{in}$$

(c)

Figure 14.7   Cells for systolic array II: (a) boundary cell; (b) internal cell; (c) final cell. *Note*: The stored valve x is initialized to be zero (i.e., real). For the boundary cell, it always remains real. Hence, the formulas for the rotation parameters c and s computed by the boundary cell can be simplified considerably, as shown in part (a). Note also that in parts (a) and (b), the values x stored in the array are elements of $\mathbf{R}^H$; hence, $r^* = x$ for all elements of the array.

In a corresponding way, the last cell in the row of internal cells appended to the triangular section produces an output equal to $\xi(n)\gamma^{1/2}(n)$.

The structure of Fig. 14.6 includes a new element referred to as the *final processing cell*, which is indicated by a small circle. This cell produces an output simply by multiplying its two inputs. Thus, with inputs equal to $\gamma^{1/2}(n)$ and $\xi(n)\gamma^{1/2}(n)$, the final processing cell in Fig. 14.6 produces an output equal to the *a posteriori* estimation error $e(n)$, in accordance with the relation [see Eq. (13.38)]

$$
\begin{aligned}
e(n) &= \xi(n)\gamma(n) \\
&= (\xi(n)\gamma^{1/2}(n))(\gamma^{1/2}(n))
\end{aligned}
\tag{14.57}
$$

As the time-skewed input data vectors enter the systolic array of Fig. 14.6, we find that updated estimation errors are produced at the output of the array at the rate of one every clock cycle. The estimation error produced on a given clock cycle corresponds, of course, to the particular element of the desired response vector $d(n)$ that entered the array M clock cycles previously.

It is noteworthy that the *a priori* estimation error $\xi(n)$ may be obtained by *dividing* the output that emerges from the last cell in the appended (bottom) row of internal cells by the output from the last boundary cell. Also, the conversion factor $\gamma(n)$ may be obtained simply by squaring the output that emerges from the last boundary cell.

Figure 14.8 summarizes, in a diagrammatic fashion, the flow of signals in the systolic array of Fig. 14.6. The figure includes the external inputs $\mathbf{u}(n)$ and $d(n)$, the resulting

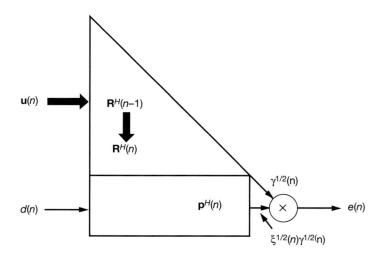

Figure 14.8   Diagrammatic representation of the flow of signals in systolic array II of Fig. 14.6.

transformations in the internal states of the triangular section and appended row of internal cells, the respective outputs of these two sections, and the overall output of the complete processor. Note that the systolic processor I of Fig. 14.1 and the systolic processor II of Fig. 14.6 are mathematically equivalent; however, they have different numerical properties.

A distinctive feature of the systolic structure shown in Fig. 14.6 is that, unlike the systolic structure of Fig. 14.1, computation of the *a posteriori* estimation error does not require knowledge of the weight vector $\hat{\mathbf{w}}(n)$. Clearly, the structure of Fig. 14.6 may be extended to include a linear systolic section (as in Fig. 14.1) so as to compute $\hat{\mathbf{w}}(n)$, if so required. However, there is a simpler method of computing the weight vector $\hat{\mathbf{w}}(n)$ as a useful *by-product* of the direct error (residual) extraction capability inherent in the systolic process of Fig. 14.6. The method for extracting the weight vector $\hat{\mathbf{w}}(n)$ is referred to as *serial weight flushing* (Ward et al., 1986; Shepherd and McWhirter, 1991). To explain the method, let $\mathbf{u}(n)$ denote the input vector and $d(n)$ denote the desired response, both at time n. Given that the weight vector at this time is $\hat{\mathbf{w}}(n)$, the corresponding *a posteriori* estimation error is

$$e(n) = d(n) - \hat{\mathbf{w}}^H(n)\mathbf{u}(n) \tag{14.58}$$

Suppose that the state of the array is *frozen* at time n_+, immediately *after* the systolic computation at time n is completed. Specifically, any update of stored values in the array is suppressed; otherwise, it is permitted to function normally in all other respects. At time n_+, we also set the desired response $d(n)$ equal to zero. We now define an input vector that consists of a string of zeros, except for the ith element that is set equal to unity, as shown by

$$\mathbf{u}^H(n_+) = [0 \ldots 010 \ldots 0]$$
$$\uparrow \tag{14.59}$$
$$i\text{th element}$$

Then, substituting these values in Eq. (14.58), we get

$$e(n_+) = -\hat{w}_i^*(n_+) \tag{14.60}$$

In other words, except for a trivial sign change, we may compute the ith element of the M-by-1 weight vector $\mathbf{w}^H(n)$ by freezing the state of the processor at time n, and subsequently setting the desired response equal to zero and feeding the processor with an input vector whose ith element is unity and the remaining $M - 1$ elements are all zero. The essence of all of this is that the Hermitian-transposed weight vector $\hat{\mathbf{w}}^H(n)$ may be viewed as the *impulse response* of the *nonadaptive* (i.e., frozen) form of the systolic array processor in the sense that it can be generated as the system output produced by inputting an $(M - 1)$-by-$(M - 1)$ identity matrix to the main triangular array and a zero vector to the bottom row of the array in Fig. 14.6 (Shepherd and McWhirter, 1991). To "flush" the entire M-by-1 weight vector $\hat{\mathbf{w}}^H(n)$ out of the systolic processor in Fig. 14.6, the procedure is therefore simply to halt the update of all stored values and input a data matrix that consists of a unit diagonal matrix (i.e., identity matrix) of dimension M.

14.4 EXTENDED QR–RLS ALGORITHM

The systolic array structure of Fig. 14.6 is suitable for adaptive filtering applications such as adaptive beamforming and acoustic echo cancelation, where the primary function is to compute the *a posteriori* estimation error without explicit knowledge of the least-squares weight vector. However, in other adaptive filtering applications such as system identification and spectrum analysis, knowledge of the weight vector on a *continuing* basis is a necessary requirement. Although, indeed, we may cater to this requirement by appending a linear section to the systolic array structure of Fig. 14.6 in the manner described in Fig. 14.1, the use of such a procedure is computationally inefficient because the data throughput of the linear section is lower than that of the triangular section. A preferable approach is to modify the QR–RLS algorithm so as to avoid the need for the cumbersome method of back substitution. This should be possible in light of what we know about the extended square-root information filter, which computes the state estimate directly from the postarray without invoking back substitution. The modified form of QR–RLS algorithm derived in this way is called the *extended QR–RLS algorithm* (Hudson et al., 1989; Yang and Böhme, 1992).

To be specific, consider the transformation of Eq. (14.39) that pertains to the extended square-root information filter. We may formulate the prearray-to-postarray transformation for the extended QR–RLS algorithm by using the one-to-one correspondences that exist between the Kalman variables and the RLS variables, and so write the following (Sayed and Kailath, 1994):

$$
\begin{bmatrix}
\lambda^{1/2}\boldsymbol{\Phi}^{1/2}(n-1) & \mathbf{u}(n) \\
\lambda^{1/2}\mathbf{p}^{H}(n-1) & d(n) \\
\mathbf{0}^{T} & 1 \\
\lambda^{-1/2}\boldsymbol{\Phi}^{-H/2}(n-1) & \mathbf{0}
\end{bmatrix}
\boldsymbol{\Theta}(n) =
\begin{bmatrix}
\boldsymbol{\Phi}^{1/2}(n) & \mathbf{0} \\
\mathbf{p}^{H}(n) & \xi(n)\gamma^{1/2}(n) \\
\mathbf{u}^{H}(n)\boldsymbol{\Phi}^{-H/2}(n) & \gamma^{1/2}(n) \\
\boldsymbol{\Phi}^{-H/2}(n) & -\mathbf{k}(n)\gamma^{-1/2}(n)
\end{bmatrix}
\tag{14.61}
$$

Moreover, we readily see from the second line of Eq. (14.40) that the updated least-squares value of the weight vector is computed using the recursion:

$$
\hat{\mathbf{w}}(n) = \hat{\mathbf{w}}(n-1) + \mathbf{k}(n)\xi^{*}(n)
\tag{14.62}
$$

$$
= \hat{\mathbf{w}}(n-1) + (\mathbf{k}(n)\gamma^{-1/2}(n))\,(\xi(n)\gamma^{1/2}(n))^{*}
$$

where the quantities $\mathbf{k}(n)\gamma^{-1/2}(n)$ and $\xi(n)\lambda^{1/2}(n)$ are read directly from the postarray in Eq. (14.61). Note, however, unlike the QR–RLS algorithm of Section 14.3, both $\boldsymbol{\Phi}^{1/2}$ and $\boldsymbol{\Phi}^{-H/2}$ are propagated in the extended QR–RLS algorithm. Accordingly, these two algorithms may behave differently in finite-precision arithmetic; we will have more to say on this issue in Chapter 17.

A summary of the extended QR–RLS algorithm in presented in Table 14.2.

Systolic Array Implementation

Figure 14.9 (drawn for the case of filter order $M = 3$) presents a systolic array implementation of the extended QR–RLS algorithm (Yang and Böhme, 1992; Sayed and Kailath, 1994). This structure consists of two triangular sections appended to each other:

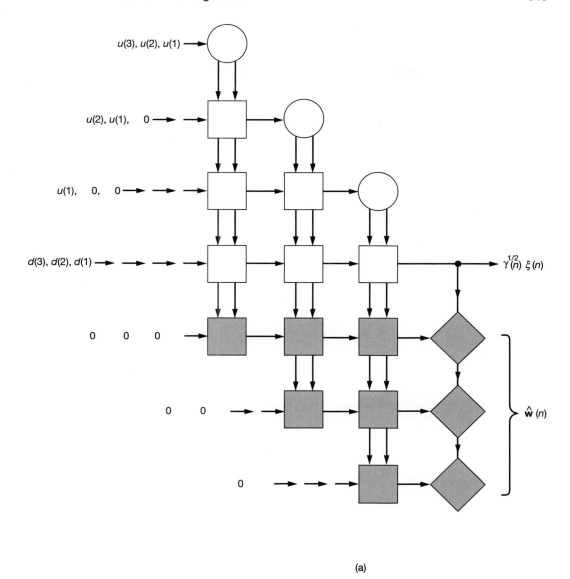

(a)

Figure 14.9   (a) Systolic array implementation of the extended QR–RLS algorithm. (b) Cells for the lower triangular section (unshaded). (c) Cells for the upper triangular section (shaded). Parts (b) and (c) of the figure are shown on the next page.

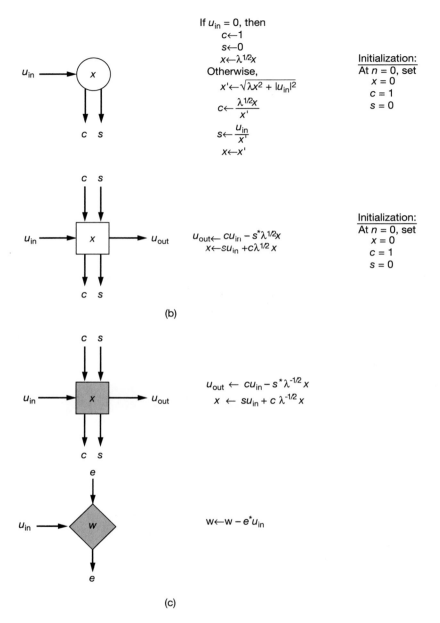

If $u_{in} = 0$, then
$$c \leftarrow 1$$
$$s \leftarrow 0$$
$$x \leftarrow \lambda^{1/2} x$$
Otherwise,
$$x' \leftarrow \sqrt{\lambda x^2 + |u_{in}|^2}$$
$$c \leftarrow \frac{\lambda^{1/2} x}{x'}$$
$$s \leftarrow \frac{u_{in}}{x'}$$
$$x \leftarrow x'$$

Initialization:
At $n = 0$, set
$$x = 0$$
$$c = 1$$
$$s = 0$$

$$u_{out} \leftarrow cu_{in} - s^* \lambda^{1/2} x$$
$$x \leftarrow su_{in} + c\lambda^{1/2} x$$

Initialization:
At $n = 0$, set
$$x = 0$$
$$c = 1$$
$$s = 0$$

(b)

$$u_{out} \leftarrow cu_{in} - s^* \lambda^{-1/2} x$$
$$x \leftarrow su_{in} + c \lambda^{-1/2} x$$

$$w \leftarrow w - e^* u_{in}$$

(c)

Figure 14.9   (*Cont.*)

- The top triangular section, shown unshaded in Fig. 14.9(a), operates in exactly the same way as the triangular section of Fig. 14.1. The computations performed by the boundary and internal cells of this section are described in Fig. 14.2; they are reproduced in Fig. 14.9(b) for convenience of presentation. The function of the top triangular section is to compute the quantity $\xi(n)\gamma^{1/2}(n)$.

- The bottom triangular section, shown lightly shaded in Fig. 14.9(a), consists of its own boundary and internal cells. The computations performed by the latter cells are described in Fig. 14.9(c). This second triangular section rotates stored values of $\lambda^{-1/2}\boldsymbol{\Phi}^{-H/2}(n-1)$ and an externally applied vector of zeros, yielding the updated $\boldsymbol{\Phi}^{-H/2}(n)$ and the desired quantity $\mathbf{k}(n)\gamma^{-1/2}(n)$.

The correction needed to update the weight vector is obtained simply by complex conjugating the top triangular section's output $\xi(n)\gamma^{1/2}(n)$ and then multiplying it by the bottom triangular section's output $\mathbf{k}(n)\gamma^{-1/2}(n)$, in accordance with Eq. (14.62). These computations are performed in the diamond-shaped boundary cells of the bottom triangular section.

14.5 ADAPTIVE BEAMFORMING

From previous discussions of adaptive beamforming, we recall that the objective of this spatial form of adaptive filtering is to modify the individual outputs of an array of sensors so as to produce an overall far-field pattern that optimizes, in some statistical sense, the reception of a target signal along a direction of interest. As with any adaptive filter, this optimization is achieved by suitable modifications of a set of weights built into the construction of the array. However, unlike other adaptive filtering applications, adaptive beamforming does not require explicit knowledge of the weights. This suggests a possible area of application for the QR–RLS algorithm implemented in the form of a systolic array, particularly the structure described in Fig. 14.6.

In this section we focus on an important type of adaptive beamforming known as *minimum variance distortionless response (MVDR) beamforming*. The key question, of course, is how to formulate the QR–RLS algorithm, and therefore the triangular systolic array of Fig. 14.6, so as to perform the MVDR beamforming task.

The MVDR Problem

Consider a linear array of M uniformly spaced sensors whose outputs are individually weighted and then summed to produce the beamformer output

$$e(i) = \sum_{l=1}^{M} w_l^*(n)u_l(i) \tag{14.63}$$

where $u_l(i)$ is the output of sensor l at time i, and $w_l(n)$ is the associated (complex) weight. To simplify the mathematical presentation, we consider the simple case of a single look direction. Let $s_1(\phi), s_2(\phi), \ldots, s_M(\phi)$ be the elements of a prescribed *steering vector* $\mathbf{s}(\phi)$; the electrical angle ϕ is determined by the look direction of interest. In particular, the element $s_l(\phi)$ is the output of sensor l of the array under the condition that there is no signal other than that due to a source of interest. We may thus state the MVDR problem as follows:

Minimize the cost function

$$\mathscr{E}(n) = \sum_{i=1}^{n} \lambda^{n-i} |e(i)|^2 \tag{14.64}$$

subject to the constraint

$$\sum_{l=1}^{M} w_l^*(n) s_l(\phi) = 1 \qquad \text{for all } n \tag{14.65}$$

Using matrix notation, we may redefine the cost function $\mathscr{E}(n)$ of Eq. (14.64) as

$$\mathscr{E}(n) = \boldsymbol{\epsilon}^H(n) \boldsymbol{\Lambda}(n) \boldsymbol{\epsilon}(n) \tag{14.66}$$

where $\boldsymbol{\Lambda}(n)$ is the exponential weighting matrix, and $\boldsymbol{\epsilon}(n)$ is the vector of constrained beamformer outputs. According to Eq. (14.63), the beamformer output vector $\boldsymbol{\epsilon}(n)$ is related to the data matrix $\mathbf{A}(n)$ by

$$\begin{aligned} \boldsymbol{\epsilon}(n) &= [e(1), e(2), \ldots, e(n)]^H \\ &= \mathbf{A}(n)\mathbf{w}(n) \end{aligned} \tag{14.67}$$

where $\mathbf{w}(n)$ is the weight vector, and the data matrix $\mathbf{A}(n)$ is defined in terms of the *snapshots* $\mathbf{u}(1), \mathbf{u}(2), \ldots, \mathbf{u}(n)$ by

$$\begin{aligned} \mathbf{A}^H(n) &= [\mathbf{u}(1), \mathbf{u}(2), \ldots, \mathbf{u}(n)] \\ &= \begin{bmatrix} u_1(1) & u_1(2) & \cdots & u_1(n) \\ u_2(1) & u_2(2) & \cdots & u_2(n) \\ \bullet & \bullet & & \bullet \\ \bullet & \bullet & & \bullet \\ \bullet & \bullet & & \bullet \\ u_M(1) & u_M(2) & \cdots & u_M(n) \end{bmatrix} \end{aligned} \tag{14.68}$$

We may now restate the MVDR problem in matrix terms as follows:

Given the data matrix $\mathbf{A}(n)$ and the exponential weighting matrix $\boldsymbol{\Lambda}(n)$, minimize the cost function

$$\mathscr{E}(n) = \| \boldsymbol{\Lambda}^{1/2}(n) \mathbf{A}(n) \mathbf{w}(n) \|^2 \tag{14.69}$$

with respect to the weight vector $\mathbf{w}(n)$, subject to the constraint

$$\mathbf{w}^H(n)\mathbf{s}(\phi) = 1 \qquad \text{for all } n$$

where $\mathbf{s}(\phi)$ is the steering vector for a prescribed electrical angle ϕ.

The solution to this constrained optimization problem is described by the MVDR formula (see Section 10.9)

$$\hat{\mathbf{w}}(n) = \frac{\boldsymbol{\Phi}^{-1}(n)\mathbf{s}(\phi)}{\mathbf{s}^H(\phi)\boldsymbol{\Phi}^{-1}(n)\mathbf{s}(\phi)} \tag{14.70}$$

where $\boldsymbol{\Phi}(n)$ is the M-by-M correlation matrix of the exponentially weighted sensor outputs averaged over n snapshots, and which is related to the data matrix $\mathbf{A}(n)$ as follows:

$$\boldsymbol{\Phi}(n) = \mathbf{A}^H(n)\boldsymbol{\Lambda}(n)\mathbf{A}(n) \tag{14.71}$$

Systolic MVDR Beamformer

Let the correlation matrix $\boldsymbol{\Phi}(n)$ be expressed in its factored form:

$$\boldsymbol{\Phi}(n) = \boldsymbol{\Phi}^{1/2}(n)\boldsymbol{\Phi}^{H/2}(n) \tag{14.72}$$

Correspondingly, we may rewrite Eq. (14.70) as follows:

$$\hat{\mathbf{w}}(n) = \frac{\boldsymbol{\Phi}^{-H/2}(n)\boldsymbol{\Phi}^{-1/2}(n)\mathbf{s}(\phi)}{\mathbf{s}^H(\phi)\boldsymbol{\Phi}^{-H/2}(n)\boldsymbol{\Phi}^{-1/2}(n)\mathbf{s}(\phi)} \tag{14.73}$$

To simplify matters, we define the auxiliary vector:

$$\mathbf{a}(n) = \boldsymbol{\Phi}^{-1/2}(n)\mathbf{s}(\phi) \tag{14.74}$$

We now note that the denominator of Eq. (14.73) is a real-valued scalar equal to the squared Euclidean norm of the auxiliary vector $\mathbf{a}(n)$. As for the numerator, it is equal to the Hermitian-transposed square root $\boldsymbol{\Phi}^{-H/2}(n)$ postmultiplied by the auxiliary vector $\mathbf{a}(n)$. We may thus simplify Eq. (14.73) to

$$\hat{\mathbf{w}}(n) = \frac{\boldsymbol{\Phi}^{-H/2}(n)\mathbf{a}(n)}{\|\mathbf{a}(n)\|^2} \tag{14.75}$$

The MVDR beamformer output, or in adaptive filtering terminology, the *a posteriori* estimation error $e(n)$ produced at time n in response to the snapshot $\mathbf{u}(n)$ is given by

$$e(n) = \hat{\mathbf{w}}^H(n)\mathbf{u}(n) \tag{14.76}$$

$$= \frac{\mathbf{a}^H(n)\boldsymbol{\Phi}^{-1/2}(n)\mathbf{u}(n)}{\|\mathbf{a}(n)\|^2}$$

Let $e'(n)$ denote a new estimation error, defined by

$$e'(n) = \mathbf{a}^H(n)\boldsymbol{\Phi}^{-1/2}(n)\mathbf{u}(n) \tag{14.77}$$

We may then reduce Eq. (14.76) to

$$e(n) = \frac{e'(n)}{\|\mathbf{a}(n)\|^2} \tag{14.78}$$

This equation shows that the MVDR beamformer output $e(n)$ is uniquely defined by two quantities: $e'(n)$ and $\mathbf{a}(n)$.

At this point in the discussion, we find it informative to recall the formula for the *a posteriori* estimation error $e(n)$ actually computed by the QR–RLS algorithm. By definition, we have

$$e(n) = d(n) - \hat{\mathbf{w}}^H(n)\mathbf{u}(n) \tag{14.79}$$

where $d(n)$ is the desired response, $\hat{\mathbf{w}}(n)$ is the least-squares weight vector, and $\mathbf{u}(n)$ is the input data vector. Substituting Eq. (14.50) in (14.79) yields

$$e(n) = d(n) - \mathbf{p}^H(n)\mathbf{\Phi}^{-1/2}(n)\mathbf{u}(n) \tag{14.80}$$

Thus, comparing Eqs. (14.77) and (14.80), we readily deduce the correspondences between the QR–RLS adaptive filtering and MVDR beamforming variables listed in Table 14.4.

The stage is now set for a recasting of the QR–RLS algorithm to suit the MVDR beamforming problem. First of all, we reformulate the prearray in Eq. (14.49) in light of the correspondences in Table 14.4 and so express it as

$$\begin{bmatrix} \lambda^{1/2}\mathbf{\Phi}^{1/2}(n-1) & \mathbf{u}(n) \\ \lambda^{1/2}\mathbf{a}^H(n-1) & 0 \\ \mathbf{0}^T & 1 \end{bmatrix}$$

Next, we determine the postarray that goes with this prearray by proceeding in the same manner as that described in Section 14.3. We may thus write

$$\begin{bmatrix} \lambda^{1/2}\mathbf{\Phi}^{1/2}(n-1) & \mathbf{u}(n) \\ \lambda^{1/2}\mathbf{a}^H(n-1) & 0 \\ \mathbf{0}^T & 1 \end{bmatrix} \mathbf{\Theta}(n) = \begin{bmatrix} \mathbf{\Phi}^{1/2}(n) & \mathbf{0} \\ \mathbf{a}^H(n) & -e'(n)\gamma^{-1/2}(n) \\ \mathbf{u}^H(n)\mathbf{\Phi}^{-H/2}(n) & \gamma^{1/2}(n) \end{bmatrix} \tag{14.81}$$

We now see that the two quantities of interest to the MVDR problem may be obtained from the postarray of Eq. (14.81) as follows:

- The updated auxiliary vector $\mathbf{a}(n)$ is read directly from the second row of the postarray.
- The estimation error $e'(n)$ is given by

$$e'(n) = (e'(n)\gamma^{-1/2}(n))(\gamma^{1/2}(n)) \tag{14.82}$$

where $e'(n)\gamma^{-1/2}(n)$ and $\gamma^{1/2}(n)$ are read directly from the nonzero entries of the second column of the postarray.

TABLE 14.4 CORRESPONDENCES BETWEEN THE QR–RLS ADAPTIVE FILTERING
AND MVDR BEAMFORMING VARIABLES

QR–RLS adaptive filtering	MVDR beamforming	Description
$e(n)$	$-e'(n)$	Estimation error
$d(n)$	0	Desired response
$\mathbf{p}(n)$	$\mathbf{a}(n)$	Auxiliary vector
$\mathbf{u}(n)$	$\mathbf{u}(n)$	Snapshot

Finally, we may implement the MVDR beamformer using the systolic array structure shown in Fig. 14.10, which is basically the same as that of Fig. 14.6 except for some minor changes (McWhirter and Shepherd, 1989). Specifically, $d(n)$ is set equal to zero for all n. With this change in place, we note the following from Fig. 14.10:

- The auxiliary vector $\mathbf{a}(n)$ is generated and stored in the bottom row of cells.
- The output of the final cell is identically equal to $-e'(n)$.

With a continuing sequence of snapshots $\mathbf{u}(n)$, $\mathbf{u}(n + 1)$, . . . , applied to the systolic array processor in Fig. 14.10, a corresponding sequence of estimation errors $e(n)$, $e(n + 1)$, . . . is generated by the MVDR beamformer in accordance with Eq. (14.78).

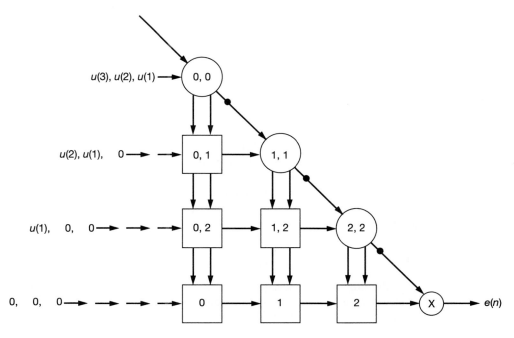

Figure 14.10 Systolic array for solving the MVDR beamforming problem.

Computer Experiment

We now illustrate the performance of the systolic array implementation of an adaptive MVDR beamformer by considering a linear array of five uniformly spaced sensors. The spacing d between adjacent elements equals one half of the received wavelength. The array operates in an environment that consists of a target signal and a single source of interference, which originate from different sources. The exponential weighting factor $\lambda = 1$.

The aims of the experiment are twofold:

1. To examine the evolution of the adapted spatial response (pattern) of the beamformer with time.
2. To evaluate the effect of varying the interference-to-target ratio on the interference-nulling capability of the beamformer.

The directions of the target and source of interference are as follows:

Excitation	Angle of incidence, θ, measured with respect to normal to the array (radians)
Target	$\sin^{-1}(0.2)$
Interference	0

The steering vector is defined by

$$\mathbf{s}^T(\phi) = [1,\, e^{-j\phi},\, e^{-j2\phi},\, e^{-j3\phi},\, e^{-j4\phi}] \tag{14.83}$$

where the electrical angle ϕ is defined in terms of the angle of incidence θ as follows:

$$\phi = \pi \sin\theta \tag{14.84}$$

The data set used for the experiment consists of three components: a target signal, elemental receiver noise, and an interfering signal. The target signal and the interfering signal originate in the far field of the array antenna and are therefore represented by plane waves impinging on the array along their respective directions. Let these directions be denoted by angles θ_1 and θ_2, measured (in radians) with respect to the normal to the array antenna. The elemental signals of the array antenna are thus expressed in baseband form as follows:

$$u(n) = A_1 \exp(jn\phi_1) + A_2 \exp(jn\phi_2 + \psi) + v(n), \qquad n = 1, 2, 3, 4, 5 \tag{14.85}$$

where A_1 is the amplitude of the target signal and A_2 is the amplitude of the interfering signal. The electrical angles ϕ_1 and ϕ_2 are related to the individual angles of arrival θ_1 and θ_2, respectively, by Eq. (14.84). Since the target and interfering signals are uncorrelated, the phase difference ψ associated with the second component in Eq. (14.85) is a random variable uniformly distributed over the interval (0, 2π]. Lastly, the additive receiver noise $v(n)$ is a complex-valued Gaussian random variable with zero mean and unit variance. The target-to-noise ratio is held constant at 10 dB; the interference-to-noise ratio is variable, assuming the values 40, 30, and 20 dB.

Figure 14.11 shows the effects of varying the target-to-interference ratio and the number of snapshots (excluding those needed for initialization) on the adapted response of the beamformer. The response is obtained by plotting $20\log_{10}|e(n)e^{j\phi}|$ versus the electrical angle ϕ; multiplication by the exponential factor $e^{j\phi}$ provides a means of spatially sampling the beamformer output. The results are presented in three parts, corresponding to 20, 100, and 200 snapshots; and each part corresponds to the three different values of interference-to-noise ratio, namely, 40, 30, and 20 dB.

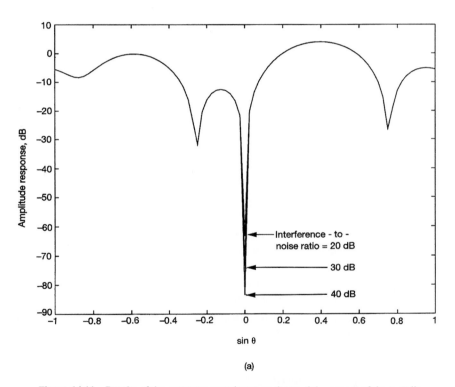

(a)

Figure 14.11   Results of the computer experiment on the spatial response of the systolic MVDR beamformer for varying interference-to-noise ratio and different number of snapshots: (a) $n = 20$; (b) $n = 100$; and (c) $n = 200$. Parts (b) and (c) of the figure are shown on the next two pages.

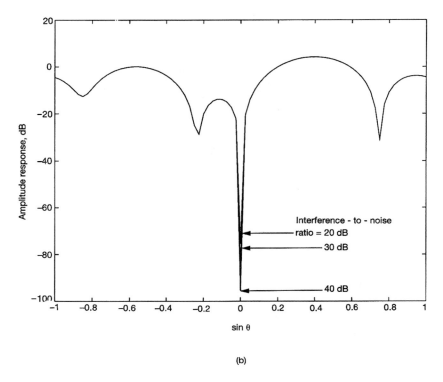

(b)

Figure 14.11 (*Cont.*)

Based on these results, we may make the following observations:

- The response of the beamformer along the target is held fixed at a value of one under all conditions, as required.
- With as few as 30 snapshots, including initialization, the beamformer exhibits a reasonably effective nulling capability, which continually improves as the beamformer processes more snapshots.
- The response of the beamformer is relatively insensitive to variations in the interference-to-target ratio.

14.6 INVERSE QR–RLS ALGORITHM

We now come to our last square-root adaptive filtering algorithm, known as the *inverse QR–RLS algorithm*. This algorithm derives its name from the fact that, instead of operating on the correlation matrix $\Phi(n)$ as in the conventional QR–RLS algorithm or extended

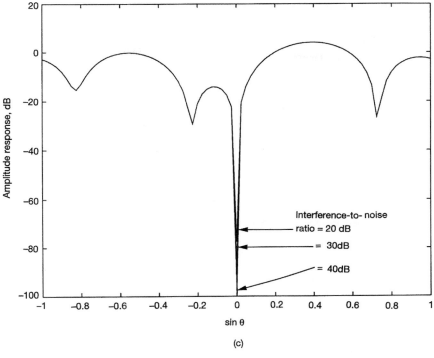

(c)

Figure 14.11   (*Cont.*)

QR–RLS algorithm, the operation is performed on the "inverse" of $\mathbf{\Phi}(n)$. In other words, the inverse QR–RLS algorithm operates on $\mathbf{P}(n) = \mathbf{\Phi}^{-1}(n)$ (Alexander and Ghirnikar, 1993; Pan and Plemmons, 1989). In light of the one-to-one correspondences between the Kalman variables and RLS variables, this means that the inverse QR–RLS algorithm is basically a reformulation of the square-root covariance (Kalman) filtering algorithm (Sayed and Kailath, 1994).

Referring to Eq. (14.26), which pertains to the square-root covariance filtering algorithm, we readily see that the corresponding prearray-to-postarray transformation for the inverse QR–RLS algorithm may be written as follows (after canceling common terms):

$$\begin{bmatrix} 1 & \lambda^{-1/2}\mathbf{u}^{H}(n)\mathbf{P}^{1/2}(n-1) \\ \mathbf{0} & \lambda^{-1/2}\mathbf{P}^{1/2}(n-1) \end{bmatrix} \mathbf{\Theta}(n) = \begin{bmatrix} \gamma^{-1/2}(n) & \mathbf{0}^{T} \\ \mathbf{k}(n)\gamma^{-1/2}(n) & \mathbf{P}^{1/2}(n) \end{bmatrix} \tag{14.86}$$

where $\mathbf{\Theta}(n)$ is a unitary rotation that operates on the block entry $\lambda^{-1/2}\mathbf{u}^{H}(n)\mathbf{P}^{1/2}(n-1)$ in the prearray by annihilating its elements, one by one, so as to produce a block zero entry in the first row of the postarray. The gain vector $\mathbf{k}(n)$ of the RLS algorithm is readily obtained from the entries in the first column of the postarray by writing

$$\mathbf{k}(n) = (\mathbf{k}(n)\gamma^{-1/2}(n))(\gamma^{-1/2}(n))^{-1} \tag{14.87}$$

TABLE 14.5 SUMMARY OF THE INVERSE QR–RLS ALGORITHM

Inputs:
    input signal vector $= \{\mathbf{u}(1), \mathbf{u}(2), \ldots, \mathbf{u}(n)\}$
    desired response $= \{d(1), d(2), \ldots, d(n)\}$

Known parameters:
    exponential weighting factor $= \lambda$

Initial conditions:
    $\mathbf{P}^{1/2}(0) = \delta^{-1/2}\mathbf{I}$, $\delta =$ small positive constant
    $\hat{\mathbf{w}}(0)$ $= \mathbf{0}$

Computations:
    For $n = 1, 2, \ldots$, compute

$$\begin{bmatrix} 1 & \lambda^{-1/2}\mathbf{u}^H(n)\mathbf{P}^{1/2}(n-1) \\ \mathbf{0} & \lambda^{-1/2}\mathbf{P}^{1/2}(n-1) \end{bmatrix} \boldsymbol{\Theta}(n) = \begin{bmatrix} \gamma^{-1/2}(n) & \mathbf{0}^T \\ \mathbf{k}(n)\gamma^{-1/2}(n) & \mathbf{P}^{1/2}(n) \end{bmatrix}$$

where $\boldsymbol{\Theta}(n)$ is a unitary rotation that produces a zero entry in the first row of the post-array.

$$\mathbf{k}(n) = (\mathbf{k}(n)\gamma^{-1/2}(n))\,(\gamma^{-1/2}(n))^{-1}$$

$$\xi(n) = d(n) - \hat{\mathbf{w}}^H(n-1)\mathbf{u}(n)$$

$$\hat{\mathbf{w}}(n) = \hat{\mathbf{w}}(n-1) + \mathbf{k}(n)\xi^*(n)$$

Hence, the least-squares weight vector may be updated in accordance with the recursion;

$$\hat{\mathbf{w}}(n) = \hat{\mathbf{w}}(n-1) + \mathbf{k}(n)\xi^*(n) \tag{14.88}$$

where the *a priori* estimation error $\xi(n)$ is defined in the usual way:

$$\xi(n) = d(n) - \hat{\mathbf{w}}^H(n-1)\mathbf{u}(n) \tag{14.89}$$

A summary of the inverse QR–RLS algorithm, including initial conditions, is presented in Table 14.5.

Here we note that since the square root $\boldsymbol{\Phi}^{1/2}(n)$ is lower triangular in accordance with Eq. (14.47), its inverse matrix $\boldsymbol{\Phi}^{-1/2}(n) = \mathbf{P}^{1/2}(n)$ is upper triangular.

The inverse QR–RLS algorithm differs from both the conventional QR–RLS algorithm and the extended QR–RLS algorithm in a fundamental way. Specifically, the input data vector $\mathbf{u}(n)$ does not appear by itself as a block entry in the prearray of the algorithm; rather, it is multiplied by $\lambda^{-1/2}\mathbf{P}^{1/2}(n-1)$. Hence, the input data vector $\mathbf{u}(n)$ has to be preprocessed prior to performing the rotations described in Eq. (14.86). The *preprocessor* to do this consists of simply computing the inner product of $\mathbf{u}(n)$ with each of the columns of the square-root matrix $\mathbf{P}^{1/2}(n-1)$ scaled by $\lambda^{-1/2}$. The preprocessor can be structured to take advantage of the upper triangular form of $\mathbf{P}^{-1/2}(n-1)$.

The inverse QR–RLS algorithm lends itself to parallel implementation in the form of two sections connected together (Alexander and Ghirnikar, 1993):

- A *triangular systolic array*, which operates on the preprocessed input vector $\lambda^{-1/2}\mathbf{P}^{H/2}(n-1)\mathbf{u}(n)$ in accordance with Eq. (14.86). Nonzero elements of the updated matrix $\mathbf{P}^{H/2}(n)$ are stored in the internal cells of the systolic array. The two other products of the systolic computation are $\gamma^{-1/2}(n)$ and $\mathbf{k}(n)\gamma^{-1/2}(n)$.

- A *linear section*, which is appended to the triangular section for the purpose of operating on the latter two products of the systolic computation to produce the elements of the updated weight vector $\hat{\mathbf{w}}(n)$ in accordance with Eqs. (14.87), (14.89), and (14.88), in that order.

The combination of these two sections is designed to operate in a completely parallel fashion.

14.7 SUMMARY AND DISCUSSION

In this chapter we discussed the derivations of three square-root adaptive filtering algorithms for exponentially weighted recursive least-squares (RLS) estimation in a unified manner. The algorithms are known as the QR–RLS algorithm, the extended QR–RLS algorithm, and the inverse QR–RLS algorithm. These algorithms bear one-to-one correspondences with the square-root information filter, the extended square-root information filter, and the square-root covariance filter, respectively, that represent square-root variants of the celebrated Kalman filter. These correspondences were exploited in the derivations of different variants of the RLS algorithm presented here.

The inverse QR–RLS algorithm is a natural extension of the standard RLS algorithm. It may therefore be legitimately referred to as the *square-root RLS algorithm*.

The QR–RLS algorithm and inverse QR–RLS algorithm propagate a *single* square root, namely, $\mathbf{\Phi}^{1/2}(n)$ and $\mathbf{P}^{1/2}(n) = \mathbf{\Phi}^{-1/2}(n)$, respectively. On the other hand, the extended QR–RLS algorithm propagates two square roots: $\mathbf{\Phi}^{1/2}(n)$ and the Hermitian transpose of $\mathbf{P}^{1/2}(n)$. This raises numerical difficulties for the extended QR–RLS algorithm, as discussed in Chapter 17.

A common feature of the QR–RLS algorithm, extended QR–RLS algorithm, and inverse QR–RLS algorithm is that, in varying degrees, they lend themselves to parallel implementation in the form of systolic arrays. Naturally, the actual details of the systolic array implementations depend on which algorithm is being considered. In particular, there are some basic differences that should be carefully noted. The conventional QR–RLS and extended QR–RLS algorithms operate directly on the input data. On the other hand, in the inverse QR–RLS algorithm the input data vector $\mathbf{u}(n)$ is transformed by the square-root matrix $\mathbf{P}^{1/2}(n) = \mathbf{\Phi}^{-1/2}(n)$ before it can be processed by the systolic array. This adds computational complexity to the parallel implementation of the inverse QR–RLS algorithm.

The parallel implementations of both the extended QR–RLS algorithm and inverse QR–RLS algorithms permit the computation of the least-squares weight vector in an efficient manner in their own individual ways. Accordingly, these two square-root adaptive filtering algorithms are well suited for applications such as system identification, spectrum estimation, and adaptive equalization, where knowledge of the weight vector is a necessary requirement. In contrast, computation of the weight vector in the conventional

QR–RLS algorithm involves the method of back substitution, which can be performed "on-the-fly" at the cost of additional computation. For this reason, the scope of practical applications for the QR–RLS algorithm is restricted to those areas such as adaptive beamforming and acoustic echo cancelation, where it is *not* necessary to have explicit knowledge of the weight vector.

Finally, the point that needs to be stressed is that all three QR–RLS algorithms preserve the desirable convergence properties of the standard RLS algorithm, namely, a fast rate of convergence and insensitivity to variations in the eigenvalue spread of the correlation matrix of incoming data.

One final comment is in order. The systolic array implementations of the QR decomposition involved in the design of the variants of the RLS algorithm described in this chapter were all based on the Givens rotation. This form of rotation provides one method for constructing the unitary rotation $\Theta(n)$. From Chapter 12 we recall that the Householder transformation (reflection) provides another method for constructing the unitary rotation $\Theta(n)$. According to an error analysis under finite-precision computations reported by Wilkinson (1965), the Householder transformation is superior to the Givens rotation. It is therefore of interest to know if a systolic implementation can be extended to the Householder transformation for QR decomposition–based RLS algorithms. Indeed, Liu et al. (1992) describe a two-level pipelined implementation of the Householder transformation on a systolic array with only local connections. The systolic array is, however, of a block-oriented kind, with the block size providing a new variable. In particular, improved numerical stability is attained by increasing the block size, but at the expense of increased latency.

PROBLEMS

1. Starting with the prearray-to-postarray transformation described in Eq. (14.32) for the square-root information filter, derive the equalities defined in Eqs. (14.33) to (14.37).

2. In this problem we revisit the square-root information filter. Specifically, the term $v(n)$ in the state-space model of Eqs. (14.2) and (14.3) is assumed to be a random variable of zero mean and variance $Q(n)$. Show that the square-root information filter may now be formulated as follows:

$$\mathbf{K}^{-1}(n) = \lambda(\mathbf{K}^{-1}(n-1) + Q^{-1}(n)\mathbf{u}(n)\mathbf{u}^H(n))$$

$$\mathbf{K}^{-1}(n)\hat{\mathbf{x}}(n+1 \mid \mathcal{Y}_n) = \lambda^{1/2}(\mathbf{K}^{-1}(n-1)\hat{\mathbf{x}}(n \mid \mathcal{Y}_{n-1}) + Q^{-1}(n)\mathbf{u}(n)y(n))$$

which includes Eqs. (14.29) and (14.30) as a special case.

3. Justify the validity of the prearray-to-postarray transformation described in Eq. (14.39) for the extended square-root information filter.

4. Let the n-by-n unitary matrix $\mathbf{Q}(n)$ involved in the QR-decomposition of the data matrix $\mathbf{A}(n)$ be partitioned as follows

$$\mathbf{Q}(n) = \begin{bmatrix} \mathbf{Q}_1(n) \\ \mathbf{Q}_2(n) \end{bmatrix}$$

where $\mathbf{Q}_1(n)$ has the same number of rows as the upper triangular matrix $\mathbf{R}(n)$ in the QR-decomposition of $\mathbf{A}(n)$. Assume that the exponential weighting factor $\lambda = 1$.

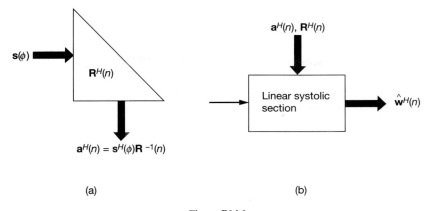

$$\mathbf{a}^H(n) = \mathbf{s}^H(\phi)\mathbf{R}^{-1}(n)$$

(a)   (b)

Figure P14.1

According to the method of least squares presented in Chapter 11, the projection operator is

$$\mathbf{P}(n) = \mathbf{A}(n)(\mathbf{A}^H(n)\mathbf{A}(n))^{-1}\mathbf{A}^H(n)$$

Show that for the problem at hand:

$$\mathbf{P}(n) = \mathbf{Q}_1^H(n)\mathbf{Q}_1(n)$$

How is the result modified for the case where $0 < \lambda \le 1$?

5. In the context of the systolic array in Fig. 14.1, explain the reasons for the following:
   (a) A total of $2M$ clock cycles are required to compute the Givens rotations in the triangular section of the systolic array in Fig. 14.1.
   (b) The latency for the entire array is $4M$ clock cycles.

6. Explain the way in which the systolic array structure of Fig. 14.6 may be used to operate as a prediction-error filter.

7. Discuss the use of a linear section, based on forward substitution, for solving the equation $\mathbf{R}^H(n)\mathbf{a}(n) = \mathbf{s}$ for the vector $\mathbf{a}(n)$; the matrix $\mathbf{R}$ is an M-by-M upper triangular matrix and $\mathbf{s}$ is an M-by-1 vector.

8. Figure P14.1 depicts a block diagram representation of an MVDR beamforming algorithm. Specifically, the triangular array in part (a) of this figure is frozen at time n and the steering vector $\mathbf{s}(\phi)$ is input into the array. The stored $\mathbf{R}^H(n)$ of the array and its output $\mathbf{a}^H(n)$ are applied to a linear systolic section as in part (b) of the figure. Do the following:
   (a) Show that the output of the triangular array is

   $$\mathbf{a}^H(n) = \mathbf{s}^H(\phi)\mathbf{R}^{-1}(n)$$

   (b) Using the method of back substitution, show that the linear systolic array produces the Hermitian transposed weight vector $\mathbf{w}^H(n)$ as its output.

9. Referring to the systolic MVDR beamformer described in Section 14.5, show that

$$\lambda\|\mathbf{a}(n-1)\|^2 = \|\mathbf{a}(n)\|^2 + |\epsilon(n)|^2$$

where $\|\mathbf{a}(n)\|$ is the Euclidean norm of the auxiliary vector $\mathbf{a}(n)$, λ is the exponential weighting factor, and $\epsilon(n)$ is some estimation error.

CHAPTER

── 15 ──

Order-Recursive Adaptive Filters

In this chapter we develop another important class of adaptive filters, the design of which is based on algorithms that involve both *order-update* and *time-update recursions*. The algorithms are rooted in recursive least-squares estimation theory[1] and therefore retain two unique attributes of the RLS algorithm: A fast rate of convergence, and insensitivity to variations in the eigenvalue spread of the underlying correlation matrix of the input data. However, unlike the RLS algorithm, the computational complexity of the algorithms considered in this chapter increases *linearly* with the number of adjustable filter parameters. This highly desirable property is a direct result of order recursiveness, which gives the adaptive filter a *computationally efficient, modular, latticelike structure*. In particular, as the filter order is increased from m to $m + 1$, say, the lattice filter permits us to carry over certain information gathered from the previous computations pertaining to the filter order m.

In deriving the order-recursive adaptive filters considered herein, we follow the same approach that we pursued in the previous chapter dealing with square-root adaptive filters. Specifically, we start from a state-space model of lattice filtering, which makes it possible to exploit the relevant aspects of Kalman filter theory. In so doing, we further consolidate the unification of adaptive filters along the lines described by Sayed and Kailath

---

[1] There is another type of order-recursive adaptive filtering algorithm, called the *gradient adaptive lattice (GAL) algorithm*, which is rooted in stochastic approximation. The derivation of GAL algorithms follows an approach similar to that of the least-mean-square (LMS) algorithm; for details, see Appendix G.

(1994). Before embarking on this development, however, we first present some background material relating to the forward and backward predictions of input data, which is fundamental to the underlying theory of order-recursive adaptive filters.

15.1 ADAPTIVE FORWARD LINEAR PREDICTION

Consider a forward linear predictor of order m, depicted in Fig. 15.1(a), whose tap-weight vector $\hat{\mathbf{w}}_{f,m}(n)$ is optimized in the least-squares sense over the entire observation interval $1 \le i \le n$. Let $f_m(n)$ denote the forward prediction error produced by the predictor at time n in response to the tap-input vector $\mathbf{u}_m(n-1)$ of size m, as shown by

$$f_m(n) = u(n) - \hat{\mathbf{w}}_{f,m}^H(n)\mathbf{u}_m(n-1) \tag{15.1}$$

According to this definition, $u(n)$ plays the role of "desired response" for forward linear prediction. The compositions of input vector $\mathbf{u}_m(n-1)$ and weight vector $\hat{\mathbf{w}}_m(n)$ are as follows, respectively:

$$\mathbf{u}_m(n-1) = [u(n-1), u(n-2), \ldots, u(n-m)]^T$$

$$\hat{\mathbf{w}}_{f,m}(n) = [w_{f,m,1}(n), w_{f,m,2}(n), \ldots, w_{f,m,m}(n)]^T$$

We refer to $f_m(n)$ as the *forward a posteriori prediction error* since its computation is based on the current value of the forward predictor's tap-weight vector, $\hat{\mathbf{w}}_{f,m}(n)$. Correspondingly, we may define the *forward a priori prediction error* as

$$\eta_m(n) = u(n) - \hat{\mathbf{w}}_{f,m}^H(n-1)\mathbf{u}_m(n-1) \tag{15.2}$$

the computation of which is based on the past value of the forward predictor's tap-weight vector, $\hat{\mathbf{w}}_{f,m}(n-1)$. In effect, $\eta_m(n)$ represents a form of innovation.

In Table 15.1 are listed the correspondences between the various quantities characterizing linear estimation in general and those characterizing forward linear prediction in particular, with the RLS algorithm in mind. With the aid of this table, it is a straightforward matter to modify the RLS algorithm developed in Sections 13.3 and 13.4 to write the recursions for adaptive forward linear prediction. Specifically, we deduce the following recursion for updating the tap-weight vector of the forward predictor:

$$\hat{\mathbf{w}}_{f,m}(n) = \hat{\mathbf{w}}_{f,m}(n-1) + \mathbf{k}_m(n-1)\eta_m^*(n) \tag{15.3}$$

where $\eta_m(n)$ is the forward *a priori* prediction error defined in Eq. (15.2), and $\mathbf{k}_m(n-1)$ is the past value of the *gain vector* defined by

$$\mathbf{k}_m(n-1) = \mathbf{\Phi}_m^{-1}(n-1)\mathbf{u}_m(n-1) \tag{15.4}$$

The matrix $\mathbf{\Phi}_m^{-1}(n-1)$ is the inverse of the correlation matrix $\mathbf{\Phi}_m(n-1)$ of the input data, with the latter matrix being defined by

$$\mathbf{\Phi}_m(n-1) = \sum_{i=1}^{n-1} \lambda^{n-1-i}\mathbf{u}_m(i)\mathbf{u}_m^H(i) \tag{15.5}$$

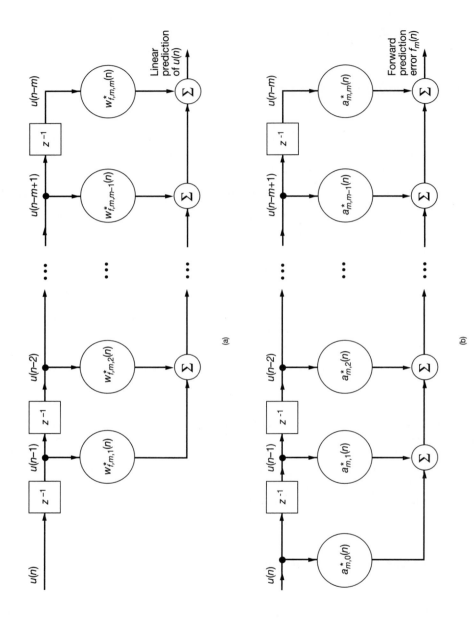

Figure 15.1 (a) Forward predictor of order m; (b) corresponding prediction-error filter.

TABLE 15.1 SUMMARY OF CORRESPONDENCES BETWEEN LINEAR ESTIMATION, FORWARD PREDICTION, AND BACKWARD PREDICTION

Quantity	Linear estimation (general)	Forward linear prediction of order m	Backward linear prediction of order m
Tap-input vector	$\mathbf{u}(n)$	$\mathbf{u}_m(n-1)$	$\mathbf{u}_m(n)$
Desired response	$d(n)$	$u(n)$	$u(n-m)$
Tap-weight vector	$\hat{\mathbf{w}}(n)$	$\hat{\mathbf{w}}_{f,m}(n)$	$\hat{\mathbf{w}}_{b,m}(n)$
A *posteriori* estimation error	$e(n)$	$f_m(n)$	$b_m(n)$
A *priori* estimation error	$\xi(n)$	$\eta_m(n)$	$\beta_m(n)$
Gain vector	$\mathbf{k}(n)$	$\mathbf{k}_m(n-1)$	$\mathbf{k}_m(n)$
Minimum value of sum of weighted error squares	$\mathscr{E}_{\min}(n)$	$\mathscr{F}_m(n)$	$\mathscr{B}_m(n)$

The use of subscript m is intended to signify the order of the prediction process. We follow this practice here and in the rest of the chapter, since some of the recursions to be developed involve an order update.

The adaptive forward linear prediction problem just described is in terms of a predictor characterized by the tap-weight vector $\hat{\mathbf{w}}_{f,m}(n)$. Equivalently, we may describe the problem by specifying a *forward prediction-error filter*, as depicted in Fig. 15.1(b). Let $\mathbf{a}_m(n)$ denote the $(m+1)$-by-1 tap-weight vector of the prediction-error filter of order m. This tap-weight vector is related to that of the forward predictor in Fig. 15.1(a) by

$$\mathbf{a}_m(n) = \begin{bmatrix} 1 \\ -\hat{\mathbf{w}}_{f,m}(n) \end{bmatrix} \tag{15.6}$$

Then we may redefine the forward *a posteriori* prediction and forward *a priori* prediction errors as follows, respectively:

$$f_m(n) = \mathbf{a}_m^H(n)\mathbf{u}_{m+1}(n) \tag{15.7}$$

and

$$\eta_m(n) = \mathbf{a}_m^H(n-1)\mathbf{u}_{m+1}(n) \tag{15.8}$$

where the input vector $\mathbf{u}_{m+1}(n)$ of size $m+1$ is partitioned in the following way:

$$\mathbf{u}_{m+1}(n) = \begin{bmatrix} u(n) \\ \mathbf{u}_m(n-1) \end{bmatrix} \tag{15.9}$$

The tap-weight vector $\hat{\mathbf{w}}_{f,m}(n)$ of the forward predictor is the solution obtained by minimizing the sum of weighted forward *a posteriori* prediction-error squares for $1 \le i \le n$,

$$\mathscr{F}_m(n) = \sum_{i=1}^{n} \lambda^{n-i}|f_m(i)|^2 \tag{15.10}$$

Equivalently, the tap-weight vector $\mathbf{a}_m(n)$ of the prediction-error filter is the solution to the same minimization problem, subject to the constraint that the first element of $\mathbf{a}_m(n)$ equals unity, in accordance with Eq. (15.6).

Finally, using (13.35), we get the following recursion for updating the minimum value of the sum of weighted forward prediction-error squares (i.e., forward prediction-error energy):

$$\mathcal{F}_m(n) = \lambda \mathcal{F}_m(n-1) + \eta_m(n) f_m^*(n) \tag{15.11}$$

where the product term $\eta_m(n) f_m^*(n)$ is real valued.

15.2 ADAPTIVE BACKWARD LINEAR PREDICTION

Consider next the *backward linear predictor of order m*, depicted in Fig. 15.2(a), whose tap-weight vector $\hat{\mathbf{w}}_{b,m}(n)$ is optimized in the least-squares sense over the entire observation interval $1 \le i \le n$. Let $b_m(n)$ denote the backward prediction error produced by this predictor at time n in response to the tap-input vector $\mathbf{u}_m(n)$ of size m, as shown by

$$b_m(n) = u(n-m) - \hat{\mathbf{w}}_{b,m}^H(n) \mathbf{u}_m(n) \tag{15.12}$$

According to this definition, $u(n-m)$ plays the role of desired response for backward linear prediction, and

$$\mathbf{u}_m(n) = [u(n), u(n-1), \dots, u(n-m+1)]^T$$

$$\hat{\mathbf{w}}_{b,m}(n) = [\hat{w}_{b,m,1}(n), \hat{w}_{b,m,2}(n), \dots, \hat{w}_{b,m,m}(n)]^T$$

We refer to $b_m(n)$ as the *a posteriori backward prediction error* since its computation is based on the current value of the backward predictor's tap-weight vector, $\hat{\mathbf{w}}_{b,m}(n)$. Correspondingly, we may define the *backward a priori prediction error* as

$$\beta_m(n) = u(n-m) - \hat{\mathbf{w}}_{b,m}^H(n-1) \mathbf{u}_m(n) \tag{15.13}$$

the computation of which is based on the past value of the backward predictor's tap-weight vector, $\hat{\mathbf{w}}_{b,m}(n-1)$.

In Table 15.1 are also listed the correspondences between the quantities characterizing linear estimation in general and those characterizing backward linear prediction in particular. To write the recursions for adaptive backward linear prediction, we may again modify the RLS algorithm developed in Sections 13.3 and 13.4 in light of these correspondences. Thus, we deduce the following recursion for updating the tap-weight vector of the backward predictor:

$$\hat{\mathbf{w}}_{b,m}(n) = \hat{\mathbf{w}}_{b,m}(n-1) + \mathbf{k}_m(n) \beta_m^*(n) \tag{15.14}$$

where $\beta_m(n)$ is the backward *a priori* prediction error defined in Eq. (15.12), and $\mathbf{k}_m(n)$ is the current value of the *gain vector* defined by

$$\mathbf{k}_m(n) = \boldsymbol{\Phi}_m^{-1}(n) \mathbf{u}_m(n) \tag{15.15}$$

The matrix $\mathbf{\Phi}_m^{-1}(n)$ is the inverse of the correlation matrix $\mathbf{\Phi}_m(n)$ of the input data, with the latter matrix being defined by

$$\mathbf{\Phi}_m(n) = \sum_{i=1}^{n} \lambda^{n-i}\mathbf{u}_m(i)\mathbf{u}_m^H(i) \tag{15.16}$$

The description of the backward linear prediction problem just presented is in terms of a backward predictor characterized by the tap-weight vector $\hat{\mathbf{w}}_{b,m}(n)$. Equivalently, we may describe the problem in terms of a *backward prediction-error filter*, as depicted in Fig. 15.2(b). Let the prediction-error filter of order m be characterized by a tap-weight vector $\mathbf{c}_m(n)$, which is related to that of the backward predictor in Fig. 15.2(a) as follows:

$$\mathbf{c}_m(n) = \begin{bmatrix} -\hat{\mathbf{w}}_{b,m}(n) \\ 1 \end{bmatrix} \tag{15.17}$$

Thus, with an input vector $\mathbf{u}_{m+1}(n)$ of size $m + 1$, the backward *a posteriori* prediction and backward *a priori* prediction errors may be rewritten as follows, respectively:

$$b_m(n) = \mathbf{c}_m^H(n)\mathbf{u}_{m+1}(n) \tag{15.18}$$

and

$$\beta_m(n) = \mathbf{c}_m^H(n-1)\mathbf{u}_{m+1}(n) \tag{15.19}$$

In this case, the input vector $\mathbf{u}_{m+1}(n)$ is partitioned in the following way:

$$\mathbf{u}_{m+1}(n) = \begin{bmatrix} \mathbf{u}_m(n) \\ u(n-m) \end{bmatrix} \tag{15.20}$$

The tap-weight vector $\hat{\mathbf{w}}_{b,m}(n)$ of the backward predictor is obtained by minimizing the sum of weighted backward *a posteriori* prediction-error squares for $1 \leq i \leq n$,

$$\mathcal{B}_m(n) = \sum_{i=1}^{n} \lambda^{n-i}|b_m(i)|^2 \tag{15.21}$$

Equivalently, the tap-weight vector $\mathbf{c}_m(n)$ of the backward prediction-error filter is the solution to the same minimization problem, subject to the constraint that the last element of $\mathbf{c}_m(n)$ equals unity, in accordance with Eq. (15.17).

Also, using Eq. (13.35), we get the following recursion for updating the minimum value of the sum of weighted backward prediction-error squares (i.e., backward prediction-error energy):

$$\mathcal{B}_m(n) = \lambda\mathcal{B}_m(n-1) + \beta_m(n)b_m^*(n) \tag{15.22}$$

where the product term $\beta_m(n)b_m^*(n)$ is real valued.

In closing this discussion of the recursive least-squares prediction problem, it is of interest to note that in the case of backward prediction, the input vector $\mathbf{u}_{m+1}(n)$ is partitioned with the desired response $u(n-m)$ as the last entry, as shown in Eq. (15.20). On the other hand, in the case of forward linear prediction the input vector $\mathbf{u}_{m+1}(n)$ is parti-

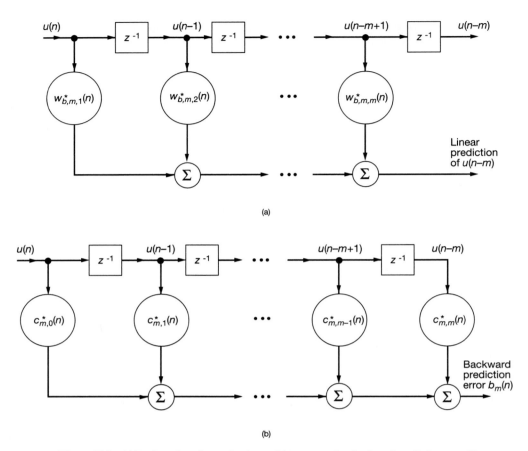

Figure 15.2   (a) Backward predictor of order m; (b) corresponding backward prediction-error filter.

tioned with the desired response $u(n)$ as the leading entry, as shown in Eq. (15.9). Note also that the update recursion for the tap-weight vector $\hat{\mathbf{w}}_{b,m}(n)$ of the backward linear predictor in Eq. (15.14) requires knowledge of the current value $\mathbf{k}_m(n)$ of the gain vector. On the other hand, the update recursion for the tap-weight vector $\hat{\mathbf{w}}_{f,m}(n)$ of the forward linear predictor in Eq. (15.3) requires knowledge of the old value $\mathbf{k}_m(n-1)$ of the gain vector.

15.3 CONVERSION FACTOR

The definition of the m-by-1 vector,

$$\mathbf{k}_m(n) = \boldsymbol{\Phi}_m^{-1}(n)\mathbf{u}_m(n)$$

may also be viewed as the solution of a special case of the normal equations for least-squares estimation. To be specific, the gain vector $\mathbf{k}_m(n)$ defines the tap-weight vector of a transversal filter that contains m taps and that operates on the input data $u(1)$, $u(2)$, ..., $u(n)$ to produce the least-squares estimate of a special desired response that equals

$$d(i) = \begin{cases} 1, & i = n \\ 0, & i = 1, 2, \ldots, n - 1 \end{cases} \tag{15.23}$$

The n-by-1 vector whose elements equal the $d(i)$ of Eq. (15.23) is called the *first coordinate vector*. This vector has the property that its inner product with any time-dependent vector reproduces the upper or "most recent" element of that vector.

Substituting Eq. (15.23) in (13.10), we find that the m-by-1 cross-correlation vector $\mathbf{z}_m(n)$ between the m tap inputs of the transversal filter and the desired response equals $\mathbf{u}_m(n)$. This therefore confirms the gain vector $\mathbf{k}_m(n)$ as the special solution of the normal equations that arises when the desired response is defined by Eq. (15.23).

For the problem described here, define the *estimation error*

$$\begin{aligned} \gamma_m(n) &= 1 - \mathbf{k}_m^H(n)\mathbf{u}_m(n) \\ &= 1 - \mathbf{u}_m^H(n)\mathbf{\Phi}_m^{-1}(n)\mathbf{u}_m(n) \end{aligned} \tag{15.24}$$

The estimation error $\gamma_m(n)$ represents the output of a transversal filter whose tap-weight vector equals the gain vector $\mathbf{k}_m(n)$ and which is excited by the tap-input vector $\mathbf{u}_m(n)$, as depicted in Fig. 15.3. Since the filter output has the structure of a Hermitian form, it follows that the estimation error $\gamma_m(n)$ is a real-valued scalar. Moreover, $\gamma_m(n)$ has the important property that it is bounded by zero and one; that is

$$0 < \gamma_m(n) \le 1 \tag{15.25}$$

This property is readily proved by substituting the recursion of Eq. (13.16) for the inverse matrix $\mathbf{\Phi}_m^{-1}(n-1)$ in Eq. (15.24), and then simplifying to obtain the result

$$\gamma_m(n) = \frac{1}{1 + \lambda^{-1}\mathbf{u}_m^H(n)\mathbf{\Phi}_m^{-1}(n-1)\mathbf{u}_m(n)} \tag{15.26}$$

The Hermitian form $\mathbf{u}_m^H(n)\mathbf{\Phi}_m^{-1}(n-1)\mathbf{u}_m(n) \ge 0$. Consequently, the estimation error $\gamma_m(n)$ is bounded as in (15.25).

It is noteworthy that $\gamma_m(n)$ also equals the sum of weighted error squares resulting from use of the transversal filter in Fig. 15.3, whose tap-weight vector equals the gain vector $\mathbf{k}_m(n)$, to obtain the least-squares estimate of the first coordinate vector (see Problem 1).

Other Useful Interpretations of $\gamma_m(n)$

Depending on the approach taken, the parameter $\gamma_m(n)$ may be given three other entirely different interpretations:

1. The parameter $\gamma_m(n)$ may be viewed as a *likelihood variable* (Lee et al., 1981). This interpretation follows from a statistical formulation of the tap-input vector in terms of its log-likelihood function, under the assumption that the tap-inputs have a joint Gaussian distribution (see Problem 11).

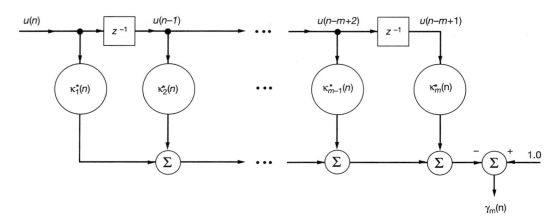

Figure 15.3   Transversal filter for defining the estimation error $\gamma_m(n)$.

2. The parameter $\gamma_m(n)$ may be interpreted as an *angle variable* (Lee et al., 1981; Carayannis et al., 1983). This interpretation follows from Eq. (15.24). In particular, following the discussion presented in Chapter 14, we may express the (positive) square root of $\gamma_m(n)$ as

$$\gamma_m^{1/2}(n) = \prod_{i=1}^{m} \cos\phi_i(n)$$

where $\phi_i(n)$ represents the angle of a plane (Givens) rotation [see Eq. (14.56)].

3. The parameter $\gamma_m(n)$ may be interpreted as a *conversion factor* (Carayannis et al., 1983). According to this interpretation, the availability of $\gamma_m(n)$ helps us determine the value of an *a posteriori* estimation error, given the value of the corresponding *a priori* estimation error.

It is this third interpretation that we pursue here. Indeed, it is because of this interpretation that we have adopted the terminology "conversion factor" as a description for $\gamma_m(n)$.

Three Kinds of Estimation Error

In linear least-squares estimation theory, there are three kinds of estimation error to be considered: the ordinary estimation error (involved in the estimation of some desired response), the forward prediction error, and the backward prediction error. Correspondingly, $\gamma_m(n)$ has three useful interpretations as a conversion factor, as described next.

1. For recursive least-squares estimation, we have

$$\gamma_m(n) = \frac{e_m(n)}{\xi_m(n)} \tag{15.27}$$

where $e_m(n)$ is the *a posteriori* estimation error and $\xi_m(n)$ is the *a priori* estimation error. This relation is readily proved by postmultiplying the Hermitian transposed sides of Eq. (13.25) by $\mathbf{u}_m(n)$, using Eq. (13.26) for the *a priori* estimation error $\xi_m(n)$, Eq. (13.27) for the *a posteriori* estimation error $e_m(n)$, and the first line of Eq. (15.24) for the variable $\gamma_m(n)$. Equation (15.27) states that given the *a priori* estimation error $\xi_m(n)$ as computed in the RLS algorithm, we may determine the corresponding value of the *a posteriori* estimation error $e_m(n)$ by multiplying $\xi_m(n)$ by $\gamma_m(n)$. We may therefore view $\xi_m(n)$ as a tentative value of the estimation error $e_m(n)$ and $\gamma_m(n)$ as the multiplicative correction.

2. For adaptive forward linear prediction, we have

$$\gamma_m(n-1) = \frac{f_m(n)}{\eta_m(n)} \tag{15.28}$$

This relation is readily proved by postmultiplying the Hermitian transposed sides of Eq. (15.3) by $\mathbf{u}_m(n-1)$, and then using the definitions of Eqs. (15.1), (15.2), (15.4), and (15.24). Equation (15.28) states that given the forward *a priori* prediction error $\eta_m(n)$, we may compute the forward *a posteriori* prediction error $f_m(n)$ by multiplying $\eta_m(n)$ by the delayed estimation error $\gamma_m(n-1)$. We may therefore view $\eta_m(n)$ as a tentative value for the forward *a posteriori* prediction error $f_m(n)$ and $\gamma_m(n-1)$ as the multiplicative correction.

3. For adaptive backward linear prediction, we have

$$\gamma_m(n) = \frac{b_m(n)}{\beta_m(n)} \tag{15.29}$$

This third relation is readily proved by postmultiplying the Hermitian transposed sides of Eq. (15.14) by $\mathbf{u}_m(n)$, and then using the definitions of Eqs. (15.12), (15.13), (15.15), and (15.24). Equation (15.29) states that given the backward *a priori* prediction error $\beta_m(n)$, we may compute the backward *a posteriori* prediction error $b_m(n)$ by multiplying $\beta_m(n)$ by the estimation error $\gamma_m(n)$. We may therefore view $\beta_m(n)$ as a tentative value for the backward prediction error $b_m(n)$ and $\gamma_m(n)$ as the multiplicative correction.

The discussion above points out the unique role of the variable $\gamma_m(n)$ in that it is the *common* factor (either in its regular or delayed form) in the conversion of an *a priori* estimation error into the corresponding *a posteriori* estimation error, be it in the context of ordinary estimation, forward prediction, or backward prediction. Accordingly, we may refer to $\gamma_m(n)$ as a *conversion factor*. Indeed, it is remarkable that through the use of this conversion factor we are able to compute the *a posteriori* errors $e_m(n), f_m(n),$ and $b_m(n)$ at time n before the tap-weight vectors of the pertinent filters that produce them have been actually computed (Carayannis et al., 1983).

15.4 LEAST-SQUARES LATTICE PREDICTOR

Returning to the time-shifting property of the input data, we note from Eq. (15.20) that the input vector $\mathbf{u}_m(n)$ for a backward linear predictor of order m and the input vector $\mathbf{u}_{m+1}(n)$ for a backward linear predictor of order $m + 1$ have exactly the same first m entries. Likewise, we note from Eq. (15.9) that the input vector $\mathbf{u}_m(n - 1)$ for a forward linear predictor of order m and the input vector $\mathbf{u}_{m+1}(n)$ for a forward linear predictor of order $m + 1$ have exactly the same m last entries. These observations prompt us to raise the following fundamental question: In the course of increasing the prediction order from $m - 1$ to m, say, is it possible to carryover information gathered from previous computations pertaining to the prediction order $m - 1$? The answer to this question is an emphatic yes, and it is embodied in a modular filtering structure known as the *least-squares lattice predictor*.

To derive this important filtering structure and its algorithmic design, we propose to proceed as follows. In this section, we use the *principle of orthogonality* to derive the basic equations that characterize the least-squares lattice predictor. Then, under the unifying umbrella of *Kalman filter theory*, we derive various algorithms for its design in subsequent sections of the chapter.

To begin with, consider the situation depicted in Fig. 15.4, involving a pair of forward and backward prediction-error filters of order $m - 1$. They are both fed by the same input vector $\mathbf{u}_m(i)$. The forward prediction-error filter, characterized by the tap-weight vector $\mathbf{a}_{m-1}(n)$, produces $f_{m-1}(i)$ at its output. The backward prediction-error filter, characterized by the tap-weight vector $\mathbf{c}_{m-1}(n)$, produces $b_{m-1}(i)$ at its output. The input data $u(i)$ occupy the observation interval $1 \leq i \leq n$. The problem we wish to address may be stated as follows:

- Given the forward prediction error $f_{m-1}(i)$ and backward prediction error $b_{m-1}(i)$, determine their order-updated values $f_m(i)$ and $b_m(i)$, respectively, in a computationally efficient manner.

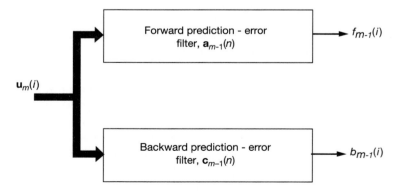

Figure 15.4   Setting the stage for formulating the least-squares lattice predictor.

By "computationally efficient," we mean the following: The input vector in Fig. 15.4 is enlarged by adding the past sample $u(i - m)$ and the prediction order is thereby increased by one; yet the computations involved in evaluating $f_{m-1}(i)$ and $b_{m-1}(i)$ remain completely intact.

The forward prediction error $f_{m-1}(i)$ is determined by the tap inputs $u(i)$, $u(i - 1)$, ..., $u(i - m + 1)$. The order-updated forward prediction error $f_m(i)$ requires knowledge of the additional tap input (i.e., past sample) $u(i - m)$. The backward prediction error $b_{m-1}(i)$ is determined by the same tap inputs as those involved in $f_{m-1}(i)$. If therefore we were to delay $b_{m-1}(i)$ by one time unit, the additional past sample $u(i - m)$ needed for computing $f_m(i)$ would be found in the composition of the *delayed* backward prediction error $b_{m-1}(i - 1)$. Thus, treating $b_{m-1}(i - 1)$ as the input to a *one-tap least-squares filter*, $f_{m-1}(i)$ as the desired response and $f_m(i)$ as the residual resulting from the least-squares estimation, we may write (see Fig. 15.5(a))

$$f_m(i) = f_{m-1}(i) + \kappa_{f,m}^*(n)\, b_{m-1}(i-1), \qquad i = 1, 2, \ldots, n \tag{15.30}$$

where $\kappa_{f,m}(n)$ is the filter's scalar coefficient to be determined. The format of Eq. (15.30) is similar to that of the corresponding order-update derived in Chapter 6 for a lattice predictor operating on stationary inputs. However, the formula for $\kappa_{f,m}(n)$ is different. For the determination of this coefficient, we turn to the principle of orthogonality discussed in Chapter 11 in the context of linear least-squares estimation. According to this principle, the estimation error (i.e., residual) produced by a linear least-squares filter in response to a set of inputs is orthogonal to each of those inputs in a time-averaged sense over the entire observation interval of interest. Thus, applying the principle of orthogonality to the input $b_{m-1}(i - 1)$ and residual $f_m(i)$ of the linear forward prediction problem postulated in Eq. (15.30), we get

$$\sum_{i=1}^{n} \lambda^{n-i} f_m^*(i)\, b_{m-1}(i-1) = 0 \tag{15.31}$$

Hence, substituting Eq. (15.30) in (15.31) and then solving for $\kappa_{f,m}(n)$, we get

$$\kappa_{f,m}(n) = -\,\frac{\displaystyle\sum_{i=1}^{n} \lambda^{n-i} f_{m-1}^*(i)\, b_{m-1}(i - 1)}{\displaystyle\sum_{i=1}^{n} \lambda^{n-i} \left| b_{m-1}(i - 1) \right|^2} \tag{15.32}$$

The denominator of this formula is the sum of weighted backward prediction-error squares for order $m - 1$:

$$\mathcal{B}_{m-1}(n - 1) = \sum_{i=1}^{n-1} \lambda^{n-1-i} \left| b_{m-1}(i) \right|^2$$

$$= \sum_{i=1}^{n} \lambda^{n-i} \left| b_{m-1}(i - 1) \right|^2 \tag{15.33}$$

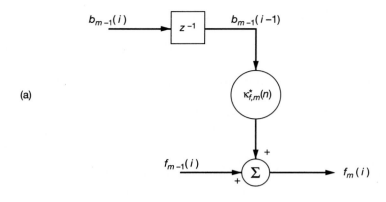

(a)

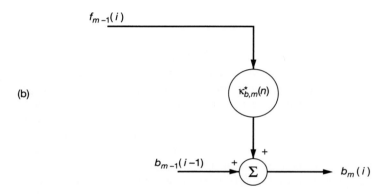

(b)

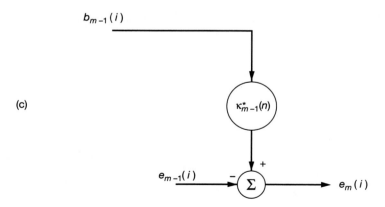

(c)

Figure 15.5 Single-coefficient linear combiners for (a) forward prediction, (b) backward prediction, and (c) joint-process estimation.

where, in the last line, we have used the fact that

$$b_{m-1}(0) = 0 \qquad \text{for all } m \geq 1$$

by virtue of prewindowing the input data. For the numerator of Eq. (15.32), we introduce a new definition:

$$\Delta_{m-1}(n) = \sum_{i=1}^{n} \lambda^{n-i} f_{m-1}^{*}(i) \, b_{m-1}(i-1) \tag{15.34}$$

Using the definitions of Eqs. (15.33) and (15.34) in Eq. (15.32), the formula for the scalar coefficient $\kappa_{f,m}(n)$ takes on the compact form:

$$\kappa_{f,m}(n) = -\frac{\Delta_{m-1}(n)}{\mathcal{B}_{m-1}(n-1)} \tag{15.35}$$

Consider next the issue of computing the order-updated backward prediction error $b_m(i)$. As before, we may work with the forward prediction error $f_{m-1}(i)$ and delayed backward prediction error $b_{m-1}(i-1)$, except for the fact that their filtering roles are now interchanged. Specifically, we have a one-tap least-squares filter with $f_{m-1}(i)$ acting as the input, $b_{m-1}(i-1)$ as the desired response, and $b_m(i)$ as the residual of the filtering process. That is, we write (see Fig. 15.5(b))

$$b_m(i) = b_{m-1}(i-1) + \kappa_{b,m}^{*}(n) \, f_{m-1}(i), \qquad i = 1, 2, \ldots, n \tag{15.36}$$

where $\kappa_{b,m}(n)$ is the filter's scalar coefficient to be determined. The format of Eq. (15.36) is also similar to the corresponding order update derived in Chapter 6 for a lattice predictor operating on stationary inputs. However, the formula for $\kappa_{b,m}(n)$ is different. In particular, we determine $\kappa_{b,m}(n)$ by applying the principle of orthogonality to the input $f_{m-1}(i)$ and residual $b_m(i)$ in the backward prediction problem postulated in Eq. (15.36), and thus write

$$\sum_{i=1}^{n} \lambda^{n-i} b_m^{*}(i) \, f_{m-1}(i) = 0 \tag{15.37}$$

Hence, substituting Eq. (15.36) in (15.37) and then solving for $\kappa_{b,m}(n)$, we get

$$\kappa_{b,m}(n) = -\frac{\displaystyle\sum_{i=1}^{n} \lambda^{n-i} b_{m-1}^{*}(i-1) \, f_{m-1}(i)}{\displaystyle\sum_{i=1}^{n} \lambda^{n-i} |f_{m-1}(i)|^{2}} \tag{15.38}$$

The numerator of this formula is the complex conjugate of the quantity $\Delta_{m-1}(n)$ defined in Eq. (15.34). The denominator is recognized as the sum of weighted forward prediction-error squares for order $m-1$:

$$\mathcal{F}_{m-1}(n) = \sum_{i=1}^{n} \lambda^{n-i} |f_{m-1}(i)|^{2} \tag{15.39}$$

Accordingly, we may recast the formula of Eq. (15.38) in the compact form

$$\kappa_{b,m}(n) = -\frac{\Delta_{m-1}^*(n)}{\mathcal{F}_{m-1}(n)} \tag{15.40}$$

The results described in Eqs. (15.30) and (15.36) are basic to the least-squares lattice predictor. For their physical interpretation, define the n-by-1 prediction-error vectors:

$$\mathbf{f}_m(n) = [f_m(1), f_m(2), \ldots, f_m(n)]^T$$

$$\mathbf{b}_m(n) = [b_m(1), b_m(2), \ldots, b_m(n)]^T$$

$$\mathbf{b}_m(n-1) = [0, b_m(1), \ldots, b_m(n-1)]^T$$

where prediction order $m = 0, 1, 2, \ldots$. Then, on the basis of Eqs. (15.30) and (15.36) we may make the following statements in the terminology of *projection* theory:

- The result of projecting the vector $\mathbf{f}_{m-1}(n)$ onto $\mathbf{b}_{m-1}(n-1)$ is represented by the residual vector $\mathbf{f}_m(n)$; the *forward reflection coefficient* $\kappa_{f,m}(n)$ is the parameter needed to do this projection.
- The result of projecting the vector $\mathbf{b}_{m-1}(n-1)$ onto $\mathbf{f}_{m-1}(n)$ is represented by the residual vector $\mathbf{b}_m(n)$; the *backward reflection coefficient* $\kappa_{b,m}(n)$ is the parameter needed to do this second projection.

To put the finishing touch to this part of the discussion, we evaluate Eqs. (15.30) and (15.36) for the end of the observation interval $i = n$. We thus have the following pair of interrelated order-update recursions:

$$f_m(n) = f_{m-1}(n) + \kappa_{f,m}^*(n)b_{m-1}(n-1) \tag{15.41}$$

$$b_m(n) = b_{m-1}(n-1) + \kappa_{b,m}^*(n)f_{m-1}(n) \tag{15.42}$$

where $m = 1, 2, \ldots, M$, and M is the *final prediction order*. When $m = 0$ there is no prediction being preformed on the input data; this corresponds to the *initial conditions* described by

$$f_0(n) = b_0(n) = u(n) \tag{15.43}$$

where $u(n)$ is the input datum at time n. Thus, as we vary the prediction order m from zero all the way up to the final value M, we get the multistage least-squares lattice predictor of Fig. 15.6, whose number of stages equals M. An important feature of the least-squares lattice predictor described herein is its *modular* structure, the implication which is that the computational complexity scales *linearly* with the prediction order.

Least-Squares Lattice Version of the Levinson–Durbin Recursion

The forward prediction error $f_m(n)$ and backward prediction error $b_m(n)$ are defined by Eqs. (15.7) and (15.18), reproduced for convenience of presentation on the next page:

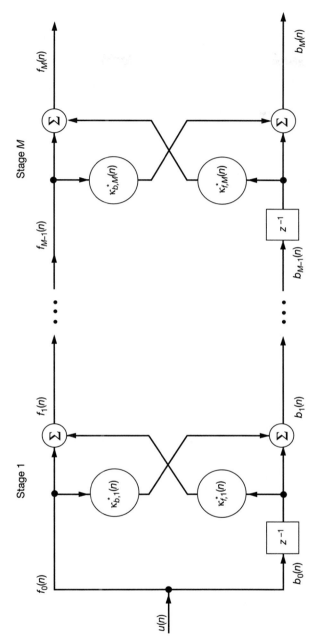

Figure 15.6 Multistage lattice predictor.

$$f_m(n) = \mathbf{a}_m^H(n)\,\mathbf{u}_{m+1}(n)$$

$$b_m(n) = \mathbf{c}_m^H(n)\,\mathbf{u}_{m+1}(n)$$

where $\mathbf{a}_m(n)$ and $\mathbf{c}_m(n)$ are the tap-weight vectors of the corresponding forward and backward prediction-error filters, respectively. The forward prediction error $f_{m-1}(n)$ and delayed backward prediction error $b_{m-1}(n-1)$, pertaining to a lower prediction order, are defined as follows, respectively:

$$f_{m-1}(n) = \mathbf{a}_{m-1}^H(n)\,\mathbf{u}_m(n)$$

$$= \begin{bmatrix} \mathbf{a}_{m-1}(n) \\ 0 \end{bmatrix}^H \begin{bmatrix} \mathbf{u}_m(n) \\ u(n-m) \end{bmatrix}$$

$$= \begin{bmatrix} \mathbf{a}_{m-1}(n) \\ 0 \end{bmatrix}^H \mathbf{u}_{m+1}(n)$$

$$b_{m-1}(n-1) = \mathbf{c}_{m-1}^H(n-1)\,\mathbf{u}_m(n-1)$$

$$= \begin{bmatrix} 0 \\ \mathbf{c}_{m-1}(n-1) \end{bmatrix}^H \begin{bmatrix} u(n) \\ \mathbf{u}_m(n-1) \end{bmatrix}$$

$$= \begin{bmatrix} 0 \\ \mathbf{c}_{m-1}(n-1) \end{bmatrix}^H \mathbf{u}_{m+1}(n)$$

The four prediction errors defined above share a common input vector, namely $\mathbf{u}_{m+1}(n)$. Therefore, substituting their defining equations into Eqs. (15.41) and (15.42) and then comparing terms on the two sides of the resultants, we deduce the following pair of order updates:

$$\mathbf{a}_m(n) = \begin{bmatrix} \mathbf{a}_{m-1}(n) \\ 0 \end{bmatrix} + \kappa_{f,m}(n) \begin{bmatrix} 0 \\ \mathbf{c}_{m-1}(n-1) \end{bmatrix} \tag{15.44}$$

$$\mathbf{c}_m(n) = \begin{bmatrix} 0 \\ \mathbf{c}_{m-1}(n) \end{bmatrix} + \kappa_{b,m}(n) \begin{bmatrix} \mathbf{a}_{m-1}(n) \\ 0 \end{bmatrix} \tag{15.45}$$

where $m = 1, 2, \ldots, M$. Equations (15.44) and (15.45) may be viewed as the least-squares version of the *Levinson–Durbin recursion* discussed in Chapter 6. Recognizing that the last element of $\mathbf{c}_{m-1}(n-1)$ and the first element of $\mathbf{a}_{m-1}(n)$ equal unity, by definition, we readily see from Eqs. (15.44) and (15.45) that

$$\kappa_{f,m}(n) = a_{m,m}(n) \tag{15.46}$$

$$\kappa_{b,m}(n) = c_{m,0}(n) \tag{15.47}$$

where $a_{m,m}(n)$ is the last element of the vector $\mathbf{a}_m(n)$, and $c_{m,0}(n)$ is the first element of the vector $\mathbf{c}_m(n)$. Unlike the situation described in Chapter 6 for a stationary environment, we generally find that in a least-squares lattice predictor:

$$\kappa_{f,m}(n) \neq \kappa_{b,m}^*(n)$$

In any event, the order updates of Eqs. (15.44) and (15.49) reveal a remarkable property of a least-squares lattice predictor of final order M: In an implicit sense, it embodies a chain of forward prediction-error filters of order $1, 2, \ldots, M$, and a chain of backward prediction-error filters $1, 2, \ldots, M$, all in one modular structure.

A Time-Update Recursion for $\Delta_{m-1}(n)$

From Eqs (15.35) and (15.40) we see that the reflection coefficients $\kappa_{f,m}(n)$ and $\kappa_{b,m}(n)$ of the least-squares lattice predictor are uniquely determined by three quantities: $\Delta_{m-1}(n)$, $\mathcal{F}_{m-1}(n)$, and $\mathcal{B}_{m-1}(n-1)$. Equations (15.11) and (15.22) provide us with the means to time-update the latter two quantities. We need the corresponding time-update recursion for $\Delta_{m-1}(n)$. A derivation of this recursion is presented in what follows.

Let $\boldsymbol{\Phi}_{m+1}(n)$ denote the $(m+1)$-by-$(m+1)$ correlation matrix of the tap-input vector $\mathbf{u}_{m+1}(i)$ applied to the forward prediction-error filter of order m, where $1 \le i \le n$. Let $\mathbf{a}_m(n)$ denote the tap-weight vector of this filter, and $\mathcal{F}_m(n)$ denote the corresponding sum of weighted prediction-error squares. We may characterize this filter by the *augmented normal equations* (see Chapter 11)

$$\boldsymbol{\Phi}_{m+1}(n)\mathbf{a}_m(n) = \begin{bmatrix} \mathcal{F}_m(n) \\ \mathbf{0}_m \end{bmatrix} \tag{15.48}$$

where $\mathbf{0}_m$ is the m-by-1 null vector. The correlation matrix $\boldsymbol{\Phi}_{m+1}(n)$ may be partitioned in two different ways, depending on how we interpret the first or last element of the tap-input vector $\mathbf{u}_{m+1}(i)$. The form of partitioning that we like to use first is the one that enables us to relate the tap-weight vector $\mathbf{a}_m(n)$, pertaining to prediction order m, to the tap-weight vector $\mathbf{a}_{m-1}(n)$, pertaining to prediction order $m-1$. This aim is realized by using

$$\boldsymbol{\Phi}_{m+1}(n) = \left[\begin{array}{c|c} \boldsymbol{\Phi}_m(n) & \boldsymbol{\phi}_2(n) \\ \hline \boldsymbol{\phi}_2^H(n) & \mathcal{U}_2(n) \end{array} \right] \tag{15.49}$$

where $\boldsymbol{\Phi}_m(n)$ is the m-by-m correlation matrix of the tap-input vector $\mathbf{u}_m(i)$, $\boldsymbol{\phi}_2(n)$ is the m-by-1 cross-correlation vector between $\mathbf{u}_m(i)$ and $u(i-m)$, and $\mathcal{U}_2(n)$ is the sum of weighted squared values of the input $u(i-m)$ for $1 \le i \le n$. Note that $\mathcal{U}_2(n)$ is zero for $n - m \le 0$. We postmultiply both sides of Eq. (15.49) by an $(m+1)$-by-1 vector whose first m elements are defined by the vector $\mathbf{a}_{m-1}(n)$ and whose last element equals zero. We may thus write

$$\boldsymbol{\Phi}_{m+1}(n) \begin{bmatrix} \mathbf{a}_{m-1}(n) \\ \hline 0 \end{bmatrix} = \left[\begin{array}{c|c} \boldsymbol{\Phi}_m(n) & \boldsymbol{\phi}_2(n) \\ \hline \boldsymbol{\phi}_2^H(n) & \mathcal{U}_2(n) \end{array} \right] \begin{bmatrix} \mathbf{a}_{m-1}(n) \\ \hline 0 \end{bmatrix}$$

$$= \begin{bmatrix} \boldsymbol{\Phi}_m(n)\mathbf{a}_{m-1}(n) \\ \boldsymbol{\phi}_2^H(n)\mathbf{a}_{m-1}(n) \end{bmatrix} \tag{15.50}$$

Both $\boldsymbol{\Phi}_m(n)$ and $\mathbf{a}_{m-1}(n)$ have the same time argument n. Furthermore, in the first line of Eq. (15.50), they are both positioned in such a way that when the matrix multiplication is performed, $\boldsymbol{\Phi}_m(n)$ becomes postmultiplied by $\mathbf{a}_{m-1}(n)$. For a forward prediction-error filter of order $m - 1$, evaluated at time n, the set of augmented normal equations defined in Eq. (15.48) takes the form

$$\boldsymbol{\Phi}_m(n)\mathbf{a}_{m-1}(n) = \begin{bmatrix} \mathcal{F}_{m-1}(n) \\ \mathbf{0}_{m-1} \end{bmatrix}$$

Define the scalar

$$\Delta_{m-1}(n) = \boldsymbol{\phi}_2^H(n)\mathbf{a}_{m-1}(n) \tag{15.51}$$

which is the same parameter defined previously in Eq. (15.34); see Problem 2. Accordingly, we may rewrite Eq. (15.50) as

$$\boldsymbol{\Phi}_{m+1}(n)\begin{bmatrix} \mathbf{a}_{m-1}(n) \\ 0 \end{bmatrix} = \begin{bmatrix} \mathcal{F}_{m-1}(n) \\ \mathbf{0}_{m-1} \\ \Delta_{m-1}(n) \end{bmatrix} \tag{15.52}$$

Consider next the backward prediction-error filter of order m. Let $\mathbf{c}_m(n)$ denote its tap-weight vector, and $\mathcal{B}_m(n)$ denote the corresponding sum of weighted prediction-error squares. This filter is characterized by the augmented normal equations written in the matrix form:

$$\boldsymbol{\Phi}_{m+1}(n)\mathbf{c}_m(n) = \begin{bmatrix} \mathbf{0}_m \\ \mathcal{B}_m(n) \end{bmatrix} \tag{15.53}$$

where $\boldsymbol{\Phi}_{m+1}(n)$ is as defined previously, and $\mathbf{0}_m$ is the m-by-1 null vector. This time we use the other partitioned form of the correlation matrix $\boldsymbol{\Phi}_m(n)$, as shown by

$$\boldsymbol{\Phi}_{m+1}(n) = \left[\begin{array}{c:c} \mathcal{U}_1(n) & \boldsymbol{\phi}_1^H(n) \\ \hdashline \boldsymbol{\phi}_1(n) & \boldsymbol{\Phi}_m(n-1) \end{array} \right] \tag{15.54}$$

where $\mathcal{U}_1(n)$ is the sum of weighted squared values of the input $u(i)$ for the time interval $1 \le i \le n$, $\boldsymbol{\phi}_1(n)$ is the m-by-1 cross-correlation vector between $u(i)$ and the tap-input vector $\mathbf{u}_m(i - 1)$, and $\boldsymbol{\Phi}_m(n - 1)$ is the m-by-m correlation matrix of $\mathbf{u}_m(i - 1)$. Correspondingly, we postmultiply $\boldsymbol{\Phi}_{m+1}(n)$ by an $(m + 1)$-by-1 vector whose first element is zero and whose m remaining elements are defined by the tap-weight vector $\mathbf{c}_{m-1}(n - 1)$ that pertains to a backward prediction-error filter of order $m - 1$. We may thus write

$$\boldsymbol{\Phi}_{m+1}(n)\begin{bmatrix} 0 \\ \mathbf{c}_{m-1}(n - 1) \end{bmatrix} = \begin{bmatrix} \mathcal{U}_1(n) & \boldsymbol{\phi}_1^H(n) \\ \boldsymbol{\phi}_1(n) & \boldsymbol{\Phi}_m(n - 1) \end{bmatrix}\begin{bmatrix} 0 \\ \mathbf{c}_{m-1}(n - 1) \end{bmatrix}$$

$$= \begin{bmatrix} \boldsymbol{\phi}_1^H(n)\mathbf{c}_{m-1}(n - 1) \\ \boldsymbol{\Phi}_m(n - 1)\mathbf{c}_{m-1}(n - 1) \end{bmatrix} \tag{15.55}$$

Both $\boldsymbol{\Phi}_m(n - 1)$ and $\mathbf{c}_{m-1}(n - 1)$ have the same time argument, $n - 1$. Also, they are both positioned in the first line of Eq. (15.55) in such a way that, when the matrix multiplication is performed, $\boldsymbol{\Phi}_{m-1}(n - 1)$ becomes postmultiplied by $\mathbf{c}_{m-1}(n - 1)$. For a backward prediction-error filter of order $m - 1$, evaluated at time $n - 1$, the set of augmented normal equations in Eq. (15.53) takes the form

$$\boldsymbol{\Phi}_m(n-1)\mathbf{c}_{m-1}(n-1) = \begin{bmatrix} \mathbf{0}_{m-1} \\ \mathcal{B}_{m-1}(n-1) \end{bmatrix}$$

Define the second scalar

$$\Delta'_{m-1}(n) = \boldsymbol{\phi}_1^H(n)\mathbf{c}_{m-1}(n-1) \tag{15.56}$$

where the prime is intended to distinguish this new parameter from $\Delta_{m-1}(n)$. Accordingly, we may rewrite Eq. (15.55) as

$$\boldsymbol{\Phi}_{m+1}(n)\begin{bmatrix} 0 \\ \mathbf{c}_{m-1}(n-1) \end{bmatrix} = \begin{bmatrix} \Delta'_{m-1}(n) \\ \mathbf{0}_{m-1} \\ \mathcal{B}_{m-1}(n-1) \end{bmatrix} \tag{15.57}$$

The parameters $\Delta_{m-1}(n)$ and $\Delta'_{m-1}(n)$, defined by Eqs. (15.51) and (15.56), respectively, are in actual fact the complex conjugate of one another; that is,

$$\Delta'_{m-1}(n) = \Delta^*_{m-1}(n) \tag{15.58}$$

where $\Delta^*_{m-1}(n)$ is the complex conjugate of $\Delta_{m-1}(n)$. We prove this relation in three stages:

1. We premultiply both sides of Eq. (15.52) by the row vector

$$[0, \mathbf{c}^H_{m-1}(n-1)]$$

where the superscript H denotes Hermitian transposition. The result of this matrix multiplication is the scalar

$$[0, \mathbf{c}^H_{m-1}(n-1)]\boldsymbol{\Phi}_{m+1}(n)\begin{bmatrix} \mathbf{a}_{m-1}(n) \\ 0 \end{bmatrix} = [0, \mathbf{c}^H_{m-1}(n-1)]\begin{bmatrix} \mathcal{F}_{m-1}(n) \\ \mathbf{0}_{m-1} \\ \Delta_{m-1}(n) \end{bmatrix} \tag{15.59}$$

$$= \Delta_{m-1}(n)$$

where we have used the fact that the last element of $\mathbf{c}_{m-1}(n-1)$ equals unity.

2. We apply Hermitian transposition to both sides of Eq. (15.57), and use the Hermitian property of the correlation matrix $\boldsymbol{\Phi}_{m+1}(n)$, thereby obtaining

$$[0, \mathbf{c}^H_{m-1}(n-1)]\boldsymbol{\Phi}_{m+1}(n) = [\Delta'^*_{m-1}(n), \mathbf{0}^T_{m-1}, \mathcal{B}_{m-1}(n-1)]$$

where $\Delta'^*_{m-1}(n)$ is the complex conjugate of $\Delta'_{m-1}(n)$, and $\mathcal{B}_{m-1}(n-1)$ is real valued. Next we use this relation to evaluate the scalar

$$[0, \mathbf{c}^H_{m-1}(n-1)]\boldsymbol{\Phi}_{m+1}(n)\begin{bmatrix} \mathbf{a}_{m-1}(n) \\ 0 \end{bmatrix}$$

$$= [\Delta'^*_{m-1}(n), \mathbf{0}^T_{m-1}, \mathcal{B}_{m-1}(n-1)]\begin{bmatrix} \mathbf{a}_{m-1}(n) \\ 0 \end{bmatrix} \tag{15.60}$$

$$= \Delta'^*_{m-1}(n)$$

where we have used the fact that the first element of $\mathbf{a}_{m-1}(n)$ equals unity.

3. Comparison of Eqs. (15.59) and (15.60) immediately yields the relation of Eq. (15.58) between the parameters $\Delta_{m-1}(n)$ and $\Delta'_{m-1}(n)$.

We are now equipped with the relations needed to derive the desired time-update for recursive computation of the parameter $\Delta_{m-1}(n)$.

Consider the m-by-1 tap-weight vector $\mathbf{a}_{m-1}(n-1)$ that pertains to a forward prediction-error filter of order $m-1$, evaluated at time $n-1$. The reason for considering time $n-1$ will become apparent presently. Since the leading element of the vector $\mathbf{a}_{m-1}(n-1)$ equals unity, we may express $\Delta_{m-1}(n)$ as follows [see Eqs. (15.58) and (15.60)]:

$$\Delta_{m-1}(n) = [\Delta_{m-1}(n), \mathbf{0}^T_{m-1}, \mathcal{B}_{m-1}(n-1)]\begin{bmatrix} \mathbf{a}_{m-1}(n-1) \\ 0 \end{bmatrix} \tag{15.61}$$

Taking the Hermitian transpose of both sides of Eq. (15.57), recognizing the Hermitian property of $\mathbf{\Phi}_{m+1}(n)$, and using the relation of Eq. (15.58), we get

$$[0, \mathbf{c}^H_{m-1}(n-1)]\,\mathbf{\Phi}_{m+1}(n) = [\Delta_{m-1}(n), \mathbf{0}^T_{m-1}, \mathcal{B}_{m-1}(n-1)] \tag{15.62}$$

Hence, substitution of Eq. (15.62) in (15.61) yields

$$\Delta_{m-1}(n) = [0, \mathbf{c}^H_{m-1}(n-1)]\mathbf{\Phi}_{m+1}(n)\begin{bmatrix} \mathbf{a}_{m-1}(n-1) \\ 0 \end{bmatrix} \tag{15.63}$$

But the correlation matrix $\mathbf{\Phi}_{m+1}(n)$ may be time-updated as follows [see Eq. (13.12)]:

$$\mathbf{\Phi}_{m+1}(n) = \lambda\mathbf{\Phi}_{m+1}(n-1) + \mathbf{u}_{m+1}(n)\mathbf{u}^H_{m+1}(n) \tag{15.64}$$

Accordingly, we may use this relation for $\mathbf{\Phi}_{m+1}(n)$ to rewrite Eq. (15.63) as

$$\Delta_{m-1}(n) = \lambda[0, \mathbf{c}^H_{m-1}(n-1)]\mathbf{\Phi}_{m+1}(n-1)\begin{bmatrix} \mathbf{a}_{m-1}(n-1) \\ 0 \end{bmatrix}$$
$$+ [0, \mathbf{c}^H_{m-1}(n-1)]\mathbf{u}_{m+1}(n)\mathbf{u}^H_{m+1}(n)\begin{bmatrix} \mathbf{a}_{m-1}(n-1) \\ 0 \end{bmatrix} \tag{15.65}$$

Next we recognize from the definition of forward *a priori* prediction error that

$$\mathbf{u}^H_{m+1}(n)\begin{bmatrix} \mathbf{a}_{m-1}(n-1) \\ 0 \end{bmatrix} = [\mathbf{u}^H_m(n), u*(n-m)]\begin{bmatrix} \mathbf{a}_{m-1}(n-1) \\ 0 \end{bmatrix}$$
$$= \mathbf{u}^H_m(n)\mathbf{a}_{m-1}(n-1) \tag{15.66}$$
$$= \eta^*_{m-1}(n)$$

and from the definition of the backward *a posteriori* prediction error that

$$[0, \mathbf{c}^H_{m-1}(n-1)]\mathbf{u}_{m+1}(n) = [0, \mathbf{c}^H_{m-1}(n-1)]\begin{bmatrix} u(n) \\ \mathbf{u}_m(n-1) \end{bmatrix}$$
$$= \mathbf{c}^H_{m-1}(n-1)\mathbf{u}_m(n-1) \tag{15.67}$$
$$= b_{m-1}(n-1)$$

Also, by substituting $n - 1$ for n in Eq. (15.52), we have

$$\Phi_{m+1}(n - 1)\begin{bmatrix} \mathbf{a}_{m-1}(n - 1) \\ 0 \end{bmatrix} = \begin{bmatrix} \mathscr{F}_{m-1}(n - 1) \\ \mathbf{0}_{m-1} \\ \Delta_{m-1}(n - 1) \end{bmatrix}$$

Hence, using this relation and the fact that the last element of the tap-weight vector $\mathbf{c}_{m-1}(n - 1)$, pertaining to the backward prediction-error filter, equals unity, we may write the first term on the right-hand side of Eq. (15.65), except for λ, as

$$[0, \mathbf{c}_{m-1}^H(n - 1)]\Phi_{m+1}(n - 1)\begin{bmatrix} \mathbf{a}_{m-1}(n - 1) \\ 0 \end{bmatrix}$$

$$= [0, \mathbf{c}_{m-1}^H(n - 1)]\begin{bmatrix} \mathscr{F}_{m-1}(n - 1) \\ \mathbf{0}_{m-1} \\ \Delta_{m-1}(n - 1) \end{bmatrix} \qquad (15.68)$$

$$= \Delta_{m-1}(n - 1)$$

Finally, substituting Eqs. (15.66), (15.67), and (15.68) in (15.65), we may express the time-update recursion for $\Delta_{m-1}(n)$ simply as

$$\Delta_{m-1}(n) = \lambda\Delta_{m-1}(n - 1) + b_{m-1}(n - 1)\eta_{m-1}^*(n) \qquad (15.69)$$

which is the desired result. Note that Eq. (15.69) for $\Delta_{m-1}(n)$ is similar to Eq. (15.11) for $\mathscr{F}_m(n)$ and Eq. (15.22) for $\mathscr{B}_m(n)$ in that in each of these three updates, the correction term involves the product of an *a posteriori* and an *a priori* prediction error.

Exact Decoupling Property of the Least-Squares Lattice Predictor

Another important property of a least-squares lattice predictor consisting of m stages is that the backward prediction errors $b_0(n), b_1(n), \ldots, b_m(n)$ produced at the various stages of the predictor are *uncorrelated* (orthogonal) with each other in a time-averaged sense at all instants of time. In other words, the least-squares lattice predictor transforms a correlated input data sequence $\{u(n), u(n - 1), \ldots, u(n - m)\}$ into a new sequence of uncorrelated backward prediction errors, as shown by

$$\{u(n), u(n-1), \ldots, u(n - m)\} \rightleftharpoons \{b_0(n), b_1(n), \ldots, b_m(n)\} \qquad (15.70)$$

The transformation shown here is *reciprocal*, which means that the least-squares lattice predictor preserves the full information content of the input data.

Consider a backward prediction-error filter of order m. Let the $(m + 1)$-by-1 tap-weight vector of the filter, optimized in the least-squares sense over the time interval $1 \leq i \leq n$, be denoted by $\mathbf{c}_m(n)$. In expanded form, we have

$$\mathbf{c}_m(n) = [c_{m,m}(n), c_{m,m-1}(n), \ldots, 1]^T$$

Let $b_m(i)$ denote the backward *a posteriori* prediction error produced at the output of the filter in response to the $(m + 1)$-by-1 input vector $\mathbf{u}_{m+1}(i)$. The expanded form of the input vector is shown by

$$\mathbf{u}_{m+1}(i) = [u(i), u(i - 1), \ldots, u(i - m)]^T, \qquad i > m$$

We may thus express the error $b_m(i)$ as

$$b_m(i) = \mathbf{c}_m^H(n)\mathbf{u}_{m+1}(i)$$

$$= \sum_{k=0}^{m} c_{m,k}^*(n)u(i - m + k) \qquad \begin{matrix} m < i \leq n \\ m = 0, 1, 2, \ldots \end{matrix} \qquad (15.71)$$

Let $\mathbf{b}_{m+1}(i)$ denote the $(m + 1)$-by-1 backward *a posteriori* prediction-error vector, defined by

$$\mathbf{b}_{m+1}(i) = [b_0(i), b_1(i), \ldots, b_m(i)]^T, \qquad \begin{matrix} m < i \leq n \\ m = 0, 1, 2, \ldots \end{matrix}$$

Substituting Eq. (15.71) into this vector, we may express the transformation of the input data into the corresponding set of backward *a posteriori* prediction errors as follows:

$$\mathbf{b}_{m+1}(i) = \mathbf{L}_m(n)\mathbf{u}_{m+1}(i) \qquad (15.72)$$

where the $(m + 1)$-by-$(m + 1)$ *transformation matrix* $\mathbf{L}_m(n)$ is defined by the lower triangular matrix:

$$\mathbf{L}_m(n) = \begin{bmatrix} 1 & 0 & \cdots & 0 \\ c_{1,1}^*(n) & 1 & \cdots & 0 \\ \cdot & \cdot & \cdot & \cdot \\ \cdot & \cdot & \cdot & \cdot \\ \cdot & \cdot & \cdot & \cdot \\ c_{m,m}^*(n) & c_{m,m-1}^* & \cdots & 1 \end{bmatrix} \qquad (15.73)$$

The subscript m in the symbol $\mathbf{L}_m(n)$ refers to the highest order of backward prediction-error filter involved in its constitution. Note also the following points:

- The nonzero elements of row l of matrix $\mathbf{L}_m(n)$ are defined by the tap weights of a backward prediction-error filter of order $(l - 1)$.
- The diagonal elements of matrix $\mathbf{L}_m(n)$ equal unity; this follows from the fact that the last tap weight of a backward prediction-error filter equals unity.
- The determinant of matrix $\mathbf{L}_m(n)$ equals one for all m; hence, the inverse matrix $\mathbf{L}_m^{-1}(n)$ exists, and the reciprocal nature of the transformation in Eq. (15.70) is confirmed.

By definition, the correlation between the backward prediction errors pertaining to different orders, k and m, say, is given by the exponentially weighted time average

$$\phi_{km}(n) = \sum_{i=1}^{n} \lambda^{n-i} b_k(i) b_m^*(i)$$

$$= \sum_{i=1}^{n} \lambda^{n-i} \mathbf{c}_k^H(n) \mathbf{u}_k(i) b_m^*(i) \qquad (15.74)$$

$$= \mathbf{c}_k^H(n) \sum_{i=1}^{n} \lambda^{n-i} \mathbf{u}_k(i) b_m^*(i)$$

where $\mathbf{c}_k(n)$ is the tap-weight vector of the backward prediction-error filter of order k that is responsible for generating the error $b_k(n)$. Without loss of generality, we may assume that $m > k$. Then, recognizing that the elements of the input vector $\mathbf{u}_k(n)$ are involved in generating the backward prediction represented by the error $b_m(n)$, optimized in the least-squares sense, we readily deduce from the principle of orthogonality that the correlation $\phi_{km}(n)$ is zero for $m > k$. In other words, for $m \neq k$, the backward prediction errors $b_k(n)$ and $b_m(n)$ are uncorrelated with each other in a time-averaged sense.

This remarkable property makes the least-squares lattice predictor an ideal device for *exact least-squares joint-process estimation*. Specifically, we may exploit the sequence of backward prediction errors produced by the lattice structure of Fig. 15.6 to perform the least-squares estimation of a desired response in the order-recursive manner described in Fig. 15.7. In particular, for order (stage) m we may write

$$e_m(n) = e_{m-1}(n) - \kappa_{m-1}^*(n) b_{m-1}(n), \qquad m = 1, 2, \ldots, M + 1 \qquad (15.75)$$

For the *initial condition* of joint-process estimation, we have

$$e_0(n) = d(n) \qquad (15.76)$$

The parameters $\kappa_{m-1}(n)$, $m = 1, 2, \ldots, M + 1$ are called *joint-process regression coefficients*. Thus, the least-squares estimation of desired response $d(n)$ may proceed on a stage-by-stage basis, alongside the linear prediction process.

Equation (15.75) represents a *single-order linear combiner*, as depicted in Fig. 15.5(c). Here we have purposely used the time index i in place of n for the estimation variables in order to be consistent with the notations used in parts (a) and (b) of the figure. The point to note here is that $b_{m-1}(i)$ may be viewed as the input and $e_{m-1}(i)$ as the desired response for $1 \leq i \leq n$. (The symbol used to denote the joint-process regression coefficient in Fig. 15.5(c) should not be confused with that used in Chapter 6 to denote the reflection coefficient of a lattice predictor.)

15.5 ANGLE-NORMALIZED ESTIMATION ERRORS

The formulation of the least-squares lattice predictor presented in the previous section was based on the forward *a posteriori* prediction error $f_m(n)$ and the backward *a posteriori* prediction error $b_m(n)$. The resulting order-recursive relations are defined in terms of the cur-

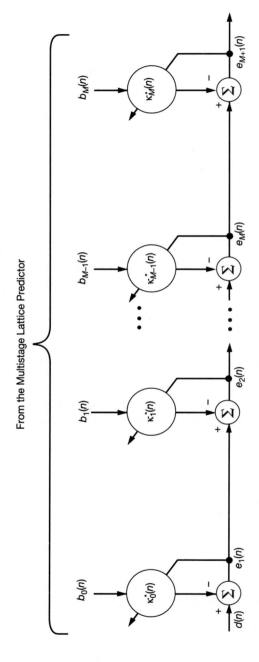

Figure 15.7 Joint-process estimation using a sequence of backward prediction errors.

rent value of forward reflection coefficient $\kappa_{f,m}(n)$ and the current value of backward reflection coefficient $\kappa_{b,m}(n)$. We may equally well formulate the least-squares lattice predictor in terms of the forward *a priori* prediction error $\eta_m(n)$ and the backward *a priori* prediction error $\beta_m(n)$. In the latter case, the order-recursive relations are defined in terms of the past value of the forward reflection coefficient $\kappa_{f,m}(n - 1)$ and the past value of the backward reflection coefficient $\kappa_{b,m}(n - 1)$.

From a developmental point of view, it would be highly desirable to formulate the least-squares lattice prediction problem in a way that is invariant to the choice of *a posteriori* or *a priori* prediction errors. This objective is attained by introducing the notion of *angle-normalized estimation errors*. Specifically, with three different forms of estimation in mind, we define the following set of angle-normalized estimation errors for a least-squares lattice predictor of order m:

- *Angle-normalized forward prediction error*:

$$\epsilon_{f,m}(n) = \gamma_m^{1/2}(n - 1)\eta_m(n) = \frac{f_m(n)}{\gamma_m^{1/2}(n - 1)} \tag{15.77}$$

  where $\gamma_m(n - 1)$ is the past value of the conversion factor.

- *Angle-normalized backward prediction error*:

$$\epsilon_{b,m}(n) = \gamma_m^{1/2}(n)\beta_m(n) = \frac{b_m(n)}{\gamma_m^{1/2}(n)} \tag{15.78}$$

  where $\gamma_m(n)$ is the current value of the conversion factor.

- *Angle-normalized joint-process estimation error*:

$$\epsilon_m(n) = \gamma_m^{1/2}(n)\xi_m(n) = \frac{e_m(n)}{\gamma_m^{1/2}(n)} \tag{15.79}$$

where $e_m(n)$ and $\xi_m(n)$ are the *a posteriori* and *a priori* values of the joint-process estimation error, respectively.

The term "angle" used in these definitions refers to an interpretation of the conversion factor as the cosine of an angle; see Section 15.3 for more details. In any event, the important point to note here is that in basing the algorithmic development of least-squares lattice filtering on angle-normalized estimation errors, we no longer have to distinguish between the *a posteriori* and *a priori* versions of the different estimation errors.

15.6 FIRST-ORDER STATE-SPACE MODELS FOR LATTICE FILTERING

With the background material presented in the previous sections at our disposal, we are ready to embark on the derivation of algorithms for the design of order-recursive adaptive filters based on least-squares estimation. The approach used here builds on the one-to-one correspondences between RLS variables and Kalman variables. To do so, we clearly need to formulate state-space representations for least-squares prediction using a lattice structure and its extension for joint-process estimation. For reasons explained in the previous section, we wish to formulate the state-space models in terms of angle-normalized estimation errors.

Consider the following two n-by-1 vectors of angle-normalized prediction errors of order $m - 1$:

$$\boldsymbol{\epsilon}_{f,m-1}(n) = \begin{bmatrix} \epsilon_{f,m-1}(1) \\ \epsilon_{f,m-1}(2) \\ \cdot \\ \cdot \\ \cdot \\ \epsilon_{f,m-1}(n) \end{bmatrix}, \, \boldsymbol{\epsilon}_{b,m-1}(n-1) = \begin{bmatrix} 0 \\ \epsilon_{b,m-1}(1) \\ \cdot \\ \cdot \\ \cdot \\ \epsilon_{b,m-1}(n-1) \end{bmatrix} \quad (15.80)$$

For initialization of the least-squares lattice predictor, we typically set

$$\mathscr{F}_{m-1}(0) = \mathscr{B}_{m-1}(-1) = \delta$$

$$\Delta_{m-1}(0) = 0$$

The constant δ is usually small enough to have a negligible effect on $\mathscr{F}_{m-1}(n)$ and $\mathscr{B}_{m-1}(n-1)$, and so it may be ignored. Accordingly, we may draw the following conclusions from Eqs. (15.11), (15.22), and (15.69), respectively:

1. The sum of weighted forward prediction-error squares $\mathscr{F}_{m-1}(n)$ is equal to the exponentially weighted squared norm of the corresponding angle-normalized vector $\boldsymbol{\epsilon}_{f,m-1}(n)$:

$$\mathscr{F}_{m-1}(n) = \boldsymbol{\epsilon}_{f,m-1}^H(n) \, \boldsymbol{\Lambda}(n) \, \boldsymbol{\epsilon}_{f,m-1}(n) \quad (15.81)$$

where $\boldsymbol{\Lambda}(n)$ is the n-by-n *exponential weighting matrix*:

$$\boldsymbol{\Lambda}(n) = \text{diag}[\lambda^{n-1}, \lambda^{n-2}, \dots, 1]$$

2. The sum of weighted backward prediction-error squares $\mathscr{B}_{m-1}(n-1)$ is equal to the exponentially weighted squared norm of the corresponding angle-normalized vector $\boldsymbol{\epsilon}_{b,m-1}(n-1)$:

$$\mathscr{B}_{m-1}(n-1) = \boldsymbol{\epsilon}_{b,m-1}^H(n-1) \, \boldsymbol{\Lambda}(n) \, \boldsymbol{\epsilon}_{b,m-1}(n-1) \quad (15.82)$$

3. The parameter $\Delta_{m-1}^*(n)$, representing a form of cross-correlation, is equal to the exponentially weighted inner product of the angle-normalized vectors $\boldsymbol{\epsilon}_{b,m-1}(n-1)$ and $\boldsymbol{\epsilon}_{f,m-1}(n)$:

$$\Delta_{m-1}^*(n) = \boldsymbol{\epsilon}_{b,m-1}^H(n-1) \, \boldsymbol{\Lambda}(n) \, \boldsymbol{\epsilon}_{f,m-1}(n) \quad (15.83)$$

These facts mean that the forward reflection coefficient

$$\kappa_{f,m}^*(n) = - \frac{\Delta_{m-1}^*(n)}{\mathscr{B}_{m-1}(n-1)}$$

is also equal to

$$\kappa^*_{f,m}(n) = - \frac{\boldsymbol{\epsilon}^H_{b,m-1}(n-1)\Lambda(n)\boldsymbol{\epsilon}_{f,m-1}(n)}{\boldsymbol{\epsilon}^H_{b,m}(n-1)\Lambda(n)\boldsymbol{\epsilon}_{b,m-1}(n-1)} \tag{15.84}$$

In other words, the scalar $\kappa^*_{f,m}(n)$ can be interpreted as the coefficient that we need in order to project $\boldsymbol{\epsilon}_{f,m-1}(n)$ onto $\boldsymbol{\epsilon}_{b,m-1}(n-1)$. This suggests that we may replace the problem of projecting the *a posteriori* prediction vector $\mathbf{f}_{m-1}(n)$ onto the *a posteriori* prediction vector $\mathbf{b}_{m-1}(n-1)$ by the equivalent problem of projecting the angle-normalized prediction vector $\boldsymbol{\epsilon}_{f,m-1}(n)$ onto the angle-normalized prediction vector $\boldsymbol{\epsilon}_{b,m-1}(n-1)$. A similar conclusion follows for the problem of projecting $\mathbf{b}_{m-1}(n-1)$ onto $\mathbf{f}_{m-1}(n)$ and also for the joint-process estimation problem.

Accordingly, referring to the signal-flow-graphs of Figs. 15.5(a), 15.5(b), and 15.5(c), we may formulate a combination of *three first-order state-space models* for stage m of the least-squares lattice filtering process, based on three projections:

1. For forward linear prediction, $\boldsymbol{\epsilon}_{f,m-1}(n)$ is projected onto $\boldsymbol{\epsilon}_{b,m-1}(n-1)$.
2. For backward linear prediction, $\boldsymbol{\epsilon}_{b,m-1}(n-1)$ is projected onto $\boldsymbol{\epsilon}_{f,m-1}(n)$.
3. For joint-process linear estimation, $\boldsymbol{\epsilon}_{m-1}(n)$ is projected onto $\boldsymbol{\epsilon}_{b,m-1}(n)$.

Thus, bearing in mind the one-to-one correspondences between the Kalman variables and RLS variables that were established in Chapter 13, the state-space characterization of stage m of the least-squares lattice filtering process may be described in three parts as follows (Sayed and Kailath, 1994):

1. *Forward prediction*:

$$x_1(n+1) = \lambda^{-1/2}x_1(n) \tag{15.85}$$

$$y_1(n) = \boldsymbol{\epsilon}^*_{b,m-1}(n-1)x_1(n) + v_1(n) \tag{15.86}$$

where $x_1(n)$ is the state-variable, and the reference signal $y_1(n)$ is defined by

$$y_1(n) = \lambda^{-n/2}\boldsymbol{\epsilon}^*_{f,m-1}(n) \tag{15.87}$$

The scalar measurement noise $v_1(n)$ is a random variable with zero mean and unit variance.

2. *Backward prediction*:

$$x_2(n+1) = \lambda^{-1/2}x_2(n) \tag{15.88}$$

$$y_2(n) = \boldsymbol{\epsilon}^*_{f,m-1}(n)x_2(n) + v_2(n) \tag{15.89}$$

where $x_2(n)$ is the second state-variable, and the second reference signal (observation) $y_2(n)$ is defined by

$$y_2(n) = \lambda^{-n/2}\boldsymbol{\epsilon}^*_{b,m-1}(n) \tag{15.90}$$

As with $v_1(n)$, the scalar measurement noise $v_2(n)$ is a random variable with zero mean and unit variance.

3. *Joint-process estimation*:

$$x_3(n + 1) = \lambda^{-1/2}x_3(n) \tag{15.91}$$

$$y_3(n) = \epsilon^*_{b,m-1}(n)x_3(n) + v_3(n) \tag{15.92}$$

where $x_3(n)$ is the third and final state-variable, and the corresponding reference signal (observation) $y_3(n)$ is defined by

$$y_3(n) = \lambda^{-n/2}\epsilon^*_{m-1}(n) \tag{15.93}$$

As before, the scalar measurement noise $v_3(n)$ is a random variable with zero mean and unit variance. The noise variables $v_1(n)$, $v_2(n)$, and $v_3(n)$ are all independent of each other.

On the basis of the state-space models described above, we may formulate the list of one-to-one correspondences, shown in Table 15.2, between Kalman variables and three sets of least-squares lattice (LSL) variables, assuming a prediction order of $m - 1$. The three sets of LSL variables refer to forward prediction, backward prediction, and joint-process estimation.

The first four lines of Table 15.2 follow readily from the state-space models of Eqs. (15.85) to (15.93) and Table 13.2 listing the one-to-one correspondences between Kalman

TABLE 15.2 SUMMARY OF ONE-TO-ONE CORRESPONDENCES BETWEEN KALMAN VARIABLES AND LSL VARIABLES IN STAGE m OF THE LATTICE PREDICTOR

Kalman Variable	LSL Variable		
	Forward Prediction	Backward Prediction	Joint-process Estimation
$y(n)$	$\lambda^{-n/2}\epsilon^*_{f,m-1}(n)$	$\lambda^{-n/2}\epsilon^*_{b,m-1}(n-1)$	$\lambda^{-n/2}\epsilon^*_{m-1}(n)$
$\mathbf{u}^H(n)$	$\epsilon^*_{b,m-1}(n-1)$	$\epsilon^*_{f,m-1}(n)$	$\epsilon^*_{b,m-1}(n)$
$\hat{\mathbf{x}}(n\|\mathcal{Y}_{n-1})$	$-\lambda^{-n/2}\kappa_{f,m}(n-1)$	$-\lambda^{-n/2}\kappa_{b,m}(n-1)$	$\lambda^{-n/2}\kappa_{m-1}(n-1)$
$\mathbf{K}(n-1)$	$\lambda^{-1}\mathcal{B}^{-1}_{m-1}(n-2)$	$\lambda^{-1}\mathcal{F}^{-1}_{m-1}(n-1)$	$\lambda^{-1}\mathcal{B}^{-1}_{m-1}(n-1)$
$r(n)$	$\dfrac{\gamma_{m-1}(n-1)}{\gamma_m(n-1)}$	$\dfrac{\gamma_{m-1}(n-1)}{\gamma_m(n)}$	$\dfrac{\gamma_{m-1}(n)}{\gamma_m(n)}$
$\alpha(n)$	$\lambda^{-n/2}\gamma^{1/2}_{m-1}(n-1)\,\eta^*_m(n)$	$\lambda^{-n/2}\gamma^{1/2}_{m-1}(n-1)\,\beta^*_m(n)$	$\lambda^{-n/2}\gamma^{1/2}_{m-1}(n)\,\xi^*_m(n)$

variables and RLS variables. To verify the remaining two lines of correspondences in Table 15.2, we may proceed as follows for the case of forward linear prediction:

1. According to Kalman filter theory, the innovation $\alpha(n)$ is defined by

$$\alpha(n) = y(n) - \mathbf{u}^H(n)\hat{\mathbf{x}}(n|\mathcal{Y}_{n-1}) \tag{15.94}$$

From the first three lines of Table 15.2, we have the following one-to-one correspondences for forward prediction:

$$y(n) \leftrightarrow \lambda^{-n/2}\, \epsilon^*_{f,m-1}(n)$$

$$\mathbf{u}^H(n) \leftrightarrow \epsilon^*_{b,m-1}(n-1)$$

$$\hat{\mathbf{x}}(n|\mathcal{Y}_{n-1}) \leftrightarrow -\lambda^{-n/2}\, \kappa_{f,m}(n-1)$$

where the double-headed arrows signify one-to-one correspondences. Therefore, substituting these correspondences into the right-hand side of Eq. (15.94), we get

$$\alpha(n) \leftrightarrow \lambda^{-n/2}(\epsilon^*_{f,m-1}(n) + \kappa_{f,m}(n)\, \epsilon^*_{b,m-1}(n-1))$$

Next, using the relations of Eqs. (15.77) and (15.78), we may equivalently write

$$\alpha(n) \leftrightarrow \lambda^{-n/2}\gamma^{1/2}_{m-1}(n-1)(\eta^*_{m-1}(n) + \kappa_{f,m}(n-1)\beta^*_{m-1}(n-1)) \tag{15.95}$$

We now recognize the following input–output relations for stage m of a lattice predictor described in terms of *a priori* prediction errors:

$$\eta_m(n) = \eta_{m-1}(n) + \kappa^*_{f,m}(n-1)\beta_{m-1}(n-1) \tag{15.96}$$

$$\beta_m(n) = \beta_{m-1}(n-1) + \kappa^*_{b,m}(n-1)\eta_{m-1}(n) \tag{15.97}$$

The fundamental difference between Eqs. (15.96) and (15.97) and their counterparts, Eqs. (15.41) and (15.42), is that the former pair is based on past values $\kappa_{f,m}(n-1)$ and $\kappa_{b,m}(n-1)$ of the forward and backward reflection coefficients, whereas the latter pair is based on their current values $\kappa_{f,m}(n)$ and $\kappa_{b,m}(n)$. Thus, in light of Eq. (15.96) we conclude from (15.95) that for forward prediction of order $m-1$:

$$\alpha(n) \leftrightarrow \lambda^{-n/2}\gamma^{1/2}_{m-1}(n-1)\, \eta^*_m(n) \tag{15.98}$$

2. According to Kalman filter theory, the filtered estimation error is given by

$$e(n) = y(n) - \mathbf{u}^H(n)\hat{\mathbf{x}}(n|\mathcal{Y}n) \tag{15.99}$$

where the filtered state vector is defined by

$$\hat{\mathbf{x}}(n|\mathcal{Y}_n) = \mathbf{F}(n, n+1)\hat{\mathbf{x}}(n+1)|\mathcal{Y}_n)$$

For the problem at hand, the transition matrix is [see Eq. (15.85)]

$$\mathbf{F}(n+1, n) = \lambda^{-1/2}$$

Since

$$\mathbf{F}(n+1, n)\mathbf{F}(n, n+1) = \mathbf{I}$$

and

$$\hat{\mathbf{x}}(n+1|\mathcal{Y}_n) \leftrightarrow -\lambda^{-(n+1)/2}\kappa_{f,m}(n)$$

it follows that

$$\hat{\mathbf{x}}(n|\mathcal{Y}_n) \leftrightarrow -\lambda^{-n/2}\kappa_{f,m}(n)$$

We may therefore use Eq. (15.99) to write for forward linear prediction:

$$e(n) \leftrightarrow \lambda^{-n/2}\,(\epsilon^*_{f,m-1}(n) + \kappa_{f,m}(n)\,\epsilon^*_{b,m-1}(n-1))$$

Again, using the relations of Eqs. (15.77) and (15.78), we may equivalently write

$$e(n) \leftrightarrow \lambda^{-n/2}\,\gamma^{-1/2}_{m-1}\,(n-1)\,(f^*_{m-1}(n) + \kappa_{f,m}(n)\,b^*_{m-1}(n-1)) \qquad (15.100)$$

But, from Eq. (15.41) we have

$$f_m(n) = f_{m-1}(n) + \kappa^*_{f,m}(n)\,b_{m-1}(n-1)$$

in light of which, we conclude from Eq. (15.100) that for forward prediction of order $m-1$:

$$e(n) \leftrightarrow \lambda^{-n/2}\,\gamma^{-1/2}_{m-1}\,(n-1)\,f^*_m(n) \qquad (15.101)$$

3. In Kalman filter theory, the conversion factor $r^{-1}(n)$ is defined by the ratio $e(n)/\alpha(n)$. We may therefore use (15.98) and (15.101) to write the following one-to-one correspondence for forward prediction of order $m-1$:

$$r(n) \leftrightarrow \frac{\gamma_{m-1}(n-1)}{\gamma_m(n-1)} \qquad (15.102)$$

Thus, Eqs. (15.102) and (15.98) provide the basis for the last two lines of correspondences between the Kalman and LSL variables for forward prediction listed in Table 15.2. By proceeding in a manner similar to that described above, we may fill in the remaining correspondences pertaining to backward prediction and joint-process estimation; this is left as an exercise for the reader.

15.7 QR-DECOMPOSITION–BASED LEAST-SQUARES LATTICE FILTERS

Equipped with the state-space models for least-squares lattice filtering described in Section 15.6 and the square-root information (Kalman) filter developed in Chapter 14, we are at long last positioned to get on with the derivation of our first and most important order-recursive adaptive filter. The writings of arrays for the filter and their expansions are presented in three parts, dealing with adaptive forward prediction, adaptive backward prediction, and adaptive joint-process estimation, in that order.

Array for Adaptive Forward Prediction

Adapting Eq. (14.39) to suit the forward prediction model described by the state-space equations (15.85) to (15.87), with the aid of Table 15.2 listing the one-to-one correspondences between the Kalman variables and LSL variables (for forward prediction), we may write the following array for stage m of the least-squares lattice predictor (Sayed and Kailath, 1994):

$$
\begin{bmatrix}
\lambda^{1/2}\mathcal{B}_{m-1}^{1/2}(n-2) & \epsilon_{b,m-1}(n-1) \\
\lambda^{1/2}p_{f,m-1}^{*}(n-1) & \epsilon_{f,m-1}(n) \\
0 & \gamma_{m-1}^{1/2}(n-1)
\end{bmatrix}
\Theta_{b,m-1}(n-1)
\tag{15.103}
$$

$$
=
\begin{bmatrix}
\mathcal{B}_{m-1}^{1/2}(n-1) & 0 \\
p_{f,m-1}^{*}(n) & \epsilon_{f,m}(n) \\
b_{m-1}^{*}(n-1)\mathcal{B}_{m-1}^{-1/2}(n-1) & \gamma_{m}^{1/2}(n-1)
\end{bmatrix}
$$

where we have done the following things. First, the common factors $\lambda^{1/2}$ and $\lambda^{-n/2}$ have been cancelled from the pre- and postarrays in the first and second rows, respectively. Second, we have multiplied the pre- and postarray in the third row by $\gamma_{m-1}^{1/2}(n-1)$. The scalar quantities $\mathcal{B}_{m-1}(n-1)$ and $p_{f,m-1}(n)$ appearing in the postarray are defined as follows:

1. The real-valued quantity $\mathcal{B}_{m-1}(n-1)$ is the autocorrelation of the delayed, angle-normalized backward prediction error $\epsilon_{b,m-1}(n-1)$ for a lag of zero:

    $$
    \mathcal{B}_{m-1}(n-1) = \sum_{i=1}^{n-1} \lambda^{n-1-i}\epsilon_{b,m-1}(i-1)\epsilon_{b,m-1}^{*}(i-1)
    \tag{15.104}
    $$

    $$
    = \lambda\mathcal{B}_{m-1}(n-2) + \epsilon_{b,m-1}(n-1)\epsilon_{b,m-1}^{*}(n-1)
    $$

    The quantity $\mathcal{B}_{m-1}(n-1)$ may also be interpreted as the minimum value of the sum of weighted backward *a posteriori* prediction-error squares, which is defined in accordance with RLS theory as follows [see Eq. (15.22)]:

    $$
    \mathcal{B}_{m-1}(n-1) = \lambda\mathcal{B}_{m-1}(n-2) + \beta_{m-1}(n-1)b_{m-1}^{*}(n-1) \tag{15.105}
    $$

    Note that the product term $\beta_{m-1}(n-1)b_{m-1}^{*}(n-1)$ is always real, in that we can write

    $$
    \beta_{m-1}(n-1)b_{m-1}^{*}(n-1) = \beta_{m-1}^{*}(n-1)b_{m-1}(n-1)
    $$

2. The complex-valued quantity $p_{f,m-1}(n)$ is, except for the factor $\mathcal{B}_{m-1}^{-1/2}(n-1)$, the cross-correlation between the angle-normalized forward and backward prediction errors, as shown by

    $$
    p_{f,m-1}(n) = \frac{\Delta_{m-1}(n)}{\mathcal{B}_{m-1}^{1/2}(n-1)}
    \tag{15.106}
    $$

where

$$\Delta_{m-1}(n) = \sum_{i=1}^{n} \lambda^{n-i} \epsilon_{b,m-1}(i-1) \epsilon_{f,m-1}^{*}(i) \tag{15.107}$$

$$= \lambda \Delta_{m-1}(n-1) + \epsilon_{b,m-1}(n-1) \epsilon_{f,m-1}^{*}(n)$$

Indeed $p_{f,m-1}(n)$ is related to the forward reflection coefficient $\kappa_{f,m}(n)$ for prediction order m as follows (Haykin, 1991):

$$\kappa_{f,m}(n) = -\frac{\Delta_{m-1}(n)}{\mathcal{B}_{m-1}(n-1)}$$

$$= -\frac{p_{f,m-1}(n)}{\mathcal{B}_{m-1}^{1/2}(n-1)} \tag{15.108}$$

The 2-by-2 matrix $\Theta_{b,m-1}(n-1)$ in Eq. (15.103) is any unitary rotation that reduces the (1,2) entry in the postarray to zero; that is, it annihilates the entry $\epsilon_{b,m-1}(n-1)$ in the prearray. This requirement is readily satisfied by using a Givens rotation, as described here:

$$\Theta_{b,m-1}(n-1) = \begin{bmatrix} c_{b,m-1}(n-1) & -s_{b,m-1}(n-1) \\ s_{b,m-1}^{*}(n-1) & c_{b,m-1}(n-1) \end{bmatrix} \tag{15.109}$$

where the cosine and sine parameters are themselves defined by

$$c_{b,m-1}(n-1) = \frac{\lambda^{1/2} \mathcal{B}_{m-1}^{1/2}(n-2)}{\mathcal{B}_{m-1}(n-1)} \tag{15.110}$$

$$s_{b,m-1}(n-1) = \frac{\epsilon_{b,m-1}(n-1)}{\mathcal{B}_{m-1}^{1/2}(n-1)} \tag{15.111}$$

Thus, using Eq. (15.109) in (15.103), we get the following update relations in addition to that of Eq. (15.104):

$$p_{f,m-1}^{*}(n) = c_{b,m-1}(n-1)\lambda^{1/2} p_{f,m-1}^{*}(n-1) + s_{b,m-1}^{*}(n-1)\epsilon_{f,m-1}(n) \tag{15.112}$$

$$\epsilon_{f,m}(n) = c_{b,m-1}(n-1)\epsilon_{f,m-1}(n) - s_{b,m-1}(n-1)\lambda^{1/2} p_{f,m-1}^{*}(n-1) \tag{15.113}$$

$$\gamma_{m}^{1/2}(n-1) = c_{b,m-1}(n-1)\gamma_{m-1}^{1/2}(n-1) \tag{15.114}$$

Equations (15.104) and (15.110) to (15.114) constitute the set of relations for a square-root information filtering solution to the adaptive forward linear problem in a least-squares lattice sense.

Array for Adaptive Backward Prediction

Consider next the adaptive backward prediction model described by the state-space equations (15.88) to (15.90). Then, with the aid of the one-to-one correspondences between the Kalman variables and LSL variables (for backward prediction) listed in Table 15.2, we may use Eq. (14.39) to write the following array for stage m of the least-squares lattice predictor (Sayed and Kailath, 1994):

$$\begin{bmatrix} \lambda^{1/2}\mathscr{F}^{1/2}_{m-1}(n-1) & \epsilon_{f,m-1}(n) \\ \lambda^{1/2}p^{*}_{b,m-1}(n-1) & \epsilon_{b,m-1}(n-1) \\ 0 & \gamma^{1/2}_{m-1}(n-1) \end{bmatrix} \Theta_{f,m-1}(n) = \begin{bmatrix} \mathscr{F}^{1/2}_{m-1}(n) & 0 \\ p^{*}_{b,m-1}(n) & \epsilon_{b,m}(n) \\ f^{*}_{m-1}(n)\mathscr{F}^{-1/2}_{m-1}(n) & \gamma^{1/2}_{m}(n) \end{bmatrix}$$

(15.115)

The two new scalar quantities $\mathscr{F}_{m-1}(n)$, and $p_{b,m-1}(n)$ appearing in the postarray of Eq. (15.115) are defined as follows:

1. The real-valued quantity $\mathscr{F}_{m-1}(n)$ is the autocorrelation of the angle-normalized forward prediction error $\epsilon_{f,m-1}(n)$ for a lag of zero:

$$\mathscr{F}_{m-1}(n) = \sum_{i=1}^{n} \lambda^{n-i}\epsilon_{f,m-1}(i)\epsilon^{*}_{f,m-1}(i)$$

$$= \lambda\mathscr{F}_{m-1}(n-1) + \epsilon_{f,m-1}(n)\epsilon^{*}_{f,m-1}(n)$$

(15.116)

It may also be interpreted as the minimum value of the sum of forward prediction-error squares, which is defined in accordance with the RLS theory as follows [see Eq. (15.11)]:

$$\mathscr{F}_{m-1}(n) = \lambda\mathscr{F}_{m-1}(n-1) + \eta_{m-1}(n)f^{*}_{m-1}(n)$$

(15.117)

As in the case of Eq. (15.105), the product $\eta_{m-1}(n)f^{*}_{m-1}(n)$ is always real-valued.

2. The complex-valued quantity $p_{b,m-1}(n)$ is, except for the factor $\mathscr{F}^{-1/2}_{m-1}(n)$, the complex conjugate of the cross-correlation between the angle-normalized forward and backward prediction errors, defined in Eq. (15.107); that is,

$$p_{b,m-1}(n) = \frac{\Delta^{*}_{m-1}(n)}{\mathscr{F}^{1/2}_{m-1}(n)}$$

(15.118)

The quantity $p_{b,m-1}(n)$ is also related to the backward reflection coefficient for prediction order m by the formula (Haykin, 1991):

$$\kappa_{b,m}(n) = -\frac{\Delta^{*}_{m-1}(n)}{\mathscr{F}_{m-1}(n)}$$

$$= -\frac{p_{b,m-1}(n)}{\mathscr{F}^{1/2}_{m-1}(n)}$$

(15.119)

The 2-by-2 matrix $\Theta_{f,m-1}(n)$ is a unitary rotation that reduces the (1,2) entry of the postarray in Eq. (15.115) to zero; that is, it annihilates the entry $\epsilon_{f,m-1}(n)$ in the prearray of this same equation. This requirement may be satisfied by using a Givens rotation described as follows:

$$\Theta_{f,m-1}(n) = \begin{bmatrix} c_{f,m-1}(n) & -s_{f,m-1}(n) \\ s^{*}_{f,m-1}(n) & c_{f,m-1}(n) \end{bmatrix}$$

(15.120)

where the cosine and sine parameters are themselves defined by

$$c_{f,m-1}(n) = \frac{\lambda^{1/2}\mathcal{F}_{m-1}^{1/2}(n-1)}{\mathcal{F}_{m-1}^{1/2}(n)} \tag{15.121}$$

$$s_{f,m-1}(n) = \frac{\epsilon_{f,m-1}(n)}{\mathcal{F}_{m-1}^{1/2}(n)} \tag{15.122}$$

Thus, substituting Eq. (15.120) in (15.115), we readily deduce the following recursions:

$$p_{b,m-1}^*(n) = c_{f,m-1}(n)\lambda^{1/2}p_{b,m-1}^*(n-1) + s_{f,m-1}^*(n)\epsilon_{b,m-1}(n-1) \tag{15.123}$$

$$\epsilon_{b,m}(n) = c_{f,m-1}(n)\epsilon_{b,m-1}(n-1) - s_{f,m-1}(n)\lambda^{1/2}p_{b,m-1}^*(n-1) \tag{15.124}$$

$$\gamma_m^{1/2}(n) = c_{f,m-1}(n)\gamma_{m-1}^{1/2}(n-1) \tag{15.125}$$

Equations (15.115) and (15.121) to (15.125) constitute the set of recursions for a square-root information filtering solution to the adaptive backward problems in a least-squares lattice sense.

Array for Joint-Process Estimation

Finally, consider the joint-process estimation problem described by the state-space equations (15.91) to (15.93), pertaining to stage m of the least-squares lattice filtering process. Thus, with the aid of the one-to-one correspondences between the Kalman variables and LSL variables (for joint-process estimation) listed in Table 15.2, we may use the array of Eq. (14.39) to write (Sayed and Kailath, 1994)

$$\begin{bmatrix} \lambda^{1/2}\mathcal{B}_{m-1}^{1/2}(n-1) & \epsilon_{b,m-1}(n) \\ \lambda^{1/2}p_{m-1}^*(n-1) & \epsilon_{m-1}(n) \\ 0 & \gamma_{m-1}^{1/2}(n) \end{bmatrix} \Theta_{b,m-1}(n) = \begin{bmatrix} \mathcal{B}_{m-1}^{1/2}(n) & 0 \\ p_{m-1}^*(n) & \epsilon_m(n) \\ b_{m-1}^*(n)\mathcal{B}_{m-1}^{-1/2}(n) & \gamma_m^{1/2}(n) \end{bmatrix} \tag{15.126}$$

In Eq. (15.126), there is only one new quantity that we have to describe, namely, $p_{m-1}(n)$. This new quantity is, except for the factor $\mathcal{B}_{m-1}^{-1/2}(n)$, the cross-correlation between the angle-normalized backward prediction error and angle-normalized joint-estimation error, as shown by

$$p_{m-1}(n) = \frac{1}{\mathcal{B}_{m-1}^{1/2}(n)} \sum_{i=1}^{n} \lambda^{n-i}\epsilon_{b,m-1}(i)\epsilon_{b,m-1}^*(i) \tag{15.127}$$

The regression coefficient $\kappa_{m-1}(n)$ for prediction order $m-1$ is correspondingly defined by (Haykin, 1991)

$$\kappa_{m-1}(n) = \frac{p_{m-1}(n)}{\mathcal{B}_{m-1}^{1/2}(n)} \tag{15.128}$$

The 2-by-2 matrix $\Theta_{b,m-1}(n)$ is a unitary rotation designed to annihilate the entry $\epsilon_{b,m-1}(n)$ in the prearray of Eq. (15.126). To do this, we may use the same Givens rotation as that in Eq. (15.109), except for a shift in time by one unit. Specifically, we may write

$$\Theta_{b,m-1}(n) = \begin{bmatrix} c_{b,m-1}(n) & -s_{b,m-1}(n) \\ s^*_{b,m-1}(n) & c_{b,m-1}(n) \end{bmatrix} \tag{15.129}$$

where

$$c_{b,m-1}(n) = \frac{\lambda^{1/2}\mathcal{B}^{1/2}_{m-1}(n-1)}{\mathcal{B}^{1/2}_{m-1}(n)} \tag{15.130}$$

$$s_{b,m-1}(n) = \frac{\epsilon_{b,m-1}(n)}{\mathcal{B}^{1/2}_{m-1}(n)} \tag{15.131}$$

Hence, substituting Eq. (15.129) in (15.126), we get the following new recursions

$$p^*_{m-1}(n) = c_{b,m-1}(n)\lambda^{1/2}p^*_{m-1}(n-1) + s^*_{b,m-1}(n)\epsilon_{m-1}(n) \tag{15.132}$$

$$\epsilon_m(n) = c_{b,m-1}(n)\epsilon_{m-1}(n) - s_{b,m-1}(n)\lambda^{1/2}p^*_{m-1}(n-1) \tag{15.133}$$

Equations (15.104), (15.110), and (15.111) with time $n-1$ replaced with n, together with Eqs. (15.132) and (15.133), constitute the set of recursions for a square-root information filtering solution to the joint-process estimation problem in a least-squares lattice sense.

Summary of the QRD–LSL Algorithm

Table 15.3 presents a summary of the angle-normalized QRD–LSL algorithm,[2] based on the arrays of Eqs. (15.103), (15.115), and (15.126). Note that the forward and backward predictions are performed for $m = 1, 2, \ldots, M$, whereas those for the joint-process estimation are performed for $m = 1, 2, \ldots, M + 1$, where M is the final prediction order.

[2]The idea of a fast QR-decomposition–based algorithm for recursive least-squares estimation was first presented by Cioffi (1988). A detailed derivation of the algorithm is described in Cioffi (1990). In the latter paper, Cioffi presents a geometric approach for the derivation that is reminiscent of his earlier work on fast transversal filters. The algorithm derived by Cioffi is of a Kalman or matrix-oriented type. Several other authors have presented seemingly simple algebraic derivations and other versions of the QRD-fast RLS algorithm (Bellanger, 1988; Proudler et al., 1988, 1989; Regalia and Bellanger, 1991). The paper by Proudler et al. (1989) is of particular interest in that it develops a novel implementation of the QRD-RLS algorithm using a lattice structure. A similar fast algorithm has been derived independently by Ling (1989) using the modified Gram–Schmidt orthogonalization procedure. The connection between the modified Gram–Schmidt orthogonalization and QR-decomposition is discussed in Shepherd and McWhirter (1991).

Haykin (1991) presented a development of the QRD–LSL algorithm that is based on a hybridization of ideas due to Proudler et al. (1989) and Regalia and Bellanger (1991). Specifically, the development followed Proudler et al. in deriving QR-decomposition–based solutions to forward and backward linear prediction problems, and Regalia and Bellanger in solving the joint-process estimation problem. By so doing, the complications in the procedure by Proudler et al. that uses forward linear prediction errors for joint-process estimation are avoided. The structure of the QRD–LSL algorithm derived by Haykin follows a philosophy directly analogous to that described for conventional LSL algorithms later in the chapter.

TABLE 15.3   SUMMARY OF THE QRD–LSL ALGORITHM

1. *Computations*
   (a) *Predictions*: For each time instant $n = 1, 2, \ldots$, perform the following computations and repeat for each prediction order $m = 1, 2, \ldots, M$, where M is the final prediction order:

$$\begin{bmatrix} \lambda^{1/2}\mathcal{B}_{m-1}^{1/2}(n-2) & \epsilon_{b,m-1}(n-1) \\ \lambda^{1/2}p_{f,m-1}^{*}(n-1) & \epsilon_{f,m-1}(n) \\ 0 & \gamma_{m-1}^{1/2}(n-1) \end{bmatrix} \Theta_{b,m-1}(n-1) = \begin{bmatrix} \mathcal{B}_{m-1}^{1/2}(n-1) & 0 \\ p_{f,m-1}^{*}(n) & \epsilon_{f,m}(n) \\ b_{m-1}^{*}(n-1)\mathcal{B}_{m-1}^{-1/2}(n-1) & \gamma_{m}^{1/2}(n-1) \end{bmatrix}$$

$$\begin{bmatrix} \lambda^{1/2}\mathcal{F}_{m-1}^{1/2}(n-1) & \epsilon_{f,m-1}(n) \\ \lambda^{1/2}p_{b,m-1}^{*}(n-1) & \epsilon_{b,m-1}(n-1) \end{bmatrix} \Theta_{f,m-1}(n) = \begin{bmatrix} \mathcal{F}_{m-1}^{1/2}(n) & 0 \\ p_{b,m-1}^{*}(n) & \epsilon_{b,m}(n) \end{bmatrix}$$

   (b) *Filtering*: For each time instant $n = 1, 2, \ldots$, perform the following computations and repeat for each prediction order $m = 1, 2, \ldots, M + 1$, where M is the final prediction order:

$$\begin{bmatrix} \lambda^{1/2}\mathcal{B}_{m-1}^{1/2}(n-1) & \epsilon_{b,m-1}(n) \\ \lambda^{1/2}p_{m-1}^{*}(n-1) & \epsilon_{m-1}(n) \end{bmatrix} \Theta_{b,m-1}(n) = \begin{bmatrix} \mathcal{B}_{m-1}^{1/2}(n) & 0 \\ p_{m-1}^{*}(n) & \epsilon_{m}(n) \end{bmatrix}$$

2. *Initialization*
   (a) *Auxiliary parameter initialization*: For order $m = 1, 2, \ldots, M$, set

$$p_{f,m-1}(0) = p_{b,m-1}(0) = 0$$

   and for order $m = 1, 2, \ldots, M + 1$, set

$$p_{m-1}(0) = 0$$

   (b) *Soft-constraint initialization*: For order $m = 0, 1, \ldots, M$, set

$$\mathcal{B}_{m}(-1) = \mathcal{B}_{m}(0) = \delta$$

$$\mathcal{F}_{m}(0) = \delta$$

   where δ is a small positive constant.
   (c) *Data initialization*: For $n = 1, 2, \ldots$, compute

$$\epsilon_{f,0}(n) = \epsilon_{b,0}(n) = u(n)$$

$$\epsilon_{0}(n) = d(n)$$

$$\gamma_{0}(n) = 1$$

   where $u(n)$ is the input and $d(n)$ is the desired response at time n.

That is, the joint-process estimation involves one final set of computations pertaining to $m = M + 1$. Note also that in the second and third arrays of Table 15.3, we have omitted the particular rows that involve updating the conversion factor. This requirement is taken care of by the first array.

Table 15.3 includes the *initialization* procedure, which is of a *soft-constraint* form. This form of initialization is consistent with that adopted for the conventional RLS algorithm, as described in Chapter 13. The algorithm proceeds with a set of *initial values* determined by the input datum $u(n)$ and desired response $d(n)$, as shown by

$$\epsilon_{f,0}(n) = \epsilon_{b,0}(n) = u(n)$$

and

$$\epsilon_0(n) = d(n)$$

The initial value for the conversion factor is chosen as

$$\gamma_0(n) = 1$$

15.8 FUNDAMENTAL PROPERTIES OF THE QRD–LSL FILTER

The QRD–LSL algorithm summarized in Table 15.3 is so called in recognition of three facts. First, the unitary transformations in Eqs. (15.103), (15.115), and (15.130), in which the (1,2) entry of each postarray is reduced to zero, are all examples of the *QR-decomposition*. Second, the algorithm is rooted in *recursive least-squares estimation*. Third, the computations performed by the algorithm proceed on a stage-by-stage fashion, with each stage having the form of a *lattice*. By virtue of these facts, the QRD–LSL algorithm is endowed with a highly desirable set of operational and implementational characteristics:

- *Good numerical properties*, which are inherited from the QR-decomposition part of the algorithm
- *Good convergence properties* (i.e., fast rate of convergence, and insensitivity to variations in the eigenvalue spread of the underlying correlation matrix of the input data), which are due to the recursive least-squares nature of the algorithm.
- A *high level of computational efficiency*, which results from the modular, lattice-like structure of the prediction process

The unique combination of these characteristics makes the QRD–LSL algorithm a powerful adaptive filtering algorithm.

The latticelike structure of the QRD–LSL algorithm, using a sequence of Givens rotations, is clearly illustrated by the multistage signal-flow graph of Fig. 15.8. In particular, we see that stage m of the predictor section of the algorithm involves the computations of the angle-normalized prediction errors: $\epsilon_{f,m}(n)$ and $\epsilon_{b,m}(n)$, where the prediction order $m = 1, 2, \ldots, M$. On the other hand, the filtering section of the algorithm involves the computation of the angle-normalized joint-process estimation error $\epsilon_m(n)$, where $m = 1, 2, \ldots, M + 1$. The details of these computations are depicted in the signal-flow graphs shown in Fig. 15.9, which further emphasize the inherent lattice nature of the QRD–LSL algorithm.

The boxes labeled $z^{-1}\mathbf{I}$ in Fig. 15.8 signify *storage*, which is needed to accommodate the fact that the Givens rotations involved in the adaptive forward prediction process are delayed with respect to those involved in the joint-process estimation process by one time unit. Note, however, that the joint-process estimation process involves one last Givens rotation, all by itself.

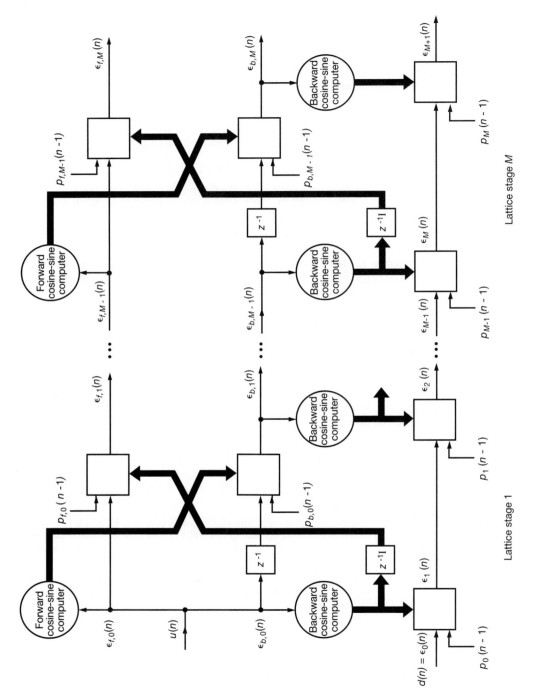

Figure 15.8 Signal-flow graph of the QRD–LSL algorithm.

From the signal-flow graph of Fig. 15.8 we clearly see that the total number of Givens rotations needed for the computation of $\epsilon_{M+1}(n)$ is $2M + 1$, which increases linearly with the final prediction order M. However, the price paid for the high level of computational efficiency of the fast algorithm described herein, compared to the conventional RLS algorithm described in Chapter 13, is that of having to write a more elaborate set of instructions.

Examining the signal-flow graphs of Figs. 15.8 and 15.9, we may identify two different sets of recursions in the formulation of the QRD–LSL algorithm:

1. *Order updates (recursions).* At each stage of the algorithm, *order updates* are performed on the angle-normalized estimation errors. Specifically, a series of m order updates applied to the initial values $\epsilon_{f,0}(n)$ and $\epsilon_{b,0}(n)$ yields the final values $\epsilon_{f,M}(n)$ and $\epsilon_{b,M}(n)$, respectively, where M is the final prediction order. To compute the final value of the angle-normalized joint-process estimation error $\epsilon_{M+1}(n)$, another series of $M + 1$ order updates are applied to the initial value $\epsilon_0(n)$. This latter set of updates includes the use of the sequence of angle-normalized backward prediction errors $\epsilon_{b,0}(n)$, $\epsilon_{b,1}(n)$, . . . , $\epsilon_{b,M}(n)$. The final order update pertains to the computation of the square root of the conversion factor, $\gamma_{M+1}^{1/2}(n)$, which involves the application of $M + 1$ order updates to the initial value $\gamma_0^{1/2}(n)$. The availability of the final values $\epsilon_{M+1}(n)$ makes it possible to compute the final value $e_{M+1}(n)$ of the joint-process estimation error.

2. *Time updates (recursions).* The computations of $\epsilon_{f,m}(n)$ and $\epsilon_{b,m}(n)$ as outputs of predictor stage m of the algorithm involve the use of the auxiliary parameters $p_{f,m-1}(n)$ and $p_{b,m-1}(n)$, respectively, for $m = 1, 2, \ldots, M$. Similarly, the computation of $\epsilon_m(n)$ involves the auxiliary parameter $p_{m-1}(n)$ for $m = 1, 2, \ldots,$ $M + 1$. The computations of these three auxiliary parameters themselves have the following common features:

   • They are all governed by *first-order difference equations.*
   • The coefficients of the equation are *time varying.* For exponential weighting (i.e., $\lambda \leq 1$), the coefficients are bounded in absolute value by one. Hence, the solution of the equation is convergent.
   • The term playing the role of "excitation" is represented by some form of an estimation error.
   • In the prewindowing method, all three auxiliary parameters are equal to zero for $n \leq 0$.

   Consequently, the auxiliary parameters $p_{f,m-1}(n)$ and $p_{b,m-1}(n)$ for $m = 1, 2,$ . . . , M, and $p_m(n)$ for $m = 0, 1, \ldots, M$, may be computed recursively in time.

The auxiliary parameters $p_{f,m-1}(n)$, $p_{b,m-1}(n)$, and $p_{m-1}(n)$ in the QRD–LSL algorithm perform functions that are analogous to those of the forward reflection coefficient $\kappa_{f,m}(n)$, the backward reflection coefficient $\kappa_{b,m}(n)$, and the joint-process regression coefficient $\kappa_{m-1}(n)$, for prediction order m, respectively. Indeed, these three sets of parameters

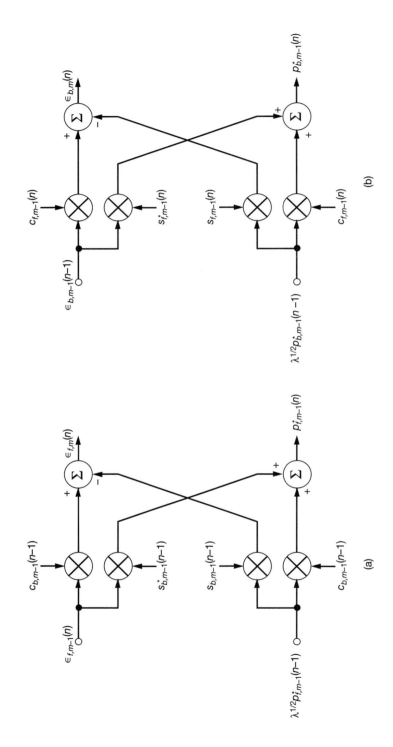

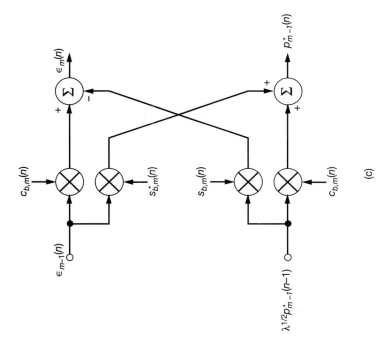

Figure 15.9 Signal-flow graphs for computing normalized variables in the QRD–LSL algorithm: (a) Angle-normalized forward prediction error $\epsilon_{f,m}(n)$. (b) Angle-normalized backward prediction error $\epsilon_{b,m}(n)$. (c) Angle-normalized joint-process estimation error $\epsilon_m(n)$.

(c)

are related to each other by Eqs. (15.108), (15.118), and (15.128), respectively, which enables us to compute $\kappa_{f,m}(n)$, $\kappa_{b,m}(n)$, and $\kappa_{m-1}(n)$ *indirectly*, if so desired.

Finally, and perhaps most importantly, the angle-normalized QRD–LSL algorithm summarized in Table 15.3 plays a central role in the derivation of the whole family of recursive least-squares lattice (LSL) algorithms. We say so because all other existing recursive LSL algorithms using *a posteriori* estimation errors or *a priori* estimation errors (or combinations thereof) may be viewed as rewritings of the QRD–LSL algorithm (Sayed and Kailath, 1994). The validity of this statement is demonstrated in Sections 15.10 and 15.11.

15.9 COMPUTER EXPERIMENT ON ADAPTIVE EQUALIZATION

In this computer experiment, we study the use of the QRD–LSL algorithm for *adaptive equalization* of a linear channel that produces unknown distortion. The parameters of the channel are the same as those used to study the RLS algorithm in Section 13.8 for a similar application. The results of the experiment should therefore help us to make an assessment of the performance of this order-recursive algorithm compared to the standard RLS algorithm.

The parameters of the QRD–LSL algorithm studied here are identical to those used for the RLS algorithm in Section 13.8:

Exponential weighting factor: $\lambda = 1$ (for stationary data)
Prediction order: $M = 10$
Number of equalizer taps: $M + 1 = 11$
Initializing constant: $\delta = 0.004$

The computer simulations were run for four different values of the channel parameter W defined in Eq. (9.105), namely, $W = 2.9, 3.1, 3.3,$ and 3.5. These values of W correspond to the following eigenvalue spreads of the underlying correlation matrix $\mathbf{R}$ of the channel output (equalizer input): $\chi(\mathbf{R}) = 6.0782, 11.1238, 21.7132,$ and 46.8216, respectively. The signal-to-noise ratio measured at the channel output was 30 dB. For more details of the experimental setup, the reader is referred to Sections 9.16 and 13.8.

Learning Curves

Figure 15.10 presents the superposition of learning curves of the QRD–LSL algorithm for the exponential weighting factor $\lambda = 1$, and four different values of the channel parameter $W = 2.9, 3.1, 3.3,$ and 3.5. Each learning curve was obtained by ensemble-averaging the squared value of the final *a priori* estimation error (i.e., the innovation) $\xi_{M+1}(n)$ over 200 independent trials of the experiment for a final prediction order $M = 10$. To compute the *a priori* estimation error $\xi_{M+1}(n)$, we first recognize that

$$\xi_{M+1}(n) = \frac{e_{M+1}(n)}{\gamma_{M+1}(n)}$$

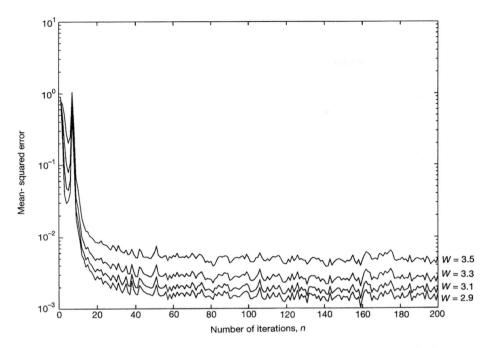

Figure 15.10   Learning curves of the QRD–LSL algorithm for the adaptive equalization experiment.

Equivalently, we may write

$$\xi_{M+1}(n) = \frac{\epsilon_{M+1}(n)}{\gamma_{M+1}^{1/2}(n)}$$

where $\epsilon_{M+1}(n)$ is the final value of the angle-normalized joint-process estimation error, and γ_{M+1} is the associated conversion factor. Note that for a final prediction order M, the number of taps involved in the joint-process estimation is $M + 1$.

For each eigenvalue spread, the learning curve of the QRD–LSL algorithm follows a path practically identical to that of the RLS algorithm, once the initialization is completed. This is readily confirmed by comparing the plots of Fig. 15.10 with those of Fig. 13.6. In both cases, double-precision arithmetic was used so that finite-precision effects are negligible.

Conversion Factor

In Fig. 15.11, we show the superposition of four ensemble-averaged plots of the conversion factor $\gamma_{M+1}(n)$ (for the final stage) versus the number of iterations n, corresponding to the four different values of the eigenvalue spread $\chi(\mathbf{R})$ as defined above. The curves plotted here are obtained by ensemble-averaging $\gamma_{M+1}(n)$ over 200 independent trials of

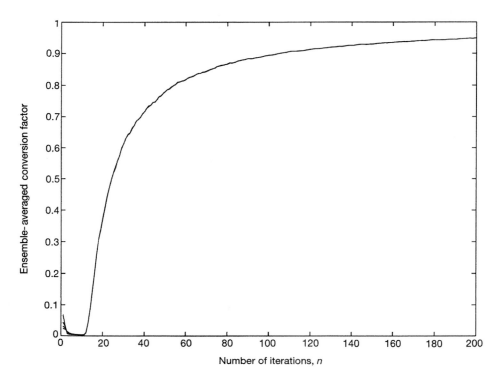

Figure 15.11   Ensemble-averaged conversion factor $\gamma_{M+1}(n)$ for varying eigenvalve spread.

the experiment. It is noteworthy that the time variation of this ensemble-averaged conversion factor $E[\gamma_m(n)]$, follows an *inverse* law, as shown by

$$E[\gamma_m(n)] \simeq 1 - \frac{m}{n} \qquad \text{for } m = 1, 2, \ldots, M + 1, \text{ and } n \geq m$$

This equation provides a good fit to the experimentally computed curve shown in Fig. 15.11 particularly for n large compared to the predictor order $m = M + 1$. The reader is invited to check the validity of this fit. Note that the experimental plots of the conversion factor $\gamma_{M+1}(n)$ are insensitive to variations in the eigenvalue spread of the correlation matrix of the equalizer input for $n \geq 10$.

Impulse Response

In Fig. 15.12 we have plotted the ensemble-averaged impulse response of the adaptive equalizer after $n = 500$ iterations for each of the four eigenvalue spreads. As before, this ensemble-averaging was performed over 200 independent trials of the experiment. The

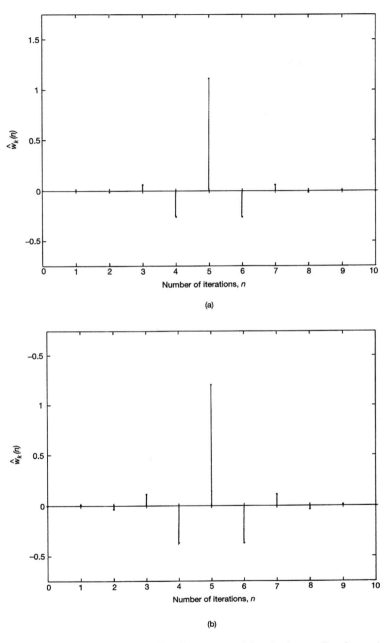

Figure 15.12   Ensemble-averaged impulse response of the adaptive equalizer for varying eigenvalue spread: (a) $W = 2.9$; $\chi(\mathbf{R}) = 6.0782$. (b) $W = 3.1$; $\chi(\mathbf{R}) = 11.1238$. (c) $W = 3.3$; $\chi(\mathbf{R}) = 21.7132$. (d) $W = 3.5$; $\chi(\mathbf{R}) = 46.8216$. Parts (c) and (d) of the figure are presented on the next page.

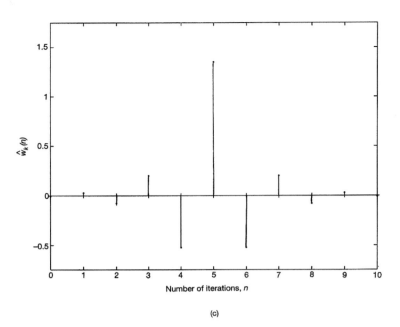

(c)

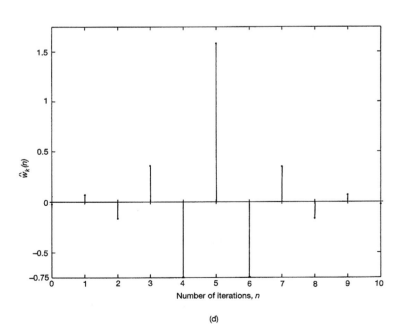

(d)

Figure 15.12   (*contd.*)

results of Fig. 15.12 for the QRD–LSL algorithm are, for all practical purposes, indistinguishable from the corresponding results of Fig. 9.22 for the LMS algorithm for a similar application, except for a scale change.

15.10 EXTENDED QRD–LSL ALGORITHM

The QRD–LSL algorithm summarized in Table 15.3 permits computations of the forward and backward reflection coefficients and joint-process regression coefficients in an indirect manner; see Eqs. (15.108), (15.118), and (15.128). If *direct* computations of these coefficients are desired, we may use the *extended QRD–LSL algorithm* that follows from the use of the extended square-root information filter. Specifically, we may expand the three arrays of Table 15.3 as described here.

1. *Adaptive forward prediction*

$$
\begin{bmatrix}
\lambda^{1/2}\mathscr{B}_{m-1}^{1/2}(n-2) & \epsilon_{b,m-1}(n-1) \\
\lambda^{1/2}p_{f,m-1}^*(n-1) & \epsilon_{f,m-1}(n) \\
0 & \gamma_{m-1}^{1/2}(n-1) \\
\lambda^{-1/2}\mathscr{B}_{m-1}^{-1/2}(n-2) & 0
\end{bmatrix} \mathbf{\Theta}_{b,m-1}(n-1)
$$

$$
=
\begin{bmatrix}
\mathscr{B}_{m-1}^{1/2}(n-1) & 0 \\
p_{f,m-1}^*(n) & \epsilon_{f,m}(n) \\
b_{m-1}^*(n-1)\mathscr{B}_{m-1}^{-1/2}(n-1) & \gamma_m^{1/2}(n-1) \\
\mathscr{B}_{m-1}^{-1/2}(n-1) & -\overline{k}_{b,m-1}(n-1)
\end{bmatrix}
$$

(15.134)

The complex-valued quantity $\overline{k}_{b,m-1}(n-1)$ in the last row of the postarray is the *normalized gain factor*, which is determined by backward prediction variables; it is defined by

$$
\overline{k}_{b,m-1}(n-1) = \gamma_m^{-1/2}(n-1)k_{b,m-1}(n-1)
$$
(15.135)

The normalized gain factor $\overline{k}_{b,m-1}(n-1)$ may be used to update the forward reflection coefficient in accordance with the (Kalman filtering) formula:

$$
\begin{aligned}
\kappa_{f,m}(n) &= \kappa_{f,m}(n-1) + k_{b,m-1}(n-1)\eta_m^*(n) \\
&= \kappa_{f,m}(n-1) + (\gamma_m^{-1/2}(n-1)k_{b,m-1}(n-1))\,(\gamma_m^{1/2}(n-1)\eta_m(n))^* \\
&= \kappa_{f,m}(n-1) + \overline{k}_{b,m-1}(n-1)\epsilon_{f,m}^*(n)
\end{aligned}
$$
(15.136)

where the entries $\overline{k}_{b,m-1}(n-1)$ and $\epsilon_{f,m}(n)$ are read directly from the last and second rows of the postarray in Eq. (15.134).

2. *Adaptive backward prediction*

$$
\begin{bmatrix}
\lambda^{1/2}\mathscr{F}_{m-1}^{1/2}(n-1) & \epsilon_{f,m-1}(n) \\
\lambda^{1/2}p_{b,m-1}^{*}(n-1) & \epsilon_{b,m-1}(n-1) \\
0 & \gamma_{m-1}^{1/2}(n-1) \\
\lambda^{-1/2}\mathscr{F}_{m-1}^{-1/2}(n-1) & 0
\end{bmatrix}
\Theta_{f,m-1}(n)
$$

$$
=
\begin{bmatrix}
\mathscr{F}_{m-1}^{1/2}(n) & 0 \\
p_{b,m-1}^{*}(n) & \epsilon_{b,m}(n) \\
f_{m-1}^{*}(n)\mathscr{F}_{m-1}^{-1/2}(n) & \gamma_{m}^{1/2}(n) \\
\mathscr{F}_{m-1}^{-1/2}(n) & -\overline{k}_{f,m-1}(n)
\end{bmatrix}
$$

$$\tag{15.137}$$

The complex-valued quantity $\overline{k}_{f,m-1}(n)$, appearing in the last row of the postarray, is the normalized gain factor determined by forward prediction variables; it is defined by

$$
\overline{k}_{f,m-1}(n) = \gamma_{m}^{-1/2}(n)k_{f,m-1}(n) \tag{15.138}
$$

As such, it may be used to update the backward reflection coefficient in accordance with the (Kalman filtering) formula:

$$
\begin{aligned}
\kappa_{b,m}(n) &= \kappa_{b,m}(n-1) + k_{f,m-1}(n)\beta_{m}^{*}(n) \\
&= \kappa_{b,m}(n-1) + (\gamma_{m}^{-1/2}(n)k_{f,m-1}(n))\,(\gamma_{m}^{1/2}(n)\beta_{m}(n))^{*} \quad (15.139) \\
&= \kappa_{b,m}(n-1) + \overline{k}_{f,m-1}(n)\epsilon_{b,m}^{*}(n)
\end{aligned}
$$

where the entries $\overline{k}_{f,m-1}(n)$ and $\epsilon_{b,m}(n)$ are read directly from the last and second rows of the postarray in Eq. (15.137).

3. *Joint-process estimation*

$$
\begin{bmatrix}
\lambda^{1/2}\mathscr{B}_{m-1}^{1/2}(n-1) & \epsilon_{b,m-1}(n) \\
\lambda^{1/2}p_{m-1}^{*}(n-1) & \epsilon_{m-1}(n) \\
0 & \gamma_{m-1}^{1/2}(n) \\
\lambda^{-1/2}\mathscr{B}_{m-1}^{-1/2}(n-1) & 0
\end{bmatrix}
\Theta_{b,m-1}(n)
$$

$$
=
\begin{bmatrix}
\mathscr{B}_{m-1}^{1/2}(n) & 0 \\
p_{m-1}^{*}(n) & \epsilon_{m}(n) \\
b_{m-1}^{*}(n)\mathscr{B}_{m-1}^{-1/2}(n) & \gamma_{m}^{1/2}(n) \\
\mathscr{B}_{m-1}^{-1/2}(n) & -\overline{k}_{b,m-1}(n)
\end{bmatrix}
$$

$$\tag{15.140}$$

The regression coefficient $\kappa_{m-1}(n)$ is computed directly in terms of the normalized gain factor $\overline{k}_{b,m-1}(n)$, appearing in the last row of the postarray, as follows:

$$\kappa_{m-1}(n) = \kappa_{m-1}(n-1) + k_{b,m-1}(n)\xi_m^*(n)$$

$$= \kappa_{m-1}(n-1) + (k_{b,m-1}(n)\gamma_m^{-1/2}(n))\,(\xi_m(n)\gamma_m^{1/2}(n))^* \quad (15.141)$$

$$= \kappa_{m-1}(n-1) + \bar{k}_{b,m-1}(n)\epsilon_m^*(n)$$

where the entries $\bar{k}_{b,m-1}(n)$ and $\epsilon_m(n)$ are read directly from the last and second rows of the postarray of Eq. (15.140).

Despite the fact that the extended QRD–LSL algorithm is able to compute the forward and backward reflection coefficients and joint-process regression coefficients directly, the QRD–LSL algorithm of Table 15.3 is to be preferred over it for two practical reasons:

1. An inherently simpler structure
2. A potentially better numerical behavior

The first point is obvious in light of what we have already said. The second point needs to be explained. Each array of the QRD–LSL algorithm propagates a *single* square root, namely, $\mathcal{B}_{m-1}^{1/2}(n-2)$ or $\mathcal{F}_{m-1}^{1/2}(n-1)$. On the other hand, each array of the extended QRD–LSL algorithm propagates one or the other of these two square roots and, in addition, the inverse of that particular square root. Therefore, finite-precision effects may cause a numerical discrepancy to arise between each of these two square roots and its inverse, unless proper precautions are taken.

15.11 RECURSIVE LEAST-SQUARES LATTICE FILTERS USING *A POSTERIORI* ESTIMATION ERRORS

In a generic sense, the family of recursive LSL algorithms may be divided into two subgroups: those that involve the use of unitary rotations, and those that do not. A well-known algorithm that belongs to the latter subgroup is the standard *recursive LSL algorithm using a posteriori estimation errors* (Morf, 1974; Morf and Lee, 1978; Lee et al., 1981). The estimation errors used in this algorithm are represented by the *a posteriori* forward prediction error $f_m(n)$, the *a posteriori* backward prediction error $b_m(n)$, and the *a posteriori* joint-process estimation error $e_m(n)$, where $m = 0, 1, 2, \ldots, M$, as shown in the signal-flow graphs of Figs. 15.6 and 15.7.

We may derive this algorithm (and for that matter, other recursive LSL algorithms) from the QRD–LSL algorithm of Table 15.3 by using a simple two-step procedure:

- The three arrays of the QRD–LSL algorithm, dealing with adaptive forward prediction, adaptive backward prediction, and adaptive joint-process estimation are squared; the effects of unitary rotations are thereby completely removed from the algorithm.

- Certain terms (depending on the algorithm of interest) on both sides of the resultant arrays are retained, and then compared.

Examples of this procedure were presented in Chapter 14 dealing with square-root adaptive filters.

Applying this procedure to the three arrays of Table 15.3 that describe the angle-normalized QRD–LSL algorithm, and expressing the results in terms of the *a posteriori* estimation errors, we get the following three sets of recursions:

1. *Adaptive forward prediction*:

$$\mathscr{B}_{m-1}(n-1) = \lambda \mathscr{B}_{m-1}(n-2) + \frac{|b_{m-1}(n-1)|^2}{\gamma_{m-1}(n-1)} \tag{15.142}$$

$$\Delta_{m-1}(n) = \lambda \Delta_{m-1}(n-1) + \frac{b_{m-1}(n-1)f_{m-1}^*(n)}{\gamma_{m-1}(n-1)} \tag{15.143}$$

$$\kappa_{f,m}(n) = - \frac{\Delta_{m-1}(n)}{\mathscr{B}_{m-1}(n-1)} \tag{15.144}$$

$$f_m(n) = f_{m-1}(n) + \kappa_{f,m}^*(n)b_{m-1}(n-1) \tag{15.145}$$

$$\gamma_m(n-1) = \gamma_{m-1}(n-1) - \frac{|b_{m-1}(n-1)|^2}{\mathscr{B}_{m-1}(n-1)} \tag{15.146}$$

where in the second line we have made use of the relation between $\Delta_{m-1}(n)$ and $p_{f,m-1}(n)$ given in Eq. (15.106), and in the third line we have made use of the definition of the forward reflection coefficient $\kappa_{f,m}(n)$ given in Eq. (15.108).

2. *Adaptive backward prediction*:

$$\mathscr{F}_{m-1}(n) = \lambda \mathscr{F}_{m-1}(n-1) + \frac{|f_{m-1}(n)|^2}{\gamma_{m-1}(n-1)} \tag{15.147}$$

$$\kappa_{b,m}(n) = - \frac{\Delta_{m-1}^*(n)}{\mathscr{F}_{m-1}(n)} \tag{15.148}$$

$$b_m(n) = b_{m-1}(n-1) + \kappa_{b,m}^*(n)f_{m-1}(n) \tag{15.149}$$

where in the second line we have made use of the relation between $\Delta_{m-1}(n)$ and $p_{b,m-1}(n)$ given in Eq. (15.118), and also the definition of the backward reflection coefficient $\kappa_{b,m}(n)$ given in Eq. (15.119).

3. *Adaptive joint-process estimation*:

$$\pi_{m-1}(n) = \lambda \pi_{m-1}(n-1) + \frac{b_{m-1}(n)e_{m-1}^*(n)}{\gamma_{m-1}(n)} \tag{15.150}$$

$$\kappa_{m-1}(n) = \frac{\pi_{m-1}(n)}{\mathscr{B}_{m-1}(n)} \tag{15.151}$$

$$e_m(n) = e_{m-1}(n) - \kappa_{m-1}^*(n)b_{m-1}(n) \tag{15.152}$$

where $\pi_{m-1}(n)$ is defined in terms of $p_{m-1}(n)$ by

$$\pi_{m-1}(n) = \mathscr{B}_{m-1}^{1/2}(n)p_{m-1}(n) \tag{15.153}$$

and the joint-process regression coefficient $\kappa_{m-1}(n)$ is defined in terms of $p_{m-1}(n)$ by

$$\kappa_{m-1}(n) = \frac{p_{m-1}(n)}{\mathscr{B}_{m-1}^{1/2}(n)} \tag{15.154}$$

Summary of the Recursive LSL Algorithm Using *A Posteriori* Estimation Errors

The complete list of order- and time-update recursions constituting the recursive LSL algorithm (based on *a posteriori* estimation errors) is summarized in Table 15.4. This summary includes the recursions and relations of Eqs. (15.143), (15.142), (15.147), (15.144), (15.148), (15.145), (15.149), (15.146), (15.150), (15.151), and (15.152), in that order.

Since the LSL algorithm summarized in Table 15.4 involves division by updated parameters at some of the steps, care must be taken to ensure that these values are not allowed to become too small. Unless a high-precision computer is used, selection of the constant δ [determining the initial values $\mathscr{F}_0(0)$ and $\mathscr{B}_0(0)$] may have a severe effect on the initial transient performance of the LSL algorithm. Friedlander (1982) suggests using some form of *thresholding*, in that if the divisor (in any computation of LSL algorithm) is less than this preassigned threshold, the corresponding term involving that divisor is set to be zero. This remark also applies to other versions of the recursive LSL algorithm, e.g., those summarized in Table 15.5.

Initialization of the Recursive LSL Algorithm

To initialize the recursive LSL algorithm using *a posteriori* estimation errors, we start with the elementary case of zero prediction order, for which we have [see Eq. (15.43)]

$$f_0(n) = b_0(n) = u(n)$$

where $u(n)$ is the lattice predictor input at time n.

The remaining set of initial values pertain to the sums of weighted *a posteriori* prediction-error squares for zero prediction order. Specifically, setting $m - 1 = 0$ in Eq. (15.117) yields

$$\mathscr{F}_0(n) = \lambda\mathscr{F}_0(n - 1) + |u(n)|^2 \tag{15.155}$$

Similarly, setting $m - 1 = 0$ and replacing $n - 1$ with n in Eqs. (15.105) yields

$$\mathscr{B}_0(n) = \lambda\mathscr{B}_0(n - 1) + |u(n)|^2 \tag{15.156}$$

With the conversion factor $\gamma_m(n - 1)$ bounded by zero and 1, a logical choice for the zeroth-order value of this parameter is

$$\gamma_0(n - 1) = 1 \tag{15.157}$$

TABLE 15.4 SUMMARY OF THE RECURSIVE LSL ALGORITHM USING *A POSTERIORI* ESTIMATION ERRORS

Predictions:
For $n = 1, 2, 3, \ldots$, compute the various order updates in the sequence $m = 1, 2, \ldots, M$, where M is the final order of the least-squares lattice predictor:

$$\Delta_{m-1}(n) = \lambda \Delta_{m-1}(n-1) + \frac{b_{m-1}(n-1)f^*_{m-1}(n)}{\gamma_{m-1}(n-1)}$$

$$\mathcal{B}_{m-1}(n-1) = \lambda \mathcal{B}_{m-1}(n-2) + \frac{|b_{m-1}(n-1)|^2}{\gamma_{m-1}(n-1)}$$

$$\mathcal{F}_{m-1}(n) = \lambda \mathcal{F}_{m-1}(n-1) + \frac{|f_{m-1}(n)|^2}{\gamma_{m-1}(n-1)}$$

$$\kappa_{f,m}(n) = -\frac{\Delta_{m-1}(n)}{\mathcal{B}_{m-1}(n-1)}$$

$$\kappa_{b,m}(n) = -\frac{\Delta^*_{m-1}(n)}{\mathcal{F}_{m-1}(n)}$$

$$f_m(n) = f_{m-1}(n) + \kappa^*_{f,m}(n)b_{m-1}(n-1)$$

$$b_m(n) = b_{m-1}(n-1) + \kappa^*_{b,m}(n)f_{m-1}(n)$$

$$\gamma_m(n-1) = \gamma_{m-1}(n-1) - \frac{|b_{m-1}(n-1)|^2}{\mathcal{B}_{m-1}(n-1)}$$

Filtering:
For $n = 1, 2, 3, \ldots$ compute the various order updates in the sequence $m = 1, 2, \ldots, M + 1$:

$$\pi_{m-1}(n) = \lambda \pi_{m-1}(n-1) + \frac{b_{m-1}(n)e^*_{m-1}(n)}{\gamma_{m-1}(n)}$$

$$\kappa_{m-1}(n) = \frac{\pi_{m-1}(n)}{\mathcal{B}_{m-1}(n)}$$

$$e_m(n) = e_{m-1}(n) - \kappa^*_{m-1}(n)b_{m-1}(n)$$

Initialization:
1. To initialize the algorithm, at time $n = 0$ set

$$\Delta_{m-1}(0) = 0$$

$$\mathcal{F}_{m-1}(0) = \delta \qquad \delta = \text{small positive constant}$$

$$\mathcal{B}_{m-1}(-1) = \delta$$

$$\gamma_0(0) = 1$$

2. At each instant $n \geq 1$, generate the various zeroth-order variables as follows:

$$f_0(n) = b_0(n) = u(n)$$

$$\mathcal{F}_0(n) = \mathcal{B}_0(n) = \lambda \mathcal{F}_0(n-1) + |u(n)|^2$$

$$\gamma_0(n-1) = 1$$

3. For joint-process estimation, initialize the algorithm by setting at time $n = 0$

$$\pi_{m-1}(0) = 0$$

At each instant $n \geq 1$, generate the zeroth-order variable

$$e_0(n) = d(n)$$

Note: For prewindowed data, the input $u(n)$ and desired response $d(n)$ are both zero for $n \leq 0$.

We complete the initialization of the algorithm for forward and backward predictions by using the following conditions at time $n = 0$:

$$\Delta_{m-1}(0) = 0 \tag{15.158}$$

and

$$\mathcal{F}_{m-1}(0) = \mathcal{B}_{m-1}(-1) = \delta \tag{15.159}$$

where δ is a small positive constant. The constant δ is used to ensure nonsingularity of the correlation matrix $\mathbf{\Phi}_m(n)$.

Turning finally to the initialization of the joint-estimation process, we see (for zero-prediction order)

$$e_0(n) = d(n)$$

where $d(n)$ is the desired response. Thus, to initiate this part of the computation, we generate $e_0(n)$ for each instant n. To complete the initialization of the recursive LSL algorithm for joint-process estimation, at time $n = 0$ we set

$$\pi_{m-1}(0) = 0 \qquad \text{for } m = 1, 2, \ldots, M + 1$$

Table 15.4 includes the initialization of the recursive LSL algorithm as described above.

15.12 RECURSIVE LSL FILTERS USING *A PRIORI* ESTIMATION ERRORS WITH ERROR FEEDBACK

To add further support to the statement made in Section 15.9 that the QRD–LSL algorithm is fundamental to the derivation of all other recursive LSL algorithms, we now pursue the derivation of another recursive LSL algorithm. Specifically, the algorithm considered in this section differs from the recursive LSL algorithm of the previous section in two respects. First, it is based on *a priori* estimation errors. Second, the reflection and regression coefficients of the algorithm are all derived *directly*.

Following the two-step procedures described previously (i.e., squaring and then comparing terms), but this time using the arrays of Eqs. (15.134), (15.137), and (15.140), we get the following three sets of results expressed in terms of *a priori* estimation errors:

1. *Adaptive Forward Prediction*:

$$\mathcal{B}_{m-1}(n-1) = \lambda\mathcal{B}_{m-1}(n-2) + \gamma_{m-1}(n-1)|\beta_{m-1}(n-1)|^2 \tag{15.160}$$

$$\gamma_m(n-1)\eta_m(n) = \gamma_{m-1}(n-1)\eta_{m-1}(n) + \gamma_{m-1}(n-1)\kappa^*_{f,m}(n)\beta_{m-1}(n-1) \tag{15.161}$$

$$\kappa_{f,m}(n) = \kappa_{f,m}(n-1) + k_{b,m-1}(n-1)\eta^*_m(n) \tag{15.162}$$

$$\gamma_m(n-1) = \gamma_{m-1}(n-1) - \frac{\gamma^2_{m-1}(n-1)}{\mathcal{B}_{m-1}(n-1)}|\beta_{m-1}(n-1)|^2 \tag{15.163}$$

where in the second line we have made use of the definition given in Eq. (15.108) for the forward reflection coefficient $\kappa_{f,m}(n)$, and in the third line we have made use of Eqs. (15.77) and (15.135). The order update of Eq. (15.161) is not quite in the right form yet. To put it in the right form, we substitute Eq. (15.163) in (15.161), cancel the common conversion factor $\gamma_{m-1}(n-1)$, and rearrange terms. We thus obtain

$$\eta_m(n) = \eta_{m-1}(n) + [\kappa_{f,m}^*(n) + \frac{\gamma_{m-1}(n-1)\beta_{m-1}^*(n-1)}{\mathscr{B}_{m-1}(n-1)}\eta_m(n)]\beta_{m-1}(n-1)$$

$$(15.164)$$

Define the gain factor $k_{b,m-1}(n-1)$ in terms of backward prediction variables as[3]

$$k_{b,m-1}(n-1) = -\frac{\gamma_{m-1}(n-1)\beta_{m-1}(n-1)}{\mathscr{B}_{m-1}(n-1)}$$

$$(15.165)$$

so that we may rewrite Eq. (15.162) as

$$\kappa_{f,m}(n) = \kappa_{f,m}(n-1) - \frac{\gamma_{m-1}(n-1)\beta_{m-1}(n-1)}{\mathscr{B}_{m-1}(n-1)}\eta_m^*(n)$$

$$(15.166)$$

Accordingly, we may recast the order update of Eq. (15.164) into the desired form:

$$\eta_m(n) = \eta_{m-1}(n) + \kappa_{f,m}^*(n-1)\beta_{m-1}(n-1)$$

$$(15.167)$$

where $\kappa_{f,m}(n-1)$ is the *past* value of the forward reflection coefficient for stage m of the lattice predictor; see Eq. (15.96).

2. *Adaptive Backward Prediction*:

$$\mathscr{F}_{m-1}(n) = \lambda\mathscr{F}_{m-1}(n-1) + \gamma_{m-1}(n-1)|\eta_{m-1}(n)|^2$$

$$(15.168)$$

$$\gamma_m(n)\beta_m(n) = \gamma_{m-1}(n-1)\beta_{m-1}(n-1) + \gamma_{m-1}\kappa_{b,m}^*(n)\eta_{m-1}(n)$$

$$(15.169)$$

$$\kappa_{b,m}(n) = \kappa_{b,m}(n-1) + k_{f,m-1}(n)\beta_m^*(n)$$

$$(15.170)$$

$$\gamma_m(n) = \gamma_{m-1}(n-1) - \frac{\gamma_{m-1}^2(n-1)|\eta_{m-1}(n)|^2}{\mathscr{F}_{m-1}(n)}$$

$$(15.171)$$

where in the second line we have made use of the definition given in Eq. (15.119) for the backward reflection coefficient $\kappa_{b,m}(n)$ of stage m in the least-squares lattice predictor, and in the third line we have made use of Eqs. (15.78) and (15.138). Here again, we find that the order update of Eq. (15.169) is note quite in the right form. To put it in the right form, we substitute Eq. (15.171) in

---

[3]The formula of Eq. (15.165), defining the gain factor $k_{b,m-1}(n-1)$, is in perfect accord with that used in the traditional approach to the derivation of recursive least-squares lattice filters (Haykin, 1991).

(15.169), cancel the common conversion factor $\gamma_{m-1}(n-1)$, and rearrange terms. We may thus write

$$\beta_m(n) = \beta_{m-1}(n-1) + \left[\kappa_{b,m}^*(n) + \frac{\gamma_{m-1}(n-1)|\eta_{m-1}^*(n)|^2}{\mathcal{F}_{m-1}(n)} \beta_m(n) \right] \eta_{m-1}(n)$$

(15.172)

Define the gain factor $k_{f,m-1}(n)$ in terms of forward prediction variables as[4]

$$k_{f,m-1}(n) = -\frac{\gamma_{m-1}(n-1)\eta_{m-1}(n)}{\mathcal{F}_{m-1}(n)}$$

(15.173)

so that we may rewrite the order update of Eq. (15.170) as

$$k_{b,m}(n) = \kappa_{b,m}(n-1) - \frac{\gamma_{m-1}(n-1)\eta_{m-1}(n)}{\mathcal{F}_{m-1}(n)} \beta_m^*(n)$$

(15.174)

Now we may put the order update of Eq. (15.172) in the desired form:

$$\beta_m(n) = \beta_{m-1}(n-1) + \kappa_{b,m}^*(n-1)\eta_{m-1}(n)$$

(15.175)

where $\kappa_{b,m}(n-1)$ is the *past* value of the backward reflection coefficient for stage m of the least-squares lattice predictor; see Eq. (15.97).

3. *Adaptive Joint-process Estimation*:

$$\gamma_m(n)\xi_m(n) = \gamma_{m-1}(n)\xi_{m-1}(n) - \gamma_{m-1}(n)\kappa_{m-1}^*(n)\beta_{m-1}(n)$$

(15.176)

$$\kappa_{m-1}(n) = \kappa_{m-1}(n-1) + k_{b,m-1}(n)\xi_m^*(n)$$

(15.177)

where in the first line we have made use of the definition given in Eq. (15.128) for the joint-process regression coefficient $\kappa_{m-1}(n)$, and in the second line we have made use of Eqs. (15.79) and (15.135). Here also we find that the order update of Eq. (15.176) is not quite in the right form. To put it in the right form, we substitute the following time update [obtained by replacing $n-1$ with n in Eq. (15.163)]

$$\gamma_m(n) = \gamma_{m-1}(n) - \frac{\gamma_{m-1}^2(n)}{\mathcal{B}_{m-1}(n)} |\beta_{m-1}(n)|^2$$

in Eq. (15.176), cancel the common conversion factor $\gamma_{m-1}(n)$, and rearrange terms. We may thus write

$$\xi_m(n) = \xi_{m-1}(n) + \left(\frac{\gamma_{m-1}(n)\beta_{m-1}^*(n)}{\mathcal{B}_{m-1}(n)} \xi_m(n) - \kappa_{m-1}^*(n) \right) \beta_{m-1}(n)$$

(15.178)

---

[4]The definition given in Eq. (15.173) for the gain factor $k_{f,m-1}(n)$ is of exactly the same form as that used in the traditional approach to the derivation of recursive least-squares lattice filters (Haykin, 1991).

Using the definition of the gain factor $k_{b,m-1}(n)$, obtained by replacing $n - 1$ in Eq. (15.165) with n, we may rewrite the time update of Eq. (15.177) as

$$\kappa_{m-1}(n) = \kappa_{m-1}(n - 1) + \frac{\gamma_{m-1}(n)\beta_{m-1}(n)}{\mathcal{B}_{m-1}(n)} \, \xi_m^*(n) \tag{15.179}$$

Accordingly, we may recast Eq. (15.178) into the desired order update form:

$$\xi_m(n) = \xi_{m-1}(n) - \kappa_{m-1}^*(n-1)\beta_{m-1}(n) \tag{15.180}$$

where $\kappa_{m-1}(n - 1)$ is the past value of the regression coefficient for prediction order $m - 1$.

Summary of the Recursive LSL Algorithm Using *A Priori* Estimation Errors with Error Feedback

Eqs. (15.168), (15.160), (15.167), (15.175), (15.166), (15.174), and (15.163), in that order, define the computations involved in the forward and backward predictions of the recursive LSL algorithm using *a priori* estimation errors with error feedback. Equations (15.180) and (15.179) define the computations involved in the joint-process estimation part of the algorithm. A complete summary of the algorithm, including initial conditions of the soft-constraint form, is presented in Table 15.5.

Figure 15.13 presents a signal-flow graph of this new algorithm, emphasizing that order updating of the variables of interest (i.e., *a priori* forward prediction, backward prediction and joint-process estimation errors) at iteration n requires knowledge of the forward reflection coefficients, backward reflections coefficients, and regression coefficients at the previous iteration $n - 1$.

An important difference between the two recursive LSL algorithms summarized in Tables 15.4 and 15.5 is the way in which the reflection coefficients and regression coefficients are updated. In the case of Table 15.4, the updating is performed *indirectly*. We first compute the cross-correlation between forward and delayed backward prediction errors and the cross-correlation between backward prediction errors and joint-process estimation errors. Next, we compute the sum of weighted forward prediction-error squares and the sum of weighted backward-error squares. Finally, we compute the reflection and regression coefficients by dividing a cross-correlation by a sum of weighted prediction-error squares. On the other hand, in Table 15.5, the updating of the reflection and regression coefficients is performed *directly*. The differences between indirect and direct forms of updating, as described herein, have an important bearing on the numerical behavior of these recursive LSL algorithms; this issue is discussed in detail in Chapter 17.

15.13 COMPUTATION OF THE LEAST-SQUARES WEIGHT VECTOR

In solving the joint-process estimation problem with an order-recursive adaptive filter, we have shown how the least-squares predictor can be expanded to include the estimation of a desired response. The solution to this problem encompasses the computation of a set of

TABLE 15.5 SUMMARY OF THE RECURSIVE LSL ALGORITHM USING *A PRIORI* ESTIMATION ERRORS WITH ERROR FEEDBACK

Predictions:

For $n = 1, 2, 3, \ldots$, compute the various order updates in the sequence $m = 1, 2, \ldots, M$, where M is the final order of the least-squares predictor:

$$\mathscr{F}_{m-1}(n) = \lambda \mathscr{F}_{m-1}(n-1) + \gamma_{m-1}(n-1)|\eta_{m-1}(n)|^2$$

$$\mathscr{B}_{m-1}(n-1) = \lambda \mathscr{B}_{m-1}(n-2) + \gamma_{m-1}(n-1)|\beta_{m-1}(n-1)|^2$$

$$\eta_m(n) = \eta_{m-1}(n) + \kappa^*_{f,m}(n-1)\beta_{m-1}(n-1)$$

$$\beta_m(n) = \beta_{m-1}(n-1) + \kappa^*_{b,m}(n-1)\eta_{m-1}(n)$$

$$\kappa_{f,m}(n) = \kappa_{f,m}(n-1) - \frac{\gamma_{m-1}(n-1)\beta_{m-1}(n-1)}{\mathscr{B}_{m-1}(n-1)}\eta^*_m(n)$$

$$\kappa_{b,m}(n) = \kappa_{b,m}(n-1) - \frac{\gamma_{m-1}(n-1)\eta_{m-1}(n)}{\mathscr{F}_{m-1}(n)}\beta^*_m(n)$$

$$\gamma_m(n-1) = \gamma_{m-1}(n-1) - \frac{\gamma^2_{m-1}(n-1)|\beta_{m-1}(n-1)|^2}{\mathscr{B}_{m-1}(n-1)}$$

Filtering:

For $n = 1, 2, 3, \ldots$, compute the various order updates in the sequence $m = 1, 2, \ldots, M+1$:

$$\xi_m(n) = \xi_{m-1}(n) - \kappa^*_{m-1}(n-1)\beta_{m-1}(n)$$

$$\kappa_{m-1}(n) = \kappa_{m-1}(n-1) + \frac{\gamma_{m-1}(n)\beta_{m-1}(n)}{\mathscr{B}_{m-1}(n)}\xi^*_m(n)$$

Initialization:

1. To initialize the algorithm, at time $n = 0$ set

$$\mathscr{F}_{m-1}(0) = \delta, \qquad \delta = \text{small positive constant}$$

$$\mathscr{B}_{m-1}(-1) = \delta$$

$$\kappa_{f,m}(0) = \kappa_{b,m}(0) = 0$$

$$\gamma_0(0) = 1$$

2. For each instant $n \geq 1$, generate the zeroth-order variables:

$$\eta_0(n) = \beta_0(n) = u(n)$$

$$\mathscr{F}_0(n) = \mathscr{B}_0(n) = \lambda \mathscr{F}_0(n-1) + |u(n)|^2$$

$$\gamma_0(n-1) = 1$$

3. For joint-process estimation, at time $n = 0$ set

$$\kappa_{m-1}(0) = 0$$

At each instant $n \geq 1$, generate the zeroth-order variable

$$\xi_0(n) = d(n)$$

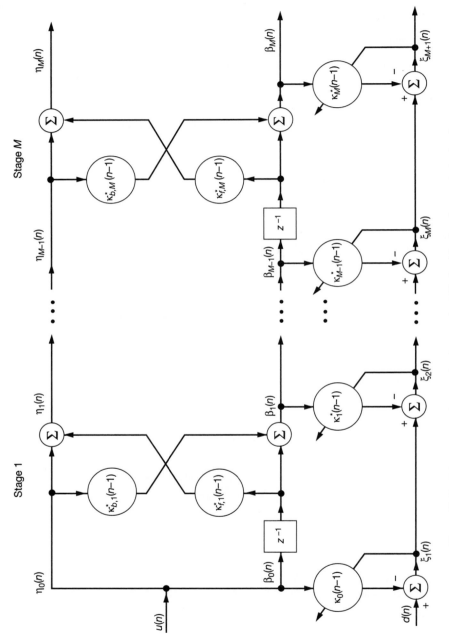

Figure 15.13 Joint process estimator using the recursive LSL algorithm based on *a priori* estimation errors.

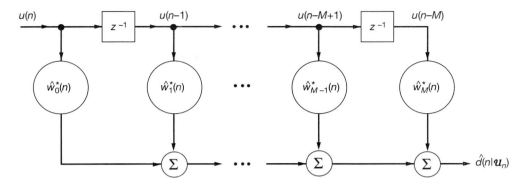

Figure 15.14   Conventional transversal fiter.

regression coefficient $\{\kappa_0(n), \kappa_1(n), \ldots, \kappa_M(n)\}$ that is fed with a corresponding set of inputs represented by the backward prediction errors $\{b_0(n), b_1, \ldots, b_M(n)\}$; see Fig. 15.7. Recognizing that there is a one-to-one correspondence between this set of backward prediction errors and the set of tap inputs $\{u(n), u(n-1), \ldots, (n-M)\}$, as shown in Eq. (15.70), we expect to find a corresponding relationship between the sequence of regression coefficients and the set of least-squares tap weights $\{\hat{w}_0(n), \hat{w}_1(n), \ldots, \hat{w}_M(n)\}$. The purpose of this section is to formally derive this relationship.

Consider the conventional tapped-delay-line or transversal filter structure shown in Fig. 15.14, where the tap inputs $u(n), u(n-1), \ldots, u(n-m)$ are derived directly from the process $u(n)$ and the tap weights $\hat{w}_0(n), \hat{w}_1(n), \ldots, \hat{w}_m(n)$ are used to form respective scalar inner products. From Chapter 11, we recall that the least-squares solution for the $(m-1)$-by-1 tap vector $\hat{\mathbf{w}}_m(n)$, consisting of the elements $\hat{w}_0(n), \hat{w}_1(n), \ldots, \hat{w}_m(n)$, is defined by

$$\boldsymbol{\Phi}_{m+1}(n)\,\hat{\mathbf{w}}_m(n) = \mathbf{z}_{m+1}(n) \tag{15.181}$$

where $\boldsymbol{\Phi}_{m-1}(n)$ is the $(m+1)$-by-$(m+1)$ correlation matrix of the tap inputs, and $\mathbf{z}_{m+1}(n)$ is the $(m+1)$-by-1 cross-correlation vector between the tap inputs and desired response. We modify Eq. (15.181) in two ways: (1) we premultiply both sides of the equation by the $(m+1)$-by-$(m+1)$ lower triangular transformation matrix $\mathbf{L}_m(n)$, and (2) we interject the $(m+1)$-by-$(m+1)$ identity matrix $\mathbf{I} = \mathbf{L}_m^H(n)\mathbf{L}_m^{-H}(n)$ between the matrix $\boldsymbol{\Phi}_{m+1}(n)$ and the vector $\hat{\mathbf{w}}_m(n)$ on the left-hand side of the equation. The matrix $\mathbf{L}_m(n)$ is defined in terms of the tap weights of backward prediction-error filters of orders $0, 1, 2, \ldots, m$, as in Eq. (15.73). The symbol $\mathbf{L}_m^{-H}(n)$ denotes the Hermitian transpose of the inverse matrix $\mathbf{L}_m^{-1}(n)$. We may thus write

$$\mathbf{L}_m(n)\boldsymbol{\Phi}_{m-1}(n)\mathbf{L}_m^H(n)\mathbf{L}_m^{-H}(n)\,\hat{\mathbf{w}}_m(n) = \mathbf{L}_m(n)\mathbf{z}_{m+1}(n) \tag{15.182}$$

Let the product $\mathbf{L}_m(n)\boldsymbol{\Phi}_{m+1}(n)\mathbf{L}_m^H(n)$ on the left-hand side of Eq. (15.182) be denoted by

$$\mathbf{D}_{m+1}(n) = \mathbf{L}_m(n)\boldsymbol{\Phi}_{m+1}(n)\mathbf{L}_m^H(n) \tag{15.183}$$

Using the formula for the augmented normal equations for backward linear prediction, we may show that the product $\boldsymbol{\Phi}_{m+1}(n)\mathbf{L}_m^H(n)$ consists of a lower triangular matrix whose diagonal elements equal the various sums of weighted backward *a posteriori* prediction-error squares, that is, $\mathscr{B}_0(n)$, $\mathscr{B}_1(n)$, . . . , $\mathscr{B}_m(n)$ (see Problem 10). The matrix $\mathbf{L}_m(n)$ is, by definition, a lower triangular matrix whose diagonal elements are all equal to unity. Hence, the product of $\mathbf{L}_m(n)$ and $\boldsymbol{\Phi}_{m+1}(n)\mathbf{L}_m^H(n)$ is a lower triangular matrix. We also know that $\mathbf{L}_m^H(n)$ is an upper triangular matrix, and so is the matrix product $\mathbf{L}_m(n)\boldsymbol{\Phi}_{m+1}(n)$. Hence, the product of $\mathbf{L}_m(n)\boldsymbol{\Phi}_{m+1}(n)$ and $\mathbf{L}_m^H(n)$ is an upper triangular matrix. In other words, the matrix $\mathbf{D}_{m+1}(n)$ is both *upper* and *lower* triangular, which can only be satisfied if it is *diagonal*. Accordingly, we may write

$$\mathbf{D}_{m+1}(n) = \mathbf{L}_m(n)\boldsymbol{\Phi}_{m+1}(n)\mathbf{L}_m^H(n) \tag{15.184}$$

$$= \text{diag}[\mathscr{B}_0(n), \mathscr{B}_1(n), \ldots, \mathscr{B}_m(n)]$$

Equation (15.184) is further proof that the backward *a posteriori* prediction errors $b_0(n)$, $b_1(n)$, . . . , $b_m(n)$ produced by the various stages of the least-squares lattice predictor are uncorrelated (in a time-averaged sense) at all instants of time.

The product $\mathbf{L}_m(n)\mathbf{z}_{m+1}(n)$ on the right-hand side of Eq. (15.182) equals the cross-correlation vector between the backward prediction errors and the desired response. Let $\mathbf{t}_{m+1}(n)$ denote this cross-correlation vector, as shown by

$$\mathbf{t}_{m-1}(n) = \sum_{i=1}^{n} \lambda^{n-i}\mathbf{b}_{m+1}(i)d^*(i) \tag{15.185}$$

where $d(i)$ is the desired response. Substituting Eq. (15.72) in (15.185), we thus get

$$\mathbf{t}_{m+1}(n) = \sum_{i=1}^{n} \lambda^{n-i}\mathbf{L}_m(n)\mathbf{u}_{m+1}(i)d^*(i)$$

$$= \mathbf{L}_m(n)\sum_{i=1}^{n} \lambda^{n-i}\mathbf{u}_{m+1}(i)d^*(i) \tag{15.186}$$

$$= \mathbf{L}_m(n)\mathbf{z}_{m+1}(n)$$

which is the desired result. Accordingly, the combined use of Eqs. (15.183) and (15.186) in Eq. (15.182) yields the *transformed RLS solution*:

$$\mathbf{D}_{m+1}(n)\mathbf{L}_m^{-H}(n)\hat{\mathbf{w}}_m(n) = \mathbf{t}_{m+1}(n) \tag{15.187}$$

Thus far we have considered how the application of lower triangular matrix $\mathbf{L}_m(n)$ transforms the RLS solution for the tap-weight vector of the conventional transversal structure shown in Fig. 15.14. We next wish to consider the RLS solution represented by the regression coefficient vector $\boldsymbol{\kappa}_m(n)$, which is denoted by

$$\boldsymbol{\kappa}_m(n) = [\kappa_0(n), \kappa_1(n), \ldots, \kappa_m(n)]^T \tag{15.188}$$

The regression coefficient vector $\boldsymbol{\kappa}_m(n)$ may be viewed as the solution that minimizes the index of performance

$$\sum_{i=1}^{n} \lambda^{n-i} |d(i) - \mathbf{b}_{m+1}^{T}(i)\boldsymbol{\kappa}_{m}^{*}(n)|^{2}$$

where $\boldsymbol{\kappa}_{m}(n)$ is held constant for $1 \leq i \leq n$. The resulting solution to this RLS problem is defined by

$$\mathbf{D}_{m+1}(n)\boldsymbol{\kappa}_{m}(n) = \mathbf{t}_{m+1}(n) \tag{15.189}$$

where, as defined before, $\mathbf{D}_{m+1}(n)$ is the $(m + 1)$-by-$(m + 1)$ correlation matrix of the backward *a posteriori* prediction errors used as inputs to the regression coefficients, and $\mathbf{t}_{m+1}(n)$ is the $(m + 1)$-by-1 cross-correlation vector between these tap inputs and the desired response.

By comparing the transformed RLS solution of Eq. (15.187) and the RLS solution of Eq. (15.189), we immediately deduce the following simple relationship between the tap-weight vector $\hat{\mathbf{w}}_{m}(n)$ in the structure of Fig. 15.14 and the corresponding regression coefficient vector $\boldsymbol{\kappa}_{m}(n)$ computed by a recursive LSL filter:

$$\boldsymbol{\kappa}_{m}(n) = \mathbf{L}_{m}^{-H}(n)\hat{\mathbf{w}}_{m}(n) \tag{15.190}$$

or, equivalently,

$$\hat{\mathbf{w}}_{m}(n) = \mathbf{L}_{m}^{H}(n)\boldsymbol{\kappa}_{m}(n) \tag{15.191}$$

We thus see that the lower triangular transformation matrix $\mathbf{L}_{m}(n)$ represents the connecting link between the regression coefficient vector $\boldsymbol{\kappa}_{m}(n)$ in Fig. 15.7 and the least-squares tap-weight vector $\hat{\mathbf{w}}_{m}(n)$ in Fig. 15.14.

15.14 COMPUTER EXPERIMENT ON ADAPTIVE PREDICTION

In this second computer experiment of the chapter, we use a first-order autoregressive (AR) process $u(n)$ to study *adaptive prediction*, with two objectives in mind:

- To evaluate the performance of the recursive LSL algorithm using *a posteriori* estimation errors
- To compare the performance of this algorithm with that of the LMS algorithm for a similar application

The evaluations are to be made for the same two sets of conditions described in Section 9.6:

1. AR parameter : $a = -0.99$
   variance of AR process $u(n)$: $\sigma_{u}^{2} = 0.93627$
2. AR parameter : $a = +0.99$
   variance of AR process $u(n)$: $\sigma_{u}^{2} = 0.995$

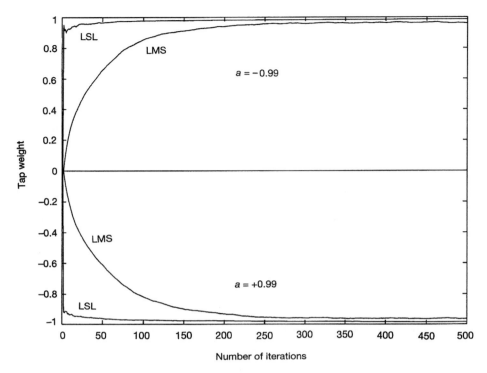

Figure 15.15   Comparison of the convergence behavior of the recursive LSL algorithm
and LMS algorithm for autoregressive modeling of order 1.

Figure 15.15 shows the results of this experiment for the recursive LSL algorithm
assuming that the exponential weighting factor $\lambda = 1$. This figure also includes the corresponding results of the experiment described in Section 9.6, using the LMS algorithm with
step-size parameter $\mu = 0.05$. In both cases, the ensemble-averaged estimate of the AR
parameter a is plotted versus the number of iterations n. The ensemble averaging was performed over 100 independent trials of the particular experiment.

The results of Fig. 15.15 show that:

- The recursive LSL algorithm converges to its steady-state condition much faster
  than the LMS algorithm
- After 500 iterations, the ensemble-averaged estimate of the AR parameter a produced by the recursive LSL algorithm is more accurate than that produced by the
  LMS algorithm.

15.15 OTHER VARIANTS OF LEAST-SQUARES LATTICE FILTERS

In Sections 15.11 and 15.12 we derived two recursive LSL filters (algorithms), one using *a posteriori* estimation errors and the other using *a priori* estimation errors with error feedback. The derivations were presented as special cases of the angle-normalized QRD–LSL algorithm, demonstrating the fundamental importance of this algorithm. There are, of course, many other recursive LSL algorithms that could also be derived from the angle-normalized QRD–LSL algorithm. Two other variants that immediately come to mind are the recursive LSL algorithm using *a posteriori* estimation errors with error feedback, and the recursive LSL algorithm using *a priori* estimation errors without error feedback. Hybrid combinations of the four recursive LSL algorithms mentioned here are obvious candidates that could be considered too.

The recursive LSL algorithm summarized in Table 15.4 is designed to propagate two reflection coefficients, one for solving the least-squares forward prediction problem and the other for solving the least-squares backward prediction problem. By employing a proper normalization of the forward and backward reflection coefficients, it is indeed possible to reformulate the recursive LSL algorithm of Table 15.4 so as to propagate a single reflection coefficient. The resulting algorithm is called the *normalized least-squares lattice algorithm* (Lee et al., 1981).

Yet another variant is the so-called *hybrid QR/lattice least-squares algorithm*, derived by Bellanger and Regalia (1991). This algorithm combines the good numerical properties of QR decomposition and the desirable order-recursive properties of least-squares lattice predictors. As shown by Sayed and Kailath (1994), the hybrid QR/lattice least-squares algorithm is essentially a rewriting of the angle-normalized QRD–LSL algorithm, whereby a certain collection of rows in the three arrays of the QRD–LSL algorithm is combined with the order updates:

$$\mathscr{F}_m(n) = \mathscr{F}_{m-1}(n) - \frac{|\Delta_{m-1}(n)|^2}{\mathscr{B}_{m-1}(n-1)} \tag{15.192}$$

$$\mathscr{B}_m(n) = \mathscr{B}_{m-1}(n-1) - \frac{|\Delta_{m-1}(n)|^2}{\mathscr{F}_{m-1}(n)} \tag{15.193}$$

These two order updates follow naturally from the augmented normal equations for least-squares estimation; see Problem 6.

The important point to note here is that by virtue of the following:

- The diversity of state-space models for characterizing the least-squares lattice filtering process
- The many variants of square-root Kalman filtering algorithms available for use
- The numerous ways in which *a priori* and *a posteriori* estimation errors can be hybridized

there is a large variety of recursive least-squares lattice algorithms that is essentially a matter of taste and patience, and that these algorithms may all be viewed as alternative rewritings of the angle-normalized QRD–LSL algorithm in exact arithmetic.

15.16 SUMMARY AND DISCUSSION

In this chapter we further consolidated the intimate relationship between Kalman filter theory and the family of adaptive linear filters that is rooted in least-squares estimation. In particular, we demonstrated how the square-root information filter, which is a variant of the Kalman filter, can be used to derive the QR–decomposition–based least-squares lattice (QRD–LSL) algorithm, which represents the most fundamental form of an order-recursive adaptive filter. We also demonstrated how other order-recursive adaptive filtering algorithms, such as the recursive LSL algorithm using *a posteriori* estimation errors and the recursive LSL algorithm using *a priori* estimation errors with error feedback, are in fact rewritings of the QRD–LSL algorithm.

The QRD–LSL algorithm combines highly desirable features of recursive least-squares estimation, QR-decomposition, and a lattice structure. Accordingly, it offers a unique set of operational and implementational advantages:

- The QRD–LSL algorithm has a *fast rate of convergence*, which is inherent in recursive least-squares estimation.

- The QRD–LSL algorithm can be implemented using a sequence of Givens rotations, which represent a form of QR decomposition. Moreover, the good numerical properties of the QR decomposition mean that the QRD–LSL algorithm is *numerically stable*.

- The QRD–LSL algorithm offers a *high level of computational efficiency*, in that its complexity is on the order of M, where M is the final prediction order (i.e., the number of available degrees of freedom).

- The lattice structure of the QRD–LSL algorithm is *modular* in nature, which means that the prediction order can be increased without having to recalculate all previous values. This property is particularly useful when there is no prior knowledge as to what the final value of the prediction order should be.

- Another implication of the modular structure of the QRD–LSL algorithm is that it lends itself to the use of *very large-scale integration (VLSI)* technology for its hardware implementation. Of course, the use of this sophisticated technology can only be justified if the application of interest calls for the use of VLSI chips in large numbers.

- The QRD–LSL algorithm includes an *integral set of desired variables and parameters* that are useful to have in signal-processing applications. Specifically, it offers the following three sets of useful by-products:
  - Angle-normalized forward and backward prediction errors

- Auxiliary parameters that can be used for the *indirect* computation of the forward and backward reflection coefficients and the regression coefficients (i.e., tap weights)

The recursive LSL algorithms enjoy many of the properties of the QRD–LSL algorithm, namely, fast convergence, modularity, and an integral set of useful parameters and variables for signal processing applications. However, the numerical properties of recursive LSL algorithms depend on whether error feedback is included or not in their composition; this issue is discussed in Chapter 17.

The order-recursive adaptive filters considered in this chapter have a computational advantage over the square-root adaptive filters considered in the previous chapter. In the former case, the computational cost increases linearly with the number of adjustable parameters, whereas in the latter case it increases as the square of the number of adjustable parameters. However, the use of order-recursive adaptive filters is limited to *temporal* signal processing applications that permit the exploitation of the time-shifting property of the input data. On the other hand, square-root adaptive filters can be used for both *temporal and spatial* signal-processing applications.

Traditionally, the derivations of these filters have been rather ad hoc, laborious, and certainly lacking a strong sense of unity. In contrast, the adoption of a Kalman filtering approach pioneered by Sayed and Kailath (1994), which we have followed in this book, not only overcomes these shortcomings of the traditional approach, but also offers the following advantages and possibilities:

- A *compact* representation of the adaptive filtering algorithms in the form of *arrays*, made up of prearrays, unitary rotations, and postarrays; the arrays propagate *all* the quantities needed to update the adaptive filtering algorithms.
- An opening to exploit the vast literature on Kalman filters, not so much to know how to develop new algorithms (as we already have enough of them) but rather to explore how to further improve the properties of adaptive filters.

In the context of the latter point, for example, we could mention that much has been written on the design of smoothing filters using Kalman filter theory (Gelb, 1974; Sorenson, 1985). It would therefore be enlightening to use this theory to derive latticelike realizations of smoothing filters based on recursive least-squares estimation. This would provide a framework for comparing notes with work done by Yuan and Stuller (1995), who have derived an algorithm for the design of order-recursive lattice smoothers that use past, present, and future data. Naturally, an adaptive smoother would outperform an adaptive filter, but would require the provision of an overall delay for physical realizeability in a real-time sense.

In closing this discussion, we should mention that the family of order-recursive adaptive filtering algorithms (including the QRD–LSL algorithm) is part of a larger class of adaptive filtering algorithms known collectively as *fast algorithms*. In the context of recursive least-squares estimation, an algorithm is said to be "fast" if its *computational complexity increases linearly with the number of adjustable parameters.* A fast algorithm is therefore similar to the LMS algorithm in its computational requirement.

The class of order-recursive adaptive filtering algorithms, based on the use of reflection coefficients or their counterparts, is well suited for such adaptive signal-processing applications as predictive modeling, noise cancelation, and equalization. For fast RLS algorithms needed for other applications such as system identification and spectrum analysis where the emphasis is on the direct computation of adaptive transversal filter coefficients, we may look to the following alternatives:

- *Fast transversal filters (FTF) algorithm,* involving the combined use of four transversal filters for forward and backward predictions, gain vector computation, and joint-process estimation (Cioffi and Kailath, 1984). This is an elegant algorithm; unfortunately, it has a tendency to become numerically unstable when it is implemented in finite-precision arithmetic. To stabilize it, the algorithm has to be modified in a certain way, as explained in Chapter 17.
- *Fast QR-decomposition-based recursive least-squares estimation,* in which the triangularization of the data matrix and back-solving for the transversal filter's coefficient vector are *performed together* rather than separately (Liu, 1995). To account for this requirement, the Householder transformation used for the triangularization is modified in a special way. Unlike the FTF algorithm, the algorithm derived by Liu appears to be numerically stable.

PROBLEMS

1. Show that the parameter $\gamma_m(n)$ defined by

$$\gamma_m(n) = 1 - \mathbf{k}_m^H(n)\mathbf{u}_m(n)$$

equals the sum of weighted error squares resulting from use of the transversal filter in Fig. 15.3. The tap-weight vector of this filter equals the gain vector $\mathbf{k}_m(n)$, and the tap-input vector equals $\mathbf{u}_m(n)$. The filter is designed to produce the least-squares estimate of a desired response that equals the *first coordinate vector.*

2. The parameter $\Delta_{m-1}(n)$ is defined in Eq. (15.34). It is also defined in Eq. (15.51). Show that these two definitions are equivalent.

3. Let $\mathbf{\Phi}_m(n)$ denote the time-averaged correlation matrix of the tap-input vector $\mathbf{u}_m(n)$ at time n; likewise for $\mathbf{\Phi}_m(n-1)$. Show that the conversion factor $\gamma_m(n)$ is related to the determinants of these two matrices as follows:

$$\gamma_m(n) = \lambda \frac{\det[\mathbf{\Phi}_m(n-1)]}{\det[\mathbf{\Phi}_m(n)]}$$

where λ is the exponential weighting factor. *Hint:* Use the identity

$$\det(\mathbf{I}_1 + \mathbf{AB}) = \det(\mathbf{I}_2 + \mathbf{BA})$$

where $\mathbf{I}_1$ and $\mathbf{I}_2$ are identity matrices of appropriate dimensions, and $\mathbf{A}$ and $\mathbf{B}$ are matrices of compatible dimensions.

4. (a) Show that the inverse of the correlation matrix $\mathbf{\Phi}_{m-1}(n)$ may be expressed as follows:

$$\mathbf{\Phi}_{m+1}^{-1}(n) = \begin{bmatrix} 0 & \mathbf{0}_m^T \\ \mathbf{0}_m & \mathbf{\Phi}_m^{-1}(n-1) \end{bmatrix} + \frac{1}{\mathcal{F}_m(n)} \mathbf{a}_m(n)\mathbf{a}_m^H(n)$$

where $\mathbf{0}_m$ is the M-by-1 null vector, $\mathbf{0}_m^T$ is its transpose, $\mathcal{F}_m(n)$ is the minimum sum of weighted forward prediction-error squares, and $\mathbf{a}_m(n)$ is the tap-weight vector of forward prediction-error filter. Both $\mathbf{a}_m(n)$ and $\mathcal{F}_m(n)$ refer to prediction order m.

(b) Show that the inverse of $\mathbf{\Phi}_{m+1}(n)$ may also be expressed in the form

$$\mathbf{\Phi}_{m+1}^{-1}(n) = \begin{bmatrix} \mathbf{\Phi}_m^{-1}(n) & \mathbf{0}_m \\ \mathbf{0}_m^T & 0 \end{bmatrix} + \frac{1}{\mathcal{B}_m(n)} \mathbf{c}_m(n)\mathbf{c}_m^H(n)$$

where $\mathcal{B}_m(n)$ is the minimum sum of weighted backward *a posteriori* prediction-error squares, and $\mathbf{c}_m(n)$ is the tap-weight vector of the backward prediction-error filter. Both $\mathcal{F}_m(n)$ and $\mathbf{c}_m(n)$ refer to prediction order m.

5. Derive the following update formulas:

$$\gamma_{m+1}(n) = \gamma_m(n-1) - \frac{|f_m(n)|^2}{\mathcal{F}_m(n)}$$

$$\gamma_{m+1}(n) = \gamma_m(n) - \frac{|b_m(n)|^2}{\mathcal{B}_m(n)}$$

$$\gamma_{m+1}(n) = \lambda \frac{\mathcal{F}_m(n-1)}{\mathcal{F}_m(n)} \gamma_m(n-1)$$

$$\gamma_{m+1}(n) = \lambda \frac{\mathcal{B}_m(n-1)}{\mathcal{B}_m(n)} \gamma_m(n)$$

6. Using Eqs. (15.52) and (15.57), derive the following order-update recursions involving the sums of forward and backward prediction-error squares, respectively:

$$\mathcal{F}_m(n) = \mathcal{F}_{m-1}(n) - \frac{|\Delta_{m-1}(n)|^2}{\mathcal{B}_{m-1}(n-1)}$$

$$\mathcal{B}_m(n) = \mathcal{B}_{m-1}(n-1) - \frac{|\Delta_{m-1}(n)|^2}{\mathcal{F}_{m-1}(n)}$$

7. In this problem we show how the various quantities of the fast prediction equations relate to each other, and the parametric redundancy that they contain.[5]

(a) By combining parts (a) and (b) from Problem 4, show that

$$\begin{bmatrix} \mathbf{\Phi}_m^{-1}(n) & \mathbf{0}_m \\ \mathbf{0}_m^T & 0 \end{bmatrix} - \begin{bmatrix} 0 & \mathbf{0}_m^T \\ \mathbf{0}_m & \mathbf{\Phi}_m^{-1}(n-1) \end{bmatrix} = \frac{\mathbf{a}_m(n)\mathbf{a}_m^H(n)}{\mathcal{F}_m(n)} - \frac{\mathbf{c}_m(n)\mathbf{c}_m^H(n)}{\mathcal{B}_m(n)}$$

(b) From the recursive equations of Chap. 13, plus Eq. (15.26), show that the time update for $\mathbf{\Phi}_m^{-1}$ may be rewritten as

$$\mathbf{\Phi}_m^{-1}(n) = \lambda^{-1}\mathbf{\Phi}_m^{-1}(n-1) - \frac{\mathbf{k}_m(n)\mathbf{k}_m^H(n)}{\gamma_m(n)}$$

where $\mathbf{k}_m(n)$ is the gain vector and $\gamma_m(n)$ is the conversion factor.

[5]This problem was originally formulated by P. Regalia, private communication, 1995.

(c) By eliminating $\Phi_m^{-1}(n-1)$ from the above two expressions, show that all the variables may be reconciled as

$$\begin{bmatrix} \Phi_m^{-1}(n) & \mathbf{0}_m \\ \mathbf{0}_m^T & 0 \end{bmatrix} - \lambda \begin{bmatrix} 0 & \mathbf{0}_m^T \\ \mathbf{0}_m & \Phi_m^{-1}(n) \end{bmatrix} = \frac{\mathbf{a}_m(n)\mathbf{a}_m^H(n)}{\mathcal{F}_m(n)} + \lambda \begin{bmatrix} 0 \\ \mathbf{k}_m(n) \end{bmatrix} \frac{[0, \mathbf{k}_m^H(n)]}{\gamma_m(n)} - \frac{\mathbf{c}_m(n)\mathbf{c}_m^H(n)}{\mathcal{B}_m(n)}$$

in which all the variables have a common time index n, and a common order index m. The left-hand side is called a *displacement residue* of $\Phi_m^{-1}(n)$ and the right-hand side, being the sum and difference of three vector dyads, has rank not exceeding three. In matrix theory, $\Phi_m^{-1}(n)$ is said to have *displacement rank* three as a result of the special structure of the data matrix exposed in Chapter 11. Note that this structure holds irrespective of the sequence $u(n)$ that builds the data matrix.

(d) Suppose we multiply the result of part (c) from the left by the row vector $[1, z/\sqrt{\lambda}, \ldots, (z/\sqrt{\lambda})^m]$, and from the right by the column vector $[1, w\sqrt{\lambda}, \ldots, (w/\sqrt{\lambda})^m]^H$, where z and w are two complex variables. Show that the result of part (c) is equivalent to the two-variable polynomial equation

$$(1 - zw^*)P(z, w^*) = A(z)A^*(w) + K(z)K^*(w) - C(z)C^*(w) \qquad \text{for all } z, w$$

provided that we make the correspondences

$$P(z, w^*) = [1, z/\sqrt{\lambda}, \ldots, (z/\sqrt{\lambda})^{m-1}]\Phi_m^{-1}(n) \begin{bmatrix} 1 \\ w^*/\sqrt{\lambda} \\ \cdot \\ \cdot \\ \cdot \\ (w^*/\sqrt{\lambda})^{M-1} \end{bmatrix}$$

and

$$A(z) = [1, z/\sqrt{\lambda}, \ldots, (z/\sqrt{\lambda})^m] \frac{\mathbf{a}_m(n)}{\sqrt{\mathcal{F}_m(n)}}$$

$$K(z) = [1, z/\sqrt{\lambda}, \ldots, (z/\sqrt{\lambda})^m] \sqrt{\frac{\lambda}{\gamma_m(n)}} \begin{bmatrix} 0 \\ \mathbf{k}_m(n) \end{bmatrix}$$

$$C(z) = [1, z/\sqrt{\lambda}, \ldots, (z/\sqrt{\lambda})^m] \frac{\mathbf{c}_m(n)}{\sqrt{\mathcal{B}_m(n)}}$$

Similarly, $A^*(w) = [A(w)]^*$, and so on.

(e) Set $z = w = e^{j\omega}$ in the result of part (d) to show that

$$|A(e^{j\omega})|^2 + |K(e^{j\omega})|^2 = |C(e^{j\omega})|^2 \qquad \text{for all } \omega$$

that is, the three polynomials $A(z)$, $K(z)$, and $C(z)$ are power complementary along the unit circle $|z| = 1$.

(f) Show that, because $\Phi_m^{-1}(n)$ is positive definite, the following system of inequalities necessarily results:

$$|A(z)|^2 + |K(z)|^2 - |C(z)|^2 = \begin{cases} <0, & |z| > 1 \\ =0, & |z| = 1 \\ >0, & |z| < 1 \end{cases}$$

Hint: Set $w^* = z^*$ in part (d) and note that, if $\Phi_m^{-1}(n)$ is positive definite, the inequality

$$P(z,z^*) > 0 \qquad \text{for all } z$$

must result. Note that the center equality is equivalent to the result of part (e).

(g) Deduce from the first inequality of part (f) that $C(z)$ must be devoid of zeros in $|z| > 1$, and hence given $A(z)$ and $K(z)$, the polynomial $C(z)$ is uniquely determined from part (e) via spectral factorization. This shows that, once the forward prediction and gain quantities are known, the backward prediction variables contribute nothing further to the solution, and hence are theoretically redundant.

8. Justify the following relationships:

(a) Joint-process estimation errors:

$$|\epsilon_m(n)| = \sqrt{|e_m(n)| \cdot |\xi_m(n)|}$$

$$\text{ang}[\epsilon_m(n)] = \text{ang}[e_m(n)] + \text{ang}[\xi_m(n)]$$

(b) Backward prediction errors:

$$|\epsilon_{b,m}(n)| = \sqrt{|b_m(n)| \cdot |\beta_m(n)|}$$

$$\text{ang}[\epsilon_{b,m}(n)] = \text{ang}[b_m(n)] + \text{ang}[\beta_m(n)]$$

(c) Forward prediction errors:

$$|\epsilon_{f,m}(n)| = \sqrt{|f_m(n)| \cdot |\eta_m(n)|}$$

$$\text{ang}[\epsilon_{f,m}(n)] = \text{ang}[f_m(n)] + \text{ang}[\eta_m(n)]$$

9. Suppose that we have computed the cosine parameters $c_{f,m}(n)$ and $c_{b,m-1}(n-1)$ pertaining to the Givens transformations $\Theta_{f,m}(n)$ and $\Theta_{b,m-1}(n-1)$, respectively. Hence, show that the conversion factor $\gamma_{m+1}(n)$ may be updated in both time and order as follows:

$$\gamma_{m+1}^{1/2}(n) = c_{f,m}(n) c_{b,m-1}(n-1) \gamma_{m-1}^{1/2}(n-1)$$

10. The correlation matrix $\Phi_{m+1}(n)$ is postmultiplied by the Hermitian transpose of the lower triangular matrix $\mathbf{L}_m(n)$, where $\mathbf{L}_m(n)$ is defined by Eq. (15.73). Show that the product $\Phi_{m+1}(n)\mathbf{L}_m^H(n)$ consists of a lower triangular matrix whose diagonal elements equal the various sums of weighted backward prediction-error squares, $\mathcal{B}_0(n)$, $\mathcal{B}_1(n)$, ..., $\mathcal{B}_m(n)$. Hence, show that the product $\mathbf{L}_m(n)\Phi_{m-1}(n)\mathbf{L}_m^H(n)$ is a diagonal matrix, as shown by

$$\mathbf{D}_{m+1}(n) = \text{diag}[\mathcal{B}_0(n), \mathcal{B}_1(n), \ldots, \mathcal{B}_m(n)]$$

11. Consider the case where the input samples $u(n), u(n-1), \ldots, u(n-M)$ have a *joint Gaussian distribution with zero mean*. Assume that, within a scaling factor, the ensemble-averaged correlation matrix $\mathbf{R}_{M+1}$ of the input signal is equal to its time-averaged correlation matrix $\Phi_{M+1}(n)$ for time $n \geq M$. Show that the log-likelihood function for this input includes a term equal to the parameter $\gamma_M(n)$ associated with the recursive LSL algorithm. For this reason, the parameter $\gamma_M(n)$ is sometimes referred to as a *likelihood variable*.

12. Let $\hat{d}(n|\mathcal{U}_{n-m+1})$ denote the least-squares estimate of the desired response $d(n)$, given the inputs $u(n-m+1), \ldots, u(n)$ that span the space $\mathcal{U}_{n-m+1}$. Similarly, let $\hat{d}(n|\mathcal{U}_{n-m})$ denote the least-squares estimate of the desired response, given the inputs $u(n-m), u(n-m+1), \ldots, u(n)$ that span the space $\mathcal{U}_{n-m}$. In effect, the latter estimate exploits an additional piece of information represented by the input $u(n - m)$. Show that this new information is represented by the corresponding backward prediction error $b_m(n)$. Also, show that the two estimates are related by the recursion

$$\hat{d}(n|\mathcal{U}_{n-m}) = \hat{d}(n|\mathcal{U}_{n-m+1}) + \kappa_m^*(n)b_m(n)$$

where $\kappa_m(n)$ denotes the pertinent regression coefficient in the joint-process estimator. Compare this result with that of Section 7.1 dealing with the concept of innovations.

13. Let $\mathbf{\Phi}(n)$ denote the $(M + 1)$-by-$(M + 1)$ correlation matrix of the input data $u(n)$. Show that the change of variables to backward prediction errors brought about by using a lattice predictor achieves exactly the Cholesky decomposition of the matrix $\mathbf{P}(n) = \mathbf{\Phi}^{-1}(n)$.

14. Expand the joint-process estimator of Fig. 15.7 so as to include (in modular form) the least-squares estimate of the desired response $d(n)$ for increasing prediction order m.

15. In Section 15.11 we discussed a modification of the *a priori* error LSL algorithm by using a form of error feedback. In this problem we consider the corresponding modified version of the *a posteriori* LSL algorithm. In particular, show that

$$\kappa_{f,m}(n) = \frac{\gamma_m(n - 1)}{\gamma_{m-1}(n - 1)}\left[\kappa_{f,m}(n - 1) - \frac{1}{\lambda}\frac{b_{m-1}(n - 1)f_{m-1}^*(n)}{\mathcal{B}_{m-1}(n - 2)\gamma_{m-1}(n - 1)}\right]$$

$$\kappa_{b,m} = \frac{\gamma_m(n)}{\gamma_{m-1}(n - 1)}\left[\kappa_{b,m}(n - 1) - \frac{1}{\lambda}\frac{f_{m-1}(n)b_{m-1}^*(n - 1)}{\mathcal{F}_{m-1}(n - 1)\gamma_{m-1}(n - 1)}\right]$$

16. The accompanying table is a summary of the *normalized LSL algorithm*. The normalized parameters are defined by

$$\bar{f}_m(n) = \frac{f_m(n)}{\mathcal{F}_m^{1/2}(n)\gamma_m^{1/2}(n - 1)}$$

$$\bar{b}_m(n) = \frac{b_m(n)}{\mathcal{B}_m^{1/2}(n)\gamma_m^{1/2}(n)}$$

$$\bar{\Delta}_m(n) = \frac{\Delta_m(n)}{\mathcal{F}_m^{1/2}(n)\mathcal{B}_m^{1/2}(n - 1)}$$

Hence, derive the steps summarized in the table.

$$\bar{\Delta}_{m-1}(n) = \bar{\Delta}_{m-1}(n - 1)[1 - |\bar{f}_{m-1}(n)|^2]^{1/2}[1 - |\bar{b}_{m-1}(n - 1)|^2]^{1/2} + \bar{b}_{m-1}(n - 1)\bar{f}_{m-1}^*(n)$$

$$\bar{b}_m(n) = \frac{\bar{b}_{m-1}(n - 1) - \bar{\Delta}_{m-1}(n)\bar{f}_{m-1}(n)}{[1 - |\bar{\Delta}_{m-1}(n)|^2]^{1/2}[1 - |\bar{f}_{m-1}(n)|^2]^{1/2}}$$

$$\bar{f}_m(n) = \frac{\bar{f}_{m-1}(n) - \bar{\Delta}_{m-1}^*(n)\bar{b}_{m-1}(n - 1)}{[1 - |\bar{\Delta}_{m-1}(n)|^2]^{1/2}[1 - |\bar{b}_{m-1}(n - 1)|^2]^{1/2}}$$

CHAPTER

16

Tracking of Time-Varying Systems

In this second part of the book we have described two families of adaptive filtering algorithms, namely, the LMS family and the RLS family. Specifically, in Chapters 9 and 13 we considered the *average* behavior of the standard LMS and RLS algorithms operating in a stationary environment. For such an environment, the error-performance surface is fixed and the essential requirement is to seek, in a step-by-step fashion, the minimum point of that surface and thereby assure optimum or near optimum performance.

In this chapter we examine the operation of these two algorithms in a *nonstationary environment*, for which the optimum Wiener solution takes on a time-varying form. The net result is that the minimum point of the error-performance surface is no longer fixed. Consequently, the adaptive filtering algorithm now has the added task of *tracking* the minimum point of the error-performance surface. In other words, the algorithm is required to continuously track the statistical variations of the input, the occurrence of which is assumed to be "slow" enough for tracking to be feasible.

Tracking is a *steady-state phenomenon*. This is to be contrasted with convergence, which is a transient phenomenon. It follows therefore that for an adaptive filter to exercise its tracking capability, it must first pass from the transient mode to the steady-state mode of operation, and there must be provision for continuous adjustment of the free parameters of the filter. Moreover, we may state that, in general, the rate of convergence and tracking capability are two different properties of the algorithm. In particular, an adaptive filtering algorithm with good convergence properties does *not* necessarily possess a fast tracking capability, and vice versa.

We begin the discussion by describing a particular time-varying model for system identification, which is subsequently used as the basis for evaluating the tracking performances of the standard LMS and RLS algorithms operating in a nonstationary environment.

16.1 MARKOV MODEL FOR SYSTEM IDENTIFICATION

Nonstationarity of an environment may arise in practice in one of two basic ways:

1. *The frame of reference provided by the desired response may be time varying.* Such a situation arises, for example, in system identification when an adaptive transversal filter is used to model a time-varying system. In this case, the correlation matrix of the tap inputs of the adaptive transversal filter remains fixed (as in a stationary environment), whereas the cross-correlation vector between the tap inputs and the desired response assumes a time-varying form.

2. *The stochastic process supplying the tap inputs of the adaptive filter is nonstationary.* This situation arises, for example, when an adaptive transversal filter is used to equalize a time-varying channel. In this second case, both the correlation matrix of the tap inputs in the adaptive transversal filter and the cross-correlation vector of the tap inputs and the desired response assume time-varying forms.

Thus, the tracking details of a time-varying system are not only dependent on the type of adaptive filter employed but are also *problem specific*.

In this chapter we will focus on a popular time-varying model for system identification, which is depicted in Fig. 16.1 (Widrow et al., 1976). The model is governed by three basic equations, as described next.

1. *First-order Markov process.* The unknown dynamic system is modeled as a transversal filter whose tap-weight vector $\mathbf{w}_o(n)$ (i.e., impulse response) undergoes a *first-order Markow process* written in vector form as follows [see Fig. 16.1(a)]:

$$\mathbf{w}_o(n + 1) = a\mathbf{w}_o(n) + \boldsymbol{\omega}(n) \tag{16.1}$$

where a is a fixed parameter of the model, and $\boldsymbol{\omega}(n)$ is the *process noise vector* assumed to be of zero mean and correlation matrix $\mathbf{Q}$. In physical terms, the tap-weight vector $\mathbf{w}_o(n)$ may be viewed as originating from a source of random vector $\boldsymbol{\omega}(n)$, whose individual elements are applied to a bank of one-pole low-pass filters. Each such filter has a transfer function equal to $1/(1 - az^{-1})$, where z^{-1} is the unit-delay operator. It is assumed that the value of parameter a is very close to 1. The significance of this assumption is that the bandwidth of the low-pass filters is very much smaller than the incoming data rate. Equivalently, we may say that many iterations of the model in Fig. 16.1(a) are required to produce a significant change in the tap-weight vector $\mathbf{w}_o(n)$.

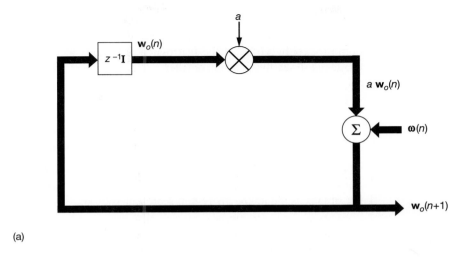

(a)

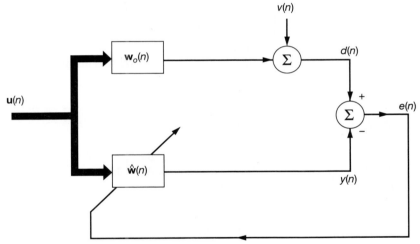

(b)

Figure 16.1   (a) Model of first-order Markov process. (b) System identification using adaptive filter.

2. *Desired response.* The desired response $d(n)$, providing a frame of reference for the adaptive filter, is defined by

$$d(n) = \mathbf{w}_o^H(n)\mathbf{u}(n) + v(n) \tag{16.2}$$

where $\mathbf{u}(n)$ is the input vector, which is common to both the unknown system and the adaptive filter; and $v(n)$ is the *measurement noise* assumed to be white with zero mean and variance σ_v^2. Thus, even though both $\mathbf{u}(n)$ and $v(n)$ may be stationary random processes, the desired response $d(n)$ is a *nonstationary* random process by virtue of the fact that $\mathbf{w}_o(n)$ varies with time. Herein lies the challenge posed to the adaptive filter.

3. *Error signal.* The error signal $e(n)$, involved in the adaptive process, is defined by [see Fig. 16.1(b)]

$$
\begin{aligned}
e(n) &= d(n) - y(n) \\
&= \mathbf{w}_o^H(n)\mathbf{u}(n) + v(n) - \hat{\mathbf{w}}^H(n)\mathbf{u}(n)
\end{aligned}
\tag{16.3}
$$

where $\hat{\mathbf{w}}(n)$ is the tap-weight vector of the adaptive filter assumed to have a transversal structure. It is also assumed that the adaptive filter has the same number of taps as the unknown system represented by $\mathbf{w}_o$.

The tap-weight vector $\mathbf{w}_o(n)$ of the unknown system represents the "target" to be tracked by the adaptive filter. Whenever the tap-weight vector $\hat{\mathbf{w}}(n)$ of the adaptive filter equals $\mathbf{w}_o(n)$, the minimum mean-square error produced by the adaptive filter equals σ_v^2.

According to Eq. (16.2), the desired response $d(n)$ applied to the adaptive filter equals the overall output of the unknown system. Since this system is time-varying, the desired response is correspondingly nonstationary. Accordingly, with the correlation matrix of the tap inputs having the fixed value $\mathbf{R}$, we find that the adaptive filter has a quadratic bowl-shaped error-performance surface whose position is in a permanent state of motion.

Assumptions

Typically, the variations represented by the process noise vector $\boldsymbol{\omega}(n)$ in the Markov model of Fig. 16.1(a) are slow (i.e., bounded). This makes it possible for the adaptive transversal filter using the LMS or RLS algorithm to *track* the statistical variations in the dynamic behavior of the unknown system in Fig. 16.1(b).

To proceed with a tracking analysis of the LMS and RLS algorithms in the environment described in Fig. 16.1, the following conditions are assumed throughout the chapter (Macchi, 1995):

1. The process noise vector $\boldsymbol{\omega}(n)$ is independent of both the input vector $\mathbf{u}(n)$ and measurement noise $v(n)$.

2. The input vector $\mathbf{u}(n)$ and measurement noise $v(n)$ are independent of each other.

3. The measurement noise $v(n)$ is white with zero mean and variance $\sigma_v^2 < \infty$.

Assumptions 2 and 3 have been made previously (see Chapters 9 and 13). Assumption 1, pertaining to the unknown system itself, is new to this chapter. Thus, there are three sources of randomness corresponding to physical phenomena to be considered: the input vector $\mathbf{u}(n)$, the measurement noise $v(n)$, and the process noise vector $\boldsymbol{\omega}(n)$. In practice, they do *not* usually relate to each other in the unrealistic manner described under assumptions 1 to 3. Nevertheless, these assumptions are commonly made in the literature on adaptive filters for the sake of mathematical tractability. Collectively, they are referred to as the *independence assumption.*

16.2 DEGREE OF NONSTATIONARITY

In order to provide a clear definition of the rather ambiguous notion of what is meant by "slow" or "fast" statistical variations of the model, we may introduce the *degree of nonstationarity* (Macchi, 1986, 1995). In the context of the Markov model described in Fig. 16.1, the degree of nonstationarity, denoted by α, is formally defined as the square root of the average noise power due to the process noise vector $\boldsymbol{\omega}(n)$ to the average noise power due to the measurement noise $v(n)$, both of which refer to the output end of the time-varying system. That is,

$$\alpha = \left(\frac{E[|\boldsymbol{\omega}^H(n)\mathbf{u}(n)|^2]}{E[|v(n)|^2]} \right)^{1/2} \tag{16.4}$$

The degree of nonstationarity is therefore a characteristic of the time varying system alone; as such, it has nothing to do with the adaptive filter.

The numerator in Eq. (16.4) may be rewritten as follows [in light of the assumption that $\boldsymbol{\omega}(n)$ is independent of $\mathbf{u}(n)$]:

$$\begin{aligned} E[|\boldsymbol{\omega}^H(n)\mathbf{u}(n)|^2] &= E[\boldsymbol{\omega}^H(n)\mathbf{u}(n)\mathbf{u}^H(n)\boldsymbol{\omega}(n)] \\ &= \mathrm{tr}\{E[\boldsymbol{\omega}^H(n)\mathbf{u}(n)\mathbf{u}^H(n)\boldsymbol{\omega}(n)]\} \\ &= E\{\mathrm{tr}[\boldsymbol{\omega}^H(n)\mathbf{u}(n)\mathbf{u}^H(n)\boldsymbol{\omega}(n)]\} \\ &= E\{\mathrm{tr}[\boldsymbol{\omega}(n)\boldsymbol{\omega}^H(n)\mathbf{u}(n)\mathbf{u}^H(n)]\} \\ &= \mathrm{tr}\{E[\boldsymbol{\omega}(n)\boldsymbol{\omega}^H(n)\mathbf{u}(n)\mathbf{u}^H(n)]\} \\ &= \mathrm{tr}\{E[\boldsymbol{\omega}(n)\boldsymbol{\omega}^H(n)]E[\mathbf{u}(n)\mathbf{u}^H(n)]\} \\ &= \mathrm{tr}[\mathbf{QR}] \end{aligned} \tag{16.5}$$

where $\mathrm{tr}[\cdot]$ denotes the *trace* of the matrix enclosed inside the square brackets, $\mathbf{R}$ is the correlation matrix of the input vector $\mathbf{u}(n)$, and $\mathbf{Q}$ is the correlation matrix of the process noise vector $\boldsymbol{\omega}(n)$. The denominator in Eq. (16.4) is simply the variance σ_v^2 of the zero-mean measurement noise $v(n)$. Accordingly, we may reformulate the degree of nonstationary for the Markov model of Fig. 16.1 simply as

$$\alpha = \frac{1}{\sigma_v} (\mathrm{tr}[\mathbf{RQ}])^{1/2} = \frac{1}{\sigma_v} (\mathrm{tr}[\mathbf{QR}])^{1/2} \tag{16.6}$$

The degree of nonstationarity α bears a useful relation to the misadjustment $\mathcal{M}$ of the adaptive filter, as explained here. First, we note that the minimum mean-squared error $J_{\min}$

that the adaptive filter in Fig. 16.1(b) can ever attain is equal to the variance σ_v^2 of the measurement noise $v(n)$. Next, we note that the best that the adaptive filter can ever do in tracking the time-varying system of Fig. 16.1(a) is to produce a *weight-error vector* $\boldsymbol{\epsilon}(n)$ that is equal to the process noise vector $\boldsymbol{\omega}(n)$. Then, following the terminology introduced in Section 9.4, we may set the weight-error correlation matrix $\mathbf{K}(n)$ equal to the correlation matrix $\mathbf{Q}$ of $\boldsymbol{\omega}(n)$. Hence, recalling from that chapter that the excess mean-squared error $J_{\text{ex}}(n)$ is equal to tr$[\mathbf{RK}(n)]$, we may state that the excess mean-squared error attained by the adaptive filter can never be less than tr$[\mathbf{RQ}]$. Thus, in light of the definition of misadjustment as the ratio of the excess mean-squared error to the minimum mean-squared error [see Eq. (9.67)], we may write

$$\mathcal{M} = \frac{J_{\text{ex}}}{J_{\min}} \geq \frac{\text{tr}[\mathbf{RQ}]}{\sigma_v^2} = \alpha^2 \tag{16.7}$$

In other words, the square root of the misadjustment $\mathcal{M}$ places an upper bound on the degree of nonstationarity.

We will have more to say on misadjustment as a measure of tracking performance in the next section. For now, we may use Eq. (16.7) to make two noteworthy remarks:

1. For slow statistical variations, α is small. This, in turn, means that it should be possible to build an adaptive filter that can track the time-varying system.
2. When the statistical variations of the environment are too fast, α may be greater than 1. In such a case, the misadjustment produced by the adaptive filter exceeds 100 percent, which means that there is no advantage to be gained in building an adaptive filter to solve the tracking problem.

16.3 CRITERIA FOR TRACKING ASSESSMENT

With the unknown dynamic system in Fig. 16.1 modeled as a transversal filter whose tap-weight vector is denoted by $\mathbf{w}_o(n)$, and with the tap-weight vector of the adaptive transversal filter denoted by $\hat{\mathbf{w}}(n)$, we may define the *weight-error vector* as

$$\boldsymbol{\epsilon}(n) = \hat{\mathbf{w}}(n) - \mathbf{w}_o(n) \tag{16.8}$$

On the basis of $\boldsymbol{\epsilon}(n)$, we may go on to define two figures of merit for assessing the tracking capability of an adaptive filter.

1. *Mean-square Deviation*

A commonly used figure of merit for tracking assessment is the *mean-square deviation* (MSD) between the actual weight vector $\mathbf{w}_o(n)$ of the unknown dynamic system and the adjusted weight vector $\hat{\mathbf{w}}(n)$ of the adaptive filter, defined by (Benveniste and Ruget, 1982; Macchi, 1986; Benveniste, 1987):

$$\mathcal{D}(n) = E[\|\hat{\mathbf{w}}(n) - \mathbf{w}_o(n)\|^2]$$

$$= E[\|\boldsymbol{\epsilon}(n)\|^2] \tag{16.9}$$

where the number of iterations n is assumed to be large enough for the adaptive filter's transient mode of operation to have finished. Equation (16.9) may be reformulated in the equivalent form (following steps similar to those presented in Eq. (16.5)]:

$$\mathcal{D}(n) = \text{tr}[\mathbf{K}(n)] \tag{16.10}$$

where $\mathbf{K}(n)$ is the correlation matrix of the error vector $\boldsymbol{\epsilon}(n)$:

$$\mathbf{K}(n) = E[\boldsymbol{\epsilon}(n)\boldsymbol{\epsilon}^H(n)] \tag{16.11}$$

Clearly, the mean-square deviation $\mathcal{D}(n)$ should be small for a good tracking performance.

The weight-error vector $\boldsymbol{\epsilon}(n)$ may be expressed as the sum of two components, as shown by

$$\boldsymbol{\epsilon}(n) = \boldsymbol{\epsilon}_1(n) + \boldsymbol{\epsilon}_2(n) \tag{16.12}$$

where $\boldsymbol{\epsilon}_1(n)$ is *weight vector noise* defined by

$$\boldsymbol{\epsilon}_1(n) = \hat{\mathbf{w}}(n) - E[\hat{\mathbf{w}}(n)] \tag{16.13}$$

where $E[\hat{\mathbf{w}}(n)]$ is the ensemble-averaged value of the tap-weight vector; and $\boldsymbol{\epsilon}_2(n)$ is the *weight vector lag* defined by

$$\boldsymbol{\epsilon}_2(n) = E[\hat{\mathbf{w}}(n)] - \mathbf{w}_o(n) \tag{16.14}$$

Invoking the assumptions made in Section 16.1, we have

$$E[\boldsymbol{\epsilon}_1^H(n)\,\boldsymbol{\epsilon}_2(n)] = E[\boldsymbol{\epsilon}_2^H(n)\boldsymbol{\epsilon}_1(n)] = 0 \tag{16.15}$$

Accordingly, we may express the mean-square deviation $\mathcal{D}(n)$ as the sum of two components, as shown by

$$\mathcal{D}(n) = \mathcal{D}_1(n) + \mathcal{D}_2(n) \tag{16.16}$$

The first term $\mathcal{D}_1(n)$ is called the *estimation variance*, which is due to the weight vector noise $\boldsymbol{\epsilon}_1(n)$; it is defined by

$$\mathcal{D}_1(n) = E[\|\boldsymbol{\epsilon}_1(n)\|^2] \tag{16.17}$$

The estimation variance $\mathcal{D}_1(n)$ is always present, even in the stationary case. The second term $\mathcal{D}_2(n)$ is called the *lag variance*, which is due to the weight vector lag; it is defined by

$$\mathcal{D}_2(n) = E[\|\boldsymbol{\epsilon}_2(n)\|^2] \tag{16.18}$$

The presence of $\mathcal{D}_2(n)$ is testimony to the nonstationary nature of the environment. The decomposition of the mean-square deviation $\mathcal{D}(n)$ into estimation variance $\mathcal{D}_1(n)$ and lag variance $\mathcal{D}_2(n)$, as described in Eq. (16.16), is called the *decoupling property* (Macchi, 1986a,b).

2. Misadjustment

Another commonly used figure of merit for assessing the tracking capability of an adaptive filter is the *misadjustment*, which is defined by (Widrow et al., 1976)

$$\mathcal{M}(n) = \frac{J_{\text{ex}}(n)}{\sigma_v^2} \tag{16.19}$$

where $J_{\text{ex}}(n)$ is the excess (residual) mean-squared error of the adaptive filter measured with respect to the variance σ_v^2 of the white noise component $v(n)$ at the output of the transversal model in Fig. 16.1(b). Here again it is assumed that the number of iterations n is large enough for the transient period to have ended. In light of Eq. (9.65), justified under the independence assumption, we may express the excess mean-squared error in terms of $\mathbf{K}(n)$, the correlation matrix of the weight error vector $\boldsymbol{\epsilon}(n)$, as follows:

$$J_{\text{ex}}(n) = \text{tr}[\mathbf{R}\mathbf{K}(n)] \tag{16.20}$$

where $\mathbf{R}$ is the correlation matrix of the input vector $\mathbf{u}(n)$. Accordingly, we may reformulate Eq. (16.19) in the equivalent form

$$\mathcal{M}(n) = \frac{\text{tr}[\mathbf{R}\mathbf{K}(n)]}{\sigma_v^2} \tag{16.21}$$

For a good tracking performance, it is apparent that the misadjustment $\mathcal{M}(n)$ should be small compared to unity.

As with the mean-square deviation, the excess mean-squared error $J_{\text{ex}}(n)$ may be expressed as the sum of two components, $J_{\text{ex1}}(n)$ and $J_{\text{ex2}}(n)$, by virtue of the assumptions made in Section 16.1. The first component $J_{\text{ex1}}(n)$ is due to the weight vector noise $\boldsymbol{\epsilon}_1(n)$; it is called the *estimation noise*. The second component $J_{\text{ex2}}(n)$ is due to the weight vector lag $\boldsymbol{\epsilon}_2(n)$; it is called the *lag noise*. The presence of the latter term is attributed directly to the nonstationary nature of the environment. Correspondingly, we may express the misadjustment $\mathcal{M}(n)$ as

$$\mathcal{M}(n) = \mathcal{M}_1(n) + \mathcal{M}_2(n) \tag{16.22}$$

where $\mathcal{M}_1(n) = J_{\text{ex1}}(n)/\sigma_v^2$ and $\mathcal{M}_2(n) = J_{\text{ex2}}(n)/\sigma_v^2$. The first term $\mathcal{M}_1(n)$ is called the *noise misadjustment* and the second term $\mathcal{M}_2(n)$ is called the *lag misadjustment*. Thus, the decoupling property is true for the misadjustment too, in that the estimation noise and lag noise are decoupled in power.

In general, both figures of merit, $\mathcal{D}(n)$ and $\mathcal{M}(n)$, depend on the number if iterations (time) n. Moreover, they highlight different aspects of the tracking problem in a complementary way, as subsequent analysis will reveal.

16.4 TRACKING PERFORMANCE OF THE LMS ALGORITHM

To proceed with a study of the tracking problem, consider the system model of Fig. 16.1 in which the adaptive (transversal) filter is implemented using the LMS algorithm. According to this algorithm, the tap-weight vector of the adaptive filter is updated as follows:

$$\hat{\mathbf{w}}(n + 1) = \hat{\mathbf{w}}(n) + \mu\mathbf{u}(n)e^*(n) \tag{16.23}$$

where μ is the step-size parameter. Substituting Eq. (16.3) for the error signal $e(n)$ in Eq. (16.23), we may reformulate the LMS algorithm in the expanded form:

$$\hat{\mathbf{w}}(n+1) = [\mathbf{I} - \mu\mathbf{u}(n)\mathbf{u}^H(n)]\hat{\mathbf{w}}(n) + \mathbf{u}(n)\mathbf{u}^H(n)\mathbf{w}_o(n) + \mu\mathbf{u}(n)v^*(n) \quad (16.24)$$

Next, using the definition of Eq. (16.7) for the weight-error vector $\boldsymbol{\epsilon}(n)$ and the description of a first-order Markov model given in Eq. (16.1), we may write (after combining terms)

$$\boldsymbol{\epsilon}(n+1) = \hat{\mathbf{w}}(n+1) - \mathbf{w}_o(n+1)$$
$$= [\mathbf{I} - \mu\mathbf{u}(n)\mathbf{u}^H(n)]\boldsymbol{\epsilon}(n) + (1-a)\,\mathbf{w}_o(n) + \mu\mathbf{u}(n)v^*(n) - \boldsymbol{\omega}(n) \quad (16.25)$$

where $\mathbf{I}$ is the identity matrix. The linear stochastic difference equation (16.25) provides a complete description of the LMS algorithm embedded in the system model of Fig. 16.1. A general tracking theory of the Markov model based on Eq. (16.25) is yet to be developed. The approach usually taken is to assume that the model parameter a is very close to 1, so that we may ignore the term $(1-a)\mathbf{w}_o$. In so doing, we are in fact developing a tracking theory for the *random walk model* (Macchi and Turki, 1992; Macchi, 1995). Thus, with $a = 1$, Eq. (16.25) reduces to

$$\boldsymbol{\epsilon}(n+1) = [\mathbf{I} - \mu\mathbf{u}(n)\mathbf{u}^H(n)]\boldsymbol{\epsilon}(n) + \mu\mathbf{u}(n)v^*(n) - \boldsymbol{\omega}(n) \quad (16.26)$$

Typically, the step-size parameter μ is assigned a small value in order to realize a good tracking performance.[1] Then under this condition, we may solve Eq. (16.26) for the weight-error vector $\boldsymbol{\epsilon}(n)$ by invoking the *direct-averaging method* (Kushner, 1984), which was discussed previously in Chapter 9. Specifically, we may state that for a small μ, the solution $\boldsymbol{\epsilon}(n)$ to the linear stochastic difference equation (16.26) is "close" to the solution of another linear stochastic difference equation that is obtained by replacing the system matrix $[\mathbf{I} - \mu\mathbf{u}(n)\mathbf{u}^H(n)]$ with its ensemble average $(\mathbf{I} - \mu\mathbf{R})$, where $\mathbf{R}$ is the correlation matrix of the input vector $\mathbf{u}(n)$. We may thus write the new stochastic difference equation as follows:

$$\boldsymbol{\epsilon}(n+1) = (\mathbf{I} - \mu\mathbf{R})\boldsymbol{\epsilon}(n) + \mu\mathbf{u}(n)v^*(n) - \boldsymbol{\omega}(n) \quad (16.27)$$

By right, we should have used a different notation for the weight-error vector in Eq. (16.27) to distinguish it from that used in Eq. (16.26). We have opted not to do so merely for convenience of presentation. To evaluate the correlation matrix of $\boldsymbol{\epsilon}(n+1)$ given in Eq. (16.27), we invoke the independence assumption described earlier. Under this assump-

---

[1] In practical situations, we often find that μ is assigned a large value that may lie outside the scope of stochastic approximation theory; this is usually done in order to obtain a fast rate of convergence. To deal with this dilemma, Perrier et al. (1994) use the perturbation expansion method, first proposed by Solo (1992), to investigate the steady-state performance of the LMS algorithm. In particular, it is shown how to explicitly integrate the correlation coefficients of the input signal up to the second order in μ.

tion, the correlation matrix of the weight-error vector $\boldsymbol{\epsilon}(n)$ is readily determined from Eq. (16.27) to be

$$
\begin{aligned}
\mathbf{K}(n+1) &= E[\boldsymbol{\epsilon}(n+1)\boldsymbol{\epsilon}^H(n+1)] \\
&= (\mathbf{I} - \mu\mathbf{R})\mathbf{K}(n)\,(\mathbf{I} - \mu\mathbf{R}) + \mu\sigma_v^2\mathbf{R} + \mathbf{Q}
\end{aligned}
\tag{16.28}
$$

For a *steady-state solution* of the difference equation (16.28), for which n is large, we may legitimately set $\mathbf{K}(n+1) = \mathbf{K}(n)$. Furthermore, assuming that the step-size parameter μ is small enough to justify ignoring the term $\mu^2\mathbf{R}\mathbf{K}(n)\mathbf{R}$ in comparison with the identity matrix $\mathbf{I}$, we may approximate Eq. (16.28) as follows (after rearranging terms):

$$
\mathbf{R}\mathbf{K}(n) + \mathbf{K}(n)\mathbf{R} \simeq \mu\sigma_v^2\mathbf{R} + \frac{1}{\mu}\mathbf{Q}
\tag{16.29}
$$

This is the equation for assessing the tracking capability of the LMS algorithm applied to the system model of Fig. 16.1, under the assumption that a is close to unity.

Mean-square Deviation of the LMS Algorithm

To proceed with the evaluation of the mean-square deviation of the LMS algorithm, we premultiply both sides of Eq. (16.29) by the inverse matrix $\mathbf{R}^{-1}$, and then take the trace of the resultant matrices. We thus obtain

$$
\text{tr}[\mathbf{K}(n)] + \text{tr}[\mathbf{R}^{-1}\mathbf{K}(n)\mathbf{R}] \simeq \mu M\sigma_v^2 + \frac{1}{\mu}\text{tr}[\mathbf{R}^{-1}\mathbf{Q}]
\tag{16.30}
$$

Next, we recognize that since $\mathbf{K}(n)$ and $\mathbf{R}$ have the same dimensions, then

$$
\begin{aligned}
\text{tr}[\mathbf{R}^{-1}\mathbf{K}(n)\mathbf{R}] &= \text{tr}[\mathbf{K}(n)\mathbf{R}\mathbf{R}^{-1}] \\
&= \text{tr}[\mathbf{K}(n)]
\end{aligned}
$$

Accordingly, we may use Eqs. (16.10) and (16.30) to evaluate the mean-square deviation of the LMS algorithm, as shown by

$$
\mathcal{D}(n) \simeq \frac{\mu}{2} M\sigma_v^2 + \frac{1}{2\mu}\text{tr}[\mathbf{R}^{-1}\mathbf{Q}], \qquad n \text{ large}
\tag{16.31}
$$

The first term, $\mu M\sigma_v^2/2$, is the estimation variance due to the measurement noise $v(n)$; it varies *linearly* with the step-size parameter μ. The second term, $\text{tr}[\mathbf{R}^{-1}\mathbf{Q}]/2\mu$, is the lag variance due to the process noise vector $\boldsymbol{\omega}(n)$; it varies *inversely* with the step-size parameter μ, thereby permitting a faster tracking speed.

Let μ_{opt} denote the optimum value of the step-size parameter for which the mean-square deviation attains its minimum value $\mathcal{D}_{\min}$. This optimum condition is realized when the estimation variance and lag variance contribute equally to the mean-square deviation. From Eq. (16.31) we thus readily find that

$$
\mu_{\text{opt}} \simeq \frac{1}{\sigma_v\sqrt{M}}\,(\text{tr}\,\mathbf{R}^{-1}\mathbf{Q})^{1/2}
\tag{16.32}
$$

and

$$D_{\min} \simeq \sigma_v \sqrt{M}\, (\mathrm{tr}[\mathbf{R}^{-1}\mathbf{Q}])^{1/2} \tag{16.33}$$

Misadjustment of the LMS Algorithm

To evaluate the misadjustment of the LMS algorithm for the system identification scenario described in Fig. 16.1, we take the trace of the matrix quantities on both sides of Eq. (16.29), and so write

$$\mathrm{tr}[\mathbf{R}\mathbf{K}(n)] + \mathrm{tr}[\mathbf{K}(n)\mathbf{R}] \simeq \mu\sigma_v^2\,\mathrm{tr}[\mathbf{R}] + \frac{1}{\mu}\,\mathrm{tr}[\mathbf{Q}] \tag{16.34}$$

Next, recognizing that the traces of $\mathbf{R}\mathbf{K}(n)$ and $\mathbf{K}(n)\mathbf{R}$ are equal, we may apply the formula of Eq. (16.21) to the problem at hand, and so express the misadjustment of the LMS algorithm as

$$\mathcal{M}(n) \simeq \frac{\mu}{2}\,\mathrm{tr}[\mathbf{R}] + \frac{1}{2\mu\sigma_v^2}\,\mathrm{tr}[\mathbf{Q}], \qquad n \text{ large} \tag{16.35}$$

The first term, $\mu\mathrm{tr}[\mathbf{R}]/2$, is the noise misadjustment caused by the measurement noise $v(n)$; this term is of the same form as in a stationary environment, which is not surprising. The second term, $\mathrm{tr}[\mathbf{Q}]/2\mu\sigma_v^2$, is the lag misadjustment caused by the process noise vector $\boldsymbol{\omega}(n)$, which is representative of nonstationarity in the environment.

The noise misadjustment varies *linearly* with the step-size parameter μ, whereas the lag misadjustment varies *inversely* with μ. The optimum value of the step-size parameter, μ_{opt}, for which the misadjustment attains its minimum value, $\mathcal{M}_{\min}$, occurs when the estimation noise and lag noise are equal. We thus readily find from Eq. (16.35) that

$$\mu_{\mathrm{opt}} \simeq \frac{1}{\sigma_v}\left(\frac{\mathrm{tr}[\mathbf{Q}]}{\mathrm{tr}[\mathbf{R}]}\right)^{1/2} \tag{16.36}$$

and

$$\mathcal{M}_{\min} \simeq \frac{1}{\sigma_v}\,(\mathrm{tr}[\mathbf{R}]\mathrm{tr}[\mathbf{Q}])^{1/2} \tag{16.37}$$

Equations (16.33) and (16.37) indicate that, in general, optimization of the two figures of merit, the mean-square deviation and misadjustment, leads to different values for the optimum setting of the step-size parameter μ. This should not be surprising since these two figures of merit emphasize different aspects of the tracking problem. However the choice is made, it is presumed that the optimum μ satisfies the condition for convergence of the LMS algorithm in the mean square.

16.5 TRACKING PERFORMANCE OF THE RLS ALGORITHM

Consider next the RLS algorithm used to implement the adaptive filter in the system model of Fig. 16.1. From Chapter 13 we recall that the corresponding update equation for the weight vector $\hat{\mathbf{w}}(n)$ of the adaptive transversal filter may be written in the form

$$\hat{\mathbf{w}}(n) = \hat{\mathbf{w}}(n-1) + \boldsymbol{\Phi}^{-1}(n)\mathbf{u}(n)\xi^*(n) \tag{16.38}$$

where $\mathbf{\Phi}(n)$ is the correlation matrix of the input vector $\mathbf{u}(n)$:

$$\mathbf{\Phi}(n) = \sum_{i=1}^{n} \lambda^{n-i} \mathbf{u}(i)\mathbf{u}^H(i) \tag{16.39}$$

and $\xi(n)$ is the *a priori* estimation error:

$$\xi(n) = d(n) - \hat{\mathbf{w}}^H(n-1)\mathbf{u}(n) \tag{16.40}$$

To accommodate the slight change in the notation for the weight vector in Eq. (16.38) compared to that in Eq. (16.23), we modify the first-order Markov model of Eq. (16.1) and the desired response $d(n)$ of Eq. (16.2) as follows, respectively:

$$\mathbf{w}_o(n) = a\mathbf{w}_o(n-1) + \boldsymbol{\omega}(n) \tag{16.41}$$

and

$$d(n) = \mathbf{w}_o^H(n-1)\mathbf{u}(n) + v(n) \tag{16.42}$$

Accordingly, using Eqs. (16.38), (16.41), and (16.42), we may express the update equation for the weight-error vector in the RLS algorithm as shown by

$$\boldsymbol{\epsilon}(n) = [\mathbf{I} - \mathbf{\Phi}^{-1}(n)\mathbf{u}(n)\mathbf{u}^H(n)]\boldsymbol{\epsilon}(n-1) + \mathbf{\Phi}^{-1}(n)\mathbf{u}(n)v^*(n) + (1-a)\mathbf{w}_o(n-1) - \boldsymbol{\omega}(n) \tag{16.43}$$

where $\mathbf{I}$ is the identity matrix. The linear stochastic difference equation (16.43) provides a complete description of the RLS algorithm embedded in the system model of Fig. 16.1, bearing in mind the aforementioned minor change in notation. As with the LMS algorithm, we assume that the model parameter a is very close to 1, so that we may ignore the term $(1-a)\mathbf{w}_o(n-1)$. That is, the process equation is described essentially by a random-walk model, for which Eq. (16.43) reduces to

$$\boldsymbol{\epsilon}(n) = [\mathbf{I} - \mathbf{\Phi}^{-1}(n)\mathbf{u}(n)\mathbf{u}^H(n)]\boldsymbol{\epsilon}(n-1) + \mathbf{\Phi}^{-1}(n)\mathbf{u}(n)v^*(n) - \boldsymbol{\omega}(n) \tag{16.44}$$

Before proceeding further, it is instructive to find an approximation for the inverse matrix $\mathbf{\Phi}^{-1}(n)$ that makes the tracking analysis of the RLS algorithm mathematically trackable in a meaningful manner. To do so, we first take the expectation of both sides of Eq. (16.39), obtaining

$$E[\mathbf{\Phi}(n)] = \sum_{i=1}^{n} \lambda^{n-i} E[\mathbf{u}(i)\mathbf{u}^H(i)]$$

$$= \sum_{i=1}^{n} \lambda^{n-i} \mathbf{R} \tag{16.45}$$

$$= \mathbf{R}(1 + \lambda + \lambda^2 + \ldots + \lambda^{n-1})$$

where $\mathbf{R}$ is the ensemble-averaged correlation matrix of the input vector $\mathbf{u}(n)$. The series inside the parentheses on the right-hand side of Eq. (16.45) represents a geometric series with the following description: a first term equal to unity, geometric ratio equal to λ, and length equal to n. Assuming that n is large enough for us to treat the geometric series to be essentially of infinite length, we may use the formula for the sum of such a series to rewrite Eq. (16.45) in the compact form

$$E[\mathbf{\Phi}(n)] = \frac{\mathbf{R}}{1 - \lambda}, \qquad n \text{ large} \tag{16.46}$$

Equation (16.46) defines the expectation (ensemble average) of $\mathbf{\Phi}(n)$, on the basis of which we may express $\mathbf{\Phi}(n)$ itself as follows (Eleftheriou and Falconer, 1986):

$$\mathbf{\Phi}(n) = \frac{\mathbf{R}}{1 - \lambda} + \tilde{\mathbf{\Phi}}(n), \qquad n \text{ large} \tag{16.47}$$

where $\tilde{\mathbf{\Phi}}(n)$ is a Hermitian *perturbation matrix* whose individual entries are represented by zero-mean random variables that are statistically independent from the input vector $\mathbf{u}(n)$. Assuming a slow adaptive process (i.e., the exponential weighting factor λ is close to unity), we may view the $\mathbf{\Phi}(n)$ in Eq. (16.47) as a *quasi-deterministic* matrix, in the sense that for large n we have[2]

$$E[\|\tilde{\mathbf{\Phi}}(n)\|^2] << E[\|\mathbf{\Phi}(n)\|^2]$$

where $\|\cdot\|$ denotes matrix norm. Under this condition, we may go one step further by ignoring the perturbation matrix $\tilde{\mathbf{\Phi}}(n)$, and so approximate the correlation matrix $\mathbf{\Phi}(n)$ as

$$\mathbf{\Phi}(n) \simeq \frac{\mathbf{R}}{1 - \lambda}, \qquad n \text{ large} \tag{16.48}$$

This approximation is crucial to the tracking analysis of the RLS algorithm presented herein. In a corresponding way to Eq. (16.48), we may express the inverse matrix $\mathbf{\Phi}^{-1}(n)$ as

$$\mathbf{\Phi}^{-1}(n) \simeq (1 - \lambda)\mathbf{R}^{-1}, \qquad n \text{ large} \tag{16.49}$$

where $\mathbf{R}^{-1}$ is the inverse of the ensemble-averaged correlation matrix $\mathbf{R}$.

Returning to Eq. (16.44) and using the approximation of (16.49) for $\mathbf{\Phi}^{-1}(n)$, we may now write

$$\begin{aligned}
\boldsymbol{\epsilon}(n) \simeq\ & [\mathbf{I} - (1 - \lambda)\mathbf{R}^{-1}\mathbf{u}(n)\mathbf{u}^H(n)]\boldsymbol{\epsilon}(n-1) \\
& + (1 - \lambda)\mathbf{R}^{-1}\mathbf{u}(n)v^*(n) - \boldsymbol{\omega}(n), \qquad n \text{ large}
\end{aligned} \tag{16.50}$$

---

[2]A completely general proof that the correlation matrix $\mathbf{\Phi}(n)$ is quasi-deterministic is yet to be presented in the literature. This issue was apparently first discussed in Eleftheriou and Falconer (1986) using heuristic arguments. It is also discussed in Macchi and Bershad (1991), where in Appendix II of that paper a proof is presented for the case of a nonstationary signal, namely, a noisy chirped sinusoid. This signal includes the commonly encountered example of a pure sinusoid in additive white Gaussian noise as a special case, which validates the proof for a stationary environment, too. However, a limitation of the proof presented by Macchi and Bershad is that it hinges on the unrealistic assumption that successive input vectors are statistically independent.

Typically, the exponential weighting factor λ is close to unity so that $1 - \lambda$ has a small value. Then, invoking the direct-averaging method, we may state that the solution $\epsilon(n)$ is "close" to the solution of the new stochastic difference equation:

$$\epsilon(n) \simeq \lambda \, \epsilon(n-1) + (1-\lambda)\mathbf{R}^{-1}\mathbf{u}(n)v^*(n) - \omega(n), \qquad n \text{ large} \qquad (16.51)$$

which is obtained by replacing the system matrix $[\mathbf{I} - (1-\lambda)\mathbf{R}^{-1}\mathbf{u}(n)\mathbf{u}^H(n)]$ in Eq. (16.50) by its ensemble average $\lambda\mathbf{I}$. For convenience of presentation, we have again retained the same notation for the weight-error vector in Eq. (16.51) as that used previously. Finally, evaluating the correlation matrix of $\epsilon(n)$ in Eq. (16.51) and invoking the independence assumption, we obtain

$$\mathbf{K}(n) \simeq \lambda^2 \, \mathbf{K}(n-1) + (1-\lambda)^2\sigma_v^2\mathbf{R}^{-1} + \mathbf{Q}, \qquad n \text{ large} \qquad (16.52)$$

Equation (16.52) for the RLS algorithm has a form that is dramatically different from that of Eq. (16.28) for the LMS algorithm, which, of course, is to be expected.

For a *steady-state solution* of the difference equation (16.52), for which n is large, we may legitimately set $\mathbf{K}(n - 1) = \mathbf{K}(n)$. Under this condition, Eq. (16.52) takes on the simplified form

$$(1-\lambda^2)\mathbf{K}(n) \simeq (1-\lambda)^2\sigma_v^2\mathbf{R}^{-1} + \mathbf{Q}, \qquad n \text{ large} \qquad (16.53)$$

For λ close to unity, we may approximate $1-\lambda^2$ as follows:

$$\begin{aligned} 1 - \lambda^2 &= (1-\lambda)(1+\lambda) \\ &\simeq 2(1-\lambda) \end{aligned} \qquad (16.54)$$

Accordingly, we may further simplify the correlation matrix $\mathbf{K}(n)$ for the RLS algorithm as follows:

$$\mathbf{K}(n) \simeq \frac{1-\lambda}{2} \, \sigma_v^2\mathbf{R}^{-1} + \frac{1}{2(1-\lambda)} \, \mathbf{Q}, \qquad n \text{ large} \qquad (16.55)$$

This is the equation for evaluating the tracking capability of the RLS algorithm for the system identification problem described in Fig. 16.1, subject to the condition that a is close to unity.

Mean-square Deviation of the RLS Algorithm

Applying the formula of Eq. (16.10) to (16.55), we readily find that the mean-square deviation of the RLS algorithm is defined by

$$\mathscr{D}(n) \simeq \frac{1-\lambda}{2} \, \sigma_v^2 \text{tr}[\mathbf{R}^{-1}] + \frac{1}{2(1-\lambda)} \, \text{tr}[\mathbf{Q}], \qquad n \text{ large} \qquad (16.56)$$

The first term, $(1 - \lambda)\sigma_v^2\text{tr}[\mathbf{R}^{-1}]/2$, is the estimation variance due to the measurement noise $v(n)$. The second term, $\text{tr}[\mathbf{Q}]/2(1 - \lambda)$, is the lag variance due to the process noise vector ω. These two contributions vary in proportion to $(1 - \lambda)$ and $(1 - \lambda)^{-1}$, respectively. The optimum value of the forgetting factor, λ_{opt}, occurs when these two contributions are equal. Thus, from Eq. (16.56) we readily find that

$$\lambda_{opt} \simeq 1 - \frac{1}{\sigma_v} \left(\frac{\text{tr}[\mathbf{Q}]}{\text{tr}[\mathbf{R}^{-1}]} \right)^{1/2} \tag{16.57}$$

Correspondingly, the minimum mean-square deviation of the RLS algorithm is given by

$$\mathcal{D}_{min} \simeq \sigma_v \, (\text{tr}[\mathbf{R}^{-1}]\text{tr}[\mathbf{Q}])^{1/2} \tag{16.58}$$

Misadjustment of the RLS Algorithm

Multiplying both sides of Eq. (16.55) by the correlation matrix $\mathbf{R}$, we get

$$\mathbf{R}\mathbf{K}(n) \simeq \frac{1 - \lambda}{2} \sigma_v^2 \mathbf{I} + \frac{1}{2(1 - \lambda)} \mathbf{R}\mathbf{Q}, \qquad n \text{ large} \tag{16.59}$$

The identity matrix $\mathbf{I}$ is of size M-by-M, where M is the number of taps in the adaptive transversal filter. Hence, taking the trace of the two sides of Eq. (16.59) yields

$$\text{tr}[\mathbf{R}\mathbf{K}(n)] \simeq \frac{1 - \lambda}{2} \sigma_v^2 M + \frac{1}{2(1 - \lambda)} \text{tr}[\mathbf{R}\mathbf{Q}], \qquad n \text{ large} \tag{16.60}$$

Finally, using the formula of Eq. (16.21), we readily find that the misadjustment of the RLS algorithm is given by

$$\mathcal{M}(n) \simeq \frac{1 - \lambda}{2} M + \frac{1}{2(1 - \lambda)\sigma_v^2} \text{tr}[\mathbf{R}\mathbf{Q}], \qquad n \text{ large} \tag{16.61}$$

The first term on the right-hand side of Eq. (16.61) represents the noise misadjustment of the RLS algorithm due to the measurement noise $v(n)$. It varies linearly with $1 - \lambda$; note also that it depends on the number of taps M in the adaptive transversal filter. The second term represents the lag misadjustment of the RLS algorithm due to the process noise vector $\boldsymbol{\omega}(n)$; it varies inversely with $1 - \lambda$. The optimum value of the forgetting factor, λ_{opt}, occurs when these two contributions are equal. We thus find from Eq. (16.61) that

$$\lambda_{opt} \simeq 1 - \frac{1}{\sigma_v} \left(\frac{1}{M} \text{tr}[\mathbf{R}\mathbf{Q}] \right)^{1/2} \tag{16.62}$$

Correspondingly, the minimum value of the misadjustment produced by the RLS algorithm is given by

$$\mathcal{M}_{min} \simeq \frac{1}{\sigma_v} (M \, \text{tr}[\mathbf{R}\mathbf{Q}])^{1/2} \tag{16.63}$$

Here also we find that the two criteria, minimum misadjustment and minimum mean-square deviation, lead to different values for λ_{opt}. For these values to be meaningful, we must have $0 < \lambda_{opt} < 1$.

We now have all the tools we need to make a quantitative comparison between the LMS and RLS algorithms in the context of the system model depicted in Fig. 16.1.

16.6 COMPARISON OF THE TRACKING PERFORMANCE OF LMS AND RLS ALGORITHMS

Realizing that the LMS and RLS algorithms are formulated in entirely different ways, it is only natural to find that they exhibit not only different convergence properties but also different tracking properties. The difference in their tracking behavior may be traced back to the stochastic difference equations (16.26) and (16.50). In the RLS algorithm, the input vector $\mathbf{u}(n)$ is premultiplied by the inverse matrix $\mathbf{R}^{-1}$, wherein lies the fundamental difference between the LMS and RLS algorithms. Moreover, comparing Eqs. (16.26) and (16.50) on which the tracking analysis presented in the previous two sections is based, we see that $1 - \lambda$ in the RLS algorithm plays an analogous role to that of μ in the LMS algorithm. In making this analogy, however, we should try to be more precise. In particular, the exponential weighting factor λ is dimensionless, whereas the step-size parameter μ has the inverse dimension of power. To correct for this dimensional discrepancy, we do the following:

- For the LMS algorithm, we define the *normalized step-size parameter*

$$\nu = \mu \sigma_u^2 \tag{16.64}$$

  where σ_μ^2 is the variance of the zero-mean tap input $u(n)$.

- For the RLS algorithm, we define the *forgetting rate*

$$\beta = 1 - \lambda \tag{16.65}$$

Moving into the main issue of interest, we may use Eqs. (16.33) and (16.58), and Eqs. (16.37) and (16.63) to formulate a corresponding pair of ratios for comparing the "optimum" tracking performance of the LMS and RLS algorithms for the system identification problem at hand; one ratio is based on the mean-square deviation and the other is based on misadjustment as the figure of merit. Specifically, we may write

$$\frac{\mathcal{D}_{\min}^{\text{LMS}}}{\mathcal{D}_{\min}^{\text{RLS}}} \simeq \left(\frac{M \text{tr}[\mathbf{R}^{-1}\mathbf{Q}]}{\text{tr}[\mathbf{R}^{-1}]\text{tr}[\mathbf{Q}]} \right)^{1/2} \tag{16.66}$$

and

$$\frac{\mathcal{M}_{\min}^{\text{LMS}}}{\mathcal{M}_{\min}^{\text{RLS}}} \simeq \left(\frac{\text{tr}[\mathbf{R}]\text{tr}[\mathbf{Q}]}{M \text{tr}[\mathbf{RQ}]} \right)^{1/2} \tag{16.67}$$

where $\mathbf{R}$ is the correlation matrix of the input vector $\mathbf{u}(n)$, $\mathbf{Q}$ is the correlation matrix of the process noise vector $\boldsymbol{\omega}(n)$, and M is the number of taps in the adaptive transversal filter of Fig. 16.1. Clearly, whatever comparison we make between the LMS and RLS algorithms on the basis of Eqs. (16.66) and (16.67), the result depends on the prevalent environmental conditions and, in particular, on how the correlation matrices $\mathbf{Q}$ and $\mathbf{R}$ are defined. In what follows, we consider three specific examples.[3]

---

[3]Example 1 is discussed in Widrow and Walach (1984) and Eleftheriou and Falconer (1986); Examples 2 and 3 are discussed in Benveniste et al. (1987) and Slock and Kailath (1993).

Example 1: $Q = \sigma_\omega^2 I$

Consider first the case of process noise vector $\omega(n)$ in the first-order Markov model of Eq. (16.1) originating from a white noise source of zero mean and variance σ_ω^2. We may thus express the correlation matrix Q of $\omega(n)$ as

$$Q = \sigma_\omega^2 I \qquad (16.68)$$

where I is the M-by-M identity matrix. Then, using Eq. (16.68) in (16.66) and (16.67), we get the following respective results (after canceling common terms):

$$\mathscr{D}_{min}^{LMS} \simeq \mathscr{D}_{min}^{RLS}, \qquad Q = \sigma_\omega^2 I \qquad (16.69)$$

and

$$\mathscr{M}_{min}^{LMS} \simeq \mathscr{M}_{min}^{RLS}, \qquad Q = \sigma_\omega^2 I \qquad (16.70)$$

Accordingly, we may state that the LMS and RLS algorithms produce essentially the same minimum levels of misadjustment and mean-square deviation for the case of a process noise vector $\omega(n)$ drawn from a white-noise source.

Example 2: $Q = c_1 R$

Consider next the example when the correlation matrix Q of the process noise vector $\omega(n)$ in the first-order Markov model of Eq. (16.1) equals a constant c_1 times the correlation matrix R of the input vector $u(n)$. The scaling factor c_1 is introduced here for two reasons:

1. To account for the fact that the process noise vector $\omega(n)$ and the input vector $u(n)$ are ordinarily measured in different units.
2. To ensure that the optimum μ for the LMS algorithm in Eq. (16.32) or (16.36), and the optimum λ for the RLS algorithm in Eq. (16.57) or (16.62) assume meaningful values.

Thus, putting $Q = c_1 R$ in Eqs. (16.66) and (16.67) and canceling the scaling factor c_1, we get the two comparative yardsticks listed under $Q = c_1 R$ in Table 16.1. Before commenting on these results, it is instructive to go on and consider the complementary example described next.

TABLE 16.1   COMPARATIVE YARDSTICKS FOR LMS AND RLS ALGORITHMS FOR EXAMPLES 2 AND 3

	$Q = c_1 R$	$Q = c_2 R^{-1}$
$\dfrac{\mathscr{D}_{min}^{LMS}}{\mathscr{D}_{min}^{RLS}}$	$\dfrac{M}{(\mathrm{tr}[R^{-1}]\,\mathrm{tr}[R])^{1/2}}$	$\dfrac{(M\,\mathrm{tr}[R^{-2}])^{1/2}}{\mathrm{tr}[R^{-1}]}$
$\dfrac{\mathscr{M}_{min}^{LMS}}{\mathscr{M}_{min}^{RLS}}$	$\dfrac{\mathrm{tr}[R]}{(M\,\mathrm{tr}[R^2])^{1/2}}$	$\dfrac{1}{M}(\mathrm{tr}[R]\mathrm{tr}[R^{-1}])^{1/2}$

Example 3. $Q = c_2 R^{-1}$

In this final example, the correlation matrix $\mathbf{Q}$ of the process noise vector $\boldsymbol{\omega}(n)$ is equal to a constant c_2 times the inverse of the correlation matrix $\mathbf{R}$ of the input vector $\mathbf{u}(n)$. The scaling factor c_2 is used here for exactly the same reasons explained in Example 2. Thus, putting $\mathbf{Q} = c_2\mathbf{R}^{-1}$ in Eqs. (16.66) and (16.67) and again canceling the scaling factor c_2, we get the remaining two comparative yardsticks listed under $\mathbf{Q} = c_2\mathbf{R}^{-1}$ in Table 16.1.

The 2-by-2 array of entries shown in this table exhibits a useful property, namely, that of *reciprocal symmetry*. The significance of this property in the context of tracking will become apparent presently.

The cross-diagonal terms in the array of Table 16.1 lend themselves to the application of the *Cauchy–Schwarz inequality*, which results in the following two useful bounds (see Problem 5):

$$\mathcal{M}_{\min}^{\text{LMS}} \leq \mathcal{M}_{\min}^{\text{RLS}} \qquad \text{for } \mathbf{Q} = c_1\mathbf{R} \tag{16.71}$$

and

$$\mathcal{D}_{\min}^{\text{RLS}} \leq \mathcal{D}_{\min}^{\text{LMS}} \qquad \text{for } \mathbf{Q} = c_2\mathbf{R}^{-1} \tag{16.72}$$

These two bounds are indeed manifestations of the property of reciprocal symmetry, on the basis of which we may make the following two statements:

1. For $\mathbf{Q} = c_1\mathbf{R}$, the LMS algorithm performs better than the RLS algorithm, in that it produces a minimum level of misadjustment $\mathcal{M}_{\min}$ that is smaller than the corresponding value produced by the RLS algorithm.

2. For $\mathbf{Q} = c_2\mathbf{R}^{-1}$, the RLS algorithm performs better than the LMS algorithm, in that it produces a minimum mean-square deviation $\mathcal{D}_{\min}$ that is smaller than the corresponding value produced by the LMS algorithm.

However, in general, we cannot be as conclusive on the implications of the diagonal entries in Table 16.1. Nevertheless, in light of the aforementioned property of reciprocal symmetry, we can say the following. If, for $\mathbf{Q} = c_1\mathbf{R}$, the minimum mean-square deviation $\mathcal{D}_{\min}$ produced by the LMS algorithm is smaller than the corresponding value produced by the RLS algorithm, then it is true that for $\mathbf{Q} = c_2\mathbf{R}^{-1}$ the minimum misadjustment $\mathcal{M}_{\min}$ produced by the RLS algorithm is smaller than the corresponding value produced by the LMS algorithm.

To illustrate the validity of this latter statement, consider the special case of an adaptive filter with $M = 2$, for which the 2-by-2 correlation matrix of the input vector $\mathbf{u}(n)$ is denoted by

$$\mathbf{R} = \begin{bmatrix} r_{11} & r_{21} \\ r_{21} & r_{22} \end{bmatrix}$$

For this specification of $\mathbf{R}$, the 2-by-2 array of Table 16.1 takes on the particular form presented in Table 16.2. Next, recognizing that since any 2-by-2 correlation matrix satisfies the condition

$$(r_{11} - r_{22})^2 + (2r_{21})^2 \geq 0,$$

TABLE 16.2   COMPARATIVE YARDSTICKS FOR LMS AND RLS
ALGORITHMS FOR EXAMPLES 2 AND 3, ASSUMING $M = 2$

	$\mathbf{Q} = c_1 \mathbf{R}$	$\mathbf{Q} = c_2 \mathbf{R}^{-1}$
$\dfrac{\mathscr{D}_{min}^{LMS}}{\mathscr{D}_{min}^{RLS}}$	$\dfrac{2\sqrt{r_{11}r_{22}-r_{21}^2}}{r_{11} + r_{22}}$	$\dfrac{\sqrt{2(r_{11}^2 + 2r_{21}^2 + r_{22}^2)}}{r_{11} + r_{22}}$
$\dfrac{\mathscr{M}_{min}^{LMS}}{\mathscr{M}_{min}^{RLS}}$	$\dfrac{r_{11} + r_{22}}{\sqrt{2(r_{11}^2 + 2r_{21}^2 + r_{22}^2)}}$	$\dfrac{r_{11} + r_{22}}{2\sqrt{r_{11}r_{22}-r_{21}^2}}$

then Table 16.2 leads us to make the following statements encompassing all the four entries of the array:

1. For $\mathbf{Q} = c_1\mathbf{R}$, the LMS algorithm performs better than the RLS algorithm, in that it yields smaller values for both $\mathscr{D}_{min}$ and $\mathscr{M}_{min}$.
2. For $\mathbf{Q} = c_2\mathbf{R}^{-1}$, the RLS algorithm performs better than the LMS algorithm, in that it yields smaller values for both $\mathscr{D}_{min}$ and $\mathscr{M}_{min}$.

Examples 2 and 3 clearly illustrate that neither the LMS algorithm nor the RLS algorithm has a complete monopoly over a good tracking behavior. Rather, we find that one or the other of these two adaptive filtering algorithms is the preferred algorithm for tracking a nonstationary environment, depending on the prevalent environmental conditions.

16.7 ADAPTIVE RECOVERY OF A CHIRPED SINUSOID IN NOISE

Up to this point in our discussion of using an adaptive filter to track a time-varying system, we have focused our attention on the performance of LMS and RLS algorithms in the context of system identification. As mentioned previously in Section 16.1, in such a scenario, only the cross-correlation vector between the input vector $\mathbf{u}(n)$ and the desired response $d(n)$ is time varying. In this section we briefly consider a more difficult problem: the *adaptive recovery of a chirped sinusoid* (tone) buried in additive white Gaussian noise (Macchi and Bershad, 1991; Bershad and Macchi, 1991; Macchi, 1995).

To perform such a task, we may use an *adaptive line enhancer* (ALE) that consists of a one-step predictor, configured as shown in Fig. 16.2. The received (input) signal $u(n)$ consists of two components:

$$u(n) = s(n) + v(n) \qquad (16.73)$$

where $s(n)$ is the desired signal and $v(n)$ is the additive noise component. Typically, the desired signal $s(n)$ has a much narrower bandwidth than the noise $v(n)$. This property is exploited by the ALE to produce an output signal $\hat{s}(n)$ that represents an estimate of the desired signal $s(n)$. The difference between the input signal $u(n)$ and the output signal $\hat{s}(n)$ defines the error signal $e(n)$, which is used to adjust the tap weights of the adaptive filter.

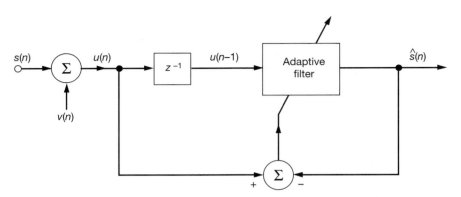

Figure 16.2   Adaptive line enhancer.

The adaptive recovery of a chirped sinusoid in noise is of special interest for several reasons:

- The chirped sinusoid represents a well-defined form of nonstationarity that is of a deterministic nature.
- The chirped sinusoid, characterized by a linear shift in frequency, may be used as a first-order model of the Doppler effect encountered in mobile communications.
- In adaptive prediction, both the correlation matrix of the input vector $\mathbf{u}(n)$ and the cross-correlation vector between $\mathbf{u}(n)$ and the desired response represented by the first element of $\mathbf{u}(n)$ are time varying. Accordingly, a mathematical analysis of tracking a chirped sinusoid in noise is more difficult than the system identification problem.

The chirped sinusoid, denoted by $s(n)$, is defined by the complex exponential

$$s(n) = \sqrt{P_s}\, \exp\!\left(j2\pi f_c n + j\frac{\psi}{2}\, n^2 + \varphi\right) \tag{16.74}$$

where $\sqrt{P_s}$ is the *signal amplitude* assumed to be constant, f_c is the *center frequency*, ψ is the chirp rate, and φ is an arbitrary phase shift. The signal $s(n)$ is deterministic but nonstationary because of the chirping. The instantaneous angular frequency of the chirped sinusoid $s(n)$ is defined by the derivative

$$\frac{d}{dn}\left(2\pi f_c n + \frac{\psi}{2}\, n^2 + \varphi\right) = 2\pi f_c + \psi n$$

The angular frequency deviation, measured inside a time interval τ, is $\psi\tau$. Equation (16.74) may therefore be viewed as a narrow-band model, provided that we have

$$\psi\tau \ll 2\pi f_c$$

The complex chirped sinusoid $s(n)$ is corrupted by additive white Gaussian noise $v(n)$ of zero mean and variance σ_v^2, as indicated in Eq. (16.73). The signal-to-noise ratio at the ALE input is thus defined by

$$\rho = \frac{P_s}{\sigma_v^2} \tag{16.75}$$

A detailed mathematical analysis of recovery of the chirped sinusoid $s(n)$ from the noisy received signal $u(n)$ using an ALE based on the LMS algorithm, is presented in (Bershad and Macchi, 1991). The corresponding analysis for the case of an ALE based on the RLS algorithm is presented in (Macchi and Bershad, 1991). In order to obtain functionally similar parameters for both the LMS and RLS algorithms, and for reasons discussed in Section 16.6, the following *normalized adaptation constants* are introduced for these two algorithms (Macchi et al., 1991):

$$v = \begin{cases} \mu \sigma_u^2 (1 + \rho) & \text{for LMS} \\ \beta = 1 - \lambda & \text{for RLS} \end{cases} \tag{16.76}$$

The definition of v for the RLS algorithm is the same as before; but for the LMS algorithm, it now incorporates the factor $(1 + \rho)$, where ρ is the signal-to-noise ratio. For both the LMS and RLS algorithms, highlights of the findings reported in (Macchi and Bershad, 1991; Bershad and Macchi, 1991) may be summarized as follows:

- The noise misadjustment due to the weight vector noise $\epsilon_1(n)$ is dominated by a term of order v, as shown by

$$\mathcal{M}_1 = \frac{MJ_{\min}}{2} v \tag{16.77}$$

where $J_{\min}$ is the minimum mean-squared error produced by the Wiener filter; for the chirp input it is given by

$$J_{\min} = \frac{M + 1}{M} \sigma_v^2 \tag{16.78}$$

- The lag misadjustment due to the mean lag vector $\epsilon_2(n)$ is dominated by a term of order v^{-2}, as shown by

$$\mathcal{M}_2 = \frac{\psi^2}{v^2} C(\rho, M) \sigma_v^2 \tag{16.79}$$

where $C(\rho, M)$ is a proportionality factor depending on the algorithm used. Specifically, we have

$$C(\rho, M) = \begin{cases} \dfrac{1}{12}(M + 1)\left(1 - \dfrac{1}{M}\right)(1 + \rho)^2 & \text{for LMS} \\[2ex] \left(\dfrac{M + 1}{2}\right)^2 \rho & \text{for RLS} \end{cases} \tag{16.80}$$

The overall misadjustment is equal to the sum of $\mathcal{M}_1$ and $\mathcal{M}_2$ for both algorithms:

$$\mathcal{M} = \frac{MJ_{\min}}{2} v + \frac{\psi^2}{v^2} C(\rho, M)\sigma_v^2 \tag{16.81}$$

The overall misadjustment $\mathcal{M}$ is minimized by setting the normalized adaptation constant ν equal to the optimum value:

$$
\nu_{\text{opt}} = \begin{cases} \left(\dfrac{\psi^2}{3}\left(1 - \dfrac{1}{M}\right)(1 + \rho)^2 \right)^{1/3} & \text{for LMS} \\[2em] (\rho\psi^2(M + 1))^{1/3} & \text{for RLS} \end{cases} \tag{16.82}
$$

For both algorithms, the minimum misadjustment is

$$
\mathcal{M}_{\text{min}} = \frac{3}{4} MJ_{\text{min}}\nu_{\text{opt}} \tag{16.83}
$$

The relative tracking performance of the LMS and RLS algorithms for the recovery of a chirped sinusoid in noise may be determined by considering the ratio:

$$
\frac{\mathcal{M}_{\text{min}}^{\text{LMS}}}{\mathcal{M}_{\text{min}}^{\text{RLS}}} = \frac{\nu_{\text{opt}}^{\text{LMS}}}{\nu_{\text{opt}}^{\text{RLS}}}
$$

$$
\simeq \left(\frac{(1 + \rho)^2}{3\rho M} \right)^{1/3}, \qquad M \text{ large} \tag{16.84}
$$

where, in the last line, the approximation is taken for large M. Based on the result of Eq. (16.84), we readily find that for $\rho < 3M$ the LMS algorithm has a smaller misadjustment than the RLS algorithm, and for $\rho > 3M$ the RLS algorithm has a smaller misadjustment than the LMS algorithm. In other words, for small signal-to-noise ratios the LMS algorithm has a better tracking performance than the RLS algorithm, and for large signal-to-noise ratios the reverse is true.

Sensitivity to the Choice of Adaptation Rate

The minimum misadjustment of Eq. (16.83) for the chirped sinusoid in noise is only 50 percent greater than the misadjustment of Eq. (16.77) for the stationary case of an ordinary sinusoid in noise, providing that the optimum adaptation constant of Eq. (16.82) is utilized. The main difference between the stationary and chirped sinusoidal cases is that, for the stationary case, the misadjustment can be made arbitrarily small by decreasing ν as indicated by Eq. (16.77). For the chirped sinusoidal case, on the other hand, the penalty incurred by setting ν too small is an unbounded increase in the lag misadjustment $\mathcal{M}_2$, as indicated by Eq. (16.79). By using the optimum value of ν specified in Eq. (16.82), the best compromise between the weight vector noise $\boldsymbol{\epsilon}_1(n)$ and the mean lag vector $\boldsymbol{\epsilon}_2(n)$ is achieved. This is illustrated further in what follows for the RLS algorithm.[4]

The sensitivity in tracking performance to the choice of the adaptation constant $\nu = (1 - \beta) = 1 - \lambda$ for the RLS algorithm increases with increasing chirp rate ψ. This can be illustrated by evaluating the reflection coefficients of the recursive least-squares lattice (LSL) algorithm (i.e., lattice implementation of the RLS algorithm). For stationary

---

[4]The material presented in this subsection is based on Zeidler, J.R., private communication, 1995.

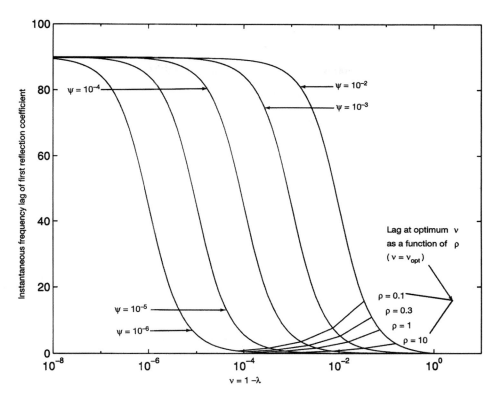

Figure 16.3   The instantaneous frequency lag as a function of the adaptation constant ν. The frequency lag for $\nu = \nu_{\text{opt}}$ is indicated for $\rho = 0.1$ to $\rho = 10$ on the curves.

inputs, the derivations of these reflection coefficients were presented in Chapter 15. The analysis of that chapter has been extended to the nonstationary case of a chirped sinusoid in noise in (Soni et al., 1995). Figure 16.3 shows the expected value of the frequency lag derived by Soni et al. for the first reflection coefficient of a lattice predictor of order $M = 3$, plotted as a function of ν for a wide range of chirp rates between $\psi = 10^{-2}$ and $\psi = 10^{-6}$. The *frequency lag* is defined as the steady-state difference in phase lag between the reflection coefficient of a particular stage in the lattice predictor and the corresponding reflection coefficient of the "optimum" lattice predictor that *works just as well with a chirped sinusoid as with a pure sinusoid.*[5] It is thus a good measure for assessing the tracking performance of the lattice predictor.

---

[5]This optimum time-varying predictor is in perfect accord with an idea first described in (Macchi and Bershad, 1991) for a transversal filter implementation of the RLS algorithm.

The magnitude of the optimum v specified by Eq. (16.82) is also overlaid on the frequency lag versus v plots for values of the signal-to-noise ratio between $\rho = 0.1$ and $\rho = 10$ in order to illustrate the frequency lag associated with the optimum adaptation constant v_{opt}. Based on the results presented in Fig. 16.3, we note the following:

- For slow chirp rates, the instantaneous frequency lag associated with the optimum adaptation constant v_{opt} is less than one degree; and there is a wide range of values for v in the vicinity of v_{opt} that would provide a negligible frequency lag.

- For a fixed signal-to-noise ratio ρ, the tracking performance of the predictor deteriorates with increasing chirp rate ψ. For example, for the extreme case of $\psi = 10^{-2}$ and $\rho = 0.1$ shown in Fig. 16.3, the frequency lag associated with the optimum adaptation constant v_{opt} increases to about 20 degrees. Moreover, at this point, the slope of the frequency lag curve versus v is high, with the result that the frequency lag changes rapidly with variations in v.

- For a fixed chirp rate ψ, the tracking performance of the predictor improves with increasing signal-to-noise ratio ρ. For example, for the fast chirp rate $\psi = 10^{-2}$ and high signal-to-noise ratio $\rho = 10$, the frequency lag associated with the optimum adaptation constant v_{opt} decreases to less than 5 degrees. Also, at this point, the slope of the frequency lag versus v curve decreases considerably relative to its value at $\rho = 0.1$; in other words, the sensitivity of the tracking performance to variations in v is considerably reduced with increasing ρ.

The above results clearly illustrate that by selecting an adaptation constant v close to its optimum value v_{opt} defined by Eq. (16.82), the lattice predictor can effectively track a chirped sinusoidal signal in noise over a wide range of chirp rates and with an acceptably small frequency lag, provided that the signal-to-noise ratio is high. The results also indicate that the sensitivity in tracking performance to the selection of v increases nonlinearly with increasing chirp rate ψ and decreasing signal-to-noise ratio, which is intuitively satisfying.

Tracking of a Chirped Nonzero-bandwidth Signal in Noise

The desired signal in the study reported by Macchi and Bershad (1991) consists of a chirped sinusoid (tone), which is deterministic and therefore has no information content. In a subsequent study reported by Wei et al. (1994), the previous theory of Macchi and Bershad is extended to the more *general* case of a chirped signal buried in noise. In effect, the desired signal assumes a *finite bandwidth*, which makes it resemble a communication signal more closely.

To compare the tracking performances of the LMS and RLS algorithms for a chirped nonzero-bandwidth signal, Wei et al. (1994) consider an autoregressive (AR) process of order one that may be used to model many narrow-band signals. Specifically, the baseband signal of interest is modeled by the recursive equation:

$$s(n) = as(n - 1) + v(n) \qquad (16.85)$$

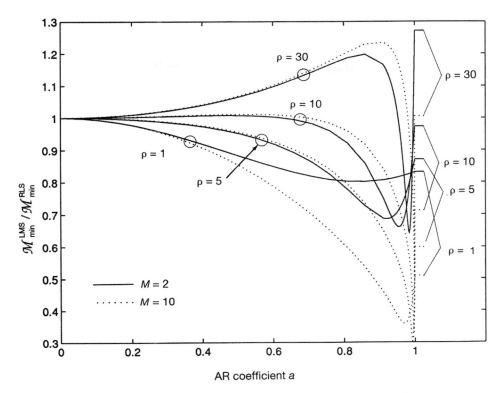

Figure 16.4   Ratio of LMS misadjustment versus RLS misadjustment for an AR process of order one.

where a is the AR coefficient, and $\nu(n)$ is a white noise process of zero mean and variance

$$\sigma_\nu^2 = \sigma_s^2 \, (1 - a^2) \tag{16.86}$$

where σ_s^2 is the variance of $s(n)$. In a mobile communications environment, for example, the baseband signal $u(n)$ is modulated onto a carrier wave for transmission over the channel, in the course of which it may also be Doppler-shifted by some relative motion between the transmitter and receiver. Accordingly, the baseband form of the received signal may be modeled as a chirped AR process of order one, as shown by (Wei et al., 1994)

$$s(n) = a\Omega\psi^{-1/2}\psi^n s(n - 1) + \nu(n) \tag{16.87}$$

where $\nu(n)$ is a white-noise process of variance $\sigma_s^2(1 - a^2)$.

Figure 16.4 plots the ratio $\mathcal{M}_{\min}^{\text{LMS}}/\mathcal{M}_{\min}^{\text{RLS}}$ versus the AR coefficient a for varying signal-to-noise ratio ρ and for two different values of filter length M. Based on this figure, we may make the following observations (Wei et al., 1994):

- At low signal-to-noise ratios ($\rho \leq 10$ dB), the LMS algorithm tracks the chirped AR process better than the RLS algorithm.

- For narrow-band signals ($a \simeq 0.95$), the LMS algorithm also performs better than the RLS algorithm.
- For high signal-to-noise ratios ($\rho \geqslant 10$ dB), the RLS algorithm performs better than the LMS algorithm.

These observations reinforce the earlier conclusions reached by Macchi and Bershad, in that the LMS algorithm or the RLS algorithm has a better tracking performance, depending on the signal-to-noise ratio.

16.8 HOW TO IMPROVE THE TRACKING BEHAVIOR OF THE RLS ALGORITHM

For a given "linear" model of a time-varying system, it is well known that the Kalman filter is the optimum tracking algorithm, assuming Gaussian statistics. In light of the material presented in Chapters 13 to 15, we know that there are one-to-one correspondences between the Kalman filter and the RLS algorithm. Simply put, the RLS algorithm is a special case of the Kalman filter. Given this intimate relationship between the RLS algorithm and Kalman filter, how then is it that the RLS algorithm and its variants have not fully inherited the good tracking properties of the Kalman filter? Before proceeding to offer a possible explanation for this dilemma, it is informative to recall the underlying state-space model of the standard RLS algorithm, which has the form (see Chapter 13)

$$\mathbf{x}(n + 1) = \lambda^{-1/2}\mathbf{x}(n) \tag{16.88}$$

$$y(n) = \mathbf{u}^{H}(n)\mathbf{x}(n) + v(n) \tag{16.89}$$

Comparing this state-space model with the first-order Markov model described in Eqs. (16.1) and (16.2), we immediately see that there is a serious *model mismatch* between these two situations. Specifically, the state equation (16.88) has zero process noise, which is in direct violation of what the Markov model described in Eq. (16.1) would imply. Herein lies the root of the problem.

The message that we wish to convey here is that if the system designer has prior knowledge of the underlying physical model of a particular task, then that knowledge should be exploited in formulating an appropriate state-space model for the RLS algorithm. In the sequel, we describe two examples illustrating how this objective can be accomplished.

Model I Incorporating a Process (State) Noise Vector

In the system identification problem described by Eqs. (16.1) and (16.2), the underlying Markov model has a nonzero process noise vector $\boldsymbol{\omega}(n)$. The state-space model for the RLS algorithm is therefore formulated as follows:

$$\mathbf{x}(n + 1) = a\mathbf{x}(n) + \mathbf{v}_1(n) \tag{16.90}$$

$$y(n) = \mathbf{u}^{H}(n)\mathbf{x}(n) + v_2(n) \tag{16.91}$$

The process (state) noise vector $\mathbf{v}_1(n)$ in Eq. (16.90) is modeled as having zero mean and correlation matrix $\mathbf{Q}(n)$, the same as that of the process noise vector $\boldsymbol{\omega}(n)$ in the Markov model of Eq. (16.2). For example, we may have $\mathbf{Q}(n) = q\mathbf{I}$, where $\mathbf{I}$ is the identity matrix. In such a case, the elements of $\mathbf{v}(n)$ constitute a set of independent white noise sources, each having zero mean and variance q. The implication of using the state equation (16.90) in place of that of (16.88) is that, in the language of Kalman filter theory, the filtered state-error correlation matrix $\mathbf{K}(n)$ and the predicted state-error correlation matrix $\mathbf{K}(n + 1, n)$ are no longer equal. In particular, in view of Eq. (16.90), and assuming that $\mathbf{Q}(n) = q\mathbf{I}$, the RLS algorithm is modified as follows (Haykin et al., 1995):

Starting with $\hat{\mathbf{w}}(0|-1) = E[\mathbf{w}_o(0)]$ and $\mathbf{P}(0,-1) = \boldsymbol{\Pi}_0$, compute for $n \geq 0$:

$$\mathbf{k}(n) = a\mathbf{P}(n, n - 1)\mathbf{u}(n) \, [\mathbf{u}^H(n)\mathbf{P}(n, n - 1)\mathbf{u}(n) + 1]^{-1}$$

$$\xi(n) = y(n) - \hat{\mathbf{w}}^H(n|n - 1)\mathbf{u}(n)$$

$$\hat{\mathbf{w}}(n + 1|n) = a\hat{\mathbf{w}}(n|n - 1) + \mathbf{k}(n)\xi^*(n)$$

$$\mathbf{P}(n) = \mathbf{P}(n, n - 1) - \frac{\mathbf{P}(n, n - 1)\mathbf{u}(n)\mathbf{u}^H(n)\mathbf{P}(n, n - 1)}{\mathbf{u}^H(n)\mathbf{P}(n, n - 1)\mathbf{u}(n) + 1}$$

$$\mathbf{P}(n + 1, n) = a^2\mathbf{P}(n) + q\mathbf{I}$$

This algorithm is hereafter referred to as the *extended RLS algorithm—version* 1 (ERLS-1).

In effect, the last computation step in the summary of the standard RLS algorithm presented in Table 13.1 has been replaced by the last two steps of ERLS-1. Note also that putting $q = 0$ and $a = 1$ reduces ERLS 1 to its standard form with $\lambda = 1$.

Model II Incorporating a Nonconstant Transition Matrix

The presence of a process noise vector, as in Eq. (16.90), describes one way in which non-stationarity can be accounted for. Another way in which nonstationarity can be represented is to have a transition matrix that is *not* a constant, as described here:

$$\mathbf{x}(n + 1) = \mathbf{F}(n + 1, n)\mathbf{x}(n) \tag{16.92}$$

$$y(n) = \mathbf{u}^H(n)\mathbf{x}(n) + v(n) \tag{16.93}$$

For example, in tracking a chirped signal in noise, the transition matrix $\mathbf{F}(n + 1, n)$ may consist of a diagonal matrix whose entries depend on the chirp rate and are unknown. Pursuing this example further, let $p_1, p_2, \ldots, p_M$ denote the *unknown* diagonal elements of the transition matrix, as shown by

$$\mathbf{F}(n + 1, n) = \mathrm{diag}[p_1, p_2, \ldots, p_M] \tag{16.94}$$

We may then define a new state vector:

$$\mathbf{x}'(n) = \begin{bmatrix} \mathbf{x}(n) \\ \mathbf{p} \end{bmatrix} \tag{16.95}$$

where

$$\mathbf{p} = [p_1, p_2, \ldots, p_M]^T \tag{16.96}$$

The original state-space model of Eqs. (16.92) and (16.93) may now be rewritten in the nonlinear form:

$$\mathbf{x}'(n+1) = \begin{bmatrix} \mathbf{x}(n+1) \\ \mathbf{p} \end{bmatrix}$$

$$= \begin{bmatrix} \mathrm{diag}[p_1, p_2, \ldots, p_M] & \mathbf{O} \\ \mathbf{O} & \mathbf{I} \end{bmatrix} \begin{bmatrix} \mathbf{x}(n) \\ \mathbf{p} \end{bmatrix} \qquad (16.97)$$

$$= \begin{bmatrix} \mathrm{diag}[p_1, p_2, \ldots, p_M] & \mathbf{O} \\ \mathbf{O} & \mathbf{I} \end{bmatrix} \mathbf{x}'(n)$$

$$y(n) = [\mathbf{u}^H(n), \mathbf{O}] \begin{bmatrix} \mathbf{x}(n) \\ \mathbf{p} \end{bmatrix} + v(n)$$

$$= [\mathbf{u}^H(n), \mathbf{O}] \, \mathbf{x}'(n) + v(n) \qquad (16.98)$$

In other words, we have a nonlinear state-space model on our hands, which may be described in words as follows:

- The modified state vector $\mathbf{x}'(n+1)$ at time $n+1$ is a nonlinear function of its value $\mathbf{x}'(n)$ at time n.
- The observation $y(n)$ consists of another nonlinear function of $\mathbf{x}'(n)$ plus a noise component $v(n)$.

The important point to note, however, is that the mathematical forms of both nonlinear functions, and the ways in which they depend on the unknown parameters $p_1, p_2, \ldots, p_M$, are *known*. This, in turn, means that we may compute the gradients of $\mathbf{x}'(n)$ and $y(n)$ with respect to these unknown parameters, and thereby set the stage for applying the RLS version of the extended Kalman filter. The *extended Kalman filter* provides a device for tracking a nonstationary system whose underlying state-space model is nonlinear. Its derivation rests on a "linearization" procedure applied to the nonlinear state-space model of the system, as discussed in Chapter 7. Basically, this procedure results in a linear "approximation" of the model, whereafter we may proceed with the application of Kalman filter theory (its RLS version in our case) to estimate both the unknown state vector $\mathbf{x}(n)$ and the unknown parameter vector $\mathbf{p}$ in the usual way. The resulting algorithm, referred to as the *extended RLS algorithm—version 2* (ERLS-2), proceeds as follows[6]

Starting with $\hat{\mathbf{w}}(0|-1) = E[\mathbf{w}_o(0)]$, $\hat{\boldsymbol{\psi}}_{|-1} = E[\boldsymbol{\psi}]$, and

$$\mathbf{P}(0, -1) = \begin{bmatrix} \Pi_0 & \mathbf{0} \\ \mathbf{0} & \pi_0 \end{bmatrix},$$

---

[6]The algorithm described here is due to A. H. Sayed; for more details, see Haykin et al. (1995).

compute for $n \geq 0$:

$$\mathbf{k}(n) = \mathbf{P}(n, n - 1) \begin{bmatrix} \mathbf{u}(n) \\ 0 \end{bmatrix} \left([\mathbf{u}^H(n)\ \ 0]\, \mathbf{P}(n, n - 1) \begin{bmatrix} \mathbf{u}(n) \\ 0 \end{bmatrix} + \sigma^2(n) \right)^{-1}$$

$$\xi(n) = y(n) - \hat{\mathbf{w}}^H(n|n - 1)\mathbf{u}(n)$$

$$\begin{bmatrix} \hat{\mathbf{w}}(n|n) \\ \hat{\boldsymbol{\psi}}_{|n} \end{bmatrix} = \begin{bmatrix} \hat{\mathbf{w}}(n|n - 1) \\ \hat{\boldsymbol{\psi}}_{|n-1} \end{bmatrix} + \mathbf{k}(n)\xi^*(n),$$

$$\hat{\mathbf{w}}(n + 1|n) = \mathbf{F}(\hat{\boldsymbol{\psi}}_{|n})\hat{\mathbf{w}}(n|n),$$

$$\mathbf{P}(n + 1, n) = \mathbf{F}(n + 1, n)\mathbf{P}(n, n)\mathbf{F}^H(n + 1, n),$$

$$\mathbf{P}(n, n) = (\mathbf{I} - \mathbf{k}(n)\, [\mathbf{u}^H(n)\ \ 0])\, \mathbf{P}(n, n - 1).$$

The important point that we are trying to make here, in presenting the ERLS-1 and ERLS-2 algorithms, is that by virtue of the one-to-one correspondences between RLS variables and Kalman variables, we have the vast literature on Kalman filter theory to build improved versions of the RLS algorithm so as to properly handle the tracking of nonstationary systems.

16.9 COMPUTER EXPERIMENT ON SYSTEM IDENTIFICATION

The subject matter of the computer experiment on system identification described in this section builds on the results of Examples 2 and 3 of Section 16.6 and material presented in Section 16.8 for an adaptive transversal filter with $M = 2$ taps. The purpose of the experiment is twofold:

- To demonstrate that it is possible for the LMS algorithm to outperform the standard RLS algorithm, and vice versa.
- To demonstrate the tracking optimality of the ERLS-1 algorithm compared with the standard RLS and LMS algorithms.

In the experiment, there are three sets of parameters to be considered:

1. *Basic parameters*, made up of the following:
   - The parameter a and the variance σ_v^2 of the zero-mean measurement noise $v(n)$ in the first-order Markov model of Eq. (16.1).
   - The elements r_{11}, r_{21}, and r_{22} of the correlation matrix $\mathbf{R}$ of the zero-mean input vector $\mathbf{u}(n)$
2. *Auxiliary parameters*, made up of the following:
   - The scaling factor c_1 for the case of $\mathbf{Q} = c_1\mathbf{R}$ (Example 2)
   - The scaling factor c_2 for the case of $\mathbf{Q} = c_2\mathbf{R}^{-1}$ (Example 3)

where $\mathbf{Q}$ is the correlation matrix of the zero-mean process noise vector $\boldsymbol{\omega}(n)$ in the first-order Markov model.

3. *Frame of reference*, chosen arbitrarily as

$$\delta = \mathcal{D}_{min}^{RLS}$$

The input vector $\mathbf{u}(n)$ is drawn from a two-dimensional Gaussian process of zero mean and correlation matrix

$$\mathbf{R} = 10^{-4} \begin{bmatrix} 1 & -0.75 \\ -0.75 & 1 \end{bmatrix}$$

The model parameters are

$$a = 0.9998$$

$$\sigma_v^2 = 0.04$$

This completes the specification of the basic parameters of the experiment. For the auxiliary parameters, we have

$$c_1 = 2.7344 \times 10^{-4}$$

$$c_2 = 0.160 \times 10^{-4}$$

Finally, the frame of reference is

$$\delta = 0.01$$

To initialize the experiments, we set

$$E[\mathbf{w}_o(0)] = \mathbf{0}$$

Under the assumption of ergodicity, the instantaneous weight-error vector

$$\boldsymbol{\epsilon}(n) = \hat{\mathbf{w}}(n|n-1) - \mathbf{w}_o(n)$$

is measured by time-averaging over one simulation run of $N = 50,000$ iterations in the steady state (i.e., after all transients have essentially dissipated). In the simulations, this is taken to occur at the iteration index $n = 50,000$. The values of n and N so chosen can be justified by noting that plots of the simulated quantities as a function of N show no discernible change by that point.

Table 16.3 presents the simulation results for the two cases:

Case 1: $\mathbf{Q} = c_1 \mathbf{R}$
Case 2: $\mathbf{Q} = c_2 \mathbf{R}^{-1}$

In each case, the experimental values included in the table pertain to the following:

- Minimum mean-square deviation $\mathcal{D}_{min}$ and minimum misadjustment $\mathcal{M}_{min}$ produced by the standard RLS algorithm

TABLE 16.3   SIMULATION RESULTS
FOR TWO DIFFERENT CASES

	Case 1 $Q = c_1 \, R$	Case 2 $Q = c_2 \, R^{-1}$
$\mathcal{D}_{\min}^{\text{RLS}}$	0.0105	0.0103
$\mathcal{D}_{\min}^{\text{LMS}}$	0.0071	0.0135
$\mathcal{D}^{\text{ERLS-1}}$	0.0069	0.0102
$\mathcal{M}_{\min}^{\text{RLS}}$	0.0842	0.0423
$\mathcal{M}_{\min}^{\text{LMS}}$	0.0698	0.0666
$\mathcal{M}^{\text{ERLS-1}}$	0.0673	0.0419

- Minimum mean-square deviation $\mathcal{D}_{\min}$ and minimum misadjustment $\mathcal{M}_{\min}$ produced by the LMS algorithm
- Mean-square deviation $\mathcal{D}$ and misadjustment $\mathcal{M}$ produced by the ERLS-1 algorithm.

The simulation results clearly demonstrate the superiority of the LMS algorithm over the standard RLS algorithm for Case 1, and vice versa for Case 2. Moreover, they show that the ERLS-1 algorithm performs better (though only marginally) than the optimal LMS/RLS algorithm in each case. Most likely, the marginal improvement is an artifact of the choice of experimental parameters; the choice makes both the relative mean-square weight deviation and relative mean-square misadjustment sufficiently small, so that differences between the performances of the algorithms are not easily discernible over what passes as normal simulation variance and numerical noise.

16.10 AUTOMATIC TUNING OF THE ADAPTATION CONSTANTS

Returning to the model of a time-varying system described in Fig. 16.1, by now we know how to calculate optimum values for the step-size parameter μ in the LMS algorithm using Eq. (16.32) or (16.36), and how to calculate optimum values for the exponential weighting factor λ in the RLS algorithm using Eq. (16.57) or (16.62). Irrespective of the optimality criterion used to assess the tracking performance of the adaptive filter employed to track the system, these calculations require knowledge of the correlation matrix Q of the process noise vector $\omega(n)$ in the model of Fig. 16.1(a) and the correlation matrix R of the input vector $u(n)$ in the model of Fig. 16.1(b). This observation prompts us to raise a basic question that lies at the heart of what adaptive filtering is all about:

- How is the optimum value of the step-size parameter μ in the LMS algorithm or that of the exponential weighting factor λ in the RLS algorithm to be chosen, when details of the underlying physical model of the system and its variability with time are not known?

In Benveniste et al. (1990), certain modifications to the LMS and RLS algorithms are proposed by superimposing adaptive schemes on their respective formulations for the purpose of tuning μ in the LMS algorithm and λ in the RLS algorithm. The practical validity of the idea described therein has been fully supported by (1) signal-processing applications on adaptive equalization and phase-locked loop presented in Brossier (1992), and (2) proof of convergence based on a fairly strong result rooted in stochastic approximation theory that is presented in Kushner and Yang (1995).

LMS Algorithm with Adaptive Gain

The purpose of the adaptive scheme suggested in Benveniste et al. (1990) and elaborated on in Kushner and Yang (1995) is to find an estimate for the particular value of the step-size parameter μ that minimizes the ensemble-averaged cost function

$$J(n) = \frac{1}{2} E[|e(n)|^2] \tag{16.99}$$

where $e(n)$ is the estimation error:

$$e(n) = d(n) - \mathbf{w}^H(n)\mathbf{u}(n) \tag{16.100}$$

Differentiating the cost function $J(n)$ with respect to the step-size parameter μ yields the scalar gradient

$$\nabla_\mu(n) = \frac{\partial J(n)}{\partial \mu}$$

$$= \frac{1}{2} E\left[\frac{\partial e(n)}{\partial \mu} e*(n) + \frac{\partial e*(n)}{\partial \mu} e(n) \right] \tag{16.101}$$

From Eq. (16.100) we readily find that

$$\frac{\partial e(n)}{\partial \mu} = -\boldsymbol{\psi}^H(n)\,\mathbf{u}(n) \tag{16.102}$$

where the vector $\boldsymbol{\psi}(n)$ denotes the gradient of the tap-weight vector $\mathbf{w}(n)$ with respect to the step-size parameter μ:

$$\boldsymbol{\psi}(n) = \frac{\partial \mathbf{w}(n)}{\partial \mu} \tag{16.103}$$

Accordingly, we may redefine the scalar gradient $\nabla_\mu(n)$ as

$$\nabla_\mu(n) = -\frac{1}{2} E[\boldsymbol{\psi}^H(n)\mathbf{u}(n)e*(n) + \mathbf{u}^H(n)\boldsymbol{\psi}(n)e(n)] \tag{16.104}$$

Let $\mu(n)$ and $\hat{\mathbf{w}}(n)$ denote the actual sequences of step-sizes and estimates of $\mathbf{w}(n)$, respectively, resulting from the operation of the adaptive scheme. The adaptation proceeds in an iterative manner. Computation of the estimate $\hat{\mathbf{w}}(n)$ follows the LMS algorithm described in Chapter 9. In a manner similar to that computation, we may formulate the recursion for updating $\mu(n)$ as follows:

$$\mu(n + 1) = \mu(n) - \alpha\hat{\nabla}_\mu(n) \tag{16.105}$$

where α is a small, positive *learning-rate parameter*, and $\hat{\nabla}_\mu(n)$ is an estimate of the scalar gradient $\nabla_\mu(n)$.

Define $\hat{\boldsymbol{\psi}}(n) = \partial\hat{\mathbf{w}}(n)/\partial\mu(n)$ as the estimate of the derivative $\boldsymbol{\psi}(n)$. Then, on the basis of Eq. (16.104) we may formulate the *instantaneous estimate*

$$\begin{aligned}
\hat{\nabla}_\mu(n) &= -\frac{1}{2}[\hat{\boldsymbol{\psi}}^H(n)\mathbf{u}(n)e^*(n) + \mathbf{u}^H(n)\hat{\boldsymbol{\psi}}(n)e(n)] \\
&= -\operatorname{Re}[\hat{\boldsymbol{\psi}}^H(n)\mathbf{u}(n)e^*(n)]
\end{aligned} \tag{16.106}$$

where Re signifies the real-part operator, and $e(n)$ is defined by Eq. (16.100) with the estimate $\hat{\mathbf{w}}(n)$ used in place of $\mathbf{w}(n)$.

We are now ready to describe the two-step adaptive scheme for tuning the step-size parameter in the LMS algorithm:

1. Given the old value $\mu(n)$ of the step-size parameter, its updated value is computed using the recursion:

$$\mu(n + 1) = \mu(n) + \alpha\operatorname{Re}[\hat{\boldsymbol{\psi}}^H(n)\mathbf{u}(n)e^*(n)] \tag{16.107}$$

2. Starting with the usual recursion for updating the tap-weight vector:

$$\hat{\mathbf{w}}(n + 1) = \hat{\mathbf{w}}(n) + \mu(n)\mathbf{u}(n)e^*(n) \tag{16.108}$$

and differentiating it with respect to $\mu(n)$, we get the recursion for updating the estimate $\hat{\boldsymbol{\psi}}(n)$, as described here

$$\begin{aligned}
\hat{\boldsymbol{\psi}}(n + 1) &= \hat{\boldsymbol{\psi}}(n) + \mathbf{u}(n)e^*(n) + \mu(n)\mathbf{u}(n)\frac{\partial e^*(n)}{\partial\mu(n)} \\
&= \hat{\boldsymbol{\psi}}(n) + \mathbf{u}(n)e^*(n) - \mu(n)\mathbf{u}(n)\mathbf{u}^H(n)\hat{\boldsymbol{\psi}}(n)
\end{aligned} \tag{16.109}$$

In the last line of Eq. (16.109), we have adapted the use of Eq. (16.102) for the problem at hand, with $\hat{\boldsymbol{\psi}}(n)$ used in place of $\boldsymbol{\psi}(n)$.

We may now summarize the *LMS algorithm with adaptive gain* as follows:
Starting with some initial values $\hat{\mathbf{w}}(0)$, $\mu(0)$, and $\hat{\boldsymbol{\psi}}(0)$, compute for $n > 0$:

$$e(n) = d(n) - \hat{\mathbf{w}}^H(n)\mathbf{u}(n)$$

$$\hat{\mathbf{w}}(n + 1) = \hat{\mathbf{w}}(n) + \mu(n)\mathbf{u}(n)e^*(n)$$

$$\mu(n + 1) = [\mu(n) + \alpha\operatorname{Re}[\hat{\boldsymbol{\psi}}^H(n)\mathbf{u}(n)e^*(n)]]_{\mu_-}^{\mu_+}$$

$$\hat{\boldsymbol{\psi}}(n + 1) = [\mathbf{I} - \mu(n)\mathbf{u}(n)\mathbf{u}^H(n)]\boldsymbol{\psi}(n) + \mathbf{u}(n)e^*(n)$$

In the third line of this summary, the bracket with μ_- and μ_+ indicates *truncation*. According to simulations reported in Kushner and Yang (1995), the lower level of truncation, μ_-, appears to play a relatively insignificant role; it may be set equal to zero or some small number. On the other hand, the upper level of truncation, μ_+, is highly crucial for good behavior of the algorithm. Typically, the optimum value of the step-size parameter for a good tracking behavior is near the point of instability, so the assignment of too large a value to μ_+ may cause the LMS algorithm to become unstable. In any application, the user usually develops enough experience to see how to set μ_+.

RLS Algorithm with Adaptive Memory

Consider next the RLS algorithm equipped with an adaptive scheme for tuning the exponential weighting factor λ. In this case, the objective is to find the particular value of λ that optimizes the cost function

$$J'(n) = \frac{1}{2} E[|\xi(n)|^2] \tag{16.110}$$

where $\xi(n)$ is the *a priori* estimation error defined by

$$\xi(n) = d(n) - \hat{\mathbf{w}}^H(n-1)\mathbf{u}(n) \tag{16.111}$$

Differentiating the cost function $J'(n)$ with respect to λ yields

$$\nabla_\lambda(n) = \frac{\partial J'(n)}{\partial \lambda}$$

$$= \frac{1}{2} E\left[\frac{\partial \xi(n)}{\partial \lambda} \xi^*(n) + \frac{\partial \xi^*(n)}{\partial \lambda} \xi(n) \right] \tag{16.112}$$

Define

$$\boldsymbol{\psi}(n) = \frac{\partial \mathbf{w}(n)}{\partial \lambda} \tag{16.113}$$

We may then, with the aid of Eq. (16.111), redefine the scalar gradient $\nabla_\lambda(n)$ as

$$\nabla_\lambda(n) = -\frac{1}{2} E[\boldsymbol{\psi}^H(n-1)\,\mathbf{u}(n)\xi^*(n) + \mathbf{u}^H(n)\,\boldsymbol{\psi}(n-1)\xi(n)] \tag{16.114}$$

The updating of the tap-weight vector in the RLS algorithm involves the gain vector $\mathbf{k}(n) = \mathbf{P}(n)\mathbf{u}(n)$, as shown by [see Eq. (13.25)]

$$\mathbf{w}(n) = \mathbf{w}(n-1) + \mathbf{P}(n)\mathbf{u}(n)\xi^*(n) \tag{16.115}$$

Let $\mathbf{S}(n)$ denote the derivative of the inverse correlation matrix $\mathbf{P}(n)$ with respect to λ:

$$\mathbf{S}(n) = \frac{\partial \mathbf{P}(n)}{\partial \lambda} \tag{16.116}$$

Then, using Eqs. (16.111), (16.115) and (16.116) in Eq. (16.113) yields

$$\boldsymbol{\psi}(n) = [\mathbf{I} - \mathbf{k}(n)\mathbf{u}^H(n)] \, \boldsymbol{\psi}(n-1) + \mathbf{S}(n)\mathbf{u}(n)\xi^*(n) \qquad (16.117)$$

For the recursion to compute $\mathbf{S}$, we first use Eqs. (13.18) and (13.19) to write

$$\mathbf{P}(n) = \lambda^{-1}\,\mathbf{P}(n-1) - \frac{\lambda^{-2}\,\mathbf{P}(n-1)\,\mathbf{u}(n)\mathbf{u}^H(n)\mathbf{P}(n-1)}{1 + \lambda^{-1}\,\mathbf{u}^H(n)\,\mathbf{P}(n-1)\mathbf{u}(n)} \qquad (16.118)$$

Hence, differentiating Eq. (16.118) with respect to λ and then collecting terms, we get

$$\mathbf{S}(n) = \lambda^{-1}\,[\mathbf{I} - \mathbf{k}(n)\mathbf{u}^H(n)]\,\mathbf{S}(n-1)\,[\mathbf{I} - \mathbf{u}(n)\,\mathbf{k}^H(n)]$$
$$+ \lambda^{-1}\mathbf{k}(n)\mathbf{k}^H(n) - \lambda^{-1}\mathbf{P}(n) \qquad (16.119)$$

We are now ready to formulate the *RLS algorithm with adaptive memory*. Let $\lambda(n)$, $\hat{\mathbf{w}}(n)$, and $\hat{\boldsymbol{\psi}}(n)$ denote the actual values of the exponential weighting factor, tap-weight vector $\mathbf{w}(n)$, and gradient $\boldsymbol{\psi}(n)$ computed by the algorithm at iteration n. Then, using the instantaneous estimate $-\mathrm{Re}[\boldsymbol{\psi}^H(n-1)\mathbf{u}(n)\xi^*(n)]$ for the scalar gradient $\nabla_\lambda(n)$, based on Eq. (16.114) we may adaptively compute the exponential weighting factor using the recursion:

$$\lambda(n) = \lambda(n-1) - \alpha\hat{\nabla}_\lambda(n)$$
$$= \lambda(n-1) + \alpha\,\mathrm{Re}[\boldsymbol{\psi}^H(n-1)\mathbf{u}(n)\xi^*(n)] \qquad (16.120)$$

where α is a small, positive learning-rate parameter. Thus, incorporating this recursion into the standard RLS algorithm, we may summarize the RLS algorithm with adaptive memory as follows:

Starting with the initial values $\hat{\mathbf{w}}(0)$, $\mathbf{P}(0)$, $\lambda(0)$, $\mathbf{S}(0)$, and $\hat{\boldsymbol{\psi}}(0)$, compute for $n > 0$:

$$\mathbf{k}(n) = \frac{\lambda^{-1}(n-1)\mathbf{P}(n-1)\mathbf{u}(n)}{1 + \lambda^{-1}(n-1)\mathbf{u}^H(n)\mathbf{P}(n-1)\mathbf{u}(n)}$$

$$\xi(n) = d(n) - \hat{\mathbf{w}}^H(n-1)\mathbf{u}(n)$$

$$\hat{\mathbf{w}}(n) = \hat{\mathbf{w}}(n-1) + \mathbf{k}(n)\xi^*(n)$$

$$\mathbf{P}(n) = \lambda^{-1}(n-1)\mathbf{P}(n-1) - \lambda^{-1}(n-1)\mathbf{k}(n)\mathbf{u}^H(n)\mathbf{P}(n-1)$$

$$\lambda(n) = [\lambda(n-1) + \alpha\,\mathrm{Re}[\hat{\boldsymbol{\psi}}^H(n-1)\mathbf{u}(n)\xi^*(n)]]_{\lambda_-}^{\lambda_+}$$

$$\mathbf{S}(n) = \lambda^{-1}(n)[\mathbf{I} - \mathbf{k}(n)\mathbf{u}^H(n)]\,\mathbf{S}(n-1)\,[\mathbf{I} - \mathbf{u}(n)\mathbf{k}^H(n)]$$
$$+ \lambda^{-1}(n)\mathbf{k}(n)\mathbf{k}^H(n) - \lambda^{-1}(n)\mathbf{P}(n)$$

$$\hat{\boldsymbol{\psi}}(n) = [\mathbf{I} - \mathbf{k}(n)\mathbf{u}^H(n)]\,\hat{\boldsymbol{\psi}}(n-1) + \mathbf{S}(n)\mathbf{u}(n)\xi^*(n)$$

As with the LMS algorithm with adaptive gain, the bracket with λ_- and λ_+ in the fifth line of this summary indicates truncation. The upper level of truncation, λ_+, may be set close to (but less than) unity. The lower level of truncation, λ_-, plays a more crucial role, and its value may be determined by the user through experimentation.

16.11 SUMMARY AND DISCUSSION

In this chapter we studied the tracking performance of adaptive filters when operating in a nonstationary environment. First and foremost, tracking is *problem-specific*. Thus, the tracking performance of an adaptive filter, used in system identification is quite different from that in channel equalization or the adaptive recovery of a desired signal in additive noise.

To assess the tracking capability of an adaptive filter, we may use the mean-square deviation $\mathcal{D}(n)$ or the misadjustment $\mathcal{M}(n)$. These two figures of merit highlight the tracking performance of the adaptive filter in their own individual ways. The important point to note is that in either case, we may identify two contributions: one being representative of the stationary case, and the other being attributed to nonstationarity of the environment.

Based on the mean-square deviation and misadjustment, we find that, in general, the LMS algorithm exhibits a more robust tracking behavior than the RLS algorithm. Therefore, it should not be surprising to find that, in data transmission over time-varying communication channels, the LMS algorithm is preferred over the RLS algorithm, not only because of its simplicity but also because of its better tracking capability.

This latter remark pertains to the conventional formulation of the RLS algorithm, as described in Chapter 13. However, we are not restricted exclusively to this version of the RLS algorithm. Rather, in light of the one-to-one correspondences that exist between RLS variables and Kalman variables, we may draw upon the vast literature on Kalman filter theory to build other versions of the RLS algorithm to suit the application. In particular, we may mention two possible routes:

1. The underlying state-space model of the RLS algorithm is formulated as in Eqs. (16.90) and (16.91), with the process noise vector $\mathbf{v}_1(n)$ included in the model to account for the nonstationary behavior of the environment. Such an approach is well suited for a system identification problem with a Markovian description, as in Eq. (16.1).

2. The nonstationarity is accounted for by including in the state-space model of the RLS algorithm a transition matrix $\mathbf{F}(n + 1, n)$ that is not a constant or even unknown. The state-space model of the RLS algorithm is then formulated in nonlinear terms, and the application of extended Kalman filter formalism is invoked. This latter approach is, for example, appropriate for the tracking of a chirped signal in additive noise.

The conclusion to be drawn from this discussion is that whatever prior knowledge is available about the task at hand, it should be exploited, so as to *minimize the mismatch* between the state-space model of the RLS algorithm and the mathematical model for the problem of interest, and thereby improve the tracking performance of the RLS algorithm in a nonstationary environment.

PROBLEMS

1. In qualitative terms, describe how the error-performance surface of an adaptive equalizer for a time-varying communication channel differs from that of an adaptive equalizer for a communication channel of fixed characteristics.

2. In prediction, the present value of a signal constitutes the desired response and a finite set of its past values constitutes the input vector. Describe how the error-performance surface of an adaptive predictor for a nonstationary process differs from that for a stationary process.

3. The weight-error vector $\boldsymbol{\epsilon}(n)$ may be expressed as the sum of the weight vector noise $\boldsymbol{\epsilon}_1(n)$ and weight vector lag $\boldsymbol{\epsilon}_2(n)$. Show that

$$E[\boldsymbol{\epsilon}_1^H(n)\boldsymbol{\epsilon}_2(n)] = E[\boldsymbol{\epsilon}_2^H(n)\boldsymbol{\epsilon}_1(n)] = 0$$

Under the assumption that $\hat{\mathbf{w}}(n)$ and $\mathbf{w}_o(n)$ are statistically independent in Fig. 16.1, show that

$$E[\|\boldsymbol{\epsilon}(n)\|^2] = E[\|\boldsymbol{\epsilon}_1(n)\|^2] + E[\|\boldsymbol{\epsilon}_2(n)\|^2]$$

4. Continuing with Problem 3, use the independence assumption to show that

$$E[\boldsymbol{\epsilon}_1^H(n)\mathbf{u}(n)\mathbf{u}^H(n)\boldsymbol{\epsilon}_1(n)] = \text{tr}[\mathbf{R}\mathbf{K}_1(n)]$$

$$E[\boldsymbol{\epsilon}_2^H(n)\mathbf{u}(n)\mathbf{u}^H(n)\boldsymbol{\epsilon}_2(n)] = \text{tr}[\mathbf{R}\mathbf{K}_2(n)]$$

$$E[\boldsymbol{\epsilon}_1^H(n)\mathbf{u}(n)\mathbf{u}^H(n)\boldsymbol{\epsilon}_2^H(n)] = E[\boldsymbol{\epsilon}_2^H(n)\mathbf{u}(n)\mathbf{u}^H(n)\boldsymbol{\epsilon}_2(n)] = 0$$

where $\mathbf{u}(n)$ is the input vector assumed to be of zero mean, $\mathbf{R}$ is the correlation matrix of $\mathbf{u}(n)$, and $\mathbf{K}_1(n)$ and $\mathbf{K}_2(n)$ are the correlation matrices of $\boldsymbol{\epsilon}_1(n)$ and $\boldsymbol{\epsilon}_2(n)$, respectively. How is the correlation matrix $\mathbf{K}(n)$ of $\boldsymbol{\epsilon}(n)$ related to $\mathbf{K}_1(n)$ and $\mathbf{K}_2(n)$?

5. Given the vectors $\mathbf{x}$ and $\mathbf{y}$, both assumed to have the same dimension, the *Cauchy–Schwarz inequality* states that

$$|\mathbf{x}^H\mathbf{y}|^2 \leq \|\mathbf{x}\|^2 \|\mathbf{y}\|^2$$

Applying this inequality to the cross-diagonal terms of the 2-by-2 array in Table 16.1, derive the results presented in Eqs. (16.71) and (16.72).

6. Continuing with the results presented in Table 16.2 for an adaptive filter with $M = 2$, do the following:
   (a) For the LMS algorithm, determine the minimum mean-square deviation $\mathcal{D}_{\min}$ and the minimum misadjustment $\mathcal{M}_{\min}$, and the corresponding optimum values of the step-size parameter μ.
   (b) For the RLS algorithm, determine the minimum mean-square deviation $\mathcal{D}_{\min}$ and the minimum misadjustment $\mathcal{M}_{\min}$, and the corresponding optimum values of the exponential weighting factor λ.

   Hence, verify the entries presented in Table 16.2.

CHAPTER

— 17 —

Finite-Precision Effects

A study of adaptive filters would be incomplete without some discussion of the effects of *quantization* or *round-off errors* that arise when they are implemented digitally.

The theory of adaptive filtering developed in previous chapters assumes the use of an *analog model* (i.e., infinite precision) for the samples of input data as well as the internal algorithmic calculations. This assumption is made in order to take advantage of well-understood continuous mathematics. Adaptive filter theory, however, cannot be applied to the construction of an adaptive filter directly; rather it provides an *idealized framework* for such a construction. In particular, in a *digital* implementation of an adaptive filtering algorithm as encountered in practice, the input data and internal calculations are all quantized to a *finite precision* that is determined by design and cost considerations. Consequently, the quantization process has the effect of causing the performance of a digital implementation of the algorithm to deviate from its theoretical value. The nature of this deviation is influenced by a combination of factors:

- The type of design details of the adaptive filtering algorithm employed
- The degree of ill-conditioning (i.e., the eigenvalue spread) in the underlying correlation matrix that characterizes the input data
- The form of numerical computation (fixed-point or floating-point) employed

It is important for us to understand the numerical properties of adaptive filtering algorithms, as it would obviously help us in meeting design specifications. Moreover, the cost of a digital implementation of an algorithm is influenced by the *number of bits* (i.e., precision) available for performing the numerical computations associated with the algorithm. Generally speaking, the cost of implementation increases with the number of bits employed. There is therefore practical motivation for using the minimum number of bits possible.

We begin our study of the numerical properties of adaptive filtering algorithms by examining the sources of quantization error and the related issues of numerical stability and accuracy.

17.1 QUANTIZATION ERRORS

In a digital implementation of an adaptive filter, there are essentially two sources of quantization error to be considered as described here.

1. *Analog-to-digital conversion.* Given that the input data are in analog form, we may use an analog-to-digital converter for their numerical representation. For our present discussion, we assume a quantization process with a *uniform step size* δ and a set of *quantizing levels* positioned at $0, \pm \delta \pm 2\delta, \cdots$. Figure 17.1 illustrates the input-output characteristic of a typical uniform quantizer. Consider a particular sample at the quantizer input, with an amplitude that lies in the range $i\delta - (\delta/2)$ to $i\delta + (\delta/2)$, where i is an integer (positive or negative, including zero) and $i\delta$ defines the *quantizer output*. The quantization process thus described introduces a region of uncertainty of width δ, centered on $i\delta$. Let η denote the quantization error. Correspondingly, the quantizer input is $i\delta + \eta$, where η is bounded as $-(\delta/2) \leq \eta \leq (\delta/2)$. When the quantization is fine enough (say, the number of quantizing levels is 64 or more), and the signal spectrum is sufficiently rich, the distortion produced by the quantizing process may be modeled as an additive independent source of white noise with zero mean and variance determined by the quantizer step size δ (Gray, 1990). It is customary to assume that the quantization error η is *uniformly distributed* over the range $-\delta/2$ to $\delta/2$. The variance of the quantization error is therefore given by

$$\sigma^2 = \int_{-\delta/2}^{\delta/2} \frac{1}{\delta} \eta^2 \, d\eta \tag{17.1}$$

$$= \frac{\delta^2}{12}$$

We assume that the quantizer input is properly scaled, so that it lies inside the interval $(-1, +1]$. With each quantizing level represented by B bits plus sign, the quantizer step size is

$$\delta = 2^{-B} \tag{17.2}$$

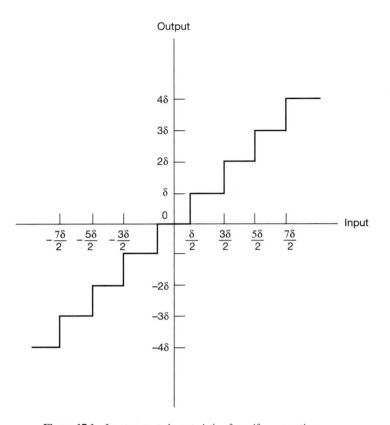

Figure 17.1 Input–output characteristic of a uniform quantizer.

Substituting Eq. (17.2) in (17.1), we find that the quantization error resulting from the digital representation of input analog data has the variance

$$\sigma^2 = \frac{2^{-2B}}{12} \tag{17.3}$$

2. *Finite word-length arithmetic.* In a digital machine, a finite word length is commonly used to store the result of internal arithmetic calculations. Assuming that *no* overflow takes place during the course of computation, additions do not introduce error (if fixed-point arithmetic is used), whereas each multiplication introduces an error after the product is quantized. The statistical characterization of finite word-length arithmetic errors may be quite different from that of analog-to-digital conversion errors. Finite word-length arithmetic errors may have a *nonzero mean*, which results from either rounding off or truncating the output of a multiplier so as to match the prescribed word length.

The presence of finite word-length arithmetic raises serious concern in the digital implementation of an adaptive filter, particularly when the tap weights (coefficients) of the filter are updated on a continuous basis. The digital version of the filter exhibits a specific *response* or *propagation* to such errors, causing its performance to deviate from the ideal (i.e., infinite-precision) form of the filter. Indeed, it is possible for the deviation to be of a catastrophic nature in the sense that the errors resulting from the use of finite-precision arithmetic may accumulate without bound. If such a situation is allowed to persist, the filter is ultimately driven into an overflow condition, and the algorithm is said to be *numerically unstable*. Clearly, for an adaptive filter to be of practical value, it has to be numerically stable. An adaptive filter is said to be *numerically stable* if the use of finite-precision arithmetic results in deviations from the infinite-precision form of the filter that are *bounded*. It is important to recognize that numerical stability is an inherent characteristic of an adaptive filter. In other words, if an adaptive filter is numerically unstable, then increasing the number of bits used in a digital implementation of the filter will not change the stability condition of that implementation.

Another issue that requires attention in a digital implementation of an adaptive filter is that of *numerical accuracy*. Unlike numerical stability, however, the numerical accuracy of an adaptive filter is determined by the number of bits used to implement the internal calculations of the filter. The larger the number of bits used, the smaller the deviation from ideal performance, and the more accurate therefore would be the digital implementation of the filter. In practical terms, it is only meaningful to speak of the numerical accuracy of an adaptive filter if it is numerically stable.

For the remainder of this chapter, we discuss the numerical properties of adaptive filtering algorithms and related issues. We begin with the LMS algorithm and then move on to RLS adaptive filtering algorithms, presented in the same order as in previous chapters of the book.

17.2 LEAST-MEAN-SQUARE ALGORITHM

In order to simplify the discussion of finite-precision effects on the performance of the LMS algorithm,[1] we will depart from the practice followed in previous chapters, and assume that the input data and therefore the filter coefficients are all *real valued*. This

---

[1] The first treatment of finite-precision effects in the LMS algorithm was presented by Gitlin et al. (1973). Subsequently, more detailed treatments of these effects were presented by Weiss and Mitra (1979), Caraiscos and Liu (1984), and Alexander (1987). The paper by Caraiscos and Liu considers steady-state conditions, whereas the paper by Alexander is broader in scope, in that it considers transient conditions. The problem of finite-precision effects in the LMS algorithm is also discussed in Cioffi (1987) and Sherwood and Bershad (1987). Another problem encountered in the practical use of the LMS algorithm is that of parameter drift, which is discussed in detail in Sethares et al. (1986). The material presented in this section is very much influenced by the contents of these papers. In our presentation, we assume the use of *fixed-point arithmetic*. Error analysis of the LMS algorithm for *floating-point arithmetic* is discussed in Caraiscos and Liu (1984).

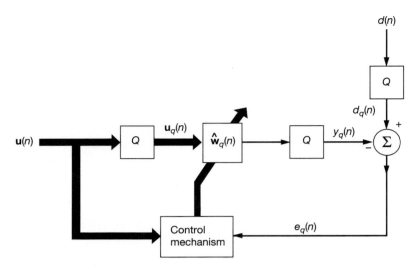

Figure 17.2   Block diagram representation of the finite-precision form of the LMS algorithm.

assumption, made merely for convenience of presentation, will in no way affect the validity of the findings presented in this section.

A block diagram of the *finite-precision least-mean-square (LMS) algorithm* is depicted in Fig. 17.2. Each of the blocks (operators) labeled Q represents a *quantizer*. Each one introduces a *quantization*, or *round-off error* of its own. Specifically, we may describe the input-output relations of the quantizers operating in Fig. 17.2 as follows:

1. For the input quantizer connected to $\mathbf{u}(n)$ we have

$$\mathbf{u}_q(n) = Q[\mathbf{u}(n)]$$
$$= \mathbf{u}(n) + \boldsymbol{\eta}_u(n) \tag{17.4}$$

where $\boldsymbol{\eta}_u(n)$ is the *input quantization error vector*.

2. For the quantizer connected to the desired response $d(n)$, we have

$$d_q(n) = Q[d(n)]$$
$$= d(n) + \eta_d(n) \tag{17.5}$$

where $\eta_d(n)$ is the *desired response quantization error*.

3. For the quantized tap-weight vector $\hat{\mathbf{w}}_q(n)$, we write

$$\hat{\mathbf{w}}_q(n) = Q[\hat{\mathbf{w}}(n)]$$
$$= \hat{\mathbf{w}}(n) + \Delta\hat{\mathbf{w}}(n) \tag{17.6}$$

where $\hat{\mathbf{w}}(n)$ is the tap-weight vector in the infinite-precision LMS algorithm, and $\Delta\hat{\mathbf{w}}(n)$ is the *tap-weight error vector* resulting from quantization.

4. For the quantizer connected to the output of the transversal filter represented by the quantized tap-weight vector $\hat{\mathbf{w}}_q(n)$, we write

$$
\begin{aligned}
y_q(n) &= Q[\mathbf{u}_q^T(n)\hat{\mathbf{w}}_q(n)] \\
&= \mathbf{u}_q^T(n)\hat{\mathbf{w}}_q(n) + \eta_y(n)
\end{aligned}
\tag{17.7}
$$

where $\eta_y(n)$ is the *filtered output quantization error*.

The finite-precision LMS algorithm is described by the following pair of relations:

$$
e_q(n) = d_q(n) - y_q(n) \tag{17.8}
$$

$$
\hat{\mathbf{w}}_q(n + 1) = \hat{\mathbf{w}}_q(n) + Q[\mu e_q(n)\mathbf{u}_q(n)] \tag{17.9}
$$

where $y_q(n)$ is itself defined in Eq. (17.7). The quantizing operation indicated on the right-hand side of Eq. (17.9) is not shown explicitly in Fig. 17.2; nevertheless, it is basic to the operation of the finite-precision LMS algorithm. The use of Eq. (17.9) has the following practical implication. The product $\mu e_q(n)\mathbf{u}_q(n)$, representing a scaled version of the gradient vector estimate, is quantized *before* addition to the contents of the *tap-weight accumulator*. Because of hardware constraints, this form of digital implementation is preferred to the alternative method of operating the tap-weight accumulator in double precision and then quantizing the tap weight to single precision at the accumulator output.

In a statistical analysis of the finite-precision LMS algorithm, it is customary to make the following assumptions:

1. The input data are properly scaled so as to *prevent overflow* of the elements of the quantized tap-weight vector $\hat{\mathbf{w}}_q(n)$ and the quantized output $y_q(n)$ during the filtering operation.

2. Each data sample is represented by B_D bits plus sign, and each tap weight is represented by B_W bits plus sign. Thus, the quantization error associated with a B_D-plus-sign bit number (i.e., data sample) has the variance

$$
\sigma_D^2 = \frac{2^{-2B_D}}{12} \tag{17.10}
$$

Similarly, the quantization error associated with a B_W-plus-sign bit number (i.e., tap weight) has the variance

$$
\sigma_W^2 = \frac{2^{-2B_W}}{12} \tag{17.11}
$$

3. The elements of the input quantization error vector $\boldsymbol{\eta}_u(n)$ and the desired response quantization error $\eta_d(n)$ are *white-noise* sequences, independent of the signals and from each other. Moreover, they have zero mean and variance σ_D^2.

4. The output quantization error $\eta_y(n)$ is a white-noise sequence, independent of the input signals and other quantization errors. It has a mean of zero and a variance

equal to $c\sigma_D^2$, where c is a constant that depends on the way in which the inner product $\mathbf{u}_q^T(n)\hat{\mathbf{w}}_q(n)$ is computed. If the individual scalar products in $\mathbf{u}_q^T(n)\hat{\mathbf{w}}_q(n)$ are all computed without quantization, then summed, and the final result is quantized in B_D bits plus sign, the constant c is unity and the variance of $\eta_y(n)$ is σ_D^2 as defined in Eq. (17.10). If, on the other hand, the individual scalar products in $\mathbf{u}^T(n)\hat{\mathbf{w}}_q(n)$ are quantized and then summed, the constant c is M and the variance of $\eta_y(n)$ is $M\sigma_D^2$, where M is the number of taps in the transversal filter implementation of the LMS algorithm.

5. The independence theory of Section 9.4, dealing with the infinite-precision LMS algorithm, is invoked.

Total Output Mean-squared Error

The filtered output $y_q(n)$, produced by the finite-precision LMS algorithm, presents a *quantized estimate* of the desired response. The *total output error* is therefore equal to the difference $d(n) - y_q(n)$. Using Eq. (17.7), we may therefore express this error as

$$
\begin{aligned}
e_{\text{total}}(n) &= d(n) - y_q(n) \\
&= d(n) - \mathbf{u}_q^T(n)\hat{\mathbf{w}}_q(n) - \eta_y(n)
\end{aligned}
\tag{17.12}
$$

Substituting Eqs. (17.4) and (17.6) in Eq. (17.12), and ignoring all quantization error terms higher than first order, we get

$$
e_{\text{total}}(n) = [d(n) - \mathbf{u}^T(n)\hat{\mathbf{w}}(n)] - [\Delta\hat{\mathbf{w}}^T(n)\mathbf{u}(n) + \eta_u^T(n)\hat{\mathbf{w}}(n) + \eta_y(n)] \tag{17.13}
$$

The term inside the first set of square brackets on the right-hand side of Eq. (17.13) is the estimation error $e(n)$ in the infinite-precision LMS algorithm. The term inside the second set of square brackets is entirely due to quantization errors in the finite-precision LMS algorithm. Because of assumptions 3 and 4 (i.e., the quantization errors η_u and η_y are independent of the input signals and of each other), the quantization error-related terms $\Delta\hat{\mathbf{w}}^T(n)\mathbf{u}(n)$, and $\eta_y(n)$ are uncorrelated with each other. Basically, for the same reason, the infinite-precision estimation error $e(n)$ is uncorrelated with both $\eta_u^T(n)\hat{\mathbf{w}}(n)$ and $\eta_y(n)$. By invoking the independence assumption of Chapter 9, we may write

$$
E[e(n)\Delta\hat{\mathbf{w}}^T(n)\mathbf{u}(n)] = E[\Delta\hat{\mathbf{w}}^T(n)]E[e(n)\mathbf{u}(n)]
$$

Moreover, by invoking this same independence assumption, we may show that the expectation $E[\Delta\hat{\mathbf{w}}(n)]$ is zero (see Problem 2). Hence, $e(n)$ and $\Delta\hat{\mathbf{w}}^T(n)u(n)$ are also uncorrelated. In other words, the infinite-precision estimation error $e(n)$ is uncorrelated with all the three quantization-error-related terms, $\Delta\hat{\mathbf{w}}^T(n)\mathbf{u}(n)$, $\eta_u^T(n)\hat{\mathbf{w}}(n)$, and $\eta_y(n)$ in Eq. (17.13).

Using these observations, and assuming that the step-size parameter μ is small, it is shown in Caraiscos and Liu (1984) that the *total output mean-squared error* produced in the finite-precision algorithm has the following *steady-state* structure:

$$
E[e_{\text{total}}^2(n)] = J_{\min}(1 + \mathcal{M}) + \xi_1(\sigma_w^2, \mu) + \xi_2(\sigma_D^2) \tag{17.14}
$$

The first term $J_{\min}(1 + \mathcal{M})$ on the right-hand side of Eq. (17.14) is the mean-squared error of the infinite-precision LMS algorithm. In particular, $J_{\min}$ is the *minimum mean-squared error* of the optimum Wiener filter, and $\mathcal{M}$ is the *misadjustment* of the infinite-precision LMS algorithm. The second term $\xi_1(\sigma_w^2, \mu)$ arises because of the error $\Delta\hat{\mathbf{w}}(n)$ in the quantized tap-weight vector $\hat{\mathbf{w}}_q(n)$. This contribution to the total output mean-squared error is *inversely proportional to the step-size parameter* μ. The third term $\xi_2(\sigma_D^2)$ arises because of two quantization errors: the error $\boldsymbol{\eta}_u(n)$ in the quantized input vector $\mathbf{u}_q(n)$ and the error $\eta_y(n)$ in the quantized filter output $y_q(n)$. However, unlike $\xi_1(\sigma_w^2, \mu)$, this final contribution to the total output mean-squared error is, to a first order of approximation, independent of the step-size parameter μ.

From the infinite-precision theory of the LMS algorithm presented in Chapter 9, we know that decreasing μ reduces the misadjustment $\mathcal{M}$ and thus leads to an improved performance of the algorithm. In contrast, the inverse dependence of the contribution $\xi_1(\sigma_w^2, \mu)$ on μ in Eq. (17.14) indicates that decreasing μ has the effect of increasing the deviation from infinite-precision performance. In practice, therefore, the step-size parameter μ may only be decreased to a level at which the degrading effects of quantization errors in the tap weights of the finite-precision LMS algorithm become significant.

Since the misadjustment $\mathcal{M}$ decreases with μ, and the contribution $\xi_1(\sigma_w^2, \mu)$ increases with reduced μ, we may (in theory) find an optimum value of μ for which the total output mean-squared error in Eq. (17.14) is minimized. However, it turns out that this minimization results in an optimum value μ_o for the step-size parameter μ that is too small to be of practical value. In other words, it does not permit the LMS algorithm to converge completely. Indeed, Eq. (17.14) for calculating the total output mean-squared error is valid only for a μ that is well in excess of μ_o. Such a choice of μ is necessary so as to prevent the occurrence of a phenomenon known as stalling, described later in the section.

Deviations during the Convergence Period

Equation (17.14) describes the general structure of the total output mean-squared error of the finite-precision LMS algorithm, assuming that the algorithm has reached steady state. During the convergence period of the algorithm, however, the situation is more complicated.

A detailed treatment of the transient adaptation properties of the finite-precision LMS algorithm is presented in Alexander (1987). In particular, a general formula is derived for the tap-weight misadjustment, or *perturbation*, of the finite-precision LMS algorithm, which is measured with respect to the tap-weight solution computed from the infinite-precision form of the algorithm. The tap-weight misadjustment is defined by

$$\mathcal{W}(n) = E[\Delta\hat{\mathbf{w}}^T(n)\Delta\hat{\mathbf{w}}(n)] \tag{17.15}$$

where the *tap-weight error vector* $\Delta\hat{\mathbf{w}}(n)$ is itself defined by [see Eq. (17.6)]

$$\Delta\hat{\mathbf{w}}(n) = \hat{\mathbf{w}}_q(n) - \hat{\mathbf{w}}(n) \tag{17.16}$$

The tap-weight vectors $\hat{\mathbf{w}}_q(n)$ and $\hat{\mathbf{w}}(n)$ refer to the finite-precision and infinite-precision forms of the LMS algorithm, respectively. To determine $\mathcal{W}(n)$, the weight up-date equation (17.9) is written as

$$\hat{\mathbf{w}}_q(n + 1) = \hat{\mathbf{w}}_q(n) + \mu e_q(n)\mathbf{u}_q(n) + \boldsymbol{\eta}_w(n) \tag{17.17}$$

where $\boldsymbol{\eta}_w(n)$ is the *gradient quantization error vector*; it results from quantizing the product $\mu e_q(n)\mathbf{u}_q(n)$ that represents a scaled version of the gradient vector estimate. The individual elements of $\boldsymbol{\eta}_w(n)$ are assumed to be uncorrelated in time and with each other, and assumed to have a common variance σ_w^2. For this assumption to be valid, the step-size parameter μ must be large enough to prevent the stalling phenomenon from occurring; this phenomenon is described later in the section.

Applying an orthogonal transformation to $\hat{\mathbf{w}}_q(n)$ in Eq. (17.17) in a manner similar to that described in Section 5.6, the propagation characteristics of the tap-weight misadjustment $\mathcal{W}(n)$ during adaptation and steady state may be studied. Using such an approach, Alexander (1987) has derived some important theoretical results, supported by computer simulation. These results are summarized here.

1. The tap weights in the LMS algorithm are the *most sensitive* of all parameters to quantization. For the case of *uncorrelated input data*, the variance σ_w^2 [that enters the statistical characterization of the tap-weight update equation (17.17)] is proportional to the reciprocal of the product $r(0)\mu$, where $r(0)$ is the average input power and μ is the step-size parameter. For the case of *correlated input data*, the variance σ_w^2 is proportional to the reciprocal of $\mu\lambda_{min}$, where λ_{min} is the smallest eigenvalue of the correlation matrix $\mathbf{R}$ of the input data vector $\mathbf{u}(n)$.

2. For uncorrelated input data, the adaptation time constants of the tap-weight misadjustment $\mathcal{W}(n)$ are heavily dependent on the step-size parameter μ.

3. For correlated input data, the adaptation time constants of $\mathcal{W}(n)$ are heavily dependent on the interaction between μ and the minimum eigenvalue λ_{min}.

From a design point of view, it is thus important to recognize that the step-size parameter μ cannot be chosen too small; this is in spite of the infinite-precision theory of the LMS algorithm that advocates a small value for μ. Moreover, the more ill-conditioned the input process $u(n)$ is, the more pronounced the finite-precision effects in a digital implementation of the LMS algorithm will be.

Leaky LMS Algorithm

To further stabilize the digital implementation of the LMS algorithm, we may use a technique known as *leakage*.[2] Basically, leakage prevents the occurrence of overflow in a limited-precision environment by providing a compromise between minimizing the mean-squared error and containing the energy in the impulse response of the adaptive filter. However, the prevention of overflow is attained at the expense of an increase in hardware cost and at the expense of a degradation in performance compared to the infinite-precision form of the conventional LMS algorithm.

---

[2]Leakage may be viewed as a technique for increasing algorithm robustness (Ioannou and Kokotovic, 1983; Ioannou, 1990). For a historical account of the leakage technique in the context of adaptive filtering, see Cioffi (1987). For discussions of the leakage LMS algorithm, see Widrow and Stearns (1985) and Cioffi (1987).

In the leaky LMS algorithm, the cost function

$$J(n) = e^2(n) + \alpha\|\hat{\mathbf{w}}(n)\|^2 \tag{17.18}$$

is minimized with respect to the tap-weight vector $\hat{\mathbf{w}}(n)$, where α is a positive control parameter. The first term on the right-hand side of Eq. (17.18) is the squared estimation error, and the second term is the energy in the tap-weight vector $\hat{\mathbf{w}}(n)$. The minimization described herein (for real data) yields the following time update for the tap-weight vector (see Problem 7, Chapter 9)

$$\hat{\mathbf{w}}(n + 1) = (1 - \mu\alpha)\hat{\mathbf{w}}(n) + \mu e(n)\mathbf{u}(n) \tag{17.19}$$

where α is a constant that satisfies the condition

$$0 \le \alpha < \frac{1}{\mu}$$

Except for the *leakage factor* $(1 - \mu\alpha)$ associated with the first term on the right-hand side of Eq. (17.19), the algorithm is of the same mathematical form as the conventional LMS algorithm.

Note that the inclusion of the leakage factor $(1 - \mu\alpha)$ in Eq. (17.19) has the equivalent effect of adding a white-noise sequence of zero mean and variance α to the input process $u(n)$. This suggests another method for stabilizing a digital implementation of the LMS algorithm. Specifically, a relatively weak white-noise sequence (of variance α), known as *dither*, is added to the input process $u(n)$, and samples of the combination are then used as tap inputs (Werner, 1983).

Stalling Phenomenon

There is another phenomenon, known as the *stalling* or *lock-up phenomenon*, not evident from Eq. (17.14), which may arise in a digital implementation of the LMS algorithm. This phenomenon occurs when the gradient estimate is *not* sufficiently noisy. To be specific, a digital implementation of the LMS algorithm *stops adapting* or *stalls*, whenever the correction term $\mu e_q(n)u_q(n - i)$ for the ith tap weight in the update equation (17.9) is smaller in magnitude than the *least significant bit (LSB)* of the tap weight, as shown by (Gitlin et al., 1973)

$$\left|\mu e_q(n_0)u_q(n_0 - i)\right| \le \text{LSB} \tag{17.20}$$

Here, n_0 is the time at which the ith tap weight stops adapting. Suppose that the condition of Eq. (17.20) is first satisfied for the ith tap eight. To a first order of approximation, we may replace $u_q(n_0 - i)$ by its *root-mean-square* (rms) value, A_{rms}. Accordingly, using this value in Eq. (17.20), we get the following relation for the rms value of the quantized estimation error when adaptation in the digitally implemented LMS algorithm stops:

$$\left|e_q(n)\right| \le \frac{\text{LSB}}{\mu A_{\text{rms}}} = e_D(\mu) \tag{17.21}$$

The quantity $e_D(\mu)$, defined on the right-hand side of (17.21), is called the *digital residual error*.

To prevent the algorithm-stalling phenomenon due to digital effects, the digital residual error $e_D(\mu)$ must be made as small as possible. According to the definition of Eq. (17.21), this requirement may be satisfied in one of two ways:

1. The least significant bit (LSB) is reduced by picking a sufficiently large number of bits for the digital representation of each tap weight.
2. The step-size parameter μ is made as large as possible, while still guaranteeing convergence of the algorithm.

Another method of preventing the stalling phenomenon is to insert *dither* at the input of the quantizer that feeds the tap-weight accumulator (Sherwood and Bershad, 1987). Dither is a random sequence that essentially "linearizes" the quantizer. In other words, the addition of dither guarantees that the quantizer input is *noisy* enough for the gradient quantization error vector $\boldsymbol{\eta}_w$ to be again modeled as white noise (i.e., the elements of $\boldsymbol{\eta}_w$ are uncorrelated in time and with each other, and have a common variance σ_w^2). When dither is used in the manner described here, it is desirable to minimize its effect on the overall operation of the LMS algorithm. This is commonly achieved by shaping the power spectrum of the dither so that it is effectively rejected by the algorithm at its output.

Parameter Drift

In addition to the numerical problems associated with the LMS algorithm, there is one other rather subtle problem that is encountered in practical applications of the algorithm. Specifically, certain classes of input excitation can lead to *parameter drift*; that is, parameter estimates or tap weights in the LMS algorithm attain arbitrarily large values despite bounded inputs, bounded disturbances, and bounded estimation errors (Sethares et al., 1986). Although such an unbounded behavior may be unexpected, it is possible for the parameter estimates to drift to infinity while all the signals observable in the algorithm converge to zero. Parameter drift in the LMS algorithm may be viewed as a *hidden form of instability*, since the tap weights represent "internal" variables of the algorithm. As such, it may result in new numerical problems, increased sensitivity to unmodeled disturbances, and degraded long-term performance.

In order to appreciate the subtleties of the parameter drift problem, we need to introduce some new concepts relating to the parameter space. We therefore digress briefly from the issue at hand to do so.

A sequence of information-bearing tap-input vectors $\mathbf{u}(n)$ for varying time n may be used to partition the *real M-dimensional parameter space* $\mathbb{R}^M$ into orthogonal subspaces, where M is the number of tap weights (i.e., the available number of degrees of freedom). The aim of this partitioning is to convert the stability analysis of an adaptive filtering algo-

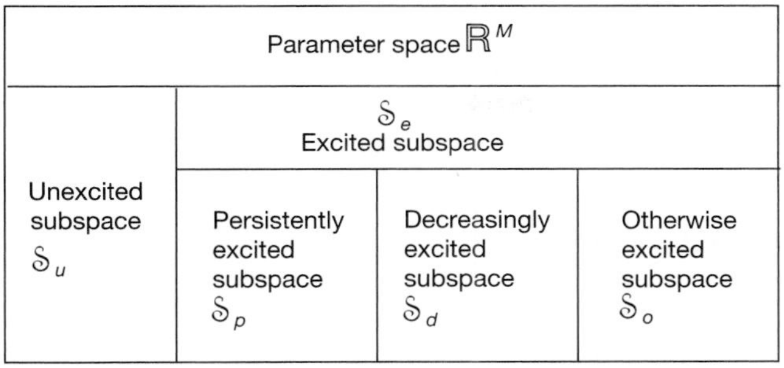

Parameter space $\mathbb{R}^M$			
Unexcited subspace $\mathcal{S}_u$	$\mathcal{S}_e$ Excited subspace		
	Persistently excited subspace $\mathcal{S}_p$	Decreasingly excited subspace $\mathcal{S}_d$	Otherwise excited subspace $\mathcal{S}_o$

Figure 17.3 Decomposition of parameter space $\mathbb{R}^M$, based on excitation.

rithm (e.g., the LMS algorithm) into simpler subsystems and thereby provide a closer linkage between the transient behavior of the parameter estimates and the filter excitations. The partitioning we have in mind is depicted in Fig. 17.3. In particular, we may identify the following subspaces of $\mathbb{R}^M$.

1. *The unexcited subspace.* Let the M-by-1 vector $\mathbf{z}$ be any element of the parameter space $\mathbb{R}^M$, which satisfies two conditions:
   - The Euclidean norm of the vector $\mathbf{z}$ is 1; that is,

$$\|\mathbf{z}\| = 1$$

   - The vector $\mathbf{z}$ is orthogonal to the tap-input vector $\mathbf{u}(n)$ for all but a finite number of n; that is,

$$\mathbf{z}^T\mathbf{u}(n) \neq 0, \qquad \text{only finitely often} \qquad (17.22)$$

   Let $\mathcal{S}_u$ denote the subspace of $\mathbb{R}^M$ that is spanned by the set of all such vectors $\mathbf{z}$. The subspace $\mathcal{S}_u$ is called the *unexcited subspace* in the sense that it spans those *directions* in the parameter space $\mathbb{R}^M$ that are excited only *finitely often*.

2. *The excited subspace.* Let $\mathcal{S}_e$ denote the *orthogonal complement* of the unexcited subspace $\mathcal{S}_u$. Clearly, $\mathcal{S}_e$ is also a subspace of the parameter space $\mathbb{R}^M$. It contains those directions in the parameter space $\mathbb{R}^M$ that are excited *infinitely often*. Thus, except for the null vector, every element $\mathbf{z}$ *belonging to the subspace $\mathcal{S}_e$* satisfies the condition

$$\mathbf{z}^T\mathbf{u}(n) \neq 0, \qquad \text{infinitely often} \qquad (17.23)$$

   The subspace $\mathcal{S}_e$ is called the *excited subspace*.

   The subspace $\mathcal{S}_e$ may itself be decomposed into three orthogonal subspaces of its own, depending on the effects of different types of excitation on the behavior of the adaptive filtering algorithm. Specifically, three subspaces of $\mathcal{S}_e$ may be identified as follows (Sethares et al., 1986):

- *The persistently excited subspace.* Let $\mathbf{z}$ be any vector of unit norm that lies in the excited subspace $\mathcal{S}_e$. For any positive integer m and any $\alpha > 0$, choose the vector $\mathbf{z}$ such that we have

$$\mathbf{z}^T \mathbf{u}(i) > \alpha \qquad \text{for } n \leq i \leq n + m \text{ and for all but a finite number of } n$$
$$(17.24)$$

Given the integer m and the constant α, let $\mathcal{S}_p(m, \alpha)$ be the subspace spanned by all such vectors $\mathbf{z}$ that satisfy the condition of (17.24). There exist a finite m_0 and a positive α_0 for which *the subspace $\mathcal{S}_p(m_0, \alpha_0)$ is* *maximal*. In other words, $\mathcal{S}_p(m_0, \alpha_0)$ contains $\mathcal{S}_p(m, \alpha)$ for all $m > 0$ and for all $\alpha > 0$. The subspace $\mathcal{S}_p \equiv \mathcal{S}_p(m_0, \alpha_0)$ is called the *persistently excited subspace*; and m_0 is called the *interval of excitation*. For every "direction" $\mathbf{z}$ that lies in the persistently excited subspace $\mathcal{S}_p$, there is an excitation of level α_0 at least once in all but a finite number of intervals of length m_0. In the persistently excited subspace, we are therefore able to find a tap-input vector $\mathbf{u}(n)$ *rich* enough to excite all the internal modes that govern the transient behavior of the adaptive filtering algorithm being probed (Narendra and Annaswamy, 1989).

- *The subspace of decreasing excitation.* Consider a sequence $u(i)$ for which we have

$$\left(\sum_{i=1}^{\infty} |u(i)|^p \right)^{1/p} < \infty \qquad (17.25)$$

Such a sequence is said to be an element of the *normed linear space* l^p for $1 < p < \infty$. The norm of this new space is defined by

$$\|\mathbf{u}\|_p = \left(\sum_{i=1}^{\infty} |u(i)|^p \right)^{1/p} \qquad (17.26)$$

Note that if the sequence $u(i)$ is an element of the normed linear space l^p for $1 < p < \infty$, then

$$\lim_{n \to \infty} u(n) = 0 \qquad (17.27)$$

Let $\mathbf{z}$ be any unit-norm vector $\mathbf{z}$ that lies in the excited subspace $\mathcal{S}_e$ such that for $1 < p < \infty$, the sequence $\mathbf{z}^T \mathbf{u}(n)$ lies in the normed linear space l^p. Let $\mathcal{S}_d$ be the subspace that is spanned by all such vectors $\mathbf{z}$. The subspace $\mathcal{S}_d$ is called the *subspace of decreasing excitation* in the sense that each direction of $\mathcal{S}_d$ is decreasingly excited. For any vector $\mathbf{z} \neq \mathbf{0}$, the two conditions

$$|\mathbf{z}^T \mathbf{u}(n)| = \alpha > 0, \qquad \text{infinitely often}$$

and

$$\lim_{n \to \infty} \mathbf{z}^T \mathbf{u}(n) = 0$$

cannot be satisfied simultaneously. In actual fact, we find that the subspace of decreasing excitation $\mathcal{S}_d$ is orthogonal to the subspace of persistent excitation $\mathcal{S}_p$.

- *The otherwise excited subspace.* Let $\mathcal{S}_p \cup \mathcal{S}_d$ denote the *union* of the persistently excited subspace $\mathcal{S}_p$ and the subspace of decreasing excitation $\mathcal{S}_d$. Let $\mathcal{S}_o$ denote the orthogonal complement of $\mathcal{S}_p \cup \mathcal{S}_d$ that lies in the excited subspace $\mathcal{S}_e$. The subspace $\mathcal{S}_o$ is called the *otherwise excited subspace.* Any vector that lies in the subspace $\mathcal{S}_o$ is not unexciting, not persistently exciting, and not in the normal linear space l^p for any finite p. An example of such a signal is the sequence

$$\mathbf{z}^T \mathbf{u}(n) = \frac{1}{\ln(1+n)}, \qquad n = 1, 2, \ldots \qquad (17.28)$$

Returning to our discussion of the parameter drift problem in the LMS algorithm, we find that for bounded excitations and bounded disturbances, in the case of unexcited and persistently exciting subspaces the parameter estimates resulting from the application of the LMS algorithm are indeed bounded. However, in the decreasing and otherwise excited cases, parameter drift may occur (Sethares et al., 1986). A common method of counteracting the parameter drift problem in the LMS algorithm is to introduce leakage into the tap-weight update equation of the algorithm. Here is another reason for using the leaky LMS algorithm that was described previously.

17.3 RECURSIVE LEAST-SQUARES ALGORITHM

The recursive least-squares (RLS) algorithm offers an alternative to the LMS algorithm as a tool for the solution of adaptive filtering problems. From the discussion presented in Chapter 13, we know that the RLS algorithm is characterized by a fast rate of convergence that is relatively insensitive to the eigenvalue spread of the underlying correlation matrix of the input data, and a negligible misadjustment (zero for a stationary environment without disturbances). Moreover, although it is computationally demanding (in the sense that its computational complexity is on the order of M^2, where M is the dimension of the tap-weight vector), the mathematical formulation and therefore implementation of the RLS algorithm is relatively simple. However, there is a numerical instability problem to be considered when the RLS algorithm is implemented in finite-precision arithmetic.

Basically, *numerical instability* or *explosive divergence* of the RLS algorithm is of a similar nature to that experienced in Kalman filtering, of which the RLS algorithm is a special case. Indeed, the problem may be traced to the fact that the time-updated matrix $\mathbf{P}(n)$ in the Riccati equation is computed as the difference between two nonnegative definite matrices, as indicated in Eq. (13.19). Accordingly, explosive divergence of the algorithm occurs when the matrix $\mathbf{P}(n)$ loses the property of positive definiteness or Hermitian sym-

TABLE 17.1 SUMMARY OF A COMPUTATIONALLY EFFICIENT SYMMETRY-PRESERVING
VERSION OF THE RLS ALGORITHM

Initialize the algorithm by setting

$$\mathbf{P}(0) = \delta^{-1}\mathbf{I}, \qquad \delta = \text{small positive constant}$$

$$\hat{\mathbf{w}}(0) = \mathbf{0}$$

For each instant of time, $n = 1, 2, \ldots$, compute

$$\boldsymbol{\pi}(n) = \mathbf{P}(n - 1)\mathbf{u}(n)$$

$$r(n) = \frac{1}{\lambda + \mathbf{u}^H(n)\boldsymbol{\pi}(n)}$$

$$\mathbf{k}(n) = r(n)\boldsymbol{\pi}(n)$$

$$\xi(n) = d(n) - \hat{\mathbf{w}}^H(n - 1)\mathbf{u}(n)$$

$$\hat{\mathbf{w}}(n) = \hat{\mathbf{w}}(n - 1) + \mathbf{k}(n)\xi^*(n)$$

$$\mathbf{P}(n) = \text{Tri}\{\ \lambda^{-1}[\mathbf{P}(n - 1) - \mathbf{k}(n)\boldsymbol{\pi}^H(n)]\}$$

---

metry. This is precisely what happens in the usual formulation of the RLS algorithm described in Table 13.1 (Verhaegen, 1989).

How then can the RLS algorithm be formulated so that the Hermitian symmetry of the matrix $\mathbf{P}(n)$ is preserved despite the presence of numerical errors? For obvious practical reasons, it would also be satisfying if the solution to this fundamental question can be attained in a computationally efficient manner. With these issues in mind, we present in Table 17.1 a particular version of the RLS algorithm from Yang (1994), which describes a computationally efficient procedure[3] for preserving the Hermitian symmetry of $\mathbf{P}(n)$ by design. The improved computational efficiency of this algorithm is achieved because it computes simply the upper/lower triangular part of the matrix $\mathbf{P}(n)$, as signified by the operator Tri{ }, and then fills in the rest of the matrix to preserve Hermitian symmetry. Moreover, division by λ is replaced by multiplication with the precomputed value of λ^{-1}.

Error Propagation Model

According to the algorithm of Table 17.1, the recursions involved in the computation of $\mathbf{P}(n)$ proceed as follows:

$$\boldsymbol{\pi}(n) = \mathbf{P}(n - 1)\mathbf{u}(n) \tag{17.29}$$

$$r(n) = \frac{1}{\lambda + \mathbf{u}^H(n)\boldsymbol{\pi}(n)} \tag{17.30}$$

---

[3]Verhaegen (1989) describes another symmetry-preserving version of the RLS algorithm; Verhaegen's version is less efficient than Yang's version in computational terms. However, both versions exhibit the same numerical behavior.

$$\mathbf{k}(n) = r(n)\boldsymbol{\pi}(n) \tag{17.31}$$

$$\mathbf{P}(n) = \text{Tri}\{\lambda^{-1}[\mathbf{P}(n-1) - \mathbf{k}(n)\boldsymbol{\pi}^{H}(n)]\} \tag{17.32}$$

where λ is the exponential weighting factor. Consider the propagation of a *single* quantization error at time $n - 1$ to subsequent recursions, under the assumption that no other quantization errors are made. In particular, let

$$\mathbf{P}_{q}(n-1) = \mathbf{P}(n-1) + \boldsymbol{\eta}_{p}(n-1) \tag{17.33}$$

where the *error matrix* $\boldsymbol{\eta}_{p}(n-1)$ arises from the quantization of $\mathbf{P}(n-1)$. The corresponding quantized value of $\boldsymbol{\pi}(n)$ is

$$\boldsymbol{\pi}_{q}(n) = \boldsymbol{\pi}(n) + \boldsymbol{\eta}_{p}(n-1)\mathbf{u}(n) \tag{17.34}$$

Let $r_{q}(n)$ denote the quantized value of $r(n)$. Using the defining equation (17.30), we may write

$$\begin{aligned}
r_{q}(n) &= \frac{1}{\lambda + \mathbf{u}^{H}(n)\boldsymbol{\pi}_{q}(n)} \\
&= \frac{1}{\lambda + \mathbf{u}^{H}(n)\boldsymbol{\pi}(n) + \mathbf{u}^{H}(n)\,\boldsymbol{\eta}_{p}(n-1)\mathbf{u}(n)} \\
&= \frac{1}{\lambda + \mathbf{u}^{H}(n)\boldsymbol{\pi}(n)}\left(1 + \frac{\mathbf{u}^{H}(n)\boldsymbol{\eta}_{p}(n-1)\mathbf{u}(n)}{\lambda + \mathbf{u}^{H}(n)\boldsymbol{\pi}(n)}\right)^{-1} \\
&= \frac{1}{\lambda + \mathbf{u}^{H}(n)\boldsymbol{\pi}(n)} - \frac{\mathbf{u}^{H}(n)\boldsymbol{\eta}_{p}(n-1)\mathbf{u}(n)}{(\lambda + \mathbf{u}^{H}(n)\boldsymbol{\pi}(n))^{2}} + O(\boldsymbol{\eta}_{p}^{2}) \\
&= r(n) - \frac{\mathbf{u}^{H}(n)\boldsymbol{\eta}_{p}(n-1)\mathbf{u}(n)}{(\lambda + \mathbf{u}^{H}(n)\boldsymbol{\pi}(n))^{2}} + O(\boldsymbol{\eta}_{p}^{2})
\end{aligned} \tag{17.35}$$

where $O(\boldsymbol{\eta}_{p}^{2})$ denotes the order of magnitude $\|\boldsymbol{\eta}_{p}\|^{2}$.

In an ideal situation, the infinite-precision scalar quantity $r(n)$ is *nonnegative*, taking on values between zero and $1/\lambda$. On the other hand, if $\mathbf{u}^{H}(n)\boldsymbol{\pi}(n)$ is small compared to λ and λ itself is small enough compared to 1, then according to Eq. (17.35), in a finite-precision environment it is possible for the quantized quality $r_{q}(n)$ to take on a *negative* value larger in magnitude than $1/\lambda$. When this happens, the RLS algorithm exhibits explosive divergence (Bottomley and Alexander, 1989).[4]

The quantized value of the gain vector $\mathbf{k}(n)$ is written as

$$\begin{aligned}
\mathbf{k}_{q}(n) &= r_{q}(n)\boldsymbol{\pi}_{q}(n) \\
&= \mathbf{k}(n) + \boldsymbol{\eta}_{k}(n)
\end{aligned} \tag{17.36}$$

where $\boldsymbol{\eta}_{k}(n)$ is the *gain vector quantization error*, defined by

$$\boldsymbol{\eta}_{k}(n) = r(n)\,(\mathbf{I} - \mathbf{k}(n)\mathbf{u}^{H}(n))\,\boldsymbol{\eta}_{p}(n-1)\mathbf{u}(n) + O(\boldsymbol{\eta}_{p}^{2}) \tag{17.37}$$

---

[4]According to Bottomley and Alexander (1989), the evolution of the factor $r(n)$ provides a good indication of explosive divergence as this factor grows large, then suddenly becomes negative.

Finally, using Eq. (17.32), we find that the quantization error incurred in computing the updated inverse-correlation matrix $\mathbf{P}(n)$ is

$$\boldsymbol{\eta}_p(n) = \lambda^{-1}(\mathbf{I} - \mathbf{k}(n)\mathbf{u}^H(n))\,\boldsymbol{\eta}_p(n-1)(\mathbf{I} - \mathbf{k}(n)\mathbf{u}^H(n))^H \tag{17.38}$$

where the term $O(\boldsymbol{\eta}_p^2)$ has been ignored.

On the basis of Eq. (17.38), it would be tempting to conclude that $\boldsymbol{\eta}_p^H(n) = \boldsymbol{\eta}_p(n)$ and therefore the RLS algorithm of Table 17.1 is *Hermitian-symmetry preserving*, if we can assume that the condition $\boldsymbol{\eta}_p^H(n-1) = \boldsymbol{\eta}_p(n-1)$ holds at the previous iteration. We are justified in making this assertion by virtue of the fact there is *no* blow-up in this formulation of the RLS algorithm, as demonstrated in what follows (it is also assumed that there is no stalling).

Equation (17.38) defines the *error propagation mechanism* for the RLS algorithm summarized in Table 17.1 on the basis of a single quantization error in $\mathbf{P}(n-1)$. The matrix $\mathbf{I} - \mathbf{k}(n)\mathbf{u}^H(n)$ plays a crucial role in the way in which the single quantization error $\boldsymbol{\eta}_p(n-1)$ propagates through the algorithm. Using the original definition given in Eq. (13.22) for the gain vector, namely,

$$\mathbf{k}(n) = \boldsymbol{\Phi}^{-1}(n)\mathbf{u}(n) \tag{17.39}$$

we may write

$$\mathbf{I} - \mathbf{k}(n)\mathbf{u}^H(n) = \mathbf{I} - \boldsymbol{\Phi}^{-1}(n)\mathbf{u}(n)\mathbf{u}^H(n) \tag{17.40}$$

Next, from Eq. (13.12) we have

$$\boldsymbol{\Phi}(n) = \lambda\boldsymbol{\Phi}(n-1) + \mathbf{u}(n)\mathbf{u}^H(n) \tag{17.41}$$

Multiplying both sides of Eq. (17.41) by the inverse matrix $\boldsymbol{\Phi}^{-1}(n)$ and rearranging terms, we get

$$\mathbf{I} - \boldsymbol{\Phi}^{-1}(n)\mathbf{u}(n)\mathbf{u}^H(n) = \lambda\boldsymbol{\Phi}^{-1}(n)\boldsymbol{\Phi}(n-1) \tag{17.42}$$

Comparing Eqs. (17.40) and (17.42), we readily deduce that

$$\mathbf{I} - \mathbf{k}(n)\mathbf{u}^H(n) = \lambda\boldsymbol{\Phi}^{-1}(n)\boldsymbol{\Phi}(n-1) \tag{17.43}$$

Suppose now we consider the effect of the quantization error $\boldsymbol{\eta}_p(n_0)$ induced at time $n_0 \leq n$. When the RLS algorithm of Table 17.1 is used and the matrix $\mathbf{P}(n)$ remains Hermitian, then according to the error-propagation model of Eq. (17.38), the effect of the quantization error $\boldsymbol{\eta}_p(n_0)$ becomes modified at time n as follows:

$$\boldsymbol{\eta}_p(n) = \lambda^{-(n-n_0)}\,\boldsymbol{\varphi}(n, n_0)\boldsymbol{\eta}_p(n_0)\boldsymbol{\varphi}^H(n, n_0),\ n \geq n_0 \tag{17.44}$$

where $\boldsymbol{\varphi}(n, n_0)$ is a *transition matrix* defined by

$$\boldsymbol{\varphi}(n, n_0) = (\mathbf{I} - \mathbf{k}(n)\mathbf{u}^H(n)) \cdots (\mathbf{I} - \mathbf{k}(n_0 + 1)\,\mathbf{u}^H(n_0 + 1)) \tag{17.45}$$

The repeated use of Eq. (17.43) in (17.45) leads us to express the transition matrix in the equivalent form

$$\varphi(n, n_0) = \lambda^{n-n_0} \boldsymbol{\Phi}^{-1}(n)\boldsymbol{\Phi}(n_0) \tag{17.46}$$

The correlation matrix $\boldsymbol{\Phi}(n)$ is defined by [see Eq. (13.9)]

$$\boldsymbol{\Phi}(n) = \sum_{i=1}^{n} \lambda^{n-i}\mathbf{u}(i)\mathbf{u}^{H}(i) \tag{17.47}$$

On the basis of this definition, the tap-input vector $\mathbf{u}(n)$ is said to be *uniformly persistently exciting* for sufficiently large n if there exist some $a > 0$ and $n > 0$ such that the following condition is satisfied (Ljung and Ljung, 1985):

$$\boldsymbol{\Phi}(n) \geq a\mathbf{I} \qquad \text{for } n \geq N \tag{17.48}$$

The notation used in Eq. (17.48) is shorthand for saying that the matrix $\boldsymbol{\Phi}(n)$ is positive definite. The condition for persistent excitation not only guarantees the positive definiteness of $\boldsymbol{\Phi}(n)$, but also guarantees its matrix norm to be uniformly bounded for $n \geq N$, as shown by

$$\|\boldsymbol{\Phi}^{-1}(n)\| \leq \frac{1}{a} \qquad \text{for } n \geq N \tag{17.49}$$

Returning to the transition matrix $\varphi(n, n_0)$ of Eq. (17.46) and invoking the *mutual consistency*[5] property of a matrix norm, we may write

$$\|\varphi(n, n_0)\| \leq \lambda^{n-n_0} \|\boldsymbol{\Phi}^{-1}(n)\| \cdot \|\boldsymbol{\Phi}(n_0)\| \tag{17.50}$$

Next, invoking the inequality of (17.49), we may rewrite that of Eq. (17.50) as

$$\|\varphi(n, n_0)\| \leq \frac{\lambda^{n-n_0}}{a} \|\boldsymbol{\Phi}(n_0)\| \tag{17.51}$$

Finally, we may use the error propagation equation (17.44) to express the vector norm of $\boldsymbol{\eta}_p(n)$ as

$$\|\boldsymbol{\eta}_p(n)\| \leq \lambda^{-(n-n_0)} \|\varphi(n, n_0)\| \cdot \|\boldsymbol{\eta}_p(n-1)\| \cdot \|\varphi^{H}(n, n_0)\|$$

which, in light of (17.51), may be rewritten as

$$\|\boldsymbol{\eta}_p(n)\| \leq \lambda^{n-n_0} M, \qquad n \geq n_0 \tag{17.52}$$

where M is a positive number defined by

$$M = \frac{1}{a^2} \|\boldsymbol{\Phi}(n_0)\|^2 \|\boldsymbol{\eta}_p(n-1)\| \tag{17.53}$$

---

[5]Consider two matrices $\mathbf{A}$ and $\mathbf{B}$ of compatible dimensions. The mutual consistency property states that (see page 168)

$$\|\mathbf{AB}\| \leq \|\mathbf{A}\| \cdot \|\mathbf{B}\|$$

Equation (17.53) states that the RLS algorithm of Table 17.1 is *exponentially stable* in the sense that a single quantization error $\boldsymbol{\eta}_p(n_0)$ occurring in the inverse correlation matrix $\mathbf{P}(n_0)$ at time n_0 decays exponentially provided that $\lambda < 1$ (i.e., the algorithm has *finite memory*).[6] In other words, the propagation of a single error through this formulation of the standard RLS algorithm with finite memory is *contractive*. Computer simulations validating this result are presented in Verhaegen (1989).

However, the single-error propagation for the case of *growing memory* (i.e., $\lambda = 1$) is *not* contractive. The reason for saying so is that when $\lambda = 1$, neither $\boldsymbol{\varphi}(n, n_0) \leq \mathbf{I}$ nor $\|\boldsymbol{\varphi}(n, n_0)\| \leq 1$ holds, even if the input vector $\mathbf{u}(n)$ is persistently exciting. Consequently, the accumulation of numerical errors may cause the algorithm to be divergent (Yang, 1994). In an independent study, Slock and Kailath (1991) also point out that the error propagation mechanism in the RLS algorithm with $\lambda = 1$ is unstable and of a random walk type. Moreover, there is experimental evidence for this numerical divergence, which is reported in (Ardalan and Alexander, 1987).

Stalling Phenomenon

As with the LMS algorithm, a second form of divergence, referred to as the *stalling phenomenon*, occurs when the tap weights in the RLS algorithm stop adapting. In particular, this phenomenon occurs when the quantized elements of the matrix $\mathbf{P}(n)$ become very small, such that multiplication by $\mathbf{P}(n)$ is equivalent to multiplication by a zero matrix (Bottomley and Alexander, 1989). Clearly, the stalling phenomenon may arise no matter how the RLS algorithm is implemented.

The stalling phenomenon is directly linked to the exponential weighting factor λ and the variance σ_u^2 of the input data $u(n)$. Assuming that λ is close to unity, we find from the definition of the correlation matrix $\boldsymbol{\Phi}(n)$ that the expectation of $\boldsymbol{\Phi}(n)$ is given by [see Eq. 16.46]

$$E[\boldsymbol{\Phi}(n)] \simeq \frac{\mathbf{R}}{1 - \lambda}, \qquad \text{large } n \tag{17.54}$$

For λ close to unity, we have

$$E[\mathbf{P}(n)] = E[\boldsymbol{\Phi}^{-1}(n)] \simeq (E[\boldsymbol{\Phi}(n)])^{-1} \tag{17.55}$$

Hence, using Eq. (17.54) in Eq. (17.55), we get

$$E[\mathbf{P}(n)] \simeq (1 - \lambda)\mathbf{R}^{-1}, \qquad \text{large } n \tag{17.56}$$

---

[6]The first rigorous proof that single-error propagation in the RLS algorithm is exponentially stable for $\lambda < 1$ was presented in (Ljung and Ljung, 1984). This was followed by a more detailed investigation in (Verhaegen, 1989). Reconfirmation that the error propagation mechanism in the RLS algorithm is exponentially stable was subsequently presented in (Slock and Kailath, 1991; Yang, 1994).

where $\mathbf{R}^{-1}$ is the inverse of matrix $\mathbf{R}$. Assuming that the tap-input vector $\mathbf{u}(n)$ is drawn from a wide-sense stationary process with zero mean, we may write

$$\mathcal{R} = \frac{1}{\sigma_u^2}\mathbf{R} \tag{17.57}$$

where $\mathcal{R}$ is a *normalized correlation matrix* with diagonal elements equal to 1 and off-diagonal elements less than or equal to 1 in magnitude, and σ_u^2 is the variance of an input data sample $u(n)$. We may therefore rewrite Eq. (17.56) as

$$E[\mathbf{P}(n)] \simeq \left(\frac{1-\lambda}{\sigma_u^2}\right)\mathcal{R}^{-1} \qquad \text{for large } n \tag{17.58}$$

Equation (17.58) reveals that the RLS algorithm may stall if the exponential weighting factor λ is close to 1 and/or the input data variance σ_u^2 is large. Accordingly, we may prevent stalling of the standard RLS algorithm by using a sufficiently large number of accumulator bits in the computation of the inverse correlation matrix $\mathbf{P}(n)$.

17.4 SQUARE-ROOT ADAPTIVE FILTERS

In Chapter 14 we discussed three particular forms of square-root adaptive filters, namely, the *QR-decomposition–based recursive least-squares (QR–RLS)* algorithm, the *extended QR–RLS algorithm*, and the *inverse QR–RLS algorithm*. The QR–RLS and extended QR–RLS algorithms are special cases of the square-root information filter, whereas the inverse QR–RLS algorithm is a special case of the square-root covariance (Kalman) filter.

QR–RLS Algorithm

It is generally agreed that the QR decomposition is one of the best numerical procedures for solving the recursive least-squares estimation problem because of two important properties:

1. The QR decomposition operates on the input data directly.
2. The QR decomposition involves the use of only numerically well-behaved unitary rotations (e.g., Givens rotations).

In particular, the QR–RLS algorithm propagates the square root of the correlation matrix $\mathbf{\Phi}(n)$ rather than $\mathbf{\Phi}(n)$ itself. Hence, the condition number of $\mathbf{\Phi}^{1/2}(n)$ equals the square root of the condition number of $\mathbf{\Phi}(n)$. This, in turn, results in a significant reduction in the dynamic range of data handled by QR-decomposition–based algorithms, and therefore a more accurate computation than the standard RLS algorithm that propagates $\mathbf{\Phi}(n)$. Moreover, the finite-precision form of the QR–RLS algorithm is *stable in a bounded input–bounded output (BIBO) sense* (Leung and Haykin, 1989; Liu et al., 1991). However, it must be stressed that the BIBO stability of the QR–RLS algorithm does not guarantee that the various quantities computed by the algorithm remain meaningful in any sense when operating in a finite-precision environment (Yang and Böhme, 1992). In particular, a unitary rotation (e.g., sequence of Givens rotations) is used to annihilate a certain vector in the prearray, and the unitary rotation then operates on other related entries in the prearray. A perturbation in internal computations may produce a corresponding perturbation in rotation angles, which introduces yet another source of numerical error in the rotated entries of the postarray. These errors, in turn, produce further perturbations of their own in subsequent computations of the rotation angles, and the process goes on. The net result is that we have a *complicated parametric feedback system*, and it is not entirely clear whether this feedback system is in fact numerically stable.

Yang and Böhme (1992) present experimental results that demonstrate the numerical stability of the QR–RLS algorithm for $\lambda < 1$; they used this algorithm to perform adaptive prediction of an autoregressive (AR) process. All the computer simulations reported in that paper were performed on a personal computer (PC) using floating-point arithmetic. In order to observe finite-precision effects in a reasonable simulation time, the effective number of mantissa bits in the floating-point representation was reduced. This was done by truncating the mantissa at a predefined position without affecting the exponent. In the experiments reported by Yang and Böhme, the mantissa length took on the values 52, 12, and 5 bits; the resulting word-length changes were found to have only a minor effect on the convergence behavior of the QR–RLS algorithm. Yang and Böhme (1992) also show that the QR–RLS algorithm diverges when $\lambda = 1$.

The numerical stability of the QR–RLS algorithm for $\lambda < 1$ has also been demonstrated experimentally by Ward et al. (1986) in the context of adaptive beamforming. In particular, they show that for the same number of bits of arithmetic precision, the QR–RLS algorithm offers a significantly better performance than the sample matrix inversion algorithm described in Reed et al. (1974).

Extended QR–RLS Algorithm

In comparing the QR–RLS and extended QR–RLS algorithms summarized in Table 14.2, we see that these two algorithms differ in the quantities that they propagate. In particular, the QR–RLS algorithm propagates $\mathbf{\Phi}^{1/2}(n)$, whereas the extended QR–RLS algorithm propagates both $\mathbf{\Phi}^{1/2}(n)$ and its Hermitian inverse $\mathbf{\Phi}^{-H/2}(n)$, independently from each other. Accordingly, in a finite-precision environment, the extended QR–RLS algorithm behaves in a profoundly different way from the QR–RLS algorithm.

Let $\mathbf{X}(n-1)$ denote the quartized version of the Hermitian inverse matrix $\mathbf{\Phi}^{-H/2}(n-1)$, stored at time $n-1$, as shown by

$$\mathbf{X}(n-1) = \mathbf{\Phi}^{-H/2}(n-1) + \mathbf{\eta}_X(n-1) \qquad (17.59)$$

where the component $\mathbf{\eta}_X(n-1)$ represents the effect of round-off errors. Then, assuming that there are no additional local errors introduced at time n, the recursion pertaining to the bottom parts of the prearray and postarray of Eq. (14.61) takes on the following form:

$$[\lambda^{-1/2}\mathbf{X}(n-1) \qquad 0]\mathbf{\Theta}(n) = [\mathbf{X}(n) \qquad -\mathbf{y}(n)] \tag{17.60}$$

where $\mathbf{\Theta}(n)$ is a unitary rotation. The vector $\mathbf{Y}(n)$ is the quantized version of $\mathbf{k}(n)\gamma^{-1/2}(n)$ as shown by

$$\mathbf{y}(n) = \mathbf{k}(n)\ \gamma^{-1/2}(n) + \mathbf{\eta}_y(n) \tag{17.61}$$

where $\mathbf{k}(n)$ is the gain vector, and $\gamma(n)$ is the conversion factor. In a corresponding way to Eq. (17.59), the updated matrix $\mathbf{X}(n)$ may be expressed as

$$\mathbf{X}(n) = \mathbf{\Phi}^{-H/2}(n) + \mathbf{\eta}_X(n) \tag{17.62}$$

Hence, substituting Eqs. (17.59), (17.61), and (17.62) in Eq. (17.60), we get

$$[\lambda^{-1/2}\mathbf{\Phi}^{-H/2}(n-1) + \lambda^{-1/2}\mathbf{\eta}_X(n-1) \qquad 0]\mathbf{\Theta}(n)$$
$$= [\mathbf{\Phi}^{-H/2}(n) + \mathbf{\eta}_X(n) \qquad -\mathbf{k}(n)\gamma^{-1/2}(n) - \mathbf{\eta}_y(n)] \tag{17.63}$$

But, under infinite-precision arithmetic, we have

$$[\lambda^{-1/2}\mathbf{\Phi}^{-H/2}(n-1) \qquad 0]\mathbf{\Theta}(n) = [\mathbf{\Phi}^{-H/2}(n) \qquad -\mathbf{k}(n)\gamma^{-1/2}(n)]$$

Accordingly, we may simplify Eq. (17.63) to

$$[\lambda^{-1/2}\mathbf{\eta}_X(n-1) \qquad 0]\mathbf{\Theta}(n) = [\mathbf{\eta}_x(n) \qquad -\mathbf{\eta}_y(n)] \tag{17.64}$$

Equation (17.64) reveals that the error propagation due to $\mathbf{\eta}_X(n-1)$ is *not* necessarily stable, in that local errors tend to grow unboundedly (Moonen and Vandewalle, 1990). The unlimited error growth is due to (1) the amplification produced by the factor $\lambda^{-1/2}$ for $\lambda < 1$, and (2) the fact that the unitary rotation $\mathbf{\Theta}(n)$ is independent of the error $\mathbf{\eta}_X(n)$. Consequently, as the recursion progresses, the stored values of $\mathbf{\Phi}^{1/2}$ and $\mathbf{\Phi}^{-H/2}$ deviate more and more from each other's Hermitian inverse, thereby contradicting the very hypothesis on which the recursion in the extended QR–RLS algorithm is based. This is indeed another example of *numerical inconsistency*, to which all forms of numerical divergence in RLS algorithms can be traced. Another manifestation of this phenomenon is that the quantity $-\mathbf{k}(n)\gamma^{-1/2}(n)$, which is the by-product of updating $\mathbf{\Phi}^{-H/2}(n-1)$, deviates more and more from its infinite-precision value; this, in turn, causes the weight updating to produce unreliable results.

To combat this problem in the extended QR–RLS algorithm, Moonen and Vandewalle (1990) describe a procedure that forces $\mathbf{\Phi}^{-H/2}(n)$ to be the exact Hermitian inverse of $\mathbf{\Phi}^{1/2}(n)$ at every iteration of the algorithm by using Jacobi-type rotations. However, the price paid for achieving this requirement is increased computational complexity.

Inverse QR–RLS Algorithm

The inverse QR–RLS algorithm propagates $\mathbf{P}^{1/2}(n)$, which is the square root of the inverse correlation matrix $\mathbf{P}(n) = \mathbf{\Phi}^{-1}(n)$. Thus, although the inverse QR–RLS algorithm differs from the QR–RLS algorithm that propagates $\mathbf{\Phi}^{1/2}(n)$, the two algorithms do share a common feature: they both avoid the propagation of the Hermitian inverse of their respective matrix quantities. Accordingly, the inverse QR–RLS algorithm is able to exploit the good numerical properties of the QR decomposition in a manner similar to the QR–RLS algorithm.

For $\lambda < 1$, the propagation of a single error in the inverse QR–RLS algorithm (and, for that matter, in the QR–RLS algorithm) is *exponentially stable*. The rationale for this statement follows the numerical stability analysis of the RLS algorithm presented in Section 17.3.

However, for $\lambda = 1$, the single-error propagation is *not* contractive. It may therefore be conjectured that the accumulation of quantization errors can cause the inverse QR–RLS algorithm to be numerically divergent; this phenomenon has been confirmed experimentally, using computer simulations.[7] A similar remark applies to the QR–RLS algorithm, for which experimental validation is presented in Yang and Böhme (1992).

Summarizing Comments

In summary, we may say the following on the requirement to operate a square-root adaptive filter in a finite-precision environment[8]:

- Use of the extended QR–RLS algorithm is not recommended.
- If only the estimation error $e(n)$ is required, the QR–RLS algorithm is the preferred choice.
- If the weight vector $\hat{\mathbf{w}}(n)$ is required, both the QR–RLS algorithm (in conjunction with back-substitution) and the inverse QR–RLS algorithms are good candidates.
- If, in addition, a systolic implementation of the filter is desired, the inverse QR–RLS algorithm is the best choice.

17.5 ORDER-RECURSIVE ADAPTIVE FILTERS

In order-recursive adaptive filtering algorithms (and, for that matter, in all fast RLS algorithms known to date), the particular section responsible for joint-process estimation is subordinate to the section responsible for performing the forward and backward linear predictions. Accordingly, the numerical stability of this class of adaptive filtering algorithms is critically dependent on how the prediction section performs its computations, as discussed in the sequel.

QRD–LSL Algorithm

In the *QR-decomposition–based least-squares lattice (QRD–LSL) algorithm* summarized in Table 15.3, the prediction section consists of $M-1$ lattice stages, where $M-1$ is the pre-

---

[7]Yang, B., private communication, 1995.

[8]The summarizing comments presented herein are based on Yang, B., private communication, 1995.

diction order. Each stage of the prediction section uses QR decomposition in the form of Givens rotations to perform its computations. The net result is that the sequence of input data $u(n)$, $u(n-1)$, ..., $u(n-M+1)$ is transformed into a corresponding sequence of angle-normalized backward prediction errors $\epsilon_{b,0}(n)$, $\epsilon_{b,1}(n)$, ..., $\epsilon_{b,m-1}(n)$. Given this latter sequence, the joint-process estimation section also uses QR decomposition, on a stage-by-stage basis, to perform its computations; the final product is a least-squares estimate of some desired response $d(n)$. In other words, all the computations throughout the algorithm are performed using QR decomposition.

From a numerical point of view, the QRD–LSL algorithm has some desirable properties:

1. The sines and cosines involved in applying the Givens rotations are all numerically *well behaved*.

2. The algorithm is *numerically consistent* in that, from one iteration to the next, each section of the algorithm propagates the *minimum* possible mumber of parameters needed for a satisfactory operation; that is, the propagation of related parameters is avoided. In particular, Table 15.3 shows that the parameters propagated by the three sections of the algorithm are as follows:

Section	Parameters propagated
Forward prediction	$\mathcal{B}^{1/2}_{m-1}(n-2)$, $p_{f,m-1}(n-1)$, $\gamma^{1/2}_{m-1}(n-1)$
Backward prediction	$\mathcal{F}^{1/2}_{m-1}(n-1)$, $p_{b,m-1}(n-1)$
Joint-process estimation	$p_{m-1}(n-1)$

3. The auxiliary parameters $p_{f,m-1}(n-1)$, $p_{b,m-1}(n-1)$, and $p_{m-1}(n-1)$, which are involved in the order updating of the angle-normalized estimation errors $\epsilon_{f,m-1}(n)$, $\epsilon_{b,m-1}(n)$, and $\epsilon_m(n)$, respectively, are all computed *directly*. That is, *local error feedback* is involved in the time-update recursions used to compute each of these auxiliary parameters. This form of feedback is another factor in assuring numerical stability of the algorithm.

There is experimental evidence for numerical stability of the QRD–LSL algorithm. In this context, we may mention computer simulation studies reported by Ling (1989), Yang and Böhme (1992), McWhirter and Proudler (1993), and Levin and Cowan (1994), all of which have demonstrated the numerical robustness of variants of the QRD–LSL algorithm. Of particular interest is the paper by Levin and Cowan (1994), in which the performance of eight different adaptive filtering algorithms of the RLS family was evaluated in a finite-precision environment. The results presented therein demonstrate the superior performance of algorithms belonging to the square-root information domain (exemplified by the QRD–LSL algorithm) over those belonging to the covariance domain. Moreover, of the eight algorithms considered in the study, the QRD–LSL algorithm appeared to be the least affected by numerical imprecision.

Further evidence for numerical robustness of the QRD–LSL is presented in Capman et al. (1995). An acoustic echo canceler is described in that paper by combining a multirate scheme with a variant of the QRD–LSL algorithm. Simulation results presented therein demonstrate the numerical robustness of this solution to the echo cancelation problem.

Recursive LSL Algorithms

Turning next to *recursive least-squares lattice (LSL) algorithms*, we note from the discussion presented in Chapter 15 that these algorithms are special cases of the QRD–LSL algorithm. Indeed, they are derived by squaring the arrays of the QRD–LSL algorithm and then comparing terms. Recognizing that, in the context of numerical behavior of an algorithm, "squaring" has an opposite effect to that of "square-rooting," we may state that the performance of recursive LSL algorithms in a limited-precision environment is always *inferior* to that of the QRD–LSL algorithm from which they are derived.

A recursive LSL algorithm provides a "fast" solution to the recursive least-squares problem by employing a multistage lattice predictor for transforming the input data into a corresponding sequence of backward prediction errors. This transformation may be viewed as a form of the classical Gram–Schmidt orthogonalization procedure. The Gram–Schmidt orthogonalization is known to be *numerically inaccurate* (Stewart, 1973). Correspondingly, a conventional form of the recursive LSL algorithm (be it based on *a posteriori* or *a priori* prediction errors) has poor numerical behavior. The key to a practical method of overcoming the numerical accuracy problem in a recursive LSL algorithm is to update the forward and backward reflection coefficients *directly*, rather than first computing the individual sums of weighted forward and backward prediction errors and their cross-correlations and then taking ratios of the appropriate quantities (as in a conventional LSL algorithm). This is precisely what is done in a recursive LSL algorithm *with error feedback* (Ling and Proakis, 1984), exemplified by the algorithm summarized in Table 15.5. For a prescribed fixed-point representation, a recursive LSL algorithm with error feedback works with much more accurate values of the forward and backward reflection coefficients; these two coefficients are the key parameters in any recursive LSL algorithm. The direct computation of the forward and backward reflection coefficients therefore has the overall effect of preserving the positive definiteness of the underlying inverse correlation matrix of the input data, despite the presence of quantization errors due to finite-precision effects. Therefore, insofar as numerical performance is concerned, recursive LSL algorithms with error feedback are preferred to their conventional forms.[9]

Gradient Adaptive Lattice Algorithm

Order-recursive adaptive filtering algorithms include the (inexact) *gradient adaptive lattice (GAL) algorithm*, in which the reflection coefficient (one per stage of the algorithm) is also updated directly (see Appendix G). In other words, there is a form of error feedback built into the operation of the GAL algorithm. In light of what has been said above, we expect the GAL algorithm to exhibit a good numerical behavior too. This is indeed borne out by the results of a computer simulation reported in (Satorius et al., 1983).

The GAL algorithm has the advantage of simplicity over recursive LSL algorithms (with or without error feedback). On the other hand, recursive LSL algorithms have an advantage over the GAL algorithm in that they provide a faster rate of convergence, because *no* approximations are made in their derivations.

17.6 FAST TRANSVERSAL FILTERS

We conclude our discussion of the numerical behavior of adaptive filtering algorithms by considering the *fast transversal filters (FTF) algorithm*. As with order-recursive adaptive filters, the FTF algorithm solves the recursive least-squares problem by exploiting the time-shift invariance property of the input data.

The FTF algorithm uses four separate transversal filters that share a common input, as indicated in the block diagram of Fig. 17.4. The four filters have distinct tasks:

- recursive forward linear prediction
- recursive backward linear prediction
- recursive computation of the gain vector
- recursive estimation of the desired response.

A summary of the FTF algorithm[10] is presented in Table 17.2; the notations used herein are the same as those described in Chapter 15. An attractive feature of this algorithm is that it permits direct computation of the coefficients of a transversal filter model.

---

[9]North et al. (1993) present computer simulations (using floating-point arithmetic), comparing the numerical behavior of a 32-bit directly updated recursive LSL algorithm (i.e., with error feedback) with a 32-bit indirectly updated recursive LSL algorithm. The study involved adaptive interference cancelation. In the recursive LSL algorithm with indirect updating, it was found that after about 10^5 iterations the accumulation of numerical errors resulted in a degradation of approximately 20 dB in interference cancelation, compared to the directly updated recursive LSL algorithm.

[10]For a derivation of the FTF algorithm, see Cioffi and Kailath (1984), Haykin (1991), Slock and Kailath (1993), and Sayed and Kailath (1994).

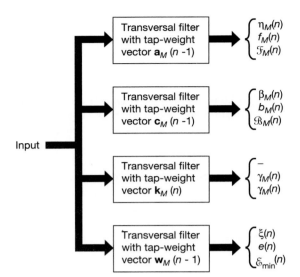

Figure 17.4 Block diagram highlighting the computations involved in the FTF algorithm, assuming a prediction order M.

Unfortunately, when the FTF algorithm is implemented in finite-precision arithmetic, numerical errors may cause the algorithm to diverge. The numerical divergence is necessarily preceded by the algorithm losing its least-squares character. In particular, the FTF algorithm contains unstable modes of propagation that are never excited in exact arithmetic, but do manifest themselves in finite-precision arithmetic (Slock, 1992; Regalia, 1992).

Rescue Variable

Experimentation with the FTF algorithm indicates that a certain, positive quantity in the algorithm becomes negative due to the accumulated effect of numerical errors just before divergence of the algorithm occurs (Lin, 1984; Cioffi and Kailath, 1984). The quantity in question equals the ratio of the conversion factors for prediction errors M and $M + 1$, as shown by

$$\zeta_M(n) = \frac{\gamma_{M+1}(n)}{\gamma_M(n)} \tag{17.65}$$

We refer to $\zeta_M(n)$ as the *rescue variable*. For the ideal case of infinite prediction, we have $0 \le \zeta_M(n) \le 1$. A violation of this restriction on the permissible value of $\zeta_M(n)$ is due to finite-precision effects.

The rescue variable $\zeta_M(n)$ may be expressed in the equivalent form (see Problem 8)

$$\zeta_M(n) = 1 - \beta_M^*(n)\, \gamma_{M+1}(n)\, \bar{k}_{M+1,M+1}(n) \tag{17.66}$$

TABLE 17.2 SUMMARY OF THE FTF ALGORITHM

Predictions

$$\eta_M(n) = \mathbf{a}_M^H(n-1)\mathbf{u}_{M+1}(n)$$

$$f_M(n) = \gamma_M(n-1)\eta_M(n)$$

$$\mathscr{F}_M(n) = \lambda\mathscr{F}_M(n-1) + \eta_M(n)f_M^*(n)$$

$$\gamma_{M+1}(n) = \lambda\frac{\mathscr{F}_M(n-1)}{\mathscr{F}_M(n)}\gamma_M(n-1)$$

$$\overline{\mathbf{k}}_{M+1}(n) = \begin{bmatrix} 0 \\ \overline{\mathbf{k}}_M(n-1) \end{bmatrix} + \lambda^{-1}\frac{\eta_M(n)}{\mathscr{F}_M(n-1)}\mathbf{a}_M(n-1)$$

$$\mathbf{a}_M(n) = \mathbf{a}_M(n-1) - f_M^*(n)\begin{bmatrix} 0 \\ \overline{\mathbf{k}}_M(n-1) \end{bmatrix}$$

$$\beta_M(n) = \lambda\mathscr{B}_M(n-1)\overline{k}_{M+1,M+1}(n)$$

$$\gamma_M(n) = [1 - \beta_M^*(n)\gamma_{M+1}(n)\overline{k}_{M+1,M+1}(n)]^{-1}\gamma_{M+1}(n)$$

Rescue[a] variable $= [1 - \beta_M^*(n)\gamma_{M+1}(n)\overline{k}_{M+1,M+1}(n)]$

$$b_M(n) = \gamma_M(n)\beta_M(n)$$

$$\mathscr{B}_M(n) = \lambda\mathscr{B}_M(n-1) + \beta_M(n)b_M^*(n)$$

$$\begin{bmatrix} \overline{\mathbf{k}}_M(n) \\ 0 \end{bmatrix} = \overline{\mathbf{k}}_{M+1}(n) - \overline{k}_{M+1,M+1}(n)\mathbf{c}_M(n-1)$$

$$\mathbf{c}_M(n) = \mathbf{c}_M(n-1) - b_M^*(n)\begin{bmatrix} \overline{\mathbf{k}}_M(n) \\ 0 \end{bmatrix}$$

Filtering

$$\xi_M(n) = d(n) - \hat{\mathbf{w}}_M^H(n-1)\mathbf{u}_M(n)$$

$$e_M(n) = \gamma_M(n)\xi_M(n)$$

$$\hat{\mathbf{w}}_M(n) = \hat{\mathbf{w}}_M(n-1) + \overline{\mathbf{k}}_M(n)e_M^*(n)$$

Note: $\overline{k}_{M+1,M+1}(n)$ is the last element of the normalized gain vector $\overline{\mathbf{k}}_{M+1}(n)$.

[a]Rescue if variable is negative: Save $\hat{\mathbf{w}}_M(n)$ as initial condition and reuse in the augmented cost function:

$$\mathscr{E}'(n) = \mu\lambda^n \|\mathbf{w}_M(n) - \hat{\mathbf{w}}_M(n)\|^2 + \sum_{i=1}^{n}\lambda^{n-i}|e(i)|^2$$

where the factor $\mu > 0$ ensures a quadratic cost function.

where $\beta_M(n)$ is the backward *a priori* prediction error, and $\overline{k}_{M+1,M+1}(n)$ is the last element of the normalized gain vector $\overline{\mathbf{k}}_{M+1}(n)$. Equation (17.66) indicates that tight control has to be exercised on the computation of $\beta_M(n)$, $\gamma_{M+1}(n)$ [or, equivalently, $\gamma_M^{-1}(n)$], and $\overline{k}_{M+1,M+1}(n)$ in order to ensure numerical stability of the FTF algorithm.

Error Feedback

An important characteristic of the FTF algorithm is that the three quantities just mentioned [i.e., $\beta_M(n)$, $\gamma_M^{-1}(n)$, and $\bar{k}_{M+1,M+1}(n)$] may each be computed in two different ways:

- The transversal filtering of an input data vector
- The manipulation of scalar quantities

To be specific, consider the computation of the backward *a priori* prediction error $\beta_M(n)$. Obviously, we may compute $\beta_M(n)$ as

$$\beta_M^{(f)}(n) = \mathbf{c}_M^H(n-1)\mathbf{u}_{M+1}(n) \tag{17.67}$$

where $\mathbf{c}_M(n-1)$ is the tap-weight vector of the backward prediction-error filter at iteration $n-1$, and $\mathbf{u}_{M+1}(n)$ is the tap-input vector. The superscript (f) in the symbol $\beta_M^{(f)}(n)$ denotes "filtering." From the FTF algorithm summarized in Table 17.2, we see that $\beta_M(n)$ may also be computed as

$$\beta_M^{(s)}(n) = \lambda \mathcal{B}_M(n-1)\bar{k}_{M+1,M+1}(n) \tag{17.68}$$

where λ is the exponential weighting factor, $\mathcal{B}_M(n-1)$ is the minimum sum of backward *a posteriori* prediction-error squares, and $\bar{k}_{M+1,M+1}(n)$ is the last element of the normalized gain vector $\bar{\mathbf{k}}_{M+1}(n)$. The superscript (s) in the symbol $\beta_M^{(s)}(n)$ denotes "scalar manipulation."

When infinite-precision arithmetic is used, the two different ways of computing $\beta_M(n)$ described in Eqs. (17.67) and (17.68) will yield identical results. However, the results will differ from each other when finite-precision arithmetic is used. Similar remarks apply to $\gamma_M^{-1}(n)$ and $\bar{k}_{M+1,M+1}(n)$.

The difference signals resulting from the different methods of computing $\beta_M(n)$, $\gamma_M^{-1}(n)$, and $\bar{k}_{M+1,M+1}(n)$ may be viewed as output signals of the error propagation mechanism. Most importantly, they may be used in a *feedback* scheme designed for the purpose of influencing the error propagation responsible for their generation in the first place. This is indeed the idea behind *error feedback* as a method for stabilizing the FTF algorithm (Slock and Kailath, 1988, 1991; Botto and Moustakides, 1989). In particular, the difference signals may be fed back to any point in the FTF algorithm without altering the true RLS character of the algorithm, since they would vanish when exact arithmetic is used. In the error feedback scheme proposed by Slock and Kailath, the difference signals are fed back into the computation of the specific quantities they are actually associated with.

Consider, for example, the feedback stabilization of $\beta_M(n)$. Given the two finite-precision values $\beta_M^{(f)}(n)$ and $\beta_M^{(s)}(n)$ computed in accordance with Eqs. (17.67) and (17.68), respectively, we use a convex combination of these two values to define the final value of $\beta_M(n)$ as described in Slock and Kailath (1991):

$$\begin{aligned}\beta_M(n) &= \beta_M^{(s)}(n) + K(\beta_M^{(f)}(n) - \beta_M^{(s)}(n)) \\ &= K\beta_M^{(f)}(n) + (1-K)\,\beta_M^{(s)}(n)\end{aligned} \tag{17.69}$$

where K is a *feedback constant*. A similar approach may be used to stabilize the other two variables of concern: $\gamma_M^{-1}(n)$ and $\bar{k}_{M+1,M+1}(n)$.

The quantity $\beta_M(n)$ is used in several places in the FTF algorithm. This motivates the use of different values for the feedback constant K in those different plans, thereby providing more freedom in affecting the error propagation in the FTF algorithm (Slock and Kailath, 1991). Such a choice of feedback mechanism is intuitively appealing; it will make it possible to stabilize all unstable modes in error propagation, which is the ultimate goal of error feedback. The cost of stabilization is a modest increase in computational complexity, from $7M$ to $(7M + M)$, where M is the filter length.

The operation of the *stabilized FTF algorithm* may thus be summarized as follows (Slock and Kailath, 1991, 1993; Slock, 1992):

- Forward and backward linear predictions are used to make the computational complexity of the algorithm linear in the filter length M. In so doing, however, the algorithm becomes potentially unstable in a finite-precision environment.

- Error feedback is used to reintroduce redundancy of order M into the algorithm, the purpose of which is to fortify the algorithm against numerical errors incurred in the recursions.

- The error feedback mechanism is, however, only able to stabilize the FTF algorithm for a restricted range of the exponential weighting factor λ, defined by:

$$1 - \frac{1}{2M} < \lambda < 1 \tag{17.70}$$

where M is the length of the transversal filter. The permissible range of values of λ described in (17.70) covers some useful choices of interest in practice.

17.7 SUMMARY AND DISCUSSION

In this chapter we discussed the numerical stability of the LMS and RLS families of adaptive filtering algorithms.

The LMS algorithm is *numerically robust*. When operating in a limited-precision environment, the point to note is that the step-size parameter μ may only be decreased to a level at which the degrading effects of round-off errors in the tap weights of the finite-precision LMS algorithm become significant. Moreover, a finite-precision implementation of the LMS algorithm may be improved by incorporating *leakage* into the algorithm.

As for the RLS family, the use of square-root filtering provides a powerful technique for realizing a robust numerical behavior. For adaptive beamforming, which is spatial in nature, a good choice is the QR–RLS algorithm, which is the RLS version of the square-root (Kalman) information filter. On the other hand, for temporal applications of adaptive filters where we can exploit the time-shift invariance property of the input data, a good choice is the order-recursive QRD–LSL algorithm, which is also rooted in square-root information filtering. In light of the numerical stability analysis of the RLS algorithm, we may say the following in the context of the QR–RLS, inverse QR–RLS, and QRD–LSL algorithms:

- The propagation of a single error is exponentially stable, provided that the exponential weighting factor λ is less than unity. This behavior appears to guarantee a "small" overall accumulation of error in the algorithm.
- When $\lambda = 1$, the error propagation is not contractive, with the result that the accumulation of numerical errors may lead to divergence of the algorithm.

Having said this, however, we lack a unified treatment of finite-precision effects in the RLS family of adaptive filtering algorithms that goes beyond a single error propagation.

In the literature, it is sometimes argued that square-root filters are (1) expensive and (2) awkward to calculate, constituting a bottleneck for overall performance. For these reasons, square-root free versions of the QR–RLS and QRD–LSL algorithms have been formulated, using special methods for performing Givens rotations without square roots, the idea for which dates back to Gentleman (1973). In the case of Kalman filtering, the method of UD-factorization (Bierman, 1977) was developed specifically for avoiding the actual use of square roots. It now appears that square-root free algorithms actually introduce a number of problems (Stewart and Chapman, 1990):

- Square-root free algorithms may become numerically unstable, and potentially suffer from serious overflow/underflow problems.
- Based on the knowledge that square roots are simpler (or even equally as complex) as divider arrays, the reformulation of standard adaptive filtering algorithms in square-root free form may actually increase arithmetic complexity.

Another noteworthy point is that RLS adaptive filtering algorithms requiring the use of square roots can be programmed very efficiently on *CORDIC processors*, in which square root operations make no explicit appearance. Here we recognize that most of these algorithms are decomposable in the two basic operations of rotation and vectoring. These two operations are indeed fundamental to a CORDIC processor (Volder, 1959):

1. *Rotation.* In the rotation mode, the CORDIC processor is given the coordinates of a two-element vector and an angle of rotation. The processor then computes the coordinate components of the original vector after rotation through the desired angle.
2. *Vectoring.* In the vectoring mode, the coordinates of a two-element vector are given. The CORDIC processor then rotates the vector until the angular argument is zero. The angle of rotation in this second mode is therefore the negative of the original angular argument. In other words, the vectoring operation is equivalent to the annihilation of an element of a two-element vector.

Apparently, a CORDIC processor yields the fastest implementation of these two basic operations. Moreover, its integral circuit implementation in the form of silicon chips makes its use for the implementation of RLS algorithms involving square roots all the more attractive.[11]

PROBLEMS

1. Consider the digital implementation of the LMS algorithm using fixed-point arithmetic, as discussed in Section 17.2. Show that the M-by-1 error vector $\Delta\hat{\mathbf{w}}(n)$, incurred by quantizing the tap-weight vector, may be updated as follows:

$$\Delta\hat{\mathbf{w}}(n + 1) = \mathbf{F}(n)\Delta\mathbf{w}(n) + \mathbf{t}(n), \qquad n = 0, 1, 2, \ldots$$

   where $\mathbf{F}(n)$ is an M-by-M matrix and $\mathbf{t}(n)$ is an M-by-1 vector. Hence, define $\mathbf{F}(n)$ and $\mathbf{t}(n)$. Base your analysis on real-valued data.

2. Using the results of Problem 1, and invoking the independence assumption of Chapter 9, show that

$$E[\Delta\hat{\mathbf{w}}(n)] = \mathbf{0}$$

3. Consider two transversal filters I and II, both of length M. Filter I has all of its tap inputs as well as tap weights represented in infinite-precision form. Filter II is identical to filter I, except for the fact that its tap weights are represented in finite-precision form. Let $y_I(n)$ and $y_{II}(n)$ denote the respective filter outputs for the tap-inputs $u(n), u(n - 1), \ldots, u(n - M - 1)$. Define the error

$$\epsilon(n) = y_I(n) - y_{II}(n)$$

   Assuming that the inputs $u(n)$ are independent random variables with a common rms value equal to A_{rms}, show that the mean-square value of the error $\epsilon(n)$ is

$$E[\epsilon^2(n)] = A_{\text{rms}}^2 \sum_{i=0}^{M-1} (w_i - w_{iq})^2$$

   where w_{iq} is the quantized version of the tap weight w_i.

4. Consider an LMS algorithm with 17 taps and a step-size parameter $\mu = 0.07$. The input data stream has an rms value of unity.
   (a) Given the use of a quantization process with 12-bit accuracy, calculate the corresponding value of the digital residual error.
   (b) Suppose the only source of output error is that due to quantization of the tap weights. Using the result of Problem 3, calculate the rms value of the resulting measurement error in the output. Compare this error with the digital residual error calculated in part (a).

5. Demonstrate the Hermitian symmetry-preserving property of the RLS algorithm summarized in Table 17.1. Assume that a single quantization error is made at time $n - 1$, as shown by

$$\mathbf{P}_q(n - 1) = \mathbf{P}(n - 1) + \boldsymbol{\eta}_p(n - 1)$$

[11]Rader (1990) describes a linear systolic array for adaptive beamforming based on the Cholesky factorization, which uses the CORDIC processor for its hardware implementations in VSLI form.

where $\mathbf{P}_q(n-1)$ is the quantized value of the matrix $\mathbf{P}(n-1)$ and $\boldsymbol{\eta}_p(n-1)$ is the quantization error matrix.

6. In Eqs. (17.24) and (17.48) we presented two different ways of defining the condition for a tap-output vector $\mathbf{u}(n)$ to be persistently exciting. Reconcile these two conditions.

7. It may be argued that the extended QRD–LSL algorithm described in Section 15.10 is potentially unstable in a limited-precision environment. Yet the recursive LSL algorithm with error feedback derived from the extended QRD–LSL algorithm in Section 15.12 has good numerical properties. Explain the rationale for these two statements.

8. Equation (17.66) defines a formula for the rescue variable $\zeta_M(n)$ that arises in the FTF algorithm. Derive the formula.

PART 4
Nonlinear Adaptive Filtering

The last part of the book is devoted to some aspects of nonlinear adaptive filtering. In particular, in Chapter 18 we study the blind deconvolution problem, the solution of which may require the use of higher-order statistics and therefore some form of nonlinearity built into the design of the adaptive filtering algorithm. The use of cyclostationarity for solving the blind equalization problem is also discussed in Chapter 18.

In the remaining two chapters we consider two important types of feedforward multilayer neural networks, the design of which relies on some form of supervised learning. Chapter 19 discusses the back-propagation learning algorithm for the training of multilayer perceptrons; this algorithm may be viewed as a generalization of the LMS algorithm. Chapter 20 discusses radial basis-function networks that operate in a manner entirely different from multilayer perceptrons.

CHAPTER

18

Blind Deconvolution

Deconvolution is a signal processing operation that ideally unravels the effects of convolution performed by a linear time-invariant system operating on an input signal. More specifically, in deconvolution, the output signal and the system are both known, and the requirement is to reconstruct what the input signal must have been. In *blind deconvolution*, or in more precise terms, *unsupervised deconvolution*, only the output signal is known (both the system and the input signal are unknown), and the requirement is to find both the input signal and the system itself. Clearly, blind deconvolution is a more difficult signal-processing task than ordinary deconvolution.

We may identify two broadly defined families of blind deconvolution algorithms, depending on the additional information that is used by the algorithm to make up for the unavailability of the system (channel) input:

1. *Higher-order statistics (HOS)-based algorithms:* This family of blind deconvolution algorithms may itself be subdivided into two groups:
   • *Implicit HOS-based algorithms,* which exploit higher-order statistics of the received signal in an implicit sense; this group of blind deconvolution algorithms includes *Bussgang algorithms,* so called because the deconvolved sequence assumes Bussgang statistics when the algorithm converges in the mean value.

- *Explicit HOS-based algorithms,* which explicitly use *higher-order cumulants* or their discrete Fourier transforms known as *polyspectra;* the property of polyspectra to preserve phase information makes them well suited for blind deconvolution.

2. *Cyclostationary statistics-based algorithms,* which exploit the second-order cyclostationary statistics of the received signal; the property of cyclostationarity is known to arise in a modulated signal that results from varying the amplitude, phase, or frequency of a sinusoidal carrier, which is basic to the electrical communications process.

We begin our study of the blind deconvolution problem by discussing its theoretical implications and practical importance, which we do in the next section.

18.1 THEORETICAL AND PRACTICAL CONSIDERATIONS

Consider an *unknown* linear time-invariant system $\mathscr{L}$ with input $x(n)$ as depicted in Fig. 18.1. The input data (information-bearing) sequence $x(n)$ is assumed to consist of *independently and identically distributed (iid) symbols;* the only thing known about the input is its probability distribution. The problem is to restore $x(n)$ or equivalently, *to identify the inverse $\mathscr{L}^{-1}$ of the system $\mathscr{L}$, given the observed sequence $u(n)$ at the system output.*

If the system $\mathscr{L}$ is *minimum-phase* (i.e., the transfer function of the system has all of its poles and zeros confined to the interior of the unit circle in the z-plane), then not only is the system $\mathscr{L}$ stable, but so is the inverse system $\mathscr{L}^{-1}$. In this case, we may view the input sequence $x(n)$ as the "innovation" of the system output $u(n)$, and the inverse system $\mathscr{L}^{-1}$ is just a *whitening filter;* with it, the blind deconvolution problem is solved. These observations follow from the study of linear prediction presented in Chapter 6.

In many practical situations, however, the system $\mathscr{L}$ may *not* be minimum phase. A system is said to be *nonminimum phase* if its transfer function has any of its zeros located outside the unit circle in the z-plane; exponential stability of the system dictates that the poles must be located inside the unit circle. Practical examples of a nonminimum phase system include a telephone channel and a fading radio channel. In this situation, the restoration of the input sequence $x(n)$, given the channel output, is a more difficult problem.

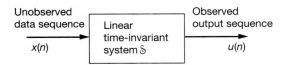

Unobserved
data sequence | Linear
time-invariant
system $\mathcal{S}$ | Observed
output sequence

$x(n)$ $u(n)$

Figure 18.1   Setting the stage for blind deconvolution.

Typically, adaptive equalizers used in digital communications require an initial *training period*, during which a *known* data sequence is transmitted. A replica of this sequence is made available at the receiver in proper synchronism with the transmitter, thereby making it possible for adjustments to be made to the equalizer coefficients in accordance with the adaptive filtering algorithm employed in the equalizer design. When the training is completed, the equalizer is switched to its *decision-directed mode*, and normal data transmission may then commence. (These two modes of operation of an adaptive equalizer were discussed in Section 7 of the introductory chapter.) However, there are practical situations where it would be highly desirable for a receiver to be able to achieve complete adaptation *without* the cooperation of the transmitter. For example, in a *multipoint data network* involving a *control unit* connected to several *data terminal equipments* (DTEs), we have a "master-slave" situation, in that a DTE is permitted to transmit only when its modem is polled by the modem of the control unit. A problem peculiar to these networks is that of retraining the receiver of a DTE unable to recognize data and polling messages, due to severe variations in channel characteristics or simply because that particular receiver was not powered on during initial synchronization of the network. Clearly, in a large or heavily loaded multipoint network, data throughput is increased and the burden of monitoring the network is eased if some form of *blind equalization* is built into the receiver design (Godard, 1980).

Another class of communication systems that may need blind equalization is *wireless communication systems* using digital technology. In particular, in a *mobile communications channel* it is impractical to employ a training sequence of long duration for two reasons:

- The system *cost* involved in the repeated transmission of a known sequence to train the equalizer at the receiving end of the system is typically too *high*.
- The unavoidable presence of *multipath fading* makes it difficult (if not impossible) to establish data transmission over the channel when outage in the system occurs; the fading phenomenon arises because the transmitted signal tends to propogate along several paths, each of different electrical length.

In reflection seismology, the traditional method of removing the source waveform from a seismogram is to use *linear-predictive deconvolution* (see Section 7 of the introductory chapter). The method of predictive deconvolution is derived from four fundamental assumptions (Gray, 1979):

1. The reflectivity series is *white*. This assumption is, however, often violated by reflection seismograms as the reflectivities result from a differential process applied to acoustic impedances. In many sedimentary basins there are thin beds that cause the reflectivity series to be correlated in sign.

2. The source signal is *minimum phase*, in that its z-transform has all of its zeros confined to the interior of the unit circle in the z-plane; here, it is presumed that

the source signal is in discrete-time form. This assumption is valid for several explosive sources (e.g., dynamite), but it is only approximate for more complicated sources such as those used in marine exploration.

3. The reflectivity series and noise are statistically independent and stationary in time. The stationarity assumption, however, is violated because of spherical divergence and attenuation of seismic waves. To cope with nonstionarity of the data, we may use adaptive deconvolution, but such a method often destroys primary events of interest.

4. The *minimum mean-square error criterion* is used to solve the linear prediction problem. This criterion is appropriate only when the prediction errors (the reflectivity series and noise) have a Gaussian distribution. Statistical tests performed on reflectivity series, however, show that their kurtosis is much higher than that expected from a Gaussian distribution. The *skewness* and *kurtosis* of a distribution function are defined as follows, respectively:

$$\gamma_1 = \frac{\mu_3}{\sigma^3}$$

and

$$\gamma_2 = \frac{\mu_4}{\sigma^4} - 3$$

where σ^2 is the variance of the distribution, and μ_3 and μ_4 are its third- and fourth-order central moments, respectively.

Assumptions 1 and 2 were explicitly mentioned in the presentation of the method of predictive deconvolution in the introductory chapter. Assumptions 3 and 4 are implicit in the application of Wiener filtering that is basic to the solution of the linear prediction problem, as presented in Chapter 6. The main point of the discussion here is that valuable phase information contained in a reflection seismogram is ignored by the method of predictive deconvolution. This limitation is overcome by using *blind deconvolution* (Godfrey and Rocca, 1981).

Blind equalization in digital communications and blind deconvolution in reflection seismology are examples of a special kind of adaptive inverse filtering that operate in an *unsupervised* manner (i.e., without access to a desired response). Only the received signal and some additional information in the form of a *probabilistic source model* are provided. In the case of equalization for digital communications, the model describes the statistics of the transmitted data sequence. In the case of seismic deconvolution, the model describes the statistics of the earth's reflection coefficients.

Having clarified the framework within which the use of blind deconvolution is feasible, we are ready to undertake a detailed study of its operation. Specifically, we begin by considering the Bussgang family of blind deconvolution algorithms in the context of equalization for digital communications.

18.2 BUSSGANG ALGORITHM FOR BLIND EQUALIZATION OF REAL BASEBAND CHANNELS

Consider the *baseband model* of a digital communications system, depicted in Fig. 18.2. The model consists of the cascade connection of a *linear communication channel* and a *blind equalizer*.

The channel includes the combined effects of a transmit filter, a transmission medium, and a receive filter. It is characterized by an impulse response h_n that is *unknown*; it may be time varying, albeit slowly. The nature of the impulse response h_n (i.e., whether it is real or complex valued) is determined by the type of modulation employed. To simplify the discussion, for the present, we assume that the impulse response is real, which corresponds to the use of *multilevel pulse-amplitude modulation (M-ary PAM)*; the case of a complex impulse response is considered in the next section. We may thus describe the sampled input-output relation of the channel by the *convolution sum*

$$u(n) = \sum_{k=-\infty}^{\infty} h_k x(n - k), \qquad n = 0, \pm 1, \pm 2, \ldots \tag{18.1}$$

where $x(n)$ is the *data (message) sequence* applied to the channel input, and $u(n)$ is the resulting *channel output*. For this introductory treatment of blind deconvolution, the effect of receiver noise is ignored in Eq. (18.1). We are justified to do so, because degradation in the performance of data transmission (over a voice-grade telephone channel, say) is usually dominated by *intersymbol interference* due to channel dispersion. We further assume that

$$\sum_{k} h_k^2 = 1 \tag{18.2}$$

Equation (18.2) implies the use of *automatic gain control* (AGC) that keeps the variance of the channel output $u(n)$ essentially constant. Also, in general, the channel is *noncausal*, which means that

$$h_n \neq 0 \qquad \text{for } n < 0 \tag{18.3}$$

The problem we wish to solve is the following:

Given the received signal $u(n)$, reconstruct the original data sequence $x(n)$ applied to the channel input.

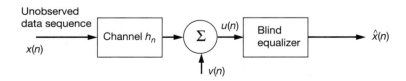

Figure 18.2   Cascade connection of an unknown channel and blind equalizer.

Equivalently, we may restate the problem as follows:

Design a blind equalizer that is the inverse of the unknown channel, with the channel input being unobservable and with no desired response available.

To solve the blind equalization problem, we need to prescribe a *probabilistic model* for the data sequence $x(n)$. For the problem at hand, we assume the following (Bellini, 1986, 1994):

1. The data sequence $x(n)$ is *white*; that is, the data symbols are *iid random variables*, with zero mean and unit variance, as shown by

$$E[x(n)] = 0 \tag{18.4}$$

   and

$$E[x(n)x(k)] = \begin{cases} 1, & k = n \\ 0, & k \neq n \end{cases} \tag{18.5}$$

   where E is the statistical expectation operator.

2. The *probability density function* of the data symbol $x(n)$ is *symmetric* and *uniform*; that is (see Fig. 18.3),

$$f_X(x) = \begin{cases} 1/2\sqrt{3}, & -\sqrt{3} \leq x < \sqrt{3} \\ 0, & \text{otherwise} \end{cases} \tag{18.6}$$

   This distribution has the merit of being independent of the number M of amplitude levels employed in the modulation process.

Note that Eq. (18.4) and the first line of Eq. (18.5) follow from Eq. (18.6).

With the distribution of $x(n)$ assumed to be symmetric, as in Fig. 18.3, we find that the whole data sequence $-x(n)$ has the same law as $x(n)$. Hence we cannot distinguish the desired inverse filter $\mathcal{L}^{-1}$ (corresponding to $x(n)$) from the opposite one $-\mathcal{L}^{-1}$ (corre-

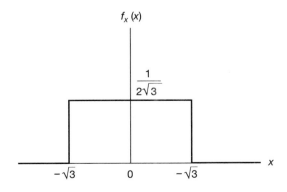

Figure 18.3   Uniform distribution.

sponding to $-x(n)$). We may overcome this *sign ambiguity problem* by *initializing* the deconvolution algorithm such that there is a single nonzero tap weight with the desired algebraic sign (Benveniste et al., 1980).

Iterative Deconvolution: The Objective

Let w_i denote the impulse response of the *ideal inverse filter*, which is related to the impulse response h_i of the channel as follows:

$$\sum_i w_i h_{l-i} = \delta_l \tag{18.7}$$

where δ_l is the *Kronecker delta*:

$$\delta_l = \begin{cases} 1, & l = 0 \\ 0, & l \neq 0 \end{cases} \tag{18.8}$$

An inverse filter defined in this way is "ideal" in the sense that it reconstructs the transmitted data sequence $x(n)$ *correctly*. To demonstrate this, we first write

$$\sum_i w_i u(n-i) = \sum_i \sum_k w_i h_k x(n-i-k) \tag{18.9}$$

Let

$$k = l - i$$

Making this change of indices in Eq. (18.9), and interchanging the order of summation, we get

$$\sum_i w_i u(n-i) = \sum_l x(n-l) \sum_i w_i h_{l-i} \tag{18.10}$$

Hence, using Eq. (18.7) in (18.10) and then applying the definition of Eq. (18.8), we get

$$\sum_i w_i u(n-i) = \sum_l \delta_l x(n-l) \tag{18.11}$$

$$= x(n)$$

which is the desired result.

For the situation described herein, the impulse response h_n is unknown. We cannot therefore use Eq. (18.7) to determine the inverse filter. Instead, we use an *iterative deconvolution procedure* to compute an *approximate inverse filter* characterized by the impulse response $\hat{w}_i(n)$. The index i refers to the *tap-weight number* in the *transversal filter* realization of the approximate inverse filter, as indicated in Fig. 18.4. The index n refers to the *iteration number*; each iteration corresponds to the transmission of a data symbol. The computation is performed iteratively in such a way that the convolution of the impulse response $\hat{w}(n)$ with the received signal $u(n)$ results in the complete or partial removal of the

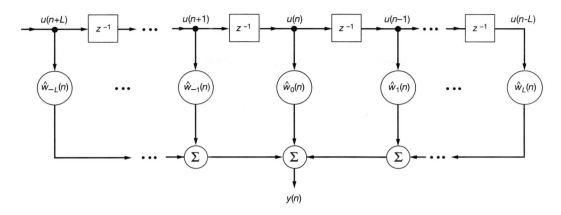

Figure 18.4 Transversal filter realization of approximate inverse filter; use of real data is assumed.

intersymbol interference (Bellini, 1986). Thus, at the nth iteration we have an approximately deconvolved sequence

$$y(n) = \sum_{i=-L}^{L} \hat{w}_i(n) u(n - i) \tag{18.12}$$

where $2L + 1$ is the truncated *length* of the impulse response $\hat{w}_i(n)$ (see Fig. 18.4). For the sake of simplicity, it is customary to assume that the transversal filter (equalizer) is symmetric about the midpoint $i = 0$ but this assumption is not required yet.

The convolution sum on the left-hand side of Eq. (18.11), pertaining to the ideal inverse filter, is *infinite* in extent, in that the index i ranges from $-\infty$ to ∞. In this case, we speak of a *doubly infinite filter (equalizer)*. On the other hand, the convolution sum on the right-hand side of Eq. (18.12) pertaining to the approximate inverse filter is *finite* in extent, in that i extends from $-L$ to L. In this latter case, which is how it usually is in practice, we speak of a *finitely parameterized filter (equalizer)*. Clearly, we may rewrite Eq. (18.12) as follows:

$$y(n) = \sum_{i} \hat{w}_i(n) u(n - i), \qquad \hat{w}_i(n) = 0 \text{ for } |i| > L$$

or, equivalently,

$$y(n) = \sum_{i} w_i u(n - i) + \sum_{i} [\hat{w}_i(n) - w_i] u(n - i) \tag{18.13}$$

Let

$$v(n) = \sum_{i} [\hat{w}_i(n) - w_i] u(n - i), \qquad \hat{w}_i = 0 \text{ for } |i| > L \tag{18.14}$$

Then, using the ideal result of Eq. (18.11) and the definition of Eq. (18.14), we may simplify Eq. (18.13) as follows:

$$y(n) = x(n) + v(n) \tag{18.15}$$

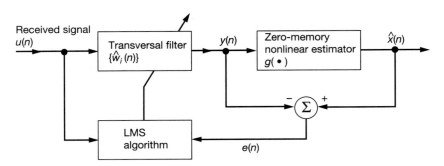

Figure 18.5   Block diagram of blind equalizer.

The term $v(n)$ is called the *convolutional noise*, representing the *residual* intersymbol interference that results from the use of an approximate inverse filter.

The inverse filter output $y(n)$ is next applied to a *zero-memory nonlinear estimator*, producing the estimate $\hat{x}(n)$ for the data symbol $x(n)$. This operation is depicted in the block diagram of Fig. 18.5. We may thus write

$$\hat{x}(n) = g(y(n)) \tag{18.16}$$

where $g(\bullet)$ is some nonlinear function. The issue of nonlinear estimation is discussed in the next subsection.

Ordinarily, we find that the estimate $\hat{x}(n)$ at iteration n is not reliable enough. Nevertheless, we may use it in an *adaptive* scheme to obtain a "better" estimate at the next iteration, $n + 1$. Indeed, we have a variety of *linear* adaptive filtering algorithms (discussed in previous chapters) at our disposal that we can use to perform this adaptive parameter estimation. In particular, a simple and yet effective scheme is provided by the LMS algorithm. To apply it to the problem at hand, we note the following:

1. The ith tap input of the transversal filter at iteration (time) n is $u(n - i)$.
2. Viewing the nonlinear estimate $\hat{x}(n)$ as the "desired" response [since the transmitted data symbol $\hat{x}(n)$ is unavailable to us], and recognizing that the corresponding transversal filter output is $y(n)$, we may express the *estimation error* for the iterative deconvolution procedure as

$$e(n) = \hat{x}(n) - y(n) \tag{18.17}$$

3. The ith tap weight $\hat{w}_i(n)$ at iteration n represents the "old" parameter estimate.

Accordingly, the "updated" value of the ith tap weight at iteration $n + 1$ is computed as follows:

$$\hat{w}_i(n + 1) = \hat{w}_i(n) + \mu u(n - i)e(n), \qquad i = 0, \pm 1, \dots, \pm L \tag{18.18}$$

where μ is the *step-size parameter*. Note that for the situation being considered here, the data are all *real* valued.

Equations (18.12), (18.16), (18.17), and (18.18) constitute the iterative deconvolution algorithm for the blind equalization of a real baseband channel (Bellini, 1986). As remarked earlier, each iteration of the algorithm corresponds to the transmission of a data symbol. It is assumed that the symbol duration is known at the receiver.

A block diagram of the blind equalizer is shown in Fig. 18.5. The idea of generating the estimation error $e(n)$, as detailed in Eqs. (18.16) and (18.17), is similar in philosophy to the decision-directed mode of operating an adaptive equalizer. More will be said on this issue later in the section.

Nonconvexity of the Cost Function

The ensemble-averaged cost function corresponding to the tap-weight update equation (18.18) is defined by

$$J(n) = E[e^2(n)]$$
$$= E[(\hat{x}(n) - y(n))^2] \tag{18.19}$$
$$= E[(g(y(n)) - y(n))^2]$$

where $y(n)$ is defined by Eq. (18.12). In the LMS algorithm, the cost function is a quadratic (convex) function of the tap weights and therefore has a well-defined minimum point. By contrast, the cost function $J(n)$ of Eq. (18.19) is a *nonconvex* function of the tap weights. This means that, in general, the error-performance surface of the iterative deconvolution procedure described here may have *local minima* in addition to *global minima*. More than one global minimum may exist, corresponding to data sequences that are equivalent under the chosen blind deconvolution criterion (e.g., sign ambiguity). The nonconvexity of the cost function $J(n)$ may arise because of the fact that the estimate $\hat{x}(n)$, performing the role of an internally generated "desired response," is produced by passing the linear combiner output $y(n)$ through a zero-memory nonlinearity, and also because $y(n)$ is itself a function of the tap weights.

In any event, the nonconvex form of the cost function $J(n)$ may result in *ill-convergence* of the iterative deconvolution algorithm described by Eqs. (18.12) and (18.16) to (18.18). The important issue of convergence is considered in some greater detail later in this section.

Statistical Properties of Convolutional Noise

The additive convolutional noise $v(n)$ is defined in Eq. (18.14). To develop a more refined formula for $v(n)$, we note that the tap input $u(n - i)$ involved in the summation on the right-hand side of this equation is given by [see Eq. (18.1)]

$$u(n - i) = \sum_k h_k x(n - i - k) \tag{18.20}$$

We may therefore rewrite Eq. (18.14) as a double summation:

$$v(n) = \sum_i \sum_k h_k[\hat{w}_i(n) - w_i]x(n - i - k) \tag{18.21}$$

Let

$$n - i - k = l$$

Hence, we may also write

$$v(n) = \sum_l x(l)\nabla(n - l) \tag{18.22}$$

where

$$\nabla(n) = \sum_k h_k[\hat{w}_{n-k}(n) - w_{n-k}] \tag{18.23}$$

The sequence $\nabla(n)$ is a sequence of small numbers, representing to the *residual impulse response* of the channel due to imperfect equalization. We imagine the sequence $\nabla(n)$ as a *long and oscillatory wave* that is convolved with the transmitted data sequence $x(n)$ to produce the convolutional noise sequence $v(n)$, as indicated in Eq. (18.22).

The definition of Eq. (18.22) is basic to the statistical characterization of the convolution noise $v(n)$. The mean of $v(n)$ is zero, as shown by

$$E[v(n)] = E\left[\sum_l x(l)\nabla(n - l)\right]$$

$$= \sum_l \nabla(n - l)E[x(l)] \tag{18.24}$$

$$= 0$$

where in the last line we have made use of Eq. (18.4). Next, the autocorrelation function of $v(n)$ for a lag j is given by

$$E[v(n)v(n - j)] = E\left[\sum_l x(l)\nabla(n - l)\sum_m x(m)\nabla(n - m - j)\right]$$

$$= \sum_l \sum_m \nabla(n - l)\nabla(n - m - j)E[x(l)x(m)] \tag{18.25}$$

$$= \sum_l \nabla(n - l)\nabla(n - l - j)$$

where in the last line we have made use of Eq. (18.5). Since $\nabla(n)$ is a long and oscillatory waveform, the sum on the right-hand side of Eq. (18.25) is nonzero only for $j = 0$, obtaining

$$E[v(n)v(n - j)] = \begin{cases} \sigma^2, & j = 0 \\ 0, & j \neq 0 \end{cases} \tag{18.26}$$

where

$$\sigma^2(n) = \sum_l \nabla^2(n - l) \tag{18.27}$$

Based on Eqs. (18.24) and (18.26), we may thus describe the convolutional noise process $v(n)$ as a *zero-mean white-noise process of time-varying variance equal to* $\sigma^2(n)$, defined by Eq. (18.27).

According to the model of Eq. (18.22), the convolutional noise $v(n)$ is a weighted sum of iid variables representing different transmissions of data symbols. If, therefore, the residual impulse response $\nabla(n)$ is long enough, the *central limit theorem* makes a *Gaussian* model for $v(n)$ to be plausible.

Having characterized the convolutional noise $v(n)$ by itself, all that remains for us to do is to evaluate the *cross-correlation* between it and the data sample $x(n)$. These two random variables are certainly *correlated* with each other, since $v(n)$ is the result of convolving the residual impulse response $\nabla(n)$ with $x(n)$, as shown in Eq. (18.22). However, the cross-correlation between $v(n)$ and $x(n)$ is negligible compared to the variance of $v(n)$. To demonstrate this, we write

$$
\begin{aligned}
E[x(n)v(n-j)] &= E\Big[x(n)\sum_l x(l)\nabla(n-l-j)\Big] \\
&= \sum_l \nabla(n-l-j)E[x(n)x(l)] \\
&= \nabla(-j)
\end{aligned}
\tag{18.28}
$$

where, in the last line, we have made use of Eq. (18.5). Here again, using the assumption that $\nabla(n)$ is a long and oscillatory waveform, we deduce that the variance of $v(n)$ is large compared to the magnitude of the cross-correlation $E[x(n)v(n-j)]$.

Since the data sequence $x(n)$ is white by assumption and the convolutional noise sequence $v(n)$ is approximately white by deduction, and since these two sequences are essentially uncorrelated, it follows that their sum $y(n)$ is approximately white too. This suggests that $x(n)$ and $v(n)$ may be taken to be essentially independent. We may thus model the convolutional noise $v(n)$ as an *additive, zero-mean, white Gaussian noise process that is statistically independent of the data sequence* $x(n)$.

Because of the approximations made in deriving the model described herein for the convolutional noise, its use in an iterative deconvolution process yields a *suboptimal* estimator for the data sequence. In particular, given that the iterative deconvolution process is convergent, the intersymbol interference (ISI) during the latter stages of the process may be small enough for the model to be applicable. In the early stages of the iterative deconvolution process, however, the ISI is typically large with the result that the data sequence and the convolutional noise are strongly correlated, and the convolutional noise sequence is more uniform than Gaussian (Godfrey and Rocca, (1981)).

Zero-Memory Nonlinear Estimation of the Data Sequence

We are now ready to consider the next important issue, namely, that of estimating the data sequence $x(n)$, given the deconvolved sequence $y(n)$ at the transversal filter output. Specifically, we may formulate the estimation problem as follows: We are given a (filtered) observation $y(n)$ that consists of the sum of two components (see Fig. 18.6):

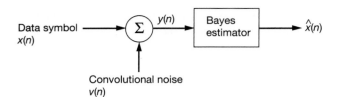

Figure 18.6 Estimation of the data symbol $x(n)$, given the observation $y(n)$.

1. A uniformly distributed data symbol $x(n)$ with zero mean and unit variance
2. A white Gaussian noise $v(n)$ with zero mean and variance $\sigma^2(n)$, which is statistically independent of $x(n)$

The requirement is to derive a *Bayes estimate* of $x(n)$, optimized in a statistical sense.

Before proceeding with this classical estimation problem, two noteworthy observations are in order. First, the estimate is naturally a *conditional estimate* that depends on the optimization criterion. Second, although the estimate (in theory) is optimum in a mean-square error sense, in the context of our present situation, it is *suboptimum* by virtue of the approximations made in the development of the model described above for the convolutional noise $v(n)$.

An optimization criterion of particular interest is that of minimizing the mean-square value of the error between the actual transmission $x(n)$ and the estimation $\hat{x}(n)$. The choice of this optimization criterion yields a *conditional mean estimator*[1] that is both sensible and robust.

For convenience of presentation, we will supress the dependence of random variables on time n. Thus, given the observation y, the conditional mean estimate $\hat{x}$ of the random variable x is written as $E[\hat{x}|y]$, where E is the expectation operator. Let $f_X(x|y)$ denote the *conditional probability density function* of x, given y. We thus have

$$\hat{x} = E[x|y]$$

$$= \int_{-\infty}^{\infty} x f_X(x|y)\,dx \tag{18.29}$$

From *Bayes' rule*, we have

$$f_X(x|y) = \frac{f_Y(y|x)\,f_X(x)}{f_Y(y)} \tag{18.30}$$

where $f_Y(y|x)$ is the conditional probability density function of y, given x; and $f_X(x)$ and $f_Y(y)$ are the probability density functions of x and y, respectively. We may therefore rewrite the formula of Eq. (18.29) as

$$\hat{x} = \frac{1}{f_Y(y)} \int_{-\infty}^{\infty} x f_Y(y|x) f_X(x)\,dx \tag{18.31}$$

---

[1] For a derivation of the conditional mean and its relation to mean-squared-error estimation, see Appendix D.

Let the deconvolved sequence $y(n)$ be a scaled version of the original data sequence $x(n)$, except for an additive noise term $v(n)$, as shown by

$$y = c_0 x + v \tag{18.32}$$

The scaling factor c_0 is slightly smaller than unity. This factor has been included in Eq. (18.32) so as to keep $E[y^2]$ equal to 1. In accordance with the statistical model for the conventional noise v developed previously, x and v are statistically independent. With v modeled to have zero mean and variance σ^2, we readily see from Eq. (18.32) that the scaling factor c_0 is

$$c_0 = \sqrt{1 - \sigma^2} \tag{18.33}$$

Furthermore, from Eq. (18.32) it follows that

$$f_Y(y|x) = f_V(y - c_0 x) \tag{18.34}$$

Accordingly, the use of Eq. (18.34) in (18.31) yields

$$\hat{x} = \frac{1}{f_Y(y)} \int_{-\infty}^{\infty} x f_V(y - c_0 x) f_X(x) \, dx \tag{18.35}$$

The evaluation of $\hat{x}$ is straightforward but tedious. To proceed with it, we may note the following:

1. The mathematical form of the estimate $\hat{x}(n)$ produced at the output of the Bayes (conditional mean) estimator depends on the probability density function of the original data symbol $x(n)$. For the analysis presented here, we assume that the data symbol x is *uniformly distributed* with zero mean and unit variance; its probability density function is given in Eq. (18.6), which is reproduced here for convenience:

$$f_X(x) = \begin{cases} 1/2\sqrt{3}, & -\sqrt{3} \le x < \sqrt{3} \\ 0, & \text{otherwise} \end{cases} \tag{18.36}$$

2. The convolutional noise v is *Gaussian distributed* with zero mean and variance σ^2; its probability density function is

$$f_V(v) = \frac{1}{\sqrt{2\pi}\sigma} \exp\left(-\frac{v^2}{2\sigma^2}\right) \tag{18.37}$$

3. The filtered observation y is the sum of $c_0 x$ and v; its probability density function is therefore equal to the convolution of the probability density function of x with that of v, as shown by

$$f_Y(y) = \int_{-\infty}^{\infty} f_X(x) f_V(y - c_0 x) \, dx \tag{18.38}$$

Using Eqs. (18.36) to (18.38) in (18.35), we get (Bellini, 1988)

$$x = \frac{1}{c_0 y} - \frac{\sigma}{c_0} \frac{Z(y_1) - Z(y_2)}{Q(y_1) - Q(y_2)} \tag{18.39}$$

where the variables y_1 and y_2 are defined by

$$y_1 = \frac{1}{\sigma}(y + \sqrt{3}\, c_0)$$

and

$$y_2 = \frac{1}{\sigma}(y - \sqrt{3}\, c_0)$$

The function $Z(y)$ is the *standardized* Gausian probability density function

$$Z(y) = \frac{1}{\sqrt{2\pi}}\, e^{-y^2/2} \tag{18.40}$$

The function $Q(y)$ is the corresponding probability distribution function

$$Q(y) = \frac{1}{\sqrt{2\pi}} \int_y^\infty e^{-u^2/2}\, du \tag{18.41}$$

A small *gain correction* to the nonlinear estimator of Eq. (18.39) is needed in order to achieve perfect equalization[2] when the iterative deconvolution algorithm [described by Eqs. (18.16) to (18.18)] converges eventually. Perfect equalization requires that $y = x$. Under the minimum mean-square error condition, the estimation error is orthogonal to each of the tap inputs in the transversal filter realization of the approximate inverse filter. Putting all of this together, we find that the following condition must hold (Bellini, 1986, 1988):

$$E[\hat{x}\, g(\hat{x})] = 1 \tag{18.42}$$

where $g(\hat{x})$ is the nonlinear estimator $\hat{x} = g(y)$ with $y = \hat{x}$ for perfect equalization; see Problem 2.

Figure 18.7 shows the nonlinear estimate $\hat{x} = g(z)$ plotted versus $|z|$ for an eight-level PAM system (Bellini, 1986, 1988). The estimator is normalized in accordance with Eq.

---

[2] In general, for perfect equalization we require that

$$y = (x - D)e^{j\phi}$$

where D is a constant delay and ϕ is a constant phase shift. This condition corresponds to an equalizer whose transfer function has magnitude one and a linear phase response. We note that the input data sequence x_i is stationary and the channel is linear time-invariant. Hence, the observed sequence $y(n)$ at the channel output is also stationary; its probability density function is therefore invariant to the constant delay D. The constant phase shift ϕ is also of no immediate consequence when the probability density function of the input sequence remains symmetric under rotation, which is indeed the case for the assumed density function given in Eq. (18.36). We may therefore simplify the condition for perfect equalization by requiring that $y = x$.

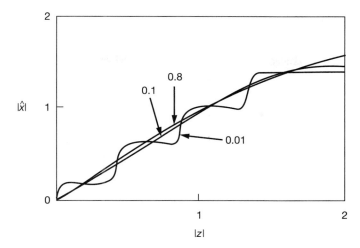

Figure 18.7   Nonlinear estimators of eight-level data in Gaussian noise with $\hat{x} = g\,(z)$. The noise-to-(signal + noise) ratios are 0.01, 0.1, and 0.8. [From Bellini (1986), with permission of the IEEE]

(18.42). In this figure, three widely different levels of convolutional noise are considered. Here we note from Eq. (18.32) that the *distortion-to (signal plus distortion) ratio* is given by

$$\frac{E[(y - x)^2]}{E[y^2]} = (1 - c_0)^2 + \sigma^2$$

$$= 2(1 - c_0) \tag{18.43}$$

where in the last line we have made use of Eq. (18.33). The curves presented in Fig. 18.7 correspond to three values of this ratio, namely, 0.01, 0.1, and 0.8. We observe the following from these curves:

1. When the convolutional noise is low, the blind equalization algorithm approaches a minimum mean-squared-error criterion.
2. When the convolutional noise is high, the nonlinear estimator appears to be independent of the fine structure of the amplitude-modulated data. Indeed, different values of amplitude modulation levels result in only very small gain differences due to the normalization defined by Eq. (18.42). This suggests that the use of a uniform amplitude distribution for multilevel modulation systems is an adequate approximation.
3. The nonlinear estimator is robust with respect to variations in the variance of the convolutional noise.

Convergence Considerations

For the iterative deconvolutional algorithm described by Eqs. (18.16) to (18.18) to converge in the mean value, we require the expected value of the tap weight $\hat{w}_i(n)$ to approach some constant value as the number of iterations n approaches infinity. Correspondingly, we find that the *condition for convergence in the mean value* is described by

$$E[u(n - i)y(n)] = E[u(n - i)g(y(n))], \qquad \text{large } n, \text{ and } i = 0, \pm 1, \ldots, \pm L$$
$$(18.44)$$

Multiplying both sides of this equation by $\hat{w}_{i-k}$ and summing over i, we get

$$E\left[y(n) \sum_{i=-L}^{L} \hat{w}_{i-k}(n)u(n - i)\right] = E\left[g(y(n)) \sum_{i=-L}^{L} \hat{w}_{i-k}(n)u(n - i)\right], \qquad \text{large } n$$
$$(18.45)$$

We next note from Eq. (18.12) that

$$y(n - k) = \sum_{i=-L}^{L} \hat{w}_i(n)u(n - k - i)$$

$$= \sum_{i=-L-k}^{L-k} \hat{w}_{i-k}(n)u(n - i), \qquad \text{large } n$$

Provided that L is large enough for the transversal equalizer to achieve perfect equalization, we may approximate the expression for $y(n - k)$ as

$$y(n - k) \simeq \sum_{i=-L}^{L} \hat{w}_{i-k}(n)u(n - i), \qquad \text{large } n \text{ and large } L \qquad (18.46)$$

Accordingly, we may use Eq. (18.46) to simplify (18.45) as follows:

$$E[y(n)y(n - k)] \simeq E[g(y(n))y(n - k)], \qquad \text{large } n \text{ and large } L \qquad (18.47)$$

We now recognize the following property. A stochastic process $y(n)$ is said to be a *Bussgang process* if it satisfies the condition

$$E[y(n)y(n - k)] = E[y(n)g(y(n - k))] \qquad (18.48)$$

where the function $g(\cdot)$ is a zero-memory nonlinearity.[3] In other words, a Bussgang process has the property that its autocorrelation function is equal to the cross-correlation between that process and the output of a zero-memory nonlinearity produced by that process, with both correlations being measured for the same lag. Note that a Bussgang process satisfies Eq. (18.48) up to a multiplicative constant; in the case discussed here, the multiplicative constant is unity by virtue of the assumption made in Eq. (18.42).

---

[3] A number of stochastic processes belong to the class of Bussgang processes. Bussgang (1952) was the first to recognize that any correlated Gaussian process has the property described in Eq. (18.48). Subsequently, Barrett and Lampard (1955) extended Bussgang's result to all stochastic processes with exponentially decaying autocorrelation functions. This includes an independent process, since its autocorrelation function consists of a delta function that may be viewed as an infinitely fast decaying exponential (Gray, 1979).

Returning to the issue at hand, we may state that the process $y(n)$ acting as the input to the zero-memory nonlinearity in Fig. 18.5 is *approximately* a Bussgang process, provided that L is large; the approximation becomes better as L is made larger. It is for this reason that the blind equalization algorithm described here is referred to as a *Bussgang algorithm* (Bellini, 1986, 1988).

In general, convergence of the Bussgang algorithm is not guaranteed. Indeed, the cost function of the Bussgang algorithm operating with a finite L is *nonconvex*; it may therefore have false minima.

For the idealized case of a doubly *infinite* equalizer, however, a rough proof of convergence of the Bussgang algorithm may be sketched as follows (Bellini, 1988). The proof relies on a theorem derived in Benveniste et al. (1980), which provides sufficient conditions for convergence.[4] Let the function $\psi(y)$ denote the dependence of the estimation error in the LMS algorithm on the transversal filter output $y(n)$. According to our terminology, we have [see Eqs. (18.16) and (18.17)]

$$\psi(y) = g(y) - y \qquad (18.49)$$

The *Benveniste–Goursat–Ruget theorem* states that convergence of the Bussgang algorithm is guaranteed if the probability distribution of the data sequence $x(n)$ is *sub-Gaussian* and the second derivative of $\psi(y)$ is negative on the interval $[0, \infty)$. In particular, we may state the following:

1. A random variable x, for example, with probability density function

$$f_X(x) = Ke^{-|x/\beta|^v}, \qquad K = \text{constant} \qquad (18.50)$$

    is sub-Gaussian when $v > 2$. For the limiting case of $v = \infty$, the probability density function of Eq. (18.50) reduces to that of a uniformly distributed random variable. Also, by choosing $\beta = \sqrt{3}$, we have $E[x^2] = 1$. Thus, the probabilistic model assumed in Eq. (18.6) satisfies the first part of the Benveniste–Goursat–Ruget theorem.

2. The second part of the theorem is also satisfied by the Bussgang algorithm, since we have

$$\frac{\partial^2 \psi}{\partial y^2} < 0 \qquad \text{for } 0 < y < \infty \qquad (18.51)$$

This is readily verified by examining the curves plotted in Fig. 18.7.

The Benveniste–Goursat–Ruget theorem exploited in this proof is based on the assumption of a doubly infinite equalizer. Unfortunately, this assumption breaks down in practice as we have to work with a finitely parameterized equalizer. To date, no zero-memory nonlinear function $g(\cdot)$ is known, which would result in global convergence of the

---

[4] Note that the function $\psi(y)$ defined in Eq. (18.49) is the negative of that defined in Benveniste et al. (1980).

blind equalizer in Fig. 18.5 to the inverse of the unknown channel (Verdu, 1984; Johnson, 1991). The global convergence of the Bussgang algorithm for an arbitrarily large but finite filter length remains an open problem. Nevertheless, there is practical evidence, supported by convergence analysis presented in Li and Ding (1995), for the conjecture that the Bussgang algorithm will converge to a desired global minimum if the transversal equalizer is long enough and initialized with a nonzero center tap, e.g., $\hat{w}_0(0) = 1$ in Fig. 18.4.

Decision-Directed Algorithm

When the Bussgang algorithm has converged and the eye pattern appears "open," the equalizer should be switched smoothly to the *decision-directed mode* of operation, and minimum mean-squared-error control of the tap weights of the transversal filter component in the equalizer is exercised, as in a conventional adaptive equalizer.

Figure 18.8 presents a block diagram of the equalizer operating in its decision-directed mode. The only difference between this mode of operation and that of blind equalization lies in the type of zero-memory nonlinearity employed. Specifically, the conditional mean estimation of the blind equalizer in Fig. 18.5 is replaced by a *threshold decision device*. Given the observation $y(n)$, that is, the equalized signal at the transversal filter output, the threshold device makes a *decision in favor of a particular value in the known alphabet of the transmitted data sequence that is closest to* $y(n)$. We may thus write

$$\hat{x}(n) = \text{dec}(y(n)) \qquad (18.52)$$

For example, in the simple case of an *equiprobable binary data sequence*, the data levels and decision levels are as follows, respectively:

$$x(n) = \begin{cases} +1 & \text{for symbol 1} \\ -1 & \text{for symbol 0} \end{cases} \qquad (18.53)$$

and

$$\text{dec}((y(n)) = \text{sgn}(y(n)) \qquad (18.54)$$

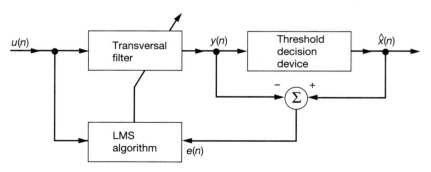

Figure 18.8   Block diagram of decision-directed mode of operation.

where sgn(·) is the *signum function* equal to $+1$ if the argument is positive, and -1 if it is negative.

The equations that govern the operation of the decision-directed algorithm are the same as those of the Bussgang algorithm, except for the use of Eq. (18.52) in place of (18.16). Herein lies an important practical advantage of a blind equalizer that is based on the Bussgang algorithm and incorporates the decision-directed algorithm: Its implementation is only slightly more complex than that of a conventional adaptive equalizer, yet it does not require the use of a training sequence.

Suppose that the following conditions are satisfied:

1. The eye pattern is open (which it should be on the completion of blind equalization).
2. The step-size parameter μ used in the LMS implementation of the decision-directed algorithm is fixed (which is a common practice).
3. The sequence of observations at the channel output, denoted by the vector $\mathbf{u}(n)$, is *ergodic* in the sense that

$$\lim_{N\to\infty} \frac{1}{N}\sum_{n=1}^{N} \mathbf{u}(n)\mathbf{u}^T(n) \to E[\mathbf{u}(n)\mathbf{u}^T(n)] \qquad \text{almost surely} \qquad (18.55)$$

Then, under these conditions, *the tap-weight vector in the decision-directed algorithm converges to the optimum (Wiener) solution in the mean-square sense* (Macchi and Eweda, 1984). This is a powerful result, making the decision-directed algorithm an important adjunct of the Bussgang algorithm for blind equalization in digital communications.

18.3 EXTENSION OF BUSSGANG ALGORITHMS TO COMPLEX BASEBAND CHANNELS

Thus far we have only discussed the use of Bussgang algorithms for the blind equalization of *M*-ary PAM systems, characterized by a real baseband channel. In this section we extend the use of this family of blind equalization algorithms to *quadrature-amplitude modulation (QAM)* systems that involve a hybrid combination of amplitude and phase modulations.

In the case of a complex baseband channel, the transmitted data sequence $x(n)$, the channel impulse response h_n, and the received signal $u(n)$ are all complex valued. We may thus write

$$x(n) = x_I(n) + jx_Q(n) \qquad (18.56)$$

$$h_n = h_{I,n} + jh_{Q,n} \qquad (18.57)$$

and

$$u(n) = u_I(n) + ju_Q(n) \qquad (18.58)$$

TABLE 18.1 SUMMARY OF BUSSGANG ALGORITHMS FOR BLIND EQUALIZATION OF COMPLEX BASEBAND CHANNELS

Initialization: Set

$$\hat{w}_i(0) = \begin{cases} 1, & i = 0 \\ 0, & i = \pm 1, \dots, \pm L \end{cases}$$

Computation: $n = 1, 2, \dots$

$$y(n) = y_I(n) + jy_Q(n)$$

$$= \sum_{i=-L}^{L} \hat{w}_i^*(n)u(n - i)$$

$$\hat{x}(n) = \hat{x}_I(n) + j\hat{x}_Q(n)$$

$$= g(y_I(n)) + jg(y_Q(n))$$

$$e(n) = \hat{x}(n) - y(n)$$

$$\hat{w}_i(n + 1) = \hat{w}_i(n) + \mu u(n - i)e^*(n), \qquad i = 0, \pm 1, \dots, \pm L$$

---

where the subscripts I and Q refer to the *in-phase* (*real*) and *quadrature* (*imaginary*) *components*, respectively. Correspondingly, the conditional mean estimate of the complex datum $x(n)$, given the observation $y(n)$ at the transversal filter output, is written as

$$\hat{x}(n) = E[x(n)|y(n)]$$

$$= \hat{x}_I(n) + j\hat{x}_Q(n) \qquad (18.59)$$

$$= g(y_I(n)) + jg(y_Q(n))$$

where $g(\cdot)$ describes a zero-memory nonlinearity. Equation (18.59) states that the in-phase and quadrature components of the transmitted data sequence $x(n)$ may be estimated separately from the in-phase and quadrature components of the transversal filter output $y(n)$, respectively. Note, however, that the conditional mean $E[x(n)|y(n)]$ can only be expressed as in Eq. (18.59) if the data transmitted in the in-phase and quadrature channels are statistically independent of each other, which is usually the case.

Clearly, Bussgang algorithms for complex baseband channels include the corresponding algorithms for real baseband channels as a special case. Table 18.1 presents a summary of Bussgang algorithms for a complex baseband channel.

18.4 SPECIAL CASES OF THE BUSSGANG ALGORITHM

The Bussgang algorithm discussed in Sections 18.2 and 18.3 is of a general formulation, in that it includes a number of blind equalization algorithms as special cases. Two special cases of the Bussgang algorithm are considered in the sequel.

Sato Algorithm

The idea of blind equalization in M-ary PAM systems dates back to the pioneering work of Sato (1975). The *Sato algorithm* consists of minimizing a *nonconvex* cost function

$$J(n) = E[(\hat{x}(n) - y(n))^2] \tag{18.60}$$

where $y(n)$ is the transversal filter output defined in Eq. (18.12), and $\hat{x}(n)$ is an estimate of the transmitted datum $x(n)$. This estimate is obtained by a zero-memory nonlinearity described as follows:

$$\hat{x}(n) = \gamma \, \text{sgn}[y(n)] \tag{18.61}$$

The constant γ sets the *gain* of the equalizer; it is defined by

$$\gamma = \frac{E[x^2(n)]}{E[|x(n)|]} \tag{18.62}$$

It is apparent that the Sato algorithm is a special (nonoptimal) case of the Bussgang algorithm, with the nonlinear function $g(y)$ defined by

$$g(y) = \gamma \, \text{sgn}(y) \tag{18.63}$$

where sgn ($\cdot$) is the signum function. The nonlinearity defined in Eq. (18.63) is similar to that in the decision-directed algorithm for binary PAM, except for the data-dependent gain factor γ.

The Sato algorithm for blind equalization was introduced originally to deal with one-dimensional multilevel (M-ary PAM) signals, with the objective of being more robust than a decision-directed algorithm. Initially, the algorithm treats such a digital signal as a "binary" signal by estimating the most significant bit; the remaining bits of the signal are treated by the algorithm as additive noise insofar as the blind equalization process is concerned. The algorithm then uses the results of this preliminary step to modify the error signal obtained from a conventional decision-directed algorithm.

The Benveniste–Goursat–Ruget theorem for convergence holds for the Sato algorithm even though the nonlinear function $\psi(\cdot)$ is not differentiable for it. According to this theorem, global convergence of the Sato algorithm can be achieved provided that the probability density function of the transmitted data sequence can be approximated by a sub-Gaussian function such as the uniform distribution [Benveniste et al. (1980)]. However, global convergence of the Sato algorithm holds only for the limiting case of a doubly infinite equalizer. Deviations from this ideal behavior have been reported in the literature:

- In Mazo (1980), Verdu (1984), and Macchi and Eweda (1985), it is shown that the Sato algorithm exhibits local minima for discrete QAM input signals.
- In Ding et al. (1989), it is shown that for finitely parameterized equalizers the Sato algorithm may converge to local minima for both discrete and sub-Gaussian inputs.

Godard Algorithm

Godard (1980) was the first to propose a family of *constant modulus* blind equalization algorithms for use in two-dimentional digital communication systems (e.g., M-ary phase-shift keying). Specifically, the *Godard algorithm* minimizes a nonconvex cost function of the form

$$J(n) = E[(|y(n)|^p - R_p)^2] \tag{18.64}$$

where p is a positive integer, and R_p is a positive real constant defined by

$$R_p = \frac{E[|x(n)|^{2p}]}{E[|x(n)|^p]} \tag{18.65}$$

The Godard algorithm is designed to penalize deviations of the blind equalizer output $x(n)$ from a constant modulus. The constant R_p is chosen in such a way that the gradient of the cost function $J(n)$ is zero when perfect equalization [i.e., $\hat{x}(n) = x(n)$] is attained.

The tap-weight vector of the equalizer is adapted in accordance with the stochastic gradient algorithm (Godard, 1980)

$$\hat{\mathbf{w}}(n + 1) = \hat{\mathbf{w}}(n) + \mu\mathbf{u}(n)e^*(n) \tag{18.66}$$

where μ is the step-size parameter, $\mathbf{u}(n)$ is the tap-input vector, and $e(n)$ is the error signal defined by

$$e(n) = y(n)|y(n)|^{p-2}(R_p - |y(n)|^p) \tag{18.67}$$

From the definition of the cost function $J(n)$ in Eq. (18.64) and from Eq. (18.67), we see that the qualizer adaptation according to the Godard algorithm does not require carrier phase recovery. The algorithm therefore tends to converge slowly. However, it offers the advantage of decoupling the ISI equalization and carrier phase recovery problems from each other.

Two cases of the Godard algorithm are of specific interest:

Case 1: $p = 1$ The cost function of Eq. (18.64) for this case reduces to

$$J(n) = E[(|y(n)| - R_1)^2] \tag{18.68}$$

where

$$R_1 = \frac{E[|x(n)|^2]}{E[|x(n)|]} \tag{18.69}$$

This case may be viewed as a modification of the Sato algorithm.

Case 2: $p = 2$ In this case, the cost function of Eq. (18.64) reduces to

$$J(n) = E[(|y(n)|^2 - R_2)^2] \tag{18.70}$$

where

$$R_2 = \frac{E[|x(n)|^4]}{E[|x(n)|^2]} \tag{18.71}$$

This second special case is referred to in the literature as the *constant modulus algorithm* (*CMA*).[5]

The Godard algorithm is considered to be the most successful among the Bussgang family of blind equalization algorithms, as demonstrated by the comparative studies reported in Shynk et al. (1991) and Jablon (1992). In particular, we may say the following (Papadias, 1995):

- The Godard algorithm is more robust than other Bussgang algorithms with respect to carrier phase offset. This important property of the Godard algorithm is due to the fact that the cost function used for its derivation is based solely on the amplitude of the received signal.

- Under steady-state conditions, the Godard algorithm attains a mean-squared error that is lower than other Bussgang algorithms.

- Last but by no means least, the Godard algorithm is often able to equalize a dispersive channel, such that the eye pattern is opened up when it is initially closed for all practical purposes.

Summary of Special Forms of the Bussgang Algorithm

The decision-directed, Sato, and Godard algorithms may be viewed as special cases of the Bussgang algorithm [Bellini (1986)]. In particular, we may use Eqs. (18.52), (18.61), and (18.67) to set up the entries shown in Table 18.2 for the special forms of the zero-memory nonlinear function $g(\cdot)$ pertaining to these three algorithms [Hatzinakos (1990)]. The entries for the decision-directed and Sato algorithms follow directly from the definition

$$\hat{x}(n) = g(y(n))$$

In the case of the Godard algorithm, we note that

$$e(n) = \hat{x}(n) - y(n)$$

or, equivalently,

$$g(y(n)) = y(n) + e(n)$$

---

[5] The constant modulus algorithm (CMA) was so named by Treichler and Agee (1983), independently of Godard's 1980 paper. It is probably the most widely investigated blind equalization algorithm and the one most widely used in practice (Treichler and Larimore, 1985a, b; Smith and Friedlander, 1985; Johnson et al., 1988).

TABLE 18.2   SPECIAL CASES OF THE BUSSGANG ALGORITHM

Algorithm	Zero-memory nonlinear function $g(\bullet)$	Definitions
Decision-directed*	$\text{sgn}(\bullet)$	
Sato	$\gamma\,\text{sgn}(\bullet)$	$\gamma = \dfrac{E[x^2(n)]}{E[\lvert x(n)\rvert]}$
Godard	$\dfrac{y(n)}{\lvert y(n)\rvert}\left(\lvert y(n)\rvert + R_p\lvert y(n)\rvert^{p-1} - \lvert y(n)\rvert^{2p-1}\right)$	$R_p = \dfrac{E[\lvert x(n)\rvert^{2p}]}{E[\lvert x(n)\rvert^{p}]}$

*The zero-memory nonlinear function sgn $(\bullet)$ for the decision-directed algorithm applies if the input data are binary; for the general case of M-ary PAM, an M-ary slicer is required.

Hence, we may use this relation and Eq. (18.67) to derive the special forms of the Godard algorithm in Table 18.2.

18.5 BLIND CHANNEL IDENTIFICATION AND EQUALIZATION USING POLYSPECTRA

The Bussgang algorithm uses the higher-order statistics of the received signal in an implicit sense. We now describe another class of blind deconvolution algorithm, which uses the higher-order statistics of the received signal in an explicit sense. For convenience of presentation, we restrict the discussion to real-valued stochastic processes.

From Chapter 3 we recall that the higher-order statistics of a stationary stochastic process are described in terms of the *cumulants* and their Fourier transforms known as *polyspectra*. Indeed, cumulants and polyspectra may be viewed as generalizations of the autocorrelation function and power spectrum, respectively. Polyspectra provide the basis for the identification (and therefore blind equalization) of a nonminimum-phase channel by virtue of their ability to preserve phase information in the channel output.

Consider then the system model described in Section 18.2 for the baseband transmission of a data sequence $x(n)$ using M-ary modulation. The probabilistic model of the sequence $x(n)$ is as described in Eqs. (18.4) to (18.6). We assume that the FIR channel transfer function $H(z)$ admits the following factorization, under the premise that $H(z)$ has no zeros on the unit circle:

$$H(z) = kI(z)O(z^{-1}) \tag{18.72}$$

where k is a scaling factor, $I(z)$ is a *minimum-phase polynomial*, and $O(z^{-1})$ is a *maximum-phase polynomial*. The polynomial $I(z)$ has all its zeros inside the unit circle in the z-plane, as shown by

$$I(z) = \prod_{l=1}^{L_1} (1 - a_l z^{-1}), \qquad |a_l| < 1 \tag{18.73}$$

The second polynomial $O(z)$ has all its zeros outside the unit circle, as shown by

$$O(z^{-1}) = \prod_{l=1}^{L_2} (1 - b_l z), \qquad |b_l| < 1 \tag{18.74}$$

According to the representation described in Eqs. (18.72) to (18.74), the channel is characterized by a finite-(length) impulse response and nonminimum-phase transfer function.

For a data sequence $x(n)$ having a symmetric uniform distribution, as described in the probabilistic model of Eq. (18.6), we have

$$E[x(n)] = 0$$

$$E[x^2(n)] = 1$$

$$E[x^3(n)] = 0$$

$$E[x^4(n)] = 9/5$$

Correspondingly, the *skewness* of $x(n)$ is $\gamma_3 = 0$, and its *kurtosis* is

$$\gamma_4 = E[x^4(n)] - 3(E[x^2(n)])^2$$

$$= \frac{9}{5} - 3 = -\frac{6}{5}$$

With $\gamma_3 = 0$, it follows that the third-order cumulant of the channel output $u(n)$ is identically zero. On the other hand, γ_4 has a nonzero value; we may therefore work in the fourth-order cumulant domain as a basis for blind equalization.

Tricepstrum

Let $c_4(\tau_1, \tau_2, \tau_3)$ denote the fourth-order cumulant of the channel output $u(n)$. We may express the trispectrum of $u(n)$ as

$$C_4(\omega_1, \omega_2, \omega_3) = F[c_4(\tau_1, \tau_2, \tau_3)] \tag{18.75}$$

where $F[\cdot]$ denotes three-dimensional discrete Fourier transformation. Define

$$\kappa_4(\tau_1, \tau_2, \tau_3) = F^{-1}[\ln C_4(\omega_1, \omega_2, \omega_3)] \tag{18.76}$$

where ln signifies the natural logarithm, and F^{-1} signifies inverse three-dimensional discrete Fourier transformation. The quantity $\kappa_4(\tau_1, \tau_2, \tau_3)$ is called the *complex cepstrum of trispectrum* or *tricepstrum* of the process $u(n)$ (Pan and Nikias, 1988; Hatzinakos and Nikias, 1989, 1991).

When a linear time-invariant system (channel) characterized by impulse response h_n is excited by a process $x(n)$ consisting of *iid* random variables, the fourth-order cumulant of the resulting output $u(n)$ is defined by (see section 3.7)

$$c_4(\tau_1, \tau_2, \tau_3) = \gamma_4 \sum_{i=0}^{\infty} h_i h_{i+\tau_1} h_{i+\tau_2} h_{i+\tau_3} \tag{18.77}$$

Note that the relation of Eq. (18.77) holds even if the linear system (channel) includes additive white Gaussian noise at its output, which is typically the case in a communications environment. In any event, taking the three-dimensional discrete Fourier transforms of both sides of Eq. (18.77), we get

$$C_4(\omega_1, \omega_2, \omega_3) = \gamma_4 H(e^{j\omega_1}) H(e^{j\omega_2}) H(e^{j\omega_3}) H(e^{-j(\omega_1+\omega_2+\omega_3)}) \tag{18.78}$$

Next, taking the natural logarithm of both sides of Eq. (18.78), we get

$$\ln C_4(\omega_1, \omega_2, \omega_3) = \ln \gamma_4 + \ln H(e^{j\omega_1}) + \ln H(e^{j\omega_2}) + \ln H(e^{j\omega_3})$$
$$+ \ln H(e^{-j(\omega_1+\omega_2+\omega_3)}) \tag{18.79}$$

The channel transfer function $H(z)$ is defined by Eqs. (18.72) to (18.74); hence, we have

$$\ln H(e^{j\omega_i}) = \ln k + \ln I(e^{j\omega_i}) + \ln O(e^{-j\omega_i})$$
$$= \ln k + \sum_{l=1}^{L_1} \ln(1 - a_l e^{-j\omega_i}) + \sum_{l=1}^{L_2} \ln(1 - b_l e^{+j\omega_i}), \qquad i = 1, 2, 3 \tag{18.80}$$

and

$$\ln H(e^{-j(\omega_1+\omega_2+\omega_3)}) = \ln k + \ln I(e^{-j(\omega_1+\omega_2+\omega_3)}) + \ln O(e^{j(\omega_1+\omega_2+\omega_3)})$$
$$= \ln k + \sum_{l=1}^{L_1} \ln(1 - a_l e^{j(\omega_1+\omega_2+\omega_3)}) + \sum_{l=1}^{L_2} \ln(1 - b_l e^{-j(\omega_1+\omega_2+\omega_3)}) \tag{18.81}$$

Thus, returning to Eq. (18.79) and taking the inverse three-dimensional discrete Fourier transform of $\ln C_4(\omega_1, \omega_2, \omega_3)$, we find that the tricepstrum has the following form:[6]

[6] To evaluate $\kappa_4(\tau_1, \tau_2, \tau_3)$, we may use the inversion formula for the three-dimensional z-transform:

$$\kappa_4(\tau_1, \tau_2, \tau_3) = \frac{1}{(2\pi j)^3} \oint_{\mathscr{C}_3} \oint_{\mathscr{C}_2} \oint_{\mathscr{C}_1} \ln C_4(z_1, z_2, z_3) z_1^{\tau_1-1} z_2^{\tau_2-1} z_3^{\tau_3-1} \, dz_1 \, dz_2 \, dz_3$$

where $C_4(z_1, z_2, z_3)$ is obtained from $C_4(\omega_1, \omega_2, \omega_3)$ by substituting z_i for $e^{j\omega i}$, where $i = 1, 2, 3$. The closed contours $\mathscr{C}_1$, $\mathscr{C}_2$, and $\mathscr{C}_3$ lie completely within the region of convergence of $\ln C_4(z_1, z_2, z_3)$. Let

$$\hat{a} = \max\{|a_l|\}, \qquad 1 \le l \le L_1$$
$$\hat{b} = \max\{|b_l|\}, \qquad 1 \le l \le L_2$$
$$e = \max\{\hat{a}, \hat{b}\}$$

The region of convergence for $\ln C_4(z_1, z_2, z_3)$ is defined by

$$R_c = \{|z_1| > e, |z_2| > e, |z_3| > e, \text{ and } |z_1 z_2 z_3| < 1/e\}$$

The unit surface defined by $\{|z_1| = 1, |z_2| = 1, \text{ and } |z_3| = 1\}$ lies within the region of convergence R_c. Accordingly, it is permissible to use the power series expansion or inversion formula to evaluate $\kappa_4(\tau_1, \tau_2, \tau_3)$.

$$\kappa_4(\tau_1, \tau_2, \tau_3) = \begin{cases} \ln k + 3 \ln \gamma_4, & \tau_1 = \tau_2 = \tau_3 = 0 \\[1.5ex] -\dfrac{1}{\tau_1} A^{(\tau_1)}, & \tau_1 > 0, \tau_2 = \tau_3 = 0 \\[1.5ex] -\dfrac{1}{\tau_2} A^{(\tau_2)}, & \tau_2 > 0, \tau_1 = \tau_3 = 0 \\[1.5ex] -\dfrac{1}{\tau_3} A^{(\tau_3)}, & \tau_3 > 0, \tau_1 = \tau_2 = 0 \\[1.5ex] \dfrac{1}{\tau_1} B^{(-\tau_1)}, & \tau_1 < 0, \tau_2 = \tau_3 = 0 \\[1.5ex] \dfrac{1}{\tau_2} B^{(-\tau_2)}, & \tau_2 < 0, \tau_1 = \tau_3 = 0 \\[1.5ex] \dfrac{1}{\tau_3} B^{(-\tau_3)}, & \tau_3 < 0, \tau_1 = \tau_2 = 0 \\[1.5ex] -\dfrac{1}{\tau_2} B^{(\tau_2)}, & \tau_1 = \tau_2 = \tau_3 > 0 \\[1.5ex] \dfrac{1}{\tau_2} A^{(\tau_2)}, & \tau_1 = \tau_2 = \tau_3 < 0 \\[1.5ex] 0, & \text{otherwise} \end{cases} \qquad (18.82)$$

where

$$A^{(m)} = \sum_{l=1}^{L_1} a_l^m \qquad (18.83)$$

and

$$B^{(m)} = \sum_{l=1}^{L_2} b_l^m \qquad (18.84)$$

The $A^{(m)}$ and $B^{(m)}$ contain *minimum-phase* and *maximum-phase* information about the channel, respectively; that is, they correspond to $I(z)$ and $O(z^{-1})$, respectively.

The *differential cepstrum paramaters* $A^{(m)}$ and $B^{(m)}$ exhibit the following properties (Hatzinakos and Nikias, 1994):

1. The differential cepstrum parameters decay exponentially at least as fast as (for positive integer m)

$$|A^{(m)}| < c_1 \alpha^m$$

and

$$|B^{(m)}| < c_2 \beta^m$$

where $\max|a_l| < \alpha < 1$ and $\max|b_l| < \beta < 1$ and c_1 and c_2 are constants.

2. The tricepstrum is invariant under a time shift (i.e., a linear phase shift).

3. Let the *minimum-phase time sequence* $i(n)$ denote the inverse z-transform of the polynomial $I(z)$. Then, $i(n)$ is related to the corresponding differential cepstrum parameter $A^{(m)}$ by

$$i(n) = -\frac{1}{n} \sum_{m=1}^{n} A^{(m)} i(n - m), \qquad n = 1, 2, \ldots, L_1 \qquad (18.85)$$

For other values of n, we have

$$i(n) = \begin{cases} 1, & n = 0 \\ 0, & n < 0 \end{cases} \qquad (18.86)$$

Next, let the *maximum-phase time sequence* $o(n)$ denote the inverse z-transform of the polynomial $O(z^{-1})$. Then, $o(n)$ is related to the corresponding differential cepstrum parameter $B^{(m)}$ by

$$o(n) = \frac{1}{n} \sum_{m=n}^{-1} B^{(-m)} o(n - m), \qquad n = -1, -2, \ldots, -L_2 \qquad (18.87)$$

For other values of n, we have

$$o(n) = \begin{cases} 1, & n = 0 \\ 0, & n > 0 \end{cases} \qquad (18.88)$$

Blind Channel Estimation and Equalization

The fourth-order cumulant $c_4(\tau_1, \tau_2, \tau_3)$ and the tricepstrum $\kappa_4(\tau_1, \tau_2, \tau_3)$ are related as follows (Pan and Nikias, 1988):

$$\sum_{r=-\infty}^{\infty} \sum_{s=-\infty}^{\infty} \sum_{t=-\infty}^{\infty} r\kappa_4(r, s, t) c_4(\tau_1 - r, \tau_2 - s, \tau_3 - t) = -\tau_1 c_4(\tau_1, \tau_2, \tau_3) \qquad (18.89)$$

The linear convolution formula of Eq. (18.89) is of fundamental importance to the solution of the blind channel estimation/equalization problem. Specifically, substituting Eq. (18.82) into (18.89), we obtain (after some algebra) the following *tricepstral equation*:

$$\sum_{m=1}^{p} (A^{(m)} [c_4(\tau_1 - m, \tau_2, \tau_3) - c_4(\tau_1 + m, \tau_2 + m, \tau_3 + m)])$$

$$+ \sum_{m=1}^{q} (B^{(m)} [c_4(\tau_1 - m, \tau_2 - m, \tau_3 - m) - c_4(\tau_1 + m, \tau_2, \tau_3)]) \qquad (18.90)$$

$$= -\tau_1 c_4(\tau_1, \tau_2, \tau_3)$$

In theory, the parameters p and q are infinitely large. In practice, however, they can both be approximated by finite (arbitrarily large) values, because $A^{(m)}$ and $B^{(m)}$ decay exponentially as m increases (Hatzinakos and Nikias, 1991, 1994). Assuming that suitable values have been assigned to p and q, we may define

$$\alpha_1 = \max(p, q)$$

$$\alpha_2 \leq \frac{\alpha_1}{2}$$

$$\alpha_3 \leq \alpha_2$$

and choose

$$\tau_1 = -\alpha_1, \ldots, -1, 1, \ldots, \alpha_1$$

$$\tau_2 = -\alpha_2, \ldots, 0, \ldots, \alpha_2$$

$$\tau_3 = -\alpha_3, \ldots, 0, \ldots, \alpha_3$$

Let

$$w = 2\alpha_1(2\alpha_2 + 1)(2\alpha_3 + 1) \qquad (18.91)$$

Accordingly, we may use Eq. (18.90) to construct the following overdetermined linear system of equations:

$$\mathbf{Ca} = \mathbf{p} \qquad (18.92)$$

where the known quantities $\mathbf{C}$ and $\mathbf{p}$ and the unknown $\mathbf{a}$ are defined as follows:

1. The matrix $\mathbf{C}$ is a w-by-$(p + q)$ matrix with entries of the form $\{c_4(\tau_1, \tau_2, \tau_3) - c_4(\tau_1', \tau_2', \tau_3')\}$; the dimension w is itself defined in Eq. (18.91).
2. The vector $\mathbf{p}$ is a w-by-1 vector with entries of the form $\{-\tau_1 c_4(\tau_1, \tau_3, \tau_3)\}$.
3. The vector $\mathbf{a}$ is a $(p + q)$-by-1 coefficient vector defined in terms of the $A^{(m)}$ and the $B^{(m)}$ by

$$\mathbf{a} = [A^{(1)}, A^{(2)}, \ldots, A^{(p)}, B^{(1)}, B^{(2)}, \ldots, B^{(q)}]^T \qquad (18.93)$$

Our main purpose is to formulate a *zero-forcing blind equalization algorithm*. The structural form of the algorithm is depicted in Fig. 18.9, which consists of two major components: a *channel estimator*, and a *channel equalizer*. Accordingly, the algorithm proceeds in two stages as follows (Hatzinakos and Nikias, 1994):

1. *Channel estimation.* Let $\hat{\mathbf{C}}$ and $\hat{\mathbf{p}}$ denote estimates of the matrix $\mathbf{C}$ and the vector $\mathbf{p}$, respectively; these estimates are themselves derived from $\hat{c}_4(\tau_1, \tau_2, \tau_3)$, a time-averaged estimate of the fourth-order cumulant that is obtained from a finite-window length of the channel output $u(n)$. Then, given $\hat{\mathbf{C}}$ and $\hat{\mathbf{p}}$, we may use the pseudoinverse matrix of $\hat{\mathbf{C}}$ to solve Eq. (18.92) for $\hat{\mathbf{a}}$ that denotes an estimate of the vector $\mathbf{a}$. The elements of $\hat{\mathbf{a}}$ define estimates of the differential cepstrum

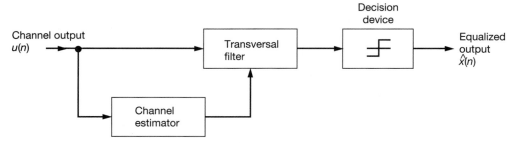

Figure 18.9   Block diagram of tricepstrum-based blind equalizer.

paramters $A^{(m)}$ and $B^{(m)}$; these estimates are denoted by $\hat{A}^{(m)}$ and $\hat{B}^{(m)}$, respectively.

2. *Channel equalization.* Using the estimates $\hat{A}^{(m)}$, and $\hat{B}^{(m)}$ of the differential cepstrum parameters in Eqs. (18.85) and (18.87), respectively, corresponding estimates of the minimum-phase sequence $i(n)$ and the maximum-phase sequence $o(n)$ are computed. Let these estimates be denoted by $\hat{i}(n)$ and $\hat{o}(n)$, respectively. Then, the impulse response of the transversal equalizer of total length $L_1 + L_2$ is obtained by convolving the inverses of $\hat{i}(n)$ and $\hat{o}(n)$. The resulting equalizer design is an approximation to the zero-forcing condition, under which the transfer function of the equalizer is the inverse of the transfer function of the channel.

The computation described under point 1 is made using block estimation approach. Alternatively, we may use an adaptive estimation approach of the LMS type. In the latter case, the recommended procedure is to permit the adaptive process to converge (i.e., reach a steady state) before proceeding with stage 2 of the estimation procedure.

18.6 ADVANTAGES AND DISADVANTAGES OF HOS-BASED DECONVOLUTION ALGORITHMS

The implicit HOS-based blind deconvolution algorithm, exemplified by Bussgang algorithms, are relatively simple to implement and generally capable of delivering good performance, as evidenced by their use in line-of-sight digital radio systems. However, they suffer from some basic limitations: (1) a potential for converging to a local minimum (Ding et al., 1991), and (2) sensitivity to timing jitter (Qureshi, 1985). In contrast, explicit HOS-based blind deconvolution algorithms, such as those that use the tricepstrum, overcome the local minimum problem by avoiding the need for minimizing a cost function, but they are computationally much more complex.

Perhaps the most serious limitation of both implicit and explicit HOS-based blind deconvolution algorithms is their slow rate of convergence (Ding, 1994). To appreciate the reason for this poor behavior, we have to recognize that time-average estimation of higher-order statistics requires a much larger sample size than is the case for second-order statistics. According to Brillinger (1975), the sample size needed to estimate the nth-order statistics of a stochastic process, subject to prescribed values of estimation bias and variance, increases almost exponentially with order n. Now, from our discussion of the tricepstrum-based deconvolution method, we know that channel identification/equalization requires at least fourth-order statistics; a similar remark also applies to Bussgang algorithms. It is therefore not surprising to find that HOS-based blind deconvolution algorithms exhibit a slow rate of convergence, compared to conventional adaptive filtering algorithms that rely on a training sequence for their operation. Thus, whereas a conventional adaptive filtering algorithm may require a few hundred iterations to converge, an existing HOS-based blind deconvolution algorithm may require several thousand iterations to converge.

The slow rate of convergence of HOS-based blind deconvolution algorithms is of no serious concern in some applications such as seismic deconvolution. However, in a more difficult environment such as mobile digital communications, the algorithm may simply not have enough time to reach a steady state, and may therefore be unable to track the statistical variations of the environment. Accordingly, this class of blind deconvolution algorithms cannot be used in applications where rapid acquisition is a necessary system requirement.

18.7 CHANNEL IDENTIFIABILITY USING CYCLOSTATIONARY STATISTICS

In HOS-based deconvolution algorithms, information about the unknown phase response of a nonminimum-phase channel is extracted by using higher-order statistics of the channel output, which is sampled at the baud rate (i.e., symbol rate). Alternatively, we may extract this phase information by exploiting another inherent characteristic of the channel output, namely, *cyclostationarity*. To explain this latter characteristic, we first rewrite the received signal in a digital communications system in its most general baseboard form as follows:

$$u(t) = \sum_{k=-\infty}^{\infty} x_k\, h(t - kT) + v(t) \qquad (18.94)$$

where a symbol x_k is transmitted every T seconds (i.e., $1/T$ is the baud rate), and t denotes continuous time; $h(t)$ is the overall impulse response of the channel (including transmit and ceive filters), and $v(t)$ is the channel noise. (The channel noise v used here should not be confused with the convolutional noise v used in the discussion on the Bussgang algorithm. All the quantities described in Eq. (18.94) are complex valued. Under the assumption that the transmitted sequence x_k and the channel noise $v(t)$ are both wide-sense stationary with zero mean, we may readily show that the received signal $u(t)$ also has zero mean, and its autocorrelation function is *periodic* in the symbol duration T (see Problem 7):

$$\begin{aligned} r_u(t_1, t_2) &= E[u(t_1)u^*(t_2)] \\ &= r_u(t_1 + T, t_2 + T) \end{aligned} \qquad (18.95)$$

That is, the received signal $u(t)$ is *cyclostationary in the wide sense*.

What makes the use of cyclostationarity as the basis of an alternative approach to blind deconvolution particularly attractive is the fact that it only uses *second-order statistics*, thereby overcoming the "slow-to-converge" limitation of HOS-based algorithms.

Apparently, Gardner (1991) was the first to recognize that cyclostationary characteristics of modulated signals permit the recovery of a communication channel's amplitude and phase responses using second-order statistics only. However, the idea of blind channel identification and equalization using cyclostationary statistics is attributed to Tong et el. (1991). Indeed, the ability to solve the difficult problem of blind deconvolution on the sole basis of second-order statistics deserves to be viewed as a major technical breakthrough.

The original idea proposed by Tong et al. relies on the use of *temporal diversity* (i.e., oversampling the received signal). Ordinarily, this operation is performed in a digital communications system for the specific purpose of timing and phase recovery. However, in the context of our present discussion, the use of oversampling leads to *fractionally-spaced equalization*, which is so called because the equalization taps are spaced closer than the reciprocal of the incoming symbol rate.

Among the many fractionally-spaced blind channel identification/equalization techniques that have been proposed to date, we have picked the *subspace decomposition method*[7] described in Moulines et al. (1995). This approach bears a close relationship to the *mu*ltiple *si*gnal *c*lassification (MUSIC) algorithm originally proposed by Schmidt (1979) for angle of arrival estimation. Thus, the material presented in the next section points to the fact that much can be gained from the extensive literature on statistical array signal processing for solving the blind deconvolution problem.

18.8 SUBSPACE DECOMPOSITION FOR FRACTIONALLY-SPACED BLIND *IDENTIFICATION*

In what follows, we assume that the channel is modeled as an FIR filter, and several measurements are made during each sampling period T. The latter requirement can be satisfied in the following ways:

1. The received signal is *oversampled*.
2. *Multiple sensors* are used, with their individual outputs sampled at the symbol rate $1/T$.
3. A combination of techniques 1 and 2 is used.

The material presented in this section focuses on the first technique.

---

[7] Another interesting approach for blind identification is based on linear prediction theory. Such an approach was first studied by Slock (1994), and has been elaborated on by Slock (1995) and Abed Meriam et al. (1995). The basic premise of time-domain blind identification using linear prediction is presented as Problem 9.

Suppose then the received signal $u(t)$ is oversampled by setting

$$t = \frac{iT}{L} \tag{18.96}$$

where L is a positive integer. Thus Eq. (18.94) takes on the discrete form

$$u\left(\frac{iT}{L}\right) = \sum_{k=-\infty}^{\infty} x_k\, h\left(\frac{iT}{L} - kT\right) + v\left(\frac{iT}{L}\right) \tag{18.97}$$

Let

$$i = nL + l, \qquad l = 0, 1, \ldots, L - 1 \tag{18.98}$$

We may then rewrite Eq. (18.97) as

$$u\left(nT + \frac{lT}{L}\right) = \sum_{k=-\infty}^{\infty} x_k\, h\left((n - k)T + \frac{lT}{L}\right) + v\left(nT - \frac{lT}{L}\right) \tag{18.99}$$

For convenience of presentation, let

$$h_n^{(l)} = h\left(nT + \frac{lT}{L}\right)$$

$$u_n^{(l)} = u\left(nT + \frac{lT}{L}\right)$$

$$v_n^{(l)} = v\left(nT + \frac{lT}{L}\right)$$

Correspondingly, we may describe the oversampled channel in the simplified form

$$u_n^{(l)} = \sum_{k=-\infty}^{\infty} x_k\, h_{n-k}^{(l)} + v_n^{(l)}, \qquad l = 0, 1, \ldots, L - 1 \tag{18.100}$$

With the channel modeled as an FIR filter, we may write

$$h_k^{(l)} = 0 \qquad \begin{array}{l} \text{for } k < 0 \text{ or } k > M, \\ \text{and all } l \end{array} \tag{18.101}$$

That is, the channel is causal and has finite time support. Furthermore, we assume that, at time n, the processing involves the use of a transmitted signal vector consisting of $(M + N)$ symbols, as shown by

$$\mathbf{x}_n = [x_n, x_{n-1}, \ldots, x_{n-M-N+1}]^T \tag{18.102}$$

At the receiving end, we find that each block consists of NL samples. Depending on how these samples are grouped together, we may distinguish two different matrix representations for the oversampled channel, as described here.

1. *Single input-multiple output (SIMO) model.* This model consists of L *virtual channels* (subchannels) fed from a common input, as depicted in Fig. 18.10 (Moulines et al., 1995; Duhamel, 1995). Each virtual channel has the same time

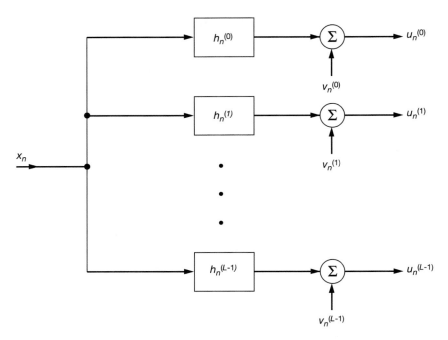

Figure 18.10   Representation of an oversampled channel as a single input–multiple output model.

support and a noise contribution of its own. Let the *l*th virtual channel be characterized with the following:

- An $(M + 1)$-by-1 tap-weight (coefficient) vector

$$\mathbf{h}^{(l)} = [h^{(l)}_0, h^{(l)}_1, \ldots, h^{(l)}_M]^T$$

- An N-by-1 received signal vector

$$\mathbf{u}^{(l)}_n = [u^{(l)}_n, u^{(l)}_{n-1}, \ldots, u^{(l)}_{n-N+1}]^T$$

- An N-by-1 noise vector

$$\mathbf{v}^{(l)}_n = [v^{(l)}_n, v^{(l)}_{n-1}, \ldots, v^{(l)}_{n-N+1}]^T$$

We may then represent Eq. (18.100), written for N successive received samples, in the compact form

$$\mathbf{u}^{(l)}_n = \mathbf{H}^{(l)}\mathbf{x}_n + \mathbf{v}^{(l)}_n \qquad l = 0, 1, \ldots, L - 1 \qquad (18.103)$$

where the transmitted signal vector $\mathbf{x}_n$ is defined in Eq. (18.102). The N-by-$(M + N)$ matrix $\mathbf{H}^{(l)}$, termed a *filtering matrix*, has a Toeplitz structure as shown by

$$
\mathbf{H}^{(l)} =
\begin{bmatrix}
h_0^{(l)} & h_1^{(l)} & \cdots & h_M^{(l)} & 0 & \cdots & 0 \\
0 & h_0^{(l)} & \cdots & h_{M-1}^{(l)} & h_M^{(l)} & \cdots & 0 \\
\cdot & \cdot & \cdot & \cdot & \cdot & \cdot & \cdot \\
\cdot & \cdot & \cdot & \cdot & \cdot & \cdot & \cdot \\
\cdot & \cdot & \cdot & \cdot & \cdot & \cdot & \cdot \\
0 & 0 & \cdots & h_0^{(l)} & h_1^{(l)} & \cdots & h_M^{(l)}
\end{bmatrix}
\tag{18.104}
$$

Finally, combining the set of L equations (18.103) into a single relation, we may write

$$
\boldsymbol{u}_n = \mathcal{H}\, \mathbf{x}_n + \boldsymbol{v}_n
\tag{18.105}
$$

where $\boldsymbol{u}_n$ is the LN-by-1 multichannel received signal vector

$$
\boldsymbol{u}_n =
\begin{bmatrix}
\mathbf{u}_n^{(0)} \\
\mathbf{u}_n^{(1)} \\
\cdot \\
\cdot \\
\cdot \\
\mathbf{u}_n^{(L-1)}
\end{bmatrix}
$$

and $\boldsymbol{v}_n$ is the LN-by-1 multichannel noise vector

$$
\boldsymbol{v}_n =
\begin{bmatrix}
\mathbf{v}_n^{(0)} \\
\mathbf{v}_n^{(1)} \\
\cdot \\
\cdot \\
\cdot \\
\mathbf{v}_n^{(L-1)}
\end{bmatrix}
$$

and $\mathcal{H}$ is the LN-by-$(M + N)$ *multichannel filtering matrix*:

$$
\mathcal{H} =
\begin{bmatrix}
\mathbf{H}^{(0)} \\
\mathbf{H}^{(1)} \\
\cdot \\
\cdot \\
\cdot \\
\mathbf{H}^{(L-1)}
\end{bmatrix}
\tag{18.106}
$$

where the individual matrix entries are themselves defined in Eq. (18.104).

2. *Sylvester matrix representation.* In this second model, the L virtual channel coefficients having the same delay index are all grouped together. Specifically, we write

$$
\mathbf{h}_k' = [h_k^{(0)}, h_k^{(1)}, \ldots, h_k^{(L-1)}]^T, \qquad k = 0, 1, \ldots, M
$$

Correspondingly, we define an L-by-1 received signal vector

$$\mathbf{u}'_n = [u^{(0)}_n, u^{(1)}_n, \ldots, u_n^{(L-1)}]^T$$

and an L-by-1 noise vector

$$\mathbf{v}'_n = [v^{(0)}_n, v^{(1)}_n, \ldots, v_n^{(L-1)}]^T$$

Then on this basis, we may use Eq. (18.100) to group the NL received samples as follows:

$$\boldsymbol{u}'_n = \begin{bmatrix} \mathbf{u}'_n \\ \mathbf{u}'_{n-1} \\ \cdot \\ \cdot \\ \cdot \\ \mathbf{u}'_{n-N+1} \end{bmatrix} \tag{18.107}$$

$$= \mathcal{H}'\mathbf{x}_n + \boldsymbol{o}'_n$$

where the transmitted signal vector $\mathbf{x}_n$ is as previously defined in Eq. (18.102). The LN-by-1 noise vector $\boldsymbol{o}'_n$ is defined by

$$\boldsymbol{o}'_n = \begin{bmatrix} \mathbf{v}'_n \\ \mathbf{v}'_{n-1} \\ \cdot \\ \cdot \\ \cdot \\ \mathbf{v}'_{n-N+1} \end{bmatrix}$$

The LN-by-$(M + N)$ matrix $\mathcal{H}'$ is defined by:

$$\mathcal{H}' = \begin{bmatrix} \mathbf{h}'_0 & \mathbf{h}'_1 & \cdots & \mathbf{h}'_M & \mathbf{0} & \cdots & \mathbf{0} \\ \mathbf{0} & \mathbf{h}'_0 & \cdots & \mathbf{h}'_{M-1} & \mathbf{h}'_M & \cdots & \mathbf{0} \\ \cdot & \cdot & \cdot & \cdot & \cdot & \cdot & \cdot \\ \cdot & \cdot & \cdot & \cdot & \cdot & \cdot & \cdot \\ \cdot & \cdot & \cdot & \cdot & \cdot & \cdot & \cdot \\ \mathbf{0} & \mathbf{0} & \cdots & \mathbf{h}'_0 & \mathbf{h}'_1 & \cdots & \mathbf{h}'_M \end{bmatrix} \tag{18.108}$$

The block-Toeplitz matrix $\mathcal{H}'$ is called a *Sylvester resultant matrix* (Rosenbrock, 1970; Tong et al., 1993), hence the terminology used to refer to this second matrix representation of an oversampled channel.

Filtering-matrix Rank Theorem

The matrices $\mathcal{H}$ and $\mathcal{H}'$, defined in Eqs. (18.106) and (18.108), respectively, differ primarily in the way in which their individual rows are arranged; they contain the same information about the channel but display it differently. Most importantly, the spaces spanned by the columns of $\mathcal{H}$ and $\mathcal{H}'$ are *canonically equivalent*. From here on, we therefore restrict the discussion to the single input-multiple output model of Fig. 18.10.

The multichannel filtering matrix $\mathcal{H}$ plays a central role in the blind identification problem. In particular, the problem is solvable if and only if the matrix $\mathcal{H}$ is of full column rank. This requirement is covered by a crucial theorem due to Tong et al. (1993), which may be stated as follows:

- The *LN*-by-$(M + N)$ multichannel filtering matrix $\mathcal{H}$ is of full column rank, that is, rank$(\mathcal{H}) = M + N$, provided that the following three conditions are satisfied:
  1. The polynomials

$$H^{(l)}(z) = \sum_{m=0}^{M} h_m^{(l)} z^{-m} \quad \text{for} \quad l = 0, 1, \ldots, L - 1$$

  have no common zeros.
  2. At least one of the polynomials $H^{(l)}(z)$, $l = 0, 1, \ldots, L - 1$, has the maximum possible degree M.
  3. The size N of the received signal vector $\mathbf{u}_n^{(l)}$ for each virtual channel is greater than M.

Equipped with this theorem, hereafter referred to as the *filtering-matrix rank theorem*, we are ready to describe the subspace decomposition-based procedure for blind identification, which we do next.

Blind Identification

The basic equation (18.106) provides a matrix description of an oversampled channel. A block diagram representation of this equation is shown in Fig. 18.11, which may be viewed as a condensed version of the single input-multiple output model of Fig. 18.10. To proceed with a statistical characterization of the channel, we make the following assumptions:

- The transmitted signal vector $\mathbf{x}_n$ and multichannel noise vector $\mathbf{v}_n$ originate from wide-sense stationary processes that are statistically independent.
- The $(M + N)$-by-1 transmitted signal vector $\mathbf{x}_n$ has zero mean and correlation matrix

$$\mathbf{R}_x = E[\mathbf{x}_n \mathbf{x}_n^H]$$

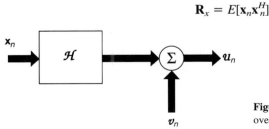

Figure 18.11   Matrix representation of an oversampled channel.

The $(M + N)$-by-$(M + N)$ matrix $\mathbf{R}_x$ has full column rank; otherwise, it is unknown.

- The N-by-1 noise vector $\mathbf{v}_n$ has zero mean and correlation matrix

$$\mathbf{R}_v = E[\mathbf{v}_n \mathbf{v}_n^H]$$
$$= \sigma^2 \mathbf{I}$$

The noise variance σ^2 is known.

Accordingly, the LN-by-1 received signal vector $\boldsymbol{u}_n$ has zero mean and a correlation matrix defined by

$$
\begin{aligned}
\mathbf{R} &= E[\boldsymbol{u}_n \boldsymbol{u}_n^H] \\
&= E[(\mathcal{H}\mathbf{x}_n + \boldsymbol{v}_n)(\mathcal{H}\mathbf{x}_n + \boldsymbol{v}_n)^H] \\
&= E[\mathcal{H}\mathbf{x}_n \mathbf{x}_n^H \mathcal{H}^H] + E[\boldsymbol{v}_n \boldsymbol{v}_n^H] \\
&= \mathcal{H} \mathbf{R}_x \mathcal{H}^H + \mathbf{R}_v
\end{aligned}
\tag{18.109}
$$

To gain some insight into the blind identification problem, we cast it in a geometrical framework first proposed by Schmidt (1979, 1981). First, we invoke the spectral theorem of Chapter 4 to describe the LN-by-LN correlation matrix $\mathbf{R}$ in terms of its eigenvalues and associated eigenvectors as follows:

$$\mathbf{R} = \sum_{k=1}^{LN} \lambda_k \mathbf{q}_k \mathbf{q}_k^H \tag{18.110}$$

where the eigenvalues are arranged in decreasing order:

$$\lambda_0 \geq \lambda_1 \geq \cdots \geq \lambda_{LN-1}$$

Next, we invoke the filtering-matrix rank theorem to divide these eigenvalues into two groups:

1. $\lambda_k > \sigma^2$, $k = 0, 1, \ldots, M + N - 1$
2. $\lambda_k = \sigma^2$, $k = M + N, M + N + 1, \ldots, LN - 1$

Correspondingly, the space spanned by the eigenvectors of matrix $\mathbf{R}$ is divided into two subspaces:

1. *Signal subspace $\mathcal{S}$*, spanned by the eigenvectors associated with the eigenvalues $\lambda_0, \lambda_1, \ldots, \lambda_{M+N-1}$. These eigenvectors are written as

$$\mathbf{s}_k = \mathbf{q}_k, \qquad k = 0, 1, \ldots, M + N - 1$$

2. *Noise subspace $\mathcal{N}$*, spanned by the eigenvectors associated with the remaining eigenvalues $\lambda_{M+N}, \lambda_{M+N+1}, \ldots, \lambda_{LN-1}$. These eigenvectors are written as

$$\mathbf{g}_k = \mathbf{q}_{M+N+k}, \qquad k = 0, 1, \ldots, LN - M - N - 1$$

The noise subspace is the orthogonal complement of the signel subspace.

By definition, we have

$$\mathbf{R}\mathbf{g}_k = \sigma^2 \mathbf{g}_k, \qquad k = 0, 1, \ldots, LN - M - N - 1 \qquad (18.111)$$

Substituting Eq. (18.109), with $\mathbf{R}_v = \sigma^2 \mathbf{I}$, in Eq. (18.111) and then simplifying, we get

$$\mathcal{H} \mathbf{R}_x \mathcal{H}^H \mathbf{g}_k = 0, \qquad k = 0, 1, \ldots, LN - M - N - 1$$

Since both matrices $\mathcal{H}$ and $\mathbf{R}_x$ are of full column rank, it follows that we must have

$$\mathcal{H}^H \mathbf{g}_k = \mathbf{0}, \qquad k = 0, 1, \ldots, LN - M - N - 1 \qquad (18.112)$$

Equation (18.112) provides the theoretical framework of the *subspace decomposition procedure for blind identification* described in Moulines et al. (1995). Specifically, it builds on two items:

- Knowledge of the eigenvectors associated with the $LN - M - N$ smallest eigenvalues of the correlation matrix $\mathbf{R}$ of the received signal vector $\mathbf{u}_n$.
- Orthogonality of the columns of the unknown multichannel filtering matrix $\mathcal{H}$ to the noise subspace $\mathcal{N}$.

In other words, the cyclostationary statistics of the received signal $\mathbf{u}_n$, exemplified by the correlation matrix $\mathbf{R}$, are indeed sufficient for blind identification of the channel to within a multiplicative constant.

Alternative Formulation of the Orthogonality Condition

From a computational point of view, we find it more convenient to work with an alternative formulation of the *orthogonality condition* described in Eq. (18.112). To begin with, we rewrite this condition in the equivalent scalar form

$$\|\mathcal{H}^H \mathbf{g}_k\|^2 = \mathbf{g}_k^H \mathcal{H} \mathcal{H}^H \mathbf{g}_k = 0, \qquad k = 0, 1, \ldots, LN - M - N - 1 \qquad (18.113)$$

Recognizing the partitioned structure of the multiparameter filtering matrix $\mathcal{H}$ displayed in Eq. (18.106), in a corresponding way we may partition the LN-by-1 eigenvector $\mathbf{g}_k$ as follows:

$$\mathbf{g}_k = \begin{bmatrix} \mathbf{g}_k^{(0)} \\ \mathbf{g}_k^{(1)} \\ \cdot \\ \cdot \\ \cdot \\ \mathbf{g}_k^{(L-1)} \end{bmatrix} \qquad (18.114)$$

where $\mathbf{g}_k^{(l)}$, $l = 0, 1, \ldots, L - 1$, is an N-by-1 vector. Next, guided by the composition of matrix $\mathbf{H}^{(l)}$ given in Eq. (18.104), we formulate the $(M + 1)$-by-$(M + N)$ matrix:

$$
\mathbf{G}_k^{(l)} =
\begin{bmatrix}
g_{k,0}^{(l)} & g_{k,1}^{(l)} & \cdots & g_{k,N-1}^{(l)} & 0 & \cdots & 0 \\
0 & g_{k,0}^{(l)} & \cdots & g_{k,N-2}^{(l)} & g_{k,N-1}^{(l)} & \cdots & 0 \\
\vdots & \vdots & \ddots & \vdots & \vdots & \ddots & \vdots \\
& & & & & & \\
0 & 0 & \cdots & g_{k,0}^{(l)} & g_{k,1}^{(l)} & \cdots & g_{k,N-1}^{(l)}
\end{bmatrix}
\tag{18.115}
$$

Finally, in light of Eq. (18.106) describing the multichannel filtering matrix $\mathcal{H}$, we use the matrices defined in Eq. (18.115) for $l = 0, 1, \ldots, L - 1$ to set up the $L(M + 1)$-by-$(M + N)$ matrix

$$
\mathcal{G}_k =
\begin{bmatrix}
\mathbf{G}_k^{(0)} \\
\mathbf{G}_k^{(1)} \\
\vdots \\
\mathbf{G}_k^{(L-1)}
\end{bmatrix},
\qquad k = 0, 1, \ldots, LN - M - N - 1
\tag{18.116}
$$

Given the $\mathcal{G}_k$ as defined here, it may be shown that (Moulines et al., 1995)

$$
\mathbf{g}_k^H \mathcal{H}\, \mathcal{H}^H \mathbf{g}_k = \mathbf{h}^H \mathcal{G}_k \mathcal{G}_k^H \mathbf{h}
\tag{18.117}
$$

where $\mathbf{h}$ is an $L(M + 1)$-by-1 vector defined in terms of the multichannel coefficients by

$$
\mathbf{h} =
\begin{bmatrix}
\mathbf{h}^{(0)} \\
\mathbf{h}^{(1)} \\
\vdots \\
\mathbf{h}^{(L-1)}
\end{bmatrix}
$$

Accordingly, we may reformulate the orthogonality condition of Eq. (18.113) in the equivalent form

$$
\mathbf{h}^H \mathcal{G}_k \mathcal{G}_k^H \mathbf{h} = 0, \qquad k = 0, 1, \ldots, LN - M - N - 1
\tag{18.118}
$$

which is the desired relation. In Eq. (18.118) the unknown multichannel coefficients feature in the simple form of vector $\mathbf{h}$, whereas in Eq. (18.113) they feature in the highly elaborate structure of matrix $\mathcal{H}$.

Estimation of the Channel Coefficients

In practice, we have to work with estimates of the eigenvectors $\mathbf{g}_k$. Let these estimates be denoted by $\hat{\mathbf{g}}_k$, $k = 0, 1, \ldots, LN - M - N - 1$. To derive a corresponding estimate of the multichannel coefficient vector $\mathbf{h}$, we use the orthogonality condition of Eq. (18.118) to define the cost function

$$\mathscr{E}(\mathbf{h}) = \mathbf{h}^H \boldsymbol{\mathcal{2}} \, \mathbf{h} \tag{18.119}$$

where $\boldsymbol{\mathcal{2}}$ is an $L(M+1)$-by-$L(M+1)$ matrix defined by

$$\boldsymbol{\mathcal{2}} = \sum_{k=0}^{LN-M-N-1} \hat{\mathcal{G}}_k \hat{\mathcal{G}}_k^H \tag{18.120}$$

The estimated matrix $\hat{\mathcal{G}}_k$ is itself defined by Eqs. (18.115) and (18.116) with $\hat{\mathbf{g}}_k$ used in place of $\mathbf{g}_k$. In the ideal case of a true correlation matrix $\mathbf{R}$, the true multichannel coefficient vector $\mathbf{h}$ is uniquely defined (except for a multiplicative constant) by the condition $\mathscr{E}(\mathbf{h}) = 0$. Working with the matrix $\boldsymbol{\mathcal{2}}$ based on the estimates $\hat{\mathbf{g}}_k$, a least-squares estimate of the vector $\mathbf{h}$ is computed by minimizing the cost function $\mathscr{E}(\mathbf{h})$. However, this minimization would have to be performed subject to a properly chosen constraint, so as to avoid the trivial solution $\mathbf{h} = \mathbf{0}$. Moulines et al. (1995) suggest two possible optimization criteria:

1. *Linear constraint.* Minimize the cost function $\mathscr{E}(\mathbf{h})$ subject to $\mathbf{c}^H \mathbf{h} = 1$, where $\mathbf{c}$ is some $L(M+1)$-by-1 vector.
2. *Quadratic constraint.* Minimize the cost function $\mathscr{E}(\mathbf{h})$ subject to $\|\mathbf{h}\| = 1$.

The first criterion requires the prescription of an arbitrary vector $\mathbf{c}$, whereas the second criterion appears to be more natural but computationally more demanding.

A successful use of the subspace-decomposition method for blind identification rests on the premise that the transfer functions of the virtual channels have no common zeros. A test would therefore have to be performed to satisfy this requirement. Such a test would, in turn, require exact knowledge of the channel model order M. The important point to note here is that, given that these requirements are satisfied, it is feasible to perform the blind equlization of a communication channel using cyclostationary second-order statistics.

18.9 SUMMARY AND DISCUSSION

Blind deconvolution is an example of *unsupervised learning* in the sense that it identifies the inverse of an unknown linear time-invariant (possibly nonminimum-phase) system *without* having access to a training sequence (i.e., desired response). This operation requires the identification of both the magnitude and phase of the system's transfer function. To identify the magnitude component, we only need second-order statistics of the received signal (i.e., system output). However, to identify the phase component is a more difficult task.

One class of procedures for blind deconvolution relies on higher-order statistics of the received signal in an implicit or explicit sense. This, in turn, requires the use of some form of nonlinearity. Most, importantly, for higher-order statistics-based approaches to blind deconvolution to succeed, the received signal must be non-Gaussian. In this chapter, we described two such procedures, one called the Bussgang algorithm and the other the tricepstrum-based identification algorithm.

The Bussgang algorithm, using higher-order statistics in an implicit sense, performs blind equalization by subjecting the received signal to an iterative deconvolution process. When the algorithm has converged in the mean value, the deconvolved sequence assumes Bussgang statistics, hence the name of the algorithm. The distinguishing features of the Bussgang algorithm are as follows:

- The minimization of a nonconvex cost function, and therefore the potential likelihood of being trapped in a local minimum
- A low computational complexity, which is slightly greater than that of a conventional adaptive equalizer having access to a training sequence.

The tricepstrum-based blind identification algorithm explicitly exploits the inherent ability of the fourth-order cumulant of the received signal to extract phase information about the channel. This second algorithm has the following characteristics:

- Channel estimation by identifying the minimum-phase and maximum-phase parts of the channel transfer function; this is done without involving the use of a cost function and thereby avoiding the local minimum problem
- A high computational complexity

A limitation common to both of these approaches to blind channel identification and equalization, based on higher-order statistics, is a slow rate of convergence, which may inhibit their use in a difficult environment that requires rapid acquisition. This limitation may be overcome by using cyclostationary second-order statistics rather than higher-order statistics of the channel output. In this chapter, we have shown that it is indeed feasible to identify an unknown channel solely on the basis of cyclostationary statistics of the received signal, as exemplified by the subspace decomposition-based blind identification procedure. However, the use of cyclostationarity for blind identification and equalization is in its early stages of development, and its commercial use is yet to be demonstrated.

PROBLEMS

1. Equation (18.39) defines the conditional mean estimate of the datum x, assuming that the convolutional noise v is additive, white, Gaussian, and statistically independent of x. Derive this formula.

2. For perfect equalization, we require that the equalizer output $y(n)$ be exactly equal to the transmitted datum $x(n)$. Show that when the Bussgang algorithm has converged in the mean value and perfect equalization has been attained, the nonlinear estimator must satisfy the condition

$$E[\hat{x}g(\hat{x})] = 1$$

where $\hat{x}$ is the conditional mean estimate of x.

3. Equation (18.18) provides an adaptive method for finding the tap weights of the transversal filter in the Bussgang algorithm for performing the iterative deconvolution. Develop an alternative method for doing this computation, assuming the availability of an overdetermined system of equations and the use of the method of least squares.

4. Derive the linear convolution formula given in Eq. (18.89), which defines the relation between the fourth-order cumulant $c_4(\tau_1, \tau_2, \tau_3)$ and the trispectrum $\kappa_4(\tau_1, \tau_2, \tau_3)$.

5. Derive the tricepstral equation (18.90) that relates the fourth-order cumulant $c_4(\tau_1, \tau_2, \tau_3)$, to the $A^{(m)}$ and the $B^{(m)}$ that contain minimum-phase and maximum-phase information about the channel.

6. Formulate an adaptive estimation approach of the LMS type for solving the overdetermined system of equations (18.92). Your formulation should also include an upper bound on the step-size parameter $\mu(n)$ used in the LMS algorithm.

7. In this problem we explore the possibility of extracting the phase response of an unknown channel using cyclostationary statistics (Ding, 1993).

   (a) Using Eq. (18.94) for the received signal $u(t)$ in a digital communications system, and invoking the assumptions made in Section 18.8 on the transmitted signal x_k and channel noise $v(t)$, show that the autocorrelation function of $u(t)$ evaluated at the two time instants t_1 and t_2 is given by

$$r_u(t_1, t_2) = E[u(t_1)u^*(t_2)]$$

$$= \sum_{k=-\infty}^{\infty} \sum_{l=-\infty}^{\infty} r_x(kT - lT)h(t_1 - kT)h^*(t_2 - lT) + \sigma_v^2 \delta(t_1 - t_2)$$

where $r_x(kT)$ is the autocorrelation function of the transmitted signal for lag kT and σ_v^2 is the noise variance. Hence, demonstrate that $u(t)$ is cyclostationary in the wide sense.

   (b) The cyclic autocorrelation function and spectral density of a cyclostationary process $u(t)$ are defined by, respectively (see Chapter 3)

$$r_u^\alpha(\tau) = \frac{1}{T} \int_{-T/2}^{T/2} r_u\left(t + \frac{\tau}{2}, t - \frac{\tau}{2}\right) \exp(j2\pi\alpha t)dt$$

$$S_u^\alpha(\omega) = \int_{-\infty}^{\infty} r_u^\alpha(\tau)\exp(-j2\pi f\tau)d\tau, \qquad \omega = 2\pi f$$

where

$$\alpha = \frac{k}{T}, \qquad k = 0, \pm 1, \pm 2, \ldots$$

Let $\Psi_k(\omega)$ denote the phase response of $S_u^{k/T}(\omega)$ and $\Phi(\omega)$ denote the phase response of the channel. Show that

$$\Psi_k(\omega) = \Phi\left(\omega + \frac{k\pi}{T}\right) - \Phi\left(\omega - \frac{k\pi}{T}\right), \qquad k = 0, \pm 1, \pm 2, \ldots$$

(c) Let $\psi_k(\tau)$ and $\phi(\tau)$ denote the inverse Fourier transforms of $\Psi_k(\omega)$ and $\Phi(\omega)$, respectively. Using the result of part (b), show that

$$\psi_k(\tau) = -2j\phi(\tau)\sin\left(\frac{\pi k\tau}{T}\right), \qquad k = 0, \pm 1, \pm 2, \ldots$$

What conclusions can you draw from this relation with regard to the possibility of extracting the phase response $\Phi(\omega)$ from $\psi_k(\tau)$?

8. Suppose that the multichannel filtering matrix $\mathcal{H}$ of the SIMO model depicted in Fig. 18.10 has been estimated using the subspace-decomposition procedure described in Section 18.8.

   Show that, in the noise-free case, perfect equalization is achieved by using a multichannel structure whose own filtering matrix is defined by the pseudoinverse of $\mathcal{H}$.

9. The use of linear prediction provides the basis of other procedures for blind identification (Slock, 1994, 1995; Abed Meriam et al., 1995). The basic idea behind these procedures resides in the *generalized Bezout identity* (Kailath, 1980). Define the L-by-1 polynomial vector

$$\mathbf{H}(z) = [H^{(0)}(z), H^{(1)}(z), \ldots, H^{(L-1)}(z)]^T$$

where $H^{(l)}(z)$ is the transfer function of the lth virtual channel. Under the condition that $\mathbf{H}(z)$ is irreducible, the generalized Bezout identity states that there exists a 1-by-L polynomial vector

$$\mathbf{G}(z) = [G^{(0)}(z), G^{(1)}(z), \ldots, G^{(L-1)}(z)]$$

such that

$$\mathbf{G}(z)\mathbf{H}(z) = 1,$$

that is,

$$\sum_{l=0}^{L-1} G^{(l)}(z)H^{(l)}(z) = 1$$

The implication of this identity is that a set of moving average processes described in terms of a white noise process $v(n)$ by the operation $\mathbf{y}(n) = \mathbf{H}(z)[v(n)]$ may also be represented by an autoregressive process of finite order.

   Consider the ideal case of a noiseless channel, for which the received signal of the lth virtual channel is defined by

$$u_n^{(l)} = \sum_{m=0}^{M} h_m^{(l)} x_{n-m}, \qquad l = 0, 1, \ldots, L - 1$$

where x_n is the transmitted symbol, and $h_n^{(l)}$ is the impulse response of the lth virtual channel. Using the generalized Bezout identity, show that

$$\sum_{l=0}^{L-1} G^{(l)}(z)[u_n^{(l)}] = x_n$$

and x_n is thus reproduced exactly; in this relation, $G^{(l)}(z)$ acts as an operator. How would you interpret this result in light of linear prediction?

19

Back-Propagation Learning

In this chapter and the next, we consider a class of *neural networks* that are quite different from the adaptive filtering structures considered in previous chapters of the book. A neural network is made up of the interconnection of a large number of nonlinear processing units referred to as *neurons*. The internal structure of the neural network may involve feedforward paths only, or feedforward as well as feedback paths. In this book we will confine our attention to the class of *feedforward* neural networks.

From a signal-processing perspective, interest in neural networks is motivated by the following important properties (Haykin, 1994):

- *Nonlinearity*. This property, attributed to the nonlinear nature of neurons in the network, is particularly useful if the underlying physical mechanism responsible for the generation of an input signal (e.g., speech signal) is inherently nonlinear.

- *Weak statistical assumptions*. A neural network relies on the availability of training data for its design; it is therefore able to capture the statistical characteristics of the environment in which it operates, provided the training data are large enough to be "representative" of the environment. In other words, a neural network permits "the dataset to speak for itself."

- *Learning*. A neural network has a built-in capability to learn from its environment by undergoing a training session for the purpose of adjusting its free parameters.

We begin our discussion of neural networks by describing the different models of a neuron that constitutes the basic processing unit of a neural network.

19.1 MODELS OF A NEURON

Figure 19.1 shows the *model* of an artificial neuron referred to hereafter simply as a neuron; we have labeled it as neuron *i* for the purpose of reference. The model consists of a *linear combiner* followed by a *nonlinear unit*. The linear combiner itself consists of a set of *synaptic weights* (adjustable parameters) connected to respective input terminals, and whose weighted outputs are combined in a *summing* junction. An external bias plus the linear combiner output constitute the net input of the nonlinear unit, which is denoted by net_i in Fig. 19.1.

We may distinguish four basic types of neuron models, depending on the exact description of the nonlinear unit:

1. **Linear model.** In this model the nonlinear unit is replaced by a *direct connection*, with the result that the output of the neuron is a weighted sum of its inputs. This special form of a neuron is basic to the operation of linear adaptive filters on which the material presented in Part III of this book is based.

2. **McCulloch–Pitts model.** In this second model of a neuron the nonlinear unit is characterized by a *threshold function* as depicted in Fig. 19.2(a); it is so named in recognition of the pioneering work done by McCulloch and Pitts on neural networks that dates back to 1943.

3. **Piecewise linear model.** The input–output characteristic of the nonlinear unit for this model of a neuron is described in Fig. 19.2(b). The piecewise linear model includes the linear model and the McCulloch–Pitts model as special cases. If the linear region of the input-output characteristic in Fig. 19.2(b) is made infinitely wide, we get the linear model. If, on the other hand, it is made infinitely narrow (i.e., the slope of the linear region is made infinitely large), we get the McCulloch–Pitts model.

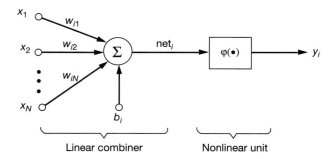

Linear combiner   Nonlinear unit

Figure 19.1 Simplified model of a neuron.

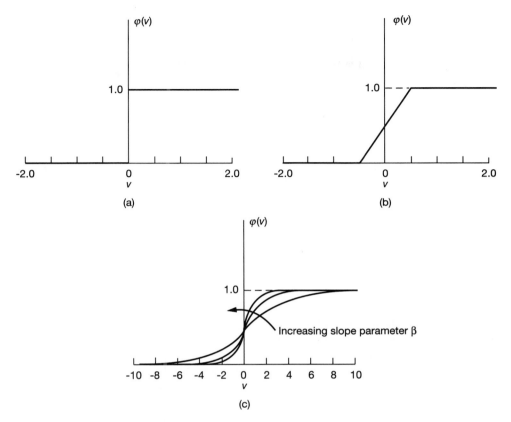

Figure 19.2 Unipolar activation functions: (a) Threshold function; (b) piecewise linear function; (c) sigmoid function.

4. **Sigmoidal model.** This last model of a neuron is so called because the *activation function* that defines the input–output characteristic of the nonlinear unit is *S-shaped*. Let the activation function be denoted by $\varphi(\cdot)$. We may then write

$$\varphi(\text{net}) = \begin{cases} 1, & \text{net} = \infty \\ \dfrac{1}{2}, & \text{net} = 0 \\ 0, & \text{net} = -\infty \end{cases} \tag{19.1}$$

where net is the sum of the linear combiner output plus the bias. A highly popular form of sigmoidal nonlinearity is the *logistic function*, defined by

$$\varphi(\text{net}) = \frac{1}{1 + e^{-\beta \, \text{net}}} \tag{19.2}$$

where β is the *slope parameter*. Figure 19.2(c) shows a depiction of the sigmoidal nonlinearity for varying β. The derivative of this nonlinearity with respect to its input is given by

$$\varphi'(\text{net}) = \frac{d\varphi}{d\text{net}} \tag{19.3}$$

$$= \beta\varphi(\text{net})\,(1 - \varphi(\text{net}))$$

where, in the first line, the prime denotes differentiation; this practice is followed in the material that follows. The maximum slope of the logistic function of Eq. (19.2) equals $\beta/4$. When the slope parameter β is made infinitely large, the sigmoidal model of a neuron reduces to the McCulloch–Pitts neuron.

In practical terms, the sigmoidal model of a neuron is by far the most widely used of all models for two reasons. First, it introduces a well-defined form of nonlinearity into the operation of a neuron. Second, it is *differentiable*. Indeed, the sigmoidal model of a neuron is basic to the construction of an important neural network structure called a multilayer perceptron using the back-propagation algorithm for training; more will be said on this important neural network in the next two sections.

The activation functions described in Fig. 19.2 are all of a *unipolar kind*, in that in each case the model's output is always nonnegative, regardless of the polarity of the input. Alternatively, the model's output is permitted to assume both positive and negative values, in which case the activation function is said to be of a *bipolar kind*. Figure 19.3 shows three examples of a bipolar activation function. Of particular interest is the sigmoid function shown in Fig. 19.3(c), an example of which is the *hyperbolic tangent function* defined by

$$\varphi(\text{net}_i) = \tanh\left(\frac{1}{2}\,\text{net}_i\right)$$

$$= \frac{1 - \exp(-\text{net}_i)}{1 + \exp(-\text{net}_i)} \tag{19.4}$$

Another way of distinguishing between the activation functions of Fig. 19.2 and those of Fig. 19.3 is to note that the former are asymmetric, whereas the latter are antisymmetric.

Example 1. Conditional mean estimator.

In Section 18.2 we derived a zero-memory nonlinear estimator as an integral part of a blind equalizer of the Bussgang type. The input–output characteristic of this estimator is plotted in Fig. 18.7. A point of particular interest is that for high levels of convolutional noise, the input–output characteristic of this nonlinear estimator is closely approximated by the bipolar sigmoidal nonlinearity:

$$x = a_1\tanh\left(\frac{a_2 y}{2}\right)$$

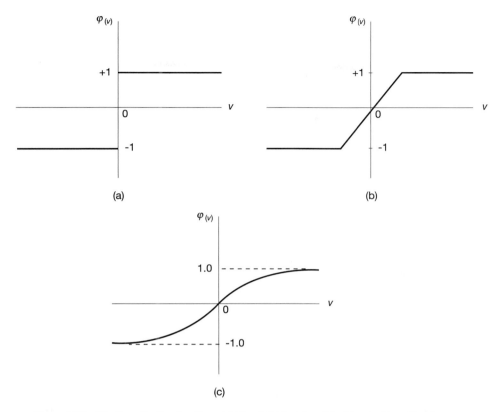

Figure 19.3   Bipolar activation functions: (a) Threshold function; (b) piecewise linear function; (c) sigmoid function.

For the situation described in Fig. 18.7, the following values for the constants a and b provide a good fit:

$$a_1 = 1.945$$
$$a_2 = 1.25$$

The resulting input–output characteristic shown in Fig. 19.4.

    Accordingly, we may view the blind equalizer of the Bussgang type depicted in Fig. 18.5 as being essentially a single neuron with its linear combiner and sigmoidal nonlinearity represented by the transversal filter and zero-memory nonlinear estimator, respectively. The error signal for adjusting the synaptic weights of the neuron is obtained by comparing the input and output signals of the nonlinear unit in the neuron.

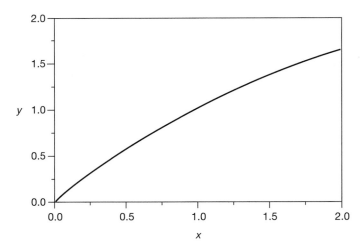

Figure 19.4   Input–output characteristic of sigmoid nonlinearity for a blind equalizer of the Bussgang type.

19.2 MULTILAYER PERCEPTRON

The *multilayer perceptron (MLP)* is a neural network that consists of an *input layer* of source nodes, one or more *hidden layers* of computational nodes (neurons), and an *output layer* also made up of computational nodes (neurons). The source nodes provide physical access points for the application of input signals. The neurons in the hidden layers act as "feature detectors"; these neurons are referred to as "hidden" neurons because they are physically inaccessible from the input end or output end of the network. Finally, the neurons in the output layer present to a user the conclusions reached by the network in response to the input signals.

Figure 19.5 depicts a multilayer perceptron with a pair of input nodes, a single layer of five hidden neurons, and a single output neuron. Two features of such a structure are immediately apparent from this figure:

1. A multilayer perceptron is a *feedforward network*, in the sense that the input signals produce a response at the output(s) of the network by propagating in the forward direction only. Simply put, there is *no* feedback in the network.

2. The network may be *fully connected*, as shown in Fig. 19.5, in that each node in a layer of the network is connected to every other node in the layer adjacent to it. Alternatively, the network may be *partially connected* in that some of the synaptic links may be missing. Locally connected networks represent an important type of partially connected networks; the term "local" refers to the connectivity of a neuron in a layer of the network only to a subset of possible inputs.

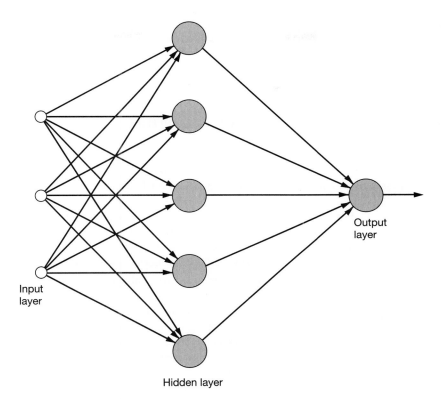

Figure 19.5   Multilayer perceptron with a single hidden layer.

The number of source nodes in the input layer is determined by the dimensionality of the observation space that is responsible for the generation of the input signals. The number of computational nodes in the output layer is determined by the required dimensionality of the desired response. Thus, the design of a multilayer perceptron requires that we address three issues:

1. The determination of the number of hidden layers
2. The determination of the number of neurons in each of the hidden layers
3. The specification of the synaptic weights that interconnect the neurons in the different layers of the network

Issues 1 and 2 relate to *neural (model) complexity*. Unfortunately, these two issues represent the weakest link in our present knowledge of how to design a multilayer perceptron. More will be said on network complexity later in the chapter. To resolve issue 3, we may use the back-error propagation algorithm, also referred to in the literature as the *backpropagation algorithm*, or simply the backprop.

In the next section, we present a derivation of the complex back-propagation algorithm, which is designed to handle complex signals. The derivation of the back-propagation algorithm for real signals follows immediately as a special case of the complex back-propagation algorithm; the latter case is treated in Section 19.4.

19.3 COMPLEX BACK-PROPAGATION ALGORITHM

The *complex back-propagation (BP) algorithm* is a generalization of the complex LMS algorithm, with an appropriately chosen nonlinear activation function. There are two passes of signals in the implementation of the BP algorithm.

1. **Forward pass.** In the forward pass, also termed the *function level adaptation*, the synaptic weights are *fixed*, and the response of the network is computed by subjecting it to a prescribed set of input signals. The forward pass in the BP algorithm is analogous to the filtering process in the LMS algorithm.

2. **Backward pass.** In the backward pass, also termed the *parameter level adaptation*, the adjustments to the synaptic weights are computed for the purpose of minimizing a cost function defined as the sum of error squares. In particular, we start by computing the error signals in the output layer, and then work *backwards* through the network, layer by layer, until the complete network is covered. The BP algorithm derives its name from the backward nature of the error computations involved in its implementation. Note also that the backward pass in the BP algorithm is analogous to the adaptive process in the LMS algorithm.

The derivation of the BP algorithm is usually presented for real-valued data (Rumelhart et al., 1986; Werbos, 1993; Haykin, 1994). However, we will pursue a different course here by first deriving the complex form of the algorithm. In this context, the key question is: How do we handle the use of complex-valued data? This need often arises in processing coherent data, for example, those found in radar, sonar, and communications fields. We may accommodate the use of complex data in either one of two ways:

1. The real and imaginary parts of each member of the input set of data are treated as two separate entities; similarly, the real and imaginary parts of each member of the network output are treated as two separate entities. The synaptic weights of the network are then computed in accordance with the *real* (conventional) form of the BP algorithm.

2. The synaptic weights are assigned complex values and their computations are performed using the *complex* form of the BP algorithm.

We can show that the two approaches are equivalent (Haykin and Ukrainec, 1993). Let the desired mapping be written as

$$\mathbf{z} = \mathbf{z}_I + j\mathbf{z}_Q \rightarrow \varphi(\mathbf{z}) = u(\mathbf{z}_I, \mathbf{z}_Q) + jv(\mathbf{z}_I, \mathbf{z}_Q) \qquad (19.5)$$

where

$$(\mathbf{z}_I, \mathbf{z}_Q) \rightarrow u(\mathbf{z}_I, \mathbf{z}_Q) \text{ and } (\mathbf{z}_I, \mathbf{z}_Q) \rightarrow v(\mathbf{z}_I, \mathbf{z}_Q) \qquad (19.6)$$

and where u and v are real functions of the complex input vector $\mathbf{z}$; the subscripts I and Q refer to the in-phase and quadrature components (i.e., real and imaginary parts), respectively. Two real-valued feedforward networks (or one real-valued feedforward network with two outputs) can thus be used to compute the resultant mapping, one giving the real part of the mapping, the other the imaginary part of the mapping. There is, however, an advantage in using a network with complex weights. Referring to the linear combiner section of Fig. 19.1, its scalar output can be written in vector notation as follows (dropping subscript i for convenience of notation):

$$\mathbf{x}^H \mathbf{w} = (\mathbf{x}_I + j\mathbf{x}_Q)^H (\mathbf{w}_I + j\mathbf{w}_Q) \qquad (19.7)$$

$$= (\mathbf{x}_I^T \mathbf{w}_I + \mathbf{x}_Q^T \mathbf{w}_Q) + j(-\mathbf{x}_Q^T \mathbf{w}_I + \mathbf{x}_I^T \mathbf{w}_Q)$$

where $\mathbf{x} = \mathbf{x}_I + j\mathbf{x}_Q$ is a complex input vector; likewise $\mathbf{w} = \mathbf{w}_I + j\mathbf{w}_Q$ is the corresponding complex weight vector. The equivalent real-valued combiner can also be constructed using only real-valued vectors, so that

$$[\mathbf{x}_I^T \quad \mathbf{x}_Q^T] \begin{bmatrix} \mathbf{u}_I & \mathbf{v}_I \\ \mathbf{u}_Q & \mathbf{v}_Q \end{bmatrix} = [\mathbf{x}_I^T \mathbf{u}_I + \mathbf{x}_Q^T \mathbf{u}_Q \quad \mathbf{x}_I^T \mathbf{v}_I + \mathbf{x}_Q^T \mathbf{v}_Q] \qquad (19.8)$$

where the weight matrix consists of real vectors $\mathbf{u}_I$, $\mathbf{u}_Q$, $\mathbf{v}_I$, and $\mathbf{v}_Q$. The resultant real vector in Eq. (19.8) contains the real and imaginary components of the complex output in Eq. (19.7). Comparing Eqs. (19.7) and (19.8), we may readily see that

$$\mathbf{u}_I = \mathbf{w}_I, \qquad \mathbf{v}_I = \mathbf{w}_Q \qquad (19.9)$$

$$\mathbf{u}_Q = \mathbf{w}_Q, \qquad \mathbf{v}_Q = -\mathbf{w}_I$$

and therefore

$$\mathbf{u}_I = -\mathbf{v}_Q \qquad \mathbf{u}_Q = \mathbf{v}_I \qquad (19.10)$$

It is apparent that a network with real-valued weights has more degrees of freedom than absolutely necessary to solve the complex mapping problem. In general, the real-valued learning algorithm treats all the weights as independent parameters, adjusting them to decrease the cost function. In the case of a complex-valued mapping, symmetries exist that are not taken advantage of by the learning algorithm.[1] In other words, the network

---

[1] It is possible to constrain the weights so that the above mentioned symmetry exists between the real-valued weights. However, the usual gradient descent algorithm does not make use of this information, which ends up being lost.

with complex-valued weights gives a parsimonious solution as compared to the network with real-valued weights. Other considerations include those of convergence. It has been shown (Horowitz and Senne, 1981) that for the LMS algorithm, superior performance is achieved for the complex LMS algorithm over that of the real version of the LMS algorithm. The algorithm is more stable, and the rate of mean-squared convergence is almost twice that of the real LMS algorithm. Since the BP algorithm is a generalization of the LMS algorithm, we may conjecture that this behavior carries over to feedforward neural networks as well.

Derivation of the complex back-propagation algorithm

We now present a detailed derivation of the complex form of the back-propagation algorithm. The complex algorithm was developed independently by several researchers (Clarke, 1990; Kim and Guest,1990; Hensler and Braspenning, 1990; Leung and Haykin, 1991; Georgiou and Koustougeras, 1992; Birx and Pipenberg, 1992; Benvenuto and Piazza, 1992). All of the approaches are fundamentally a generalization of the complex least-mean square (LMS) algorithm to a network with multiple layers of multiple linear combiners with nonlinearities. The introduction of nonlinearity into the network raises the basic question: What form does the complex activation function take? The answer to this question requires a consideration of the nature of differentiable functions of complex variables.

A multilayer perceptron, shown in Fig. 19.6, consists of many adaptive linear combiners with a nonlinearity at the output; such a combiner is shown in Fig. 19.1. The input–output relationship of such a unit in layer l of the network is characterized by the nonlinear difference equation

$$x_i^{(l+1)} = \varphi\left(\sum_{p=1}^{N} w_{ip}^{*(l)}x_p^{(l)} + b_i^{*(l)}\right) \tag{19.11}$$

with the output being the ith node in the $(l + 1)$th layer. The parameter b is a bias term, equivalent to a weight with a constant $+ 1$ input. Equation (19.11) is generalized to all units in the multilayer perceptron as shown in Fig. 19.6.

The error signal is defined to be the difference between some desired response and the actual output of the network. Specifically, for the ith output neuron we may write

$$e_i(n) = d_i - y_i(n), \qquad i = 1, 2, \ldots, N_M \tag{19.12}$$

where d_i is the desired response at the ith node of the output layer, $y_i(n)$ is the output at the ith node of the output layer, and N_M is the number of neurons in the output layer of the neural network, referred to hereafter as the Mth layer; n refers to the number of iterations of the algorithm. The sum of error squares produced by the network defines the cost function

$$\mathscr{E}(n) = \frac{1}{2}\sum_{i=1}^{N_M} e_i(n)e_i^*(n) = \frac{1}{2}\sum_{i=1}^{N_M} |e_i(n)|^2 \tag{19.13}$$

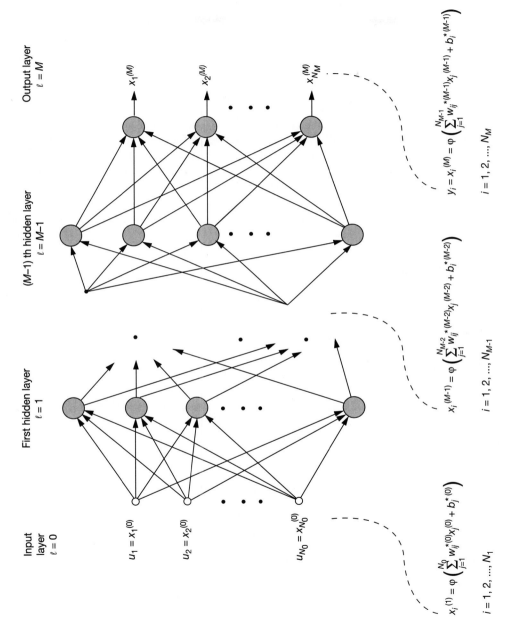

Figure 19.6 Definitions of signals in various layers of multilayer perceptron.

The BP algorithm minimizes the cost function $\mathscr{E}(n)$ by recursively adjusting the complex weights of the multilayer perceptron, using an approximation to the gradient descent technique. The weight update equation is

$$w_{ip}^{(l)}(n + 1) = w_{ip}^{(l)}(n) + \Delta w_{ip}^{(l)}(n) \tag{19.14}$$

The weights are changed in proportion to the negative of the gradient. The update term is defined to be

$$\Delta w_{ip}^{(l)}(n) = -\mu \nabla_{w_{ip}}^{(l)} \mathscr{E}(n) \tag{19.15}$$

where μ is a learning-rate parameter and $\nabla_{w_{ip}}^{(l)} \mathscr{E}(n)$ is the gradient of the cost function $\mathscr{E}(n)$ with respect to the weight $w_{ip}^{(l)}$. We must first find the partial derivative of $\mathscr{E}(n)$ with respect to the complex weights of the $(M - 1)$th layer, and then extend it to the coefficients of all the hidden layers. The gradient of the cost function $\mathscr{E}(n)$ with respect to the complex weights in the $(M - 1)$th layer is defined as

$$\nabla_{w_{ip}}^{(M-1)} \mathscr{E}(n) = \frac{\partial \mathscr{E}(n)}{\partial w_{I,ip}^{(M-1)}(n)} + j \frac{\partial \mathscr{E}(n)}{\partial w_{Q,ip}^{(M-1)}(n)} \tag{19.16}$$

where the complex weight connecting the pth node to the ith node for layer $(M - 1)$ at iteration n is given by

$$w_{ip}^{(M-1)}(n) = w_{I,ip}^{(M-1)}(n) + jw_{Q,ip}^{(M-1)}(n) \tag{19.17}$$

where the subscripts I and Q signify the real and imaginary components of the complex weight in question, respectively. The output $y_i(n)$ is therefore

$$y_i = x_i^{(M)} = \varphi(\text{net}_i^{(M-1)}) \tag{19.18}$$

where

$$\text{net}_i^{(M-1)} = \text{net}_{I,i}^{(M-1)} + j\text{net}_{Q,i}^{(M-1)} \tag{19.19}$$

$$= \sum_{p=1}^{N_{M-1}} w_{ip}^{*(M-1)} x_p^{(M-1)} + b_i^{*(M-1)}$$

Assume that $\varphi(\text{net}_i^{(M-1)})$ is a suitable complex activation function; hence, let

$$\varphi(\text{net}_i) = \varphi(\text{net}_{I,i} + j\text{net}_{Q,i}) \tag{19.20}$$

$$= u(\text{net}_{I,i}, \text{net}_{Q,i}) + jv(\text{net}_{I,i}, \text{net}_{Q,i})$$

where u and v are real functions. The derivative of the activation function, if it exists, is defined by

$$\varphi'(\text{net}_i) = \frac{d\varphi(\text{net}_i)}{d\text{net}_i} \tag{19.21}$$

The partial derivatives of u and v with respect to the real and imaginary parts of the internal signal net_i are defined by

$$u'_{I,i} = \frac{\partial u_i}{\partial net_{I,i}}$$

$$u'_{Q,i} = \frac{\partial u_i}{\partial net_{Q,i}} \qquad (19.22)$$

$$v'_{I,i} = \frac{\partial v_i}{\partial net_{I,i}}$$

$$v'_{Q,i} = \frac{\partial v_i}{\partial net_{Q,i}}$$

Next, we need to find expressions for $\partial \mathscr{E}(n)/\partial w_{I,ij}^{(M-1)}$ and $\partial \mathscr{E}(n)/\partial w_{Q,ij}^{(M-1)}$. Using the chain rule of calculus, we may write

$$\frac{\partial \mathscr{E}}{\partial w_{I,ip}^{(M-1)}} = \frac{\partial \mathscr{E}}{\partial u_i} \left(\frac{\partial u_i}{\partial net_{I,i}} \frac{\partial net_{I,i}}{\partial w_{I,ip}} + \frac{\partial u_i}{\partial net_{Q,i}} \frac{\partial net_{Q,i}}{\partial w_{I,ip}} \right)$$

$$+ \frac{\partial \mathscr{E}}{\partial v_i} \left(\frac{\partial v_i}{\partial net_{I,i}} \frac{\partial net_{I,i}}{\partial w_{I,ip}} + \frac{\partial v_i}{\partial net_{Q,i}} \frac{\partial net_{Q,i}}{\partial w_{I,ip}} \right) \qquad (19.23)$$

$$\frac{\partial \mathscr{E}}{\partial w_{Q,ip}^{(M-1)}} = \frac{\partial \mathscr{E}}{\partial u_i} \left(\frac{\partial u_i}{\partial net_{I,i}} \frac{\partial net_{I,i}}{\partial w_{Q,ip}} + \frac{\partial u_i}{\partial net_{Q,i}} \frac{\partial net_{Q,i}}{\partial w_{Q,ip}} \right)$$

$$+ \frac{\partial \mathscr{E}}{\partial v_i} \left(\frac{\partial v_i}{\partial net_{I,i}} \frac{\partial net_{I,i}}{\partial w_{Q,ip}} + \frac{\partial v_i}{\partial net_{Q,i}} \frac{\partial net_{Q,i}}{\partial w_{Q,ip}} \right)$$

Evaluating the partial derivative $\partial net_{I,i}/\partial w_{I,ip}$, we may write

$$\frac{\partial net_{I,i}}{\partial w_{I,ip}} = \frac{\partial}{\partial w_{I,ip}} [w_{I,ip}x_{I,p} + w_{Q,ip}x_{Q,p} + b_{I,i}] = x_{I,p} \qquad (19.24)$$

The other partial derivatives of Eqs. (19.23) can be found in a similar manner. Summarizing our findings up to this point in the discussion:

$$\frac{\partial net_{I,i}}{\partial w_{I,ip}} = x_{I,p} \qquad (19.25)$$

$$\frac{\partial net_{I,i}}{\partial w_{Q,ip}} = x_{Q,p} \qquad (19.26)$$

$$\frac{\partial net_{Q,i}}{\partial w_{I,ip}} = x_{Q,p} \qquad (19.27)$$

$$\frac{\partial net_{Q,i}}{\partial w_{Q,ip}} = -x_{I,p} \qquad (19.28)$$

Substituting Eqs. (19.22) and (19.25) to (19.28) into Eq. (19.23), the partial derivatives in the two lines of the latter equation can be expressed respectively as

$$\frac{\partial \mathscr{E}(n)}{\partial w_{I,ip}^{(M-1)}} = \frac{\partial \mathscr{E}(n)}{\partial u_i^{(M-1)}} (u_{I,i}' x_{I,p}^{(M-1)} + u_{Q,i}' x_{Q,p}^{(M-1)}) \tag{19.29}$$

$$+ \frac{\partial \mathscr{E}(n)}{\partial v_i^{(M-1)}} (v_{I,i}' x_{I,p}^{(M-1)} + v_{Q,i}' x_{Q,p}^{(M-1)})$$

$$\frac{\partial \mathscr{E}(n)}{\partial w_{Q,ip}^{(M-1)}} = \frac{\partial \mathscr{E}(n)}{\partial u_i^{(M-1)}} (u_{I,i}' x_{Q,p}^{(M-1)} - u_{Q,i}' x_{I,p}^{(M-1)}) \tag{19.30}$$

$$+ \frac{\partial \mathscr{E}(n)}{\partial v_i^{(M-1)}} (v_{I,i}' x_{Q,p}^{(M-1)} - v_{Q,i}' x_{I,p}^{(M-1)})$$

where, as mentioned previously, primes indicate differentiation. For the weights belonging to layer $(M - 1)$ in the network, the partial derivatives of the cost function can be readily found as follows:

$$\frac{\partial \mathscr{E}(n)}{\partial u_i^{(M-1)}} = -[d_{I,i} - y_{I,i}(n)] = -e_{I,i}(n) \tag{19.31}$$

$$\frac{\partial \mathscr{E}(n)}{\partial v_i^{(M-1)}} = -[d_{Q,i} - y_{Q,i}(n)] = -e_{Q,i}(n) \tag{19.32}$$

Substituting Eqs. (19.29) and (19.30) into (19.16) and simplifying, we get

$$\nabla_{w_{ip}}^{(M-1)}\mathscr{E}(n) = x_p^{(M-1)}(n) \left(\frac{\partial \mathscr{E}(n)}{\partial u_i^{(M-1)}} (u_{I,i}'^{(M-1)} - j u_{Q,i}'^{(M-1)}) \right. \tag{19.33}$$

$$\left. + \frac{\partial \mathscr{E}(n)}{\partial v_i^{(M-1)}} (v_{I,i}'^{(M-1)} - j v_{Q,i}'^{(M-1)}) \right)$$

Then, using Eqs. (19.31) and (19.32):

$$\nabla_{w_{ip}}^{(M-1)}\mathscr{E}(n) = -x_p^{(M-1)}(n)[e_{I,i}(n)(u_{I,i}' - j u_{Q,i}') + e_{Q,i}(n)(v_{I,i}' - j v_{Q,i}')] \tag{19.34}$$

Hence, the weight update rule of Eq. (19.14) becomes

$$w_{ip}^{(M-1)}(n+1) = w_{ip}^{(M-1)}(n) + \mu x_p^{(M-1)}(n)[e_{I,i}(n)(u_{I,i}' - j u_{Q,i}') \tag{19.35}$$

$$+ e_{Q,i}(n)(v_{I,i}' - j v_{Q,i}')]$$

The update rule for the bias term b can be derived in a similar manner. We will just state it here to be

$$b_i^{(M-1)}(n+1) = b_i^{(M-1)}(n) + \mu[e_{I,i}(n)(u_{I,i}' - j u_{Q,i}') \tag{19.36}$$

$$+ e_{Q,i}(n)(v_{I,i}' - j v_{Q,i}')]$$

Pattern classification tasks often require a mapping from a multidimensional feature space to a class label. Feature data belonging to a class $\mathscr{C}_k$ are trained to map to a constant value at an output node k in the neural network. For this application, a bounded, nonlinear activation function at the output is desirable. However, if a continuous mapping is required, as in a nonlinear prediction problem that is an example of nonlinear regression, it is necessary to remove the nonlinearity from the output unit and thereby have the final layer operate as a linear combiner; this modification allows the output to vary in an unbounded fashion. In this latter case, we let

$$\varphi(\text{net}_i) = \text{net}_i \tag{19.37}$$

The partial derivatives of this function reduce to

$$u'_{I,i} = 1$$

$$u'_{Q,i} = 0$$

$$v'_{I,i} = 0$$

$$v'_{Q,i} = 1$$

Substituting these values into the weight update rule of Eq. (19.35), we get

$$w_{ip}^{(M-1)}(n+1) = w_{ip}^{(M-1)}(n) + \mu x_p^{(M-1)}(n) e_i^*(n) \tag{19.38}$$

which corresponds to the familiar complex LMS algorithm.

We have now shown the process for updating the $(M-1)$th layer (i.e., output layer) of weights. The next step is to derive the relations necessary to update the weights in the hidden layers of the multilayer perceptron. The main idea is to find expressions that relate the error in the lth layer to the $(l-1)$th layer. In this way we may *back-propagate* the error, stepping from the output layer back towards the input layer in a layer-by-layer manner.

Restating Eqs. (19.29) and (19.30) in terms of a hidden layer of the multilayer perceptron, the expressions for the partial derivatives of the cost function $\mathscr{E}(n)$ with respect to the weights in layer $(M-2)$ are as follows:

$$\frac{\partial \mathscr{E}(n)}{\partial w_{I,ip}^{(M-2)}} = \frac{\partial \mathscr{E}(n)}{\partial u_i^{(M-2)}} \left(u'_{I,i} x_{I,p}^{(M-2)} + u'_{Q,i} x_{Q,p}^{(M-2)} \right) \tag{19.39}$$

$$+ \frac{\partial \mathscr{E}(n)}{\partial v_i^{(M-2)}} \left(v'_{I,i} x_{I,p}^{(M-2)} + v'_{Q,i} x_{Q,p}^{(M-2)} \right)$$

$$\frac{\partial \mathscr{E}(n)}{\partial w_{Q,ip}^{(M-2)}} = \frac{\partial \mathscr{E}(n)}{\partial u_i^{(M-2)}} \left(u'_{I,i} x_{Q,p}^{(M-2)} - u'_{Q,i} x_{I,p}^{(M-2)} \right) \tag{19.40}$$

$$+ \frac{\partial \mathscr{E}(n)}{\partial v_i^{(M-2)}} \left(v'_{I,i} x_{Q,p}^{(M-2)} - v'_{Q,i} x_{I,p}^{(M-2)} \right)$$

Using the chain rule, we may now write

$$\frac{\partial \mathscr{E}(n)}{\partial u_i^{(M-2)}} = \sum_k \frac{\partial \mathscr{E}(n)}{\partial u_k^{(M-1)}} \left(\frac{\partial u_k^{(M-1)}}{\partial \text{net}_{I,k}} \frac{\partial \text{net}_{I,k}}{\partial u_i^{(M-2)}} + \frac{\partial u_k^{(M-1)}}{\partial \text{net}_{Q,k}} \frac{\partial \text{net}_{Q,k}}{\partial u_i^{(M-2)}} \right)$$

$$+ \sum_k \frac{\partial \mathscr{E}(n)}{\partial v_k^{(M-1)}} \left(\frac{\partial v_k^{(M-1)}}{\partial \text{net}_{I,k}} \frac{\partial \text{net}_{I,k}}{\partial u_i^{(M-2)}} + \frac{\partial v_k^{(M-1)}}{\partial \text{net}_{Q,k}} \frac{\partial \text{net}_{Q,k}}{\partial u_i^{(M-2)}} \right) \tag{19.41}$$

$$\frac{\partial \mathcal{E}(n)}{\partial v_i^{(M-2)}} = \sum_k \frac{\partial \mathcal{E}(n)}{\partial u_k^{(M-1)}} \left(\frac{\partial u_k^{(M-1)}}{\partial \text{net}_{I,k}} \frac{\partial \text{net}_{I,k}}{\partial v_i^{(M-2)}} + \frac{\partial u_k^{(M-1)}}{\partial \text{net}_{Q,k}} \frac{\partial \text{net}_{Q,k}}{\partial v_i^{(M-2)}} \right)$$

$$+ \sum_k \frac{\partial \mathcal{E}(n)}{\partial v_k^{(M-1)}} \left(\frac{\partial v_k^{(M-1)}}{\partial \text{net}_{I,k}} \frac{\partial \text{net}_{I,k}}{\partial v_i^{(M-2)}} + \frac{\partial v_k^{(M-1)}}{\partial \text{net}_{Q,k}} \frac{\partial \text{net}_{Q,k}}{\partial v_i^{(M-2)}} \right) \tag{19.42}$$

The partial derivatives of the cost function $\mathcal{E}(n)$ in Eqs. (19.41) and (19.42) are expressed in terms of the partial derivatives pertaining to the previous layer $(M - 1)$. We now only need to determine the partial derivatives of

$$\text{net}_k^{(M-1)} = \sum_i w_{ki}^{*(M-1)} \varphi(\text{net}_i^{(M-2)}) + b_k^{*(M-1)}$$

$$= \sum_i (u_i^{(M-2)} w_{I,ki}^{(M-1)} + v_i^{(M-2)} w_{Q,ki}^{(M-1)} + b_{I,k}^{(M-1)}) \tag{19.43}$$

$$+ j(v_i^{(M-2)} w_{I,ki}^{(M-1)} - u_i^{(M-2)} w_{Q,ki}^{(M-1)} - b_{Q,k}^{(M-1)})$$

with respect to the u and the v of the previous layer $(M - 2)$. Summarizing these partial derivatives, we have

$$\frac{\partial \text{net}_{I,k}^{(M-1)}}{\partial u_i^{(M-2)}} = w_{I,ki}^{(M-1)} \tag{19.44}$$

$$\frac{\partial \text{net}_{I,k}^{(M-1)}}{\partial v_i^{(M-2)}} = w_{Q,ki}^{(M-1)} \tag{19.45}$$

$$\frac{\partial \text{net}_{Q,k}^{(M-1)}}{\partial u_i^{(M-2)}} = -w_{Q,ki}^{(M-1)} \tag{19.46}$$

$$\frac{\partial \text{net}_{Q,k}^{(M-1)}}{\partial v_i^{(M-2)}} = w_{I,ki}^{(M-1)} \tag{19.47}$$

Substituting Eqs. (19.44) to (19.47) into (19.41) and (19.42), we may express the partial derivatives of interest as

$$\frac{\partial \mathcal{E}(n)}{\partial u_i^{(M-2)}} = \sum_k \frac{\partial \mathcal{E}(n)}{\partial u_k^{(M-1)}} (u_{I,k}'^{(M-1)} w_{I,ki}^{(M-1)} - u_{Q,k}'^{(M-1)} w_{Q,ki}^{(M-1)}) \tag{19.48}$$

$$+ \sum_k \frac{\partial \mathcal{E}(n)}{\partial v_k^{(M-1)}} (v_{I,k}'^{(M-1)} w_{I,ki}^{(M-1)} - v_{Q,k}'^{(M-1)} w_{Q,ki}^{(M-1)})$$

$$\frac{\partial \mathcal{E}(n)}{\partial v_i^{(M-2)}} = \sum_k \frac{\partial \mathcal{E}(n)}{\partial u_k^{(M-1)}} (u_{I,k}'^{(M-1)} w_{Q,ki}^{(M-1)} + u_{Q,k}'^{(M-1)} w_{I,ki}^{(M-1)}) \tag{19.49}$$

$$+ \sum_k \frac{\partial \mathcal{E}(n)}{\partial v_k^{(M-1)}} (v_{I,k}'^{(M-1)} w_{Q,ki}^{(M-1)} + v_{Q,k}'^{(M-1)} w_{I,ki}^{(M-1)})$$

Combining these results as the real and imaginary parts of a complex partial derivative, we may simplify matters by writing

$$\frac{\partial \mathscr{E}(n)}{\partial u_i^{(M-2)}} + j \frac{\partial \mathscr{E}(n)}{\partial v_i^{(M-2)}}$$

$$= \sum_k w_{ki}^{(M-1)}(n) \left(\frac{\partial \mathscr{E}(n)}{\partial u_k^{(M-1)}} (u'_{I,k} + ju'_{Q,k}) + \frac{\partial \mathscr{E}(n)}{\partial v_k^{(M-1)}} (v'_{I,k} + jv'_{Q,k}) \right) \qquad (19.50)$$

where the primed variables refer to layer $M - 1$.

Using induction, we can extend this relationship to the other hidden layers of the multilayer perceptron. Equation (19.50) gives us the means to back-propagate the error from the output layer, (M), to the input layer (0). After the values for $\partial \mathscr{E}(n)/\partial u$ and $\partial \mathscr{E}(n)/\partial v$ for the particular layer have been determined, Eq. (19.33) gives the gradient, and hence the weight update values.

Complex-valued activation function

One of the difficulties encountered in extending the real BP algorithm to the complex domain involves the appropriate choice of activation function. The straightforward extension of the sigmoidal from the real domain to the complex domain is inadequate, due to the fact that it has singularities, such that

$$\frac{1}{1 + e^{-z}} \to \infty \qquad \text{for } z = \pm j(2k + 1)\pi, \qquad k \text{ any integer} \qquad (19.51)$$

For a practical implementation of the complex multilayer perceptron, it is necessary that the activation function be bounded. Without such a guarantee, there is a risk of arithmetic overflow in software implementation of multilayer perceptron; hardware implementation would suffer in an analogous manner, with unbounded outputs resulting in possible clipping at node outputs. Singularities in an activation function must therefore be avoided.

Georgiou and Koutsougeras (1992) have developed a set of properties, which a complex activation function must satisfy in order to be useful in a multilayer perceptron trained with the back-propagation algorithm. These properties are summarized here:

1. The activation function $\varphi(z)$ should be nonlinear in both z_I and z_Q, which denote the real and imaginary parts of the argument z; otherwise, there is no advantage in having a multilayer perceptron. A multilayer perceptron that is linear may always be collapsed to an equivalent single-layer network. The motivation here is to have a nonlinear network that can compute a more general set of functions than is possible with a linear network.

2. The function $\varphi(z)$ should be bounded. The computation of the forward pass of the multilayer perceptron is required to be bounded; otherwise, clipping or numerical overflow can occur.

3. The partial derivatives of $\varphi(z)$ should exist and be bounded. The learning phase updates the complex weights of the multilayer perceptron by amounts proportional to the partial derivatives, so they also need to be bounded.

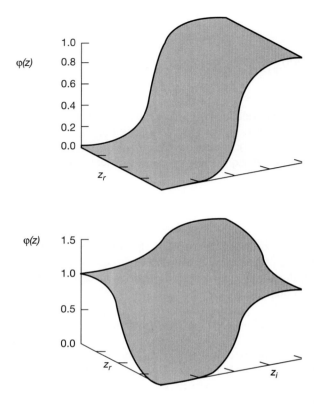

$\varphi(z)$

z_r

$\varphi(z)$

z_r z_i

$z = z_r + jz_i$

Figure 19.7   Real part (top) and magnitude (bottom) of the activation function $\varphi(z) = c/(1 + e^{-kz_r}) + jc/(1 + e^{-kz_i})$

4. The function $\varphi(z)$ should not be an entire function. *Entire functions* are defined as complex functions that are analytic everywhere in the complex domain. A function is defined to be *analytic* at some point z_0 if it is differentiable in some neighborhood of z_0. By *Liouville's theorem*,[2] we know that a bounded, entire function is constant. Clearly, a function that is entire is not a suitable choice for an activation function for the reasons stated in Property 1.

5. The partial derivatives of the cost function $\mathscr{E}(n)$ should satisfy the condition:

$$\frac{\partial \mathscr{E}}{\partial u} + \frac{j\partial \mathscr{E}}{\partial v} \neq 0$$

---

[2] A review of complex variable theory, including Liouville's theorem, is presented in Appendix A.

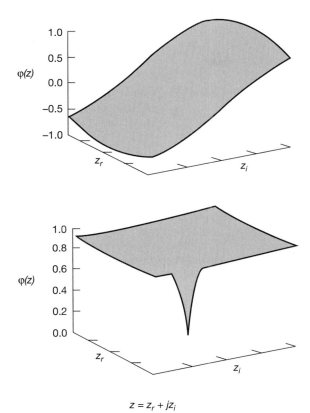

$$z = z_r + jz_i$$

Figure 19.8 Real part (top) and magnitude (bottom) of the activation function $\varphi(z) = z/[c + (1/r)\,|z|]$.

For this condition to hold, the partial derivatives of $\varphi(z)$ should not satisfy the relation $u_I v_Q = u_Q v_I$. This relationship can be satisfied by simultaneously setting the real and imaginary parts of Eq. (19.33) equal to zero, for $x_p(n) \neq 0$. Should the partial derivatives of the activation function satisfy the above relation, this would imply that in the presence of both nonzero input and error, it would be possible that $\nabla_w \mathscr{E} = 0$, and therefore a stationary point would be reached. No further learning could then take place, since the weight update is proportional to the gradient.

Figures 19.7 and 19.8 show two possible choices for the complex activation function. Figure 19.7 shows the complex activation function suggested by Benvenuto and Piazza (1992). The function is a superposition of real and imaginary sigmoids, as shown by

$$\varphi(z) = \frac{c}{1 + \exp(-kz_r)} + j\,\frac{c}{1 + \exp(-kz_i)} \tag{19.52}$$

where z_r and z_i are the real and imaginary parts of z, respectively.

TABLE 19.1 SUMMARY OF THE COMPLEX BACK-PROPAGATION ALGORITHM

1. *Initialization*
   Set all weights and biases to small complex random values
2. *Present input and desired outputs*
   Present input vector $\mathbf{x}(1)$, $\mathbf{x}(2)$, ..., $\mathbf{x}(N)$ and desired response $\mathbf{d}(1)$, $\mathbf{d}(2)$, ..., $\mathbf{d}(N)$, where N is the total number of training patterns
3. *Calculate actual outputs*
   Use the formulas in Fig. 19.6 to calculate the output signals $y_1, y_2, \ldots, y_{N_M}$
4. *Adapt weights and biases*

$$\Delta w_{ip}^{(l-1)}(n) = -\mu x_p^{(l-1)}(n) \left(\frac{\partial \mathscr{E}(n)}{\partial u_i^{(l-1)}} (u'_{I,i} - ju'_{Q,i}) + \frac{\partial \mathscr{E}(n)}{\partial v_i^{(l-1)}} (v'_{I,i} - jv'_{Q,i}) \right)$$

$$\Delta b_i^{(l-1)}(n) = -\mu \left(\frac{\partial \mathscr{E}(n)}{\partial u_i^{(l-1)}} (u'_{I,i} - ju'_{Q,i}) + \frac{\partial \mathscr{E}(n)}{\partial v_i^{(l-1)}} (v'_{I,i} - jv'_{Q,i}) \right)$$

where

$$\frac{\partial \mathscr{E}(n)}{\partial u_i^{(l-1)}} + j \frac{\partial \mathscr{E}(n)}{\partial v_i^{(l-1)}} = \left\{ \begin{array}{ll} -[d_i - y_i(n)] & \text{for } l = M \\[2ex] \sum_k w_{ki}^{(l)} \left(\frac{\partial \mathscr{E}(n)}{\partial u_k^{(l)}} (u'_{I,k} + ju'_{Q,k}) + \frac{\partial \mathscr{E}(n)}{\partial v_k^{(l)}} (v'_{I,k} + jv'_{Q,k}) \right) & \text{for } 1 \leq l < M \end{array} \right\}$$

where $x_p(n) =$ output of node p or input to node i at iteration n

---

Another possible activation function is suggested by Georgiou and Koutsougeras (1992); it is also a sigmoidlike function, as shown in Fig. 19.8, with

$$\varphi(z) = \frac{z}{c + (1/r)|z|} \tag{19.53}$$

This function maps the z-domain to an open disk $|z| < r$; hence, the activation function effectively squashes the range of $|\varphi(z)|$ to the interval $[0,r]$.

Now that we have identified suitable choices for the complex activation functions, we may finish our discussion of the complex back-propagation algorithm by summarizing the important steps involved in its application, as outlined in Table 19.1.

Incorporation of a Momentum Term

The back-propagation learning process may be accelerated by incorporating a *momentum* term (Rumelhart et al., 1986). Specifically, the correction $\Delta w_{ip}^{(l)}(n)$ applied to the synaptic $w_{ip}^{(l)}(n)$ in layer l of the network, defined in Eq. (19.15), is modified as follows:

$$\Delta w_{ip}^{(l)}(n) = \alpha \Delta w_{ip}^{(l)}(n-1) - \frac{1}{2} \mu \nabla w_{ip}^{(l)} \mathscr{E}(n) \tag{19.54}$$

where α is called the *momentum constant*, and $\Delta w_{ip}^{(l)}(n-1)$ is the previous value of the correction. As before, μ is the learning-rate parameter.

The use of momentum introduces a feedback loop around $\Delta w_{ip}^{(l)}(n)$. As such, it can have a highly beneficial effect on learning behavior of the back-propagation algorithm. In particular, it may have the benefit of preventing the learning process from being stuck at a local minimum on the error-performance surface of the multilayer perceptron (Rumelhart et al., 1986; Haykin, 1994).

19.4 BACK-PROPAGATION ALGORITHM FOR REAL PARAMETERS

The common development of the back-propagation algorithm is for real-valued data and parameters. We now show that this is merely a special case of the more general, complex back-propagation algorithm developed in the previous section.

We proceed by considering all the parameters to be real-valued, including the input and desired output data. In terms of the complex-valued neural network, the quadrature components are all set equal to zero. Applying this principle to Eq. (19.23), we readily observe that only the first term survives, so that

$$\frac{\partial \mathcal{E}}{\partial w_{ip}} = \frac{\partial \mathcal{E}}{du_i} \frac{\partial u_i}{\partial \text{net}_i} \frac{\partial \text{net}_i}{\partial w_{ip}} \tag{19.55}$$

Note that the in-phase designation, I, is dropped from the variables in this equation; since there is no longer a quadrature signal component to consider, it is a redundant notation. We also replace all occurrences of u with

$$\varphi(\text{net}_i) = u(\text{net}_i, 0) \tag{19.56}$$

and

$$\varphi'(\text{net}_i) = u'_{I,i} = \frac{\partial u_i}{\partial \text{net}_{I,i}} \tag{19.57}$$

Equation (19.55) can now be rewritten as

$$\frac{\partial \mathcal{E}}{\partial w_{ip}} = \frac{\partial \mathcal{E}}{\partial \varphi(\text{net}_i)} \frac{\partial \varphi(\text{net}_i)}{\partial \text{net}_i} \frac{\partial \text{net}_i}{\partial w_{ip}} \tag{19.58}$$

The activation function, $\varphi(\text{net})$, can be any bounded, differentiable, monotonically increasing function. The sigmoid function is often the function of choice.

We now define a new variable

$$\delta_i^{(l-1)}(n) = -\frac{\partial \mathcal{E}(n)}{\partial \text{net}_i^{(l-1)}} \tag{19.59}$$

For the case of the output layer of the multilayer perceptron, that is, $l = M$, we may write

$$\delta_i^{(M-1)}(n) = -\frac{\partial \mathcal{E}}{\partial \varphi(\text{net}_i)} \frac{\partial \varphi(\text{net}_i)}{\partial \text{net}_i} = \varphi'(\text{net}_i^{(M-1)})e_i(n) \tag{19.60}$$

TABLE 19.2   SUMMARY OF THE REAL BACK-PROPAGATION ALGORITHM

1. *Initialization*
   Set all weights and biases to small real random values

2. *Present input and desired outputs*
   Present input vector $\mathbf{x}(1)$, $\mathbf{x}(2)$, . . . , $\mathbf{x}(N)$ and desired response $\mathbf{d}(1)$, $\mathbf{d}(2)$, . . . , $\mathbf{d}(N)$, where N is the number of training patterns

3. *Calculate actual outputs*
   Use the formulas in Fig. 19.6 to calculate the output signals y_1, y_2, . . . , y_{N_M}

4. *Adapt weights and biases*

$$\Delta w_{ij}^{(l-1)}(n) = \mu x_j(n)\delta_i^{(l-1)}(n)$$

$$\Delta b_i^{(l-1)}(n) = \mu\, \delta_i^{(l-1)}(n)$$

where

$$\delta_i^{(l-1)}(n) = \begin{cases} \varphi'(\text{net}_i^{(l-1)})\, [d_i - y_i(n)], & l = M \\[2mm] \varphi'(\text{net}_i^{(l-1)}) \displaystyle\sum_k w_{ki}\, \delta_k^{(l)}(n), & 1 \le l < M \end{cases}$$

where $x_j(n)$ = output of node j or input to node i at iteration n

---

where the crime on the right-hand side of the equation signifies differentiation. For a hidden layer of the network, that is, $1 \le l < M$, we have

$$\delta_i^{(l-1)}(n) = \varphi'(\text{net}_i^{(l-1)}) \sum_k w_{ki}(n)\delta_k^l(n) \tag{19.61}$$

The variable δ is interpreted as a back-propagated error term, which can be recursively computed for each layer of the multilayer perceptron, starting from the output layer.

A summary of the real-back-propagation algorithm is presented in Table 19.2. Note that a momentum term can also be added here in a manner similar to that described for the complex back-propagation algorithm.

19.5 UNIVERSAL APPROXIMATION THEOREM

A multilayer perceptron trained with the back-propagation algorithm provides a powerful device for approximating a *nonlinear input–output mapping* of a general nature. In this context, a key question needs to be considered: What is the number of hidden layers that would be needed in the design of the multilayer perceptron to do the approximation in a uniform manner? The answer to this fundamental question lies in the *universal approximation theorem*, which was developed independently by Cybenko (1989), Funahashi

(1989), and Hornik et al. (1989). The universal approximation theorem may be stated as follows:

Let $f(\cdot)$ be a nonconstant, bounded, and monotone-increasing continuous function. Let I_{N_0} denote the N_0-dimensional unit hypercube. The space of continuous functions on I_{N_0} is denoted by $C(I_{N_0})$. Then, given any function $f \in C(I_{N_0})$ and $\varepsilon > 0$, there exist an integer N_1 and sets of real constants α_i, b_i, and w_{ip}, where $i = 1, 2, \ldots, N_1$ and $p = 1, 2, \ldots, N_0$ such that we may define

$$F(u_1, u_2, \ldots, u_{N_0}) = \sum_{i=1}^{N_1} \alpha_i \, \varphi \left(\sum_{p=1}^{N_0} w_{ip} x_p + b_i \right) \qquad (19.62)$$

as an approximate realization of the function $f(\cdot)$; that is, the absolute value of the approximation satisfies the condition

$$\left| F(u_1, u_2, \ldots, u_{N_0}) - f(u_1, u_2, \ldots, u_{N_0}) \right| < \varepsilon$$

for all $\{ u_1, u_2, \ldots, u_{N_0} \} \in I_{N_0}$.

The universal approximation theorem is directly applicable to a multilayer perceptron having the following description:

- An input layer of N_0 nodes, whose individual inputs are denoted by $x_1, x_2, \ldots, x_{N_0}$
- A single hidden layer of N_1 sigmoidal neurons, with the synaptic weights of the ith hidden neuron being denoted by $w_{i1}, w_{i2}, \ldots, w_{iN_0}$
- An output layer consisting of a single linear neuron

It should be emphasized that the universal approximation theorem is an existence theorem, in the sense that it provides the mathematical justification for the approximation of an arbitrary continuous function as opposed to exact representation. Equation (19.62), which is the backbone of the theorem, merely generalizes approximations by finite Fourier series. In effect, the theorem states that a single hidden layer is sufficient for a multilayer perceptron to compute a uniform ε approximation into a given training set represented by the sets of inputs $x_1, x_2, \ldots, x_{N_0}$ and a desired (target) output denoted by $f(x_1, x_2, \ldots, x_{N_0})$. From a theoretical viewpoint, the universal approximation theorem is therefore important. Without such a theorem we could be conceivably searching for a solution that cannot exist. However, the theorem does not say that a single hidden layer is optimum in the sense of learning time or ease of implementation.

From a practical perspective, the problem with multilayer perceptrons using a single hidden layer is that the hidden neurons tend to interact with each other. In complex situations, this interaction makes it difficult to improve the approximation at one point without worsening it at some other point. On the other hand, with two hidden or more layers the approximation (curve-fitting) process may become more manageable (Chester, 1990). It is for this reason that we find in solving large-scale problems, the recommended procedure is to use a multilayer perceptron with two (or possibly more) hidden layers.

19.6 NETWORK COMPLEXITY

To solve real-world problems with multilayer perceptrons, we usually require the use of highly structured networks of a rather large size. A practical issue that arises in this context is that of minimizing the size of the network and yet maintaining good performance. As a general rule, a neural network with minimum size is less likely to learn the idiosyncrasies or noise in the training data, and may therefore generalize better to new data. *Generalization*, a term borrowed from psychology, refers to the ability of a neural network, having learned the essential information content of training data, to achieve "reasonable" performance for test data drawn from the same input space but not seen before.

We may achieve the design objective of minimum network size by proceeding in either one of the following two ways:

- *Network growing*, in which we start with a multilayer perceptron that is small for accomplishing the task at hand, and then add a new neuron or a new layer of hidden neurons only when we are unable to meet the design specification.
- *Network pruning*, in which we start with a large multilayer perceptron with an adequate performance for the task at hand, and then prune it by eliminating "unreliable" synaptic weights in a selective and orderly manner.

Although both of these approaches are used in practice, it is safe to say that in current practice, network pruning is the more popular one of the two.

In this section we describe the so-called *weight-eliminating procedure* (Weigend et al., 1991), the objective of which is to find a weight vector $\mathbf{w}$ that minimizes the total *risk*

$$\mathcal{R}(\mathbf{w}) = \mathcal{E}_s(\mathbf{w}) + \lambda \mathcal{E}_c(\mathbf{w}) \tag{19.63}$$

The first term, $\mathcal{E}_s(\mathbf{w})$, is the *standard* performance measure that depends on both the network (model) and the input data. In back-propagation learning, $\mathcal{E}_s(\mathbf{w})$ is typically defined as a mean-squared error whose evaluation extends over the output neurons of the network, and which is carried out for all the training data. The second term, $\mathcal{E}_c(\mathbf{w})$, is the *complexity penalty*, which depends on the network (model) alone. The evaluation of $\mathcal{E}_c(\mathbf{w})$ is confined to the synaptic connections of the network. In the weight-elimination procedure, the complexity penalty is defined by

$$\mathcal{E}_c(\mathbf{w}) = \sum_{i \in l} \frac{(w_i/w_o)^2}{1 + (w_i/w_o)^2} \tag{19.64}$$

where w_o is a prescribed free parameter of the procedure, and w_i refers to the weight of synapse i in the network. The set l refers to all the synaptic connections in the network. An

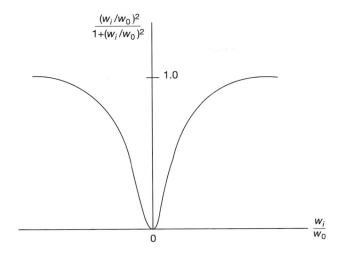

Figure 19.9   Complex penalty function.

individual penalty term varies with w_i/w_o in a symmetric fashion, as shown in Fig. 19.9. We may identify two limiting conditions:

- When $|w_i| << w_o$, the complexity penalty (cost) for that weight approaches zero. The implication of this condition is that insofar as learning from training data is concerned, the ith synaptic weight is unreliable and should therefore be eliminated from the network.
- When $|w_i| >> w_o$, the complexity penalty (cost) for that weight approaches the maximum value of unity, which means that w_i is important to the back-propagation learning process.

The parameter λ in Eq. (19.63) plays the role of a *regularization parameter*. When λ is zero, the back-propagation learning process is unconstrained, in which case the network is completely determined by the training data in the manner described in Section 19.3 for complex data and Section 19.4 for real data. When λ is made infinitely large, on the other hand, the implication is that the constraint imposed by the complexity penalty is by itself sufficient to specify the network. This is another way of saying that the training data are unreliable and should therefore be ignored. In practical applications of the weight-elimination procedure, the regularization parameter λ is assigned a value somewhere between these two limiting cases.

Thus, starting with a multilayer perceptron designed by means of the back-propagation algorithm, and having chosen a value for the regularization parameter λ that is appropriate for the particular situation under study, the network is pruned by minimizing the total risk $\mathcal{R}(\mathbf{w})$ defined in Eq. (19.63). Clearly, the computational effort involved in this minimization is highly dependent on the size of the network. The ultimate aim is, of course, to *make the network complexity a close match to the complexity of the data used to train the network.*

19.7 FILTERING APPLICATIONS

Before proceeding to describe some filtering applications of multilayer perceptrons, it is instructive to distinguish between learning and adaptation, which go on, for example, in the back-propagation (BP) and the least-mean-square (LMS) algorithms, respectively. In the LMS algorithm, adjustments to the tap weights of a transversal filter are made while at the same time the input signal is being processed. This kind of a process is an example of *continuous learning*, which never stops. In contrast, in the BP algorithm, the synaptic weights of the multilayer perceptron are adjusted during the training phase; and once steady-state conditions are reached, all the synaptic weights in the network are fixed thereafter. In other words, in a multilayer perceptron, learning precedes signal processing. Clearly then, signal-processing (filtering) applications of multilayer perceptrons have to take full account of the way in which this class of neural networks learns from its environment.

For our present discussion, we have chosen three applications of multilayer perceptrons: the first relating to system identification, the second involving the time-delay neural network for temporal signal processing, and the third dealing with target detection. In the sequel, these three applications are described in that order. In all three cases the emphasis is on nonlinear signal processing in one form or another.

System Identification

In light of what we have just said about back-propagation learning, the multilayer perceptron is basically a *static* network. We may extend its use for the identification of a nonlinear dynamic system by the incorporation of unit-delay elements at its input, output, or both, as described next.

For the identification of a nonlinear dynamic system, we may formulate four different models, depending on how the output of the system is defined in terms of past values of the output and past values of the input. Specifically, we may describe the models in terms of nonlinear difference equations as described in Narendra and Pasthasarathy (1990).

Model I.   The output $y(n + 1)$ at time $n + 1$ depends linearly on N past values of the output, $y(n), \ldots, y(n - N + 1)$, and nonlinearly on M past values of the input, $u(n)$, $\ldots, u(n - M + 1)$, as shown by

$$y(n + 1) = \sum_{i=0}^{N-1} \alpha_i \, y(n-i) + g(u(n), u(n - 1), \ldots, u(n - M + 1)) \qquad (19.65)$$

where $g(\cdot, \cdot, \cdots, \cdot)$ is a nonlinear function that is differentiable with respect to its arguments. It is assumed that $M \leq N$ for all four models.

Model II.   The output $y(n + 1)$ at time $n + 1$ depends nonlinearly on N past values of the output and linearly on M past values of the input, as shown by

$$y(n + 1) = f(y(n), y(n - 1), \ldots, y(n - N + 1)) + \sum_{i=0}^{M-1} \beta_i u(n-i) \qquad (19.66)$$

where $f(\cdot, \cdot, \cdots, \cdot)$ is another nonlinear function that is also differentiable with respect to its arguments.

Model III.   The output $y(n + 1)$ at time $n + 1$ depends nonlinearly on past values of both the output and the input in a separable manner, as shown by

$$\begin{aligned} y(n + 1) = f(y(n), y(n-1), \ldots, y(n-N+1)) \\ + g(u(n), u(n - 1), \ldots, u(n-M+1)) \end{aligned} \qquad (19.67)$$

Model IV.   The output at time $n + 1$ depends nonlinearly on past values of both the output and the input in a nonseparable manner, as shown by

$$y(n+1) = f(y(n), y(n-1), \ldots, y(n-N+1); u(n), u(n - 1), \ldots, u(n-M+1)) \qquad (19.68)$$

Clearly, Model IV is the most general one, in that it includes the other three models as special cases. However, in spite of its generality, model IV is the least tractable in analytic terms, which makes the other three models more attractive for practical applications (Narendra and Pasthasarathy, 1990).

Figure 19.10 presents block diagram descriptions of the four models. The elements labeled z^{-1} at the input and output ends of each model represent unit-delay elements.

In light of the universal approximation theorem described in Section 19.5, we may say that under fairly weak conditions on the nonlinear function f and/or g in Eqs. (19.65) to (19.68), multilayer perceptrons can indeed be designed using the back-propagation algorithm to approximate the input-output mapping described by models I to IV over compact sets (Narendra and Pasthasarathy, 1990). Thus, although the multilayer perceptron is a static network by itself, it assumes a dynamic behavior by embedding it in models I through IV, described in Fig. 19.10. The choice of a particular model is dictated by the application of interest.

It is of interest to note that if the coefficients, α_i, $i = 0, 1, \ldots, N - 1$, were all to be reduced to zero, then Eq. (19.65) takes the form

$$y(n)+1) = g(u(n), u(n-1), \ldots, u(n-M+1)) \qquad (19.69)$$

The model output $y(n+1)$ is now recognized as the one-step prediction

$$y(n+1) = \hat{u}(n+1 \,|\, \mathcal{U}_n) \qquad (19.70)$$

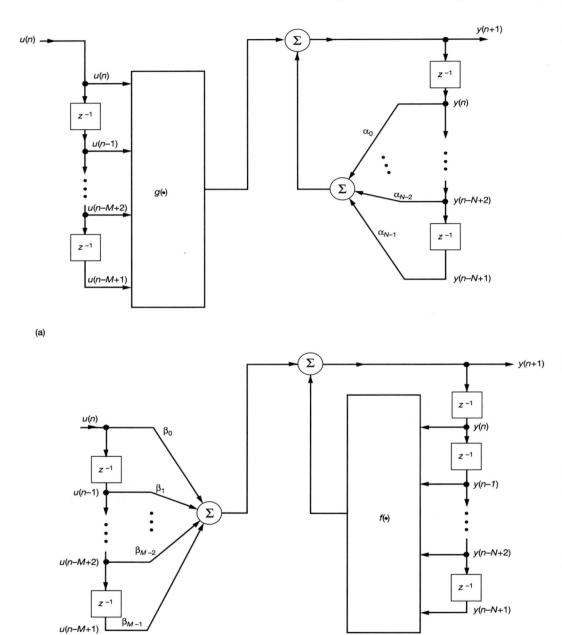

(a)

(b)

Figure 19.10   Four different models for system identification. Parts (c) and (d) of the figure are presented on the next two pages.

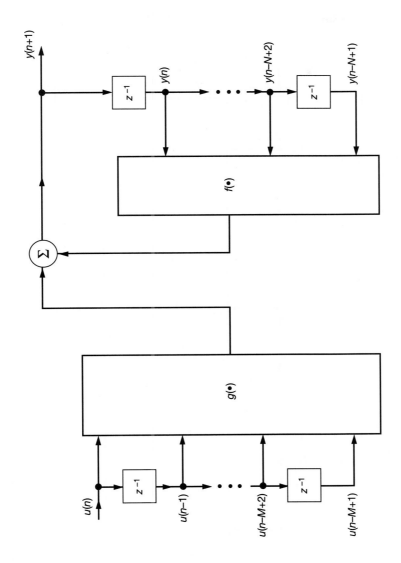

(c)

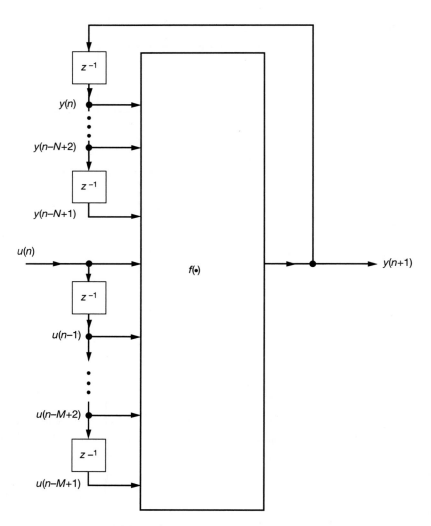

(d)

Figure 19.10   (concluded)

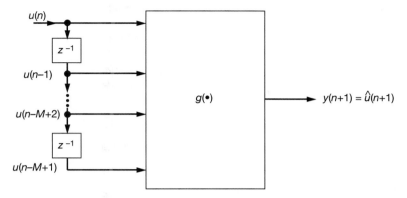

Figure 19.11   Nonlinear one-step predictor.

where $\mathcal{U}_n$ denotes the M-dimensional space defined by past values of the input, $u(n)$, $u(n-1), \ldots, u(n-M+1)$. Accordingly, model II includes the nonlinear predictor as a special case, as depicted in the block diagram of Fig. 19.11.

Time-Delay Neural Network

The models described in Fig. 19.10 describe a straightforward method for extending the use of a multilayer perceptron to account for *time*, which is accomplished by the inclusion of *memory* in the form of unit-delay elements outside of the network. In another structure known as the *time-delay neural network (TDNN)*, time delays are incorporated inside the network, as depicted in Fig. 19.12. The TDNN consists of a multilayer feedforward network, whose hidden neurons and output neurons are all *replicated across time*. It was originally devised by Lang and Hinton (1988) to capture explicitly the notion of time symmetry as encountered in the recognition of an isolated word (phoneme) using a *spectrogram*. The *spectrogram* is a two-dimensional image in which the vertical dimension corresponds to frequency and the horizontal dimension corresponds to time; the intensity (darkness) of the image corresponds to signal energy (Rabiner and Schafer, 1978). In effect, the spectrogram provides a method for making speech "visible."

Figure 19.12(a) illustrates a single-layer hidden version of the TDNN. For the example considered here (Lang and Hinton, 1988), the input layer consists of $16 \times 12 = 192$ sensory nodes encoding the spectrogram. The hidden layer consists of 10 copies of 8 hidden neurons. The output layer consists of 6 copies of 4 output neurons. The various replicas of a hidden neuron apply the same set of synaptic weights to narrow (3 time-step) windows of the spectrogram. Similarly, the various replicas of an output neuron apply the same set of synaptic weights to narrow (5 time-step) windows of the pseudospectrogram computed by the hidden layer. Figure 19.12(b) presents a time-delay interpretation of the replicated neural network of Fig. 19.12(a), hence the name "time-delay neural network." For the example of a single hidden layer considered here, the TDNN has a total of 544 synaptic weights. In a more elaborate structure described by Waibel et al. (1989), the

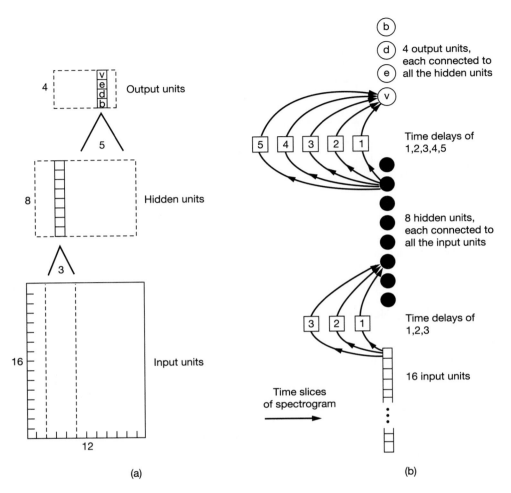

Figure 19.12   (a) A three-layer network whose hidden units and output units are replicated across time. (b) Time-delay neural network (TDNN) representation. (From K.J. Lang and G.E. Hinton, 1988 with permission.)

TDNN is expanded to include two hidden layers. In any case, the standard back-propagation algorithm may be used to train the TDNN.

The TDNN has been used by several investigators for speech recognition (Lang and Hinton, 1988; Waibel et al., 1989). In this context, it appears that the temporal processing power of the TDNN lies in its ability to develop shift-invariant internal representations of speech and to use them for making "optimal" classifications. Another useful application of the TDNN for acoustic echo cancelation is described in Birkett and Goubran (1995).

Specifically, the TDNN is used to model the system nonlinearities and acoustic path in a "hands-free telephone" environment. Simulations are presented therein, demonstrating that such a nonlinear model can provide a significant improvement in system performance over a linear acoustic echo canceler using the normalized LMS algorithm.

The TDNN topology is in fact embodied in a multilayer perceptron in which each synapse is represented by a finite-duration impulse response (FIR) filter. In such a generalization, known as the *FIR multilayer perceptron*, time takes on a "distributed" representation at the synaptic level, thereby enhancing the temporal signal-processing power of the multilayer perceptron in a significant way. To train the FIR multilayer perceptron, we may *unfold it in time* and thereby develop an equivalent structure in the form of a static multilayer perceptron of much larger size, to which the standard back-propagation algorithm may be applied in the usual way. However, such a procedure is highly inefficient. A more practical approach is to use a *temporal back-propagation algorithm* devised by Wan (1990), which works directly with the FIR multilayer perceptron.[3]

Target Detection in Clutter

For our third application we have chosen the detection of a radar target signal buried in a background of clutter. In radar terminology, *clutter* refers to reflections (echoes) of the transmitted signal produced by unwanted objects. In such a situation, the clutter is typically dominant, not only overpowering the receiver noise but also the wanted target signal. We thus have a binary hypothesis testing problem that may be described essentially as follows:

- *Hypothesis that a target is present*, in which the received signal $u(n)$ at time n consists of a target signal $s(n)$ plus clutter $c(n)$, as shown by

$$u(n) = s(n) + c(n)$$

- *Null hypothesis*, in which the received signal $u(n)$ consists of clutter alone, as shown by

$$u(n) = c(n)$$

In the traditional approach to the detection problem described here, a parametric model is formulated for the clutter process and a detection strategy (e.g., Neyman–Pearson criterion) is used to solve the problem. However, with such an approach it is difficult to account fully for an inherent characteristic of radar clutter, that it is in reality the product of a *nonlinear dynamical process*. Indeed, a detailed experimental study reported in Haykin and Li (1995), using real-life radar data, has shown that sea clutter (i.e., radar backscatter from an ocean surface) is largely chaotic. A *chaotic process* is the result of a

---

[3] For a detailed discussion of temporal processing using the multilayer perceptron and other neural networks, see Haykin (1994).

deterministic mechanism, but it exhibits many of the characteristics ordinarily associated with a stochastic process. The important point to note here is that radar clutter is deterministically predictable in a short-term sense.

Recognizing that learning is a natural attribute of neural networks, we may propose a new strategy for the detection of a radar target signal in clutter as follows (Li and Haykin, 1993; Haykin and Li, 1995):

- Starting with actual clutter data that are representative of the environment of interest, a neural network such as a multilayer perceptron is trained (using the back-propagation algorithm) as a *one-step predictor*. Provided that the network is of the right size and the training data set is large enough, the prediction error produced at the output of the network, under the null hypothesis should closely approximate the sample function of a white Gaussian noise process. In effect, the network is trained to perform the function of a clutter model.

- When the network is fed with a received signal that consists of a target signal plus clutter, the presence of the target signal in the input causes a corresponding perturbation at the output of the network. That is, the network tends to preserve the essential characteristics of the target signal at its output. Thus, under the hypothesis that a target is present, the output signal consists essentially of a component identifiable with the target of interest, superimposed on a white Gaussian noise background.

The novelty of the detection strategy described here lies in the fact that a difficult signal detection problem is transformed into the detection of an unknown signal in additive white Gaussian noise, which may be viewed as the communication theorist's dream. The complete receiver thus consists of a one-step nonlinear predictor followed by a conventional constant false-alarm rate (CFAR) processor, as depicted in Fig. 19.13. The important advantages of this receiver include the following:

- Weak statistical assumptions about the environment in which the radar operates
- Inherent ability to account for nonlinear characteristics of the received radar signal

Most importantly, in a clutter dominated environment, the receiver of Fig. 19.13 has the potential to outperform a conventional radar receiver.

Figure 19.14, taken from Haykin et al. (1995b), shows the results of applying the neural-network approach described herein to an operational marine environment using a noncoherent radar. Specifically, Fig. 19.14(a) shows a sample azimuthal time series taken along a range ring containing a 10 square-meter target. The corresponding output of the neural network is shown in Fig. 19.14(b). As can be readily observed, the neural network has captured the dynamics of the clutter, such that the learned clutter component has been effectively removed, yet the target gives a significant response. Fig. 19.14 thus clearly

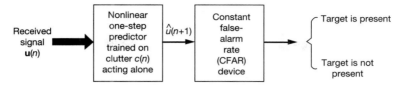

Figure 19.13   Neural network-based radar receiver.

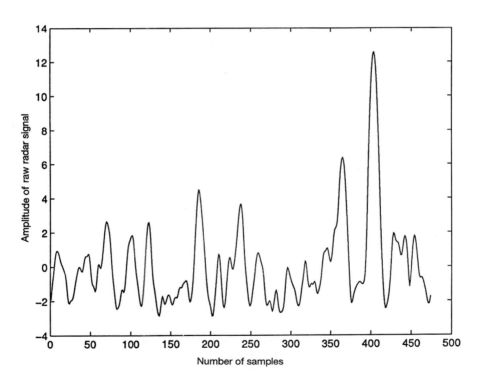

Figure 19.14   (a) Azimuthal time series consisting of target signal plus clutter.

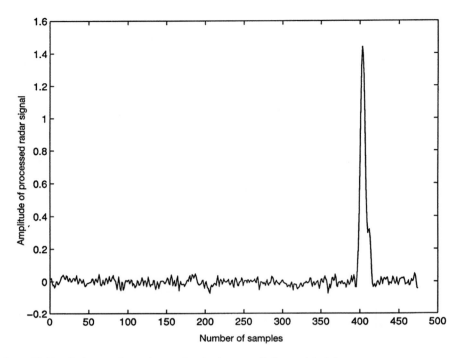

Figure 19.14   (b) Output of neural network trained as a predictive model of clutter.

illustrates the action of the neural network based predictive model as a radar clutter canceler.

19.8 SUMMARY AND DISCUSSION

Just as the LMS algorithm has established itself as the workhorse of linear adaptive filters, so it is with the back-propagation algorithm in the context of neural networks. The back-propagation algorithm is relatively simple to implement, which has made it the most popular algorithm in use today for the design of neural networks. In particular, *it provides a powerful device for storing the information content of the training data in the synaptic weights of the network.* As long as the dataset used to train the neural network is large enough to be representative of the environment in which the network is embedded, the network develops the capability to *generalize*. Specifically, the network delivers a "satisfactory" performance when it is fed with test data drawn from the same input space as the training data but not previously seen by the network.

The multilayer perceptron, consisting of a feedforward network with one or more hidden layers, is the neural network structure commonly used in conjunction with the back-propagation algorithm. With terminology in mind, it is wrong to speak of a "back-propagation network." Rather, we have a multilayer perceptron as the neural network, which is trained with the back-propagation algorithm.

Multilayer perceptrons have been applied successfully in a variety of diverse areas. In terms of functional tasks, the applications may be categorized as follows:

- Pattern classification (recognition)
- Control
- Signal processing

The ability of the multilayer perceptron to *learn* from its environment befits its use for these tasks, each one in its own specific way.

Back-propagation learning is an example of *supervised learning*, so called by virtue of the fact that the desired response (target signal) in the training data plays the role of a "teacher." The important issue to note here is that the learning process is *statistical* in nature. The reason for stochasticity is rooted in the environment in which the neural network is embedded. The net result is that the network is merely one form in which "empirical" knowledge about the environment is represented (White, 1989). The difficulties encountered in a study of the learning process are twofold (Haykin, 1994):

1. A neural network is *nonlinear*, which makes a detailed statistical analysis of the learning process to be a challenging undertaking.

2. In a neural network, *knowledge* about the environment is represented by the values taken on by the free parameters (synaptic weights and biases) of the network; the distributed nature of the knowledge stored in this manner in the network makes for a difficult interpretation.

Finally, from a computational point of view we should stress that the back-propagation algorithm is characterized by a slow rate of convergence. The problem becomes particularly serious when the requirement is to solve a large-scale task. In this context, we are once again reminded of analogy with the LMS algorithm, which is known for its own slow rate of convergence. Various procedures have been devised to accelerate the application of the back-propagation algorithm through the use of learning-rate adaptation (Jacobs, 1988). Alternatively, we may resort to the use of other supervised learning algorithms rooted in nonlinear system identification (Palmieri et al., 1991; Feldkamp, 1994) or function optimization theory (Battiti, 1992; Johansson et al., 1992). For a discussion of the issues raised herein, the reader is referred to the book by Haykin (1994).

PROBLEMS

1. A neuron j receives inputs from four other neurons whose activity levels are 10, -20, 4, and -2. The respective synaptic weights of neuron j are 0.8, 0.2, -1.0, and -0.9. Calculate the output of neuron j for the following two situations:
   (a) The neuron is linear.
   (b) The neuron is represented by the McCulloch–Pitts model.

2. Repeat Problem 1 for a neuron j whose model is based on the logistic function

$$\varphi(\text{net}) = \frac{1}{1 + \exp(-\text{net})}$$

3. Consider a multilayer feedforward network, all the neurons of which operate in their linear regions. Justify the statement that such a network is equivalent to a feedforward network with a single layer of computation nodes.

4. Consider the following nonlinear functions:

   (a) $\varphi(x) = \dfrac{1}{\sqrt{2\pi}} \displaystyle\int_{-\infty}^{x} \exp\left(-\frac{t^2}{2}\right) dt$

   (b) $\varphi(x) = \dfrac{2}{\pi} \tan^{-1}(x)$

   Explain why both of these functions satisfy the properties of an activation function fitting the requirements of the universal approximation theorem. How do these two activation functions differ from each other?

5. The momentum constant α is normally assigned a positive value in the range $0 \le \alpha < 1$. Justify the fact that α may also be assigned a negative value in the range $-1 < \alpha \le 0$.

6. A time series is created using a discrete *Volterra model* of the form

$$u(n) = \sum_i a_i v(n-i) + \sum_i \sum_j a_{ij} v(n-i)\, v(n-j) + \cdots$$

   where $a_i, a_{ij}, \ldots$ are the Volterra coefficients, the $v(n)$ are samples of a white Gaussian noise sequence, and $u(n)$ is the model output. Using a neural network, construct an implemenation of this Volterra model made up as follows:
   (a) The linear term has coefficients corresponding to $i = 1, 2, 3$.
   (b) The quadratic term has coefficients corresponding to $i, j = 1, 2$.
   (c) The cubic and all higher-order terms are zero.

7. The risk $\mathcal{R}$ defined in Eq. (19.63) has a form similar to that for the minimum-description length (MDL) criterion for stochastic model complexity. Discuss how these criteria are related.

8. Construct an FIR multilayer perceptron equivalent of the TDNN described in Fig. 19.12, in which each synapse consists of a simple FIR filter with a single coefficient and a single delay element.

9. In neural network terminology, a *recurrent network* is a network whose output is a function of both its input samples and past samples of the output. With this definition in mind, which of the networks described in Fig. 19.10 would qualify as a recurrent network?
   Can we refer to the TDNN as a recurrent network? Why?

CHAPTER

20

Radial Basis Function Networks

The training process of a neural network may be viewed as one of *curve fitting*. In particular, we are given a set of data points in the observation space defined by specified values of the input signal and a desired response (target signal), and the requirement is to find an input–output mapping that passes through these points. In a corresponding way, the generalization process may be viewed as one of *interpolation*, in that the network is called upon to express its response to test data never seen before. This viewpoint is exploited in the design of another important type of neural network known as a *radial basis function (RBF) network* (Broomhead and Lowe, 1988). The RBF network is a multilayer feedforward network with a single layer of hidden units which operate as "kernel" nodes. As such, it represents an alternative to the multilayer perceptron. Advantages of RBF networks over multilayer perceptrons trained with the back-propagation algorithm include a more straightforward training process, and a simpler network structure.

Ordinarily, the development of RBF networks is pursued assuming real data and real free parameters. In the study of RBF networks presented in this chapter, we will consider the more general case of *complex RBF networks*, which maintains the precise formulation and elegant structure of complex signals as encountered in radar, sonar, and communication systems (Chen et al., 1994). Naturally, the treatment presented herein includes real RBF networks as a special case.

We begin the discussion by considering the structure of RBF networks, emphasizing the features that distinguish them from multilayer perceptrons.

20.1 STRUCTURE OF RBF NETWORKS

A radial basis-function (RBF) network consists of an input layer of source nodes, a single hidden layer of nonlinear processing units, and an output layer of linear weights, as depicted in Fig. 20.1. Using the outputs computed by the hidden layer in response to an input vector in combination with a desired response presented to the output layer, the weights are trained in a supervised fashion using an appropriate linear filtering method, thereby providing a bridge between linear adaptive filters and neural networks.

RBF networks differ from multilayer perceptrons in the following structural/operational respects:

- RBF networks have a single hidden layer, whereas multilayer perceptrons may have one or more hidden layers.
- In RBF networks, the transfer functions connecting the input layer to the hidden layer are nonlinear and those connecting the hidden layer to the output layer are linear. In multilayer perceptrons, the transfer functions of each hidden layer con-

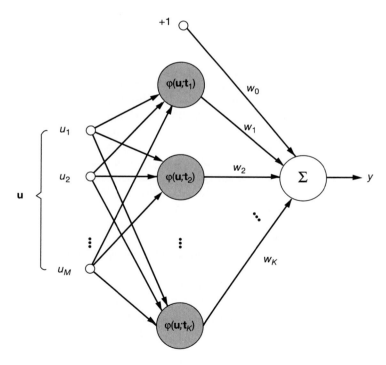

Figure 20.1   RBF network.

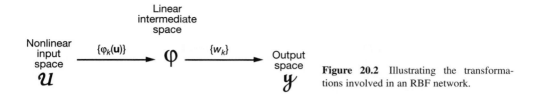

Figure 20.2 Illustrating the transformations involved in an RBF network.

necting it to the previous layer are all nonlinear, and the transfer functions of the output layer may be nonlinear or linear, depending on the application of interest.

- Each hidden unit of an RBF network computes a distance function between the input vector and the center of a *radial basis function* characterizing that particular unit. On the other hand, each neuron of a multilayer perceptron computes the inner product (dot product) of the input vector applied to that neuron and the vector of associated synaptic weights.

RBF networks and multilayer perceptrons do, however, share a common property: they are both universal approximators of the feedforward type. Naturally, they perform their input–output mapping in different ways, as explained later.

Without loss of generality, the RBF network of Fig. 20.1 is shown to have a single output node. Using the terminology of this figure, we may describe the input–output mapping performed by the RBF network as follows:

$$y = \sum_{k=1}^{K} w_k \, \varphi(\mathbf{u}; \mathbf{t}_k) + w_0 \tag{20.1}$$

The term $\varphi(\mathbf{u}; \mathbf{t}_k)$ is the kth radial-basis function (kernel) that computes the "distance" between an input vector $\mathbf{u}$ and its own center $\mathbf{t}_k$; the output signal produced by the kth hidden unit (also referred to as the kernel node) is a *nonlinear function* of that distance. The scaling factor w_k in Eq. (20.1) represents a *complex weight* that connects the kth hidden node to the output node of the network. Finally, the constant term w_0 in Eq. (20.1) represents a *bias* that may be complex.

The input–output mapping performed by the RBF network is accomplished in two stages, as depicted in Fig. 20.2:

- A *nonlinear transformation*, which maps the complex-valued input space $\mathcal{U}$ onto a real-valued intermediate space φ
- A *linear transformation*, which maps the intermediate space φ onto the complex-valued output space $\mathcal{Y}$

The nonlinear transformation is defined by the set of radial-basis functions ϕ_k, and the linear transformation is defined by the set of weights w_k, $k = 1, 2, \ldots, K$.

For the second transformation to be effective, the vector of random variables produced by the hidden layer should desirably represent a "linear process." How to test for the linearity of this process is an open problem.[1]

For the present it suffices to say that "linearization" of the input space is highly likely if the dimensionality of the intermediate space φ (i.e., the number of radial-basis functions) is large enough compared to the dimensionality of the input space. This observation is made in view of an earlier result by Cover (1965) on nonlinear separability of patterns in the context of pattern classification.

20.2 RADIAL-BASIS FUNCTIONS

At the heart of an RBF network is the hidden layer that is defined by a set of *radial-basis functions*, from which the network derives its name. Typical examples of real-valued radial-basis functions are the following (Broomhead and Lowe, 1988; Poggio and Girosi, 1990):

1. *Thin-plate-spline function*:

$$\varphi(r) = \left(\frac{r}{\sigma}\right)^2 \ln\left(\frac{r}{\sigma}\right) \qquad \text{for some } \sigma > 0, \text{ and } r \geq 0 \qquad (20.2)$$

2. *Gaussian function*:

$$\varphi(r) = \exp\left(-\frac{r^2}{2\sigma^2}\right) \qquad \text{for some } \sigma > 0, \text{ and } r \geq 0 \qquad (20.3)$$

Of these two examples, the Gaussian function is the one most commonly used in practice. In the remainder of this chapter, we will confine our discussion to the use of Gaussian functions.

The selection of a radial-basis function for a complex-valued RBF network is in fact the same as that for a real-valued RBF network, with some minor modifications. Specifically, given an input vector $\mathbf{u}$, the kth Gaussian radial-basis function of the RBF network is defined by

$$\varphi(\mathbf{u}; \mathbf{t}_k) = \exp[-(\mathbf{u} - \mathbf{t}_k)^H \mathbf{C}_k(\mathbf{u} - \mathbf{t}_k)], \qquad k = 1, 2, \ldots, K \qquad (20.4)$$

---

[1]Tugnait (1994), building on earlier work by Subba Rao and Gabr (1980, 1984), presents an approach based solely on the bispectrum of the input data to test for linearity of a stationary time series. The stochastic model used for this approach assumes the possible presence of an additive noise component. In the case of an RBF network acting on "noisy" data, the noise appearing at the output of the hidden layer may be of a multiplicative kind due to the highly nonlinear nature of the processing units in that layer. The applicability of Tugnait's method to the hidden layer of an RBF network for noisy input data may therefore be somewhat uncertain.

where the vector $\mathbf{t}_k$ defines the center of the kth radial basis function and the matrix $\mathbf{C}_k$ defines its *width* or *smoothing factor*;[2] the superscript H denotes Hermitian transposition. Using the concept of the *Mahalanobis metric (distance)*, we may rewrite Eq. (20.4) in the compact form

$$\varphi(\mathbf{u}; \mathbf{t}_k) = \exp(-\|\mathbf{u} - \mathbf{t}_k\|^2_{Mc}), \qquad k = 1, 2, \ldots, K \tag{20.5}$$

A simple choice for the matrix $\mathbf{C}_k$ is the diagonal matrix

$$\mathbf{C}_k = \frac{1}{\sigma^2_k} \mathbf{I} \tag{20.6}$$

where $\mathbf{I}$ is the identity matrix. On this basis, we may redefine the kth radial-basis function as follows:

$$\varphi(\mathbf{u}; \mathbf{t}_k) = \exp\left(-\frac{1}{\sigma^2_k}\|\mathbf{u} - \mathbf{t}_k\|^2\right), \qquad k = 1, 2, \ldots, K \tag{20.7}$$

where $\mathbf{t}_k$ is the center, σ_k is the width, and $\|\mathbf{u} - \mathbf{t}_k\|$ denotes the Euclidean distance between $\mathbf{u}$ and $\mathbf{t}_k$. Note that $\varphi(\mathbf{u}; \mathbf{t}_k)$ is *radially symmetric* in the sense that

$$\varphi(\mathbf{u}_i; \mathbf{t}_k) = \varphi(\mathbf{t}_k; \mathbf{u}_i) \qquad \text{for all } i \text{ and } k$$

Thus, substituting Eq. (20.7) in (20.1), we may formulate the input–output mapping realized by a Gaussian RBF network as follows:

$$y = \sum_{k=1}^{K} w_k \exp(-\frac{1}{\sigma^2_k}\|\mathbf{u} - \mathbf{t}_k\|^2) \tag{20.8}$$

From a design point of view, the requirement is to select suitable values for the parameters of each of the K Gaussian radial-basis functions, namely σ_k and $\mathbf{t}_k$, $k = 1, 2, \ldots, K$, and solve for the weights of the output layer. In the sequel, we describe three different procedures for the design of a Gaussian RBF network, each with its own merit.

20.3 FIXED CENTERS SELECTED AT RANDOM

The simplest approach for the design of an RBF network involves selecting a set of *fixed* radial basis functions for the hidden units of the network. In particular, the locations of the centers may be chosen *randomly* from the training data set. Such an approach, first described by Broomhead and Lowe (1988), is considered to be "sensible," since random sampling would distribute the centers according to the probability density function of the

---

[2]From analogy with a multivariate complex Gaussian distribution (Miller, 1974), the vector $\mathbf{t}_k$ and the matrix $\mathbf{C}_k$ in Eq. (20.4) play the roles of a mean vector and the inverse of a covariance matrix. A discussion of complex Gaussian functions is presented in Chapter 2.

training data. This assumes that the training data are distributed in a representative manner for the problem at hand. We may thus write

$$\varphi(\mathbf{u}; \mathbf{t}_k) = \exp\left(-\frac{K}{d_{\max}^2} \|\mathbf{u} - \mathbf{t}_k\|^2\right), \qquad k = 1, 2, \ldots, K \tag{20.9}$$

where K is the number of centers, and $d_{\max}$ is the maximum distance between the chosen centers. In effect, the width σ_k for each Gaussian radial-basis function is fixed at the common value

$$\sigma = \frac{d_{\max}}{\sqrt{K}} \tag{20.10}$$

This formula ensures that the individual functions are not too peaked or too flat; clearly both of these two extremes are to be avoided. Alternatively, we may use individually scaled centers with broader widths in areas of lower data density.

Having fixed the radial-basis functions, we then move on to compute the weights in the output layer of the RBF network. For this computation, we may use the *method of least squares* (described in Chapter 11), which is of a *block (batch) processing* kind. Let the training set be denoted by $\{\mathbf{u}_i, d_i\}$, where $\mathbf{u}_i$ denotes the input vector and d_i denotes the desired response belonging to the ith example, with $i = 1, 2, \ldots, N$. We may then define the following matrix and vector quantities:

$$\Phi = \begin{bmatrix} 1 & \varphi(\mathbf{u}_1; \mathbf{t}_1) & \varphi(\mathbf{u}_1; \mathbf{t}_2) & \cdots & \varphi(\mathbf{u}_1; \mathbf{t}_K) \\ 1 & \varphi(\mathbf{u}_2; \mathbf{t}_1) & \varphi(\mathbf{u}_2; \mathbf{t}_2) & \cdots & \varphi(\mathbf{u}_2; \mathbf{t}_K) \\ \cdot & \cdot & \cdot & \cdot & \cdot \\ \cdot & \cdot & \cdot & \cdot & \cdot \\ \cdot & \cdot & \cdot & \cdot & \cdot \\ 1 & \varphi(\mathbf{u}_N; \mathbf{t}_1) & \varphi(\mathbf{u}_N; \mathbf{t}_2) & \cdots & \varphi(\mathbf{u}_N; \mathbf{t}_K) \end{bmatrix} \tag{20.11}$$

$$\mathbf{d} = [d_1, d_2, \ldots, d_N]^T \tag{20.12}$$

The real-valued matrix Φ, called an *interpolation matrix*, is of size N-by-$(K + 1)$, where N is the number of training examples and K is the number of radial-basis functions; the first column of unity terms is included in this matrix to account for the use of a bias. The *desired response vector $\mathbf{d}$ is of size N-by-1*.

Evaluating Eq. (20.1) for each of the N examples in the training set, we may write

$$y_i = \sum_{k=1}^{K} w_k \, \varphi(\mathbf{u}_i; \mathbf{t}_k) + w_0, \qquad i = 1, 2, \ldots, N \tag{20.13}$$

Using the matrix notation of Eq. (20.11), we may rewrite the set of N equations (20.13) in the compact form:

$$\mathbf{y} = \Phi \, \mathbf{w} \tag{20.14}$$

where $\mathbf{y}$ is the N-by-1 *output vector*:

$$\mathbf{y} = [y_1, y_2, \ldots, y_N]^T \tag{20.15}$$

and $\mathbf{w}$ is the $(K + 1)$-by-1 weight vector:

$$\mathbf{w} = [w_0, w_1, w_2, \ldots, w_K]^T \tag{20.16}$$

According to the definitions of Eqs. (20.11) and (20.16), the bias term may be viewed as a weight w_0 connected to an input φ_0 fixed at $+1$, as indicated in Fig. 20.1.

Suppose that during training, the RBF network output vector is constrained to equal the desired response vector:

$$\mathbf{y} = \mathbf{d} \tag{20.17}$$

We may then rewrite Eq. (20.14) as[3]

$$\mathbf{d} = \mathbf{\Phi} \, \mathbf{w} \tag{20.18}$$

With $N > K$, Eq. (20.18) represents an *overdetermined* system of equations in that we have more equations than unknowns. To solve Eq. (20.18) for the weight vector $\mathbf{w}$, we may use the method of least squares. In particular, a robust solution for $\mathbf{w}$ is provided by the *minimum norm solution*, written as follows (see Eq. (11.13)):

$$\mathbf{w} = \mathbf{\Phi}^+ \tag{20.19}$$

where $\mathbf{\Phi}^+$ is the pseudoinverse of the interpolation matrix $\mathbf{\Phi}$. The recommended procedure for computing the pseudoinverse matrix $\mathbf{\Phi}+$ is to use the method of singular value decomposition (SVD) described in Chapters 11 and 12.

Summarizing, the *method of fixed centers* based on batch (block) processing proceeds as follows:

1. For a specified number of radial-basis factors, K, select the centers (and therefore their widths) randomly from the training data. Hence, using a Gaussian model, define the radial-basis functions in accordance with Eq. (20.9).

2. Use Eq. (20.11) to determine the interpolation matrix $\mathbf{\Phi}$ for the given set of N training examples.

3. Compute the weight vector of the output layer using Eq. (20.19).

As an alternative to the block processing method used to compute the weight vector $\mathbf{w}$, we may use an iterative procedure such as the LMS algorithm described in Chapter 9 or the RLS algorithm described in Chapter 13.

---

[3]For *strict interpolation*, we should have $K + 1 = N$. In this case, $\mathbf{\Phi}$ assumes the form of a square matrix. Equation (20.18) may then be solved for $\mathbf{w}$, as shown by

$$\mathbf{w} = \mathbf{\Phi}^{-1} \, \mathbf{d}$$

where $\mathbf{\Phi}^{-1}$ is the inverse of matrix $\mathbf{\Phi}$. Although, in theory, we are always assured of a solution to the strict interpolation problem, in practice we cannot always solve for $\mathbf{w}$ particularly when the matrix $\mathbf{\Phi}$ is arbitrarily close to singular. Moreover, for large N, the size of the RBF network becomes prohibitively expensive to implement. Both of these problems are overcome by choosing the number of centers K small compared to the size N of the training data, in which case $\mathbf{\Phi}$ assumes the form of a rectangular matrix. In a strict sense, when $K < N$ the matrix $\mathbf{\Phi}$ is no longer an interpolation matrix.

20.4 RECURSIVE HYBRID LEARNING PROCEDURE

The main problem with the use of fixed centers just described for the design of an RBF network is the fact it may require a large dataset for a prescribed level of performance. One way of overcoming this limitation is to use a *hybrid learning procedure*, which combines the following:

- *Self-organized learning algorithm* for the selection of the centers of the radial-basis functions in the hidden layer
- *Supervised learning algorithm* for the computation of the weights in the output layer

Although block (batch) processing can be used to implement these two operations, it is particularly advantageous to take an adaptive (iterative) approach. For example, we may use the k-means clustering algorithm[4] (among others) for the self-organized learning part of the hybrid procedure. As for the supervised learning part, we may use the RLS or LMS algorithm, depending on complexity requirements.

The *k-means clustering algorithm* computes k centers and thereby partitions the input data into k clusters (Duda and Hart, 1973). Specifically, it places the centers of the radial-basis functions in only those regions of the input space $\mathcal{U}$ where significant data are present. Let K denote the number of radial-basis functions; the determination of a suitable value for K may require experimentation. Let $\mathbf{t}_k(n)$, $k = 1, 2, \ldots, K$, denote the centers of the radial-basis functions at iteration n. Then, the k-means clustering algorithm may proceed as follows:[5]

1. *Initialization.* Choose random values for the initial centers $\mathbf{t}_k(0)$; the only restriction here is that the $\mathbf{t}_k(0)$ be different for $k = 1, 2, \ldots, K$. It may also be desirable to keep the Euclidean norm of the centers small.

2. *Sampling.* Draw a sample vector $\mathbf{u}$ from the input space $\mathcal{U}$ with a certain probability. The vector $\mathbf{u}$ represents the input applied to the RBF network.

3. *Similarity matching.* Find the best-matching (winning) center $\tilde{k}(\mathbf{u})$ at iteration n, using the minimum-distance Euclidean criterion:

$$\tilde{k}(\mathbf{u}) = \arg \min_k \|\mathbf{u}(n) - \mathbf{t}_k(n)\|, \qquad k = 1, 2, \ldots, K \tag{20.20}$$

4. *Updating.* Adjust the locations of the centers, using the update rule

$$\mathbf{t}_k(n+1) = \begin{cases} \mathbf{t}_k(n) + \eta[\mathbf{u}(n) - \mathbf{t}_k(n)], & k = \tilde{k}(\mathbf{u}) \\ \mathbf{t}_k(n), & \text{otherwise} \end{cases} \tag{20.21}$$

where η is the *learning-rate parameter* that lies in the range $0 < \eta < 1$.

---

[4]The use of the k-means clustering algorithm for the design of RBF network was first proposed by Moody and Darken (1989). Its use for complex RBF networks is discussed in Chen et al. (1994).

[5]The procedure described herein is a special case of a more general self-organized learning algorithm known as the *self-organizing feature map (SOFM)*, originally developed by Kohonen (1982, 1990).

A limitation of the conventional *k*-means algorithm described above is that it can only achieve a local optimum solution that depends on the initial choice of cluster centers. Consequently, computing resources may be wasted in that some initial centers get stuck in regions of the input space $\mathcal{U}$ with a scarcity of data points and therefore never move to new locations where they are needed. The net result is an unnecessarily large network. To overcome this limitation, Chen (1995) proposes the use of an *enhanced k-means clustering algorithm* due to Chirungrueng and Sequin (1994), which is based on a cluster variation-weighted measure that enables the algorithm to converge to an optimum or near-optimum configuration, independent of the initial center locations.

In any event, having identified the individual centers of the Gaussian radial-basis functions and their common width using the *k*-means algorithm or its enhanced version, we may move onto the output layer. If computational complexity is of no particular concern here, we may use the RLS algorithm or one of its variants to compute the weight vector **w** in the output layer. If, on the other hand, the requirement is to minimize computational complexity, the recommended procedure is to use the LMS algorithm. For a complex RBF network, the complex form of the RLS or LMS algorithm would naturally be used, with one important modification. Specifically, the vector of output signals produced by the hidden layer, which constitutes the input vector for the RLS or LMS algorithm, is *real-valued* in the present scenario. However, the weight vector **w** is complex-valued, since the RBF network is required to produce a complex-valued overall output to approximate the complex-valued desired response. Note also that the *k*-means clustering algorithm for the hidden layer and the RLS or LMS algorithm for the output layer can proceed with their own individual computations concurrently.

The hybrid approach described in this section and the method of fixed centers described in the previous section share a common feature: In both cases, the selection of centers in the hidden layer is decoupled from the design of linear weights in the output layer, which makes a theoretical understanding of what goes on inside the network somewhat difficult. This observation leads us to consider a fully supervised learning procedure, described next.

20.5 STOCHASTIC GRADIENT APPROACH

In the stochastic gradient approach for the design of an RBF network, the centers of the radial-basis functions and all other free parameters of th network undergo a *supervised learning process* (Lowe, 1989). In other words, the RBF network design takes on its more generalized form. A natural candidate for such a process is *error-correction learning*, which is most conveniently implemented using a stochastic gradient descent of the error criterion (Poggio and Girosi, 1990; Kassam and Cha, 1993), and whose basic concept is similar to the LMS algorithm.

The first step in the development of this supervised learning procedure is to define the instantaneous value of the cost function

$$\mathcal{E}(n) = \frac{1}{2} |e(n)|^2, \quad n = 1, 2, \ldots, N \tag{20.22}$$

TABLE 20.1 SUMMARY OF THE STOCHASTIC GRADIENT ALGORITHM FOR THE DESIGN OF RBF NETWORKS USING COMPLEX-VALUED DATA

$$y(n) = \sum_{k=1}^{K} w_k(n)\phi(\mathbf{u}(n); \mathbf{t}_k(n))$$

$$e(n) = d(n) - y(n)$$

$$w_k(n+1) = w_k(n)\ \mu_w e^*(n)\ \varphi(\mathbf{u}(n); \mathbf{t}_k(n))$$

$$\mathbf{t}_k(n+1) = \mathbf{t}_k(n) + 2\mu_t e^*(n)\ w_k(n)\varphi(\mathbf{u}(n); \mathbf{t}_k(n))\frac{\mathbf{u}(n) - \mathbf{t}_k(n)}{\sigma_k^2(n)}$$

$$\sigma_k^2(n+1) = \sigma_k^2(n) + \mu_\sigma\, e^*(n)\, w_k(n)\varphi(\mathbf{u}(n); \mathbf{t}_k(n))\frac{\|\mathbf{u}(n) - \mathbf{t}_k(n)\|^2}{\sigma_k^2(n)}$$

where

$$\varphi(\mathbf{u}(n); \mathbf{t}_k(n)) = \exp\left(-\frac{1}{\sigma_k^2(n)}\|\mathbf{u}(n) - \mathbf{t}_k(n)\|^2\right)$$

where $e(n)$ is the error signal produced in response to the nth example, and N is the total number of examples in the training set. The error signal is defined by

$$e(n) = d(n) - y(n)$$
$$= d(n) - \sum_{k=1}^{K} w_k(n) \exp\left(-\frac{1}{\sigma_k^2(n)}\|\mathbf{u}(n) - \mathbf{t}_k(n)\|^2\right) \tag{20.23}$$

where $\mathbf{t}_k(n)$ is the center and $\sigma_k^2(n)$ is the squared width of the kth radial-basis function for example n, and $w_k(n)$ is the corresponding value of the kth weight in the output layer. The objective is to find the values of these free parameters that minimize $\mathscr{E}(n)$. The results of a stochastic gradient procedure aimed at this minimization are summarized in Table 20.1; the derivations of these results are presented as an exercise to the reader as Problem 3.

The following points are noteworthy in Table 20.1:

1. The cost function $\mathscr{E}(N)$, averaged over the entire set of N training examples, is convex with respect to the weights $w_k(n)$ of the output layer, but nonconvex with respect to the centers $\mathbf{t}_k(n)$ and squared widths $\sigma_k^2(n)$ of the radial-basis functions in the hidden layer; in the latter case, the search for optimality may get stuck at a local minimum of the error-performance surface.

2. The update rules for $\mathbf{t}_k(n)$, $\sigma_k^2(n)$, and $w_k(n)$ are (in general) assigned different learning-rate parameters μ_t, μ_σ, and μ_w, respectively.

3. Unlike the back-propagation algorithm, the stochastic gradient descent for the RBF network described herein does *not* involve the back-propagation of errors.

4. The gradient vector $\partial\mathscr{E}/\partial\mathbf{t}_k$ has an effect similar to a clustering effect that is task-dependent (Poggio and Girosi, 1990).

For the *initialization* of the stochastic gradient algorithm, the free parameters of the RBF network may be assigned a subset of values drawn at random from the training set. In so doing, we are building on an idea described in Section 20.3. In particular, the search in parameter space begins from a *structured* initial condition, in which case the likelihood of converging to an undesirable local minimum on the error-performance surface is reduced.

20.6 UNIVERSAL APPROXIMATION THEOREM (REVISITED)

In the previous chapter we presented a form of the universal approximation theorem that applies directly to multilayer perceptrons. RBF networks, however, differ from multilayer perceptrons, in that the activation functions of their hidden units (i.e., the radial-basis functions) have an argument that is *nonlinearly* dependent on the input vector **u**. Hence, the universal approximation theorem as stated in Chapter 19 is not applicable to RBF networks.

In this section, we consider another form of the *universal approximation theorem* that is directly applicable to RBF networks. This issue, in the context of Gaussian hidden units, was apparently first considered by Brown.[6] Then it was reconsidered independently by Hartman et al. (1990), and in a broader setting by Park and Sandberg (1991).

Let $\mathcal{U}$ be any convex compact subset of $\mathbb{R}^M$. Let $\mathbf{u} \in \mathcal{U}$, and $\mathbf{t}_k \in \mathcal{U}$ for $k = 1, 2, \ldots, K$. Consider then a two-parameter family $\mathcal{F}$ of restricted Gaussian functions for real-valued data:

$$\varphi(\mathbf{u}; \mathbf{t}_k) = \exp\left(\frac{-\|\mathbf{u} - \mathbf{t}_k\|^2}{2\sigma_k^2}\right) \qquad \sigma_k > 0 \qquad (20.24)$$

Let $\mathcal{L}$ be the set of all finite linear combinations of elements (with real coefficients) drawn from $\mathcal{F}$. Then, we may state the following theorem (Hartman et al., 1990; Park and Sandberg, 1991):

Any function in the algebra $C(\mathcal{U})$ of all continuous functions on $\mathcal{U}$ with the supremum norm can be uniformly approximated to an arbitrary accuracy by elements of $\mathcal{L}$.

In other words, RBF networks with Gaussian hidden units are universal approximators. The universal approximation theorem as stated herein is formulated in the context of real RBF networks. Its extension to complex RBF networks is intuitively obvious.

In light of the version of the universal approximation theorem stated here and the version of it stated in the previous chapter, we may now justifiably say that RBF networks and multilayer perceptrons are both universal approximators. Accordingly, it is not sur-

---

[6]In an appendix to a chapter contribution by Powell (1992), based on a lecture presented in 1990, credit is given to a result due to A. L. Brown. The result, apparently obtained in 1981, states that an RBF network can map an arbitrary function from a *closed* domain in $\mathbb{R}^M$ to $\mathbb{R}$.

prising to find that there always exists an RBF network capable of accurately mimicking a specified multilayer perceptron, or vice versa. However, these two neural networks perform their individual approximation tasks in entirely different ways. Multilayer perceptrons construct *global* approximations to nonlinear input–output mapping. Consequently, they are capable of extrapolation in regions of the input space where there is a scarcity of training data. In contrast, RBF networks construct *local* approximations to nonlinear input–output mapping, with the result that these networks are capable of fast learning and reduced sensitivity to the order of presentation of training data. In many cases, however, we find that in order to represent a mapping to some desired degree of smoothness, the number of radial-basis functions required to span the input space adequately may have to be very large. This problem, largely due to the number of available data points, becomes particularly acute in trying to solve large-scale problems such as image processing and speech recognition.

20.7 FILTERING APPLICATIONS

In light of the universal approximation theorem just discussed, it is apparent that many, if not all, of the signal processing applications that befit the use of multilayer perceptrons would befit RBF networks just as well, and vice versa. System identification and target detection, considered as possible applications of multilayer perceptrons in Section 19.7, may be equally served by means of RBF networks. By the same token, the first application of RBF networks selected for discussion in this section, namely, adaptive equalization, qualifies equally well for the use of multilayer perceptrons.[7]

Adaptive Equalization

In the conventional form of an adaptive equalizer, based on the linear adaptive filter theory presented in Part III of the book, the equalizer operates as an inverse model. In particular, in the absence of noise and in the case of a minimum-phase channel, the cascade combination of the channel and the equalizer provides distortionless transmission. When, however, noise is present and/or the channel is nonminimum phase, the use of an inverse-model is no longer optimum.

An alternative viewpoint to that of inverse modeling is to approach the equalization process as a *pattern classification* problem (Theodoridis et al., 1992). For the simple case of bipolar data transmission, the received samples, corrupted by intersymbol interference (ISI) and noise, would have to be classified as -1 and $+1$. The equalizer now has the function of assigning each received sample to the correct decision region. According to this

---

[7]For the application of multilayer perceptrons (trained with the back-propagation algorithm) to simultaneous equalization and decoding of severe intersymbol interference due to data transmission over a communication channel, see Al-Mashouq and Reed (1994); the experimental results presented in that paper show a substantial improvement over conventional methods for equalization and decoding.

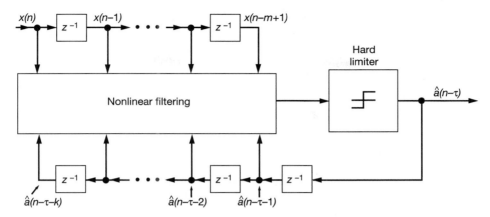

Figure 20.3   Block diagram of decision-feedback equalizer using a nonlinear filter.

viewpoint, a linear equalizer is equivalent to a linear pattern classifier. However, in realistic situations where noise is present in the received signal and/or the channel is nonminimum phase, the optimum pattern classifier is in fact nonlinear (Gibson and Cowan, 1990), which may therefore benefit from the use of a neural network.

In order to see how the design of a nonlinear adaptive equalizer can be based on neural networks, consider first the block diagram of Fig. 20.3, which depicts a *nonlinear decision-feedback equalizer*. The equalizer consists of a nonlinear filter with feedforward inputs denoted by $u(n)$, $u(n-1)$, . . . , $u(n - M)$ and feedback inputs denoted by $a(n - \tau - 1)$, $a(n - \tau - 2)$, . . . , $a(n - \tau - k)$, where $u(n)$ is the channel output (received signal) at time n, and $a(n - \tau)$ is the equalizer output representing an estimate of the transmitted symbol $a(n)$ delayed by τ seconds. The equalizer output is produced by passing the output of a nonlinear filter inside the equalizer through a hard limiter, whereby decisions are made on a symbol-by-symbol basis. Using Bayesian considerations, Chen et al. (1992a, b, 1994), have shown that the optimum form of the nonlinear filter in the decision-feedback equalizer of Fig. 20.3 has an identical structure to that of the RBF network.

Chen et al. have also used computer simulation to investigate the performance of an RBF decision-feedback equalizer and compared it with that of (1) a standard decision-feedback equalizer using a transversal filter, and (2) a maximum-likelihood sequential estimator known as the Viterbi algorithm. The investigations were carried out for both stationary and nonstationary channels; the highly nonstationary channel considered in the study was chosen to be representative of a mobile radio environment. The results of the investigations reported by Chen et al. may be summarized as follows:

- The maximum-likelihood sequential estimator provides the best attainable performance for the case of a stationary channel; the corresponding performance of the RBF decision-feedback equalizer is worse by about 2 dB, but better than that of the standard decision-feedback equalizer by roughly an equal amount.

- In the case of a highly nonstationary channel, the RBF decision-feedback equalizer outperforms the maximum-likelihood sequential estimator; in the latter case, the degradation in performance is attributed to the accumulation of errors.

The results of the study of Chen et al. appear to show that the RBF decision-feedback equalizer is robust, and that it provides a viable solution for the equalization of highly nonstationary communication channels.

Dynamic RBF Network for Time Series Prediction

For the second application, we consider a simplified implementation of the *dynamic Gaussian RBF network* used as a nonlinear predictor, in which learning takes place on a *continuous* basis, hence the description of the network as "dynamic." There are three key parameters of the network to be determined: The *centers*, the common *width*, and the linear *weighting coefficients*. Collectively, these parameters completely define the prediction of an input sample $u(n + 1)$, which is computed at time step n as

$$\hat{u}(n + 1) = F(\mathbf{u}(n))$$

$$= \sum_{k=1}^{K} w_k(n) \exp\left(-\frac{1}{2\sigma^2(n)} \|\mathbf{u}(n) - \mathbf{t}_{n-k}\|^2\right) \tag{20.25}$$

where it is assumed that the input data are real valued. Specifically, the input vector $\mathbf{u}(n) = [u(n), u(n - 1), \ldots, u(n - M + 1)]^T$ is available for processing at time step n, where M is the prediction order. The network parameters are described as follows:

1. The centers $\mathbf{t}_{n-k}$, $k = 1, 2, \ldots, K$, constitute the set of *process-state vectors*.
2. The width $\sigma(n)$ is typically computed as a function of the empirical covariance of the time series data, and is common to all centers; this forces the interpolation matrix $\mathbf{\Phi}(n)$ to be symmetric, thereby improving numerical stability.
3. The coefficient vector $\mathbf{w}(n) = [w_1(n), w_2(n), \ldots, w_K(n)]^T$ satisfies the *strict interpolation (SI) condition:*

$$[\mathbf{\Phi}(n) + \lambda(n)\,\mathbf{I}]\,\mathbf{w}(n) = \mathbf{d}(n) \tag{20.26}$$

where the interpolation matrix $\mathbf{\Phi}(n)$ is defined by

$$\mathbf{\Phi}(n) = \left\{\exp\left(-\frac{1}{2\sigma^2(n)} \|\mathbf{u}_{n-i} - \mathbf{u}_{n-k}\|^2\right)\right\}, \qquad (i,k) = 1, 2, \ldots, K \quad (20.27)$$

and $\lambda(n)$ is the *regularization parameter* at time step n, typically estimated from a window of the available time series data or fixed *a priori*. The desired response vector $\mathbf{d}(n)$ is defined by

$$\mathbf{d}(n) = [u(n), u(n - 1), \ldots, u(n - K + 1)]^T \tag{20.28}$$

From this formulation, we see that as new time series data become available, the dynamic Gaussian RBF predictor naturally evolves from time step k to $k + 1$ via a shift in the predictor centres and the subsequent recomputation of $\mathbf{\Phi}(n)$, $\sigma(n)$, $\lambda(n)$, and $\mathbf{w}(n)$. The recomputation of $\mathbf{w}(n)$ is $O(K^3)$ in general but it can be shown that if $\lambda(n)$ is fixed and $\mathbf{\Phi}(n)$ is only partially updated by a low-rank matrix, then the complexity can be reduced to $O(K^2)$. For further discussion on this reduced complexity algorithm and its experimental effects, the reader is referred to Yee and Haykin (1995).

We shall illustrate the essential characteristics of the network by way of a nonlinear time series prediction experiment. This experiment involves the prediction of a reasonably noise-free male speech signal represented by a total of 10,000 samples at 8kHz and 8 bits per sample. For the purposes of the experiment, the speech signal is shifted to zero mean and scaled to unit total amplitude for both training and testing.

Intuitively, we would expect that in the case of a (strictly) stationary time series, there would exist an optimum set of dynamic Gaussian RBF predictor parameters that (at least, in principle) could be learned by some appropriate means as in the case of the LMS algorithm for linear processes. Where the interest lies, however, is in the prediction of a nonstationary, nonlinear process. Here the current state of the art revolves around the use of local *linear* approximations to the process state-space mapping, leading to algorithms such as the *extended Kalman Filter (EKF)* and its generalized counterparts. The *dynamic* component of the Gaussian RBF predictor extends this idea to a series of local *nonlinear* approximations to the process state-space mapping via continuously updated Gaussian RBF curve fits over the most currently available windows of time series data. Again, we may expect this dynamic updating to yield improved tracking for a significantly nonstationary process. Indeed, in Fig. 20.4, we compare the performance of a static RBF predictor to a dynamic one over the speech signal. By "static Gaussian RBF predictor", we mean a Gaussian RBF predictor with $K = 250$ centers trained once with a given initial window of time series data and then frozen thereafter. In contrast, the dynamic predictor with $K = 100$ centers has its parameters updated once per time step according to the scheme previously outlined. Both predictors use a state-space order of $M = 50$ and a fixed regularization parameter $\lambda = 0.01$. The plot in Fig. 20.4(a) clearly demonstrates how the static predictor tracks well over those segments of the speech signal that are similar to the initial segment (unvoiced speech) upon which it was trained, but is unable to track the middle segment of more quickly varying speech (voiced speech). On the other hand, the plot in Fig. 20.4(b) shows that the dynamic predictor, despite having fewer than half the number of centers of the static predictor, is able to adapt to and maintain tracking over all of the speech signal.

The use of regularization in the solution of interpolation and approximation problems is well-established (Morozov, 1993). Roughly speaking, regularization stabilizes the solution of *ill-posed* problems (of which nonparametric curve fitting is one), in the sense that it can make the solution less sensitive to noise and errors in the given data. As a simple example, the interpolation matrix $\mathbf{\Phi}(n)$ for the dynamic Gaussian RBF predictor should be, in principle, nonsingular for any distinct choice of centers drawn from the time series data; in practice, however, we observe that the likelihood of ill-conditioning in $\mathbf{\Phi}(n)$

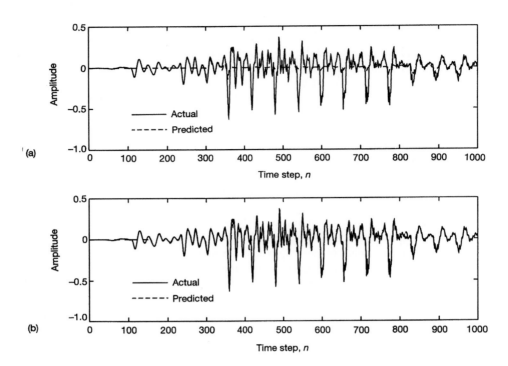

Figure 20.4   Tracking ability of (a) static nonlinear predictor versus (b) dynamic nonlinear predictor.

increases as the size K increases. The stabilizing effect of regularization on the dynamic predictor in this context can be seen in Fig. 20.5, where a 100-point segment of the speech signal shown along with two regularized dynamic Gaussian RBF predictors with $K = 100$ centers of state-space order $M = 2$. The essentially "nonregularized" predictor, which uses only a minimal regularization parameter $\lambda = 0.01$ to avoid singularity in the interpolation matrix, exhibits numerical instability near time step $n = 50$; on the other hand, the regularized predictor with $\lambda = 0.1$ has no such difficulty. Note also that even where the nonregularized predictor achieves some degree of tracking, the regularized one tracks the speech signal more closely.

As a final note, we should mention that when compared with the pseudo-linear adaptive predictor specified in CCITT Recommendation G.726 operating at 32kbits/s, a dynamic Gaussian RBF predictor with K = 100 centers shows a nontrivial improvement of approximately 4 dB in prediction SNR, as measured both segmentally and completely over the entire speech signal (Yee and Haykin, 1995). This result suggests that nonlinear

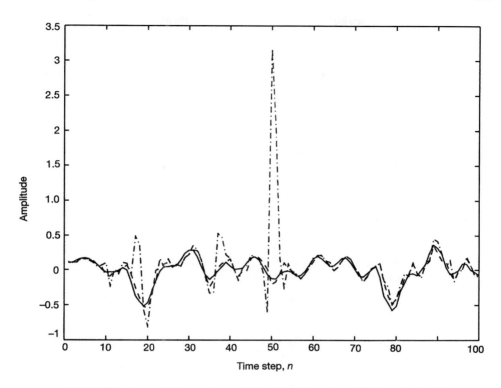

Figure 20.5   Regularized versus nonregularized predictor, $K = 100$, $M = 2$: ('—' is actual, '- - -' is nonlinear prediction for regularization parameter $\lambda = 0.1$, and '-.-.-' is nonlinear prediction for $\lambda = 0.01$)

methods, as opposed to the current linear ones, of predicting speech can yield yet better performance than previously thought possible.

20.8 SUMMARY AND DISCUSSION

The structure of an RBF network is unusual in that the constitution of its hidden units is entirely different from that of its output units. The theory of RBF networks is linked intimately with that of radial-basis functions, which is nowadays one of the main fields of study in numerical analysis (Singh, 1992). Another interesting point to note is that with linear weights of the output layer providing a set of adjustable parameters, much can be gained by studying the well-developed theory of linear adaptive filters presented in Part III of the book.

RBF networks have a rich theory of their own, as summarized here:

- With appropriate extensions, RBF networks are an important case in *regularization theory* (Poggio and Girosi, 1990). In the context of approximation problems, the basic idea of regularization is to *stabilize* the solution by means of some auxiliary nonnegative functional that embeds prior knowledge (e.g., smoothness constraints on the input–output mapping), and thereby makes an ill-posed problem into a well-posed one.

- For function approximation, an RBF network has been shown (under certain conditions) to provide the minimum variance approximation to a function when the input data are corrupted by additive noise (Webb, 1994). In this case, the nonlinear activation functions are determined by the probability density function of the additive noise in the input signal. The necessary conditions are that the standard deviation of the noise be large compared to the sample spacing of the data points. This form of an RBF network solution for a finite number of training examples is not imposed *a priori*; rather it follows naturally as a direct consequence of a least-squares approach to the problem of function approximation and generalization.

- The input–output mapping of a Gaussian RBF network, described by Eq. (20.8), is closely related to *mixture models*, that is, mixtures of Gaussian distributions. Mixtures of probability distributions, in particular, Gaussian distributions, have been used extensively as models in a wide variety of applications where the data of interest arise from two or more populations mixed together in some varying properties (Tetterington et al., 1985; McLachlan and Basford, 1988).

- RBF networks are closely related to *kernel-based methods*, on which there is a large amount of literature (Duda and Hart, 1973; Davijver and Kittler, 1982; Fukuraga, 1990). A *kernel* is a function $K(\mathbf{x}, \mathbf{x}_j)$ that attains its maximum value at the point $\mathbf{x} = \mathbf{x}_j$ and decreases monotonically as the distance between the vectors $\mathbf{x}$ and $\mathbf{x}_j$ increases, subject to the condition that

$$\int K(\mathbf{x}, \mathbf{x}_j)\, d\mathbf{x} = 1 \qquad \text{for all } \mathbf{x}_j$$

The kernel function $K(\mathbf{x}, \mathbf{x}_j)$ provides information about the unknown conditional probability density function $f(\mathbf{x}|\omega_i)$ of a p-dimensional vector (query point) $\mathbf{x}$, given that a particular class ω_i is true; the information is built up by using a set of training examples $\mathbf{x}_j$, $j = 1, 2, \ldots, N$, that belong to class ω_i. In the well-known *Parzen density estimator* (Parzen, 1962), a Gaussian kernel of fixed width σ is commonly used, and an estimate of the unknown probability density function is given by

$$\hat{f}(\mathbf{x}|\omega_i) = \frac{1}{N} \sum_{j=1}^{N} K(\mathbf{x}, \mathbf{x}_j)$$

where (assuming real-valued data)

$$K(\mathbf{x}, \mathbf{x}_j) = \frac{1}{(2\pi\sigma^2)^{p/2}} \exp\left(-\frac{1}{2\sigma^2}(\mathbf{x} - \mathbf{x}_j)^T(\mathbf{x}-\mathbf{x}_j)\right)$$

The estimate $\hat{f}(\mathbf{x}|\omega_i)$ is simply the sum of a number of multivariate Gaussian distributions that are centered on the particular set of training examples. The sum so defined can approximate any smooth density function. An important advantage of the Parzen density estimator is the fact that it is probably *consistent*, which means that the estimate $\hat{f}(\mathbf{x}|\omega_i)$ approaches the optimal Bayes estimate for class ω_i as the size N of the training set for that class approaches infinity. However, it has the disadvantage of requiring a very large amount of storage. The *probabilistic neural network* described in Specht (1990) is an implementation of the Parzen window estimator.

Perhaps the most interesting aspect of RBF networks is that the basic expansion described in Eq. (20.1) with the output y viewed as a probability density function over a space of real numbers, the $\phi(\mathbf{u}; \mathbf{t}_k)$ viewed as a set of expansion functions, and the w_k viewed as the corresponding amplitudes, may be given a *neurobiological interpretation* (Anderson and Van Essen, 1994). According to such an interpretation, the amplitudes are represented physically as neuronal firing rates, while the functions forming the mathematical basis of the representation provide a convenient set of rules by which the amplitudes are manipulated.

In conclusion, multilayer perceptrons (trained with the back-propagation algorithm) and RBF networks constitute the backbone of supervised neural networks in their own individual ways. They are both examples of multilayer feedforward networks that are universal approximators. The basic difference between them is that Gaussian RBF networks provide local approximation, whereas multilayer perceptrons provide global approximation. This, in turn, means that for the approximation of a nonlinear input–output mapping, the multilayer perceptron may be more *parimoneous* (i.e., require a smaller number of scalar coefficients) than the RBF network for a prescribed degree of accuracy. On the other hand, when continuous learning is required as in the tracking of a time-varying environment, the use of nested nonlinearities in a multilayer perceptron makes it difficult to evolve the network in a dynamic fashion. That is, if we want to include a new example in the training set or enlarge the multilayer perceptron by adding new synaptic weights, then the whole network must be retrained all over again. In contrast, the structure of an RBF network permits it to operate dynamically, such that the centers of the radial-basis functions in the hidden layer and the linear weights of the output layer may be updated without having to recompute them from scratch (Yee and Haykin, 1995).

PROBLEMS

1. It may be argued, by virtue of the central limit theorem, that for a Gaussian RBF network the output produced by the network in response to a random input vector may be approximated by a Gaussian distribution, and that the approximation gets better as the number of centers in the network is increased. Rationalize the validity of this statement.

2. In describing the recursive hybrid learning procedure for the design of a complex RBF network presented in Section 20.5, we mentioned the RLS algorithm and the LMS algorithm as possible candidates for computing the weight vector of the output layer. Formulate the algorithm for performing this computation, using:

(a) The RLS algorithm

(b) The LMS algorithm

3. Table 20.1 presents a summary of the stochastic gradient algorithm for computing the centers and widths of the Gaussian hidden units and the linear weights in the output layer of a complex RBF network. Present detailed derivations of the results summarized in Table 20.1.

4. A requirement exists for the design of an RBF network to perform interference cancelation, in which the reference signal and interference are nonlinearly correlated. Describe how this requirement can be achieved.

5. Investigate the possible use of a Gaussian RBF network to perform the blind equalization of a nonminimum-phase communication channel.

6. A *normalized* Gaussian basis function is defined in Moody and Darken (1989) as follows:

$$\varphi_i = \frac{\exp\left(-\dfrac{1}{\sigma_i^2}\|\mathbf{u} - \mathbf{t}_i\|^2\right)}{\displaystyle\sum_{k=1}^{K} \exp\left(-\dfrac{1}{\sigma_k^2}\|\mathbf{u} - \mathbf{t}_k\|^2\right)}, \qquad i = 1, 2, \ldots, K$$

(a) On this basis, φ_i may be viewed as the probability that the hidden neuron with center $\mathbf{t}_i$ is the "winning" neuron (i.e., the neuron closest to the input vector $\mathbf{u}$ in Euclidean norm). Explain the rationale for this statement.

(b) In what way is the φ_i for $i = 1$ or $i = K$ different from other values of φ_i?

APPENDIX

A

Complex Variables

This Appendix presents a brief review of the functional theory of complex variables. In the context of the material considered in this book, a complex variable of interest is the variable z associated with the z-transform. We begin the review by defining analytic functions of a complex variable, and then derive the important theorems that make up the important subject of complex variables[1].

A.1 CAUCHY–REIMANN EQUATIONS

Consider a complex variable z defined by

$$z = x + jy$$

where $x = \text{Re}[z]$, and $y = \text{Im}[z]$. We speak of the plane in which the complex variable z is represented as the *z-plane*. Let $f(z)$ denote a *function of the complex variable z*, written as

$$w = f(z) = u + jv$$

The function $w = f(z)$ is *single-valued* if there is only one value of w for each z in a given region of the z-plane. If, on the other hand, more than one value of w corresponds to z, the function $w = f(z)$ is said to be *multiple-valued*.

---

[1]For a detailed treatment of the functional theory of complex variables, see Guillemin (1949), Levinson and Redheffer (1970), and Wylie and Barrett (1982).

875

We say that a point $z = x + jy$ in the z-plane approaches a fixed point $z_0 = x_0 + jy_0$ if $x \to x_0$ and $y \to y_0$. Let $f(z)$ denote a single-valued function of z that is defined in some neighborhood of the point $z = z_0$. The *neighborhood* of z_0 refers to the set of all points in a sufficiently small circular region centered at z_0. Let

$$\lim_{z \to z_0} f(z) = w_0$$

In particular, if $f(z_0) = w_0$, then the function $f(z)$ is said to be *continuous* at $z = z_0$.

Let $f(z)$ be written in terms of its real and imaginary parts as

$$f(z) = u(x, y) + jv(x, y)$$

Then, if $f(z)$ is continuous at $z_0 = x_0 + jy_0$, its real and imaginary parts $u(x, y)$ and $v(x, y)$ are continuous functions at (x_0, y_0), and vice versa.

Let $w = f(z)$ be continuous at each point of some region of interest in the z-plane. The complex quantities w and z may then be represented on separate planes of their own, referred to as the w- and z-planes, respectively. In particular, a point (x, y) in the z-plane corresponds to a point (u, v) in the w-plane by virtue of the relationship $w = f(z)$.

Consider an incremental change Δz such that the point $z_0 + \Delta z$ may lie anywhere in the neighborhood of z_0, and throughout which the function $f(z)$ is defined. We may then define the *derivative* of $f(z)$ with respect to z at $z = z_0$ as

$$f'(z_0) = \lim_{\Delta z \to 0} \frac{f(z_0 + \Delta z) - f(z_0)}{\Delta z} \tag{A.1}$$

Clearly, for the derivative $f'(z_0)$ to have a unique value, the limit in Eq. (A.1) must be independent of the way in which Δz approaches zero.

For a function $f(z)$ to have a unique derivative at some point $z = x + jy$, it is necessary that its real and imaginary parts satisfy certain conditions, as shown next. Let

$$w = f(z) = u(x, y) + jv(x, y)$$

With $\Delta w = \Delta u + j\Delta v$ and $\Delta z = \Delta x + j\Delta y$, we may write

$$
\begin{aligned}
f'(z) &= \lim_{\Delta z \to 0} \frac{\Delta w}{\Delta z} \\
&= \lim_{\substack{\Delta x \to 0 \\ \Delta y \to 0}} \frac{\Delta u + j\Delta v}{\Delta x + j\Delta y}
\end{aligned}
\tag{A.2}
$$

Suppose that we let $\Delta z \to 0$ by first letting $\Delta y \to 0$ and then $\Delta x \to 0$, in which case Δz is purely real. We then deduce from Eq. (A.2) that

$$
\begin{aligned}
f'(z) &= \lim_{\Delta x \to 0} \frac{\Delta u}{\Delta x} + j \frac{\Delta v}{\Delta x} \\
&= \frac{\partial u}{\partial x} + j \frac{\partial v}{\partial x}
\end{aligned}
\tag{A.3}
$$

Suppose next that we let $\Delta z \to 0$ by first letting $\Delta x \to 0$ and then $\Delta y \to 0$, in which case Δz is purely imaginary. This time we deduce from Eq. (A.2) that

$$f'(z) = \lim_{\Delta y \to 0} \frac{\Delta v}{\Delta y} - j \frac{\Delta u}{\Delta y}$$

$$= \frac{\partial v}{\partial y} - j \frac{\partial u}{\partial y} \tag{A.4}$$

If the derivative $f'(z)$ is to exist, it is necessary that the two expressions in Eqs. (A.3) and (A.4) be one and the same. Hence, we require

$$\frac{\partial u}{\partial x} + j \frac{\partial v}{\partial x} = \frac{\partial v}{\partial y} - \frac{\partial u}{\partial y}$$

Accordingly, equating real and imaginary parts, we get the following pair of relations, respectively:

$$\frac{\partial u}{\partial x} = \frac{\partial v}{\partial y} \tag{A.5}$$

$$\frac{\partial v}{\partial x} = -\frac{\partial u}{\partial y} \tag{A.6}$$

Equations (A.5) and (A.6), known as the *Cauchy–Riemann equations*, were derived from a consideration of merely two of the infinitely many ways in which Δz can approach zero. For $\Delta w/\Delta z$ evaluated along these other paths to also approach $f'(z)$, we need only make the additional requirement that the partial derivatives in Eqs. (A.5) and (A.6) are continuous at the point (x, y). In other words, provided that the real part $u(x, y)$ and the imaginary part $v(x, y)$ together with their first partial derivatives are continuous at the point (x, y), the Cauchy–Riemann equations are not only necessary but also sufficient for the existence of a derivative of the complex function $w = u(x, y) + jv(x, y)$ at the point (x, y).

A function $f(z)$ is said to be *analytic*, or *homomorphic*, at some point $z = z_0$ in the z-plane if it has a derivative at $z = z_0$ and at every point in the neighborhood of z_0; the point z_0 is called a *regular point* of the function $f(z)$. If the function $f(z)$ is *not* analytic at a point z_0, but if every neighborhood of z_0 contains points at which $f(z)$ is analytic, the point z_0 is referred to as a *singular point* of $f(z)$.

A.2 CAUCHY'S INTEGRAL FORMULA

Let $f(z)$ be any continuous function of the complex variable z, analytic or otherwise. Let $\mathscr{C}$ be a sectionally smooth path joining the points $A = z_0$ and $B = z_n$ in the z plane. Suppose that the path $\mathscr{C}$ is divided into n segments Δs_k by the points z_k, $k = 1, 2, \ldots, n - 1$, as illustrated in Fig. A.1. This figure also shows an arbitrary point ζ_k on segment Δs_k, depicted as an elementary arc of length Δz_k. Consider then the summation $\Sigma_{k=1}^{n} f(\zeta_k)\, \Delta z_k$.

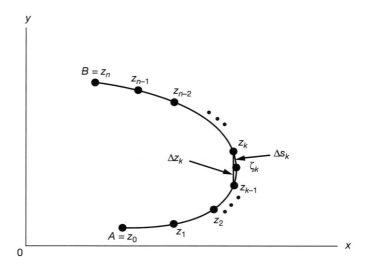

Figure A.1 Sectionally smooth path.

The *line integral* of $f(z)$ along the path $\mathscr{C}$ is defined by the limiting value of this summation as the number n of segments is allowed to increase indefinitely in such a way that Δz_k approaches zero. That is

$$\oint_{\mathscr{C}} f(z) \, dz = \lim_{n \to \infty} \sum_{k=1}^{n} f(\zeta_k) \, \Delta z_k \tag{A.7}$$

In the special case when the points A and B coincide and $\mathscr{C}$ is a closed curve, the integral in Eq. (A.7) is referred to as a *contour integral* that is written as $\oint_{\mathscr{C}} f(z) dz$. Note that, according to the notation described herein, the contour $\mathscr{C}$ is transversed in a *counterclockwise direction*.

Let $f(z)$ be an analytic function in a given region R, and let the derivative $f'(z)$ be continuous there. The line integral $\oint_{\mathscr{C}} f(z) dz$ is then independent of the path $\mathscr{C}$ that joins any pair of points in the region R. If the path $\mathscr{C}$ is a closed curve, the value of this integral is zero. We thus have *Cauchy's integral theorem*, stated as follows:

If a function $f(z)$ is analytic throughout a region R, then the contour integral of $f(z)$ along any closed path $\mathscr{C}$ lying inside the region R is zero, as shown by

$$\oint_{\mathscr{C}} f(z) \, dz = 0 \tag{A.8}$$

This theorem is of cardinal importance in the study of analytic functions.

An important consequence of Cauchy's theorem is known as *Cauchy's integral formula*. Let $f(z)$ be analytic within and on the boundary $\mathscr{C}$ of a simple connected region. Let z_0 be any point in the interior of $\mathscr{C}$. Then Cauchy's integral formula states that

$$f(z_0) = \frac{1}{2\pi j} \oint_{\mathscr{C}} \frac{f(z)}{z - z_0} \, dz \tag{A.9}$$

where the contour integration around $\mathscr{C}$ is taken in the counterclockwise direction.

Cauchy's integral formula expresses the value of the analytic function $f(z)$ at an interior point z_0 of $\mathscr{C}$ in terms of its values on the boundary of $\mathscr{C}$. Using this formula, it is a straightforward matter to express the derivative of $f(z)$ of all orders as follows:

$$f^{(n)}(z_0) = \frac{n!}{2\pi j} \oint_{\mathscr{C}} \frac{f(z)}{(z - z_0)^{n+1}} \, dz \tag{A.10}$$

where $f^{(n)}(z_0)$ is the nth derivative of $f(z)$ evaluated at $z = z_0$. Equation (A.10) is obtained by repeated differentiation of Eq. (A.9) with respect to z_0.

Cauchy's Inequality

Let the contour $\mathscr{C}$ consist of a circle of radius r and center z_0. Then, using Eq. (A.10) to evaluate the magnitude of $f^{(n)}(z_0)$, we may write

$$|f^{(n)}(z_0)| = \frac{n!}{2\pi} \left| \oint_{\mathscr{C}} \frac{f(z)}{(z - z_0)^{n+1}} \, dz \right|$$

$$\leq \frac{n!}{2\pi} \oint_{\mathscr{C}} \frac{|f(z)|}{|z - z_0|^{n+1}} \, |dz|$$

$$\leq \frac{n!}{2\pi} \frac{M}{r^{n+1}} \oint_{\mathscr{C}} |dz|$$

$$= \frac{n!}{2\pi} \frac{M}{r^{n+1}} 2\pi r$$

$$= n! \frac{M}{r^n} \tag{A.11}$$

where M is the maximum value of $f(z)$ on $\mathscr{C}$. The inequality of (A.11) is known as *Cauchy's inequality*.

A.3 LAURENT'S SERIES

Let the function $f(z)$ be analytic in the annular region of Fig. A.2, including the boundary of the region. The annular region consists of two concentric circles $\mathscr{C}_1$ and $\mathscr{C}_2$, whose common center is z_0. Let the point $z = z_0 + h$ be located inside the annular region as depicted in Fig. A.2. According to *Lauren's series*, we have

$$f(z_0 + h) = \sum_{k=-\infty}^{\infty} a_k h^k \tag{A.12}$$

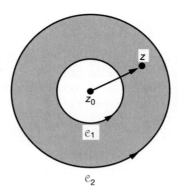

Figure A.2   Annular region.

where the coefficients a_k for varying k are given by

$$
a_k = \begin{cases} \dfrac{1}{2\pi j} \displaystyle\oint_{\mathscr{C}_2} \dfrac{f(z)\, dz}{(z - z_0)^{k+1}}, & k = 0, 1, 2, \ldots \\[3mm] \dfrac{1}{2\pi j} \displaystyle\oint_{\mathscr{C}_1} \dfrac{f(z)\, dz}{(z - z_0)^{k+1}}, & k = -1, -2, \ldots \end{cases}
\tag{A.13}
$$

Note that we may also express the Laurent expansion of $f(z)$ around the point z as

$$
f(z) = \sum_{k=-\infty}^{\infty} a_k (z - z_0)^k
\tag{A.14}
$$

When all the coefficients of negative index have the value zero, then Eq. (A.14) reduces to *Taylor's series*:

$$
f(z) = \sum_{k=0}^{\infty} a_k (z - z_0)^k
\tag{A.15}
$$

In light of Eq. (A.10) and the first line of Eq. (A.13), we may define the coefficient a_k as

$$
a_k = \frac{f^{(k)}(z_0)}{k!}, \qquad k = 0, 1, 2, \ldots
\tag{A.16}
$$

Taylor's series provides the basis of Liouville's theorem, considered next.

Liouville's Theorem

Let a function $f(z)$ of the complex variable z be bounded and analytic for all values of z. Then, according to *Liouville's theorem*, $f(z)$ is simply a constant.

To prove this theorem, we first note that since $f(z)$ is analytic everywhere inside the z-plane, we may use Taylor's series to expand $f(z)$ about the origin:

$$
f(z) = \sum_{k=0}^{\infty} \frac{f^{(k)}(0)}{k!} z^k
\tag{A.17}
$$

The power series of Eq. (A.17) is convergent, and therefore provides a valid representation of $f(z)$. Let contour $\mathscr{C}$ consist of a circle of radius r and origin as center. Then, invoking Cauchy's inequality of (A.11), we may write

$$\left|f^{(k)}(0)\right| \leq \frac{k!\, M_c}{r^k} \tag{A.18}$$

where M_c is the maximum value of $f(z)$ on $\mathscr{C}$. Correspondingly, the value of the kth coefficient in the power series expansion of Eq. (A.17) is bounded as

$$|a_k| = \frac{\left|f^{(k)}(0)\right|}{k!} \leq \frac{M_c}{r^k} \leq \frac{M}{r^k} \tag{A.19}$$

where M is the bound on $|f(z)|$ for all values of z. Since, by hypothesis, M does exist, it follows from (A.19) that for an arbitrarily large r:

$$a_k = \begin{cases} f(0), & k=0 \\ 0, & k = 1, 2, \ldots \end{cases} \tag{A.20}$$

Accordingly, Eq. (A.17) reduces to

$$f(z) = f(0) = \text{constant}$$

which proves Liouville's theorem.

A function $f(z)$ that is analytic for all values of z is said to be an *entire function*. Thus, Liouville's theorem may be restated as follows: *An entire function that is bounded for all values of z is a constant* (Wylie and Barrett, 1982).

A.4 SINGULARITIES AND RESIDUES

Let $z = z_0$ be a singular point of an analytic function $f(z)$. If the neighborhood of $z = z_0$ contains *no* other singular points of $f(z)$, the singularity at $z = z_0$ is said to be *isolated*. In the neighborhood of such a singularity, the function $f(z)$ may be represented by the Laurent series

$$\begin{aligned} f(z) &= \sum_{k=-\infty}^{\infty} a_k(z - z_0)^k \\ &= \sum_{k=0}^{\infty} a_k(z - z_0)^k + \sum_{k=-\infty}^{-1} a_k(z - z_0)^k \\ &= \sum_{k=0}^{\infty} a_k(z - z_0)^k + \sum_{k=1}^{\infty} \frac{a_{-k}}{(z - z_0)^k} \end{aligned} \tag{A.21}$$

The particular coefficient a_{-1} in the Laurent expansion of $f(z)$ in the neighborhood of the isolated singularity at the point $z = z_0$ is called the *residue* of $f(z)$ at $z = a$. The residue plays an important role in the evaluation of integrals of analytic functions. In particular, putting $k = -1$ in Eq. (A.13) we get the following connection between the residue a_{-1} and the integral of the function $f(z)$:

$$a_{-1} = \frac{1}{2\pi j} \oint_{\mathscr{C}} f(z)\, dz \tag{A.22}$$

There are two nontrivial cases to be considered:

1. The Laurent expansion of $f(z)$ contains *infinitely* many terms with negative powers of $z - z_0$, as in Eq. (A.21). The point $z - z_0$ is then called an *essential singular point* of $f(z)$.
2. The Laurent expansion of $f(z)$ contains at most *a finite* number of terms, m, with negative powers of $z - z_0$, as shown by

$$f(z) = \sum_{k=0}^{\infty} a_k (z - z_0)^k + \frac{a_{-1}}{z - z_0} + \frac{a_{-2}}{(z - z_0)^2} + \cdots + \frac{a_{-m}}{(z - z_0)^m} \tag{A.23}$$

According to this latter representation, $f(z)$ is said to have a *pole of order m* at $z = z_0$. The *finite sum* of all the terms containing *negative powers* on the right-hand side of Eq. (A.22) is called the *principal part* of $f(z)$ at $z = z_0$.

Note that when the singularity at $z = z_0$ is a pole of order m, the residue of the pole may be determined by using the formula

$$a_{-1} = \frac{1}{(m-1)!} \frac{d^{m-1}}{dz^{m-1}} [(z - z_0)^m f(z)]_{z=z_0} \tag{A.24}$$

In effect, by using this formula we avoid the need for the deduction of the Laurent series. For the special case when the order $m = 1$, the pole is said to be *simple*. Correspondingly, the formula of Eq. (A.24) for the residue a_{-1} of a simple pole reduces to

$$a_{-1} = \lim_{z \to z_0} (z - z_0) f(z) \tag{A.25}$$

A.5 CAUCHY'S RESIDUE THEOREM

Consider a closed contour $\mathscr{C}$ in the z-plane containing within it a number of isolated singularities of some function $f(z)$. Let $z_1, z_2, \ldots, z_n$ define the locations of these isolated singularities. Around each singular point of the function $f(z)$, we draw a circle small enough to ensure that it does not enclose the other singular points of $f(z)$, as depicted in Fig. A.3. The original contour $\mathscr{C}$ together with these small circles constitute the boundary of a *multiply connected region* in which $f(z)$ is analytic everywhere and to which Cauchy's integral theorem may therefore be applied. Specifically, for the situation described in Fig. A.3 we may write

$$\frac{1}{2\pi j} \oint_{\mathscr{C}} f(z)\, dz + \frac{1}{2\pi j} \oint_{\mathscr{C}_1} f(z)\, dz + \cdots + \frac{1}{2\pi j} \oint_{\mathscr{C}_n} f(z)\, dz = 0 \tag{A.26}$$

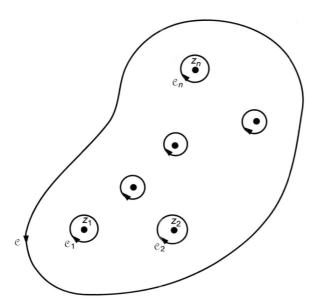

Figure A.3   Multiply connected region.

Note that in Fig. A.3 the contour $\mathscr{C}$ is traversed in the *positive* sense (i.e., counterclockwise direction), whereas the small circles are traversed in the *negative* sense (i.e., clockwise direction).

Suppose now we *reverse* the direction along which the integral around each small circle in Fig. A.3 is taken. This operation has the equivalent effect of applying a minus sign to each of the integrals in Eq. (A.26) that involve the small circles $\mathscr{C}_1, \ldots, \mathscr{C}_n$. Accordingly, for the case when *all* the integrals around the original contour $\mathscr{C}$ and the small circles $\mathscr{C}_1, \ldots, \mathscr{C}_n$ are taken in the counterclockwise direction, we may rewrite Eq. (A.26) as

$$\frac{1}{2\pi j} \oint_{\mathscr{C}} f(z)\,dz = \frac{1}{2\pi j} \oint_{\mathscr{C}_1} f(z)\,dz + \cdots + \frac{1}{2\pi j} \oint_{\mathscr{C}_n} f(z)\,dz \qquad (A.27)$$

By definition, the integrals on the right-hand side of Eq. (A.27) are the residues of the function $f(z)$ evaluated at the various isolated singularities of $f(z)$ within the contour $\mathscr{C}$. We may thus express the integral of $f(z)$ around the contour $\mathscr{C}$ simply as

$$\oint_{\mathscr{C}} f(z)\,dz = 2\pi j \sum_{k=1}^{n} \mathrm{Res}(f(z), z_k) \qquad (A.28)$$

where $\mathrm{Res}(f(z), z_k)$ stands for the residue of the function $f(z)$ evaluated at the isolated singular point $z = z_k$. Equation (A.28) is called *Cauchy's residue theorem*. This theorem is extremely important in the theory of functions in general and in evaluating definite integrals in particular.

A.6 PRINCIPLE OF THE ARGUMENT

Consider a complex function $f(z)$, characterized as follows:

1. The function $f(z)$ is analytic in the interior of a closed contour $\mathscr{C}$ in the z-plane, except at a finite number of poles.
2. The function $f(z)$ has neither poles nor zeros on the contour $\mathscr{C}$. By a "zero" we mean a point in the z-plane at which $f(z) = 0$. In contrast, at a "pole" as defined previously, we have $f(z) = \infty$. Let N be the *number of zeros* and P be the *number of poles* of the function $f(z)$ in the interior of contour $\mathscr{C}$, where each zero or pole is counted according to its *multiplicity*.

We may then state the following theorem (Levinson and Redheffer, 1970; Wylie and Barrett, 1982):

$$\frac{1}{2\pi j} \oint_{\mathscr{C}} \frac{f'(z)}{f(z)} \, dz = N - P \tag{A.29}$$

where $f'(z)$ is the derivative of $f(z)$. We note that

$$\frac{d}{dz} \ln f(z) = \frac{f'(z)}{f(z)} \, dz$$

where ln denotes the natural logarithm. Hence,

$$\oint_{\mathscr{C}} \frac{f'(z)}{f(z)} \, dz = \ln f(z)|_{\mathscr{C}}$$

$$= \ln |f(z)|_{\mathscr{C}} + j \arg f(z)|_{\mathscr{C}} \tag{A.30}$$

where $|f(z)|$ denotes the magnitude of $f(z)$, and $\arg f(z)$ denotes its argument. The first term on the right-hand side of Eq. (A.30) is zero, since the logarithmic function $\ln f(z)$ is single-valued and the contour $\mathscr{C}$ is closed. Hence,

$$\oint_{\mathscr{C}} \frac{f'(z)}{f(z)} \, dz = j \arg f(z)|_{\mathscr{C}} \tag{A.31}$$

Thus, substituting Eq. (A.31) in (A.29), we get

$$N - P = \frac{1}{2\pi} \arg f(z)|_{\mathscr{C}} \tag{A.32}$$

This result, which is a reformulation of the theorem described in Eq. (A.29), is called the *principle of the argument*.

For a geometrical interpretation of this principle, let $\mathscr{C}$ be a closed contour in the z-plane as in Fig. A.4(a). As z traverses the contour $\mathscr{C}$ in a counterclockwise direction, we find that $w = f(z)$ traces out a contour $\mathscr{C}'$ of its own in the w-plane; for the purpose of illustration, $\mathscr{C}'$ is shown in Fig. A.4(b). Suppose now a line is drawn in the w-plane from the

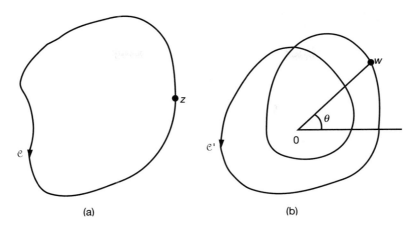

Figure A.4   (a) Contour $\mathscr{C}$ in the z-plane; (b) Contour $\mathscr{C}'$ in the w-plane, where $w = f(z)$.

origin to the point $w = f(z)$, as depicted in Fig. A.4(b). Then the angle θ which this line makes with a fixed direction (shown as the horizontal direction in Fig. A.4(b)) is $\arg f(z)$. The principle of the argument thus provides a description of the number of times the point $w = f(z)$ winds around the origin of the w-plane (i.e., the point $w = 0$) as the complex variable z traverses the contour $\mathscr{C}$ in a counterclockwise direction.

Rouché's Theorem

Let the function $f(z)$ be analytic on a closed contour $\mathscr{C}$ and in the interior of $\mathscr{C}$. Let $g(z)$ be a second function which, in addition to satisfying the same condition for analyticity as $f(z)$, also fulfills the following condition on the contour $\mathscr{C}$:

$$|f(z)| > |g(z)|$$

In other words, on the contour $\mathscr{C}$ we have

$$\left| \frac{g(z)}{f(z)} \right| < 1 \tag{A.33}$$

Define the function

$$F(z) = 1 + \frac{g(z)}{f(z)} \tag{A.34}$$

which has no poles or zeros on $\mathscr{C}$. By the principle of the argument applied to $F(z)$, we have

$$N - P = \frac{1}{2\pi} \arg F(z)\big|_{\mathscr{C}} \tag{A.35}$$

However, the implication of the condition (A.33) is that when z is on the contour $\mathscr{C}$, then

$$|F(z) - 1| < 1 \tag{A.36}$$

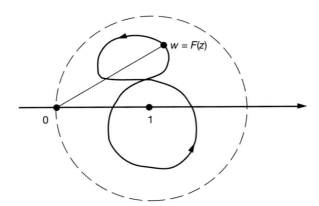

Figure A.5 Point $w = F(z)$ on a closed contour inside the unit circle.

In other words, the point $w = F(z)$ lies inside a circle with center at $w = 1$ and unit radius, as illustrated in Fig. A.5. It follows therefore that

$$\left| \arg F(z) \right| < \frac{\pi}{2} \qquad \text{for } z \text{ on } \mathscr{C} \tag{A.37}$$

Equivalently, we may write

$$\arg F(z)\big|_{\mathscr{C}} = 0 \tag{A.38}$$

Hence, from Eq. (A.38) we deduce that $N = P$, where both N and P refer to $f(z)$. From the definition of the function $F(z)$ given in Eq. (A.34) we note that the poles of $F(z)$ are the zeros of $f(z)$, and the zeros of $F(z)$ are the zeros of the sum $f(z) + g(z)$. Accordingly, the fact that $N = P$ means that $f(z) + g(z)$ and $f(z)$ have the same numbers of zeros. The result that we have just established is known as *Rouché's theorem*, which may be formally stated as follows:

Let $f(z)$ and $g(z)$ be analytic on a closed contour $\mathscr{C}$ and in the interior of $\mathscr{C}$. Let $|f(z)| > |g(z)|$ on $\mathscr{C}$. Then $f(z)$ and $f(z) + g(z)$ have the same number of zeros inside contour $\mathscr{C}$.

Example

Consider the contour depicted in Fig. A.6(a) that constitutes the boundary of a multiply connected region in the z-plane. Let $F(z)$ and $G(z)$ be two polynomials in z^{-1}, both of which are analytic on this contour and in the interior of it. Moreover, Let $|F(z)| > |G(z)|$. Then, according to Rouché's theorem, both $F(z)$ and $F(z) + G(z)$ have the same number of zeros inside the contour described in Fig. A.6(a).

Suppose now that we let the radius R of the outside circle $\mathscr{C}$ in Fig. A.6(a) approach infinity. Also, let the separation l between the two straight-line portions of the contour approach zero. Then, in the limit, the region enclosed by the contour described in Fig. A.6(a)

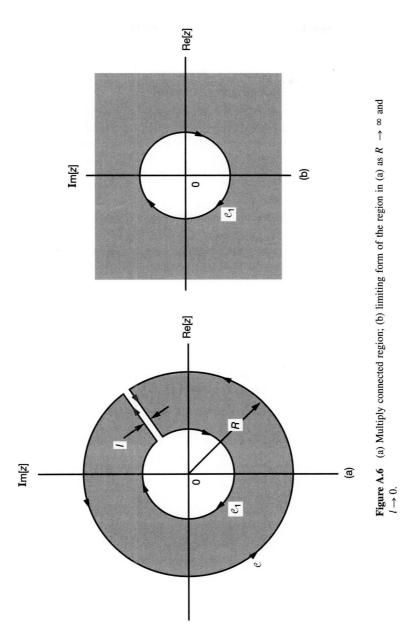

Figure A.6 (a) Multiply connected region; (b) limiting form of the region in (a) as $R \rightarrow \infty$ and $l \rightarrow 0$.

will be made up of the entire area that lies *outside* the inner circle $\mathscr{C}_1$ as depicted in Fig. A.6(b). In other words, the polynomials $F(z)$ and $F(z) + G(z)$ have the same number of zeros outside the circle $\mathscr{C}_1$, under the conditions described above. Note that the circle $\mathscr{C}_1$ is traversed in the clockwise direction (i.e., negative sense).

A.7 INVERSION INTEGRAL FOR THE z-TRANSFORM

The material presented in Sections A.1 through A.6 is applicable to functions of a complex variable in general. In this section and the next one, we consider the special case of a complex function defined as the z-transform of a sequence of samples taken in time.

Let $X(z)$ denote the z-transform of a sequence $x(n)$, which converges to an analytic function in the annular domain $R_1 < |z| < R_2$. By definition, $X(z)$ is written as the Laurent series

$$X(z) = \sum_{m=-\infty}^{\infty} x(m)z^{-m}, \qquad R_1 < |z| < R_2 \tag{A.39}$$

where, for the convenience of presentation, we have used m in place of n as the index of time. Let $\mathscr{C}$ be a closed contour that lies inside the *region of convergence* $R_1 < |z| < R_2$. Then, multiplying both sides of Eq. (A.39) by z^{n-1}, integrating around the contour $\mathscr{C}$ in a counterclockwise direction, and interchanging the order of integration and summation, we get

$$\frac{1}{2\pi j} \oint_{\mathscr{C}} X(z)z^n \frac{dz}{z} = \sum_{m=-\infty}^{\infty} x(n) \frac{1}{2\pi j} \oint_{\mathscr{C}} z^{n-m} \frac{dz}{z} \tag{A.40}$$

The interchange of integration and summation is justified here because the Laurent series that defines $X(z)$ converges uniformly on $\mathscr{C}$. Let

$$z = re^{j\theta}, \qquad R_1 < r < R_2 \tag{A.41}$$

Hence,

$$z^{n-m} = r^{n-m}e^{j(n-m)\theta}$$

and

$$\frac{dz}{z} = j\,d\theta$$

Correspondingly, we may express the contour integral on the right-hand side of Eq. (A.40) as

$$\frac{1}{2\pi j} \oint_{\mathscr{C}} z^{n-m} \frac{dz}{z} = \frac{1}{2\pi} \int_0^{2\pi} r^{n-m}e^{j(n-m)\theta}\,d\theta \tag{A.42}$$

$$= \begin{cases} 1, & m = n \\ 0, & m \neq n \end{cases}$$

Inserting Eq. (A.42) in (A.40), we get

$$x(n) = \frac{1}{2\pi j} \oint_{\mathscr{C}} X(z) z^n \frac{dz}{z} \tag{A.43}$$

Equation (A.43) is called the *inversion integral formula* for the z-transform.

A.8 PARSEVAL'S THEOREM

Let $X(z)$ denote the z-transform of the sequence $x(n)$ with the region of convergence $R_{1x} < |z| < R_{2x}$. Let $Y(z)$ denote the z-transform of a second sequence $y(n)$ with the region of convergence $R_{1y} < |z| < R_{2y}$. Then *Parseval's theorem* states that

$$\sum_{n=-\infty}^{\infty} x(n) y^*(n) = \frac{1}{2\pi j} \oint_{\mathscr{C}} X(z) Y^*\left(\frac{1}{z^*}\right) \frac{dz}{z} \tag{A.44}$$

where $\mathscr{C}$ is a closed contour defined in the overlap of the regions of convergence of $X(z)$ and $Y(z)$, both of which are analytic. The function $Y^*(1/z^*)$ is obtained from the z-transform $Y(z)$ by using $1/z^*$ in place of z, and then complex-conjugating the resulting function. Note that the function $Y^*(1/z^*)$ obtained in this way is analytic too.

To prove Parseval's theorem, we use the inversion integral of Eq. (A.43) to write

$$\sum_{n=-\infty}^{\infty} x(n) y^*(n) = \frac{1}{2\pi j} \sum_{n=-\infty}^{\infty} y^*(n) \oint_{\mathscr{C}} X(z) z^n \frac{dz}{z} \tag{A.45}$$

$$= \frac{1}{2\pi j} \oint_{\mathscr{C}} X(z) \sum_{n=-\infty}^{\infty} y^*(n) z^n \frac{dz}{z}$$

From the definition of the z-transform of $y(n)$, namely,

$$Y(z) = \sum_{n=-\infty}^{\infty} y(n) z^{-n}$$

we note that

$$Y^*\left(\frac{1}{z^*}\right) = \sum_{n=-\infty}^{\infty} y^*(n) z^n \tag{A.46}$$

Hence, using Eq. (A.46) in (A.45), we get the result given in Eq. (A.44), and the proof of Parseval's theorem is completed.

APPENDIX

B

Differentiation with Respect to a Vector

An issue commonly encountered in the study of optimization theory is that of differentiating a cost function with respect to a parameter vector of interest. In the text we used an ordinary gradient operation. The purpose of Appendix B is to address the more difficult issue of differentiating a cost function with respect to a complex-valued parameter vector. We begin by introducing some basic definitions.

B.1 BASIC DEFINITIONS

Consider a complex function $f(\mathbf{w})$ that is dependent on a parameter vector $\mathbf{w}$. When $\mathbf{w}$ is complex valued, there are two different mathematical concepts that require individual attention: (1) the vector nature of $\mathbf{w}$, and (2) the fact that each element of $\mathbf{w}$ is a complex number.

Dealing with the issue of complex numbers first, let x_k and y_k denote the real and imaginary parts of the kth element w_k of the vector $\mathbf{w}$; that is,

$$w_k = x_k + jy_k \tag{B.1}$$

We thus have a function of the real quantities x_k and y_k. Hence, we may use Eq. (B.1) to express the real part x_k in terms of the pair of *complex conjugate coordinates* w_k and w_k^* as

$$x_k = \tfrac{1}{2}(w_k + w_k^*) \tag{B.2}$$

and express the imaginary part y_k as

$$y_k = \frac{1}{2j}(w_k - w_k^*) \tag{B.3}$$

where the asterisk denotes complex conjugation. The real quantities x_k and y_k are functions of both w_k and w_k^*. It is only when we deal with analytic functions f that we are permitted to abandon the complex-conjugated term w_k^* by virtue of the Cauchy–Riemann equations. However, most functions encountered in physical sciences and engineering are *not* analytic.

The notion of a derivative must tie in with the concept of a differential. In particular, the chain rule of changes of variables must be obeyed. With these important points in mind, we may define certain complex derivatives in terms of real derivatives, as shown by (Schwartz, 1967)

$$\frac{\partial}{\partial w_k} = \frac{1}{2}\left(\frac{\partial}{\partial x_k} - j\frac{\partial}{\partial y_k}\right) \tag{B.4}$$

and

$$\frac{\partial}{\partial w_k^*} = \frac{1}{2}\left(\frac{\partial}{\partial x_k} + j\frac{\partial}{\partial y_k}\right) \tag{B.5}$$

The derivatives defined here satisfy the following two basic requirements:

$$\frac{\partial w_k}{\partial w_k} = 1$$

$$\frac{\partial w_k}{\partial w_k^*} = \frac{\partial w_k^*}{\partial w_k} = 0$$

(An analytic function f satisfies $\partial f/\partial z^* = 0$ everywhere.)

The next issue to be considered is that of differentiation with respect to a vector. Let w_0, w_1, . . ., w_{M-1} denote the elements of an M-by-1 complex vector $\mathbf{w}$. We may extend the use of Eqs. (B.4) and (B.5) to deal with this new situation by writing (Miller, 1974)

$$\frac{\partial}{\partial \mathbf{w}} = \frac{1}{2}\begin{bmatrix} \dfrac{\partial}{\partial x_0} - j\dfrac{\partial}{\partial y_0} \\[2mm] \dfrac{\partial}{\partial x_1} - j\dfrac{\partial}{\partial y_1} \\[2mm] \bullet \\ \bullet \\ \bullet \\[2mm] \dfrac{\partial}{\partial x_{M-1}} - j\dfrac{\partial}{\partial y_{M-1}} \end{bmatrix} \tag{B.6}$$

and

$$\frac{\partial}{\partial \mathbf{w}^*} = \frac{1}{2} \begin{bmatrix} \dfrac{\partial}{\partial x_0} + j\dfrac{\partial}{\partial y_0} \\[2ex] \dfrac{\partial}{\partial x_1} + j\dfrac{\partial}{\partial y_1} \\[1ex] \bullet \\ \bullet \\ \bullet \\[1ex] \dfrac{\partial}{\partial x_{M-1}} + j\dfrac{\partial}{\partial y_{M-1}} \end{bmatrix} \qquad (B.7)$$

where we have $w_k = x_k + jy_k$ for $k = 0, 1, \ldots, M - 1$. We refer to $\partial/\partial \mathbf{w}$ as a *derivative* with respect to the vector $\mathbf{w}$, and to $\partial/\partial \mathbf{w}^*$ as a *conjugate derivative* also with respect to the vector $\mathbf{w}$. These two derivatives must be considered together. They obey the following relations:

$$\frac{\partial \mathbf{w}}{\partial \mathbf{w}} = \mathbf{I}$$

and

$$\frac{\partial \mathbf{w}}{\partial \mathbf{w}^*} = \frac{\partial \mathbf{w}^*}{\partial \mathbf{w}} = \mathbf{O}$$

where $\mathbf{I}$ is the identity matrix and $\mathbf{O}$ is the null matrix.

For subsequent use, we will adopt the definition of (B.7) as the derivative with respect to a complex-valued vector.

B.2 EXAMPLES

In this section, we illustrate some applications of the derivative defined in Eq. (B.7). The examples are taken from Chapter 5 dealing with optimum linear filtering, and Chapter 11 dealing with the method of least squares.

Example 1

Let $\mathbf{p}$ and $\mathbf{w}$ denote two complex-valued M-by-1 vectors. There are two inner products, $\mathbf{p}^H \mathbf{w}$ and $\mathbf{w}^H \mathbf{p}$, to be considered

Let $c_1 = \mathbf{p}^H \mathbf{w}$. The conjugate derivative of c_1 with respect to the vector $\mathbf{w}$ is

$$\frac{\partial c_1}{\partial \mathbf{w}^*} = \frac{\partial}{\partial \mathbf{w}^*}(\mathbf{p}^H \mathbf{w}) = \mathbf{0} \qquad (B.8)$$

where $\mathbf{0}$ is the null vector. Here we note that $\mathbf{p}^H \mathbf{w}$ is an analytic function; see Problem 1 of Chapter 5. We therefore find that the derivative of $\mathbf{p}^H \mathbf{w}$ with respect to $\mathbf{w}$ is zero, in agreement with Eq. (B.8).

Consider next $c_2 = \mathbf{w}^H\mathbf{p}$. The conjugate derivative of c_2 with respect to $\mathbf{w}$ is

$$\frac{\partial c_2}{\partial \mathbf{w}^*} = \frac{\partial}{\partial \mathbf{w}^*}(\mathbf{w}^H\mathbf{p}) = \frac{\partial}{\partial \mathbf{w}^*}(\mathbf{p}^T\mathbf{w}^*) = \mathbf{p} \tag{B.9}$$

Here we note that $\mathbf{w}^H\mathbf{p}$ is not an analytic function; see Problem 1 of Chapter 5. Hence, the derivative of $\mathbf{w}^H\mathbf{p}$ with respect to $\mathbf{w}^*$ is nonzero, as in Eq. (B.9).

Example 2

Consider next the quadratic form

$$c = \mathbf{w}^H\mathbf{R}\mathbf{w}$$

where $\mathbf{R}$ is a Hermitian matrix. The conjugate derivative of c (which is real) with respect to $\mathbf{w}$ is

$$\frac{\partial c}{\partial \mathbf{w}^*} = \frac{\partial}{\partial \mathbf{w}^*}(\mathbf{w}^H\mathbf{R}\mathbf{w})$$

$$= \mathbf{R}\mathbf{w} \tag{B.10}$$

Example 3

Consider the real-valued cost function (see Chapter 5)

$$J(\mathbf{w}) = \sigma_d^2 - \mathbf{w}^H\mathbf{p} - \mathbf{p}^H\mathbf{w} + \mathbf{w}^H\mathbf{R}\mathbf{w}$$

Using the results of Examples 1 and 2, we find that the conjugate derivative of J with respect to the tap-weight vector $\mathbf{w}$ is

$$\frac{\partial J}{\partial \mathbf{w}^*} = -\mathbf{p} + \mathbf{R}\mathbf{w} \tag{B.11}$$

Let $\mathbf{w}_o$ be the optimum value of the tap-weight vector $\mathbf{w}$ for which the cost function J is minimum or, equivalently, the derivative $(\partial J/\partial \mathbf{w}^*) = \mathbf{0}$. Hence, from Eq. (B.11) we deduce that

$$\mathbf{R}\mathbf{w}_o = \mathbf{p} \tag{B.12}$$

This is the matrix form of the Wiener–Hopf equations for a transversal filter operating in a stationary environment.

Example 4

Consider the real log-likelihood function (see Chapter 11)

$$l(\tilde{\mathbf{w}}) = F - \frac{1}{\sigma^2}\boldsymbol{\epsilon}^H\boldsymbol{\epsilon} \tag{B.13}$$

where F is a constant and

$$\boldsymbol{\epsilon} = \mathbf{b} - \mathbf{A}\tilde{\mathbf{w}} \tag{B.14}$$

Substituting Eq. (B.14) in (B.13), we get

$$l(\tilde{\mathbf{w}}) = F - \frac{1}{\sigma^2}\mathbf{b}^H\mathbf{b} + \frac{1}{\sigma^2}\mathbf{b}^H\mathbf{A}\tilde{\mathbf{w}} + \frac{1}{\sigma^2}\tilde{\mathbf{w}}^H\mathbf{A}^H\mathbf{b} - \frac{1}{\sigma^2}\tilde{\mathbf{w}}^H\mathbf{A}^H\mathbf{A}\tilde{\mathbf{w}} \tag{B.15}$$

Evaluating the conjugate derivative of l with respect to $\tilde{\mathbf{w}}$, and adapting the results of Examples 1 and 2 to fit our present situation, we get

$$\frac{\partial l}{\partial \tilde{\mathbf{w}}^*} = \frac{1}{\sigma^2}\mathbf{A}^H\mathbf{b} - \frac{1}{\sigma^2}\mathbf{A}^H\mathbf{A}\tilde{\mathbf{w}}$$

Setting $(\partial l/\partial \tilde{\mathbf{w}}^*) = \mathbf{0}$, and then simplifying, we thus get

$$\mathbf{A}^H\mathbf{b} - \mathbf{A}^H\mathbf{A}\mathbf{w}_o = \mathbf{0}$$

where $\mathbf{w}_o$ is the special value of $\tilde{\mathbf{w}}$ for which the log-likelihood function is maximum. Hence,

$$\mathbf{A}^H\mathbf{A}\mathbf{w}_o = \mathbf{A}^H\mathbf{b} \tag{B.16}$$

This is the matrix form of the normal equations for the method of least squares.

B.3 RELATION BETWEEN THE DERIVATIVE WITH RESPECT TO A VECTOR AND THE GRADIENT VECTOR

Consider the real cost function $J(\mathbf{w})$ that defines the error-performance surface of a linear transversal filter whose tap-weight vector is $\mathbf{w}$. In Chapter 5, we defined the *gradient vector* of the error-performance surface as

$$\nabla J = \begin{bmatrix} \dfrac{\partial J}{\partial x_0} + j\dfrac{\partial J}{\partial y_0} \\[2ex] \dfrac{\partial J}{\partial x_1} + j\dfrac{\partial J}{\partial y_1} \\[1ex] \bullet \\ \bullet \\ \bullet \\[1ex] \dfrac{\partial J}{\partial x_{M-1}} + j\dfrac{\partial J}{\partial y_{M-1}} \end{bmatrix} \tag{B.17}$$

where $x_k + jy_k$ is the kth element of the tap-weight vector $\mathbf{w}$, and $k = 0, 1, \ldots, M - 1$. The gradient vector is *normal* to the error-performance surface. Comparing Eqs. (B.7) and (B.17), we see that the conjugate derivative $\partial J/\partial \mathbf{w}^*$ and the gradient vector ∇J are related by

$$\nabla J = 2\frac{\partial J}{\partial \mathbf{w}^*} \tag{B.18}$$

Thus, except for a scaling factor, the definition of the gradient vector introduced in Chapter 5 is the same as the conjugate derivative defined in Eq. (B.7).

APPENDIX

C

Method of Lagrange Multipliers

Optimization consists of determining the values of some specified variables that minimize or maximize an *index of performance* or *cost function*, which combines important properties of a system into a single real-valued number. The optimization may be *constrained* or *unconstrained*, depending on whether the variables are also required to satisfy side equations or not. Needless to say, the additional requirement to satisfy one or more side equations complicates the issue of constrained optimization. In this appendix, we derive the classical *method of Lagrange multipliers* for solving the *complex* version of a constrained optimization problem. The notation used in the derivation is influenced by the nature of applications that are of interest to us. We consider first the case when the problem involves a single side equation, followed by the more general case of multiple side equations.

C.1 OPTIMIZATION INVOLVING A SINGLE EQUALITY CONSTRAINT

Consider the minimization of a real-valued function $f(\mathbf{w})$ that is a quadratic function of a vector $\mathbf{w}$, subject to the *constraint*

$$\mathbf{w}^H \mathbf{s} = g \tag{C.1}$$

where $\mathbf{s}$ is a prescribed vector and g is a complex constant. We may redefine the constraint by introducing a new function $c(\mathbf{w})$ that is linear in $\mathbf{w}$, as shown by

$$c(\mathbf{w}) = \mathbf{w}^H \mathbf{s} - g$$
$$= 0 + j0 \tag{C.2}$$

In general, the vectors $\mathbf{w}$ and $\mathbf{s}$ and the function $c(\mathbf{w})$ are all *complex*. For example, in a beamforming application the vector $\mathbf{w}$ represents a set of complex weights applied to the individual sensor outputs, and $\mathbf{s}$ represents a steering vector whose elements are defined by a prescribed "look" direction; the function $f(\mathbf{w})$ to be minimized represents the mean-square value of the overall beamformer output. In a harmonic retrieval application, $\mathbf{w}$ represents the tap-weight vector of a transversal filter, and $\mathbf{s}$ represents a sinusoidal vector whose elements are determined by the angular frequency of a complex sinusoid contained in the filter input; the function $f(\mathbf{w})$ represents the mean-square value of the filter output. In any event, assuming that the issue is one of minimization, we may state the constrained optimization problem as follows:

$$\text{Minimize a real-valued function } f(\mathbf{w}), \\ \text{subject to the constraint } c(\mathbf{w}) = 0 + j0 \tag{C.3}$$

The *method of Lagrange multipliers* converts the problem of constrained minimization described above into one of unconstrained minimization by the introduction of *Lagrange multipliers*. First we use the real function $f(\mathbf{w})$ and the complex constraint function $c(\mathbf{w})$ to define a new real-valued function

$$h(\mathbf{w}) = f(\mathbf{w}) + \lambda_1 \, \text{Re}[c(\mathbf{w})] + \lambda_2 \, \text{Im}[c(\mathbf{w})] \tag{C.4}$$

where λ_1 and λ_2 are *real Lagrange* multipliers, and

$$c(\mathbf{w}) = \text{Re}[c(\mathbf{w})] + j \, \text{Im}[c(\mathbf{w})] \tag{C.5}$$

Define a *complex Lagrange multiplier*:

$$\lambda = \lambda_1 + j\lambda_2 \tag{C.6}$$

We may then rewrite Eq. (C.4) in the form

$$h(\mathbf{w}) = f(\mathbf{w}) + \text{Re}[\lambda^* c(\mathbf{w})] \tag{C.7}$$

where the asterisk denotes complex conjugation.

Next, we minimize the function $h(\mathbf{w})$ with respect to the vector $\mathbf{w}$. To do this, we set the conjugate derivative $\partial h/\partial \mathbf{w}^*$ equal to the null vector, as shown by

$$\frac{\partial f}{\partial \mathbf{w}^*} + \frac{\partial}{\partial \mathbf{w}^*}(\text{Re}[\lambda^* c(\mathbf{w})]) = \mathbf{0} \tag{C.8}$$

The system of simultaneous equations, consisting of Eq. (C.8) and the original constraint given in Eq. (C.2), define the optimum solutions for the vector $\mathbf{w}$ and the Lagrange multiplier λ. We call Eq. (C.8) the *adjoint equation* and Eq. (C.2) the *primal equation* (Dorny, 1975).

C.2 OPTIMIZATION INVOLVING MULTIPLE EQUALITY CONSTRAINTS

Consider next the minimization of a real function $f(\mathbf{w})$ that is a quadratic function of the vector $\mathbf{w}$, subject to a set of *multiple linear constraints*

$$\mathbf{w}^H\mathbf{s}_k = g_k, \qquad k = 1, 2, \ldots, K \tag{C.9}$$

where the number of constraints, K, is less than the dimension of the vector $\mathbf{w}$, and the g_k are complex constants. We may state the multiple-constrained optimization problem as follows:

> Minimize a real function $f(\mathbf{w})$, subject to the
> constraints $c_k(\mathbf{w}) = 0 + j0$ for $k = 1, 2, \ldots, K$ $\qquad$ (C.10)

The solution to this optimization problem is readily obtained by generalizing the previous results of Section C.1. Specifically, we formulate a system of simultaneous equations, consisting of the adjoint equation

$$\frac{\partial f}{\partial \mathbf{w}^*} + \sum_{k=1}^{K} \frac{\partial}{\partial \mathbf{w}^*}(\mathrm{Re}[\lambda_k^* c_k(\mathbf{w})]) = \mathbf{0} \tag{C.11}$$

and the primal equation

$$c_k(\mathbf{w}) = 0^* + j0, \qquad k = 1, 2, \ldots, K \tag{C.12}$$

This system of equations defines the optimum solutions for the vector $\mathbf{w}$ and the set of complex Lagrange multipliers $\lambda_1, \lambda_2, \ldots, \lambda_K$.

C.3 Example

By way of an example, consider the problem of finding the vector $\mathbf{w}$ that minimizes the function

$$f(\mathbf{w}) = \mathbf{w}^H\mathbf{w} \tag{C.13}$$

and which satisfies the constraint

$$c(\mathbf{w}) = \mathbf{w}^H\mathbf{s} - g = 0 + j0 \tag{C.14}$$

The adjoint equation for this problem is

$$\frac{\partial}{\partial \mathbf{w}^*}(\mathbf{w}^H\mathbf{w}) + \frac{\partial}{\partial \mathbf{w}^*}(\mathrm{Re}[\lambda^*(\mathbf{w}^H\mathbf{s} - g)]) = \mathbf{0} \tag{C.15}$$

Using the rules for differentiation developed in Appendix B, we have

$$\frac{\partial}{\partial \mathbf{w}^*}(\mathbf{w}^H\mathbf{w}) = \mathbf{w}$$

and

$$\frac{\partial}{\partial \mathbf{w}^*}(\mathrm{Re}[\lambda^*(\mathbf{w}^H\mathbf{s} - g)]) = \lambda^*\mathbf{s}$$

Substituting these results in Eq. (C.15), we get

$$\mathbf{w} + \lambda^*\mathbf{s} = \mathbf{0} \tag{C.16}$$

or, equivalently,

$$\mathbf{w}^H + \lambda\mathbf{s}^H = \mathbf{0}^T \tag{C.17}$$

Next, postmultiplying both sides of Eq. (C.17) by $\mathbf{s}$ and then solving for the unknown λ, we obtain

$$\begin{aligned} \lambda &= -\frac{\mathbf{w}^H\mathbf{s}}{\mathbf{s}^H\mathbf{s}} \\ &= -\frac{g}{\mathbf{s}^H\mathbf{s}} \end{aligned} \tag{C.18}$$

Finally, substituting Eq. (C.18) in (C.16) and solving for the optimum value $\mathbf{w}_o$ of the weight vector $\mathbf{w}$, we get

$$\mathbf{w}_o = \left(\frac{g^*}{\mathbf{s}^H\mathbf{s}}\right)\mathbf{s} \tag{C.19}$$

This solution is optimum in the sense that $\mathbf{w}_o$ satisfies the constraint of Eq. (C.14) and has minimum length.

APPENDIX

D

Estimation Theory

Estimation theory is a branch of probability and statistics that deals with the problem of deriving information about properties of random variables and stochastic processes, given a set of observed samples. This problem arises frequently in the study of communications and control systems. *Maximum likelihood* is by far the most general and powerful method of estimation. It was first used by the famous statistician R. A. Fisher in 1906. In principle, the method of maximum likelihood may be applied to any estimation problem with the proviso that we formulate the joint probability density function of the available set of observed data. As such, the method yields almost all the well-known estimates as special cases.

D.1 LIKELIHOOD FUNCTION

The method of maximum likelihood is based on a relatively simple idea: Different populations generate different data samples and any given data sample is more *likely* to have come from some population than from others (Kmenta, 1971).

Let $f_{\mathbf{U}}(\mathbf{u}|\boldsymbol{\theta})$ denote the *conditional joint probability density function* of the *random vector* $\mathbf{U}$ represented by the observed *sample* vector $\mathbf{u}$, where the sample vector $\mathbf{u}$ has u_1, $u_2, \ldots, u_M$ for its elements, and $\boldsymbol{\theta}$ is a *parameter vector* with $\theta_1, \theta_2, \ldots, \theta_K$ as elements.

The method of maximum likelihood is based on the principle that we should estimate the parameter vector $\boldsymbol{\theta}$ by its most *plausible values*, given the observed sample vector $\mathbf{u}$. In other words, the maximum-likelihood estimators of $\theta_1, \theta_2, \ldots, \theta_K$ are those values of the parameter vector for which the conditional joint probability density function $f_{\mathbf{U}}(\mathbf{u}|\boldsymbol{\theta})$ is at maximum.

The name *likelihood function*, denoted by $l(\boldsymbol{\theta})$, is given to the conditional joint probability density function $f_{\mathbf{U}}(\mathbf{u}|\boldsymbol{\theta})$, viewed as a function of the parameter vector $\boldsymbol{\theta}$. We thus write

$$l(\boldsymbol{\theta}) = f_{\mathbf{U}}(\mathbf{u}|\boldsymbol{\theta}) \tag{D.1}$$

Although the conditional joint probability density function and the likelihood function have exactly the same formula, nevertheless, it is vital that we appreciate the physical distinction between them. In the case of the conditional joint probability density function, the parameter vector $\boldsymbol{\theta}$ is fixed and the observation vector $\mathbf{u}$ is variable. On the other hand, in the case of the likelihood function, the parameter vector $\boldsymbol{\theta}$ is variable and the observation vector $\mathbf{u}$ is fixed.

In many cases, it turns out to be more convenient to work with the natural logarithm of the likelihood function rather than with the likelihood itself. Thus, using $L(\boldsymbol{\theta})$ to denote the *log-likelihood function*, we write

$$\begin{aligned} L(\boldsymbol{\theta}) &= \ln[l(\boldsymbol{\theta})] \\ &= \ln[f_{\mathbf{U}}(\mathbf{u}|\boldsymbol{\theta})] \end{aligned} \tag{D.2}$$

The logarithm of $l(\boldsymbol{\theta})$ is a *monotonic transformation* of $l(\boldsymbol{\theta})$. This means that whenever $l(\boldsymbol{\theta})$ decreases, its logarithm $L(\boldsymbol{\theta})$ also decreases. Since $l(\boldsymbol{\theta})$, being a formula for conditional joint probability density function, can never become negative, it follows that there is no problem in evaluating its logarithm $L(\boldsymbol{\theta})$. We conclude therefore that the parameter vector for which the likelihood function $l(\boldsymbol{\theta})$ is at maximum is exactly the same as the parameter vector for which the log-likelihood function $L(\boldsymbol{\theta})$ is at its maximum.

To obtain the ith element of the maximum-likelihood estimate of the parameter vector $\boldsymbol{\theta}$, we differentiate the log-likelihood function with respect to θ_i and set the result equal to zero. We thus get a set of first-order conditions:

$$\frac{\partial L}{\partial \theta_i} = 0, \qquad i = 1, 2, \ldots, K \tag{D.3}$$

The first derivative of the log-likelihood function with respect to parameter θ_i is called the *score* for that parameter. The vector of such parameters is known as the *scores vector* (i.e., the gradient vector). The scores vector is identically zero at the maximum-likelihood estimates of the parameters, that is, at the values of $\boldsymbol{\theta}$ that result from the solutions of Eq. (D.3).

To find how effective the method of maximum likelihood is, we can compute the *bias* and *variance* for the estimate of each parameter. However, this is frequently difficult to do. Rather than approach the computation directly, we may derive a *lower bound* on the variance of any *unbiased* estimate. We say an estimate is unbiased if the average value of the estimate equals the parameter we are trying to estimate. Later we show how the variance of the maximum-likelihood estimate compares with this lower bound.

D.2 CRAMÉR–RAO INEQUALITY

Let $\mathbf{U}$ be a random vector with conditional joint probability density function $f_{\mathbf{U}}(\mathbf{u}|\boldsymbol{\theta})$, where $\mathbf{u}$ is the observed sample vector with elements $u_1, u_2, \ldots, u_M$ and $\boldsymbol{\theta}$ is the parameter vector with elements $\theta_1, \theta_2, \ldots, \theta_K$. Using the definition of Eq. (D.2) for the log-likelihood function $L(\hat{\mathbf{u}})$ in terms of the conditional joint probability density function $f_{\mathbf{U}}(\mathbf{u}|\boldsymbol{\theta})$, we form the K-by-K matrix:

$$
\mathbf{J} = -
\begin{bmatrix}
E\left[\dfrac{\partial^2 L}{\partial\theta_1^2}\right] & E\left[\dfrac{\partial^2 L}{\partial\theta_1\partial\theta_2}\right] & \cdots & E\left[\dfrac{\partial^2 L}{\partial\theta_1\partial\theta_K}\right] \\[2ex]
E\left[\dfrac{\partial^2 L}{\partial\theta_2\partial\theta_1}\right] & E\left[\dfrac{\partial^2 L}{\partial\theta_2^2}\right] & \cdots & E\left[\dfrac{\partial^2 L}{\partial\theta_2\partial\theta_K}\right] \\[2ex]
\cdot & \cdot & & \cdot \\
\cdot & \cdot & & \cdot \\
\cdot & \cdot & & \cdot \\
E\left[\dfrac{\partial^2 L}{\partial\theta_K\partial\theta_1}\right] & E\left[\dfrac{\partial^2 L}{\partial\theta_K\partial\theta_2}\right] & \cdots & E\left[\dfrac{\partial^2 L}{\partial\theta_K^2}\right]
\end{bmatrix}
\tag{D.4}
$$

The matrix $\mathbf{J}$ is called *Fisher's information matrix.*

Let $\mathbf{I}$ denote the inverse of Fisher's information matrix $\mathbf{J}$. Let I_{ii} denote the ith diagonal element (i.e., the element in the ith row and ith column) of the inverse matrix $\mathbf{I}$. Let $\hat{\theta}_i$ be *any* unbiased estimate of the parameter θ_i, based on the observed sample vector $\mathbf{u}$. We may then write (Van Trees, 1968; Nahi, 1969)

$$
\mathrm{var}[\hat{\theta}_i] \geq I_{ii}, \qquad i = 1, 2, \ldots, K
\tag{D.5}
$$

Equation (D.5) is called the *Cramér–Rao inequality*. This theorem enables us to construct a lower limit (greater than zero) for the variance of any unbiased estimator, provided, of course, that we know the functional form of the log-likelihood function. The lower limit specified in the theorem is called the *Cramér–Rao lower bound* (CRLB).

If we can find an unbiased estimator whose variance equals the Cramér–Rao lower bound, then according to the theorem of Eq. (D.5) there is no other unbiased estimator with a smaller variance. Such an estimator is said to be *efficient*.

D.3 PROPERTIES OF MAXIMUM-LIKELIHOOD ESTIMATORS

Not only is the method of maximum likelihood based on an intuitively appealing idea (that of choosing those parameters from which the actually observed sample vector is most likely to have come), but also the resulting estimates have some desirable properties.

Indeed, under quite general conditions, the following *asymptotic* properties may be proved (Kmenta, 1971):

1. Maximum-likelihood estimators are *consistent*. That is, the value of θ_i for which the score $\partial L/\partial \theta_i$ is identically zero *converges in probability* to the true value of the parameter θ_i, $i = 1, 2, \ldots, K$, as the *sample size M* approaches infinity.

2. Maximum-likelihood estimators are *asymptotically efficient*; that is,

$$\lim_{M \to \infty} \left\{ \frac{\text{var}[\theta_{i,\text{ml}} - \theta_i]}{I_{ii}} \right\} = 1, \qquad i = 1, 2, \ldots, K$$

   where $\theta_{i,\text{ml}}$ is the maximum-likelihood estimate of parameter θ_i, and I_{ii} is the ith diagonal element of the inverse of Fisher's information matrix.

3. Maximum-likelihood estimators are *asymptotically Gaussian*.

In practice, we find that the large-sample (asymptotic) properties of maximum-likelihood estimators hold rather well for sample size $M \geq 50$.

D.4 CONDITIONAL MEAN ESTIMATOR

Another classic problem in estimation theory is that of the *Bayes estimation of a random parameter*. There are different answers to this problem, depending on how the *cost function* in the Bayes estimation is formulated (Van Trees, 1968). A particular type of the Bayes estimator of interest to us in this book is the so-called *conditional mean estimator*. We now wish to do two things: (1) derive the formula for the conditional mean estimator from first principles, and (2) show that such an estimator is the same as a minimum mean-squared-error estimator.

Consider a *random parameter x*. We are given an *observation y* that depends on x, and the requirement is to estimate x. Let $\hat{x}(y)$ denote an *estimate* of the parameter x; the symbol $\hat{x}(y)$ emphasizes the fact that the estimate is a function of the observation y. Let $C(x, \hat{x}(y))$ denote a *cost function*. Then, according to Bayes estimation theory, we may write an expression for the *risk* as follows (Van Trees, 1968):

$$\mathcal{R} = E[C(x, \hat{x}(y))]$$
$$= \int_{-\infty}^{\infty} dx \int_{-\infty}^{\infty} C(x, \hat{x}(y)) f_{X,Y}(x, y) \, dy \tag{D.6}$$

where $f_{X,Y}(x, y)$ is the joint probability density function of x and y. For a specified cost function $C(x, \hat{x}(y))$, the *Bayes estimate* is defined as the estimate $\hat{x}(y)$ that *minimizes* the risk $\mathcal{R}$.

A cost function of particular interest (and which is very much in the spirit of the material covered in this book) is the *mean-squared error*. In this case, the cost function is specified as the square of the *estimation error*. The estimation error is itself defined as the difference between the actual parameter value x and the estimate $\hat{x}(y)$, as shown by

$$\epsilon = x - \hat{x}(y) \tag{D.7}$$

Correspondingly, the cost function is defined by

$$C(x, \hat{x}(y)) = C(x - \hat{x}(y)) \tag{D.8}$$

or, more simply,

$$C(\epsilon) = \epsilon^2 \tag{D.9}$$

Thus, the cost function varies with the estimation error ϵ in the manner indicated in Fig. D.1. It is assumed here that x and y are both real. Accordingly, for the situation at hand, we may rewrite Eq. (D.6) as follows:

$$\mathcal{R}_{ms} = \int_{-\infty}^{\infty} dx \int_{-\infty}^{\infty} [x - \hat{x}(y)]^2 f_{X,Y}(x, y)\, dy \tag{D.10}$$

where the subscripts in the risk $\mathcal{R}_{ms}$ indicate the use of mean-squared error as its basis.

Using *Bayes' rule*, we have

$$f_{X,Y}(x, y) = f_X(x|y) f_Y(y) \tag{D.11}$$

where $f_X(x|y)$ is the conditional probability density function of x, given y, and $f_Y(y)$ is the (marginal) probability density function of y. Hence, using Eq. (D.11) in (D.10), we have

$$\mathcal{R}_{ms} = \int_{-\infty}^{\infty} dy f_Y(y) \int_{-\infty}^{\infty} [x - \hat{x}(y)]^2 f_X(x|y)\, dx \tag{D.12}$$

We now recognize that the inner integral and $f_Y(y)$ in Eq. (D.12) are both nonnegative. We may therefore minimize the risk $\mathcal{R}_{ms}$ by simply minimizing the inner integral. Let the estimate so obtained be denoted by $\hat{x}_{ms}(y)$. We find $\hat{x}_{ms}(y)$ by differentiating the inner integral with respect to $\hat{x}(y)$ and then setting the result equal to zero.

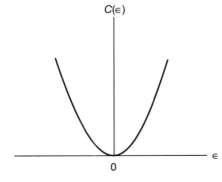

Figure D.1   Mean-squared error as the cost function.

To simplify the presentation, let I denote the inner integral in Eq. (D.12). Then differentiating I with respect to $\hat{x}(y)$ yields

$$\frac{dI}{d\hat{x}} = -2 \int_{-\infty}^{\infty} x f_X(x|y) \, dx + 2\hat{x}(y) \int_{-\infty}^{\infty} f_X(x|y) \, dx \tag{D.13}$$

The second integral on the right-hand side of Eq. (D.13) represents the total area under a probability density function and therefore equals 1. Hence, setting the derivative $dI/d\hat{x}$ equal to zero, we obtain

$$\hat{x}_{ms}(y) = \int_{-\infty}^{\infty} x f_X(x|y) \, dx \tag{D.14}$$

The solution defined by Eq. (D.14) is a unique minimum.

The estimator $\hat{x}_{ms}(y)$ defined in Eq. (D.14) is naturally a *minimum mean-squared-error estimator*, hence the use of the subscripts "ms." For another interpretation of this estimator, we recognize that the integral on the right-hand side of Eq. (D.14) is just the *conditional mean* of the parameter x, given the observation y.

We therefore conclude that the minimum mean-squared error estimator and the conditional mean estimator are indeed one and the same. In other words, we have

$$\hat{x}_{ms}(y) = E[x|y] \tag{D.15}$$

Substituting Eq. (D.15) for the estimate $\hat{x}(y)$ in Eq. (D.12), we find that the inner integral is just the *conditional variance* of the parameter x, given y. Accordingly, the minimum value of the risk $\mathcal{R}_{ms}$ is just the average of this conditional variance over all observations y.

APPENDIX

E

Maximum-Entropy Method

The *maximum-entropy method (MEM)* was originally devised by Burg (1967, 1975) to overcome fundamental limitations of Fourier-based methods for estimating the power spectrum of a stationary stochastic process. The basic idea of MEM is to choose the particular spectrum that corresponds to the most *random* or the most *unpredictable* time series whose autocorrelation function agrees with a set of known values. This condition is equivalent to an extrapolation of the autocorrelation function of the available time series by *maximizing* the *entropy* of the process, hence the name of the method. Entropy is a measure of the average information content of the process (Shannon, 1948). Thus, MEM bypasses the problems that arise from the use of window functions, a feature that is common to all Fourier-based methods of spectrum analysis. In particular, MEM avoids the use of a periodic extension of the data (as in the method based on smoothing the periodogram and its computation using the fast Fourier transform algorithm) or of the assumption that data outside the available record length are zero (as in the Blackman–Tukey method based on the sample autocorrelation function). An important feature of the MEM spectrum is that it is *nonnegative at all frequencies*, which is precisely the way it should be.

E.1 MAXIMUM-ENTROPY SPECTRUM

Suppose that we are given $2M + 1$ values of the autocorrelation function of a stationary stochastic process $u(n)$ of zero mean. We wish to obtain the special value of the power spectrum of the process that corresponds to the most random time series whose autocorrelation function is consistent with the set of $2M + 1$ known values. In terms of information theory, this statement corresponds to the *principle of maximum entropy* (Jaynes, 1982).

In the case of a set of Gaussian-distributed random variables of zero mean, the entropy is given by (Middleton, 1960)

$$H = \tfrac{1}{2}\ln[\det(\mathbf{R})] \tag{E.1}$$

where $\mathbf{R}$ is the correlation matrix of the process. When the process is of infinite duration, however, we find that the entropy H diverges, and so we cannot use it as a measure of information content. To overcome this divergence problem, we may use the *entropy rate* defined by

$$h = \lim_{M \to \infty} \frac{H}{M + 1}$$

$$= \lim_{M \to \infty} \tfrac{1}{2}\ln[\det(\mathbf{R})]^{1/(M+1)} \tag{E.2}$$

Let $S(\omega)$ denote the power spectrum of the process $u(n)$. The limiting form of the determinant of the correlation matrix $\mathbf{R}$ is related to the power spectrum $S(\omega)$ as follows (see Problem 14 of Chapter 4):

$$\lim_{M \to \infty} [\det(\mathbf{R})]^{1/(M+1)} = \exp\left\{\frac{1}{2\pi} \int_{-\pi}^{\pi} \ln S(\omega)\, d\omega\right\} \tag{E.3}$$

Hence, substituting Eq. (E.3) in (E.2), we get

$$h = \frac{1}{4\pi} \int_{-\pi}^{\pi} \ln[S(\omega)]d\omega \tag{E.4}$$

Although this relation was derived on the assumption that the process $u(n)$ is Gaussian, nevertheless, the form of the relation is valid for any stationary process.

We may now restate the MEM problem in terms of the entropy rate. We wish to find a real positive-valued power spectrum characterized by entropy rate h, satisfying two simultaneous requirements:

1. The entropy rate h is *stationary* with respect to the *unknown* values of the autocorrelation function of the process.
2. The power spectrum is *consistent* with respect to the *known* values of the autocorrelation function of the process.

We will address these two requirements in turn.

Since the autocorrelation sequence $r(m)$ and power spectrum $S(\omega)$ of a stationary process $u(n)$ form a discrete-time Fourier-transform pair, we write

$$S(\omega) = \sum_{m=-\infty}^{\infty} r(m) \exp(-jm\omega) \tag{E.5}$$

Equation (E.5) assumes that the sampling period of the process $u(n)$ is normalized to unity. Substituting Eq. (E.5) in (E.4), we get

$$h = \frac{1}{4\pi} \int_{-\pi}^{\pi} \ln\left[\sum_{m=-\infty}^{\infty} r(m) \exp(-jm\omega) \right] d\omega \tag{E.6}$$

We extrapolate the autocorrelation sequence $r(m)$ outside the range of known values, $-M \leq m \leq M$, by choosing the unknown values of the autocorrelation function in such a way that no information or entropy is added to the process. That is, we impose the condition

$$\frac{\partial h}{\partial r(m)} = 0, \qquad |m| \geq M + 1 \tag{E.7}$$

Hence, differentiating Eq. (E.6) with respect to $r(m)$ and setting the result equal to zero, we find that the conditions for *maximum entropy* are as follows:

$$\int_{-\pi}^{\pi} \frac{\exp(-jm\omega)}{S_{\mathrm{MEM}}(\omega)} d\omega = 0, \qquad |m| \geq M + 1 \tag{E.8}$$

where $S_{\mathrm{MEM}}(\omega)$ is the special value of the power spectrum resulting from the imposition of the condition in Eq. (E.7). Equation (E.8) implies that the power spectrum $S_{\mathrm{MEM}}(\omega)$ is expressible in the form of a truncated Fourier series:

$$\frac{1}{S_{\mathrm{MEM}}(\omega)} = \sum_{k=-M}^{M} c_k \exp(-jk\omega) \tag{E.9}$$

The complex Fourier coefficient c_k of the expansion satisfies the Hermitian condition

$$c_k^* = c_{-k} \tag{E.10}$$

so as to ensure that $S_{\mathrm{MEM}}(\omega)$ is real for all ω.

The next requirement is to make the power spectrum $S_{\mathrm{MEM}}(\omega)$ consistent with the set of known values of the autocorrelation function $r(m)$ for the interval $-M \leq m \leq M$. Since $r(m)$ is a Hermitian function, we need only concern ourselves with $0 \leq m \leq M$. Accordingly, $r(m)$ must equal the inverse discrete-time Fourier transform of $S_{\mathrm{MEM}}(\omega)$ for $0 \leq m \leq M$, as shown by

$$r(m) = \frac{1}{2\pi} \int_{-\pi}^{\pi} S_{\mathrm{MEM}}(\omega) \exp(jm\omega) \, d\omega, \qquad 0 \leq m \leq M \tag{E.11}$$

Therefore, substituting Eq. (E.9) in (E.11), we get

$$r(m) = \frac{1}{2\pi} \int_{-\pi}^{\pi} \frac{\exp(jm\omega)}{\displaystyle\sum_{k=-M}^{M} c_k \exp(-jk\omega)} \, d\omega, \qquad 0 \leq m \leq M \tag{E.12}$$

Clearly, in the set of complex Fourier coefficients $\{c_k\}$, we have the available degrees of freedom needed to satisfy the conditions of Eq. (E.12).

To proceed with the analysis, however, we find it convenient to use z-transform notation by changing from the variable ω to z. Define

$$z = \exp(j\omega) \tag{E.13}$$

Hence,

$$d\omega = \frac{1}{j}\frac{dz}{z}$$

and so we rewrite Eq. (E.12) in terms of the variable z as the contour integral

$$r(m) = \frac{1}{j2\pi} \oint \frac{z^{m-1}}{\displaystyle\sum_{k=-M}^{M} c_k z^{-k}}\, dz, \qquad 0 \leq m \leq M \tag{E.14}$$

The contour integration in Eq. (E.14) is performed on the unit circle in the z-plane in a counterclockwise direction. Since the complex Fourier coefficient c_k satisfies the Hermitian condition of Eq. (E.10), we may express the summation in the denominator of the integral in Eq. (E.14) as the product of two polynomials, as follows:

$$\sum_{k=-M}^{M} c_k z^{-k} = G(z)G^*\!\left(\frac{1}{z^*}\right) \tag{E.15}$$

where

$$G(z) = \sum_{k=0}^{M} g_k z^{-k} \tag{E.16}$$

and

$$G^*\!\left(\frac{1}{z^*}\right) = \sum_{k=0}^{M} g_k^* z^{k} \tag{E.17}$$

We choose the first polynomial $G(z)$ to be minimum phase, in that its zeros are all located inside the unit circle in the z-plane. Correspondingly, we choose the second polynomial $G^*(1/z^*)$ to be maximum phase, in that its zeros are all located outside the unit circle in the z-plane. Moreover, the zeros of these two polynomials are the inverse of each other with respect to the unit circle. Thus, substituting Eq. (E.15) in (E.14), we get

$$r(m) = \frac{1}{j2\pi} \oint \frac{z^{m-1}}{G(z)G^*(1/z^*)}\, dz, \qquad 0 \leq m \leq M \tag{E.18}$$

We next form the summation

$$\sum_{k=0}^{M} g_k r(m - k) = \frac{1}{j2\pi} \oint \frac{z^{m-1} \sum_{k=0}^{M} g_k z^{-k}}{G(z)G^*(1/z^*)} \, dz \tag{E.19}$$

$$= \frac{1}{j2\pi} \oint \frac{z^{m-1}}{G^*(1/z^*)} \, dz, \qquad 0 \le m \le M$$

where in the first line we have used Eq. (E.18), and in the second line we have used Eq. (E.16).

To evaluate the contour integral of Eq. (E.19), we use *Cauchy's residue theorem* of complex variable theory (see Appendix A). According to this theorem, the contour integral equals $2\pi j$ times the sum of *residues* of the poles of the integral $z^{m-1}/G^*(1/z^*)$ that lie inside the unit circle used as the contour of integration. Since the polynomial $G^*(1/z^*)$ is chosen to have no zeros inside the unit circle, it follows that the integral in Eq. (E.19) is analytic on and inside the unit circle for $m \ge 1$. For $m = 0$ the integral has a simple pole at $z = 0$ with a *residue* equal to $1/g_0^*$. Hence, application of Cauchy's residue theorem yields

$$\oint \frac{z^{m-1}}{G^*(1/z^*)} \, dz = \begin{cases} \dfrac{2\pi j}{g_0^*}, & m = 0 \\[2mm] 0, & m = 1, 2, \ldots, M \end{cases} \tag{E.20}$$

Thus, substituting Eq. (E.20) in (E.19), we get

$$\sum_{k=0}^{M} g_k r(m - k) = \begin{cases} \dfrac{1}{g_0^*}, & m = 0 \\[2mm] 0, & m = 1, 2, \ldots, M \end{cases} \tag{E.21}$$

We recognize that the set of $(M + 1)$ equations in (E.21) has a mathematical form similar to that of the augmented Wiener–Hopf equations for forward prediction of order M (see Chapter 6). In particular, by comparing Eqs. (E.21) and (6.16), we deduce that

$$g_k^* = \frac{1}{g_0 P_M} a_{M,k}, \qquad 0 \le k \le M \tag{E.22}$$

where the $a_{M,k}$ are coefficients of a prediction-error filter of order M, and P_M is the average output power of the filter. Since $a_{M,0} = 1$ for all M, by definition, we find from Eq. (E.22) that for $k = 0$:

$$|g_0|^2 = \frac{1}{P_M} \tag{E.23}$$

Finally, substituting Eqs. (E.15), (E.22), and (E.23) in (E.9) with $z = \exp(j\omega)$, we get

$$S_{\text{MEM}}(\omega) = \frac{P_M}{\left|1 + \displaystyle\sum_{k=1}^{M} a_{M,k} e^{-jk\omega}\right|^2} \tag{E.24}$$

We refer to the formula of Eq. (E.24) as the *MEM spectrum*.

E.2 COMPUTATION OF THE MEM SPECTRUM

The formula for the MEM spectrum given in Eq. (E.24) may be recast in the alternative form

$$S_{\text{MEM}}(\omega) = \frac{1}{\displaystyle\sum_{k=-M}^{M} \psi(k) e^{-j\omega k}} \tag{E.25}$$

where $\psi(k)$ is defined in terms of the prediction-error filter coefficients as follows:

$$\psi(k) = \begin{cases} \dfrac{1}{P_M} \displaystyle\sum_{i=0}^{M-k} a_{M,i}\, a^*_{M,i+k} & \text{for } k = 0, 1, \ldots, M \\[2ex] \psi^*(-k) & \text{for } k = -M, \ldots, -1 \end{cases} \tag{E.26}$$

The parameter $\psi(k)$ may be viewed as some form of a *correlation coefficient for prediction-error filter coefficients*.

Examination of the denominator polynomial in Eq. (E.25) reveals that it represents the *discrete Fourier transform* of the sequence $\psi(k)$. Accordingly, we may use the *fast Fourier transform (FFT)* algorithm (Oppenheim and Schafer, 1989) for the efficient computation of the denominator polynomial and therefore the MEM spectrum. Given the autocorrelation sequence $r(0), r(1), \ldots, r(M)$, pertaining to a wide-sense stationary stochastic process $u(n)$, we may now summarize an efficient procedure for computing the MEM spectrum:

Step 1: Levinson–Durbin Recursion.

Initialize the algorithm by setting

$$a_{0,0} = 1$$

$$P_0 = r(0)$$

For $m = 1, 2, \ldots, M$, compute

$$\kappa_m = -\frac{1}{P_{m-1}} \sum_{i=0}^{M-1} r(i-m)a_{m-1,i}$$

$$a_{m,i} = \begin{cases} 1 & \text{for } i = 0 \\ a_{m-1,i} + \kappa_m a_{m-1,m-i}^* & \text{for } i = 1, 2, \ldots, m-1 \\ \kappa_m & \text{for } i = m \end{cases}$$

$$P_m = P_{m-1}(1 - |\kappa_m|^2)$$

Step 2: Correlation for Prediction-Error Filter Coefficients.

Compute the correlation coefficient

$$\psi(k) = \begin{cases} \dfrac{1}{P_M} \sum_{i=0}^{M-k} a_{M,i}\, a_{M,i+k}^* & \text{for } k = 0, 1, \ldots, M \\ \psi^*(-k) & \text{for } k = -M, \ldots, -1 \end{cases} \tag{E.26}$$

Step 3: MEM Spectrum.

Use the fast Fourier transform algorithm to compute the MEM spectrum for varying angular frequency:

$$S_{\text{MEM}}(\omega) = \frac{1}{\displaystyle\sum_{k=-M}^{M} \psi(k)e^{-j\omega k}}$$

APPENDIX

F

Minimum-Variance Distortionless Response Spectrum

In Section 5.8, we derived the formula for the *minimum-variance distortionless response (MVDR) spectrum* for a wide-sense stationary stochastic process. In this appendix we do two things. First, we develop a fast algorithm for computing the MVDR spectrum, given the ensemble-averaged correlation matrix of the process (Musicus, 1985); the algorithm exploits the Toeplitz property of the correlation matrix. Second, in deriving the algorithm, we develop an insightful relationship between the MVDR and MEM spectra.

F.1 FAST MVDR SPECTRUM COMPUTATION

Consider a zero-mean wide-sense stationary stochastic process $u(n)$ characterized by an $(M + 1)$-by-$(M + 1)$ ensemble-averaged correlation matrix $\mathbf{R}$. The *minimum-variance distortionless response (MVDR) spectrum* for such a process is defined in terms of the inverse matrix $\mathbf{R}^{-1}$ by

$$S_{\text{MVDR}}(\omega) = \frac{1}{\mathbf{s}^H(\omega)\mathbf{R}^{-1}\mathbf{s}(\omega)} \tag{F.1}$$

where

$$\mathbf{s}(\omega) = [1, e^{-j\omega}, e^{-j2\omega}, \dots, e^{-jM\omega}]^T$$

Let $R_{l,k}^{-1}$ denote the (l, k)th element of $\mathbf{R}^{-1}$. Then, we may rewrite Eq. (F.1) in the form

$$S_{\text{MVDR}}(\omega) = \frac{1}{\displaystyle\sum_{k=-M}^{M} \mu(k)e^{-j\omega k}} \tag{F.2}$$

where

$$\mu(k) = \sum_{l=\max(0,k)}^{\min(M-k,M)} R_{l,l+k}^{-1} \tag{F.3}$$

We recognize that the correlation matrix $\mathbf{R}$ is Toeplitz. We may therefore use the *Gohberg–Semencul formula* (Kailath et al., 1979) to express the (l,k)th element of the inverse matrix $\mathbf{R}^{-1}$ as follows:

$$R_{l,k}^{-1} = \frac{1}{P_M} \sum_{i=0}^{l} (a_{M,i}a_{M,i+k-l}^* - a_{M,M+1-i}^* a_{M,M+1-i-k+l}), \qquad k \geq l \tag{F.4}$$

where $1, a_{M,1}, \ldots, a_{M,M}$ are the coefficients of a prediction-error filter of order M, and P_M is the average prediction-error power. Substituting Eq. (F.4) in (F.3) and confining attention to $k \geq 0$, we get

$$\mu(k) = \frac{1}{P_M} \sum_{l=0}^{M-k} \sum_{i=0}^{l} a_{M,i}a_{M,i+k}^* - \frac{1}{P_M} \sum_{l=0}^{M-k} \sum_{i=0}^{l} a_{M,M+1-i}^* a_{M,M+1-i-k} \tag{F.5}$$

Interchanging the order of summations and setting $j = M + 1 - i - k$, we may rewrite $\mu(k)$ as

$$\mu(k) = \frac{1}{P_M} \sum_{i=0}^{M-k} \sum_{l=i}^{M-k} a_{M,i} a_{M,i+k}^* - \frac{1}{P_M} \sum_{j=1}^{M+1-k} \sum_{l=M+1-j-k}^{M-k} a_{M,j+k}^* a_{M,j} \tag{F.6}$$

The terms that do not involve the index l permit us to collapse the summation over l into a multiplicative integer constant. We may thus combine the two summations in Eq. (F.6). Moreover, we may use the Levinson–Durbin recursion for computing the prediction-error filter coefficients. Given the autocorrelation sequence $r(0), r(1), \ldots, r(M)$, we may now formulate a fast algorithm for computing the MVDR spectrum as follows (Musicus, 1985):

Step 1: Levinson–Durbin Recursion.

Initialize the algorithm by setting

$$a_{0,0} = 1$$

$$P_0 = r(0)$$

Hence, compute for $m = 1, 2, \ldots, M$:

$$\kappa_m = -\frac{1}{P_{m-1}} \sum_{i=0}^{m-1} r(i - m)a_{m-1,i}$$

$$a_{m,i} = \begin{cases} 1 & \text{for } i = 0 \\ a_{m-1,i} + \kappa_m a_{m-1,m-i}^* & \text{for } i = 1, 2, \ldots, m - 1 \\ \kappa_m & \text{for } i = m \end{cases}$$

$$P_m = P_{m-1}(1 - |\kappa_m|^2)$$

Step 2: Correlation of the Predictor Coefficients.

Compute the parameter $\mu(k)$ for varying k:

$$\mu(k) = \begin{cases} \dfrac{1}{P_M} \displaystyle\sum_{i=0}^{M-k} (M + 1 - k - 2i)a_{M,i}a_{M,i+k}^* & \text{for } k = 0, \ldots, M \\ \mu^*(-k) & \text{for } k = -M, \ldots, -1 \end{cases} \tag{F.7}$$

Step 3: MVDR Spectrum Computation.

Use the fast Fourier transform algorithm to compute the MVDR spectrum for varying angular frequency:

$$S_{\text{MVDR}}(\omega) = \frac{1}{\displaystyle\sum_{k=-M}^{M} \mu(k)e^{-j\omega k}} \tag{F.8}$$

F.2 COMPARISON OF MVDR AND MEM SPECTRA

Comparing the formula for computing the MVDR spectrum with that for computing the MEM spectrum, we see that the only difference between the MVDR formula in Eq. (F.8) and the MEM formula in Eq. (E.25) lies in the definitions of their respective correlations of predictor coefficients. In particular, a *linear taper* is used in the definition of $\mu(k)$ given in Eq. (F.7) for the MVDR formula. On the other hand, the definition of the corresponding parameter $\psi(k)$ given in Eq. (E.26) for the MEM formula does *not* involve a taper. This means that for a large-enough model order M, such that $a_{M,i} = 0$ for $i > M/2$, the linear taper involved in the computation of $\mu(k)$ acts like a triangular window on the product terms $a_{M,i}a_{M,i+k}^*$. This has the effect of deemphasizing higher-order terms with large i for large values of lag k (Musicus, 1985). Accordingly, for a given process, an MVDR spectrum is *smoother* in appearance than the corresponding MEM spectrum.

APPENDIX

G

Gradient Adaptive Lattice Algorithm

The adaptive lattice filtering algorithms considered in Chapter 15 are all *exact* manifestations of recursive least-squares estimation, exact in the sense that no approximations are made in their derivations. In this appendix we derive another adaptive lattice filtering algorithm known as the *gradient adaptive lattice (GAL) algorithm* (Griffiths, 1977, 1978), which is a natural extension of the least-mean-square (LMS) algorithm.

Consider a single-stage lattice structure the input–output relation of which is characterized by a single parameter, namely, the *reflection coefficient* κ_m. We assume that the input data are wide-sense stationary and that κ_m is complex valued. Define a cost function for this stage as

$$J_m = E[|f_m(n)|^2 + |b_m(n)|^2] \tag{G.1}$$

where $f_m(n)$ is the forward prediction error and $b_m(n)$ is the backward prediction error, both measured at the output of the stage; E is the statistical expectation operator. The input–output relations of the lattice stage under consideration are described by

$$f_m(n) = f_{m-1}(n) + \kappa_m^* b_{m-1}(n-1)$$

$$b_m(n) = b_{m-1}(n-1) + \kappa_m f_{m-1}(n)$$

The gradient of the cost function J_m with respect to the real and imaginary parts of the reflection coefficient κ_m is given by

$$\nabla J_m = 2E[f_m^*(n)b_{m-1}(n-1) + b_m(n)f_{m-1}^*(n)] \tag{G.2}$$

where $f_{m-1}(n)$ is the forward prediction error and $b_{m-1}(n-1)$ is the delayed backward prediction error, both measured at the input of the lattice stage; the other two prediction errors in Eq. (G.2) refer to the output of the stage. Following the development of the LMS algorithm as presented in Chapter 9, we may use instantaneous estimates of the expectations in Eq. (G.2) and thus write

$$E[f_m^*(n)b_{m-1}(n-1)] \simeq f_m^*(n)b_{m-1}(n-1)$$

$$E[b_m(n)f_{m-1}^*(n)] \simeq b_m(n)f_{m-1}^*(n)$$

Correspondingly, we may express the *instantaneous estimate* of the gradient $\nabla_m J$ as

$$\hat{\nabla}_m J(n) = 2[f_m^*(n)b_{m-1}(n-1) + b_m(n)f_{m-1}^*(n)] \qquad \text{(G.3)}$$

Let $\hat{\kappa}_m(n-1)$ denote the *old estimate* of the reflection coefficient κ_m of the mth lattice stage. Let $\hat{\kappa}_m(n)$ denote the *updated estimate* of this reflection coefficient. We may compute this updated estimate by adding to the old estimate $\kappa_m(n-1)$ a *correction* term proportional to the gradient estimate $\hat{\nabla}_m J(n)$, as shown by

$$\hat{\kappa}_m(n) = \hat{\kappa}_m(n-1) - \frac{1}{2}\mu_m(n)\hat{\nabla}_m J(n) \qquad \text{(G.4)}$$

where μ_m denotes a *time-varying step-size parameter* associated with the mth lattice stage. Substituting Eq. (G.3) in (G.4), we thus get

$$\hat{\kappa}_m(n) = \hat{\kappa}_m(n-1) - \mu_m(n)[f_m^*(n)b_{m-1}(n-1) + b_m(n)f_{m-1}^*(n)] \qquad \text{(G.5)}$$

The adaptation parameter $\mu_m(n)$ is chosen as

$$\mu_m(n) = \frac{\tilde{\mu}}{\mathcal{E}_{m-1}(n)} \qquad \text{(G.6)}$$

where

$$\mathcal{E}_{m-1}(n) = \sum_{i=1}^{n} [|f_{m-1}(i)|^2 + |b_{m-1}(i-1)|^2]$$
$$= \mathcal{E}_{m-1}(n-1) + |f_{m-1}(n)|^2 + |b_{m-1}(n-1)|^2] \qquad \text{(G.7)}$$

For a well-behaved convergence of the algorithm, we usually set $\tilde{\mu} < 0.1$. The parameter $\mathcal{E}_{m-1}(n)$ represents the total energy of both the forward and backward prediction errors at the input of the mth stage, measured up to and including time n.

In practice, a minor modification is made to the *energy estimator* of Eq. (G.7) by writing it in the form of a *single-pole average* of squared data, as shown by (Griffiths, 1977, 1978)

$$\mathcal{E}_{m-1}(n) = \beta\mathcal{E}_{m-1}(n-1) + (1-\beta)[|f_{m-1}(n)|^2 + |b_{m-1}(n-1)|^2] \qquad \text{(G.8)}$$

where $0 < \beta < 1$. The introduction of the parameter β in Eq. (G.8) provides the GAL algorithm with a finite *memory*, which helps it deal better with statistical variations when operating in a nonstationary environment.

TABLE G.1   SUMMARY OF THE GAL ALGORITHM

Parameters: M = final prediction order
β = constant, lying in the range $0 < \beta < 1$
$\tilde{\mu} < 0.1$

Initialization: For prediction order $m = 1, 2, \ldots, M$, put

$$f_m(0) = b_m(0) = 0$$

$$\mathscr{E}_{m-1}(0) = \delta, \qquad \delta = \text{small constant}$$

$$\hat{\kappa}_m(0) = 0$$

For time $n = 1, 2, \ldots$, put

$$f_0(n) = b_0(n) = u(n), \qquad u(n) = \text{lattice predictor input}$$

Prediction: For prediction order $m = 1, 2, \ldots, M$ and time $n = 1, 2, \ldots$, compute

$$f_m(n) = f_{m-1}(n) + \hat{\kappa}_m^*(n)b_{m-1}(n-1)$$

$$b_m(n) = b_{m-1}(n-1) + \hat{\kappa}_m(n)f_{m-1}(n)$$

$$\mathscr{E}_{m-1}(n) = \beta\mathscr{E}_{m-1}(n-1) + (1-\beta)(|f_{m-1}(n)|^2 + |b_{m-1}(n-1)|^2)$$

$$\hat{\kappa}_m(n) = \hat{\kappa}_m(n-1) - \frac{\tilde{\mu}}{\mathscr{E}_{m-1}(n)} [f_{m-1}^*(n)b_m(n) + b_{m-1}(n-1)f_m^*(n)]$$

A summary of the GAL algorithm is presented in Table G.1.

Properties of the GAL Algorithm

The use of time-varying step-size parameter $\mu_m(n) = \tilde{\mu}/\mathscr{E}_{m-1}(n)$ in the update equation for the reflection coefficient $\hat{\kappa}_m(n)$ introduces a form of *normalization* similar to that in the normalized LMS algorithm. From Eq. (G.8) we see that for small magnitudes of the prediction errors $f_{m-1}(n)$ and $b_{m-1}(n)$ the value of the parameter $\mathscr{E}_{m-1}(n)$ is correspondingly small or, equivalently, the step-size parameter $\mu_m(n)$ has a correspondingly large value. Such a behavior is desirable from a practical point of view. Basically, a small value for the prediction errors means that the adaptive lattice predictor is providing an accurate model of the external environment in which it is operating. Hence, if there is any increase in the prediction errors, it should be due to variations in the external environment, in which case it is highly desirable for the adaptive lattice predictor to respond rapidly to such variations. This objective is indeed realized by having the step-size parameter $\mu_m(n)$ assume a large value, which makes it possible for the GAL algorithm to provide an initially rapid convergence to the new environmental conditions. If, on the other hand, the input data applied to the adaptive lattice predictor are too noisy (i.e., they contain a strong white-noise component in addition to the signal of interest), we find that the prediction errors produced by the adaptive lattice predictor are correspondingly large. In such a situation, the parameter $\mathscr{E}_{m-1}(n)$ has a large value or, equivalently, the step-size parameter $\mu_m(n)$ has a small value. Accordingly, the GAL algorithm does *not* respond rapidly to variations in the external environment, which is precisely the way we would like the algorithm to behave (Alexander, 1986a).

Another point of interest is that the convergence behavior of the GAL algorithm is somewhat more rapid than that of the LMS algorithm, but inferior to that of exact recursive LSL algorithms.

APPENDIX

H

Solution of the Difference Equation (9.75)

In this appendix we fill in the mathematical details concerning the mean-squared error analysis of the LMS algorithm. We begin by reproducing Eq. (9.75):

$$\mathbf{x}(n+1) = \mathbf{B}\mathbf{x}(n) + \mu^2 J_{\min}\boldsymbol{\lambda} \tag{H.1}$$

where $\mathbf{B}$ is a real, positive, and symmetric matrix; $\boldsymbol{\lambda}$ is a vector of eigenvalues pertaining to an ensemble-averaged correlation matrix $\mathbf{R}$ of size M-by-M.

Equation (H.1) is a difference equation of order 1 in the vector $\mathbf{x}(n)$. Therefore, assuming an initial value $\mathbf{x}(0)$, the solution to this equation is[1]

$$\mathbf{x}(n) = \mathbf{B}^n\mathbf{x}(0) + \mu^2 J_{\min}\sum_{i=0}^{n-1} \mathbf{B}^i\boldsymbol{\lambda} \tag{H.2}$$

By analogy with the formula for the sum of a geometric series, we may express the finite sum $\sum_{i=0}^{n-1} \mathbf{B}^i$ as follows:

$$\sum_{i=0}^{n-1} \mathbf{B}^i = (\mathbf{I} - \mathbf{B}^n)(\mathbf{I} - \mathbf{B})^{-1} \tag{H.3}$$

where $\mathbf{I}$ is the identity matrix. Substituting Eq. (H.3) in (H.2), we thus get

$$\mathbf{x}(n) = \mathbf{B}^n[\mathbf{x}(0) - \mu^2 J_{\min}(\mathbf{I} - \mathbf{B})^{-1}\boldsymbol{\lambda}] + \mu^2 J_{\min}(\mathbf{I} - \mathbf{B})^{-1}\boldsymbol{\lambda} \tag{H.4}$$

---

[1] The approach we follow here is adapted from Mazo (1979). However, we differ from Mazo in that our analysis is for complex data, whereas that of Mazo is for real data.

The first term on the right-hand side of Eq. (H.4) is the *transient* component of the vector $\mathbf{x}(n)$, and the second term is the *steady-state* component. Since the matrix $\mathbf{B}$ is symmetric, we may apply to it an orthogonal similarity transformation. We may thus write

$$\mathbf{G}^T \mathbf{B} \mathbf{G} = \mathbf{C} \tag{H.5}$$

The matrix $\mathbf{C}$ is a diagonal matrix with elements $c_i = 1, 2, \ldots, M$, which are the eigenvalues of $\mathbf{B}$. The matrix $\mathbf{G}$ is an *orthonormal matrix* whose ith column is the eigenvector $\mathbf{g}_i$ of $\mathbf{B}$, associated with eigenvalue c_i. Because of the property

$$\mathbf{G}\mathbf{G}^T = \mathbf{I} \tag{H.6}$$

we find that

$$\mathbf{B}^n = \mathbf{G}\mathbf{C}^n\mathbf{G}^T \tag{H.7}$$

Hence, we may rewrite Eq. (H.4) in the form

$$\mathbf{x}(n) = \mathbf{G}\mathbf{C}^n\mathbf{G}^T[\mathbf{x}(0) - \mu^2 J_{\min}(\mathbf{I} - \mathbf{B})^{-1}\boldsymbol{\lambda}] + \mu^2 J_{\min}(\mathbf{I} - \mathbf{B})^{-1}\boldsymbol{\lambda} \tag{H.8}$$

Since $\mathbf{C}$ is a diagonal matrix, we have

$$\mathbf{C}^n = \mathrm{diag}[c_1^n, c_2^n, \ldots, c_M^n] \tag{H.9}$$

It follows therefore that the solution defined by Eq. (H.8) is stable if and only if the eigenvalues of matrix $\mathbf{B}$ all have a magnitude less than 1. The eigenvalues of matrix $\mathbf{B}$ are all positive, since the matrix $\mathbf{B}$ is positive definite. For stability, we therefore require the condition

$$0 < c_i < 1 \qquad \text{for all } i \tag{H.10}$$

When this condition is satisfied, the transient component in Eq. (H.8) decays to zero as the number of iterations, n, approaches infinity. This would then leave the steady-state component as the only component. We may thus write

$$\mathbf{x}(\infty) = \mu^2 J_{\min}(\mathbf{I} - \mathbf{B})^{-1}\boldsymbol{\lambda} \tag{H.11}$$

Substituting Eq. (H.11) in (H.8), we may rewrite the solution as

$$\mathbf{x}(n) = \mathbf{G}\mathbf{C}^n\mathbf{G}^T[\mathbf{x}(0) - \mathbf{x}(\infty)] + \mathbf{x}(\infty) \tag{H.12}$$

In view of the diagonal nature of matrix $\mathbf{C}^n$, and since the orthonormal matrix $\mathbf{G}$ consists of the eigenvectors of $\mathbf{B}$ as its columns, we may express the matrix product $\mathbf{G}\mathbf{C}^n\mathbf{G}^T$ as follows:

$$\mathbf{G}\mathbf{C}^n\mathbf{G}^T = \sum_{i=1}^{M} c_i^n \mathbf{g}_i \mathbf{g}_i^T \tag{H.13}$$

Accordingly, we may rewrite Eq. (H.12) one more time in the equivalent form

$$\mathbf{x}(n) = \sum_{i=1}^{M} c_i^n \mathbf{g}_i \mathbf{g}_i^T[\mathbf{x}(0) - \mathbf{x}(\infty)] + \mathbf{x}(\infty) \tag{H.14}$$

This is the desired solution to the difference equation (H.1).

APPENDIX

I

Steady-State Analysis of the LMS Algorithm Without Invoking the Independence Assumption

In this Appendix, we revisit the steady-state analysis of the LMS algorithm by taking an iterative approach that avoids the independence assumption (Butterweck, 1995a). The theory applies to small values of the step-size parameter. It proceeds in two stages. First, a power series solution is derived for the weight-error vector in terms of the step-size parameter. The result so obtained is next used to derive a corresponding expansion for the weight-error correlation matrix.

I.1 ITERATIVE SOLUTION FOR THE WEIGHT-ERROR VECTOR

The weight-error vector $\boldsymbol{\epsilon}(n)$ computed by the LMS algorithm is defined by the stochastic difference equation (9.55), reproduced here for convenience of presentation:

$$\boldsymbol{\epsilon}(n+1) = [\mathbf{I} - \mu\mathbf{u}(n)\mathbf{u}^H(n)]\boldsymbol{\epsilon}(n) + \mu\mathbf{u}(n)e_o^*(n) \tag{I.1}$$

where $\mathbf{u}(n)$ is the tap-input vector, μ is the step-size parameter, and $e_o(n)$ is the estimation error produced by the Wiener solution. Under the condition that μ is small, the direct-averaging method leads us to say that the solution of this equation is approximately the same as that of Eq. (9.56), reproduced here in the form

$$\boldsymbol{\epsilon}_0(n+1) = (\mathbf{I} - \mu\mathbf{R})\boldsymbol{\epsilon}_0(n) + \mu\mathbf{u}(n)e_o^*(n) \tag{I.2}$$

where $\mathbf{R} = E[\mathbf{u}(n)\mathbf{u}^H(n)]$. For reasons that will become apparent presently, we have used a different symbol for the weight-error vector in Eq. (I.2). Note that the solutions of Eqs. (I.1) and (I.2) become equal for the limiting case of a vanishing step-size parameter μ.

In the iterative procedure described by Butterweck (1995a), the solution of Eq. (I.2) is used as a starting point for generating a whole set of solutions of the original stochastic difference equation (I.1). The accuracy of the solution so obtained improves with increasing iteration order. Thus, starting with the solution $\boldsymbol{\epsilon}_0(n)$, the solution of Eq. (I.1) is expressed as a sum of partial functions, as shown by

$$\boldsymbol{\epsilon}(n) = \boldsymbol{\epsilon}_0(n) + \boldsymbol{\epsilon}_1(n) + \boldsymbol{\epsilon}_2(n) + \cdots \tag{I.3}$$

Define the zero-mean difference matrix:

$$\mathbf{P}(n) = \mathbf{u}(n)\mathbf{u}^H(n) - \mathbf{R} \tag{I.4}$$

Then, substituting Eq. (I.4) in (I.1) yields

$$\boldsymbol{\epsilon}_0(n + 1) + \boldsymbol{\epsilon}_1(n + 1) + \boldsymbol{\epsilon}_2(n + 1) + \cdots$$
$$= (\mathbf{I} - \mu\mathbf{R})[\boldsymbol{\epsilon}_0(n) + \boldsymbol{\epsilon}_1(n) + \boldsymbol{\epsilon}_2(n) + \cdots$$
$$- \mu\mathbf{P}(n)[\boldsymbol{\epsilon}_1(n) + \boldsymbol{\epsilon}_2(n) + \cdots] + \mu\mathbf{u}(n)e_o^*(n)$$

from which we readily deduce that

$$\boldsymbol{\epsilon}_i(n+1) = (\mathbf{I} - \mu\mathbf{R})\boldsymbol{\epsilon}_i(n) + \mathbf{f}_i(n), \qquad i = 0, 1, 2, \cdots \tag{I.5}$$

where the subscript i refers to the iteration order. The "driving force" $\mathbf{f}_i(n)$ for the difference equation (I.5) is defined by

$$\mathbf{f}_i(n) = \begin{cases} \mu\mathbf{u}(n)e_o^*(n), & i = 0 \\ -\mu\mathbf{P}(n)\boldsymbol{\epsilon}_{i-1}(n), & i = 1, 2, \cdots \end{cases} \tag{I.6}$$

Thus, a time-varying system characterized by the stochastic difference equation (I.1) is transformed into a set of equations having the same basic format as described in (I.5), such that the solution to the ith equation in the set (i.e., step i in the iterative procedure) follows from the $(i-1)$th equation. In particular, the problem is reduced to a study of the transmission of a stationary stochastic process through a low-pass filter with an extremely low cutoff frequency.

I.2 SERIES EXPANSION OF THE WEIGHT-ERROR CORRELATION MATRIX

On the basis of Eq. (I.3), we may express the weight-error correlation matrix in the form of a corresponding series as follows:

$$\mathbf{K}(n) = E[\boldsymbol{\epsilon}(n)\boldsymbol{\epsilon}^H(n)]$$
$$= \sum_i \sum_k E[\boldsymbol{\epsilon}_i(n)\boldsymbol{\epsilon}_k^H(n)], \qquad (i,k) = 0, 1, 2, \cdots \tag{I.7}$$

Expanding this series in light of the definitions given in Eqs. (I.5) and (I.6), and then grouping equal-order terms in the step-size parameter μ, we get the following series expansion:

$$\mathbf{K}(n) = \mathbf{K}_0(n) + \mu\mathbf{K}_1(n) + \mu^2\mathbf{K}_2(n) + \cdots \tag{I.8}$$

where the various matrix coefficients are themselves defined as follows:

$$\mathbf{K}_j(n) = \begin{cases} E[\boldsymbol{\epsilon}_0(n)\boldsymbol{\epsilon}_0^H(n)] & \text{for } j = 0 \\[2mm] \displaystyle\sum_i \sum_k E[\boldsymbol{\epsilon}_i(n)\boldsymbol{\epsilon}_k^H(n)] & \begin{array}{l}\text{for all } (i,k) \geq 0 \\ \text{such that } i + k = 2j-1,\ 2j\end{array} \end{cases} \tag{I.9}$$

These matrix coefficients are defined, albeit in a rather complex fashion, by the spectral and probability distribution of the environment in which the LMS algorithm operates. In a general setting with arbitrarily colored signals, the calculation of $\mathbf{K}_j(n)$ for $j \geq 1$ can be rather tedious, except in some special cases (Butterweck, 1995a).

The zero-order term $\mathbf{K}_0(n)$ in Eq. (I.8) is of special interest for two reasons. First, for a small μ it may be used as an approximation to the actual $\mathbf{K}(n)$, as discussed in Section 9.4. Second, it lends itself to examination without any statistical assumptions concerning the environment in which the LMS algorithm operates. In particular, we find that under steady-state conditions (i.e., large n), $\mathbf{K}_0(n)$ is determined as the solution to the equation (Butterweck, 1995b):

$$\mathbf{R}\mathbf{K}_0(n) + \mathbf{K}_0(n)\mathbf{R} = \mu \sum_l J_{\min}^{(l)}\ \mathbf{R}^{(l)}, \qquad \text{large } n \tag{I.10}$$

where

$$J_{\min}^{(l)} = E[e_o(n)\ e_o^*(n - l)], \qquad l = 0, 1, 2, \ldots \tag{I.11}$$

$$\mathbf{R}^{(l)} = E[\mathbf{u}(n)\mathbf{u}^H(n - l)], \qquad l = 0, 1, 2, \ldots \tag{I.12}$$

Note that for $l = 0$, we have $J_{\min}^{(0)} = J_{\min}$ and $\mathbf{R}^{(0)} = \mathbf{R}$.

The steady-state value of the misadjustment $\mathcal{M}$ derived in Chapter 9 under the independence assumption corresponds to setting $l = 0$ in Eq. (I.10) and ignoring all higher-order terms. This special case corresponds to the assumption that the estimation error $e(n)$ produced by the LMS algorithm is drawn from a white noise process. Thus, Eq. (I.10) is approximated by

$$\mathbf{R}\mathbf{K}_0(n) + \mathbf{K}_0(n)\mathbf{R} \simeq \mu J_{\min}\mathbf{R}, \qquad \text{large } n$$

from which we readily find that the misadjustment is

$$\mathcal{M} = \frac{\text{tr}[\mathbf{R}\mathbf{K}_0(n)]}{J_{\min}}$$

$$\simeq \frac{\mu}{2}\ \text{tr}[\mathbf{R}]$$

$$= \frac{\mu}{2} \sum_{i=1}^{M} \lambda_i$$

This is indeed the result derived in Eq. (9.95).

APPENDIX

$$\boxed{\text{J}}$$

The Complex Wishart Distribution

The Wishart distribution plays an important role in statistical signal processing. In this appendix we present a summary of some important properties of the Wishart distribution for complex-valued data. In particular, we derive a result that is pivotal to a rigorous analysis of the convergence behavior of the standard RLS algorithm, presented in Chapter 13. We begin the discussion with a definition of the complex Wishart distribution.

J.1 DEFINITION

Consider an M-by-M time-averaged (sample) correlation matrix $\boldsymbol{\Phi}(n)$, defined by

$$\boldsymbol{\Phi}(n) = \sum_{i=1}^{n} \mathbf{u}(i)\mathbf{u}^{H}(i) \tag{J.1}$$

where

$$\mathbf{u}(i) = [u_1(i), u_2(i), \ldots, u_M(i)]^{T}$$

In what follows, we assume that $\mathbf{u}(1), \mathbf{u}(2), \ldots, \mathbf{u}(n)$ $(n > M)$ are *independently and identically distributed*. We may then formally define the *complex Wishart distribution* as follows (Muirhead, 1982):

If $\{u_1(i), u_2(i), \ldots, u_M(i) \mid i = 1, 2, \ldots, n\}, n \geq M$, is a sample from the M-dimensional Gaussian distribution $\mathcal{N}(\mathbf{0}, \mathbf{R})$, and if $\mathbf{\Phi}(n)$ is the time-averaged correlation matrix defined in Eq. (J.1), then the elements of $\mathbf{\Phi}(n)$ have the complex Wishart distribution $\mathcal{W}_M(n, \mathbf{R})$, which is characterized by the parameters M, n, and $\mathbf{R}$.

In specific terms, we may say that if matrix $\mathbf{\Phi}$ is $\mathcal{W}_M(n, \mathbf{R})$, then the probability density function of $\mathbf{\Phi}$ is

$$f(\mathbf{\Phi}) = \frac{1}{2^{Mn/2} \Gamma_M\left(\frac{1}{2}n\right)(\det(\mathbf{R}))^{n/2}} \, \mathrm{etr}\left(-\frac{1}{2}\mathbf{R}^{-1}\mathbf{\Phi}\right)(\det(\mathbf{\Phi}))^{(n-M-1)/2} \tag{J.2}$$

where $\det(\cdot)$ denotes the determinant of the enclosed matrix, $\mathrm{etr}(\cdot)$ denotes the exponential raised to the trace of the enclosed matrix, and $\Gamma_M(a)$ is the *multivariate gamma function* defined by

$$\Gamma_M(a) = \int_{\mathbf{A}} \mathrm{etr}(-\mathbf{A})(\det(\mathbf{A}))^{a-(M+1)/2}d\mathbf{A} \tag{J.3}$$

where $\mathbf{A}$ is a positive definite matrix.

J.2 *THE* CHI-SQUARE DISTRIBUTION AS A SPECIAL CASE

For the special case of a univariate distribution, that is, $M = 1$, Eq. (J.1) reduces to the scalar form:

$$\varphi(n) = \sum_{i=1}^{n} |u(i)|^2 \tag{J.4}$$

Correspondingly, the correlation matrix $\mathbf{R}$ reduces to the variance σ^2. Let

$$\chi^2(n) = \frac{\varphi(n)}{\sigma^2} \tag{J.5}$$

Then, using Eq. (J.2) we may define the normalized probability density function of the normalized random variable $\chi^2(n)$ as

$$f(\chi^2) = \frac{\left(\dfrac{\chi^2}{2}\right)^{n/2-1} e^{-\chi^2/2}}{2^{n/2}\Gamma\left(\dfrac{1}{2}n\right)} \tag{J.6}$$

where $\Gamma(1/2n)$ is the (scalar) *gamma function.*[1] The variable $\chi^2(n)$, defined above, is said to have a *chi-square distribution with n degrees of freedom.* We may thus view the complex Wishart distribution as a generalization of the univariate chi-square distribution.

A useful property of a chi-square distribution with n degrees of freedom is the fact that it is *reproductive with respect to 1/2n* (Wilks, 1962). That is, the rth moment of $\chi^2(n)$ is

$$E[\chi^{2r}(n)] = \frac{2^r \Gamma\left(\dfrac{n}{2} + r\right)}{\Gamma\left(\dfrac{n}{2}\right)} \qquad (J.7)$$

Thus, the mean, mean-square, and variance of $\chi^2(n)$ are as follows, respectively:

$$E[\chi^2(n)] = n \qquad (J.8)$$

$$E[\chi^4(n)] = n(n + 2) \qquad (J.9)$$

$$\text{var}[\chi^2(n)] = n(n + 2) - n^2 = 2n \qquad (J.10)$$

Moreover, from Eq. (J.7) we find that the mean of the reciprocal of $\chi^2(n)$ is

$$E\left[\frac{1}{\chi^2(n)}\right] = \frac{1}{2}\frac{\Gamma\left(\dfrac{n}{2} - 1\right)}{\Gamma\left(\dfrac{n}{2}\right)}$$

$$\qquad (J.11)$$

$$= \frac{1}{2}\frac{\Gamma\left(\dfrac{n}{2} - 1\right)}{\left(\dfrac{n}{2} - 1\right)\Gamma\left(\dfrac{n}{2} - 1\right)} = \frac{1}{n - 2}$$

---

[1]For the general case of a complex number g whose real part is positive, the *gamma function* $\Gamma(g)$ is defined by the definite integral (Wilks, 1962)

$$\Gamma(g) = \int_0^\infty x^{g-1} e^{-x} \, dx$$

Integrating it by parts, we readily find that

$$\Gamma(g) = (g - 1)\Gamma(g - 1)$$

For the case when g is a positive integer, we may thus express the gamma function $\Gamma(g)$ as the factorial

$$\Gamma(g) = (g - 1)!$$

When $g > 0$, but not an integer, we have

$$\Gamma(g) = (g - 1)\,\Gamma(\delta)$$

where $0 < \delta < 1$. For the particular case of $\delta = 1/2$, we have $\Gamma(\delta) = \sqrt{\pi}$.

J.3 PROPERTIES OF THE COMPLEX WISHART DISTRIBUTION

Returning to the main theme of this appendix, the complex Wishart distribution has some important properties of its own, which are summarized as follows (Muirhead, 1982; Anderson, 1984):

1. If Φ is $\mathcal{W}_M(n, \mathbf{R})$ and $\mathbf{a}$ is any M-by-1 random vector distributed independently of Φ with $P(\mathbf{a} = \mathbf{0}) = 0$ (i.e., the probability that $\mathbf{a} = \mathbf{0}$ is zero), then $\mathbf{a}^H \Phi \mathbf{a} / \mathbf{a}^H \mathbf{R} \mathbf{a}$ is chi-square distributed with n degrees of freedom, and is independent of $\mathbf{a}$.

2. If Φ is $\mathcal{W}_M(n, \mathbf{R})$ and $\mathbf{Q}$ is a matrix of dimensions M-by-k and rank k, then $\mathbf{Q}^H \Phi \mathbf{Q}$ is $\mathcal{W}_k(n, \mathbf{Q}^H \mathbf{R} \mathbf{Q})$.

3. If Φ is $\mathcal{W}_M(n, \mathbf{R})$ and $\mathbf{Q}$ is a matrix of dimensions M-by-k and rank k, then $(\mathbf{Q}^H \Phi^{-1} \mathbf{Q})^{-1}$ is $\mathcal{W}_k(n - M + k, (\mathbf{Q}^H \mathbf{R}^{-1} \mathbf{Q})^{-1})$.

4. If Φ is $\mathcal{W}_M(n, \mathbf{R})$ and $\mathbf{a}$ is any M-by-1 random vector distributed independently of Φ with $P(\mathbf{a} = \mathbf{0}) = 0$, then $\mathbf{a}^H \mathbf{R}^{-1} \mathbf{a} / \mathbf{a}^H \Phi^{-1} \mathbf{a}$ is chi-square distributed with $n - M + 1$ degrees of freedom.

5. Let Φ and $\mathbf{R}$ be partitioned into p and $M - p$ rows and columns, as shown by

$$\Phi = \begin{bmatrix} \Phi_{11} & \Phi_{12} \\ \Phi_{21} & \Phi_{22} \end{bmatrix}$$

$$\mathbf{R} = \begin{bmatrix} \mathbf{R}_{11} & \mathbf{R}_{12} \\ \mathbf{R}_{21} & \mathbf{R}_{22} \end{bmatrix}$$

If Φ is distributed according to $\mathcal{W}_M(n, \mathbf{R})$, then Φ_{11} is distributed according to $\mathcal{W}_p(n, \mathbf{R}_{11})$.

J.4 EXPECTATION OF THE INVERSE CORRELATION MATRIX $\Phi^{-1}(n)$

Property 4 of the complex Wishart distribution may be used to find the expectation of the inverse correlation matrix $\Phi^{-1}(n)$, which is associated with the convergence of the RLS algorithm in the mean square. Specifically, for any fixed and nonzero $\boldsymbol{\alpha}$ in $\mathbb{R}^M$, we know from Property 4 described above that $\boldsymbol{\alpha}^H \mathbf{R}^{-1} \boldsymbol{\alpha} / \boldsymbol{\alpha}^H \Phi^{-1} \boldsymbol{\alpha}$ is chi-square distributed with $n - M + 1$ degrees of freedom. Let $\chi^2(n - M + 1)$ denote this ratio. Then, using the result described in Eq. (J.11), we may write

$$E[\boldsymbol{\alpha}^H \Phi^{-1}(n) \boldsymbol{\alpha}] = \boldsymbol{\alpha}^H \mathbf{R}^{-1} \boldsymbol{\alpha} E\left[\frac{1}{\chi^2(n - M + 1)} \right]$$

$$= \frac{1}{n - M - 1} \boldsymbol{\alpha}^H \mathbf{R}^{-1} \boldsymbol{\alpha}, \qquad n > M + 1$$

which, in turn, implies that

$$E[\Phi^{-1}(n)] = \frac{1}{n - M - 1} \mathbf{R}^{-1}, \qquad n > M + 1 \tag{J.12}$$

Glossary

TEXT CONVENTIONS

1. Boldfaced lowercase letters are used to denote column vectors. Boldfaced uppercase letters are used to denote matrices.

2. The estimate of a scalar, vector, or matrix is designated by the use of a hat (ˆ) placed over the pertinent symbol.

3. The symbol | | denotes the magnitude or absolute value of a complex scalar enclosed within. The symbol ang[ ] or arg[ ] denotes the phase angle of the scalar enclosed within.

4. The symbol ‖ ‖ denotes the Euclidean norm of the vector or matrix enclosed within.

5. The symbol det( ) denotes the determinant of the square matrix enclosed within.

6. The open interval (a, b) of the variable x signifies that $a < x < b$. The closed interval $[a, b]$ signifies that $a \leq x \leq b$, and $(a, b]$ signifies that $a < x \leq b$.

7. The inverse of nonsingular (square) matrix $\mathbf{A}$ is denoted by $\mathbf{A}^{-1}$.

8. The pseudoinverse of matrix $\mathbf{A}$ (not necessarily square) is denoted by $\mathbf{A}^{+}$.

9. Complex conjugation of a scalar, vector, or matrix is denoted by the use of an asterisk as superscript. Transposition of a vector or matrix is denoted by superscript T. Hermitian transposition (i.e., complex conjugation and transposition

combined) of a vector or matrix is denoted by superscript H. Backward rearrangement of the elements of a vector is denoted by superscript B.

10. The symbol $\mathbf{A}^{-H}$ denotes the Hermitian transpose of the inverse of a nonsingular (square) matrix $\mathbf{A}$.

11. The square root of a square matrix $\mathbf{A}$ is denoted by $\mathbf{A}^{1/2}$.

12. The symbol $\text{diag}[\lambda_1, \lambda_2, \ldots, \lambda_M]$ denotes a diagonal matrix whose elements on the main diagonal equal $\lambda_1, \lambda_2, \ldots, \lambda_M$.

13. The order of linear predictor or the order of autoregressive model is signified by a subscript added to the pertinent scalar or vector parameter.

14. The statistical expectation operator is denoted by $E[\cdot]$, where the quantity enclosed is the random variable or random vector of interest. The variance of a random variable is denoted by $\text{var}[\cdot]$, where the quantity enclosed is the random variable.

15. The conditional probability density function of random variable U, given that hypothesis H_i is true, is denoted by $f_U(u|H_i)$, where u is the sample value of random variable U.

16. The inner product of two vectors $\mathbf{x}$ and $\mathbf{y}$ is defined as $\mathbf{x}^H\mathbf{y} = \mathbf{y}^T\mathbf{x}^*$. Another possible inner product is $\mathbf{y}^H\mathbf{x} = \mathbf{x}^T\mathbf{y}^*$. These two inner products are the complex conjugate of each other. The outer product of the vectors $\mathbf{x}$ and $\mathbf{y}$ is defined as $\mathbf{x}\mathbf{y}^H$. The inner product is a scalar, whereas the outer product is a matrix.

17. The trace of a square matrix $\mathbf{R}$ is denoted by $\text{tr}[\mathbf{R}]$; it is defined as the sum of the diagonal elements of $\mathbf{R}$. The exponential raised to the trace of matrix $\mathbf{R}$ is denoted by $\text{etr}[\mathbf{R}]$.

18. The autocorrelation function of stationary discrete-time stochastic process $u(n)$ is defined by

$$r(k) = E[u(n)u^*(n - k)]$$

The cross-correlation function between two jointly stationary discrete-time stochastic process $u(n)$ and $d(n)$ is defined by

$$p(-k) = E[u(n - k)d^*(n)]$$

19. The ensemble-averaged correlation matrix of a random vector $\mathbf{u}(n)$ is defined by

$$\mathbf{R} = E[\mathbf{u}(n)\mathbf{u}^H(n)]$$

20. The ensemble-averaged cross-correlation vector between a random vector $\mathbf{u}(n)$ and a random variable $d(n)$ is defined by

$$\mathbf{p} = E[\mathbf{u}(n)d^*(n)]$$

21. The time-averaged (sample) correlation matrix of a vector $\mathbf{u}(i)$ over the observation interval $1 \leq i \leq n$ is defined by

$$\mathbf{\Phi}(n) = \sum_{i=1}^{n} \mathbf{u}(i)\mathbf{u}^H(i)$$

The exponentially weighted version of $\boldsymbol{\Phi}(n)$ is

$$\boldsymbol{\Phi}(n) = \sum_{i=1}^{n} \lambda^{n-i} \, \mathbf{u}(i)\mathbf{u}^{H}(i)$$

22. The time-averaged cross-correlation vector between a vector $\mathbf{u}(i)$ and a scalar $d(i)$ over the observation interval $1 \leq i \leq n$ is defined by

$$\mathbf{z}(n) = \sum_{i=1}^{n} \mathbf{u}(i)d^*(i)$$

Its exponentially weighted version is

$$\mathbf{z}(n) = \sum_{i=1}^{n} \lambda^{n-i} \, \mathbf{u}(i)d^*(i)$$

23. The discrete-time Fourier transform of a time function $u(n)$ is denoted by $F[u(n)]$. The inverse discrete-time Fourier transform of a frequency function $U(\omega)$ is denoted by $F^{-1}[U(\omega)]$.

24. In constructing block diagrams (signal-flow graphs) involving scalar quantities, the following symbols are used. The symbol

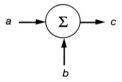

denotes an adder with $c = a + b$. The same symbol with algebraic signs added as in the following

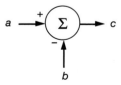

denotes a substractor with $c = a - b$. The symbol

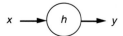

denotes a multiplier with $y = hx$. This multiplication is also represented as

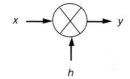

The unit-sample (delay) operator is denoted by

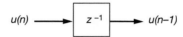

25. In constructing block diagrams (signal-flow graphs) involving matrix quantities, the following symbols are used. The symbol

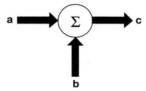

denotes summation with $\mathbf{c} = \mathbf{a} + \mathbf{b}$. The symbol

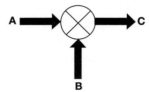

denotes multiplication with $\mathbf{C} = \mathbf{AB}$. The symbol

denotes a branch having transmittance $\mathbf{H}$, with $\mathbf{y} = \mathbf{Hx}$. The unit-sample operator is denoted by the symbol

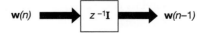

ABBREVIATIONS

ADPCM	Adaptive differential pulse-code modulation
AGC	Automatic gain control
AIC	An information-theoretic criterion
ALE	Adaptive line enhancer
AR	Autoregressive
ARMA	Autoregressive-moving average
as	Almost surely
BIBO	Bounded input-bounded output
b/s	Bits per second
BLP	Backward linear prediction
BP	Back-propagation
BPSK	Binary phase-shift keying
CMA	Constant modulus adaptive
DCT	Discrete cosine transform
DFT	Discrete Fourier transform
DPCM	Differential pulse-code modulation
DTE	Data terminal equipment
EKF	Extended Kalman filter
FBLP	Forward-backward linear prediction
FDAF	Frequency-domain adaptive filter
FFT	Fast Fourier transform
FIR	Finite-duration impulse response
FLP	Forward linear prediction
FTF	Fast transversal filters ((algorithm)
GAL	Gradient adaptive lattice
GSLC	Generalized sidelobe canceler
HF	High frequency
HOS	Higher-order statistics
Hz	Hertz
IFFT	Inverse fast Fourier transform
iid	Independent and identically distributed
IIR	Infinite-duration impulse response
INR	Interference-to-noise ratio
kb/s	Kilobits per second
kHz	Kilohertz
LCMV	Linearly constrained minimum variance

LMS	Least-mean-square
LBS	Least significant bit
LSL	Least-squares lattice
LPC	Linear predictive coding
MA	Moving average
MDL	Minimum description length (criterion)
MEM	Maximum entropy method
MN	Minimum norm
MLP	Multilayer perceptron
MSE	Mean-squared error
MVDR	Minimum-variance distortionless response
PAM	Pulse amplitude modulation
PARCOR	Partial correlation
PCM	Pulse-code modulation
PN	Pseudonoise
QAM	Quadrature amplitude modulation
QPSK	Quadrature phase-shift keying
QR-RLS	QR-decomposition–based recursive least squares
QRD-LSL	QR-decomposition–based least-squares lattice
RBF	Radial basis function
RLS	Recursive least squares
rms	root-mean-square
ROC	Region of convergence (z-transform)
ROC	Rate of convergence (adaptive filter)
s	Second
SIMO	Single input–multiple output
SNR	Signal-to-noise ratio
SOBAF	Self-orthogonalizing block adaptive filter
SRF	Square-root filtering
SVD	Singular value decomposition
TDNN	Time-delay neural network
wp1	With probability one

PRINCIPAL SYMBOLS

$a_{M,k}(n)$	kth tap weight of forward prediction-error filter of order M (at iteration n), with $k = 0, 1, \ldots, M$; note that $a_{m,0}(n) = 1$

$\mathbf{a}_M(n)$	Tap-weight vector of forward prediction-error filter of order M (at iteration n)
$\mathbf{A}$	Data matrix in the covariance method
$\mathbf{A}(n)$	Data matrix in the pre-windowing method
$b_M(n)$	Backward (*a posteriori*) prediction error produced at iteration n by prediction-error filter of order M
$\mathbf{b}(n)$	Backward (*a posteriori*) prediction-error vector representing sequence of errors produced by backward prediction-error filters of orders $0, 1, \ldots, M$
$\mathcal{B}_M(n)$	Sum of weighted backward prediction error squares produced by backward prediction-error filter of order M
$c_{M,k}(n)$	kth tap weight of backward prediction-error filter of order M (at iteration n), with $k = 0, 1, \ldots, M$; note that $c_{M,M}(n) = 1$
$\mathbf{c}_M(n)$	Tap-weight vector of backward prediction-error filter of order M (at iteration n)
$\mathbf{c}(n)$	Weight-error vector in steepest-descent algorithm
$c_k(\tau_1, \tau_2, \ldots, \tau_k)$	kth-order cumulant
$C_k(\omega_1, \omega_2, \ldots, \omega_k)$	kth-order polyspectrum
$\mathcal{C}$	Contour
$\mathcal{C}^M$	Complex M-dimensional parameter space
$\mathcal{C}(n)$	Convergence ratio
$\det()$	Determinant of the enclosed matrix
$\operatorname{diag}()$	Diagonal matrix
$d(n)$	Desired response
$\mathbf{d}$	Desired response vector in the covariance method
$\mathbf{d}(n)$	Desired response vector in the pre-windowing method
D	Unit-delay operator (same as z^{-1})
$\mathbf{D}_{m+1}(n)$	Correlation matrix of backward prediction errors
$\mathcal{D}$	Mean-square deviation
$\operatorname{dec}()$	Function describing the decision performed by a threshold device
$e(n)$	*A posteriori* estimation error
$e_m(n)$	*A posteriori* estimation error at the output of stage m in the joint-process estimator using the recursive LSL algorithm or QRD–LSL algorithm
e	Base of natural logarithm
$\operatorname{etr}()$	Exponential raised to the trace of the enclosed matrix
$\exp$	Exponential

E	Expectation operator
$\mathscr{E}(\mathbf{w}, n)$	Cost function defined as the sum of weighted error squares expressed as a function of iteration n
$\mathscr{E}(\mathbf{w})$	Cost function defined as the sum of error squares, expressed as a function of the tap-weight vector $\mathbf{w}$
$\mathscr{E}_{\min}$	Minimum value of $\mathscr{E}(\mathbf{w})$
$\mathscr{E}(n)$	Cost function defined as the sum of weighted error squares, expressed as a function of iteration n
$f_M(n)$	Forward (*a posteriori*) prediction error produced at iteration n by forward prediction-error filter of order M
$\mathbf{f}(n)$	Forward (*a posteriori*) prediction error vector representing sequence of errors produced by forward prediction-error filters of orders $0, 1, \ldots, M$
$f_U(n)$	Probability density function of random variable U, whose sample value equals u
$f_{\mathbf{U}}(\mathbf{u})$	Joint probability density function of the elements of random vector $\mathbf{U}$, whose sample value equals $\mathbf{u}$
$F_M(z)$	z-transform of sequence of forward prediction errors produced by forward prediction-error filter of order M
$\mathbf{F}(n + 1, n)$	Transition matrix
$\mathscr{F}_M(n)$	Weighted sum of forward prediction-error squares produced by forward prediction-error filter of order M
$F[\ ]$	Fourier transform operator
$F^{-1}[\ ]$	Inverse Fourier transform operator
$g(\cdot)$	Nonlinear function used in blind equalization
$\mathbf{G}(n)$	Kalman gain
h_k	Kth regression coefficient of joint-process estimation based on lattice predictor for stationary impulse
h_n	Minimum-phase polynomial used in blind equalization
H_i	ith hypothesis
$H(z)$	Transfer function of discrete-time linear filter
I	Subscript for signifying the in-phase (real) component of a complex baseband signal
$\mathbf{I}$	Identity matrix
$\mathbf{I}$	Inverse of Fisher's information matrix $\mathbf{J}$
j	Square root of -1
$J(\mathbf{w})$	Cost function used to formulate the Wiener filtering problem, expressed as a function of the tap-weight vector $\mathbf{w}$

$\mathbf{J}$	Fisher's information matrix
$\mathbf{k}(n)$	Gain vector in the RLS algorithm
K	Final order of moving average model
$\mathbf{K}(n)$	Correlation matrix of weight-error vector $\boldsymbol{\epsilon}(n)$
$\ln$	Natural logarithm
$\mathbf{L}(n)$	Transformation matrix in the form of lower triangular matrix
m	Variable order of linear predictor or autoregressive model
M	Final order of linear predictor or autoregressive model
M, K	Final order of autoregressive-moving average model
$\mathcal{M}$	Misadjustment
n	Discrete-time or number of iterations applied to recursive algorithm
N	Data length
$\mathcal{N}$	Symbol signifying the Gaussian (normal) distribution
$O(z)$	Maximum phase polynomial
$O(M^k)$	Order of M^k
$p(-k)$	Element of cross-correlation vector $\mathbf{p}$ for lag k
$\mathbf{p}$	Cross-correlation vector between tap-input vector $\mathbf{u}(n)$ and desired response $d(n)$
P_M	Average value of (forward or backward) prediction-error power for prediction order M for stationary inputs
$\mathbf{P}(n)$	Matrix equal to the inverse of the time-averaged correlation matrix $\boldsymbol{\Phi}(n)$ used in formulating the RLS algorithm
q_{ki}	ith element of kth eigenvector
$\mathbf{q}_k$	kth eigenvector
Q	Subscript for signifying the quadrature (imaginary) component of a complex baseband signal
$\mathbf{Q}$	Unitary matrix that consists of normalized eigenvectors in the set $\{\mathbf{q}_k\}$ used as columns
$Q(y)$	Probability distribution function of standardized Gaussian random variable
$r(k)$	Element of (ensemble-averaged) correlation matrix $\mathbf{R}$ for lag k
$\mathbf{R}$	Ensembled-average correlation matrix of stationary discrete-time process $u(n)$
$\mathbb{R}^M$	Real M-dimensional parameter space

s	Signal vector; steering vector
sgn()	Signum function
$S(\omega)$	Power spectral density
$S_{\text{AR}}(\omega)$	Autoregressive spectrum
$S_{\text{MEM}}(\omega)$	MEM (maximum entropy method) spectrum
$S_{\text{MVDR}}(\omega)$	Minimum variance distortionless response spectrum
$\mathscr{S}$	System
$\mathscr{S}_d$	Decreasingly excited subspace
$\mathscr{S}_o$	Otherwise excited subspace
$\mathscr{S}_p$	Persistently excited subspace
$\mathscr{S}_u$	Unexcited subspace
t	Time
t	Vector arising in joint-process estimation for nonstationary inputs
$\mathbf{t}_k$	Vector defining the center of the kth kernel in RBF network
$u(n)$	Sample value of tap input in transversal filter at time n
$\mathbf{u}(n)$	Tap-input vector consisting of $u(n)$, $u(n-1)$, . . . , as elements
$u_I(n)$	In-phase component of $u(n)$
$u_Q(n)$	Quadrature component of $u(n)$
$\mathbf{u}_k$	kth left-singular vector of data matrix **A**
U	Matrix of left-singular vectors of data matrix **A**
$\mathscr{U}_n$	Space spanned by tap inputs $u(n)$, $u(n-1)$, . . .
$\mathscr{U}(n)$	Sum of weighted squared values of tap inputs $u(i)$, $i = 1, 2, \ldots, n$
$v(n)$	Sample value of white-noise process of zero mean
$\mathbf{v}_1(n)$	Process noise vector
$\mathbf{v}_2(n)$	Measurement noise vector
$\mathbf{v}(n)$	Process noise vector in random-walk state model
$\mathbf{v}_k(n)$	kth right-singular vector of data matrix **A**
V	Matrix of right singular vectors of data matrix **A**
$w_k(n)$	kth tap weight of transversal filter at time n
$w_{b,m,k}(n)$	kth tap weight of backward predictor of order m at iteration n
$w_{f,m,k}(n)$	kth tap weight of forward predictor of order m at iteration n
$\mathbf{w}(n)$	Tap-weight vector of transversal filter at time n

$\mathbf{w}_{b,m}(n)$	Tap-weight vector of backward predictor of order m at iteration n
$\mathbf{w}_{f,m}(n)$	Tap-weight vector of forward predictor of order m at iteration n
$\mathcal{W}$	Symbol signifying the Wishart distribution
$\mathbf{x}(n)$	State vector
$\mathbf{y}(n)$	Observation vector used in formulating Kalman filter theory
$\mathcal{Y}_n$	Vector space spanned by $y(n), y(n-1), \ldots$
z^{-1}	Unit-sample (delay) operator used in defining the z-transform of a sequence
$\mathbf{z}$	Time-averaged cross-correlation vector between tap-input vector $\mathbf{u}(i)$ and desired response $d(i)$
$Z(y)$	Standardized Gaussian probability density function
$\alpha(n)$	Innovation at time n
$\boldsymbol{\alpha}(n)$	Innovations vector
β	Constant used in the DCT–LMS algorithm
β	Constant used in the GAL algorithm
$\beta_m(n)$	Backward prediction error of order m
$\gamma(n)$	Conversion factor used in the FTF algorithm, recursive LSL algorithm, and recursive QRD–LSL algorithm
$\Gamma(g)$	Gamma function of g
γ_3	Skewness of a random variable
γ_4	Kurtosis of a random variable
δ	Constant used in the initialization of the RLS family of algorithms
$\boldsymbol{\delta}$	First coordinator vector
δ_l	Kronecker delta, equal to 1 for $l = 0$ and zero for $l \neq 0$
Δ_m	Cross-correlation between forward prediction error $f_m(n)$ and delayed backward prediction error $b_m(n-1)$
$\Delta_m(n)$	Parameter in recursive LSL algorithm
$\epsilon_m(n)$	Angle-normalized joint-process estimation error for prediction order m
$\epsilon_{b,m}(n)$	Angle-normalized backward prediction error for prediction order m
$\epsilon_{f,m}(n)$	Angle-normalized forward prediction error for prediction order m
$\boldsymbol{\epsilon}(n)$	Weight-error vector
$\boldsymbol{\epsilon}$	Estimation error vector in the covariance method

$\boldsymbol{\epsilon}(n)$	Estimation error vector in the prewindowed method
$\eta(n)$	Forward (*a priori*) prediction error
$\boldsymbol{\theta}$	Parameter vector
$\boldsymbol{\Theta}$	Unitary rotation
κ_m	mth reflection of a lattice predictor for stationary environment
$\kappa_{b,m}(n)$	mth backward reflection coefficient of a least-squares lattice for a non-stationary environment
$\kappa_{f,m}(n)$	mth forward reflection coefficient of a least squares lattice for a non-stationary environment
$\kappa_m(n)$	mth joint-process regression coefficient in the recursive LSL and QLD-LSL algorithms at iteration n
$\kappa_4(\tau_1, \tau_2, \tau_3)$	Tricepstrum
λ	Exponential weighting vector in RLS, FTF, LSL, QR -RLS, and QRD–LSL algorithms
λ_k	kth eigenvalue of correlation matrix $\mathbf{R}$
$\lambda_{\max}$	Maximum eigenvalue of correlation matrix $\mathbf{R}$
$\lambda_{\min}$	Minimum eigenvalue of correlation matrix $\mathbf{R}$
Λ	Likelihood ratio
$\ln \Lambda$	Log-likelihood ratio
$\boldsymbol{\Lambda}(n)$	Diagonal matrix of exponential weighting factors
μ	Mean value
μ	Step-size parameter in steepest-descent algorithm or LMS algorithm
μ	Constant used in the FTF algorithm with soft constraint
$\boldsymbol{v}(n)$	Normalized weight-error vector in steepest-descent algorithm
$\boldsymbol{\pi}$	Vector in RLS algorithm
$\pi_m(n)$	mth parameter in QRD–LSL algorithm
$\xi(n)$	*Apriori* estimation error
$\phi(t, k)$	t, kth element of time-averaged correlation matrix $\boldsymbol{\Phi}$
$\boldsymbol{\varphi}(n, n_0)$	Transition matrix arising in finite-precision analysis of RLS algorithms
$\boldsymbol{\Phi}$	Interpolation matrix in RBF network
$\boldsymbol{\Phi}$	Time-averaged correlation matrix
$\boldsymbol{\Phi}(n)$	Time-averaged correlation matrix expressed as a function of the observation interval n
$\chi^2(n)$	Chi-square distributed random variable with n degrees of freedom

$\varphi(v)$	Activation function of a neuron, expressed as a function of input v
$\chi(\mathbf{R})$	Eigenvalue spread (i.e., ratio of maximum eigenvalue to minimum eigenvalue) of correlation matrix $\mathbf{R}$
ω	Normalized angular frequency; $0 < \omega \le 2\pi$
$\boldsymbol{\omega}(n)$	Process noise vector in Markov model
ρ_m	Correlation coefficient or normalized value of autocorrelation function for lag m
σ^2	Variance
τ_k	Time constant of kth natural mode of steepest-descent algorithm
$\tau_{\text{mse,av}}$	Time constant of a single decaying exponential that approximates the learning curve of LMS algorithm
$\nabla(n)$	Convolutional noise in blind equalization
$\boldsymbol{\nabla}$	Gradient vector

Bibliography

ABRAHAM, J. A., ET AL. (1987). "Fault tolerance techniques for systolic arrays," *Computer,* vol. 20, pp. 65–75.

AKAIKE, H. (1973). "Maximum likelihood identification of Gaussian autoregressive moving average models," *Biometrika*, vol. 60, pp. 255–265.

AKAIKE, H. (1974). "A new look at the statistical model identification," *IEEE Trans. Autom. Control,* vol. AC-19, pp. 716–723.

AKAIKE, H. (1977). "An entropy maximisation principle," in *Proceedings of Symposium on Applied Statistics*, ed. P. Krishnaiah, North-Holland, Amsterdam.

ALBERT, A. E., and L. S. GARDNER, JR. (1967). *Stochastic Approximation and Nonlinear Regression*, MIT Press, Cambridge, Mass.

AL-MASHOUQ, K.A., and I. S. REED (1994). "The use of neural nets to combine equalization with decoding for severe intersymbol interference channels," *IEEE Trans. Neural Networks*, vol. 5, pp. 982–988.

ALEXANDER, S. T. (1986a). *Adaptive Signal Processing: Theory and Applications*, Springer-Verlag, New York.

ALEXANDER, S. T. (1986b). "Fast adaptive filters: a geometrical approach," *IEEE ASSP Mag.*, pp. 18–28.

ALEXANDER, S. T. (1987). "Transient weight misadjustment properties for the finite precision LMS algorithm," *IEEE Trans. Acoust. Speech Signal Process.*, vol. ASSP-35, pp. 1250–1258.

ALEXANDER, S. T., and A. L. GHIRNIKAR (1993). "A method for recursive least-squares filtering based upon an inverse QR decomposition," *IEEE Trans. Signal Process.*, vol. 41, pp. 20–30.

ANDERSON, T. W. (1963). "Asymptotic theory for principal component analysis," *Ann. Math. Stat.*, vol. 34, pp. 122–148.

ANDERSON, T. W. (1984). *An Introduction to Multivariate Statistical Analysis*, 2nd ed., Wiley, New York.

ANDERSON, B. D. O., and J. B. MOORE (1979). *Linear Optimal Control*, Prentice-Hall, Englewood Cliffs, N.J.

ANDERSON, C. H., and D. C. VAN ESSEN (1994). "Neurobiological computational systems," presented at the *IEEE World Congress on Computational Intelligence*, Orlando, Fla., June 26–July 1, 11 pages.

ANDREWS, H. C., and C. L. PATTERSON (1975). "Singular value decomposition and digital image processing," *IEEE Trans. Acoust. Speech Signal Process.*, vol. ASSP-24, pp. 26–53.

APPLEBAUM, S. P. (1966). "Adaptive arrays," Syracuse University Research Corporation, Rep. SPL TR 66-1.

APPLEBAUM, S. P., and D. J. CHAPMAN (1976). "Adaptive arrays with main beam constraints," *IEEE Trans. Antennas Propag.*, vol. AP-24, pp. 650–662.

ARDALAN, S. H. (1986). "Floating-point error analysis of recursive least-squares and least-mean-squares adaptive filters," *IEEE Trans. Circuits Syst.*, vol. CAS-33, pp. 1192–1208.

ARDALAN, S. H., and S. T. ALEXANDER (1987). "Fixed-point roundoff error analysis of the exponentially windowed RLS algorithm for time varying systems," *IEEE Trans. Acoust. Speech Signal Process.*, vol. ASSP-35, pp. 770–783.

ÅSTRÖM, K. J., and P. EYKHOFF (1971). "System identification—a survey," *Automatica*, vol. 7, pp. 123–162.

ATAL, B. S., and S. L. HANAUER (1971). "Speech analysis and synthesis by linear prediction of the speech wave," *J. Acoust. Soc. Am.*, vol. 50, pp. 637–655.

ATAL, B. S., and M. R. SCHROEDER (1970). "Adaptive predictive coding of speech signals," *Bell Syst. Tech. J.*, vol. 49, pp. 1973–1986.

AUSTIN, M. E. (1967). *"Decision-feedback equalization for digital communication over dispersive channels,"* Tech. Rep. 437, MIT Lincoln Laboratory, Lexington, Mass.

AUTONNE, L. (1902). "Sur les groupes linéaires, réels et orthogonaux," *Bull. Soc. Math., France*, vol. 30, pp. 121–133.

BARRETT, J. F. and D. G. LAMPARD (1955). "An expansion for some second-order probability distributions and its application to noise problems," *IRE Trans. Information Theory*, vol. IT-1, pp. 10–15.

BARRON, A. R. (1993). "Universal approximation bounds for superpositions of a sigmoidal function," *IEEE Trans. Information Theory*, vol. 39, pp. 930–945.

BATTITI, R. (1992). "First- and second-order methods for learning: between steepest descent and Newton's method," *Neural Computation*, vol. 4, pp. 141–166.

BEAUFAYS, F. (1995a). "Transform-domain adaptive filters: an analytical approach," *IEEE Trans. Signal Process.*, vol. 43, pp. 422–431.

BEAUFAYS, F. (1995b). "Two-layer linear structures for fast adaptive filtering," Ph.D. dissertation, Stanford University, Stanford, Calif.

BEAUFAYS, F., and B. WIDROW (1994) "Two-layer linear structures for fast adaptive filtering," *World Congress on Neural Networks*, vol. III, San Diego, Calif., pp. 87–93.

BELFIORE, C. A., and J. H. PARK, JR. (1979). "Decision feedback equalization," *Proc. IEEE*, vol. 67, pp. 1143–1156.

BELLANGER, M. G. (1988a). *Adaptive Filters and Signal Analysis*, Dekker, New York.

BELLANGER, M. G. (1988b). "The FLS-QR algorithm for adaptive filtering," *Signal Process.*, vol. 17, pp. 291–304.

BELLINI, S. (1986). "Bussgang techniques for blind equalization," in GLOBECOM, Houston, Tex., pp. 1634–1640.

BELLINI, S. (1988). "Blind equalization," *Alta Freq.*, vol. 57, pp. 445–450.

BELLINI, S. (1994). "Bussgang techniques for blind deconvolution and equalization," in *Blind Deconvolution*, ed. S. Haykin, Prentice-Hall, Englewood Cliffs, N.J.

BELLINI, S., and F. ROCCA (1986). "Blind deconvolution: polyspectra or Bussgang techniques?" in *Digital Communications*, ed. E. Biglieri and G. Prati, North-Holland, Amsterdam, pp. 251–263.

BELLMAN, R. (1960). *Introduction to Matrix Analysis*, McGraw-Hill, New York.

BENVENISTE, A. (1987). "Design of adaptive algorithms for the tracking of time-varying systems," *Int. J. Adaptive Control Signal Proc.*, vol. 1, pp. 3–29.

BENVENISTE, A., and M. GOURSAT (1984). "Blind equalizers," *IEEE Trans. Commun.*, vol. COM-32, pp. 871–883.

BENVENISTE, A., M. GOURSAT, and G. RUGET (1980). "Robust identification of a nonminimum phase system: blind adjustment of a linear equalizer in data communications," *IEEE Trans. Autom. Control*, vol. AC-25, pp. 385–399.

BENVENISTE, A., M. MÉTIVIER, and P. PRIOURET (1987). *Adaptive Algorithms and Stochastic Approximations*, Springer-Verlag, New York.

BENVENUTO, N., ET AL. (1986). "The 32 kb/s ADPCM coding standard," *AT&T J.*, vol. 65, pp. 12–22.

BENVENUTO, N., and F. PIAZZA (1992). "On the complex backpropagation algorithm," *IEEE Trans. Signal Process.*, vol. 40, pp. 967–969.

BERGMANS, J. W. M. (1990). "Tracking capabilities of the LMS adaptive filter in the presence of gain variations," *IEEE Trans. Acoust. Speech Signal Process.* vol. 38, pp. 712–714.

BERKHOUT, A. J., and P. R. ZAANEN (1976). "A comparison between Wiener filtering, Kalman filtering, and deterministic least squares estimation," *Geophysical Prospect.*, vol. 24, pp. 141–197.

BERSHAD, N. J. (1986). "Analysis of the normalized LMS algorithm with Gaussian inputs," *IEEE Trans. Acoust. Speech Signal Process.*, vol. ASSP-34, pp. 793–806.

BERSHAD, N. J., and P. L. FEINTUCH (1986). "A normalized frequency domain LMS adaptive algorithm," *IEEE Trans. Acoust. Speech Signal Process.*, vol. ASSP-34, pp. 452–461.

BERSHAD, N. J., and L. Z. QU (1989). "On the probability density function of the LMS adaptive filter weights," *IEEE Trans. Acoust. Speech Signal Process.*, vol. ASSP-37, pp. 43–56.

BERSHAD, N. J., and O. MACCHI (1991). "Adaptive recovery of a chirped sinusoid in noise, Part 2: Performance of the LMS algorithm," *IEEE Trans. Acoust. Speech Signal Process.*, vol. 39, pp. 595–602.

BIERMAN, G. J. (1977). *Factorization Methods for Discrete Sequential Estimation*, Academic Press, New York.

BIERMAN, G. J., and C. L. THORNTON (1977). "Numerical comparison of Kalman filter algorithms: orbit determination case study," *Automatica*, vol. 13, pp. 23–35.

BIRKETT, A. N., and R. A. GOUBRAN, 1995. "Acoustic echo cancellation using NLMS-neural network structures," in *Proc. ICASSP*, Detroit, Michigan. vol. 5, pp. 3035–3038.

BIRX, D. L., and S. J. PIPENBERG (1992). "Chaotic oscillators and complex mapping feed forward networks (CMFFNS) for signal detection in noisy environments," in *International Joint Conference on Neural Networks*, Baltimore, MD, vol. 2, pp. 881–888.

BITMEAD, R. R. and B. D. O. ANDERSON (1980a). "Lyapunov techniques for the exponential stability of linear difference equations with random coefficients," *IEEF Trans. Autom. Control*, vol. AC-25, pp. 782–787.

BITMEAD, R. R., and B. D. O. ANDERSON (1980b). "Performance of adaptive estimation algorithms in dependent random environments," *IEEE Trans. Autom. Control,* vol. AC-25, pp. 788–794.

BITMEAD, R. P., and B. D. O. ANDERSON (1981). "Adaptive frequency sampling filters," *IEEE Trans. Circuits Syst.*, vol. CAS-28, pp. 524–535.

BJÖRCK, A. (1967). "Solving linear least squares problems by Gram-Schmidt orthogonalization," *BIT*, vol. 7, pp. 1–21.

BODE, H. W., and C. E. SHANNON (1950). "A simplified derivation of linear least square smoothing and prediction theory," *Proc. IRE*, vol. 38, pp. 417–425.

BOJANCZYK, A. W., and F. T. LUK (1990). "A unified systolic array for adaptive beamforming," *J. Parallel Distrib. Comput.*, vol. 8, pp. 388–392.

BORAY, G. K., and M. D. SRINATH (1992). "Conjugate gradient techniques for adaptive filtering," *IEEE Trans. Circuits Syst. Fundam. Theory Appli.*, vol. 39, pp. 1–10.

BOTTO, J. L., and G. V. MOUSTAKIDES (1989). "Stabilizing the fast Kalman algorithms," *IEEE Trans. Acoust., Speech Signal Process.*, vol. ASSP-37, pp. 1342–1348.

BOTTOMLEY, G. E., and S. T. ALEXANDER (1989). "A theoretical basis for the divergence of conventional recursive least squares filters," in *Proc. ICASSP*, Glasgow, Scotland, pp. 908–911.

BOX, G. E. P., and G. M. JENKINS (1976). *Time Series Analysis: Forecasting and Control*, Holden-Day, San Francisco.

BRACEWELL, R. N. (1978). *The Fourier Transform and Its Applications*, 2nd ed., McGraw-Hill, New York.

BRADY, D. M. (1970). "An adaptive coherent diversity receiver for data transmission through dispersive media," in *Conf. Rec. ICC 70*, pp. 21-35–21-40.

BRENT, R. P., F. T. LUK, and C. VAN LOAN (1983). "Decomposition of the singular value decomposition using mesh-connected processors," *J. VLSI Comput. Syst.*, vol. 1, pp. 242–270.

BRILLINGER, D. R. (1974). *Time Series: Data Analysis and Theory*, Holt, Rinehart, and Winston, New York.

BROGAN, W. L. (1985). *Modern Control Theory*, 2nd ed., Prentice-Hall, Englewood Cliffs, N.J.

BROOKS, L. W., and I. S. REED (1972). "Equivalence of the likelihood ratio processor, the maximum signal-to-noise ratio filter, and the Wiener filter," *IEEE Trans. Aerospace Electron. Syst.*, vol. AES-8, pp. 690–692.

BROOMHEAD, D. S., and D. LOWE (1988). "Multi-variable functional interpolation and adaptive networks," *Complex Syst.*, vol. 2, pp. 269–303.

BROSSIER, J. M. (1992). "Egalisation adaptive et estimation de phase: Application aux communications sous-marines," These de Docteur, del'Institut National Polytechnique de Grenoble, France.

BRUCKSTEIN, A., and T. KAILATH (1987). "An inverse scattering framework for several problems in signal processing," *IEEE ASSP Mag.*, vol. 4, pp. 6–20.

BUCKLEW, J. A., T. KURTZ, and W. A. SETHARES (1993). "Weak convergence and local stability properties of fixed stepsize recursive algorithms," *IEEE Trans. Information Theory*, vol. 39, pp. 966–978.

BUCKLEY, K. M., and L. J. GRIFFITHS (1986). "An adaptive generalized sidelobe canceller with derivative constraints," *IEEE Trans. Antennas Propag.*, vol. AP-34, pp. 311–319.

BUCY, R. S. (1994). *Lectures on Discrete Time Filtering*, Springer-Verlag, New York.

BURG, J. P. (1967). "Maximum entropy spectral analysis," in *37th Ann. Int. Meet., Soc. Explor. Geophys.*, Oklahoma City, Okla.

BURG, J. P. (1968). "A New Analysis Technique for Time Series Data," NATO Advanced Study Institute on Signal Processing, Enschede, The Netherlands.

BURG, J. P. (1972). "The relationship between maximum entropy spectra and maximum likelihood spectra," *Geophysics*, vol. 37, pp. 375–376.

BURG, J. P. (1975). "Maximum Entropy Spectral Analysis," Ph.D. dissertation, Stanford University, Stanford, Calif.

BUSSGANG, J. J. (1952). Cross Correlation Functions of Amplitude-Distorted Gaussian Signals, Tech. Rep. 216, MIT Research Laboratory of Electronics, Cambridge, Mass.

BUTTERWECK, H. J. (1995a). "A steady-state analysis of the LMS adaptive algorithm without use of the independence assumption," in *Proc. ICASSP*, Detroit, Michigan, pp. 1404–1407.

BUTTERWECK, H. J. (1995b). "Iterative analysis of the steady-state weight fluctuations in LMS-type adaptive filters," private communication.

CAPMAN, F., J. BOUDY, and P. LOCKWOOD (1995). "Acoustic echo cancellation using a fast QR-RLS algorithm and multirate schemes," in *Proc. ICASSP*, Detroit, Michigan, pp. 969–971.

CAPON, J. (1969). "High-resolution frequency-wavenumber spectrum analysis," *Proc. IEEE*, vol. 57, pp. 1408–1418.

CARAISCOS, C., and B. LIU (1984). "A roundoff error analysis of the LMS adaptive algorithm," *IEEE Trans. Acoust. Speech Signal Process.*, vol. ASSP-32, pp. 34–41.

CARAYANNIS, G., D. G. MANOLAKIS, and N. KALOUPTSIDIS (1983). "A fast sequential algorithm for least-squares filtering and prediction," *IEEE Trans. Acoust. Speech Signal Process.*, vol. ASSP-31, pp. 1394–1402.

CHANG, R. W. (1971). "A new equalizer structure for fast start-up digital communications," *Bell Syst. Tech. J.*, vol. 50, pp. 1969-2014.

CHAO, J., H. PEREZ, and S. TSUJII (1990). "A fast adaptive filter algorithm using eigenvalue reciprocals as step sizes," *IEEE Trans. Acoust. Speech Signal Process.* vol. ASSP-38, pp. 1343–1352.

CHAZAN, D., Y. MEDAN, and U. SHVADRON (1988). "Noise cancellation for hearing aids," *IEEE Trans. Acoust. Speech Signal Process.*, vol. ASSP-36, pp. 1697–1705.

CHEN, S. (1995). "Nonlinear time series modelling and prediction using Gaussian RBF networks with enhanced clustering and RLS learning," *Electronics Letters*, vol. 31, no. 2, pp. 117–118.

CHEN, S., S. MCLAUGHLIN, and B. MULGREW, (1994a). "Complex-valued radial basis function network, Part I: Network architecture and learning algorithms," *Signal Proc.* vol. 35, pp. 19–31.

CHEN, S., S. MCLAUGHLIN, and B. MULGREW, (1994b). "Complex-valued radial basis function network, Part II: Application to digital communications channel equalisation," *Signal Proc.*, vol. 36, pp. 175–188.

CHESTER, D. L. (1990). "Why two hidden layers are better than one," *International Joint Conference on Neural Networks*, Washington, D.C., vol. 1, pp. 265–268.

CHILDERS, D. G., ed. (1978). *Modern Spectrum Analysis*, IEEE Press, New York.

CHINRUNGRUENG, C., and C. H. SÉQUIN (1995). "Optimal adaptive k-means algorithm with dynamic adjustment of learning rate," *IEEE Trans. Neural Networks*, vol. 6, pp. 157–169.

CHUI, C. K., and G. CHEN (1987). *Kalman Filtering with Real-time Application*, Springer-Verlag, New York.

CIOFFI, J. M. (1987). "Limited-precision effects in adaptive filtering," *IEEE Trans. Circuits Syst.*, vol. CAS-34, pp. 821–833.

CIOFFI, J. M. (1988). "High speed systolic implementation of fast QR adaptive filters," in *Proc ICASSP*, New York, pp. 1584–1588.

CIOFFI, J. M. (1990). "The fast adaptive rotor's RLS algorithm," *IEEE Trans. Acoust. Speech Signal Process.*, vol. ASSP-38, pp. 631–653.

CIOFFI, J. M., and T. KAILATH (1984). "Fast, recursive-least-squares transversal filters for adaptive filtering," *IEEE Trans. Acoust. Speech Signal Process.*, vol. ASSP-32, pp. 304–337.

CLASSEN, T. A. C. M., and W. F. G. MECKLANBRÄUKER (1985). "Adaptive techniques for signal processing in communications," *IEEE Commun.*, vol. 23, pp. 8–19.

CLARK, G. A., S. K. MITRA, and S. R. PARKER (1981). "Block implementation of adaptive digital filters," *IEEE Trans. Circuits Syst.*, vol. CAS-28, pp. 584–592.

CLARK, G. A., S. R. PARKER, and S. K. MITRA (1983). "A unified approach to time- and frequency-domain realization of FIR adaptive digital filters," *IEEE Trans. Acoust. Speech Signal Process.*, vol. ASSP-31, pp. 1073–1083.

CLARKE, T. L. (1990). "Generalization of neural networks to the complex plane," *International Joint Conference on Neural Networks*, San Diego, Calif., vol. II, pp. 435–440.

CLARKSON, P. M. (1993). *Optimal and Adaptive Signal Processing*, CRC Press, Boca Raton, Fla.

COHN, A. (1922). "Über die Anzahl der Wurzeln einer algebraischen Gleichung in einem Kreise," *Math. Z.*, vol. 14, pp. 110–148.

COMPTON, R. T. (1988). *Adaptive Antennas: Concepts and Performance*, Prentice-Hall, Englewood Cliffs, N.J.

COWAN, C. F. N. (1987). "Performance comparisons of finite linear adaptive filters," *IEE Proc. (London)*, part F, vol. 134, pp. 211–216.

COWAN, C. F. N., and P. M. GRANT (1985). *Adaptive Filters*, Prentice-Hall, Englewood Cliffs, N. J.

COWAN, J. D. (1990). "Neural networks: the early days." In *Advances in Neural Information Processing Systems* 2, ed. D. S. Touretzky, pp. 828–842, Morgan Kaufman, San Mateo, Calif.

COX, H., R. M. ZESKIND, and M. M. OWEN (1987). "Robust adaptive beamforming," *IEEE Trans. Acoust. Speech Signal Process.*, vol. ASSP-35, pp. 1365–1376.

CROCHIERE, R. E., and L. R. RABINER (1983). *Multirate Digital Signal Processing*, Prentice-Hall, Englewood Cliffs, N.J.

CUTLER, C. C. (1952). *Differential Quantization for Communication Signals*, U. S. Patent 2,605,361.

CYBENKO, G. (1989). *Approximation by superpositions of a sigmoidal function,"* Mathematics of Control, Signals, and Systems, vol. 2, pp. 303–314.

DE COURVILLE, M., and P. DUHAMEL (1995). "Adaptive filtering in subbands using a weighted criterion," in *Proc. ICASSP*, Detroit, Michigan, vol. 2, pp. 985–988.

DEIFT, P. J., DEMMEL, C. TOMAL, and L.-C. LI, (1989). The Bidiagonal Singular Value Decomposition and Hamiltonian Mechanics, Rep. 458, Department of Computer Science, Courant Institute of Mathematical Sciences, New York University, New York.

DEMMEL, J. (1994). "Designing High Performance Linear Algebra Software for Parallel Computers," CS Division and Math Dept., UC Berkeley, December 9.

DEMMEL, J., and W. KAHAN (1990). "Accurate singular values of bidiagonal matrices," *SIAM J. Sci. Stat. Comp.*, vol. 11, pp. 873–912.

DEMMEL, J., and K. VESELIĆ (1989). *Jacobi's Method Is More Accurate than QR*, Tech. Rep. 468, Department of Computer Science, Courant Institute of Mathematical Sciences, New York University, New York.

DEMOOR, B. L. R., and G. H. GOLUB (1989). "Generalized Singular Value Decompositions: A Proposal for a Standardized Nomenclature," Manuscript NA-89-05, Numerical Analysis Project, Computer Science Department, Standard University, Stanford, Calif.

DENTINO, M., J. MCCOOL, and B. WIDROW (1978). "Adaptive filtering in the frequency domain," *Proc. IEEE*, vol. 66, no. 12, pp. 1658–1659.

DEPRETTERE, E. F., ed. (1988). *SVD and Signal Processing: Algorithms, Applications, and Architectures*, North-Holland, Amsterdam.

DEVIJVER, P. A., and J. KITTLER (1982). *Pattern Recognition: A Statistical Approach*, Prentice-Hall International, London.

DEWILDE, P. (1969). "Cascade Scattering Matrix Synthesis," Ph.D. dissertation, Stanford University, Stanford, Calif.

DEWILDE, P., A. C. VIEIRA, and T. KAILATH (1978). "On a generalized Szegö–Levinson realization algorithm for optimal linear predictors based on a network synthesis approach," *IEEE Trans. Circuits Syst.*, vol. CAS-25, pp. 663–675.

DHRYMUS, P. J. (1970). *Econometrics: Statistical Foundations and Applications*, Harper & Row, New York.

DING, Z. (1994). "Blind channel identification and equalization using spectral correlation measurements, Part I: Frequency-domain approach," in *Cyclostationarity in Communications and Signal Processing*," ed. W. A. Gardner, IEEE Press, New York, pp. 417–436.

DING, Z., and Z. MAO (1995). "Knowledge based identification of fractionally sampled channels," in *Proc. ICASSP*, Detroit, Michigan, vol. 3, pp. 1996–1999.

DING, Z., C. R. JOHNSON, JR., and R. A. KENNEDY (1994). "Global convergence issues with linear blind adaptive equalizers," in *Blind Deconvolution*, ed. S. Haykin, Prentice-Hall, Englewood Cliffs, N.J.

DINIZ, P. S. R., and L. W. P. BISCAINHO (1992). "Optimal variable step size for the LMS/Newton algorithm with application to subband adaptive filtering," *IEEE Trans. Signal Process.*, vol. 40, pp. 2825–2829.

DITORO, M. J. (1965). "A new method for high speed adaptive signal communication through any time variable and dispersive transmission medium," in *1st IEEE Annu. Commun. Conf.*, pp. 763–767.

DONGARRA, J. J., ET AL. (1979). *LINPACK User's Guide*, Society for Industrial and Applied Mathematics, Philadelphia.

DONOHO, D. L. (1981). "On minimum entropy deconvolution," in *Applied Time Series Analysis II*, ed. D. F. Findlay, Academic Press, New York.

DOOB, L. J., (1953). *Stochastic Processes*, Wiley, New York.

DORNY, C. N. (1975). *A Vector Space Approach to Models and Optimization*, Wiley-Interscience, New York.

DOUGLAS, S. C. (1994). "A family of normalized LMS algorithms," *IEEE Signal Processing Letters*, vol. 1, pp. 49–51.

DOUGLAS, S. C., and T. H.-Y. MENG (1994). "Normalized data nonlinearities for LMS adaptation," *IEEE Trans. Acoust. Speech Signal Process.*, vol. 42, pp. 1352–1365.

DOYLE, J. C., K. GLOVER, P. KHARGONEKAR, and B. FRANCIS (1989). "State-space solutions to standard H_2 and H_∞ control problems," *IEEE Trans. Autom. Control*, vol. AC-34, pp. 831–847.

DUGARD, L., M. M'SAAD, and I. D. LANDAU (1993). *Adaptive Systems in Control and Signal Processing*, Pergamon Press, Oxford, United Kingdom.

DUDA, R. O., and P. E. HART (1973). *Pattern Classification and Scene Analysis*, Wiley, New York.

DUHAMEL, P. (1995). "Tutorial: Blind equalization," *The 1995 IEEE International Conference on Acoustics, Speech, and Signal Processing*, Detroit, Michigan.

DURBIN, J. (1960). "The fitting of time series models," *Rev. Int. Stat. Inst.*, vol. 28, pp. 233–244.

DUTTWEILER, D. L., and Y. S. CHEN (1980). "A single-chip VLSI echo canceler," *Bell Syst. Tech. J.*, vol. 59, pp. 149–160.

ECKART, G., and G. YOUNG (1936). "The approximation of one matrix by another of lower rank," *Psychometrika*, vol. 1, pp. 211–218.

ECKART, G., and G. YOUNG (1939). "A principal axis transformation for non-Hermitian matrices," *Bull. Am. Math. Soc.*, vol. 45, pp. 118–121.

EDWARDS, A. W. F. (1972). *Likelihood*, Cambridge University Press, New York.

ELEFTHERIOU, E., and D. D. FALCONER (1986). "Tracking properties and steady state performance of RLS adaptive filter algorithms," *IEEE Trans. Acoust. Speech Signal Process.*, vol. ASSP-34, pp. 1097–1110.

EWEDA, E. (1994). "Comparison of RLS, LMS, and sign algorithms for tracking randomly time-varying channels," *IEEE Trans. Signal Process.*, vol. 42, pp. 2937–2944.

EWEDA, E., and O. MACCHI (1985). "Tracking error bounds of adaptive nonstationary filtering," *Automatica*, vol. 21, pp. 293–302.

EWEDA, E., and O. MACCHI (1987). "Convergence of the RLS and LMS adaptive filters," *IEEE Trans. Circuits Syst.*, vol. CAS-34, pp. 799–803.

FALCONER, D. D., and L. LJUNG (1978). "Application of fast Kalman estimation to adaptive equalization," *IEEE Trans. Commun.*, vol. COM-26, pp. 1439–1446.

FARDEN, D. C. (1981a). "Stochastic approximation with correlated data," *IEEE Trans. Information Theory*, vol. IT-27, pp. 105–113.

FARDEN, D. C. (1981b). "Tracking properties of adaptive signal processing algorithms," *IEEE Trans. Acoust. Speech Signal Process.*, vol. ASSP-29, pp. 439–446.

FELDKAMP, L. A., G. V. PUSKORIUS, L. I. DAVIS, JR., and F. YUAN (1994). "Enabling concepts for application of neurocontrol," in *Proc. Eighth Yale Workshop on Adaptive and Learning Systems*, Yale University, New Haven, Conn., pp. 168–173.

FERNANDO, K. V., and B. N. PARLETT (1994). "Accurate singular values and differential qd algorithms," *Numerische Mathematik*, vol. 67, pp. 191–229.

FERRARA, E. R., JR., (1980). "Fast implementation of LMS adaptive filters," *IEEE Trans. Acoust. Speech Signal Process.*, vol. ASSP-28, pp. 474–475.

FERRARA, E. R., JR., (1985). "Frequency-domain adaptive filtering," in *Adaptive Filters,* ed. C. F. N. Cowan and P. M. Grant, pp. 145–179, Prentice-Hall, Englewood Cliffs, N.J.

FIJALKOW, I., J. R. TREICHLER, and C. R. JOHNSON JR. (1995). "Fractionally spaced blind equalization: loss of channel disparity," in *Proc. ICASSP*, Detroit, Michigan, pp. 1988–1991.

FISHER, B., and N. J. BERSHAD (1983). "The complex LMS adaptive algorithm—transient weight mean and covariance with applications to the ALE," *IEEE Trans. Acoust. Speech Signal Process.*, vol. ASSP-31, pp. 34–44.

FLANAGAN, J. L. (1972). *Speech Analysis, Synthesis and Perception*, 2nd ed., Springer-Verlag, New York.

FLANAGAN, J. L., ET AL. (1979). "Speech coding," *IEEE Trans. Commun.*, vol. COM-27, pp. 710–737.

FOLEY, J. B., and F. M. BOLAND (1987). "Comparison between steepest descent and LMS algorithms in adaptive filters," *IEE Proc. (London), Part F*, vol. 134, pp. 283–289.

FOLEY, J. B., and F. M. BOLAND (1988). "A note on the convergence analysis of LMS adaptive filters with Gaussian data," *IEEE Trans Acoust Speech Signal Process.*, vol. 36, pp. 1087–1089.

FORNEY, G. D. (1972). "Maximum-likelihood sequence estimation of digital sequence in the presence of intersymbol interference," *IEEE Trans. Information Theory*, vol. IT-18, pp. 363–378.

FORSYTHE, G. E., and P. HENRICI (1960). "The cyclic Jacobi method for computing the principal values of a complex matrix," *Trans. Am. Math. Soc.*, vol. 94, pp. 1–23.

FOSCHINI, G. J. (1985). "Equalizing without altering or detecting data," *AT&T Tech. J.*, vol. 64, pp. 1885–1911.

FRANKS, L. E. (1969). *Signal Theory*. Prentice-Hall, Englewood Cliffs, N.J.

FRANKS, L. E., ed. (1974). *Data Communication: Fundamentals of Baseband Transmission*, Benchmark Papers in Electrical Engineering and Computer Science, Dowden, Hutchinson & Ross, Stroudsburg, Pa.

FRASER, D. C. (1967). "A new technique for the optimal smoothing of data," Sc.D. thesis, Massachusetts Institute of Technology, Cambridge, Mass.

FRIEDLANDER, B. (1982). "Lattice filters for adaptive processing," *Proc. IEEE*, vol. 70, pp. 829–867.

FRIEDLANDER, B. (1988). "A signal subspace method for adaptive interference cancellation," *IEEE Trans. Acoust. Speech Signal Process.*, vol. ASSP-36, pp. 1835–1845.

FRIEDLANDER, B., and B. PORAT (1989). "Adaptive IIR algorithm based on high-order statistics," *IEEE Trans. Acoust. Speech Signal Process.*, vol. ASSP-37, pp. 485–495.

FRIEDRICHS, B. (1992). "Analysis of finite-precision adaptive filters. I. Computation of the residual signal variance," *Frequenz*, vol. 46, pp. 218–223.

FROST III, O. L. (1972). "An algorithm for linearly constrained adaptive array processing," *Proc. IEEE*, vol. 60, pp. 926–935.

FUKUNAGA, K. (1990). *Statistical Pattern Recognition*, 2nd ed., Academic Press, New York.

FUNAHASHI, K. (1989). "On the approximate realization of continuous mappings by neural networks," *Neural Networks*, vol. 2, pp. 183–192.

GABOR, D., W. P. L. WILBY, and R. WOODCOCK (1961). "A universal non-linear filter, predictor and simulator which optimizes itself by a learning process," *IEE Proc. (London)*, vol. 108, pt. B, pp. 422–438.

GABRIEL, W. F. (1976). "Adaptive arrays: an introduction," *Proc. IEEE*, vol. 64, pp. 239–272.

GALLIVAN, K. A., and C. E. LEISERSON (1984). "High-performance architectures for adaptive filtering based on the Gram–Schmidt algorithm," in *Proc. SPIE*, vol. 495, Real Time Signal Processing VII, pp. 30–38.

GARBOW, B. S., ET AL. (1977). *Matrix Eigensystem Routines—EISPACK Guide Extension*, Lecture Notes in Computer Science, vol. 51, Springer-Verlag, New York.

GARDNER, W. A. (1984). "Learning characteristics of stochastic-gradient-descent algorithms: a general study, analysis and critique," *Signal Process.*, vol. 6, pp. 113–133.

GARDNER, W. A. (1987). "Nonstationary learning characteristics of the LMS algorithm," *IEEE Trans. Circuits Syst.*, vol. CAS-34, pp. 1199–1207.

GARDNER, W. A. (1990). *Introduction to Random Processes with Applications to Signals and Systems*, McGraw-Hill, New York.

GARDNER, W. A. (1991). "A new method of channel identification," *IEEE Trans. Commun.*, vol. COM-39, pp. 813–817.

GARDNER, W. A. (1993). "Cyclic Wiener filtering: Theory and method", *IEEE Trans. Signal Process.*, vol. 41, pp. 151–163.

GARDNER, W. A. ed. (1994a). *Cyclostationarity in Communications and Signal Processing*, IEEE Press, New York.

GARDNER, W. A. (1994b). "An introduction to cyclostationary signals," in *Cyclostationarity in Communications and Signal Processing*, ed., W. A. Gardner, IEEE Press, New York, pp. 1–90.

GARDNER, W. A., and L. E. FRANKS (1975). "Characterization of cyclostationary random signal processes," *IEEE Trans. Information Theory*, vol. IT-21, pp. 4–14.

GARDNER, W. A., and C. M. SPOONER (1994). "The cumulant theory of cyclostationary time-series, Part I: foundation," *IEEE Trans. Signal Process.*, vol. 42, pp. 3387–3408.

GAUSS, C. F. (1809). *Theoria motus corporum coelestium in sectionibus conicus solem ambientum*, Hamburg (translation: Dover, New York, 1963).

GELB, A., ed. (1974). *Applied Optimal Estimation*, MIT Press, Cambridge, Mass.

GENTLEMAN, W. M. (1973). "Least squares computations by Givens transformations without square-roots," *J. Inst. Math. Its Appl.*, vol. 12, pp. 329–336.

GENTLEMAN, W. M., and H. T. KUNG (1981). "Matrix triangularization by systolic arrays," in *Proc. SPIE*, vol. 298, Real Time Signal Processing IV, pp. 298–303.

GEORGIOU, G. N., and C. KOUTSOUGERAS (1992). "Complex domain backpropagation," *IEEE Trans. Circuits Syst. Part II: Analog and Digital Signal Processing*, vol. 39, pp. 330–334.

GERSHO, A. (1968). "Adaptation in a quantized parameter space," in *Proc. Allerton Conf. on Circuit and System Theory*, Urbana, Il., pp. 646–653.

GERSHO, A. (1969). "Adaptive equalization of highly dispersive channels for data transmission," *Bell Syst. Tech. J.*, vol. 48, pp. 55–70.

GERSHO, A., B. GOPINATH, and A.M. OLDYZKO (1979). "Coefficient inaccuracy in transversal filtering," *Bell Syst. Tech. J.*, vol. 58, pp. 2301–2316.

GHOLKAR, V. A. (1990). "Mean square convergence analysis of LMS algorithm (adaptive filters)," *Electron Letters*, vol. 26, pp. 1705–1706.

GIANNAKIS, G. B., and S. D. HALFORD (1995). "Blind fractionally-spaced equalization of noisy FIR channels: adaptive and optimal solutions," in *Proc. ICASSP*, Detroit, Michigan, pp. 1972–1975.

GIBSON, J. D. (1980). "Adaptive prediction in speech differential encoding systems," *Proc. IEEE*, vol. 68, pp. 488–525.

GIBSON, G. J. and C. F. N. COWAN (1990). "On the decision regions of multilayer perceptrons," *Proc. IEEE*, vol. 78, pp. 1590–1599.

GILLOIRE, A., and M. VETTERLI, 1992. "Adaptive filtering in subbands with critical sampling: analysis, experiments, and applications to acoustic echo cancellation," *IEEE Trans. Circuits Syst.*, vol. 40, pp. 1862–1875.

GILL, P. E., G. H. GOLUB, W. MURRAY, and M. A. SAUNDERS (1974). "Methods of modifying matrix factorizations," *Math. Comput.*, vol. 28, pp. 505–535.

GITLIN, R. D., and F. R. MAGEE, Jr., (1977). "Self-orthogonalizing adaptive equalization algorithms," *IEEE Trans. Commun.*, vol. COM-25, pp. 666–672.

GITLIN, R. D., and S. B. WEINSTEIN (1979). "On the required tap-weight precision for digitally implemented mean-squared equalizers," *Bell Syst. Tech. J.*, vol. 58, pp. 301–321.

GITLIN, R. D., and S. B. WEINSTEIN (1981). "Fractionally spaced equalization: an improved digital transversal equalizer," *Bell Syst. Tech. J.*, vol. 60, pp. 275–296.

GITLIN, R. D., J. E. MAZO, and M. G. TAYLOR (1973). "On the design of gradient algorithms for digitally implemented adaptive filters," *IEEE Trans. Circuit Theory*, vol. CT-20, pp. 125–136.

GIVENS, W. (1958). "Computation of plane unitary rotations transforming a general matrix to triangular form," *J. Soc. Ind. Appl. Math*, vol. 6, pp. 26–50.

GLASER, E. M. (1961). "Signal detection by adaptive filters," *IRE Trans. Information Theory*, vol. IT-7, pp. 87–98.

GLOVER, J. R., JR. (1977). "Adaptive noise cancelling applied to sinusoidal interferences," *IEEE Trans. Acoust. Speech Signal Process.*, vol. ASSP-25, pp. 484–491.

GODARA, L. C., and A. CANTONI (1986). "Analysis of constrained LMS algorithm with application to adaptive beamforming using perturbation sequences," *IEEE Trans. Antennas Propag.*, vol. AP-34, pp. 368–379.

GODARD, D. N. (1974). "Channel equalization using a Kalman filter for fast data transmission," *IBM J. Res. Dev.*, vol. 18, pp. 267–273.

GODARD, D. N. (1980). "Self-recovering equalization and carrier tracking in a two-dimensional data communication system," *IEEE Trans. Commun.*, vol. COM-28, pp. 1867–1875.

GODFREY R., and F. ROCCA (1981). "Zero memory non-linear deconvolution," *Geophys. Prospect.*, vol. 29, pp. 189–228.

GOLD, B. (1977). "Digital speech networks," *Proc. IEEE*, vol. 65, pp. 1636–1658.

GOLOMB, S. W., ed. (1964). *Digital Communications with Space Applications*, Prentice-Hall, Englewood Cliffs, N.J.

GOLUB, G. H., (1965). "Numerical methods for solving linear least squares problems," *Numer. Math.*, vol. 7, pp. 206–216.

GOLUB, G. H., and W. KAHAN (1965). "Calculating the singular values and pseudo-inverse of a matrix," *J. SIAM Numer. Anal. B.*, vol. 2, pp. 205–224.

GOLUB, G. H., and C. REINSCH (1970). "Singular value decomposition and least squares problems," *Numer. Math.*, vol. 14, pp. 403–420.

GOLUB, G. H., and C. F. VAN LOAN (1989). *Matrix Computations*, 2nd ed., The Johns Hopkins University Press, Baltimore, Md.

GOLUB, G. H., F. T. LUK, and M. L. OVERTON (1981). "A block Lanczos method for computing the singular values and corresponding singular vectors of a matrix," *ACM Trans. Math. Software*, vol. 7, pp. 149–169.

GOODWIN, G. C., and R. L. PAYNE (1977). *Dynamic System Identification: Experiment Design and Data Analysis*, Academic Press, New York.

GOODWIN, G. C., and K. S. SIN (1984). *Adaptive Filtering, Prediction and Control*, Prentice-Hall, Englewood Cliffs, N.J.

GRAY, R. M. (1972). "On the asymptotic eigenvalue distribution of Toeplitz matrices," *IEEE Trans. Information Theory*, vol. IT-18, pp. 725–730.

GRAY, R. M. (1977). Toeplitz and Circulant Matrices: II, Tech. Rep. 6504-1, Information Systems Laboratory, Stanford University, Stanford, Calif.

GRAY, R. M. (1990). "Quantization noise spectra," *IEEE Trans. Inform Theory*, vol. 36, pp. 1220–1244.

GRAY, R. M., and L. D. DAVISSON (1986). *Random Processes: A Mathematical Approach for Engineers*, Prentice-Hall, Englewood Cliffs, N.J.

GRAY, W. (1979). "Variable Norm Deconvolution," Ph.D. dissertation, Department of Geophysics, Stanford University, Stanford, Calif.

GREEN, M., and D. J. N. LIMEBEER (1995). *Linear Robust Control*, Prentice-Hall, Englewood Cliffs, N.J.

GRENANDER, U., and G. SZEGÖ (1958). *Toeplitz Forms and Their Applications*, University of California Press, Berkeley, Calif.

GRIFFITHS, L. J. (1975). "Rapid measurement of digital instantaneous frequency," *IEEE Trans. Acoust. Speech Signal Process.*, vol. ASSP-23, pp. 207–222.

GRIFFITHS, L. J. (1977). "A continuously adaptive filter implemented as a lattice structure," in *Proc. ICASSP*, Hartford, Conn., pp. 683–686.

GRIFFITHS, L. J. (1978). "An adaptive lattice structure for noise-cancelling applications," in *Proc. ICASSP*, Tulsa, Okla., pp. 87–90.

GRIFFITHS, L. J., and C. W. JIM (1982). "An alternative approach to linearly constrained optimum beamforming," *IEEE Trans. Antennas Propag.*, vol. AP-30, pp. 27–34.

GRIFFITHS, L. J., and R. PRIETO-DIAZ (1977). "Spectral analysis of natural seismic events using autoregressive techniques," *IEEE Trans. Geosci. Electron.*, vol. GE-15, pp. 13–25.

GRIFFITHS, L. J., F. R. SMOLKA, and L. D. TREMBLY (1977). "Adaptive deconvolution: a new technique for processing time-varying seismic data," *Geophysics*, vol. 42, pp. 742–759.

GU, M., and S. C. EISENSTAT (1994). "A Divide-and-Conquer Algorithm for the Bidiagonal SVD," Research Report YALEU/DCS/RR-933, UC Berkeley, Calif., April 4.

GU, M., J. DEMMEL, and I. DHILLON (1994). "Efficient Computation of the Singular Value Decomposition with Applications to Least Squares Problems," Department of Mathematics, UC Berkeley, Calif., September 29.

GUNNARSSON, S., and L. LJUNG (1989). "Frequency domain tracking characteristics of adaptive algorithms," *IEEE Trans. Signal Process.*, vol. 37, pp. 1072–1089.

GUILLEMIN, E. A. (1949). *The Mathematics of Circuit Analysis*, Wiley, New York.

GUPTA, I. J., and A. A. KSIENSKI (1986). "Adaptive antenna arrays for weak interfering signals," *IEEE Trans. Antennas Propag.*, vol. AP-34, pp. 420–426.

GUTOWSKI, P. R., E. A. ROBINSON, and S. TREITEL (1978). "Spectral estimation: fact or fiction," *IEEE Trans. Geosci. Electron.*, vol. GE-16, pp. 80–84.

HADHOUD, M. M., and D. W. THOMAS (1988). "The two-dimensional adaptive LMS (TDLMS) algorithm," *IEEE Trans. Circuits Syst.*, vol. CAS-35, pp. 485–494.

HAMPEL, F. R., ET AL. (1986). *Robust Statistics: The Approach Based on Influence Functions*, Wiley, New York.

HANSON, R. J., and C. L. LAWSON (1969). "Extensions and applications of the Householder algorithm for solving linear least squares problems," *Math. Comput.*, vol. 23, pp. 787–812.

HARTMAN, E. J., J. D. KEELER, and J. M. KOWALSKI (1990). "Layered neural networks with Gaussian hidden units as universal approximations," *Neural Computation*, vol. 2, pp. 210–215.

HASSIBI, B., A.H. SAYED, and T. KAILATH (1996). "H^∞ optimality of the LMS algorithm," IEEE Trans. Signal Process., vol. 44, pp. 267–280.

HASTINGS-JAMES, R., and M. W. SAGE (1969). "Recursive generalized-least-squares procedure for online identification of process parameters," *IEE Proc. (London)*, vol. 116, pp. 2057–2062.

HATZINAKOS, D. (1990). "Blind equalization based on polyspectra," Ph.D. thesis, Northeastern University, Boston, Mass.

HATZINAKOS, D., and C. L. NIKIAS (1989). "Estimation of multipath channel response in frequency selective channels," *IEEE J. Sel. Areas Commun.*, vol. 7, pp. 12–19.

HATZINAKOS, D., and C. L. NIKIAS (1991). "Blind equalization using a tricepstrum based algorithm," *IEEE Trans. Commun.*, vol. COM-39, pp. 669–682.

HATZINAKOS, D., and C. L. NIKIAS (1994). "Blind equalization based on higher-order statistics (HOS)," in *Blind Deconvolution*, ed. S. Haykin, Prentice-Hall, Englewood Cliffs, N.J.

HAYKIN, S., ed. (1983). *Nonlinear Methods of Spectral Analysis*, 2nd ed., Springer-Verlag, New York.

HAYKIN, S., ed. (1984). *Array Signal Processing*, Prentice-Hall, Englewood Cliffs, N.J.

HAYKIN, S. (1989a). *Modern Filters*, Macmillan, New York.

HAYKIN, S. (1989b). "Adaptive filters: past, present, and future," *Proc. IMA Conf. Math. Signal Process.*, Warwick, England.

HAYKIN, S. (1991). *Adaptive Filter Theory*, 2nd ed., Prentice-Hall, Englewood Cliffs, NJ.

HAYKIN, S. (1994a). *Communication Systems*, 3rd ed., Wiley, New York.

HAYKIN, S. (1994b). *Neural Networks: A Comprehensive Foundation*, Macmillan, New York.

HAYKIN, S., and X. B. LI (1995). "Detection of signals in chaos," *Proc. IEEE*, vol. 83, pp. 95–122.

HAYKIN, S., and A. UKRAINEC (1993). "Neural networks for adaptive signal processing," in *Adaptive System Identification and Signal Processing Algorithms,* ed. N. Kalouptsidis and S. Theodoridis, pp. 512–553, Prentice-Hall, Englewood Cliffs, N.J.

HAYKIN, S., A. H. SAYED, J. R. ZEIDLER, P. YEE, and P. WEI (1995a). "Tracking of linear time-variant systems," MILCOM 95, San Diego, CA.

HAYKIN, S., A.UKRAINEC, B. CURRIE, B. LI, AND M. AUDETTE (1995b). A neural network-based non-coherent radar processor for a chaotic ocean environment," ANNIE, St. Louis, Missouri.

HENSELER, J., and P. J. BRASPENNING (1990). *Training Complex Multi-Layer Neural Networks*, Tech. Rep. CS90-02, University of Limburg, Maastricht, Department of Computer Science, The Netherlands.

HERZBERG, H., and R. HAIMI-COHEN (1992). "A systolic array realization of an LMS adaptive filter and the effects of delayed adaptation," *IEEE Trans. Signal Process.* vol. 40, pp. 2799–2803.

HO, Y. C. (1963). "On the stochastic approximation method and optimal filter theory," *J. Math. Anal. Appl.*, vol. 6, pp. 152–154.

HODGKISS, W. S., JR., and D. ALEXANDROU (1983). "Applications of adaptive least-squares lattice structures to problems in underwater acoustics," in *Proc. SPIE*, vol. 431, Real Time Signal Processing VI, pp. 48–54.

HONIG, M. L., and D. G. MESSERSCHMITT (1981). "Convergence properties of an adaptive digital lattice filter," *IEEE Trans. Acous. Speech Signal Process.*, vol. ASSP-29, pp. 642–653.

HONIG, M. L., and D. G. MESSERSCHMITT (1984). *Adaptive Filters: Structures, Algorithms and Applications*, Kluwer, Boston, Mass.

HORNIK, K., M. STINCHCOMBE, and H. WHITE (1989). "Multilayer feedforward networks are universal approximators," *Neural Networks*, vol. 2, pp. 359–366.

HOROWITZ, L. L., and K. D. SENNE (1981). "Performance advantage of complex LMS for controlling narrow-band adaptive arrays," *IEEE Trans. Acoust. Speech Signal Process.*, vol. ASSP-29, pp. 722–736.

HOUSEHOLDER, A. S. (1958a). "Unitary triangularization of a nonsymmetric matrix," *J. Assoc. Comput. Mach.*, vol. 5, pp. 339–342.

HOUSEHOLDER, A. S. (1958b). "The approximate solution of matrix problems," *J. Assoc. Comput. Mach.*, vol. 5, pp. 204–243.

HOUSEHOLDER, A. S. (1964). *The Theory of Matrices in Numerical Analysis*, Blaisdell, Waltham, Mass.

HOWELLS, P. W. (1965). *Intermediate Frequency Sidelobe Canceller*, U.S. Patent 3,202,990, August 24.

HOWELLS, P. W. (1976). "Explorations in fixed and adaptive resolution at GE and SURC," *IEEE Trans. Antennas Propag.*, vol. AP-24, Special Issue on Adaptive Antennas, pp. 575–584.

HSIA, T. C. (1983). "Convergence analysis of LMS and NLMS adaptive algorithms," in *Proc. ICASSP*, Boston, Mass., pp. 667–670.

HSU, F. M. (1982). "Square root Kalman filtering for high-speed data received over fading dispersive HF channels," *IEEE Trans. Information Theory*, vol. IT-28, pp. 753–763.

HU, Y. H. (1992). "CORDIC-based VLSI architectures for digital signal processing," *IEEE Signal Process Magazine*, vol. 9, pp. 16–35.

HUBER, P. J. (1981). *Robust Statistics*, Wiley, New York.

HUBING, N. E., and S. T. ALEXANDER (1990). "Statistical analysis of the soft constrained initialization of recursive least squares algorithms," in *Proc. ICASSP*, Albuquerque, N. Mex.

HUDSON, J. E. (1981). *Adaptive Array Principles*, Peregrinus, London.

HUDSON, J. E., and T. J. SHEPHERD (1989). "Parallel weight extraction by a systolic least squares algorithm," in *Proc. SPIE, Advanced Algorithms and Architectures for Signal Processing IV*, vol. 1152, pp. 68–77.

HUHTA, J. C., and J. G. WEBSTER (1973). "60-Hz interference in electrocardiography," *IEEE Trans. Biomed. Eng.*, vol. BME-20, pp. 91–101.

IOANNOU, P. A., (1990). "Robust adaptive control," in *Proc. Sixth Yale Workshop on Adaptive and Learning Systems*, Yale University, New Haven, Conn., pp. 32–39.

IOANNOU, P. A., and P. V. KOKOTOVIC (1983). *Adaptive Systems with Reduced Models*, Springer-Verlag, New York.

ITAKURA, F., and S. SAITO (1971). "Digital filtering techniques for speech analysis and synthesis," in *Proc. 7th Int. Conf. Acoust.*, Budapest, vol. 25-C-1, pp. 261–264.

ITAKURA, F., and S. SAITO (1972). "On the optimum quantization of feature parameters in the PARCOR speech synthesizer," in *IEEE 1972 Conf. Speech Commun. Process.*, New York, pp. 434–437.

JABLON, N. K. (1986). "Steady state analysis of the generalized sidelobe canceller by adaptive noise canceling techniques," *IEEE Trans. Antennas Propag.*, vol. AP-34, pp. 330–337.

JABLON, N. K. (1991). "On the complexity of frequency-domain adaptive filtering", *IEEE Trans. Signal Process.*, vol. 39, pp. 2331–2334.

JABLON, N. K. (1992). "Joint blind equalization, carrier recovery, and timing recovery for high-order QAM constellations," *IEEE Trans. Signal Process.*, vol. 40, pp. 1383–1398.

JACOBI, C. G. J. (1846). "Über ein leichtes verfahren, die in der theorie der säkularstörungen vorkommenden gleichungen numerisch aufzulösen," *J. Reine Angew. Math.* vol. 30, pp. 51–95.

JACOBS, R. A. (1988). "Increased rates of convergence through learning rate adaptation," *Neural Networks*, vol. 1, pp. 295–307.

JAYANT, N. S., and P. NOLL (1984). *Digital Coding of Waveforms*, Prentice-Hall, Englewood Cliffs, N.J.

JAYANT, N. S. (1986). "Coding speech," *IEEE Spectrum*, vol. 23, pp. 58–63.

JAYNES, E. T. (1982). "On the rationale of maximum-entropy methods," *Proc. IEEE*, vol. 70, pp. 939–952.

JAZWINSKI, A. H. (1969). "Adaptive filtering," *Automatica*, vol. 5, pp. 475–485.

JAZWINSKI, A. H. (1970). *Stochastic Processes and Filtering Theory*, Academic Press, New York.

JOHANSSON, E. M., F. U. DOWLA, and D. M. GOODMAN (1990) "Back-propagation learning for multi-layer feed-forward networks using the conjugate gradient method," Report UCRL-JC-104850, Lawrence Livermore National Laboratory, Livermore, Calif.

JOHNSON, C. R., JR. (1984). "Adaptive IIR filtering: current results and open issues," *IEEE Trans. Information Theory*, vol. IT-30, Special Issue on Linear Adaptive Filtering, pp. 237–250.

JOHNSON, C. R., JR. (1988). *Lectures on Adaptive Parameter Estimation*, Prentice-Hall, Englewood Cliffs, N.J.

JOHNSON, C. R., JR. (1991). "Admissibility in blind adaptive channel equalization: a tutorial survey of an open problem," *IEEE Control Systems Magazine*, vol. 11, pp. 3–15.

JOHNSON, C. R., JR., S. DASGUPTA, and W. A. SETHARES (1988). "Averaging analysis of local stability of a real constant modulus algorithm adaptive filter," *IEEE Trans. Acoust. Speech Signal Process.*, vol. ASSP-36, pp. 900–910.

JOHNSON, C. R., JR., B. EGARDT, and G. KUBIN. (1995). "Frequency-domain interpretation of LMS performance," *IEEE Trans. Signal Process.*, (submitted).

JOHNSON, D. H., and P. S. RAO (1990). "On the existence of Gaussian noise," in The 1990 Digital Signal Processing Workshop, New Paltz, NY, Sponsored by IEEE Signal Processing Society, pp. 8.14.1–8.14.2.

JONES, S. K., R. K. CAVIN III, and W. M. REED (1982). "Analysis of error-gradient adaptive linear equalizers for a class of stationary-dependent processes," *IEEE Trans. Information Theory*, vol. IT-28, pp. 318–329.

JOU, J.-Y., and A. ABRAHAM (1986). "Fault-tolerant matrix arithmetic and signal processing on highly concurrent computing structures," *Proc. IEEE*, Special Issue on Fault Tolerance in VLSI, vol. 74, pp. 732–741.

JUSTICE, J. H. (1985). "Array processing in exploration seismology," in *Array Signal Processing*, ed. S. Haykin, pp. 6–114, Prentice-Hall, Englewood Cliffs, N.J.

KAILATH, T. (1960). Estimating Filters for Linear Time-Invariant Channels, Quarterly Progress Rep. 58, MIT Research Laboratory for Electronics, Cambridge, Mass., pp. 185–197.

KAILATH, T. (1968). "An innovations approach to least-squares estimation: Part 1. Linear filtering in additive white noise," *IEEE Trans. Autom. Control*, vol. AC-13, pp. 646–655.

KAILATH, T. (1969). "A generalized likelihood ratio formula for random signals in Gaussian noise," *IEEE Trans. Information Theory*, vol. IT-15, pp. 350–361.

KAILATH, T. (1970). "The innovations approach to detection and estimation theory," *Proc. IEEE*, vol. 58, pp. 680–695.

KAILATH, T. (1974). "A view of three decades of linear filtering theory," *IEEE Trans. Information Theory*, vol. IT-20, pp. 146–181.

KAILATH, T., ed. (1977). *Linear Least-Squares Estimation*, Benchmark Papers in Electrical Engineering and Computer Science, Dowden, Hutchinson & Ross, Stroudsburg, Pa.

KAILATH, T. (1980). *Linear Systems*, Prentice-Hall, Englewood Cliffs, N.J.

KAILATH, T. (1981). *Lectures on Linear Least-Squares Estimation*, Springer-Verlag, New York.

KAILATH, T. (1982). "Time-variant and time-invariant lattice filters for nonstationary processes," in *Outils et Modèles Mathématique pour l'Automatique, l'Analyse de Systèms et le Traitement du Signal*, vol. 2, ed. I. Laudau, CNRS, Paris, pp. 417–464.

KAILATH, T., and P. A. FROST (1968). "An innovations approach to least-squares estimation: Part 2. Linear smoothing in additive white noise," *IEEE Trans. Autom. Control*, vol. AC-13, pp. 655–660.

KAILATH, T., and R. A. GEESEY (1973). "An innovations approach to least-squares estimation: Part 5. Innovation representations and recursive estimation in colored noise," *IEEE Trans. Autom. Control*, vol. AC-18, pp. 435–453.

KAILATH, T., A. VIEIRA, and M. MORF (1978). "Inverses of Toeplitz operators, innovations, and orthogonal polynomials," *SIAM Rev.*, vol. 20, pp. 106–119.

KALLMANN, H. J. (1940). "Transversal filters," *Proc. IRE*, vol. 28, pp. 302–310.

KALMAN, R. E. (1960). "A new approach to linear filtering and prediction problems," *Trans. ASME, J. Basic Eng.*, vol. 82, pp. 35–45.

KALMAN, R. E., and R. S. BUCY (1961). "New results in linear filtering and prediction theory," *Trans. ASME, J. Basic Eng.*, vol. 83, pp. 95–108.

KALOUPTSIDIS, N., and S. THEODORIDIS (1987). "Parallel implementation of efficient LS algorithms for filtering and prediction," *IEEE Trans. Acoust. Speech Signal Process.*, vol. ASSP-35, pp. 1565–1569.

KALOUPTSIDIS, N., and S. THEODORIDIS, eds. (1993). *Adaptive System Identification and Signal Processing Algorithms*, Prentice-Hall, Englewood Cliffs, N.J.

KAMINSKI, P. G., A. E. BRYSON, and S. F. SCHMIDT (1971). "Discrete square root filtering: A survey of current techniques," *IEE Trans. Autom. Control*, vol. AC-16, pp.727–735.

KANG, G. S., and L. J. Fransen (1987). "Experimentation with an adaptive noise-cancellation filter," *IEEE Trans. Circuits Syst.*, vol. CAS-34, pp. 753–758.

KASSAM, S. A., and I. CHA (1993). "Radial basis functions networks in nonlinear signal processing applications," in Conf. Rec. Asilomar Conference on Signals, Systems, and Computers, Pacific Grove, Calif., pp. 1021–1025.

KAY, S. M. (1988). *Modern Spectral Estimation: Theory and Application*, Prentice-Hall, Englewood Cliffs, N.J.

KAY, S. M., and L. S. MARPLE, JR. (1981). "Spectrum analysis—a modern perspective," *Proc. IEEE*, vol. 69, pp. 1380–1419.

KELLY, J. L, JR., and R. F. LOGAN (1970). *Self-Adaptive Echo Canceller*, U.S. Patent 3,500,000, March 10.

KELLY, E. J., I. S. REED, and W. L. ROOT (1960). "The detection of radar echoes in noise: I," *J. SIAM*, vol. 8, pp. 309–341.

KHARGONEKAR, P. P., and K. M. NAGPAL (1991). "Filtering and smoothing in an H^∞-setting," *IEEE Trans. Autom. Control*, vol. AC-36, pp. 151–166.

KIM, M. S., and C. C. GUEST (1990). "Modification of backpropagation networks for complex-valued signal processing in frequency domain," *International Joint Conference on Neural Networks*, San Diego, Calif., vol. III, pp. 27–31.

KIMURA, H. (1984). "Robust realizability of a class of transfer functions," *IEEE Trans. Autom. Control*, vol. AC-29, pp. 788–793.

KLEMA, V. C., and A.J. LAUB (1980). "The singular value decomposition: Its computation and some applications," *IEEE Trans Autom. Control*, vol. AC-25, pp. 164–176.

KMENTA, J. (1971). *Elements of Econometrics*, Macmillan, New York.

KNIGHT, W. C., R. G. PRIDHAM, and S. M. KAY (1981). "Digital signal processing for sonar," *Proc. IEEE*, vol. 69, pp. 1451–1506.

KOH, T., and E. J. POWERS (1985). "Second-order Volterra filtering and its application to nonlinear system identification," *IEEE Trans. Acoust. Speech Signal Process.*, vol. ASSP-33, pp. 1445–1455.

KOHONEN, T. (1982). "Self-organized formation of topologically correct feature maps," *Biological Cybernetics*, vol. 43, pp. 59–69.

KOHONEN, T. (1990). "The self-organizing map," *Proc. IEEE*, vol. 78, pp. 1464–1480.

KOLMOGOROV, A. N. (1939). "Sur l'interpolation et extrapolation des suites stationaries," *C.R. Acad. Sci.*, Paris, vol. 208, pp. 2043–2045. [English translation reprinted in Kailath, 1977.]

KOLMOGOROV, A. N. (1968). "Three approaches to the quantitative definition of information," *Probl. Inf. Transm. USSR*, vol. 1, pp. 1–7.

KREIN, M. G. (1945). "On a problem of extrapolation of A. N. KOLMOGOROV," *C. R. (Dokl.) Akad. Nauk SSSR*, vol. 46, pp. 306–309. [Reproduced in Kailath, 1977.]

KULLBACK, S., and R. A. LEIBLER (1951). "On information and sufficiency," *Ann. Math. Statist.*, vol. 22, pp. 79–86.

KUMAR, R. (1983). "Convergence of a decision-directed adaptive equalizer," in *Proc. Conf. Decision Control*, vol. 3, pp. 1319–1324.

KUNG, H. T. (1982). "Why systolic architectures?" *Computer*, vol. 15, pp. 37–46.

KUNG, H. T., and C. E. LEISERSON (1978). "Systolic arrays (for VLSI)," *Sparse Matrix Proc. 1978*, *Soc. Ind. Appl. Math.*, 1978, pp. 256–282. [A version of this paper is reproduced in Mead and Conway, 1980.]

KUNG, S. Y. (1988). *VLSI Array Processors*, Prentice-Hall, Englewood Cliffs, N.J.

KUNG, S. Y., H. J. WHITEHOUSE, and T. KAILATH, eds. (1985). *VLSI and Modern Signal Processing*, Prentice-Hall, Englewood Cliffs, N.J.

KUNG, S. Y., ET AL. (1987). "Wavefront array processors—concept to implementation," *Computer*, vol. 20, pp. 18–33.

KUSHNER, H. J. (1984). *Approximation and Weak Convergence Methods for Random Processes with Applications to Stochastic System Theory*, MIT Press, Cambridge, Mass.

KUSHNER, H. J., and D. S. CLARK (1978). *Stochastic Approximation Methods for Constrained and Unconstrained Systems*, Springer-Verlag, New York.

KUSHNER, H. J., and J. YANG (1995). "Analysis of adaptive step size SA algorithms for parameter tracking," *IEEE Trans. Autom. Control*, Vol. 40, pp. 1403–1410.

LANDAU, I. D. (1984). "A feedback system approach to adaptive filtering," *IEEE Trans. Information Theory*, vol. IT-30, Special Issue on Linear Adaptive Filtering, pp. 251–262.

LANG, K. J., and G. E. HINTON (1988). "The development of the time-delay neural network architecture for speech recognition," Technical Report, CMU-CS-88-152, Carnegie-Mellon University, Pittsburgh, Pa.

LANG, S. W., and J. H. MCCLELLAN (1979). "A simple proof of stability for all-pole linear prediction models," *Proc. IEEE*, vol. 67, pp. 860–861.

LAWRENCE, R. E., and H. KAUFMAN (1971). "The Kalman filter for the equalization of a digital communication channel," *IEEE Trans. Commun. Technol.*, vol. COM-19, pp. 1137–1141.

LAWSON, C. L., and R. J. HANSON (1974). *Solving Least Squares Problems*, Prentice-Hall, Englewood Cliffs, N.J.

LEE, D. T. L. (1980). "Canonical ladder form realizations and fast estimation algorithms," Ph.D. dissertation, Stanford University, Stanford, Calif.

LEE, D. T. L., M. MORF, and B. FRIEDLANDER (1981). "Recursive least-squares ladder estimation algorithms," *IEEE Trans. Circuits Syst.*, vol. CAS-28, pp. 467–481.

LEE, J. C., and C. K. UN (1986). "Performance of transform-domain LMS adaptive algorithms," *IEEE Trans. Acoust. Speech Signal Process.*, vol. ASSP-34, pp. 499–510.

LEGENDRE, A. M. (1810). "Méthode des moindres quarrés, pour trouver le milieu le plus probable entre les résultats de différentes observations," *Mem. Inst. France*, pp. 149–154.

LEHMER, D. H. (1961). "A machine method for solving polynomial equations," *J. Assoc. Comput. Mach.*, vol. 8, pp. 151–162.

LEUNG, H., and S. HAYKIN (1989). "Stability of recursive QRD-LS algorithms using finite-precision systolic array implementation," *IEEE Trans. Acoust. Speech Signal Process.*, vol. ASSP-37, pp. 760–763.

LEUNG, H., and S. HAYKIN (1991). "The complex backpropagation algorithm," *IEEE Trans. Acoust. Speech Signal Process.*, vol. ASSP-39, pp. 2101–2104.

LEV-ARI, H., T. KAILATH, and J. CIOFFI (1984). "Least-squares adaptive lattice and transversal filters: A unified geometric theory," *IEEE Trans. Information Theory*, vol. IT-30, pp. 222–236.

LEVIN, M. D., and C. F. N. COWAN (1994). "The performance of eight recursive least squares adaptive filtering algorithms in a limited precision environment," in *Proc. European Signal Process. Conf.*, Edinburgh, Scotland, pp. 1261–1264.

LEVINSON, N. (1947). "The Wiener RMS (root-mean-square) error criterion in filter design and prediction," *J. Math Phys.*, vol. 25, pp. 261–278.

LEVINSON, N., and R. M. REDHEFFER (1970). *Complex Variables*, Holden-Day, San Francisco.

LEWIS, A. (1992). "Adaptive filtering-applications in telephony," *BT Technol. J.*, vol. 10, pp. 49–63.

LI, Y., and Z. DING (1995). "Convergence analysis of finite length blind adaptive equalizers," *IEEE Trans. Signal Process.*, vol. 43, pp. 2120–2129.

LIAPUNOV, A. M. (1966). *Stability of Motion*, trans. F. Abramovici and M. Shimshoni, Academic Press, New York.

LII, K. S. and M. ROSENBLATT (1982). "Deconvolution and estimation of transfer function phase and coefficients for non-Gaussian linear processes," *Ann. Stat.*, vol. 10, pp. 1195–1208.

LILES, W. C., J. W. DEMMEL, and L. E. BRENNAN (1980). Gram–Schmidt Adaptive Algorithms, Tech. Rep. RADC-TR-79-319, RADC, Griffiss Air Force Base, N. Y.

LIN, D. W. (1984). "On digital implementation of the fast Kalman algorithm," *IEEE Trans. Acoust. Speech Signal Process.*, vol. ASSP-32, pp. 998–1005.

LING, F. (1989). "Efficient least-squares lattice algorithms based on Givens rotation with systolic array implementations," in *Proc. ICASSP*, Glasgow, Scotland, pp. 1290–1293.

LING, F., 1991. "Givens rotation based least-squares lattice and related algorithms," *IEEE Trans. Signal Process.*, vol. 39, pp. 1541–1551.

LING, F., and J. G. PROAKIS (1984a). "Numerical accuracy and stability: Two problems of adaptive estimation algorithms caused by round-off error," in *Proc. ICASSP*, San Diego, Calif., pp. 30.3.1–30.3.4.

LING, F., and J. G. PROAKIS (1984b). "Nonstationary learning characteristics of least squares adaptive estimation algorithms," in *Proc. ICASSP*, San Diego, Calif., pp. 3.7.1–3.7.4.

LING, F., and J. G. PROAKIS (1986). "A recursive modified Gram–Schmidt algorithm with applications to least squares estimation and adaptive filtering," *IEEE Trans. Acoust. Speech Signal Process.*, vol. ASSP-34, pp. 829–836.

LING, F., D. MANOLAKIS, and J. G. PROAKIS (1985). "New forms of LS lattice algorithms and an analysis of their round-off error characteristics," in *Proc. ICASSP*, Tampa, Fla., pp. 1739–1742.

LING, F., D. MANOLAKIS, and J. G. PROAKIS, (1986). "Numerically robust least-squares lattice-ladder algorithm with direct updating of the reflection coefficients," *IEEE Trans. Acoust. Speech Signal Process.*, vol. ASSP-34, pp. 837–845.

LIPPMANN, R. P. (1987). "An introduction to computing with neural nets," *IEEE ASSP Magazine*, vol. 4, pp. 4–22.

LITTLE, G. R., S. C. GUSTAFSON, and R. A. SENN (1990). "Generalization of the backpropagation neural network learning algorithm to permit complex weights," *Appl. Opt.*, vol. 29, pp. 1591–1592.

LIU, K. J. R., S.-F. HSIEH, and K. YAO (1992). "Systolic block Householder transformation for RLS algorithm with two-level pipelined implementation," *IEEE Trans. Signal Process.*, vol. 40, pp. 946–958.

LIU, Z.-S. (1995). "QR methods of O(N) complexity in adaptive parameter estimation," *IEEE Trans. Signal Process.*, vol. 43, pp. 720–729.

LJUNG, L. (1977). "Analysis of recursive stochastic algorithms," *IEEE Trans. Autom. Control.*, vol. AC-22, pp. 551–575.

LJUNG, L. (1984). "Analysis of stochastic gradient algorithms for linear regression problems," *IEEE Trans. Information Theory*, vol. IT-30, Special Issue on Linear Adaptive Filtering, pp. 151–160.

LJUNG, L. (1987). *System Identification: Theory for the User*, Prentice-Hall, Englewood Cliffs, N.J.

LJUNG, L., and T. SÖDERSTRÖM (1983). *Theory and Practice of Recursive Identification*, MIT Press, Cambridge, Mass.

LJUNG, L., M. MORF, and D. FALCONER (1978). "Fast calculation of gain matrices for recursive estimation schemes," *Int. J. Control*, vol. 27, pp. 1–19.

LJUNG, L, and S. GUNNARSSON (1990). "Adaptation and tracking in system identification—A survey," *Automatica*, vol. 26, pp. 7–21.

LJUNG, S., and L. LJUNG (1985). "Error propagation properties of recursive least-squares adaptation algorithms," *Automatica*, vol. 21, pp. 157–167.

LORD, RAYLEIGH. (1879). "Investigations in optics with special reference to the spectral scope," *Philos. Mag.*, vol. 8, pp. 261–274.

LORENZ, H., G. M. RICHTER, M. CAPACCIOLI, and G. LONGO (1993). "Adaptive filtering in astronomical image processing. I. Basic considerations and examples," *Astron. Astrophys.*, vol. 277, pp. 321–330.

LOWE, D. (1989). "Adaptive radial basis function nonlinearities and the problem of generalization," in *First IEE Int. Conf. Artif. Neural Networks*, London, pp. 171–175.

LUCKY, R. W. (1965). "Automatic equalization for digital communication," *Bell Syst. Tech. J.*, vol. 44, pp. 547–588.

LUCKY, R. W. (1966). "Techniques for adaptive equalization of digital communication systems," *Bell Syst. Tech. J.*, vol. 45, pp. 255–286.

LUCKY, R. W. (1973). "A survey of the communication literature: 1968–1973," *IEEE Trans. Information Theory*, vol. IT-19, pp. 725–739.

LUCKY, R. W., J. SALZ, and E. J. WELDON, JR. (1968). *Principles of Data Communication*, McGraw-Hill, New York.

LUENBERGER, D. G. (1969). *Optimization by Vector Space Methods*, Wiley, New York.

LUK, F. T. (1986). "A triangular processor array for computing singular values," *Linear Algebra Applications*, vol. 77, pp. 259–273.

LUK, F. T., and H. PARK (1989). "A proof of convergence for two parallel Jacobi SVD algorithms," *IEEE Trans. Comput.*, vol. 38, pp. 806–811.

LUK, F. T., and S. QIAO (1989). "Analysis of a recursive least-squares signal-processing algorithm," *SIAM J. Sci. Stat. Comput.*, vol. 10, pp. 407–418.

LYNCH, M. R., and P. J. RAYNER (1989), "The properties and implementation of the non-linear vector space connectionist model," in *Proc. First IEE Int. Conf. Artif. Neural Networks*, London, pp. 186–190.

MACCHI, O. (1986a). "Advances in Adaptive Filtering," in *Digital Communications*, ed. E. Biglieri and G. Prati, North-Holland, Amsterdam, pp. 41–57.

MACCHI, O. (1986b). "Optimization of adaptive identification for time-varying filters," *IEEE Trans. Autom. Control*, vol. AC-31, pp. 283–287.

MACCHI, O. (1995). *Adaptive Processing: The LMS Approach with Applications in Transmission*, Wiley, New York.

MACCHI, O., and N. J. BERSHAD (1991). "Adaptive recovery of a chirped sinusoid in noise, Part I: Performance of the RLS algorithm," *IEEE Trans. Acoust. Speech Signal Process.*, vol. 39, pp. 583–594.

MACCHI, O., and M. TURKI (1992). "The nonstationarity degree: can an adaptive filter be worse than no processing?" in *Proc. IFAC International Symposium on Adaptive Systems in Control and Signal Processing*, Grenoble, France, pp. 743–747.

MACCHI, O., N. J. BERSHAD, and M. M-BOUP (1991). "Steady-state superiority of LMS over LS for time-varying line enhancer in noisy environment," *IEE Proc. (London), part F*, vol. 138, pp. 354–360.

MACCHI, O., and E. EWEDA (1984). "Convergence analysis of self-adaptive equalizers," *IEEE Trans. Information Theory*, vol. IT-30, Special Issue on Linear Adaptive Filtering, pp. 161–176.

MACCHI, O., and M. JAIDANE-SAIDNE (1989). "Adaptive IIR filtering and chaotic dynamics: application to audio-frequency coding," *IEEE Trans. Circuits Syst.*, vol. 36, pp. 591–599.

MAKHOUL, J. (1975). "Linear prediction: A tutorial review," *Proc. IEEE*, vol. 63, pp. 561–580.

MAKHOUL, J. (1977). "Stable and efficient lattice methods for linear prediction," *IEEE Trans. Acoust. Speech Signal Process.*, vol. ASSP-25, pp. 423–428.

MAKHOUL, J. (1978). "A class of all-zero lattice digital filters: properties and applications," *IEEE Trans. Acoust. Speech Signal Process.*, vol. ASSP-26, pp. 304–314.

MAKHOUL, J. (1981). "On the eigenvectors of symmetric Toeplitz matrices," *IEEE Trans. Acoust. Speech Signal Process.*, vol. ASSP-29, pp. 868–872.

MAKHOUL, J., and L. K. COSSELL (1981), "Adaptive lattice analysis of speech," *IEEE Trans. Circuits Syst.*, vol. CAS-28, pp. 494–499.

MANOLAKIS, D., F. LING, and J. G. PROAKIS (1987). "Efficient time-recursive least-squares algorithms for finite-memory adaptive filtering," *IEEE Trans. Circuits Syst.*, vol. CAS-34, pp. 400–408.

MANSOUR, D., and A. H. GRAY, JR. (1982). "Unconstrained frequency-domain adaptive filters," *IEEE Trans. Acoust. Speech Signal Process.*, vol. ASSP-30, pp. 726–734.

MARCOS, S., and O. MACCHI (1987). "Tracking capability of the least mean square algorithm: Application to an asynchronous echo canceller," *IEEE Trans. Acoust. Speech Signal Process.*, vol. ASSP-35, pp. 1570–1578.

MARDEN, M. (1949). "The geometry of the zeros of a polynomial in a complex variable," *Am. Math. Soc. Surveys*, no. 3, chap. 10, American Mathematical Society, New York.

MARKEL, J. D., and A. H. GRAY JR. (1976). *Linear Prediction of Speech*, Springer-Verlag, New York.

MARPLE, S. L., JR. (1980). "A new autoregressive spectrum analysis algorithm," *IEEE Trans. Acoust. Speech Signal Process.*, vol. ASSP-28, pp. 441–454.

MARPLE, S. L., JR. (1981). "Efficient least squares FIR system identification," *IEEE Trans. Acoust. Speech Signal Process.*, vol. ASSP-29, pp. 62–73.

MARPLE, S. L., JR. (1987). *Digital Spectral Analysis with Applications*, Prentice-Hall, Englewood Cliffs, N.J.

MARSHALL, D. F., W. K. JENKINS, and J. J. MURPHY (1989). "The use of orthogonal transforms for improving performance of adaptive filters," *IEEE Trans. Circuits Syst.*, vol. 36, pp. 474–484.

MASON, S. J. (1956). "Feedback theory; further properties of signal flow graphs," *Proc. IRE*, vol. 44, pp. 920–926.

MATHEWS, V. J., and Z. XIE (1993). "A stochastic gradient adaptive filter with gradient adaptive step size," *IEEE Trans. Signal Process.*, vol. 41, pp. 2075–2087.

MATHIAS, R. (1995). "Accurate eigen system computation by Jacobi methods," *SIMAX*, vol. 16, pp. 977–1003.

MAYBECK, P. S. (1979). *Stochastic Models, Estimation, and Control*, vol. 1, Academic Press, New York.

MAYBECK, P. S. (1982). *Stochastic Models, Estimation, and Control*, vol. 2, Academic Press, New York.

MAZO, J. E. (1979). "On the independence theory of equalizer convergence," *Bell Syst. Tech. J.*, vol. 58, pp. 963–993.

MAZO, J. E. (1980). "Analysis of decision-directed equalizer convergence," *Bell Syst. Tech. J.*, vol. 59, pp. 1857–1876.

McCanny, J. V., and J. G. McWhirter (1987). "Some systolic array developments in the United Kingdom," *Computer*, vol. 2, pp. 51–63.

McCulloch, W. S., and W. Pitts (1943). "A logical calculus of the ideas immanent in nervous activity," *Bulletin of Mathematical Biophysics*, vol. 5, pp. 115–133.

McCool, J. M., et al. (1980). *Adaptive Line Enhancer*, U.S. Patent 4,238,746, December 9.

McCool, J. M., et al. (1981). *An Adaptive Detector*, U.S. Patent 4,243,935, January 6.

McDonald, R. A. (1966). "Signal-to-noise performance and idle channel performance of differential pulse code modulation systems with particular applications to voice signals," *Bell Syst. Tech. J.*, vol. 45, pp. 1123–1151.

McGee, W. F. (1971). "Complex Gaussian noise moments," *IEEE Trans. Information Theory*, vol. IT-17, pp. 149–157.

McLachlan, G. J., and K. E. Basford (1988). *Mixture Models: Inference and Applications to Clustering*, Dekker, New York.

McWhirter, J. G. (1983). "Recursive least-squares minimization using a systolic array," *Proc. SPIE, Real-Time Signal Processing VI*, vol. 431, San Diego, Calif., pp. 105–112.

McWhirter, J. G. (1989). "Algorithmic engineering-an emerging technology," *Proc. SPIE, Real-Time Signal Processing VI*, vol. 1152, San Diego, Calif.

McWhirter, J. G., and I. K. Proudler (1993). "The QR family," in *Adaptive System Identification and Signal Processing Algorithms*, ed. N. Kalouptsidis and S. Theodoridis, pp. 260–321, Prentice-Hall, Englewood Cliffs, N.J.

McWhirter, J. G., and T. J. Shepherd (1989). "Systolic array processor for MVDR beamforming," *IEE Proc. (London)*, part F, vol. 136, pp. 75–80.

Mead, C., and L. Conway (1980). *Introduction to VLSI Systems*, Addison-Wesley, Reading, Mass.

Medaugh, R. S., and L. J. Griffiths (1981). "A comparison of two linear predictors," in *Proc. ICASSP*, Atlanta, Ga., pp. 293–296.

Mehra, R. K. (1972). "Approaches to adaptive filtering," *IEEE Trans. Autom. Control*, vol. AC-17, pp. 693–698.

Mendel, J. M. (1973). *Discrete Techniques of Parameter Estimation: The Equation Error Formulation*, Dekker, New York.

Mendel, J. M. (1974). "Gradient estimation algorithms for equation error formulations," *IEEE Trans. Autom. Control*, vol. AC-19, pp. 820–824.

Mendel, J. M. (1986). "Some modeling problems in reflection seismology," *IEEE ASSP Mag.*, vol. 3, pp. 4–17.

Mendel, J. M. (1990a). *Maximum-Likelihood Deconvolution: A Journey into Model-Based Signal Processing*, Springer-Verlag, New York.

Mendel, J. M. (1990b). "Introduction," *IEEE Trans. Autom. Control*, vol. AC-35, Special Issue on Higher Order Statistics in System Theory and Signal Processing, p. 3.

Mendel, J. M. (1995). *Lessons in Digital Estimation Theory*, 2nd ed., Prentice-Hall, Englewood Cliffs, N.J.

Meriam, K. A. et al. (1995). "Prediction error methods for time-domain blind identification of multichannel FIR filters," in *Proc. ICASSP*, Detroit, Michigan, vol. 3, pp. 1968–1971.

Mermoz, H. F. (1981). "Spatial processing beyond adaptive beamforming," *J. Acoust. Soc. Am.*, vol. 70, pp. 74–79.

MESSERCHMITT, D. G. (1984). "Echo cancellation in speech and data transmission," *IEEE J. Sel. Areas Commun.*, vol. SAC-2, pp. 283–297.

METFORD, P. A. S., and S. Haykin (1985). "Experimental analysis of an innovations-based detection algorithm for surveillance radar," *IEE Proc. (London),* vol. 132, part F, pp. 18–26.

MIDDLETON, D. (1960). *An Introduction to Statistical Communication Theory*, McGraw-Hill, New York.

MILLER, K. S. (1974). *Complex Stochastic Processes: An Introduction to Theory and Application*, Addison-Wesley, Reading, Mass.

MINSKY, M. L., and S. A. PAPPERT (1969). *Perceptrons*, MIT Press, Cambridge, Mass.

MONSEN, P. (1971). "Feedback equalization for fading dispersive channels," *IEEE Trans. Information Theory*, vol. IT-17, pp. 56–64.

MONZINGO, R. A., and T. W. MILLER (1980). *Introduction to Adaptive Arrays*, Wiley-Interscience, New York.

MOODY, J. E., and C. J. DARKEN (1989). "Fast learning in networks of locally-tuned processing units," *Neural Computation*, vol. 1, pp. 281–294.

MOONEN, M., and J. VANDEWALLE (1990). "Recursive least squares with stabilized inverse factorization," *Signal Process.*, vol. 21, pp. 1–15.

MORF, M. (1974). "Fast algorithms for multivariable systems," Ph.D. dissertation, Stanford University, Stanford, Calif.

MORF, M., and D. T. LEE (1978). "Recursive least squares ladder forms for fast parameter tracking," in *Proc. 1978 Conf. Decision Control*, San Diego, Calif., pp. 1362–1367.

MORF, M., and T. KAILATH (1975). "Square-root algorithms for least-squares estimation," *IEEE Trans. Autom. Control*, vol. AC-20, pp. 487–497.

MORF, M., T. KAILATH, and L. LJUNG (1976). "Fast algorithms for recursive identification," in *Proc. 1976 Conf. Decision Control*, Clearwater Beach, Fla., pp. 916–921.

MORF, M., A. VIEIRA, and D. T. LEE (1977). "Ladder forms for identification and speech processing," in *Proc. 1977 IEEE Conf. Decision Control*, New Orleans, pp. 1074–1078.

MORONEY, P. (1983). *Issues in the Implementation of Digital Feedback Compensators*, MIT Press, Cambridge, Mass.

MOROZOV, V. A. (1993). *Regularization Methods for Ill-posed Problems*, CRC Press, Boca Raton, Fla.

MORSE, P. M., and H. FESHBACK (1953). *Methods of Theoretical Physics*, Pt. I, McGraw-Hill, New York.

MOSCHNER, J. L. (1970). Adaptive Filter with Clipped Input Data, Tech. Rep. 6796-1, Stanford University Center for Systems Research, Stanford, Calif.

MOULINES, E., P. DUHAMEL, J.-F. CARDOSO, and S. MAYRARGUE (1995). "Subspace methods for blind identification of multichannel FIR filters," *IEEE Trans. Signal Process.*, vol. 43, pp. 516–525.

MUELLER, M. S. (1981a). Least-squares algorithms for adaptive equalizers," *Bell Syst. Tech. J.*, vol. 60, pp. 1905–1925.

MUELLER, M. S. (1981b). "On the rapid initial convergence of least-squares equalizer adjustment algorithms," *Bell Syst. Tech. J.*, vol. 60, pp. 2345–2358.

MUIRHEAD, R. J. (1982). *Aspects of Multivariate Statistical Theory*, Wiley, New York.

MULGREW, B. (1987). "Kalman filter techniques in adaptive filtering," *IEE Proc. (London)*, part F, vol. 134, pp. 239–243.

MULGREW, B., and C. F. N. COWAN (1987). "An adaptive Kalman equalizer: structure and performance," *IEEE Trans. Acoust. Speech Signal Process.*, vol. ASSP-35, pp. 1727–1735.

MULGREW, B., and C. F. N. COWAN (1988). *Adaptive Filters and Equalizers*, Kluwer, Boston, Mass.

MURANO, K., ET AL. (1990). "Echo cancellation and applications," *IEEE Commun.*, vol. 28, pp. 49–55.

MUSICUS, B. R. (1985). "Fast MLM power spectrum estimation from uniformly spaced correlations," *IEEE Trans. Acoust. Speech Signal Process.*, vol. ASSP-33, pp. 1333–1335.

NAGUMO, J. I., and A. NODA (1967). "A learning method for system identification," *IEEE Trans. Autom. Control*, vol. AC-12, pp. 282–287.

NAHI, N. E. (1969). *Estimation Theory and Applications*, Wiley, New York.

NARAYAN, S. S., A. M. PETERSON, and M. J. NARASHIMA (1983). "Transform domain LMS algorithm," *IEEE Trans. Acoust. Speech Signal Process.*, vol. ASSP-31, pp. 609–615.

NARENDRA, K. S., and A. M. ANNASWAMY (1989). *Stable Adaptive Systems*, Prentice-Hall, Englewood Cliffs, N.J.

NARENDRA, K. S., and K. PARTHASARATHY (1990). "Identification and control of dynamical systems using neural networks," *IEEE Trans. Neural Networks*, vol. 1, pp. 4–27.

NAU, R. F., and R. M. OLIVER (1979). "Adaptive filtering revisited," *J. Oper. Res. Soc.*, vol. 30, pp. 825–831.

NIELSEN, P. A., and J. B. THOMAS (1988). "Effect of correlation on signal detection in arctic under-ice noise," in Conf. Rec. Twenty-Second Asilomar Conference on Signals, Systems and Computers," Pacific Grove, Calif., pp. 445–450.

NIKIAS, C. L. (1991). "Higher-order spectral analysis," in *Advances in Spectrum Analysis and Array Processing*, vol. 1, ed. S. Haykin, Prentice-Hall, Englewood Cliffs, N.J.

NIKIAS, C. L., and M. R. RAGHUVEER (1987). "Bispectrum estimation: A digital signal processing framework," *Proc. IEEE*, vol. 75, pp. 869–891.

NISHITANI, T., ET AL. (1987). "A CCITT standard 32 kbits/s ADPCM LSI codec," *IEEE Trans. Acoust. Speech Signal Process.*, vol. ASSP-35, pp. 219–225.

NORMILE, J. O. (1983). "Adaptive filtering with finite wordlength constraints," *IEE Proc. (London)*, part E vol. 130, pp. 42–46.

NORTH, D. O. (1963). "An analysis of the factors which determine signal/noise discrimination in pulsed carrier systems," *Proc. IEEE*, vol. 51, pp. 1016–1027.

NORTH, R. C., J. R. ZEIDLER, W. H. KU, and T. R. ALBERT, 1993. "A floating-point arithmetic error analysis of direct and indirect coefficient updating techniques for adaptive lattice filters," *IEEE Trans. Signal Process.*, vol. 41, pp. 1809–1823.

NUTTAL, A. H. (1976). Spectral Analysis of a Univariate Process with Bad Data Points via Maximum Entropy and Linear Predictive Techniques, Naval Underwater Systems Center (NUSC) Scientific and Engineering Studies, New London, Conn.

OPPENHEIM, A. V., and J. S. LIM (1981). "The importance of phase in signals," *Proc. IEEE*, vol. 69, pp. 529–541.

OPPENHEIM, A. V., and R. W. SCHAFER (1989). *Discrete-Time Signal Processing*, Prentice-Hall, Englewood Cliffs, N.J.

OWSLEY, N. L. (1973). "A recent trend in adaptive spatial processing for sensor arrays: constrained adaptation," in *Signal Processing*, ed. J. W. R. Griffiths et al., Academic Press, New York, pp. 591–604.

OWSLEY, N. L. (1985). "Sonar array processing," in *Array Signal Processing*, ed. S. Haykin, Prentice-Hall, Englewood Cliffs, N.J., pp. 115–193.

PALMIERI, F., and S. A. SHAH (1990). "Fast training of multi-layer perceptrons using multilinear parameterization," in *International Joint Conference on Neural Networks*, Washington, D.C., vol. 1, pp. 696–699.

PAN, C. T., and R. J. PLEMMONS (1989). "Least squares modifications with inverse factorizations: Parallel implications," *J. Comput. Appl. Math.*, vol. 27, pp. 109–127.

PAN, R., and C. L. NIKIAS (1988). "The complex cepstrum of higher order cumulants and nonminimum phase identification," *IEEE Trans. Acoust. Speech Signal Process.*, vol. ASSP-36, pp. 186–205.

PANDA, G., B. MULGREW, C. F. N. COWAN, and P. M. GRANT (1986). "A self-orthogonalizing efficient block adaptive filter," *IEEE Trans. Acoust. Speech Signal Process.*, vol. ASSP-34, pp. 1573–1582.

PAPADIAS, C. (1995). "Methods for blind equalization and identification of linear channels," Ph. D. thesis, Ecole Nationale Supérieure des Télécommunications, Paris, France.

PAPOULIS, A. (1984). *Probability, Random Variables, and Stochastic Processes*, 2nd ed., McGraw-Hill, New York.

PARK, J., and I. W. SANDBERG (1991). "Universal approximation using radial-basis-function networks," *Neural Computation*, vol. 3, pp. 246–257.

PARLETT, B. N. (1980). *The Symmetric Eigenvalue Problem*, Prentice-Hall, Englewood Cliffs, N.J.

PARZEN, E. (1962). "On the estimation of a probability density function and mode," *Ann. Math. Stat.*, vol. 33, pp. 1065–1076.

PATRA, J. C., and G. PANDA (1992). "Performance evaluation of finite precision LMS adaptive filters using probability density approach," *J. Inst. Electron. Telecommun. Eng.*, vol. 38, pp. 192–195.

PEACOCK, K. L., and S. TREITEL (1969). "Predictive deconvolution: theory and practice," *Geophysics*, vol. 34, pp. 155–169.

PERRIER, A., B. DELYON, and E. MOULINES (1994). "On the validity of the independence assumption for stochastic gradient identification algorithm," submitted for publication.

PETRAGLIA, M. R., and S. K. MITRA, 1993. "Performance analysis of adaptive filter structures based on subband decomposition," in *Proceedings of International Symposium on Circuits and Systems*, pp. I.60–I.63, Chicago, Illinois.

PICCHI, G., and G. PRATI (1984). "Self-orthogonalizing adaptive equalization in the discrete frequency domain," *IEEE Trans. Commun.*, vol. COM-32, pp. 371–379.

PICCHI, G., and G. PRATI (1987). "Blind equalization and carrier recovery using a 'stop-and-go' decision-directed algorithm," *IEEE Trans. Commun.*, vol. COM-35, pp. 877–887.

PLACKETT, R. L. (1950). "Some theorems in least squares," *Biometrika*, vol. 37, p. 149.

POGGIO, T., and F. GIROSI (1990). "Networks for approximation and learning," *Proc. IEEE*, vol. 78, pp. 1481–1497.

PORAT, B., and T. KAILATH (1983). "Normalized lattice algorithms for least-squares FIR system identification," *IEEE Trans. Acoust. Speech Signal Process.*, vol. ASSP-31, pp. 122–128.

PORAT, B., B. FRIEDLANDER, and M. MORF (1982). "Square root covariance ladder algorithms," *IEEE Trans. Autom. Control.* vol. AC-27, pp. 813–829.

POTTER, J. E. (1963). "New Statistical Formulas," Instrumentation Laboratory, MIT, Cambridge, Mass., Space Guidance Analysis Memo No. 40.

POWELL, M. J. D., 1992. "The theory of radial basis function approximation in 1990," in *Advances in Numerical Analysis*, Vol. II: *Wavelets, Subdivision Algorithms, and Radial Basis Functions*, ed. W. Light, pp. 105–210, Oxford Science Publications, Oxford, United Kingdom.

PRESS, W. H., ET AL. (1988). *Numerical Recipes in C*, Cambridge University Press, Cambridge, United Kingdom.

PRIESTLEY, M. B. (1981). *Spectral Analysis and Time Series*, vols. 1 and 2, Academic Press, New York.

PROAKIS, J. G. (1975). "Advances in equalization for intersymbol interference," in *Advances in Communication Systems*, ed. A. V. Balakrishnan, vol. 4, Academic Press, New York, pp. 123–198.

PROAKIS, J. G. (1989). *Digital Communications*, 2nd ed., McGraw-Hill, New York.

PROAKIS, J. G. (1991). "Adaptive equalization for TDMA digital mobile radio," *IEEE Trans. Vehicular Technol.*, vol. 40, pp. 333–341.

PROAKIS, J. G., and J. H. MILLER (1969). "An adaptive receiver for digital signaling through channels with intersymbol interference," *IEEE Trans. Information Theory*, vol. IT-15, pp. 484–497.

PROUDLER, I. K., J. G. McWHIRTER, and T. J. SHEPHERD (1988). "Fast QRD-based algorithms for least squares linear prediction," in *Proc. IMA Conf. Math. Signal Process.*, Warwick, England.

PROUDLER, I. K., J. G. McWHIRTER, and T. J. SHEPHERD (1991). "Computationally efficient, QR decomposition approach to least squares adaptive filtering," *IEE Proc. (London)*, part F, vol. 138, pp. 341–353.

QURESHI, S. (1982). "Adaptive equalization," *IEEE Commun. Soc. Mag.*, vol. 20, pp. 9–16.

QURESHI, S. U. H. (1985). "Adaptive equalization," *Proc. IEEE*, vol. 73, pp. 1349–1387.

RABINER, L. R., and B. GOLD (1975). *Theory and Application of Digital Signal Processing*, Prentice-Hall, Englewood Cliffs, N.J.

RABINER, L. R., and R. W. SCHAFER (1978). *Digital Processing of Speech Signals*, Prentice-Hall, Englewood Cliffs, N.J.

RADER, C. M. (1990). "Linear systolic array for adaptive beamforming," in The 1990 Digital Signal Processing Workshop, New Paltz, N.Y., Sponsored by IEEE Signal Processing Society, pp. 5.2.1–5.2.2.

RADER, C. M., and A. O. STEINHARDT (1986). "Hyperbolic householder transformations", *IEEE Trans. Acoust., Speech, and Signal Process.*, vol. ASSP-34, pp. 1589–1602.

RALSTON, A. (1965). *A First Course in Numerical Analysis*, McGraw-Hill, New York.

RAO, C. R., (1973). *Linear Statistical Inference and its Applications*, 2nd ed., Wiley, New York.

RAO, S. K., and T. KAILATH (1986). "What is a systolic algorithm?" in *Proc. SPIE, Highly Parallel Signal Processing Architectures*, San Diego, Calif., vol. 614, pp. 34–48.

RAO, K. R., and P. YIP (1990). *Discrete Cosine Transform: Algorithms, Advantages, Applications*, Academic Press, San Diego, Calif.

RAYNER, P. J. W., and M. F. LYNCH (1989), "A new connectionist model based on a non-linear adaptive filter," in *Proc. ICASSP*, Glasgow, Scotland, pp. 1191–1194.

REDDI, S. S. (1979). "Multiple source location—A digital approach," *IEEE Trans. Aerospace Electron. Syst.*, vol. AES-15, pp. 95–105.

REDDI, S. S. (1984). "Eigenvector properties of Toeplitz matrices and their application to spectral analysis of time series," *Signal Process.*, pp. 45–56.

REDDY, V. U., B. EGARDT, and T. KAILATH (1981). "Optimized lattice-form adaptive line enhancer for a sinusoidal signal in broad-band noise," *IEEE Trans. Acoust. Speech Signal Process.*, vol. ASSP-29, pp. 702–710.

REDDY, V. U., and A. NEHORAI (1981). "Response of adaptive line enhancer to a sinusoid in lowpass noise," *IEE Proc. (London)*, part F, vol. 128, no. 3, pp. 6–66.

REED, I. S. (1962). "On a moment theorem for complex Gaussian processes," *IRE Trans. Information Theory*, vol. IT-8, pp. 194–195.

REED, I. S., J. D. MALLET, and L. E. BRENNAN (1974). "Rapid convergence rate in adaptive arrays," *IEEE Trans. Aerospace Electron. Syst.*, vol. AES-10, pp. 853–863.

REEVES, A. H. (1975). "The past, present, and future of PCM," *IEEE Spectrum*, vol. 12, pp. 58–63.

REGALIA, P. A. (1992). "Numerical stability issues in fast least-squares adaptation algorithms," *Optical Engineering*, vol. 31, pp. 1144–1152.

REGALIA, P. A. (1993). "Numerical stability properties of a QR-based fast least squares algorithm," *IEEE Trans. Signal Process.*, vol. 41, pp. 2096–2109.

REGALIA, P. A. (1994). *Adaptive IIR Filtering in Signal Processing and Control*, Dekker, New York.

REGALIA, P. A., and G. BELLANGER (1991). "On the duality between fast QR methods and lattice methods in least squares adaptive filtering," *IEEE Trans. Signal Process.*, vol. 39, pp. 879–891.

RICKARD, J. T., ET AL. (1981). "A performance analysis of adaptive line enhancer-augmented spectral detectors," *IEEE Trans. Circuits Sys.*, vol. CAS-28, pp. 534–541.

RICKARD, J. T., and J. R. ZEIDLER (1979). "Second-order output statistics of the adaptive line enhancer," *IEEE Trans. Acoust. Speech Signal Process.*, vol. ASSP-27, pp. 31–39.

RICKARD, J. T., J. R. ZEIDLER, M. J. DENTINO, and M. SHENSA (1981). "A performance analysis of adaptive line enhancer-augmented spectral detectors," *IEEE Trans. Circuits Syst.*, vol. CAS-28, no. 6, pp. 534–541.

RIDDLE, A. (1994). "Engineering software: Mathematical power tools," *IEEE Spectrum*, vol. 31, pp. 35–47, 95, November.

RISSANEN, J. (1978). "Modelling by shortest data description," *Automatica*, vol. 14, pp. 465–471.

RISSANEN, J. (1986). "Stochastic complexity and modeling," *Ann. Stat.*, vol. 14, pp. 1080–1100.

RISSANEN, J. (1989). *Stochastic complexity in statistical enquiry*, Series in Computer Science, vol. 15, World Scientific, Singapore.

ROBBINS, H., and S. MONRO (1951). "A stochastic approximation method," *Ann. Math. Stat.*, vol. 22, pp. 400–407.

ROBINSON, E. A. (1954). "Predictive decomposition of time series with applications for seismic exploration," Ph.D. thesis, Massachusetts Institute of Technology, Cambridge, Mass.

ROBINSON, E. A. (1982). "A historical perspective of spectrum estimation," *Proc. IEEE*, vol. 70, Special Issue on Spectral Estimation, pp. 885–907.

ROBINSON, E. A. (1984). "Statistical pulse compression," *Proc. IEEE*, vol. 72, pp. 1276–1289.

ROBINSON, E. A., and T. DURRANI (1986). *Geophysical Signal Processing*, Prentice-Hall, Englewood Cliffs, N.J.

ROBINSON, E. A., and S. TREITEL (1980). *Geophysical Signal Analysis*, Prentice-Hall, Englewood Cliffs, N.J.

ROSENBLATT, M. (1985). *Stationary Sequences and Random Fields*, Birkhäuser, Stuttgart.

ROSENBROCK, H. H., 1970. *State-space and Multivariable Theory*, Wiley, New York.

Ross, F. J. (1989). "Blind equalization for digital microwave radio," Masters thesis, McMaster University, Hamilton, Ontario, Canada.

Rumelhart, D. E., and J. L. McClelland, eds. (1986). *Parallel Distributed Processing*, vol. 1. Foundations, MIT Press, Cambridge, Mass.

Rumelhart, D. E., G. E. Hinton, and R. J. Williams (1986). "Learning representations by back-propogating errors," *Nature*, vol. 323, pp. 533–536.

Saito, S., and F. Itakura (1966). *The Theoretical Consideration of Statistically Optimum Methods for Speech Spectral Density*, Rep. 3107, Electrical Communication Laboratory, N. T. T., Tokyo (in Japanese).

Sakrison, D. (1966). "Stochastic approximation: a recursive method for solving regression problems," in *Advances in Communication Systems*, vol. 2, ed. A. V. Balakrishnan, pp. 51–106, Academic Press, New York.

Sambur, M. R. (1978). "Adaptive noise cancelling for speech signals," *IEEE Trans. Acoust. Speech Signal Process.*, vol. ASSP-26, pp. 419–423.

Samson, C. (1982). "A unified treatment of fast Kalman algorithms for identification," *Int. J. Control*, vol. 35, pp. 909–934.

Sanders, J. A., and F. Verhulst (1985). *Averaging Methods in Nonlinear Dynamical Systems*, Springer-Verlag, New York.

Sari, H. (1992). "Adaptive equalization of digital line-of-sight radio systems," in *Adaptive Systems in Control and Signal Processing 1992*, ed. L. Dugard, M. M'Saad, and I. D. Landau, Pergamon Press, Oxford, United Kingdom, pp. 505–510.

Sato, Y. (1975). "Two extensional applications of the zero-forcing equalization method," *IEEE Trans. Commun.*, vol. COM-23, pp. 684–687.

Sato, Y., 1994. "Blind equalization and blind sequence estimation," *IEICE Trans Commun.*, vol. E77-B, pp. 545–556.

Satorius, E. H., and S. T. Alexander (1979). "Channel equalization using adaptive lattice algorithms," *IEEE Trans. Commun.*, vol. COM-27, pp. 899–905.

Satorius, E.H., and J. D. Pack (1981). "Application of least squares lattice algorithms to adaptive equalization," *IEEE Trans. Commun.*, vol. COM-29, pp. 136–142.

Satorius, E. H., et al. (1983), "Fixed-point implementation of adaptive digital filters," in *Proc. ICAASP*, Boston, Mass., pp. 33–36.

Sayed, A. H., and T. Kailath (1994). "A state-space approach to adaptive RLS filtering," *IEEE Signal Process. Mag.*, vol. 11, pp. 18–60.

Sayed, A. H., and M. Rupp, 1994. "Local and global optimality criteria for gradient-type algorithms," *IEEE Trans. Signal Process.* (submitted).

Scharf, L. L., and D. W. Tufts (1987). "Rank reduction for modeling stationary signals," *IEEE Trans. Acoust. Speech Signal Process.*, vol. ASSP-35, pp. 350–355.

Scharf, L. L., and L. T. McWhorter (1994). "Quadratic estimators of the correlation matrix," in IEEE ASSP Workshop on Statistical Signal and Array Processing, Quebec City, Quebec, June 27–29.

Scharf, L. L., and J. K. Thomas (1995). "Data adaptive low rank modelling," in *National Radio Science Meeting*, Boulder, Colorado, p. 200.

Schell, S. V., and W. A. Gardner (1993). "Spatio-temporal filtering and equalization for cyclostationary signals," in *Control and Dynamic Systems*, ed. C. T. Leondes, vol. 66, Academic Press, New York, pp. 1–85.

SCHETZEN, M. (1981). "Nonlinear system modeling based on the Wiener theory," *Proc. IEEE*, vol. 69, pp. 1557–1572.

SCHMIDT, R. O. (1979). "Multiple emitter location and signal parameter estimation," in *Proc. RADC Spectral Estimation Workshop*, pp. 243–258, Griffith AFB, Rome, N.Y.

SCHMIDT, R. O. (1981). "A signal subspace approach to multiple emitter location and spectral estimation," Ph.D. dissertation, Stanford University, Stanford, Calif.

SCHREIBER, R. J. (1986). "Implementation of adaptive array algorithm," *IEEE Trans. Acoust. Speech Signal Process.*, vol. ASSP-34, pp. 1038–1045.

SCHROEDER, M. R. (1966). "Vocoders: analysis and synthesis of speech," *Proc. IEEE*, vol. 54, pp. 720–734.

SCHROEDER, M. R. (1985). "Linear predictive coding of speech: review and current directions," *IEEE Commun. Mag.*, vol. 23, pp. 54–61.

SCHUR, I. (1917). "Über Potenzreihen, die im Innern des Einheitskreises beschränkt sind," *J. Reine Angew. Math.*, vol. 147, pp. 205–232; vol. 148, pp. 122–145.

SCHUSTER, A. (1898). "On the investigation of hidden periodicities with applications to a supposed 26-day period of meterological phenomena," *Terr. Magn. Atmos. Electr.*, vol. 3, pp. 13–41.

SCHWARTZ, G. (1978). "Estimating the dimension of a model," *Ann. Stat.*, vol. 6, pp. 461–464.

SCHWARTZ, L. (1967). *Cours d'Analyse*, vol. II. Hermann, Paris, pp. 271–278.

SENNE, K. D. (1968). *Adaptive Linear Discrete-Time Estimation*, Tech. Rep., 6778-5, Stanford University Center for Systems Research, Stanford, Calif.

SETHARES, W. A. (1993). "The least mean square family," in *Adaptive System Identification and Signal Processing Algorithms*, ed. N. Kalouptsidis and S. Theodoridis, pp. 84–122, Prentice-Hall, Englewood Cliffs, N.J.

SETHARES, W. A., D. A. LAWRENCE, C. R. JOHNSON, JR., and R. R. BITMEAD (1986). "Parameter drift in LMS adaptive filters," *IEEE Trans. Acous. Speech Signal Process.*, vol. ASSP-34, pp. 868–879.

SHALVI, O., and E. WEINSTEIN (1990). "New criteria for blind equalization of non-minimum phase systems (channels)," *IEEE Trans. Inf. Theory*, vol. 36, pp. 312–321.

SHAN, T.-J., and T. KAILATH (1985). "Adaptive beamforming for coherent signals and interference," *IEEE Trans. Acoust. Speech Signal Process.*, vol. ASSP-33, pp. 527–536.

SHANBHAG, N. R., and K. K. PARHI, 1994. *Pipelined Adaptive Digital Filters*, Kluwer, Boston, Mass.

SHANNON, C. E. (1948). "The mathematical theory of communication," *Bell Syst. Tech. J.*, vol. 27, pp. 379–423, 623–656.

SHARPE, S. M., and L. W. NOLTE (1981). "Adaptive MSE estimation," in *Proc. ICASSP*, Atlanta, Ga., pp. 518–521.

SHENSA, M. J. (1980). "Non-Wiener solutions of the adaptive noise canceller with a noisy reference," *IEEE Trans. Acoust. Speech Signal Process.* vol. ASSP-28, pp. 468–473.

SHEPHERD, T. J., and J. G. MCWHIRTER (1993). "Systolic adaptive beamforming," in *Radar Array Processing*, ed. S. Haykin, J. Litva, and T. J. Shepherd, pp. 153–243, Springer-Verlag, New York.

SHERWOOD, D. T., and N. J. BERSHAD (1987). "Quantization effects in the complex LMS adaptive algorithm: linearization using dither-theory," *IEEE Trans. Circuits Systems*, vol. CAS-34, pp. 848–854.

SHI, K. H., and F. KOZIN (1986). "On almost sure convergence of adaptive algorithms," *IEEE Trans. Autom. Control*, vol. AC-31, pp. 471–474.

SHICHOR, E. (1982). "Fast recursive estimation using the lattice structure," *Bell Syst. Tech. J.*, vol. 61, pp. 97–115.

SHYNK, J. J. (1989). "Adaptive IIR filtering," *IEEE ASSP Mag.*, vol. 6, pp. 4–21.

SHYNK, J. J. (1992). "Frequency-domain and multirate adaptive filtering," *IEEE Signal Process. Mag.*, vol. 9, no. 1, pp. 14–37.

SHYNK, J. J., R. P. GOOCH, G. KRISHNAMURTHY, and C. K. CHAN (1991). "A comparative performance study of several blind equalization algorithms," in *Proc. SPIE, Adaptive Signal Processing*, vol. 1565, pp. 102–117, San Diego, Calif.

SIBUL, L. H. (1984). "Application of singular value decomposition to adaptive beamforming," in *Proc. ICASSP*, San Diego, Calif., vol. 2, pp. 33.11/1–4.

SICURANZA, G. L. (1985). "Nonlinear digital filter realization by distributed arithmetic," *IEEE Trans. Acoust. Speech Signal Process.*, vol. ASSP-33, pp 939–945.

SICURANZA, G. L., and G. RAMPONI (1986). "Adaptive nonlinear digital filters using distributed arithmetic," *IEEE Trans. Acoust. Speech Signal Process.*, vol. ASSP-34, pp. 518–526.

SINGH, S. P., ed. (1992). *Approximation Theory, Spline Functions and Applications*, Kluwer, The Netherlands.

SKOLNIK, M. I. (1982). *Introduction to Radar Systems*, 2nd ed., McGraw-Hill, New York.

SLEPIAN, D. (1978). "Prolate spheroidal wave functions, Fourier analysis, and uncertainty-V: The discrete case," *Bell Syst. Tech. J.*, vol. 57, pp. 1371–1430.

SLOCK, D. T. M. (1989). "Fast algorithms for fixed-order recursive least-squares parameter estimation," Ph.D. dissertation, Stanford University, Stanford, Calif.

SLOCK, D. T. M. (1992). "The backward consistency concept and roundoff error propagation dynamics in RLS algorithms," *Optical Engineering*, vol. 31, pp. 1153–1169.

SLOCK, D. T. M. (1994). "Blind fractionally-spaced equalization, perfect-reconstruction filter banks and multichannel linear prediction," in *Proc. ICASSP*, Adelaide, Australia, vol. 4, pp. 585–588.

SLOCK, D. T. M., and T. KAILATH (1991). "Numerically stable fast transversal filters for recursive least squares adaptive filtering," *IEEE Trans. Signal Process.* vol. 39, pp. 92–114.

SLOCK, D. T. M., and T. KAILATH (1993). "Fast transversal RLS algorithms," in *Adaptive System Identification and Signal Processing Algorithms*, eds. N. Kalouptsidis and S. Theodoridis, pp. 123–190, Prentice-Hall, Englewood Cliffs, N.J.

SLOCK, D. T. M., and C. B. PAPADIAS (1995). "Further results on blind identification and equalization of multiple FIR channels," in *Proc. ICASSP*, Detroit, Michigan, vol. 3, pp. 1964–1967.

SOLO, V. (1989). "The limiting behavior of LMS," *IEEE Trans. Acoust. Speech Signal Process.*, vol. 37, pp. 1909–1922.

SOLO, V. (1992). "The error variance of LMS with time-varying weights," *IEEE Trans. Signal Process.*, vol. 40, pp. 803–813.

SOLO, V., and X. KONG (1995). *Adaptive Signal Processing Algorithms*, Prentice-Hall, Englewood Cliffs, N.J.

SOMMEN, P. C. W., and J. A. K. S. JAYASINGHE (1988). "On frequency-domain adaptive filters using the overlap-add method," in *Proc. IEEE Int. Symp. Circuits Systems*, Espoo, Finland, pp. 27–30.

SOMMEN, P. C. W., P. J. VAN GERWEN, H. J. KOTMANS, and A. E. J. M. JANSEN (1987). "Convergence analysis of a frequency-domain adaptive filter with exponential power averaging and generalized window function," *IEEE Trans. Circuits Syst.*, vol. CAS-34, pp. 788–798.

SONDHI, M. M., (1967). "An adaptive echo canceller," *Bell Syst. Tech. J.*, vol. 46, pp. 497–511.

SONDHI, M. M. (1970). *Closed Loop Adaptive Echo Canceller Using Generalized Filter Networks*, U.S. Patent, 3,499,999, March 10.

SONDHI, M., and D. A. BERKLEY (1980). "Silencing echoes in the telephone network," *Proc. IEEE*, vol. 68, pp. 948–963.

SONDHI, M. M., and A. J. PRESTI (1966). "A self-adaptive echo canceller," *Bell Syst. Tech. J.*, vol. 45, pp. 1851–1854.

SONI, T., J. R. ZEIDLER, and W. H. KU, 1995. "Behavior of the partial correlation coefficients of a least squares lattice filter in the presence of a nonstationary chirp input," *IEEE Trans. Signal Process.*, vol. 43, pp. 852–863.

SOO, J.-S., and K. K. PAGN (1991). "A multistep size (MSS) frequency domain adaptive filter," *IEEE Trans. Signal Process.*, vol. 39, pp. 115–121.

SUBBA RAO, T., and M. M. GABR (1980). "A test for linearity of stationary time series," *J. Time Series Analysis*, vol. 1, pp. 145–158.

SORENSON, H. W. (1967). "On the error behavior in linear minimum variance estimation problems," *IEEE Trans. Autom. Control*, vol. AC-12, pp. 557–562.

SORENSON, H. W. (1970). "Least-squares estimation: from Gauss to Kalman," *IEEE Spectrum*, vol. 7, pp. 63–68.

SORENSON, H. W., ed. (1985). Kalman Filtering: Theory and Application, IEEE Press, New York.

Special Issue on Adaptive Antennas (1976). *IEEE Trans. Antennas Propaga.*, vol. AP-24, September.

Special Issue on Adaptive Arrays (1983). *IEE Proc. Commun. Radar Signal Process.*, London, vol. 130, pp. 1–151.

Special Issue on Adaptive Filters (1987). *IEE Proc. Commun. Radar Signal Process.*, London, vol. 134, pt. F.

Special Issue on Adaptive Processing Antenna Systems (1986). *IEEE Trans. Antennas Propag.*, vol. AP-34, pp. 273–462.

Special Issue on Adaptive Signal Processing (1981). *IEEE Trans. Circuits Syst.*, vol. CAS-28, pp. 465–602.

Special Issue on Adaptive Systems (1976). *Proc. IEEE*, vol. 64, pp. 1123–1240.

Special Issue on Adaptive Systems and Applications (1987). *IEEE Trans. Circuits Syst.*, vol. CAS-34, pp. 705–854.

Special Issue on Higher Order Statistics in System Theory and Signal Processing (1990). *IEEE Trans. Autom. Control*, vol. AC-35, pp. 1–56.

Special Issue on Linear Adaptive Filtering (1984). *IEEE Trans. Information Theory*, vol. IT-30, pp. 131–295.

Special Issue on Linear-Quadratic-Gaussian Problem (1971). *IEEE Trans. Autom. Control*, vol. AC-16, December.

Special Issue on Neural Networks (1990). *Proc. IEEE*, vol. 78: Neural Nets I, September; Neural Nets II, October.

Special Issue on Spectral Estimation (1982). *Proc. IEEE*, vol. 70, pp. 883–1125.

Special Issue on System Identification and Time-series Analysis (1974). *IEEE Trans. Autom. Control*, vol. AC-19, pp. 638–951.

Special Issue on Systolic Arrays (1987). *Computer*, vol. 20, No. 7.

SPECHT, D. F. (1990). "Probabilistic neural networks and the polynomial Adaline as complementary techniques for classification," *IEEE Trans. Neural Networks*, vol. 1, pp. 111–121.

STEINHARDT, A. O. (1988). "Householder transforms in signal processing," *IEEE ASSP Mag.*, vol. 5, pp. 4–12.

STEWART, G. W. (1973). *Introduction to Matrix Computations*, Academic Press, New York.

STEWART, R. W., and R. CHAPMAN (1990). "Fast stable Kalman filter algorithms utilizing the square root," in *Proc. ICASSP*, Albuquerque, N. Mexico, pp. 1815–1818.

STOER, J., and BULLIRSCH (1980). *Introduction to Numerical Analysis*, Springer-Verlag, New York.

STRANG, G. (1980). *Linear Algebra and Its Applications*, 2nd ed., Academic Press, New York.

STROBACH, P. (1990). *Linear Prediction Theory*, Springer-Verlag, New York.

SUZUKI, H. (1994). "Adaptive signal processing for optimal transmission in mobile radio communications," *IEICE Trans. Communi.*, vol. E77-B, pp. 535–544.

SWAMI, A., and J. M. MENDEL (1990). "Time and lag recursive computation of cumulants from a state-space model," *IEEE Trans. Autom. Control*, vol AC-35, pp. 4–17.

SWERLING, P. (1958). A Proposed Stagewise Differential Correction Procedure for Satellite Tracking and Prediction, Rep. P-1292, Rand Corporation.

SWERLING, P. (1963). "Comment on 'A statistical optimizing navigation procedure for space flight,'" *AIAA J.*, vol. 1, p. 1968.

SZEGÖ, G. (1939). *Orthogonal polynomials*, Colloquium Publications, no. 23, American Mathematical Society, Providence, R.I.

TARRAB, M., and A. FEUER (1988). "Convergence and performance analysis of the normalized LMS algorithm with uncorrelated Gaussian data," *IEEE Trans. Information Theory*, vol. IT-34, pp. 680–691.

TETTERINGTON, D. M., A. F. M. SMITH, and U. E. MAKOV (1985). *Statistical Analysis of Finite Mixture Distributions*, Wiley, New York.

THAKOR, N. V., and Y.-S. ZHU (1991). "Applications of adaptive filtering to ECG analysis: noise cancellation and arrhythmia detection," *IEEE Trans. Biomed. Eng.*, vol. 38, pp. 785–794.

THEODORIDIS, S., C. M. S. SEE, and C. F. N. COWAN, 1992. "Nonlinear channel equalization using clustering techniques," in *ICC*, Chicago, Il., vol. 3, pp. 1277–1279.

THOMSON, D. J. (1982). "Spectral estimation and harmonic analysis," *Proc. IEEE*, vol. 70, pp. 1055–1096.

THOMSON, W. T. (1950). "Transmission of elastic waves through a stratified solid medium," *J. Appl. Phys.*, vol. 21, pp. 89–93.

TONG, L., G. XU, and T. KAILATH, 1993. "Fast blind equalization via antenna arrays," in *Proc. ICASSP*, Minneapolis, Minnesota, vol. 4. 272–275.

TONG, L., G. XU, and T. KAILATH (1994a). "Blind identification and equalization based on second-order statistics: a time-domain approach," *IEEE Trans. Information Theory*, vol. 40, pp. 340–349.

TONG, L., G. XU, and T. KAILATH (1994b). "Blind channel identification and equalization using spectral correlation measurements, Part II: A time-domain approach," in *Cyclostationarity in Communications and Signal Processing*, ed. W. A. Gardner, IEEE Press, New York, pp. 437–454.

TREICHLER, J. R. (1979). "Transient and convergent behavior of the adaptive line enhancer," *IEEE Trans. Acoust. Speech Signal Process.*, vol. ASSP-27, pp. 53–62.

TREICHLER, J. R. and B. G. AGEE (1983). "A new approach to multipath correction of constant modulus signals," *IEEE Trans. Acoust. Speech Signal Process.*, vol. ASSP-31, pp. 459–471.

TREICHLER, J. R., and M. G. LARIMORE (1985a). "New processing techniques based on the constant modulus adaptive algorithm," *IEEE Trans. Acoust. Speech Signal Process.*, vol. ASSP-33, pp. 420–431.

TREICHLER, J. R., and M. G. LARIMORE (1985b). "The tone capture properties of CMA-based interference suppressions," *IEEE Trans. Acoust. Speech Signal Process.*, vol. ASSP-33, pp. 946–958.

TREICHLER, J. R., C. R. JOHNSON, JR., and M. G. LARIMORE (1987). *Theory and Design of Adaptive Filters*, Wiley-Interscience, New York.

TRETTER, S. A. (1976). *Introduction to Discrete-Time Signal Processing*, Wiley, New York.

TUGNAIT, J. K. (1994). "Testing for linearity of noisy stationary signals," *IEEE Trans. Signal Process.*, vol. 42, pp. 2742–2748.

TUGNAIT, J. K. (1995). "On fractionally-spaced blind adaptive equalization under symbol timing offsets using Godard and related equalizers," in *Proc. ICASSP*, Detroit, Michigan, vol. 3, pp. 1976–1979.

UNGERBOECK, G. (1972). "Theory on the speed of convergence in adaptive equalizers for digital communication," *IBM J. Res. Dev.*, vol. 16, pp. 546–555.

UNGERBOECK, G. (1976). "Fractional tap-spacing equalizer and consequences for clock recovery in data modems," *IEEE Trans. Commun.*, vol. COM-24, pp. 856–864.

ULRYCH, T. J., and R. W. CLAYTON, (1976). "Time series modelling and maximum entropy," *Phys. Earth Planet. Inter.*, vol. 12, pp. 188–200.

ULRYCH, T. J. and M. OOE (1983). "Autoregressive and mixed autoregressive-moving average models and spectra," in *Nonlinear Methods of Spectral Analysis*, ed. S. Haykin, pp. 73–125, Springer-Verlag, New York.

VAIDYANATHAN, P. P. (1987). "Quadrature mirror filter bands, M-band extensions and perfect reconstruction techniques," *IEEE ASSP Magazine*, vol. 4, pp. 4–20.

VAIDYANATHAN, P. P. (1993). *Multirate Systems and Filter Banks*, Prentice-Hall, Englewood Cliffs, N.J.

VALENZUELA, R. A. (1989). "Performance of adaptive equalization for indoor radio communications," *IEEE Trans. Commun.*, vol. 37, pp. 291–293.

VAN DE KERKHOF, L. M., and W. J. W. KITZEN (1992). "Tracking of a time-varying acoustic impulse response by an adaptive filter," *IEEE Trans. Signal Process.*, vol. 40, pp. 1285–1294.

VAN DEN BOS, A. (1971). "Alternative interpretation of maximum entropy spectral analysis," *IEEE Trans. Information Theory*, vol. IT-17, pp. 493–494.

VAN HUFFEL, S., J. VANDEWALLE, and A. HAEGEMANS (1987). "An efficient and reliable algorithm for computing the singular subspace of a matrix, associated with its smallest singular values," *J. Comput. Appl. Math.*, vol. 19, pp. 313–330.

VAN HUFFEL, S., and J. VANDEWALLE (1988). "The partial total least squares algorithm," *J. Comput. Appl. Math.*, vol. 21, pp. 333–341.

VAN LOAN, C. (1989). "Matrix computations in signal processing," in *Selected Topics in Signal Processing*, ed. S. Haykin, Prentice-Hall, Englewood Cliffs, N.J.

VAN TREES, H. L. (1968). *Detection, Estimation and Modulation Theory*, part I, Wiley, New York.

VAN VEEN, B. (1992). "Minimum variance beamforming," in *Adaptive Radar Detection and Estimation*, ed. S. Haykin and A. Steinhardt, Wiley-Interscience, New York.

VAN VEEN, B. D., and K. M. BUCKLEY (1988). "Beamforming: a versatile approach to spatial filtering," *IEEE ASSP Mag.*, vol. 5, pp. 4–24.

VARVITSIOTIS, A. P., S. THEODORIDIS, and G. MOUSTAKIDES (1989). "A new novel structure for adaptive LS FIR filtering based on QR decomposition," in *Proc. ICASSP*, Glasgow, Scotland, pp. 904–907.

VEMBU, S., S. VERDÚ, R. A. KENNEDY, and W. SETHARES (1994). "Convex cost functions in blind equalization," *IEEE Trans. Signal Process.*, vol. 42, pp. 1952–1960.

VERDÚ, S. (1984). "On the selection of memoryless adaptive laws for blind equalization in binary communications," in *Proc. 6th Intern. Conference on Analysis and Optimization of Systems*, Nice, France, pp. 239–249.

VERHAEGEN, M. H. (1989). "Round-off error propagation in four generally-applicable, recursive, least-squares estimation schemes," *Automatica*, vol. 25, pp. 437–444.

VERHAEGEN, M. H., and P. VAN DOOREN (1986). "Numerical aspects of different Kalman filter implementations," *IEEE Trans. Autom. Control*, vol. AC-31, pp. 907–917.

VETTERLI, M., and J. KOVAČCEVIĆ (1995). *Wavelets and Subband Coding*, Prentice-Hall, Englewood Cliffs, N.J.

VOLDER, J. E. (1959). "The CORDIC trigonometric computing technique, *IEEE Trans. Electron. Comput.*, vol. EC-8, pp. 330–334.

WAKITA, H. (1973). "Direct estimation of the vocal tract shape by inverse filtering of acoustic speech waveforms," *IEEE Trans. Audio Electroacoust.*, vol. AU-21, pp. 417–427.

WALACH, E., and B. WIDROW (1984). "The least mean fourth (LMF) adaptive algorithm and its family," *IEEE Trans. Information Theory*, vol. IT-30, Special Issue on Linear Adaptive Filtering, pp. 275–283.

WALKER, G. (1931). "On periodicity in series of related terms," *Proc. Royal Soc.*, vol. A131, pp. 518–532.

WALZMAN, T., and M. SCHWARTZ (1973). "Automatic equalization using the discrete frequency domain," *IEEE Trans. Information Theory*, vol. IT-19, pp. 59–68.

WAN, E. (1990). "Temporal backpropagation for FIR neural networks," *IEEE International Joint Conference on Neural Networks*, San Diego, Calif., vol. 1, pp. 575–580.

WARD, C. R., ET AL. (1984). "Application of a systolic array to adaptive beamforming," *IEE Proc. (London)*, pt. F, vol. 131, pp. 638–645.

WARD, C. R., P. H. HARGRAVE, and J. G. MCWHIRTER (1986). "A novel algorithm and architecture for adaptice digital beamforming," *IEEE Trans. Antennas Propag.*, vol. AP-34, pp. 338–346.

WAX, M., 1995. "Model based processing in sensor arrays," in *Advances in Spectrum Analysis and Array Processing*, vol. 3, ed. S. Haykin, pp. 1–47, Prentice-Hall, Englewood Cliffs, N. J.

WAX, M., and T. KAILATH (1985). "Detection of signals by information theoretic criteria," *IEEE Trans. Acoust. Speech Signal Process.*, vol. ASSP-33, pp. 387–392.

WAX, M., and I. ZISKIND (1989). "Detection of the number of coherent signals by the MDL principle," *IEEE Trans. Acoust. Speech Signal Process.*, vol. ASSP-37, pp. 1190–1196.

WEBB, A. R. (1994). "Functional approximation by feed-forward networks: a least-squares approach to generalisation," *IEEE Trans. Neural Networks*, vol. 6, pp. 363–371.

WEI, P., J. R. ZEIDLER, and W. H. KU (1994). "Adaptive recovery of a Doppler-shifted mobile communications signal using the RLS algorithm," in *Conf. Rec. Asilomar Conference on Signals, Systems, and Computers*, Pacific Groves, Calif., vol. 2, pp. 1180–1184.

WEIGEND, A. S., D. E. RUMELHART, and B. A. HUBERMAN (1991). "Generalization by weight elimination with application to forecasting," in *Advances in Neural Information Processing Systems 3*, pp. 875–882, Morgan Kaufman, San Mateo, Calif.

WEISBERG, S. (1980). *Applied Linear Regression*, Wiley, New York.

WEISS, A., and D. MITRA (1979). "Digital adaptive filters: conditions for convergence, rates of convergence, effects of noise and errors arising from the implementation," *IEEE Trans. Information Theory*, vol. IT-25, pp. 637–652.

WELLSTEAD, P. E., G. R. WAGNER, and J. R. CALDAS-PINTO (1987). "Two-dimensional adaptive prediction, smoothing and filtering," *IEE Proc. (London)*, part F, vol. 134, pp. 253–268.

WERBOS, P. J. (1974). "Beyond regression: new tools for prediction and analysis in the behavioral sciences," Ph.D. dissertation, Harvard University, Cambridge, Mass.

WERBOS, P. J. (1993). *The Roots of Backpropagation: From Ordered Derivatives to Neural Networks and Political Forecasting*, Wiley-Interscience, New York.

WERNER, J. J. (1983). *Control of Drift for Fractionally Spaced Equalizers,* U.S. Patent 438 4355.

WHEELWRIGHT, S. C. and S. MAKRIDAKIS. (1973). "An examination of the use of adaptive filtering in forecasting," *Oper. Res Q.*, vol. 24, pp. 55–64.

WHITTAKER, E. T., and G. N. WATSON (1965). *A Course of Modern Analysis*, Cambridge University Press, Cambridge, United Kingdom.

WHITTLE, P. (1963). "On the fitting of multivariate autoregressions and the approximate canonical factorization of a spectral density matrix," *Biometrika*, vol. 50, pp. 129–134.

WIDROW, B. (1966). *Adaptive Filters I: Fundamentals*, Rep. SEL-66-126 (TR 6764-6), Stanford Electronics Laboratories, Stanford, Calif.

WIDROW, B. (1970). "Adaptive filters," in *Aspects of Network and System Theory*, ed. R. E. Kalman and N. DeClaris, Holt, Rinehart and Winston, New York.

WIDROW, B., and M. E. HOFF, JR. (1960). "Adaptive switching circuits," *IRE WESCON Conv. Rec.*, pt. 4, pp. 96–104.

WIDROW, B. and M. LEHR (1990). "30 years of adaptive neural networks: Perceptron, madaline, and backpropagation," *Proc IEEE*, Special Issue on Neural Networks I, vol. 78, September.

WIDROW, B. and S. D. STEARNS (1985). *Adaptive Signal Processing*, Prentice-Hall, Englewood Cliffs, N.J.

WIDROW, B., and E. WALACH (1984). "On the statistical efficiency of the LMS algorithm with nonstationary inputs," *IEEE Trans. Information Theory*, vol. IT-30, Special Issue on Linear adaptive Filtering, pp. 211–221.

WIDROW, B., et al. (1967). "Adaptive antenna systems," *Proc. IEEE*, vol. 55, pp. 2143–2159.

WIDROW, B., J. McCOOL, and M. BALL (1975a). "The complex LMS algorithm," *Proc. IEEE*, vol. 63, pp. 719–720.

WIDROW, B., et al. (1975b). "Adaptive noise cancelling: principles and applications," *Proc. IEEE*, vol. 63, pp. 1692–1716.

WIDROW, B., et al. (1976). "Stationary and nonstationary learning characteristics of the LMS adaptive filter," *Proc. IEEE*, vol. 64, pp. 1151–1162.

WIDROW, B., K. M. DUVALL, R. P. GOOCH, and W. C. Newman (1982). "Signal cancellation phenomena in adaptive antennas: Causes and cures," *IEEE Trans. Antennas Propag.*, vol. AP-30, pp. 469–478.

WIDROW, B., et al. 1987. "Fundamental relations between the LMS algorithm and the DFT," *IEEE Trans. Circuits Syst.*, vol. CAS. 34, pp. 814–819.

WIENER, N. (1949). *Extrapolation, Interpolation, and Smoothing of Stationary Time Series, with Engineering Applications*, MIT Press, Cambridge, Mass. (originally issued as a classified National Defense Research Report in February 1942).

WIENER, N. (1958). *Nonlinear Problems in Random Theory*, Wiley, New York.

WIENER, N., and E. HOPF (1931). "On a class of singular integral equations," *Proc. Prussian Acad. Math-Phys. Ser.*, p. 696.

WILKINSON, J. H. (1963). *Rounding Errors in Algebraic Processes*, Prentice-Hall, Englewood Cliffs, N.J.

WILKINSON, J. H. (1965). *The Algebraic Eigenvalue Problem*, Oxford University Press, Oxford, United Kingdom.

WILKINSON, J. H., and C. REINSCH, eds. (1971). *Handbook for Automatic Computation*, vol. 2, *Linear Algebra*, Springer-Verlag, New York.

WILKS, S. S. (1962). *Mathematical Statistics*, Wiley, New York.

WILLIAMS, J. R., and G. G. RICKER (1972). "Signal detectability performance of optimum Fourier receivers," *IEEE Trans. Audio and Electroacoustics*, vol. AU-20, pp. 254–270.

WILSKY, A. S. (1979). *Digital Signal Processing and Control and Estimation Theory: Points of Tangency, Areas of Intersection, and Parallel Directions*, MIT Press, Cambridge, Mass.

WOLD, H. (1938). *A Study in the Analysis of Stationary Time Series*, Almqvist and Wiksell, Uppsala, Sweden.

WOODBURY, M. (1950). Inverting Modified Matrices, Mem. Rep. 42, Statistical Research Group, Princeton University, Princeton, N.J.

WOZENCRAFT, J. M., and I. M. JACOBS (1965). *Principles of Communications Engineering*, Wiley, New York.

WYLIE, C. R., and L. C. BARRETT (1982). *Advanced Engineering Mathematics*, 5th ed., McGraw-Hill, New York.

YANG, B. (1994). "A note on the error propagation analysis of recursive least squares algorithms," *IEEE Trans. Signal Process.*, vol. 42, pp. 3523–3525.

YANG, V., and J. F. BÖHME (1992). "Rotation-based RLS algorithms: unified derivations, numerical properties and parallel implementations," *IEEE Trans. Signal Process.*, vol. 40, pp. 1151–1167.

YASSA, F. F. (1987). "Optimality in the choice of the convergence factor for gradient-based adaptive algorithms," *IEEE Trans. Acoust. Speech Signal Process.*, vol. ASSP-35, pp. 48–59.

YEE, P., and S. HAYKIN (1995). "A dynamic regularized Gaussian radial basis function network for nonlinear, nonstationary time series prediction," in *Proc. ICASSP*, Detroit, Michigan, vol. 5, pp. 3419–3422.

YOGANANDAM, Y., V. U. REDDY, and T. KAILATH (1988). "Performance analysis of the adaptive line enhancer for sinusoidal signals in broad-band noise," *IEEE Trans. Acoust. Speech Signal Process.*, vol. ASSP-36, pp. 1749–1757.

YOUNG, P. C. (1984). *Recursive Estimation and Time-Series Analysis*, Springer-Verlag, New York.

YUAN, J.-T., and J. A. STULLER (1995). "Least-squares order-recursive lattice smoothers," *IEEE Trans. Signal Process.*, vol. 43, pp. 1058–1067.

YULE, G. U. (1927). "On a method of investigating periodicities in disturbed series, with special reference to Wölfer's sunspot numbers," *Philos. Trans. Royal Soc. London*, vol. A226, pp. 267–298.

ZAMES, G. (1981). "Feedback and optimal sensitivity: model reference transformations, multiplicative seminorms, and approximate inverses," *IEEE Trans. Autom. Control*, vol. AC-26, pp. 301–320.

ZAMES, G., and B. A. FRANCIS (1983). "Feedback, minimax sensitivity, and optimal robustness," *IEEE Trans. Autom. Control*, vol. AC-28, pp. 585–601.

ZEIDLER, J. R. (1990). "Performance analysis of LMS adaptive prediction filters," *Proc. IEEE*, vol. 78, pp. 1781–1806.

ZEIDLER, J. R., E. H. SATORIUS, D. M. CHABRIES, and H. T. WEXLER (1978). "Adaptive enhancement of multiple sinusoids in uncorrelated noise," *IEEE Trans. Acous. Speech Signal Process.*, vol. ASSP-26, pp. 240–254.

ZHANG, QI-TU, and S. HAYKIN (1983). "Tracking characteristics of the Kalman filter in a nonstationary environment for adaptive filter applications," in *Proc. ICASSP*, Boston, pp. 671–674.

ZHANG, Q-T., S. HAYKIN, and P. YIP (1989). "Performance limits of the innovations-based detection algorithm," *IEEE Trans. Information Theory*, vol. IT-35, pp. 1213–1222.

ZIEGLER, R. A., and J. M. CIOFFI (1989). "A comparison of least squares and gradient adaptive equalization for multipath fading in wideband digital mobile radio," in *GLOBECOM*, vol. 1, New York, pp. 102–106.

ZIEGLER, R. A., and J. M. CIOFFI (1992). "Adaptive equalization for digital wireless data transmission," in *Virginia Tech Second Symposium on Wireless Personal Communications Proceedings*, pp. 5/1–5/12.

Index

A

Acoustic noise reduction, 54
Activation function of neuron, 833
Adaptive autoregressive spectrum analysis,
45
Adaptive beamforming, 59, 388, 617
historical notes, 76
Adaptation in beam space, 63
Adaptation in data space, 63
Adaptive differential pulse-code modulation,
42
Adaptive equalization, 34, 71, 866
historical notes, 71
Adaptive filtering algorithms
classification of, 477
complex form, 14
factors in choice of, 3
finite-precision effects (see Finite-precision
effects)
historical notes, 67
real form, 15
(see also names of specific algorithms)
Adaptive filter applications, 18
identification, 18
interference canceling, 20
inverse modeling, 20
prediction, 20

Adaptive filters, 2
algorithms (see Adaptive filter algorithms)
applications of (see Adaptive filter applications)
filter structures, 4
historical notes, 69
how to choose, 14
linear versus nonlinear, 3
Adaptive filter theory, development
approaches, 9
least-squares estimation, 12
stochastic gradient approach, 11
Adaptive line enhancer (ALE), 49, 385, 719
Adaptive noise canceling, 50, 377
historical notes, 75
Adaptive speech enhancement, 54
All-pass filters, 86
All-pole filters, 83, 110
All-zero filters, 83, 110
Analog-to-digital conversion, 739
Angle-normalized backward prediction error,
655
Angle-normalized forward prediction error,
655
Angle-normalized joint-process estimation
error, 655
Angle variable (see Conversion factor)

An information-theoretic criterion (AIC), 128
Array pattern, 65
Autocorrelation function, 97
Autocorrelation method of data windowing,
    486
Autocovariance function, 97
Automatic tuning of adaptation constants, 731
Autoregressive (AR) models, 109
    asymptotic stationarity of autoregressive
        process, 116
    autoregressive process of order 2, 120
    least-squares estimation, 506
    model order of, 109
    relation between linear prediction and, 245
Autoregressive power spectrum, 275
Autoregressive-moving average (ARMA)
    models, 112

B
Back-propagation algorithm, 817
    complex, 824
    real, 837
Back substitution, 605
Backward prediction, 248
    augmented Wiener-Hopf equations for, 253
    Cholesky factorization and, 276
    fast recursive algorithms using, 634
    Givens rotation and, 663
    relation between forward prediction and,
        251
Backward prediction error, 250
Backward reflection coefficients, 643, 659
Bartlett window, 138
Baseband, 14
Bayes' risk, 902
Beamforming, 59
Benveniste-Goursat-Ruget theorem, 789
Best linear unbiased estimate (BLUE), 504
Bezout identity, generalized, 816
Bispectrum, 152
Blind deconvolution, 772
    Bussgang algorithm for (see Bussgang
        algorithm)
    tricepstrum algorithm for (see Tricepstrum
        algorithm)
    using cyclostationary statistics (see Sub-
        space decomposition)
Blind equalization, 39, 776

Block adaptive filter, 446
Block estimation, 290
Block LMS algorithm (see Block adaptive
    filters)
Bootstrap technique, 73
Burg formula, 292
Bussgang algorithm, 776
    advantages and disadvantages of, 802
    convergence considerations, 788
    decision-directed mode, 790
    extension to complex baseband channels,
        791
    nonconvexity of the cost function, 781
    special cases of, 792
    statistical properties of convolutional noise,
        781
    zero-memory nonlinear estimation of data
        sequence, 783

C
Canonical form of error-performance surface,
    209
Canonical model of complex LMS algorithm,
    372
Cauchy-Riemann equations, 877
Cauchy-Schwarz inequality, 430, 737
Cauchy's inequality, 879
Cauchy's integral formula, 877
Cauchy's residue theorem, 882
Causality, 83
Characteristic equation, 121
Chi-square distribution, 925
Chirped sinusoid, 719
Cholesky factorization, 276
Circular convolution, 88
Circularly complex Gaussian process, 131
Complementary function, 116
Complex variables, theory of, 875
Conditional mean estimator, 784, 820, 902
Condition number, 168
Constant modulus algorithm (CMA), see
    Godard algorithm
Constrained optimization, 895
Conversion factor, 318, 636
Convolution, 6
Convolution sum, 81
Convolutional noise, 780
    properties of, 781

CORDIC processors, 768
Correlation coefficient, 119
Correlation matrix, ensemble averaged
    defined, 100
    eigenvalues and eigenvectors of (see
        Eigenvalues; Eigenvectors)
    properties of, 101
Correlation matrix, time averaged
    defined, 495
    properties of, 495
Correspondences between Kalman and LSL
    variables, 658
Correspondences between Kalman and RLS
    variables, 585
Covariance (Kalman) filter, 324
    square-root, 591
Covariance method of windowing, 486
Cramér-Rao inequality, 901
Cramér spectral representation for a station-
    ary process, 144
Cross-correlation vector, ensembled averaged,
    205
Cross-correlation vector, time-averaged, 493
Cumulants, 151
Cyclic autocorrelation function, 155
Cyclic Jacobi algorithm, 544
Cycloergodic process, 155
Cyclostationary statistics, 150
    channel identifiability using, 803

D
DCT-LMS algorithm, 462
    experiment on adaptive equalization, 469
    summary of, 470
Data matrix, 497
Data terminal equipment, 774
Data windowing, 486
Decision-directed learning method, 37, 790
Decision-feedback equalizer, 72, 867
Deconvolution, 32
    (see also Blind deconvolution)
Decorrelation parameter (see Prediction
    depth)
Decoupling property (see Orthogonality of
    backward prediction errors)
Degenerate eigenvalues, 161

Degree of nonstationarity, 705
    relation to misadjustment, 706
Delay-and-sum beamformer, 60
Diagonalization, 185
Differential pulse-code modulation, 43
Differentiation with respect to a vector, 890
    relation to gradient vector, 894
Digital residual error, 748
Dynamical system, linear discrete-time, 306
Direct-averaging method, 391
Dirichlet kernel, 145
Discrete cosine transform, 93
    relation to discrete Fourier transform, 93
    sliding, 462
Discrete Fourier transform, 87
    inverse, 87
    implementing convolutions using, 87
Discrete-time wide-sense stationary stochastic
        process, 96
    autoregressive modeling of, 273
    complex Gaussian process, 130
    correlation matrix (see Correlation matrix)
    defined, 96
    eigenanalysis (see Eigenanalysis)
    mean ergodic theorem, 98
    partial characterization of, 97
    power spectral density of (see Power spec-
        tral density)
    stochastic models and (see Stochastic
        models)
    strictly stationary, 97
    transmission through linear filters, 140
    wide-sense stationary, 98
Discrete-time signal processing, 79
Displacement rank, 698
Dither, 747

E
Echo cancelation, 56
Eigenanalysis, 160
    characteristic equation, 161
    eigenfilters, 181
    eigenvalue computations, 184
    eigenvalue problem, 160
    properties of eigenvalues and eigenvectors,
        162
    (see also Singular-value decomposition)

Eigenfilters, 181
Eigenvalues, 161
    computations, 184
    degenerate, 161
Eigenvalue spread, 169
Eigenvectors, 161
Einstein-Wiener-Khintchine relations, 139
Electrical angle, 62
Entropy, 300
Equalization of communication channel, 217
Entire function, 881
Error feedback, 761, 762, 766
Error-performance surface, 206
Error-propagation model, 752
Estimate and plug-in procedure, 2
Estimation theory, 899
Euclidian norm or length, 169
Excess mean-squared error, 395, 397
Exchange matrix, 190
Excited subspace, 749
Exponential weighting factor, 564
Exponential weighting matrix, 599
Extended Kalman filter, 328
Extended square-root information filter, 596
Extended QR-RLS algorithm, 614
    systolic array implementation of, 614
Eye pattern, 38

F
Fast convolution, 93
Fast (recursive) algorithms, 13, 695
    adaptive backward linear prediction, 634
    adaptive forward linear prediction, 631
    conversion factor (angle variable), 636
    fast transversal filters (see Fast transversal
        filters algorithm)
    lattice predictor-based (see Recursive least-
        squares lattice algorithms)
    QR-decomposition-based (see QR-decom-
        position-based least-squares algo-
        rithms)
Fast transversal filters (FTF) algorithm, 696,
    763
    finite-precision effects on, 764
    rescue variable, 764
    summary of, 765
Filtered state estimate, 317

Filtered state-error correlation matrix, 318
Filtering, 1
Filtering matrix rank theorem, 808
Filters
    defined, 1
    linear time-invariant, 81
    linear versus nonlinear, 1
Filtering structures, 4
Finite-duration impulse response (FIR) filters,
    9
Finite-precision effects, 738
    error-propagation model, 752
    extended QRD-LSL algorithm, 758
    fast transversal filters algorithm, 764
    GAL algorithm, 763
    inverse QR-RLS algorithm, 759
    LMS algorithm, 741
    numerical accuracy, 741
    numerical stability, 741
    QRD-LSL algorithm, 760
    QR-RLS algorithm, 757
    quantization errors, 739
    recursive LSL algorithm, 762
    RLS algorithm, 751
First coordinate vector, 637
Fisher's information matrix, 901
Forgetting factor (see Exponential weighting
    factor)
Forward and backward linear prediction
    (FBLP) algorithm, 506
Forward prediction, 242
    augmented Wiener-Hopf equations for, 246
    fast recursive algorithms using, 631
    Givens rotation and, 663
    relation between backward prediction and,
        251
Forward prediction error, 242
Forward reflection coefficients, 641, 659
Fractionally spaced equalizer (FSE), 72
    subspace decomposition for, 74
Fredholm integral equation of the first kind,
    146
Frequency-domain adaptive filters (FDAF),
    445
    block, 446
    fast, 451
    unconstrained, 457

FTF algorithm (see Fast transversal filters algorithm)
Full column rank, defined, 500
Fundamental equation of power spectrum analysis, 146

G
Gain vector, 567
Gaussian moment factoring theorem, 132, 441
Gaussian process, 130
Generalized sidelobe canceler, 227
Givens rotations, 537, 602, 662, 663
Godard algorithm, 794
Gohberg-Semencul formula, 913
Golub-Kahan algorithm, 554
Gradient adaptive lattice (GAL) algorithm, 763, 915
  properties of, 917
Gradient-based adaptation (see Steepest descent, method of; Stochastic gradient-based algorithms)
Gradient noise, 366
Gradient vector, 342
Gram-Schmidt orthogonalization, 277

H
H^∞ criterion, 430
Hadamard theorem, 560
Hermitian matrix, 101
Higher-order statistics (HOS), 150
  blind deconvolution using, 772, 802
Householder bidiagonalization, 552
Householder transformation, 548
  properties of, 549
Hypothesis testing, 46

I
Ill-conditioned matrix, 167
Independence assumption (theory), 392, 704
Infinite-duration impulse response (IIR) filters, 9
Information (Kalman) filter, 324
  square root, 593
Innovations process, 303, 307
  correlation matrix of, 308
  properties of, 303

Instantaneous frequency measurement, 373
Interpolation matrix, 860
Intersymbol interference (ISI), 34, 217
Inverse correlation matrix, 567
Inverse filtering, 283
Inverse Levinson-Durbin algorithm, 261
Inverse QR-RLS algorithm, 624
  finite-precision effects, 759
  summary of, 626
  systolic implementation of, 626
Inversion integral for the z-transform, 80, 888
Iterative deconvolution, 778

J
Jacobi algorithm
  cyclic, 544
  two-sided, for real data, 538
Jacobi rotations (see Givens rotations)
Joint-process estimation, 6, 286, 653
Kalman-Bucy filter, 69
Kalman filters, 302
  block diagram of, 322
  conversion factor, 318
  covariance filter, 324
  correspondences between Kalman and RLS variables, 585
  correspondence between Kalman and LSL variables, 658
  extended, 328
  filtering operation, 317
  information filtering, 324
  innovations process, 303, 307
  Kalman gain, 311
  measurement equation, 307
  measurement matrix, 307
  problem statement, 306
  process equation, 307
  recursive minimum mean-square estimation for scalar random variables, 303
  Riccati equation, 312
  square-root covariance filter, 591
  square-root filtering, 326, 589
  square-root information filter, 593
  state transition matrix, 307
  summary based on one-step prediction algorithm, 320

summary of variables (see also State-space
   model), 321
UD-factorization, 327
unforced dynamics model, 323
variants of, 322
Kalman gain, 311
Karhunen-Loève expansion, 175, 460
k-means clustering algorithm, 862
   enhanced, 863
Kullback-Leibler mean information, 129
Kurtosis, 775, 797

L

Lag misadjustment, 708
Lag variance, 707
Lagrange multipliers, method of, 895
Lattice predictor, 280
   block estimation of reflection coefficients,
      290
   correlation properties, 300
   decoupling property, 277
   defined, 283
   exact least-squares, 640
   inverse filtering, 283
   joint-process estimation, 286, 653
   normalized, 298
   order-update recursions for prediction
      errors, 282
Laurent's series, 879
Layered earth modeling, 22
Leaky LMS algorithm, 441, 746
Learning, 817
   supervised, 842
   unsupervised, 772
Least-mean-square (LMS) algorithm, 367
   adaptive process, 366
   application examples, 372
   average time constant, 403
   compared with method of steepest descent,
      404
   compared with RLS algorithm for tracking
      nonstationarity, 716
   computer experiment on adaptive beam-
      forming, 421
   computer experiment on adaptive equaliza-
      tion, 413

computer experiment on adaptive prediction, 406
convergence analysis (see stability
   analysis)
convergence criteria, 393
DCT-LMS algorithm, 462
direct-averaging method applied to, 391
directionality of convergence, 425
estimation of gradient vector, 370
excess mean-squared error, 395, 397
fast, 451
filtering process, 365
finite-precision effects on, 741
independence theory and, 392
leaky, 441
misadjustment, 402
normalized, 432
operation in nonstationary environment,
   708
overview of structure and operation of, 365
robustness of, 427
signal-flow graph representation, 371
simple working rules, 402
stability analysis of, 390
vs. steepest-descent algorithm, 404
steady-state analysis without invoking
   independence assumption, 921
summary of, 405
transform domain, 480
transient behavior of mean-squared error,
   399
weight-error correlation matrix, 394
with adaptive gain, 732
Least significant bit (LSB), 747
Least squares, method of (see Least-squares
   estimation)
Least-squares estimation
   autoregressive spectrum estimation, 506
   correlation matrix, 495
   data windowing, 486
   fast recursive algorithms (see Fast recur-
      sive algorithms)
   forward-backward linear prediction (FBLP)
      method, 506
   minimum sum of error squares, 491, 494
   minimum variance distortionless response
      spectrum estimation, 512

Least-squares estimation (*cont.*):
  normal equations, 492
  orthogonal complement projector, 498
  orthogonality principle, 487
  parametric spectrum estimation, 506
  problem statement, 483
  projection operator, 498
  properties of estimates, 502
  relation to LMS algorithm, 530
  singular-value decomposition (see Singular-value decomposition)
  uniqueness theorem, 500
Least-squares lattice predictor (exact), 640
  decoupling property, 651
  finite-order state-space model, 655
  orthogonality principle, 641
  reflection coefficients, 659
Levinson-Durbin algorithm, 254
  inverse, 261
  least-squares version, 644
Likelihood function, 899
Likelihood ratio, 240
Likelihood variable (see also Conversion factor), 637, 699
Linearly constrained minimum variance (LCMV) filters, 220
Linearly constrained minimum variance (LCMV) beamforming, 222
Linear prediction, 241
  backward (see Backward prediction)
  block estimation, 290
  Cholesky factorization, 276
  eigenvector representations of prediction-error filters, 269
  forward (see Forward prediction)
  lattice predictors (see Lattice predictors)
  relation between autoregressive modeling and, 245
Linear predictive coding (LPC), 39
Linear time-invariant filters, 81
Liouville's theorem, 880
LMS algorithm (see Least-mean-square algorithm)
Lock-up phenomenon (see Stalling phenomenon)
Logistic function, 819
Low-rank modeling, 176

LSL algorithm (see Recursive Least-squares lattice algorithm)

M

Markov model, first order, 702
Matrix-factorization lemma, 590
Matrix-inversion lemma, 565
Maximin theorem, 175
Maximum entropy method (MEM), 905
Maximum entropy power spectrum, 910
  fast computation of, 910
Maximum-likelihood estimation, 900
  properties of, 901
McCulloch-Pitts model of neuron, 818
Mean, convergence of the, 393
Mean, ergodic theorem, 98
Mean square, convergence in the, 394
Mean-squared error criterion, 199
Mean-square value, 98
Mean-value function, 97
Measurement equation, 307
Measurement error, 484
Measurement matrix, 307
Mercer's theorem (see Spectral theorem)
Minimax theorem, 171
Minimum description length (MDL) criterion, 129
Minimum mean-squared error, 201
Minimum-norm solution to the linear least-squares problem, 526
Minimum-phase filters, 86
Minimum sum of error squares, 491, 494
Minimum-variance distortionless response (MVDR) beamforming, 60, 225, 617
Minimum variance distortionless response (MVDR) spectrum, 226, 912
  fast algorithm for computing, 913
Misadjustment, 367
  LMS algorithm, 402
  RLS algorithm, 402
Mixture models, Gaussian, 872
Model order, selection of, 128
Modem, 38
Momentum, for backpropagation algorithm, 836
Moore-Penrose generalized inverse (see Pseudoinverse)

Moving average (MA) models, 112
Multichannel filtering matrix, 807
Multilayer perceptron, 822
   network complexity, 840
   system identification using, 842
Multipath fading, 774
Multiple linear regression model, 484, 574
Multiple sidelobe canceler, 63
Multiple windows, method of, 148
Mutual consistency, 168, 755

N
Neural networks, 17, 71
   fault tolerance, 71
   feedforward, 71
   generalization, 71
   historical notes, 71
   learning, 71
Neyman-Pearson criterion, 47
Noise misadjustment, 708
Nonlinear adaptive filters, 15
Noise subspace, 148, 810
Nonnegative-definite correlation matrix, 102
Nonminimum-phase filters, 86
Nonsingular matrix, 103
Norm, of matrix, 169
Normal equations, 492
Normalized least-mean-square algorithm, 432
Numerical accuracy, 741
Numerical stability, 741

O
Observation vector, 306
Optimum linear discrete-time filters (see
   Kalman filters; Linear prediction;
   Wiener filters)
Order-recursive adaptive filters, 630
   adaptive backward linear prediction, 634
   adaptive forward linear prediction, 631
   conversion factor, 636
   least-squares lattice predictor, 640
   (see also Lattice predictor)
Otherwise excited subspace, 751
Orthogonality of backward prediction errors,
   277
Orthogonality principle, 197
   corollary to, 200

time-averaged form of, 487
Overdetermined system, 519, 524
Overlap-add method, 89
Overlap-save method, 90

P
Parameter drift, 748
Parseval's theorem, 889
Partial correlation (PARCOR) coefficients,
   259
Particular solution, 116
Parzen density estimator, 872
Periodogram, 138
Perron's theorem, 401
Phase-shift keying, 38
Piecewise-linear model of neuron, 818
Plane rotations (see Givens rotations)
Polyspectra, 150
Positive-definite matrix, 102
Postwindowing method, 487
Power spectral density, 136
   Cramér spectral representation for a sta-
      tionary process, 144
   defined, 138
   estimation of, 146
   fundamental equation, 146
   properties of, 138
   transmission of a stationary process
      through a linear filter, 140
Power spectrum (see Power spectral density)
Power spectrum analyzer, 142
Predicted state-error correlation matrix, 310
Predicted state-error vector, 309
Prediction, linear (see Linear prediction)
Prediction depth, 49
Prediction-error filter, backward, 252
   maximum-phase property, 267
Prediction-error filter, forward, 246
   eigenrepresentation, 269
   minimum-phase property, 265, 297
   relation between autocorrelation function
      and reflection coefficients, 262
   transfer function, 264
   whitening property, 268
Predictive deconvolution, 31
Prewindowing method, 487
Principle of the argument, 884

Principle of minimal disturbance, 436
Probabilistic neural network, 873
Projection operator, 498
Pseudoinverse, 524
Pulse-amplitude modulation (PAM) system,
    34, 776
Pulse-code modulation, 42

Q
QL algorithm, for eigen-computation, 186
QR algorithm, for SVD computation, 551
QR-decomposition-based least-squares lattice
    (QRD-LSL) algorithm, 660
  array for adaptive backward prediction,
      662
  array for adaptive forward prediction, 661
  array for adaptive joint-process estimation,
      664
  computer experiment on adaptive equaliza-
      tion, 672
  extended, 677
  finite-precision effects on, 760
  properties of, 667
  relationships between conventional LSL
      algorithms and, 679, 683
  summary of, 666
QR-decomposition-based recursive least-
    squares (QR-RLS) algorithm, 598
  extended, 614
  finite-precision effects on, 757
  implementation considerations, 600
  serial weight flushing, 613
QR-RLS algorithm (see QR-decomposition-
    based recursive least-squares algorithm)
QRD-LSL algorithm (see QR-decomposition-
    based least-squares lattice algorithm)
Quantization errors, 739
Quiescent weight vector, 234

R
Radial basis functions, 858
  Gaussian, 858
  thin-plate spline, 858
Radial-basis function (RBF) networks, 855
  applications of, 866
  comparison with multilayer perceptrons,
      873

  dynamic, 868
  fixed centers selected at random, 859
  hybrid learning procedure, 862
  stochastic gradient approach, 863
  structure of, 856
  universal approximation theorem using,
      865
Random processes (see Discrete-time wide-
    sense stationary stochastic processes)
Rank deficient matrix, 523
Rank determination, 523
Rayleigh quotient, 164
Recursive algorithms (see also Adaptive filter
    algorithms)
Recursive least-squares estimation (see
    Recursive least-squares algorithm; Order
    recursive adaptive filters; Square-root
    adaptive filters)
Recursive least-squares lattice (LSL) algo-
    rithms
  computer experiment on adaptive predic-
      tion using, 691
  finite-precision effects on, 762
  initialization of, 681
  normalized, 700
  summary of, 682, 687
  using *a posteriori* errors, 679
  using *a priori* errors with error feedback,
      683
Recursive least-squares (RLS) algorithm, 70,
    566
  compared with LMS algorithm for tracking
      nonstationarity, 716
  computer experiment on adaptive equaliza-
      tion, 580
  convergence analysis of, 573
  exponentially weighted, 566
  fast (see Fast recursive least-squares algo-
      rithm)
  finite-precision effects on, 751
  how to improve tracking performance of,
      726
  initialization of, 569
  Kalman filter theory and, 585
  learning curve of, 578
  matrix inversion lemma, 565
  operation in nonstationary environment,
      711

Riccati equation for, 567

single-weight adaptive noise canceler
using, 572

state-space formulation of, 583

summary of, 569

update recursion for the sum of weighted
error squares, 571

with adaptive memory, 734

Recursive minimum mean-square estimation
for scalar random variables, 303

Reflection coefficients, 6, 258, 659

Regression coefficients (joint-process), 289

Regularization, 869

Rescue variable, 764

Residue, 881

Riccati equation, 312, 567

RLS algorithm (see Recursive least-squares
algorithm)

Robustness, 3

Rouche's theorem, 885

Round-off errors (see Quantization errors)

S

Sample correlation matrix, 468 (see also
Time-averaged correlation matrix)

Sampling theorem, 2

Sato algorithm, 793

Scalar random variables, recursive minimum
mean-square estimation for, 303

Scanning vector, 226

Schur-Cohn test, 271

Seismic deconvolution, 32, 774

Self-orthogonalizing adaptive filters, 458

Self-orthogonalizing block adaptive filter
(SOBAF), 478

Serial weight flushing, 613

Sigmoidal model of neuron, 819

Signal detection, 46

Signal subspace, 148, 810

Signal-to-noise ratio, 107, 182

Sine wave plus noise, correlation matrix of,
106

Single input-multiple output (SIMO) model,
805

Singular-value decomposition (SVD), 517

applications of, 532

cyclic Jacobi algorithm for computing, 544

interpretation of singular values and singular vectors, 525

minimum norm solution to the linear least-
squares problem, 526

pseudoinverse, 524

QR algorithm for computing, 551

terminology and relation to eigenanalysis,
522

Singularities, 881

Skewness, 775, 797

Slepian sequences, 148

Smoothing, 1

Soft-constrained initialization, 570

Spectral-correlation density, 154, 815

Spectral norm, 169

Spectral theorem, 167

Spectrum analysis, 136

historical notes (see also Power spectral
density) 74

Spectrum estimation, nonparametric methods,
148

method of multiple windows, 148

periodogram-based methods, 148

Spectrum estimation, parametric methods,
146

eigendecomposition-based methods, 147

minimum variance distortionless response
method, 147

model identification procedures, 146

Square-root adaptive filters, 597

Square-root information filter, 593

extended, 596

Square-root Kalman (covariance) filter, 591

Square-root RLS algorithm (see inverse
QR-RLS algorithm)

Square root vs. square-root free Kalman
filtering, 328

Squared error deviation, 394

Stability, bounded input-bounded output
criterion, 83

Stalling phenomenon
LMS algorithm, 747
RLS algorithm, 756

State-space model, 306

State transition matrix (see Transition matrix)

State vector, 306

Stationary processes (see Discrete-time wide-
sense stationary stochastic processes)

Steepest descent, method of, 339
  effects of eigenvalue spread and step-size
    parameter, 350
  feedback model, 343
  vs. least-mean square algorithm, 404
  stability of, 343
  transient behavior of mean-squared error,
    349
  transversal filter structure, 340
Step-size parameter
  least-mean-square algorithm and, 370
  steepest descent algorithm and, 342
Stochastic gradient algorithms, 11
  gradient adaptive lattice (GAL) algorithm,
    915
  least-mean-square (LMS) algorithm, 367
Stochastic models, 108
  autoregressive (see Autoregressive models)
  autoregressive-moving average, 112
  moving average, 112
  selection of model order, 128
Stochastic processes (see Discrete-time wide-
    sense stationary stochastic processes)
Subspace of decreasing excitation, 750
Subspace decomposition, 177
Subspace decomposition method for fraction-
    ally spaced blind identification, 804
  orthogonality condition, 811
Sum of error squares, 486
  minimum, 491, 494
Super-resolution spectra, 515
Sylvester resultant matrix, 808
System identification, 20, 702, 842
Systolic arrays, 8, 600
Szegö's theorem, 190

T
Tapped-delay line filters (see Transversal
    filters)
Target detection in (radar) clutter, 849
Time-averaged autocorrelation function, 492
Time averaged correlation matrix, 493
  properties of, 495
Time-delay neural network (TDNN), 847
Toeplitz matrix, 101
Trace, of matrix, 167

Tracking of time-varying systems, 701
  assessment criteria, 706
  computer experiment on system identifica-
    tion, 702
  degree of nonstationarity, 705
  tracking behavior of LMS algorithm, 708
  tracking behavior of RLS algorithm, 711
Transfer function, defined, 82
Transition matrix, state, 307
Transversal filter, 5
  backward predictor using, 248
  channel equalizer using, 217
  fast (see Fast transversal filters algorithm)
  forward predictor using, 242
  least-mean-square algorithm and, 365
  least-squares estimation and, 486
  method of steepest descent and, 339
Triangle inequality, 168
Triangularization, 186
Tricepstrum algorithm, 797
  advantages and disadvantages of, 802
  channel estimation using, 800
Trispectrum, 152, 797

U
UD-factorization, 327, 768
Underdetermined system, 520, 525
Unexcited subspace, 749
Unforced dynamics, 323, 657
Uniform distribution, 777
Uniqueness theorem, 500
Unitary matrix, 166
Unitary similarity transformation, 165
Universal approximation theorem
  for multilayer perceptrons, 837
  for radial-basis function networks, 865
Unit-delay operator, 5
Unvoiced speech sound, 41

V
Vandermonde matrix, 163
Variance, 98
Voiced speech sound, 41
Voltera-based nonlinear adaptive filters, 16